MARE MAIUS
PONTUS ET
GALATIA
BITHINIA
CAPPADOCIA
ASIA PROPIE
ARMENIA MINOR
PAMPHILIA
CILICIA
LICIA
CIPRUS
SIRIACUM
SIRIA
IUDEA
MARE AEGITIACU
ARABIA PETRAEA
AEGIPTUS

THE MEDITERRANEAN SEA:
A Natural Sedimentation Laboratory

THE MEDITERRANEAN SEA:
A Natural Sedimentation Laboratory

Edited by DANIEL J. STANLEY
Smithsonian Institution

Assisted by GILBERT KELLING and YEHEZKIEL WEILER
University of Wales *The Hebrew University*

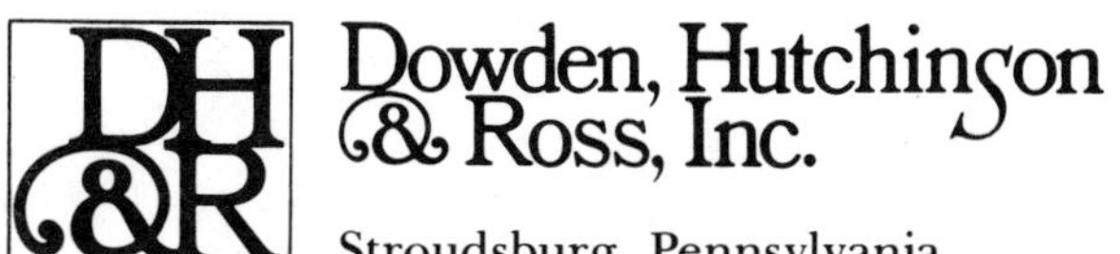

Dowden, Hutchinson & Ross, Inc.
Stroudsburg, Pennsylvania

Library of Congress Catalog Card Number: 72–88984
ISBN: 0–87933–010–4

Library of Congress Cataloging in Publication Data

Symposium on the Mediterranean Sea, Heidelberg, 1971.
The Mediterranean Sea.

English or French; each paper with English and French résumés.
The majority of the papers were presented at the Symposium on the Mediterranean Sea, held Aug. 30-Sept. 1, 1971, during the 8th International Sedimentological Congress.
1. Marine sediments—Mediterranean Sea--Congresses.
2. Geology--Mediterranean Sea--Congresses. I. Stanley, Daniel J., ed. II. International Sedimentological Congress, 8th, Heidelberg, 1971.
GC389.S95 1971 551.4'62 72-88984
ISBN 0-87933-010-4

Manufactured in the United States of America.

Exclusive distributor outside the United States and Canada:
John Wiley & Sons, Inc.

Contributors

TORBJÖRN ALEXANDERSSON
Department of Historical Geology
University of Uppsala
SWEDEN

GENEVIEVE ALLA
Musée Océanographique de Monaco
MONACO

GILBERT BELLAICHE
Centre de Recherches Géodynamiques
Alpes Maritimes, FRANCE

RICHARD H. BENSON
Department of Paleobiology
Smithsonian Institution
Washington, D.C., U.S.A.

PIERRE E. BISCAYE
Lamont-Doherty Geological Observatory
of Columbia University
Palisades, New York, U.S.A.

JEAN J. BLANC
Laboratoire de Géologie Marine
Centre Universitaire de Luminy
Marseille, FRANCE

LAURE BLANC-VERNET
Laboratoire de Géologie Marine
Centre Universitaire de Luminy
Marseille, FRANCE

ENRICO BONATTI
School of Marine and Atmospheric Sciences
University of Miami
Miami, Florida U.S.A.

ANTONIO BRAMBATI
Istituto di Geologia e Paleontologia
dell'Università degli Studi
Trieste, ITALY

LLOYD R. BRESLAU
Coast Guard Headquarters
Washington, D.C., U.S.A.

PIERRE F. BUROLLET
Compagnie Français des Pétroles
Puteaux, FRANCE

TERENCE G. CARTER
U.S. Naval Oceanographic Office
Washington, D.C., U.S.A.

JEAN PIERRE CAULET
Laboratoire de Géologie
Muséum National d'Histoire Naturelle
Paris, FRANCE

GUSTAVE A. CAUWET
Centre de Recherches de Sédimentologie Marine
Université de Montpellier
Perpignan, FRANCE

HERVE CHAMLEY
Laboratoire de Géologie Marine
Centre d'Océanographie, Luminy
Marseille, FRANCE

HENRY CHARNOCK
National Institute of Oceanography
Wormley, Godalming
Surrey, ENGLAND

MARIA B. CITA
Istituto di Geologia
Università degli Studi di Milano
Milano, ITALY

CLAUDE DEGIOVANNI
Faculté des Sciences, Alger et
Centre Universitaire de Marseille-Luminy
Marseille, FRANCE

DANIEL DESSOLIN
Musée Océanographique de Monaco
MONACO

PHILIPPE DUFAURE
Compagnie Française des Pétroles
Puteaux, FRANCE

PAULIAN DUMITRICA
Geological Institute
Bucarest, RUMANIA

SOLANGE DUPLAIX
Laboratoire de Géologie Dynamique
Université Paris VI
Paris, FRANCE

HAROLD E. EDGERTON
Massachusetts Institute of Technology
Cambridge, Massachusetts, U.S.A.

E.M. EMELYANOV
P.P. Shirshov Institute of Oceanology
Atlantic Department
Kaliningrad, U.S.S.R.

AUGUSTO FABBRI
Laboratorio di Geologia Marina
Consiglio Nazionale Delle Ricerche
Bologna, ITALY

FRANK FABRICIUS
Institut für Geologie
Technische Hochschule
München, WEST GERMANY

RHODES W. FAIRBRIDGE
Department of Geology
Columbia University
New York, New York, U.S.A.

GUILIANO FIERRO
Istituto di Geologia dell'Università
Genova, ITALY

JOSEPH P. FLANAGAN
U.S. Naval Oceanographic Office
Washington, D.C., U.S.A.

NICHOLAS C. FLEMMING
National Institute of Oceanography
Wormley, Godalming
Surrey, ENGLAND

JEAN-CHARLES FONTES
Laboratoire de Géologie Dynamique
Université de Paris VI
Paris, FRANCE

FRANÇOIS Y. GADEL
Centre de Recherches de Sédimentologie Marine
Université de Montpellier
Perpignan, FRANCE

THOMAS R. GOEDICKE
Department of Geology
American University of Beirut
Beirut, LEBANON

NORMAN HAMILTON
Department of Geology
The University of Southampton
Southampton, ENGLAND

YVONNE HERMAN
Department of Geology
Washington State University
Pullman, Washington, U.S.A.

J. BRACKETT HERSEY
Office of Naval Research
Arlington, Virginia, U.S.A.

REINHARD HESSE
Department of Geological Sciences
McGill University
Montreal, Quebec, CANADA

JOSE HONNOREZ
School of Marine and Atmospheric Sciences
University of Miami
Miami, Florida, U.S.A.

KENNETH J. HSU
Geologisches Institut
Eidg. Technische Hochschule
Zürich, SWITZERLAND

TER-CHIEN HUANG
Graduate School of Oceanography
University of Rhode Island
Kingston, Rhode Island, U.S.A.

OIVA JOENSUU
School of Marine and Atmospheric Sciences
University of Miami
Miami, Florida, U.S.A.

C. REED JONES
U.S. Naval Oceanographic Office
Washington, D.C., U.S.A.

AMITAI KATZ
Department of Geology
The Hebrew University
Jerusalem, ISRAEL

GEORGE H. KELLER
National Oceanic and Atmospheric Administration
Atlantic Oceanographic Laboratories
Miami, Florida, U.S.A.

GILBERT KELLING
Department of Geology
University of Wales
Swansea, WALES

MAHMOUD M. KHOLIEF
Earth Sciences Laboratory
National Research Centre
Dokki, Cairo, U.A.R. EGYPT

HENRI LACOMBE
Laboratoire d'Océanographie Physique
Muséum National d'Histoire Naturelle
Paris, FRANCE

DOUGLAS N. LAMBERT
National Oceanic and Atmospheric Administration
Atlantic Oceanographic Laboratories
Miami, Florida, U.S.A.

LUCIEN LECLAIRE
Laboratoire de Géologie
Muséum National d'Histoire Naturelle
Paris, FRANCE

OLIVIER LEENHARDT
Musée Océanographique de Monaco
MONACO

RENE LETOLLE
Laboratoire de Géologie Dynamique
Université de Paris VI
Paris, FRANCE

R. MICHAEL LLOYD
Shell Development Co.
Houston, Texas, U.S.A.

JENNIFER M. LORT
University of Cambridge
Cambridge, ENGLAND

FRANCIS L. MARCHANT
U.S. Naval Oceanographic Office
Washington, D.C., U.S.A.

WOLF MAYNC
Geological Consulting Service
Bern, SWITZERLAND

ARTHUR R. MILLER
Woods Hole Oceanographic Institution
Woods Hole, Massachusetts, U.S.A.

JOHN D. MILLIMAN
Woods Hole Oceanographic Institution
Woods Hole, Massachusetts, U.S.A.

ANDRE A. MONACO
Centre de Recherches de Sédimentologie Marine
Université de Montpellier
Perpignan, FRANCE

ROBERT R. MURCHISON
U.S. Naval Oceanographic Office
Washington, D.C., U.S.A.

BRUCE W. NELSON
University of South Carolina
Columbia, South Carolina, U.S.A.

WLADIMIR D. NESTEROFF
Laboratoire de Géologie Dynamique
Université de Paris VI
Paris, FRANCE

GUY PAUTOT
Centre Océanologique de Bretagne
Brest, FRANCE

GIOVANNI B. PIACENTINO
Istituto de Geologia dell' Università
Genova, ITALY

JACK W. PIERCE
Division of Sedimentology
Smithsonian Institution
Washington, D.C., U.S.A.

SERGE PIERROT
Musée Océanographique de Monaco
MONACO

ULRICH von RAD
Bundesanstalt für Bodenforschung
Hannover-Buchholz, GERMANY

JACK H. REBMAN
U.S. Naval Oceanographic Office
Washington, D.C., U.S.A.

ANTHONY I. REES
Natural Environment Research Council
London, ENGLAND

GRAZIELLA RICCIARDI
Istituto di Geologia dell' Università
Genova, ITALY

WILLIAM B.F. RYAN
Lamont-Doherty Geological Observatory
of Columbia University
Palisades, New York, U.S.A.

HAROLD RYDELL
School of Marine and Atmospheric Sciences
University of Miami
Miami, Florida, U.S.A.

EYTAN SASS
Department of Geology
The Hebrew University
Jerusalem, ISRAEL

PAUL SCHMIDT-THOME
Institut für Geologie
Technische Hochschule
München, GERMANY

RAIMONDO SELLI
Laboratorio di Geologia Marina
Consiglio Nazionale delle Ricerche
Bologna, ITALY

K.M. SHIMKUS
P.P. Shirshov Institute of Oceanology
Atlantic Department
Gelendjik, U.S.S.R.

DANIEL J. STANLEY
Division of Sedimentology
Smithsonian Institution
Washington, D.C., U.S.A.

L.M.J.U. van STRAATEN
Geologisch Instituut
Rijksuniversiteit
Groningen, THE NETHERLANDS

HERBERT STRADNER
Geologische Bundestalt
Vienna, AUSTRIA

JOHN C. SYLVESTER
U.S. Naval Oceanographic Office
Washington, D.C., U.S.A.

PAUL TCHERNIA
Laboratoire d'Océanographie Physique
Muséum National d'Histoire Naturelle
Paris, FRANCE

KOLLA VENKATARATHNAM
Lamont-Doherty Geological Observatory
of Columbia University
Palisades, New York, U.S.A.

CLAUDIO VITA-FINZI
Department of Geography
University College, University London
London, ENGLAND

YEHEZKIEL WEILER
Department of Geology
The Hebrew University
Jerusalem, ISRAEL

FORESE C. WEZEL
Università di Catania
Catania, ITALY

JOSEPH C. WHITNEY
U.S. Naval Oceanographic Office
Washington, D.C., U.S.A.

Foreword

For over a century, much of sedimentary and structural geology has been founded on investigations of the region bordering the Mediterranean Sea. Nevertheless, until little more than two decades ago, Mediterranean waters have maintained a complete air of mystery about what lay beneath them. The mystery began to be dispelled by the echo soundings made prior to 1939. In the interval 1946 to 1960, several surveys of coastal regions provided the first detailed knowledge of its continental shelves and slopes. This period was also one of sketchy scientific forays over its deep waters. Samples of sediment were taken, their seismic properties were measured in a scattering of observations, and we began to form an increasingly reliable notion of the morphology of the deep basins. Widely spaced measurements of gravity had been made during the 1930's. By 1960 we had enough information to make useful speculations about both sediment and basement, but not nearly enough for understanding.

Meanwhile, various new methods of study had been developed and used with increasing intensity in the Mediterranean and elsewhere. For the sedimentologist, these include the piston corer, the deep-sea camera, the seismic profiler, the precision echo sounder, developments in clay and heavy mineral analysis, and so on. In a series of rather major campaigns of the late 50's and early 60's, these techniques were extensively employed in the Mediterranean. The results of these early campaigns have convinced many scientists that the undersea Mediterranean is at least as fruitful a natural laboratory for studying sedimentary processes as its borderlands had already proved to be. Since 1964 a bewilderingly complex pattern of investigations of Mediterranean sediments and sedimentary rock has been made. The full complement of tools has been employed, including lately the deep-sea drill on GLOMAR CHALLENGER. Several of the expeditions forming part of this pattern are so recent that their full significance is scarcely available.

With that caveat, this book constitutes as complete an account as it is practical to give of our knowledge of sedimentary processes and the entire geologic history of the Mediterranean region. It is obvious that our methods and our ability to interpret data are turbulently evolutionary. This state is represented in the discussions to follow by conflicting interpretations of data now in hand. It will also be obvious that we need to know much more than we do now in order to decide about several of the broader, more fundamental aspects of process and history of the Mediterranean region. We still stand in the shadow of an ancient admonition—a little knowledge is a dangerous thing!

J. B. HERSEY
Office of Naval Research
Arlington, Virginia

Preface

notre vieille Méditerranée, si étroitement mêlée, non seulement à l'histoire de nos terres, mais même à notre propre histoire . . . elle semble bien délaissée aujourd'hui

Jacques Bourcart, 1949
Géographie du Fond des Mers

Contemplation of natural phenomena in the Mediterranean Sea is an old preoccupation and it is not surprising that this, the cradle of western civilization, should be the first ocean to receive serious attention by those intrigued by the sea. Among the first to consider problems of sedimentation were the Greek scientists who by the sixth century B.C. were already providing rational explanations for the operation of physical causes. Coastal and near-shore processes continued to retain the attention of philosophers, navigators and naturalists intermittently from the time of Herodotus to the Rennaissance. At the beginning of what may appropriately be called the era of modern oceanography Count Luigi Ferdinand Marsigli in his *Histoire Physique de la Mer* (1725, Amsterdam) recorded that sediment locally covered rock outcrops on the continental margin off the French coast in the Gulf of Lion. His observation that rocks dredged offshore are similar to those on land served to reinforce a fundamental concept: the key to understanding land geology lies in the adjacent submarine realm—and vice versa.

Yet, some two and one-half centuries later it is clear that the advancement of oceanography in this region has not kept pace with its progress elsewhere, and in many respects the Mediterranean has been bypassed. Jacques Bourcart, who perhaps more than any other in recent years was instrumental in calling the attention of geologists to the Mediterranean, forcefully repeated that this area must once again receive the emphasis it deserves.

The picture is changing. Impetus has been provided by military considerations and economic pressures (petroleum and fisheries), and more recently, by concern with pollution. Oceanographic techniques developed in other oceans are now applied to this almost totally enclosed and, in many respects, unique body of water. In the geological sciences there are exciting developments. The results of the intensified geophysical research of the 1960's contributed significantly to the recent geodynamic concepts of sea floor spreading and oceanization, and the first results of the August 1970 deep-sea drilling cruise (JOIDES) have provided some startling pieces of evidence.

The present volume arose out of a series of frustrations which I experienced while researching a sedimentological project in the Alboran Sea. The first of these, apparent to anyone who has made a bibliographic search, is the logistical difficulty in finding sedimentological information on this region. There is no consolidated source summarizing modern sediment studies. Data are spread across hundreds of sources. Even information about who has worked on what, or where, is not easily come by.

A second observation is that many studies tend to be provincial, *i.e.*, unrelated to other regions and prepared by workers of one school who often tend to neglect the results of other groups. The isolation is, in part, related to language obstacles. Moreover, there has been very little effort spent in comparing modern Mediterranean sediments with those exposed in the adjacent land areas.

I felt that some of these deficiencies could be repaired by the preparation of a multi-authored, multidisciplinary, and bilingual reference volume. At the very least, such a venture should highlight some of the exciting ongoing sedimentological studies in the Mediterranean Sea. And it seems that the venture is a timely one. In planning such a volume I was impressed by the success of recent joint geological, biological and hydrological projects, such as the MEDOC group cruises in 1969. Most colleagues were enthusiastic when contacted about initiating the volume and agreed that sedimentology, which to a large degree still enjoys the individualistic and generalist approach, could well benefit from this form of team effort.

THE MEDITERRANEAN SEA AS A NATURAL SEDIMENTATION LABORATORY

The major theme of this volume is that Mediterranean basins are particularly well suited as natural laboratories for the study of marine sedimentation processes. Here the term laboratory can be used in the sense given in Webster's *Third New International Dictionary, Unabridged* (1967), i.e., *an environment that provides opportunity for systematic observation, experimentation, or practice.* A number of distinctive aspects enables the Mediterranean Sea to serve such a function and a few of these are summarized below.

Logistics

The *mare internum* is a narrow, east-west trending land-enclosed depression, small (about 2,500,000 km^2) in comparison to the major world ocean basins, and formed above a transitional crust. It extends almost 4000 km from the Strait of Gibraltar to the foot of the Lebanon Mountains, yet one is never further than 370 km from shore and, more often, considerably less. The slopes, readily visible on the new chart at the back of the book, are generally short and steep, locally exceeding 10°, and bathyal to abyssal depths are attained in one to a few hours steaming time. In addition to its "compact" configuration, two characteristics make the Mediterranean an ideal region to study marine sediments: its enclosure and virtual isolation from the Atlantic Ocean and Black Sea, and the extremely rapid lateral geographical and geological changes. Immediately obvious is the high variability of sea floor topography and relief and that of the adjacent borderland, and diversity of climate, coastline, and geometry of the deep sea basins and trenches. It is convenient to distinguish a western, central and eastern province, each of which is further subdivided into distinct depositional traps by submarine ridges, tectonic blocks and the major peninsulas (Iberian, Apennine and Hellenic). The western province, for example, includes the Alboran Sea and the Balearic (or Algéro-Provençal) Basin, the largest of the depressions, and submarine (Alboran Ridge, for example) and partially exposed (Balearic) blocks. The Corso-Sardinian block, Sicily and the Italian peninsula proper serve to isolate the topographically complex and deep Tyrrhenian Sea from the shallow, elongate Adriatic in the central Mediterranean. The Aegean and Ionian seas and the Levantine Basin, with its trenches, Mediterranean Ridge, and island arc configuration south of the Aegean, comprise the eastern province. The isolation of these distinct basins, each with its own fluvial drainage and source areas, and affected by a distinct climate and hydrography, insures that the key factors affecting sedimentation in each region can be investigated. Each basin can be examined as a semi-closed system. The marine scientist combining effort with the land geologist thus can isolate those factors which are truly significant from background 'noise.'

Regional Diversity

The most attractive aspect of the Mediterranean is that it has well defined limits. This possibility of isolating the assemblage of factors controlling provenance, dispersal, and depositional patterns, and pin-pointing dominant phenomena in each region, means that we can deal with laboratory-like conditions. The lateral variability of the basin margins is an example. The coastline totals 22,500 km, but its distribution is irregular, with almost 13,000 km along the European margin contrasting with only about 5000 km along the African front and the remainder along the Levantine margin. Shelf development is generally poor. The only neritic platforms of any appreciable width lie off the Po Delta, forming the northern half of the Adriatic, and off Tunisia. Lesser shelfdepth margins also form the Strait of Sicily, the Gulf of Valencia–Ebro Delta region, the Gulf of Lion, and the Elba region in the eastern Ligurian Sea, while narrower platforms occur off Sirtica and the Nile and the Seyhan-Ceyhan rivers of Turkey. It is noteworthy that shelf platforms are in most cases related to the few major rivers—Nile, Rhône, Po, and Ebro—which drain geological source-terrains far removed from the present coastline. Although few in number, these major fluvial sources account for an enormous volume of sediment injected into the system. On the other hand, the highly irregular coast, much of which is fronted by high mountain barriers, is distinguished by short, often torrential, rivers draining small areas on a highly seasonal basis. It is on these smaller rivers that we must focus attention in order to understand pulsating sedimentation patterns in deep sea basins, for they flood in winter and dry up during the spring-to-autumn droughts. These rivers flow in narrow gorges and emerge on narrow coastal plains, debouching their petrographically identifiable load as deltas or sand bars; the periodic flooding of coastal plains by such rivers is of historical record. By-passing of the narrow coastal margin by fluvial sediment and spill-over directly onto the slope and basin plains beyond is not uncommon. The origin of the coarse fraction, as well as the clays, in cores collected on basin plains in many cases can be traced to specific stream outlets.

Provenance and dispersal is affected not only by diverse source terrains, but also by climatic differences on the northern and southern coasts. The northern sectors of the basin, lying within the zone of prevailing westerlies, are distinguished by spring and autumn showers that curtail the summer drought, and have a moderate climate. The southeastern Iberian peninsula and most of the North African margin receives an annual rainfall of less than 50 cm, or considerably less than half of the rainfall on much of the Italian peninsula and Adriatic region. Rainfall intensity and duration decrease both from west to east and from north to south. Temperature, on the other hand, increases from north to south and from west to east. Rainfall throughout the region is seasonal, with a marked minimum in summer. Wind force, temperature, and humidity are variable, drying the vegetation in summer and bringing cold rain and snow in winter. The mountains, acting as barriers to rain-bearing wind, produce their own local climatic belts. The amount and duration of rainfall results in an arid eastern zone, a less arid western province and a generally moist northern province north of the 40th parallel. Seasonality accounts for more than episodic fluvial input along the margins of the basin. As a factor of the first order, it affects the water masses that are responsible for the transport of suspensates. For example, surface cooling during the late winter in the northwestern Mediterranean results in the sinking of large masses of water to considerable depths.

Submarine or emergent topography can also exert a profound effect on sedimentary processes. Narrow or shallow straits, for instance, accelerate erosion, and zones of high relief, such as the Mediterranean Ridge, modify deep-water flow and produce a *barrière-en-creux* effect that isolates sediments from basin to basin. The relative influence of Atlantic and Black Sea water on sedimentation of plankton and suspensates can also be tested; microfauna and clay mineralogy provide a means to evaluate the relative importance of these allochthonous water masses as opposed to the fluvial influence exerted along the basin margins. They also serve to gauge the role of the sills at both ends of, and within, the depression. Topographic highs are found throughout the region and the compartmentation of the sea floor promotes sedimentary ponding, a phenomenon that varies in importance from basin to basin across the Mediterranean.

Changes with Time

Modern Mediterranean sedimentation is significantly affected by tectonics and climate, and this was also the case in the past (probably even more so than at present). The recent geologic evolution of the Tethys makes this sea particularly worthy of study. Sub-bottom records coupled with cores provide a basis for evaluating the fourth dimension—time. The late Miocene–Pliocene "revolution" produced a change in source terrains and hydrography of such magnitude that its impact must be preserved in the sedimentary record. The uplift of selected circum-Mediterranean margins, the *effondrement* of the slopes with associated canyon cutting and tectonic damming, and the formation of diapirs and the Mediterranean Ridge are all related to the probable movement of Africa relative to southern Europe and have directly influenced the styles and rates of sedimentation. The Pliocene and Quaternary volcanism (sea mounts, volcanic ridges, and tephra) in the Alboran, Tyrrhenian and Aegean seas, as well as earthquake activity, reflect the intensity of these earth movements. The recent recovery of shallow-water Miocene carbonates and evaporites (sabkha?) from the deep sea floor is further indication of the radical physiographic changes that have affected the region in the recent geological past. Superimposed on these tectonic modifications are the changes in climate and sea level which have modified coastline, source terrains, fluvial drainage, and even water-mass circulation. The JOIDES cores of Leg 13 have provided a limited window to the past and furnish a unique tool for reconstructing paleogeography. Whether or not the sea-floor spreading–plate movement hypothesis is finally accepted will depend, to a large degree, on the careful interpretation of such data.

But all changes affecting sedimentation in the Mediterranean are not due to the natural evolution of geographic and geological conditions alone. The decline in vegetation and agriculture, long recorded by archeologists is a result, to a large degree, of man's direct influence, *i.e.*, denudation of soil and deforestation. The acceleration of this process should be recognized in cores of the enclosed basins, and one might expect that this man-made progradation would vary in distance from centers of population and also with the type of human activity, and type of terrain involved. For instance, limestone, once deforested, does not recover in arid areas. Surveys of the specific man-made tracers, now more and more frequently injected by industrial complexes along the shores of the Mediterranean, could serve to define dispersal paths and rates of sedimentation in a quantitative manner.

In sum, the almost closed physical-chemical-biological systems that are represented in Mediterranean basins provide the opportunity for observing sedimentation under controlled conditions. Thus, these natural laboratories will provide a better means to understand Mesozoic–Tertiary sedimentary units exposed in the circum-Mediterranean land region. Moreover, knowledge of these systems hopefully will lead in the future to the proper control of man's own rapidly growing influence in this internal sea.

EVOLUTION OF THE VOLUME

During the summer of 1969 marine scientists who had in one way or another worked in the Mediterranean Sea were contacted and asked to submit contributions in their area of specialization. A large majority responded rapidly to the invitation which also involved participation in the Symposium on the Mediterranean Sea (to be held two years later, in 1971, at the VIIIth International Sedimentological Congress at Heidelberg). This first group of participants was also asked to provide the names of other workers who originally may have been overlooked and who could contribute to this international and multi-disciplinary teamwork effort.

The 85 contributors comprise a mix of established senior and younger scientists interested in various aspects of the Mediterranean. The list of contributors reveals the wide diversity of interests and parent organizations of the authors. These specialists from national and private oceanographic institutions, the academic community, and industry represent 15 countries, and only 35 (or 44%) have English as their mother tongue. It was apparent from the start of the project that the bulk of the work, and certainly that pertaining to the western Mediterranean, has been done by French and French-speaking workers. It thus follows that a comprehensive volume claiming to serve as a basic reference on this region should include chapters in French as well as English. My bilingual talents have been put to the test: over a quarter of the volume is in French. All papers have an English abstract and a French résumé, and if all language barriers are not completely resolved the doors are at least open to anyone interested in the subject.

The production of a reference compendium useful to any investigator of the Mediterranean Sea regardless of his own speciality required, first of all, that the volume be more than a loose collection of individual papers—too often synonymous with the term symposium volume. Accordingly, a book plan was established after initial titles had been received from potential contributors, and several additional chapters were assigned to complete the geographic and historical setting of the Mediterranean. Moreover, the symposium at Heidelberg was held *after* the papers had been prepared so that contributors could actually meet, exchange ideas and incorporate in their chapters any last-minute changes made on the basis of these discussions.

Perhaps the major responsibilities of an editor are to insure that the quality of papers be maintained at a high level and that these manuscripts be dovetailed into a reasonably cohesive volume. Although quite tolerant of the subject matter accepted, I insisted that each manuscript had to be reviewed by two, and in some cases three and even four specialists. The reviewers were selected, whenever possible, from the list of authors but the talents of outside specialists were also called upon. As a result of this critique system several chapters were rejected and fell by the wayside. Before final acceptance, all papers were once again reviewed by Drs. G. Kelling and Y. Weiler and edited at my end. In several instances we took considerable liberty in polishing up the English, but more often we have tried to preserve the author's style at the expense of a uniform text. I also have left the door open to some ideas which are either contradictory to my own or which, at this time, frankly appear eccentric. It was felt that the volume should be an accurate record of the state of the art at the beginning of the 1970's and as such will be that much more interesting as a document in years to come.

The style and treatment of subjects is admittedly uneven, and there are some overlaps and, more important, some obvious gaps. These "holes," some of them quite significant, are discussed in the Epilogue chapter at the end of the book. The focus provided on these problems could serve to prompt new lines of investigations and reorient present research programs of workers who, like ourselves, have caught a serious case of the *virus mediterraneum*.

THE SYMPOSIUM AT HEIDELBERG

In order to increase the level of interaction among the contributors of the volume and between them and other specialists of Mediterranean geology and related disciplines, a symposium entitled *Sedimentation in the Mediterranean Sea* was convened at the VIIIth International Sedimentological Congress in Heidelberg. The majority of papers in this volume were presented orally during a three-day period (30 August to 1 September, 1971). Formal and informal discussions arising from the presentations enabled authors to share experiences, coordinate their efforts and make necessary last-minute changes in their manuscripts. Forty-four papers were included in the Symposium (abstracts are published in *Program with Abstracts, 1971, VIIIth IAS,* Laboratorium für Sedimentforschung, Heidelberg).

Thus, the symposium served as unique means for bringing together the majority of sedimentologists and scientists in related fields active in the Mediterranean and circum-Mediterranean Sea. The papers were grouped as follows: the geological setting, coastal and nearshore sedimentation, regional sedimentation problems, suspended sediment transport, and geochemistry. At the end of the symposium an attempt was made to summarize work accomplished to date and, more important, to focus on major problems that need the attention of marine scientists. Those areas in which collaboration on research projects would prove most fruitful are discussed in the Epilogue section of the present volume.

ACKNOWLEDGMENTS

A venture of this kind clearly requires a team effort and could only have been achieved by a broadly based authorship and international cooperation. We are indebted to many persons and organizations. First in the list are the authors themselves, who are thanked for their effort and willingness to participate in the venture. Their promptness in submission of manuscripts, corrections, and galleys, their willingness to accept the suggestions of outside critics and modify their papers accordingly, and their patience with their not always patient editor is to be commended. It obviously would not have been possible to orchestrate a work of this scope without considerable flexibility and good will on their part. Particular thanks are due to two of the contributors: Drs. Gilbert Kelling and Yehezkiel Weiler. In addition to preparing their own manuscripts for the volume and fulfulling their already heavy professional commitments, they accepted several months of thankless editorial responsibilities during the final stages of book preparation. They also served as an informal consulting board. Publication forcibly would have been delayed without their help.

To my parent organization, the Smithsonian Institution, goes the credit of bearing the secretarial burden and support for the one thousand and one needs that face an editor, but most of all for providing the most important ingredient—the unique scholarly climate needed to complete such a task. Financial aid for some of the editorial needs were provided by Smithsonian Research Foundation Grant No. 427235, and the Department of Paleobiology of the National Museum of Natural History. We thank Mrs. E. Doane and Mrs. G. Smith for their secretarial help.

My wife, Anne Marie, willingly accepted the task of translating English abstracts into French résumés as well as helping to review the French manuscripts—no small effort. But I owe her an even greater debt for her understanding during these past two years of editorial duties and not-infrequent burning of the midnight oil.

The "unsung heroes" are those scientists who accepted the role of outside critic, and in many cases helped beyond the call of duty. They supplemented the authors in reviewing manuscripts and have served the oceanographic community by insuring that professional standards were maintained. Our appreciation is expressed to: T. F. Anderson, W. A. Berggren, K. Boström, K. W. Butzer, R. Cifelli, J. R. Curray, E. J. Dasch, C. G. Day, J. van Donk, K. O. Emery, C. E. Emiliani, M. Felsher, G. M. Friedman, M. Gennesseaux, N. Güven, J. B. Hersey, T. J. Hirst, G. deV. Klein, D. C. Krause, I. G. Macintyre, F. T. Manheim, C. H. Moore, D. Neev, F. L. Parker, R. Passega, J. Paul, J. Phillips, A. F. Richards, K. S. Rodolfo, G. Ruggieri, F. L. Sayles, T. E. Simkin, D. J. P. Swift, P. C. Sylvester-Bradley, R. Todd, J. N. Valette, and T. R. Waller.

Particular thanks are also due to N. Rupke and P. Hearn for their help in preparing the subject index.

Appreciation is expressed to the Naval Oceanographic Office, and Mr. John C. Sylvester in particular, for their efforts in making available the bathymetric chart (N.O. 310) available for inclusion in this volume. The Apollo 9 space photograph used on the jacket was provided by Dr. R. L. Stevenson, and authorization to reproduce it was obtained from the National Aeronautics and Space Administration, Washington, D.C. The Portolano maps used as end papers were furnished by the Map Section of the Library of Congress, Washington, D.C.

Publication of a specialized volume of this type is admittedly a gamble. All of us involved with this project extend our thanks to our publisher, Dowden, Hutchinson, and Ross, Inc., and its president Mr. C. Hutchinson, for their willingness to ensure that these many pages finally appear in print.

DANIEL J. STANLEY
On the Mall
Washington, D.C.
December 1972

Contents

PART 6. Coastal and Shallow Water Sedimentation: Detrital Sediments

PART 7. General Aspects of Recent Deep-sea Sedimentation

PART 8. Suspended Sediments and Regional Dispersal Patterns

PART 9. Sedimentation in the Western Mediterranean

PART 10. Sedimentation in the Central and Eastern Mediterranean

PART 11. Geochemical Aspects of Mediterranean Deposits

PART 12. Epilogue

PART 1. The Geographic Setting and Hydrography

A New Bathymetric Chart and Physiography of the Mediterranean Sea

Terence G. Carter, Joseph P. Flanagan
C. Reed Jones, Francis L. Marchant
Robert R. Murchison, Jack H. Rebman
John C. Sylvester, and Joseph C. Whitney

U.S. Naval Oceanographic Office
Washington, D.C.

ABSTRACT

A new bathymetric chart of the Mediterranean Sea has been compiled at the U.S. Naval Oceanographic Office from a larger amount of precisely controlled bathymetric data than has previously been available. A series of seven physiographic maps and descriptive texts were prepared to accompany the chart. The physiographic diagrams depart from traditional presentations in that they include some isobaths to more clearly define the provinces, and are for the most part quite detailed. Shelf widths and shelf edge depths are determined primarily from close examination of existing nautical charts. The construction of the chart and the physiographic diagrams emphasize some previously undiscussed or unknown aspects of the Mediterranean Sea floor: (1) a series of NW-SE trending ridges and valleys on the Nile Fan which were once thought to be abyssal hills; (2) the Mediterranean Ridge may be more properly called the Mediterranean Rise; (3) the Mediterranean Rise does not reach the foot of Italy and shows a very abrupt change in trend southwest of Crete; and (4) shelf edge depths are found to be quite variable and dependent primarily on shelf width.

RESUME

Le U.S. Naval Oceanographic Office a établi une carte bathymétrique à partir de données dont le nombre et la précision dépassent tout ce qui a été rassemblé jusqu'à présent. Cette carte est accompagnée de sept diagrammes physiographiques et de leur texte descriptif. Les diagrammes s'écartent de la forme traditionelle car on y a ajouté des isobathes afin de mieux délimiter les différentes zones et sont, de façon générale, très détaillés. L'étude approfondie des cartes marines déjà établies a permis de déterminer la largeur du plateau et la profondeur de son rebord. La carte et les diagrammes mettent en évidence des aspects du fond de la Méditerranée qui étaint jusque là inconnus ou négligés: (1) des lignes de relief orientées N.O.-S.E. sur le glacis du Nil, qu'on croyait être des collines abyssales, constituent en réalité une série de dorsales et de vallées; (2) il serait plus approprié de donner à la dorsale méditerranéenne le nom de *Mediterranean Rise;* (3) ce Mediterranean Rise ne s'étend pas jusqu'au pied de l'Italie, et présente un changement brusque de direction au Sud-Ouest de la Crète; (4) la profondeur du rebord du plateau est inégale et sa variation est en grande partie fonction de la largeur du plateau.

INTRODUCTION

(John C. Sylvester)

The Mediterranean Sea is unique in many ways. It has a complex geological history, the resolution of which has fascinated scientists for years. Through the history of man numerous and diverse cultures and empires have risen and fallen on its shores. It has been the scene of much warfare. Trading and commerce have been carried out on this waterway probably from the time men began to build ships. It is safe to assume that as sailing developed in this area so did man's need to know about the depths of the Sea.

Today a precise and detailed bathymetric chart has been compiled of the Mediterranean Sea and its Atlantic approaches. This chart is included in the sleeve at the end of this volume.

The density diagram (Figure 1) and track diagram (Figure 2) portray the coverage of sounding data utilized in the primary interpretation of the bottom topography. These data were all collected by survey ships utilizing precision depth recorders and more highly accurate navigational control than was heretofore available. The greatest density of soundings occur in the Alboran Sea, Ligurian Sea, Adriatic Sea, and portions of the Tyrrhenian Sea. Well over two-thirds of the Mediterranean Sea is covered by high quality, precisely controlled sounding lines 20 km or less apart. Only portions of the Levantine Sea are relatively sparsely covered.

These data formed the essential framework for the evaluation of other sounding data collected by ships (U.S. Navy, Merchant Marine and others) utilizing less precise depth recorders or less precise navigational control. Based on this evaluation the topography shown on the lesser quality data was compared to the basic framework. Depending on the fit the data from less precise sources were either rejected or utilized to enhance and extend the interpretation of bottom topography.

In areas where the precise data did not cover the shelf, an extensive examination of existing and available nautical charts was conducted to determine shelf widths, shelf edge depths, and the relationship of bathymetric features to land features. The reliability of these charts varied according to the extent of coastal hydrographic surveys conducted in the area.

Many descriptive texts and good to excellent charts of the Mediterranean Sea such as those by Mikhaylov (1965) and Ryan *et al.* (1970) were consulted. All references consulted cannot be cited specifically in the limited space available. Among others, however, are the bathymetric charts published by the Musée Océanographique de Monaco (Bourcart, 1958a, b, 1960a, b) which were especially useful for nearshore bathymetry in portions of the Balearic Sea. Charts published by the Italian Hydrographic Office were helpful for interpretation in the Adriatic Sea and portions of the Tyrrhenian Sea (Angrisano and Segre, 1969; Anonymous, 1967a, b). The Spanish Oceanographic Institution charts (Oliver, 1960, 1961, 1968a, b; Massuti, 1967) proved useful around the Balearic Islands and the coast of Spain.

Other published bathymetric charts cover the Algerian continental margin (Rosfelder, 1955); the Ligurian, Tyrrhenian and Adriatic Seas (Debrazzi and Segre, 1959a, b); portions of the western Mediterranean (Giermann, 1962, 1968); and the eastern Mediterranean (Pfannenstiel, 1960).

Twelve charts (Figure 1) were compiled at the Naval Oceanographic Office on a mercator projection at a scale of 1° longitude = 4 inches (10.2 cm). From these charts a composite was constructed at a scale of 1° longitude = 1 inch, or 2.54 cm (see map in sleeve). The contour interval is 200 m assuming 1500 m per second sound velocity. The bathymetry of the Black Sea is from a Woods Hole Oceanographic Institution (Ross *et al.*, 1972) compilation which was corrected for sound velocity according to Matthews Tables.

The physiographic maps constructed for this paper are based almost exclusively on the large bathymetric chart. Each map is accompanied by a discussion which emphasizes the highlights of that area.

As a result of the large amount of precisely collected and controlled data, certain previously unknown or undiscussed characteristics of the Mediterranean sea floor become apparent. These include: variations in shelf and slope characteristics; newly discovered features in some areas, particularly the Tyrrhenian Sea; a different concept and configuration of the Mediterranean Ridge; and ridges and valleys on the Nile Fan which were previously thought to be abyssal hills.

Nomenclature and terminology utilized on the physiographic maps and in the discussions adhere to names and definitions approved by the U.S. Board on Geographic Names (Anonymous, 1969). *Names printed within quotation marks have not yet received approval by this group and are subject to change or rejection.* For the purpose of this discussion, we have subdivided the Mediterranean into seven parts which coincide approximately with recognized major physiographic provinces, or seas.

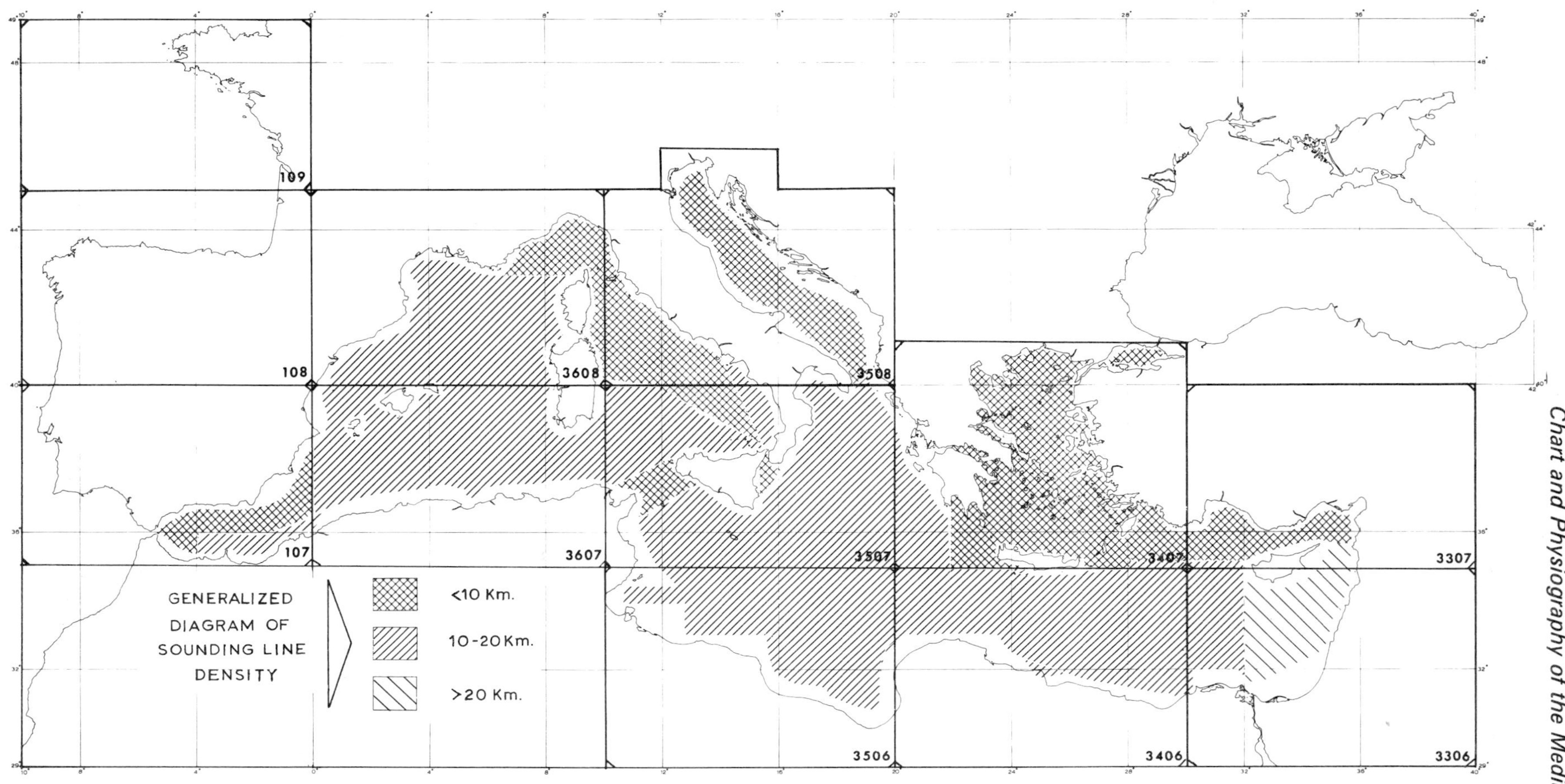

Figure 1. Diagram showing Mediterranean Sea chart coverage and sounding line density (chart of the Mediterranean in sleeve at end of volume).

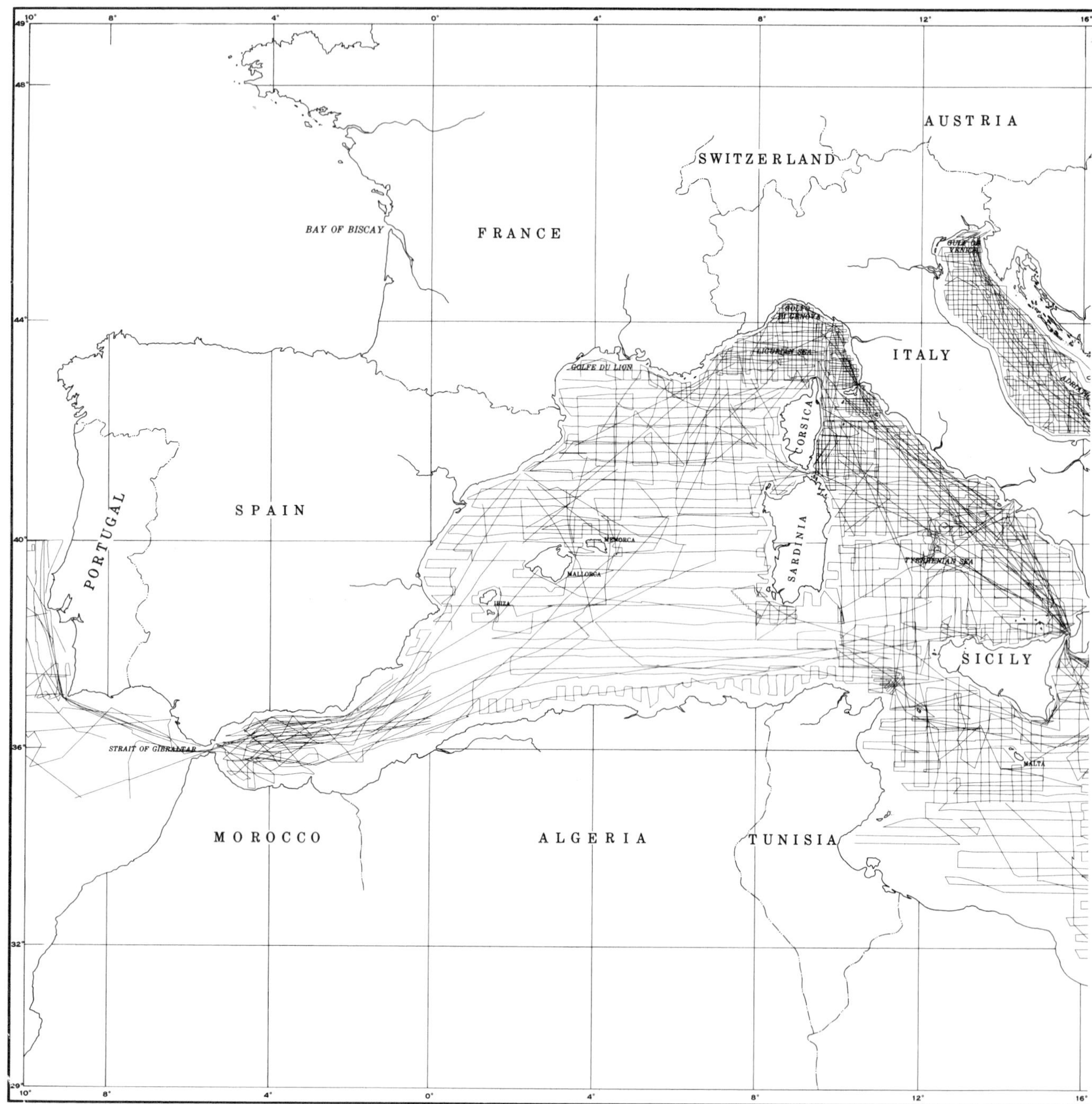

Figure 2. Track chart of precision sounding data in the Mediterranean Sea. Sources of tracks shown: U.S. Naval Oceanographic Office, Woods Hole Oceanographic Institution and Saclant ASW Research Center, La Spezia, Italy. All data reduced to uncorrected soundings in meters, assumed velocity 1500 m/sec.

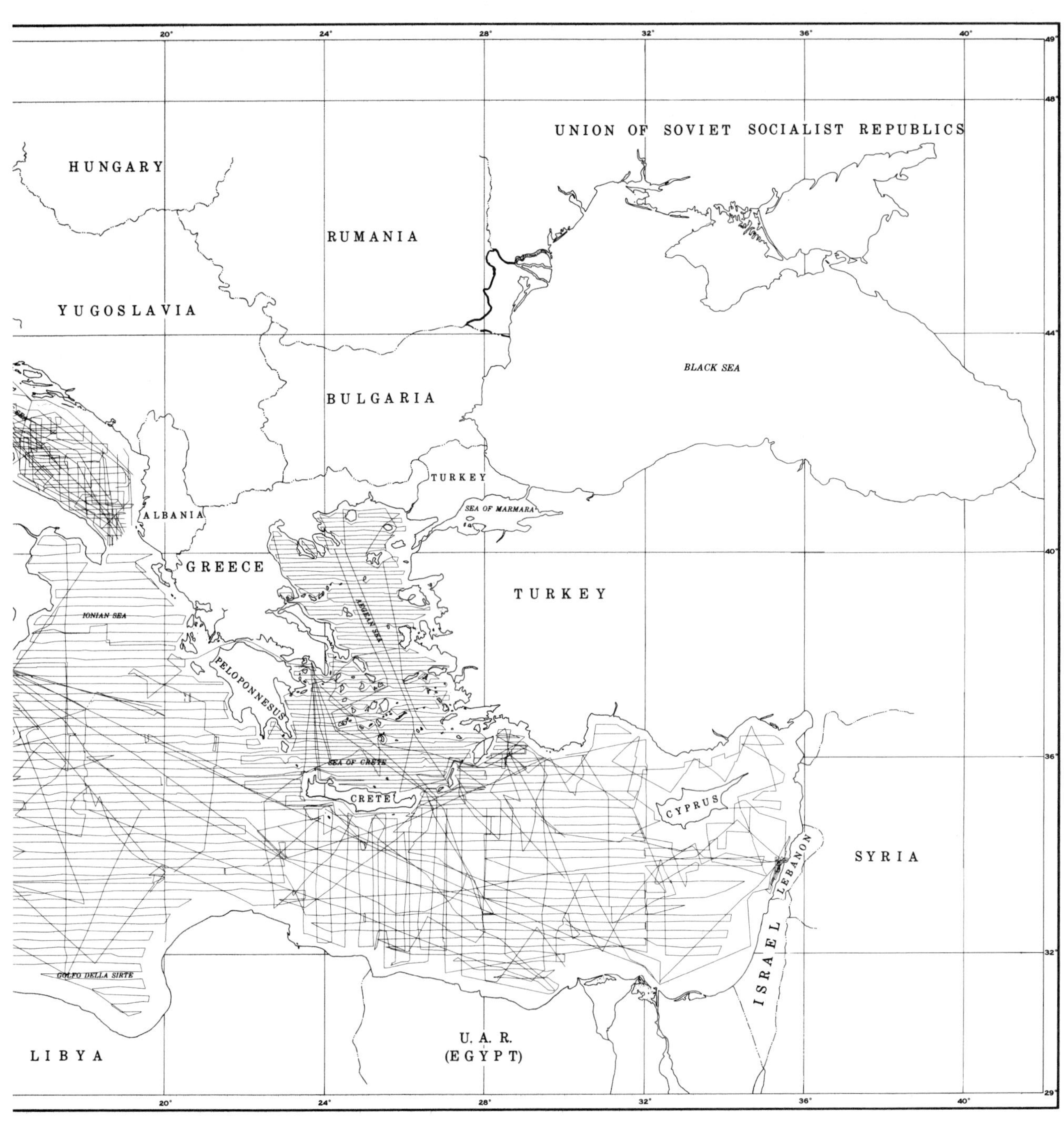

20°
24°
28°
32°
36°
40°
49°
48°
44°
40°
36°
32°
29°
UNION OF SOVIET SOCIALIST REPUBLICS
HUNGARY
RUMANIA
YUGOSLAVIA
BULGARIA
BLACK SEA
TURKEY
SEA OF MARMARA
ALBANIA
GREECE
TURKEY
IONIAN SEA
PELOPONNESUS
SEA OF CRETE
CRETE
CYPRUS
SYRIA
LEBANON
ISRAEL
GOLFO DELLA SIRTE
LIBYA
U. A. R.
(E G Y P T)

PHYSIOGRAPHY OF THE WESTERN MEDITERRANEAN SEA

The Alboran Sea (Joseph P. Flanagan)

The Alboran Sea is located in the westernmost sector of the Mediterranean Sea between the Strait of Gibraltar on the west and the Balearic Sea on the east (Figure 3). Approximately 54,000 square km in area with a maximum depth slightly less than 1500 m, it is bordered on the north by Spain and on the south by Morocco and western Algeria. Geologically it is situated between the Betic Cordillera on the Iberian Peninsula and the Moroccan Rif and Atlas Mountains of North Africa. The boundaries of the Alboran Sea as defined by the International Hydrographic Organization are the eastern limits of the Strait of Gibraltar and a line joining Cape de Gata, Spain to Cape Fegalo, Algeria. A section of the continental rise associated with the Algerian Basin falls within these limits and will be included here. A description of the Strait of Gibraltar is discussed elsewhere in this volume (Kelling and Stanley, this volume).

Continental Shelf

The continental shelf around the Alboran Sea can be divided into an upper shelf out to a first shelf break and a lower shelf-like area out to the continental slope.

The upper shelf may also be divided into four sections of contrasting width. From the Strait of Gibraltar northeastward to Cape de Gata the shelf width varies from a maximum of 10 km off Malaga to a minimum of 2 km off Cape Sacratif with 5 km as an average. The depth of the shelf break is about 100 m. Along the coast of North Africa from the Strait of Gibraltar to Point Busicur, Morocco the shelf is uniformly narrow (an average width of 6 km). The depth of the shelf break is 100 m. From Point Busicur to Cape de Tres Forcas the shelf width fluctuates from a minimum of 3 km to a maximum of 18 km. The depth of the shelf break is also 100 m. From Cape de Tres Forcas to the eastern limit of the Alboran Sea at Cape Fegalo the shelf is wide and nearly uniform with an average width of 15 km. The depth of the shelf break here is 150 m.

Marginal Ridges and Plateaus

The lower shelf-like area off the North African coast contains an extensive area of varied topography. Between Point Almina and Xauen Bank there is an elongate, gently sloping marginal plateau. This marginal plateau is approximately 92 km long and 27 km wide and extends down to 600 m in depth. Between Xauen Bank and Cape de Tres Forcas there is a small marginal plateau and a large ridge containing Xauen and Tofiño Banks and Alboran Island. This ridge, also called the "seuil d'Alboran" (Giermann *et al.*, 1968), is oriented in a northeast-southwest direction

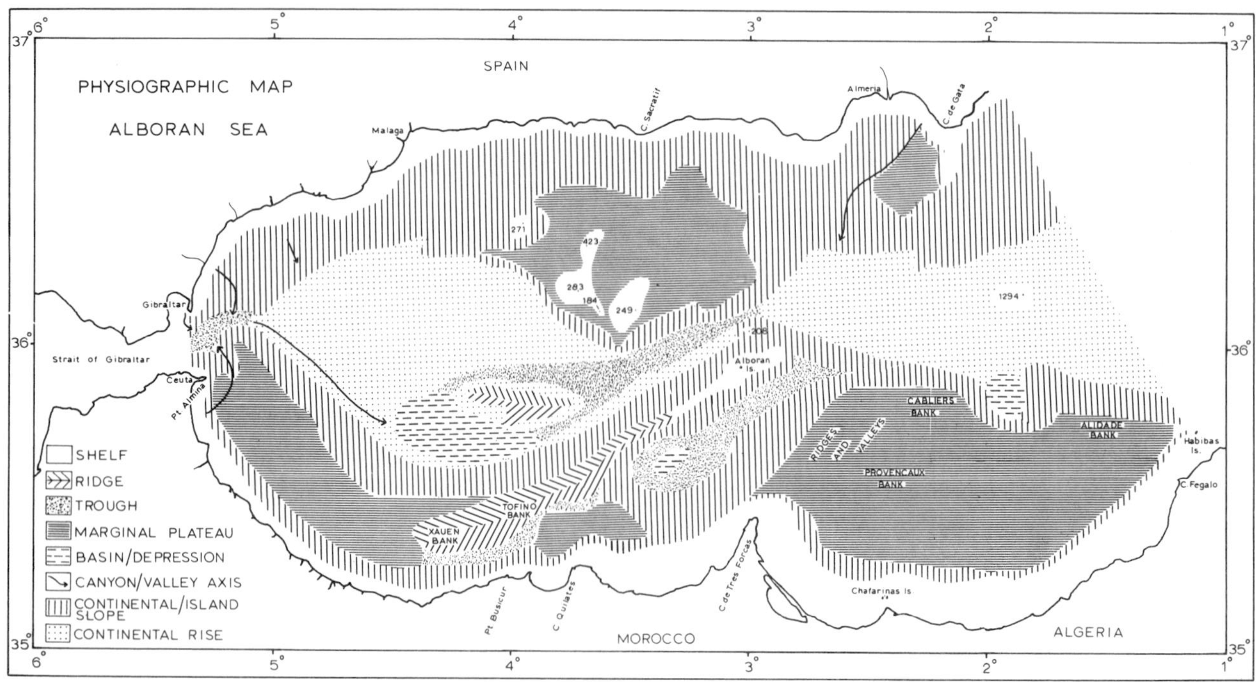

Figure 3. Physiographic map of the Alboran Sea and westernmost Mediterranean.

and is approximately 160 km in length. The southwestern extremity of the ridge, with the banks, is separated from the Alboran Island platform by a saddle, and is separated from the North African shelf by a narrow trough. This trough, termed the "South Alboran Graben" by Giermann (1962), is over 400 m in depth and parallels the shoreline for a short distance before turning northeast to follow the trend of the Alboran Ridge. The northeast-southwest trend of the Ridge is indicative of one of the faults postulated for the western Mediterranean Sea by Bourcart (1960).

Between Cape de Tres Forcas and Cape Fegalo there is a large marginal plateau. This plateau, approximately 135 km by 63 km, can be divided into a western and eastern section. The western section consists of several ridges and valleys that trend almost north-south and plunge toward the continental slope.

Provençaux Bank and Cabliers Bank are found on the easternmost ridge. The eastern section is, except for two banks in the northeast corner, a featureless area that grades gently downward from the shelf edge to its termination at about the 1000 m contour. Two marginal plateaus also exist off the coast of Spain. There is a small plateau southwest of Cape de Gata that is bordered on the west by the Almeria Canyon. Between 3° and 4° west longitude there is a large plateau, about 80 km by 55 km in dimensions, that contains several peaks of volcanic origin (Giermann et al., 1968). The lower boundary of the plateau is defined approximately by the 900 m contour. Below is a steep lower continental slope that leads into a large trough, the Alboran Trough, which connects the western Alboran Basin[1] with an area of the Algerian Basin to the east. This trough is the "Strait of Alboran" of Giermann (1962).

Continental Slope

The gradient of the continental slope off Spain is relatively constant with 1:30 being a representative value. The slope between 3° and 4°W consists of a steep upper and a very steep lower slope interrupted by the previously mentioned marginal plateau.

The continental slope and Alboran Island slope off North Africa are narrow. The gradients are steep and vary from about 1:15 north of Xauen Bank to 1:9 north of Alidade Bank. An anomalously steep slope east of Cabliers Bank leads from a marginal plateau down to a small depression that has a maximum depth of about 2400 m. This depression, believed to be a caldera, has an open silled northern side. There is also a sill in the substratum and the depression contains a thick accumulation of sediments (Giermann *et al.*, 1968).

Basins

The Alboran Sea actually includes an eastern and a western depression divided by the northeast-southwest trending Alboran Ridge (Giermann, 1962). The western section, called the "Western Alboran Basin" by Stanley *et al.* (1970), is the larger of the two and is defined approximately by the 1400 m contour. This basin is 40 km in length and 22 km in width and its maximum depth is slightly greater than 1500 m. The floor of the basin has been smoothed by sediments derived from many different sources. An important northwest source has been suggested by the presence of the Gibraltar and Ceuta Canyons as well as the Gibraltar Valley (Stanley *et al.*, 1970). Additional sources of sediment are the coasts of Spain and Morocco, Xauen and Tofiño Banks, and to the northeast, the Alboran Ridge (Stanley *et al.*, 1970). The western basin is separated from the Alboran Trough to the northeast by a small east-west trending ridge. This ridge is approximately 32 km in length and consists of several separate peaks. The western basin is connected to the Alboran Trough both north and south of the ridge. The trough continues eastward down to a depth of 1800 m due north of Alboran Island where it joins the Algerian Basin.

The eastern section, called the "Eastern Alboran Basin" by Stanley *et al.* (1970), is a small graben-like feature (Giermann, 1962) and is defined by the 1200 m contour. The basin is approximately 13km long and 6 km wide and the maximum depths are slightly greater than 1200 m. This basin, like the "Western Alboran Basin," has a northeast-southwest trending trough adjoining it. This trough is terminated at the continental slope that leads down to the Algerian Basin.

Alboran Island

Alboran Island is a small volcanic island, 600 m long and 250 m wide at its widest point. The coast of the island is steep with cliffs rising as high as 12 to 13 m above sea level. The base of a Quaternary beach is noticeable in several places and the island is surrounded by a wide intertidal plateau about 25 to 30 m in width. The surface is gently inclined in a south-north direction and the configuration is smooth and represents a typical marine abrasional platform (Hernandez-Pacheco and Asensio Amor 1968).

[1] The term *Basin*, when capitalized, refers to a specific, generally (but not always) large, morphological depression. There may be several such basins within a single sea (example: the "Western Alboran Basin" and "Eastern Alboran Basin" in the Alboran Sea). The Algerian Basin is the largest such feature in the Mediterranean, and forms at least half of the sea floor in the Balearic Sea.

Balearic Sea and the Algerian Basin (Terence G. Carter)

The largest physiographic province of the Western Mediterranean Sea is the vast, featureless area known as the Algerian Basin in the Balearic Sea (Figure 4). The basin is also called the Algero-Provençal Basin. The 2600 m isobath rather markedly delineates the Basin limit. The area of the Basin and related physiography discussed in this section is approximately 240,000 square km with a maximum depth of 2801 m. The configuration of this area is roughly triangular terminating at the continental rise in the southwest, the base of the Algerian-Tyrrhenian Trough in the southeast, and the Ligurian Canyon in the northeast. The projection of the Balearic Island Chain trending southwest-northeast interrupts this triangular uniformity.

Continental Shelf

The continental shelves in this area vary greatly in width and character. Off eastern Spain the shelves underlying the Gulf of Alicante and the Gulf of Valencia are broad and vary from 27 to 63 km (near the mouth of the Ebro River) in width. Northeast of Cabo Tortosa the shelf narrows to an average width of 18 km for some 120 km, then widens into a series of lobate structures averaging 36 km in width which have been created by canyon incisions. The shelf attains its

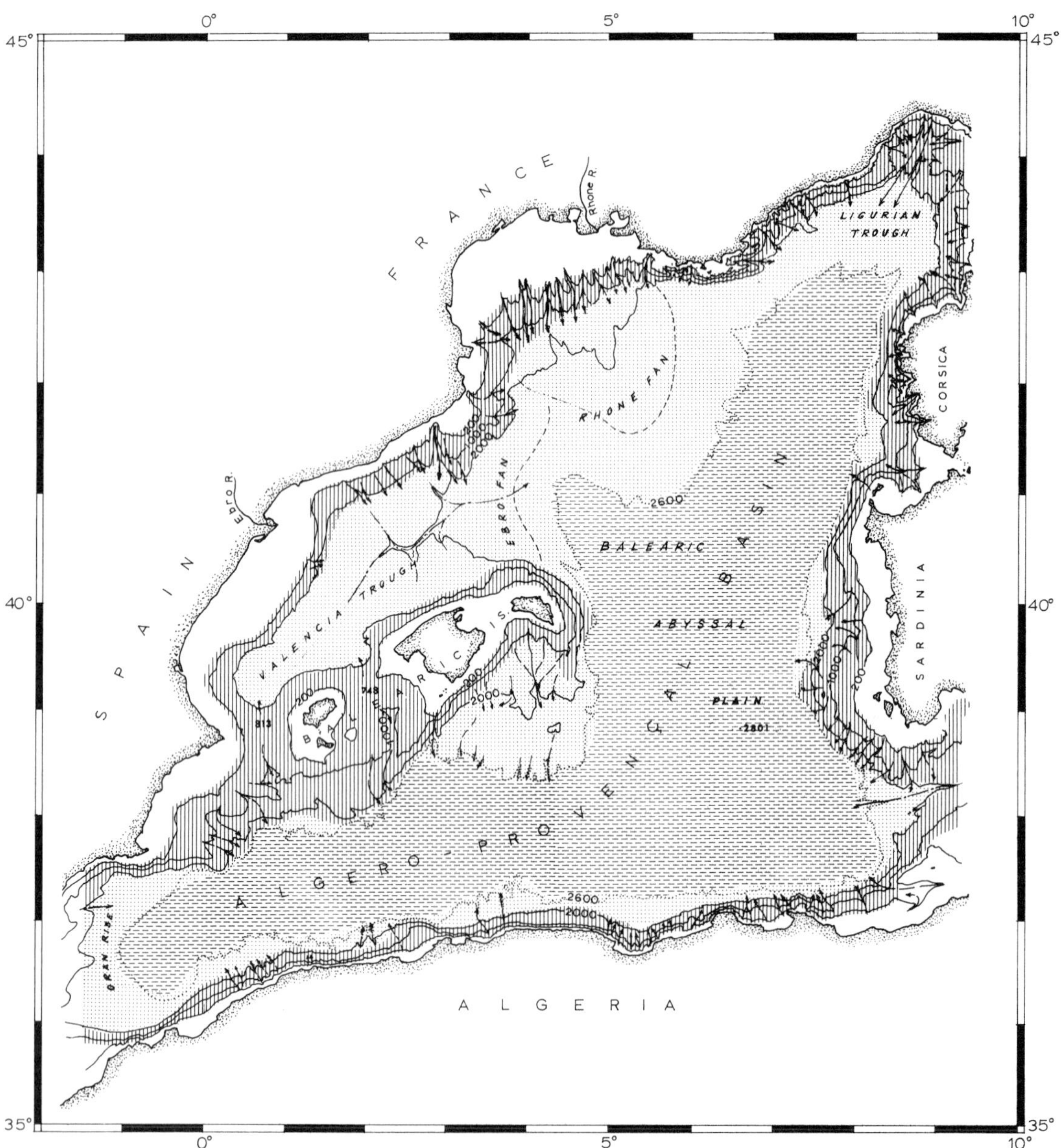

Figure 4. Physiographic map of the Balearic Sea and Algerian Basin.

maximum width, 72 km, in the region of the Gulf of Lion. Bourcart (1959) has compared this shelf with that of the eastern United States between New York and Florida, noting that both have undergone subsidence since Mesozoic time. From Toulon to Genoa the shelf is extremely narrow averaging 3 to 9 km in width and incised by numerous canyons.

With the exception of the region between Cape Corsica and Valinco, the island shelves of western Corsica and Sardinia are relatively broad averaging 30 to 50 km in width.

The shelf along the African coast varies in this region from a maximum width of 54 km at Cape Serratt to progressively narrower widths of 4 to 20 km west to the continental rise.

The islands of Mallorca and Menorca have a common shelf. The island of Ibiza has its own shelf and is separated from the other islands and the Spanish shelf by broad sills reaching depths of 748 m and 813 m on the eastern and western flanks respectively.

Continental Margins, Slopes, and Rises

Overall, continental slopes around the periphery of the Algerian Basin are steep and deeply dissected. In some areas, for example, west of Corsica and Sardinia and southeast of the sill between the islands of Mallorca and Ibiza, there is no apparent continental rise; rather an abrupt transition from basin floor to continental slope. Also in evidence on the slopes are terraces, prominent spurs and numerous canyons.

Slopes adjacent to the young mountain ranges of the Pyrenees and Maritime Alps, the Atlas Mountains of North Africa, and southeast of the Balearic Island Chain are the steepest and most complex.

The continental rise forms the southwestern terminus of the Algerian Basin. Clockwise from the region, south of the Spanish coast from Cartagena to Cabo de Palos, the slope is precipitous trending east-west. It is called the "Mazarron Escarpment" by Ryan *et al.* (1970). The continental margin from Cabo de Palos north to the Gulf of Valencia is the region of the Balearic Island Ridge. This Ridge was probably created by a rotational shifting from the Spanish mainland (Vogt *et al.,* 1971). The majority of slopes on the Ridge have rather moderate gradients with the exception of the sectors east and northeast of the island of Menorca and south of Mallorca, called the "Emile Baudot Escarpment" by Ryan *et al.* (1970).

In the region of the Gulf of Lion, the continental slope is indistinct, incised by numerous canyons and valleys extending for many kilometers seaward into the basin proper. The gradient of the slope is the most moderate of all continental/island slopes in the region, reflecting the high rates of sedimentation in front of the mouth of the Rhône River. The morphology of the Rhône Fan on the northern margin of the Algerian Basin is detailed in Menard *et al.* (1965).

From Toulon to Genoa the slope is steep and incised by numerous canyons. The continental rise is broad and of low gradient.

The continental slope around the Ligurian Canyon is complex and of low gradient. Two very prominent parallel canyons, whose axes are about 9 km apart, head at the shelf south of Genoa and are traceable into the deep basin.

The island slopes off western Corsica and Sardinia are of lesser gradient than the typically steep slopes bordering most of the region. Canyons, though numerous, are less prominent and of shorter length. Some are continuations of terrestrial valleys (*e.g.,* the Ajaccio Valley). Island rises are often absent and the abyssal plain abuts the slope for many kilometers.

The African continental slope rising from the Algerian-Tyrrhenian Trough is of low gradient. However, the gradient increases to the west and the slope is incised by numerous canyons. The continental rise is restricted.

Basins and Troughs

Centrally located within the Algerian Basin is the Balearic Plain, a region of virtually no slope (1:3000). The floor of the Basin genetically constitutes a plain of sediment accumulation, analogous to the deep water basins of the Black Sea, Southern Okhotsk Sea, and the Japan Sea (Mikhaylov, 1965).

Although the contour interval (200 m) of the bathymetric chart precludes their portrayal, numerous knolls of low relief and small areal extent (1 to 3 km in diameter) are widely distributed on the basin floor. Seismic reflection profiles recorded by various ships traversing the area indicate similar structures buried by sediments. These have been identified as diapiric structures (Hersey, 1965b). The properties of these structures have been described (Ryan *et al.,* 1970).

Between the Balearic Islands and the Spanish mainland the sea floor is characterized by the broad U-shaped Valencia Trough, the axis of which is inclined to the northeast. A dendritic-like channel system is in evidence near the trough terminus, the Ebro Fan, suggesting active bottom scour possibly by turbidity currents.

The sill between southern Sardinia and Tunisia separates the Algerian and Tyrrhenian basins. A conspicuous V-shaped channel trending east-west lies roughly midway between the aforementioned land masses. Sill depths along the channel axis are slightly less than 2000 m (based on the prime sounding data used to compile the bathymetric chart).

PHYSIOGRAPHY OF THE CENTRAL MEDITERRANEAN SEA

The Tyrrhenian Sea (C. Reed Jones)

The Tyrrhenian Sea is bounded on the west by the Corsica-Sardinia Island chain; on the north and the east by the mainland of Italy; and on the south by Sicily and the Afro-Sicilian borderland (Figure 5).

The sea is approximately 231,000 square km in area, has a maximum depth of 3557 m and is enclosed except for four passageways to other Mediterranean waters. The main passage, south of Sardinia, leads to the Algerian Basin and is a narrow, deep trough with a minimum depth slightly less than 2000 m. Another passage, the Strait of Messina, with a sill depth of about 100 m (Leroy, 1966), leads to the Ionian Sea. The northern passage, the Tuscany Trough, leads to the Ligurian Sea and has a sill depth between 300 and 400 m. Between Corsica and Sardinia, the Strait of Bonifacio gives access to the Algerian Basin and has a minimum depth less than 40 m.

The structures in the Tyrrhenian Basin have been described as resulting from extensive foundering of continental crust during middle Pliocene (Heezen *et al.,* 1971; Zarudzki, 1971). Zarudzki describes circum-Tyrrhenian normal faulting and tilting of blocks, resulting in a basin and range topography that controls the flow of terrigenous sediments to the abyssal plain. Sediment ponding varies from individual, small basins to large elongate terraces, depending on the original structural configuration and subsequent sediment fill.

The Tyrrhenian Basin can be subdivided on the basis of morphology into a northwestern and a southeastern half. The dividing line between the sections extends southwesterly from Cape Circeo to the Algerian-Tyrrhenian Trough.

The northwestern Tyrrhenian Basin is characterized by broad, continuous, multi-level, marginal terraces, troughs and ridges, and by the scarcity of obvious volcanics.

The southeastern Tyrrhenian Basin is characterized by complex, broad slopes with volcanic structures and small, discontinuous areas of sediment ponding. The basin plains are located in this area. Along the southern margins are barrier volcanic structures which give rise to fairly extensive upper slope terraces.

The type and distribution of topographic features in the Tyrrhenian Basin are best seen by studying the physiographic diagram (Figure 5). To enhance the understanding of the height and depth of many features, the boundary lines have been annotated with depth values.

Shelf and Borderland

The continental shelf is normally defined as the relatively shallow sea floor adjacent to a land mass where the gradient is less than 1:40. The shelf edge, or break, is the point at which the gradient steepens and becomes greater than 1:40. Around the Tyrrhenian Basin, the equivalent of the shelf is often steeper than 1:40 gradient and the equivalent of the shelf break may be absent. There are no established terms for the shelf and shelf break when the usual gradient is exceeded, so the term 'shelf' and 'shelf break' will be used to denote this condition. On the physiographic diagram (Figure 5), the 'shelf break' is indicated by a dotted line. The figures for shelf width and depth given here will generally represent the average rather than the extremes seen at headlands and bays.

In general, the east Sardinian island shelf is separated into two parts by the Gulf of Orosei. The southern part of the shelf has a relatively consistent width of 4 km and a break at a depth of 100 m. The shelf off Corsica and northern Sardinia varies in width from 5 to 25 km, with a break at depths of 60 to 180 m. Along the Italian coast the shelf narrows in width from 65 km around the Tuscany Islands to 5 to 10 km at the Gulf of Policastro and southward.

There are no normal shelf developments or obvious gradient changes noted south of the Island of Montecristo, around the Gulf of Policastro, and south of the Gulf of Sant'Eufemia. The transition from broad to narrow shelves takes place rather abruptly at Cape Palinuro. From La Spezia down to Cape Miseno the shelf break varies in depth from 100 to 200 m. South of Cape Miseno to Cape Vaticano, the shelf break ranges from 100 to 400 m in depth.

Along northern Sicily, the shelf is generally 3 to 6 km wide and the shelf break is at depths of 100 to 140 m. Off the western end of Sicily, the shelf becomes very broad where it merges with the borderland sometimes called the "Afro-Sicilian Platform". This borderland extends toward Africa and Sardinia and is a complex area of shoals, ridges, canyons, and small basins.

Continental and Island Slopes

The slopes around the Tyrrhenian Basin are complex and best understood by studying the physiographic diagram (Figure 5). The slope along Corsica and Sardinia consists of a southern half typified by large, north-south terraces and a northern half in which north-south ridges and troughs are more prominent. This transition in morphology takes place

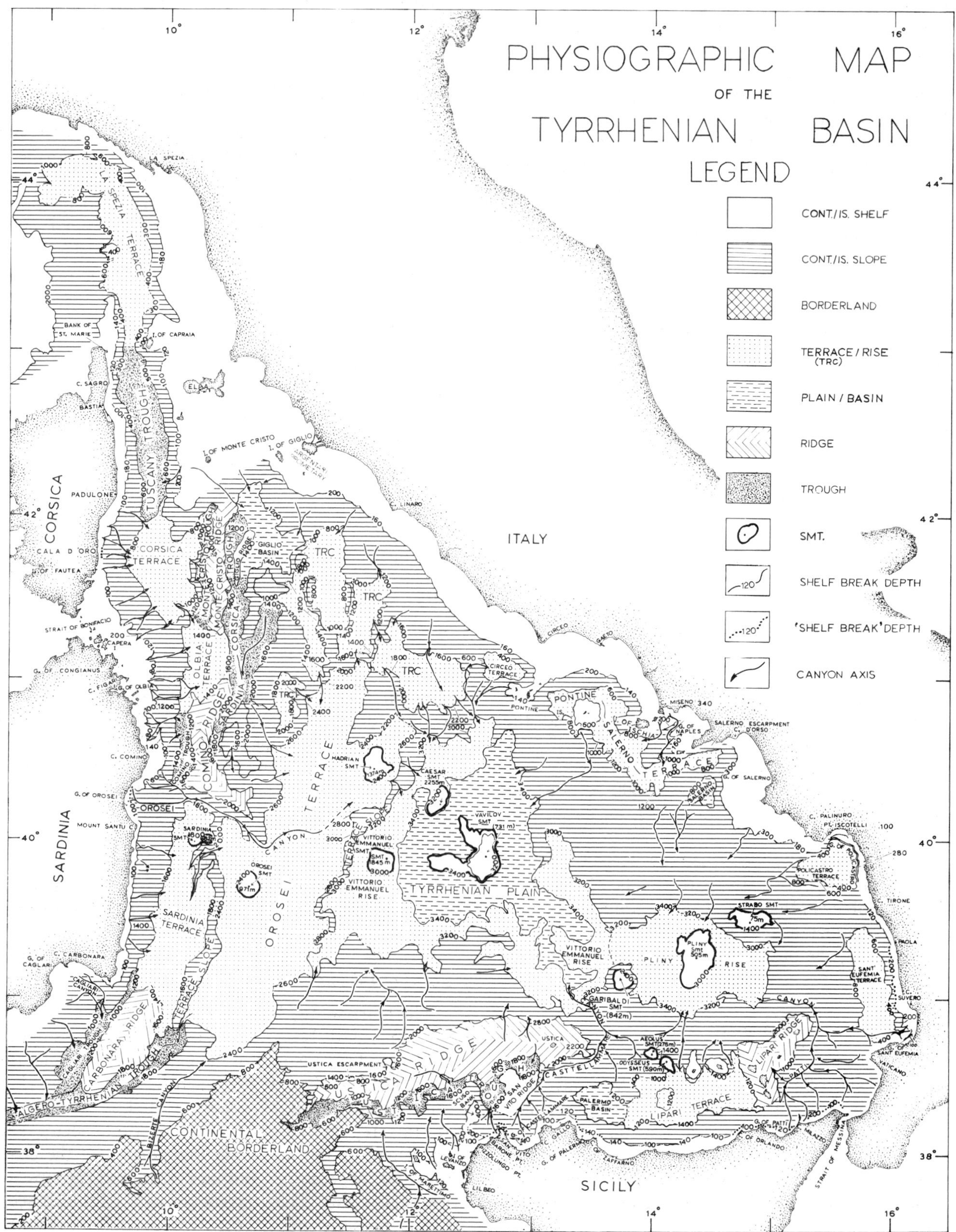

Figure 5. Physiographic map of the Tyrrhenian Sea.

principally across the Orosei Canyon axis. The location of Orosei Canyon may be tectonically significant, since the shelf character off Sardinia also changes in line with the canyon axis.

The complex slope off Corsica and Sardinia is about 180 km wide, or nearly one-third the width of the Tyrrhenian Basin. The northward trending ridges and troughs of this slope continue into the slope off the Italian mainland. Thus, while one can gain a concept of a basin margin existing on the Corsica-Sardinia side, this concept is lost where the oblique trends merge with the Italian basin margin.

Southward, along the Italian mainland, the presence of a shelf above, and a rise and plain below, delineates the complex slope. Along northern Sicily, terraces, ridges, and troughs trend east-west, paralleling the coast.

Rises and Plains

The upper limit of the continental or island rise in the Tyrrhenian Basin is delineated by large terraces and complex slopes. Generally, the rise begins at depths of 3000 to 3200 m and extends down to 3400 m.

The rise area may be divided into an eastern and a western part by the presence of a north-south sill just north of Garibaldi Seamount. The sill depth here is 3283 m. The eastern section, contains a broad, shallow trough which encircles the western margin of Pliny Seamount. The western section, encircles the Tyrrhenian Plain on three sides and includes two large seamounts, the Vittorio Emmanuel and the Caesar.

The Tyrrhenian Plain surrounds the Vavilov Seamount, is delineated here by the 3400 m contour, and has a maximum depth of 3557 m in its western part.

A detailed survey described by Hersey (1965a) shows that the northern part of the Tyrrhenian Plain is bilobate in character. Contours at 5 m intervals show a channel a few meters deep connecting a northeastern lobe of the plain with the deeper northwestern lobe.

Adriatic Sea
(Jack H. Rebman)

The Adriatic Basin, an arm of the Mediterranean Sea with an area of approximately 135,000 square km and a maximum depth of 1230 m, is situated between the Italian Peninsula and the Dalmatian coast of Albania and Yugoslavia. It is contiguous to the south with the Ionian Basin at the Strait of Otranto.

On the basis of the sea floor morphology, the Adriatic Plain can be divided into two provinces: a northern and central continental shelf and a southern basin. The physiographic diagram (Figure 6) portrays this distinction.

North and Central Shelf Province

A continental shelf underlies the whole of the northern and central Adriatic Sea. The northern continental shelf extends from the Gulf of Venice 330 km southeastward to the Mid-Adriatic depression. Shelf gradients along this section range from 1:3000 between the Gulf of Venice shoreline and the 100 m isobath, to 1:500 between the 100 and 150 m isobaths. Precision echosounders (Figure 6, profiles A-B and A-C) show a smooth terrace trending northwest-southeast along the Italian coast. This terrace, composed of mud, can be traced from 44°N to 41°50′N. The smooth terrace is in contrast to the uneven sandy bottom farther seaward (Figure 6, profiles A-B and A-C).

The central continental shelf includes the Mid-Adriatic depression at the north end, a broad shallow channel connecting the depression with a southern basin, and knolls and islands scattered about on the shelf on each side of the connecting channel.

The Mid-Adriatic depression (Figure 6, profile I-I′) trending northeast-southwest, is bounded on the northwest by a 1:100 gradient slope 90 to 110 m high and is divided into three separate depressions. A northwest-trending ridge has 58 to 75 m of relief between the middle (266 m deep) and the southwestern depression (248 m deep). The middle and northeastern depression (241 m deep) are separated by a knoll with a minimum depth of 198 m (Figure 6, profile E-E′). This knoll is in line, to the southeast, with Jabuka Reef and the islands of Sveti Andrija, Vis, and Korcula.

The broad shallow channel connecting the Mid-Adriatic depression with the southern basin is defined by the 160 m isobath; a sill depth of 163 m is located at 42°46′N and 15°39′E.

Southern Basin

The morphology beneath the southeastern Adriatic Sea consists of broad low relief, continental shelves, slopes, and rises. South of the Gargano Peninsula, the shelf has a width of 80 km, narrows to 20 km at 41°N and 17°30′E, then broadens to 30 km at 40°40′N. In this region the shelf break is quite sharp and consistent at 180 m. Along this portion of the coast the 100 m contour outlines scattered groups of irregular crests. These crests consist of lithified sediments, possibly ancient coral reefs partially covered by recent terrigenous materials (Giorgetti and Mosetti, 1969).

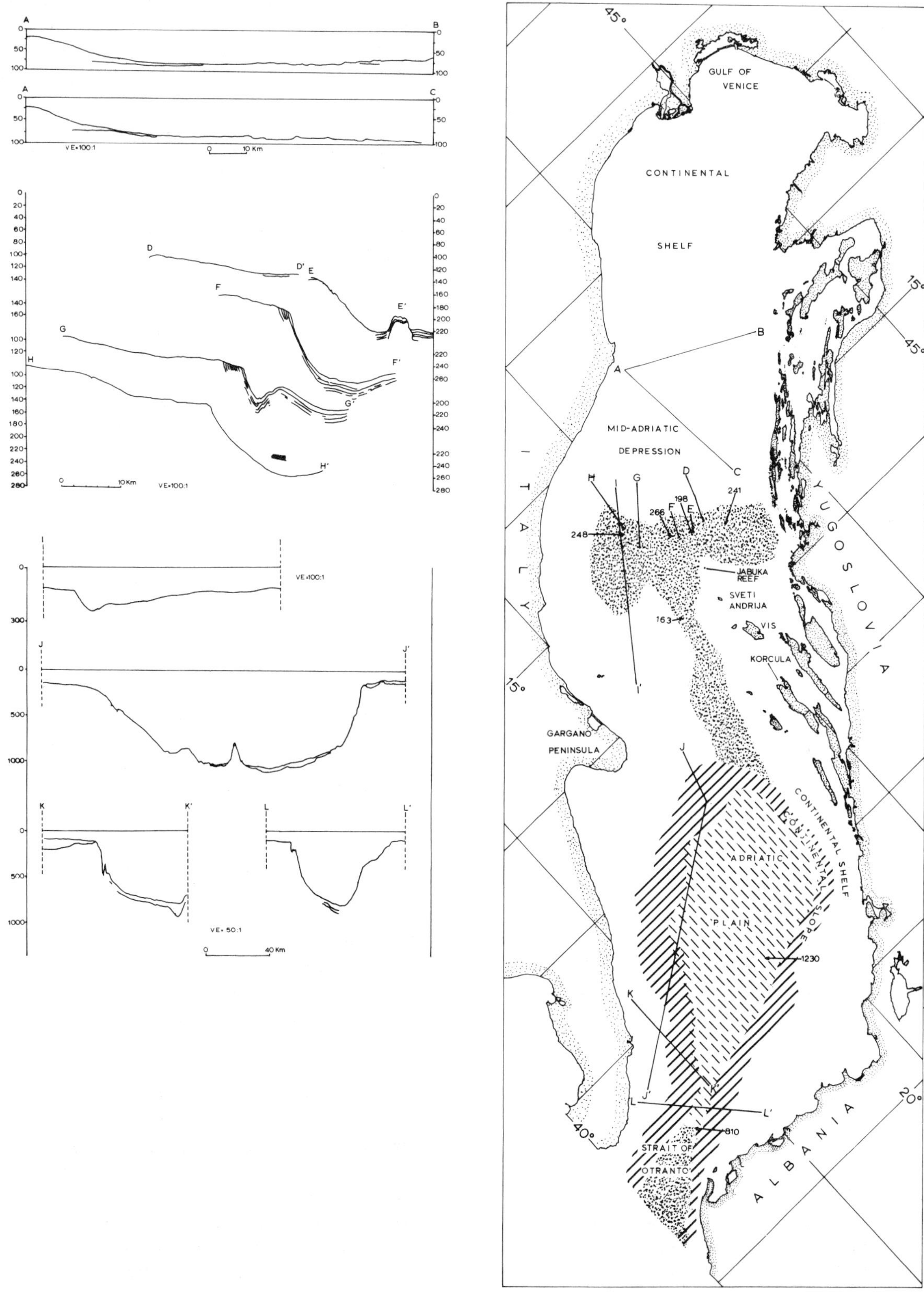

Figure 6. Physiographic map and profiles of the Adriatic Sea. Sections A through H reproduced from van Straaten, 1965; sections I through L from Giorgetti and Mosetti, 1969.

Along the Albanian and southeastern Yugoslavian coast, the shelf has a maximum width of 75 km at 41°50′N narrowing to 20 km at 42°40′N. The shelf break is at 400 m and the gradient is 1:60 to 1:100 on the shelf between 100 and 400 m. On the continental slope the gradient is 1:20 to 1:30 between 400 and 1000 m. The 1000 m isobath defines what has been called the Adriatic Plain. The lowest gradients of this plain are not sufficiently flat to be consistent with the term "abyssal plain" as defined by Heezen (Heezen *et al.*, 1959). The maximum depth of the basin is 1230 m. The Adriatic Basin is separated from the Ionian Sea by a sill depth of 810 m at 40°35′N and 18°54′E.

Ionian Sea
(Francis L. Marchant)

The Ionian Sea is located in the central part of the Mediterranean Sea. It extends north to Sicily, Italy and Greece, connecting to the Adriatic Sea by the Strait of Otranto; east to a line from Akra Kios south to 34° North and southwest from there to Ras Aamer; south to Libya and Tunisia; and on the west includes the Straits of Sicily. (Figure 7). It is approximately 616,000 square km in area and has a maximum depth of 5093 m, which is the greatest recorded depth in the Mediterranean Sea (Hersey, 1965).

Continental Margins

A large, shallow sea floor between Tunisia and Sicily is a type of continental borderland. Its northwestern limit is a line across the Straits of Sicily from the west side of Cap Bon to Adventure Bank.

Adventure Bank, the Malta Rise, and an intermediate section of narrow shelf are adjacent to the borderland on the Sicilian side. Along this narrow shelf are breaks at various depths between 80 and 190 m. The break is progressively deeper to the east as shown on the physiographic diagram (Figure 7).

The widest and largest shelf in the Mediterranean exists along the coast of Tunisia. Gradient changes equivalent to the shelf break occur between 90 m and 130 m. Below these breaks the bottom gradient again flattens into broad shelf-like areas down to approximately the 240 m isobath. There are several troughs in this area.

The borderland is terminated on the south and southeast by the westward extension of a broad depression related to the Surt (Sirte) Rise. This boundary varies from one based on a change in contour direction near the Gulf of Gabes to a well defined slope break south of the Malta Rise. Near the Malta Rise the borderland terminates along the southward extension of the Malta Escarpment.

The graben-like troughs in the borderland trend northwest-southeast. Additional smaller depressions elsewhere in the borderland area trend in the same direction suggesting widespread distribution of the basic structural trend beneath a sediment cover. The Malta Trough has been sounded to a depth of 1702 m and is the deepest and largest trough in the borderland.

Two broad shelf breaks are often observed off the Libyan coast. The shelf break as normally defined (greater than 1:40) lies at a depth of 250 to 300 m. Between Tripoli and Bengasi a distinct slope break at less than 1:40 gradient is noted at depths of 80 to 110 m. Furthermore, a number of smaller terraces and breaks in slope are present between each major break, particularly near Tripoli.

The shelf is extremely narrow off Southern Italy. Mikhaylov (1965) describes the shelf as 16 km wide and the break at 100 m in the Gulf of Taranto. This study shows a shelf varying from 2 to 7km wide and from 30 to 130 m deep.

Continental Slope and Rise

The Surt Rise occupies theentire continental margin below the shelf in the Gulf of Surt (Sidra). The lower limit of the rise is generally well delineated by the 3600 m contour. A northeast-southwest trending slope separates the rise from part of the borderland to the northwest. The east-west Malta Ridge separates this slope from the Malta Escarpment. The Herodotus Seamount occupies a somewhat similar position relative to the rise margin near Bengasi. There are several northwest-southeast trending gradient changes at the margin of the rise and shelf around 15° East longitude, which are independent of gradient changes related to sediment distribution. These changes align with the south edge of the depression extending toward the Gulf of Gabes and the southeastern edge of the Gulf of Surt suggesting possible fault control.

A highly dissected, low gradient rise is present below the shelf along the southeastern coast of Italy. The section near the Strait of Messina is called the "Messina-Cone" by Ryan and Heezen (1965); however, there is little to differentiate the dissected topography here from that to the northeast. Ryan and Heezen (1965, p. 916) note that profiles across the "Messina Cone" reveal broken relief "suggesting either recent and/or present tectonic deformation".

The canyons on the "Messina Cone" section of the rise follow a relatively direct southeasterly course down to the Sicilia Plain. Canyons below the Gulf of Taranto follow the same orientation down to just below the 3000 m contour and then head more southerly to reach the plain. Canyons originating south of the Strait of Taranto and near the Greek

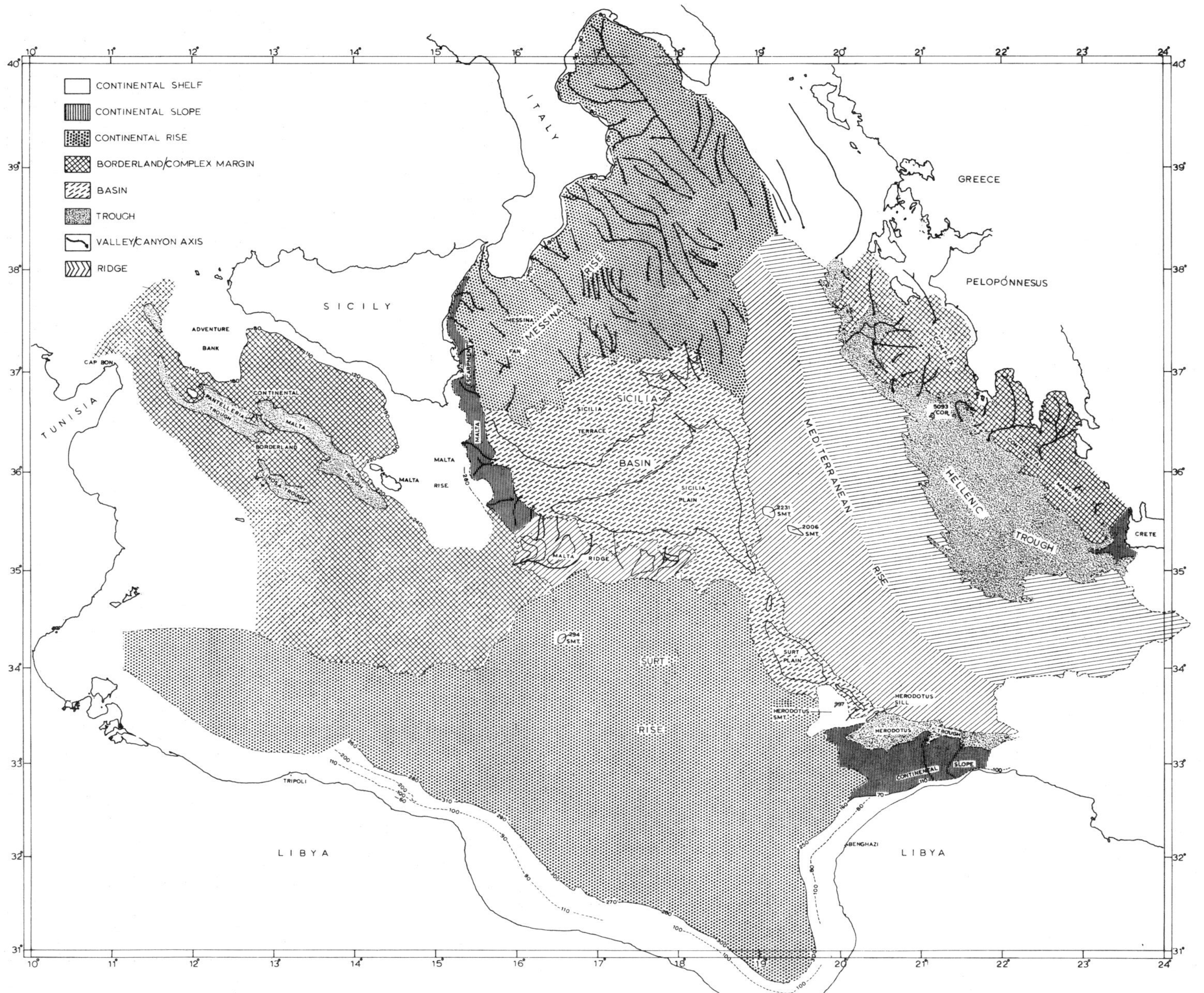

Figure 7. Physiographic map of the Ionian Sea.

mainland begin on a southeasterly course, change to southwesterly, and then swing back southeasterly to terminate in the "Hellenic Trough."

There is a considerable flattening in gradient on the "Messina Cone" between the 3200 and 3400 m contours. This area has been delineated as a terrace on the physiographic diagram, but it should be specified that the surface is quite irregular because of the numerous canyons traversing the area.

The canyon heading into the Gulf of Taranto is the largest and longest off this section of the Italian coast. There is a graben on shore at the end of the Canyon indicating possible structural control (Anonymous, 1964).

The continental margin of the Ionian Basin off Greece is highly complex. Large depressions and valleys are common suggesting block faulting and extension.

The dominant trend of depressions is northwest, or parallel to the general trend of the continental margin. The valleys have a northerly trend but consistently change to other directions down slope.

Basins and Troughs

The Sicilia Basin is delineated in general by the 3600 m contour. The basin consists of a northern section bordered by the "Messina Cone," Malta Ridge, and "Mediterranean Rise," and a southern section bordered by the Surt Rise and "Mediterranean Rise."

The northern section includes the Sicilia Plain (Ryan *et al.,* 1970) and reaches a depth of 4103 m. The southern section includes the Surt Plain (Ryan *et al.,* 1970) and reaches a depth of 3847 m.

The "Hellenic Trough" is a low area on the east flank of the "Mediterranean Rise" and at the base of the complex continental margin off Greece. The trough has been delineated by the 3000 m contour and is about 330 km in length. The floor of this trough combines the complex structure of both bordering features but the northwest trending small ridges and depressions of the "Mediterranean Rise" predominate. Many of the depressions within the trough are separated and in échelon. Sediments trapped in individual depressions form small sediment ponds at differing levels. The deepest of these depressions at 5093 m (corrected) is the deepest recorded point in the Mediterranean (Hersey, 1965). The Herodotus Trough is a small narrow depression at the foot of the continental margin of Libya. The trough defined by the 3000 m contour essentially incorporates the lower part of the continental slope, "Mediterranean Rise" and Herodotus Seamount. An additional small trough is seen on the northeast side of the Herodotus Seamount. This feature, delineated by the 3400 m contour, is separated from the Herodotus Trough by the Herodotus Sill.

Ridges and Rises

One of the most prominent features in the Mediterranean Sea is the "Mediterranean Rise" which occupies a significant portion of the floor of the Ionian Sea (as well as the Levantine Sea). The eastern margin is essentially the base of the complex continental margin, but as indicated, the portion below the 3000 m line has been included in the Hellenic Trough. The western margin of the "Mediterranean Rise" is delineated by the 3600 m contour opposite the Sicilia Basin, and by the 3000 m contour opposite the Herodotus Trough. These contours are located within an area of steeper gradient on the flanks of the rise.

The "Mediterranean Rise" crest plunges northward from 2000 m to about 3300 m over a distance of 530 km in the Ionian Basin. The surface area of the ridge is made up of many individual small ridges and depressions, which trend northwest-southeast paralleling the main ridge crest. Contrary to some previous representations (Mikhaylov, 1965; Ryan *et al.,* 1970) the rise is not a continuous, arcuate feature. There is a severe change in the direction of the rise axis southwest of Crete. Additionally, the rise does not run up to the tip of southern Italy, but appears to terminate where it meets the Messina Rise.

PHYSIOGRAPHY OF THE EASTERN MEDITERRANEAN SEA

The Aegean and Related Seas (Joseph C. Whitney)

The Aegean Basin, which constitutes the floor of most of the Aegean Sea's approximately 181,000 square km between Greece and Turkey, is characterized by a borderland with more than 200 islands and a series of northeast-southwest trending troughs. This basin is connected with the Ionian Basin through the Gulf of Corinth which has a 56 m sill depth west of it and an 8 m deep canal at its eastern end (Heezen *et al.,* 1966).

The Crete basin is located south of the Aegean basin and their boundary is delineated by a line that traces the gradient change between the two areas. The bottom gradient increases to the south and the inflection point varies from 500 to 1000 m in depth. The Cretan island arc, containing Kithira, Crete, and Rhodes, forms the southern boundary of the basin. There are several passages between the Crete basin and the Mediterranean to the South. The main passages are Andikithira Strait, Kasos Strait, Karpathos Strait, and the strait between Rhodes and Turkey with sill depths of 827, 808, 887, and 343 m, respectively. The prominent feature of the basin is the Cretan Trough which extends from the Gulf of Argolis eastward in a large arc, convex to the south, and ends at the western slope of Rhodes. The maximum depth of the Aegean Sea, 2509 m, is in this trough northwest of Karpathos.

The Marmara basin underlies the approximately 10,350 square km of the Sea of Marmara, located northeast of the Aegean basin, and is joined to the Aegean by the Dardanelles. Its main feature is a trough associated with the Anatolian Trough which cuts across the northern Aegean Borderland.

The bathymetric chart (insert) and the physiographic map of the Aegean area (Figure 8) provide details not included in the text.

Aegean Basin

Generally, the shelf surrounding the deeper parts of the basin is wide with a deep shelf break in embayment areas, and is narrow with a shallow shelf break off headlands.

The shelf at the northern end of the Aegean basin is 25 to 95 km wide except where a series of northwest trending ridges and valleys, related to continental structures, interrupt it. In the Thermaic Gulf, the shelf break is at 130 m but additional low gradient bottom extends to about 400 m and forms a marginal plateau at the edge of the Anatolian Trough. To the east, the shelf break is at about 150 m, and the descent into the trough takes place abruptly beyond the shelf edge.

Along the Turkish coast, sections of shelf extend out to connect islands to the mainland. There is usually a northeast-southwest trending trough between adjacent island and shelf sections. An example of one of these sections is the shelf around Limnos and Imroz which is about 85 km wide and attains a depth of 265 m.

The large shelf section containing Andros. Naxos, and Thira and the associated topography in the rest of the Cyclades area maintains a width and trend similar to that of the Greek mainland. The shelf edge depths are variable, but most occur between 130 and 170 m.

Immediately off the coasts of many islands in the Aegean basin, there are steep, rudimentary slopes approaching 1:20 gradients that end at about 90 m in depth. From this point, a normal shelf gradient is developed out to the shelf edge.

Below the shelf edge, the bottom gradient remains low over most of the basin. The term continental borderland is used to describe this area because of the large number of troughs and basins separated by ridges and shelflike areas.

The Aegean borderland has a general northwest-southeast trend and is divided into two parts by the Cyclades Islands. The eastern half consists primarily of segments less than 400 m deep, broken up by island shelves, banks, and depressions. The western half lacks large troughs and its low gradient segments increase in depth toward the Cretan Trough. Many of the smaller troughs and basins are less than 30 km wide and 1000 m deep.

The Anatolian Trough, located at the northern end of the Aegean borderland, extends from the Gulf of Saros to the Magnesia Peninsula. This northeast-southwest trending graben is a continuation of the North Anatolian Fault that also forms the Marmara Trough (Vogt and Higgs, 1969; Mikhaylov, 1965). The trough may be divided into two segments at about 25° longitude where there is a slight change in trend and a sill depth of 518 m. The broad western half has a flat floor averaging 1050 m except for a deep rent in the southeast corner with a maximum depth of 1468 m. The narrow eastern half, trending more sharply to the east, has a maximum depth of 1549 m in a cleft at its southwest end.

South of the Anatolian Trough there are many smaller troughs that appear as tears in the basin floor with trends parallel to adjacent shelf edges. The maximum depths range from 656 m to 1219 m.

It is noteworthy that the major troughs within the borderland have parallel trends and are evenly spaced at 90 km.

Crete Basin

The shelf areas bordering the Crete basin are generally poorly developed. Along the east coast of Peloponnesus the shelf is only 2 to 3 km wide and the shelf break occurs at 40 m. The width increases to about 6 km off the northern coast of Crete with the shelf edge averaging 120 m. The western shelf of Kasos is 6 km wide and breaks at about 120 m. Karpathos' western shelf is 2 km wide and breaks at 30 m. Off the west coast of Rhodes the shelf width averages 4 km and the shelf edge depth averages 70 m.

Northwest of Crete there is a narrow and featureless slope that extends from the shelf edge to the trough margin at 1200 m. North of Crete the slope includes a series of valleys and ridges that grade into the trough from both sides. The slope along the Aegean borderland is precipitous in some areas and there is an escarpment that extends to 2400 m off the west coasts of Kasos and Karpathos. Associated with the basin's slope is a marginal plateau that

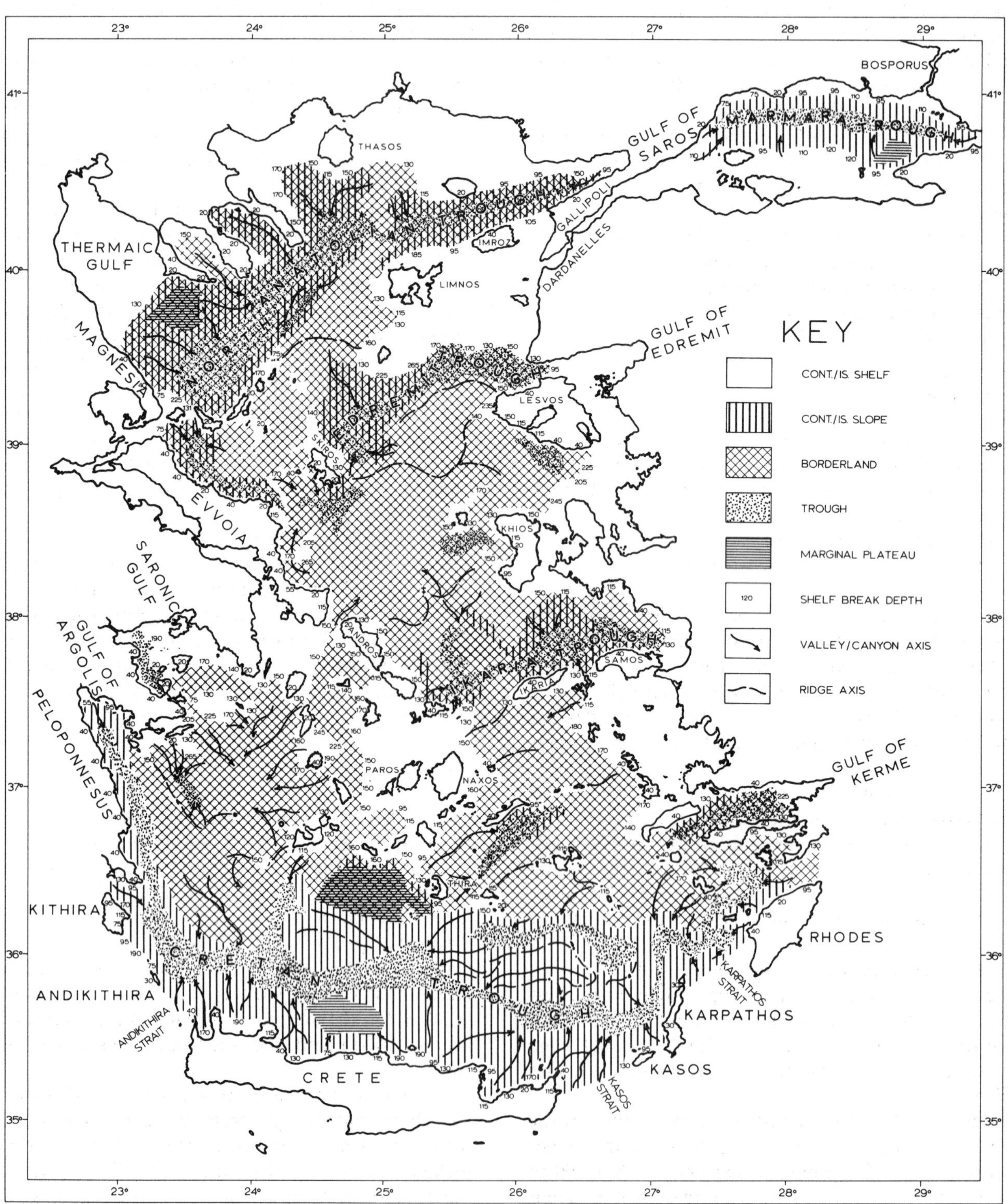

Figure 8. Physiographic map of the Aegean Sea and related basins.

connects the slope with the trough and lies between 800 and 1200 m in depth.

The Cretan Trough is a large arcuate feature associated with the Cretan island arc system which also includes the "Mediterranean Rise," the "Hellenic Trough" north of the ridge, the islands of Kithira, Crete, and Rhodes, and the inner volcanic island arc 150 to 200 km north of Crete. This system resembles a typical island arc (Vogt and Higgs, 1969; Maley and Johnson, 1971).

The trough contains many small depressions that generally become deeper to the east. The shallowest basin is in the Gulf of Argolis at 803 m and the deepest, at 2509 m, lies off the west slope of Karpathos. Mikhaylov (1965) suggests the deep narrow portion of the trough west of Karpathos may delineate a fault. Maley and Johnson (1971) offer profiles of the same area that indicate step faulting.

North of the Cretan Trough there is a subordinate trough separated from the larger one by a U-shaped ridge on which numerous small islands are located. Over most of its length the trough is delineated by the 800 m contour. Its maximum depth of 1294 m is in a small basin at the eastern end of the trough.

Marmara Basin

Over half of the area occupied by this region is continental shelf. The shelf along the northern coast varies in width from 2 to 13 km with shelf edge depths at 20 to 110 m. The southern half of the shelf is broad, 33 km wide, with a break at 115 m, except along the southeast coast where the shelf is 2 to 3 km wide and breaks at 20 to 30 m.

The continental slope, extending to 1200 m, completely encloses the Marmara Trough. The slope is interrupted southeast of the trough by a marginal plateau that lies between 200 and 400 m in depth.

The Marmara Trough is divided into 3 small depressions that have maximum depths from west to east of 1097, 1389, and 1238 m. They are separated by low sills with depths of approximately 720 m. The trough is a continuation of the North Anatolian Fault.

Levantine Sea
(Robert R. Murchison)

The eastern Mediterranean generally has been divided into two major basins, the Ionian and the Levantine (Sverdrup *et al.,* 1942). The boundary between basins formerly was placed at "the ridge extending from Greece to Africa." Recent bathymetric data, however, show that the two continents are not connected by a ridge. A better western boundary is a line from Ras el-Hilal on the Libyan coast thence to the island of Crete by way of Gavdhos Island (Figure 9). The area covered by the Sea is approximate-

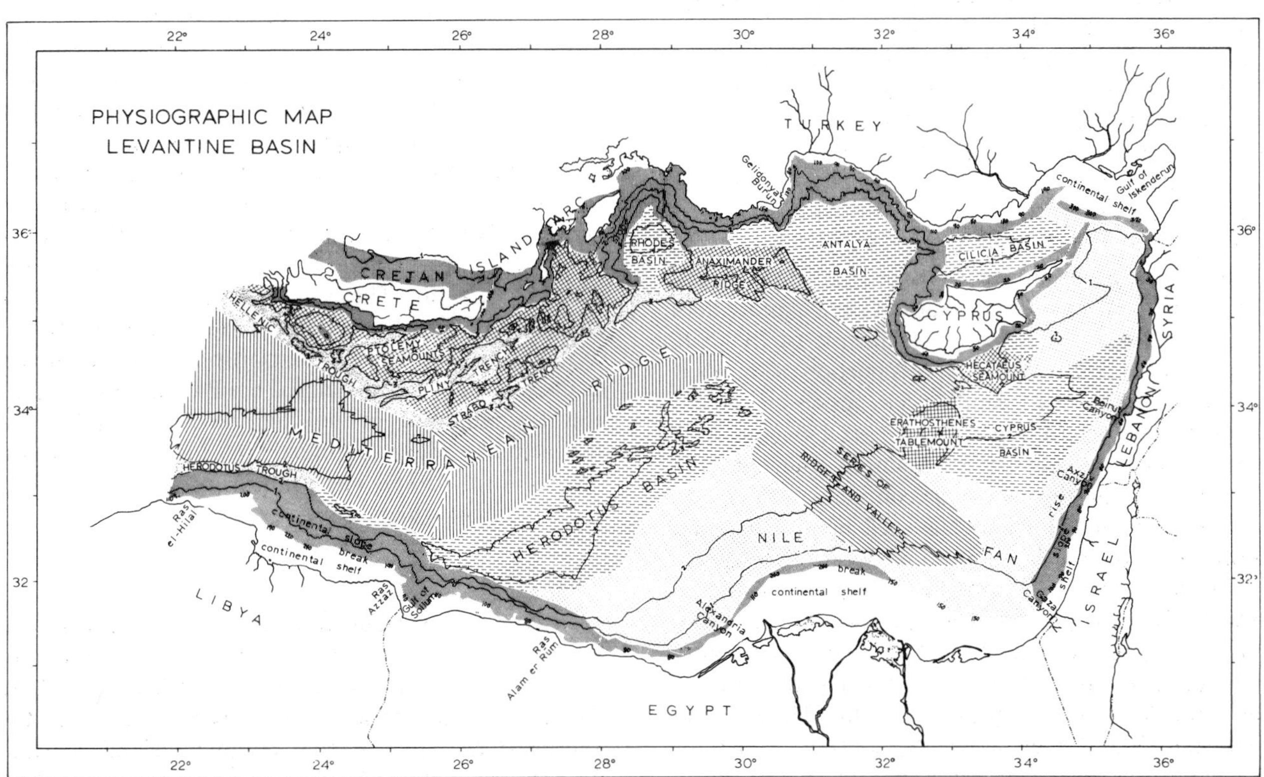

Figure 9. Physiographic map of the Levantine Sea and easternmost Mediterranean.

ly 320,000 square km with a greatest depth of 4384 m. The Levantine Sea is bordered on the north by Crete, the Dodecanese Islands and Turkey. On the east it is bordered by Syria, Lebanon and Israel and to the south by Egypt and Libya. The island of Cyprus is located in the northeast quadrant of the basin. The margin of the basin includes an island arc from Crete through the Dodecanese Island to Rhodes. The remaining basin margins are of the less complex types consisting of continental shelves, slopes, rises, and associated features. The predominant topographic feature of this region is the "Mediterranean Rise."

Continental Shelf

There is little information on the shelf depths around the eastern Mediterranean, possibly because of the frequent variation in depths. One exception is the Israeli continental shelf that has been studied in detail (Emery and Bentor, 1960). Depths for other areas were obtained by studying large scale nautical charts. These depths have been indicated along the shelf edge on the physiographic map (Figure 9). Note that the shelf edge depths generally increase with an increase in shelf width.

Along the Turkish coast from the Rhodes channel eastward to Gelidonya Burun there is no significant continental shelf. From Gelidonya Burun to the northeastern corner of the Levantine Basin the average width is 4.5 km. Along this shelf the break varies from 40 m to 130 m in depth. Off the Gulf of Iskenderun a composite delta of several rivers has built a shelf 70 km wide. The shelf break is about 300 m deep along this portion of the shelf. The shelf is very narrow or absent at the Turkish-Syrian border. From this area southward to the Gaza Strip the continental shelf gradually widens to about 18 km. There is an accompanying depth increase of the shelf break from 20 m to 120 m north to south. A shelf of more than 70 km has been built by the sediments of the Nile River (Emery *et al.*, 1966). On this portion of the shelf the break is found at depths of as much as 260 m. Beyond the western edge of the Nile Fan and to a point midway between Ras 'Alam er Rum and the Gulf of Sollum the irregular continental shelf has an average width of 7 km. The shelf break is located approximately at a depth of 80 m. West of this point to Ras Azzaz the shelf widens to 25 km. This increase in width is accompanied by an increase in shelf break depth ranging from 80 to 240 m. The continental shelf off Ras el-Hilal is about 4 km wide and the shelf break is approximately 110 m. There is no significant island shelf along the southern coasts of the islands of the Cretan island arc. On the northern side of Cyprus the shelf averages 1 km wide and the shelf break is about 20 m deep. The southern coast has a shelf width averaging 2 km and a shelf break at 30 m.

Continental Slope

Along the southern coast of Turkey the continental slope has a gradient ranging from 1:24 to 1:10. The slope is modified in the Gulf of Iskenderun by sediments from the adjacent rivers. Along the Lebanon-Israel coast the gradient ranges from 1:7 at the north to 1:21 at the south (Emery *et al.*, 1966). Three canyons, Gaza, Akziv and Beirut incise this eastern end of the Levantine Basin. These canyons were named for the geographic locations situated at their heads. The continental margin along the Egyptian coast is occupied by the Nile Fan. This fan is approximately 320 km wide and extends from the shelf 160 km out to depths of 2800 m on the western end and 1600 m on the east end. The Nile Fan can be divided into two fans. The larger, occupying the western part of the fan has several canyons, the largest being the Alexandria Canyon. A northwest trending series of ridges and valleys, previously described as abyssal hills (Emery *et al.*, 1966), are seen along the eastern margin of the fan. These trends are parallel to a ridge which connects the island of Cyprus and the Anaximander Mountain south of the Turkish coast. A postulated transcurrent fault passing between Cyprus and Erathosthenes Tablemount (Wong *et al.*, 1971) also shows a similar trend. This series of ridges and valleys would seem to be structurally controlled. Along the Libyan coast there is a well developed continental slope incised by several canyons. The gradient along this coast ranges from 1:35 to 1:10. The steepest portion is near the Nile Fan. The slope off the Cretan island arc is a complex island margin bisected by several deep depressions. These depressions trend northeast and parallel the trench and "Mediterranean Rise" topography farther east. A portion of this complex island margin has been named the Ptolemy Seamounts and is composed of uplifted blocks of indurated sedimentary rocks bounded by high angle faults (Wong *et al.*, 1971). The island slope east of Rhodes has a gradient which approaches 1:8.

Trenches

A pair of parallel trenches, Pliny Trench and Strabo Trench, are situated between the Ptolemy Seamounts and the "Mediterranean Rise." The Pliny Trench, the deeper at 4384 m, is located along the southeastern edge of this complex island margin. This trench is about 200 km long and has many small sediment filled basins along its course (Wong *et al.*,

1971). The trend of the west end of this trench is interrupted by a southward extension of the complex island margin. Depressions parallel the base of the island margin. The relationship between the trench and these depressions cannot be determined from the soundings. The Strabo Trench is approximately 40 km south and parallel to the Pliny Trench. Most of the trench floor is at a depth of 3000 m to 3200 m, but a depth of 3720 m does occur at its northeast end just south of the Rhodes Basin. The Strabo Trench is separated from the Pliny Trench by a high ridge made up of several individual segments. The Strabo Trench goes beyond the extension of the complex island margin which terminates the Pliny Trench.

A southeast trending series of unnamed depressions connects with the east end of the Strabo Trench. These depressions often consist of a series of individual small troughs. The similarity in size and distribution of relief between these small southeast trending troughs and the northeast trending trough in the "Mediterranean Rise" causes considerable difficulty in determining the point of termination of the "Mediterranean Rise" from soundings. These depressions are bordered on the northeast by the Anaximander Mountains, the island margin of Cyprus and a low ridge joining the two.

Rises

The "Mediterranean Rise" is a broad low arch averaging 150 km in width, located midway between the Cretan Island Arc and the North African continental mass. From its western end south of the boot of Italy it trends southeast to its high point of 1330 m located north of Ras el-Hilal, Libya. At this point the ridge abruptly changes trend to east-northeast and terminates west of Cyprus. The crest of the rise is about 700 m above the abyssal plains and trenches which border it (Emery *et al.*, 1966). The topography in general is rough with many low hills and small depressions. Recent geophysical evidence indicates that this system is not typical of other mid-ocean ridge or rise systems and it does not have a spreading origin (Ryan *et al.*, 1970; Wong *et al.*, 1971).

Basins

There are several minor basins within the Levantine Basin. Most of these minor basins are not distinct entities but the low areas of segments of several features, such as the continental margin and "Mediterranean Rise." This leads to difficulty in delineating the basin limits consistently as can be seen in the varying limits shown in the literature. The limits shown here will be primarily depth contours. Preference is given to those contours that mark some gradient change in the surrounding features.

The Herodotus Basin is delineated here by the 2800 m line. This depth approximately marks a flattening in gradient on the "Mediterranean Rise" and the Nile Fan. The maximum depth of this basin is 3156 m near its southwest end. The "Herodotus Plain" occupies a narrow section of the center of the basin. The plain is bordered by irregular "Mediterranean Rise" type topography on the northwest side, and by smooth Nile Fan type topography on the southeast side. A trough situated between the base of the African continental slope and the "Mediterranean Rise" extends westward from the Herodotus Basin into the Ionian Basin.

The Cyprus Basin is delineated by the 1600 m line along the coast of Lebanon and Israel and to the north of the Nile Fan. South of the island margin off Cyprus this basin is delineated by the 2000 m line. This basin slopes to the northwest having a depth of approximately 2300 m between the Cyprus island margin and Erathosthenes Tablemount. The Cyprus island margin includes Hecataeus Seamount, minimum depth 242 m, which is believed to be a submerged part of that island (Wong *et al.*, 1971). Erathosthenes Tablemount is a conspicuous single mass rising to 649 m from the surface. It is believed to be a non-volcanic uplifted block and the western extension of Cyprus before their separation along a postulated transcurrent fault (Wong *et al.*, 1971).

The Antalya Basin is located between Antalya Bay, southern Turkey, and the Anaximander Mountain-Cyprus ridge complex. This basin is delineated here by the 2200 m contour line. This basin has a westward slope deepening to approximately 2600 m. This Antalya Basin is connected to the Rhodes Basin to its west by a trough.

The Rhodes Basin located southeast of the island of Rhodes is delineated here by the 3600 m line. The maximum depth of this basin is 4307 m. The "Rhodes Plain" occupies the center portion of this basin.

The "Cilicia Plateau" is located north of Cyprus and south of Turkey. This feature is delineated here by the 800 m contour on the Turkish continental margin, and the 600 m contour on the Cyprus island margin. To the east the limit is arbitrarily placed along the trend of the 800 m line.

SUMMARY
(John C. Sylvester)

The amount and quality of the data utilized in compiling the bathymetric chart have enabled the authors to compile detailed physiographic maps of the seven major provinces of the Mediterranean Sea.

The bathymetric chart itself presents considerable detail and hopefully will be a useful tool for future work in the area.

A number of aspects of this land-locked sea were discovered. Among the more significant are: (1) The extensive regional variation in shelf width and shelf edge depths, and considerable deviation from normally accepted shelf gradient (often in excess of 1:40) and relief. (2) Newly-discovered features or characteristics discussed in the text, particularly in the complex Tyrrhenian Sea but also in other areas, are noteworthy. The Aegean Sea, for instance, has many small troughs not previously delineated. Furthermore, measured sill depths in certain areas appear to differ from earlier reported depths. (3) The marked feature commonly known as the Mediterranean Ridge may be more properly called the "Mediterranean Rise", even though it has rather rough surface topography. It is not a consistently arcuate feature as portrayed or described in much of the literature. It shows an abrupt change in trend southwest of Crete and, according to the bathymetric data, does not continue to the foot of Italy, but terminates south of it. (4) A series of northwest-southeast trending ridges and valleys, previously described as abyssal hills, occur on the Nile Fan.

There is no intention on the part of the authors to present this chart as the final definitive configuration of the Mediterranean Sea floor. There are areas where additional surveys should be conducted, and given the information supplied in the text and on the chart, future exploration in the Mediterranean can be more efficiently planned.

REFERENCES

Angrisano, G., and A. G. Segre 1969. La carta batimetrica del Mediterraneo nord occidentale. Carta No. 1501. *Istituto Idrografico della Marina, Genova.*

Anonymous 1964. *Fizico-Geograficheskiy Atlas Mira,* Akademiya Nauk SSSR i Glavnoe Upraylinic Geodezii i Kartografii GGK SSSR. (U.S.S.R. Academy of Sciences, Moscow) 298 p.

Anonymous 1967a. Carta batimetrica del Tirreno meridionale, Carta No. 1502. *Istituto Idrografico della Marina, Genova.*

Anonymous 1967b. Carta batimetrica del Adriatico Settentrionale, Carta No. 1505. *Istituto Idrografico della Marina, Genova.*

Anonymous 1969. *Undersea Features.* United States Board on Geographic Names, Gazateer No. 111. 142 p.

Allan, T. D., and C. Morelli 1970. Bathymetry (Maps No. 1, 2, 3, 4, 5, 6, 7, 13, and 14): Alboran Sea, Balearic Basin—West, Balearic Basin—North, Ligurian Sea, Balearic Basin—East, Tyrrhenian Sea, Strait of Sicily, Aegean Sea, Ionian Sea—East. *NATO Sub-committee on Oceanographic Research.*

Bourcart, J. (Editor) 1958a. Topographic map: Précontinent entre Antibes et Gênes (1:200,000). *Musée Océanographique de Monaco.*

Bourcart, J. (Editor) 1958b. Topographic map: Précontinent entre Marseille et Antibes (1:200,000). *Musée Océanographique de Monaco.*

Bourcart, J. 1959. Le Plateau continental de la Méditerranée occidentale. *Comptes Rendus des Séances de l'Académie des Sciences,* Paris 249:1380–1382.

Bourcart, J. 1960. Carte topographique du fond de la Méditerranée occidentale. *Bulletin de l'Institut Océanographique,* 1163:20 p.

Bourcart, J. (Editor) 1960a. Topographic map: Précontinent sous-marin corse de Porto aux bouches de Bonifacio (1:200,000). *Musée Océanographique de Monaco.*

Bourcart, J. (Editor) 1960b. Topographic map: Précontinent sous-marin corse du nord de Porto au Cap Corse (1:200,000). *Musée Océanographique de Monaco.*

Debrazzi, E., and A. G. Segre 1959a. Carta batimetrica del Mediterraneo centrale, Mare Ligure e Tirreno Settentrionale, Carta No. 1250. *Istituto Idrografico della Marina, Genova.*

Debrazzi, E., and A. G. Segre 1959a. Carta batimetrica del Mediterraneo centrale, Mare Adriatico. Carta No. 1253. *Istituto Idrografico della Marina, Genova.*

Emery, K. O., and Y. K. Bentor 1960. The continental shelf of Israel. *Bulletin of the Geological Survey, Israel,* 26:25–41.

Emery, K. O., B. C. Heezen and T. D. Allan 1966. Bathymetry of the eastern Mediterranean Sea. *Deep-Sea Research,* 13:173–192.

Giermann, G. 1962. Erlauterungen zur bathymetrischen Karte des westlichen Mittelmeers (zwischen 6°40′W. L. und 1°0′E. L.). *Bulletin de l'Institut Océanographique,* 1254:24 p.

Giermann, G., M. Pfannenstiel and W. Wimmenauer 1968. Relations entre morphologie, tectonique et volcanisme en mer d'Alboran (Méditerranée occidentale). Résultats préliminaires de la campagne JEAN-CHARCOT (1967). *Comptes Rendus Sommaire de Séances de la Société Géologique de France,* 4:116–117.

Giorgetti, F. and F. Mosetti 1969. General morphology of the Adriatic Sea. *Bollettino di Geofiscia,* 1:49–56.

Heezen, B. C., M. Ewing and G. L. Johnson 1966. The Gulf of Corinth floor. *Deep-Sea Research,* 13:381–411.

Heezen, B. C., C. Gray, A. G. Segre and E. F. K. Zarudzki 1971. Evidence of foundered continental crust beneath the central Tyrrhenian Sea. *Nature,* 229:327–329.

Heezen, B. C., M. Tharp and M. Ewing 1959. The floor of the oceans, I. The North Atlantic. *Geological Society of America Special Paper,* 65:112 p.

Hernandez-Pacheco, F. and I. Asensio Amor 1968. Depositos cuartenarios de la isla de Alboran. *Boletino Real Sociedad Española de Historia Natural,* 66:381–392.

Hersey, J. B. 1965a. Sediment ponding in the deep sea. *Geological Society of American Bulletin,* 76:1251–1260.

Hersey, J. B. 1965b. Sedimentary basins of the Mediterranean Sea. In: *Submarine Geology and Geophysics,* eds. Whittard, W. F. and R. Bradshaw, Butterworths, London, 75–91.

Leroy, C. C. 1966. Sound propagation in the Mediterranean Sea. *Underwater Acoustics,* 2:203–241.

Maley, T. S. and G. L. Johnson 1971. Morphology and structure of the Aegean Sea. *Deep-Sea Research,* 18:109–122.

Massuti, M. 1967. Carta de pesca de la región surmediterranea española. (Desde Estepona a Adra). *Trabajos Instituto Español de Oceanografia,* 33:24 p.

Menard, H. W., S. M. Smith and R. M. Pratt 1965. The Rhone deep-sea fan. In: *Submarine Geology and Geophysics,* eds. Whittard, W. F. and R. Bradshaw, Butterworths, London, 271–285.

Mikhaylov, O. V. 1965. The relief of the Mediterranean Sea bottom. In: *Basic Features of the Geological Structure of the Hydrological Regime and Biology of the Mediterranean Sea,* ed. Fomin, L. M. Nauka, Moscow, 224 p. (translation).

Oliver, M. 1960. Carta de pesca de las Baleares. II.—Norte de Mallorca y Menorca y este de Mallorca. *Trabajos Instituto Español de Oceanografia,* 29:12 p.

Oliver, M. 1961. Carta de pesca de Cataluña. I.–Desde el paralelo de Cabo Bear a Palamos. *Trabajos Instituto Español de Oceanografia,* 30:9 p.

Oliver, M. 1968a. Carta de pesca de Cataluña. I.–Desde el Cabo San Sebastian a Barcelona. *Trabajos Instituto Español de Oceanografia,* 35:11 p.

Oliver, M. 1968b. Carta de pesca de Cataluña. III.–Desde Barcelona a Cabo Tortosa. *Trabajos Instituto Español de Oceanografia,* 36:11 p.

Ross, D. A., E. Uchupi, K. E. Prada and J. A. MacIlvaine 1972. Bathymetry and microtopography of the Black Sea. In: *Black Sea: Its Geology, Chemistry and Biology,* eds. Degens, E. T. and Ross, D. A. Tulsa, Oklahoma, American Association of Petroleum Geologists, (in press).

Pfannenstiel, M. 1960. Erlauterungen zu den bathymetrischen Karten des östlichen Mittelmeeres. *Bulletin de l'Institut Océanographique,* 1192:60 p.

Rosfelder, A. 1955. Carte provisoire au 1:500,000 de la marge continentale Algérienne. *La Carte Géologique de l'Algérie (Nouvelle Série), Bulletin,* 5:57–106.

Ryan, W. B. F. and B. C. Heezen 1965. Ionian Sea submarine canyons and the 1908 Messina Turbidity Current. *Geological Society of America Bulletin,* 76:915–932.

Ryan, W. B. F., D. J. Stanley, J. B. Hersey, D. A. Fahlquist and T. D. Allan 1970. The tectonics and geology of the Mediterranean Sea. In: *The Sea,* ed. Maxwell, A. E. Wiley Interscience, New York, 4 (II):387–492.

Stanley, D. J., C. E. Gehin and C. Bartolini 1970. Flysch-type sedimentation in the Alboran Sea, western Mediterranean. *Nature,* 228:979–983.

van Straaten, L. M. J. U. 1965. Sedimentation in the northwestern part of the Adriatic Sea. In: *Submarine Geology and Geophysics,* eds. Whittard, W. F. and R. Bradshaw, Butterworths, London, 143–162.

Sverdrup, H. U., M. W. Johnson and R. H. Fleming 1942. The Earth and the Ocean Basins. In: *The Oceans: Their Physics, Chemistry, and General Biology.* Prentice Hall, New York, 34–35.

Trotti, L. 1968. A bathymetric and geological survey in the middle Adriatic Sea. *International Hydrographic Review,* 45:59–71.

Vogt, P. R. and R. H. Higgs 1969. An aeromagnetic survey of the eastern Mediterranean Sea and its interpretation. *Earth and Planetary Science Letters,* 5:439–448.

Vogt, P. R., R. H. Higgs and G. L. Johnson 1971. Hypotheses on the origin of the Mediterranean Basin: Magnetic data. *Journal of Geophysical Research,* 76:3207–3228.

Wong, H. K., E. F. K. Zarudzki, J. D. Phillips and G. K. F. Giermann 1971. Some geophysical profiles in the eastern Mediterranean. *Geological Society of America Bulletin,* 82:91–100.

Zarudzki, E. F. K. 1971. Regional NW-SE seismic reflection profile across Tyrrhenian (abstract). *Transactions of the American Geophysical Union* 52:357.

Caractères Hydrologiques et Circulation des Eaux en Méditerranée

Henri Lacombe et Paul Tchernia

Muséum National d'Histoire Naturelle, Paris

RESUME

L'océanographie physique de la Méditerranée est déterminée par sa climatologie; les pertes d'eau par évaporation sont plus grandes que les apports par précipitations et débit des fleuves: la Méditerranée est un bassin de concentration. Elle emprunte à l'Atlantique l'eau en déficit. Son contenu en sel est aussi constant—à l'échelle humaine des temps. Pour équilibrer à la fois son bilan d'eau et son bilan de sel, la Méditerranée fonctionne comme une "machine" qui transforme l'eau atlantique entrante, concentrée et refroidie par l'évaporation, diluée par les précipitations et les apportes des fleuves, en une eau dense et salée, typiquement méditerranéenne, qui, finalement, s'écoule dans l'Atlantique dans les parties profondes du Détroit de Gibraltar. La formation de cette eau typique a principalement lieu en hiver dans un petit nombre de régions de dimensions limitées. Le fonctionnement de la "machine" détermine aussi bien les caractères physiques des eaux de la mer que leurs principaux mouvements.

ABSTRACT

The physical oceanography of the Mediterranean is controlled by the climatology of the region. Water loss by evaporation is greater than the gain resulting from precipitation and river discharge. Thus, the Mediterranean, a concentration basin, receives the water in deficit from the Atlantic Ocean. The salt content is also constant—at least on the usual human time scale. To achieve the balance of both its water and salt contents, the Mediterranean functions as a "machine." It transforms the incoming water from the Atlantic, concentrated and cooled near the surface by evaporation and diluted by rain and river outflow, into typical dense and salted Mediterranean water mass which ultimately flows back into the Atlantic via the deeper stretches of the Strait of Gibraltar. The formation of this typical water mainly takes place in winter in a few selected areas of limited extent in the Mediterranean. The mechanism of this "machine" determines not only the physical properties of the Mediterranean water masses, but their main flow patterns as well.

CARACTERES GENERAUX

Les Seuils

Si la Méditerranée, dans son ensemble, est un bassin à seuil séparé de l'Océan Atlantique par le Détroit et le seuil de Gibraltar, elle est elle-même formée par une *série de bassins à seuil*:

1. La Mer Egée communique avec la Mer Noire par deux seuils (Bosphore, Dardanelles), entre lesquels s'étend la Mer de Marmara;

2. En Mer Egée, deux seuils sont présents: l'un, au Nord du débouché des Dardanelles, supporte les îles d'Imbros et de Lemnos, l'autre, sépare le bassin Nord de la Mer Egée de la Mer de Crète, dans une région à topographie compliquée, entre les Cyclades;

3. La Mer de Crète ouvre sur le bassin oriental de la mer par des seuils en arc, du Péloponèse à l'Anatolie, supportant les îles de Cérigo, Cérigotto, la Crète, Casso, Scarpanto, Rhodes;

4. Un seuil profond (2400 m) et qui ne semble pas avoir d'effet hydrologique important occupe le fond du Bassin Levantin; il est orienté vers l'OSO à partir de l'Ouest de Chypre;

5. Le seuil et le Détroit d'Otrante gisent au Sud de l'Adriatique;

6. Plus à l'Ouest, se trouvent le seuil siculo-tunisien, les seuils de Messine, de l'île d'Elbe, de Bonifacio;

puis les seuils de l'éperon des Baléares, le seuil, peu marqué, de la Mer d'Alboran. Enfin le Détroit et le seuil de Gibraltar limitent la Méditérranée à l'Ouest.

Le Climat

Le climat de la Méditerranée et de son bassin versant est caractérisé par sa diversité: des terres désertiques occupent une grande partie de la région Est de son périmètre méridional. Au contraire, le bassin versant qui l'alimente au NO est boisé et humide, particulièrement en hiver. Le débit des fleuves qui s'y jettent est surtout important dans le Nord de la Mer et en Mer Noire; le seul grand fleuve du rivage Sud est le Nil. Mais si les débits du Rhône, du Pô, du Danube sont relativement réguliers au cours de l'année, ceux des fleuves russes vers la Mer Noire présentent un débit maximum à la fin du printemps et au début de l'été.

Pour l'ensemble du bassin versant, les étés sont secs, les hivers sont humides. En été, l'anticyclone des Açores s'avance vers l'Europe occidentale; l'échauffement solaire sur les grandes étendues continentales à l'Est et au Sud de la Mer y engendre des pressions relativement faibles: en sorte qu'il y a tendance à l'établissement, sur la Mer, d'un régime de vents anticycloniques, soufflant du Nord dans l'Est de la Mer, et soufflant de l'Est sur la partie occidentale des côtes d'Afrique du Nord. En hiver, les dépressions météorologiques venant de l'Atlantique balayent le bassin d'Ouest en Est mais tendent fréquemment à stationner et à se creuser dans certaines régions comme le Golfe de Gênes, l'Adriatique, Chypre. Aussi la circulation des vents tend elle à être cyclonique et à provoquer des précipitations particulièrement fortes sur les côtes exposées à l'Ouest. Mais ces situations entraînent, derrière les dépressions, la présence de vents violents soufflant du NO ou du Nord, particulièrement forts dans le Golfe du Lion (*Tramontane*, *Mistral*) et dans le Nord de l'Adriatique (*Bora*). En hiver aussi, les régions Nord des bassins, occidental et oriental, peuvent être soumises à l'influence d'arrivées d'air polaire continental, sec et froid, lorsque le versant Sud d'anticyclones situés sur l'Europe centrale et orientale couvre les rivages septentrionaux de la mer.

Un caractère météorologique particulier est la présence de phénomènes transitoires, à petite échelle, mais très violents, pour les vents comme pour les précipitations. Celles-ci, en particulier, sont brusques, diluviennes et courtes. Des masses d'eau considérables, chargées en matériaux solides, sont alors déversées à la mer par des fleuves côtiers à crues soudaines. Des courants de turbidité, chargés en sédiments et issus de ces fleuves sont capables de jouer un grand rôle dans l'apport et la dispersion des sédiments, même à grande profondeur.

BILANS D'EAU ET DE SEL DU BASSIN

La Méditerranée, profondément enfoncée dans de grandes masses continentales au climat sec et dont une part au moins ne reçoit que fort peu de précipitations même en hiver, est un bassin à bilan propre négatif en eau: les pertes d'eau par évaporation, E, excèdent les gains, G, dûs aux précipitations et au débit des fleuves. $G - E$ est négatif pour l'ensemble de la Mer: la Méditerranée est un *bassin de concentration*; c'est là un élément fondamental pour l'étude de cette mer. Par contre, la Mer Noire est, à elle seule, un bassin de *dilution*; $G - E$ est positif.

Le temps de renouvellement de l'eau de la Méditerranée proprement dite est de l'ordre de 100 ans (*cf.*, ci-dessous *in fine*). Donc un changement du signe de $G - E$, du fait, par exemple, d'une augmentation des apports d'eau douce (fonte des glaces) peut, s'il dure pendant une période de l'ordre du siècle, inverser le sens des échanges avec l'océan et, éventuellement,—comme pour la Mer Noire actuellement—les couches marines profondes peuvent devenir anoxiques; la nature de la sédimentation peut alors être très différente selon les époques. Le phénomène a pu se produire au cours de l'histoire géologique récente de la Mer.

Comme le niveau de la Méditerranée—à échelle d'une vie humaine—reste constant, c'est que la mer s'alimente en eau sur l'Atlantique, à laquelle elle emprunte un volume d'eau Vo. Comme l'eau atlantique entrante est salée ($So\% \simeq 36{,}15‰$), le sel qu'elle apporte doit, en moyenne ressortir de la Mer, puisque le *bilan de sel* en est—à échelle humaine—constant. Ainsi la double nécessité du maintien du niveau et de la salinité dans la Mer exige à la fois *une entrée* d'eau Atlantique de masse volumique ρo, de salinité So sous un volume Vo et un sortie d'eau méditerranéenne pour laquelle les grandeurs homologues sont ρm, Sm, Vm. Une telle situation exige la présence, dans le détroit de Gibraltar, de deux écoulements inverses: l'écoulement méditérranéen sortant, d'eau plus dense, se fait près du fond; l'écoulement atlantique entrant, de densité plus faible, se fait en surface. Ces deux nappes superposées, en mouvement moyen inverse, présentent une interface commune relativement bien tranchée à une immersion moyenne qui est d'environ 150m dans la partie Ouest du Détroit.

La constance du contenu d'eau du bassin exige $G - E + Vo - Vm = o$.

La constance du contenu de sel impose $\rho o \; Vo \; So = \rho m \; Vm \; Sm$.

$$\text{Soit: } \frac{Vo - Vm}{Vm} = \frac{E - G}{Vm} = \frac{\rho m\, Sm - \rho o\, So}{\rho o\, So}$$

$$\text{et } \frac{Vo - Vm}{Vo} = \frac{\rho mSm - \rho oSo}{\rho mSm}$$

Les ρ ne différant que de 2.10^{-3} environ, on peut écrire, sans erreur appréciable $\Delta V/Vm = \Delta S/So$ (I); $\Delta V/Vo = \Delta S/Sm$ avec $\Delta V = Vo - Vm = E - G$ et $\Delta S = Sm - So$.

L'estimation des trois facteurs, G, E, ΔV du bilan d'eau total de la mer, nul en moyenne, a été l'objet de nombreux travaux (Carter, 1956; Tixeront, 1970, notamment); mais elle est sujette à d'importantes incertitudes plus particulièrement celle de l'évaporation E et du bilan ΔV des échanges nets à travers le détroit de Gibraltar. L'évaporation E doit être particulièrement élevée en hiver, notamment près des bords septentrionaux de la Mer, sur lesquels arrivent des vents secs, plus froids que l'eau marine. Carter évalue E et G; il en déduit ΔV. Mais les valeurs élevées de l'évaporation marine en été sont inattendues.

On peut aussi considérer E comme l'inconnue et évaluer ΔV selon la formule (I) ci-dessus, sous réserve qu'on ait,—par des mesures convenables—évalué Vm (ou Vo) Sm et So. E résulte alors de l'annulation du bilan total d'eau.

La connaissance de Sm et So relève, en principe, de déterminations hydrologiques simples, faites dans le détroit, d'une part dans la couche méditerranéenne profonde, d'autre part dans la couche atlantique. Cependant l'évaluation est loin d'être aussi simple qu'il y paraît en raison de la présence d'une couche de transition relativement épaisse (40 à 60 m) entre les masses d'eau en question, de la complexité du régime des courants du détroit, de la présence d'ondes internes de grande amplitude dans la couche de transition, enfin de l'évolution spatiale des caractéristiques hydrologiques moyennes dans le détroit lui-même. Quant à la mesure de Vm (et de Vo), elle se heurte à des difficultés pratiques considérables, liées à la profondeur du détroit, à la nature des fonds, à la violence des courants et, plus encore, à leurs variations du fait de la marée et des conditions météorologiques.

Des mesures de courants relativement prolongées (Lacombe *et al.*, 1964) effectuées en 1960 (septembre) et 1961 (mai-juin), dans la partie ouest du détroit où la topographie des fonds définit bien l'écoulement méditerranéen sortant et lui confère une valeur moyenne autorisant une bonne précision, ont conduit à la valeur moyenne $Vm = 1{,}15 \times 10^6$ m^3/sec, soit, si on admet cette valeur pour l'année, Vm = 36260 km^3/an. So et Sm déterminées à la suite de mesures faites à l'emplacement même des mesures de courant conduisent sensiblement à So = 36,15‰; Sm = 37,90‰; soit ΔS: 1,75‰. La précision sur ΔS est faible 10%–15%; elle réagit sur celle de ΔV donc aussi sur celle de E. Par l'équation (1) on déduit Vo = 38000km^3/an ou $1{,}20 \times 10^6$ m^3/sec. Le bilan *net* à Gibraltar est donc 1740km^3/an.

Si l'on veut connaître les éléments du bilan total de la Méditerranée proprement dite, indépendamment de celui de la Mer Noire, il faut tenir compte des échanges d'eau entre la Mer Noire et la Méditerranée. La Mer Noire déverse en Méditerranée en surface environ 400km^3/an. La Méditerranée reçoit donc de l'Atlantique et de la Mer Noire un total de 38400 Km^3/an. Comme son volume est sensiblement $3{,}71 \times 10^6$ Km^3, le temps de renouvellement de l'eau de la Méditerranée

$$\text{est: } \frac{3{,}71 \times 10^6}{38.400} = 97 \text{ ans.}$$

Quant au bilan *net* à travers le Bosphore, il est évalué à 189 Km^3/an, au profit de la Méditerranée proprement dite: le déficit total de celle-ci est donc 1740 + 189 km^3/an = 1929 km^3/an. Sur une surface de $2{,}53 \times 10^6$ Km^2, la hauteur déficitaire moyenne (s'il n'y avait pas d'apport atlantique) serait donc de 0,76 cm.

On en déduit, en se fondant sur les estimations de Tixeront (1970) pour les précipitations et les apports des fleuves les éléments du bilan d'eau (Tableau I).

Rappelons que les valeurs de l'évaporation E résultent de la connaissance de ΔV et de G et de l'annulation du bilan d'eau total.

Tableau 1. Bilan d'eau en Méditerranée proprement dite et en Mer Noire.

		Méditerranée proprement dite		Mer Noire (sans mer d'Azov)		Auteur
		km³/an	*m/an*	*km³/an*	*m/an*	
Ruissellement et débit des fleuves	G	514	0,20	400	0.95	Tixeront (1970)
Précipitations	G	884	0,35	181	0.43	Tixeront (1970)
Echanges nets avec l'extérieur	ΔV	1929 (Gib + Mer Noire)	0,76 entrant (à 10–15% près)	189	0.45 (sortant)	Tixeront (1970) Mer Noire
Evaporation	E	3327	1,31	392	0.91	Tixeront (1970) Mer Noire

LE MECANISME HYDROLOGIQUE PROPRE DE LA MEDITERRANEE

Dans le Détroit de Gibraltar existent deux veines d'eau superposées très différentes, en mouvement moyen inverse; les caractères de la couche inférieure, typiquement méditerranéenne, sont le résultat en définitive, des modifications qu'a encourues, dans la mer, l'eau atlantique entrante; subissant une évaporation relativement importante et recevant l'eau des précipitations et des fleuves, elle est transformée par un mécanisme hydrologique propre à la Méditerranée et lié au climat, en une eau sortante typique. C'est le mécanisme de cette transformation qui constitue le problème océanographique majeur de la Méditerranée: il implique des mouvements verticaux descendants puisque l'eau méditerranéenne de forte densité se trouve en profondeur dans le Détroit de Gibraltar.

La façon dont l'eau atlantique entrante répond au climat méditerranéen est radicalement différente en été et en hiver.

En été, l'eau atlantique de surface subit une "prétransformation" qui l'oriente, en quelque sorte, vers les modifications plus radicales qu'elle subira en hiver. A partir du mois de mai, environ, l'échauffement des couches marines superficielles résultant de l'absorption du rayonnement solaire excède les pertes thermiques par rayonnement infra-rouge pelliculaire de surface et par évaporation. Il en résulte que, dans la couche superficielle "marquée" par l'eau atlantique (150–200 m) d'épaisseur à l'Est de Gibraltar, se forme une thermocline, vers 20–40 m d'immersion qui constitue un écran pour l'échange des caractères entre les couches marines qu'elle sépare. L'eau superficielle s'échauffe et s'évapore, sa salinité augmente, mais c'est la température qui commande la structure de densité. Au-dessus de la thermocline elle atteint à la fin de l'été 22 à 26°, plus forte à l'Est qu'à l'Ouest; la salinité peut s'élever alors à 39.50 ‰ dans les parages de Rhodes et Chypre. Immédiatement sous la thermocline, se recontre un minimum de salinité associé à la présence d'une certaine proportion d'eau atlantique; ce minimum s'atténue fortement, par suite de mélanges, au fur et à mesure de la progression générale vers l'Est de l'eau marquée par l'eau atlantique. Le minimum de salinité, dans l'Est de la Mer, dépasse 38,90‰. On suit par continuité, sur un diagramme T. S. (Figure 1), l'évolution du minimum de salinité, depuis environ 36.15‰ à Gibraltar jusqu'à cette valeur élevée de 38.90 dans l'extrême Est à la fin de l'été. Ce minimum de salinité est aussi présent en Mer de Crète où ses caractères sont sensiblement T de 17 à 21° et S. de 38,5 à 38,9‰ en fin d'été. Il se distingue nettement en température, comme en salinité, d'un autre minimum de salinité que l'on trouve aussi en Mer de Crète et qui est associé à la présence sur les bords occidentaux de la Mer Egée et jusque contre le rivage Est du Péloponèse, d'eau peu salée (26‰) issue des Dardanelles. Cette eau voit sa salinité augmenter vers le Sud jusqu'à atteindre aussi 38.90 ‰ près du Cap Maléa: mais les températures en sont plus élevées (24°–25°) que celles de l'eau du minimum de salinité "atlantique". Elle reste collée contre le rivage Ouest de la Mer sous l'effet de la force de Coriolis.

Ainsi la climatologie estivale méditerranéenne a pour effet qu'une couche de surface, chaude et salée, s'est formée au-dessus de la thermocline; sa température et sa salinité croissent nettement vers l'Est, traduisant l'effet, sur la seule couche située au-dessus de la thermocline, des conditions climatiques rencontrées. Au-dessous de la thermocline, l'effet atlantique se suit, en été, jusqu'aux rivages orientaux de la Mer. Le stade estival de la transformation subie par l'eau atlantique entrante ne joue qu'au-dessus de la thermocline; les couches profondes sont isolées des effets superficiels d'échanges avec l'air.

A l'automne et en hiver, le bilan thermique s'inverse. L'évaporation marine et le transfert de chaleur de la mer à l'air sont notablement accrus sous diverses influences: la température de la mer excède la température de l'air, parfois de plus de 10°C; il en résulte une instabilité des basses couches de l'air et une augmentation des échanges de la mer à l'air. Les vents froids, secs, violents (Tramontane, Mistral, Bora) soufflent sur la mer, particulièrement dans sa partie septentrionale. La couche superficielle de forte salinité, présente à la fin de l'été, augmente de densité et se trouve, comme l'air au-dessus d'elle, en instabilité de stratification de densité; des mouvements de convection verticale sont engendrés: ils homogénéisent la couche superficielle et parviennent, en certaines régions, à effacer la thermocline; alors d'épaisses couches homogènes se forment en surface: leur établissement est d'autant plus rapide que les vents sont plus froids, plus secs et plus violents; il suffit de quelques jours pour que la stratification de densité s'efface sur plusieurs centaines de mètres.

Dans plusieurs des bassins à seuils successifs que comporte la Méditerranée, de tels phénomènes existent ou peuvent exister certaines années particulièrement froides, donc avec des variantes quant à leur importance et à leurs caractères. Ainsi, dans le bassin Nord de la Mer Egée (Bassin du Mont-Athos) au Nord du seuil supportant les îles de Lemnos et Imbros, de l'eau de forte densité ($\sigma t = 29.40$) a été trouvée; elle est formée, par hiver froid, sur le plateau continental du Golfe de Saros (T = 12°55, S = 38.80). En été cette eau a été trouvée occupant le fond du bassin en une couche homogène de 500 à 1400 m. Cette eau semble jouer un rôle important sur les caractères des eaux

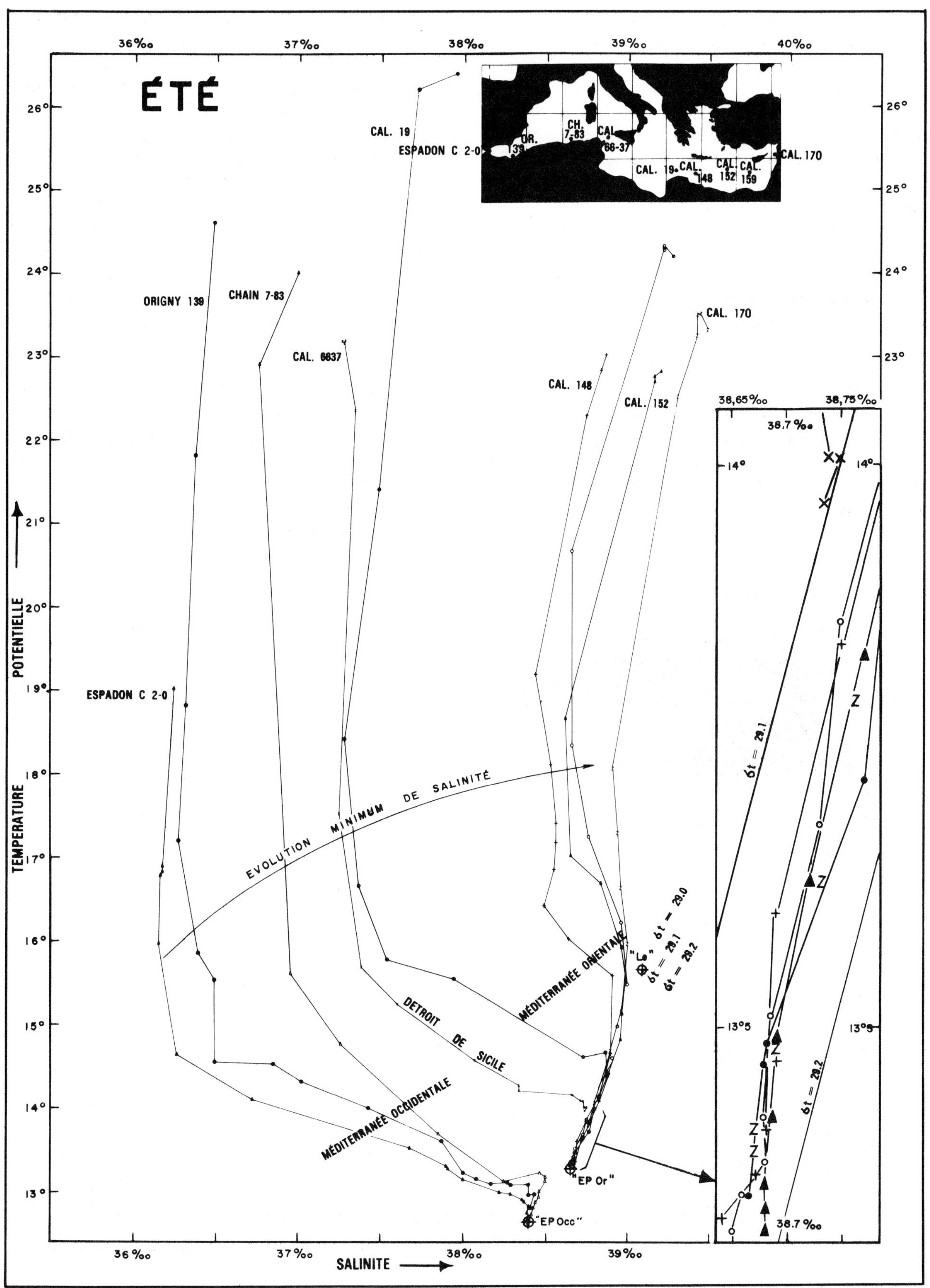

Figure 1. Diagramme de température—salinité (T.S.) en été.

profondes de la Mer Egée.

Dans d'autres bassins, ces effets ont suffisamment d'ampleur pour avoir une influence prépondérante sur une grande partie de la mer, par exemple sur la totalité d'un des deux grands bassins. Tel est le cas des trois régions de la Méditerranée où se forment des eaux qui déterminent les caractères des eaux profondes des deux bassins. Ce sont: (1) la région SE de la Mer Egée et la zone Rhodes-Chypre, (2) la Mer Adriatique, et (3) la partie Nord du Bassin Occidental.

Mer Egée du Sud-Est et Zone Rhodes—Chypre

Cette région se trouve, comme les deux autres, une de celles sur lesquelles arrivent parfois, en hiver, des masses d'air polaire continental: la densité marine superficielle excède en février 1,0290 ($\sigma t = 29.0$). Par mélange vertical, il s'y établit une couche mélangée d'une épaisseur de 100 à 150 m, dans laquelle se fondent l'eau superficielle, chaude et salée de la fin de l'été, l'eau de la couche marquée d'eau atlantique, sous la thermocline d'été; le bas de la couche rejoint la couche intermédiaire salée, reste profond de l'eau homogène de l'hiver précédent, qui est à 100–150 m en septembre.

L'eau de la couche mélangée hivernale est formée d'eau levantine hivernale "Le" (Figures 1 et 2) dont la température est d'environ 15°7 (la station 4692 de l'ATLANTIS (Figure 2) faite en avril traduit déjà un échauffement des couches superficielles), la salinité d'environ 39.10‰ avec un σt voisin de 29.0. Cette eau se répand, plus ou moins horizontalement, dans tout le bassin oriental où sa présence est décelée, en toutes saisons, par une couche de haute salinité, en équilibre de densité dans le bassin au SO et à l'Ouest de la zone de formation, à des immersions de 250 à 400 m. Elle se trouve en été sous la couche faiblement marquée par l'eau atlantique (minimum de salinité). Une partie s'en écoule, au-dessus du seuil de Sicile, vers la Méditerranée occidentale où elle se répand aussi plus ou moins horizontalement et forme la couche "d'eau intermédiaire", relativement chaude et salée et dont les caractères s'atténuent vers l'Ouest. Au large de la Côte provençale, sa température est d'environ 13°4, sa salinité 38.55 et son σt 29.06.

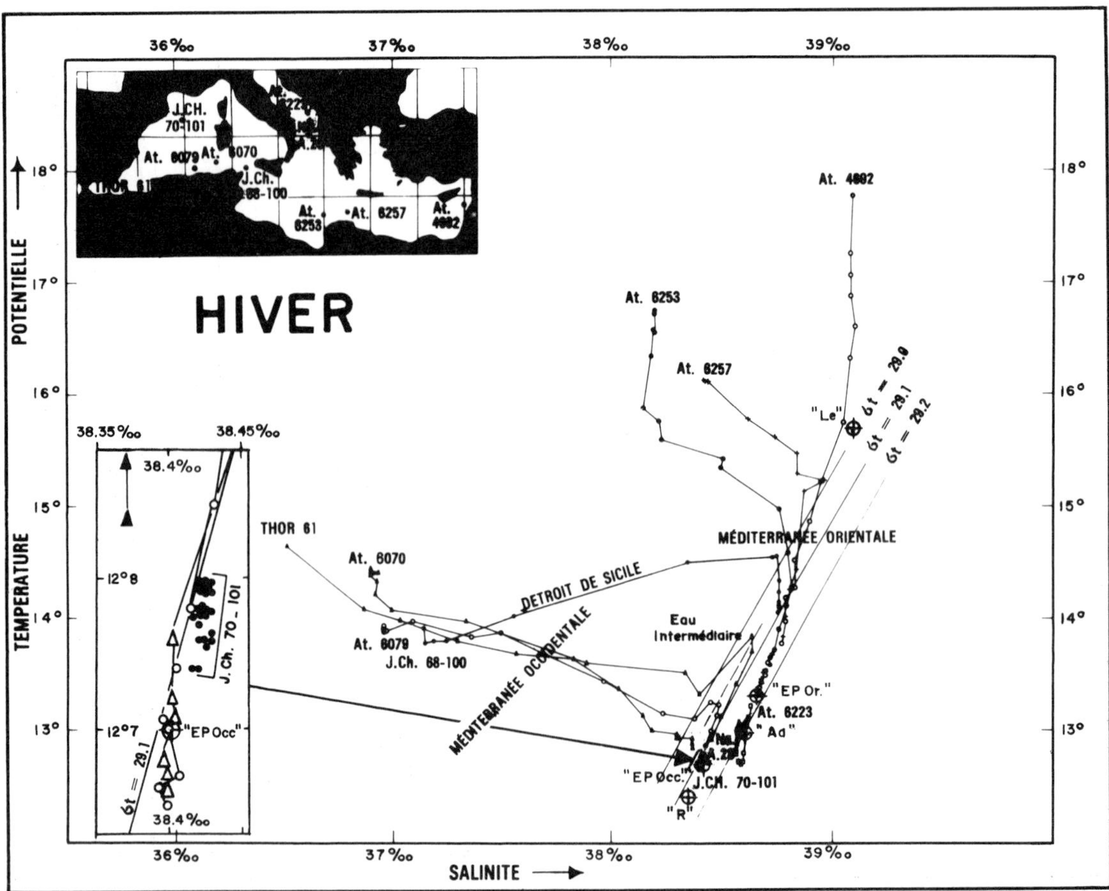

Figure 2. Diagramme de température—salinité (T.S.) en hiver.

La Mer Adriatique

Dans le Sud de l'Adriatique (Zore-Armanda, 1963), en hiver, se forme une eau, "Ad" de densité élevée ($\sigma t = 29.20$) qui résulte du mélange d'une eau froide ($T < 11°C$) existant sur les petites profondeurs du Nord de la mer et formée en présence de la Bora et d'une eau relativement chaude et salée, riche en eau Levantine "Le" qui pénètre dans l'Adriatique près du bord Est du Canal d'Otrante à des profondeurs de l'ordre de 300m. L'eau d'hiver "Ad" homogène jusqu'au fond (1400m) à une température potentielle d'environ 12°95 C une salinité d'environ 38,60‰ et un σt potentiel d'environ 29°20. Ces caractères peuvent évoluer légèrement d'une année à l'autre, en raison des variations annuelles de la rigueur des hivers et du régime des vents.

L'eau d'hiver Adriatique "Ad" s'écoule en Mer Ionnienne, se mélangeant avec l'eau d'hiver Levantine "Le", elle donne naissance à une très importante masse d'eau homogène qui est l'eau profonde "EP.Or" de la Méditerranée orientale: Tpot = 13°30, S. = 38.65, σtpot. = 29.17 (Figures 1 et 2): l'eau "Ad" y intervient sensiblement pour 7/8; l'eau "Le" pour 1/8.

La Méditerranée du Nord-Ouest

Enfin des processus, peut-être plus remarquables car on les trouve au large, se déroulent dans le Nord du bassin occidental, avec une intensité inégale de point en point, dans une bande située à environ 60–80 milles au large de la côte méridionale française et de la Riviera italienne. Les mécanismes y sont radicalement différents de ceux que l'on rencontre dans le Nord de l'Adriatique où l'eau froide qui intervient dans l'eau profonde se forme sur les petits fonds (*processus de shelf*).

En toute saison, sauf localement en hiver, précisément, on trouve dans cette région, comme dans toute la Méditerranée occidentale, un *système à trois couches*: une eau superficielle et sub-superficielle (sous la thermocline) marquée plus ou moins par l'eau atlantique, et de densité relativement faible; l'eau *intermédiaire* citée plus haut, relativement chaude et salée et qui a franchi le détroit de Sicile; eau profonde EP.Occ" de Méditerranée occidentale (Figures 1 et 2).

Au large de la côte provençale les deux premières couches sont animées d'un mouvement cyclonique: en présence de la rotation terrestre, ce fait entraîne l'existence d'une plus faible stabilité de la stratification de densité dans le centre du circuit cyclonique décrit, c'est-à-dire à 60–80 milles de la côte provençale: les eaux les plus légères sont rejetées à la périphérie du circuit, avec présence d'une *divergence* (ascendance) au centre. En outre la région est, en toute saison, le siège de vents dominants, secs et violents, Tramontane et Mistral.

A l'automne et en hiver, ces vents qui sont, surtout en hiver, franchement plus froids que l'eau (jusqu'à plus de 10°C de moins) favorisent l'évaporation et les pertes thermiques de la mer. Il en résulte un mélange vertical notable: la couche mélangée superficielle atteint sa densité maximum là où se trouvait aussi la densité maximum d'été, c'est-à-dire au centre du tourbillon. Lorsque cette eau superficielle limite de la Riviera "R" atteint la densité de la couche d'eau intermédiaire (σt 29.06), elle se mélange à elle jusqu'à des profondeurs de 500 à 800 m, formant une épaisse couche homogène, curieusement chaude et salée, relativement, car elle a emprunté chaleur et sel à l'eau intermédiaire; par exemple les température et salinité sont 12°86–38.44‰, alors que l'eau d'hiver limite de la Riviera "R" (de même densité que l'eau intermédiaire) avait les valeurs 12°40–38.26‰. Les conditions hivernales continuant à agir avec vigueur, par le vent et, parfois aussi, par l'arrivée sur la région de masses d'air continental froid et sec venant du NE sur le flanc Sud d'un anticyclone qui peut s'établir (février 1956 et 1963) sur l'Europe orientale, l'évolution des caractères de la couche homogène devient plus lente, à cause de son épaisseur même: mais elle atteint pratiquement tous les hivers le σt de 29.10, sensiblement égal à celui de l'eau profonde: alors peut se produire une homogénéisation complète en T,S‰ et σt sur toute la couche d'eau (Figure 2, Stations 70–101 du CHARCOT: l'eau profonde se renouvelle et s'enrichit en oxygène par des mouvements de convection portant sur toute l'épaisseur d'eau (2500 m). La variabilité de la rigueur des hivers successifs ne semble pas tant provoquer des valeurs extrêmes de T et S que d'étendre, par hiver froid et long, la région où l'on trouve l'épaisse couche homogène.

On peut rencontrer ces phénomènes de formation d'épaisses couches homogènes au large de la côte provençale entre 4°E et 8°E. Mais la zone ou le phénomène est le plus régulier en hiver englobe le point 42°N–5°E, situé curieusement à la convergence des directions, à terre, de la Tramontane et du Mistral, et occupe une ellipse allongée de 50 milles dans le sens E–O et 30 milles dans le sens N–S. Dans l'Est, vers 8°E, l'homogénéisation est moins profonde en général et intéresse des zones plus restreintes.

La formation de l'eau homogène est très localisée et variable: en présence de vents de NO violents (35–40 noeuds) elle peut en 4–5 jours MEDOC Group, 1970) se produire sur 1500 m. Les vitesses verticales associées aux mouvements d'homogénéisation atteignent des valeurs insoupçonnées, jusqu'à 8 cm/sec pendant quelques heures (Voorhis et Webb, 1970). Ces valeurs sont transitoires et localisées. Les zones homogènes ont au départ des dimensions de l'ordre de quelque milles seulement; elles peuvent s'étendre ensuite sur 50–80 milles, notamment pendant les hivers froids prolongés (1963), au cours desquels le stock d'eau profonde formé est probablement particulièrement important. Cependant, il se forme chaque année de l'eau profonde. Un caractère remarquable de cette eau est que, malgré le caractère localisé et transitoire des zones de formation, et la variabilité des hivers successifs, la température potentielle (12°70 env.), la salinité (38.405 env.) et le σ potentiel (29.11) de l' "E.P. occ." sont remarquablement constants. Il est probable que, dans cette eau, l'eau limite d'hiver "R" intervient pour les 4/5, l'eau intermédiaire pour 1/5.

Une eau dense se forme aussi sur le plateau continental du Golfe du Lion et s'écoule sur le talus par les vallées sous-marines. Mais il ne semble pas que, sauf hiver exceptionnel, cette eau de *shelf* se forme en quantité suffisante pour jouer un rôle important dans la formation de l'eau profonde "E.P. occ."

Nous voyons ainsi comment, par des processus qui sont liés à la climatologie et à la météorologie méditerranéennes, l'eau atlantique entrante, d'abord *conditionnée* en été par le climat local, subit dans les trois régions citées l'effet radical des conditions hivernales qui lui imposent des caractères spécifiques typiquement méditerranéens. Ceux-ci apparaissent pleinement dans les eaux profondes des bassins oriental et occidental qui occupent, à elles seules, plus de 70 % du volume marin. Les processus d'échange vertical permettent l'oxygénation des couches les plus profondes à partir de l'oxygène que l'eau superficielle collecte au contact de l'air. Les trois eaux citées, après divers mélanges profonds, donnent finalement naissance à l'eau profonde de la Méditerranée occidentale dont est essentiellement constituée l'eau de la mer d'Alboran, dans sa partie Ouest: là, en effet, la couche d'eau intermédiaire s'est effacée. C'est essentiellement de l'eau profonde de la Méditerranée occidentale qui sort de la mer dans l'Atlantique.

LA MER NOIRE

Ainsi qu'on l'a vu plus haut, la Mer Noire a les caractères d'un *bassin de dilution* (G–E positif), en raison de l'importance des apports d'eau du Danube en toutes saisons et de ceux des fleuves russes au printemps et en été. Le bilan d'eau à travers le Bosphore varie donc notablement au cours de l'année; le plus grand écoulement vers la Mer Egée se produit en été et des eaux peu salées parviennent dans la moitié Ouest de la Mer Egée près de la surface. La constance du sel en Mer Noire est assurée par un écoulement venant de la Mer Egée (*cf.* Tableau 1), à travers la Mer de Marmara, où la salinité profonde est d'environ 38,50 ‰, au-dessous de la couche superficielle dessalée. Le flux salé entrant en Mer Noire (moyenne 211 km^3/an) est sujet à d'importantes variations, en rapport, notamment, avec les situations météorologiques: les vents du Nord le font décroître ou même l'arrêtent. En toutes saisons, en Mer Noire, la densité de la couche de surface (23°, 18,5 ‰ en été) est toujours plus faible que celle de l'eau profonde; une pycnocline est donc présente en toutes saisons et fait écran pour les échanges de propriétés à travers elle: il n'y a pas de formation d'eau profonde en Mer Noire; l'eau hivernale formée sur les petites profondeurs est peu salée et n'atteint pas une densité suffisante pour pénétrer la couche profonde. Aussi n'y a-t'il pas d'oxygène au delà de 150 à 200 m de profondeur; ces couches sont, par contre, riches en hydrogène sulfuré.

La circulation des eaux de la couche de surface est constituée par deux tourbillons cycloniques dans les deux moitiés Est et Ouest de la Mer. La Mer de Marmara, en été, est une *mer à deux couches*: en surface la salinité est relativement faible (< 30‰ au-delà de 20–25 m) et au-delà elle est d'environ 38.5‰.

Ayant ainsi décrit les principales masses d'eau méditerranéenne, indiqué leurs zones de formation et leur destinée, il est possible de présenter quelques éléments de ce que l'on sait sur leurs mouvements.

CIRCULATION GENERALE DES EAUX

La force qui commande l'essentiel de la circulation des eaux méditerranéennes résulte des caractères de bassin de concentration de la mer. Par les mécanismes décrits ci-dessus, la mer *fabrique* une eau dense qui remplit plus de 70 % du bassin et s'écoule finalement vers l'Atlantique. Les deux écoulements inverses superposés dans le détroit de Gibraltar sont provoqués par un gradient horizontal de pression dirigé vers la Méditerranée dans la couche superficielle et vers l'Océan dans la couche profonde. Schématiquement, on peut dire que les pressions de part et d'autre du détroit s'équilibrent au niveau de l'interface moyenne entre les eaux présentes, soit à environ 150 m (150 dbars). La densité de l'eau marine au-dessus de cette surface étant plus grande dans la mer, cette pression de 150 dbars est provoquée par une hauteur d'eau moindre que dans l'Atlantique: le niveau moyen marin est donc plus bas en Méditerranée qu'en Atlantique. La différence de niveau est de l'ordre de 10 à 15 cm de part et d'autre des entrées Est et Ouest du détroit; elle croît vers l'Est; elle est de l'ordre de 30 cm, par rapport à l'Atlantique, sur la côte méridionale française et croît probablement vers l'Est, à mesure que la densité moyenne des eaux augmente. Cette évolution des niveaux moyens comparés provoque les gradients horizontaux de pression cités plus haut et commande la circulation d'ensemble de la mer. Celle-ci est donc liée aux effets atmosphériques méditerranéens sur la température et la salinité des eaux: la circulation correspondante est donc "thermohaline".

En outre les vents, généralement très irréguliers, mais qui peuvent être forts sont susceptibles de provoquer de notables modifications de la circulation thermohaline superficielle.

L'évolution spatiale des caractères hydrologiques montre que l'eau atlantique entrante a un mouvement général vers l'Est, qui se fait sentir jusqu'à l'extrême Est de la mer; au contraire l'eau levantine, l'eau adriatique dans le bassin oriental, puis l'eau intermédiaire et l'eau profonde de la Méditerranée occidentale ont un mouvement général vers l'Ouest; le sens des gradients horizontaux de pression indique les mêmes résultats.

Mais, en présence de la force de Coriolis, une telle orientation de ces gradients tend à provoquer dans les bassins de la mer des mouvements cycloniques dans les diverses couches marines, à partir du point d'entrée de l'eau (Détroit de Gibraltar pour l'eau atlantique) ou à partir des "sources d'eau profonde" pour les autres eaux. C'est bien le schéma général présent; mais l'existence de détroits et de seuils provoque la partition des flux en veines dans les divers bassins.

Passons en revue les principaux caractères du mouvement, dans la couche superficielle (marquée par l'eau atlantique), la couche intermédiaire du maximum de salinité, et la couche profonde.

Couche Superficielle marquée par l'influence Atlantique

Dans son mouvement vers l'Est, la *couche superficielle* suit la côte africaine contre laquelle elle est collée en *coin*, en une couche plus épaisse près de la côte qu'au large (Figure 3): 150 à 200 m d'abord (jusqu'au méridien 5°E environ), puis 100 m près du détroit de Sicile, 40 m seulement au large de la Cyrénaïque. Les plus grandes vitesses fluides se rencontrent dans une bande côtière de 30 à 60 milles de large, le long des côtes *à droite* du courant, elles décroissent le long des circuits: 0,5 à 0,7m/sec le long de la côte algérienne jusqu'à 5°E, puis 0,2 à 0,5m/sec dans le détroit de Sicile et le long du circuit cyclonique décrit au large de la côte Sud de France. En Mer Tyrrhénienne et dans le bassin oriental, la vitesse moyenne est plus petite. Le mouvement général semble plus faible en été qu'en hiver; ainsi, le long de la côte française, le volume d'eau transportée est de l'ordre de 1,5 à 2 millions de m^3/sec en été, mais 3 à 4 en hiver. Des séparations en deux branches de la veine initiale se produisent: aux environs de 5°E, dans le détroit de Sicile, sur le méridien 22°E, près du Canal d'Otrante.

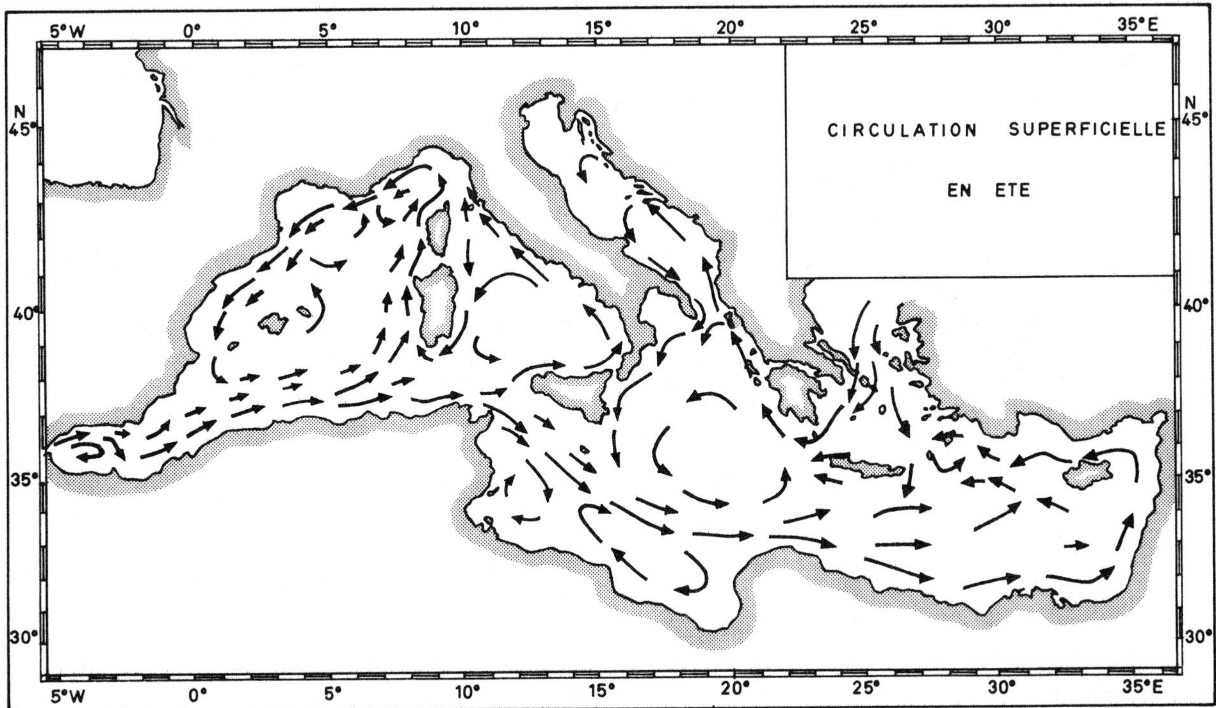

Figure 3. Circulation superficielle en été.

Au large des veines longeant les côtes, les courants sont faibles et variables; ils sont influencés par le vent: celui-ci, lorsqu'il y a une "thermocline" (en été), provoque un courant qui atteint 1% environ de la vitesse du vent, à environ 90° à droite de la direction vers laquelle souffle le vent; mais il présente de longues "oscillations d'inertie" sur une période de l'ordre de 17h, liée à la rotation terrestre. En hiver, la corrélation du courant avec le vent *local* paraît beaucoup moins nette.

Des exceptions à la loi générale de circulation cyclonique se rencontrent en deux régions au moins: dans la mer d'Alboran occidentale, l'eau atlantique entrant sous forme d'un *jet* bien différencié portant à l'ENE tourne au Sud-Est près de l'île d'Alboran et envoie une veine anticyclonique vers sa droite, dans le Sud du bassin, ce qui provoque un courant vers l'Ouest près de la côte africaine. Un cas analogue se produit après franchissement du détroit de Sicile au large de la Syrte où existent deux tourbillons anticycloniques à droite de la veine principale du mouvement.

Couche Intermédiaire de Maximum de Salinité

Le mouvement de la *couche intermédiaire de maximum de salinité subsuperficiel* (Figure 4) tend aussi à suivre des trajets cycloniques dans les bassins, avec des veines plus intenses et des *coins* d'eau à droite du mouvement. Celui-ci est relié à la valeur du maximum de salinité de l'eau *levantine* du bassin oriental et de l'eau *intermédiaire* du bassin occidental. Ce maximum s'évanouit, on l'a vu, en Mer d'Alboran.

Les vitesses de courant sont mal connues; elles sont probablement de l'ordre de 5 à 10 cm/sec dans les veines principales de courant; elles sont probablement beaucoup plus faibles au centre des bassins. Selon Wüst (1961), elles seraient plus grandes en hiver qu'en été. Sur les seuils, les vitesses sont probablement plus élevées; selon les estimations de Morel (1971), le flux d'eau orientale vers la Méditerranée occidentale serait de l'ordre de 0,7 million de m^3/sec, soit de l'ordre des 2/3 du flux sortant à Gibraltar: les flux seraient donc sensiblement proportionnels aux aires concernées puisque la Méditerranée orientale couvre environ les 2/3 de la Méditerranée proprement dite. Dans le détroit de Sicile, des vitesses de l'ordre de 50 cm/sec en direction NO ont été citées localement, près du fond.

Comme pour la couche de surface, on rencontre des exceptions à la règle de circulation cyclonique, notamment après le franchissement de seuils ou de vallées: Détroit de Sicile, Sud de la Sardaigne. Là il y a séparation de la veine en deux; dans le premier détroit cité, une des deux veines fait le tour du bassin Tyrrhénien dans le sens cyclonique et, après avoir envoyé une branche au-dessus du seuil de l'île d'Elbe

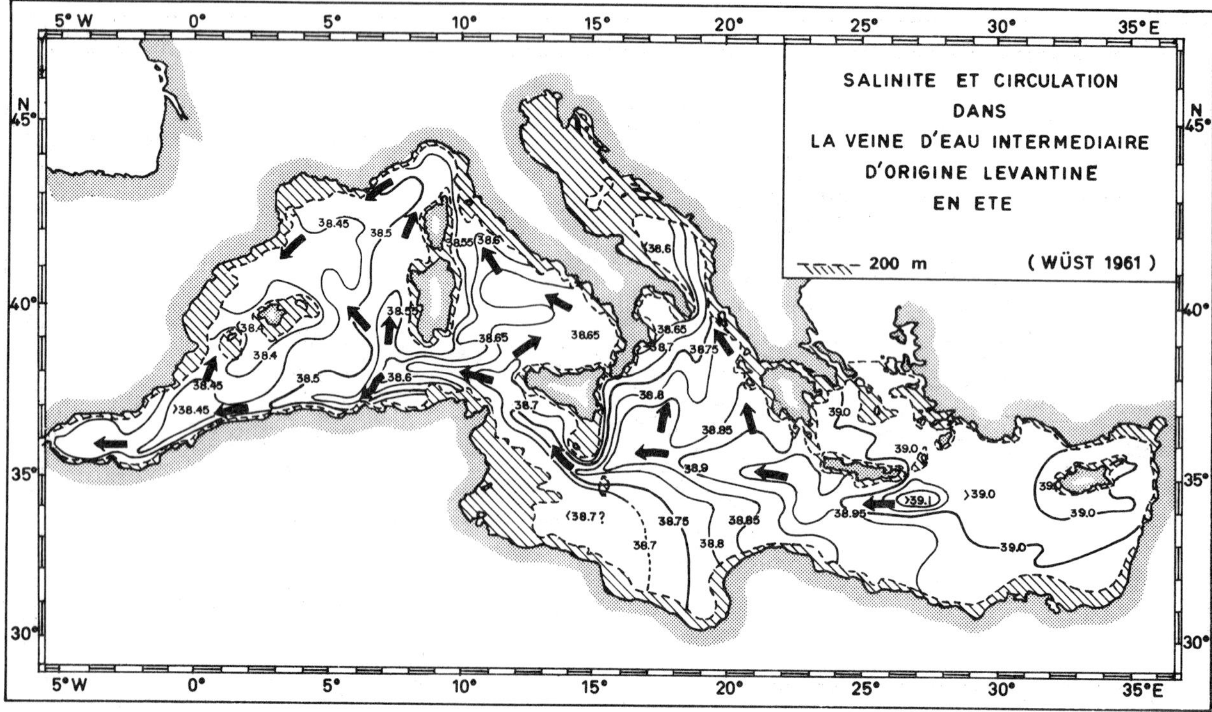

Figure 4. Salinité et circulation dans la veine d'eau intermédiaire d'origine Levantine en été (d'après Wüst, 1961).

vers le bassin algéro-provençal, elle redescend vers le Sud le long du Talus Sarde oriental. Mais la deuxième veine se dirige directement vers le canal de Sardaigne; là une autre partition se produit vers 9°E; la veine septentrionale contourne par l'Ouest le talus Sarde, où elle est animée de vitesses de l'ordre de 10 cm/sec et alimente le circuit cyclonique au Sud de la côte française; mais une autre veine porte directement vers l'OSO vers le bord du Talus Algérien à des vitesses de 3 à 6 cm/sec.

Une veine simple semble se reconstituer dans la partie orientale de la mer d'Alboran, où elle perd son maximum de salinité. Son mouvement vers l'Ouest est très faible, mais resserré par le relief du fond, il va alimenter le flux s'écoulant dans l'Atlantique.

Couche Homogène Profonde

Dans la couche homogène profonde et près du fond, les courants sont faibles et mal connus. Dans le bassin oriental, il y a tendance probable à un mouvement cyclonique; près du fond ce mouvement est issu du Canal d'Otrante, puisque l'Adriatique contribue pour une très grande part à la constitution de l'eau profonde de Méditerranée orientale. Dans les couches profondes (vers 1000 m) il faut s'attendre, dans les veines contre les talus, à de lents mouvements vers l'Ouest (2–3 cm/sec, probablement irréguliers). Dans le bassin occidental, près du fond, le mouvement cyclonique doit être issu de la zone source d'eau profonde, à 60–80 milles au SSO de Marseille. Dans les couches profondes (1000–2000 m) les courants sont de 2 à 3 cm/sec en général, mais variables dans le temps en vitesse comme en direction. L'écoulement vers l'Atlantique implique un très lent mouvement d'ensemble vers l'Ouest.

MAREES ET COURANTS DE MAREE

La marée est faible en Méditerranée, du fait que la marée océanique n'y pénètre pratiquement pas et que l'étendue restreinte de la mer limite la marée propre. L'amplitude est, en général, de 2 à 4 décimètres. La périodicité est essentiellement semi-diurne mais, par endroits, les ondes diurnes sont notables et même prépondérantes parfois (Alger notamment). La marée n'est importante que sur les rivages bordés de zones peu profondes de grandes dimensions: Golfe de Gabès (2 m), Trieste (1 m). Souvent la marée *astronomique* est bien plus faible que les oscillations du niveau marin dues aux effets météorologiques: celles-ci atteignent couramment 1 m. Sur les rivages bordés de vastes zones de petits fonds et ouverts à de grands *fetchs* de vents forts, des "storm surges" se produisent, notamment à Venise, en présence de vents forts et prolongés de SE (2m en novembre 1966).

Mais la relative faiblesse de la marée n'empêche pas que certaines aires limitées, notamment certains détroits, soient le siège de courants de marée importants (Bonifacio, Messine, Euripe, *etc*). Les phénomènes de cet ordre ont une importance particulière dans le Détroit de Gibraltar.

INDICATIONS SUR LES COURANTS DANS LE DETROIT DE GIBRALTAR ET SES ABORDS

Dans cette aire marine, existent des courants superposés inverses transportant des volumes d'eau à peu près égaux en moyenne, à 5% près. Mais la variation, en fonction de l'immersion, de la largeur du bras de mer, au voisinage du méridien du Cap Spartel affecte la forme d'un triangle, pointe en bas; il en résulte que l'aire de la section offerte au flux sortant n'est que de l'ordre du quart de celle qui s'ouvre au flux entrant; ainsi la valeur moyenne du courant entrant près de la surface—de l'ordre de 0,2 m/sec—est de l'ordre du tiers de celle dont est animée la masse d'eau sortante, 0,6 m/sec.

La différence notable des marées de part et d'autre du détroit y induit d'importants courants de marée alternatifs qui se superposent au courant moyen évoqué. Dans la partie Ouest du détroit, c'est-à-dire à l'Ouest du Seuil, la phase de ces courants, qui est à peu près la même à toutes les immersions, est calée sur la marée superficielle: le maximum de courant de marée entrant se produit sensiblement 3 heures après la pleine mer de Tarifa; le maximum de courant sortant 3 à 4 heures avant cette pleine mer.

Dans la partie Ouest du détroit (vers 5°55′O), la valeur maximale du courant de marée propre est de 0,7 à 0,8 m/sec jusqu'à environ 200 m de profondeur; à 350 m elle tombe à environ 0,4 m/sec (par fonds d'environ 400 m).

Ainsi la *part alternative* du courant de marée a dans les 200 m superficiels, une amplitude plus grande que la composante moyenne entrante du courant; le courant global, somme du courant moyen et du courant de marée, s'inverse donc au cours de la marée dans la couche atlantique. Au contraire, au delà, l'amplitude du courant de marée est plus faible que le courant moyen de sortie; alors le courant global ne s'inverse plus avec la marée au delà de 250 m environ: *il est toujours sortant* et sa vitesse oscille entre un minimum (environ 0,3 m/sec), lorsque la composante alternative du courant de marée est maximale entrante et une valeur beaucoup plus élevée (1,0 à 1,1 m/sec) lorsque la composante alternative est maximale sortante.

Les valeurs du courant maximal sortant croissent vers l'Ouest jusque vers 6°20'O (4,9 noeuds signalés) tandis que l'interface entre les deux eaux superposées se rapproche du fond, dont l'immersion varie peu dans la vallée à l'Ouest de l'entrée Ouest du détroit. Au delà, l'eau sortante atteint la pente du talus et suit des vallées sous-marines qui, bien qu'ayant un faible relief propre, canalisent l'écoulement (Madelain, 1970); celui-ci, tant qu'il reste au contact du fond, garde une remarquable individualité et forme un *courant de densité* bien différencié jusqu'au voisinage du Cap Saint Vincent, pour certaines veines.

A l'Est du Seuil du détroit, certains caractères des courants sont profondément modifiés, en raison de la formation d'un *front interne*, qui est analogue à une onde instable sur l'interface des eaux superposées. Le front prend naissance sur le Seuil (5°40'O environ), à peu près à l'heure de la pleine mer locale, et se propage vers l'Est, à une vitesse de l'ordre de 3 à 4 noeuds jusqu'en mer d'Alboran (Frassetto, 1964); il est suivi d'importantes oscillations internes formant son *sillage*. Le passage du front interne correspond à un effondrement rapide (100 m et plus) de l'interface.

L'interface, en position moyenne, remonte vers l'Est; la forme du chenal près de l'entrée Est du détroit, où les profondeurs atteignent 900 m, est telle que la section offerte à l'écoulement méditerranéen sortant est bien plus grande que celle qui est offerte à l'écoulement atlantique entrant; contrairement à ce qui se passe dans l'Ouest du détroit, les courants sont alors bien plus forts dans la couche superficielle, bien plus faibles dans la couche profonde, particulièrement dans la partie étroite du détroit, du méridien de Tarifa jusqu'à l'entrée de la baie d'Algésiras, ils atteignent 4 noeuds (2 m/sec). Il y a prédominance du courant vers l'Est et, même, la composante du courant global vers l'Ouest disparaît dans une aire assez vaste sur le méridien de Tarifa et à l'Est; ce phénomène est associé à des mouvements ascendants d'eau méditerranéenne sur le flanc Est du Seuil à certains moments de la marée (Lacombe *et al.*, 1964, Lacombe, sous presse).

Il convient d'ajouter qu'en plus des fluctuations périodiques liées directement ou indirectement à la marée, le régime des courants dans le détroit est notablement influencé par la distribution des pressions atmosphériques sur la Méditerranée occidentale. En gros, on peut dire que la diminution de ces pressions provoque une augmentation du flux entrant dans le détroit et inversement. Aussi, le flux entrant, moyenné sur une période de la marée semi-diurne, peut-il subir des fluctuations atteignant 100% de sa valeur moyenne.

Le régime des courants dans le détroit y conditionne la sédimentation locale ou plutôt l'absence de sédimentation; la présence d'eau méditerranéenne a des répercussions sur les conditions du dépôt des particules transportées jusque dans l'Ouest du détroit et, même, dans la très vaste région de l'Atlantique ou l'effet méditerranéen se fait sentir sous forme d'une couche étendue de maximum de salinité à une immersion de 1000 à 1200 m, dans la moitié Est de l'Atlantique Nord entre 20° et 55° de latitude Nord.

BIBLIOGRAPHIE

Carter, D. B. 1956. The water balance of the Mediterranean and Black seas. *Publication in Climatology*, Centerton, New-Jersey. Drexel Institute of Technology. Laboratory of Climatology. 9 (3): 123–174.

Frassetto, R. 1960. A preliminary survey of the thermal microstructure in the Straits of Gibraltar. *Deep Sea Research*, 7 (3): 152–162.

Frasetto, R. 1964. Short-period vertical displacements of the upper layer in the Straits of Gibraltar *Saclant ASW Research Center Technical Report* 30:49 p.

Lacombe, H. and P. Tchernia 1960. Quelques traits généraux de l'hydrologie Méditerranéenne. *Cahiers Océanographiques*, 12 (8): 527–547.

Lacombe, H., P. Tchernia, C. Richez and L. Gamberoni 1964. Deuxième contribution à l'étude du détroit de Gibraltar. *Cahiers Océanographiques*, 16 (4): 283–327. Résultats d'observations. (Id. Ibid.: 23–94).

Lacombe, H. 1971. Le détroit de Gibraltar; Océanographie Physique. *Mémoires du Service Géologique du Maroc*, 222 bis.

Madelain, F. 1970. Influence de la topographie du fond sur l'écoulement méditerranéen entre le détroit de Gibraltar et le Cap Saint-Vincent. *Cahiers Océanographiques*, 22 (1): 43–61. Résultats d'observations. *Cahiers Océanographiques Supplément* 1:89 p.

MEDOC Group 1970. Observations of formation of deep water in the Mediterranean Sea, 1969. *Nature*, 227: 1037–1040.

Miller, A. R., P. Tchernia and H. Charnock 1970. Mediterranean Sea Atlas. *Atlas series, Woods Hole Oceanographic Institution*, 3:190 p.

Morel, A. 1971. Caractères hydrologiques des eaux échangées entre le bassin oriental et le bassin occidental de la Méditerranée. *Cahiers Océanographiques*, 23 (4): 329–342.

Tixeront, F. 1970. Le bilan hydrologique de la Mer Noire et de la Méditerranée. *Cahiers Océanographiques*, 22 (3): 227–237.

Voorhis, A. and D. C. Webb 1970. Large vertical currents in a winter sinking region of the Northwestern Mediterranean. *Cahiers Océanographiques*, 22 (6): 571–580.

Wüst, G. 1961. On the vertical circulation of the Mediterranean Sea. *Journal of Geophysical Research*, 66 (10): 3261–3271.

Zore-Armanda, M. 1963. Les masses d'eau de la Mer Adriatique. *Acta Adriatica*, 10 (3):1–93.

Speculations Concerning Bottom Circulation in the Mediterranean Sea*

Arthur R. Miller

Woods Hole Oceanographic Institution, Woods Hole, Massachusetts

ABSTRACT

Speculative concern about bottom currents in the Mediterranean Sea builds up a case for seasonal deposition and local movements with transient features. The supply of suspended material may also have seasonal patterns with dependence upon geographical origins. The implications of stagnant deep water are examined and discussed. While there are no clear evidences of well-defined bottom water movement, neither is there any evidence of stagnancy with the sole exception of the Sea of Marmara, whose low oxygen content and high salinity values suggest isolation from both the Black and the Aegean Seas.

RESUME

L'intérêt spéculatif pour les courants de fond de la Méditerranée a conduit à dégager des hypothèses sur les dépôts saisonniers et mouvements locaux avec des traits intermédiaires. La source du matériau en suspension pourrait dépendre des saisons et des origines géographiques. Les hypothèses sur les eaux profondes stagnantes sont examinées et discutées. Bien qu'il n'y ait aucune preuve de courant de fond, il n'y a pas non plus de preuve de stagnation à l'exception de la Mer de Marmara dont la faible teneur en oxygène et haute salinité suggèrent une séparation d'avec la Mer Noire et la Mer Egée.

INTRODUCTION

The dearth of actual measurements of bottom currents in the Mediterranean Sea points up the need for definitive positioning of potentially active areas tied to known circulatory features. This scarcity should not be taken as indicative of a lack of interest or even a lack of measurable phenomena in this inland sea. It does, however, permit one to set up imaginative speculations regarding the degree of movement, the localities where movements are likely to occur, and the nature of the techniques which might be useful for their determinations. For this purpose I intend to bring in the notion of seasonal or episodic occurrences where the sedimentary geologist may ally himself with the physical oceanographer at the interfacial boundary. The purpose of this discussion is to point out areas of potential interest to both disciplines and to suggest some indirect procedures which might contribute to knowledge of bottom flow. This discussion is intended to be supplementary to the general exposition of the circulation of the Mediterranean by Lacombe and Tchernia in the previous chapter.

VARIABILITY OF BOTTOM CURRENTS

There are some features of Mediterranean circulation which lend themselves to speculation about variability of bottom currents. These features can be separated into two distinct regimes (Figure 1). The first regime is relatively shallow and concerns the

*Contribution Number 2750 from Woods Hole Oceanographic Institution, Woods Hole, Massachusetts.

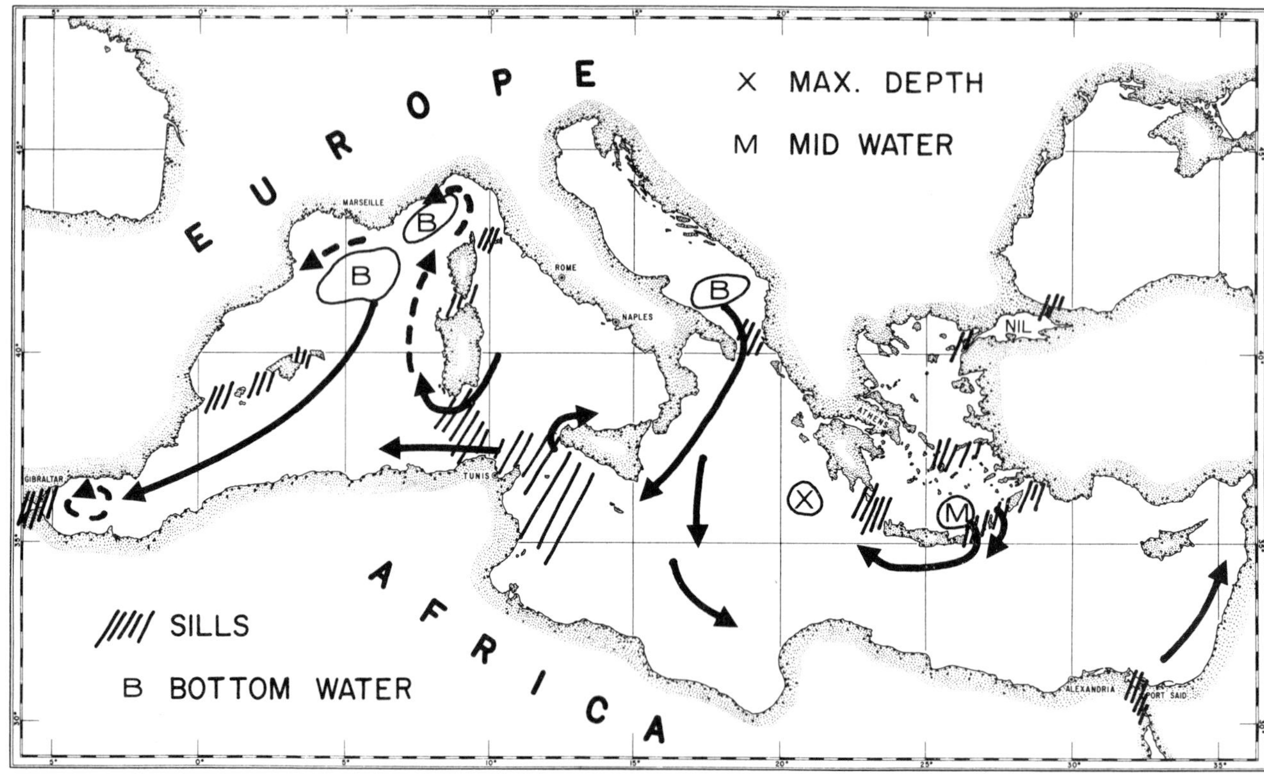

Figure 1. Chart showing potentially active bottom areas in the Mediterranean.

sill-depth phenomena. The Mediterranean is composed of a number of seas separated from each other by thresholds of limited depth which act as gates controlling the quantity and quality of interchange between seas. Although tidal flow is minimal within the Mediterranean Sea, it is known that barometric differences play a part in forcing flow across the sills. While these fluctuations in pressure are not to be compared in magnitude with those over the North Atlantic, for instance, there are sufficient east-west differences to note a three-fold greater variation in winter than in summer amounting to about a 4-millibar difference. This effect of pressure differences could be combined with the driving effect of gales and other disturbances to accelerate flow across the sills. This flow would be partially baroclinic due to the winds and partially barotropic due to the pressure and could consequently affect the average bottom current situation in a particular area. This is the first speculative element tying bottom currents to seasonal phenomena.

The second regime is quite different from the first but is, nevertheless, linked with seasonal occurrences. In the deep water there seems to be little justification for pointing to any special area of bottom activity except in the consideration of the basic exchange system within the Mediterranean, as developed by Lacombe and Tchernia (this volume). The inflow of Atlantic water is increased in density mainly by increasing its salt content by evaporation. The product of dense deep water is accumulating from winter to winter, and the balance is maintained by flow out of the Strait of Gibraltar. The interim period, that is, the residency of the dense water, is not confined to sill depth, otherwise deeper levels would stagnate. At least one area has been shown to form bottom water during periods of severe winter weather MEDOC Group, 1970).

The flow of bottom water out of the area of generation has yet to be observed, and no definitive measurable estimates have been made as to quantities of bottom water produced each year. Miller (ms, in preparation) has postulated that 45×10^3 km^3 of 38.5‰ water is required to be removed from the Mediterranean each year to maintain a steady state condition (Lacombe and Tchernia, this volume, calculate a volume of 36.2×10^3 km^3). Of course, the explicit salinity value is affected by dilution since it is never observed in this concentration at Gibraltar. The area of generation of bottom water has been observed off the coast of France. The mode in which it is removed is subject to conjecture. An enlarged spread of T-S correlation values in deep samplings (Miller, 1963) suggest that the mode of removal might be in discrete parcels with possibly random paths from here to there. The new bottom water in its most direct path to Gibraltar must travel at least 700 to 800 miles (1297–1483 km) by way of the eastern slope of the Balearic Islands and the southeastern coast of Spain. During the submersible (Alvin) search for the lost bomb off Palomares the crew of the submarine experienced significant current near the bottom and have estimated that current to be about $\frac{1}{2}$ knots (26 cm/sec), downslope (E. Hays, Personal communication). This search was conducted in the Alboran Sea where Spanish investigators have noted a semi-permanent large-scale eddy to the east of Gibraltar. It seems certain that bottom movement in this area is complex and linked to the general circulation of the area.

The two regimes, deep water bottom currents and sill depth currents, can also be inter-related. In the Eastern Mediterranean the deep water is considered to have its origins in the Adriatic (Pollak, 1951). In general, there seems to be little evidence for any significant flow along the bottom with a few exceptions. Water out of the Adriatic Sea appears to flow out of the Strait of Otranto over the sill hugging the Italian side of the Strait. There is reason to believe this flow will sink downslope past the sill, in part, and as a separate phenomenon mix with Levantine Intermediate Water at depths of 300 to 600 meters. The product of this mixing, water of 14.5°C and 38.83‰, appears to be a point-source for water making up the Intermediate supply in the Tyrrhenian Sea and it returns to the west over the Sicilian sill. With the periodic and/or aperiodic emptying of cold, saline water from the Adriatic Basin there probably is a carry-over of suspended matter which could be found along the Italian slope of the Ionian Sea. Conjecture suggests that flow out of Otranto could be sporadic and fast, diminishing in intensity with distance from origin.

A flow somewhat similar to that over the Otranto Sill can take place in the Straits of Caso and Scarpanto between Crete and Rhodes. Here, however, the water should tend to move along the southern Cretan coast at depths of 1000 to 1200 meters and not flow further downslope. Northwest of Crete there is a deep greater than 5000 meters in depth. Its position is 36°34′N, 21°04′E and it is not anoxic (Miller *et al.*, 1970). Presumably, since it is several thousand meters deeper than the mean depth around it some processes of renewal may occur, but it is difficult to conceive what mechanism could cause such renewal. Possibly the meager nutrient situation could inhibit oxygen utilisation and renewal may have little significance. In the Aegean Sea, in the deep southern part, relatively high oxygen content suggests dynamic activity throughout the water column, but, similar to the fore-mentioned deep, oxygen utilisation may be minimized because of the poor nutrient quality of the water thereby reducing the inference of deep water movement. In the far east, east of Cyprus, some deep water activity is suggested by the slope of isohalines and isotherms along the coasts of Israel and Lebanon. In the north in the Sea of Marmara, the extremely low oxygen and high salinity of the deep water point to little activity, if any, along the bottom, and this area, the channel between the Black and Aegean Seas, contrasts remarkably with either of the adjoining seas.

The speculations concerning bottom currents have taken into account some of the general features of Mediterranean circulation and, for the most part, these speculations are linked with seasonal phenomena and sporadic occurrences. It should be clear that no semi-permanent features have been identified within these conjectures and that only probable tendencies have been pointed out. If some large-scale movement were detected by observation it would be difficult to assign any more significance to it other than chance and the element of predictability would be missing. By linking the observations to the tendencies inherent to the area it might be possible to define probabilities of occurrence and repeatability. However, this requires a broad attack and is perhaps more dependent upon the results of sedimentology than on direct current measurements.

TRACERS AND BOTTOM CURRENT FLOW

There are potential tracers which can serve to define those areas of deposition or erosion by which a pattern of bottom flow might evolve (Figure 2). Fine-grained sediments will be distributed according to fluid stress along the bottom (White, 1970) and, with the assumption of lifting of loose grains where fluid flow moves from zero, the net sedimentary accumulation should be indicative of significant bottom flow whether it is transient or not. Completely stagnant areas, if existent, should have distinctive sediment distributions as opposed to non-stagnant areas. Sedimentary deposition will, in part, depend on river flow and the alluvium should have geographical import respecting the sources. But run-off in the Mediterranean with the exception of the Nile River is completely in the northern sectors. It should be pointed out, however, that the Mediterranean European coasts have a higher frequency of precipitation than the rest of continental Europe, with the Dalmatian mountains having the greatest probability of occurrence. This precipitation is concentrated seasonally and these same areas lack precipitation

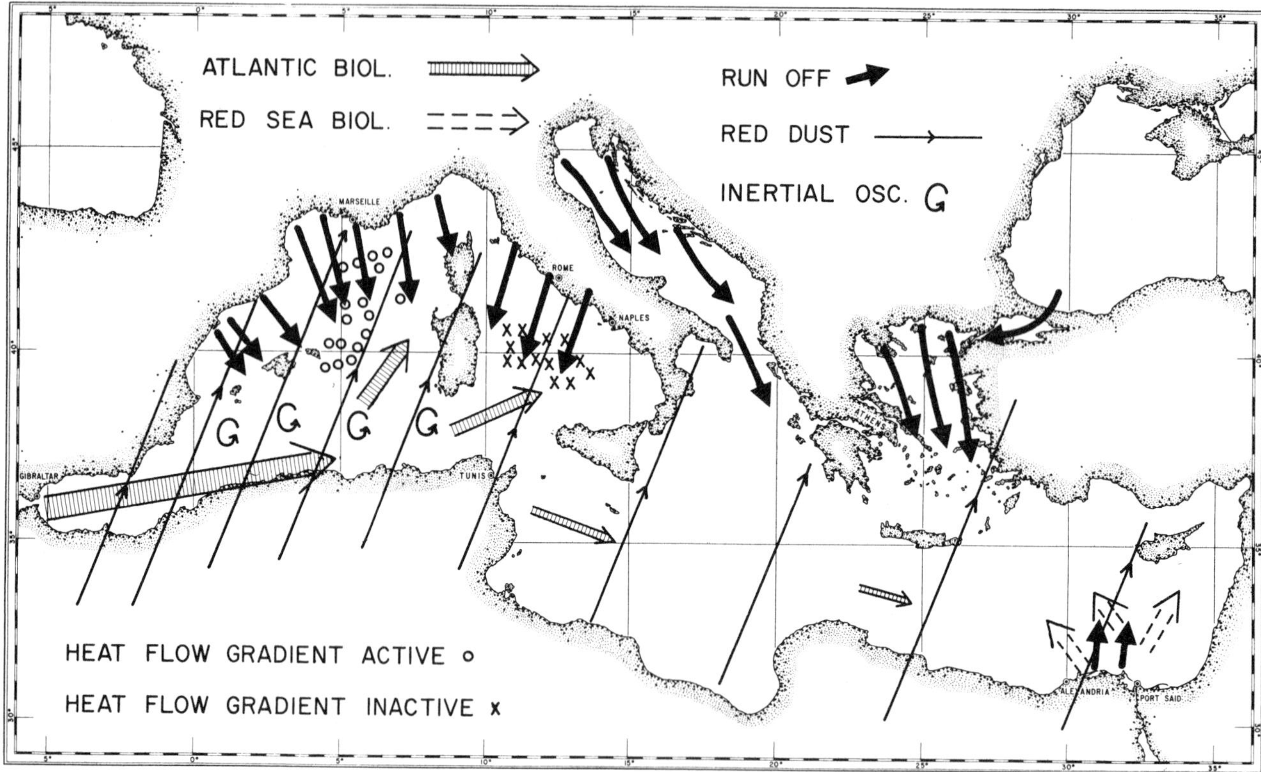

Figure 2. Chart showing potential sedimentary tracers and their movement in the Mediterranean.

in the summer. Consequently, the supply of alluvium is not constant but pulsing. The seasonal pulsations, then, of sediment in suspension would emanate from the North, exclusively, now that the Aswan Dam controls the Nile outflow.

Other suspended materials subject to deposition would be biological in origin. From west to east the nutrient content diminishes appreciably. The inflow from the Atlantic should carry with it Atlantic organisms which will be carried by surface currents to the east ultimately dying off, sinking and decomposing. The skeletal remains should have geographic distribution related to the source. Conceivably, indigenous species will be distributed according to salinity distributions. It is possible that Red Sea organisms might migrate into the Eastern Mediterranean with the Bitter Lake salt barrier in the canal no longer effective. However, these might only be swimming organisms if the net flow through the Suez Canal is southward.

One important depositional material may be eolian in origin. There is a plethora of names for the various winds of the Mediterranean usually characterizing their violence, gustiness, temperature, and origin. Those winds, important to the sedimentary geologist, are dust storms which carry aloft tons of African soil sometimes deep into Europe as far as the Alps, England, Denmark and Eastern Russia. To quote Biel (1944, p. 19: "Along a warm front from 1 to 2 million tons of African desert dust may be transported to the north and northeast." These storms are of the *sirocco* type, hot and dry, with a frequency in the western Mediterranean of about 50 siroccos per year, the number decreasing eastward. The season for siroccos is spring and they are responsible for the *blood rains* in Europe. This writer has experienced a sirocco, or *khamsin,* while hove-to aboard the R/V ATLANTIS, north of Crete. The furled sails quickly accumulated a cake of red dust and all on-deck equipment became coated. Some years earlier, on the same ship, Worthington (Personal communication) recalls a similar experience in the western Mediterranean in the Alboran Sea. "Blood rains" are frequently noted over Malta, southern Italy, and Yugoslavia.

The kinds of material that can be whirled aloft by winds should be sufficiently small in diameter to be carried about for great distances within the waters of the Mediterranean before sinking. The accumulation, or lack of accumulation, in the surficial sediments may be a tracer for the existence of bottom currents.

The MEDOC cruises of 1969 and 1970 point to a different kind of tracer, heat distribution. There is about 0.2°C temperature difference between the bottom waters of the Tyrrhenian Sea and those of the Western Basin (Algéro-Provençal Basin) (Miller,

1969). There is some speculation that the extra heat of of the Tyrrhenian basin is due to heat flow of volcanic origin. However, the Algéro-Provençal basin is the recipient of the products of deep water formation occurring off the coast of France in winter. This cold water, so dense that it sinks to the bottom, is probably responsible for the temperature differences between the two basins. Heat flow probes in the area of bottom water formation (Ryan *et al.*, 1970) could show gradients, and, possibly, seasonal changes in the gradients, which might be correlated with the fluid temperature changes. A pattern of heat-flow distribution could be inferential to the presence of bottom currents.

TRANSIENT NATURE OF BOTTOM CURRENT FLOW

The basic theme of these speculations regarding bottom currents is the probable transient nature of any currents to be observed. This transiency is for the most part related to seasonal events either in the case of direct observation or indirect methods using opportune tracers. Inertial oscillations have been found in the entire water column in the western Mediterranean with long persistence of over three weeks (Perkins, 1970). Whether this is of significance to currents along the bottom is not known. Forcing of such oscillations may have surface origins. An adequate description of bottom currents in the Mediterranean is more apt to be derived from sedimentary analysis than from direct observations where transiency seems to dominate.

On the basis that rates of deposition might vary from as little as $\frac{1}{10}$ mm per year to as much as 2 mm per year sedimentary analysis might be difficult in defining a seasonal source of sedimentary supply. Pulsations of the supply of particles in suspension could be smoothed by circulatory rates. It would be a challenge to identify a "tree-ring" history of annual deposition. The Mediterranean, with its seasonal sources of detritus and sediment load, and its seasonal dynamic activity, and its geographical complexity, poses interesting problems in paleo-oceanography.

CIRCULATION DURING THE GEOLOGICAL PAST

The above problems are related to the present. In the historical past, for instance, how sensitive was circulation of the Mediterranean related to exterior influence? If the general sea level were lowered would the greater restriction of the Gibraltar Strait slow the inward flow of Atlantic water? And, with a change in the water budget, to what degree would this affect the climatology? If the balance of precipitation and evaporation were reversed the interior circulation would be drastically changed and the distribution of sedimentary transport would be affected as a consequence (see Huang and Stanley, this volume).

In the geologic past, presumably during periods of glacial ice melt when stratification of the water column was promoted by the presence of fresh water at the surface, stagnation persisted in the eastern Mediterranean. Sapropelic layers of sediment were deposited during these periods of stagnation (Olausson, 1960). In an enclosed sea, the setting up of stratified layers in the water column could block vertical circulation and, if persistent, deep water oxygen would be entirely consumed by processes of decay. The resulting stagnant anoxic condition would then be conducive for the deposition of sapropelic muds.

The absence of bottom movement implied by the discovery of sapropelic muds in the eastern Mediterranean is not too greatly different from the implied sluggishness of present deep water activity in the Eastern Basin. A considerable difference between present conditions and those probably prevailing during the periods of Quaternary stagnation, aside from the assumption of stratification, lies in the high organic content of sapropelic muds (greater than 2% carbon). The eastern Mediterranean today is a sea of very low productivity and is relatively impoverished in nutrient chemicals. The implication that the eastern half of the Mediterranean became another Black Sea does not explain the high productivity associated with these muds. The opening at Gibraltar would, during the stagnant period, cease to provide significant nutrient material. River run-off might provide some nutrients but it seems more likely that the high productivity should have been associated with oceanic phenomena. A possible explanation for the high carbon content of the muds may be in the opening of the eastern Mediterranean to the south. It could then have had access to one of the most productive areas of the world ocean, the Arabian Sea and the Indian Ocean.

In any case, today, the Eastern Mediterranean poses no dangers of stagnation. In fact, the Aegean Sea, for example, is convectively very active and is well-oxygenated from top to bottom. The implications of the sapropelic muds with regard to paleo-circulations suggest a need for more geographical exploration to determine the bounds and continuity of stagnant regimes.

REFERENCES

Biel, E. R. 1944. Climatology of the Mediterranean area, *Miscellaneous Reports, Publications of the Institute of Meteorology of the University of Chicago,* 13:180 p.

Huang, T.-C. and Stanley, D. J. 1972. Western Alboran Sea: sediment dispersal, ponding and reversal of currents. In: *The Mediterranean Sea: A Natural Sedimentation Laboratory*, ed. Stanley, D. J., Dowden, Hutchinson and Ross, Inc., Stroudsburg, Pennsylvania, 521–559.

Lacombe, H. and Tchernia, P. 1972. Caractères hydrologiques et circulation des eaux en Méditerranée, In: *The Mediterranean Sea: A Natural Sedimentation Laboratory*, ed. Stanley, D. J., Dowden, Hutchinson and Ross, Inc., Stroudsburg, Pennsylvania, 25–36.

MEDOC Group. 1970. Observations of formation of deep water in the Mediterranean Sea, 1969, *Nature*, 227:1037–1040.

Miller, A. R. 1963. Physical oceanography of the Mediterranean Sea: a discourse. *Rapports et Procès-verbaux des réunions de la Commission Internationale pour l'Exploration Scientifique de la Mer Méditerranée*. 17(3):857–871.

Miller, A. R. 1969. La circolazione generale del mare Mediterraneo e le sue relazioni con la circolazione del mare Tirreno, *Annali dell' Istituto Universitario Navale di Napoli*, 38:5 p.

Miller, A. R. 1972. The distributions of temperature, salinity, and oxygen from the *Woods Hole Mediterranean Sea Atlas* (In preparation).

Miller, A. R., P. Tchernia, H. Charnock and D. McGill 1970. Mediterranean Sea Atlas. Temperature, Salinity, Oxygen Profiles and Data from Cruises of R. V. ATLANTIS and R. V. CHAIN, *Woods Hole Oceanographic Atlas Series*, 3:190 p.

Olausson, E. 1960. Studies of sediment cores. *Reports of the Swedish Deep-Sea Expedition, 1947–1948*, 8(6):337–391.

Perkins, H. 1970. *Inertial Oscillations in the Mediterranean*, Doctoral Dissertation; Massachusetts Institute of Technology and Woods Hole Oceanographic Institution, 155 p.

Pollak, M. J. 1951. The sources of the deep water of the Eastern Mediterranean Sea, *Journal of Marine Research*, 10(1):128–152.

Ryan, W. B. F., D. J. Stanley, J. B. Hersey, D. A. Fahlquist and T. D. Allan 1970. The tectonics and geology of the Mediterranean Sea. In: *The Sea*, ed., Maxwell, E. A., Wiley-Interscience, New York, 4(2):387–492.

White, S. J. 1970. Plane bed thresholds of fine grained sediments. *Nature*, 228:152–153.

Supply of Fluvial Sediment to the Mediterranean during the Last 20,000 Years

Claudio Vita-Finzi

University College London, London

ABSTRACT

There have been notable fluctuations in the rate of fluvial sediment in suspension supplied to the Mediterranean Sea during the last 20,000 years. The sequence of events, when considered in light of the eustatic record, suggests that within this time span the periods of high sediment supply ranged from 10,000 to 5,000 years B.P. and from 300 years B.P. to the present.

RESUME

Au cours des 20.000 dernières années des changements notables ont apparu dans la proportion des sédiments en suspension, d'origine fluviale, déversés dans la Méditerranée. La séquence, considérée en fonction des variations eustatiques, suggère que les périodes de dépôt intense se situent entre 10.000 et 5.000 ans B.P. et entre 300 ans B.P. et le présent.

INTRODUCTION

This paper deals with changes in the rate at which fluvial sediment in suspension has been supplied to the Mediterranean Sea in the course of the last 20,000 years. It is hoped that the proposed sequence will be found helpful in the interpretation of cores taken from the sea floor, if only by removing the need to make assumptions about terrigenous sedimentation in terms of climatic changes.

ALLUVIAL HISTORY OF THE MEDITERRANEAN CATCHMENT

The streams shown on Figure 1 have deposited two major alluvial fills since the Tertiary. Stratigraphic and archeological data suggest that both deposits were laid down synchronously throughout the area sampled; where radiocarbon dates are available they support this inference. It is believed that Fill I was laid down between 50,000 and 10,000 B.P., and Fill II between 2,000 and 300 B.P. (Vita-Finzi, 1969a, b; 1971a). This is summarized in tabular form (Table 1).

To judge from published accounts, both deposits are represented in parts of the Mediterranean catchment that have not been studied by the writer, including Syria (van Liere, 1960–1), the Rhône basin (Bourdier, 1961–2) and the Po valley (Gabert, 1962). In Lebanon, Wright (1962) has described a deposit which appears to correspond to Fill I. According to one interpretation of the Nile record, the Sebilian Silts (Sandford, 1934) accumulated between 25,000 and 10,000 years ago (Fairbridge, 1962; *cf.*, Berry and Whiteman, 1968, p. 31); silts deposited during early Islamic times (Butzer, 1959) have yielded a radiocarbon date of 860 B.P. (Butzer and Hansen, 1968, p. 123).

Further details are given in the references listed below the table. The industries listed under the regional representatives of Fills I and II are found within the deposits; the usual limitations of archeological dating thus apply. Those given in the middle column of the table refer both to open and to straitified sites. Figures

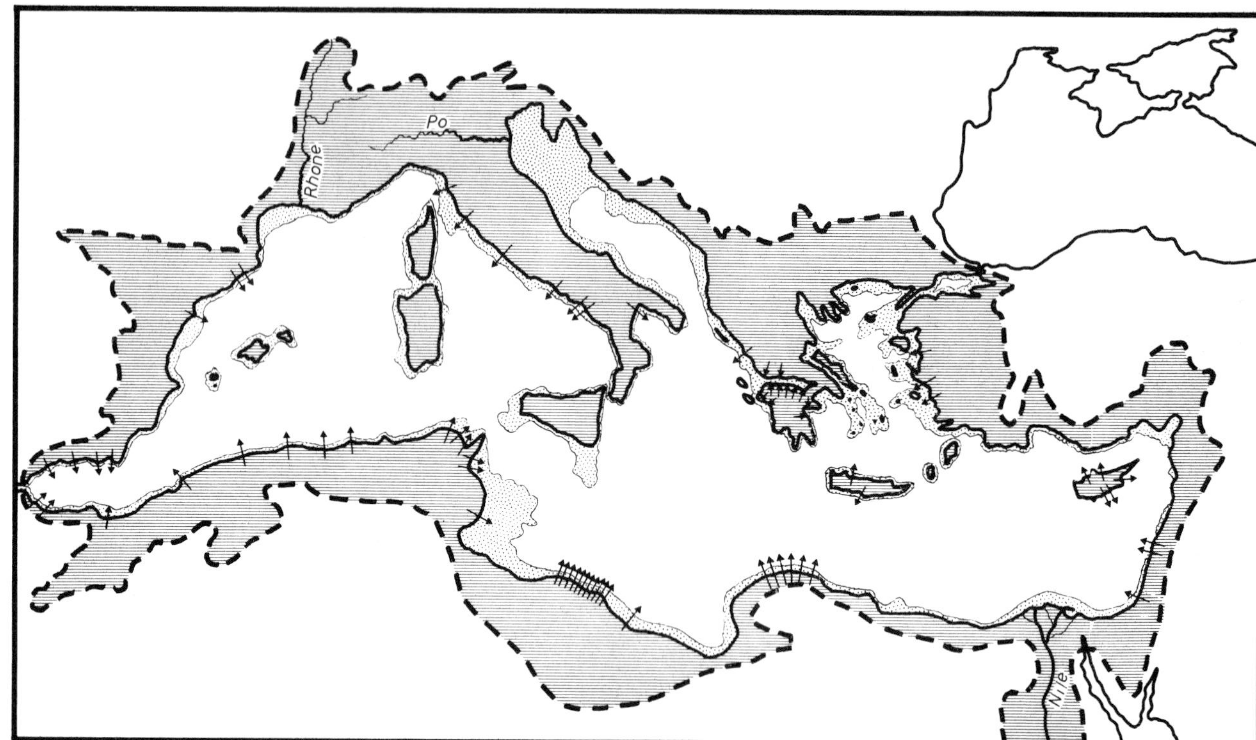

Figure 1. The Mediterranean catchment, based on Carter (1956). The position of the coastline with sea-level at about −100 m is also shown. Arrows mark streams on which the alluvial chronology outlined in this paper is based.

Table 1. Summary of fluvial deposition and erosion in the circum-Mediterranean region.

Region	Fill I	Erosion	Fill II
Cyrenaica	*Kuf Alluvium* Lev.-Moust. (45,000–40,000)		*Bel Ghadir Alluvium* R.
Tripolitania	*Gefara Alluvium* Late Ater. (30,000)	Intergétulo-Neo. (12,000–9,000)	*Lebda Alluvium* R, M. C^{14}:614
Tunisia	*Limons Rouges* Ater., Oranian		*Basse Terrasse* R.
Algeria	*Mazouna Fill* Ater., Oranian	Oranian	*Chelif Fill* R.
Morocco[1]	*Soltanian* Ater., C^{14}: 11,360	Oranian (12,000)	*Rharbian* R. C^{14}: 1900, 1950, 1530, 800, 490
Spain	*Alluvions* } *Rouges* *Limons* }		*Low Terrace* R, M.
Italy[2]	*Alluvions Terrassées* Moust.		*Historical Deposit* C^{14}: 1400, 1140, 1670, 1350
Greece[3]	*Red Beds* Moust. and Adv. Pal. (40,000–12,000)	Neo. C^{14}: 7380	*Valley-Floor Alluvium* R.
Turkey	*Sinop Alluvium* M. Pal.		*Meander Alluvium* R.
Israel[4]	*Older Fill* Pal.		*Historical Deposit* R.

Notes to Table 1: Abbreviations: Lev.-Moust. = Levalloiso-Mousterian; Ater. = Aterian; Adv./M. Pal. = Advanced/Middle Paleolithic; Pal. = Palaeolithic; Neo. = Neolithic; R = Roman; M = Medieval. The following references supplement those given in Vita-Finzi (1969a and b; 1971 a): [1]Additional C^{14} dates from Hébrard (1970); Fature and Hugot (1966); [2]Selli (1962). The last two C^{14} dates refer to the Tiber (1-4801, 1-4802, unpublished); [3]Higgs *et al.* (1967) and current work in Corfu, Crete and Cyprus; and [4]Vita-Finzi and Higgs (1970).

in brackets are C^{14} dates obtained locally for the industry in question (all dates are given in years ago). Inclusion does not signify that the author cited subscribes to the sequence proposed here.

Several writers have challenged the above scheme on the grounds that it disregards the complexity of the evidence and is based on inadequate dates. The conflicting interpretations advanced for some of the areas, notably as regards the 'Sebilian Silts' of the Nile (*e.g.*, Butzer and Hansen, 1968; Heinzelin, 1967) pose an additional problem. The twofold succession may, none the less, be found to be of some value as a working hypothesis.

SEDIMENT YIELD

Little fluvial material reached the coast during the accumulation of Fill I. Flow was predominantly

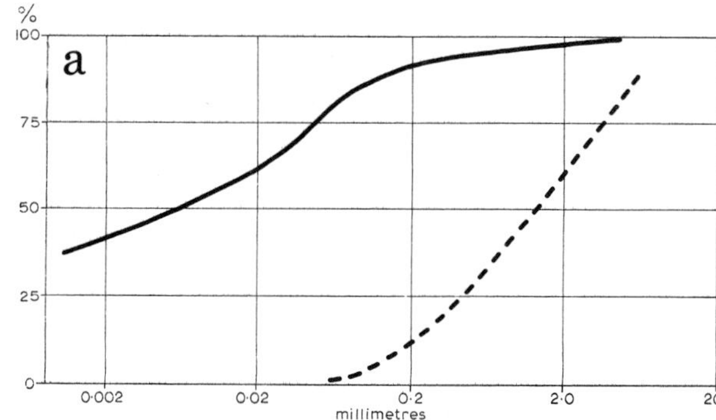

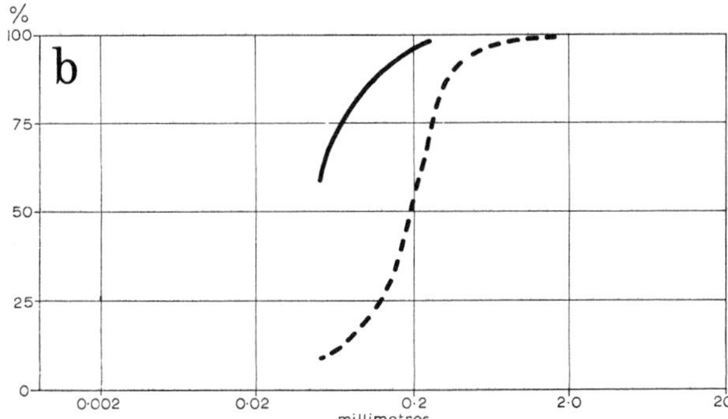

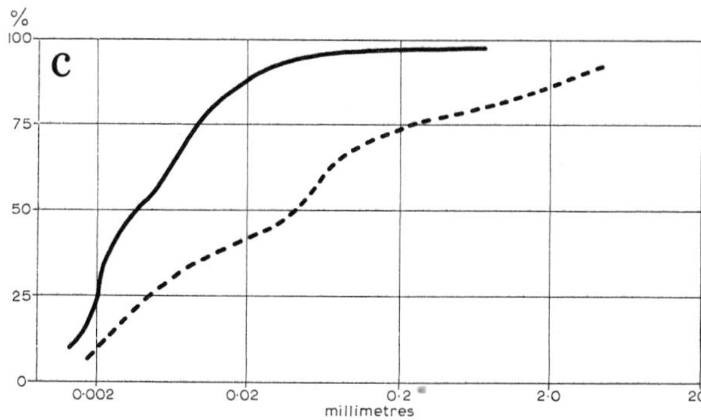

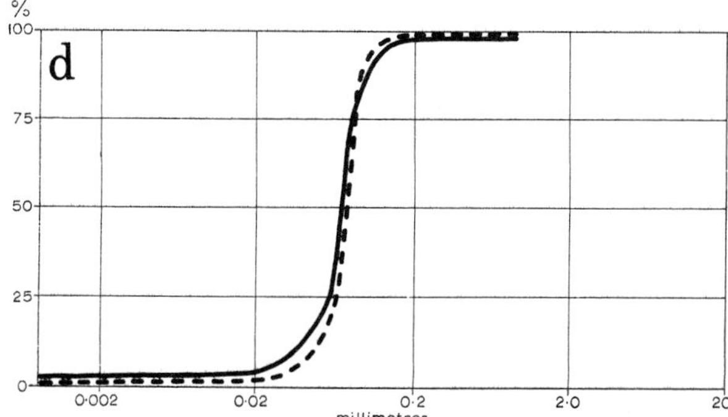

Figure 2. Grain-size distribution curves for Fill I (solid lines) and Fill II (dashed lines) at (a) Kambi (Greece), (b) Haghia Triada (Crete), (c) Boyabat (Turkey) and (d) Ganima (Libya).

ephemeral (Vita-Finzi, 1969a, b; 1971a; Fairbridge, 1963; Berry and Whiteman, 1968) and stream loads were deposited in their entirety (Vita-Finzi, 1971b; see Figure 2, solid lines). Between 10,000 and 2,000 B.P. downcutting and bank erosion gave high sediment yields. During the deposition of Fill II, flow was sufficiently sustained for the selective export of the clay and fine silt fractions (Figure 2, dashed lines). Since about 300 B.P. sediment yields have again been high (Holeman, 1968).

It is widely held that 20,000 years ago sea-level lay 100 to 120 m below present datum (Blanc, 1937; Curray, 1961; van Straaten, 1965). The onset of the Flandrian transgression is put at about 14,000 B.P. (Milliman and Emery, 1968); by 5,000 B.P. the sea had risen to within 2 m of its present position (Mörner, 1971). Deltas trap between 10 and 95 per cent of the suspended load supplied to them by their streams (Strakhov, 1967); since a rise in sea-level promotes both the growth and the erosion of deltas (Moore, 1966 p. 102–3), it was probably not until after 5,000 B.P. that storage in deltas and estuaries led to any important reduction in the volume of material carried out to sea. Delta building was reduced during deposition of Fill II (Vita-Finzi, 1969a, b; Eisma, 1962) and has accelerated since about 300 B.P. (Grabau, 1924, p. 609; Dongus, 1963; Bradford, 1957, p. 246; Houston, 1964, p. 4).

CONCLUSIONS

When combined with the eustatic record, the alluvial chronology of the Mediterranean catchment indicates the following very general pattern in the supply of suspended fluvial sediment to the sea:

20,000–10,000 B.P.	: little sediment supplied.
10,000–5,000 B.P.	: high sediment yield.
5,000–2,000 B.P.	: increasing proportion trapped in deltas.
2,000–300 B.P.	: little sediment supplied, chiefly clay and fine silt.
300–0 B.P.	: high sediment yield; much material stored in deltas.

ACKNOWLEDGMENTS

I am grateful to Dr. K. S. Sandford and Dr. A. J. Smith for their comments on the manuscript, to Mr. A. Fishwick and Mr. J. Marshall for mechanical analyses, and to Mr. K. J. Wass for drawing the figures. Much of the fieldwork to which reference is made was financed by the Nuffield Foundation.

REFERENCES

Berry, L. and A. J. Whiteman 1968. The Nile in the Sudan. *Geographical Journal,* 134:1–37.

Blanc, A. C. 1937. Low levels of the Mediterranean Sea during the Pleistocene glaciation. *Quarterly Journal of the Geological Society of London,* 93:621–651.

Bourdier, F. 1961–2. *Le bassin du Rhône au Quaternaire.* Centre National de la Recherche Scientifique, Paris, 2 volumes, 364 p.

Bradford, J. 1957. *Ancient Landscapes.* G. Bell & Sons, London, 297 p.

Butzer, K. W. 1959. Some Recent geological deposits in the Egyptian Nile Valley. *Geographical Journal,* 125:75–79.

Butzer, K. W. and Hansen, C. L. 1968. *Desert and River in Nubia.* University of Wisconsin Press, Madison, 562 p.

Carter, D. B. 1956. The water balance of the Mediterranean and Black seas. *Publications in Climatology,* 9(3):123–174.

Curray, J. R. 1961. Late Quaternary sea level: a discussion. *Geological Society of America Bulletin,* 72:1707–1712.

Dongus, H. 1963. Die Entwicklung der östlichen Po-Ebene seit frühgeschichtlicher Zeit. *Erdkunde,* 17:205–222.

Eisma, D. 1962. Beach ridges near Selçuk, Turkey. *Tijdschrift van het Koninklijk Nederlandsch Aardrijkskundig Genootschap,* 79: 234–46.

Fairbridge, R. W. 1962. New radiocarbon dates of Nile sediments. *Nature,* 196:108–110.

Fairbridge, R. W. 1963. Nile sedimentation above Wadi Halfa during the last 20,000 years. *Kush,* 11:96–107.

Faure, H. and H.-J. Hugot 1966. Chronologie absolue du Quaternaire en Afrique de l'ouest. *Bulletin de l'Institut Français d'Afrique Noire,* 28:384–97.

Gabert, P. 1962. *Les Plaines Occidentales du Pô et leur Piedmonts.* Imprimerie Louis-Jean, Gap, 531 p.

Grabau, A. W. 1924. *Principles of Stratigraphy* (Reprinted 1960). Dover, New York, 2 volumes, 1185 p.

Hébrard, L. 1970. Fichier des âges absolus du Quaternaire d'Afrique au Nord de l'équateur, 6me Série. *Bulletin de l'Association Sénégalaise pour l'Etude du Quaternaire de l'Ouest Africain* (ASÉQUA), 27–28: 39–59.

Heinzelin, J. de 1967. Pleistocene sediments and events in Sudanese Nubia. In: *Background to Evolution in Africa,* eds. Bishop, W. W. and J. D. Clark, The University of Chicago Press, Chicago, 313–328.

Higgs, E. S., C. Vita-Finzi, D. R. Harris and A. E. Fagg 1967. The climate, environment and industries of Stone Age, Greece: Part III. *Proceedings of the Prehistoric Society of London,* 33: 1–29.

Holeman, J. 1968. The sediment yield of major rivers of the world. *Water Resources Research,* 4:737–47.

Houston, J. M. 1964. *The Western Mediterranean World.* Longmans, London, 800 p.

van Liere, W. J. 1960–1. Observations on the Quaternary of Syria. *Berichten van de Rijksdienst voor het Oudheidkundig Bodemonderzoek,* 10–11: 7–69.

Milliman, J. D. and K. O. Emery 1968. Sea levels during the last 35,000 years. *Science,* 162:1121–1123.

Moore, D. 1966. Deltaic sedimentation. *Earth-Science Reviews,* 1:87–104.

Mörner, N.-A. 1971. Eustatic changes during the last 20,000 years and a method of separating the isostatic and eustatic factors in an uplifted area. *Palaeogeography, Palaeoclimatology, Palaeoecology,* 9:153–181.

Sandford, K. S. 1934. *Paleolithic Man and the Nile Valley in Upper and Middle Egypt.* University of Chicago Press, Chicago, 131 p.

Selli, R. 1962. Le Quaternaire marin du versant Adriatique Ionien de la Péninsule italienne. *Quaternaria,* 6:391–413.

Strakhov, N. M. 1967. *Principles of Lithogenesis.* Oliver and Boyd, Edinburgh, 1:245 p.

van Straaten, L. M. J. U. 1965. Sedimentation in the north-western part of the Adriatic Sea. In: *Submarine Geology and Geophysics,* eds. Whittard, W. F. and R. Bradshaw, Butterworths, London, 143–160.

Vita-Finzi, C. 1969a. *The Mediterranean Valleys.* Cambridge University Press, Cambridge, 140 p.

Vita-Finzi, C. 1969b. Late Quaternary continental deposits of central and western Turkey. *Man,* 4:605–619.

Vita-Finzi, C. 1971a. Alluvial history of northern Libya since the last interglacial. In: *Symposium on the Geology of Libya,* ed. Gray, C., Imprimerie Catholique, Beirut, 409–429.

Vita-Finzi, C. 1971b. Heredity and environment in clastic sediments: silt/clay depletion. *Geological Society of America Bulletin,* 82:187–190.

Vita-Finzi, C. and E. S. Higgs 1970. Prehistoric economy in the Mount Carmel area of Palestine: site catchment analysis. *Proceedings of the Prehistoric Society of London,* 36:1–37.

Wright, H. E., Jr. 1962. Late Pleistocene geology of coastal Lebanon. *Quaternaria,* 6:525–539.

PART 2. The Geological Setting

Evolution de la Sédimentation Pendant le Néogène en Méditerranée d'après les Forages JOIDES-DSDP

Wladimir D. Nesteroff[1], William B. F. Ryan[2], Kenneth J. Hsü[3], Guy Pautot[4], Forese C. Wezel[5], Jennifer M. Lort[6], Maria B. Cita[7], Wolf Maync[8], Herbert Stradner[9], et Paulian Dumitrica[10]

RESUME

Les forages profonds en Méditerranée (Leg 13) du GLOMAR CHALLENGER permettent de dégager, dans les séries néogènes, trois grandes unités lithologiques. L'unité inférieure est formée de marnes bleues à faune planctonique pauvre. Elles sont d'âge langhien, serravalien et tortonien. Elles sont très semblables aux marnes bleues de même âge connues à terre: Bassin du Rhône, de Vienne, Panonique, etc. Ceci suggère un type de sédimentation uniforme dans toute la Téthys vindobonienne.

La seconde unité est formée de séries évaporitiques: halites, anhydrites, gypses et marnes dolomitiques d'âge messinien. Elle est séparée des formations sous et sus-jacentes par des limites très nettes. Là encore, les séries messiniennes connues sur la terre ferme, en Espagne, dans les Appenins, en Sicile, etc., sont très semblables. Enfin, des vases semi-pélagiques forment l'unité supérieure, d'âge plio-quaternaire.

Une série de données: faunes saumâtres ou lagunaires, structures stromatolitiques, études isotopiques, etc., indiquent que les évaporites se sont déposées sous une très faible couche d'eau. Le modèle qui explique le mieux cette situation suppose qu'au Messinien la Méditerranée est isolée de l'Océan Mondial. Elle s'évapore alors et se transforme en un immense bassin peu profond où se déposent des sels. Au Pliocène, la mer envahit de nouveau ce bassin asséché et des conditions de sédimentation de mer ouverte s'établissent.

La fermeture au Messinien et la réouverture au Pliocène de la Méditerranée ne peut être que tectonique. Ainsi dans cette mer intracontinentale c'est la tectonique qui a contrôlé, à chaque période du Néogène, la circulation océanique et par voie de conséquence le type de sédimentation. Dans ces conditions très particulières les grandes formations marines doivent leur unité lithologique aux évenements tectoniques.

ABSTRACT

Examination of the GLOMAR CHALLENGER cores in the Mediterranean Sea (Leg 13) enables us to distinguish three main lithological units in the Neogene series. The lower unit is composed of blue marls with a poor, or dwarfed, fauna. The marls are of Langhian, Serravalian and Tortonian age, and are very similar to the blue marls of the same age known on land (in the Rhône basin, Vienna Basin, the Panonic Basins, etc.). This suggests that the Vindobonian Tethys received a uniform type of sedimentation. The second unit comprises evaporitic series: halites, anhydrites, gypsum and dolomitic marls of Messinian age. This unit is interbedded with sharp contacts separating lower and upper units. It is also very similar to the Messinian evaporitic series on land around the Mediterranean. The upper and last unit is composed of hemipelagic marl-oozes of Pliocene and Quaternary age.

Distinctive characteristics (brackish and lagoonal fauna, stromatolitic structures, isotopic analysis) enable us to conclude that the evaporites were deposited in shallow water. The model which best fits this data suggests isolation of the Mediterranean from the world ocean during the Messinian. Once isolated, evaporation turned this sea into a dessication basin with salt deposition. In the Pliocene, sea water again invaded this dry basin, and open sea conditions and sedimentation were resumed.

The closing of the Mediterranean in the Messinian and its reopening in the Pliocene necessitates a structural origin. During each stage of the Neogene, the oceanic circulation—and thus the sedimentation—of the sea surrounded by continents was controlled tectonically. Under these circumstances the lithologic uniformity of the major Neogene marine formations is the result of tectonic events.

[1]Université de Paris, Paris; [2]Lamont-Doherty Geological Observatory, Palisades, New York; [3]Geologisches Institut, Zurich; [4]Centre Océanologique de Bretagne, Brest; [5]Università di Catania, Catania; [6]University of Cambridge, Cambridge; [7] Università degli Studi di Milano, Milan; [8]Geological Consulting Service, Berne; [9]Geologische Bundestalt, Vienna; [10]Geological Institute, Bucarest.

INTRODUCTION

Durant la treizième campagne du bàtiment de forage GLOMAR CHALLENGER, consacrée à la Méditerranée, nous avons effectué 27 forages profonds répartis en 14 sites (Figures 1, 2 et 3). Les emplacements ainsi que le but scientifique recherché pour les divers forages ont été décrits par Scientific Staff (1970). Les colonnes sédimentaires, bien qu'encore très disséminées dans le Bassin Méditerranéen, mais que nous espérons pouvoir compléter au cours de futures campagnes, nous permettent dès à présent de tenter un essai de reconstitution de l'histoire sédimentaire de cette mer au cours du Néogène.

LE MIOCENE MOYEN ET SUPERIEUR

Les dépôts les plus anciens forés au cours de la campagne du GLOMAR CHALLENGER en Méditerranée sont miocènes. En Mer d'Alboran (site 121), seul le Tortonien a été rencontré tandis que dans le Bassin oriental (sites 126 et 129), le Langhien et le Serravalien ont été atteints. Enfin les formations évaporitiques messiniennes furent reconnues dans de nombreux sites répartis dans toute la Méditerranée (sites 122, 124, 125, 126, 129, 132 et 134).

La Mer d'Alboran (site 121)

Le site 121 est situé dans la partie occidentale de la mer d'Alboran, près d'une crête isolant un petit bassin intérieur (Figures 1 et 4). Le forage a atteint à 864 m, un socle formé de gneiss, schistes et granodiorites.

La base de la colonne sédimentaire qui surmonte le socle est composée de marnes d'âge tortonien. Elles sont sombres: gris foncé, semi-consolidées et présentent de fines laminations qui leur confèrent une allure schisteuse. Elles sont essentiellement composées de clastiques terrigènes fins: quartz et argiles. La fraction biologique comprend du nannoplancton et des foraminifères. Cette faune, relativement abondante et diversifiée, indique une mer ouverte. La proportion de carbonates reste toujours faible (30%) sauf à l'approche du socle où elle s'élève à 40–45%. La présence de dolomie, en petite quantité, est à noter. Elle serait d'origine détritique. Les minéraux argileux sont représentés par l'illite, la chlorite et la kaolinite. Les interstratifiés et la montmorillonite sont rares.

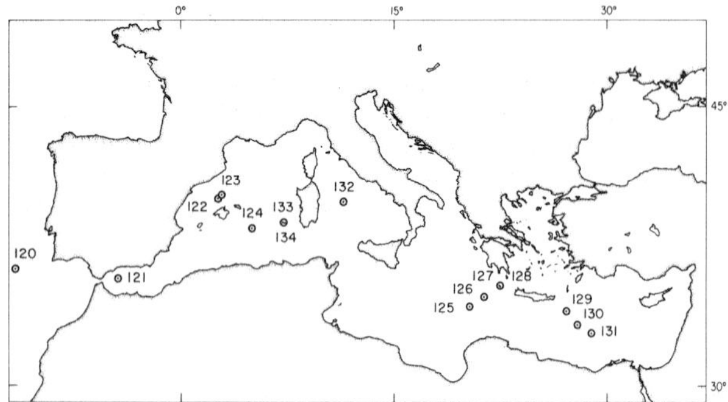

Figure 1. Position des forages du Leg 13, JOIDES-DSDP (août-octobre 1970).

Bien que la plupart des sédiments carottés dans cette série soient des marnes, 3 carottes sur 10 montrent des horizons de turbidites contenant des passages de sables grossiers. Ceci témoigne d'une certaine activité des courants de turbidité durant le dépôt de la séquence marneuse. Toutefois, le faible taux de sédimentation (2,8 cm/1000 ans), semblable à celui des dépôts pélagiques plio-quaternaires de la Méditerranée orientale, suggère que les arrivées de turbidites étaient des évènements exceptionnels.

Des phénomènes de consolidation intermittente ont été observés au milieu de ces marnes. Ainsi un mince horizon a été transformé en calcaire dolomitique et, parmi les rares passages de turbidites, deux niveaux sableux ont été cimentés en grès. Enfin un certain nombre de niveaux durs ont été traversés par le forage sans être échantillonnés. Ces horizons indurés correspondraient aux réflecteurs notés audessus du socle sur les profils sismiques (Figure 4).

Le sommet des marnes tortoniennes n'a pas été carotté. Il se situe entre 680 et 689 m et nous l'avons arbitrairement placé à 686 m. En effet, les enregistrements sismiques comportent, vers cette cote, une forte discordance angulaire et les carottes montrent que des turbidites pliocènes reposent sur les marnes tortoniennes. Une lacune, comprenant le Messinien et une partie de Pliocène inférieur, sépare les deux formations.

La base des marnes repose sur le socle qui a été pénétré sur quelques dizaines de centimètres. Les fragments remontés comprennent du schiste à biotite et des conglomérats contenant des gneiss, des granodiorites et des schistes.

Enfin, le forage ayant été implanté sur le flanc d'une crête, les profils sismiques montrent des horizons sédimentaires situés en-dessous de ceux atteints à l'extrémité du forage. Ainsi les dépôts recouvrant le socle dans le site 121 ne seraient pas les plus anciens du bassin.

La Mer Ionienne (site 126)

Le forage 126 est situé dans une vallée abrupte qui entaille la Dorsale méditerranéenne. Cette vallée a été reconnue sur plus d'une centaine de kilomètres. Les profils sismiques montrent que dans ses flancs

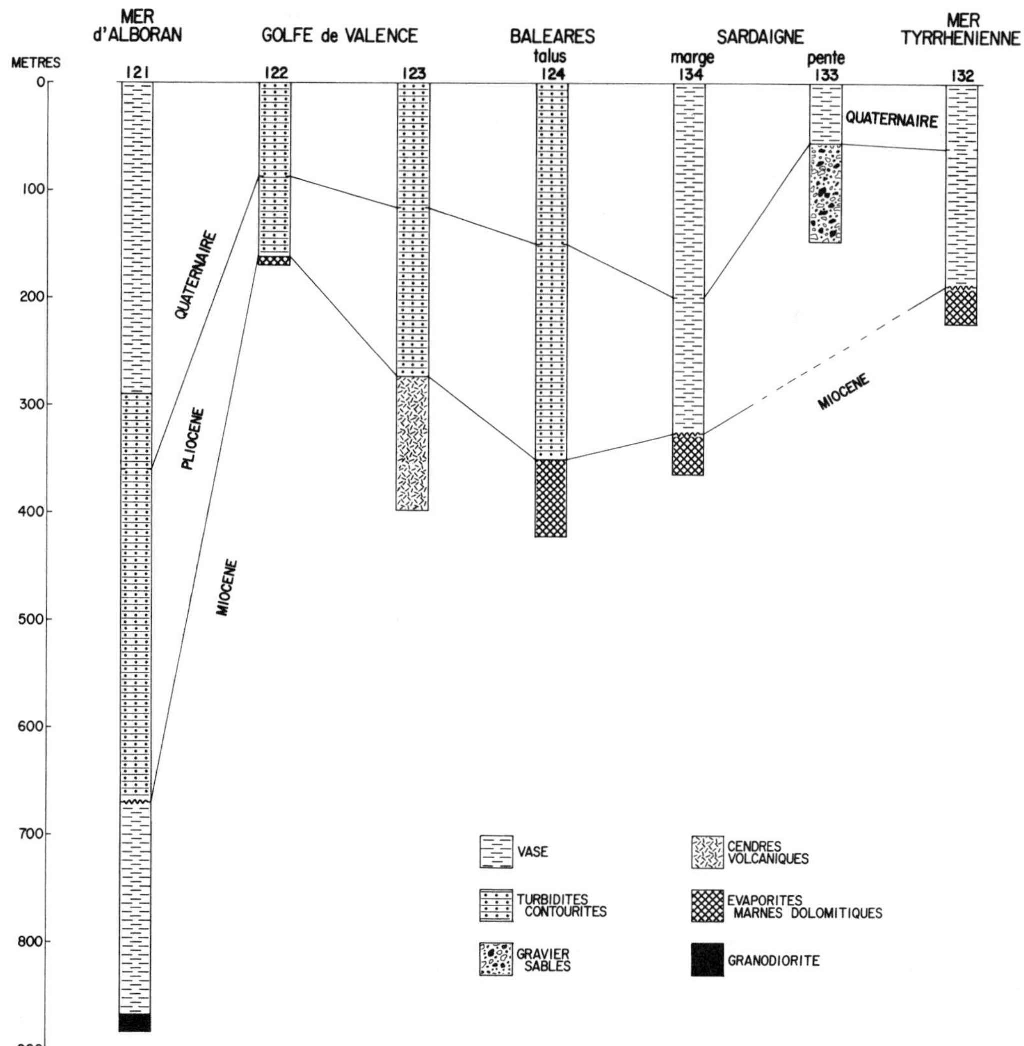

Figure 2. Logs des forages du Bassin occidental.

affleurent des horizons situés sous le *réflecteur M* (Ryan *et al.*, 1971) qui, comme nous l'avons montré, correspond au toit du Miocène. Nous pouvions donc espérer descendre assez profondément dans le Miocène.

Sous une centaine de metres de sédiments non consolidés quaternaires, le forage pénétra des marnes serravaliennes. Il s'agit de marnes bleu-noires, extrêmement indurées, fissiles et cassantes. Les rayons X montrent qu'elles sont composées de quartz, de calcite, de montmorillonite et de kaolonite. Elles contiennent une abondante faune de nannofossiles calcaires et une maigre faune de foraminifères planctoniques nains accompagnés de quelques rares formes benthiques.

Cette association suggère une mer partiellement isolée où le nannoplancton, peu sensible aux variations du régime océanique, contribue presque seul à la sédimentation biologique. Les foraminifères par contre sont affectés par ces variations. Enfin, les fines laminations des dépôts indiquent l'absence des perforants benthiques, c'est-à-dire des conditions réductrices, confirmées par l'abondance de la pyrite.

La Montagne sous-marine du Strabo (site 129)

Le Strabo, situé entre la Crète et Chypre, fait partie de l'arc Péloponèse-Crète-Chypre. Les trois forages du site 129 ont été implantés sur le flanc Nord de la montagne sous-marine, près de l'étroite fosse qui borde l'arc au Sud.

Les forages 129 et 129A ont d'abord rencontré une mince couche de vase quaternaire pélagique avant de traverser un horizon dur épais d'un mètre. En-dessous, une soixantaine de mètres de vases non

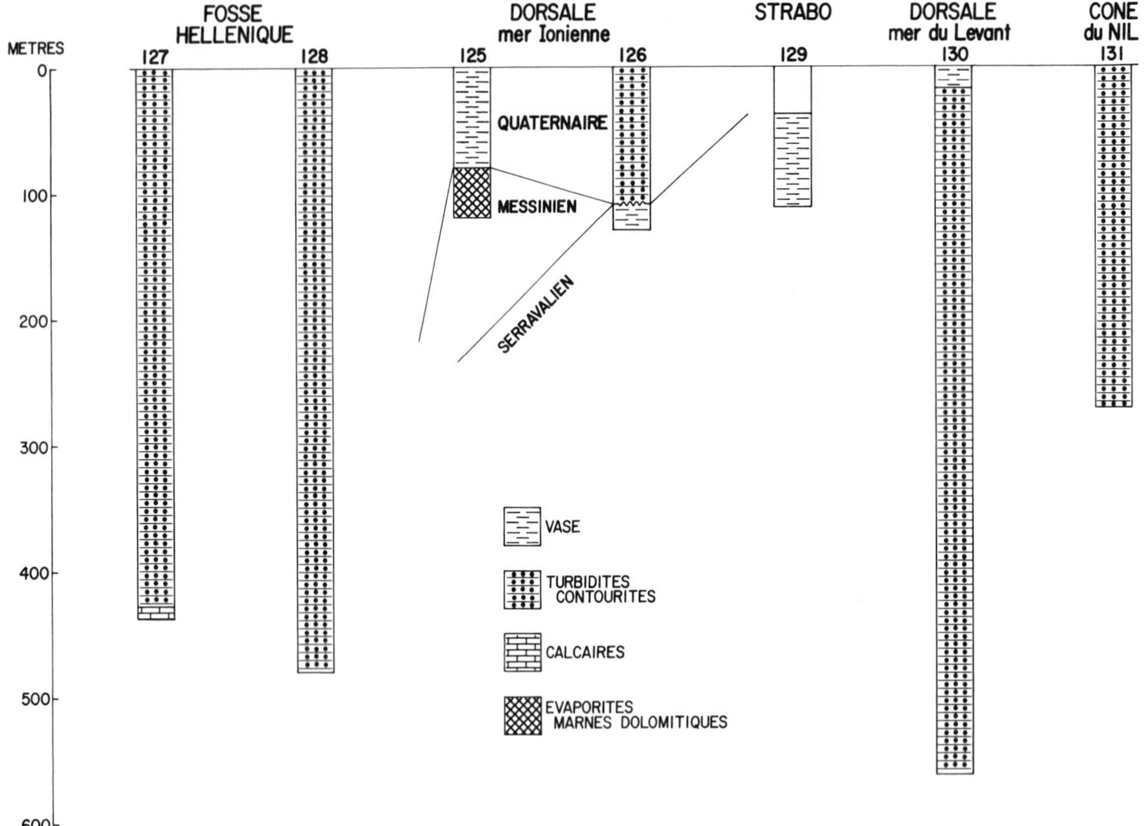

Figure 3. Logs des forages du Bassin oriental.

consolidées ont été pénétrées avant que le trépan ne butte sur un horizon dur. Le forage fut alors abandonné.

Cet horizon dur s'est révélé être le dépôt le plus ancien carotté dans toute la Méditerranée. Il s'agit d'un calcaire d'âge langhien, ancienne vase à foraminifères cimentée par la calcite. Le dépôt reflète une sédimentation pélagique avec apports terrigènes. La couleur sombre, le fin litage, la pyrite ainsi que l'absence de perforations suggèrent un milieu réducteur.

L'intervalle formé de sédiments non consolidés situé entre le calcaire langhien et la couche dure de sub-surface a livré une série de carottes difficiles à interpréter. Il s'agit de vases à nannoplancton et foraminifères. La fraction terrigène est importante, souvant dominante. La minéralogie des carbonates est curieuse. On observe un mélange de calcite et de dolomite. Souvent la dolomite subsiste seule. On note aussi des stratifications obliques avec concentration, dans les strates, de foraminifères et de sablons quartzeux suggérant l'action de courants de fond. De plus, certaines strates sont plissotées, indiquant des glissements de terrain ou de petites déformations tectoniques. Les faunes de ces dépôts sont serravalliennes et indiquent une mer partiellement fermée.

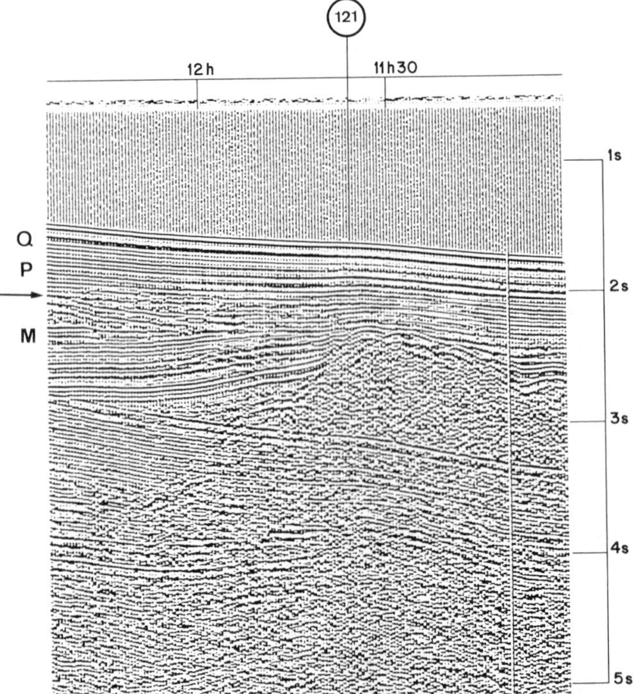

Figure 4. Profil de sismique réflexion (flexo-tir) du site 121 dans la mer d'Alboran. Notez l'épaisse série sub-horizontale (P,Q) de sédiments bien stratifiés (Plio-Quaternaire). Ils reposent avec une discordance angulaire sur les marnes tortoniennes (M).

Au-dessus des dépôts précédents, dans le forage 129A, des vases de faciès saumâtre ont été carottées. Ce faciès, caractérisé par l'association de l'ostracode *Cyprideis pannonica* et du foraminifère benthique *Ammonia beccarii tepida*, indique un âge messinien.

Enfin le niveau dur près de la surface a livré des fragments de calcaires pélagiques indurés du Pliocène inférieur.

Cette succession de dépôts peut être interprétée comme une série normale qui aurait été comprimée et déformée en un "mélange" tectonique. Certains horizons peuvent aussi correspondre à une resédimentation de foraminifères et de nannofossiles du Miocène moyen dans des formations messiniennes stériles.

De toutes façons, les faunes indiquent que le Bassin oriental était, au Miocène moyen, une mer ouverte. Vers la fin du Langhien et surtout au Serravalien, la circulation de ce bassin a subi certaines restrictions. Au Messinien, le Bassin était devenu un lac saumâtre. Enfin avec le Pliocène une sédimentation pélagique a été rétablie.

LE MESSINIEN

Des séries évaporitiques, d'âge messinien ont été rencontrées dans toute la Méditerranée, aussi bien dans le Bassin oriental que dans le Bassin occidental (sites 122, 124, 125, 126, 132 et 134). Mais la récupération a été très inégale. Certaines colonnes sont excellentes, d'autres médiocres. La pénétration s'est située entre 40 et 70 m. Comme nous savons d'après la sismique réflection que l'épaisseur moyenne de ces séries se situe, dans le Bassin Occidental, entre 300 et 500 m, nous n'avons pénétré que dans les couches supérieures et terminales de la formation. Malgré ces restrictions, nos carottes nous permettent de présenter un tableau d'ensemble de la partie supérieure des évaporites de la Méditerranée.

Les couches les plus profondes des séries évaporitiques sont formées, dans tous les sites forés, de bancs de sels interstratifiés avec des boues dolomitiques. Les bancs de sel massif sont généralement épais de 1 à 2 mètres. Selon leur disposition plus ou moins centrale dans les bassins et aussi selon les bassins, ils sont composés d'espèces minérales différentes.

En mer Tyrrhénienne (site 132), il s'agit de gypse. Les bancs les plus profonds sont colorés de couleurs vives, jaunes et rouges, et présentent les laminations ondulées caractéristiques des récifs à algues (stromatolites) des zones intercotidales des mers chaudes (Figure 5A et B). Ces bancs de gypse massif sont interstratifiés avec des lits de gypse plastique dans lesquels subsistent des traces de quartz, d'argiles et de carbonates (Figure 5C).

Dans la plaine abyssale du Bassin Algéro-Pro-

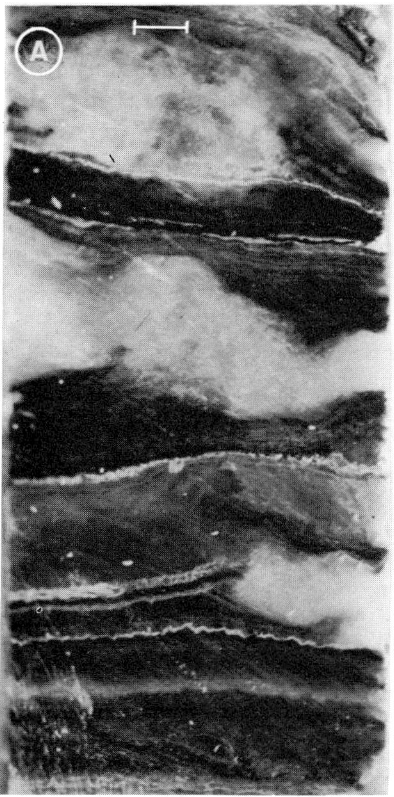

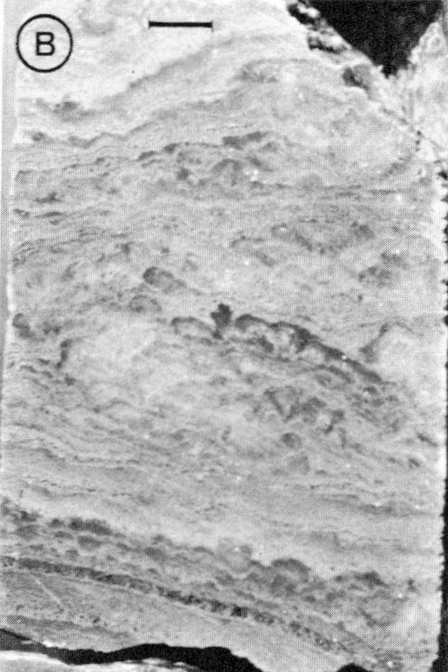

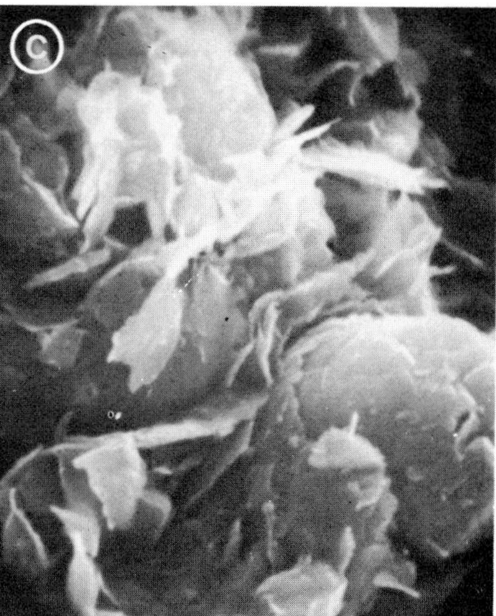

Figure 5. A, Série évaporitique de la Mer Tyrrhénienne. Banc de gypse massif (site 132–27–1–50 à 70 cm). Barre d'échelle = 1 centimètre. B, Série évaporitique de la Mer Tyrrhénienne. Banc de gypse massif. Litage ondulé caractéristique des stromatolites ou récifs à algues (site 132–27–2–130 cm). Barre d'échelle = 1 centimètre. C, Séries évaporitiques de la Mer Tyrrhénienne. Ultrastructure (X 12000) d'un lit de gypse plastique (avec trace de quartz) interstratifié entre des bancs de gypse massif (site 132–26–1–80 cm).

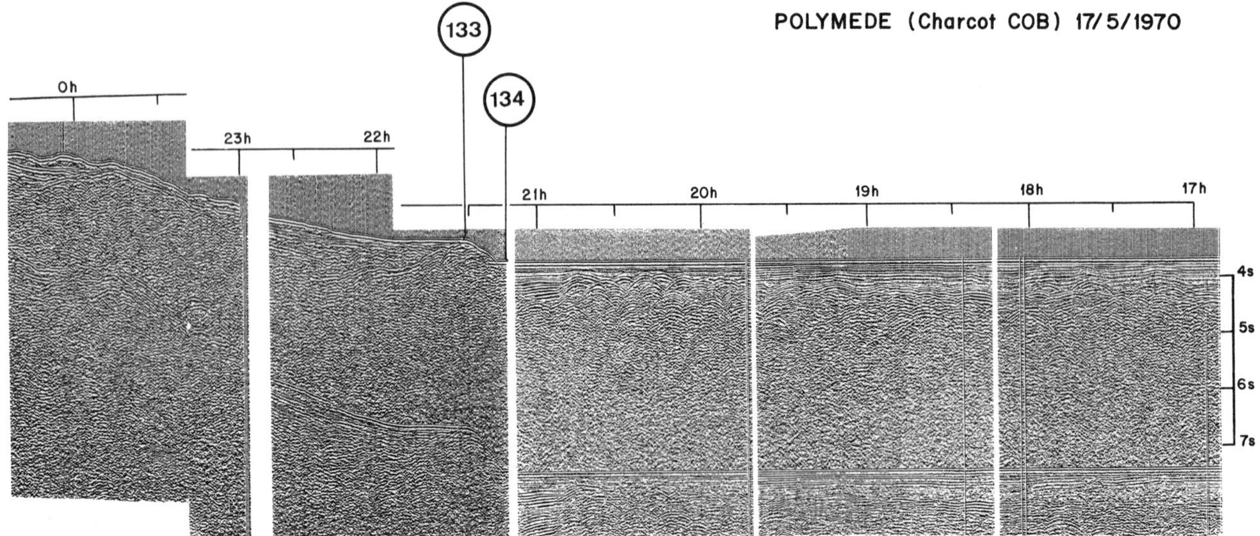

Figure 6. Marge de la Sardaigne et plaine abyssale du Bassin Algéro-Provençal (sites 133 et 134). Profil de sismique réflexion. Notez la terminaison abrupte, faillée, du socle sarde entre 133 et 134, et les dômes de sel sous la plaine abyssale.

vençal (site 134, Figure 6), il s'agit de halite et d'anhydrite. Le banc le plus profond consiste en 1,70 m de sel gemme. Celui-ci est translucide et lité (Figure 7). De plus, de fins lits millimétriques d'anhydrite s'intercalent à des intervalles de 20 cm environ. Plus haut, un banc d'anhydrite rubannée (lits clairs et sombres) a été rencontré.

De l'autre côté de cette plaine abyssale, à la base du glacis continental des Baléares (site 124), il s'agit de bancs d'anhydrite massive interstratifiés avec des boues dolomitiques. En profondeur, les anhydrites présentent un aspect en mosaïque ou alvéolaire (chicken wire) résultant de la coalescence de cristaux recristallisés (Figure 8A). On remarque d'ailleurs tous les stades intermédiaires en remontant dans la colonne sédimentaire. Le banc d'anhydrite situé le plus haut montre que la roche originale comportait de fines laminations alternées de couleurs claires et sombres. Ces "varves" sont caractéristiques des dépôts dans les sabkha (Kinsman, 1966; Rooney et French, 1968).

Dans tous les sites, des lits non consolidés sont intercalés entre les bancs de sel massif (Figure 8B). Comme nous venons de le voir, dans le site 124, ce sont des boues dolomitiques (Figure 8C) contenant encore des témoins biologiques: diatomées d'eaux saumâtres ou faunes marines naines et appauvries. Dans le site 134, un lit de boue dolomitique est intercalé entre deux bancs de halite. C'est une ancienne vase contenant une faune marine appauvrie. Des traces d'hydrocarbures migrés y ont été décelées. Enfin, dans le site 132, les lits intercalaires consistent d'abord en gypse plastique, puis, plus haut, en boues dolomitiques à faune marine abondante.

Au-dessus de la zone de bancs de sel massif, les couches terminales des séries évaporitiques sont formées de marnes plus ou moins dolomitiques. En profondeur, juste au-dessus des bancs massifs, elles sont souvent chargées de cristaux de gypse (Figure 9A et B). Plus haut, vers le toit des séries, le gypse disparaît. Lorsqu'elles sont dolomitiques, ces boues sont généralement stériles. Par contre, les marnes contiennent des faunes marines ou saumâtres appauvries.

Dans deux sites, seule cette partie terminale des séries évaporitiques a été pénétrée. Ainsi en Mer Ionienne (site 125), le forage traversa 50 mètres de boues dolomitiques contenant une faune marine naine avant de buter sur un banc massif de gypse. D'une façon analogue, dans le Golfe de Valence (site 122, Figure 10), le forage fut arrêté par un banc de gypse après avoir percé une vingtaine de mètres de vases légèrement dolomitiques.

Ainsi, dans tous les sites, les séries évaporitiques se terminent par une trentaine de mètres de marnes souvent dolomitiques. Ce nouveau faciès, succédant aux bancs de sels, annonce la fin de l'époque évaporitique et la transgression du Pliocène.

LE PLIO-QUATERNAIRE

Les dépôts plio-quaternaires méditerranéens sont franchement différents de ceux qui les précèdent, aussi bien des séries évaporitiques du Messinien, que des sédiments marins du Miocène moyen. De plus, la limite entre Pliocène et Miocène est extrêmement nette.

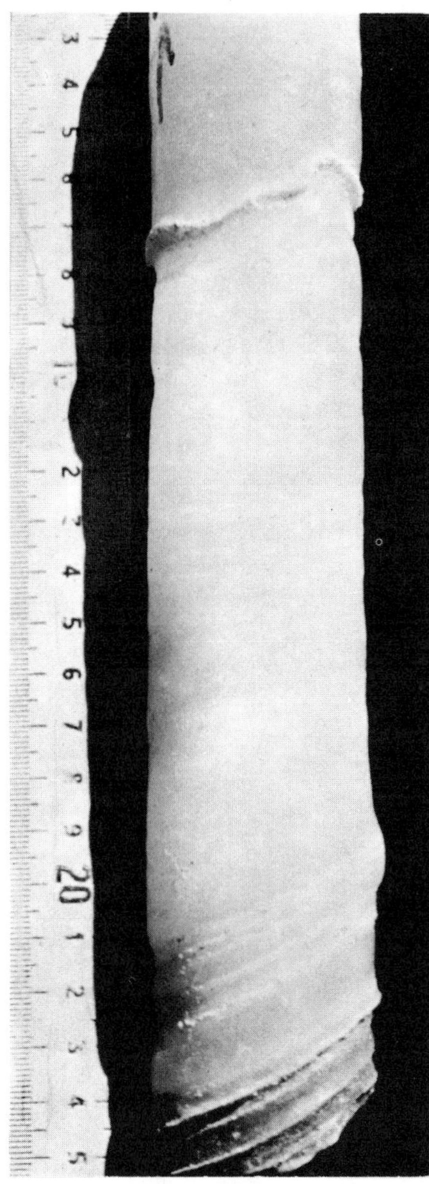

Figure 7. Séries évaporitiques du Bassin occidental. Halite transparente, litée, interstratifiée avec des lits millimétriques d'anhydrite (site 134–10–2). Echelle en centimètres.

La Limite Miocène-Pliocène

Dans les trois sites 125, 132 et 134 où elle a été carottée, la limite entre le Miocène et le Pliocène est extrêmement brutale. On observe des vases pélagiques ou des contourites de couleur claire, non consolidées et contenant une faune pélagique de mer ouverte, surmontées de marnes dolomitiques semi-indurées, de couleur gris-bleu et contenant une faune appauvrie (Figure 11). Le contact est toujours brutal et érosif En 134 et 125, une importante lacune (quelques centaines de milliers d'années en 134) sépare les deux formations. En 132, bien que le contact soit érosif, aucune solution de continuité n'a été notée dans la succession faunistique.

Dans la Mer d'Alboran, le contact n'a pas été carotté, mais un important hiatus comprenant la base du Pliocène et tout le Messinien a été observé. Au même niveau, une discordance angulaire apparaît sur les profils de sismique réflection.

Enfin, dans le Golfe de Valence (site 122), où le contact n'a pas été carotté non plus, des turbidites terrigènes de couleur claire et des cailloutis plio-quaternaires surmontent des argiles à gypse, légèrement dolomitiques, du Messinien.

Pétrographie du Plio-Quaternaire

Dans nos forages, les sédiments déposés particule par particule au Plio-Quaternaire sont des vases de couleur claire. Elles sont composées de clastiques terrigènes fins (20 à 75%) et de débris biologiques, essentiellement de nannoplancton calcaire avec quelques foraminifères. Ces derniers sont souvent peu importants ne formant que 1 à 10% du sédiment total. Les apports terrigènes comprennent du quartz et des minéraux argileux. La nature des argiles est variable et liée à la position géographique des forages, c'est-à-dire aux aires d'alimentation terrigènes dont ils dépendent.

Il s'agit de vases péri-continentales ou hémi-pélagiques, semblables à celles que l'on rencontre au Plio-Quaternaire dans les autres océans de notre globe. Elles se distinguent des vases pélagiques océaniques par une plus forte proportion d'apports terrigènes, mais leur composition varie avec leur position géographique. Certaines se rapprochent des vases terrigènes (25 à 30% de $CaCO_3$ dans la Mer d'Alboran) tandis que d'autres atteignent parfois la composition des vases pélagiques (70 à 80% de $CaCO_3$ dans le centre des bassins).

La couleur est claire (beige à blanche) contrairement aux sédiments miocènes qui sont toujours sombres: gris foncé à noir. Toutefois les dépôts plio-quaternaires à forte proportion de clastiques terrigènes fins sont plus sombres (gris clair à gris) sans atteindre les teintes miocènes.

Les dépôts quaternaires sont généralement stratifiés, présentant une succession de lits de 10 à 20 cm d'épaisseur, séparés par des limites très nettes. Les couleurs, la consistance et la proportion de foraminifères varient d'un lit à l'autre. Au fur et à mesure que l'on descend dans les colonnes sédimentaires, le litage s'atténue puis disparaît.

Enfin les dépôts plio-quaternaires ne sont pas consolidés à l'exception de quelques rares lits situés au milieu des séries de turbidites.

Figure 8. A Série évaporitique du Bassin occidental. Anyhydrite. Structure en mosaïque ou *chicken-wire* (site 124–13–2–10 cm). Barre d'échelle = 1 centimètre. B, Série évaporitique du Bassin occidental. Horizon de dolomie plastique, finement litée, interstratifiée entre des bancs d'anhydrite. La photographie montre un contact dolomie-anhydrite (site 124–11–115 à 135 cm). Barre d'échelle = 1 centimètre. C, Série évaporitique du Bassin occidental. Ultrastructure (X 5500) d'un horizon de boue dolomitique intercalée entre des bancs d'anhydrite. Dolomie dominante avec de rares quartz (124–13–2–81 cm).

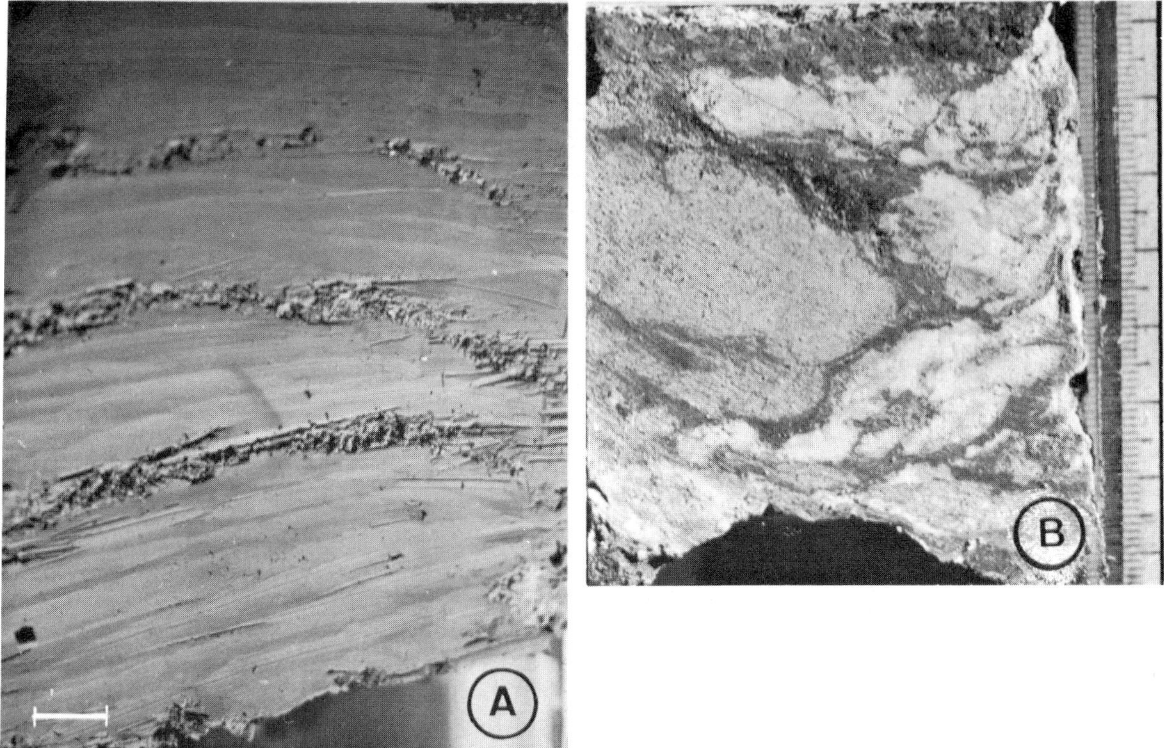

Figure 9. Série évaporitique de la Mer Tyrrhénienne. Lentilles de cristaux de gypse dans un lit de vase (site 132–25–1–0 à 10 cm). Barre d'échelle = 1 centimètre. B, Série évaporitique du Bassin occidental. Vase grise où se sont développés de larges cristaux de gypse saccaroïde (plages blanches). Site 134 D–1–1. Echelle en centimètres.

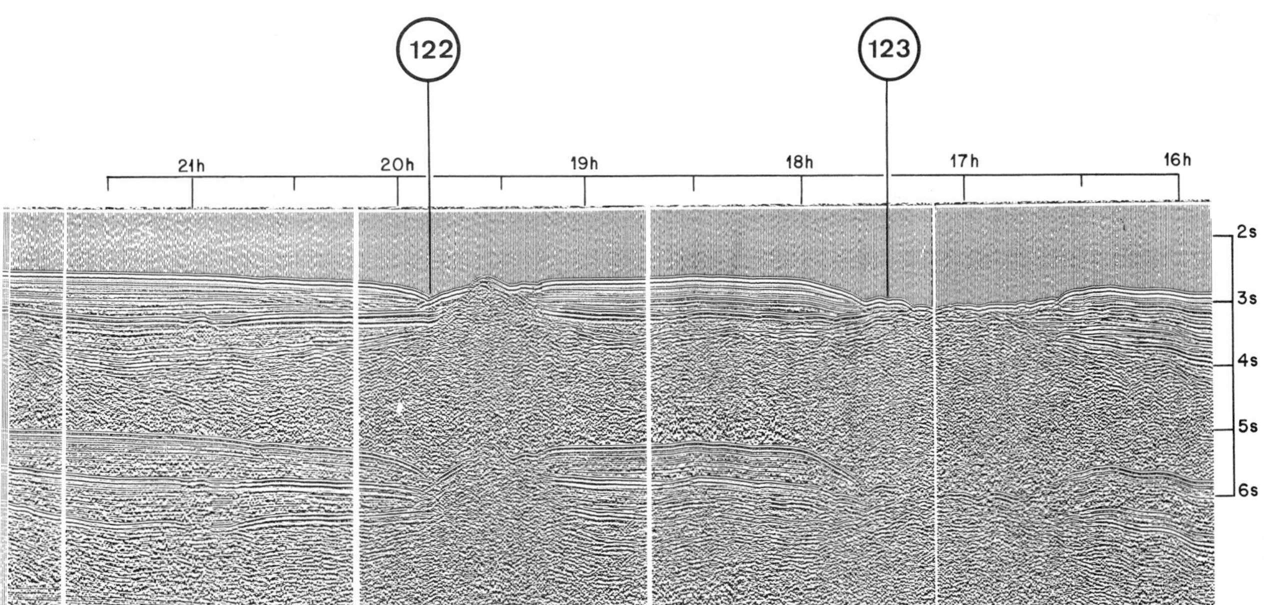

Figure 10. Golfe de Valence (sites 122 et 123). Profil de sismique réflexion. Notez les pointements du socle à travers la couverture sédimentaire et l'implantation du forage 122 dans un canyon sous-marin entaillant seulement les turbidites plio-quaternaires et respectant les séries évaporitiques sous-jacentes.

Figure 11. Limite Pliocène-Miocène dans le Bassin Algéro-Provençal. Une vase calcaire claire, d'âge pliocène, repose en discordance sur des marnes dolomitiques sombres du Miocène. Une lacune de plusieurs centaines de milliers d'années sépare les deux formations. Notez les micro-déformations dans le Miocène (site 134–7–5). Barre d'échelle = 1 centimètre.

Les vitesses de dépôt varient de 2,6 à 3,2 cm/1000 ans, vitesses très voisines de celles généralement observées pour les vases péri-continentales.

De bonnes colonnes sédimentaires, représentant d'une façon continue le Plio-Quaternaire ont été obtenues dans les forages 132 ainsi que dans le site 130, sur la Dorsale Méditerranéenne. De plus, dans ce dernier site, des séquences pélagiques quaternaires sont interstratifiées avec des turbidites du Nil. Toutes ces carottes ont permis de présenter le type pétrographique décrit ci-dessus. Ces colonnes montrent une évolution graduelle mais nette du bas (Pliocène) vers le haut (Actuel). Ainsi, le litage absent à la base du Pliocène, apparaît et devient de plus en plus net lorsqu'on remonte dans les colonnes. Les variations de composition des lits successifs semblent liées aux fluctuations climatiques mises en évidence par l'étude des faunes. On observe aussi une diminution progressive de la proportion des foraminifères. Ainsi les vases du Pliocène inférieur contiennent de nombreux foraminifères visibles à l'oeil nu sur les carottes fraîchement coupées (faciès Trubi). Ce caractère disparaît lorsqu'on remonte dans le Quaternaire.

Enfin des horizons de sapropels ont été observés dans les sites 125, 126, 127, 128, 130 et 131, tous situés en Méditerranée orientale. Deux niveaux de sapropels ont pu être identifiés et corrélés. Ils se rencontrent respectivement dans les zones de nannofossiles de *Gephyrocapsa oceanica* (NN 20) et de *Pseudomilia lacunosa* (NN 19). Les sapropels superficiels, décrits par Ryan *et al.* (1971) n'ont pas été carottés.

Dans les dépôts pélagiques (sur la Dorsale Méditerranéenne, site 125, et dans les passées pélagiques de 130 et 131), les sapropels sont des lits de couleur noire, interstratifiés avec des contacts nets avec des vases banales de couleur claire. Ils se caractérisent par une forte proportion de pyrite et de matière organique. Dans les séries de turbidites, dans la Fosse Hellénique (sites 127 et 128) et sur la Dorsale Méditerranéenne (site 126) on observe des séquences granoclassées de couleur sombre. Il s'agit de sapropels et de sédiments normaux, repris et redéposés par des courants de turbidité ou de contour.

Turbidites et contourites plio-quaternaires

Dans certaines parties de la Méditerranée, on rencontre d'importantes accumulations de turbidites ou de contourites.

Parmi les accumulations explorées par nos forages, la plus importante est celle du cône sous-marin du Nil qui s'étend au large du delta de ce fleuve. Toutes les autres turbidites se sont accumulées dans des pièges topographiques comme la Fosse Hellénique ou la Mer d'Alboran. Enfin, le forage de la marge continentale des Baléares (site 124) traversa une épaisse succession de contourites.

Un premier forage (site 131), implanté à la base du Cône du Nil par 3035 mètres de profondeur, pénétra une épaisse série (272 m) de turbidites quaternaires provenant du Nil. Près de la surface, les séquences, épaisses de 2 à 3 mètres, sont presque entièrement formées de sable quartzeux grossier surmonté d'une fine couche d'argile noire. Plus profondément, les proportions s'inversent et on observe des séquences où la couche basale de sable ne représente plus qu'un tiers ou un dizième de la séquence. Du point de vue du mode de dépôt, il s'agit de fluxoturbidites analogues à celles décrites par Dzulynski dans les Carpates et Stanley dans les Alpes.

Un second forage (site 130) fut implanté beaucoup plus au large du Delta du Nil, de l'autre côté de la plaine abyssale d'Hérodote, sur la Dorsale Méditerranéenne. Il traversa d'abord 20 mètres de vases pélagiques litées, d'âge quaternaire. Il pénétra ensuite une épaisse série (563 m) de turbidites du Nil. Ces dernières sont semblables à celles du site 131, mais de faciès beaucoup plus "distal": courtes séquences d'argiles noires du Nil comportant une fine passée de sablon ou de sable quartzeux à la base. De rares lits de vases pélagiques claires ainsi que quelques sapropels sont intercalés avec des contacts francs au milieu de séquences de turbidites.

Dans la Fosse Hellénique, un premier forage (site 127) perça 427 mètres de turbidites quaternaires avant de rencontrer des blocs de calcaires et de dolomie crétacés, interstratifiés avec du Pliocène pélagique. Ces blocs sont interprétés comme "mélange tectonique". Un second forage (site 128) pénétra 475 mètres de turbidites quaternaires avant d'être abandonné. La Fosse Hellénique est donc remplie de turbidites d'âge quaternaire, probablement interstratifiées avec quelques contourites. Les séquences sont assez classiques, granoclassées de sable à vase. Les sables sont souvent mixtes, terrigènes et bioclastiques. Enfin, si l'épaisseur moyenne des séquences est de 30 centimètres, nous avons observé quelques séquences assez spéciales épaisses de 9 mètres ou plus, et entièrement formées de vases homogènes avec un fin lit de sable à la base.

Dans le Golfe de Valence (sites 122 et 123, Figure 10), la couverture plio-quaternaire est formée de turbidites et de contourites. Les horizons de sable y sont essentiellement de composition terrigène.

Le glacis continental des Baléares (site 124) est recouvert d'une épaisse (350 mètres) couverture plio-quaternaire. Elle est formée de turbidites et de contourites. Ces dernières, déposées par des courants balayant le glacis tout en épousant la forme, dominent la formation.

Enfin la Mer d'Alboran (site 121, Figure 4) qui, à première vue, se présente comme un bon piège à turbidites, en contient relativement peu. Seule la partie inférieure des dépôts post-miocènes est formée de turbidites. Dès la fin du Calabrien, elles sont remplacées par des vases péri-continentales à fortes proportions (70%) d'apports terrigènes fins (voir aussi Huang et Stanley, ce volume).

DISCUSSION

Le Miocène Moyen et Supérieur

En Méditerranée occidentale, le Burdigalien marque une phase tectonique majeure (Durand Delga, 1969), et le début de la grande transgression miocène sur la périphérie du Bassin. Dans nos forages implantés dans le Bassin occidental, un seul, situé en Mer d'Alboran (site 121), est descendu sous les séries messiniennes (Figure 12). Le Burdigalien n'a pas été atteint, mais le Tortonien y est représenté par des marnes gris-bleues, hémi-pélagiques, à fortes proportions de clastiques terrigènes. La faune de foraminifères est normale et indique des dépôts de mer ouverte, assez profonds.

Il existe, dans les coupes géologiques périméditerranéennes, de nombreux affleurements néogènes. Mais pour pouvoir faire des comparaisons valables, il faut évidemment comparer des sédiments déposés dans des conditions similaires. Nous adoptons donc l'hypothèse que dans un bassin donné, la *sédimentation-type* caractérisant une période correspond aux sédiments pélagiques, c'est-à-dire déposés particule par particule sur le fond. Les dépôts de courants de turbidité, de courants de fond ou de contour, sont donc exclus, ainsi que les dépôts littoraux.

Dans cette optique, les dépôts profonds des assises tortoniennes des bassins périphériques présentent un faciès très semblable au Tortonien du site 121. Ce sont les marnes bleues à Pleurotomes du Bassin du Rhône, de Tortona dans l'Apennin Nord, de l'Afrique du Nord et de l'Espagne. Lorsqu'on s'enfonce plus profondément dans ces bassins, on rencontre des faciès peu profonds: sables, molasses sableuses du Bassin Suisse, *etc.* . . . Au sommet, ces séries tortoniennes passent à des dépôts plus grossiers comme par exemple les grès sableux du Lyonnais, puis sont recouvertes de nappes d'alluvions continentales d'âge pontien.

Dans le Bassin oriental, les dépôts que nous avons forés sont plus délicats à correler. Le calcaire pélagique d'âge langhien du Strabo (site 129) contient une faune de mer ouverte. Au-dessus, le Serravalien et le Tortonien montrent, par contre, des faunes pélagiques moins abondantes et moins diversifiées. De même, en Mer Ionienne (site 126), le Serravalien qui se présente toujours sous le faciès des marnes bleues contient une faune très appauvrie et beaucoup de pyrite. Ces caractères suggèrent un bassin partiellement isolé. Le nannoplancton, peu sensible aux changements des conditions hydrauliques, reste le principal agent de sédimentation biologique tandis que les foraminifères sont affectés. Leur abondance, ainsi que le nombre des espèces, diminuent. Les formes sont souvent naines. Enfin, sur le fond règnent des conditions réductrices. Seuls quelques foraminifères benthiques survivent. Les organismes fouisseurs disparaissent, laissant des dépôts finement lités où se forme de la pyrite.

Le faciès des dépôts forés dans le Bassin oriental est donc assez semblable aux marnes bleues du Bassin occidental jusqu'au Serravalien. A partir de cette époque, à l'Est, les faunes de foraminifères sont affectées. Mais comme la fraction d'origine biologique des sédiments est essentiellement composée de nannofossiles et non de foraminifères, l'aspect macroscopique des dépôts reste le même: celui des marnes bleues.

Dans toute la Méditerranée orientale, au Vindobonnien, ce faciès de marnes bleues semble caractériser les dépôts profonds. On les observe dans le Tortonien des forages implantés sur les Iles Ioniennes (Bizon, 1967), dans le centre de l'Europe, dans le Bassin de Vienne où elles forment le faciès "schlier". à l'Est de l'Europe dans le Bassin Panonique, *etc.*

Ainsi depuis le Burdigalien et jusqu'à la fin du Tortonien, des sédiments de même faciès, les marnes bleues, se déposent dans tout le bassin méditerranéen. Il s'agit ici du bassin au sens large, de la Téthys vindobonienne, comprenant l'Europe centrale, l'Europe du Sud-Est, ainsi que tous les bassins périphériques comme celui du Rhône. Ces marnes bleues forment un ensemble lithologique homogène. Il s'agit d'une *formation marine* au sens géologique du terme. Ces marnes sont caractérisées par des valeurs d'environ 30% de $CaCO_3$ et par leur couleur sombre. La proportion de carbonates montre que les apports terrigènes fins l'emportent largement sur la productivité biologique. La couleur sombre indique des conditions réductrices au niveau du fond, c'est-à-dire une circulation des eaux difficiles dans l'ensemble des bassins.

Nous ne connaissons pas la base de cette formation, car nos forages n'ont pas atteint l'Aquitanien. Il reste, en particulier, à préciser si un changement de faciès accompagne l'avènement du Burdigalien en Méditerranée profonde. Par contre, la fin du Tortonien est marquée à terre par une émersion généralisée qui semble bien correspondre à la *crise de salinité* messinienne. Ainsi, le toit de la grande formation de marnes bleues coïnciderait avec la limite supérieure du Tortonien. En Méditerranée profonde, ce toit correspond à un passage des dépôts marins à des dépôts saumâtres et évaporitiques.

La crise de salinité messinienne

Dans toute la Méditerranée, aussi bien dans le Bassin oriental qu'occidental, tous les dépôts d'âge messinien forés correspondent à des séries évaporitiques: marnes dolomitiques interstratifiées avec des horizons de gypse, d'anhydrite et de halite (Figure 12). Jusqu'à une époque récente, de telles séries auraient été considérées comme peu profondes. Toutefois, depuis la confirmation de l'existence de dômes de sel sous les plaines abyssales (Burk *et al.*, 1969), l'hypothèse d'une précipitation des sels dans des bassins profonds a été proposée (Schmalz, 1969). Nous avons donc été conduits à rechercher dans nos forages des critères permettant d'opter pour l'une ou l'autre de ces hypothèses. Nous avons noté que:

1. Des horizons d'anciennes vases marines à faciès de mer ouverte sont interstratifiés au milieu des bancs de gypse, d'anhydrite ou de halite. A l'origine ces vases étaient formées, quant à leur composition biologique, de nannoplancton et de foraminifères. Depuis, certains horizons ont été épigénisés à des degrés divers, soit par des sels, soit par de la dolomie.
2. L'étude des isotopes de l'oxygène et du carbone confirme l'origine marine de certains horizons. Mais elle montre surtout que ces lits marins sont interstratifiés avec des couches déposées lorsque le bassin était alimenté en eaux continentales (Fontes *et al.*, 1972).
3. Dans certains horizons, les associations faunistiques suggèrent des faciès littoraux ou saumâtres. Ainsi, dans la Montagne du Strabo (site 129), les premières couches messiniennes contiennent des foraminifères benthiques d'eau peu profonde et des ostrocodes d'eaux saumâtres. En Mer Ionienne (site 125), les évaporites terminales abondent en foraminifères benthiques d'eau peu profonde. Dans le Bassin Algéro-Provençal (site 124) quelques horizons de la série évaporitique contiennent des diatomites finement laminées dont la flore indique un milieu saumâtre.
4. Le faciès des épais lits d'anhydrite et de gypse est souvent semblable à celui des dépôts de sels actuels, littoraux ou supralittoraux, des régions chaudes ou tropicales comme le Golfe Persique ou les sabkha tunisiennes (Kinsman, 1966; Rooney et French, 1968). Ainsi, nous avons observé des anhydrites finement laminées avec alternance de lits clairs et sombres. Dans ces lits se forment des cristaux secondaires, nodulaires, qui en se développant aboutissent à des structures alvéolaires appelées *chicken wire* (Figure 8A). Nous avons aussi noté, dans des gypses, les laminations ondulées caractéristiques des récifs à algues littoraux ou stromatolites (Figure 5B).
5. Enfin des lits de sables terrigènes à stratification entrecroisée et contenant des débris biologiques bien roulés indiquent l'action de courants. Toutefois, ces courants peuvent aussi bien être littoraux que profonds.

Tous ces caractères nous conduisent à penser que les sels se sont déposés en eau peu profonde. Toutefois, des épisodes marins venaient interrompre, d'une façon intermittente, le dépôt des sels.

Le modèle qui répond le mieux à ces deux conditions

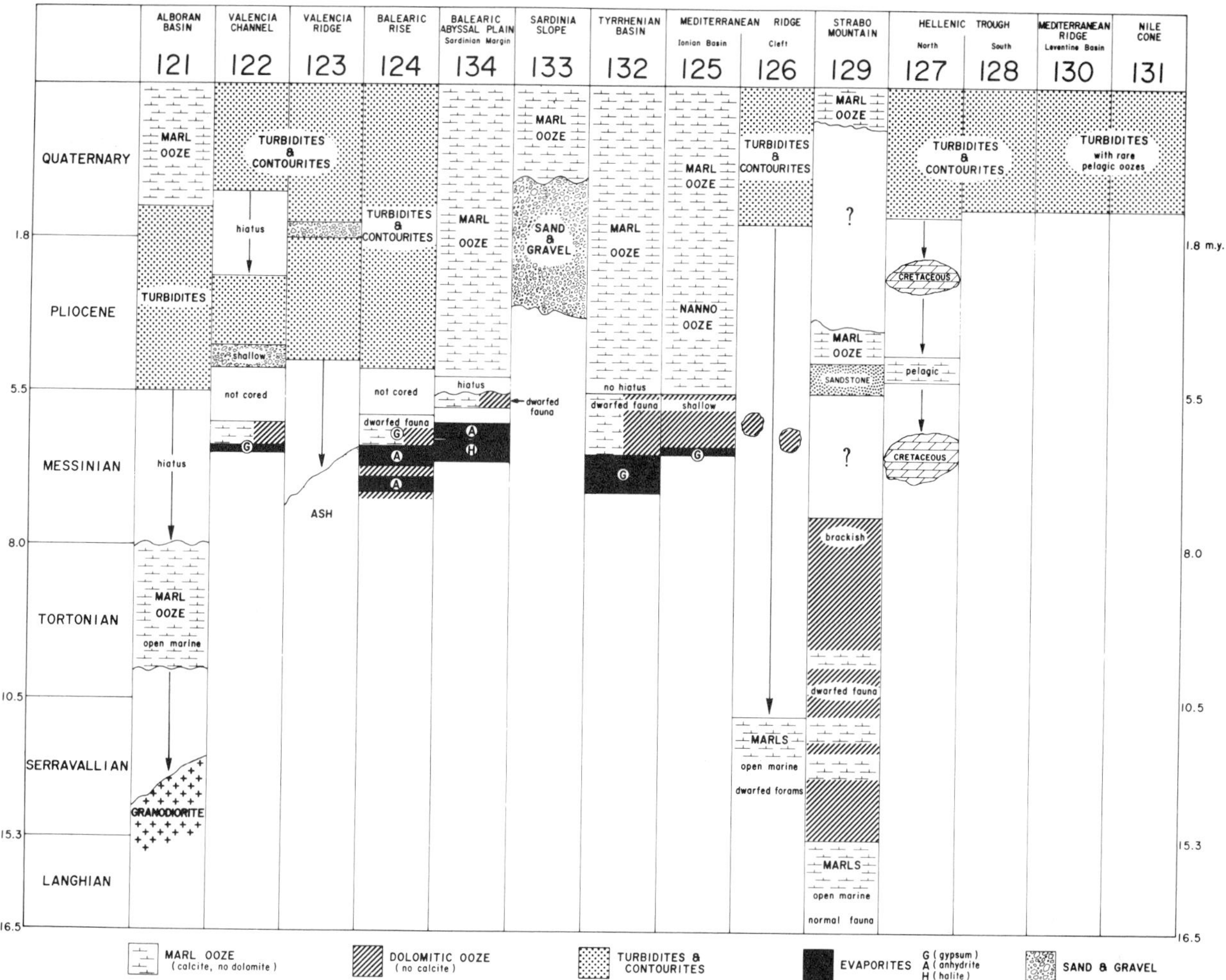

Figure 12. Lithostratigraphie des forages. Leg XIII du GLOMAR CHALLENGER (JOIDES-DSDP). D'après Nesteroff *et al.*, (1972).

est un bassin isolé, situé en contrebas du niveau de l'Océan Mondial. L'évaporation y excède les apports d'eaux continentales et conduit ainsi à un assèchement. Des lagunes peu profondes où se déposent des évaporites subsistent au fond du bassin. Mais si l'épaisseur des évaporites qui s'y accumulent atteint plusieurs centaines de mètres, il n'est guère possible d'envisager leur dépôt par simple évaporation de la colonne d'eau sus-jacente. Une alimentation limitée, mais assez continue, en eau de mer est nécessaire.

D'autre part, ce bassin est sporadiquement envahi par la mer. Il reçoit alors, pendant quelques dizaines ou centaines de milliers d'années des dépôts marins avant de s'assécher à nouveau.

Tout suggère qu'au Messinien, la Méditerranée ait été transformée en un bassin de ce type. En effet, à l'Est, les communications s'interrompent avec l'Océan Indien. Elles ne seront pas rétablies. Tout le Sud-Est de l'Europe, la Yougoslavie, la Russie du Sud, jusqu'à la mer d'Aral et la Caspienne se transforment en d'immenses mer-lacs dessalés (Gignoux, 1950). Dans ces provinces humides, situés au Nord du Bassin Méditerranéen proprement dit, les apports d'eaux continentales et les précipitations l'emportent sur l'évaporation.

Plus à l'Ouest, dans les Alpes, dans le Bassin du Rhône, c'est la régression pontienne, laissant la place aux dépôts continentaux. Dans le Golfe du Lion, le forage Mistral montre une lacune correspondant au Miocène supérieur (Burollet et Dufaure, ce volume). Dans la Mer d'Alboran, notre forage (site 121) a mis en évidence une lacune correspondant à la fin du Miocène. La Méditerranée se trouve donc isolée au Messinien, époque que les auteurs placent vers 9 ou 8 millions d'années. Il faudrait en rechercher la cause dans un rapprochement de la plaque africaine de celle de l'Eurasie. Ce rapprochement interrompt les communications à l'Est avec l'Océan Indien. Il les coupe aussi à l'Ouest en exondant les détroits Nord Bétique et Sud Riffain. Une communication sporadique subsiste toutefois avec l'Atlantique. C'est probablement par elle que se fait l'approvisionnement en eau salée nécessaire à l'alimentation des formations évaporitiques ainsi que les brutales invasions marines qui transforment de temps à autre ce bassin desséché en mer intérieure. Ainsi pendant le Messinien, des séries évaporitiques se déposent dans les parties profondes de la Méditerranée asséchée. Les études de sismique réflexion montrent en effet que les dômes de sel, expression morphologique des séries salifères, sont limités à certaines parties du bassin, aussi bien à l'Ouest (Glangeaud, 1966) qu'à l'Est (Emery *et al.,* 1966).

Des lambeaux de ces séries évaporitiques se retrouvent actuellement soulevés sur la périphérie du Bassin Méditerranéen. Les plus connus sont ceux de l'Apennin et de la Sicile (Ogniben, 1957). Ces gisements se poursuivent par des petits lambeaux en Afrique du Nord. En Espagne, des séries épaisses de 1000 mètres ont été reconnues en forage. Dans le Bassin oriental, les exemples les plus spectaculaires viennent de Chypre et de Turquie où les forages ont traversé des centaines de mètres de sel. Enfin, dans tous les gisements reconnus, le toit des évaporites et la transgression pliocène sont extrêmement nets et tranchés (Bourcart, 1960–62).

Les évaporites messiniennes du Bassin Méditerranéen constituent une grande formation lithologique dont les limites inférieures et supérieures sont très bien définies pétrographiquement. De plus, comme elles coïncident avec celles de l'étage messinien, cette unité lithologique est, en même temps, une unité *chronostratigraphique* et *biostratigraphique* (Bell *et al.*, 1961).

Cette formation représente le plus vaste système évaporitique actuellement connu. L'ensemble des divers bassins dépasse largement celui du Zechstein (700,000 km^2) ou la Mer Morte. Seul le Bassin de Tarim, dans le Sinkiang, pourrait être comparé à la Méditerranée.

Le Plio-Quaternaire

Avec le Pliocène se produit une transgression extrêmement brutale. Tous les contacts forés sont érosifs avec souvent des lacunes importantes. Les nouveaux dépôts sont, dès leur base, marins et pélagiques. De plus les associations de foraminifères indiqueraient qu'ils sont relativement profonds: plusieurs centaines de mètres (Cita, 1971). Les ostracodes témoigneraient même de profondeurs beaucoup plus importantes (Benson, ce volume).

Ces données confirment bien l'envahissement par la mer d'un bassin situé en contre-bas du niveau général. Ainsi, la transgression pliocène en Méditerranée aurait un caractère particulier qui la distingue des transgressions qui recouvrent les bordures de nos continents.

L'arrivée de la mer ne peut se faire que par l'Ouest. En effet à la fin du Miocène toutes les communications ont été interrompues du côté oriental. A l'Ouest le détroit Nord-Bétique, fonctionnel au Miocène, est émergé au Pliocène. C'est probablement la région Sud-Riffaine qui fait office de détroit, relayée plus tard par Gibraltar.

Ainsi, il y a environ 5,5 millions d'années, le détroit de Gibraltar s'ouvre et l'eau de l'Atlantique envahit le Bassin Méditerranéen asséché, situé en contrebas. Cette *mise en eau* provoque une érosion plus ou moins importante des couches terminales du Miocène. Des conditions de mer ouverte s'installent dans toute la Méditerranée et des vases semi-pélagiques, très différentes des séries évaporitiques du Messinien, commencent à s'y déposer.

Si l'on se base sur les deux colonnes sédimentaires pélagiques complètes que nous avons forées (site 132 Mer Tyrrhénienne, site 125 Mer Ionienne), la sédimentation plio-quaternaire apparaît comme continue. Toutefois, les enregistrements sismiques montrent des discordances angulaires à la base de certaines marges continentales, comme celles des Baléares (site 124). Or, ces glacis sont formés de contourites, c'est-à-dire de dépôts par courants de fond. Les discordances suggèrent donc des changements de régime des courants généraux de la Méditerranée.

Enfin, l'apparition de sapropels au Pliocène supérieur et au Quaternaire, indique des périodes de stagnation au fond des bassins, c'est-à-dire de nouveau des changements dans le régime hydraulique général. Il est intéressant de noter que les sapropels ne se rencontrent que dans le Bassin oriental.

Les lambeaux de Plio-Quaternaire actuellement émergés sur la périphérie du bassin méditerranéen correspondent bien, lorsqu'ils sont de faciès pélagique profond, aux séries-types de nos forages. En particulier, les marnes claires du faciès *trubi* qui forment la base du Pliocène dans la série Gessosso Solfifera des Apennins et de Sicile ont bien le même faciès que les vases claires à foraminifères de la base du Pliocène de nos sites 125, 132 et 134.

De même, la transgression extrêmement brutale du Pliocène avait été depuis longtemps remarquée par les géologues méditerranéens (Bourcart, 1960–1962).

Enfin, en ce qui concerne le Quaternaire supérieur, de nombreuses campagnes océanographiques ont montré la grande homogénéité des dépôts pélagiques en Méditerranée (Ryan *et al.,* 1970). D'un bassin à l'autre, les vases pélagiques ne diffèrent que par une variation lente et limitée dans les proportions de leurs principaux constituants et dans la nature des minéraux argileux.

CONCLUSIONS

L'aspect le plus intéressant de nos forages est d'avoir pu montrer, en nous basant sur les coupes-types de bassins profonds, que les affleurements néogènes éparpillés non seulement autour du Bassin-Mediterranéen, mais aussi dans tout le domaine alpin, se rangent dans trois grandes unités lithologiques. Très souvent, de par leur position marginale, il s'agit d'équivalents latéraux à faciès plus terrigène, plus grossier et plus littoral.

La plus ancienne de ces unités lithologiques est celle des marnes bleues, d'âge burdigalien-vindobonien. Elles sont sombres, pauvres en carbonates, bien litées et souvent pyriteuses. Il s'agit d'un dépôt où les apports terrigènes l'emportent sur la productivité biologique et où des conditions réductrices règnent au fond. Tous ces caractères suggèrent une circulation difficile entre les divers bassins.

La seconde unité lithologique comprend les séries évaporitiques messiniennes.

La troisième unité est celle des marnes plio-quaternaires. Elles sont plus claires que les marnes vindoboniennes et plus riches en carbonates. Elles correspondent à une plus grande productivité biologique et à des conditions oxygénées au fond. Parallèlement la proportion d'apports terrigènes fins est moins importante. Elles témoignent d'une bonne circulation océanique dans des bassins moins étendus et profonds.

Les limites entre ces trois unités lithologiques sont extrêmement nettes et de plus isochromes. Il en résulte que ces grandes formations lithologiques sont en même temps des unités chronostratigraphiques et biostratigraphiques. Nous pouvons donc proposer, pour le domaine méditerranéen, trois grandes formations au sens géologique du terme. La plus remarquable est la *formation d'évaporites messiniennes*, qu'encadrent les marnes bleues vindoboniennes et les marnes beiges plio-quaternaires.

Cette évolution sédimentaire exceptionnelle est due au caractère intracontinental des domaines méditerranéens et alpins et, en conséquence, à leur dépendance des mouvements tectoniques. Ceci est évident pour l'épisode d'évaporites durant lequel les mouvements relatifs des plaques lithosphériques Afrique et Europe-Asie ont isolé (8 millions d'années) puis réuni à nouveau (5,5 millions d'années) la Méditerranée à l'Océan Mondial. Mais ce sont aussi les mouvements tectoniques qui durant tout le Néogène ont déterminé la profondeur des bassins et leurs interconnections.

En conclusion, c'est la tectonique du domaine alpin qui, en modifiant la circulation océanique de la Téthys, a controlé pendant le Néogène le dépôt des formations lithologiques dans cette mer.

REMERCIEMENTS

L'état major scientifique du Leg 13 est profondément redevable à la National Science Foundation des Etats-Unis qui a permis la réalisation du Deep Sea Drilling Project.

Le centre National d'Exploitation des Océans, et le Centre National de la Recherche Scientifique, ont assuré la participation française à l'état-major scientifique embarqué.

BIBLIOGRAPHIE

Bell, W. C., M. Kay, G. E. Murray, H. E. Wheeler et J. A. Wilson 1961. Geochronologic and chronostratigraphic units. *American Association of Petroleum Geologists Bulletin,* 45:666–673.

Benson, R. H. 1972. Ostracods as indicators of threshold depth in the Mediterranean during the Pliocene. In: *The Mediterranean Sea: A Natural Sedimentation Laboratory,* ed. Stanley, D. J. Dowden, Hutchinson and Ross, Inc., Stroudsburg, Pennsylvania, 63–73.

Bizon, G. 1967. *Contribution à la Connaissance des Foraminifères Planctoniques d'Epire et des Iles Ioniennes (Grèce Occidentale) Depuis le Paléogène Supérieur jusqu'au Pliocène.* Technip, Publications de l'Institut Français du Pétrole, Société des Editions, Paris, 142 p.

Bourcart, J. 1960–1962. La Méditerranée et la révolution du Pliocène. Livre à la mémoire du Professeur Paul Fallot. *Société Géologique de France,* 1:103–116.

Burk, C. A., M. Ewing, J. L. Worzel, A. O. Beall, W. A. Berggren, D. Bukry, A. G. Fischer et E. A. Pessagno Jr. 1969. Deep-sea drilling into the Challenger Knoll Central Gulf of Mexico. *American Association of Petroleum Geologists Bulletin,* 53:1338–1347.

Burollet, P. F. et Ph. Dufaure 1972. The Neogene Series Drilled by the MISTRAL No. 1 Well. In: *The Mediterranean Sea: Natural Sedimentation Laboratory,* ed. Stanley, D. J., Dowden, Hutchinson and Ross, Inc., Stroudsburg, Pennsylvania, 91–98.

Cita, M. 1971. Deep Sea Neogene Stratigraphy. *5ème Congrès du Néogène Méditerranéen,* (in press).

Durand Delga, M. 1969. Mise au point sur la structure du Nord-Est de la Berbérie. *Publication du Service Géologique Algérie* (Nouvelle Série), 39:89–131.

Emery, K. O., B. C. Heezen et T. D. Allan 1966. Bathymetry of the eastern Mediterranean Sea. *Deep Sea Research,* 13:173–192.

Fontes, J. C., R. Letolle et W. D. Nesteroff 1972. Les forages DSDP en Méditerranée (Leg 13): Reconnaissance isotopique. In: *The Mediterranean Sea: A Natural Sedimentation Laboratory,* ed. Stanley, D. J. Dowden, Hutchinson and Ross, Inc., Stroudsburg, Pennsylvania, 671–680.

Glangeaud, L. 1966. Les grands ensembles structuraux de la Méditerranée occidentale d'après les données de Géomède I. *Comptes Rendus des Séances de l'Académie des Sciences, Paris,* 262:2405–2408.

Gignoux, M. 1950. *Géologie Stratigraphique*. Masson, Paris, 735p.

Huang, T. C. et D. J. Stanley 1972. Western Alboran Sea: sediment dispersal, ponding and reversal of currents. In: *The Mediterranean Sea: A Natural Sedimentation Laboratory*, ed. Stanley, D. J. Dowden, Hutchinson and Ross, Inc., Stroudsburg, Pennsylvania, 521–559.

Kinsman, D. J. J. 1966. Gypsum and anhydrite of Recent age, Trucial Coast, Persian Gulf. In: Second Symposium on salt, ed. Raup, *Northern Ohio Geological Society* 1:302–326.

Nesteroff, W. D. and others 1972. Summary of lithostratigraphical findings and problems. In: *Initial Reports of the Deep Sea Drilling Project*, National Science Foundation. Government Printing Office, Washington, D.C., 13(42).

Ogniben, L. 1957. Petrografia della seria solfifera Siciliana e considerazioni geologiche relative. *Memorie Descrittive della Carta Geologica d'Italia*, 33:1–275.

Rooney, L. F. and R. R. French 1968. Allogenic quartz and the origin of penemosaic texture in evaporites of the Detroit River Formation (Middle Devonian) in northern Indiana. *Journal of Sedimentary Petrology*, 38:755–765.

Ryan, W. B. F., D. J. Stanley, J. B. Hersey, D. A. Fahlquist et T. Allan 1970. The tectonics and geology of the Mediterranean Sea. In: *The Sea*, ed. Maxwell, A. E., John Wiley and Sons, New York, 4(2):387–492.

Schmalz, R. F. 1969. Deep water evaporite deposition: A genetic model. *American Association of Petroleum Geologists Bulletin*, 53:798–823.

Scientific Staff 1970. Deep Sea Drilling Project: Leg 13. *Geotimes*, 15(10):12–15.

Ostracodes as Indicators of Threshold Depth in the Mediterranean during the Pliocene

Richard H. Benson

Smithsonian Institution, Washington, D.C.

ABSTRACT

Three diagnostic species of "bathyal" psychrospheric (deep-sea) ostracodes were found in the Pliocene and lowermost Pleistocene strata of a core (DSDP Hole 132) obtained from the floor of the Tyrrhenian Basin. This particular deep-sea assemblage is most likely to occur living in the open ocean at depths between 1,000 and 1,500 meters (bottom temperatures between 4°C and 6°C). It is known from outcrop sections of Neogene age in Italy and Crete, and is believed to have become extinct in the Mediterranean during the early Pleistocene (Calabrian). No ostracodes, and few other benthic animal remains of Plio-Pleistocene age were present in a second core (DSDP Hole 125) examined from the eastern Ionian Basin.

The entry and maintenance of the world-ocean psychrospheric fauna within a series of partially land-locked basins requires free access to the open ocean at considerable depths. Temperatures of basin waters below the average winter atmospheric lows are possible only if these water masses are derived externally from higher (polar) latitudes. The discovery of fossil psychrospheric ostracodes and other deep meiobenthonic organisms, more than 1,000 miles from present deep-ocean conditions, suggests that in the past the opening that now forms the Gibraltar sill was much deeper and wider. The fact that the oldest of the Pliocene specimens in DSDP Hole 132 were only a few meters above a Messinian (upper Miocene) evaporite sequence suggests that radical changes in threshold depth in the western opening took place in a very brief geologic time span. The apparent absence of the benthic fauna contemporaneously in the Ionian basin suggests that the restricted hydrologic conditions that had prevailed in the late Miocene throughout the Mediterranean continued into the Pliocene in the east.

RESUME

Aux niveaux des couches du Pliocène et du Pléistocène inférieur d'une carotte prélevée au fond du Bassin Tyrrhénien (DSDP carotte No. 132) on a trouvé trois espèces d'ostracodes formant un groupe caractéristique de mer profonde (ou psychrosphère). Un tel assemblage se développe vraisemblablement en pleine mer, à des profondeurs variant entre 1000 m et 1500 m (température du fond entre 4°C et 6°C). Le même assemblage a été découvert dans des affleurements du Néogène en Italie et en Crète et a probablement disparu de la Méditerranée au début du Pléistocène (Calabrien). Dans une autre carotte (DSDP 125), prélevée au fond du Bassin Ionien, on n'a trouvé aucun ostracode et très peu d'organismes benthiques caractéristiques du Plio-Pléistocène.

La pénétration et subsistance de la faune psychrosphérique dans une série de bassins partiellement isolés nécessitent un accès libre à la pleine mer sur des profondeurs considérables. La température de l'eau de ces bassins ne peut être plus basse que la moyenne des minima d'hiver dans l'atmosphère que si des masses d'eau plus froide (de latitude polaires) y sont introduites. La présence, à plus de 1.000 miles de la configuration actuelle de l'océan, de fossiles d'ostracodes de mer profonde ainsi que d'autres organismes méiobenthoniques, semble indiquer que le passage qui forme le détroit de Gibraltar était plus large et plus profond. Dans la carotte DSDP 132 les spécimens du Pliocène les plus anciens n'étaient situés qu'à quelques mètres au-dessus d'une séquence d'évaporite messinienne (Miocène supérieur); ceci semble indiquer que les changements dans les profondeurs des seuils ont pris place dans dans un espace de temps très court à l'échelle géologique. L'absence apparente de faune benthique dans le Bassin Ionien suggère que les conditions hydrologiques qui ont prévalu au Miocène supérieur dans toute la Méditerranée se sont maintenues au Pliocène en Méditerranée occidentale.

INTRODUCTION

The fact that *deep water* facies are present in an otherwise shallow sedimentary record of the Tethys "Sea" has long been accepted. Beneo's (1956) *olistostrome* was formed in deep water. The Pindus eugeosynclinal furrow in the Hellenides of the *pre-Flysch period* was interpreted by Aubouin (1965) as the site of sedimentation of the "deep-sea" type. Glangeaud (1962) interpreted the Tyrrhenian Basin as a paleo-ocean. Gass (1968) interpreted the Troodos Mountains of Cyprus as an overthrust relic of a Mesozoic "mid-ocean ridge".

In earlier works it was not suggested that the deep water facies were necessarily contiguous with the deeper parts of the world-ocean system. Haug (1900) considered the *Aptychus* schists, among other sedimentary types characterized by pelagic fossils, as being bathyal. These sediments include radiolarites that were thought to be abyssal by Cayeux (1924). Neither worker based his judgment of depth on the presence of deep benthic fossils, but on the dominance of pelagic remains and the absence of familiar shallow forms. Ruggieri (1953, 1954) was the first to suggest that a few of the Calabrian ostracode species had deep-sea affinities. Recently, it was suggested (Benson and Sylvester-Bradley, 1971) that because these same deep-sea ostracodes could be traced back to the beginning of the Tertiary in southern Europe, the Tethys may have been an ocean in part.

Three of the four most diagnostic species of a "bathyal" psychrospheric[1] ostracode assemblage have been found in a long core (223 meters), Messinian to Quaternary in age, from one of the deeper parts (2813 meters) of the Tyrrhenian Basin (DSDP Hole 132, longitude 11°26.47′E; latitude 40°15.70′N; Figure 1). Evaporites are present in the bottom of the core (presumably the Messinian *gessoso solfifera* formation). Psychrospheric ostracodes (47 specimens of eight species) were found at five of seven intervals examined from just above the Messinian evaporite and red marl sequence in strata of Zanclian (Trubi Formation) to Calabrian age (Table 1). These specimens represent a significant part of the entire assemblage, which has been previously reported (Benson and Sylvester-Bradley, 1971; Colalongo, 1965) from rocks of the same age near Bologna, in Calabria

[1] The term *psychrosphere* was proposed by Bruun (1957) to replace the ambiguous term stratosphere (used by German oceanographers) as applied to the deeper region of the oceans with temperatures of less than 10°C. The overlying warmer waters were termed the *thermosphere*. The exact temperature boundary between thermospheric and psychrospheric ostracodes is not known, but it is suspected to be closer to 8°C than to 10°C.

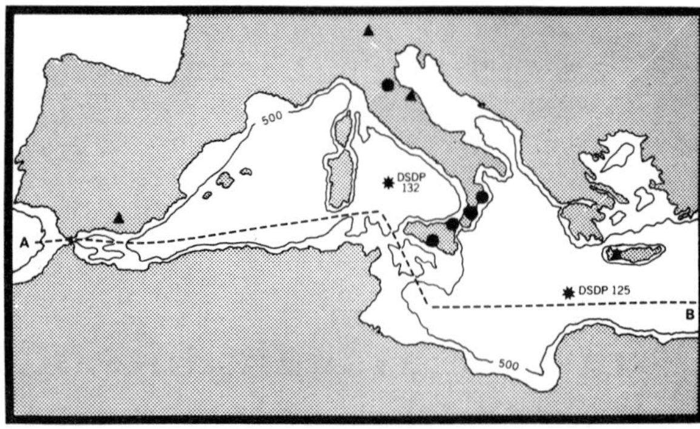

Figure 1. Sample location and gross basin bathymetry of the region of the present study also showing the general positions (A-B) of the profiles given in Figure 7. The two Deep-Sea Drilling Project Sites (125 and 132) are shown along with the general outcrop localities of Pliocene and Pleistocene (dots) and older (solid triangles) strata where psychrospheric ostracodes were found. The present 500 meter bathymetric contour outlines the two major threshold regions (Straits of Gibraltar and Sicily) responsible for thermospheric conditions throughout the Mediterranean today.

and in Sicily. Portions of the assemblage have also been reported from the Eocene of northern Italy (Ascoli, 1969), the lower Miocene of sections near Ancona, Italy, the Miocene of Gavdos Island, south of Crete, the Middle Miocene (Serravallian) of Vittoria, Sicily and the uppermost Oligocene of the same locality (Ruggieri, personal communication).

A second long core, which was found to be barren of ostracodes, was examined from the Mediterranean Ridge on the eastern side of the Ionian Basin (DSDP Site 125, 34°37.31′N, 20°25.68′E). This core (97 meters in length) also represents strata ranging in age from Messinian to Quaternary. It also penetrated the evaporite sequence and was rich in pelagic marls, yet no meiobenthonic fossil remains were found.

In a recent paper (Benson and Sylvester-Bradley, 1971), my colleague and I stated that the discovery of psychrospheric ostracodes in the Cenozoic of several regions of the Mediterranean constitutes benthic fossil evidence of a former link of Tethys with the world-ocean system. We claimed that the presence of this fauna in outcrop sections, now far removed from contact with world-ocean conditions, could only come about through a broader opening of the Iberian Portal (the gateway between the Mediterranean or western Tethys Ocean and the Atlantic; the term *Gibraltar Strait* for the early Cenozoic may be too specific). We suggested that a passage of considerable magnitude was required to permit the deeper, colder waters of the open ocean and its fauna to invade and be maintained in an otherwise restricted, east-west trending, warm water-producing intercontinental sea.

We stated that the strong development of this oceanic fauna in the central Mediterranean during the

Pliocene suggested the maintenance of this passage through late Miocene time, yet we expressed puzzlement over the obvious environmental implications of the presence and distribution of the Messinian evaporites, the *gessoso solfifera* formation. Development of evaporites would suggest an interruption in the connection with the deeper open ocean. An adequate non-catastrophic explanation for such a radical interruption is yet to be given.

Since our first study of fossil psychrospheric ostracodes in the Mediterranean, I have examined the contents of the two Deep-Sea Drilling Project cores, mentioned earlier (Figure 1), which were obtained from the Ionian (Site 125) and Tyrrhenian Basins (Site 132). I have also analyzed the world-ocean depth distribution of the four diagnostic bathyal ostracode genera. In addition, the distributions of several Foraminifera species of the Plio-Pleistocene Mediterranean psychrospheric fauna have been studied. The results of these studies, as extensions of our original study, are given here. In addition, schists (Heezen *et al.,* 1971) and evaporites (Scientific Staff, 1970), have been discovered in the floor of the Tyrrhenian Basin. If this really means that the Tyrrhenian is floored by foundered continental crust, then it must have already been ocean in Pliocene time. More probably the continental rocks have slumped into the oceanic depths as olistostromes.

I will attempt to demonstrate the likelihood of the following: (1), that during the Pliocene the pattern of water circulation in the western Mediterranean included the invasion of an intermediate level oceanic water mass (upper psychrosphere) with its included microfaunal assemblages; (2), that at least two sills probably existed in the Mediterranean during the Pliocene, one at the entrance (the Iberian Portal), and a second west of the Ionian Basin; and (3), that the Iberian Portal must have been considerably wider and deeper during the Pliocene than the Strait of Gibraltar is at present, but that the sill of the Sicilian-African marginal platform (now forming the Sicilian Straits) was deeper than at present but not as deep as the sill of the Iberian Portal.

THE IMPORTANCE OF THRESHOLDS

The threshold is the lowest point of the sill[1] across which deep waters can flow through the confining boundaries of a basin. It controls the character and structure of the water mass filling the deeper part of the basin. Depending on the density of the surface-water layers of the basin (formed by low temperatures or increased salinity), waters may be forced to flow into or out of the basin across the threshold. Thresholds act to limit the flow and level of density stratification of the basin water relative to the stratification of surrounding or adjoining water masses. The character of the environment governing the admittance of a faunal assemblage into the depths of the basin is largely a function of the control exerted by the depth of the threshold on adjacent water mass movement.

It follows, therefore, that the deepest dwelling animals that can enter a basin, like the deepest water masses, will be determined by the threshold depth of the basin. Animals outside of the basin, which are tolerant only of depths greater than the threshold depth, will be excluded, although the basin itself may be as deep as surrounding areas. In an enclosed basin, the waters below threshold depth are uniform; therefore, a species once in the basin could conceivably descend to greater depths than it was accustomed to living in the open ocean. The character of the fauna inhabiting the deeper parts of a basin reflects the greatest depth of the confining sill that allowed its member species to enter. The threshold depth may, in fact, influence the deep fauna of the basins more than do the prevailing shallow-water conditions of the basin, depending on the volume, depth and direction of flow of bottom waters through the portal.

A demonstration of this principle can be found in the nature and distribution of the fauna now inhabiting the deep basins of the Mediterranean. It is a submergent, thermospheric fauna, whose descendency into deeper levels has been determined by the formation of dense, warm surface waters (generally about 15°C to 16°C) with the consequent depression of the thermocline (the deep basin waters are about 13°C), and the restricted outflow of this basinal water mass over a shallow threshold (about 320 meters, and with a very small and turbulent portal cross-section). These waters are denser than any of the layers of the Atlantic within reach of the present Gibraltar threshold. The Atlantic now has a negligible influence on the waters of the Mediterranean (in fact, it is to the contrary). By dilution, through a higher rate of runoff or precipitation, or by lowering of the sill, the present outflow circulation could be reversed. However, even if the former occurred, the inflow across the threshold would not be psychrospheric, as these colder waters lie at considerable depths beneath the level of the sill. The 8°C isotherm is at more than 1,500 meters depth.

[1] The terms *portal, sill,* and *threshold depth* as used in this report refer to the following: *portal,* the opening between basins through which water flows; *sill,* the submerged elevation separating two basins; *threshold depth* (sometimes called *sill depth*), the greatest depth at which there is free, horizontal communication between basins.

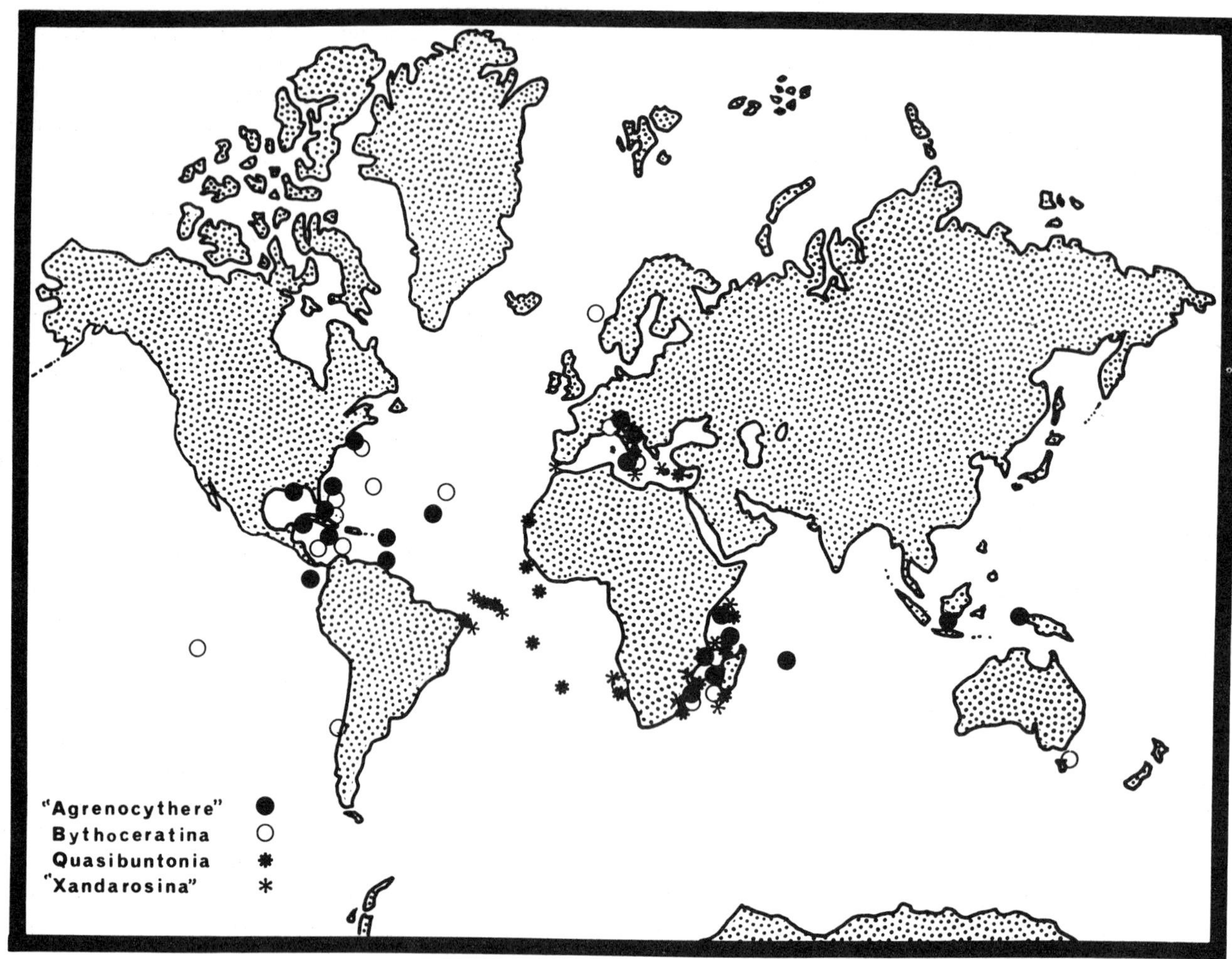

Figure 2. The Recent world ocean distribution of the four psychrospheric ostracode genera found only as fossils in the Mediterranean, as known from the collections of the author. The frequency of depth distribution is shown in Figure 4.

THE PSYCHROSPHERIC OSTRACODE FAUNA

At present the benthic ostracodes living on the continental shelves (underlying part of the thermosphere) throughout the world ocean are of a different history and assemblage composition than those found in waters generally deeper than 500 meters (generally the uppermost limit of the psychrosphere). The psychrospheric ostracode fauna, which has relatively few and long-lasting species, tends to be much more cosmopolitan than the many and more diverse thermospheric faunas. Its forms are usually larger (about 20% in length), blind (beginning at about 600 meters), more delicate, and the species with sculptured carapaces are architecturally more complex and ornate (Benson, 1969). The psychrospheric fauna seems to have been phyletically separate and distinct from the shelf fauna since before the Eocene. Some of the Recent species have a Mesozoic, rather than a modern morphologic aspect.

At present the psychrospheric ostracode fauna consists of a deeper abyssal assemblage (2,000 to 5,500 meters; water temperatures less than 4°C) and a shallower bathyal assemblage (500 to 3,000 meters; water temperatures from about 4°C to about 8°C). The depth boundary between these two groups of species is not precise and in the fossil record, it may become even less so.

The bathyal assemblage appears to have taxonomic elements that are older than those of the deeper assemblage. As suggested by Bruun (1957) this fauna may be a relic of the original ocean floor fauna; however, this is yet speculation. The terms bathyal and abyssal may lose their meanings if applied to older (early Cenozoic) fossil assemblages. What is of concern here is the likelihood of change in the depth of the boundary between these two assemblages with progressive cooling of the oceans during the Neogene. Only the shallower psychrospheric assemblage is represented in the Pliocene fossil record in the Mediterranean. Therefore, a lower limit can be placed on the depth of the western threshold at that time. If there was a progressive raising of the isotherms in the deep oceans during the Cenozoic, the zone of transition between the two assemblages during the Pliocene (if temperature is indeed the primary factor) might have been as much as 500 to 1,000 meters deeper than it is now (2–3,000 meters, including compensation for changes in sea level of about 100 meters).

Today there are few if any psychrospheric ostracodes remaining as relics in the Mediterranean. All but one species (*Bathycythere vanstraateni* Sissingh) were absent in more than 100 Recent core samples examined. The bottom waters are too warm (13°C). However, as shown on the maps of the Recent and Neogene distribution of the four characteristic psychrospheric genera found in Plio-Pleistocene deposits in the Mediterranean (Figure 2), these taxa seem to occur scattered about a Mediterranean focus. In fact, there is a remarkable similarity between some of the species of these taxa and their counterparts found in the Indian Ocean (Figure 3). These two patterns of distribution suggest that formerly the Mediterranean or Tethys constituted a link in a broader oceanic distribution. The Pliocene and earliest Pleistocene fauna represents the last remnant of that linkage, though perhaps an interrupted one.

OSTRACODA IN THE DSDP CORES

Two long cores (96.9 meters and 315 meters) obtained by the *D/S* GLOMAR CHALLENGER (Deep Sea Drilling Project, Leg XIII) were examined for ostracodes. A long core (Site 125) from the Mediterranean Ridge on the east side of the Ionian Basin (latitude 34°37.49′N and longitude 20°25.76′E; depth 2772 meters) was sampled to study the deep ostracode fauna east of the Sicilian-African marginal platform. This core included Quaternary and Pliocene marine sediment and upper Miocene (Messinian)

Figure 3. Similar psychrospheric ostracodes from the Calabrian of Italy and the Recent of the western Indian Ocean. 1, *Bythoceratina scaberrima* (Brady), from the Recent of Mozambique Channel (IIOE 407D, depth 1360 meters; USNM 168382; × 90); 2, *Quasibuntonia sulcifera* (Brady), from the Recent of Mozambique Channel (IIOE 363B, depth 2980 meters; USNM 168383; × 70); 3, *Bythoceratina scaberrima* (Brady), from the Calabrian of the Le Castella Section, Crotone Basin, Calabria (USNM 168387; × 75); and 4, *Quasibuntonia radiatopora* (Seguenza), from the Calabrian 20 kilometers west of Messina, Sicily (USNM 168385; × 70).

evaporites. The second hole (Site 132), yielding a core of similar age, was drilled in the Tyrrhenian Basin (latitude 40°15.70′N and longitude 11°26.47′E; depth 2813 meters). This core was selected to be representative of one containing the deep ostracode fauna of the western Mediterranean. It also ended in Messinian evaporites underlying a thick sequence of pelagic marine sediments.

The locations of these two cores (Figures 1 and 7) are critical in the examination of the effects of the thresholds that are now extant in terms of their earlier history. Site 125 on the Mediterranean Ridge in the Ionian Basin is well east of the present Sicilian-Tunisian divide that now physiographically and hydrologically separates the present eastern and western Mediterranean. An important question is: Did the present shallow threshold (now about 400 meters) of the Straits of Sicily exist during the Pliocene to prohibit entry of the psychrospheric fauna into the eastern basins? Site 132 in the Tyrrhenian Basin is both west of this divide and in an area considered by some to have been continental during the Miocene and to have foundered through the process of oceanization (Heezen *et al.*, 1971, and van Bemmelen, 1969). The localities also occur close to and on either side of the Sicilian and Calabrian Pliocene (Trubi Formation) and Pleistocene (Calabrian; Crotone Basin) outcrop localities.

Five of seven core samples (120 cc in size) of Site 132 in the Tyrrhenian Basin, ranging in age from early Pliocene (Core 18; below the zone of extinction of *Sphaeroidinella dehiscens* and above the first appearance of *Sphaeroidinellopsis seminuda*) to early Pleistocene (Core 6; just above the level of the first appearance of *Globorotalia truncatulinoides*), yielded 47 specimens representing eight species of Ostracoda (Table 1). *Agrenocythere pliocenica* (Seguenza) was found (six specimens) in the lower and middle Pliocene cores

Table 1. Specimen Counts in JOIDES DSDP Leg 13 Core Samples

	Pleistocene			Pliocene	
Core	6	7	8	11	18
Section	1	4	1	3	4
Interval (in centimeters)	16–24	72–82	54–63	45–54	4–13
Penetration (in meters)			70	100*	165*
Bathycythere vanstraateni	1				
Pseudocythere sp.	1				
Krithe sp.	4	8		8	
Henryhowella asperrima				5	
Cytherella volgata		3	1	3	
Agrenocythere pliocenica				4	2
Bythoceratina scaberrima			1		
Bythocypris obtusata (?)		6			

*Depths approximate.

(11 and 18, nannoplankton zones 16 and 12). These cores are stratigraphically equivalent to the Trubi Formation of southern Sicily and eastern Calabria, where this form was also found in outcrops (M. B. Cita, personal communication). *Bythoceratina scaberrima* (Brady) was found in the uppermost part of the Pliocene Section (Core 8, nannoplankton zone 18). This form is also a characteristic Mediterranean psychrospheric fossil ostracode in strata ranging in age from Miocene of Gavdos Island, Crete, to the lower Pleistocene (Le Castella Section, Calabria). *Bathycythere vanstraateni* Sissingh was found in the youngest sample studied (lower Pleistocene; nannoplankton zone 19). This psychrospheric ostracode is sometimes found as a subfossil in the deeper Recent sediments of the present Mediterranean. The other genera, *Cytherella, Krithe, Henryhowella*, and *Pseudocythere* are commonly members of deep-water ostracode assemblages, but occasionally, and commonly in the case of *Cytherella*, intrude into shallower waters. *Agrenocythere pliocenica* and *Bythoceratina scaberrima* are extinct in the present Mediterranean.

No ostracodes were found in six samples examined from Site 125 in the eastern Ionian Basin. These samples (120 cc in size) ranged in age from early Pliocene (lower core 8; equivalent to core 18 or site 132) to early Pleistocene (upper core 3; equivalent to core 6 or site 132). Very few remains of meiobenthic organisms were seen in the samples. Sapropels at several levels in the cores suggest local stagnation. The differences in the fossil content of the two cores examined strongly suggest the deeper waters of this region were not in communication with those of Site 132 or with the Plio-Pleistocene localities of the present outcrops of Sicily or Calabria.

The ostracode assemblage found in the core of Site 132 is essentially the same as that of the Zanclian (Trubi Formation) in Sicily, Calabria and central Italy. Fewer specimens were found in the core samples than have been obtained from outcrop samples, therefore it is probable that only the more abundant species of those known to occur in the complete assemblage were found. The rare large psychrospheric form of *Quasibuntonia*, often found in outcrop samples, was conspicuously absent.

MODERN DEPTH DISTRIBUTION OF PLIO-PLEISTOCENE PSYCHROSPHERIC OSTRACODA AND FORAMINIFERA

The Recent world-ocean distribution of the four most diagnostic psychrospheric ostracode genera found in deep-sea sediments in the Pliocene and lower Pleistocene of the central Mediterranean (DSDP

Site 132, Tyrrhenian Basin; Le Castella Section, Crotone Basin; San Ruffilio Section, near Bologna; Trubi Formation, near Agrigento, Sicily) have been examined and plotted to determine the likelihood of their occurrence at specific depth intervals of 500 meters over a depth range of 0 to 5,000 meters (Figure 4). Similar plots (Figure 5) of a selected Recent distribution of six of the nine species of benthic Foraminifera used by Emiliani *et al.* (1961) to estimate the depth of water in the Crotone Basin are given to compare with their estimates and those suggested by the ostracode data. The purpose of these comparisons is to test the degree of similarity between the two kinds of data; to see what probable depth ranges they suggest for the threshold; and to look for anomalies suggesting adaptive departures in the depth and temperature requirements of the psychrospheric species near the end of their existence in the Mediterranean.

The analysis of ostracode distribution showed (Figure 4) that all four genera (*Bathycythere;* a new genus (Benson, 1972) named *"Agrenocythere"*, including forms related to *"Bradleya" pliocenica; Bythoceratina,* forms related to *B. scaberrima* and *B. vandenboldi;* and *Quasibuntonia,* the larger forms of this taxon with thinner carapaces and few median ridges) individually tend to be found in depths greater than 500 meters and less than 2,000 meters (except *Bythoceratina,* whose distribution is bimodal and may be also abundant at depths between 2,500 and 3,000 meters). In combination they are most likely to occur in modern seas at depths of from 1,000 to 1,500 meters. More than thirty percent of the reported findings (10 of 31) of the genus *Agrenocythere* in the open ocean were from depths between 1,000 and 1,500 meters. No other depth interval had more than four reported occurrences. The concentration of findings of this group at this depth interval and its relative abundance in the Pliocene of the central Mediterranean is considered significant as representing the probable threshold depths of the Gibraltar sill. According to these distributions, the threshold depths could possibly have been greater, but the absence of genera characteristics of deeper levels (more than 2,500 meters) suggests otherwise.

The Recent depth distributions of six species of the

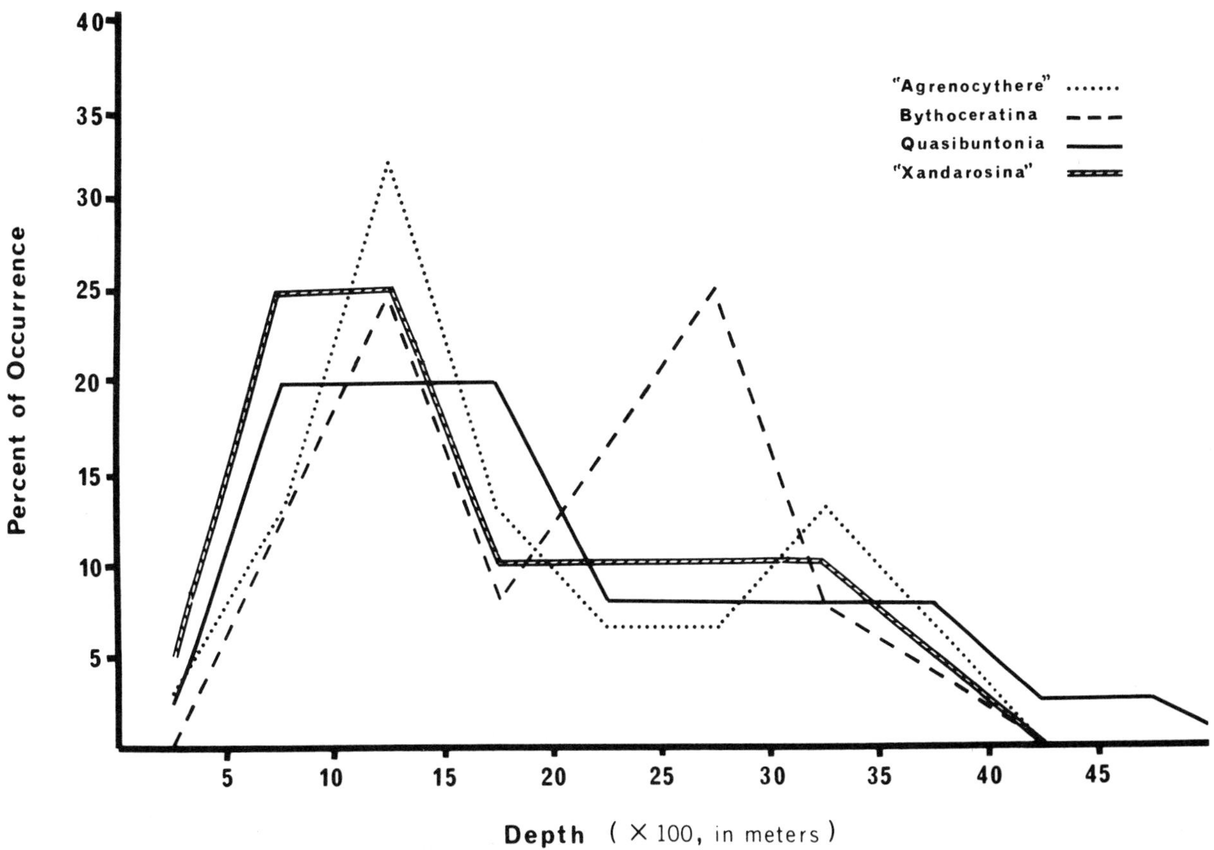

Figure 4. The Recent world-ocean depth distribution of four psychrospheric ostracode genera found in Plio-Pleistocene cores and outcrops in the Mediterranean. These frequency distributions represent the likelihood of these taxa being found at any given depth interval of 500 meters from 0 to 5,000 meters in the stations shown in Figure 2.

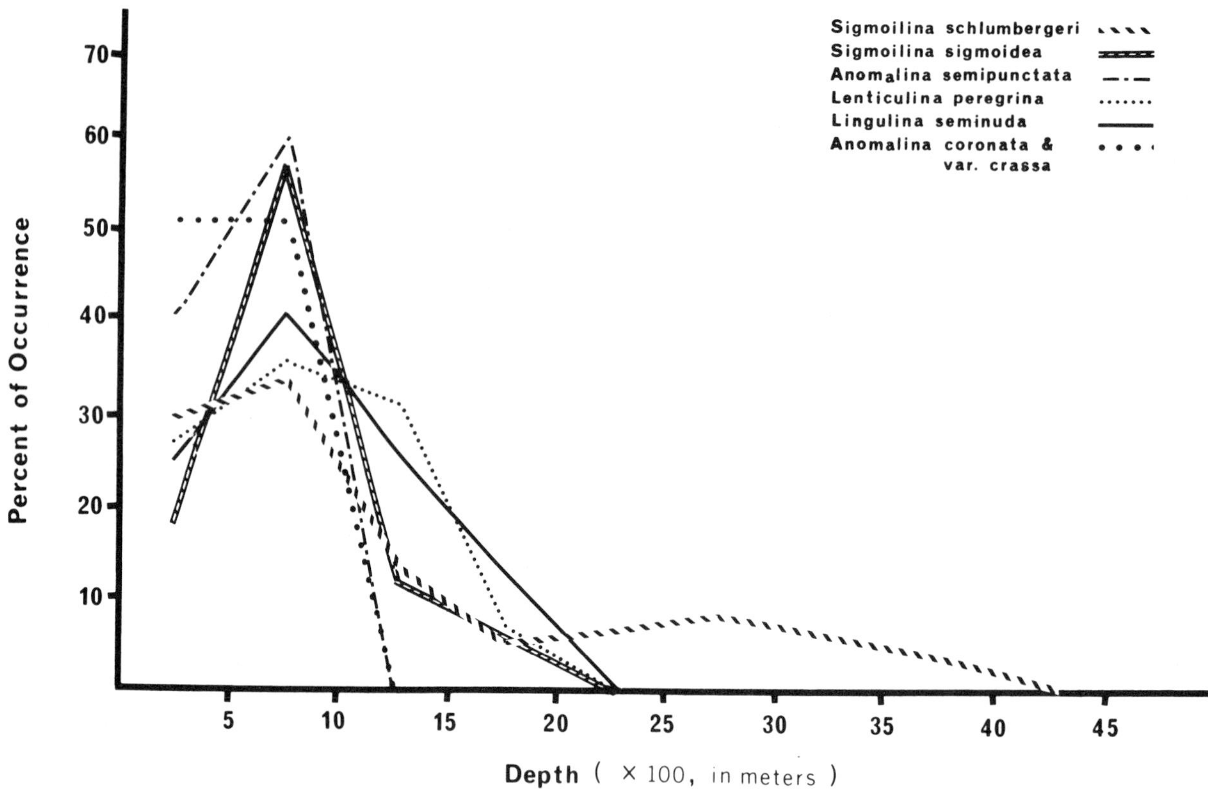

Figure 5. The Recent depth distribution of six of nine Foraminifera species claimed by Emiliani *et al.* (1961) to be representative of the benthic fauna found in the Plio-Pleistocene boundary stratotype in the Le Castella section of the Crotone Basin in Calabria. These frequency distributions here represent the likelihood of these species being found in any given depth interval of 500 meters from 0 to 4,500 meters in the stations shown in Figure 6.

nine Foraminifera identified by Emiliani *et al.* (1961) from Le Castella in Calabria (Crotone Basin) have been plotted in the same manner as were the ostracode distributions (Figure 5). These species include *Sigmoilina sigmoidea* (16 stations), *S. schlumbergeri* (58 stations), *Lenticulina peregrina* (26 stations), *Lingulina seminuda* (8 stations), *Anomalina semipunctata* (10 stations), and *A. coronata* var. *crassa* (8 stations). These data are based on identifications of specimens in the collections in the U.S. National Museum of Natural History. None were included from the Mediterranean, but from the basins of the West and East Indies (Figure 6) where the present conditions might approximate some of the conditions that could have existed during the formation of the Crotone Basin.

The results of the Foraminifera plots suggest that individually three of the six species are twice as likely to be found in depths of from 500 to 1,000 meters as in depths less than 500 meters. Two of the species are not to be expected in depths greater than 1,500 m and two more are rarely found in such depths. The likelihood of finding this assemblage, such as it occurs in the Plio-Pleistocene section of Le Castella, in depths over 1,500 meters is negligible, and in depths shallower than 500 meters is less than about thirty percent. This evidence combined with that of the ostracode distribution would suggest that at least

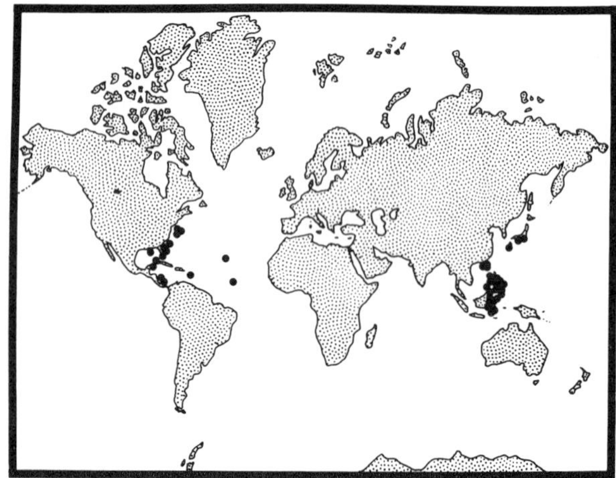

Figure 6. The collecting localities of Recent specimens of the six benthic Foraminifera species found as fossils in the Le Castella Plio-Pleistocene stratotype section (see text). For an estimate of the probable depth distribution of these species (Figure 5) data was selected from two regions (generally West and East Indies), which might now have circumstances comparable to those present in the Mediterranean during the late Pliocene and early Pleistocene.

part of the Le Castella section was deposited under psychrospheric conditions. However, the presence of shallower species in some of the samples suggests that these samples represent the uppermost range of the psychrosphere.

The Recent low temperature ranges of the Foraminifera species given by Emiliani *et al.* (1961) from the North Atlantic, but not the O^{16}/O^{18} paleotemperature ranges (Emiliani, 1971, suggests some doubt about these formerly stated absolute temperatures), were confirmed by the present data. Most of the species preferred temperatures of from 4°C to 6°C. This temperature interval occurs below 1,500 meters along the slope of northwest Africa. More interesting, however, was the discovery that several of these species are now living in the Sulu Basin (west of the Philippines) at temperatures much higher than those of open ocean (up to 17°C). Nevertheless, they are most common to depths in the 500 to 1,000 meter interval. This suggests that some Foraminifera are capable of adapting to the increased temperatures of confined basins as was shown by the work of Bandy and Chierici (1966) in their comparison of the temperature ranges of four other species common to California and the Mediterranean. Evidence of similar adaptation to temperature change in the psychrospheric ostracodes was not obvious and has yet to be demonstrated in the open ocean. Their ranges terminate in the Calabrian part of the section as shallow species become more abundant.

Of particular interest in my search for modern ranges of Foraminifera species is the fact that Cushman identified *Hyalinea* ("*Anomalina*") *baltica* from at least nine localities in the area of the Sulu Basin ranging in depth from 22 meters to 1,450 meters with temperature ranges of from 5.7°C (Albatross station 5284; depth 760 meters) to 11.3°C (Albatross station 5112; depth 318 meters). The deepest station (Albatross 5526; depth 1,450 meters) had a bottom temperature of 11.2°C as the result of a temperature-salinity inversion. The first appearance of *Hyalinea baltica* in the Crotone Basin is the basis on which the Plio-Pleistocene boundary in this section was fixed, with the theoretical assumption being that it represents the invasion of the cold-water Pleistocene shelf fauna (*nordic guests*) of the open Atlantic.

Although not immediately germane to the present estimate of possible threshold depths of Gibraltar or the Sicilian-African marginal platform during the Plio-Pleistocene, it should be noted that the psychrospheric fauna continues across the boundary determined by the first appearance and assumed invasion of *Hyalinea baltica.* It is sometimes found in the northern parts of both the Atlantic and Pacific Oceans in depths as great as 4,500 meters (although normally less than 2–300 meters). At Le Castella in Calabria (the Plio-Pleistocene boundary stratotype), it is found in a section whose ostracode assemblage changes from psychrospheric to outer-shelf (Colalongo, 1965). This is not a simple transition, however, because there are many indications of reworking and contamination by specimens from shallower waters. It is of interest to note here that elements of a shallow ostracode fauna (including *Aurila*) are mixed in with elements of the psychrospheric fauna (including *Agrenocythere pliocenica,* confined to the Pliocene; and *Bythoceratina scaberrima,* confined to the Calabrian) with the first appearance of *Hyalinea baltica.* I would suggest from the ostracode evidence that this section had an active general tectonic history of downward movement during this critical time interval, and that the first appearance of *Hyalinea baltica* in this particular section (Emiliani, 1971), may well be an erroneous recording of the actual time of invasion of this species into the Mediterranean.

A MODEL OF MEDITERRANEAN EVOLUTION

The pace of geological evolution of the Mediterranean from Tethys accelerated toward the end of the Tertiary as Africa approached closer to Europe. Compressive, gravitational, isostatic, and transcurrent movements all were integrated into a series of very complex paleogeographic changes. Their effects on the Tethyan faunas were to isolate and exclude former Indian Ocean ties, eventually to form Paratethys. Soon afterward even communication with the Atlantic was cut off to form a series of isolated lagoons setting the stage for the first of the Neogene *crises*—the Messinian evaporite crisis.

The evaporites are known from many parts of the Mediterranean from Spain to Sicily, in outcrop and now from the basin floors (Scientific Staff, 1970). The Miocene marine fauna was almost totally destroyed during this event (Ruggieri, 1967).

At the beginning of the Pliocene the gates of the Iberian Portal opened with catastrophic suddenness, allowing deeper, colder waters to pour in from the Atlantic. The psychrosphere advanced eastward to spill across parts of the central Mediterranean platform to the region of the northern Appenines, and to eastern Sicily and Calabria (Figure 7B). Meanwhile, its influence to the east was dampened by mixing with still warm and dense waters, in which the circulation was poor and sapropels became interdispersed with planktonic biogenic fallout.

With the onset of continental climatic deterioration and Alpine uplift, which culminate with the glacial

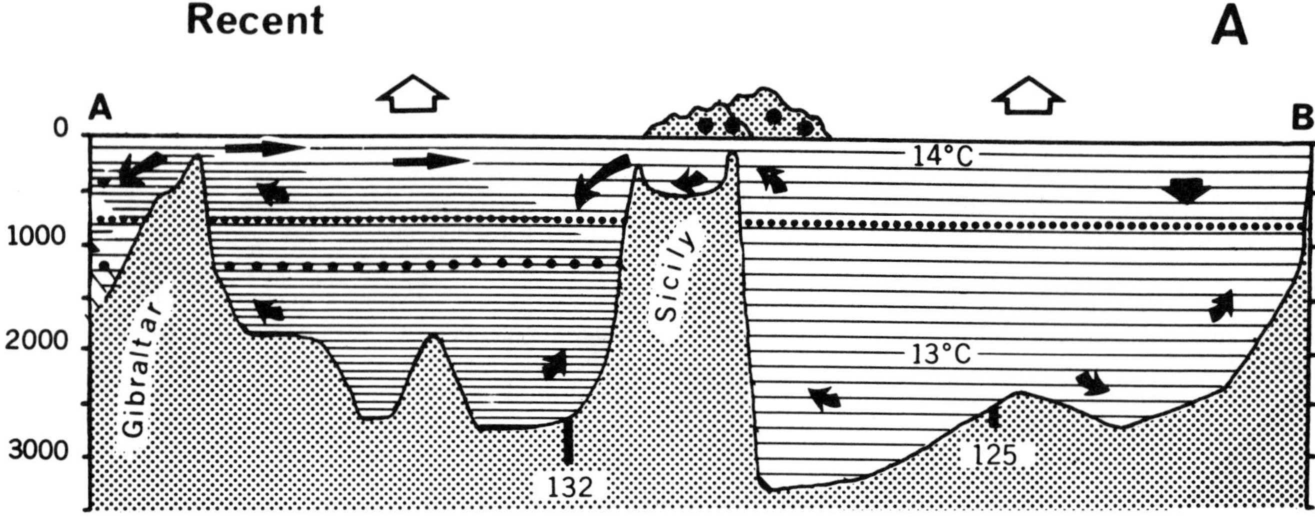

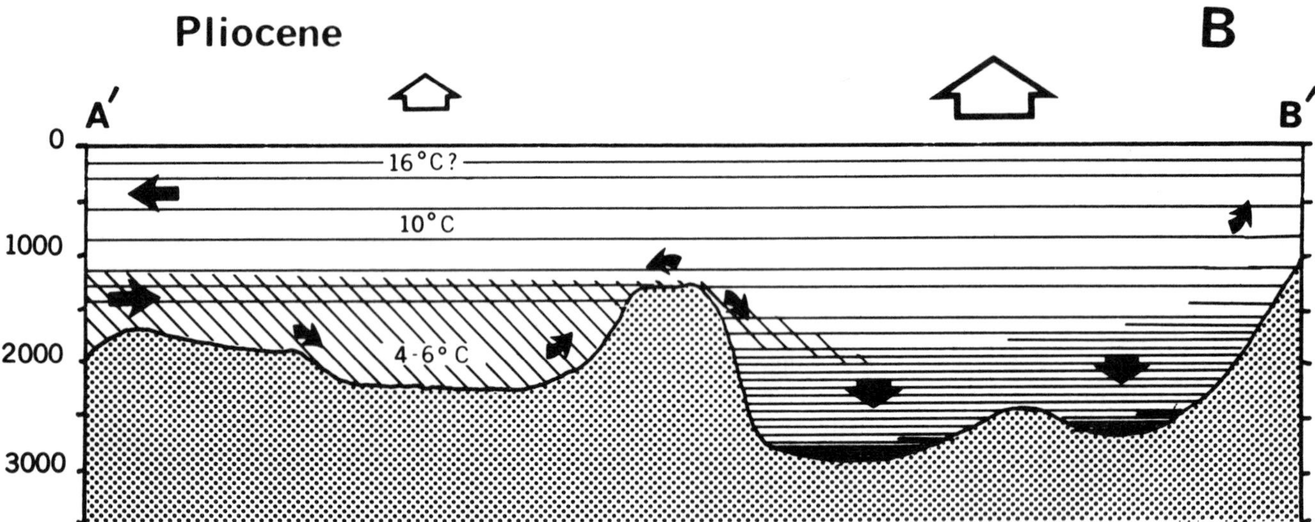

Figure 7. Water mass circulation in the Mediterranean during the Recent and the Pliocene showing the present domination by the thermosphere (horizontal lines) and the past intrusion of the psychrosphere (diagonal lines). It is postulated that the two controlling thresholds at Gilbratar and in the region of the Straits of Sicily were absent or much lower in former times, when greater evaporation in the east and warmer surface waters led to poor circulation, while cold and deep Atlantic waters spilled in from the west. Today, the thresholds have risen tectonically with some of the fossil localities (dots) to separate the deepening basins where the DSDP cores (125 and 132) were obtained. In the Recent model the dotted lines represent the approximate and postulated level of the boundary between the psychrosphere (now outside of the Mediterranean) and the thermosphere at present (upper line) and during the Pliocene (lower line).

episodes of the Pleistocene, the water masses of the basins of the western Mediterranean became "lagoonal" again, though this time they were deeper and the net evaporation loss much less than before. The restricting sills that were either suddenly removed or only partially effective during the Pliocene were raised anew to form circulation and water mass patterns approaching those of today (Figure 7A).

The faunas of this last episode—the *climatic crisis*, include the nordic guests that invaded the sanctuary of the Mediterranean, while being pressed southward during the cooling of shallow waters in the northern latitudes. They invaded and shared not only the shelf regions with relics of the former fauna, but some also descended into the warm basins, which were protected by newly developed thresholds. Meanwhile, all but a few eurythermic members of the psychrospheric assemblages became extinct.

From among the many factors that influenced this history, the development and disappearance of sills that controlled the character of the basin waters were paramount. Probably in few comparable instances have local tectonic changes had such broad environmental effects. Perhaps the importance of their formation is only equalled, in terms of biotic dispersal of both marine and terrestrial faunas in the Mediterranean, by the formation of intercontinental land bridges. The most important result of the present study is to emphasize the apparent suddenness of change that took place in the Mediterranean between the end of the Miocene and the beginning of the Pliocene. I suggest that it is simplest to explain this change in terms of threshold formation and destruction in the region of the Iberian Portal.

ACKNOWLEDGMENTS

I would like to thank P. C. Sylvester-Bradley for his guidance and support of my interest in the history of the Mediterranean and review of this manuscript, Ruth Todd for her help with the Foraminifera distribution data and review of the manuscript, G. Ruggieri for his encouragement and critique of my study, W. B. F. Ryan for making available the samples from the DSDP cores, Maria B. Cita for her helpful remarks about the previous study leading to this one, Laurie Jennings who helped me prepare the samples and illustrations, and to R. E. Grant and J. E. Hazel who also reviewed the manuscript. The study was sponsored by Smithsonian Research Foundation Grant No. SRF-43602 and a National Science Foundation Grant No. GA-17325.

REFERENCES

Aubouin, J. 1965. *Geosynclines.* Elsevier Publishing Co., Amsterdam, 335 p.

Ascoli, P. 1969. First data on the ostracod biostratigraphy of the Possagno and Brendola sections (Paleogene, Italy). *Bureau de Recherches Geologiques et Minières, France, Memoire* 69: 50–71.

Bandy, O. L. and M. A. Chierici 1966. Depth-temperature evaluation of selected California and Mediterranean bathyal Foraminifera. *Marine Geology,* 4:259–271.

van Bemmelen, R. W. 1969. Origin of the western Mediterranean Sea. In: Symposium on the Problems of Oceanization in the Western Mediterranean, *Transactions of the Royal Geological and Mining Society of the Netherlands,* 26:13–52.

Beneo, E. 1956. Accumuli terziari da risedimentazione (olitostroma) nell'Apennino centrale et franesotlomarine. *Bollettino Servizio Geologico d'Italia,* 78:291–321.

Benson, R. H. 1969. Preliminary report on the study of abyssal ostracodes. In: *Taxonomy, Morphology and Ecology of Recent Ostracoda,* ed. Neale, J. W., Oliver and Boyd, Edinburgh, 475–480.

Benson, R. H. 1972. The *Bradleya* Problem, with descriptions of two new psychrospheric genera *Agrenocythere* and *Poseidonamicus* (Ostrac., Crust.), *Smithsonian Contributions to Paleobiology,* 12:1–140.

Benson, R. H. and P. C. Sylvester-Bradley 1971. Deep-sea ostracodes and the transformation of ocean to sea in the Tethys. In: *Proceedings of the Colloquium on the Paleoecology of Ostracodes,* ed. Oertli, H. J., *Bulletin de Centre de Recherches, Pau,* 5:960 p.

Bruun, A. F. 1957. Deep sea and abyssal depths. *Geological Society of America, Memoir,* 67:641–672. [Chapter 22 In: Treatise on Marine Ecology and Paleoecology, 1, Ecology].

Cayeux, L. 1924. La question des jaspes à radiolaires. *Comptes Rendus de la Société Géologique de France,* 1924:11–12.

Colalongo, Maria L. 1965. Gli ostracodi della serie de le Castella (Calabria). *Giornale di Geologia, Annali del Museo Geologico di Bologna,* 33:83–123.

Emiliani, C. 1971. Paleotemperature variations across the Plio-Pleistocene boundary. *Science,* 171:60–62.

Emiliani, C., T. Mayeda, and R. Selli 1961. Paleotemperature analysis of the Plio-Pleistocene section at Le Castella, Calabria, southern Italy. *Geological Society of America Bulletin,* 72: 679–688.

Gass, I. G. 1968. Is the Troodos Massif of Cyprus a fragment of Mesozoic ocean floor? *Nature,* 220:39–42.

Glangeaud, L. 1962. Paléogéographie dynamique de la Méditerranée et de ses bordures. Le rôle des phases ponto-plio-quaternaires. In: *Océanographie Géologique et Géophysique de la Méditerranée occidentale,* Colloques Internationaux CNRS, Villefranche, 125–165.

Haug, E. 1900. Les géosynclinaux et les aires continentales. Contribution à l'étude des régressions et des transgressions marines. *Bulletin de la Société Géologique de France,* 28(3):617–711.

Heezen, B. C., C. Gray, A. G. Segre, and E. F. K. Zarudzki 1971. Schist and tectonized sediments beneath the central Tyrrhenian Sea: evidence of foundered continental crust. *Nature,* 229:327–329.

Ruggieri, G. 1953. Eta e faune di un terrazzo marino sulla costa ionica della Calabria. *Giornale di Geologia, Annali del Museo Geologico di Bologna,* 23:17–168.

Ruggieri, G. 1954. Iconografia degli ostracodi marini del Pliocene e del Pleistocene italiani. *Atti del Società Italiana di Scienze Naturali e del Museo Civico di Storia Naturale in Milano,* 43: 561–575.

Ruggieri, G. 1967. The Miocene and later evolution of the Mediterranean Sea. *Systematics Association,* 7:238–290.

Scientific Staff 1970. Deep-sea drilling project: Leg 13. *Geotimes,* 15:12–15.

The Structure and Stratigraphy of the Tyrrhenian Sea

Augusto Fabbri and Raimondo Selli

Consiglio Nazionale Delle Ricerche, Bologna

ABSTRACT

Some 1300 nautical miles of continuous seismic reflection profiles and numerous deep dredge samples from the Tyrrhenian Sea have elucidated the sub-bottom structure and stratigraphy of this area and show that this feature is not more than 4 million years old. Immediately prior to that time, most of the region now occupied by the Tyrrhenian Sea was emergent. Regional foundering, which commenced in the Middle Pliocene, ushered in progressively deeper marine environments, as indicated by a characteristic regional transgression sequence. The mean rate of foundering (up to the present time) is calculated at 1.1 mm per year. Thus the Tyrrhenian Sea originated in the Pliocene and may well be the youngest deep sea area in the world.

RESUME

La structure et la stratigraphie de la Mer Tyrrhénienne ont été étudiées à l'aide de profils sismiques continus sur une étendue de 1300 miles marins et de nombreux dragages en profondeur. Ces études ont révélé que l'existence de la Mer Tyrrhénienne ne date que de quatre millions d'années. La plupart de la région qu'elle occupe était constituée de terrains émergés. Un effondrement général de cette région a eu lieu au Pliocène moyen et a entrainé la formation de milieux marins profonds ainsi que l'indique une séquence de sédiments de transgression caractéristique. La vitesse moyenne d'effondrement (jusqu'à nos jours) est de 1,1mm par an. En conséquence, la Mer Tyrrhénienne est d'âge Pliocène et, probablement, la mer profonde la plus récente dans le monde entier.

INTRODUCTION

Numerous and detailed bathymetric studies have provided much information on the physiography of the Tyrrhenian Sea floor. Seven major morphologic units can be distinguished (Selli, 1970) and are shown in Figure 1. They are: *a* continental shelf, *b* upper continental slope, *c* peri-Tyrrhenian basins, *d* peri-Tyrrhenian seamounts, *e* lower continental slope, *f* bathyal plain, and *g* central-Tyrrhenian seamounts.

Features *c, d, f* and *g* are characteristic of this region. A discontinuous belt of large sedimentary peri-Tyrrhenian basins is generally bordered seawards by the peri-Tyrrhenian seamounts, which tend to dam back terrigenous sediments, which are carried into the basins by turbidity currents passing through canyons cut into the upper slope. The bathyal plain of the Tyrrhenian Sea differs from a true oceanic abyssal plain in possessing a gently concave profile, a high sedimentation rate and the fact that it is underlain by an intermediate type of crust (Fahlquist and Hersey, 1969). The numerous central-Tyrrhenian seamounts rise to a height of 2900 m above the bathyal plain or its borders.

A generalized physiographic profile (Figure 1, inset) across the Tyrrhenian Sea south of the 41°N parallel, shows the average depth, width, and gradient of the seven morphological units.

RESULTS

Reflection Profiles

Some 2400 km (1300 nautical miles) of continuous seismic reflection profiles were obtained with a 24

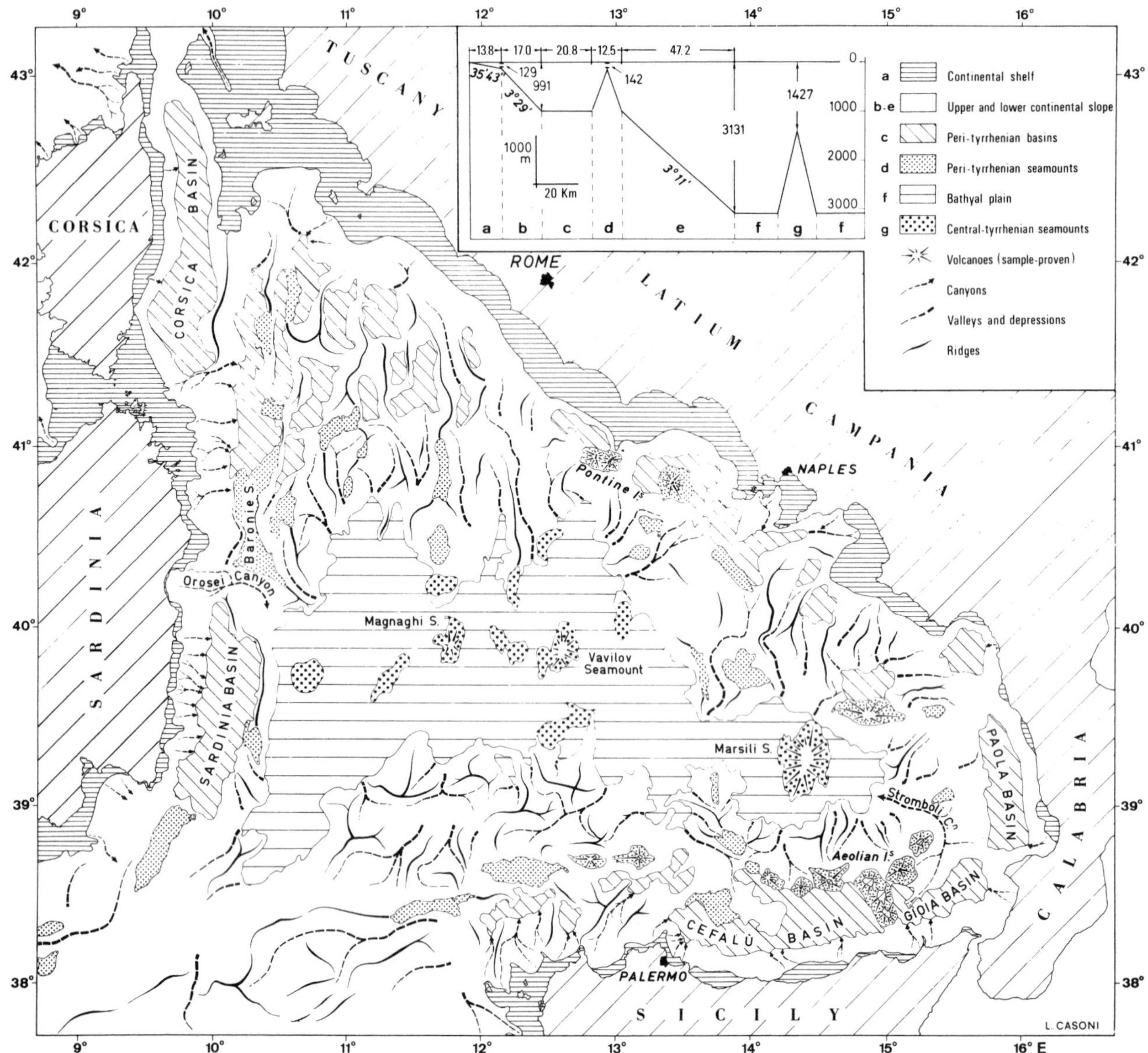

Figure 1. Chart showing principal morphologic features of the Tyrrhenian Sea. Inset: generalized profile across the central and southern Tyrrhenian Sea with average width (in km), depth (in m) and gradient of main morphologic units: *a* continental shelf, *b* upper continental slope, *c* peri-Tyrrhenian basins, *d* peri-Tyrrhenian seamounts, *e* lower continental slope, *f* bathyal plain, *g* central-Tyrrhenian seamounts. Simplified from Selli (1970).

kilojoule sparker. The profiles were concentrated in three main areas (Figure 2A) and incorporate both the marginal and central portions of the Tyrrhenian Sea. The profiles reveal three main acoustic units (Figure 3), here designated A, B and C, in descending order, which are defined on the basis of different character of seismic reflections contained within these units. The three most prominent reflectors are termed respectively x (top reflector; separates layers A and B), y (middle reflector; within layer B), and z (basal reflector; separates units B and C).

Unit A

Unit A displays many undisturbed, sharp and continuous near-horizontal reflections which can be followed over great distances. Occasionally, extensive and thick submarine slumps are present. This unit attains its greatest thickness (about 1100 meters, assuming a seismic interval velocity of 1800 m/sec) in the central zone of the sedimentary basins both on the continental slope (peri-Tyrrhenian basins) and under the bathyal plain. Unit A lies discordantly above the underlying

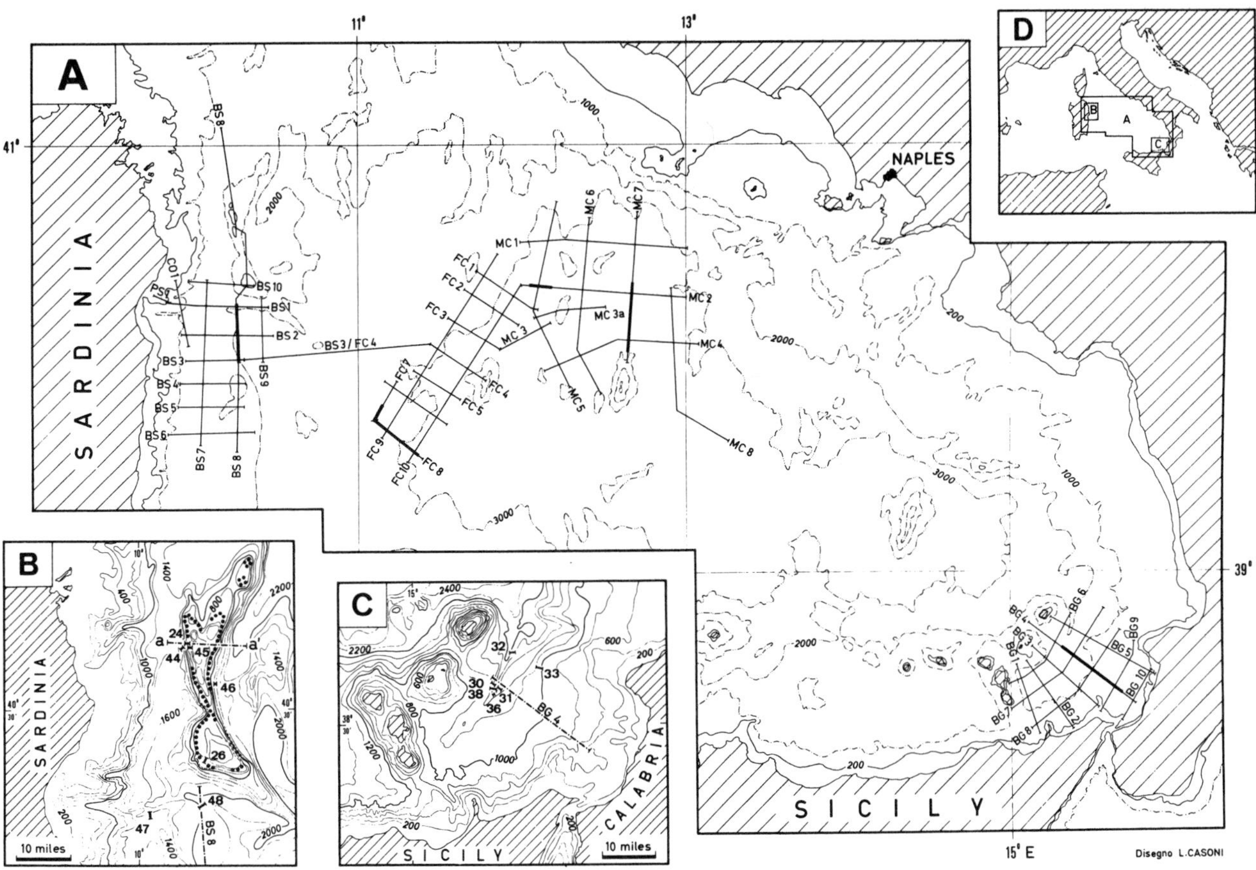

Figure 2. Maps showing location of continuous seismic reflection profiles and dredge hauls. A, General location of seismic survey; heavy lines indicate the profiles reproduced in Figures 3 and 4. B, Dredging locations (heavier numbers) on Baronie Seamount and in Orosei Canyon; heavy-dotted line indicates the step or terrace around Baronie Seamount; *a-a'* bathymetric profile of Figure 4, V. C, Dredging locations in Stromboli Canyon. D, Index map of the central Mediterranean.

units B or C near the flanks of the sedimentary basins, but Unit A appears to pass conformably down into Unit B in the deeper, central parts of the basins.

Unit B.

Unit B is characterized by reflections which become progressively less continuous and regular downward and is affected by a number of folds and faults which also become more intensely developed downward. The thickness of this unit varies from about 0 to 900 m (assuming an interval velocity of 2000 m/sec).

Unit C.

Unit C usually lacks clearly defined continuous interval reflections and forms the acoustic substratum of the Tyrrhenian Sea. Unit C probably represents the basement, while we interpret units B and A to represent two separate sedimentary and tectonic cycles within the sedimentary cover.

Dredge Samples and Stratigraphy

Dredge samples were obtained in the Stromboli and Orosei Canyons and on the Baronie Seamount (Figures 1 and 2) in order to identify and date the seismic units[1].

Large fragments of white or yellowish indurated marls were dredged frequently near the base of both flanks of Stromboli Canyon, that is within the upper part of seismic Unit B. The microfauna is assigned to the *Globorotalia margaritae* zone (Lower Pliocene) and to the *Globorotalia aemiliana* zone (the lowermost part of the Middle Pliocene). The lithologic and micropaleontologic characters of the marls correspond

[1] We can subdivide the Pliocene into three parts, following the standard bio-zonation scheme proposed by Cati *et al.*, 1968, *i.e.:*

Globorotalia inflata zone = Upper Pliocene

Globorotalia crassaformis zone + *Gl. aemiliana* zone = Middle Pliocene

Globorotalia margaritae zone = Lower Pliocene

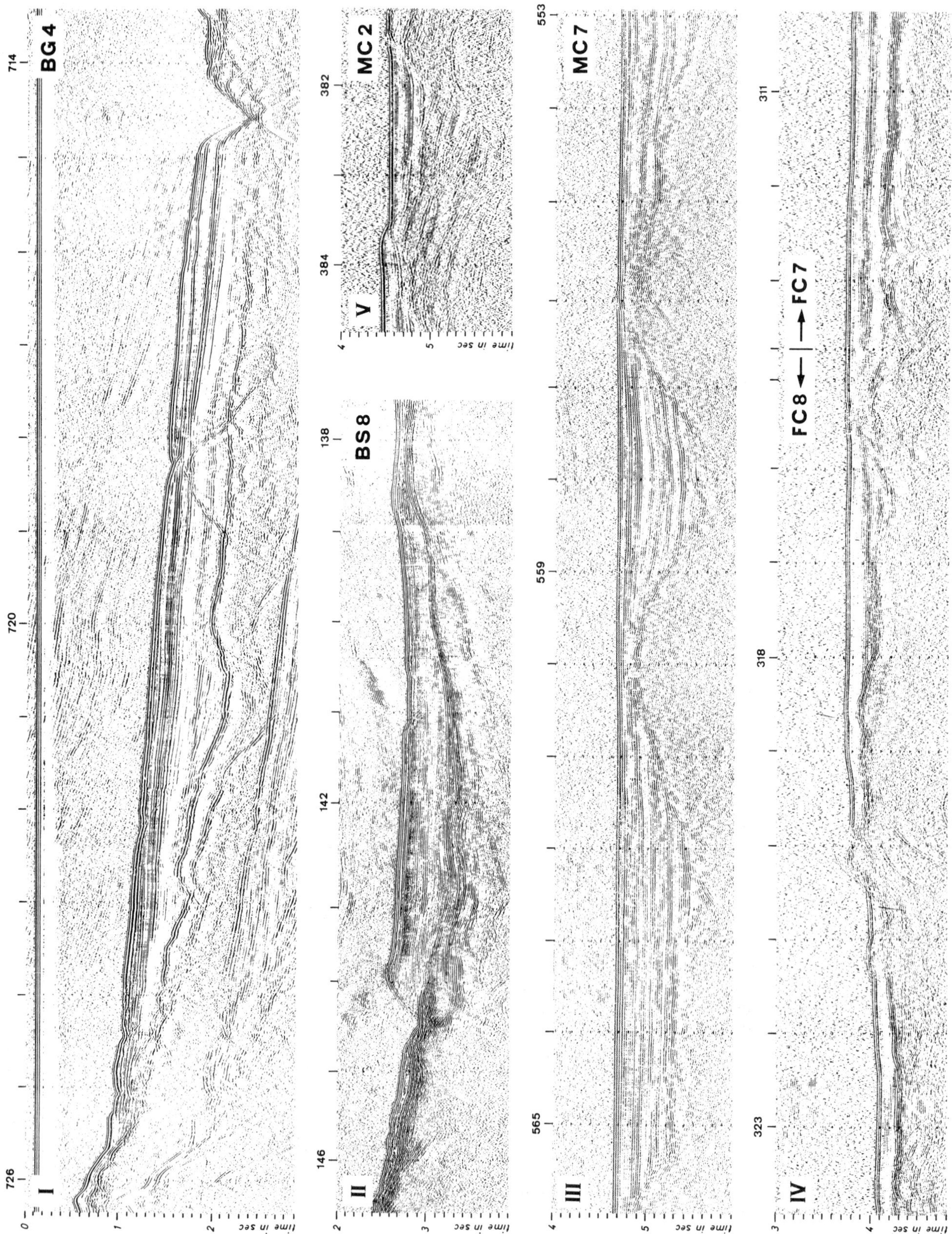

Figure 3. Typical continuous seismic reflection profiles across the Tyrrhenian area (location on Figure 2). Ordinate: two-way travel time in seconds. Interpretation, scale and orientation shown in Figure 4.

exactly to those of the "Trubi" Formation of Sicily and southern Calabria. Numerous samples of blue marly clays have been dredged from the middle and upper flanks of Stromboli Canyon, above the discordant reflector *x* (Figure 4, I), that is, from the seismic Unit A. Foraminiferal zones in these dredged clays range from the uppermost part of the Middle Pliocene (*Globorotalia crassaformis* zone) to the Lower Pleistocene (*Globigerina pachyderma* zone of Colalongo, 1968 and D'Onofrio, 1968). Deeper seismic units, not cut by the Stromboli Canyon, are correlated with formations outcropping further east in Calabria by their attitude and their seismic response. Thus, Unit C can be identified as the metamorphic and granitic basement, while the discordant reflector *z* (Figure 4, I) at the base of Unit B, may represent the bottom of the Lower Miocene transgressive sequence (Selli, 1957). The strong reflector *y* (Figure 4, I) probably marks the top of the Lower Messinian "gessoso-solfifera" Formation (Selli, 1954 and 1964).

Blocks of a conglomeratic rock were dredged from the base of the southern flank of Orosei Canyon (2200 m depth), near the outcrop of the discordant seismic horizon *x* (Figure 4, II). Pebbles of basalt, sandstone and marl of Lower Pliocene age (*Globorotalia margaritae* zone) occur in this rock. The matrix includes littoral and shallow water molluscs, which are Middle or Upper Pliocene in age (Colantoni, 1970). Another marl sample, dredged at the contact with the conglomerate, contains a rich fauna of Foraminifera

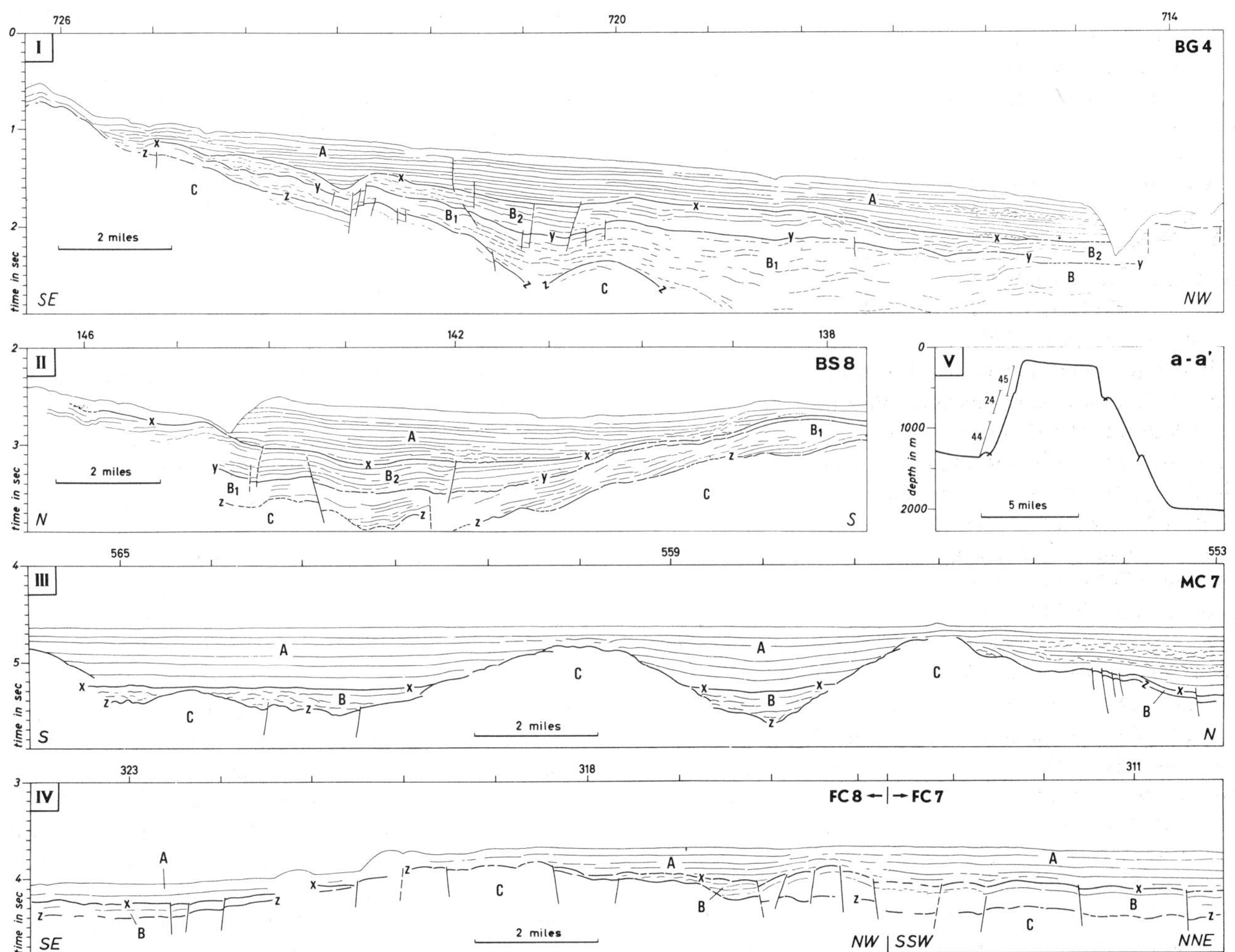

Figure 4. Geologic interpretation of the continuous seismic reflection profiles shown in Figure 3. **A**, continuous sequence from uppermost Middle Pliocene to the present (seismic Unit A); *x-x*, inferred Middle Pliocene unconformity and transgressive sequences, laterally passing into a conformable junction (dotted line); **B**, Miocene, Lower Pliocene and lowermost Middle Pliocene sequence (seismic Unit B); **B_2**, Upper Messinian, Lower Pliocene and lowermost Middle Pliocene sequence (seismic sub-Unit **B_2**); *y-y*, probable top of gypsum ("gessoso-solfifera" Formation of Lower Messinian); **B_1**, Lower and Middle Miocene (?) sequence (seismic sub-Unit **B**); *z-z*, probable Lower Miocene unconformity; **C**, pre-Miocene basement (seismic Unit C). The seismic profiles cross the Gioia Basin and Stromboli Canyon (I); the Sardinia Basin and Orosei Canyon (II); the western (III) and eastern (IV) bathyal plain. For location of profiles see Figure 2. Bathymetric profile across the Baronie Seamount (V) shows dredged horizons.

representing the upper part of the *Globorotalia crassaformis* zone, *i.e.*, the uppermost part of the Middle Pliocene. Therefore the conglomerate and the associated transgression are assigned a Middle Pliocene age. The sedimentary sequence continues with bluish and greenish clays (Unit A); the samples of these contain foraminiferal faunas representing all zones from the Upper Pliocene to the Pleistocene. The age of deeper seismic layers which do not outcrop in the Orosei Canyon is difficult to determine since it is impossible to correlate them with the Sardinian stratigraphic sequence. However, tentatively, seismic Unit C is assigned to the Paleozoic metamorphic and granitic basement, horizon z to the basal, transgressive deposits of the Lower Miocene and horizon y to the top of the Lower Messinian "gessoso solfifera" Formation.

The acoustic basement (or seismic Unit C) crops out on the Baronie Seamount, where it is capped by a very thin sedimentary cover. The flanks of the seamount are cut by a step, or an irregular narrow terrace, at depths between 550 m and 740 m (Figure 4, V), and a Middle Pliocene littoral biogenic calcarenite (*Globorotalia crassaformis* zone) has been dredged from this level. Numerous blocks of slightly tectonized and strongly cemented conglomerate have been recovered (dredge 45) from the submerged cliff above this terrace. This conglomerate is identical with the so-called Verrucano Formation of Sardinia (Lower Permian) and Tuscany (Permo-Trias). From these data we infer that the terrace around the Baronie Seamount probably represents an ancient beach of Middle Pliocene age, formed at a time when the upper 400–500 m of the seamount emerged above sealevel. Thus, foundering of the area, totalling between 550 m and 740 m, must have occurred subsequent to the Middle Pliocene.

So far, no rock samples have been collected from the bathyal plain or adjacent features. However, correlation between the Sardinia Basin and the bathyal plain by means of the seismic profile BS 3/FC4 (Figure 2, A) suggests the following conclusions (Figures 4, III and 4, IV). Seismic Unit A is Plio-Pleistocene in age; the discordant reflector at its base probably represents the Middle Pliocene transgressive surface. Unit B may be ascribed mainly to the Miocene and the basal discordance probably indicates the Lower Miocene transgressive phase. Unit C, the acoustic basement, appears to be heterogeneous. In several areas of the bathyal plain, especially in the west, it is metamorphic (Heezen *et al.*, 1971) but in some central and eastern areas the upper part of this unit is bedded and probably sedimentary (Paleogene or Mesozoic), as indicated by seismic reflections (Figure 3, V).

SUMMARY AND CONCLUSIONS

The most pertinent data and interpretations can be briefly summarized as follows:

1. In the central and southern Tyrrhenian Sea, especially in the bathyal plain and in the peri-Tyrrhenian basins, three main seismic units designated A, B, and C are recognized in the continuous seismic profiles.

2. Above the basement (Unit C) lie two well-defined sedimentary and tectonic cycles (Units A and B), both of which display a marked basal discordance, representing regional transgressive events (seismic horizons x and z).

3. The upper cycle (Unit A) comprises a continuous near-horizontal sequence of clays and sandy clays, spanning the period from the uppermost Middle Pliocene to the Present. The basal discordance (seismic horizon x) is of Middle Pliocene age and is marked by a littoral conglomerate, which outcrops in the Orosei Canyon.

4. The discordant reflector x at the base of Unit A normally truncates reflectors within the lower cycle (Unit B) and, in places, within the basement (Unit C). However, in the deepest parts of the peri-Tyrrhenian basins and above the buried depressions of the bathyal plain, the two sequences appear to be concordant in attitude. In these areas sedimentation probably was continuous throughout the interval of time represented by the two cycles.

5. The important unconformity indicated at the base of Unit A corresponds in age and in attitude to the extensive Middle Pliocene transgressive phase, which is widely developed in Italy and Sicily along the Tyrrhenian coasts, but chiefly known from the Apennine foredeep region (where this sequence extends without interruption from the Po River Basin to central Sicily).

6. The lower cycle (seismic Unit B) is affected by tectonic deformation (folds and faults) contemporaneous with the Apennine orogeny; the degree of tectonism increases downward. The upper half of this cycle is of Lower Pliocene to lowermost Middle Pliocene age, as demonstrated by dredge samples from Stromboli Canyon, while the base (seismic horizon z) probably corresponds to the transgression of the Lower Miocene, well known in Southern Italy (Selli, 1957), Sicily and Sardinia.

7. The reflector y, within the seismic Unit B, corresponds to the "gessoso-solfifera" Formation of upper Miocene (Messinian). Its presence throughout much of the Tyrrhenian area supports the hypothesis, earlier advanced by Selli, 1954 (p. 92), that evaporitic sedimentation took place during the Messinian over the entire floor of the western Mediterranean (and

therefore also in the Tyrrhenian). This is demonstrated by the evaporitic outcrops of the Italian Peninsula, Sicily, Spain, North Africa, Ionian Islands, Crete, and other areas.

8. Few data are at present available on the nature of the pre-Miocene basement (seismic Unit C). Metamorphic rocks (Heezen *et al.*, 1971) and Paleozoic rocks (Baronie Seamount) have been dredged; probably younger sedimentary rocks are present as well.

9. From the new data reported here, we infer that during the Middle Pliocene the Tyrrhenian area was largely emergent and probably resembled an archipelago consisting of numerous large islands separated by channels or sounds.

10. Foundering of the Tyrrhenian area commenced in the Middle Pliocene and has proceeded since at an average rate of about 1 to 1.1 mm per year. Thus, the Middle Pliocene surface today occurs at depths of 4500 m (in the deepest part of the bathyal plain). This foundering broadly coincides in time with the last events of the Apennine orogeny during the Plio-Pleistocene. We thus conclude that the present *Tyrrhenian Sea is Pliocene in age* and may well be the youngest deep sea environment in the world.

REFERENCES

Cati, F., M. L. Colalongo, U. Crescenti, S. D'Onofrio, U. Follador, C. Pirini Raddrizzani, A. Pomesano Cherchi, G. Salvatorini, S. Sartoni, I. Premoli Silva, C. F. Wezel, V. Bertolino, G. Bizon, H. M. Bolli, A. M. Borsetti Cati, L. Dondi, H. Feinberg, D. G. Jenkins, E. Perconig, M. Sampo, and R. Sprovieri 1968. Biostratigrafia del Neogene mediterraneo basata sui Foraminiferi planctonici. *Bollettino della Societa Geologica Italiana,* 87(3): 491–503.

Colalongo, M. L. 1968. Cenozone a Foraminiferi ed Ostracodi nel Pliocene e basso Pleistocene delle serie del Santerno e dell'Appennino Romagnolo. In: *Committee on Mediterranean Neogene Stratigraphy Proceedings of the fourth Session,* ed. Selli, R., *Giornale di Geologia,* 35(3):29–61.

Colantoni, P. 1970. Littoral Pliocene molluscs dredged at 2200 m of depth in the Tyrrhenian Sea. *Report presented to the Congress of C.I.E.S.M.M.* (Commission Internationale pour l'Exploration Scientifique de la Mer Méditerranée), Rome.

D'Onofrio, S. 1968. Biostratigrafia del Pliocene e Pleistocene inferiore nelle Marche. In: *Committee on Mediterranean Neogene Stratigraphy Proceedings of the fourth Session,* ed. Selli, R., *Giornale di Geologia,* 35(3):99–114.

Fahlquist, D. A. and J. B. Hersey 1969. Seismic refraction measurements in the western Mediterranean. *Bulletin de l'Institut Océanographique de Monaco,* 67(1386): 52 p.

Heezen, B. C., C. Gray, A. G. Segre and E. F. K. Zarudzki 1971. Evidence of foundered continental crust beneath the central Tyrrhenian Sea. *Nature,* 229(5283):327–329.

Selli, R. 1954. Il bacino del Metauro. *Giornale di Geologia,* 24: 1–268.

Selli, R. 1957. Sulla trasgressione del Miocene nell'Italia Meridionale. *Giornale di Geologia,* 26: 1–54.

Selli, R. 1964. The Mayer-Eymar Messinian 1867. Proposal for a Neostratotype. *Proceedings of XXI International Geological Congress, Nordern,* 28:311–333.

Selli, R. 1970. I. Cenni morfologici generali sul Mar Tirreno (Crociera C.S.T. 1968). In: *Ricerche Geologiche Preliminari nel Mar Tirreno,* ed. Selli, R., *Giornale di Geologia,* 37:4–24.

Age and Nature of the Pan-Mediterranean Subbottom Reflector M*

Pierre E. Biscaye[1], William B. F. Ryan[1]
and Foreze C. Wezel[2]

[1]*Lamont-Doherty Geological Observatory, Palisades, New York, and*
[2]*Università di Catania, Catania*

ABSTRACT

A subbottom reflecting horizon has been observed in seismic reflection profiles in all physiographic provinces beneath much of the Mediterranean Sea. An outcrop of the top of this horizon, Reflector M, has been sampled by a piston core in the Ionian Basin. The recovered material consists of an indurated, pelagic coccolith ooze, containing a foraminiferal assemblage identical to that of the Trubi Formation (Italy) which is basal Pliocene in age. The wide distribution of this acoustic reflector suggests its origin in a lithification process that took place at, or close to, the sediment-water interface about five million years ago.

RESUME

Des profils de réflection sismique ont révélé la présence d'un réflecteur dans la plupart du substratum de la Méditerranée. On a prélevé des échantillons d'un affleurement correspondant à la partie supérieure de ce réflecteur (*Reflector M* dans le texte) situé dans le Bassin Ionien. Le matérial recueilli, un calcaire pélagique à cocolites, contient une association de foraminifères identique à celle de la formation Trubi (Italie) qui est du Pliocène inférieur. La répartition étendue de ce réflecteur acoustique suggère qu'il résulte d'un processus de lithification ayant eu lieu il y a environ 5 millions d'années près de la surface de contact de l'eau et des sédiments.

INTRODUCTION

Discrete layering can sometimes be seen in the sediment carpet of the deep ocean basins. Surveys employing the technique of continuous seismic reflection profiling (Hersey, 1963) have demonstrated that in sedimentary provinces individual subbottom interfaces act as very strong reflectors of acoustic energy (Ewing and Ewing, 1964). Particularly strong reflecting interfaces have been identified, and their distributions have been mapped (Ewing *et al.*, 1966; Windisch *et al.*, 1968). Dredging and recent drilling of the ocean floor have confirmed that in many instances the strongly reflecting horizons occur at levels in the sediment bodies where physically hard layers exist (Ewing *et al.*, 1970). This has been emphasized recently in the discovery of widespread chert in many of the cores of the Deep-Sea Drilling Project, both in the Atlantic and Pacific (Participating Scientists, 1969–1971). In the Atlantic some of the chert layers initially described seem to be restricted to clastic sedimentary units. The first or uppermost chert layer generally is in Eocene sediment, but conclusions as to the age of the chertification process await future work on the DSDP cores. The importance, however, of diagenetic processes to certain kinds of acoustical reflectivity is appreciated.

*Lamont-Doherty Geological Observatory Contribution Number 1891

The purpose of this paper is to describe samples and analyses of a strong subsurface reflecting horizon that is widespread throughout the Mediterranean Sea. Its occurrence in each of the three major deep basins of the Mediterranean Sea has been reported by by Ryan *et al.* (1970) who have called it *Reflector M.*

In regions isolated from bottom-transported clastics, *e.g.*, on top of the Mediterranean Ridge, the thickness of sediment section above Reflector M ranges between 80 and 140 meters (Wong and Zarudzki, 1970). In consideration of relatively high rates of contemporary sediment accumulation (Parker, 1958; Olausson, 1961; Ninkovich and Heezen, 1965), the shallow subbottom depth of Reflector M suggests that it might consist of lithified sediments very much younger than Eocene. Since the circum-Mediterranean stratigraphic sections (*e.g.*, in the Apennines, the Tel-Atlas, the Hellenides, and the Alps) do not contain chert younger than Eocene, the origin of Reflector M is an intriguing question whose answer might be fundamentally different from that of deep ocean chertification.

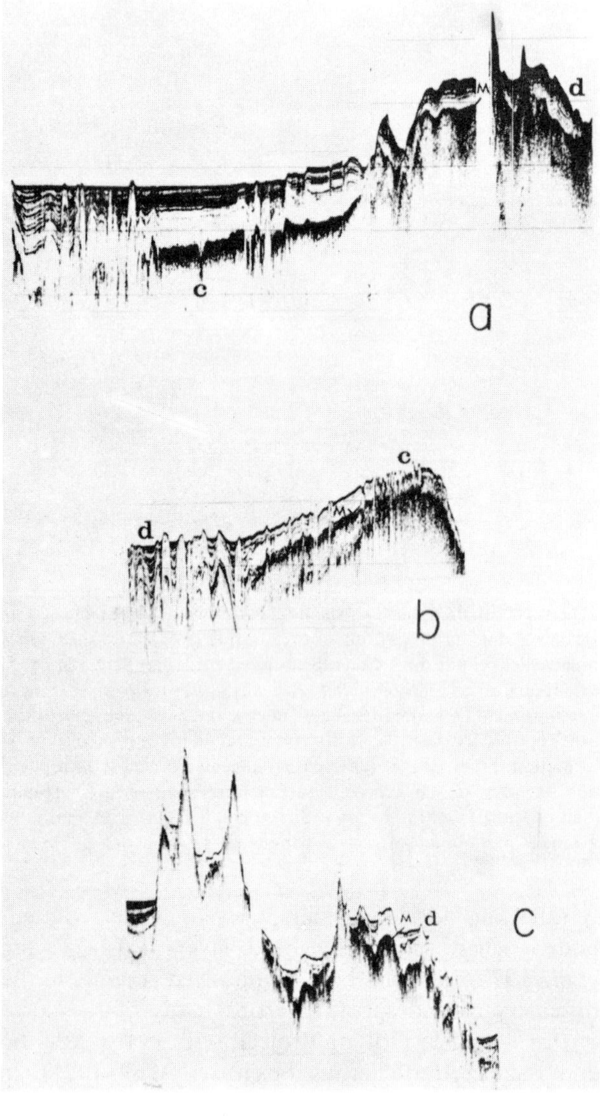

Figure 1. a, A section oriented southwest-northeast (left to right) showing layer M extending from under the Balearic Abyssal Plain up under the slope of the Balearic Island platform in the western Mediterranean. Note also the diapirs penetrating layer M and younger beds under the abyssal plain. b, East-west section of Cyprus in the eastern Mediterranean showing layer M extending across a range of topographic features, and under the abyssal plain to the east, being cut by diapiric structures. c, A section (approximately east-west) across part of the Tyrrhenian Sea showing layer M appearing under the continental slope and extending down toward the abyssal plain.

Beneath the modern abyssal plains Reflector M is in some places buried by more than 1500 meters of unconsolidated sediment. At such localities of sizable overburden (the Rhône Cone, Balearic Abyssal Plain, Tyrrhenian Abyssal Plain, Antalya Abyssal Plain, Nile Cone, and Herodotus Abyssal Plain) the reflecting horizon is pierced by dome-like structures originating in a transparent layer some 50–100 milliseconds below the uppermost surface of Horizon M. In high-resolution reflection profiles (Hersey, 1965; Glangeaud *et al.*, 1967; Mauffret, 1970; Leenhardt, 1970), the lower surface of Reflector M is characterized by numerous, small, overlapping, crescent-shaped echo sequences. (This corrugated surface is equivalent to Mauffret's Reflector K in the northern Balearic Basin.) What often appears as a single, bold subbottom reflector in low-resolution airgun profiles, is actually a series of individual reflecting horizons. As many as seven phases have been identified. Reflector M occurs in all physiographic provinces and can be traced in continuous seismic profiles beneath abyssal plains, on the flanks of seamounts, up continental slopes and onto the shelves (Figure 1a, b, c). The ubiquity of the reflector suggests that it might represent a time synchronous event over the entire Mediterranean Basin. It would be extremely interesting to the history of the Mediterranean Sea to learn the nature of this widespread reflector, to understand what event or process it represents, and what age or ages may be assigned to its level in the stratigraphic column.

SAMPLING OF MATERIAL FROM VICINITY OF REFLECTOR M

A few escarpments have been discovered where the upper surface of Reflector M actually outcrops or subcrops at very shallow depths beneath the sea floor. At these locations the proximity of the reflecting horizon to the sea floor appears to have resulted primarily from faulting and/or local erosional cuts rather than from a regional thinning of the superficial sediment cover. Exposures due to erosion are most common along the Malta Escarpment and around the periphery of the Balearic Platform.

A sample of material we believe to be from or near the upper surface of Reflector M has been obtained by coring in a narrow cleft in the Ionian Basin of the eastern Mediterranean. The seismic reflection profile from the area of this outcrop (Figure 2) extends across the southwestern flank of the Mediterranean Ridge to the Sirte Abyssal Plain and the continental rise in the Gulf of Sirte. Reflector M (indicated in the figure by an arrow) can be traced from beneath the continental rise northeastward up on Ridge. Between 0 and 20 miles on the profile, the reflector has been recorded at a two-way reflection time of 4.29 seconds. In a precision fathogram along this part of the profile (Figure 3), small hyperbolic echoes at a depth of 1700 to 1720 fathoms (3110–3148 m) are equivalent to an echo time of 4.25 seconds and mark the narrow axis of valleys incised into the ridge. These valleys are believed to be small clefts in the superficial sediments above Reflector M and apparently only extend as deep as this reflector. The valley floors are thus potential sites at which samples of the reflecting horizon might be obtained. From three lines of evidence we believe that Lamont-Doherty piston core RC9-188 did in fact sample material from the vicinity of the reflector in one of these valleys.

The fathogram in Figure 3 shows the ship's approach to the area where core RC9-188 was taken (*slow* and *stop*). From right to left, the next sections of the fathogram show the corer being lowered, the corer in the bottom and being raised again to the ship. The ship then started steaming (*underway* and *full ahead*). During the time the corer was approaching the bottom a strong echo return was visible at 1715 fathoms (3137 m) which we believe represents the floor of one of these clefts directly beneath the ship.

Two arguments are offered to demonstrate that the corer hit the bottom of the valley floor beneath the ship rather than the shallower bottom surrounding the valley.

Figure 4A illustrates a graph of wire-out in fathoms (one fathom = 6 feet) versus the echo-sounding (PDR) depth at the time of the hit in another unit of fathoms (one fathom = 1/400 second). Wire-out depths are indicated as "fathoms" in the text whereas PDR depths are designated fathoms without quotes. The data are from a series of coring stations prior to and subsequent to core RC9-188. All these stations were in areas of locally smooth relief such that the PDR depth at the moment of hit represented the local water depth for that region. At these stations the surface drift of the ship was negligible and we assume negligible catenary in the wire because these data points plot as a straight line. Because the surface at station RC9-188 was also negligible (sea state zero) we feel justified in using the relationship of Figure 4A to obtain a PDR equivalent depth from the wire out. For a wire-out value of 1797 "fathoms" for core RC9-188, the PDR depth corresponds to 1715 ± 5 fathoms, which is precisely the depth of the strong echo return at 4.28 seconds in the fathogram of Figure 3, significantly deeper than the general bottom reflections at 4.15 seconds (about 1660 fathoms).

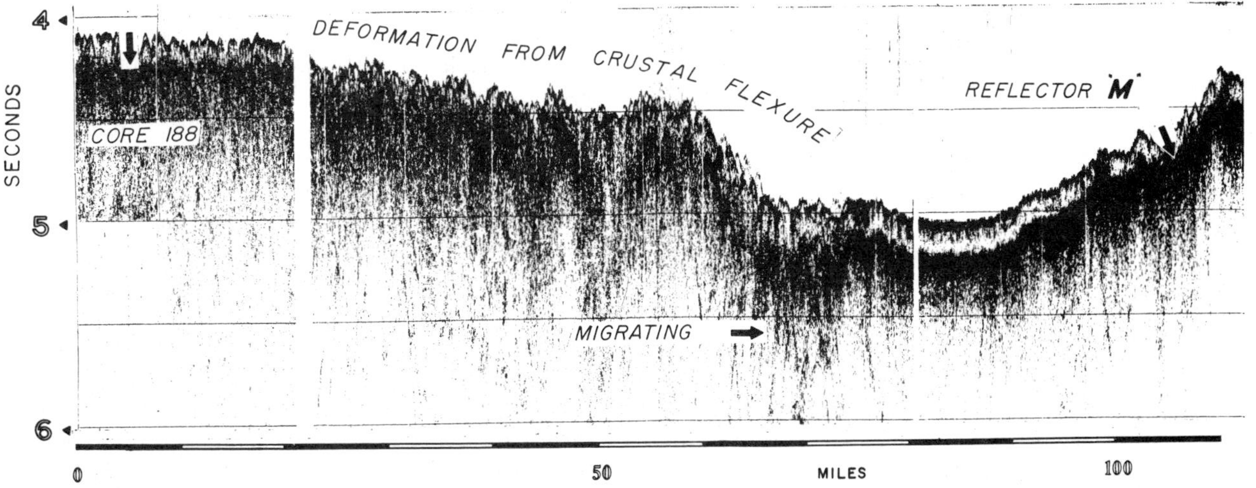

Figure 2. Seismic reflection profile across the Mediterranean Ridge and Sirte Abyssal Plain. A strong sub-bottom reflecting interface is noted, particularly beneath the abyssal plain and the continental rise off Libya. Materials from this horizon called Reflector M have been sampled in core RC9–188 at a location on the ridge, and consist of fragments of a dense cemented foraminifera ooze of Pliocene age. Reflector M corresponds to the upper surface of sedimentary layer with a compressional wave velocity of 3.4 km/sec.

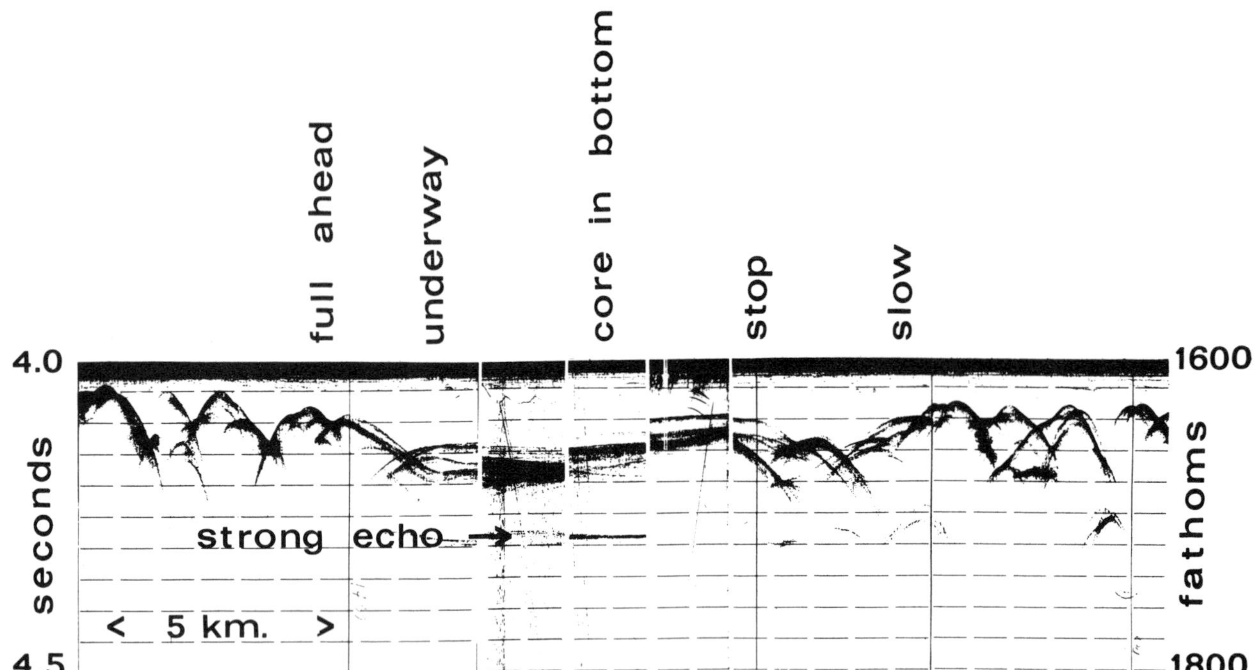

Figure 3. A fathogram in the vicinity of core site RC9-188. Small valleys, cut into the Mediterranean Ridge in this region, extend to a depth of 1700 fathoms (1 fathom = 1/400 second). This depth corresponds to the approximate depth of Reflector M. The fathogram is sectioned to show the echo traces during initial descent of the coring apparatus, the period when the core was in the bottom, and the subsequent surfacing. At the time the core was in the vicinity of the sea floor, a strong echo was recorded at 1715 fathoms, suggesting that the ship was then over the axis of one of these valleys.

Further confirmation that the corer hit the sea floor in the valley axis was made possible with a pressure gauge recording in a Thermograd instrument (Gerard *et al.*, 1962) attached to the corehead. The pressure gauge trace from this section was carefully calibrated against data for stations at core sites RC9-184 and RC9-186 (on abyssal plains) and was found to indicate an absolute depth of 3285 meters. The corrected meter equivalent of 1715 fathoms is 3290 meters (Matthews, 1939). We are confident that these independent lines of evidence indicate that the core was taken at a water depth equivalent to the level where Reflector M locally outcrops.

NATURE OF THE RECOVERED MATERIAL

The recovered core contained a microbreccia at 695 to 735 centimeters depth in the core. The microbreccia is comprised of lithified fragments of pink, yellow, blue, and orange and white foraminifera-rich lutites set in a loosely consolidated matrix. One fragment, almost pure white in color, was noted to be much harder than the others. Microscopic examination of several of the lithified lumps showed in all cases foraminifera set in a micritic matrix of very high birefringence (Figure 5a and b) which under the Scanning Electron Microscope (SEM) examination (Figure 6a and b) is shown to be coccoliths. We examined the fragments in an attempt to discern the nature of the lithification. Both the hard and less hard fragments appeared to be almost identical under SEM even up to magnification ×30,000. Both consist of foraminifera in a matrix of coccoliths (Figure 6) but in both cases the coccoliths appear equally fresh and the cause of cementation is not visible. As discussed in the next section the faunal assemblages of the various fragments indicate two different ages of deposition.

Samples of dense, less-dense fragments and matrix were analyzed for total carbonate content and were x-rayed to try to identify the cause of cementation. The most dense fragment was highest in carbonate (expressed $CaCO_3$): 89.5 ± 1% and the less-dense fragments and matrix ranged from 73.0 ± 1% to 85.9 ± 1%. X-ray diffraction revealed no difference between the fragments in the nature of the calcium carbonate using quartz as an internal peak position standard. Preferential destruction of $CaCO_3$ revealed minor dolomite, quartz and feldspars in all samples analyzed, so one cannot propose, for example, secondary dolomite cementation as the exclusive cause for induration of the most dense material.

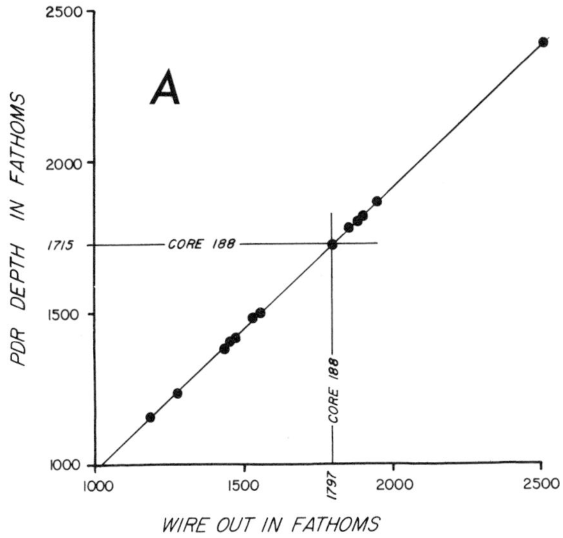

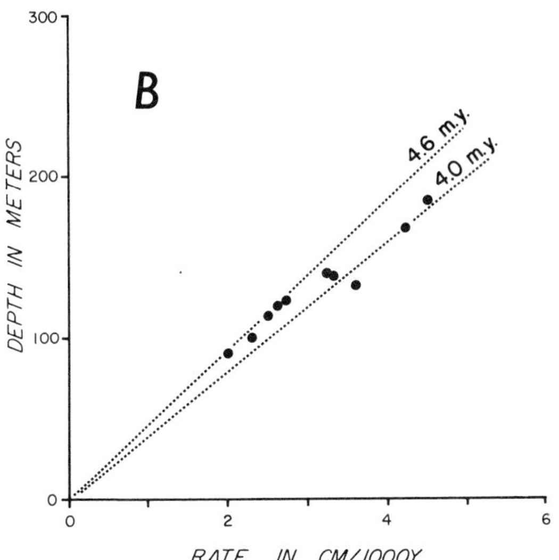

Figure 4. a, Graph of wire-out in "fathoms" on the coring winch versus sea floor PDR depth in fathoms (one fathom = 1/400 second). The points on this curve are derived using data from the coring logs at a series of core stations prior to and subsequent to core site RC9-188. At this station the wire-out value of 1797 "fathoms" at the core hit suggests that the apparatus penetrated into the sea floor at a water depth of 1715 ± 5 fathoms. This depth corresponds precisely to that of a strong echo sequence on the PDR which was evident during the period the core was in the vicinity of the bottom. b, Graph of sedimentation rate versus depth to Reflector M for 10 core sites in the eastern Mediterranean. The slope of the best fit straight line indicates an apparent age of 4.3 ± 0.3 million years for Reflector M.

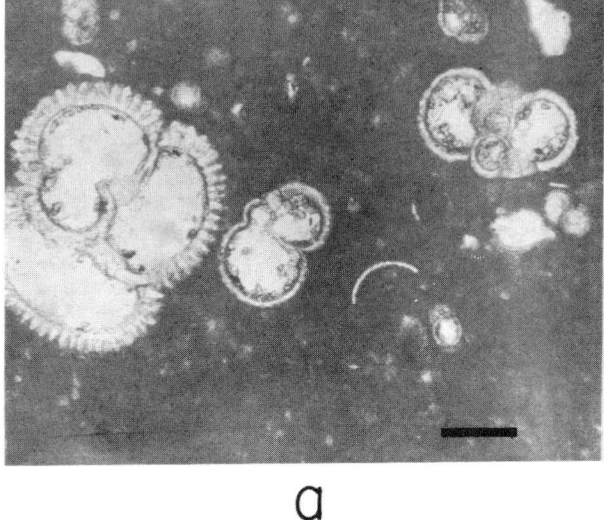

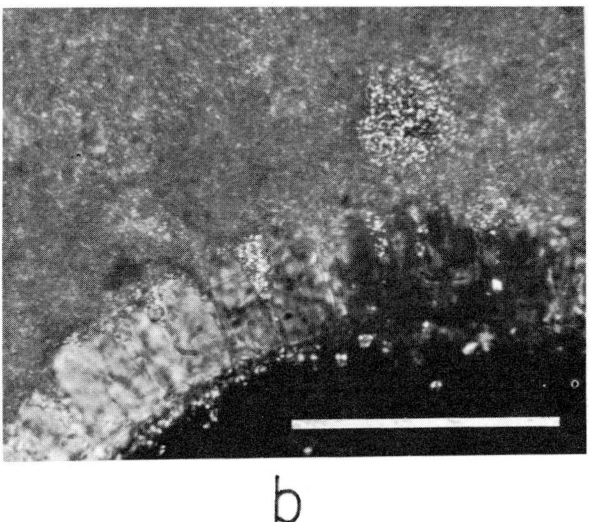

Figure 5. a, Photomicrograph of thin section of one of the dense lumps from layer M. The foraminifera are set in a very high birefringence matrix. Scale bar represents 100 microns. b, Detail of interface between foraminifera test and matrix. No evidence of recrystallization around these surfaces is obvious. Scale bar represents 100 microns.

AGE OF REFLECTOR M

The foraminifera in the particularly dense, white lump include *Globorotalia acostaensis, Globorotalia humerosa, Globorotalia merotumida, Globorotalia scitula, Sphaeroidinellopsis subdehiscens, Globigerinoides elongatus* and *Globigerinoides gomitulus* which are species diagnostic of open marine pelagic ooze at a stratigraphic level close in age to the Miocene-Pliocene boundary, possibly uppermost Messinian (Miocene) or lowermost Pliocene. This fauna is almost exactly the same as contained in the Trubi Formation of Sicily which has been studied by one of us (F.W.). No benthic foraminifera were found and the fact that the *Sphaeroidinellopsis* were not abundant indicates that our sample was not lowermost Trubi.

a

b

Figure 6. a, Scanning Electron Micrograph (X4900) of Pleistocene, unconsolidated core material showing coccoliths which comprise > 90% of the sample. This is the matrix material in which the indurated lumps of Mio-Pliocene, layer M material were found. Scale bar represents 10 microns. b, Scanning Electron Micrograph (X5100) of Miocene-Pliocene indurated material from layer M. The micrograph shows the matrix of coccoliths in which the foraminifera (upon which its age is in part based). As in the Pleistocene material above the coccoliths comprise > 90% of the sample, but display no obvious reason why this material is much harder than the Pleistocene material. Scale bar represents 10 microns.

The matrix of the microbreccia and other less-dense lumps in the microbreccia contain common *G. truncatulinoides* and indicate a Quaternary age (Calabrian or Sicilian). If the particularly dense lump is from the upper surface of Reflector M, this reasoning would indicate that the age of the reflector is lowermost Pliocene or uppermost Miocene. The Miocene-Pliocene faunal boundary corresponds to an absolute age of 5.5 m.y. as dated by paleomagnetic methods (Berggren, 1969).

Rates of sediment accumulation during the latter part of the glacial Pleistocene were calculated for a suite of ten long piston cores located on the Mediterranean Ridge at sites where the thickness of the sediment above Reflector M could be estimated (Ryan, 1970). The sediment thickness of the superficial cover above Reflector M can be derived from the reflection records and is initially recorded in units of travel time. The travel time is converted to meters by using estimates of the velocity of sound in sediment, which in this study were obtained by extrapolation from graphs by Houtz *et al.* (1968) on data from the North Atlantic. Figure 4B is a graph of recent sedimentation rate versus depth to Reflector M beneath the sea floor using the data from the suite of Mediterranean Ridge cores. The mean slope of this graph indicates an age of 4.3 ± 0.3 million years for Reflector M.

This age is younger by one million years than that of the Miocene-Pliocene boundary given by Berggren. It is not unlikely, however, that the recent rates of sediment accumulation in the glacial Pleistocene are higher than those of the earlier nonglacial Pliocene because of additional detrital influx during glacial epochs. Thus, extrapolation of these higher rates would necessarily imply an apparently young age for a given subbottom horizon. In fact the percent of biogenic material, if estimated from the bulk $CaCO_3$ content, is significantly greater for Miocene-Pliocene white colored materials from the breccia (average = 80% $CaCO_3$), than for the uppermost Pleistocene-Holocene material recovered in piston cores (average = 60% $CaCO_3$). If one were to assume that the rates of deposition of biogenic skeletal material have remained constant throughout the Pliocene and Pleistocene, and that the lower $CaCO_3$ content of the younger material represents additional influxes of terrigenous detrital components during the periods of glaciation (thus diluting the carbonate components), then a refined extrapolation of the age of Reflector M can be calculated based on sedimentation rates which are inversely proportional to to carbonate content. Based on arguments of Selli (1967) that the glacial Pleistocene was about 1 million years in duration, leaving 4.5 m.y. for the nonglacial Pleistocene and Pliocene, and that the average $CaCO_3$ content for pelagic sediments of the glacial epochs is 60%, and is 80% for older material, then the extrapolated age of the M Reflector increases to 5.4 ± 0.4 m.y.

CONCLUSIONS

Because none of the observational or analytical techniques used to date reveal any diagnostic difference between the indurated Mio-Pliocene material and the softer Pleistocene material, we cannot offer a satisfactory explanation as the nature or origins of Reflector M in the Mediterranean. From the available information we can, however, draw certain conclusions which impose boundary conditions for the genesis of Reflector M.

1. From the fact that the M-horizon occurs in seismic reflection profiles from the entire spectrum of Mediterranean physiographic provinces, we conclude that it represents a diagenetic process which took place on the Lower Pliocene-Upper Miocene sea floor *i. e.*, at or near the sediment-water interface. This process was one operative in the physiographic provinces from the continental shelf to the abyssal plain.

Fischer and Garrison (1967) also concluded that indurated Cenozoic samples which they studied from early dredgings in the Mediterranean and Barbados regions were formed by diagenesis. Their Mediterranean material is most certainly from a different part of the stratigraphic column than ours and they arrive at their conclusions from different observations. Their Mediterranean sample came from east of Rhodes and they reported a late Miocene age based on foraminifera but none of the foraminifera are common to our Trubi fauna. Other differences are that their foraminifera, in a fine grained matrix, were at least partially recrystallized and ours show little or no evidence of this; their fine grained matrix is called micritic and consists of magnesian-rich (about 7 to 10 wt. percent $MgCO_3$ by the method of Chave, 1952) calcite which they interpret to be the recrystallized phase responsible for induration. Our matrix consists of extremely fresh looking coccoliths which, together with the foraminifera, give a calcite pattern slightly less magnesian (about 4 to 6 wt. percent using Chave's method) and which show no evidence of recrystallization or even cementation under SEM. They attribute the induration to recrystallization of the micritic material that took place during a period of slow deposition at or near the sediment-water interface. We feel that the diagenetic phenomenon that produced Reflector M must have occurred at or near the sediment-water interface, but must involve some process other than the specific one postulated by Fischer and Garrison.

2. If the process that formed the reflector is due to some sort of precipitation of cement at or near the sediment-water interface, it must constitute a very small percent of the lithified material. From the fact that all observational and analytical measurements show no diagnostic differences between the indurated Mio-Pliocene and softer Pleistocene material, we conclude that the indurating cement or process must have taken place only at points of contact between the coccoliths and foraminifera. To resolve this possibility considerably more SEM work must be done at very high magnification as well as possibly microprobe work at points of contact. The SEM might reveal the presence of recrystallization and the microprobe whether or not a cementing agent, different in composition from the coccoliths, is responsible for induration.

3. Reflector M is penetrated in many areas by diapiric structures (see Figures 1a, b) which are most likely salt domes. If the age of the horizon is, as we conclude, Lower Pliocene–Upper Miocene, it is probable that the source of these diapirs is salt of Messinian age. Messinian and Tortonian evaporites are common in the Pan-Mediterranean continental sequence. Such evaporite sequences represent a special type of geomorphic-climatic situation wherein sea water can evaporate under restricted conditions but be replenished by continuous inflow. With its restricted access to the Atlantic Ocean at the west, the Mediterranean as a whole has the potential of undergoing or having undergone just such an evaporational regime. The proximity of Reflector M to a thick, diapiric salt sequence (here postulated to be of Messinian age) suggests to us the possibility of a genetic relationship. That is, for a diagenetic process to take place contemporaneously at all water depths in the Mediterranean would appear to require a change in the chemistry of the entire Mediterranean water column. We suggest that the type of tectonic-climatic controls that earlier yielded an evaporite sequence may have re-occurred either only briefly or to a lesser degree during the late Miocene–early Pliocene to produce an induration of the sediment at or near the sediment-water interface.

ACKNOWLEDGMENTS

We wish to acknowledge the National Science Foundation and the U.S. Navy Office of Naval Research for their continued support of Lamont-Doherty research vessel operations and the U.S. Navy Office of Naval Research (TO-4) and the U.S. Atomic Energy Commission (AT [30-1] 4055) grants under whose financial support this work was carried out. We thank P. Jeff Fox and Bruce Heezen of Lamont-Doherty Geological Observatory for helpful discussions and criticisms.

REFERENCES

Berggren, W. A. 1969. Cenozoic chronostratigraphy, planktonic foraminiferal zonation and the radiometric time scale. *Nature*, 224:1072–1075.

Chave, K. E. 1952. A solid solution between calcite and dolomite. *Journal of Geology*, 60:190–192.

Ewing, J., C. Windisch, and M. Ewing, 1970. Correlation of horizon A with JOIDES bore-hole results, *Journal of Geophysical Research*, 75:5645–5653.

Ewing, J., J. L. Worzel, M. Ewing, and C. Windisch, 1966. Ages of horizon A and the oldest Atlantic sediments. *Science*, 154: 1125–1132.

Ewing, M. and J. Ewing, 1964. Distribution of oceanic sediments. In: *Studies on Oceanography*, ed. K. Yoshida, University of Tokyo Press, Tokyo, 523–537.

Fischer, A. G. and R. E. Garrison, 1967. Carbonate lithification on the sea floor, *Journal of Geology*, 75:488–496.

Gerard, R., M. G. Langseth and M. Ewing, 1962. Thermal gradient measurements in the water and bottom sediment of the western Atlantic, *Journal of Geophysical Research*, 67:785–803.

Glangeaud, L., J. Alinat, J. Polveche, A. Guillaume, and O. Leenhardt, 1967. Grandes structures de la mer Ligure, leur évolution et leurs relations avec les chaînes continentales, *Bulletin de la Société Géologique de France*, 7:921–937.

Hersey, J. B. 1963. Continuous reflecting profile. In: *The Sea*, ed. Hill, M. N., Interscience Publishers, New York, 3:47–72.

Hersey, J. B. 1965. Sedimentary basins of the Mediterranean Sea. In: *Submarine Geology and Geophysics*, eds. Whittard, W. F. and R. Bradshaw, Butterworths, London, 75–91.

Houtz, R., J. Ewing, and X. LePichon, 1968. Velocity of deep-sea sediments from sonobuoy data. *Journal of Geophysical Research*, 73:2615–2641.

Leenhardt, O. 1970. Sondages sismiques continus en Méditerranée occidentale, Enregistrement, analyse, interprétation. *Mémoires de l'Institut Océanographique*, 1:120 p.

Matthews, D. J. 1939. *Tables of the Velocity of Sound in Pure Water and Sea Water*. Hydrographic Department, Admiralty, London, 52 p.

Mauffret, A. 1970. Structure des fonds marins autour des Baléares. *Cahiers Océanographiques*, 22:33–42.

Ninkovich, D. and B. C. Heezen 1965. Santorini tephra. In: *Submarine Geology and Geophysics*, eds. Whittard, W. F. and R. Bradshaw, Butterworths, London, 413–554.

Olausson, E. 1961. Studies of deep-sea cores. *Reports of the Swedish Deep-Sea Expedition, 1947–1948*, Report 8, fascicule 4(6): 3–391.

Parker, F. L. 1958. Eastern Mediterranean foraminifera. *Reports of the Swedish Deep-Sea Expedition, 1947–1948*, Report 8, fascicule 2(4):217–283.

Participating Scientists. 1969–1971. *Initial Reports of the Deep Sea Drilling Project*. National Science Foundation, Government Printing Office, Washington, D.C., 1–6.

Ryan, W. B. F. 1970. *The Floor of the Mediterranean Sea*. Ph.D. thesis, Columbia University, New York, 300 p.

Ryan, W. B. F., D. J. Stanley, J. B. Hersey, D. A. Fahlquist, and T. D. Allen 1970. The tectonics and geology of the Mediterranean Sea. In: *The Sea*, ed. Maxwell, A. E., Interscience Publishers, New York, 387–492.

Selli, R. 1967. The Pliocene-Pleistocene boundary in Italian marine sections and its relationship to continental stratigraphies. In: *Progress In Oceanography*, ed. Sears, M., Pergamon Press, London, 4:67–86.

Windisch, C. C., R. J. Leyden, J. L. Worzel, T. Saito, and J. Ewing, 1968. Investigation of horizon B, *Science*, 162(3861):1473–1479.

Wong, H. and E. F. K. Zarudzki, 1969. Thickness of unconsolidated sediments in the eastern Mediterranean Sea, *Geological Society of America Bulletin*, 80:2611–2614.

NOTE ADDED IN PROOF: Since this paper was written, Leg XIII of the JOIDES Deep-Sea Drilling Project found thick, Late Miocene evaporites in the Mediterranean Sea whose uppermost surface corresponds to Reflector M. These evaporites probably immediately underlie the cemented material we found in RC9–188 whose site is not far from DSDP site 125 which penetrated the evaporites. The abundance of *Sphaeroidinellopsis* spp. in our sample indicates that it belongs to the *Sphaeroidinellopsis* acme-zone of lowermost Pliocene age (5.2–5.4 m.y.; *in* Chapter 47, volume XIII *Initial Reports of the Deep-Sea Drilling Project*, 1972, U.S. Government Printing Office, Washington, D.C.).

The Neogene Series Drilled by the Mistral No. 1 Well in the Gulf of Lion

Pierre F. Burollet and Philippe Dufaure

Compagnie Française des Pétroles, Puteaux

ABSTRACT

The exploratory well Mistral No. 1, situated some 55 kilometers from the coast in the Gulf of Lion, was drilled to a depth of 3552 meters below sea level. Approximately 3135 meters of Miocene and Pliocene lying unconformably over a Paleozoic basement series were encountered in this well. The faunas and mineralogy of the Neogene series are described in some detail, particular attention being paid to stratigraphically diagnostic faunas. Conclusions have been drawn with regard to the depositional environments represented by the Neogene sediments and, where possible, the regional implications of these environmental interpretations are discussed. It is concluded that the area of the gulf has foundered by more than 3000 meters since Lower Miocene time.

RESUME

Le puit Mistral no. 1, situé dans le Golfe du Lion, à 55 kilomètres de la côte, foré par la Compagnie Française des Pétroles et la British Petroleum, a été arrêté à 3552 m au-dessous du niveau de la mer. Il a atteint le Paléozoïque, recouvert en discordance par une serie Miocène-Pliocène de 3135 m d'épaisseur. Le Néogène est décrit en détail. La faune caractéristique qui a servi de base pour les datations est précisée et les résultats des analyses minéralogiques des argiles et des carbonates sont donnés. Des conclusions ont été tirées en ce qui concerne les conditions de dépôt dans cette zone. La région s'est enfoncée d'environ 3000 m depuis le Miocène inférieur.

INTRODUCTION

The Mistral No. 1 well was drilled in the Gulf of Lion (also named Lions or Lyon) off the coast of French Languedoc, some 55 km SSE of Sète and 125 km WSW of Marseilles (Figure 1). The drilling location coordinates were 42°57′35″N and 3°53′54″E and the water depth 98 m. The well was drilled by Compagnie Française des Pétroles (Métropole), operator for the following groups who hold the concession: Compagnie Française des Pétroles, Société Française des développements pétroliers B.P. et Société Française des Pétroles B.P. The well was spudded on the 7th of November 1968, finishing on the 7th of February 1969. Total depth was 3552 meters below sea level or 3454 meters below the sea floor.

This deep oil exploration well is one of the first in an area previously little explored. Detailed micropalaeontological and X-ray studies of the well samples have made it possible to establish the stratigraphy and local conditions of deposition. It has also been possible to place the well in its regional setting and to compare the area with the deeper western Mediterranean basins studied by Montadert *et al.* (1970) and others.

STRATIGRAPHIC INTERPRETATION

The limits used are those corresponding to lithological breaks confirmed by the well logs. They correspond in the main to the interpretation based on micropalaeontological studies. These latter studies were made with the well cuttings and as a result the

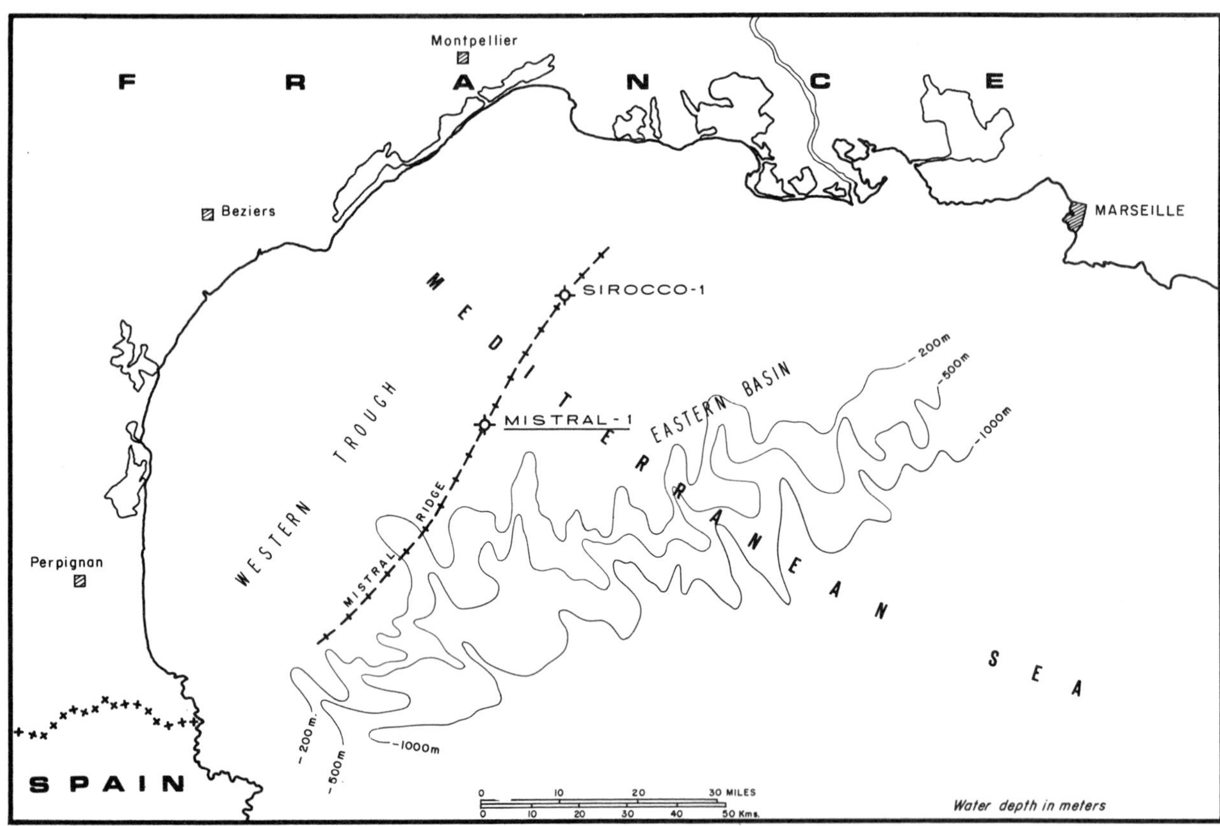

Figure 1. Chart showing position of the Mistral No. 1 well in the Gulf of Lion off the southwestern coast of France. Note the structural Mistral Ridge trending NE–SW across the shelf.

limits are not precise. Due to the evolutionary complexity of the microfauna and the sedimentary variability, it is very difficult to attribute the chosen intervals to stratigraphic zones. Furthermore, the relative imprecision and imbrication of the stages of the Aquitanian and Pleistocene of the Tethys area make it extremely hazardous to assign a definitive stratigraphic nomenclature.

The substratum, lying between 3455.5 and 3552 meters (Figure 2), consists of schists and contorted sandstone that resemble the Ordovician of the eastern Pyrenees or the "Montagne Noire". Depths of the stratigraphic horizons are given in meters below sea level.

MIOCENE

Basal Miocene (Lower Aquitanian ?)

A basal Miocene (Lower Aquitanian?) section was recovered between 3455.5 m and 3181 m. Several facies were examined.

1, From the base at 3455.5 to 3448 m: Polygenic breccia, consisting of microcrystalline beige dolomite, indurated grayish black shale, anhydritic white shale, and sericitic schist. The matrix, when studied separately (a sidewall core), seems to consist of barren shale or marl of undetermined age.

2, 3448 m to 3340 m: Black indurated silty marl. All attempts to separate the microfauna failed except for a sample from 3366 m where small Globigerinidae were found, including *Globigerinoides* aff. *primordius* Blow and Banner, which could be attributed to the base of the Miocene (Lower Aquitanian).

3, From 3340 m to 3181 m: The series consists of an alternation of indurated calcareous shale which is often silty or very silty, marl with thin beds of micritic shaly limestone, and sandstone or argillaceous glauconitic sands. There are occasionally traces of bitumen.

A thin section study showed small Globigerinidae throughout most of the interval, including the lower part (sidewall cores at 3428 m and 3400 m).

From 3322 m to 3270 m there are zones with micrite and common Lithothamnia algal debris. We note, amongst other fragments, very rare spicules, large pieces of thick-walled Bryozoan (3284 m to 3290 m), some sections of *Amphistegina* (3310 m), some Trochamminidae (3400 m) and Textularidae (3414 m).

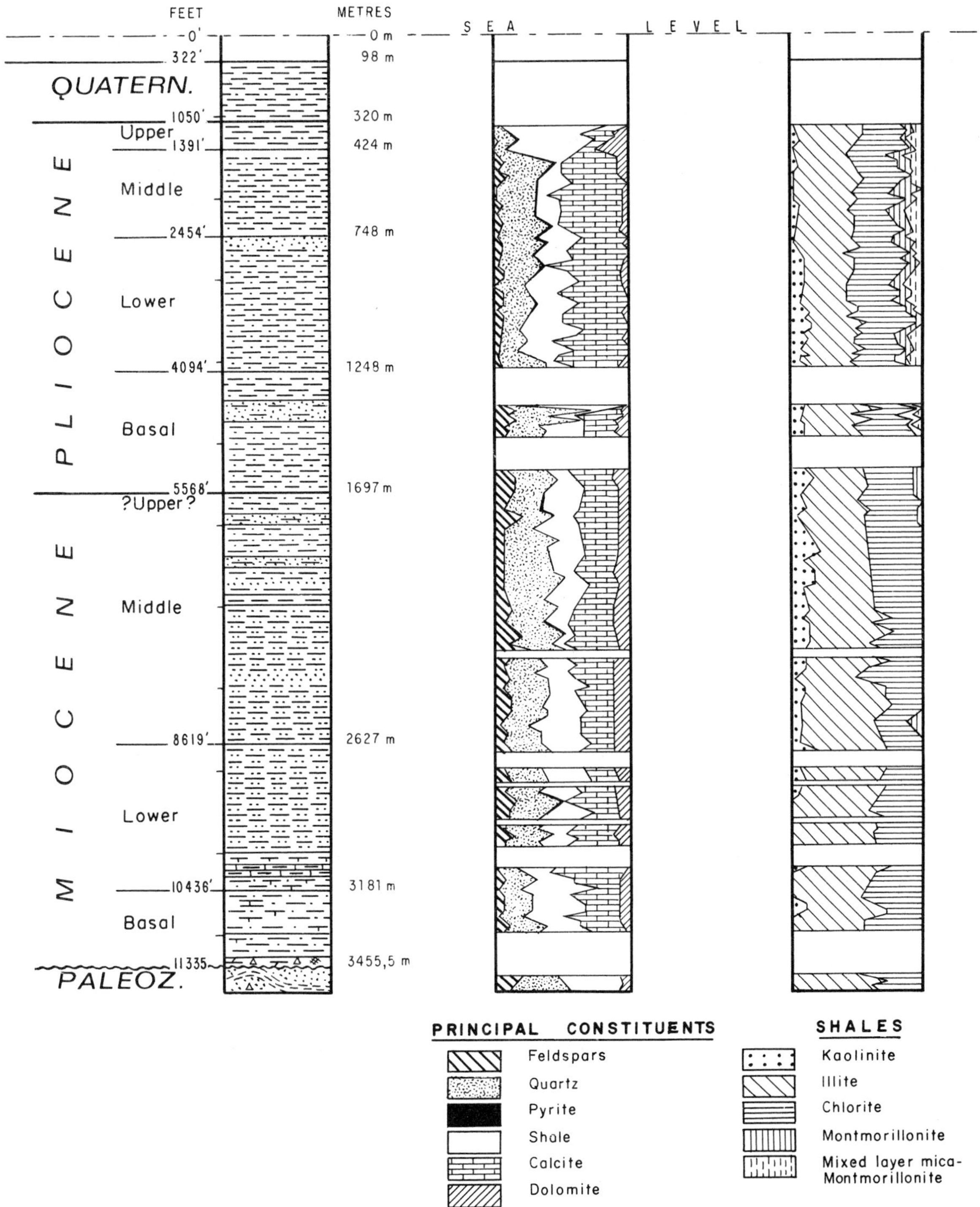

Figure 2. Stratigraphy and lithologic logs of the Mistral No. 1 well in the Gulf of Lion.

The argillaceous part of this section has been shown to contain, on average, 60% illite and 49% chlorite, with small amounts of kaolinite, especially in the middle part of the section.

This series is characterised by the presence of *Globeriginoides* aff. *primordius* Blow and Banner and the absence of *Globigerinoides trilobus* (Reuss).

Lower Miocene

A Lower Miocene section was recovered between 3181 m and 2627 meters. Two horizons can be distinguished in this section.

1. 3181 m to 3032 meters: Indurated marl and argillaceous limestone, which is slightly silty or contains very fine sand, and rare fragments of metamorphic schist or reworked quartz. The shales contain 60% illite and 40% chlorite in the lower part; with up to 10% kaolinite in the upper part, the average percentage becomes illite 60% and chlorite 30%. The carbonates form 40% of the rock, but locally reach 60%.

The microfauna was difficult to separate but the following forms were found:

Globigerinoides trilobus (Reuss)
Globorotalia gr. mayeri Cushman and Ellisor aff. Bolli (1956)
and some small *Globoquadrina* sp.

One could consider this lower part to be of Upper Aquitanian-Burdigalian age, though further studies might modify this preliminary age determination.

2. 3032 m to 2627 meters: This is a homogeneous section of silty indurated marl with small sandstone nodules that are very fine, marly and sometimes glauconitic. There are some scattered crystalline rock fragments between 3032 and 2840 meters. On the average, the rock consists of 40% detrital material (10 to 15% feldspars and 25 to 30% quartz), from 20 to 25% shale and from 35 to 40% carbonates, including 10% dolomite. The shales increase in importance towards the top and consist of about 60% illite, 30 to 40% chlorite and little or no kaolinite.

There is an abundant planktonic microfauna[1]:

Globigerinoides trilobus (Reuss), f.
Glg. bisphaericus Todd, r.
Glg. transitorius Blow, r.
Glg. sacculiferus (Brady), tr.
Globoquadrina langhiana Cita and Gelati, c.
Glq. cf. *dehiscens* Chapman, Parr and Collins, tr.
Glq. cf. *altispira* var. *globosa* Bolli, r.

Two specimens of *Globorotalia scitula* (Brady) have been found at 2762 m, and one specimen of *Orbulina suturalis* Bronnimann at 2756 m (possibly cavings).

It is common to find benthonic fauna encrusted with silt. Consequently these specimens are difficult to identify. Amongst the arenaceous foraminifera found are *Haplophragmoides* sp. towards the base, and very commonly *Sigmoilopsis* sp. Unidentified species of smooth Ostracoda were also found.

These beds represent the Burdigalian and correspond, at least partially, to the Langhian. They are characterised by the abundance and predominance of *Globigerinoides trilobus* (Reuss), and its association in the upper part with *Globigerinoides bisphaericus* Todd and *Globigerinoides sacculiferous* Brady. Also there is an absence of *Orbulina suturalis* Bronniman and *Orbulina universa* d'Orbigny whilst *Globoquadrina langhiana* Cita and Gelati is common.

Middle Miocene

A Middle Miocene section was recovered between 2627 m and about 1697 m. The sediment type is a mixture of gray to grayish blue marl, silty or with fine sand with intercalations of shaly, fine grained, glauconitic, sandstone. The lower part (2627 to 2310 meters) contains an average of 30 to 40% detrital matter (10% feldspar and 20 to 30% quartz), 20 to 30% shale and 30 to 40% carbonate. The shales are on the average 60% illite, 10% kaolinite, and 30% chlorite, which becomes more abundant towards the top. There is a little montmorillonite in the lower part.

The upper part (from 2310 m to 1697 meters) is generally richer in detrital constituents, which attain 40 to 50%. The feldspars reach 20% (including potassics). The carbonates average 40% with about 10% dolomite. The shaly portion shows a reduction in illite (40 to 50%) and the kaolinite sometimes reaches 20%, whilst chlorite remains the same at 40%. Montmorillonite is absent or rare, except in the uppermost part, above 1859 m.

The planktonic microfauna is abundant and homogeneous.

Globorotalia mayeri Cushman and Ellisor, c.
Globoquadrina aff. *dehiscens* (Chapman, Parr and Collins), f.
Glq. cf. *altispira globosa* Bolli, c.
Globigerinoides trilobus (Reuss), f.
Glg. bisphaericus Todd, af.
Glg. transitorius Blow, r.
Glg. glomerosus Blow, r.
Glg. sacculiferus (Brady), r.
Globigerina falconensis Blow, r.
Gl. gr. *trilocularis* d'Orbigny or *bulbosa* Cushman and Jarvis, r.
Orbulina universa d'Orbigny, af.
O. suturalis Bronnimann, r.

The benthonic fauna, varied and frequent, comprises many smooth *Lenticulites,* Anomalinidae with *Cibicides mexicanus* Nuttal var. *dertonensis* Ruscelli and *Cibicides* gr. *refulgens* (Montfort), as well as the

[1] The following abbreviations have been used throughout the text to describe the faunal distribution:

ab	: abundant	r	: rare
f	: frequent	tr	: very rare
af	: fairly frequent	ex	: exceptional
c	: common		

following:

Gyroidina gr. *soldani* (d'Orbigny)
Ammonia gr. *beccarii* (Linne) (flattened globular forms)
Uvigerina cf. *flinti* Cushman
Uv. tenuistriata Reuss
Uv. gr. *schwageri*, Brady
Siphonodosaria verneuili (d'Orbigny)
S. nuttali var. *gracillima* Cushman and Jarvis
S. consobrina var. *emaciata* (Reuss)
S. vertebralis (Batsch)
Arenaceous forms such as *Spiroplectammina carinata d'Orbigny*, and costulate Ostracods.

Due to the continued presence of abundant *Globorotalia mayeri* Cushman and Ellisor and common *Orbulina* together with various *Globigerinoides* and abundant *Globoquadrina* (*Glq.* cf. *altispira* var. *globosa*, *Glq.* aff. *dehiscens*, *Glq.* sp.) it is possible to date the section as Middle Miocene. This is confirmed by the absence of such forms as *Globorotalia menardii*, *Glr. praemenardii* and *Globigerina nepenthes*.

Upper Miocene

The Upper Miocene is probably absent. It is felt that the beds above 1697 m can best be interpreted as being of Lower Pliocene age, although microfossils are rare or absent.

Up to 1500 m there are few samples and they are of doubtful quality. For instance, the sample from 1555 m is mineralogically similar to the section from 1500 to 1377 m where there is abundant quartz, and the carbonates are less common than above or below. There is no longer the small percentage of altered illite in the micas that was regularly present in the Lower and Middle Miocene.

PLIOCENE

Basal Pliocene

A Basal Pliocene section was recovered between 1248 and 1697 meters. The whole of this section consists of silty shales and shaly sandstones, which are capped by sandy shales with heterogeneous pebbles. Details are lacking from the upper part of the 1697 to 1356 m section as there were no samples from 1620 to 1555 m or from 1555 to 1500 m. Several facies were recognized in the remaining section.

1. 1697 to 1620 meters: Silty marls with Middle and Lower Miocene affinities, such as equal amounts of illite and chlorite, some of the illite being altered.
2. 1620 to 1500 meters: The only samples (at 1555 m) seem to belong to the overlying section.
3. 1500 to 1356 meters: The slightly calcareous shales contain silty sandstone beds which are rich in feldspar, including some potassic feldspars. There is never more than 30% carbonate material whilst there is often 10% dolomite. In the shales, the illite percentage is greater than that of the chlorite, with about 10% kaolinite, and a small but constant proportion of montmorillonite. For the first time, 10% of the rock consists of mixed layer mica-montmorillonite minerals. This percentage is constant in the upper part of the section.
4. 1356 to 1248 meters: Here a gray sandy shale includes beds of carbonate pebbles, metamorphic or crystalline rocks (micritic glauconitic limestone, dolomitic limestone, schists with quartz veining, metamorphosed arkosic sandstone, veined quartz, opalised amphiboles, granite, etc.).

In both the shales and conglomerates the microfauna is scarce, and certain zones are barren (1330, 1340, 1486, 1540, 1646, 1670 m). The planktonic microfauna consists:

Orbulina universa d'Orbigny, tr. or af.
O. suturalis Bronnimann, tr.
O. bilobata (d'Orbigny), ex.
Globigerinoides glomerosus Blow, c.
Glg. gr. *obliquus*, Bolli, af.
Glg. sacculiferus (Brady), tr.
Glg. trilobus (Reuss), r. or c.
Globigerina pseudopachyderma Cita, Premolisilva and Rossi, r.
Globigerina falconensis Blow, r.
Gl. nepenthes Todd, tr.
Gl. cf. *picassiana* Perconig, r.
Globorotalia aff. *acostaensis* Blow, r.
Glr. acostaensis (Blow) *trochoidea* Bizon, tr.
Glr. gr. *margaritae* Bolli and Bermudez, tr. or c.
Glr. scitula (Brady), ex.
Glr. gr. *puncticulata* (Deshayes), tr. or c.
Sphaeroidinellopsis rutschi (Cushman and Renz), r.
Sph. grimsdalei (Keijzer), tr.

The benthonic microfauna is sparse and suggests a littoral marine environment with slight deltaic influences: *Elphidium* sp., *Cibicides* sp., *Lenticulina* sp. with occasional *Nonion* sp. and *Ammonia* sp., especially in the conglomeratic beds of the upper part of the section.

Initially, the pebble beds were attributed to the Pontian, and the underlying silty shales to the Upper Miocene. But the presence in the whole of the section of *Globorotalia* gr. *margaritae*, *Gl.* gr. *puncticulata* and *Sphaeroidinellopsis*, which cannot be easily explained away as caved well samples from an overlying Tabianian, suggest a lower Pliocene age. No typical Messinian foraminifera were found.

G. and J. J. Bizon (personal communication) who have examined the well samples are inclined to date the beds as Upper Miocene age, and they remark a similarity with a series known in southeast Spain. However the poor quality of the sample material does not permit a more precise datation.

Lower Pliocene (Tabianian)

The Lower Pliocene section was recovered between 1248 and 748 meters. This section consists essentially of slightly silty, bluish grey, clayey marls. From 983 to 783 m the amount of granular quartz increases. Overall, the environment is much less littoral than that of the Basal Pliocene, or the Middle Pliocene, and tends to be pelagic. The water depths may have been about 150 to 200 m.

This is a very homogeneous lithological section. Calcite is regularly abundant (an average of 40%, increasing occasionally to 60%) as is commonly the case in the lower part of the section. Only traces of dolomite exist. The feldspars are less abundant than previously and consist almost entirely of plagioclase. The shaly minerals consist of 10 to 15% kaolinite, 30 to 40% illite, 30 to 40% chlorite, and 10% montmorillonite, whilst the mixed layer mica-montmorillonite minerals can form up to 20% of the rock.

The abundant planktonic microfauna consist:

Globorotalia gr. *puncticulata* (Deshayes), ab.
Glr. margaritae Bolli and Bermudez, tr. or ab.
Glr. aff. *scitula ventricosa* Ogniben, r.
Sphaeroidinella aff. *dehiscens* (Parker and Jones) *subdehiscens* Blow
Sph. rutschi (Cushman and Renz), r.
Globigerinoides gr. *obliquus* Bolli, f., with variety *extremus*.
Glg. sacculiferus (Brady), r.
Glg. trilobus (Reuss), r.
Glg. helicinus (d'Orbigny), ex.
Globigerina cf. *bradyi* Wiesner, r.
Gl. aff. *nepenthes* Todd, r.
Orbulina universa d'Orbigny, f.
O. suturalis Bronniman, tr.

The benthonic microfauna is abundant and diversified, but consist mainly of Lagenidae, Ellipsoidinidae and Rotaliforma.

Robulus echinatus (d'Orbigny)
R. ariminensis (d'Orbigny)
R. curviseptus (Seguenza)
R. cf. *spinulosus* (Costa)
Marginulina costata Batsch var. *coarctata* Silvestri
Planularia auris (Defrance)
P. auris (Defr.) var. *cymba* (d'Orbigny)
P. cf. cassis (Fichtel and Moll)
Vaginulinopsis inversa (Costa)
V. inversa (Costa) var. *carinata* Silvestri
Lingulina costata d'Orbigny
Siphonodosaria hispida (d'Orbigny)
S. consobrina (d'Orbigny)
S. vertebralis (Batsch)
Siphonia reticulata Czjzek
Sn. planoconvexa (Fichtel and Moll)
Anomalina ornata (Costa)
Planulina (Anomalina) *helicina* (Costa)

The characteristics of the Tabianian are based on studies of samples taken from the stratotype area. There are three recognisable intervals. Starting from the base they are characterised by an abundance of *Sphaeroidinellopsis, Globorotalia margaritae* (1200 to 1100 meters), then *Glr. puncticulata* (1060 to 780 meters).

These three intervals are very similar to the Lower Pliocene micropalaeontological zones of Italy and Greece. But the well samples are too doubtful for one to be able to confirm whether these interval changes are stratigraphical successions or variations of depositional environment.

The Lower Pliocene section is characterised by several rare specimens of *Sphaeroidinella* aff. *dehiscens* Parker and Jones *subdehiscens* Blow, the presence and relative abundance of *Globigerinoides* gr. *obliquus* Bolli, and the disappearance of *Globigerinoides* aff. *nepenthes* Todd.

Middle Pliocene (Plaisancian)

The Middle Pliocene (Plaisancian) section was recovered between 748 and 424 meters. This interval consists of grayish blue, clayey sandy or silty marls. The mineralogical constituents are very similar to those in the underlying beds with over 50% of carbonates. There is invariably some pyrites and aragonite. The shaly portion has generally 40 to 50% illite, some 10% kaolinite, 20 to 30% chlorite and varying amounts of montmorillonite and illite/montmorillonite mixed layer minerals.

There are abundant planktonic foraminifera such as:

Globigerinoides gomitulus Seguenza, c.
Glg. gr. *obliquus* Bolli, f.
Glg. adriaticus (Fornasini), r.
Glg. elongatus (d'Orbigny), tr.
Glg. helicinus (d'Orbigny), tr.
Glg. sacculiferus (Brady), r.
Glg. trilobus (Reuss), r.
Globigerina falconensis Blow, r.
Gl. bulloides d'Orbigny, af.
Gl. quinqueloba Natland, tr.
Gl. aff. *microstoma* Cita and Premolisilva, r.
Gl. cf. *juvenilis* Bolli, tr.
Globorotalia cf. gr. *crassaformis* Galloway and Wissler, ex.

This association is identical to that found in the

Middle and Upper Plaisancian beds of the Po valley. The benthonic microfauna is also abundant and varied:

Dorothia gibbosa (d'Orbigny)
Marginulina costata (Batsch)
Orthomorphina tenuicostata (Costa)
O. proxima Silvestri
Bolivina punctata d'Orbigny
B. italica Cushman
Brizalina aenariensis (Costa)
Eponides umbonatus (Reuss) var. *stellatus* (Silvestri)

The presence of *Globorotalia* cf. gr. *crassaformis* Galloway and Wissler is characteristic of the Middle Pliocene.

Upper Pliocene (Astian)

The Upper Pliocene (Astian) section was recovered between 424 and 320 meters. The sediment includes a pelecypod and gastropod coquina with a grayish blue, clayey shale matrix.

Here we have a genuine shale that is only slightly dolomitic, with a major reduction of the calcite and quartz elements compared with the underlying series. Illite is the dominating mineral followed by the chlorites, with only a small amount of montmorillonite and kaolinite, and an average of 10% illite/montmorillonite mixed layer minerals. There are traces of aragonite and hemigordite.

The planktonic foraminifera are generally rare:

Globigerinoides adriaticus (Fornasini), r.
Glg. obliquus Bolli var. *extremus* Perconig
Glg. obliquus Bolli, r.
Glg. trilobus (Reuss), tr.
Globigerina bulloides d'Orbigny, r.
Globigerina pachyderma (Erhenberg), tr.

Their characteristics, as well as those of the other faunal elements, are comparable to those found in the Asti sands; some forms are common to, and typical of, both the Middle Pliocene and the Quaternary. This section has many *Globigerinoides obliquus* Bolli var. *extremus* Perconig, and *Globigerinoides obliquus* Bolli, which are not known in the Quaternary.

The benthonic specimens are common and have marked Pliocene affinities:

Bolivina aff. *italica* Cushman
Brizalina gr. *catanensis* Seguenza
Elphidium aff. *complanatum* (d'Orbigny)

QUATERNARY

Between the 320 meters level and the sea bed (at a depth of 98 m) the only samples were those recovered from the drilling bit when pulled to surface. They were grayish blue clayey shales with pelecypod, gasteropod, echinoid, and bryozoa debris, some sand and worked pebbles of sandstone and quartzite.

The planktonic fauna is common and much more abundant than in the underlying Astian:

Globorotalia inflata (d'Orbigny), af.
Glr. truncatalinoides (d'Orbigny), r. to c.
Globigerinoides conglobatus (Brady), r.
Glg. ruber (d'Orbigny), r.
Glg. obliquus Bolli, tr.
Glg. cf. *elongatus* (d'Orbigny), tr.
Globigerina aff. *quinqueloba* Natland, f.
Gl. bulloides d'Orbigny, f.
Orbulina universa d'Orbigny, r.

Globorotalia truncatalinoides (d'Orbigny) is characteristic of the Quaternary.

CONCLUSIONS

The Mistral No. 1 well shows the presence of 3135.5 m of Neogene strata in the Gulf of Lion. The major stratigraphic features observed are:

1. The transgression of a Paleozoic series by the Lower Miocene;
2. A break, or at least a considerable thinning, in the Upper Miocene series;
3. A Lower Pliocene transgression, with reworked pebbles that could represent the Pontian;
4. A development of marine Pliocene, 1377 meters thick, wholly marine in the Lower Pliocene, with more and more littoral facies as the higher beds of the series, such as Plaisancian and Astian, are reached.

The Mistral well was located on a ridge which joins the Pyrenees to the Marseilles and Maures massifs to the northeast (Figure 1). The ridge seperates a subsiding basin to the northwest from the Mediterranean Sea to the southeast. If, with the use of seismic profiles, one places the section drilled by Mistral No. 1 in its regional setting, one sees that the lower part of the Pliocene is a discordant series overlying a fairly thick Miocene section. Sometimes, as is the case to the northeast and at times near the continental slope, this lowermost Pliocene directly overlies pre-Tertiary beds. There is an almost continuous seismic horizon in the Lower Pliocene which overrides areas of infill which probably correspond to the Basal Pliocene or Upper Miocene beds.

Comparison of the Pliocene facies described for the deep water areas of the western Mediterranean and those of the Mistral area shows similar planktonic fauna. Thicknesses are less important in the bathyal zones. Nonetheless, there are up to 1000 meters of Pliocene strata with large thicknesses of terrigenous material deposited in an unstable basin surrounded by continents.

In the deep Mediterranean basin the Upper Miocene unit, which was not deposited at the Mistral site, is formed by an evaporitic series with salt dome tectonics (Glangeaud, 1966; Glangeaud and Rehault, 1968; and Leenhardt, 1969). This suggests that shallow water conditions, with a restricted environment, existed at this time. Thus, these beds have foundered 2000 meters since the end of the Miocene.

Such phenomena, which are also known at the coast (*e.g.*, in the Bizerta area of Tunisia; Burollet, 1951) are fairly general in the western Mediterranean. Although Mistral No. 1 is situated on a relatively stable ridge, when compared with the neighbouring areas, it has subsided by almost 1700 meters since the Lower Pliocene and over 3400 meters since the early Miocene. This data can be evaluated in light of recent studies in the western Mediterranean, particularly those in the Tyrrhenian Sea (see, for instance, Fabbri and Selli, this volume). In the alpine zones *sensu stricto* surrounding the Mediterranean (Andalusia, North African Atlas, Sicily, Apennines and Alps) there is evidence of deposition of Lower, and sometimes Middle, Miocene sediments prior to nappe movements; Middle and Upper Miocene beds were deposited after this tectonic phase. At the Mistral well site, the Lower and Middle Miocene beds were deposited transgressively on Paleozoic rocks forming the ridge. The absence or reduction of the Upper Miocene beds in the well probably corresponds to an epirogenic change resulting from the tectonic activity in the southern and eastern areas of the western Mediterranean, although the area of Mistral seems to be outside the areas of nappe tectonics during the Miocene.

ACKNOWLEDGMENTS

This note, written by P. F. Burollet and Ph. Dufaure, is also based on the work of J. Cravatte, L. Durand, J. Ferrat, and A. Laumondais who were responsible for the geological supervision of the well operations, and of J. Ferrero who did the X-ray analyses study of the mineral constituents. R. S. M. Templeton was kind enough to translate the paper from the French.

REFERENCES

Burollet, P. F. 1951. Etude géologique des bassins mio-pliocènes du N.E. de la Tunisie. *Annales des Mines et de la Géologie, Tunis*, 7.

Fabbri, A. and R. Selli 1972. The structure and stratigraphy of the Tyrrhenian Sea. In: *The Mediterranean Sea: A Natural Sedimentation Laboratory*, ed. Stanley, D. J., Dowden, Hutchinson and Ross, Inc., Stroudsburg, Pennsylvania, 75–81.

Glangeaud, L. 1966. Les grands ensembles structuraux de la Méditerranée occidentale d'après les données de Géomède. *Comptes Rendus de l'Académie des Sciences, Paris,* 262:2405–2408.

Glangeaud, L. and J. P. Rehault 1968. Evolution ponto-plio-quaternaire du golfe de Gênes. *Comptes Rendus de l'Académie des Sciences, Paris,* 266:60–63.

Leenhardt, O. 1969. *Sondages sismiques continus en Méditerranée Occidentale*. Thèse Sciences Appliquées, Paris 1–120.

Montadert, L., J. Sancho, J. P. Fail, J. Debyser and E. Winnock, 1970. De l'âge tertiaire de la série salifère responsable des structures diapiriques en Méditerranée Occidentale (Nord-Est des Baléares). *Comptes Rendus de l'Académie des Sciences, Paris,* 271:812–815.

PART 3. The Quaternary Record

Quaternary Sedimentation in the Mediterranean Region Controlled by Tectonics, Paleoclimates and Sea Level

Rhodes W. Fairbridge

Columbia University, New York City, New York

ABSTRACT

A systematic outline for an integration of the structural, stratigraphic, topographic and climatic framework of the Mediterranean and its Quaternary history is presented as an aid to the understanding of marine sedimentation. The sediment-controlling framework is influenced by three interacting systems:

(a) *Tectonic Change:* The present cycle is dominated by "neotectonic" vertical movements, in places exceeding 2 km during the last 2 million years, thus averaging 1 mm/yr. Subordinate strikeslip faulting is active, mainly in an E-W, sinistral sense, and not dominantly dextral, as commonly claimed. Active subduction is limited to the eastern basin.

(b) *Paleoclimatic Oscillations:* The long-established concept of pluvial maxima coinciding with glacial maxima is now positively rejected. Maximum cold is followed by maximum aridity, loss of vegetation, solifluction, stream loading and terrace building. Only the transitional phases towards the beginning and end of each glacial cycle would be marked by increased terrigenous supply to the deep-sea, facilitated by drop of sea level and turbidity currents, as well as eolian activity. Paleosols of red and brown types mark interglacial warm-humid intervals, at which time the vegetational cover would inhibit large-scale erosion and supply to the deep sea. In spite of the recognition of multiple oscillations, four major glaciations still dominate the record.

(c) *Eustatic Oscillations:* Contemporary sea-level trends are disclosed by tide-gauge studies, and interpretation is favored by the almost tideless seas and low current energies. Neotectonic uplift is confirmed in orogenic areas, and downwarping persists in basins. Glacioeustatic cycles during the Quaternary are superimposed upon a tectono-eustatic trend which has led to a mean drop of at least 100 m, believed to be related to the world trend of plate tectonics, marked especially at this time by quasicratonic collapse of marginal seas.

RESUME

On a entrepris une étude systématique de la structure, de la stratigraphie, de la topographie et du climat de la Méditerranée ainsi que de son évolution au Quaternaire afin de comprendre la sédimentation marine. La sédimentation est influencée par trois systèmes de facteurs interdépendants:

(a) *Variations tectoniques:* Le cycle actuel est caractérisé par des mouvements verticaux "néotectoniques" dépassant parfois 2 km ce qui, au cours des deux derniers millions d'années, conduit à une moyenne de 1 mm par an. Ces mouvements entrainent un relief actif de failles orientées Est-Ouest et dirigées vers la gauche et non vers la droite contrairement à ce qu'il est dit ordinairement. La subduction active est limitée au Bassin oriental.

(b) *Oscillations paléoclimatiques:* On rejette la théorie, établie depuis longtemps, qui fait coïncider maxima pluviaux et maxima glaciaires. La période la plus froide est suivie par un maximum d'aridité, la disparition de la végétation, la solifluction, l'engorgement des cours d'eau et la formation de terraces. Ce n'est que pendant les périodes intermédiaires du début et de la fin de chaque cycle glaciaire qu'on pourrait noter un accroissement des apports terrigènes vers la haute mer facilités par un abaissement du niveau de la mer, des courants de turbidité et par l'action du vent. Des paléosols marrons et rouges marquent les intervalles interglaciaires tiède-humide, pendant lesquels la végétation retarderait l'érosion massive et le transport vers la haute mer. On distingue quatre périodes glaciaires principales parmi les nombreuses oscillations.

(c) *Oscillations eustatiques:* Les mesures des variations du niveau de la mer actuel sont facilitées par la faiblesse de l'amplitude des marées et des courants. Les soulèvements néotectoniques sont localisés dans les zones orogéniques et les effondrements dans les bassins. Les cycles glacio-eustatiques du Quaternaire se sont surimposés à un mouvement tectono-eustatique ce qui a provoqué un abaissement de cent mètres du niveau de la mer. Cet abaissement.

An integration of these three systems offers a genetic approach to sedimentation problems that has not been possible before now.

sans doute relié au mouvement des plaques tectoniques, est marqué en particulier par l'effondrement des mers marginales.

Une synthèse de ces trois systèmes permet une approche génétique des problèmes de la sédimentation impossible jusqu'à présent.

INTRODUCTION

Three fundamental factors (tectonic, climatic and eustatic) have clearly controlled the nature, rate and distribution of Quaternary marine sedimentation in the Mediterranean area.

The Tectonic Factor

It was recognized by Eduard Suess nearly a century ago that the Cenozoic stratigraphic history of southern Europe is marked by a succession of almost synchronous transgressions and regressions of Mediterranean origin. They are peculiar to this Sea, because identical sequences are not duplicated around the margins of the great oceans. Some modifications to the Suess model have emerged through the years (see Gignoux, 1955, for example), but the basic generalization seems to be correct.

The deep-sea drilling results of the GLOMAR CHALLENGER in 1970 (Leg 13, Scientific Staff, 1970) provided a first clue to an explanation for Suess' transgressions and regressions. The GLOMAR CHALLENGER cores showed that in the Tertiary deep-sea sequence in the Mediterranean there were several layers of evaporites—anhydrite and halite—evidently marking episodes of isolation of the entire basin from the world ocean. Subbottom reflection profiling had already disclosed widespread areas marked by diapiric structures (Glangeaud *et al.*, 1966; Watson and Johnson, 1968). These diapirs now can be interpreted as salt domes (Burk *et al.*, 1969). It has been claimed (Schmalz, 1969) that halite can crystallize in a deep-sea setting through the sinking of heavy brines, but the cores seem to suggest total isolation and dessication. The present-day evaporation/precipitation ratio in the Mediterranean involves a loss from the net water column of nearly 1 m per year. As a result, an isolated 3000 m deep basin could be reduced to a brine-covered playa in as many years.

The Tertiary evaporites therefore constitute strong arguments for a total, possibly tectonic, isolation of the Mediterranean basin from time to time. This tectonic factor constitutes the first great variable in the control of Mediterranean sedimentation.

The methods by which the tectonic framework is studied involve several disciplines. Basically there is the traditional study based on stratigraphy and structural geology, established by field work. Cartography provides regional geotectonic syntheses such as the "Geotectonic Map of Europe".

The second procedure involves neotectonic integration which is concerned with what is happening either *now* or during recent geological millenia.

A third field of geotectonic research is in the submarine area. Information here about the structural history is derived from thickness, age and distribution of sediments. This is obtained by means of subbottom reflection profiling, monitored by selected shallow (< 30 m) gravity coring or, in exceptional instances, by deep core-drilling such as that performed recently by GLOMAR CHALLENGER.

The Climatic Factor

The second fundamental parameter controlling sedimentation is the climatic one. The Mediterranean latitudes (30–45° North) correspond to the delicate transition zone between the belt of Prevailing Westerlies and the Subtropical High Pressure belt, while the alternation here between land and sea, islands, mountains, and gulfs, provides the region with extraordinarily varied local climates.

During the repeated shifts from Pleistocene interglacials to glacial phases this region suffered enormous changes in climatic regimes. Under glacial conditions there would be a dominance of the Westerlies, with stages influenced by the cold, dry high pressure cell of the Eurasiatic continent. Even relatively low altitude mountains became glaciated and short-run rivers brought their melt-water detritus into the sea. During interglacial maxima, there would have been a northward shift of the subtropical high pressure cell, leading to climatic conditions more or less analogous to the hot-humid subtropical regime of the present-day Gulf of Mexico.

Thus the climatic factor furnishes a mechanism for supplying weathered rock detritus, which generated in landscapes ranging from those of the mountain glacier to the hyper-arid desert and the subtropical swamp.

The Eustatic Factor

The third great sedimentologic variable in the Quaternary period is the eustatic factor (Fairbridge, 1961a). Three types of eustasy have become recognized —sedimentary, tectonic and glacial.

Sedimento-eustasy:

Suess suggested that the sedimentary filling of the Mediterranean basins would lead to transgressions. Although within a narrow, shallow basin, sedimentation can be a significant factor, it cannot be responsible for major, open-sea, *marine* transgressions (Fairbridge, 1961a).

Tectono-eustasy

This type relates to the change in shape of ocean basins. Polar migration leads to drowning in one quadrant, emergence in another (Jardetzky, 1962). Tectonic opening of ocean basins, as suggested by sea floor spreading, might seem to favor a fall of water-level. Actually a spreading axis is associated with very high heat-flow due to which the oceanic crust probably is higher than normal, so that crescendos of sea floor spreading most likely lead to a *rise* of sea level, while relaxation of mid-oceanic ridge heat flow could lead to a *fall* of sea level (Brookfield, 1970), perhaps in cycles of 10^7–10^8 years (Armstrong, 1969). At the sites of deep-sea trenches, acceleration of subduction (White *et al.*, 1970) can lead to deepening and favors a *fall* of sea level, whereas orogeny, axial melting and gravitational tectogenesis can lead to filling of the trenches and therefore to a eustatic *rise*. According to the concept of plate tectonics, Moores (1970) has proposed that plate splitting and dismemberment leads to a multiplication of shallow seas, and an extension of the world coastline, so that sea level rises by *positive* tectono-eustasy. Conversely, plate collisions reduce the length of coastlines and the number of shallow seas, and may well cause *negative* eustasy.

Glacio-eustasy

Recognized as an important consequence of continental glaciation for nearly a century and a half, there is still no unanimity as regards its amplitude during the various phases of the Quaternary. However, studies in the Mediterranean and Black Sea by local specialists have disclosed a statistical pattern of emerged Quaternary shorelines that provide some support for the generalization proposed by Depéret (1906, 1918–20, 1926).

GEOTECTONIC FRAMEWORK OF THE QUATERNARY MEDITERRANEAN

Plate Motions

The present Mediterranean represents a severely restricted inland sea, a lineal descendant of a much wider and more open Tethys of the Mesozoic and Tertiary. During much of this time, it was an almost equatorial seaway, as shown by paleomagnetic data.

Evidence shows that there has been an alternation of withdrawal and collision ("epeirophoresis") between the two cratonic elements, resulting in progressive growth of orogenic belts in the Alpine system (Staub, 1928; Salomon-Calvi, 1930–33; McKenzie, 1970; Vogt *et al.*, 1971).

Paleomagnetic analyses (Irving, 1967) suggest that there also have been appreciable strike-slip displacements between the two plates, which confirm in principle the earlier concept of Carey (1955, 1958) of a *Tethyan Megashear*. The torsional couple has been widely discussed under expressions such as the "Tethyan Twist" (Tanner, 1964; Krause, 1966, 1967; Fairbridge, 1961c). During Mesozoic and Tertiary times, a complex series of these E-W strike-slip displacements took place in the Tethyan region and beyond in a sinistral (left lateral) sense (Pitman and Talwani, 1972).

The climactic Europe-Africa collision was probably that at about the Eocene/Oligocene boundary ("Pyreneen phase") and it seems probable that from Miocene to early Quaternary there was a sinistral shift along Tethys of about 500 km, at a mean rate of about 2.5 cm/yr.

Mediterranean Geotectonics

The geological evidence suggests that most of the tectonic activity associated with this region involves important strike-slip components. However, depending upon the trend of the fractures, and the interaction between E-W and N-S motions, both sinistral and dextral couples are observed.

Oceanic entrances to, and channels within, the Mediterranean have been greatly modified, or even opened and closed, periodically during the late Tertiary and Quaternary, due primarily to this tectonic activity, but they have been partially modified also by eustatic oscillations. Such features include the Strait of Gibraltar, Pyrenees foothills, Sicilian Channel, Suez Isthmus, Bosporus-Dardanelles, and Aleppo Platform. It is evident from the sedimentary record that from time to time there has been general stagnation (black muds), or almost total evaporation (anhydrite and halite).

Active orogeny, associated with overthrusting (see GLOMAR CHALLENGER results), subduction and trench development, appears to be limited in the Mediterranean area to the Hellenic Arc, south of Crete. However, during the Quaternary active volcanism has been widespread in and around the periphery of the Mediterranean from the Massif Central of France, the southern Apennines and Sicily to the southern borders of the Aegean, North Africa and the Dead Sea-Jordan-Bekaa Rift belt of the Middle East.

Current seismic activity (shallow shocks) is largely restricted to the central and eastern Mediterranean, notably Italy, Sicily, Greece and Turkey. Intermediate shocks of over 100 km depth are recorded from the eastern Tyrrhenian to Cyprus. Much of the western Mediterranean floor is aseismic. One seismic zone today corresponds to the Agadir or South Atlas Fault, *i.e.,* the southern border of the Atlas Mountains through Tunis to Morocco. Another zone seems to follow the North African shore to the Strait of Gibraltar, and may connect with the Azores.

Quaternary Activity

It is appropriate to ask here what geotectonic operations one might expect to play an active role during the Quaternary. These principles may be generalized as follows:

(1.) *Epeirogenic uplift* is to be expected in all high relief areas. All the circum-Mediterranean orogenic belts continued to rise during the Quaternary, although many strictly orogenic phases had been terminated. The Hellenic Arc and Cretan Trench now appear to form the only active orogenic island-arc sector in this part of the world. Neotectonics, seismicity, geodesy, tide gauge records, gravimetry, as well as general geomorphology and stratigraphic data suggest that geologically recent motions have been largely vertical and thus more or less independent of Alpine orogenic structures (de Booy, 1969). Isostatic reactions to continued erosion of mountainous relief, matched by subsidence in the coastal plains, deltas and basins, must continue for a lengthy period (Bourcart's "flexure continentale," 1938, 1959, 1962; see discussion by Lagrula, 1964; Glangeaud, *et al.,* 1966 and Glangeaud, 1967). Glacio-isostasy in the ice-cap affected areas such as the Alps may have played an important triggering role. The hydro-isostatic (water-loading) factor also has to be considered (Brotchie and Silvester, 1969).

2. *Strike-slip motions* on many of the major fault lines ("geosutures") have continued in the same sinistral fashion, persisting through the late Tertiary into the Quaternary. These fault lines include the Alpine-Vardar-Anatolian fault system, the Agadir-South Atlas fault, the main Tethyan Geosuture through the Strait of Gibraltar and the Sicilian Channel ("Moroccan Fault" of Carey), and the NW-SE "Teisseyre Line" (bordering the Russian Shield). There were also periodical north-south adjustments resulting in many complex structures; for example, the Vardar-Lavantal Zone and much of the Aegean-Corinth areas. Although the borders of the deep Mediterranean basins thus appear to be fault controlled, most of these faults appear to be inactive at present.

3. *Platform areas and structural blocks* ("paleoblocks") represent regions marked mainly by low relief shorelines, where little differential crustal activity is to be expected except adjacent to certain fracture-lines associated with ancient block-faulting. Such rather stable sectors are found along the North African shores in Libya and Egypt, and along the Adriatic coast of Italy, Tectonic blocks adjacent to more active block faulting offshore may present high relief sectors, *e.g.,* the coasts of Corsica, Sardinia, Catalonia, Lebanon and the Greek and Turkish shores and islands of the Aegean. Even if active tectonism is not present, the high relief of such coasts calls for isostatic adjustment under erosional unloading.

Late Cenozoic Mediterranean Barriers

Structurally each of the main barriers to the Mediterranean is controlled by major geotectonic features, as follows:

1. *Gibraltar:* key E-W lineament of the globe, the geosuture known as the Tethyan Megashear (of Carey, 1958). The principal motion in Mesozoic-Cenozoic time was left-lateral. At Gibraltar this shear is crossed by the arcuate Cenozoic orogenic belt represented by the Betic folds of southern Spain and the Rif arcs of Morocco.

2. *Pyrenees:* secondary E-W lineament, probably connected to the Tethyan shear associated with initial Cretaceous opening of the Bay of Biscay.

3. *Sicilian Channel:* transverse to the Cenozoic orogenic trends that join Sicily with Tunis (Ball, 1939) and intersected by the Tethyan Megashear, this channel was probably repeatedly opened and closed in the same manner as Gibraltar, possibly at the same times (de Booy, 1969).

4. *Suez Isthmus:* this parallels the northwestern extension of the Suez Graben and the major taphrogenic rift of the Red Sea, which was repeatedly opened, closed and reactivated during the Cenozoic (Hassan and El-Dashlouty, 1970).

5. *Bosporus*: at the present time this is a narrow N-S overflow channel leading from the Black Sea via the Sea of Marmara and the Dardanelles to the Aegean Sea and eastern Mediterranean. Its maximum depth is 40 m, so that it was closed completely during the late Quaternary glacial stages (Pfannenstiel, 1944). Earlier in the Quaternary the overflow channel —the Sakaria-Izmit "Strait"—lay farther east, and may have been blocked by motion in the active E-W tectonic belt which lies just north of the right-lateral strike-slip Anatolian Fault (Pavoni, 1961, 1964, 1969; Ketin, 1969).

6. *Aleppo Platform:* prior to the mid-Tertiary the Tethyan seaway lay open to the east. Early to middle Tertiary orogeny blocked the pre-existing central Tethyan seaway to the east which passed across Anatolia to Iran and the Indo-Pacific. However, for a while there was still a connection along the southern border of the Anatolian folds, joining the Mesopotamian Basin with the Alexandretta-Latakia region of the northeastern Mediterranean. Final closure in this region occurred before the Burdigalian stage (Lower Miocene). Although marine Pliocene deposits spread briefly into Syria, it appears that even during the highest sea-level stages of the Pliocene or the Quaternary, the sea was unable to cross the Aleppo divide (around 400 m).

Block Rotation in the Mediterranean

The work of the Utrecht school and others, summarized by de Boer (1965) and by Creer (1970; see also discussion van der Voo, 1967; Nairn and Westphal, 1968; Watkins and Richardson, 1970; Hospers and van Andel, 1969) suggests a rotation of Italy since the Carboniferous of 32°, since the Permian 26°, since the Triassic 20°, and since the Eocene 5°, with respect to Meso-Europa (see van Bemmelen, 1969; Rutten, 1969). Consideration of the timing of the orogenies in Italy and Spain, of the opening of the Tyrrhenian Sea, the western Mediterranean, the Ligurian Sea and Bay of Biscay, all suggest that the main rotation of these Tethyan blocks must have occurred during Eocene/Miocene times.

The GLOMAR CHALLENGER deep-sea Mediterranean cores (Scientific Staff, 1970), revealed bedded evaporites of mid-Tertiary age. Deep-water pelagic sediments were found directly covering the evaporites, showing that they represent desiccation of a deep basin rather than a product of super-brine formation such as that proposed by Schmalz (1969). The mean sedimentation rate in the western Mediterranean since the Miocene has averaged 15 cm/1000 yrs (see Ryan *et al.*, 1970).

Further evidence for a Miocene inception of the collapse of western Mediterranean highland blocks (the "Mediterranean Revolution"), is contained in: (a) near-vertical faulted borders (Yemelyanov *et al.*, 1964), cut by rias (Pannekoek, 1969; Schülke, 1969), and by submarine canyons containing Pliocene sediments (Bourcart, 1959); (b) sediment source studies, *e.g.*, of the early Tertiary flysch in southern France (Stanley and Mutti, 1968); (c) stream capture and reversals, *e.g.*, the headwaters of the Durance in the Maures block, which formerly flowed northwestwards into a Miocene shallow sea (Cornet, 1957 and 1965); and (d) movements of nappes and olistostromes, from the oceanside towards the land in the Apennines, Sicily, North Africa and southern Spain (see summary by Pannekoek, 1969). Off SE Spain some sediments appear to have come from the south as late as the Upper Pliocene (Völk, 1966).

It may be concluded that the high values for the geomagnetic orientation differences between Meso-Europa and Neo-Europa for the late Carboniferous to Triassic do not reflect merely the individual rotation of mid-Tethyan blocks, but the *general counterclockwise rotation* of Africa and Tethys vis-à-vis Meso-Europa (Dietz and Holden, 1970).

Dating of the sea-floor spreading history of the North Atlantic (Pitman and Talwani, 1972; Vogt *et al.*, 1971) demonstrates clearly that the separation of Africa and North America began over 100 million years earlier than the separation of Northern Europe-Greenland-North America (60 m.y. ago). GLOMAR CHALLENGER drilling (on Leg 13) further showed that Iberia began to part from North America at least 130 m.y. ago.

Geometrically the only way to accommodate such a situation, assuming that the Bullard "computer fit" of the late Paleozoic is even approximately correct, is to imagine Africa moving southeastwards with respect to Laurasia, during part of the Mesozoic and early Cenozoic, while subject to a relative rotation in a counterclockwise sense. Such geometry is indicated also by the curving traces of the North Atlantic fracture zones (Pitman and Talwani, 1972).

In considering the general tectonic framework of Quaternary Mediterranean sedimentation, a final key question is this: the latest megacycle of seafloor spreading began in the early Mesozoic (Wilson, 1966) and still continues, but at several stages the direction and rate have changed—did such a subcycle coincide with the beginning of the Quaternary (de Booy, 1968, 1969)?

PALEOCLIMATIC CONSIDERATIONS IN THE MEDITERRANEAN AREA

Geochronological Developments

New geochronological developments have contributed greatly to an understanding of the theory of climate change and provide a basis for correlation with climatically related sedimentation. Milankovitch (1941) speculated on the solar radiation control of climatic cycles which celestial mechanics data suggest are in the range 25,000 to 90,000 years (see discussion by Kukla, 1970, 1972). The writer (Fairbridge, 1961b and 1963) examined the last Milankovitch cycle and found a relationship between the last predicted solar radiative (summer) warming and the melting of

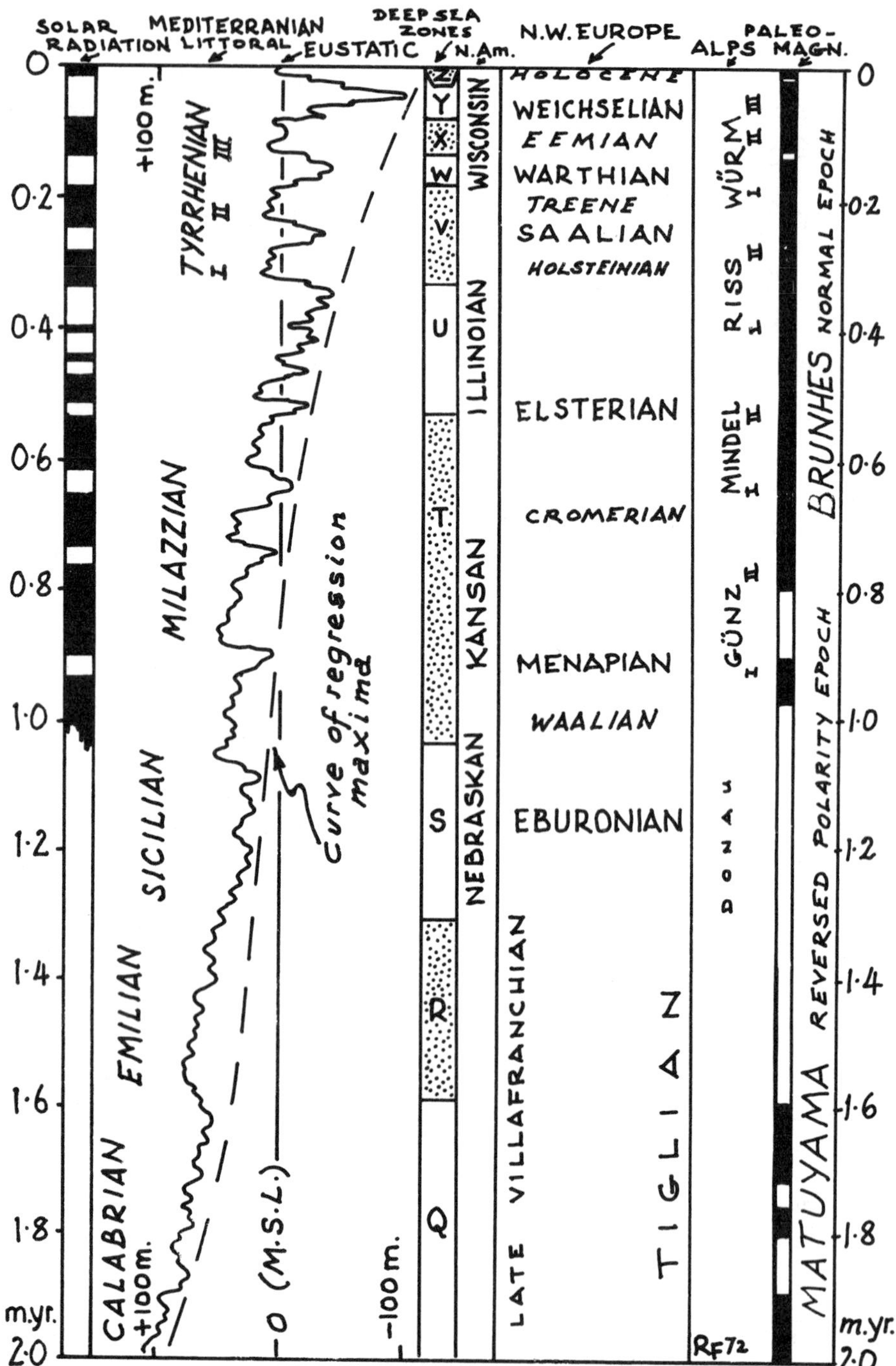

Figure 1. Tentative suggestions for the chronology of the Quaternary Period, following the indications of absolute dating, paleomagnetic reversals, deep-sea foraminiferal zonation and the Milankovitch solar radiation patterns. The eustatic curve represents an integration of the tectono-eustatic trend (progressive lowering) and the glacio-eustatic modulation that is a reciprocal of the global ice-water budget. Regression maxima become thus progressively greater and total amplitude appears to increase. Under "Solar Radiation", black indicates periods of high, and white, periods of low effective radiation received in the mid-latitudes of the northern hemisphere.

glaciers during the latest Pleistocene and early Holocene. However, there was a retardation of the rise of sea level with respect to the rise in the radiative value for 65° N by approximately 10,000 years. This figure is just about the delay that could be predicted on the basis of latent heat and associated factors related to the melting of major ice sheets.

Moreover, extension of absolute dating techniques, using uranium daughter products in aragonitic corals, by Broecker *et al.* (1968) to the last three of the Milankovitch cycles has substantiated the validity of this hypothesis (see also: Cherdyntsev *et al.*, 1965; Kind, 1968; Izett, *et al.*, 1970; Richmond, 1970).

Recently, application of the paleomagnetic technique to the Quaternary loess sequence in central Europe, notably at Červeny Kopec (Red Hill) near Brno, Moravia, by Kukla (1969; 1970) has disclosed eight complete cycles of cold loess/warm pedogenic conditions above the Matuyama-Brunhes paleomagnetic boundary (Cox, 1969) and a similar number are found in the deep-sea cores. This boundary is calculated by several different procedures to be at approximately 0.7–0.8 million years. The Milankovitch curve during the same interval also predicts eight major glacial events (see Figure 1).

As a working hypothesis, therefore, it will be assumed that solar radiation is primarily responsible for the timing of the advance and retreat of continental glaciers, although complex feedback mechanisms, positive and negative, may retard or accelerate the build-up or disintegration of the ice sheets [*cf.*, role of snow cover and resultant albedo control (Kukla, 1969, 1972)].

Anaglacial Stages

In the Mediterranean latitudes the beginning of a *glacial stage* was marked by expansion of glaciers, and accompanied by a general *lowering* of the snowline (by pleniglacial times amounting to 1000 m or more: Butzer, 1964; Messerli, 1967), a deflection of the main storm tracks of the Prevailing Westerlies to more southerly latitudes than today, and a resulting increase in precipitation over much of the Mediterranean area, including the North African littoral and inland areas of high relief. In the Hoggar, Aïr, Tibesti and other mid-Saharan mountainous areas, a typical "Mediterranean" flora (pines and scrub) developed. Mean temperatures eventually dropped by at least 6–7°C in North Africa and the Middle East (Messerli, 1967); and it seems likely from universal evidence of frost disturbed soils that mean winter temperatures at extremes dropped to 15°C below the present. Glaciers in the Sierra Nevada (Spain) extended 14 km; on Olympus (Greece) moraines are found down to 1400 m; in the Lebanon (at 33°N) they reach down to 2500 m and in the Taurus down to 2000 m above present sea level. Actual sea level dropped about 50 m during anaglacials, to a maximum regression of about 130 m at pleniglacial extremes.

Trevisan (1949) named this early cooling the "anaglacial" stage, and pointed out that it would be marked by gravel accumulations in the lower valleys, stimulated by intensive solifluction in the headwaters. In North America, a similar conclusion was reached by Schumm (1965).

During this early glacial phase there must have been long periods during which warm ocean waters would permit continued high evaporation from the mid- to high latitude sea surfaces, leading to plentiful production of moisture-bearing clouds (*cf.*, G. C. Simpson, see references *in* Fairbridge, 1961b).

As the ocean level dropped, a broad belt of unvegetated foreshore was exposed to wind action and extensive dunes developed rapidly, as can be seen around the shores of the Mediterranean in southern France, Corsica, Spain, Morocco, Algeria, Italy, Greece, Lebanon, Israel, Egypt and Libya (Butzer, 1958 and 1962; and personal observation). The eolianites alternate with paleosols and carbonate crusts, showing that rainy cycles were not infrequent (Fairbridge and Teichert, 1953). Their correlation with the anaglacial regression is particularly well seen in northwestern Corsica (Ottmann, 1958).

Pleniglacial Stages

At the maximum cold stages of glaciation, the cooling of the ocean gradually led to reduced precipitation, resulting in *cold arid climates* (Fairbridge, 1963, 1964, 1967 and 1970). The size and stability of the Siberian and Canadian High Pressure centers were greatly augmented and the northern hemisphere's global circulation reverted from a normal (interglacial) zonal flow to a blocking condition (Lamb and Woodroffe, 1970). As a result cold, dry northerly and northeasterly winds affected much of the Mediterranean. Loess deposits, laid down by these northeasterly winds bringing dust from the glacial margins and periglacial areas, extend to the western shores of France and Portugal, and cross Hungary, Yugoslavia and Bulgaria into Greece.

At critical times during the Quaternary cold, relatively arid conditions may have extended to almost the whole world, or at least there were long dry seasons and major, catastrophic droughts. In Africa (Tricart *et al.*, 1957; Fairbridge, 1964), the sand dunes of the Sahara and the Kalahari migrated vast distances, and actually crossed the Congo. Radiocarbon dates (de Ploey, 1965) show that the dunes reached the region of Kinshasa (Leopoldville) sometime between about 50,000 and 10,000 years B.P. Equatorial Guiana and Brazil were no less arid (Bakker, 1970; Bigarella *et al.*, 1965; Damuth and Fairbridge, 1970).

Palynological studies of dated lake and swamp deposits show that during the glacial stages of the last (Würm) cycle, very arid conditions existed in Spain (Menendez Amor and Florschütz, 1964), in Italy (Bonatti, 1966, 1968; Frank, 1969) and in Greece (van der Hammen *et al.*, 1967). With mean annual temperatures here 8–10°C colder (Messerli, 1967; Frenzel, 1969; Frank, 1969), the mean precipitation was extraordinarily low, perhaps averaging 200 mm. Geomorphology and fluvial sedimentology tell the same story (de Vaumas, 1964, 1969 and 1970).

In the deep-sea cores there is frequently evidence of wind-blown material. In the Swedish core 210 (SE of Spain) Eriksson (1967) found numerous microstratified layers corresponding to the middle and late Würm which he interpreted as water-laid loess.

In the middle Nile valley extensive studies have shown that there was general siltation during the last glacial stage probably because of a rise in sediment-load (due to a decrease in vegetative cover) combined with a drop in fluvial transport energy (Fairbridge, 1962), due to lowering of precipitation rates in the equatorial headwaters. Lake Victoria, according to Kendall (1969) at this time was under a low level, carbonate-rich, semi-arid regime, and was not high enough to overflow.

From the glaciated and periglacial (high solifluction) areas during the anaglacial phases and to some extent continuing seasonally during the maxima there was a great increase of stream sediment load, engendered by decreased vegetational protection, but without a corresponding rise in annual run-off. Seasonally there were increases in run-off. Vast gravel trains were carried downstream. The stream regimes were mainly of the braided, semi-arid type under which the gravels would be transported seasonally, but the general pattern led to a choking of valleys by gravel fill (Fairbridge, 1968, p. 1124). Many of these gravels never reached the ocean. The gravel terraces are particularly well seen in Corsica, where the heads can be ascribed to small valley glaciers and the distal sectors related to the contemporary sea level (Ottmann, 1958; Bourcart and Ottmann, 1957). As in the Alps there are four major cycles, divided by paleosols (representing interglacials). The most striking warm cycle is the middle one, presumed to be Mindel/Riss, which is marked by extreme weathering and "rubification" (Conchon, 1970).

Kataglacial to Early Postglacial Stages

In the climatic transition from glacial maximum to the late glacial ("kataglacial") and post-glacial warm-up, as exemplified by conditions at about the late Pleistocene/Holocene boundary, there is evidence from many parts of the world of a great increase in precipitation (see Figure 2). A pleniglacial low was reached in the Milankovitch curve (Kukla, 1972). Evidently several thousand years passed before the surface marine waters began to warm up again and increased oceanic evaporation became effective. Heavy monsoonal rains from the Indian Ocean began to affect the Ethiopian region about 13,000 B.P. or somewhat earlier. Later, Lake Victoria reached its highest level around 12,500 B.P. and began overflowing into the Nile. Sediment studies in the deep eastern Mediterranean (Herman, *et al.*, 1969) show that Nile floods into the Mediterranean were renewed at this time.

Marked changes of precipitation occur around the Holocene boundary (10,000 ± 300 B.P.) in many parts of the world (*e.g.* Lake Victoria in Africa, Lakes Bonneville/Lahontan in N. America, Lake Balaton in southeastern Europe). The Nile alluvium reveals several periods of downcutting alternating with brief stages of accumulation. Associations of mollusca and human remains have rendered it possible to work out an interesting paleoclimatic record (Fairbridge, 1962, 1968; Wendorf *et al.*, 1970) which involves not only the main glacial-interglacial climatic transition, but also the minor oscillations, long-known in northern Europe from paleobotanical data.

Interglacial Stages and Holocene Pattern

The interglacial climatic pattern in the Mediterranean region may be visualized from the post-glacial (Holocene) record which radiocarbon dating and extensive palynological analyses have now elucidated. At least a dozen different climatic oscillations can be recognized within this epoch (see "Holocene" entry, *in* Fairbridge, 1968; also 1971), but the critical climatic events in the Holocene record in Mediterranean latitudes may be simplified to three principal phases. The early Holocene is marked by a general warming trend, with increased precipitation and the widespread return of forests, even along the North African littoral. The second or mid-Holocene phase averaged 2.5°C warmer than today in northern Europe, and pollen records from southern Europe indicate that the temperature was certainly a little higher than today, but that precipitation was appreciably greater. Warm-wet conditions probably favored widespread

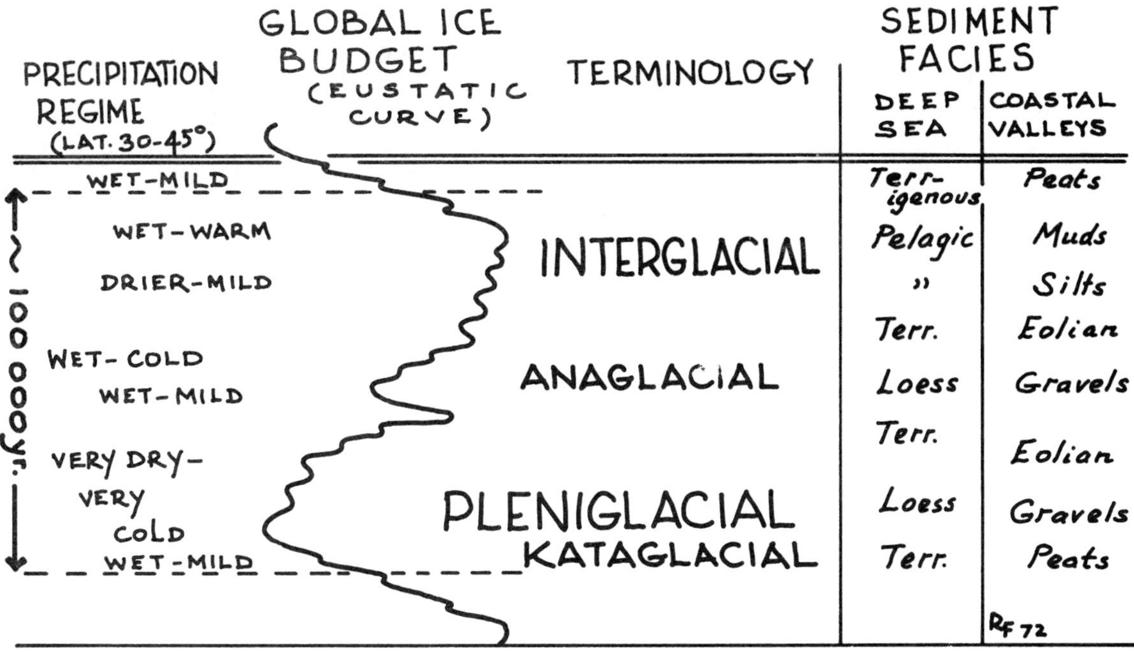

Figure 2. Diagram to illustrate a typical glacial cycle, taken from mid-point to mid-point on the eustatic curve that represents also the reciprocal of the global ice-water ratio. Note correlations between maximum cold (*Pleniglacial*) and arid climate indications (wind-blown, loess), and between transition stages with wet-mild climates and terrace dissection.

reforestation and restricted sediment loads. Primitive man (up to the Neolithic level) became widespread and in North Africa adequate monsoonal rains evidently reached right across the Sahara, forming widespread lakes there (Conrad and Conrad, 1965; Servant, *et al.*, 1969; van Zinderen Bakker, 1969).

The third major division of the Holocene began somewhere around 4500 B.P. and was marked by a world-wide cooling, but in the high pressure belts this process was reflected more by dryness and by hot (cloudless) summers. Desiccation began in North Africa, and in many parts of Spain, Italy, Greece and Anatolia. As a result of vegetative loss, and more extreme seasonality (alternating wet and dry), stream valleys began to silt up and choke (Vita-Finzi, 1969a; 1969b). For a short period, in Roman times there was a cooler and wetter period, when many of the Alpine glaciers advanced considerably, and when the Romans were able to raise wheat in North Africa. Possibly due to a climatically controlled fall of sea level, at this time the Black Sea became more toxic (Deuser, 1970). Towards the time of the fall of Rome there was another phase of desiccation which resulted in widespread loss of vegetative cover, siltation, blocking of wells, dams and irrigation systems all the way from eastern Asia to Morocco, although it has been argued that Man and his animals (goats, *etc.*) were responsible for this environmental deterioration.

EUSTATIC INFLUENCES IN THE MEDITERRANEAN

The alternation of dilatation and collision processes connected with plate tectonics in the Mediterranean and elsewhere during the Tertiary appears to have been the primary cause of major eustatic (tectono-eustatic) transgressions and regressions prior to the glacio-eustasy of the Quaternary (Fairbridge, 1961a; Jardetzky, 1954, 1962).

Stages of Low Sea Level during the Tertiary

Geomorphic studies indicate that during the late Miocene or earliest Pliocene (*ca.* 10 to 7 million years ago), many areas of the globe were affected by deep subaerial (fluvial) dissection or gradation to base-levels of the order of −200 m with respect to present mean sea level (Negris, 1912; Boillot, 1964; Galloway, 1970). Thus on Corsica which was a relatively stable area during the Quaternary, Ottmann (1958) has shown that on the west coast the nearly horizontal Pliocene marine formations were laid down within deep rias and gulfs that had been eroded since early Miocene time when the principal uplift occurred (see also Dulemba, 1970). At Aswan in the Nile Valley estuarine Plaisancian occurs on a bedrock floor at −180 m (Chumatov, 1967).

Many parts of continental shelves and submarine canyon heads were thus apparently eroded subaerially, well before the Quaternary low sea level stands, which lasted for only very brief intervals and certainly could not have persisted long enough to permit planation across a crystalline basement.

Chronology of Quaternary Eustatic Oscillations

Although some correlation problems still exist, certain limiting dates have elucidated the Quaternary eustatic history of the Mediterranean.

First, the paleomagnetic record has now furnished an approximate date for the "official" Pliocene/Pleistocene boundary at Le Castella in Calabria. By broad paleontological correlation with deep-sea sediments (Banner and Blow, 1965; Glass *et al.*, 1967), this boundary is shown to be close to the Olduvai and Gilsà paleomagnetic events, during the Matuyama Reversed Epoch and has been dated by several different procedures between 1.5 and 2.0 million years ago (Cox, 1969). Secondly, paleomagnetic studies of the Dutch Lower Pleistocene sequence (van Montfrans, 1971) permit a reasonable correlation with the European vertebrate faunal stages and thereby with the classic Alpine Günz/Mindel glacial chronology. Approximate ages for the maximal extensions of the Alpine Günz-Mindel-Riss-Würm formations may now be set approximately at 0.8, 0.5, 0.3, and 0.02 million years.

This "new" time scale indicates that the early pre-Günz Quaternary lasted a very long time, as has also long been suspected by students of the Villafranchian.

Recent studies have suggested statistical support for the generalization proposed by Depéret (1906, 1918–20; 1926), that the high level Quaternary terraces of the Mediterranean form a eustatic sequence, the highest being the oldest and the lowest being the youngest (see discussion by Morrison, 1968). While many areas of local tectonic deformation also exist, there is a fundamental nonglaciogenic eustatic control wherein a long term fall of sea-level (at least 100 m) during the Quaternary is *modulated* by many short term oscillations, corresponding to the glacial/interglacial alternation and to minor climatic cycles (bibliography: Richards and Fairbridge, 1965).

Few of the terraces in the Mediterranean region have been adequately dated. However, many have been linked paleontologically on the basis of the vertebrate faunas and human industries to the continental deposits, and thus correlated with the classical glacial and interglacial stages.

Four distinctive sets of river gravels occur in the Mediterranean region, and in certain areas these can be traced upstream to the terminal moraines of mountain and cirque glaciers. These four sets of fluvial deposits, separated by paleosols with differing degrees of weathering, presumably correspond to the four major Alpine glaciations. Locally recognized layers indicating interstadial oscillations, as well as from the early Pleistocene Donau phases, introduce complications, but they are not so widespread nor so prolonged as the major Alpine events.

These glacial river gravels were also graded to progressively lower base levels during the Pleistocene. Base level for the oldest gravel was actually higher than present MSL. This would imply that the eustatic fall after each interglacial high-stand was about 100 m, the amount corresponding to the volume of ice required to form the corresponding continental glacial sheets (Flint, 1971).

According to Bonifay (1964), Flint (1966) and Broecker *et al.* (1968), the highest interglacial sea levels probably did not reach more than 20–30 m above present MSL, the balance being explained by tectonic upwarping. However, the uniform uplift of areas in totally different structural and lithologic settings seems highly improbable while equivalent isostatic uplift is discounted because of the great differences in relief of the areas concerned. The distinctive role of solifluction must be considered; the effects of this process reach below sea level and into the warmest latitudes (de Vaumas, 1964, 1970). In many areas solifluction appears to have removed almost all traces of loose material where one might expect to find the higher and older shoreline deposits (Chamley, 1969, 1971). The solifluction debris may exceed 40 m in thickness.

It appears that during the early glacial stages the lowest sea level stands were actually higher than those of the present (or last interglacial) level.

A long-standing paradox can now be resolved. Early Pleistocene Calabrian and Sicilian faunas associated with high level deposits in the Mediterranean possess cold-water species of both shallow and deep water facies. Because of tectono-eustasy it is quite possible to find fossil indications of a cold-stage 50 m or more above present sea level. Corresponding warm-water facies of earliest interglacials may some day be found rising more than 100 m higher.

It may be noted that the Black Sea is separated from the Aegean by a relatively low barrier. The incision of the Bosporus channel was only achieved in the last glacial epoch (Pfannenstiel, 1944). The early Pleistocene Chaudian-Bakinian terraces of the Black Sea, commonly found at elevations around 100 m, are marked by a cold, relatively freshwater fauna and can be correlated with the Sicilian of the Mediterranean (Federov, 1963). In the Black Sea sequence, freshwater facies mark glacials and reflect cool air, with low evaporation, whereas more saline facies indicate warmer interglacial stages marked by higher evaporation and the influx of warm salty Mediterranean water resulting from higher sea levels. The Sicilian fauna contains *Cyprina islandica,* a boreal pelecypod today not found south of latitude 44°N. Caspian species have been identified as far south as Greece, near Corinth (Gillet, 1938). The high eustatic levels apparently led to the Caspian being connected to the Black Sea at times in the early Pleistocene. Castany and Ottmann (1957) have explained some of the "cold faunas" of Calabrian and Sicilian age as deep-water (cold) facies that have been uplifted tectonically. There is clear warping along certain well-defined belts, but elsewhere there is no substantiating evidence for tectonic uplift.

CONCLUSIONS

1. The present Mediterranean is a geotectonically restricted remnant of an immensely broader and longer seaway, the *Tethys* of Suess, a descendant of the still older *Mesogaea* of Wegener.

2. It coincides in part with a global fracture zone, the *Tethyan Megashear* of Carey, along which during the last 10^7 to 10^8 years the relative motion has been a sinistral or left-lateral strike-slip, associated with anticlockwise rotation of Iberia, Corsica-Sardinia, the Apennines and Arabia. The dextral shift deduced by many writers is contested.

3. Support is also adduced for the old concept of Suess involving repeated collision ("epeirophoresis") and withdrawals of the European and African "Forelands" (now called "plates"), along an intercontinental suture.

4. Tide gauges, geodetic relevelling and geomorphology show that the contemporary tectonic style in the Mediterranean consists largely of epeirogenic motions, usually positive in the former orogenic zones and negative in adjacent depressions (Lennon, 1966).

5. Orogenic movements are now restricted to the Hellenic Arc, which is an active subduction zone. Rather widespread volcanic activity marked almost the entire Mediterranean periphery during the Quaternary.

6. Shallow focus earthquakes characterize the great strike-slip lines, showing continued contemporary motion, mainly in a left-lateral sense (Hodgson, 1957, 1959; Hodgson and Stevens, 1958; Scheidegger, 1964; Vredenskaya and Ruprekhtova, 1961).

7. Between the Tertiary orogenic belts and beyond the tectonically active margins of major plates, there seem to be numerous minor plates or geotectonic blocks, "paleoblocks", that behave as rather stable semi-autonomous units (Hospers and van Andel, 1969).

8. Since the Eocene/Oligocene orogenic collision of the European and African plates, periodic openings and closings of the main Mediterranean seaway have have been actuated by both tectonic and eustatic processes.

9. Two general paleoclimatic trends are recognizable during the Quaternary: first, a general trend toward world cooling with progressively increasing continentality (greater temperature extremes, greater seasonality, greater aridity), and secondly a modulation in the range of 25,000–90,000 years characterized by the increasing effectiveness, frequency and severity of cold epochs. The fact that the last advance wiped out much of the record of earlier glaciations sometimes gives a (false) impression of monoglacialism (*e.g.*, Vita-Finzi, 1969c). The overall trend is attributed to world paleogeographic controls (orogenic uplift, the closing of major oceanic circulation routes, *etc.*) and the modulation to the Milankovitch astronomic control of effective solar radiation (Kukla, 1972).

10. Two distinctive, climatologically defined, parts of the Quaternary are recognized: (a) an early Quaternary period with relatively modest glacial stages in mountainous areas (Donau), typified by the Villafranchian on land (and, less distinctively, by the Calabrian and Emilian in the marine realm), which persisted up to the Matuyama/Brunhes geomagnetic boundary (0.7 to 0.8 m years B.P.); and (b) a middle to late Quaternary period characterized by marked oscillations between progressively greater continental glacial advances in mid-latitudes and extremely rapid withdrawals (interglacials).

11. Atmospheric circulation during the interglacials in the European-North African sector was marked by a dominance of zonal winds, with the principal North Atlantic storm tracks heading north of Scandinavia. During the glacials, in contrast, there was blocking action, with meridional circulation engendered by the semi-permanent high pressure system over northern Europe. During the onset of continental glaciation the westerly storm tracks passed over the Mediterranean, sweeping up into European Russia (Lamb and Woodroffe, 1970) and bringing "cold pluvial" conditions to the Mediterranean and North Africa. The extension of continental ice from Scandinavia to the foot of the Carpathians and from secondary ice centers such as the Alps and Pyrenees, gave rise to northerly and northeasterly winds which caused extreme cooling and desiccation in the Mediterranean. This was also true in North Africa generally (van Zinderen Bakker, 1969), except at high altitudes,

where the lowered snowline brought moister conditions (pine forests) and local glaciers (*e.g.*, the Atlas Mountains).

12. Following the last glacial cold peak at 17,000 years B.P., higher evaporation rates returned first to the equatorial and tropical oceans, notably in the Indian Ocean (isolated from the Arctic). Heavy rains began again in Ethiopia and East Africa to bring the Nile floods northwards in peak flows at 13,000 to 12,000 years B.P. Continental ice retreat in northern and eastern Europe had now begun to open the southern Baltic and westerly airflow over the Mediterranean brought in general pluviosity.

13. The postglacial "climatic optimum" reached its peak around 6000 B.P., when the glacioeustatic sea level regained its present state, and oceanicity became maximal. Monsoonal (summer) rains affected almost all of Africa to the Mediterranean shores, and the bordering lands received both summer and winter rains. Since that time there has been a general drop in summer solar radiation (but with minor modulations) reflected by an oscillating fall in precipitation in the region, and during the last 2000 years many parts of North Africa have become totally arid (Fairbridge, 1968, 1971). This Holocene record may serve as a "standard" for other interglacials, and the present fall in precipitation may perhaps be regarded as the initial stage of the next glacial epoch.

14. Major eustatic influences affecting the sedimentary patterns of the Mediterranean fall into two categories: *tectonoeustatic*, related to global (plate) tectonics, and *glacioeustatic*, related to withdrawal and return of water to the ocean during glaciation and deglaciation.

15. There were apparently several protracted low tectonoeustatic phases during the Tertiary (developed at a sea level of not less than −200 m), the most important being around late Miocene times (*ca.* 10 million years B.P.), during an episode of accelerated sea floor spreading and orogeny when the Mediterranean may have become isolated, as indicated by anhydrite and halite evaporites. Worldwide erosion of the more stable continental shelves took place, to a base level well below the lowest Pleistocene level.

16. There was next a late Pliocene and early Pleistocene tectonoeustatic high phase; differential uplift has modified the eustatic component, but the latter seems to be at least +100 m and possibly 200 m. In some areas solifluction has effaced almost all traces of the early Pleistocene shorelines.

17. Early Pleistocene 100 m terraces (Chaudian-Bakinian) in the Black Sea are characterized by cool, almost freshwater faunas and correlate with the Mediterranean Sicilian, also marked by cold-water faunas. These high terraces have generally been regarded as interglacial, but are here re-interpreted as low-level glacioeustatic representatives.

18. Re-study of the Pleistocene marine terraces by INQUA Subcommission on Mediterranean and Black Sea Shorelines discloses good agreement with the Depéret model of a secular decrease in the heights of the terraces. This survey rejects the alternative view that all the earlier (high) Pleistocene terraces have been tectonically uplifted.

19. Around the northern border of the Mediterranean, the older gravel trains of fluvioglacial origin are graded to base-levels *above* present mean sea level. It seems probable that only the last two (Riss, Würm) glacial stages were base-levelled to depths well below the present mean sea level.

20. Re-examination of the Quaternary chronology in the light of paleomagnetic chronology, certain absolute (K/Ar) dates, the Ericson deep-sea foraminiferal zonation, and Kukla's loess chronology, confirms the important fact that the "classic" continental glacial stages are limited to the second half of the period.

REFERENCES

Armstrong, R. L. 1969. Control of sea level relative to the continents. *Nature*, 221:1042–1043.

Bakker, J. P. 1970. Differential tectonic movements and climatic changes in the mountain area of Surinam (Guyana) during the Quaternary Period. *Acta Geographica Lodziensia*, 24:43–60.

Ball, J. 1939. *Contributions to the geography of Egypt*, Survey and Mines Department, Cairo, 308 p.

Banner, F. T. and W. H. Blow 1965. Progress in the planktonic foraminiferal biostratigraphy of the Neogene, *Nature*, 208: 1164–1166.

van Bemmelen, R. W. 1969. The Alpine loop of the Tethys zone. *Tectonophysics*, 8:107–113.

Bigarella, J. J. and G. O. de Andrade 1965. Contributions to the study of the Brazilian Quaternary. In: *International Studies on the Quaternary*, eds. Wright, H. E. and D. G. Frey, *Geological Society America Special Paper* 84:433–451.

de Boer, J. 1965. Paleomagnetic indications of megatectonic movements in the Tethys. *Journal Geophysical Research*, 70:931–945.

Boillot, G. 1964. Géologie de la Manche occidentale. Fonds rocheux, dépôts quaternaires, sédiments actuels. *Annales Institut océanographique*, France, 42 (1): 220 p.

Bonatti, E. 1966. North Mediterranean climate during the last Würm glaciation. *Nature*, 209:984–986.

Bonatti, E. 1968. Late-Pleistocene and Postglacial stratigraphy of a of a sediment core from the lagoon of Venice (Italy). *Mémoire Biogéographie Adriatique* (Venezia), 7 (supplement): 1–18.

Bonifay, E. 1964. Pliocène et Pléistocène méditerranéens: vue d'ensemble et essai de corrélations avec la chronologie glaciaire. *Annales Paléontologie Vertébrés*, 50:197–226.

de Booy, T. 1968. Mobility of the Earth's crust: a comparison between the present and the past. *Tectonophysics*, 6:177–206.

de Booy, T. 1969. Repeated disappearance of continental crust during the geological development of the western Mediterranean area. *Verhandelingen Koninklijk Nederlands Geologisch Mijnbouwkundie Genootschap*, 26:79–103.

Bourcart, J. 1938. La marge continentale. *Bulletin Société Géologique France,* ser. 5, 8:393–474.

Bourcart, J. 1959. Morphologie du précontinent des Pyrénées à la Sardaigne. *Colloque International, C.N.R.S.,* Paris, 83:33–52.

Bourcart, J. 1962. La Méditerranée et la révolution du Pliocène. *Société Géologique de France,* Livre P. Fallot, 1:103–116.

Bourcart, J. and F. Ottman 1957. Recherches de géologie marine dans la région du Cap Corse. *Revue Géographie Physique et Géologie Dynamique,* série 2, 1:65–78.

Broecker, W. S., D. L. Thurber, T. L. Ku, R. K. Matthews and K. J. Mesolella 1968. Milankovitch hypothesis supported by precise dating of coral reefs and deep sea sediments. *Science,* 159:297–300.

Brookfield, M. E. 1970, Eustatic changes of sea-level and orogeny in the Jurassic. *Tectonophysics,* 9:347–363.

Brotchie, J. F. and R. Silvester 1969. On crustal flexure. *Journal Geophysical Research,* 74:5240–5252.

Burk, C. A., M. Ewing, J. L. Worzel, A. O. Beall, Jr., W. A. Berggren, D. Bukry, A. G. Fischer, and E. A. Pessagno, Jr. 1969. Deep-sea drilling into the Challenger Knoll, central Gulf of Mexico. *American Association of Petroleum Geologists Bulletin,* 53: 1338–1347.

Butzer, K. W. 1958. Quaternary stratigraphy and climate in the Near East. *Bonner Geographische Abhandlungen,* 24:1–157.

Butzer, K. 1962. Coastal geomorphology of Majorca. *Annals Association American Geographers,* 52:191–212.

Butzer, K. 1964. *Environment and Archeology.* Aldine Publications Co., Chicago, 524 p.

Carey, S. W. 1955. The orocline concept in geotectonics. *Papers and Proceedings of the Royal Society of Tasmania,* 89: 255–288.

Carey, S. W. 1958. A tectonic approach to continental drift. In: *Continental Drift* (Symposium), ed. Carey, S. W., University of Tasmania, Hobart, 177–355.

Castany, G. and F. Ottmann 1957. Le Quaternaire marin de la Méditerranée occidentale. *Revue Géographic Physique et Géologie Dynamique* serie 2, 1:46–55.

Chamley, H. 1969. Témoins d'un niveau marin quaternaire à la côte + 4 m sur l'ensemble du littoral des Maures. *Compte Rendus de l'Académie des Sciences, Paris,* 269: 1478–1481.

Chamley, H. 1971. *Recherches sur la sédimentation argileuse en Méditerranée.* Thèses, Université d'Aix-Marseille, 401 p.

Cherdyntsev, V. V., I. V. Kazachevskiy and Ye. A. Kuzmina 1965. Dating of Pleistocene carbonate formations by the thorium and uranium isotopes. *Geochemistry International,* 2: 794–801.

Chumatov, I. S. 1967. Le Pliocène et le Pléistocène de la vallée du Nil en Nubie et en Haute Egypte. *Transactions Geological Institute, Academy Science, USSR,* Moscow, 170:115p.

Conchon, O. 1970. Précisions sur la chronologie des formations fluviatiles de Corse orientale. *Compte Rendus de l'Académie des Sciences, Paris,* 270:283–286.

Conrad, G. and J. Conrad 1965. Précisions stratigraphiques sur les dépôts holocènes du Sahara occidental grâce à la géochronologie absolue. *Compte Rendu Sommaire de la Société Géologique de France,* 7:234–236.

Cornet, C. 1957. Etude tectonique et morphologique de la région de Méounes et de la Roquebrussane (Provence calcaire). *Revue Géographie Physique et Géologie Dynamique* (2), 1:233–242.

Cornet, C. 1965. Evolution tectonique et morphologique de la Provence depuis l'Oligocène. *Mémoir de la Société Géologique de France,* Nouvelle Série, 44 (103):252p.

Cox, A. 1969. Geomagnetic reversals. *Science,* 163:237–245.

Creer, K. M. 1970. A review of palaeomagnetism. *Earth-Science Reviews,* 6:369–466.

Damuth, J. E. and R. W. Fairbridge, 1970. Equatorial Atlantic deep-sea arkosic sands and ice-age aridity in tropical South America, *Geological Society of America Bulletin,* 81:189–206.

Depéret, C. 1906. Les anciennes lignes de rivage de la côte française de la Méditerranée. *Bulletin de la Société Géologique de France,* ser. 4 (4):207–230.

Depéret, C. 1918–20. Essai de coordination chronologique générale des temps Quaternaires. *Compte Rendus de l'Académie des Sciences,* Paris, 166:884–9; 167:418–22; 167:979–984; 168: 868–873; 170:159–163; 171:212.

Depéret, C. 1926. Essai de classification générale des temps quaternaires. *Compte Rendu Congrès Géologique International* (13th, Belgium, 1922), 3:1409–1426.

Deuser, W. G. 1970. Carbon-13 in Black Sea waters and implications for the origin of hydrogen sulfide. *Science,* 168 (3939):1575–1577.

Dietz, R. S. and J. C. Holden 1970. Reconstruction of Pangaea: breakup and dispersion of continents, Permian to Present. *Journal of Geophysical Research,* 75:4939–4956.

Dulemba, J. L. 1970. Quelques remarques sur l'origine des canyons sous-marins situés au large des côtes ouest de la Corse. *Geologische Rundschau,* 59:601–604.

Eriksson, K. G. 1967. Some deep-sea sediments in the western Mediterranean Sea. In: *Progress in Oceanography,* ed. Sears, M. Pergamon Press Oxford, 4: 267–280.

Fairbridge, R. W. 1961a. Eustatic changes in sea-level. In: *Physics and Chemistry of the Earth.* eds. Ahrens, L.H., F. Press, K. Rankama and S. K. Runcorn, Pergamon Press, London, 4:99–185.

Fairbridge, R. W. 1961b. Convergence of evidence on climatic change and ice ages. *Annals New York Academy of Sciences,* 95, (1):542–579.

Fairbridge, R. W. 1961c. The Melanesian Border Plateau, a zone of crustal shearing in the S.W. Pacific. *Publication Bureau Centrale Seismologie Internationale,* ser. A, 22:137–149.

Fairbridge, R. W. 1962. New radiocarbon dates of Nile sediments. *Nature,* 196:108–110.

Fairbridge, R. W. 1963. Mean sea level related to solar radiation during the last 20,000 years. In: *Changes of Climate* (Proceedings Rome Symposium, 1960), Paris, UNESCO:229–242.

Fairbridge, R. W. 1964. African ice-age aridity. In: *Problems in Palaeoclimatology,* ed. Nairn, A. E. M., Interscience Publications, J. Wiley & Son, New York, 356–363.

Fairbridge, R. W. (ed.) 1967. *Encyclopedia of Atmospheric Sciences and Astrogeology.* Reinhold, New York, 1200 p.

Fairbridge, R. W. (ed.) 1968. *Encyclopedia of Geomorphology.* Reinhold, New York, 1295 p.

Fairbridge, R. W. 1970. World paleoclimatology of the Quaternary. *Revue Géographie Physique et Géologie Dynamique,* 2, (12):97–104.

Fairbridge, R. W. 1971. Holocene. In: *Encyclopedia Britannica,* Chicago, 11.

Fairbridge, R. W. and C. Teichert 1953. Soil horizons and marine bands in the coastal limestones of western Australia. *Journal Proceedings Royal Society New South Wales,* 86:68–87.

Federov, P. V. 1963. Stratigraphy of Quaternary sediments on the coast of the Crimea and Caucasus and some problems connected with the geological history of the Black Sea. *Trudy Geologiski Institut, Akademiya Nauk, S.S.S.R.,* 88(available in English translation, R.T.S. U.K. 2572).

Flint, R. F. 1966. Comparison of interglacial marine stratigraphy in Virginia, Alaska, and Mediterranean area. *American Journal of Science,* 264: 673–684.

Flint, R. F. 1971. *Glacial and Quaternary Geology.* John Wiley and Sons, New York, 830 p.

Frank, A. H. E. 1969. Pollen stratigraphy of the Lake of Vico (central Italy) *Palaeogeography, Palaeoclimatology, Palaeoecology,* 6:67–85.

Frenzel, B. 1969. The Pleistocene vegetation of northern Europe. *Science,* 163:637–649.

Galloway, R. W. 1970. Coastal and shelf geomorphology and late Cenozoic sea levels. *Journal of Geology,* 78:603–610.

Gignoux, M. 1955. *Stratigraphic Geology*. W.H. Freeman and Co., San Francisco, 682 p.

Gillet, S. 1938. Sur la présence d'éléments caspiques dans la faune quaternaire de Corinthe. *Compte Rendu Sommaire de la Société Géologique de France*, 163–164.

Glangeaud, L. 1967. Epirogenèses ponto-plio-quaternaires de la marge continentale franco-italienne du Rhône à Gênes. *Bulletin Société Géologique de France*, 9:426–449.

Glangeaud, L., J. Alinat, J. Polveche, A. Guillaume and O. Leenhardt 1966. Grandes structures de la mer ligure, leur évolution et leur relations avec les chaînes continentales. *Bulletin Societe Geologique de France*, ser. 7, 8:921–937.

Glangeaud, L. and P. Olive 1970. Structures mégamétriques de la Méditerranée; évolution de la mésogée de Gibraltar à l'Italie. *Compte Rendus de l'Académie Sciences de France*, série D, 271:1161–1166.

Glass, B., D. B. Ericson, B. C. Heezen, N. D. Opdyke and J. A. Glass 1967. Geomagnetic reversal and Pleistocene chronology. *Nature*, 216:437–442.

van der Hammen, T. T. A. Wijmstra and W. H. van der Molen 1967. Palynological study of a very thick peat section in Greece, and the Würm-Glacial vegetation in the Mediterranean region. *Geologie en Mijnbouw*: 44, 37–39.

Hassan, F. and S. El-Dashlouty 1970. Miocene evaporites of Gulf of Suez region and their significance. *American Association of Petroleum Geologists Bulletin*, 54:1686–1696.

Herman, Y., Y. Thommeret and C. Vergnaud-Grazzini 1969. Micropaleontology, paleotemperatures and radiocarbon dates of Quaternary Mediterranean deep-sea cores. *Paris, INQUA (VIII Congrès)*, Résumés:174(b).

Hodgson, J. H. 1957. Nature of faulting in large earthquakes. *Geological Society of America Bulletin*, 68:611–644.

Hodgson, J. H. (ed.) 1959. The mechanics of faulting, with special reference to the fault-plane work—a symposium. *Ottawa, Dominion Observatory Publications*, 20 (2): 251–418.

Hodgson, J. H. and A. Stevens 1958. Direction of faulting in some of the larger earthquakes of 1955–56. *Ottawa, Dominion Observatory Publications* 19 (8): 34 p.

Hospers, J. and S. I. van Andel 1969. Palaeomagnetism and tectonics, a review. *Earth Science Reviews*, 5:5–44.

Irving, E. 1967. Palaeomagnetic evidence for shear along the Tethys. In: *Systematics Association* (Aspects of Tethyan Biogeography), eds. Adams, C. G. and D. V. Ager 7:59–76.

Izett, G. A., R. E. Wilcox, H. A. Power, and others, 1970. The Bishop ash bed, a Pleistocene marker bed in the western United States. *Quaternary Research*; 1:121–132.

Jardetzky, W. S. 1954. Principal characteristics of the formation of the earth's crust. *Science*, 119:361–365.

Jardetzky,W. S. 1962. A periodic pole shift and deformation of the earth's crust. *Journal of Geophysical Research*, 67:4461–4472.

Kendall, R. L. 1969. An ecological history of the Lake Victoria basin. *Ecological Monographs*, 39:121–176.

Ketin, I. 1969. Uber die nordanatolische Horizontalverschiebung. *Bulletin Mineral Resources Exploration Institute Turkey*, Foreign Edition, 72:1–28.

Kind, N. V. 1968. Correlation of Late Pleistocene glaciations in the northern hemisphere by radiological data. *23rd International Geological Congress*, Prague, 115–116.

Krause, D. C. 1966. Equatorial shear zone. In: The World Rift System, *Geological Survey of Canada*, 66-14:400–443.

Krause, D. C. 1967. Bathymetry and geologic structure of the north-western Tasman Sea-Coral Sea-south Solomon Sea area of the south-western Pacific Ocean. *New Zealand Oceanographical Institute Memoir*, 41:48 p.

Kukla, J. 1969. The cause of the Holocene climate change. *Geologie en Mijnbouw*, 48:307–334.

Kukla, J. 1970. Correlations between loesses and deep-sea sediments. *Geologiska Forenings Forhandlingar*, Stockholm, 92: 148–180.

Kukla, G. J. 1972. Isolation and glacials. *Boreas*, 1:63–96.

Lagrula, J. 1964. Relations entre l'isostasie, l'érosion, la sédimentation et l'épirogenèse. *Comptes Rendus de l'Académie des Sciences, Paris*, 258:279.

Lamb, H. H. and A. Woodroffe 1970. Atmospheric circulation during the last Ice Age. *Quaternary Research* 1:29–58.

Lennon, G. W. 1966. An investigation of secular variations of sea level in European waters. *Annales Academia Fennicae*, A III, 90:225–236.

McKenzie, D. P. 1970. Plate tectonics of the Mediterranean region. *Nature*, 226:239–243.

Menendez Amor, J. and F. Florschütz 1964. Results of the preliminary palynological investigation of samples from a 50-meter boring in southern Spain. *Boletin Real Societa España Historia Natura*, Seccion Geologica 3, 62:251–255.

Messerli, B. 1967. Die eiszeitliche und die gegenwärtige Vergletscherung in Mittelmeerraum. *Geographia Helvetiae*, 3:105–228.

Milankovitch, M. 1941. Canon of insolation and the ice-age problem. *Royal Serbian Academy, Belgrade*, special publication 132, Math-Nat. 33, (Translated from German: *Israel Program Science Translations*, Jerusalem, 1969).

van Montfrans, H. M. 1971. *Paleomagnetic dating in the North Sea basin*. Thesis, University Amsterdam, Amsterdam 113 p.

Moores, E. M. 1970. Patterns of continental fragmentation and reassembly: some implications. *Geological Society of America*, Abstracts 2:629.

Morrison, R. B. 1968. Means of time-stratigraphic division and long-distance correlation of Quaternary successions: In: *Means of Correlation of Quaternary Successions*, eds. Morrison, R. B. and W. E. Wright, Jr. University Utah Press (VII INQUA Proceedings 8): 1–113.

Nairn, A. E. M. and M. Westphal 1968. Possible implications of the palaeomagnetic study of late Palaeozoic igneous rocks of northwestern Corsica. *Palaeogeography, Palaeoclimatology, Palaeoecology*, 5:179–204.

Negris, P. 1912. *La régression quaternaire*. P. D. Sakellarios, Athènes, 98 p.

Ottmann, F. 1958. Les formations Pliocènes et Quaternaires sur le littoral Corse. *Mémoires de la Société Géologique de France*, nouvelle série 37 (4), 1–176.

Pannekoek, A. J. 1966. The ria problem. *Tijdschrift Koninklijk Nederlands Aardrijkskundig Genootschap*, 83:289–297.

Pannekoek, A. J. 1969. Uplift and subsidence in and around the western Mediterranean since the Oligocene: a review. *Verhandelingen Koninklijk Nederlands Geologisch Mijnbouwkundig Genootschap*, 26:53–77.

Pavoni, N. 1961. Die nordanatolische Horizontalverschiebung. *Geologische Rundschau*, 51:122–139.

Pavoni, N. 1964. Aktive Horizontalverschiebungszonen der Erdkruste. *Bulletin Verem Schweizerische Petroleum Geologen und Ingenieuren*, 31: 54–78.

Pavoni, N. 1969. Zonen lateraler horizontaler Verschiebung in der Erdkruste und daraus ableitbare Aussagen zur globalen Tektonik. *Geologische Rundschau*, 59:56–77.

Pfannenstiel, M. 1944. Die diluvialen Entwicklungsstadien und die Urgeschichte von Dardanellen, Marmarameer und Bosporus. *Geologische Rundschau*, 34:341–434.

Pitman, W. C. III and M. Talwani 1972. Sea-floor spreading in the North Atlantic. *Geological Society of America Bulletin*, 83 (619–643).

de Ploey, J. 1965. Position géomorphologique, génèse et chronologie de certains dépôts superficiels au Congo occidental. *Quaternaria*, 7:131–154.

Richards, H. G. and R. W. Fairbridge 1965. Annotated bibliography of Quaternary shorelines (1945–1964). *Philadelphia, Academy Natural Sciences,* Special Publication 6:280 p. (also Supplement for 1965–1969, published 1970).

Richmond, G. M. 1970. Comparison of the Quaternary stratigraphy of the Alps and Rocky Mountains. *Quaternary Research,* 1:3–28.

Rutten, M. G. 1969. *The Geology of Western Europe,* Elsevier, Amsterdam, 520 p.

Ryan, W. B., D. J. Stanley, J. B. Hersey, D. A. Fahlquist and T. D. Allan 1970. The tectonics and geology of the Mediterranean Sea. In: *The Sea,* ed. Maxwell, A. E., John Wiley-Interscience, New York 4 (2): 387–492.

Salomon-Calvi, W. 1930–33. Epirophorese. (3 Parts). *Sitzungsberichte Heidelberg Akademie Wissenschaften,* Math-Nat. Kl.

Scheidegger, A. E. 1964. The tectonic stress and tectonic motion direction in Europe and western Asia as calculated from earthfault plane solutions. *Bulletin Seismological Society America,* 54:1519–1528.

Schmalz, R. F. 1969. Deep-water evaporite deposition: a genetic model. *American Association of Petroleum Geologists Bulletin,* 53:798–823.

Schülke, H. 1969. Bestimmungsversuch des Ria-Begriffes durch das Kriterium der Fluvialität. *Erdkunde,* 23:264–280.

Schumm, S. A. 1965. Quaternary paleohydrology. In: *The Quaternary of the United States,* eds. Wright, H. E. and D. G. Frey, Princeton University Press, Princeton, 783–794.

Schwarzbach, M. 1963. *Climates of the past.* Van Nostrand Co. London, 328 p.

Scientific Staff 1970. Deep Sea Drilling Project: Leg 13. *Geotimes,* 15:12–15.

Servant, M., P. Ergenzinger and Y. Coppens 1969. Datations absolues sur un delta lacustre quaternaire au sud du Tibesti (Angamma). *Compte Rendu Sommaire de la Société Géologique de France,* 8:313–314.

Stanley, D. J. and E. Mutti 1968. Sedimentological evidence for an emerged land mass in the Ligurian Sea during the Paleogene. *Nature,* 218:32–36.

Staub, R. 1928. *Der Bewegungsmechanismus der Erde.* Gebrüder Borntraeger, Berlin, 270 p.

Tanner, W. F. 1964. Unified basis for tectonic theory. *Tectonophysics,* 2:135–158.

Trevisan, L. 1949. Genèse de terrasses fluviatiles en relation avec les cycles climatiques. *Compte Rendu Congrès International Géographie,* Lisbon, 2:511–528.

Tricart, J., P. Michel and J. Vogt 1957. Oscillations climatiques quaternaires en Afrique occidentale. *Congreso Internationale INQUA* (V, Madrid), Résumés: 187–188.

de Vaumas, E. 1964. Phénomènes cryogéniques et systèmes morphogénétiques en Méditerranée orientale (Chypre, Galilée). *Revue Géographie Physique et Géologie Dynamique,* (2), 6:291–311.

de Vaumas, E. 1969. Formes de relief, consolidation et dissolution des grès littoraux du Quaternaire—Méditerranée Orientale (Syrie-Liban-Palestine-Chypre). *Recherches Méditerranéennes,* Études et Travaux, 7:151–187.

de Vaumas, E. 1970. Phénomènes cryogéniques de la côte libanaise. *Revue Géographie Physique et Géologie Dynamique* (2), 12:265–292.

Vita-Finzi, C. 1969a. *The Mediterranean Valleys.* Cambridge University Press, Cambridge, 131 p.

Vita-Finzi, C. 1969b. Late Quaternary continental deposits of central and western Turkey. *Man,* 4:605–619.

Vita-Finzi, C. 1969c. Mediterranean monoglacialism? *Nature,* 224:173.

Vogt, P. R., R. H. Higgs, and G. L. Johnson 1971. Hypotheses on the origin of the Mediterranean Basin: magnetic data. *Journal of Geophysical Research,* 76:3207–3228.

Völk, H. R. 1966. Geologische Gründe für die Existenz sialischen Krustenmaterials im Mittelmeer östlich von Vera (SE-Spanien) zur Zeit des jüngsten Pliozäns. *Verhandelingen Koninklijke Nederlandse Akademie Wetenschappen,* Amsterdam, B, 69: 446–451.

van der Voo, R. 1967. The rotation of Spain: paleomagnetic evidence from the Spanish Meseta. *Palaeogeography, Palaeoclimatology, Palaeoecology,* 34:393–416.

Vredenskaya, A. V. and L. Ruprekhtova 1961. Characteristic features of stress distribution of the foci of earthquakes at the bend of the Carpathian arc. *Izvestia Akademiya Nauk, Geofysika:* 953–965 (translated from Russian, as *Bulletin of the Academy of Sciences, USSR,* Geophysics series, 629–636).

Watkins, N. D. and A. Richardson 1970. Rotation of the Iberian Peninsula. *Science,* 167 (3915):209.

Watson, J. A. and G. L. Johnson 1968. Mediterranean diapiric structures. *American Association of Petroleum Geologists Bulletin,* 52:2247–2249.

Wendorf, F., R. Said and R. Schild 1970. Egyptian Prehistory: some new concepts. *Science,* 169:1161–1171.

White, D. A., D. H. Roeder, T. H. Nelson and J. C. Crowell 1970. Subduction. *Geological Society of America Bulletin,* 81:3431–3432.

Wilson, J. T. 1966. Did the Atlantic close and then reopen? *Nature,* 211:676–681.

Yemel'yanov, Ye. M., O. V. Mikhailov, V. N. Moskalenko, and K. M. Shimkus 1964. Main features of the tectonic structure of the floor of the Mediterranean Sea. *International Geological Congress,* 22, New Delhi, 16:97–113.

van Zinderen Bakker, E.M. (ed.) 1969. *Palaeoecology of Africa, IV* (1966–1968). Balkema, A. A., Capetown, 274 p.

Données Micropaléontologiques et Paléoclimatiques d'après des Sédiments Profonds de Méditerranée

Laure Blanc-Vernet

Centre Universitaire de Luminy, Marseille

RESUME

L'étude comparative de huit carottes de Méditerranée occidentale et orientale a permis de préciser la valeur des diverses espèces de foraminifères pélagiques comme indicateurs climatiques. Il est possible de décrire les caractères faunistiques des différents stades du Würm et du Post-glaciaire dans l'ensemble de la Méditerranée.

Les séquences climatiques mises en évidence dans les deux bassins montrent une bonne corrélation. Elles sont en accord avec l'évolution du climat établie en milieu continental dans le Sud de la France d'une part et en Europe centrale d'autre part. Le caractère plus chaud du Bassin Oriental, net dans l'actuel et dans les interstades, tend à s'estomper lors des périodes froides où la microfaune devient beaucoup plus homogène. Cependant, en Méditerranée occidentale, les stades glaciaires sont plus importants. En Méditerranée orientale, au contraire, ces épisodes froids, quoique temporairement rigoureux, *paraissent* plus brefs.

Quelques interstades majeurs ont déterminé un réchauffement important des eaux dans l'ensemble de la Méditerranée. On peut les attribuer aux périodes suivantes: interstade Würm I/II (Néotyrrhénien), interstade Würm II/III (Laufen), interstade inter-Würm III, dit de "la Salpétrière" en Provence et de "Stillfried B" en Europe, et enfin, dans le Post-glaciaire, les deux "optima climatiques" de l'Atlantique et du Sub-Atlantique.

ABSTRACT

A comparative study of planktonic foraminifera in eight cores collected in the Mediterranean Sea enables us to determine the value of different species as climatic indicators, and to establish the faunal characteristics of the various stages during Würm and postglacial time throughout the Mediterranean region.

The climatic sequences based on faunal analyses are similar in both the eastern and western Mediterranean, and are also similar to climatic sequences recognized on land in the south of France and in central Europe. The warm aspect of the microfauna assemblage in the eastern Mediterranean is especially conspicuous during interstadials and at present; this is less conspicuous during cooler intervals when assemblages are more homogeneous. In the western Mediterranean the glacial intervals were more important; in the eastern Mediterranean, cool periods, although at times quite marked, were of shorter duration.

Several major warm intervals resulted in a warming of water masses throughout the Mediterranean. These intervals are correlated with the following interstadials: Würm I/II (Neotyrrhenian), Würm II/III (Laufen), inter-Würm III (called "Salpétrière" in the Provence region and "Stillfried B" in central Europe), and two postglacial "climatic optima" (Atlantic and Sub-Atlantic).

INTRODUCTION

L'interprétation climatique des microfaunes pélagiques est une technique d'étude des carottages qui a donné de bons résultats en Méditerranée comme dans les autres mers. Ayant eu l'occasion d'examiner un certain nombre de carottes, il m'a paru intéressant de confronter les principaux résultats obtenus à partir d'échantillons pris dans des régions différentes pour tenter d'en extraire quelques données très générales. Ces dernières concernent la signification écologique des différentes espèces, les corrélations climatiques et chronologiques et la stratigraphie du Quaternaire marin profond en Méditerranée depuis le début de la dernière glaciation.

Ces remarques sont fondées essentiellement sur l'examen de huit carottages (Figure 1, Tableau 1). mais il a été tenu compte, bien entendu, des renseignements fournis éventuellement par d'autres sédiments, ainsi que des données apportées par les différents auteurs qui ont abordé ces problèmes. Citons principalement les travaux de Bartolini et Vergnaud-Grazzini (1969), Barusseau *et al.* (1966), Bottema et van Stratten (1966), Cita et d'Onofrio (1967), Eriksson (1965 et 1967), Vergnaud-Grazzini (1968), et Vergnaud-Grazzini et Herman Rosenberg (1969).

La description des quatre carottes marquées * a déjà ètè publiée (Blanc-Vernet, 1969; Blanc-Vernet, *et al.,* 1969; Blanc-Vernet et Chamley, 1971), seule leur courbe climatique estimée a été reproduite ici. Les quatre autres, inédites, sont analysées dans ce travail.

METHODE D'ETUDE DES ASSEMBLAGES DE FORAMINIFERES

Les variations de fréquence des foraminifères ont été comparées de deux manières: (1) *par établissement de graphiques* indiquant les pourcentages des différentes espèces aux différents niveaux, et (2) *par la méthode des affinités de Imbrie et Purdy* (1962). Cette méthode permet de chiffrer le degré d'analogie qui existe entre le comportement de deux espèces ou entre la composition de deux échantillons. Chaque échantillon est assimilé à un vecteur défini dans un système à *n* coordonnées (correspondant aux propriétés considérées) et la mesure du degré de similitude entre les échantillons se ramène à une mesure d'angle. Cette méthode peut être appliquée à l'étude des microfaunes pélagiques du Quaternaire.

Sans redécrire ici le principe de la méthode, rappelons en les formules: On considère une suite de constituants (1, 2, . . .i, j, . .n) présents dans une suite d'échantillons (1, 2, . . p, q, . . N). Si les espèces i et j présentent respectivement dans le niveau p les fréquences Xip et Xjp, et dans le niveau q les fréquences Xiq et Xjq, le coefficient d'affinité entre les deux espèces i et j sera:

$$\cos\theta = \frac{\sum_{p=1}^{p=N} Xip \times Xjp}{\sqrt{\sum_{p=1}^{p=N} Xip^2 \times \sum_{p=1}^{p=N} Xjp^2}}$$

tandis que le coefficient de similitude entre les niveaux p et q sera, pour l'espèce i:

$$\cos\theta = \frac{\sum_{i=1}^{i=n} Xpi \times Xqi}{\sqrt{\sum_{i=1}^{i=n} Xip^2 \times \sum_{i=1}^{i=n} Xiq^2}}$$

Dans les deux cas, cos θ variera de 0 (répulsion) à 1 (affinité maximum).

REFERENCES CHRONOLOGIQUES ET CLIMATIQUES

Les huit carottes considérées dans ce travail ont été estimées couvrir 60.000 à 70.000 ans. C'est donc l'évolution du climat pendant cette période qui est envisagée et reliée, lorsque cela est possible, à des phénomènes décrits en milieu continental: importances des phases glaciaires, degré d'humidité, séismes et phénomènes volcaniques.

Je me suis référée principalement, pour la Méditerranée occidentale, aux travaux de Bonifay (1962), Lumley-Woodyear (1965 a et b), Escalon de Fonton (1968 et 1969). Ce dernier notamment a, dans une série de publications (1968, 1969)), proposé une séquence climatique pour la période du Würm II à l'actuel. Fondée sur l'étude des sédiments des grottes du Midi méditerranéen, repérées chronolo-

Tableau 1. Localisation des carottages (voir Figure 1).

Carotte	Latitude	Longitude	Profondeur (en m)	Région
C_3*	43°14′05″N	07°17′05″E	2000	Golfe de Gênes
JC_3*	42°12′30″N	05°48′E	2400	Sud Provence
1-4-69	41°00N	05°40′E	2620	Sud Provence
2-4-49	42°07′5″N	05°43′5″E	2511	Sud Provence
1 Mo 67*	41°52′N	05°52′E	2460	Sud Provence
2 Mo 67	35°41′N	21°40′E	3710	Sud Péloponnèse
3 Mo 67	34°25′5″N	24°50′E	1950	Sud Crète
45 Mo 67*	35°53′3″N	22°24′E	4420	Sud Péloponnèse

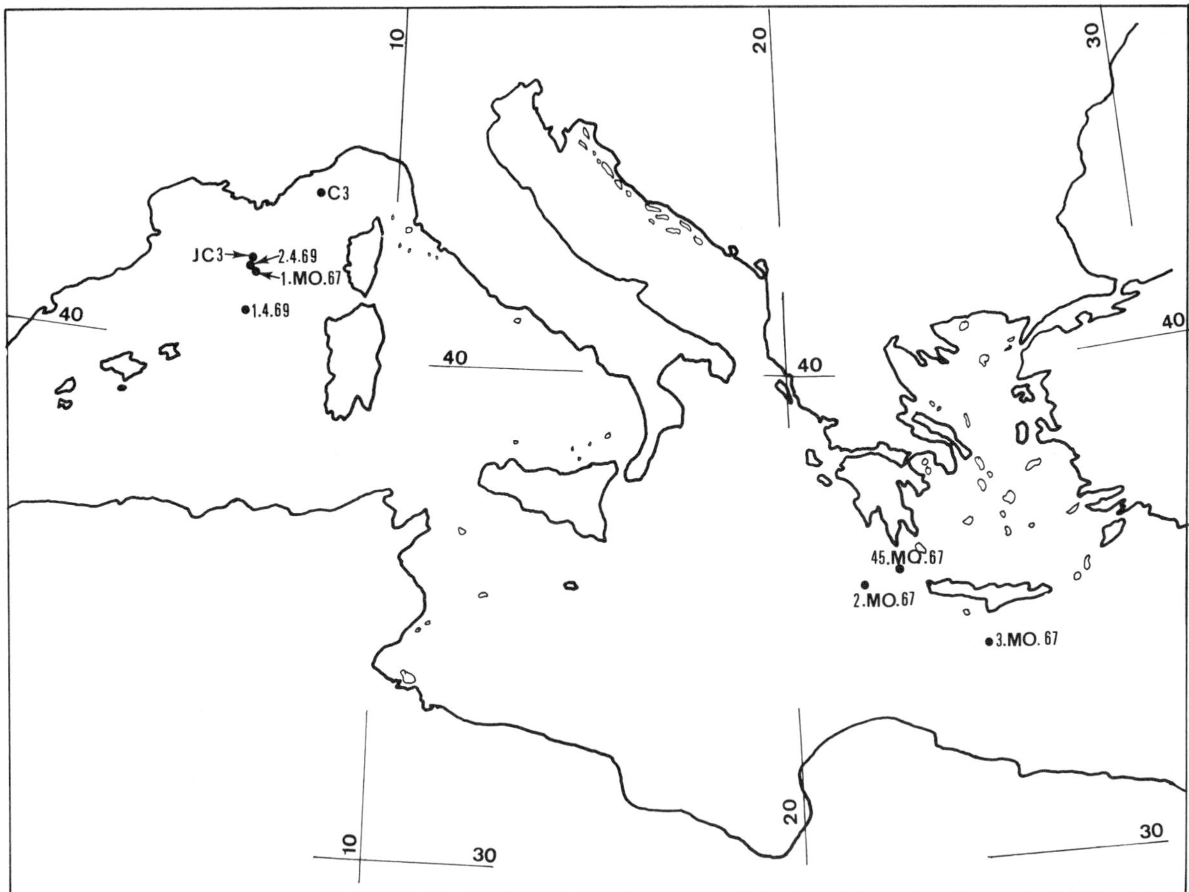

Figure 1. Localisation des carottages en Méditerranée.

giquement par rapport aux industries préhistoriques et grâce à des datations absolues par la méthode du C^{14}, cette séquence rend compte essentiellement des variations de température et d'humidité.

Pour la Méditerranée orientale, nous ne disposons malheureusement pas de renseignements aussi précis sur l'évolution du climat continental. Cependant l'interprétation chronologique des échantillons est facilitée dans cette région par l'existence de niveaux repères: cendres volcaniques issues du volcan Santorin et boues sapropéliques qui ont été étudiées et datées (Mellis, 1954; Olausson, 1961; Ninkovich et Heezen, 1966). Remarquons enfin que les interprétations avancées dans ce travail concordent de façon satisfaisantes avec les résultats de l'analyse des argiles (Blanc-Vernet et Chamley 1971, et travaux en cours). On note toutefois un décalage assez systématique entre les indications fournies par les foraminifères liés à la température des masses d'eau, et les renseignements apportés par les minéraux argileux qui suivent sans doute de plus près l'évolution climatique du continent dont ils sont issus.

INTERPRETATION CLIMATIQUE ET CHRONOLOGIQUE D'APRES LA MICROFAUNE DES CAROTTES

Les Figures 2 and 3 résument les variations de la microfaune au long des quatre carottes. Lors de l'analyse, les variations de toutes les espèces ont été considérées séparément; pour que les graphiques demeurent lisibles on a été toutefois amené à grouper certaines d'entre elles dont le comportement était très voisin, voire même identique. C'est le cas, en particulier pour *Globigerinita glutinata, Globigerina quinqueloba*, *Globorotalia scitula*, parmi les formes tolérantes au froid, et de *Globigerinella aequilateralis* et *Orbulina universa*, parmi les espèces *chaudes*.

Les variations de fréquence de *Globigerina pachyderma*, sous sa forme sénestre, ont été indiquées pour les deux premières carottes (Figure 2), Par contre, en Méditerranée orientale, cette forme est trop peu abondante dans la plupart des niveaux pour que ses fluctuations puissent être considérées comme significatives.

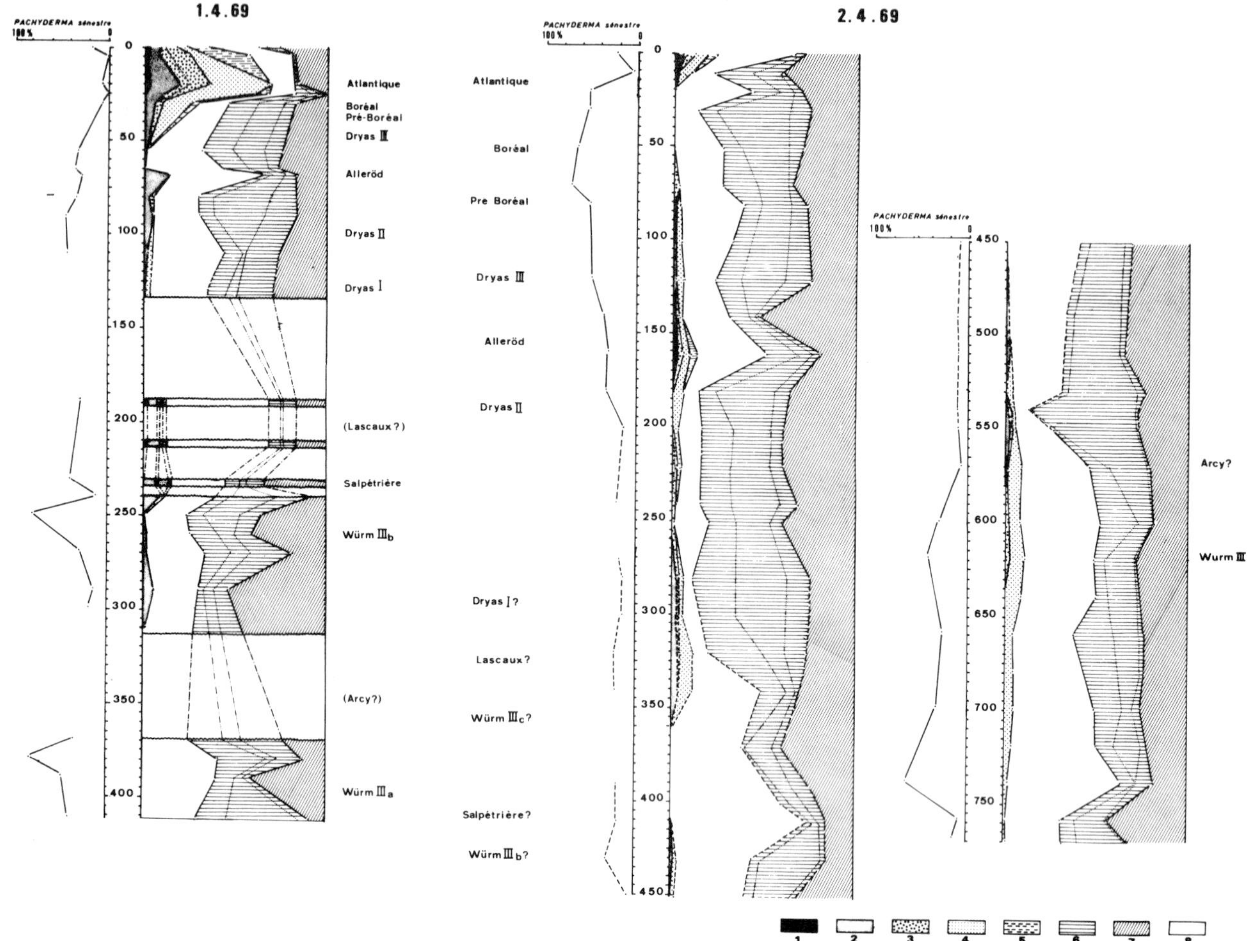

Figure 2. Successions climatiques et interprétation chronologique des carottes 1–4–69 et 2–4–69. L'echelle verticale est indiquée en cm. 1, % *Globigerinoides trilobus/sacculifer*; 2, % *G. ruber*; 3, % *Globigerinella aequilateralis* + *Orbulina universa*; 4, % *Globorotalia inflata*; 5, % *G. truncatulinoides*; 6, % *Globigerinita glutinata* + *Globigerina quinqueloba* + *Globorotalia scitula*; 7, *Globigerina pachyderma*; et 8, *autres espèces*.

La carotte 1–5–69 (Figure 2) mesure 4,30 m et comporte un certain nombre de passées sableuses (turbidites) qui interrompent la sédimentation pélagique régulière. Ces passées contiennent parfois quelques foraminifères benthiques déplacés, mais ils sont toujours peu abondants. L'apport minéral masque complètement le dépôt d'origine planctonique. De tels sédiments se rencontrent en particulier de 135 à 240 cm et de 310 à 370 cm. Lorsque les apports turbides ont été plus ou moins intermittents, on observe une alternance de vases pélagiques et de sables déplacés. C'est le cas entre 188 et 240 cm. On constate alors que la microfaune des différents lits de vase contient exactement les mêmes assemblages. Ceci indique probablement une mise en place rapide des turbidites au cours d'une période suffisamment brève pour que la microfaune n'ait présenté aucune variation. Les différents niveaux pélagiques ont été raccordés les uns aux autres et pris seuls en considération pour l'interprétation climatique et chronologique.

La carotte 2–4–69 (Figure 2), pour sa part, ne présente pas d'interruptions aussi tranchées; cependant un apport sableux plus ou moins important se rencontre tout au long. De ce fait, le taux de sédimentation est nettement plus élevé. Bien que longue de 7,82 m, il semble que cette carotte ne permette pas de remonter beaucoup plus loin dans le temps que la précédente. D'autre part, les apports en provenance des étages littoral ou bathyal supérieur entraînent des mélanges de faune importants. On remarquera les variations parfois anarchiques de la courbe de *G. pachyderma* sénestre et l'augmentation locale des "formes indéterminées" liée à la présence de nombreux individus de petite taille. Les portions correspondantes du diagramme sont tracées en pointillé et l'interprétation donnée avec réserves.

Les carottes 2 Mo 69 et 3 Mo 69 mesurent respectivement 6,60 m et 4,47 m (Figure 3). La première, prélevée au Sud du Péloponnèse, dans le secteur tourmenté des fosses de Matapan, présente un certain nombre de passées sableuses. L'autre, qui provient

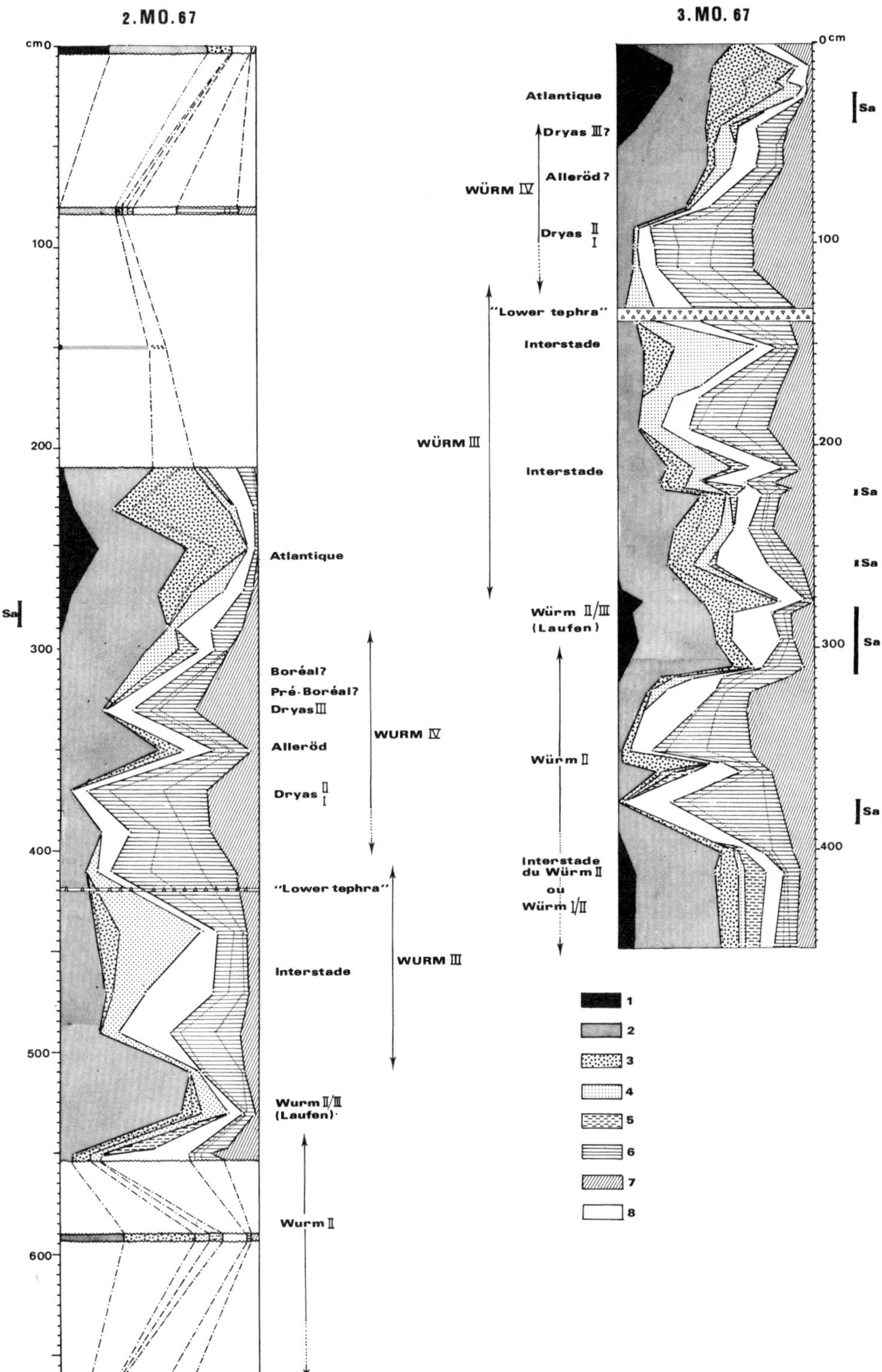

Figure 3. Successions climatiques et interprétation chronologique des carottes 2 Mo 67 et 3 Mo 67 (1–8, comme pour la Figure 2).

du Sud de la Crète, témoigne au contraire d'une sédimentation beaucoup plus régulière.

Les verres volcaniques caractéristiques de la *lower tephra* (datés d'au moins 25.000 ans) se rencontre dans la 3 Mo entre 135 et 137 cm et dans la 2 Mo à 420 cm. Une passée de vases sapropéliques présente respectivement à 26–40 cm et 75–88 cm dans ces deux carottes correspond au dernier épisode de stagnation qui se situerait entre 7.000 et 9.000 ans d'après Olson et Broecker (1961). Le maximum de la faune chaude (optimum climatique post-glaciaire) est situé légèrement au-dessus.

Dans la carotte 24 Mo, provenant de la même campagne et étudiée par Pastouret (1970), des datations absolues effectuées par la méthode du C_{14} ont donné, pour cette passée récente, un âge de −7900 au ± 170 ans BP (datation effectuée par le Centre des Faibles Radioactivités de Gyf-sur-Yvette).

D'autres passées de vases sapropéliques sont visibles dans la carotte 3 Mo à 223–225 cm, 282–313 cm et 376–387 cm. Elles n'ont pas d'équivalent dans la 2 Mo.

Un essai d'interprétation chronologique des différents stades climatiques a été indiqué pour ces quatre carottes. Il tient compte de toutes les données exposées plus haut et sera discuté par comparaison avec l'ensemble des carottages.

VALEUR DES DIFFERENTES ESPECES COMME INDICATEURS CLIMATIQUES EN MEDITERRANEE

Vaste bassin en relation avec l'Atlantique par le Détroit de Gibraltar, la Méditerranée est peuplée d'une microfaune pélagique en provenance de ce dernier. De cet ensemble faunistique sont absentes cependant les formes du groupe de *Globorotalia menardii*.

En Atlantique on met en évidence une zonation plus ou moins régulière: faunes froides de hautes latitudes, faunes tempérées de moyennes latitudes, faunes chaudes de basses latitudes. Au cours du Quaternaire les limites de ces zones variant lors des réchauffements et des refroidissements successifs, les divers assemblages se retrouveront superposés dans les carottes.

En Méditerranée, au contraire, l'ensemble pélagique apparaît beaucoup plus constant dans sa composition. Les provinces faunistiques actuelles diffèrent surtout par les proportions des différentes espèces: côtes algériennes où les variations de l'apport océanique sont trés sensibles, Golfe du Lion et de Gênes aux eaux fraîches, Bassin Oriental dans son ensemble nettement plus chaud. Au cours du Quaternaire, les variations climatiques ont superposé leur effet à celui des conditions locales. D'où l'intérêt de préciser au maximum les conditions de vie de chaque espèce pour pouvoir rendre compte des différents types de climat représentés dans l'ensemble de la Méditerranée aux diverses époques considérées.

Les Formes Tolérantes au Froid

L'étude graphique et statistique du groupe *froid* classique (*G. pachyderma, G. quinqueloba, Globigerinita glutinata, Globorotalia scitula*) montre qu'il ne s'agit pas là d'un ensemble homogène. En effet, si la présence de ces quatre espèces caractérise les périodes glaciaires, on constate que les pourcentages des trois dernières ne varient pas toujours, au sein de ces périodes, dans le même sens que *G. pachyderma*.

Ce phénomène s'observe essentiellement en Méditerranée orientale. L'analyse statistique des fréquences montre effectivement que si *G. glutinata, G. quinqueloba, G. scitula* ont entre elles un coefficient d'affinité élevé (cos θ compris entre 0,7 et 0,8 dans les différents niveaux), leur affinité vis à vis de *G. pachyderma*, espèce froide incontestable, est beaucoup plus faible: cos θ est en général voisin de 0,5, parfois inférieur.

On est amené à admettre que l'abondance des trois premières espèces caractérise des niveaux relativement frais mais qu'elle tend à diminuer sensiblement dans les épisodes de froid maximum. Ces derniers seront essentiellement caractérisés par les hautes fréquences de *G. pachyderma*, et par l'augmentation des formes sénestres de cette espèce.

Ceci est confirmé par les observations faites en Méditerranée orientale. Les maxima de froid paraissent toujours moins importants dans cette région. *G. pachyderma* est moins abondante, la forme sénestre moins répandue, et les quatre espèces précitées varient parallèlement traduisant un climat moins rigoureux. La valeur de cos θ, mesurée entre *G. pachyderma, G. glutinata* et *G. scitula* est le plus *G. scitula* est le plus souvent comprise entre 0,8 et 0,9. L'affinité de ces trois formes avec *G. quinqueloba* est, voisine de 0,7.

Remarquons que les termes de *froid* et de *chaud* sont employés pour donner une idée générale du climat mais recouvrent, en fait, un ensemble de conditions écologiques qu'il est difficile de définir avec précision, étant donné le caractère incomplet de nos connaissances sur l'écologie des foraminifères actuels de la Méditerranée.

Il est certain qu'après la température de l'eau, la salanité doit être le facteur le plus important. *G. pachyderma*, et notamment sa variété sénestre qui vit dans des eaux polaires, s'accomode aussi bien des basses températures que des faibles salinités. Quant à *Globigerinita glutinata,* elle se rencontre en Atlantique jusque dans la zone tropicale, avec le groupe de *Globigerinoides ruber*; cependant très euryhaline, elle est la seule de ce groupe à tolérer des salinités très basses (jusqu'à 32‰, Jones, 1966). Probablement indépendante de la température, son augmentation relative coïncide dans doute avec les fontes glaciaires ou avec les fortes pluviosités qui entrainent des apports d'eaux douces. Remarquons que cette espèce en Méditerranée occidentale est fréquente dans le Dryas III (froid modéré, forte humidité), tandis qu'elle est beaucoup moins représentée dans les Dryas I et II, plus froids mais surtout plus secs.

Les Formes Tolérantes au Chaud

Le groupe des espèces *chaudes* montre un phénomène inverse. On constate ici l'homogénéité du groupe en Méditerranée occidentale et une diversification au contraire en Méditerranée orientale où divers degrés pourront être observés dans le réchauffement.

En Mediterranée occidentale, les espèces chaudes sont essentiellement *Globigerinoides ruber, Globorotalia inflata, Globigerinella aequilateralis* et *Orbulina universa.* Cos θ, calculé entre ces formes prises deux à deux, est toujours supérieur à 0,8. Entre les deux dernières il atteint même 0,9. L'ensemble réagit donc de la même manière aux fluctuations du climat. La présence localement de *Globigerinoides trilobus/sacculifer*, d'ailleurs peu abondante, marque les maxima de réchauffement.

En Méditerranée orientale où l'assemblage chaud est beaucoup mieux représenté, on peut mettre en évidence une modification de ce dernier au fur et à mesure de l'adoucissement du climat. *G. ruber* se montre la plus tolérante et persiste en général tout au long des carottes. Son affinité paraît identique vis-à-vis des formes du groupe chaud et vis-à-vis des formes du groupe froid (cos $\theta = 0{,}7$ dans tous les cas). *G. inflata*, espèce tempérée, manque dans les maxima de réchauffement. Elle prend par contre un développement considérable à certains niveaux qui correspondent, sans doute, aux conditions écologiques les meilleures pour cette espèce: début du réchauffement "atlantique", interstadiaire du Würm III, par exemple. *G. aequilateralis* et *O. universa* ont toujours une grande similitude de comportement. Elles augmentent lorsque le réchauffement s'accentue, suivies par *G. trilobus/sacculifer*, puis par *"Sphaeroidinella" dehiscens* dans quelques niveaux exceptionnels (carotte 45 Mo, Blanc-Vernet et Chamley, 1971).

Le cas des orbulines doit être examiné à part. Il est en effet probable que cette forme représente un stade—probablement reproductif—susceptible de se présenter au cours du développement de plusieurs espèces. Cette distinction cependant n'intervient pas ici; les critères d'affinités montrent que le comportement de la forme "orbuline" est voisin, dans tous les cas, de celui de *Globigerinella aequilateralis.* Pour cette raison *Orbulina universa* a été considérée comme un constituant de la microfaune, sans tenir compte de son caractère polyspécifique possible.

Les variations de *Globorotalia truncatulinoides* sont également intéressantes. Cette espèce apparaît étroitement limitée à certains niveaux—en général tempérés. On remarque, en Méditerranée occidentale, *une grande majorité d'individus sénestres*, pendant la tranche de temps considérée (80 à 100 des individus). En Méditerranée orientale, la forme sénestre se rencontre surtout dans le Post-glaciaire (50 à 95% des individus), tandis que les apparitions plus anciennes montrent une majorité d'individus dextres (80 à 100%).

En ce qui concerne l'interprétation de cette dernière espèce, une grande prudence doit être observée en raison de son cycle complexe de développement qui s'effectue à divers niveaux de la tranche d'eau. L'incidence des courants, cause d'erreur toujours à craindre dans les interprétations des microfaunes pélagiques, apparaît encore plus grande dans ces conditions (Bé et Ericson, 1963).

L'essentiel de ces données, à l'exception des variations de *G. truncatulinoides*, trop épisodique, a été résumé dans la Figure 4. Il convient d'insister toutefois, sur le caractère très schématique de ce tableau, tant à propos des termes employés pour définir le climat, qu'en ce qui concerne les limites faunitisques et climatiques.

ESSAI DE CORRELATIONS: LES SEQUENCES CLIMATIQUES ET LA CHRONOLOGIE DU WURM ET DU POST-GLACIAIRE EN MEDITERRANEE

En tenant compte de tous les critères, chaque carotte peut être résumée par une courbe climatique estimée sur laquelle l'amplitude des passées *chaudes* ou *froides* a été établie en tenant compte des divers types de microfaune décrits précédemment.

Ce procédé a déjà été employé pour l'interprétation de la carotte 1Mo (Blanc-Vernet *et al.*, 1969). L'échelle stratigraphique est celle qui a été adoptée dans cette publication et qui tient compte, ainsi que je l'ai indiqué plus haut, de plusieurs travaux concernant la stratigraphie, la chronologie et la climato-

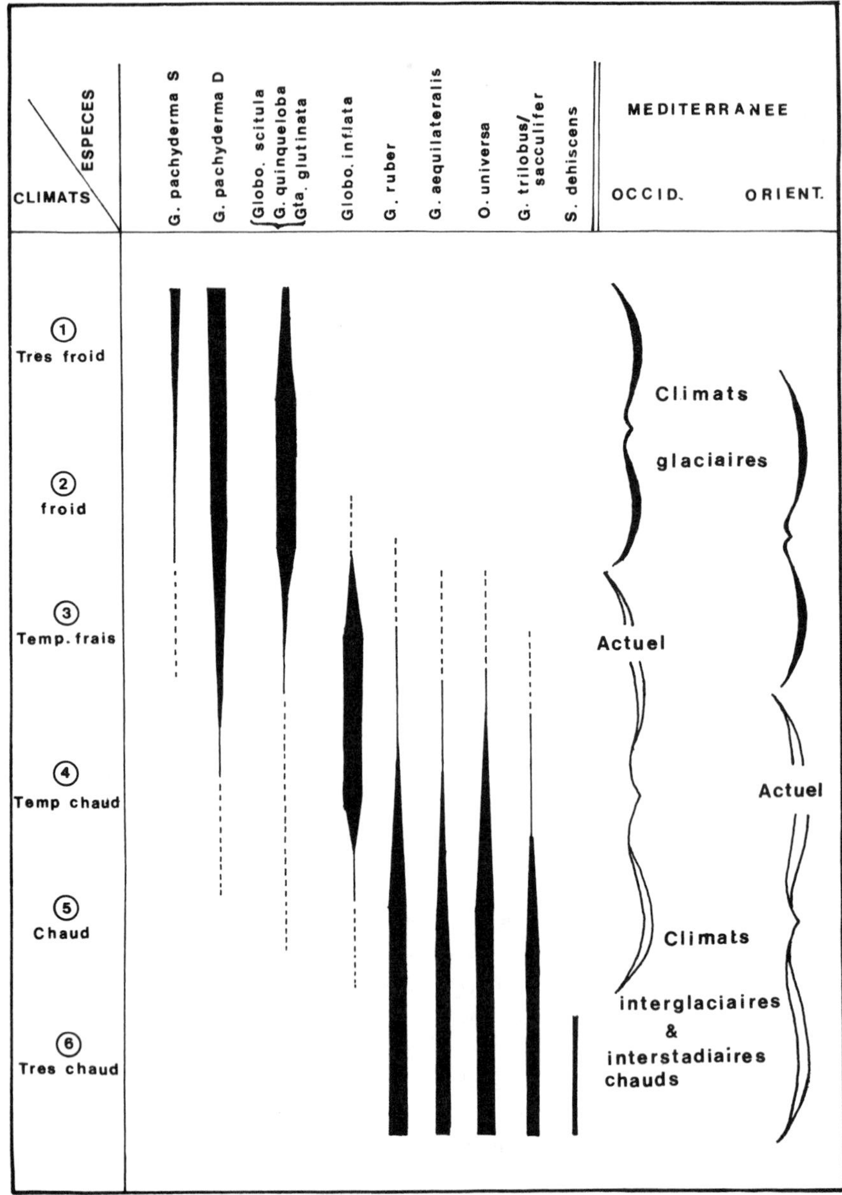

Figure 4. Tableau montrant la correspondance entre les différents types de climats et la microfaune en Méditerranée.

logie en milieu continental et marin dans la région du Sud de la France. La comparaison avec les renseignements que nous possédons sur la stratigraphie et la climatologie en milieu continental en Europe centrale sont donnés dans le texte.

La Figure 5 groupe les courbes correspondant aux huit carottes considérées. Les principales corrélations proposées sont indiquées sur la planche par des flèches. Ces graphiques permettent de comparer les différents stades plus commodément que les diagrammes d'espèces; c'est ainsi que l'interprétation de la carotte C_3 qui avait été antérieurement réalisée à partir des variations globales de la microfaune. (Blanc-Vernet, 1969), a pu être corrigée à la lumière de ces nouvelles données. La position des stades "Atlantique" et "Alleröd" a été précisée.

Les Divers Types de Climat

La Méditerranée contenant actuellement, dans son ensemble, une microfaune tempérée, nous pouvons définir les peuplements pélagiques du Bassin Occi-

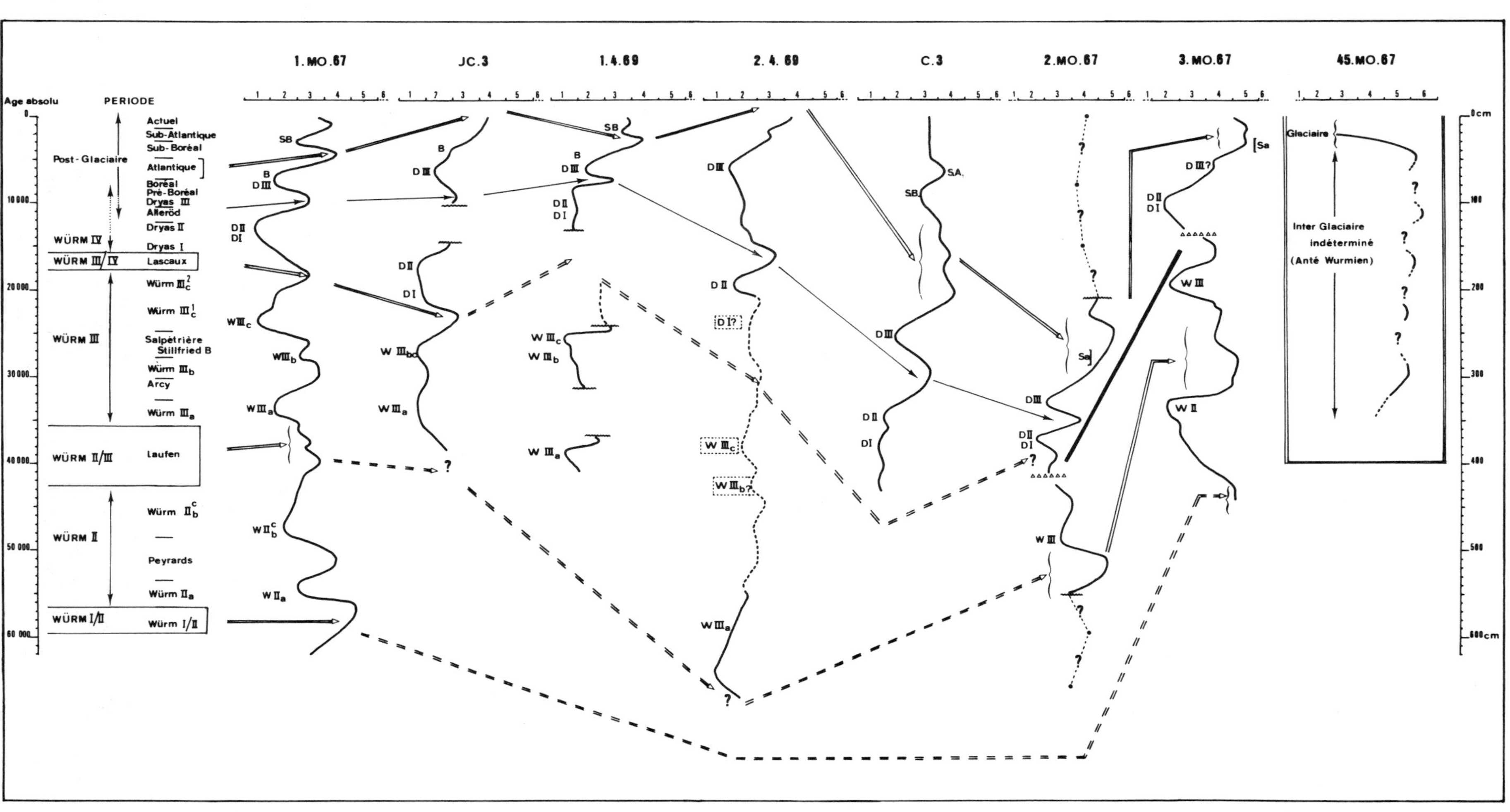

Figure 5. Courbe climatique estimée de huit carottes en Méditerranée.

dental comme *tempérés frais* et ceux du Bassin Oriental comme *tempérés chauds*. Ceci concorde avec les observations de Todd (1958) et de Parker (1958) et correspond également aux caractères de la microfaune benthique (Blanc-Vernet, 1969).

Ce décalage thermique persiste dans les stades glaciaires mais il tend toutefois à s'estomper au cours des périodes les plus froides. Il est par contre très net lors des réchauffements qui déterminent en Méditerranée orientale des microfaunes tout à fait particulières. Le stade le plus chaud a été reconnu dans la carotte 45 Mo. Cette carotte, prélevée à proximité de la 2 Mo dans le secteur faillé du Sud du Cap Matapan, paraît avoir recoupé des sédiments beaucoup plus anciens—tyrrhéniens ou anté-tyrrhéniens. Elle est représentée sur la Figure 5 en raison de son intérêt climatique mais, en l'absence de repères chronologiques certains, elle n'a pas été corrélée avec les autres échantillons.

L'Evolution Générale du Climat

Etablies à partir des données précédentes, les courbes de la Figure 5 nous montrent les successions climatiques dans l'ensemble du bassin méditerranéen au cours du Quaternaire récent.

Dans le Bassin Occidental, la séquence climatique würmienne présente les caractères suivants. Les différents stades glaciaires sont nettement définis par leur microfaune; le Würm II, relativement tempéré comporte des oscillations thermiques assez progressives. Le Würm III est franchement plus froid et les variations climatiques sont brusques lors des réchauffements interstadiaires. Au Würm IV, le froid est vif à la base; les stades du Dryas I et II ne peuvent être distingués, l'interstade dit de Bölling n'ayant probablement pas été assez important (en température et/ou en durée) pour que la microfaune marine accuse une variation. Le climat devient ensuite plus modéré au Dryas III et surtout au Boréal. La proportion des formes sénestres de *G. pachyderma* demeure faible à ce niveau.

Les interstades qui interrompent les périodes glaciaires sont en général assez courts et souvent moins chauds que le climat actuel. Le degré de similitude (cos θ) mesuré entre ces niveaux est toujours élevé, quelle qu'en soit l'amplitude ce qui montre que le caractère de la microfaune est demeuré très constant et que le réchauffement n'a pas entrainé de modifications qualitatives de l'assemblage.

En Méditerranée orientale, au contraire, les passées chaudes, mieux développées, "découpent" nettement les courbes en isolant les stades glaciaires: Würm IV, Würm III, Würm II, eux mêmes nets mais réduits (*cf.*, courbe 3 Mo).

Les Niveaux Repères

L'interprétation des carottes se fait généralement de haut en bas en partant des sédiments de surface, pris pour référence. Pour cette raison, les différents stades sont ici décrits et corrélés en partant des plus récents.

Le Post-Glaciaire: Lorsque la série est complète, le Sub-Atlantique et surtout l'Atlantique correspondent dans toute la Méditerranée à deux optima climatiques nets. Ces périodes, datées respectivement sur le continent de —2.000 à 2.500 et —5.000 à 7.500 ans BP, fournissent un bon repère pour l'interprétation des niveaux post-glaciaires. De nombreuses courbes montrent cependant une lacune ou une faible sédimentation dans les niveaux récents et la microfaune atlantique se rencontre en surface ou peu en dessous.

En Méditerranée occidentale, l'Atlantique est souligné par le développement de *Globorotalia inflata* et *G. truncatulinoides.*
En Méditerranée orientale, ces deux formes diminuent lors du maximum au profit des espèces plus chaudes. *G. inflata* réapparaît d'ailleurs de nouveau lors du rafraîchissement sub-boréal (Pastouret 1970).

Ces critères très constants font de cette époque un repère faunistique particulièrement intéressant dans l'ensemble de la Méditerranée. Dans la région Est, en outre, le réchauffement maximum de la microfaune suit de peu le dernier sapropel qui est daté de — 9.000 à — 7.000 ans environ.

Le Wurm IV: En Méditerranée occidentale, l'oscillation d'Alleröd (chaude mais surtout humide en milieu continental) constitue un jalon intéressant, bien enregistré en général par la microfaune. Son amplitude demeure cependant plus faible que celle de l'Atlantique. Par contre l'interstade Würm III/IV (Lascaux-Laugerie) est souvent masqué par des apports sableux. Nous reviendrons spécialement sur l'origine possible de ce phénomène.

Des fluctuations analogues apparaissent au sein du dernier épisode würmien dans les carottes de Méditerranée orientale. Nous manquons cependant de documents pour les rattacher avec certitude aux mêmes stades.

Le Wurm III: Deux interstades interrompent, en Méditerranée occidentale, la période glaciaire attribuée au Würm III. On peut les rapprocher probablement des interstades de la Salpétrière et d'Arcy. Ce dernier surtout semble correspondre à une variation climatique importante. Le réchauffement paraît plus accentué et la portion correspondante du diagramme est plus étalée. Ceci provient soit d'une plus grande durée, soit d'un taux de sédimentation plus élevé (lié peut-être à un adoucissement du climat et à une humidité plus grande: crues, fontes glaciaires . . .). Dans certaines carottes, un apport turbide important masque à ce niveau le stock pélagique.

En Méditerranée orientale, il existe pour cette période un repère chronologique, la *lower tephra* qui daterait, d'après les auteurs, d'environ — 25.000 ans. Les verres volcaniques caractéristiques de cette éruption sont nettement visibles dans les carottages; ils se rencontrent dans une période froide, équivalent chronologique du Würm IIIc de Méditerranée occidentale.

A quelques dizaines de centimètres sous la *lower tephra* on observe sur les carottes 2 Mo et 3 Mo un sédiment dont la microfaune, caractérisée par l'abondance de *G. inflata*, témoigne d'un climat adouci. Ce matériel occupe la place chronologique de l'interstade "Salpétrière" du Sud de la France. Ce réchauffement, intervenu environ entre 25.000 et 30.000 ans B.P. paraît donc général dans l'ensemble de la Méditerranée. On doit le rapprocher également de l'interstade dit du "Würm supérieur" ou "Stillfried B" d'Europe centale (Allemagne, Basse Autriche, Tchécoslovaquie). Cet interstade a permis l'édification dans ces régions de sols datés de — 26.000 à — 27.000 ans.

Le réchauffement en milieu marin à cette époque paraît assez modéré.

Le Wurm II/III: Cet épisode correspond dans le Sud de la France à un climat tempéré humide qui a persisté pendant un laps de temps assez important.

Les traces de ce réchauffement sont visibles sur la carotte 1 Mo pour la Méditerranée occidentale, et sur les carottes 2 Mo et 3 Mo pour la Méditerranée orientale. Il s'agit là d'un stade marquant de la chronologie du Würm. Il correspond en Europe centrale à un sol plus marqué que celui de Stillfried B. Cet interstade, connu également sous les noms de "Laufen", est appelé "interstade du Würm moyen" (Valloch, 1967).

Le Wurm I/II: Enfin, le plus ancien interstade décelable dans les carottes 1 Mo et 3 Mo a été attribué au Würm I/II. Cette attribution a été faite compte tenu de la position de ce niveau et du caractère de la microfaune. Celle ci témoigne d'un réchauffement important. Or on sait que le Würm I/II ou Néotyrrhénien a vu, en certains points de Méditerranée, la réapparition de la faune "sénégalienne" caractéristique du Tyrrhénien.

Il faut toutefois souligner que cet épisode présente un caractère moins accusé en Méditerranée orientale, tandis qu'en Méditerranée occidentale il contraste d'avantage avec le climat plus frais des niveaux environnants.

LE PROBLEME DES NIVEAUX SABLEUX

Des passées de sable interrompent fréquemment les carottes sur une longueur plus ou moins grande. Elles vont de simples apports décelables uniquement à l'examen microscopique jusqu'à des lits qui atteignent plusieurs dizaines de centimètres d'épaisseur. L'exposé des caractères granulométriques de ces sédiments et l'examen de leur faciès assimilable à des turbidites sortiraient du cadre de ce travail. Nous nous bornerons à quelques remarques concernant leur position stratigraphique en Méditerranée occidentale.

On remarquera la présence fréquente de ces sables au niveau de la limite Würm III/IV: niveaux de quelques mm à quelques cm dans la carotte JC_3 et à la base de la carotte C_3, passées plus importantes dans les carottes 1–4 et 2–4. On peut penser que l'interstade de Lascaux, phase de réchauffement et d'humidité, a dû correspondre à une augmentation du ruissellement et du débit des cours d'eau. L'accroissement de l'apport minéral détritique rejeté à la mer peut être à l'origine d'une remise en mouvement des sédiments sur le fond dans certaines zones.

Outre "Lascaux", on trouve également des turbidites au niveau de "L'Alleröd" (JC_3) et, plus fréquemment, au niveau des interstades majeurs du Würm III ("Salpétrière" et "Arcy"). Ce sont là également des périodes de réchauffement et de ruissellement. Dans la carotte 2–4 les apports sableux sont pratiquement continus depuis la fin du Würm IIIa jusqu'au Dryas I.

Signalons en outre que la Provence a été également à plusieurs reprises le siège de séismes qui ont provoqué, en particulier, des éboulements importants visibles dans les grottes. Escalon de Fonton (1968 et comm. verb.) en signale dans, le Würm III, le Dryas et dans le Post-glaciaire. Le rôle possible de ces mouvements du sol dans les remaniements de sédiments marins ne peut être exclu *a priori.*

En Méditerranée orientale, des sédiments allochtones sont décelables dans la carotte 2 Mo et probablement aussi dans la carotte 45 Mo. Ruissellement, pluviosité d'une part, phénomènes volcaniques et sismiques d'autre part, ont probablement joué un rôle dans leur mise en place. Cependant nous manquons de documents pour avancer même des hypothèses.

CONCLUSIONS

L'étude comparative d'un ensemble de carottes de Méditerranée apporte des précisions sur la composition faunistique et l'importance des différents stades climatiques du Quaternaire récent dans le bassin occidental et dans le Bassin Oriental.

On peut ainsi définir et comparer les séquences climatiques correspondant à ces deux bassins: au climat de Méditerranée occidentale, dans l'ensemble très froid, surtout durant le Würm III et le début du Würm IV, et interrompu par des réchauffements parfois peu marqués, s'opposent les stades glaciaires souvent rigoureux mais plus discontinus de Méditerranée orientale où les phases interstadiaires prennent souvent un grand développement.

Un certain nombre de ces interstades, plus longs ou plus accentués, présentent une grand importance dans l'ensemble de la Méditerranée.

1. Le Würm I/II: Dans le contexte de Méditerranée orientale, dans l'ensemble plus chaud, ce stade ne se différencie guère en intensité des autres réchauffements. En Méditerranée occidentale, par contre, il témoigne d'une élévation notable de la température de l'eau qui n'a pas eu d'équivalent ultérieur.

2. L'interstade Würm II/III: peut également être retrouvé dans les carottes.

3. Un réchauffement postérieur très constant et bien visible paraît synchrone de l'interstade de la "Salpétrière" décrit en Provence, et des sols dits de "Stillfried B" ou du Würm supérieur en Europe Centrale.

4. Enfin, dans le post-glaciaire, les deux optima climatiques de l'Atlantique" et du "Sub-Atlantique" constituent deux repères chronologiques intéressants. Le caractère plus chaud du bassin oriental s'affirme dès lors nettement et le rafraîchissement général du climat qui se manifeste ensuite amène l'établissement des deux types de climat actuels: tempéré frais en Méditerranée occidentale et tempéré chaud en Méditerranée orientale.

Les passées de sédiments déplacés qui entrecoupent les carottes et entrainent des variations considérables du taux de sédimentation, paraissent correspondre dans de nombreux cas aux interstades chauds et humides (Alleröd, Würm III/IV, inter-Würm III). Les apports dus aux fontes glaciaires et aux crues pourraient être en partie à l'origine de ces remaniements de sédiments sur le fond marin. Le rôle de mouvements sismiques peut aussi être envisagé à titre d'hypothèse.

Il serait intéressant de compléter cette étude en considérant d'autres régions de Méditerranée et surtout en l'étendant dans le temps. Néamoins, pour la période considérée ici, les corrélations entre le Bassin Occidental et le Bassin Oriental sont satisfaisantes. La correspondance entre les indications de la microfaune marine et celle des critères climatiques continentaux (concrétionnements, altérations, sols . . .) est bonne.

Ceci souligne l'intérêt de l'étude comparatives des carottes—même lorsqu'il s'agit d'échantillons provenant de régions éloignées—et justifie cette tentative de reconstitution du climat dans l'ensemble de la Méditerranée pendant le Quaternaire récent.

BIBLIOGRAPHIE

Bartolini, C., et C. Vergnaud-Grazzini 1969. Evolution paléoclimatique des sédiments post-würmiens, en Mer d'Alboran, Méditerranée occidentale. *Communication Orale, 8ème Congrès de l'INQUA*, Paris.

Barusseau, J.P., G. Bellaiche, A. Levy, A. Monaco, et G. Pautot 1966. Variations paléoclimatiques et sédimentologiques des dépôts quaternaires des rechs du Roussillon (Golfe du Lion). *Comptes Rendus des séances de l'Académie des sciences*, Paris, 263:712–715.

Bé, A.W.H. et D.B. Ericson 1963. Aspects of calcification in planktonic Foraminifera (Sarcodina). *Annals of the New York Academy of Sciences*, 109 (1): 65–81.

Blanc-Vernet, L. 1969. Contribu,ion à l'étude des Foraminifères de Méditerranée. Relations entre la microfaune et le sédiment. Biocoenoses actuelles, thanatocoenoses pliocènes et quaternaires. *Recueil des Travaux de la Station Marine d'Endoume*, 64:1–251.

Blanc-Vernet, L. et H. Chamley 1971. Sédimentation à attapulgite et *Globigerinoides trilobus* forme *dehiscens* dans une carotte profonde de Méditerranée orientale. *Deep Sea Research*, 18: 631–637.

Blanc-Vernet, L., H. Chamley et C. Froget 1969. Analyse paléoclimatique d'une carotte de Méditerranée Nord-occidentale. Comparaison entre les résultats de trois études: Foraminifères, Ptéropodes, fraction sédimentaire issue du continent. *Paleogeography, Paleoclimatology, Paleoecology*, 6 (3): 215–235.

Bonifay, E. 1962. *Les terrains quaternaires dans le Sud-Est de la France*. Travaux de l'Institut de Préhistoire de l'Université de Bordeaux, 194 p.

Bottema, S. et L. M. J. U. van Straaten 1966. Malacology and palynology of two cores from the Adriatic Sea floor. *Marine Geology*, 4 (6):553–564.

Cita, M. B. et S. d'Onofrio 1967. Climatic fluctuations in submarine cores from the Adriatic Sea (Mediterranean). In: *Progress in Oceanography*, ed. Sears, M., 4:161–178.

Eriksson, K. G. 1965. The sediment core no. 210 from the western Mediterranean Sea. *Reports of the Swedish Deep Sea Expedition, 1947–1948*, 8(7):395–594.

Eriksson, K. G. 1967. Some deep sea sediments in the western Mediterranean Sea. In: *Progress in Oceanography*, ed. Sears, M., 4:268–280.

Escalon de Fonton, M. 1968. Problèmes posés par les blocs d'effondrement des stratigraphies préhistoriques du Würm à l'Holocène dans le midi de la France. *Bulletin de l'Association Française pour l'Etude du Quaternaire*, 4: 289–296.

Escalon de Fonton, M. 1969. Les séquences sédimento-climatiques du midi Méditerranéen du Würm à l'Holocène. *Bulletin du Musée d'Anthropologie Préhistorique*, 14: 125–184.

Imbrie, J. et E. G. Purdy 1962. Classification of modern Bahamian carbonate sediments. *Classification of Carbonate Rocks*, (ed. Ham, W. E.,), *The American Association of Petroleum Geologists*, Tulsa, Oklahoma, 1: 253–272.

Jones J. 1966. Planctonic Foraminifera as indicators organisms in the eastern Atlantic equatorial current system. *Actes du Symposium sur l'Océanographie et les Ressources halieutiques de l'Atlantique Tropical*, Rapports et Communications (Abidjan, Octobre 1966), publication de l'UNESCO: 213–230.

Lumley-Woodyear, H. de 1965a. *Le Paléolithique ancien et moyen du midi méditerranéen dans son cadre géologique*. Thèse, Faculté des Sciences de Paris, 1266 p.

Lumley-Woodyear, H. de 1965b. Evolution des climats quaternaires d'après le remplissage des grottes de Provence et du Languedoc méditerranéen. *Bulletin de l'Association Française pour l'Etude du Quaternaire*, 2: 165–170.

Mellis, O. 1954. Volcanic ash horizons in deep sea sediments. *Deep-Sea Research*, 2: 89–92.

Ninkovich, D. et B.C. Heezen 1966. Santorini Tephra. In: *Submarine Geology and Geophysics* eds. Whittard, W. F. and R. Bradshaw, Butterworths, London: 413–453.

Olausson, E., 1961. Studies of deep sea cores. *Reports of the Swedish Deep Sea Expedition 1947–1948*, 8 (4): 337–391.

Olson, E. A. and W. S. Broecker, 1961. Lamont natural radiocarbon measurements, VII. *American Journal of Science, Radiocarbon Supplement*, 3: 141–175.

Parker, F. L. 1958. Eastern Mediterranean Foraminifera. *Reports of the Swedish Deep Sea Expedition 1947–1948*, 8 (2): 217–285.

Pastouret, L. 1970. Etude sédimentologique et paléoclimatique de carottes prélevées en Méditerranée orientale. *Tethys*, 2 (1): 227–266.

Todd, R. 1958. Foraminifera from western Mediterranean deep sea cores. *Reports of the Swedish Deep Sea Expedition 1947–1948*, 8 (2): 167–215.

Valloch, K. 1967. La subdivision du Pléistocène récent et l'apparition du Paléolitique supérieur en Europe centrale. *Bulletin de l'Association Française pour l'Etude du Quaternaire*, 4: 263–270.

Vergnaud-Grazzini, C. 1968. Problèmes posés par l'étude géodynamique des microfaunes actuelles et son application à la stratigraphie, la paléoécologie, la paléoclimatologie, et la paléothermomérrie. *Revue de Géographie Physique et de Géologie Dynamique*, 10 (4): 397–406.

Vergnaud-Grazzini, C. et Y. Herman-Rosenberg 1969. Etude paléoclimatique d'une carotte de Méditerranée orientale. *Revue de Géographie Physique et de Géologie Dynamique*, 11 (3): 279–292.

Quaternary Eastern Mediterranean Sediments: Micropaleontology and Climatic Record

Yvonne Herman

Washington State University, Pullman, Washington

ABSTRACT

Faunal analyses and megascopic examination of deep-sea cores coupled with paleotemperature and C^{14} age determinations have been used to reconstruct the climatic and hydrologic history of the eastern Mediterranean. Changes in planktonic population with time, in response to variations in surface water temperatures and in hydrographic conditions, took place during the late Quaternary. The time interval represented by the longest core exceeds 120,000 years. It includes the Holocene, the Last Glacial and the Last Interglacial periods. The Holocene has a fauna similar to that of the present; one short period of stagnation occurred about 7,000 years ago.

The Last Glacial began about 70,000 years ago; several mild oscillations of varying length and amplitude interrupted this cold period. Repeated evidence of stagnation of the deep water in the upper part of the Last Glacial distinguishes late from earlier Glacial deposits. One core penetrates sediments deposited during the Last Interglacial: $O^{18/16}$ measurements of planktonic foraminiferal tests suggest that mean surface water temperatures during this time were similar to those of the Holocene.

Deposition of volcanic ash layers, turbidites and slump deposits, suggests that the eastern Mediterranean was tectonically active during this period. Biostratigraphic correlations accompanied by radiometric datings indicate that climatic oscillations in the Mediterranean were synchronous with those in the Atlantic Ocean and Red Sea.

RESUME

Le climat et l'hydrologie de la Méditerranée orientale ont été reconstitués grâce à l'étude de la faune, l'analyse mégascopique d'échantillons et à la paléotempérature ainsi qu'au datage au carbone 14. Au Quaternaire supérieur, la population planctonique a subit des changements liés aux variations de la température des eaux de surface et des conditions hydrographiques. La carotte la plus longue couvre une période dépassant 120.000 ans; elle comprend l'Holocène, la dernière époque glaciaire et l'Interglaciaire supérieur. L'Holocène possède une faune analogue à la faune actuelle; une brève période de stagnation a eu lieu il y a environ 7.000 ans.

La dernière glaciation a commencé il y a à peu près 70.000 ans; plusieurs oscillations tempérées, de durées et d'amplitudes variées ont interrompu cette période froide. Plusieurs périodes de stagnation des eaux de fond dans les couches supérieures permettent de différencier celles-ci des couches inférieures. Une carotte recoupe des sédiments Interglaciaire supérieur. L'analyse isotopique des foraminifères planktoniques suggère que les moyennes des températures l'eau de surface pendant cette époque étaient comparables aux moyennes des températures pendant l'Holocène.

Le dépôt des cendres volcaniques, des turbidites et des slumpings suggère une activité tectonique en Méditerranée orientale pendant cette période. Des corrélations biostratigraphiques ainsi que des datages radiométriques indiquent que les oscillations climatiques de la Méditerranée étaient en synchronisation avec celles de l'Atlantique et de la Mer Rouge.

INTRODUCTION

With the systematic large scale sampling of the sea floor which started about two decades ago, thousands of sediment cores have been collected from major oceans and marginal and inland seas. These sediments have been the subject of investigation by a number of researchers. The results of their studies have added to our knowledge of Quaternary climates, oceanography and processes of sedimentation, and have led to modifications of traditional concepts of Quaternary climates.

As the major efforts were directed towards the study of ocean basins, the investigation of small seas has been largely neglected. More recently, students from various disciplines have undertaken a thorough study of inland seas. Quaternary climates, recorded in sediment assemblages of pteropods and planktonic foraminifers from the Mediterranean basins, present some advantages over investigations of sediments deposited on land and in major oceans. Climatic changes recorded by terrestrial plants and animals are incomplete as compared to those preserved in marine deposits. Lacunae, which include both non-depositional and erosional hiatuses, are characteristic of land deposits (Ericson and Wollin, 1964); on the other hand, a large body of water such as the Atlantic will not be affected by minor temperature oscillations of short duration. From their studies of Atlantic deep-sea cores, Ericson *et al.* (1961, p. 282) conclude that "any variation of water temperature which may have occurred between 60,000 and 11,000 years B. P. must have been of insufficient intensity or of insufficient duration to have left a discernible mark on the faunal record."

In inland seas, seasonal as well as long-range temperature variations are relatively marked because of their comparatively small size and their proximity to land. This statement is substantiated by Emiliani and Flint (1963) who state that oxygen isotopic analysis performed on the shells of planktonic foraminifera indicates that the amplitude of surface water temperature oscillations was smallest in the Pacific, larger in the Atlantic, and largest in the Mediterranean. Moreover, detailed micropaleontologic investigation of Red Sea cores indicates that the Last Glacial period embraces two long cool intervals upon which several mild oscillations of short duration were superimposed (Herman-Rosenberg, 1965). The main warm oscillation occurred about 32–36,000 years ago (C^{14} date *in* Herman, 1968). Faunal analysis of a continuous sequence of samples, accompanied by $O^{18/16}$ and C^{14} determinations in Ionian Sea core (LDGO) V10-67 also provides evidence for numerous climatic oscillations during the Last Glacial (Herman, *et al.*, 1969; Vergnaud-Grazzini and Herman-Rosenberg, 1969). The few examples given indicate that inland seas afford an excellent opportunity for a detailed reconstruction of climatic changes.

PREVIOUS INVESTIGATIONS

In view of the extensive literature on sediments and fauna, as well as on the climatic, hydrologic and geologic features of the Mediterranean region, no attempt will be made to present a complete review of the published work; instead, this section summarizes the present state of knowledge, emphasizing aspects of greatest significance to the subject discussed herein.

Hydrology—Present knowledge of the water characteristics and circulation during different seasons is due to the work of Biel (1944), Pollak (1951), Tchernia and Lacombe (1959), and Wüst (1959, 1960, 1961).

Geology—I have drawn upon available data on the bottom topography (Koczy, 1956), sediments and structural features of the Mediterranean (Emery and Bentor, 1960; Emery *et al.*, 1966; Ryan *et al.*, 1965; Ludwig *et al.*, 1965; Hersey, 1965), and distribution of volcanic ash layers (Mellis, 1954; Ninkovich and Heezen, 1965). Chemical analyses and megascopic descriptions of core sediments have been made by Olausson (1960, 1961).

Paleontology—Foraminifera from deep-sea cores in the western Mediterranean Sea have been studied by Todd (1958), the eastern Mediterranean Sea by Parker (1958), from the Adriatic Sea by Cita and Chierici (1962); pteropods, planktonic foraminifers, and the mineralogy of a western Mediterranean core by Blanc-Vernet *et al.* (1969); planktonic fauna, paleotemperatures, and absolute age determinations of Mediterranean cores by Herman (1966), Herman *et al.* (1969), Vergnaud-Grazzini and Herman-Rosenberg (1969), and Herman (1971).

Eighteen long cores (average penetration 10 m) collected by the Swedish Deep-Sea Expedition from the western part of the Mediterranean Sea west of Italy and Sicily were studied by Todd (1958). Globigerinid and pteropod ooze was the most common type of sediment encountered. Intervals of sand and volcanic glass shards occur in minor quantities and reflect local diastrophism and vulcanism. An attempt was made to recognize and delineate climatic variations within the Pleistocene on the basis of fluctuations in the percentages of planktonic foraminifera. Benthonic foraminifera, present in many of the cores, gave an indication of regional depth zonation during Late Pleistocene time; displaced faunas from shallow environments were shown to be indicative of slumping and/or turbidity currents.

Parker (1955, 1958) studied the distribution and abundance of foraminifers in sixteen long cores (penetration varies from 0.8–11.5 m) from the eastern Mediterranean. She suggested that the modern planktonic fauna is probably indigenous due to the sill barrier in the Strait of Gibraltar which restricts water exchange between the sea and the open ocean. An increase in warm-tolerant forms was observed eastward in the Mediterranean. In the Aegean Sea, the faunas reflect cooler conditions. Four depth-assemblage boundaries were recognized, based on benthonic foraminiferal species: <25 m (bay-open-ocean); 143–205 m; 500–700 m; and 1000–1300 m. With the exception of the first, these boundaries coincide with water layer boundaries.

Displacement of benthonic faunas was observed in several cores, particularly those near shallow coastal areas. Pleistocene climatic variations were discussed on the basis of fluctuations in warm and cold-tolerant species. In most cores, two long cold intervals separated by a short warm one, and preceded by a long fluctuating warm interval, were recorded. The two long cold intervals were tentatively correlated with the Last Glacial (Würm I, II, III) and the long warm interval preceding them with the Last Interglacial (Riss/Würm). Sedimentation rates were estimated to vary between 5.7 cm/1000 years to 22.1 cm/1000 years in the area studied, but these values are highly speculative.

Parker also attempted to correlate the late Quaternary eastern Mediterranean climatic fluctuations with those from other marine and land deposits. Her interpretation is at variance with that of Emiliani (1955) who bases his correlation on oxygen isotope analysis of planktonic foraminiferal shells from an eastern Mediterranean core. Inasmuch as very little information on the absolute and relative rates of sedimentation were available at the time of their studies, the interpretation of their results is speculative (Parker, 1958).

Eighteen cores (penetration varied from 0.9–8.9 m) were studied from the Adriatic Sea (Cita and Chierici, 1962). Planktonic foraminifera were found to be absent at depths less than 40 m, and present in only low percentages (*ca.* 1%) between 40–100 m. Rich planktonic faunas appear only at a depth of 166 m where they represent about 22% of the fauna; this percentage increases with depth. Fluctuations in percent-

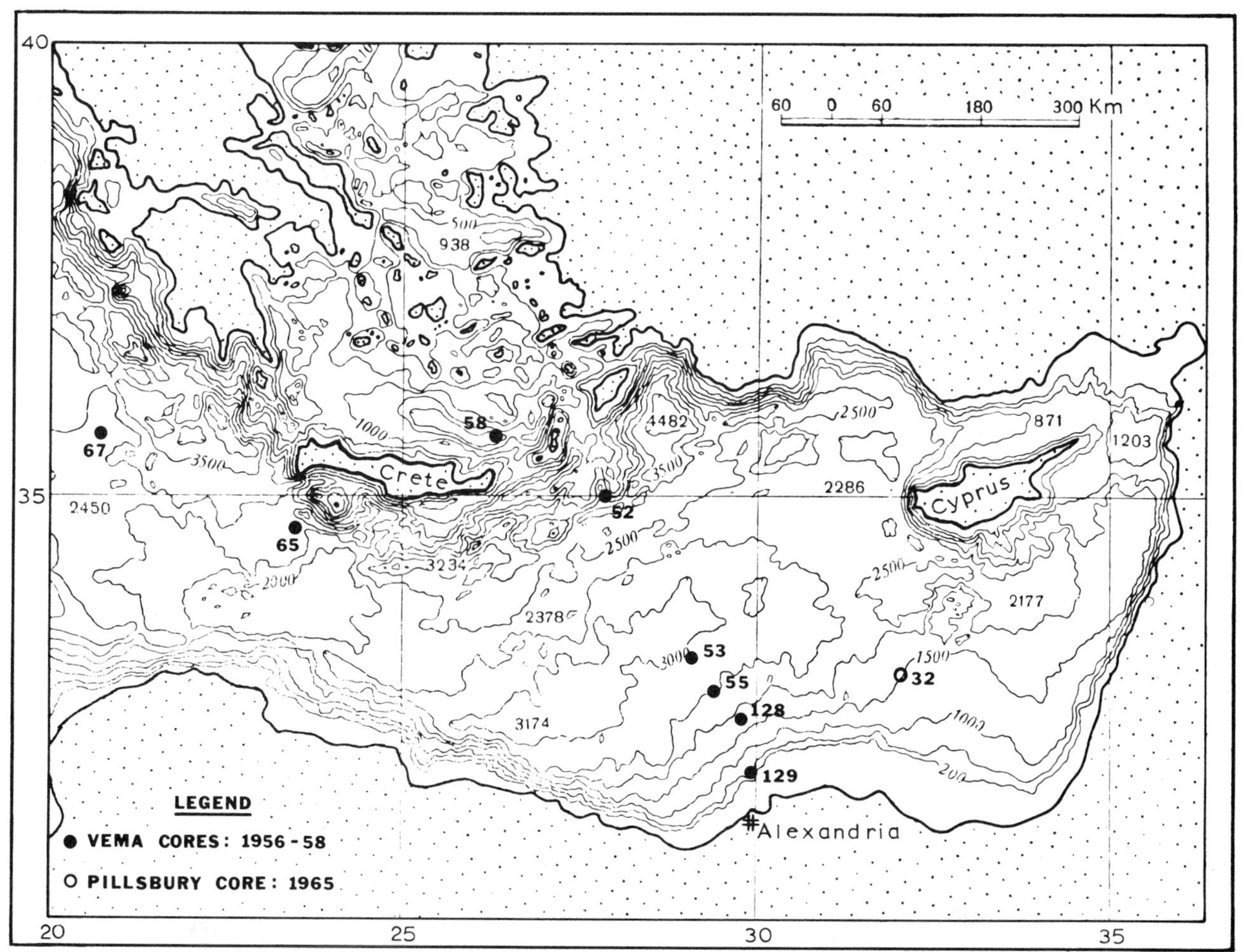

Figure 1. Physiographic map of the eastern Mediterranean (after Goncharov and Mikhailov, 1964) showing core locations.

age of planktonic foraminiferal species suggest that warm and temperate species predominate in the upper part of the deepest core (853 m) and cooler-water indicators prevail in its lower part, a reflection of Late Pleistocene climatic changes in the Adriatic Sea.

MATERIAL AND METHODS

Approximately thirty piston cores were examined megascopically for details of texture and structure; visual inspection of cores is of particular value for detecting anomalous layers such as slump and turbidite deposits and "flow-in" sections. Of these, representative cores from different physiographic provinces were selected for detailed microfaunal analyses. The location, water depths, and core lengths are given in Table 1 and Figures 1 and 2.

Equal volume samples were taken at 10–50 cm intervals or in places where lithologic changes were observed. However, the cores were sampled at closer intervals in sections of particular interest. The dried and weighed material was washed through a sieve with 74μ apertures, except for the PILLSBURY "H" cores for which a sieve with 64μ apertures was used. After drying at 100°C the coarse fraction was weighed for a second time and recorded as a percentage of the total. For faunal analyses the coarse fraction was split into sub-samples and counts totaling 300–500 specimens were made; the entire population within a sample was counted when total specimens numbered less than 300. Their frequency was recorded according to the following scale: Rare = 1–5 tests; Frequent = 6–11 tests; Common = 12–25 tests; Abundant = 26–100 tests; Very abundant = > 100 tests. It was found that each sample is characterized by the:

1. Co-occurrence of a number of species and the relative abundance of one species to the rest of the population;
2. Ratio of two species whose abundance is negatively correlated (with different ecologic requirements, Figure 4);
3. Absolute number of stenohaline and stenothermal forms (Figures 3, 5; Table 2);
4. Coiling direction of *Globorotalia truncatulinoides* (Figures 4, 6) and characteristic benthonic foraminifera.

Observations 2 and 3 are complementary inasmuch as the former eliminates the errors in interpretation introduced by variations in rates of accumulation of the fine fraction ($<64\mu$ or $<74\mu$), while the latter reflects the actual biological production.

Pteropods were counted separately from the foraminifers. For convenience, the species were grouped under two main headings, *Dominant* and *Subordinate*. As shown in a previous publication (Vergnaud-Grazzini and Herman-Rosenberg, 1969), each stage is characterized by discrete faunal associations. Moreover, the distinction between various interstadials (and stadials) is generally reflected by the relative abundance of different forms.

When several consecutive samples were counted, a certain amount of variation in the faunal composition

Table 1. Location, depth, and length of cores.

Core	Latitude	Longitude	Depth (m)	Length (cm)
H-29	33°51.00′ N	35°22.00′ E	1104	60
H-32	32°55.00′ N	32°00.00′ E	1573	120
H-37	32°11.00′ N	30°34.00′ E	873	103
H-38	32°00.00′ N	30°35.00′ E	199	123
H-39	31°48.00′ N	30°32.00′ E	51	70
H-42	31°51.00′ N	31°53.00′ E	94	115
H-44	31°50.00′ N	32°29.00′ E	186	130
H-46	32°15.00′ N	32°15.00′ E	921	130
H-48A	31°41.00′ N	33°43.00′ E	565	110
H-48	32°00.00′ N	33°30.00′ E	813	140
V10-52	35°00.30′ N	27°49.30′ E	2595	360
V10-53	33°11.30′ N	29°06.00′ E	2897	650
V10-55	32°46.30′ N	29°23.00″ E	2520	422
V10-58	35°40.30′ N	26°18.00′ E	2193	735
V10-65	34°37.00′ N	23°25.00′ E	2586	960
V10-67	35°42.00′ N	20°43.00′ E	2890	830
V14-128	32°27.00′ N	29°45.00′ E	1931	626
V14-129	31°47.00′ N	29°55.00′ E	576	650

Table 2. Foraminiferal assemblages from the Levantine Basin cores.

Postglacial	Glacial
DOMINANT (80%)	DOMINANT (95%)
Globigerinoides ruber f.A.	*Globigerinoides ruber* f.B.
Globigerinoides sacculifer	*Globigerina bulloides*
Globigerinoides ruber f.B.	*Globigerina pachyderma*
Globigerinita glutinata	*Globorotalia scitula* (dextral)
Globigerinita tenellus	*Globigerinita glutinata*
SUBORDINATE (20%)	SUBORDINATE (5%)
Globigerina quinqueloba	*Globigerina quinqueloba*
Orbulina universa	*Globigerinita uvula*
Globigerinoides ruber (pink)	*Globorotalia inflata*
Hastigerina pelagica	*Globorotalia truncatulinoides* (sinistral and dextral)
Globorotalia anfracta	
Globigerina digitata	
Globigerina bulloides	
Globigerina falconensis	
Globigerina rubescens	
Globorotalia truncatulinoides (sinistral)	
Globorotalia scitula	
Globigerinoides conglobatus	
Globigerinoides ruber f. *pyramidalis*	
Globigerinella aequilateralis	
Globigerina pachyderma (dextral)	
Globorotalia inflata	

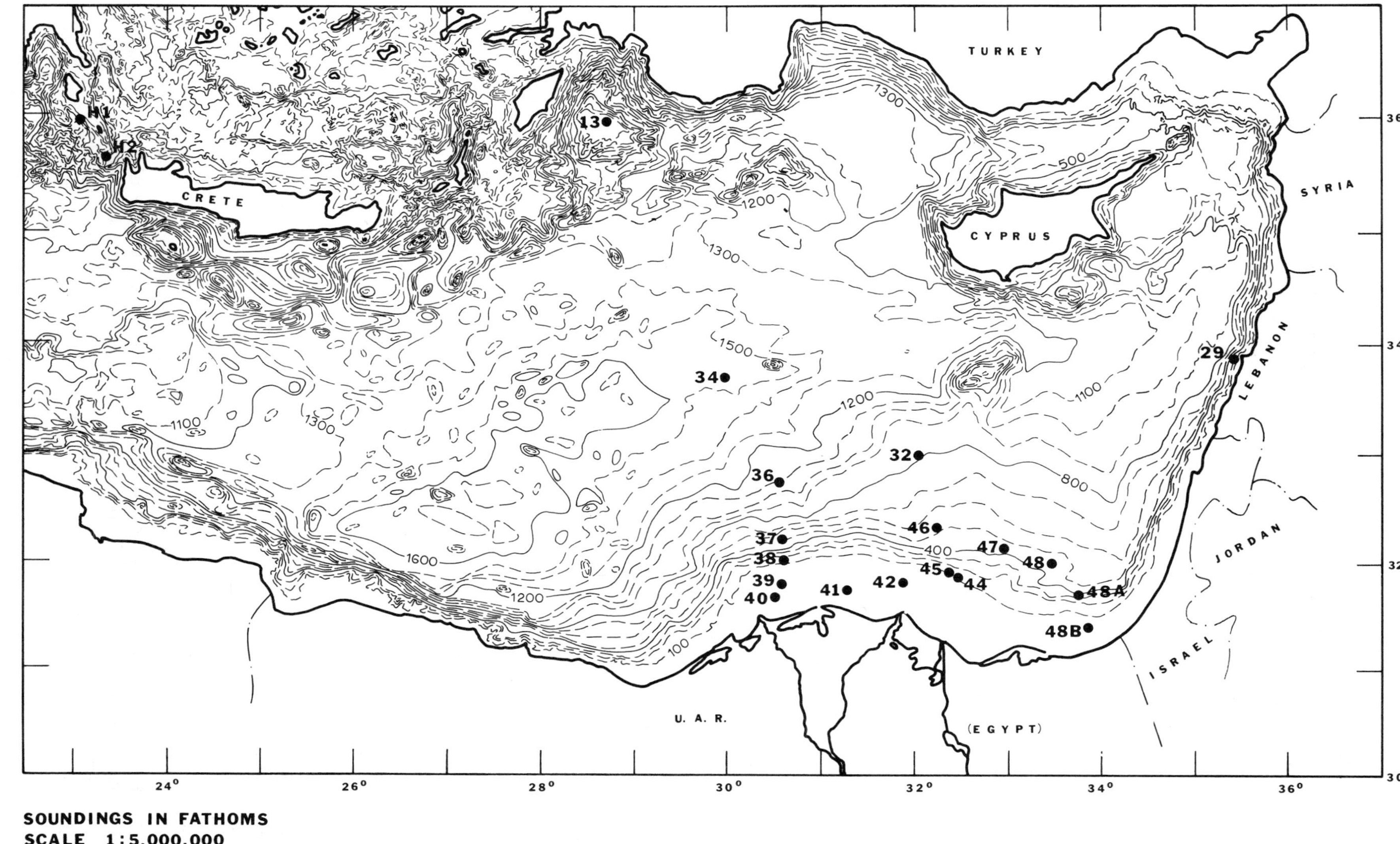

Figure 2. Physiographic map of the eastern Mediterranean compiled by the U. S. Hydrographic office, showing the location of PILLSBURY cores.

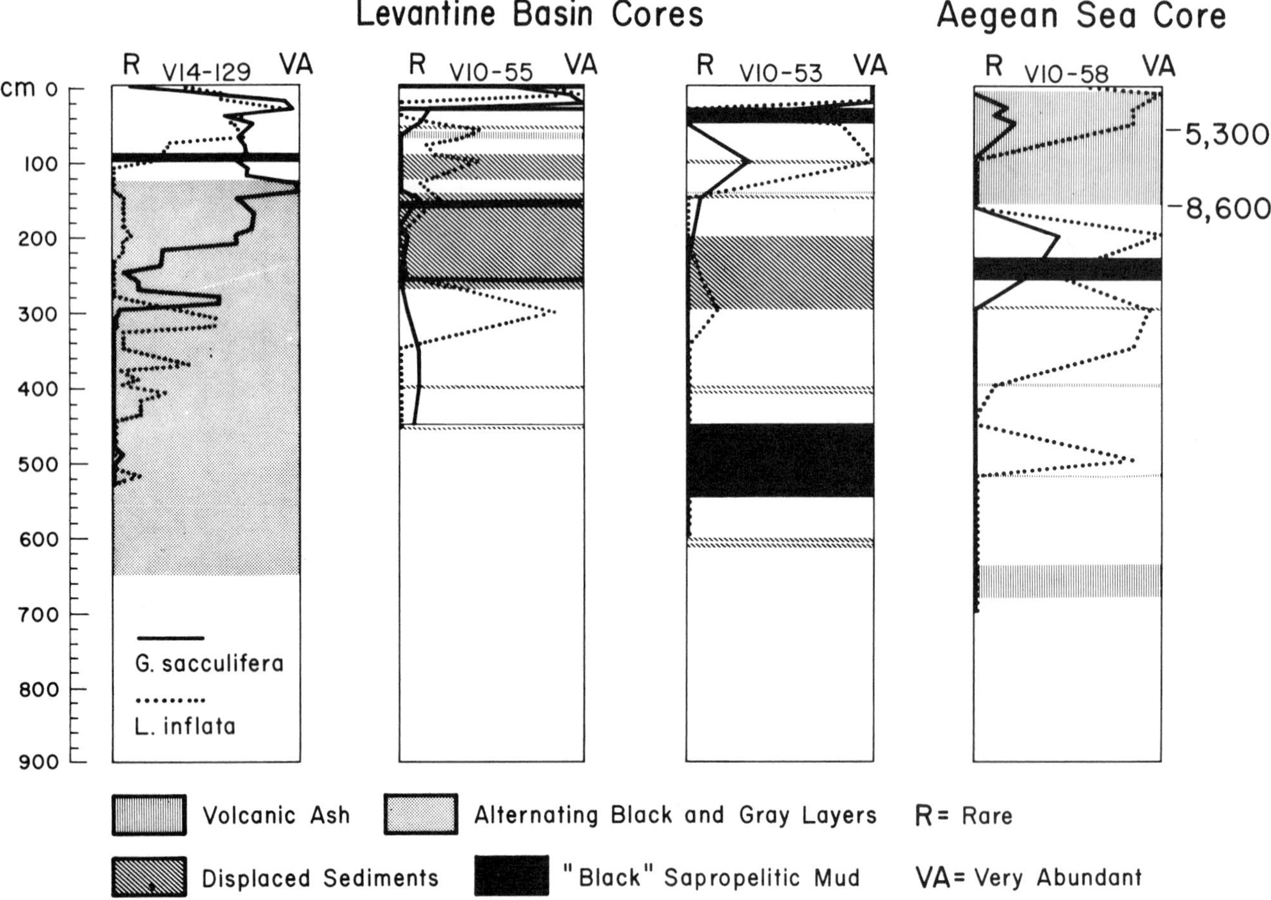

Figure 3. Levantine Basin and Aegean Sea cores showing sediment types and the variation in abundance of *Limacina inflata* and *Globigerinoides sacculifer*. R = 1–5 tests; VA >100 tests. (After Herman, 1971).

of samples belonging to the same climatic phase was observed. These variations are to be expected, and are accounted for by the random distribution and burial as well as by variations in the rates of post-depositional solution of tests. Furthermore, redistribution by currents and burrowers, and changes in accumulation rates of clastics will affect the composition of the preserved faunal assemblages. One sample count is inadequate to characterize a whole population or an entire climatic stage.

Therefore, several samples from each climatic stage were examined and counted. Moreover, several selected cores from each province were studied and the faunal associations were compared. By this method a more complete coverage of the actual population can be attained, eliminating sampling error to a great extent. The sample size is also important; it was found that approximately 500 counts per sample are statistically satisfactory. In an ideal case, the sample mean should be equal to the population mean. A deep-sea sediment assemblage, however, represents only a part of the living population for reasons mentioned in the preceding paragraph. Possibly, the closest values to those of the living population are found in assemblages preserved in sapropelic silty lutites accumulated during periods of rapid detrital sedimentation in a warm, anaerobic basin (*e. g.*, Postglacial and Last Glacial interstade deposits: Figures 3, 4, 5, 6).

PHYSIOGRAPHIC SETTING

Within the major physiographic provinces of the eastern Mediterranean, numerous smaller features are discussed in detail by Emery *et al.* (1966). For simplicity, we have combined them into three provinces:

1. *The Nile Cone Province,* extending from the shelf and slope across the continental rise, to a narrow elongate abyssal plain (Emery *et al.*, 1966). Numerous channels cut across the cone, forming a large distributary system that leads to the abyssal plain (Ryan *et al.*, 1966).

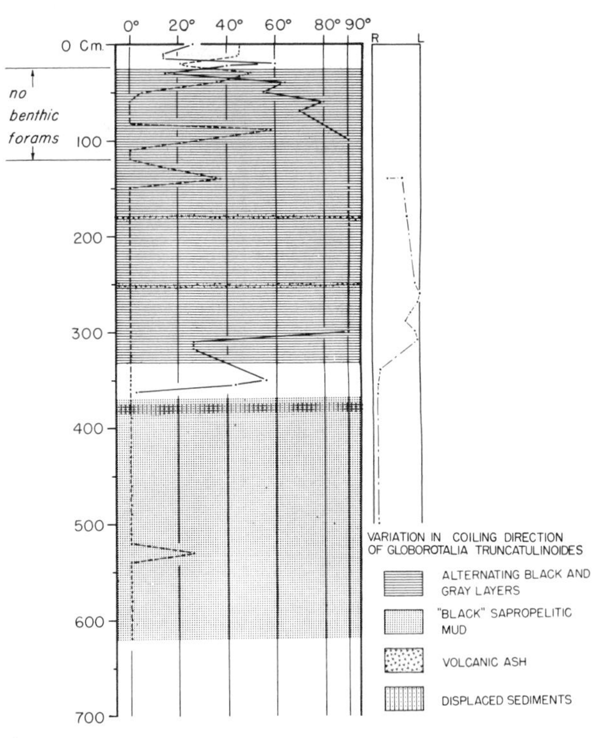

Figure 4. Levantine Basin core showing sediment types and climatic curves, based on percentages of warm and cool water microfauna. The solid line indicates the ratio of *Creseis* spp. to *Limacina inflata*, the broken line indicates the ratio of *Globigerinoides sacculifer* to *Globigerinoides ruber*. The ratios were converted into arctangent and plotted as degrees. (After Herman, 1971).

2. *The Central Levantine Province,* comprising the narrow elongate abyssal plain and numerous abyssal hills as well as the median ridge which extends from Italy, passes between Crete and Libya and curves northeastward to Cyprus. This latter major topographic feature acts as a barrier, impeding the spread of the Nile-derived sediments. (Emery *et al.*, 1966).

3. *The Aegean Province,* extending between the Ionian Sea and western Turkey; deep depressions extend from the Ionian Islands south of Crete and Rhodes and into the Gulf of Antalya, forming the Hellenic Trough. The greatest depth in the Mediterranean (5092 m) occurs in one of these depressions (Ryan *et al.*, 1966). Terrigenous sediments are derived from the Greek and Turkish mainlands and from the Aegean Islands (Wong and Zarudzki, 1969).

SEDIMENTS

During the late Quaternary, deposition of sediments in the eastern Mediterranean was controlled primarily by regional tectonic activity, as indicated by the occurrence of numerous volcanic ash layers, turbidite and slump deposits interbedded with pelagic sediments (Figures 3, 4, 5, 6). Climatic changes were of subordinate importance. Five main sediment types are encountered.

Terrigenous Clastics

Dark-gray terrigenous silty lutites predominate in the Nile Cone Province. The major supplier of sediments is the Nile, with an annual discharge estimated by Holeman (1968) to be 120 million tons. The sediment's dark gray color is mainly due to high percentages of finely disseminated organic debris. Calcareous skeletal remains (foraminifers, pteropods, and heteropods) are scarce, probably because of the high rates of clastic sedimentation. Microscopic examination of the washed samples indicates that most silt and sand size particles are composed of mica and quartz.

Displaced Sediments

Turbidite and slump deposits are widespread in the eastern Mediterranean (Figures 3, 4, 5, 6) and generally detectable with the naked eye. Graded beds containing displaced shallow water benthonic foraminifers, particles of the calcareous algae *Halimeda* and black plant detritus are most common in the Nile Cone and Central Levantine Provinces (Figures 3, 4, 5, 6).

Sapropelic Deposits

Repeatedly during late Last Glacial time and at least once during the Postglacial the water circulatory system was altered as a result of increased freshwater discharge and/or glacio-eustatic lowering of sea level, resulting in the stagnation of deeper water layers and consequent deposition of black sapropelic silty lutites (Figures 3, 4, 5, 6). The sediment's dark color is due to the high content of finely disseminated organic matter and to iron sulphide minerals. In sapropels, pteropods and planktonic foraminifers are more abundant and better preserved than in sediments laid down during aerobic conditions, whereas benthonic foraminifers are absent or scarce.

Compared to the oxidized deposits, sapropels contain 10–40 times more carbon (2-8%), 10–20 times more Ba and Mo, and 2–5 times more Ni, V, and Co (Ryan *et al.,* 1966).

Volcanic Ash Layers

Volcanic eruptions, in particular that of Santorini, have contributed widespread tephra layers to the Aegean Province (Mellis, 1954; Ninkovich and Hee-

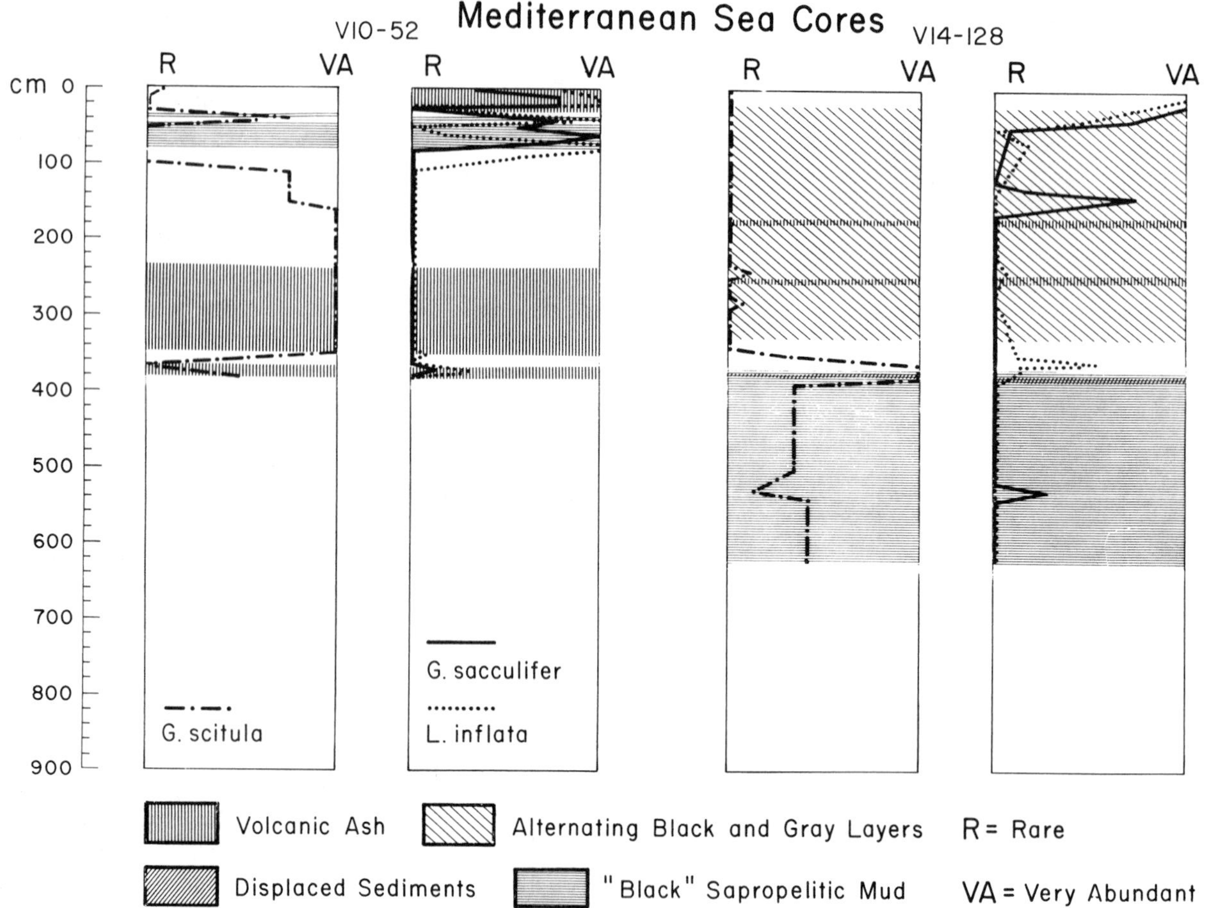

Figure 5. Eastern Mediterranean Sea cores showing sediment types and the variation in abundance of *Limacina inflata, Globigerinoides sacculifer*, and *Globorotalia scitula*. R = 1–5 tests; VA = > 100 tests (After Herman, 1971).

zen, 1965). A *lower tephra* (n = 1.521) and an *upper tephra* (n = 1.509) were identified by Mellis (1954) and Ninkovich and Heezen (1965) who suggested that the volcanic ash layers originated from Santorini eruptions and were distributed by winds. Both tephra layers are composed of a colorless fine-grained, frothy ash. The *upper tephra* was deposited in Post-glacial time, about 5,000 years ago (Ninkovich and Heezen, 1965), the *lower tephra* was deposited in late Last Glacial time, about 27,000 years ago (Vergnaud-Grazzini and Herman-Rosenberg, 1969). In addition to the colorless Santorini tephra layers, black and dark brown volcanic ash layers occur sporadically, some associated with abundant radiolarian and diatom oozes, others with anomalously high percentages of pteropods.

Calcareous Oozes

Planktonic foraminifers, pteropods, heteropods and coccolithophorids build skeletal structures of calcite and aragonite and are quantitatively the chief contributors to calcareous oozes. Benthonic foraminifers generally constitute 0.5–5% of the fauna in the coarse fraction. According to Ryan *et al.* (1966), the average carbonate content of eastern Mediterranean sediments is about 40% while that of the western Mediterranean is 30%. In Holocene sediments, pteropods and foraminifers constitute 90–95% of the planktonic fauna in the coarse fraction, the remainder being made up of heteropods. Due to the relatively higher solubility of aragonite, the original composition of the faunal assemblages is modified by post-depositional solution; consequently pteropods and heteropods are scarce in sediments older than 12–13,000 years.

RATES OF SEDIMENTATION

Among the important factors determining rates of sediment accumulation are: (a) topographic setting; (b) winnowing of sediments by bottom currents; (c) rates of clastic and biogenic supply; and (d) pre- and post-depositional solution. Seismic reflections (Wong and Zarudzki, 1969), and radiocarbon age

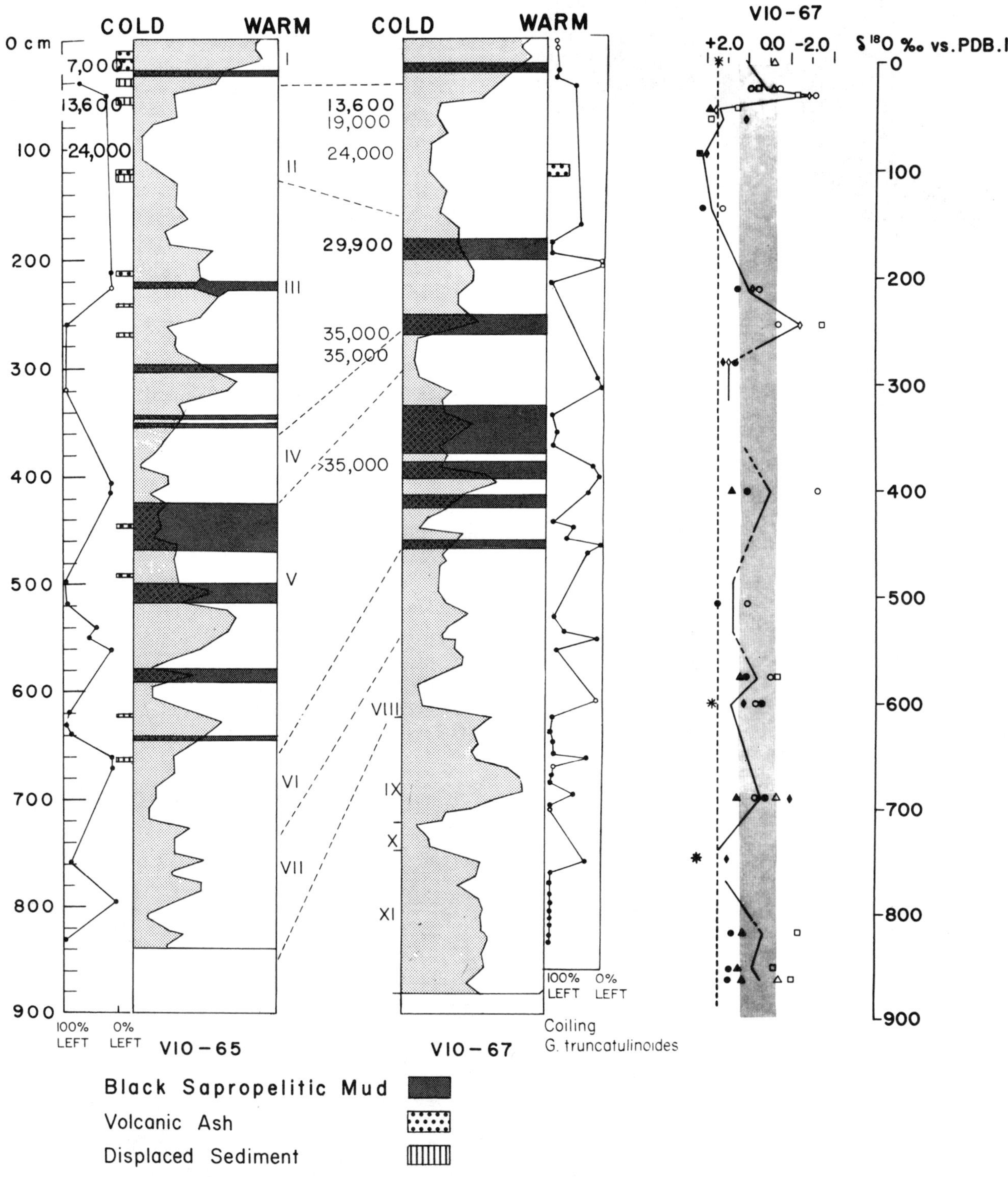

Figure 6. Core depth in centimeters. Climatic curves based on total planktonic faunal analysis. Paleotemperature curve (right hand); Solid line = Variations in average isotopic concentrations; Dashed vertical line = Average isotopic composition of living benthonic species. *G. truncatulinoides* ● > 20 specimens, ○ < 20 specimens, shaded zone = Range of isotopic variation in living planktonic foraminifers, □ *G. ruber*, △ = *G. sacculifer*, ◇ Juvenile epiplanktonic forms (< 100μ) and *Limacina* sp, ○ *O. universa*, ■ *G. pachyderma*, ▲ *G. truncatulinoides*, ◆ *G. bulloides*, ● *G. inflata*, * Benthonic species. Geologic—climate units are indicated by Roman numerals. (Modified after Vergnaud-Grazzini and Herman-Rosenberg, 1969).

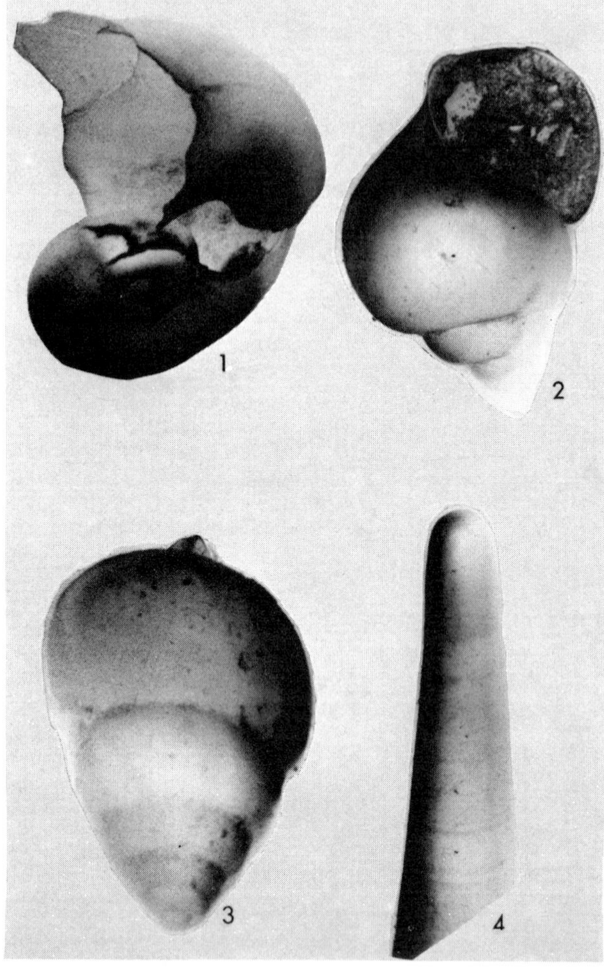

Figure 7. Scanning electron microscope (SEM) illustrations of pteropods in the Eastern Mediterranean. 1, *Limacina inflata* (d'Orbigny), spiral view ×125, from core CH 61–40–15 cm. 2, *Limacina trochiformis* (d'Orbigny), apertural view, ×100, CH 61–40–15 cm. 3, *Limacina bulimoides* (d'Orbigny), spiral view, ×100, CH 61–40–15 cm. 4, *Creseis acicula* (Rang), ×300, CH 61–40–15 cm.

determinations (Herman *et al.*, 1969; Vergnaud-Grazzini and Herman-Rosenberg, 1969) indicate that sedimentation rates were in the order of 10 to 40 cm/1000 years during late Quaternary time.

MICROPALEONTOLOGY

Species of Pteropoda

A detailed synonymy and taxonomic discussion has not been attempted here; the nomenclature used by Tesch (1946, 1948) is followed in most cases. Pteropods present in eastern Mediterranean sediments are listed in alphabetical order:

Cavolinia inflexa (Lesueur)
C. longirostris (Lesueur)
C. tridentata (Forskal)
C. uncinata (Rang)
Clio cuspidata (Bosc)
C. polita Pelseneer
C. pyramidata Linne
C. pyramidata f. pyramidata Linne
Creseis acicula (Rang) [Figure 7, *4*]
C. conica Eschscholtz
C. virgula Rang f. *constricta* Chen
Diacria quadridentata (Lesueur)
D. trispinosa (Lesueur)
Hyalocylix striata (Rang)
Limacina bulimoides (d'Orbigny) [Figure 7, *3*]
L. inflata (d'Orbigny) [Figure 7, *1*]
L. retroversa (Fleming)
L. trochiformis (d'Orbigny) [Figure 7, *2*]
Peraclis bispinosa Pelseneer
Peraclis sp.
Styliola subula (Quoy and Gaimard)

Species of Planktonic Foraminifera

Planktonic foraminifers present in eastern Mediterranean cores are listed in alphabetical order; the nomenclature used by Parker (1962) is followed in most cases:

Globigerina bulloides d'Orbigny
G. calida Parker
G. digitata Brady
G. falconensis Blow
G. pachyderma (Ehrenberg)
G. praedigitata Parker

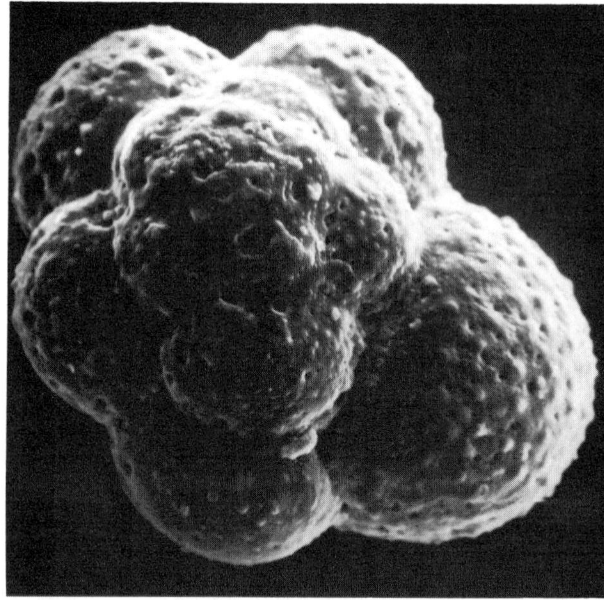

Figure 8. SEM illustration of *Globigerina quinqueloba* Natland, dorsal view, ×800, from core V10–67, 731–744 cm.

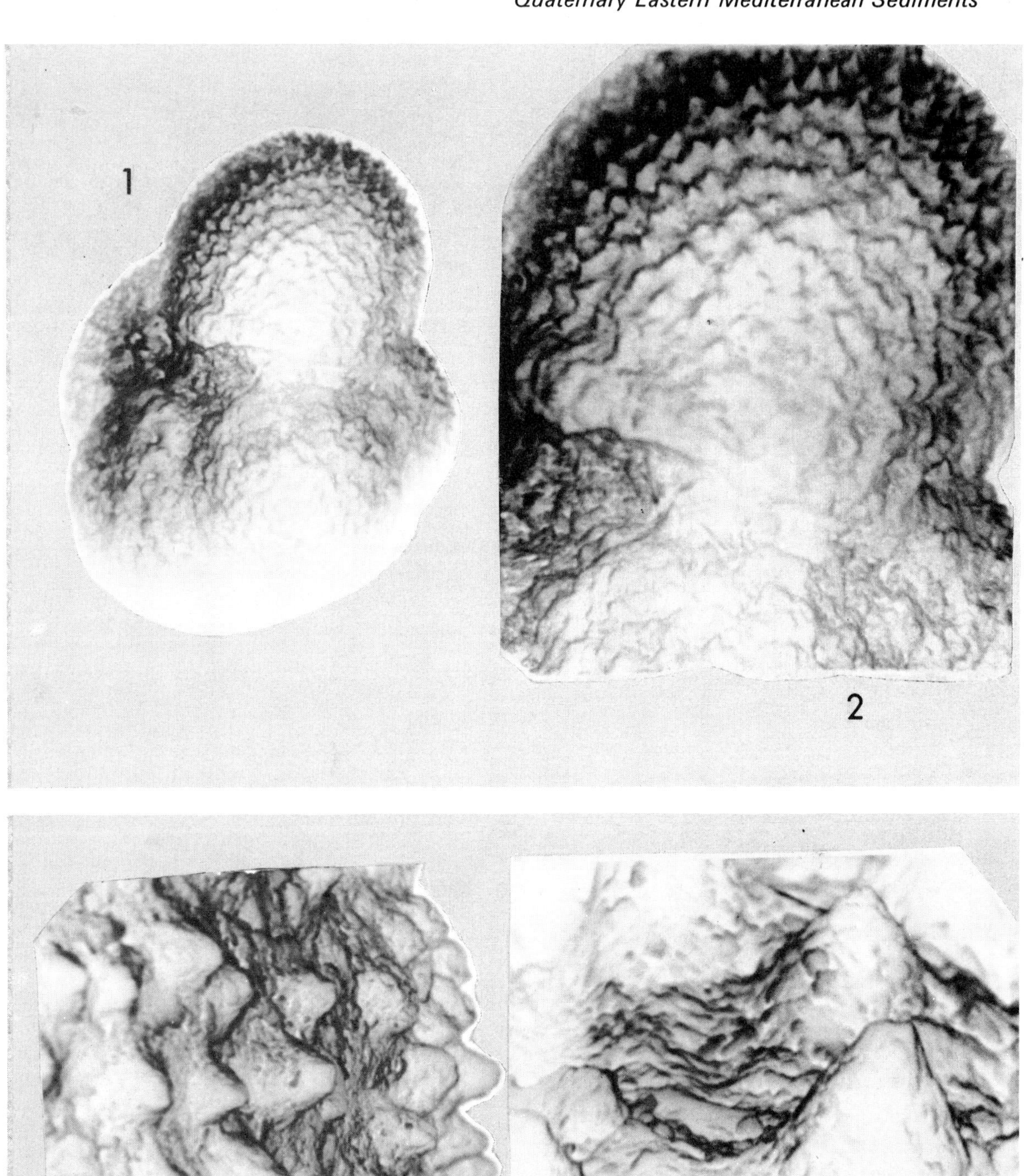

Figure 9. SEM illustration of *Globigerina quiqueloba* Natland. 1, ventral view, × 340, from core V 10–4–180 cm; 2, same specimen, × 640; 3, same specimen, × 2250; and 4, same specimen, × 6000.

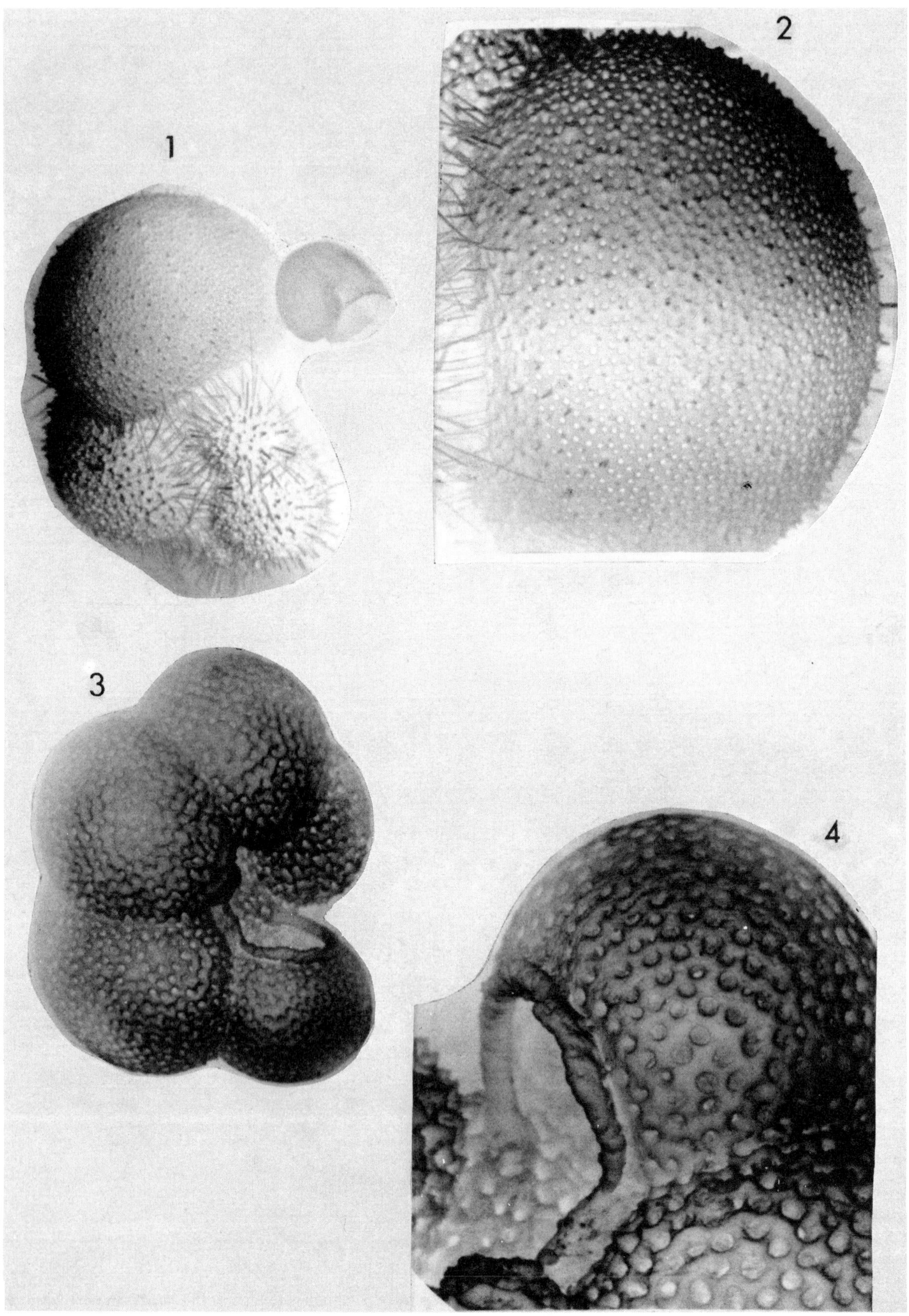

Figure 10. SEM illustrations of planktonic foraminifera. 1, *Globigerinella aequilateralis* (Brady), equatorial view, × 160, from core CH 61–40–226. 2, Same specimen, × 300, last chamber. 3, *Globoquadrina dutertrei* (d'Orbigny), ventral view, × 260, from core CH 61–40–226 cm. 4. Same specimen, ventral view, × 600, last chamber, showing the even distribution of circular pores and the smooth last chamber. Note the thick interpore ridges in penultimate chamber and the fine pores piercing the apertural lip.

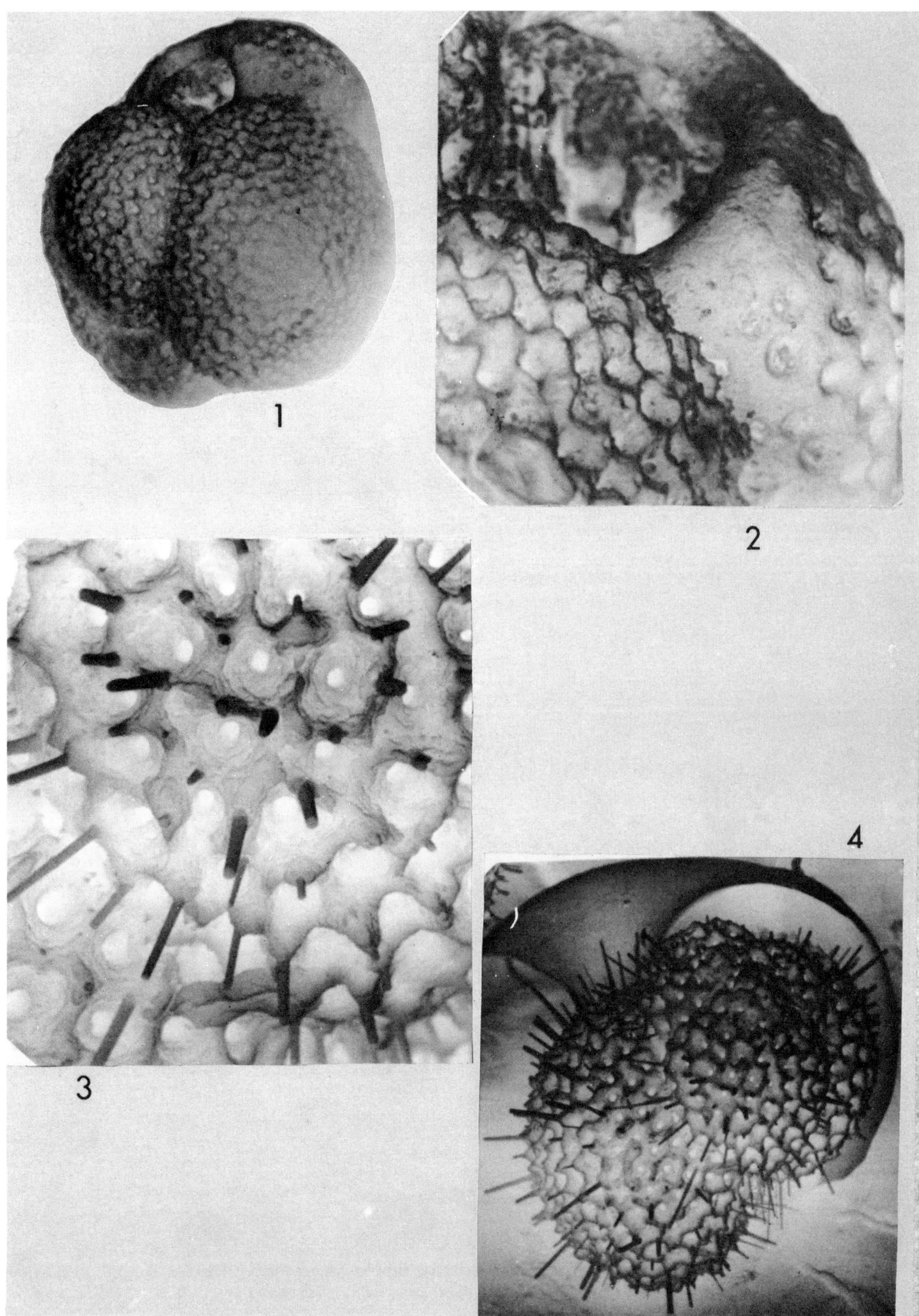

Figure 11. SEM illustrations of planktonic foraminifera in the eastern Mediterranean. 1, *Globigerinoides ruber f. B,* × 235, from core, CH 61–31–195 cm; 2, Same specimen, × 750. 3, *Globigerinoides ruber f. A,* (d'Orbigny), × 1000, from core, CH 61–40–226 cm; 4, Same specimen, × 300.

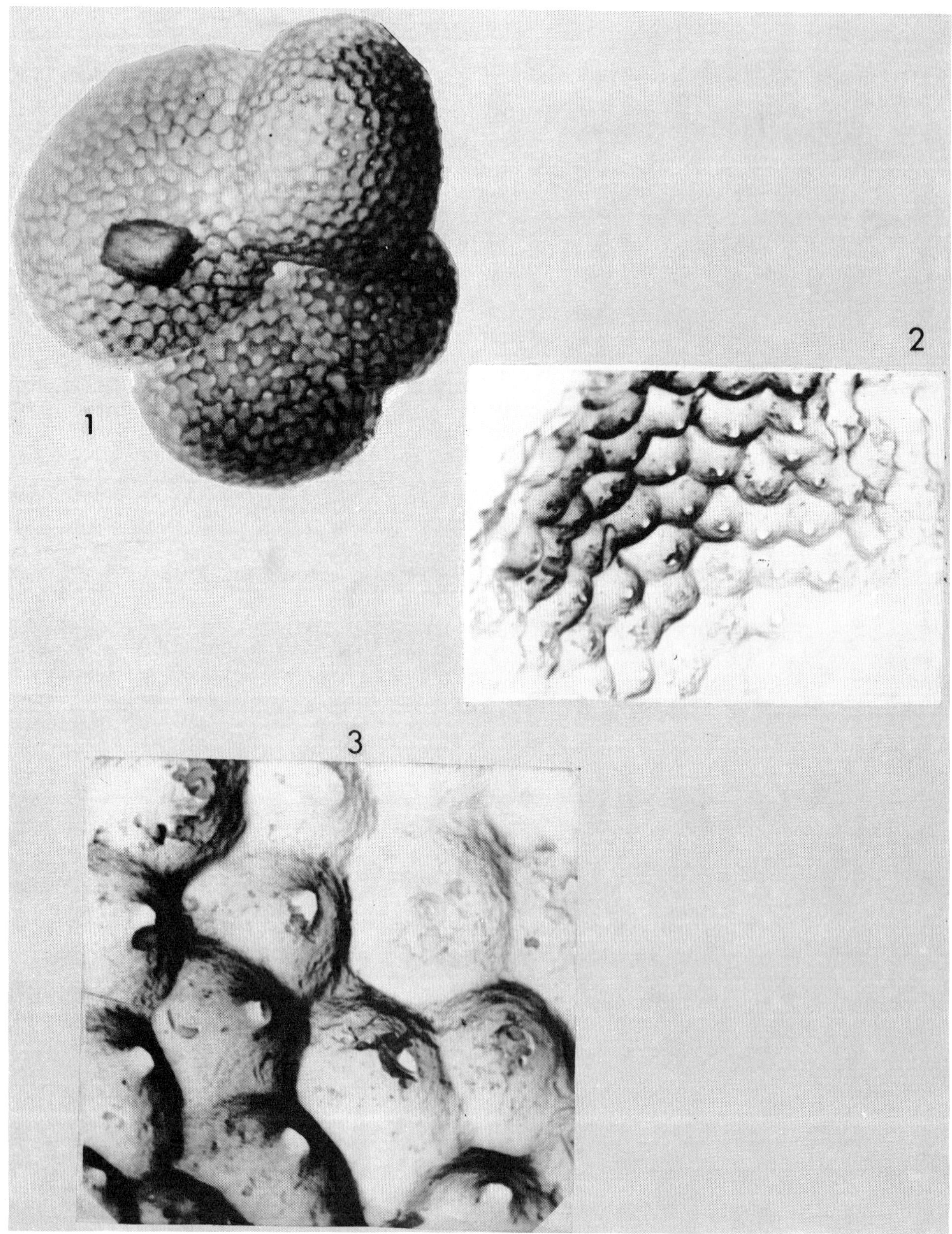

Figure 12. SEM illustrations of *Globigerinoides sacculifer* (Brady). 1, ventral view, × 200, from core CH 61–40–15 cm. 2, × 750, from core CH 61–31–195 cm; 3, Same specimen, × 2250.

G. quinqueloba Natland [Figures 8 and 9]
G. rubescens Hofker
Globigerinella aequilateralis (Brady) [Figure 10, *1–2*]
Globigerinita glutinata (Egger)
G. uvula (Ehrenberg)
Globigerinoides conglobatus (Brady)
G. obliquus (Bolli)
G. ruber f. A (d'Orbigny) [Figure 11, *3–4*][1]
G. ruber f. B [Figure 11, *1–2*][2]
G. ruber (d'Orbigny) *f. pyramidalis* (Van den Broeck)
G. sacculifer (Brady) [Figure 12]
G. tenellus Parker
Globoquadrina dutertrei (d'Orbigny) [Figure 10, *3–4*]
Globorotalia anfracta Parker
G. inflata (d'Orbigny)
G. scitula (Brady)
G. truncatulinoides (d'Orbigny)
Hastigerina pelagica (d'Orbigny)
Orbulina universa (d'Orbigny)

Faunal Analysis

Distributional patterns of living pteropods and planktonic foraminifers indicate that many species have a limited tolerance to changes in temperature and salinity. Food, light and oxygen are also known to determine their distribution and abundance. Accordingly, variations in planktonic faunal composition in consecutive sediment layers are interpreted as reflecting alterations in production rates as well as changes in climatic and hydrologic conditions at the time of, and shortly after, their burial. In addition to the above mentioned factors, other variables determine the composition of faunal remains in sediments. Important among these are redistribution by currents and burrowers, changes in accumulation rates of detrital sediments and the solution of limy tests. Dissolution of foraminiferal shells is selective (Ericson *et al.*, 1961; Berger, 1967); it varies with genera and species, the more fragile forms being eliminated first, thus changing species composition. The criteria which determine the selective solution of foraminifers most probably also apply to pteropods. For instance, pteropods and fragile, minute planktonic foraminifers abound in Holocene sediments raised in relatively shallow water but are, with few exceptions, absent or present at low frequencies in deeper water cores, particularly in sediments deposited during the Last Glacial and Last Interglacial.

A synopsis of the faunal assemblages contained in Levantine, south Aegean and southern Ionian cores is given in Tables 2–5 and *in* Vergnaud-Grazzini and Herman-Rosenberg (1969).

DISCUSSION

In Holocene sediments, pteropods and foraminifers constitute 90–95% of the planktonic fauna, the remainder being made up of heteropods. In Last Glacial and Last Interglacial sediments, planktonic foraminifers make up the bulk of the faunal remains.

Most sediments sampled from the Nile cone and its vicinity (*e.g.*, H–29, H–37, H–38, H–39, H–42, H–44, H–46, H–48, H–48A, V14–128 and V14–129, Table 1; Figures 1 and 2) contain large amounts of clay and silt-size land-derived detritus. The dark gray color of the sediment is mainly due to finely disseminated organic debris and to crystalline aggregates of pyrite. Calcareous skeletal remains are scarce, because of the high rates of clastic sedimentation. Core V14–129 (Figure 3) exemplifies these sediments: the entire core (630 cms in length) represents Holocene deposits. The composition of the benthonic fauna changes with depth. It is abundant and varied down to 130 cm; between 130 and 250 cms its relative abundance with respect to the total fauna fluctuates, and benthonic forms are virtually absent from 250 cm down to the core bottom. The absence

Table 3. Pteropodal assemblages from Levantine Basin cores.

Postglacial	Glacial
DOMINANT (> 90%)	
Creseis acicula	
Creseis spp.	
Limacina inflata	
Styliola subula	
Clio pyramidata	*Clio pyramidata* f. *pyramidata*
Limacina trochiformis	*Limacina retroversa*
SUBORDINATE (< 10%)	
Clio polita	
Clio cuspidata	
Hyalocylix striata	
Cavolinia uncinata	
Cavolinia longirostris	
Cavolinia inflexa	
Limacina bulimoides	Present day salinity: 38%
Cavolinia tridentata	Summer temperature: 29°C
Diacria trispinosa	Winter temperature: 16°C
Peraclis bispinosa	[Data from Wüst]
Peraclis sp.	

[1] *G. ruber f.* A = typical inflated chambers.
[2] *G. ruber f.* B = test more compact, aperture smaller, walls thicker than in the typical form, chambers flattened; this may be a colder water variant of form A.

(or low frequencies) of benthonic foraminifers correspond to levels rich in black plant debris, finely disseminated organic matter, and aggregates of pyrite crystals. Qualitatively, the composition of the planktonic foraminifera and pteropods remains essentially unchanged throughout the core. Variations in abundance of *Globigerinoides sacculifer* and *Limacina inflata* (Figure 3) as well as that of other forms, are believed to represent mainly changes in rates of detrital sedimentation and/or adverse environmental conditions.

In Figure 4, the two curves represent the ratios of *Creseis* spp. to *Limacina inflata* and of *Globigerinoides sacculifer* to *Globigerinoides ruber* in core V14–128. As the result of high rates of clastic sedimentation, a detailed chronological sequence of events may be followed.

The Holocene is divided into three phases. The first phase, contained in the upper 25 cm of the core, has a planktonic fauna dominated by *Limacina inflata*, *Limacina trochiformis*, *Creseis acicula*, *Styliola subula*, *Clio pyramidata*, *Globigerinoides ruber f.* A, *f.* B, *Globigerinoides sacculifer*, *Globigerinella aequilateralis*, and *Orbulina universa*. The second phase, contained between 25–120 cms, is devoid of benthonic foraminifers. The dominant pteropods are epipelagic *Creseis* spp, *Clio pyramidata* and *Diacra trispinosa*. The presence of black plant debris and iron sulfide minerals, in addition to the absence of benthonic foraminifera, suggests an euxinic episode. During the third phase, below 120 cms, there is a gradual increase in the number of cold water foraminifers. Pteropods are scarce or absent; Mn oxide minerals are present as thin coating on the foraminiferal shells. Benthonic foraminifers fluctuate in abundance, occasionally displaced shallow water forms and *Halimeda* spp. are present. The Holocene-Pleistocene boundary is at about 360 cms and it is marked by the change in coiling direction of *Globorotalia truncatulinoides* from dextral in Pleistocene, to sinistral in Holocene (Figure 4).

Central Levantine basin cores are characterized by the intercalation of normal pelagic sediments with numerous sporadically emplaced terrigenous silts and sands, some of which are graded (Figure 3, Cores V10–53 and V10–55). The close coincidence of the curves representing the variation in relative abundance of *Limacina inflata* and *Globigerinoides sacculifer* in zones representing normal sedimentation is illustrated in Figure 3.

Core V10–58 (Figure 3) raised from the vicinity of Crete exemplifies *Aegean Province* sediments: between 10 and 160 cms it contains the *upper Santorini tephra* (n = 1.509); from 635 to 680 cms, the *lower Santorini tephra* (n = 1.521), (Mellis, 1954; Ninkovich and Heezen, 1965). As indicated in Figure 3, other ash layers intervene between the calcareous holopelagic sediments. Shallow water benthonic foraminifers and *Halimeda* spp. were observed at several levels, attesting to sporadic, post-depositional sediment transport. Radiolarian and diatom-rich layers occur in sediments containing volcanic ash. The Postglacial-Last Glacial boundary is at about 320–330 cms; it is marked by the replacement of cold water forms such as *Globigerina pachyderma* (dextral) *Globigerina bulloides*, *Globorotalia scitula*, *Limacina retroversa*, and *Clio pyramidata*, by warm water *Globigerinoides sacculifer*, *Globigerinoides ruber f.* A., B., the pink variety, *Globigerinella aequilateralis*, *Limacina inflata* and *Creseis* spp., etc. The core terminates in upper late Last Glacial sediments. Another typical Aegean Province core is V10–52 (Figure 5): the Postglacial-Last Glacial boundary is between 80 and 90 cm depth in the core. This boundary is marked by the replacement of cold water *Globigerina pachyderma* (dextral), *Globigerina bulloides*,

Table 4. Pteropodal assemblages from southern Aegean Sea cores.

Postglacial	Interstadial	Glacial
DOMINANT	DOMINANT	
Limacina inflata	*Limacina inflata*	
Creseis spp.	*Creseis* spp.	
Styliola subula	*Clio pyramidata*	
	Limacina trochiformis	
SUBORDINATE	SUBORDINATE	*Limacina retroversa*
Clio pyramidata	*Diacria quadridentata*	*Clio pyramidata* f. *pyramidata*
Limacina trochiformis	*Hyalocylix striata*	
Limacina bulimoides	*Cavolinia* sp.	
Diacria trispinosa		Present day salinity: 39%
Cavolinia sp.		Summer temperature: 25°C
Hyalocylix striata		Winter temperature: 12°C
Peraclis sp.		[Data from Wüst]

Table 5. Foraminiferal assemblages from southern Aegean sea cores.

Postglacial	Glacial
DOMINANT	DOMINANT
Globigerinoides ruber f.A.	*Globigerina pachyderma* (dextral)
Globigerinella aequilateralis	*Globigerina bulloides*
Globigerinoides ruber (pink)	*Globorotalia scitula*
Globigerinoides ruber f.B.	*Globigerina quinqueloba*
Globigerinoides sacculifer	*Globigerinita glutinata*
Orbulina universa	*Globigerinoides ruber* f.B.
Globigerina quinqueloba	
Globigerina calida	
Globigerinita glutinata	
SUBORDINATE	SUBORDINATE
Globigerina bulloides	*Globorotalia truncatulinoides* (sinistral and dextral)
Globorotalia inflata	*Globoquadrina dutertrei*
Globigerinoides conglobatus	*Globigerinella aequilateralis*
Globigerina pachyderma (dextral)	*Globigerinoides tenellus*
Globigerinoides tenellus	
Globoquadrina dutertrei	
Globigerina rubescens	
Hastigerina pelagica	
Globorotalia scitula	
Globigerina digitata	
Globorotalia truncatulinoides (sinistral)	

Globorotalia inflata, and *Globorotalia scitula* by warm water forms such as *Globigerinoides sacculifer*, and *Globigerinoides ruber f.* A., B., and the pink variety, of *Globigerinoides ruber*. Pteropods abound in Postglacial deposits, constituting about 50% of the faunal assemblages, whereas they are absent or rare throughout most of the late Last Glacial sediments. The *upper Santorini* tephra is present between 20 and 30 cm depth in this core, which also contains another thick ash layer between 240 and 390 cm (Figure 5).

Figure 6 represents climatic variations in Ionian basin cores V10–65 and V10–67 as inferred from changes in warm and cold water fauna and $O^{18/16}$ determinations. The climatic curve in core V10–67 is based on faunal analysis of a continuous sequence of samples (Herman, *in* Vergnaud-Grazzini and Herman-Rosenberg, 1969) and is supplemented by $O^{18/16}$ determinations and C^{14} dates. Core V10–65 was analyzed at 10 to 50 cm intervals. Inasmuch as pteropods are scarce or absent in Last Glacial and Last Interglacial sediments, the climatic record is based principally on planktonic foraminifers and $O^{18/16}$ determinations. The time interval represented by core V10–67 is estimated to be about 120,000 years. This age has been estimated using extrapolated sedimentation rates, determined by C^{14} dating of several samples from the upper 280 cm of core V10–67 (Figure 6).

Core V10–65, located in the vicinity of Crete and containing several ash and turbidite layers, did not penetrate the sediments laid down during the last interglacial and present in core V10–67. A synopsis of assemblages contained in the eastern Mediterranean and south Aegean is given in Tables 2 to 5.

The Holocene-Pleistocene boundary is at about 40 cm depth in both cores V10–65 and V10–67; at this level *Globorotalia truncatulinoides* changes coiling direction from dextral in Pleistocene to sinistral in Holocene (Figure 6). This shift in coiling direction is accompanied by changes in the composition of faunal assemblages (Vergnaud-Grazzini and Herman-Rosenberg, 1969).

The Last Glacial began about 70,000 years ago (base of unit VIII in Figure 6); several mild oscillations of varying length and amplitude interrupted this cold period (Figure 6; also Emiliani, 1955; Parker, 1958; Vergnaud-Grazzini and Herman-Rosenberg, 1969). The last interstade commenced about 35,000 B. P. and ended 28–29,000 years ago, when the last cold stade began (Figure 6). As a consequence of temporary stagnation of the deep water, black sapropelic silty lutites were deposited repeatedly in the last 50,000 years.

Units IX, X, and XI represent a part of the Last Interglacial (Figure 6) but since the core terminates in this bed its duration is unknown. Paleotemperature ($O^{18/16}$) measurements indicate that during the Last Interglacial surface water temperatures were similar to those of the present (Vergnaud-Grazzini and Herman-Rosenberg, 1969).

In conclusion, micropaleontologic analyses and lithologic examination of eastern Mediterranean cores indicate that variations in faunal composition and sediment characteristics with time were due to climatic and hydrologic oscillations.

Radiometric age determinations suggest that the climatic oscillations in the Mediterranean were contemporaneous with those in the Red Sea and the Atlantic Ocean (Ericson *et al.,* 1961; Herman-Rosenberg, 1965; Herman, 1971).

The eastern Mediterranean Holocene fauna has close affinities with the Red Sea Holocene fauna. In contrast with the Holocene, during the Last Glacial stades the faunas of the Red Sea and the Mediterranean were totally different, suggesting that the amplitude of climatic fluctuations was greater in the Mediterranean.

Compared to the climatic fluctuations in the Atlantic Ocean (Ericson *et al.*, 1961), surface water temperature changes in the Mediterranean were more marked because of its relatively small size and proximity to land.

In addition to climatic oscillations during the Quaternary, the Mediterranean was the scene of continuing tectonic activity, as indicated by geophysical measurements (Hersey, 1965; and Emery *et al.*, 1966). The occurrence of numerous ash layers, turbidites, and slump deposits interbedded with Quaternary pelagic sediments in the eastern Mediterranean provide additional evidence of Quaternary regional tectonic movements.

ACKNOWLEDGMENTS

I thank R. E. Garrison of the University of California at Santa Cruz for making the scanning electron miscroscope (SEM) available and G. Wolery for technical assistance; the micrographs were taken with a JEOLCO J. M. S.–2 instrument; N. Schneiderman, University of Illinois, for one micrograph taken with a Cambridge Mark II A instrument at the Central Electron Microscopy Laboratory; Lamont-Doherty Geological Observatory scientists for collecting the VEMA cores; Iaakov Nir, Israel Geological Survey, for collecting the "H" cores aboard the R/V PILLSBURY; P. E. Rosenberg and J. W. Mills for critically reading the manuscript; R. Capo, LDGO, for shipping several core samples; Z. Reiss for making Laboratory facilities available during the writer's visit to the Hebrew University in Jerusalem; J. Thommeret, C^{14} Laboratory, Institute of Oceanography, Monaco, and D. Thurber and W. Broecker, LDGO, for the radiocarbon determinations.

Funds for shipboard coring operations and curatorial services of the Lamont-Doherty Geological Observatory cores were provided by grants ONR (N00014–67–A–0108–0004) and NSF–GS–10635 to Lamont-Doherty Geological Observatory. This investigation was supported by the Oceanography Section, National Science Foundation, NSF GA–16500 and by funds provided by the WSU Graduate School Development Fund (14 N–2940–0020).

REFERENCES

Berger, W. H. 1967. Foraminifera ooze: solution at depths. *Science,* 156 (3773): 383–385.

Biel, E. R. 1944. Climatology of the Mediterranean area. *A Publication of the Institute of Meteorology of the University of Chicago, Miscellaneous Reports,* 13, 180 p.

Blanc-Vernet, L., H. Chamley and C. Froget 1969. Analyse paléoclimatique d'une carotte de Méditerranée nord-occidentale. Comparaison entre les résultats de trois études: foraminifères, ptéropodes, fraction sédimentaire issue du continent. *Paleogeography, Paleoclimatology, Paleoecology,* 6 (3): 215–235.

Cita, M. B., and M. A. Chierici 1962. Crociera talassografica Adriatica 1955. V. Ricerche sui foraminiferi contenuti in 18 carote prelevate sul fondo del Mare Adriatico. *Archivio di Oceanografia e Limnologia Venezia,* 12 (3): 297–359.

Emery, K. O., and Y. K. Bentor 1960. The continental shelf of Israel, *Israel Geological Survey Bulletin,* 26: 25–41.

Emery, K. O., B. C. Heezen, and T. D. Allan 1966. Bathymetry of the Eastern Mediterranean Sea. *Deep-Sea Research,* 13: 173–192.

Emiliani, C. 1955. Pleistocene temperature variations in the Mediterranean. *Quaternaria* 2: 87–98.

Emiliani, C., and R. F. Flint 1963. The Pleistocene record. In: *The Sea,* ed. Hill, M. N., Interscience, New York, 3: 888–927.

Ericson, D. B., M. Ewing, G. Wollin and B. C. Heezen 1961. Atlantic deep-sea sediment cores. *Geological Society of America Bulletin,* 72: 193–286.

Ericson, D. B., and G. Wollin 1964. *The Deep and the Past,* A. A. Knopf, New York, 288 p.

Goncharov, V. P., and O. V. Mikhailov 1964. New data on the bottom relief of the Mediterranean. *Deep-Sea Research,* 11: 625–628.

Herman, Y. 1966. Climatic changes in Quaternary cores from the Mediterranean and Red Sea basins recorded by 1) pteropods, 2) planktonic foraminifera. *Second International Oceanographic Congress,* (abstract): 156–157.

Herman, Y. 1968. Evidence of climatic changes in Red Sea cores. *7th INQUA Congress Proceedings, Means of Correlation of Quaternary Sequences,* 8: 325–348.

Herman, Y. 1971. Vertical and horizontal distribution of pteropods in Quaternary sequences. In: *Micropaleontology of Oceans,* Funnel, B. and W. Riedel, eds., SCOR, Cambridge Proceedings, Cambridge University Press, Cambridge, 463–486.

Herman, Y., J. Thommeret and C. Grazzini 1969. Micropaleontology, paleotemperatures, and radiocarbon dates of Quaternary Mediterranean deep-sea cores. *8th INQUA Congress Proceedings,* (abstract): 174.

Herman-Rosenberg, Y. 1965. Etudes des sédiments quaternaires de la Mer Rouge. *Annales de l'Institut Océanographique,* Masson et Cie, 42 (3): 343–415.

Hersey, J. B. 1965. Sedimentary basins of the Mediterranean Sea. In: *Submarine Geology and Geophysics,* eds., Whittard, W. F. and R. Bradshaw. Butterworths, London, 75–91.

Holeman, J. N. 1968. The sediment yield of major rivers of the world. *Water Resources Research,* 4 (4): 737–741.

Koczy, F. F. 1956. Echo soundings. *Reports of the Swedish Deep-Sea Expedition, 1947–1948,* 4 (2): 99–131.

Ludwing, W. J., B. Gunturi and M. Ewing 1965. Sub-bottom reflection measurements in the Tyrrhenian and Ionian seas. *Journal of Geophysical Research,* 70 (18): 4719–4723.

Mellis, O. 1954. Volcanic ash-horizons in deep-sea sediments from the eastern Mediterranean. *Deep-Sea Research,* 2: 89–92.

Ninkovich, D., and B. C. Heezen 1965. Santorini Tephra. In: *Submarine Geology and Geophysics,* eds., Wittard, W. F. and R. Bradshaw, Butterworths, London, 413–453.

Olausson, E. 1960. Description of sediment cores from the Mediterranean and the Red Sea. *Reports of the Swedish Deep-Sea Expedition, 1947–1948,* 8 (3): 287–334.

Olausson, E. 1961. Sediment cores from the Mediterranean Sea and the Red Sea, studies of deep-sea cores. *Reports of the Swedish Deep-Sea Expedition, 1947–1948,* 8 (4): 337–387.

Parker, F. L. 1955. Distribution of planktonic foraminifera in some Mediterranean sediments. *Deep-Sea Research Supplements,* (Papers in Marine Biology and Oceanography) 3: 204–211.

Parker, F. L. 1958. Eastern Mediterranean foraminifera. *Reports of the Swedish Deep-Sea Expedition, 1947–1948,* 8 (2) :217–283.

Parker, F. L. 1962. Planktonic foraminiferal species in Pacific sediments. *Micro-paleontology,* 8 (2): 219–254.

Pollak, M. J. 1951. The source of the deep water of the eastern Mediterranean Sea. *Journal of Marine Research,* 10 (1): 128–152.

Ryan, W., F. Workum and J. B. Hersey 1965. Sediments on the Tyrrhenian abyssal plain. *Geological Society of America Bulletin,* 76: 1261–1282.

Ryan, W., E. Olausson and R. W. Fairbridge 1966. Mediterranean Sea. In: *The Encyclopedia of Oceanography,* ed.

Fairbridge, R. W., Reinhold Publishing Company, New York, 490–502.

Tchernia, P., and H. Lacombe 1959. Hydrological cycle in the Mediterranean. *International Oceanographic Congress* (abstract): 520–521.

Tesch, J.J. 1946. The Thecosomatous pteropods: The Atlantic. In: *Dana Report,* 28, Carlsberg Foundation, Copenhagen, 82 p.

Tesch, J. J. 1948. The Thecosomatous pteropods: The Indo-Pacific. In: *Dana Report*, 30, Carlsberg Foundation, Copenhagen. 45 p.

Todd, R. 1958. Foraminifera from Western Mediterranean deep-sea cores. *Reports of the Swedish Deep-Sea Expedition, 1947–1948,* 8 (3): 169–215.

Vergnaud-Grazzini, C., and Y. Herman-Rosenberg 1969. Etude paléoclimatique d'une carotte de Méditerranée orientale. *Revue de Géographie Physique et Géologie Dynamique*, 11 (3): 279–292.

Wong, H. K., and E. F. K. Zarudzki 1969. Thickness of unconsolidated sediments in the eastern Mediterranean Sea. *Geological Society of America Bulletin*, 80: 2611–2614.

Wüst, G. 1959. Remarks on the circulation of the intermediate and deep water masses. In: *The Mediterranean Seas and the Methods of their Further Exploration. Annali dell'Instituto Universitario Navale de Napoli*, 28: 3–15.

Wüst, G. 1960. Die Tiefenzirkulation des Mittelländischen Meeres in den Kernschichten des Zwischen- und des Tiefenwassers. *Deutsche Hydrographische Zeitschrift,* 13 (3): 105–130.

Wüst, G. 1961. On the vertical circulation of the Mediterranean Sea. *Journal of Geophysical Research*, 66 (10): 3261–3271.

Stratigraphy of Late Quaternary Sediments in the Eastern Mediterranean*

William B. F. Ryan

Lamont-Doherty Geological Observatory of Columbia University, Palisades, New York

ABSTRACT

The frequency abundancies of pelagic foraminifera have been quantitatively evaluated by Factor- and Vector-analysis in a sediment sequence ranging from the present back approximately 400,000 years. R-mode groupings of proportional similarity have been interpreted as indicating species of foraminifera which respond principally to temperature and salinity changes in the surface waters. Realistic climate curves can be constructed by Q–mode analysis.

The times of deposition of sapropelic mud were synchronous throughout the deep-water basins of the eastern Mediterranean. Numerous sediment layers can be cross-correlated across large distances on the basis of lithology, mineral composition, faunal zones and sapropel chronology. Climate curves inferred from faunal and O^{18}/O^{16} isotopic measurements are similar to established curves from the Caribbean.

The sapropel layers reflect stagnant conditions which always occur on the warming trend of the generalized climate curve near a maximum in sea level. The sudden onset of stagnant conditions is believed to have been caused by a marked density stratification created when the level of the eastern basin during a transgression, reached the −40 m sill depth in the Bosphorus. At this time the dense saline waters of the Aegean entered the Black Sea and triggered a turnover which effectively flushed the lighter Black Sea water back into the eastern basin of the Mediterranean, where it resided as a discrete thin film of surface water for several thousand years.

Sediment deposition rates during the "cold" W and Y faunal zones were noticeably higher in the Ionian Basin than was corresponding deposition in the Levantine Basin. Deposition was very slow in the "warm" X faunal zone of the Ionian Basin sediments. Mediterranean Ridge sedimentation rates range from approximately 2.5 to 6 cm/1000 years. On the abyssal plains accumulation has been as much as 200 cm/1000 years.

RESUME

L'analyse factorielle et vectorielle quantitative de la répartition des populations de foraminifères pélagiques a été poursuivie dans une séquence de sédiments dont l'âge remonte du présent jusqu'à 400.000 ans. Les groupements de module R de similitudes proportionnelles ont été interprétés comme étant une indication de l'existence d'espèces de foraminifères qui sont particulièrement sensibles aux variations de température et de la salinité des eaux de surface. Il est possible d'établir des courbes de climat réel à partir des données de l'analyse de module Q.

Les dépôts de boues sapropéliques sont du même âge dans tous les bassins de la Méditerranée orientale. L'étude de la lithologie, de la composition minérale, des zones de la faune et de la chronologie des sapropels permet d'établir une corrélation entre de nombreuses couches de sédiments sur de grandes distances. Les courbes climatiques déduites des mesures de la faune et des isotopes ^{18}O et ^{16}O sont similaires aux courbes climatiques des Caraïbes.

Les couches de sapropels présentent les caractères de stagnation qu'on trouve toujours sur la pente positive d'une courbe climatique théorique un en point voisin du maximum en niveau de mer. La cause de l'apparition soudaine des conditions de stagnations est attribuée à une stratification de densité marquée qui a pris place quand le bassin oriental, lors d'une transgression, a atteint le seuil de −40 m de profondeur dans le Bosphore. A cette époque les eaux salines denses de la Mer Egée ont pénétré dans la Mer Noire et déclenchèrent un retournement en repoussant les eaux de la mer Noire moins denses, dans le Bassin oriental de la Méditerranée où elles sont restées pendant plusieurs milliers d'années sous la forme d'un film d'eau de surface.

Pendant la formation des zones faunales "froides" W et Y, la vitesse de dépôt des sédiments était nettement plus grande dans le Bassin Ionien que celle de dépôts correspondants dans le Bassin du Levant. Le dépôt était très lent dans la zone faunale "tiede," X des sédiments de bassin Ionien. Les vitesses de dépôts sédimentaires

*Lamont-Doherty Geological Obervatory Contribution No. 1894

Core RC9–181 has a complete succession of sediment layers and shows steady and uniform deposition during the last 400,000 years, with a record of five brief excursions (short events) in the geomagnetic field. It is suggested that this core be considered a standard stratigraphic section for the Late Quaternary in the eastern Mediterranean Sea.

sur la dorsale méditerranéenne sont comprises entre 2,5 et 6 cm/1000 ans. Dans les plaines abyssales, la vitesse d'accumulation a atteint jusqu'à 200 cm/1000 ans.

La carotte RC9–181 présente une succession complète de couches de sédiments et révèle un dépôt uniforme pendant les 400.000 dernières années avec l'indication de 5 manifestations du champ magnétique. On suggère que cette carotte soit considérée comme étalon de section stratigraphique pour le Quaternaire supérieur dans la Méditerranée orientale.

INTRODUCTION

The stratigraphy of the Mediterranean Sea reveals many details concerning ancient climates, ancient geography, and ancient oceanic circulation. An absolute chronology has been established for the surficial section of the sedimentary series which extends back approximately one-half million years. However, the most detailed record lies in the uppermost ten meters of soft mud which have been repeatedly sampled in piston cores, involving events which occurred in this region in the last few hundred thousand years.

The first systematic examination of eastern Mediterranean sediments from a stratigraphic point of view was made on long piston cores collected by the ALBATROSS during the Swedish Deep-Sea Expedition of 1947–1948. Kullenberg (1952) investigated the interstitial salinity in the sediments and attempted correlation of a few of the cores by means of chemical analyses and also from other lithological and faunal controls. He recognized the individual beds of sapropelic mud and their rich content of organic matter and suggested that at the time of their deposition the deep water mass of this basin was completely without oxygen. He speculated that the most likely cause of stagnation would be the development of a density stratification during pluvial conditions that prevailed during a glacial maximum. He considered the following criteria to be important in producing the density stratification: higher precipitation, lower temperature, and lower eustatic sea-level. The last two would necessarily accompany the cooling stage of a glacial cycle when the lower temperature would result in lower evaporation and the lower sea-level would create a shallower sill at Gibraltar, thus restricting the two-way circulation with the Atlantic Ocean.

Earlier, Bradley (1938) predicted that stagnant conditions and the deposition of sapropelic muds would be confined exclusively to the eastern Mediterranean basin. Like Kullenberg, he emphasized the simultaneous occurrence of pluvials and low stands of sea level. The pluvials would create low salinity and a low-density film of surface water, which Bradley inferred would reverse the current direction in the Strait of Gibraltar. At that time the Atlantic water would enter the Mediterranean as an underflow and circulate freely in the Balearic and Tyrrhenian basins. However, during low sea level the very restricted and shallow passage to the eastern basin would prevent circulation and renewal there and would result in a stagnant phase. "Under these conditions the water becomes poisoned with hydrogen sulfide which kills off all the bottom fauna. With no bottom dwelling organisms to disturb the sediment and aid in the destruction of the organic matter the sediment accumulates as successive pairs of thin laminae, one of which is rich in organic matter." (Bradley, 1938, p. 377).

However, subsequent studies by Mellis (1954) on the correlation of tephra layers and by Rubin and Suess (1955), who were able to obtain radiocarbon dates of 9000 years B. P. for the sediment below the uppermost sapropel, demonstrated that the last period of oxygen deficiency resulting in complete stagnation occurred, not under low sea level conditions in the glacial period, but rather in postglacial (Holocene) times. Further radiocarbon dating on additional cores showed that this stagnant period started about 9000 years B.P. and lasted approximately 2000 to 3000 years (Olsson, 1959; Menzies *et al.*, 1961). Thus the most recent stagnant condition does not necessarily coincide with either a glacial condition or a major lowering of sea level. To determine the environment at the times of other stagnant periods, it is necessary to examine the stratigraphy and lithology of the sedimentary series of this region.

Parker (1958) carried out a study of the foraminifera in 15 of the ALBATROSS cores. She noticed that the modern planktonic fauna of the eastern Mediterranean is to a certain extent indigenous, probably because of an ecological barrier in the Strait of Gibraltar caused by a sudden temperature change and/or other factors. Species present in the adjacent Atlantic are absent from the Mediterranean. From analysis of relative faunal abundances of certain species which were identified as "warm" fauna and others which reflected "cool" conditions, Parker found that she could construct warm-cold curves which showed similar sequences in the different cores. Furthermore, her curve for core ALB 189 agreed semi-qualitatively with a paleotemperature curve based on the isotopic measurements of O^{18}/O^{16} (Emiliani, 1955).

Like Emiliani, Parker defined intervals which were predominantly cold and those which were predominantly warm. These intervals were separated by boundaries placed at the levels of transition. The resulting stages were correlated to a glacial sequence of the late Pleistocene and Recent. In subsequent studies, Olausson (1961, 1965) has analyzed Emiliani's and Parker's data and has redefined his own stages in relation to the then-known Pliestocene chronology. Olausson's provisional correlation is in general agreement with the sequences of Kullenberg (1952), and Emiliani (1955, 1958) and disagrees with that of Parker (1958) as to whether the first two cold intervals belong to the Würm (Würm I and Würm II-III respectively) or whether the entire Würm is represented in the first major cold interval.

An important contribution of Olausson (1961) was the recognition that several of the sapropels were deposited after major cool intervals. He proposed that the density stratification might be produced by meltwater from the glaciers and/or the general increase in the temperature of the water column after an ice age. The warming of the sea surface in itself would decrease the possibilities of forming surface water sufficiently dense to sink to the bottom. Another possibility, suggested by Olausson, is the contribution of water from the Black Sea, particularly the flushing of this basin at the time when sea level would rise to the –40 m sill depth of the Bosporus. This flushing would be caused by a supply of Mediterranean water to the Black Sea, which, because of its greater density, would displace the large volume of brackish meltwater of the Black Sea back into the Aegean and then on into the eastern basin. Olausson was unsure exactly which of these phenomena was the primary cause of the euxinic phase, or whether some combination of these factors was responsible. This question will be pursued in a later section.

Before one can establish the causes of the euxinic phases it is necessary to establish a precise chronology for the sedimentary succession.

CLIMATIC INFERENCES FROM PLANKTONIC FORAMINIFERA

The climate curve established by Parker (1958) contains many more secondary oscillations than the isotopic paleotemperature curve (Emiliani, 1966). It is likely that many of the species of foraminifera respond to ecological conditions in the surface water environment other than water temperature. The restructuring of the water mass during periods of decreased ventilation is apt to be accompanied by changes in salinity and even nutrient content. Therefore, before a true paleotemperature curve can be obtained from data on the frequency abundance of the planktonic foraminifera, the particular responses to temperature must be understood and must be distinguished from the responses to other influences.

Finally, in order to demonstrate that response to temperature can actually be determined, it is necessary to correlate the resulting Mediterranean climate curve to other climate curves established in cores from exterior regions such as the North Atlantic or Caribbean, where the changes in salinity or nutrient budget would not be expected to be marked.

Schott (1935), Ericson and Wollin (1956), and Ericson *et al.* (1961) have shown that *Globorotalia menardii* (d'Orbigny) and its subspecies are useful markers of climatic zones in the North Atlantic and Caribbean. *Globorotalia menardii* and similarly *Pulleniatina obliquiloculata* (Parker and Jones) are found in strata deposited beneath relatively warm water masses.

Neither of these two important species is present in the Mediterranean (Parker, 1958). The limited latitudinal range in the eastern Mediterranean does not allow for a large range of surface-water temperatures in any one season; consequently it is impossible on the basis of data on the present distribution of foraminifera alone to classify which species here are cold-water indicators and which are warm-water indicators.

Parker (1958) has noted that a few species which previous work in the North Atlantic had shown to prefer cold-water environments, appear in the Mediterranean cores and fluctuate in unison and in opposition to a few species preferring warm-water environments. However, she also states that "in the eastern Mediterranean the problem of detecting how planktonic species respond to temperature changes is complicated by the fact that some species appear to be influenced sometimes by other factors." (Parker, 1958, p. 235). That the other factors may be important is clearly demonstrated by the discovery that in certain sediment layers the tests of both the cold- and warm-water species are found together in exceedingly large abundances. At these times, for as yet unknown reasons, some of the species did not act in an easily perceptible rhythm with other species having supposedly similar temperature preferences, or conversely, with those of opposite preferences.

Warm-Cold Curves

In her study of eastern Mediterranean cores, Parker (1958) concluded that the most reliable indicators of cool-water environments are *Globigerina pachyderma* (Ehrenberg) and *Globorotalia scitula* (Brady). Those species that prefer warmer water are *Globigerinella aequilateralis* (Brady), *Globigerinoides ruber* (d'Orbigny), *Globigerinoides sacculifer* (Brady), and *Hastigerina pelagica* (d'Orbigny).

Warm-cold climate zones were identified by examining the relative abundances of these species along the lengths of many of the ALBATROSS cores. Using these inferred climate zones and the lithologic properties of the sediments, Olausson (1961) was able to establish a correlation of individual bedding sequences in Cores 189, 194 and 195; Figure 1 illustrates this correlation.

The coring process with the Kullenberg apparatus occasionally failed to recover the near-surface sediments, as shown in the correlation diagram. This failure is apparent in Cores 189 and 195. Olausson (1961) felt that except for its missing top, Core 189 contained both the longest and most complete sediment sequence in the ALBATROSS suite. Consequently this core has been chosen for the subsequent analysis.

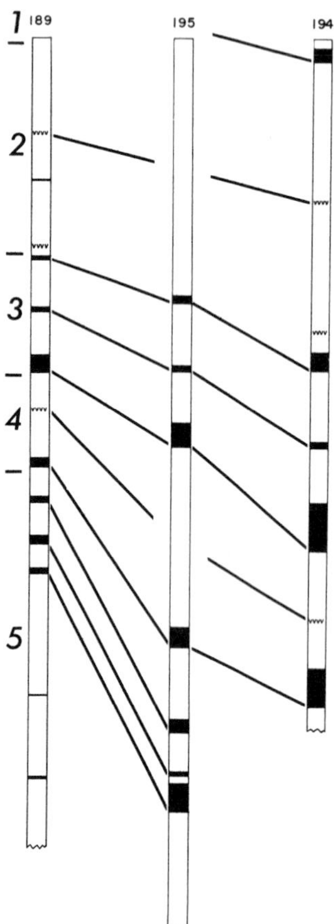

Figure 1. Cross-correlation of ALBATROSS Cores 189, 195 and 194(after Olausson, 1961). The black layers are the sapropelic muds, the v's represent beds of tephra. The numbered stages follow the usage of Olausson in defining the paleoclimate zones of the eastern Mediterranean sediment sequence. At core sites 189 and 195 the coring process did not recover the uppermost seafloor sediment.

Figure 2 is a diagram of Core 189 in which many variables are plotted against depth. The black zones in the core sketch are the sapropel muds. The upper 30 cm of Core 194 have been added to the top of Core 189 to create a generally complete sequence from the present back into the Quaternary.

The agreement in the trends of the O^{18}/O^{16} curves with those of the faunal curves is generally good. However, there are many spikes on the curves which are not found on the isotope curve and which are synchronous with the deposition of sapropelitic mud. In these muds the tests of several of the species are noted to have anomalous or "abnormal" (after Parker, 1958) abundances. It was believed that these peculiarities might furnish the additional information necessary to recognize environmental changes other than temperature; the peculiarities were examined as follows.

Factor and Vector Analysis of Faunal-Frequency Abundances

A factor and vector analysis was performed on the frequency abundance data from all species of planktonic foraminifera at all sample depths in Core 189 and from the top of Core 194. The analysis employed a technique developed by Imbrie and van Andel (1964) and used a computer program (COVAP) written by Manson and Imbrie (1964).

The factor- and vector-analysis approach was taken primarily because in the sapropelic muds the abundances of several of the species [*i.e., Globigerina eggeri* (Rhumbler), *G. inflata, Globigerinoides ruber*, and *Globigerinita glutinata* (Egger)] seem to fluctuate rather widely. Considering the enclosed nature of the eastern Mediterranean basin and its isolation during the large oscillations of the Pleistocene sea levels, it was believed the compositional changes in the seawater might affect these species. Olausson (1965) has shown that in a pronounced stagnant phase, such as that at 335 to 350 centimeters in Core 189, the isotopic "temperatures" determined from different species showed large deviations. Notably, *Globigerina eggeri* yields higher apparent temperatures in stagnant phases and lesser apparent temperatures in the cool intervals than do other species. Olausson considered that these anomalous temperatures might indicate isotopic compositional changes in the surface water caused by excess glacial run-off and precipitation at these times. This in turn suggested that fluctuations in the surface salinity might have been influential ecologically.

Because they believe that Emiliani's (1966) allow-

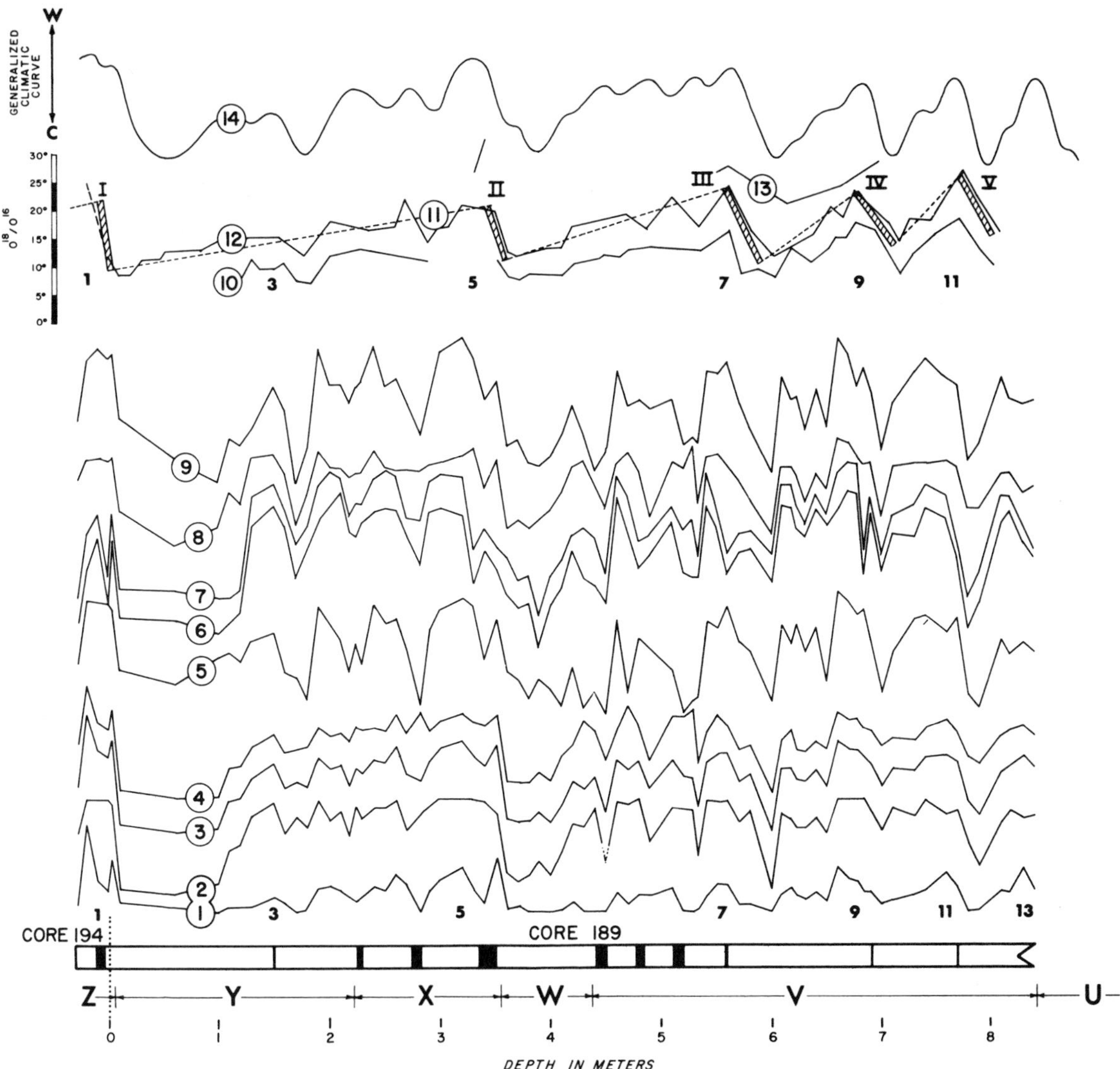

Figure 2. Climate curves in the eastern Mediterranean, based on analyses of faunal and isotopic data from ALBATROSS Core 189 and the upper 30 cm of Core 194. The black layers are the sapropelic muds.

Curve one is the sum of the frequency abundances of "warm" planktonic foraminifera selected by Parker (1958). The curve consists of the cumulative abundances of *Glogigerina aequilateralis, Globigerinoides sacculifer, Hastigerina pelagica,* and 1/10 the abundance of *Globigerinoides ruber*.

Curve two is an inverse of the sum of the frequency abundances of "cold" planktonic foraminifera selected by Parker (1958). This curve contains the cumulative abundances of *Globigerina pachyderma* and *Globorotalia scitula.*

Curve three is the sum of curves 1 and 2.

Curve four is curve 3 plus 1/5 the frequency abundance of *Globigerina eggeri.*

Curve five is the inverse of a plot of the coefficients of proportionality with respect to Vector B in the Q-mode factor- and vector analysis. The reference sample of Vector B is at 450 cm, and contains predominantly "cold" species of foraminifera.

Curve six is a plot of the coefficients of proportionality of Vector A in the Q-mode analysis. The reference sample of Vector A is at 670 cm, and contains predominantly "warm" species.

Curve seven is a plot of Vector A minus the sum of Vector B and Vector C, all in the Q-mode analysis. Curve seven is a total climate curve derived from the Q-mode analysis.

Curve eight is a climate curve based on the R-mode analysis where the frequency abundances of the temperature-sensitive species of Vector A were weighted and cumulated. This curve is considered the best estimate of *temperature* oscillations, with less disturbance from other environmental parameters.

Curve nine consists of the cumulative frequency abundance of all the "warm" species minus all the "cold" species. The "Warm" species according to Ruddiman (personal communication) are *Globigerina aequilateralis, Globigerinoides conglabatus, G. ruber, G. sacculifer, G. tenellus, Globorotalia truncatulinoides,* and *Hastigerina pelagica.* The "cold" species include *Globigerina bulloides, G. inflata, G. pachyderma, G. quinqueloba,* and *Globorotalia scitula.*

Curve ten is a paleotemperature curve after Emiliani (1955) from O^{18}/O^{16} measurements made on tests of *Globigerina inflata.*

Curve eleven is a generalized saw-toothed curve after Broecker and van Donk (1969) showing the definition of the primary glacial cycles.

Curve twelve is a paleotemperature curve after Emiliani (1955) using only the round tests of *Globigerinoides ruber*.

Curve thirteen is a paleotemperature curve using the tapered tests of *G. ruber.*

Curve fourteen is the "generalized climate curve" of Emiliani (1966) for cores from the Caribbean and North Atlantic which has been slightly stretched or squeezed to correlate as closely as possible with the *temperature* curve 8. The numbered stages follow the usage of Emiliani. The letter symbols at the bottom of the diagram are the faunal zones of Ericson *et al.* (1961). The boundaries of these zones are established by correlation of the eastern Mediterranean climate sequence with that of the Caribbean, and not by the *G. menardii* complex criteria.

ance for changing isotopic composition is much too small, several workers have doubted the validity of his assumption and, therefore, his conclusions (Olausson, 1965; Shackleton, 1967). The change in ocean isotopic composition results from the fact that the isotopic composition of atmospheric precipitation, and hence that of an ice sheet, differs from that of the ocean (Dansgaard, 1961).

The eastern Mediterranean is likely to be a reservoir for large volumes of glacial meltwater derived from the European and Scandinavian ice sheets, hence Olausson (1965) was of the opinion that contamination from meltwater is a very critical factor in evaluating the paleotemperature record from the region. It is suggested that the marine plankton have been as much influenced by changes in salinity and nutrients linked to the waxing and waning of the Pleistocene ice sheets as they have been by the surface-water temperature.

In the vector analysis both the R-mode and Q-mode tests were performed on the data from the two cores. In the Q-mode the frequency abundance values were percent-range-transformed in order to give each species equal weight. The raw data are presented in Parker (1958, her tables 10 and 15). The mean percent abundance and standard deviation from that mean for each species and all samples is tabulated in Table 1. In the R-mode test four rotated factors were selected on the basis of the list of positive eigenvalues shown in Table 2.

The subsequent reordered oblique projection matrix is shown in Table 3. In the first column of Table 3 the reference vector (called Vector A) has been chosen as *Globigerina bulloides* (d'Orbigny). A comparison of the present surface distribution of *G. bulloides*

Table 1. General statistics on the faunal abundances.

Variable		Mean	Standard Deviation
No.	Species		
1	*G. bulloides*	16.11	7.62
2	*G. digitata*	0.88	1.52
3	*G. eggeri*	14.01	15.70
4	*G. inflata*	6.91	6.49
5	*G. pachyderma*	4.22	5.82
6	*G. quinqueloba*	5.55	7.51
7	*G. radians*	0.10	0.29
8	*G. aequilateralis*	1.83	2.24
9	*G. glutinata*	4.22	4.62
10	*G. conglobatus*	0.22	0.85
11	*G. ruber*	29.34	20.27
12	*G. sacculifer*	1.11	3.24
13	*G. tenellus*	4.41	4.73
14	*G. scitula*	4.41	5.17
15	*G. truncatulinoides*	1.83	4.24
16	*G. pelagica*	0.10	0.25
17	*O. universa*	4.23	4.12

Table 2. Table of positive eigenvalues—R-Mode.

No.	Eigenvalue	Percent of Communality Over:			
		All (17) Factors		4 Rotated Factors	
1	3.953	23.3	23.3	46.5	46.5
2	1.744	10.3	33.5	20.5	67.0
3	1.496	8.8	42.3	17.6	84.5
4	1.315	7.7	50.0	15.5	100.0
5	1.158	6.8	56.9		
6	1.049	6.2	63.0		
7	1.014	6.0	69.0		
8	0.817	4.8	73.8		
9	0.763	4.5	78.3		
10	0.721	4.2	82.5		
11	0.663	3.9	86.4		
12	0.629	3.7	90.1		
13	0.560	3.3	93.4		
14	0.470	2.8	96.2		
15	0.390	2.3	98.5		
16	0.258	1.5	100.0		
17	0.001	0.0	100.0		

Trace of original matrix 17.000
Communality over 17 Factors = 17.000
4 Factors = 8.508

in the North Atlantic with surface temperature maps indicates that this species prefers cold waters in the range of 9 to 14°C (Bé and Tolderlund, 1971). Results show that *Globigerina pachyderma* and *Globorotalia scitula*, respectively, are found in the core samples in relative abundances proportionally similar to *Globigerina bulloides*. This is not surprising because *G. pachyderma* is well known as an indicator species in the Arctic Ocean and is the dominant form in the waters north of the Arctic Circle (Bé and Tolderlund, 1971).

Globigerinoides ruber, Globigerinella aequilateralis (Brady), and *Globorotalia truncatulinoides* (d'Orbigny) are found in generally inverse proportions to *Globigerina bulloides*. In fact, Bé and Tolderlund (1971, p. 132) state that "*Globigerinoides ruber* is the most successful warm water species in terms of distribution and abundance in the Atlantic Ocean." According to Tolderlund (1969) the temperature range of *Globigerinella aequilateralis* is similar to that of *Globigerinoides ruber*. *Globorotalia truncatulinoides* is predominantly a subtropical species. Consideration of these known traits suggests that Vector A associates with temperatures and exhibits a two-direction (warm-cold) polarity. The vector analysis reveals two species (*Globigerina bulloides* and *Globigerinoides ruber*) at the cold and warm extremes, respectively. The former was not used at all by Parker (1958) in the construction of her warm-cold curves.

In the second column of the oblique projection

matrix *Globigerina eggeri* (= *Globoquadrina dutertrei* (d'Orbigny) of some authors) was selected by the computations as the reference vector (called Vector B). In core sections this species is found in both cold and warm sequences and reaches very large abundances in many of the stagnant zones containing sapropelitic muds. Kullenberg (1952) and Parker (1958) discussed these abnormal accumulations of *Globigerina eggeri* and concluded that the large species population bears no relation to temperature conditions but instead might reflect reduced salinity. Bé and Tolderlund (1971, p. 122) state that "*Globoquadrina dutertrei* is a subtropical species which is especially abundant near the continental margins." The slope waters are known regions of reduced surface salinity. In the North Atlantic, Ruddiman (1969) emphasized that this species is the key indication of low salinity in the overlying surface water mass. *Globigerina quinqueloba* (Natland) has a coefficient of proportionality of 0.789, suggesting that it reacts in unison with *G. eggeri*. However, *G. quinqueloba* prefers cold water environments and is a predominantly sub-polar species. Nonetheless, *G. quinqueloba* is also found concentrated in low-salinity slope waters, particularly off the northeastern United States and Spain (Bé and Tolderlund, 1971).

The species exhibiting inverse proportionality to reference Vector B are *Globigerina radians* (Egger) and *Globigerinoides tenellus* (Parker). Tests of the latter are found in sediments beneath the south central Sargasso Sea, which is the region of maximum salinity in the North Atlantic (Ruddiman, 1969). It is suggested that the species associated with reference Vector B are influenced more by sea-surface salinity than by temperature.

In the sapropels, *Globigerinoides ruber* is occasionally found in large abundances, together with *Globigerina eggeri*. Bé and Tolderlund (1971) believe that *Globigerinoides ruber* is an euryhaline species whose maximum frequencies occur at the more extreme salinity values, *i.e.*, either above 36.0‰ or below 34.5‰. The abundance of this species during the stagnant phases is therefore further evidence that these phases are accompanied by a reduced salinity layer of surface water.

The reference species of Vector C in column three of the oblique projection matrix is *Globigerina inflata*. In the Atlantic Ocean this species is found only in transitional waters in regions of vertical mixing. *Orbulina universa* (d'Orbigny) is found in both transitional waters and subtropical waters (Bé and Tolderlund, 1971). *Globigerinita glutinata* (Egger) which occurs in a marked inverse proportionality to *Globigerina inflata* is one of the most ubiquitous species of planktonic foraminifera with a continuous distribution for subarctic to subantarctic waters (Bé and Tolderlund, 1971). Vector C seems to be associated with temperature, except that, in contrast to that of Vector A, the species have large temperature ranges and also prefer transitional regions. The species of Vector C occur most commonly during either warming or cooling trends, and are replaced by the Vector A species near the maximum warm phases or near the coolest phases. Vector C is interpreted as a temperature-gradient indicator, a first derivative of temperature with time (*i.e.*, depth in the core). Because individuals live only a very short time and are unable to perceive the general climate change, their presence is associated with phenomena that accompany long-period temperature fluctuations. The phenomenon in the eastern Mediterranean most sensitive to these changes is the rate of vertical mixing and ventilation. The overturn produced by the formation of bottom water in limited areas of the Adriatic and Aegean seas is the greatest during cooling phases and least during warming phases. The rate of vertical mixing controls both the production of the deep-water mass and the rate of outflow of the Levantine intermediate water mass. McGill (1961) has shown that the nutrient budget of the eastern Mediterranean is controlled by these exchanges of water, and that at the present time this region is greatly impoverished because the nutrients that enter this basin from rivers and the already depleted surface water flow from the North

Table 3. Reordered oblique projection matrix—R-Mode.

Name-Species	*G. bulloides*	*G. eggeri*	*G. inflata*	*H. pelagica*
	(A)			
G. bulloides	*1.000*	0.000	0.0000	0.000
G. pachyderma	0.914	0.283	−0.125	0.028
G. scitula	0.878	−0.005	−0.125	−0.180
G. truncatulinoides	−0.631	0.028	0.433	−0.263
G. aequilateralis	−0.710	−0.600	−0.108	0.040
G. ruber	−0.743	−0.600	−0.191	0.294
		(B)		
G. eggeri	0.000	*1.000*	0.000	0.000
G. quinqueloba	0.431	0.789	−0.086	0.045
G. tenellus	−0.438	−0.830	0.056	−0.584
G. radians	−0.266	−0.849	−0.437	−0.834
			(C)	
G. inflata	0.000	0.000	*1.000*	0.000
O. universa	0.047	−0.087	0.773	−0.136
G. digitata	0.268	0.087	0.741	0.537
G. sacculifer	−0.289	−0.473	−0.751	0.254
G. glutinata	0.191	−0.827	−0.831	−0.736
				(D)
H. pelagica	0.000	0.000	0.000	*1.000*
G. conglobatus	0.401	−0.050	−0.156	0.690

Atlantic are rapidly returned to this ocean before they have a chance to accumulate in the deep water mass. However, during periods of rapid cooling in the eastern Mediterranean the combination of lower evaporation and higher precipitation (glacial pluvials) might reduce the surface salinity in this area and reverse the current directions in the Strait of Gibraltar. This would result in a vast nutrient enrichment similar to present conditions in the Black Sea. Vector C is, therefore, tentatively associated with rates of vertical circulation and nutrient balance.

Vector D is partly an artifact of the vector-analysis calculation. The proportional similarity between *Hastigerina pelagica* and *Globigerinoides conglobatus* (Brady) reflects to some extent that these species only occur in very small abundances during the very warmest periods. Tolderlund (1969) has shown that *G. conglobatus* is an excellent indicator of fall conditions in subtropical waters; at that time the water temperatures are at their maximum. Vector D is possibly associated with phenomena related to the second derivative of temperature with time. *Hastigerina pelagica* occurs precisely at the temperature maxima of stages 11, 9, 7, 5, and 3 in Core 189, indicating that these limited periods are when this species finds its optimal and possibly very selective supply of nutrients. If a decrease in diversity of the phytoplankton were to accompany the temperature maximum as has been recognized in other oceans, then it would be expected that a decrease in the diversity of planktonic foraminifera would follow suit.

Climate Curves Based on R-Mode Analysis

With the oblique projection matrix as an *a priori* guide to species selection, it is possible to construct frequency abundance curves which may define either warm-cold intervals, periods of high or low salinity, or even warming or cooling trends. As an example, species associated with Vector A were chosen to create a paleotemperature curve. The frequency abundance of the individual species was weighted according to the projection of the species onto the reference vector. The resulting curve (Figure 2, Number 8), shows a marked qualitative agreement with the isotope paleotemperature curve 12. The matching of these curves is better than that of Parker's (curves 1 and 2) which used a subjective choice of species. The temperature minima between the major numbered stages are better defined and the secondary modulations of these cycles are fewer and have lower amplitudes, particularly in stage 5. The separation of stage 3 from stage 5 is very marked.

According to this temperature curve, the occurrence of sapropelic muds has a definite relationship to temperature. These layers occur near temperature highs (major or minor) in most cases along the warming part of the curve.

Ruddiman's Curve

To evaluate these curves impartially, the author requested that Ruddiman suggest weights for the various species to create a "total" climate curve, using information for all the species. Curve 9 is based on the knowledge of the present surface-sediment distribution of these species in the North Atlantic (Ruddiman, 1969). It is interesting to note that this curve is also in good qualitative agreement with the isotopic paleotemperature curves and with Curve 8.

Q-Mode Analysis

In the Q-mode, attention is focused on the 79 samples, and results follow from inspection of a 79 × 79 matrix of relationships between all pairs of samples. For this particular study three rotated factors were chosen on the basis of the list of positive eigenvalues in Table 4. The oblique projection matrix for the Q-mode analysis is given in Table 5.

Reference Vector A from this matrix is associated with a core sample taken at 670 cm depth in Core 189. Column 1 of Table 5 lists the value of proportional similarity of all the other core samples to this one particular reference sample. Curve 6 of Figure 2 is

Table 4. Table of positive eigenvalues—Q-mode.

No.	Eigenvalue	Percent of Communality Over All (79) Factors		3 Rotated Factors	
1	39.830	50.4	50.4	72.7	72.7
2	10.432	13.2	63.6	19.0	91.7
3	4.551	5.8	69.4	8.3	100.0
4	3.409	4.3	73.7		
5	3.140	4.0	77.7		
6	2.727	3.5	81.1		
7	2.122	2.7	83.8		
8	2.046	2.6	86.4		
9	1.830	2.3	88.7		
10	1.679	2.1	90.8		
11	1.430	1.8	92.7		
12	1.233	1.6	94.2		
13	1.176	1.5	95.7		
14	1.074	1.4	97.1		
15	0.865	1.1	98.2		
16	0.818	1.0	99.2		
17	0.638	0.8	100.0		

Trace of original matrix 79.000
Communality over 79 factors = 78.999
3 factors = 54.812

a plot of these values against sample depth. The striking similarity of this curve with Curves 8 and 9 suggests that this is a temperature curve. Subsequent inspection of the frequency abundance of the various species in the reference sample revealed that this sample is from a warm interval at the peak of stage 9.

Similarly, column 2 of Table 5 contained reference Vector B from 450 cm in this core. Curve 5 is an upside-down plot of the values related to this reference sample. The close inverse relationship of Vectors A and B is explained when it is noted that the fauna at 450 cm prefer cold-water environments.

Reference Vector C is located at 30 cm in Core 194, which in the graphs of Figure 2, corresponds to 1 cm left of the origin for Core 189. This sample is from very near the base of the sapropelic mud horizon in core stage 1 where there is a high concentration of both *Globigerina bulloides*, *Globigerinoides ruber* and *Globigerinita glutinata*. A curve constructed from values of proportional similarity from column 3 is not shown, but has sharp maxima in all the sapropelic mud layers as well as near the two major tephra horizons at 104 and 390 cm.

A Combined Curve from the Q-Mode Analysis

Curves 6 and 5, respectively, are similar since one is based on samples proportionally similar to a "warm" reference sample and the other to a "cold" reference sample. It is noted that the zones in the cores in which the sapropelitic muds were deposited contain faunal abundances such that these regions of the curves least resemble the isotopic temperature curves. For this reason it was decided that the coefficients of proportional similarity for Vector C would be added to those of Vector B and that this total would be subtracted from Vector A. The combined curve shows that in the eastern Mediterranean, climate curves which are not confined to any subjective *a priori* reasoning can be constructed from faunal data. The Q-Mode analysis has served to identify strong polarities or extremes in relative abundance patterns to which the individual samples are compared. It is reasonable to expect that these extremes are caused by environmental conditions and that curves based on proportional similarity would reflect oscillations in the paleoenvironment, such as climate and circulation changes accompanying the glacial cycles of the Pleistocene Epoch.

PALEOMAGNETIC MEASUREMENTS

The paleomagnetic method was chosen to correlate objectively the climatic fluctuations in the eastern

Table 5. Reordered oblique projection matrix—Q-mode.

Core Number and Depth (cm)	89–670	89–450	94–1	
	(1)			
89–670	*1.000*	0.000	0.000	(1) Vector A
89–210	0.988	0.333	−0.053	
89–678	0.978	0.111	−0.109	
89–460	0.946	0.160	0.031	
89–540	0.873	0.169	−0.133	
89–200	0.869	0.178	−0.147	
89–150	0.867	0.258	−0.016	
89–608	0.857	0.200	0.061	
89–617	0.852	0.376	−0.132	
89–740	0.837	0.117	0.118	
89–299	0.835	0.065	−0.011	
89–640	0.831	0.223	0.156	
89–250	0.830	0.181	−0.206	
89–660	0.825	−0.098	0.102	
89–260	0.821	0.141	−0.063	
89–818	0.816	0.465	−0.039	
89–290	0.806	0.242	−0.102	
89–690	0.803	0.339	0.125	
89–240	0.803	0.023	0.077	
89–320	0.773	−0.040	0.082	
89–760	0.745	0.270	0.062	
89–190	0.728	0.032	0.114	
89–510	0.721	0.607	−0.099	
89–810	0.709	0.334	−0.037	
89–229	0.708	0.550	−0.089	
89–160	0.680	0.611	−0.059	
89–130	0.679	0.334	−0.048	
89–622	0.654	0.223	0.195	
89–550	0.644	0.238	0.318	
89–218	0.638	0.616	−0.025	
89–710	0.611	0.458	0.087	
89–269	0.606	0.402	0.191	
89–224	0.601	0.362	−0.026	
94–10	0.592	−0.005	0.519	
89–730	0.587	0.149	0.180	
89–630	0.586	0.445	0.224	
89–3	0.579	0.061	0.409	
89–830	0.540	0.363	0.306	
89–580	0.482	0.448	0.239	
89–570	0.452	0.358	0.354	
89–490	0.427	0.420	0.292	
		(2)		
89–450	0.000	*1.000*	0.000	(2) Vector B
89–520	0.269	0.981	−0.503	
89–790	−0.004	0.941	0.018	
89–430	0.216	0.938	−0.556	
89–410	0.126	0.921	−0.056	
89–282	0.338	0.912	−0.550	
89–600	0.182	0.878	−0.060	
89–528	0.229	0.869	−0.628	
89–380	−0.008	0.859	0.147	
89–180	0.480	0.855	−0.235	
89–533	0.233	0.841	0.054	
89–780	−0.224	0.814	0.361	
89–470	0.612	0.804	−0.597	
89–700	0.281	0.803	0.097	
89–440	0.158	0.792	−0.286	
89–400	−0.042	0.779	0.216	
89–60	−0.151	0.726	0.390	
89–650	0.687	0.699	−0.114	

Table 5. (continued)

Core Number and Depth (cm)	89–670	89–450	94–1	
89–370	−0.049	0.686	0.358	
89–389	−0.402	0.678	0.664	
89–170	0.228	0.659	0.026	
89–360	0.065	0.617	0.132	
89–10	−0.129	0.596	0.517	
89–420	0.462	0.567	0.193	
89–340	0.466	0.495	0.153	
			(3)	
94–1	0.000	0.000	*1.000*	(3) Vector C
89–559	0.259	0.028	0.834	
89–684	0.250	0.058	0.794	
89–110	−0.177	0.448	0.768	
94–28	−0.159	0.521	0.744	
89–330	0.160	0.037	0.678	
89–770	0.405	0.143	0.665	
89–351	0.296	0.142	0.660	
89–100	−0.274	0.482	0.651	
89–120	−0.081	0.523	0.648	
94–20	0.298	−0.024	0.584	
89–840	0.411	0.435	0.530	
89–480	0.310	0.319	0.400	

Mediterranean cores with established core sequences from other oceanic regions. Kullenberg (1952), Emiliani (1955), Parker (1958), and Olausson (1961, 1965) have proposed schemes for comparing the faunally determined climate curves for Core 189 to an established Pleistocene sequence. Olausson (1965) has further extended this correlation to other ALBATROSS cores in the North Atlantic.

In the eastern Mediterranean cores there is no diagnostic species which is present or absent in any one particular warm or cold interval (such as Ericson's *G. menardii* complex) which can be used unequivocally to differentiate this interval from an adjacent one. The strongest argument used by Olausson in his correlation is that the basic sequence in the climate curve is similar to that found in North Atlantic cores. The basic approach consisted of cross-correlation on the basis of lithology and non-biogenous stratigraphy of several of the long ROBERT D. CONRAD Cruise 9 cores, and also correlation with the ALBATROSS cores, in particular Core 189 for which the climate curves were established. Because the entire sequence under investigation was probably less than $\frac{1}{2}$ million years in age, the sediments all would have been deposited within the Brunhes Normal Epoch.

The author wondered if variations in the magnetic field intensity, small excursions of the magnetic poles (called 'magnetic happenings'), or very short magnetic events could be detected in detailed correlatable lithologies of cores where generally rapid rates of deposition prevailed (3 to 10 cm/1000 years). First, magnetic correlations between the Mediterranean cores, which could be verified by other correlation techniques must be found, and then similar correlations looked for in North Atlantic and Caribbean cores for which faunal analyses and isotopic measurements had been made.

The author and J.H. Foster investigated magnetic measurements of a dozen CONRAD cores. The investigations of Atlantic, Caribbean and Indian Ocean cores has been reported on by Smith and Foster (1969).

Table 6. Core locations.

Core No.	N. Latitude	E. Longitude	Depth (m)
Mediterranean Cores			
ALB – 189	33°54′	28°29′	2664
ALB – 194	34°48′	23°29′	3000
ALB – 195	35°52′	21°53′	3665
V10 – 65	34°37′	23°25′	2586
V10 – 67	35°42′	20°43′	2890
RC9 – 174	32°58′	32°25′	1397
RC9 – 175	35°51′	32°16′	2639
RC9 – 176	36°01′	31°28′	2465
RC9 – 177	33°42′	30°05′	2820
RC9 – 178	33°44′	27°55′	2628
RC9 – 179	34°16′	27°11′	2604
RC9 – 180	34°06′	25°41′	2653
RC9 – 181	33°25′	25°01′	2286
RC9 – 182	33°48′	23°36′	1794
RC9 – 183	34°30′	23°25′	2684
RC9 – 185	34°27′	20°07′	2858
RC9 – 188	36°11′	19°30′	3290
RC9 – 189	36°59′	19°41′	3378
RC9 – 190	38°39′	19°14′	1712
RC9 – 191	38°12′	18°02′	2345
Caribbean Cores			
A179 – 4	16°36′	74°48′	2965
V12 – 122	17°00′	74°24′	2800

Methods of Sampling and Spinning

Because no small magnetic events had been detected in the Brunhes Normal Epoch, except for somewhat inconclusive evidence of a short interval of reversed polarity 7000 to 9000 years ago (Bon Hommet and Babkine, 1967), the Mediterranean cores were sampled at very close intervals. A separation of 5 centimeters was chosen, assuring one measurement for nearly every 1000 years of deposition. In areas of particular interest the interval was decreased to include overlapping samples. Nearly 2000 individual samples were examined by use of the spinner magnetometer described by Foster (1966). All samples were spun first without magnetic washing, then were washed at 50 oersted. In several cases the samples were rewashed in progressively higher magnetic fields.

Lithologic Correlations

Earlier studies of the ALBATROSS cores from the eastern Mediterranean, followed by investigations of the cores by VEMA in 1956 and 1958 (Ryan and Heezen, 1965; Emery *et al.*, 1966; Ninkovich and Heezen, 1965 and 1967) demonstrated that discrete properties of the sediments could be identified and cross-correlated with other cores. These properties included color, texture, mineral composition, kind of burrowing, bedding contacts, tephra layers, bedding sequences, discrete layers of pteropods or diatomaceous ooze, and sapropels. Ninkovich and Heezen (1965 and 1967) demonstrated that correlations could be made with tephra layers based on the color, shape, index of refraction and chemical contents of the glass shards. Ryan and Heezen (1965) established a correlation of the upper few meters of several cores in the Ionian Basin based on a sequence of light grey tephra, pteropods, multicolored laminated muds, a sapropel mud layer, a zone of light tan foraminiferal ooze, and a second layer of brownish tephra.

Nevertheless, a most important conclusion drawn from these early studies was that in the majority of cores from the Mediterranean Ridge and Hellenic Trough numerous discontinuities were found which made a complete correlation of the entire recovered sequence impossible. It was subsequently realized that the incomplete records in these cores resulted from slumping, induced by contemporary sedimentary tectonics in this region. When the CONRAD cores were taken in 1965 special care was employed to choose some coring sites that would be less subject to slumping, and might permit the recovery of complete sediment sequences. The correlations presented here are based to a great extent on notes recorded at the time of collection, and on both color and black and white photographs of the core sections.

Figures 3 and 4 illustrate some of the details of the correlations; for instance (Figure 3) a thin layer of light diatomaceous mud near the base of an individual sapropel bed can be used as diagnostic identification of this particular horizon. In other sapropels, burrowed tops, laminae, diatoms, or tephra, or a combination of these can be used. Figure 4 shows the correlation of tephra layers. The lower Santorini tephra layer (Ninkovich and Heezen, 1965) not only has a unique index of refraction (1.521 ± 3) but also occurs at precisely the same place in a sequence of layers of slightly different coloring and hue. In the lower parts of several cores between widely spaced sapropel zones a sequence of rose-pink lutites was found interbedded in a light buff chalk. Such details as burrow structures, thickness of beds, and identical color were repeated in this zone from cores taken several hundred miles apart along the Mediterranean Ridge.

The absence of changes in slope along the 3 curves,

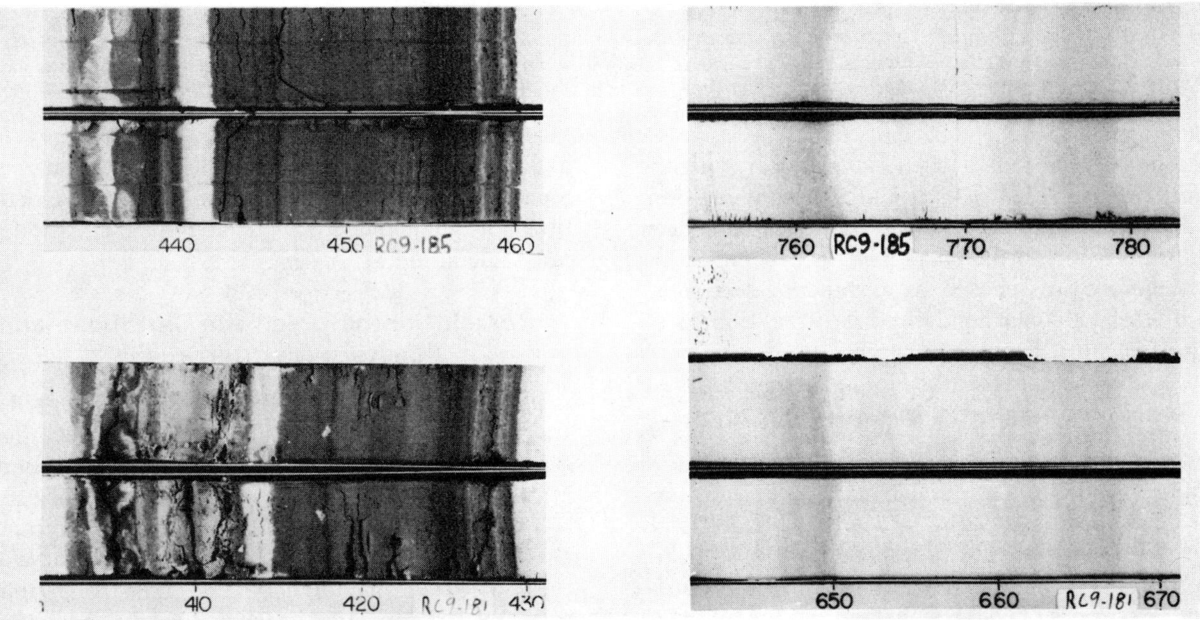

Figure 3. Correlation of lithologic units between cores RC9–181 and RC9–185, taken approximately 500 kilometers apart on the Mediterranean Ridge. In core sections at left, the sapropel layer in the upper part of the V faunal zones contains diagnostic thin layers of diatomaceous mud which are found in identical laminated sequences in both cores. In core sections at right, thin alternating layers of reddish lutites are interbedded between light buff chalks. Even the fine details of the burrow structures are identical in this sequence.

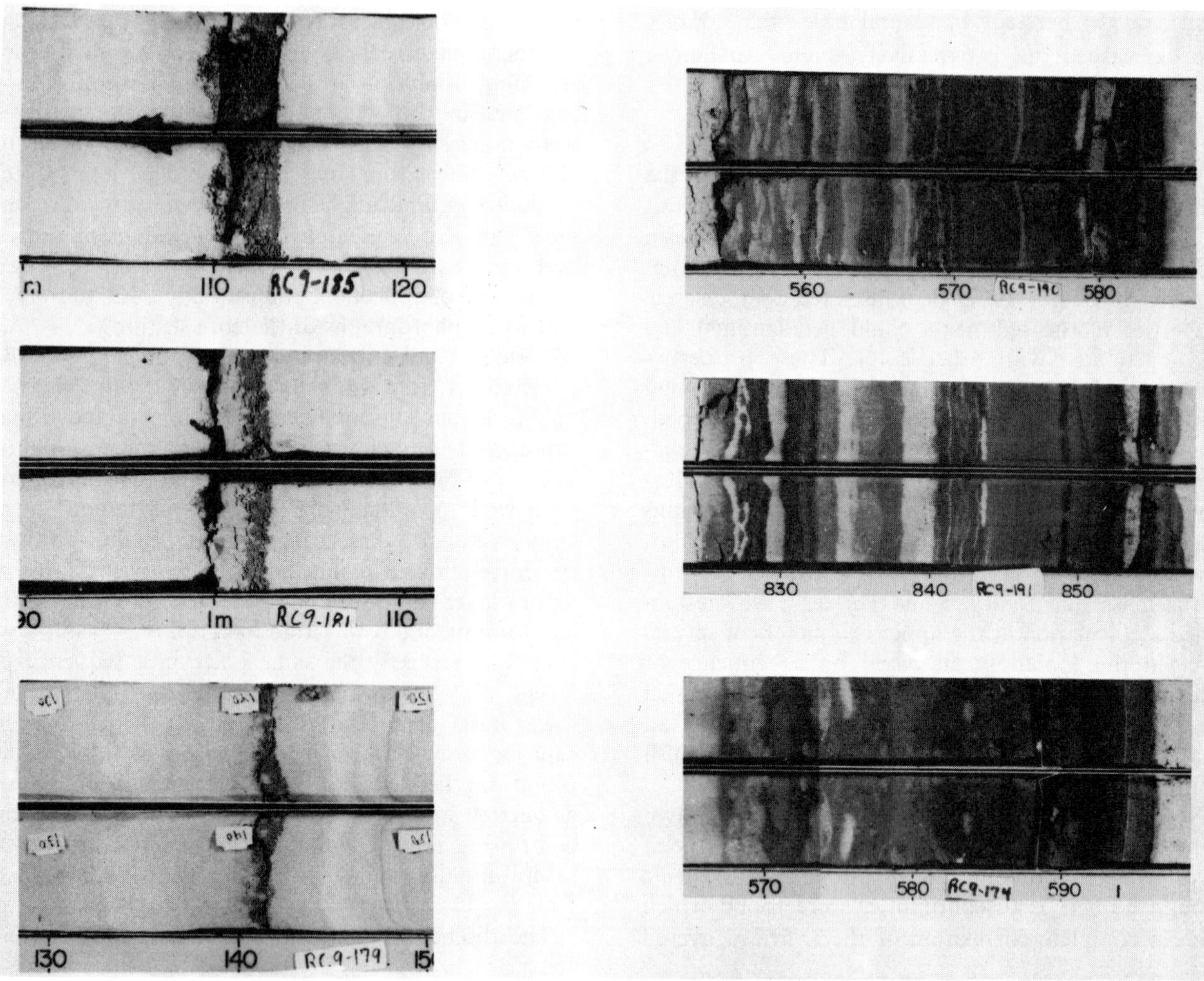

Figure 4. Other lithologic correlations in cores in the eastern Mediterranean. In cores at left, the "lower" Santorini tephra layer (Ninkovich and Heezen, 1965) is found interbedded in an identical sequence of various colored lutites. In cores at right, both the sequences of diatomaceous muds and the burrow mottling in a sapropel bed are markedly similar in Core RC9–174 on the eastern Nile Cone and in cores RC9–190 and RC9–191 south of Calabria, Italy, a separation of almost 1500 kilometers.

except at a thick tephra layer, at identical horizons implies that RC9–181 records uniform sedimentation, or that the changes in sedimentation rates have been synchronous for all the cores examined. Core RC9–181 therefore was chosen as a standard section to which the paleomagnetic studies were compared (Figure 5).

The Paleomagnetic Measurements on RC9–181 and V12–122

Figure 6 shows plots of inclination for the remnant magnetism in core RC9–181 (Latitude 33°25′N) from the eastern Mediterranean and V12–122 (Latitude 17°00′N) from the Jamaica Ridge in the Caribbean. Four zones of negative inclinations are observed in RC9–181 and are shown by shading. Three zones of negative inclination in V12–122 are also marked. The upper section of this core contains negative inclinations which are not explained. These measurements might reflect post-depositional remagnetization, although the processes of remagnetization are not understood at present.

Correlation between the Caribbean and Eastern Mediterranean

The climatic statigraphy of V12–122 has been studied by Ericson, using abundance of *Globorotalia menardii* as an indicator of surface-water temperature. Figure 7 shows a series of diagrams for Caribbean cores A179–4, V12–122 and Mediterranean cores ALB–189 and RC9–181. The faunal curves for the two Caribbean cores have been taken from publications by Ericson (1961, 1968). The oxygen isotopic measurements for Core A179–4 are from Emiliani 1955 and 1966) and for Core 12–122, from Broecker and van Donk (1969). The zones marked with letters

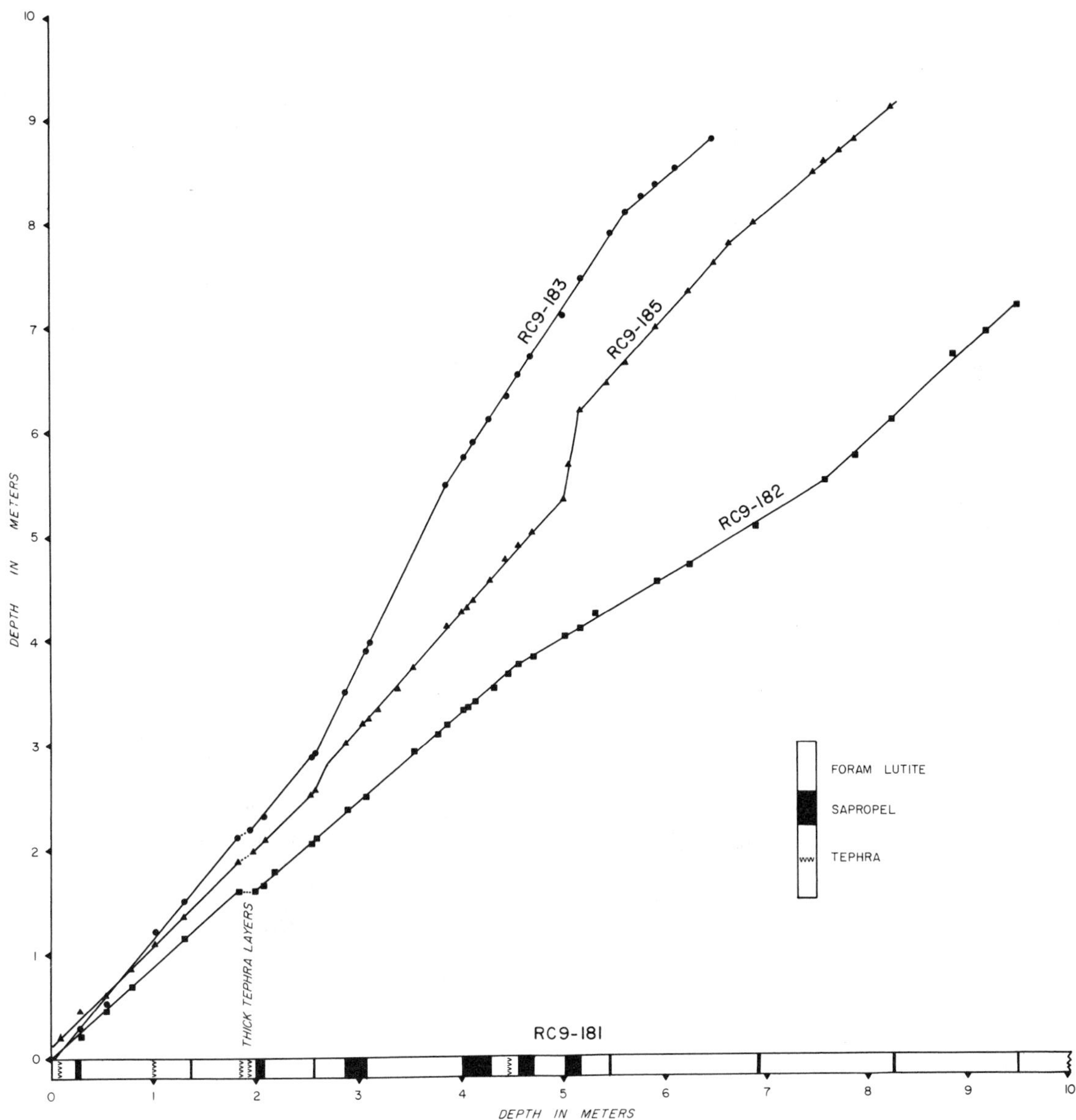

Figure 5. The established lithologic correlations between Core RC9–181 and Cores RC9–182, RC9–183, and RC9–185. Over 50 individual lithologic units were chosen in this correlation. Changes in slope of the curve indicate changes in relative depositional rates of the three individual cores with respect to Core RC9–181. None of the breaks in slope occur in the same bedding sequence of continuous uniform and uninterrupted sediment, and for this reason the succession of layers in this core is considered a standard section.

are defined by Ericson (1961) on the basis of the presence or absence of the *G. menardii* complex. The X zone in particular is based on the presence of abundant *G. m. menardii* and *G. m. flexuosa*, whereas in the W zone the *G. menardii* complex is absent or rare.

The stages on the isotope curves are numbered after Emiliani (1955 and 1966) and by correlation of V12–122 to PILLSBURY Cores P-6034–8 and P-6034–9 by Broecker. The hiatus in A179–4 between stage 7 and stage 11 has been established by Emiliani (1966) using correlations of the coiling direction of *Globorotalia crassaformis* and *Globorotalia truncatulinoides* between this core and the PILLSBURY cores (which Emiliani believes contain a complete sequence). The short magnetic event located at the base of the X zone in A179–4 was reported by Smith and Foster (1969), and was found in the same faunal zone in seven other cores from the North Atlantic and Indian Oceans.

The uppermost event (Blake Event of Smith and Foster, 1969) in V12–122 appears in the same part of the X zone on the identical first cooling trend of stage 5. This event is found in RC9–183 and in RC9–181 in an identical stratigraphic position above the lowermost sapropel of this core stage. The Blake Event is used here to establish an absolute time syn-

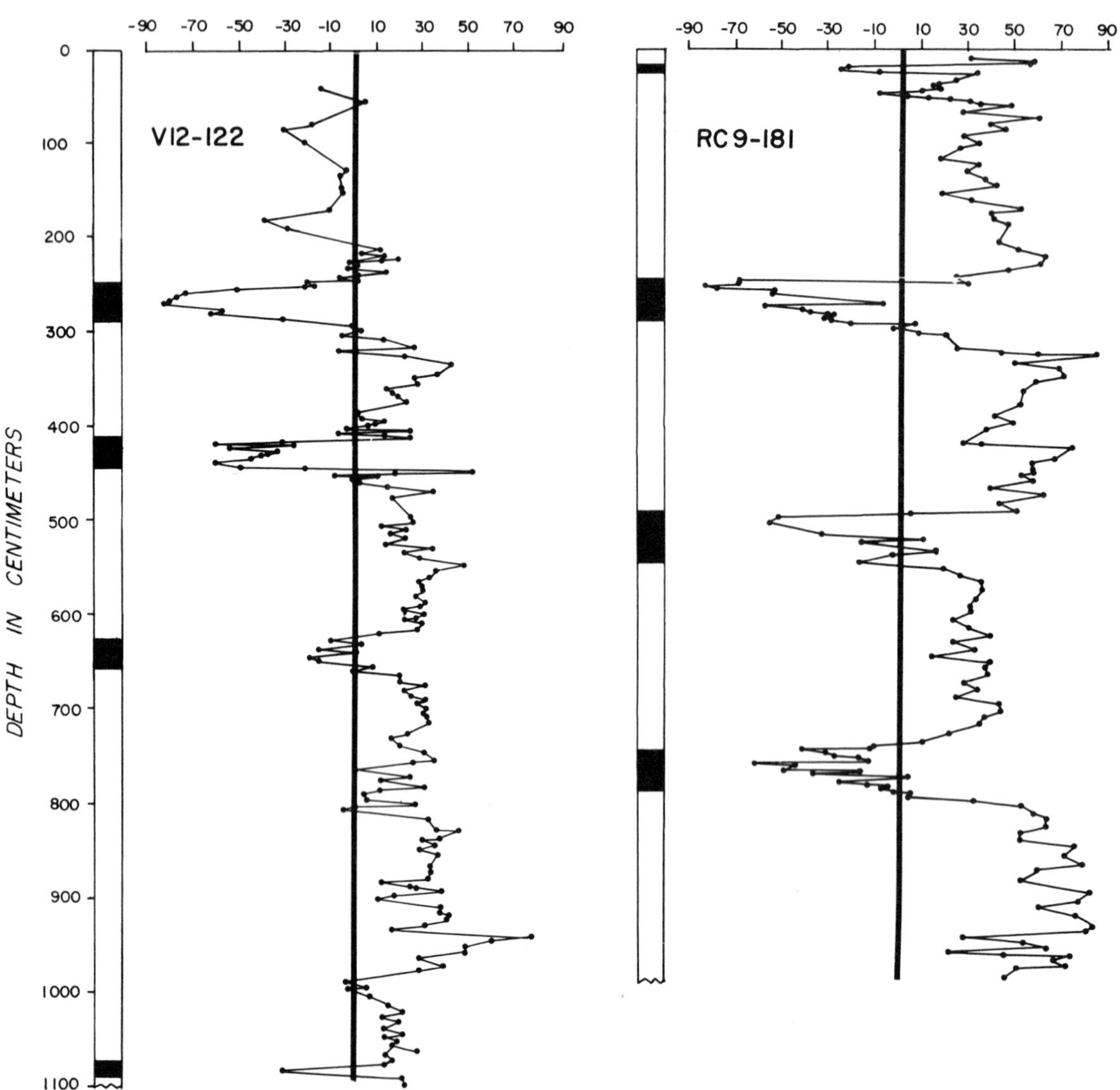

Figure 6. Plots of inclination (in degrees) of the remnant magnetism of the sediments in Core V12–122 from the Caribbean and Core RC9–181 from the eastern Mediterranean. The graphs are for the samples previously AC-demagnetized in a 50–oersted field. The periods of negative inclination indicate magnetic events of reversed polarity in the Brunhes Normal Magnetic Epoch. The identified events are shown in the accompanying diagrams.

chroneity between the Mediterranean stratigraphic successions and those of the Caribbean and North Atlantic.

Evidence for this correlation appears to be established, and because of the completeness of this sequence one might expect that the remainder of the Mediterranean sequence would also be correlatable. However, the next event in Core RC9–181 is found near the base of stage 7, followed by another short event in the middle of stage 11. In the Caribbean Core V12–122 the next event occurs in the middle of a major stage labelled III by Broecker and van Donk (1969), and the next older event is in their stage IV.

The severe secondary cooling minimum in the middle of stage II is not nearly as prominent in the other Caribbean cores examined by Emiliani (1966). Considering the hiatus already reported for A179–4, taken only 30 miles from V12–122 on the Jamaica Ridge, the author and Foster (Personal communication) suggest that the zone from 480 to 500 cm in V12–122 marks a hiatus also. If this is the case, then the upper maximum of stage III is stratigraphically equivalent to Emiliani's stage 7 and the magnetic event is time-synchronous with the same event found in the Mediterranean. This event is provisionally named the Jamaica Event from its discovery in V12–122 from the Jamaica Ridge. Having shown a hiatus

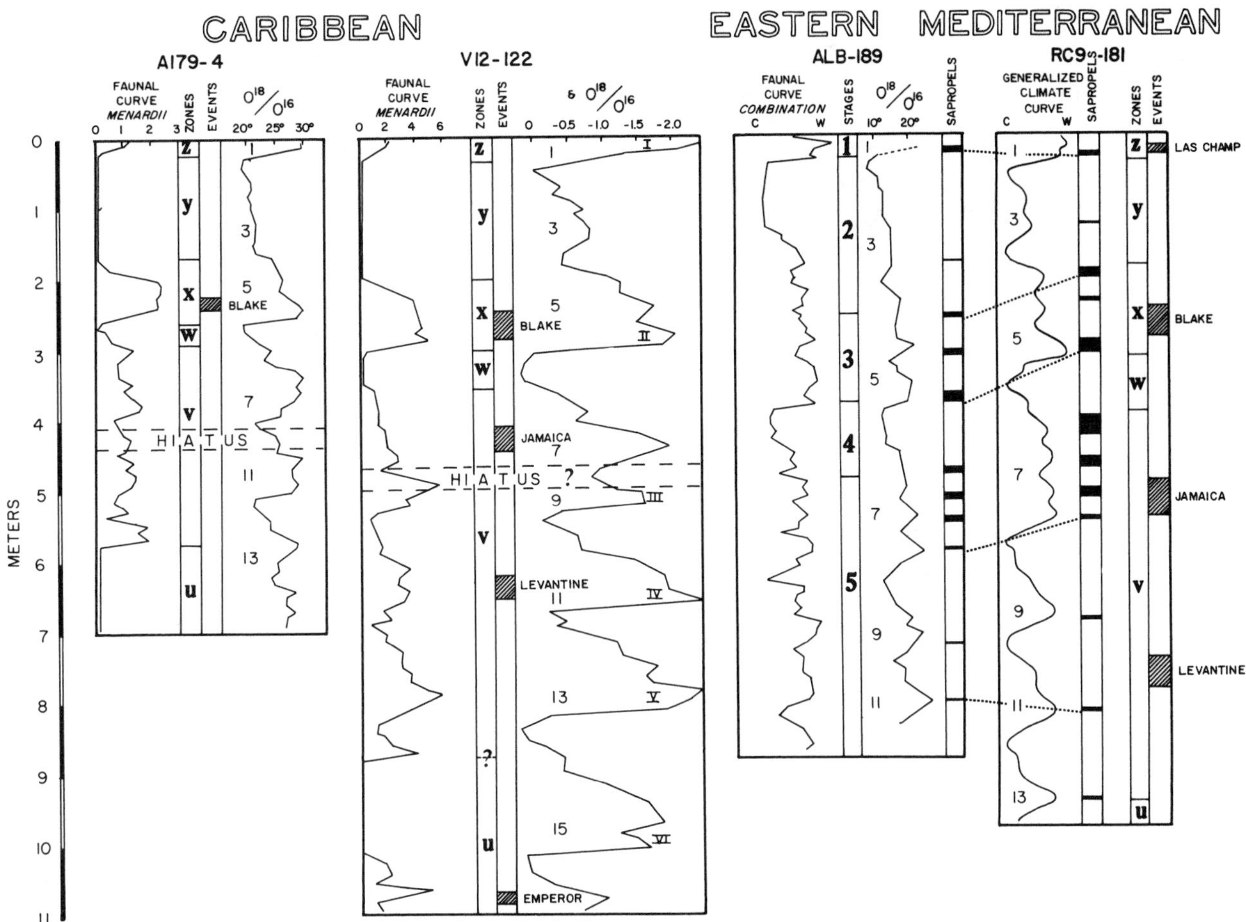

Figure 7. Faunal, isotopic O^{18}/O^{16}, and magnetic stratigraphies for Cores A179–4 and V12–122 from the Jamaica Ridge in the Caribbean and Cores ALB–189 and RC9–181 from the Mediterranean Ridge. The stage numbers under the O^{18}/O^{16} curves are from Emiliani (1955 and 1966); the stage numbers for ALB–189 are from Olausson (1961). The letters indicate faunal zones from Ericson *et al.*, (1961).

in the middle part of stage III, the lower part of this stage thus corresponds to stage 9 and stage IV to stage 11. The next lower event which is provisionally termed the Levantine Event after its discovery in cores from the Levantine Basin of the eastern Mediterranean is then found in each case in the middle of stage 11. The confirmation of this interpretation will have to await further documentation from data on the changes in coiling directions of *Globorotalia crassaformis* and *Globorotalia truncatulinoides* in Core V12–122 which are being carried out at present (Broecker, personal communication).

Some confusion arises in the correlation of faunal stages based on the *Globorotalia menardii* complex with the climate stages inferred from the oxygen isotope data because of the generally non-synchronous placing of the U/V boundary. In A179–4 this boundary occurs midway through stage 13. In A172–6 studied by Ericson (1961) and Emiliani (1964) this boundary is in the lower part of stage 13; V12–122 boundary is between stages 13 and 15. Possibly the climate curves constructed by the two approaches agree only down to the bottom of the W zone as has been previously suggested by Ericson *et al.* (1964) and Ku and Broecker (1966).

Another small magnetic event in the U zone in V12–122 has been correlated tentatively with similar inclination changes found in CONRAD cores from the Pacific. This event is provisionally called the Emperor Event. The uppermost zone of negative inclination in RC9–181 is believed to be the equivalent of the Las Champ Event reported by Bon Hommet and Babkine (1967), except that in the Mediterranean cores the Las Champ Event occurs after the deposition of the uppermost sapropel layer and therefore is younger than 7000 years.

Faunal Zones

Faunal zones from RC9–181 and the rest of the eastern Mediterranean cores examined in this study are shown in Figure 8. These zones apply to all Lamont cores in the eastern Mediterranean which can be stratigraphically correlated with RC9–181. Because of their widespread use at the Lamont-Doherty Geological Observatory and in the literature, the author has adopted the letter symbols of Ericson (1961). However, it must be emphasized that boundaries of these zones in the eastern Mediterranean have not been defined by the *Globorotalia menardii* complex criterion, but rather by time-equivalent faunal, oxygen isotope, and paleomagnetic correlation with Ericson's Caribbean and Atlantic sequences. The boundaries of Ericson's zones are identical to the boundaries of the core stages defined by Olausson (1961), the only exception being that the W/X boundary is above the uppermost sapropel of Emiliani's stage 7 whereas the 5/4 boundary is just below this sapropel (see Figures 1 and 2).

CORRELATION OF EASTERN MEDITERRANEAN CORES

Using the faunal zones established above and lithological characteristics, an east to west sediment profile has been constructed for the eastern basin (Figure 8). The profile illustrates that the sediments in the generally cool Y and W faunal zones were deposited at accelerated sedimentation rates toward the west in the Ionian Basin, whereas, during the warmer X zone sedimentation was greatly reduced and periods of nondeposition or of erosion occurred. At the eastern end of the profile on the Nile Cone, sediments in the X zone were deposited at greater rates than in the Y and W zones.

Sedimentation on the Herodotus Abyssal Plain southeast of Cyprus was very rapid as shown in RC9–175 in the Z zone, where the sediments consist of numerous fine-grained turbidite layers. Turbidites are absent in the part of the Y zone recovered.

The basal contacts of the microbreccias are often discontinuities, indicating erosion during the initial slumping processes followed by deposition of the brecciated fragments. Pliocene sediments were recovered beneath an obvious lithologic discontinuity in Core RC9–178. These Pliocene sediments consist of turbidites and are found in contact with late Pleistocene foraminiferal lutite.

Tephra layers are more abundant in the Y zone in cores south of Calabria, and were probably derived from volcanoes in this region.

An Absolute Chronology for the Eastern Mediterranean Sedimentary Sequence

Ku and Broeckner (1966) and Broecker and van Donk (1969) have obtained an isotopically dated chronology of Core V12–122, using Pa^{231} and Th^{230} measurements. The base of the X zone as calculated is approximately 126,000 years (± 6000). Recently Rona and Emiliani (1969) repeated the Th^{230} and Pa^{231} measurements on PILLSBURY cores P6304–8 and P6304–9 and obtained a date of approximately 100,000 years from the W/X boundary, in general agreement with the earlier results of Rosholt *et al.* (1961).

Broecker and van Donk failed to recognize the apparent hiatus in their stage III, which was discussed previously. Actually, in Figure 9A of their paper, which plots excess Th^{230} against depth, a better fit to their data is provided if the hiatus at 480 cm is replaced by one meter of additional sediment.

Figure 9 shows a diagram of Core RC9–181 along side a generalized climate curve showing the core stages after Emiliani (1955, 1966) and core zones after this paper. The generalized curve here was constructed from a best-fit analysis of Emiliani's (1966) generalized temperature curve for the surface water of the central Caribbean and the climate curves shown in Figure 2. A mean absolute time scale is drawn with the base of the X zone at 126,000 years. Using this time scale the U/V boundary (or peak of stage 13) falls at 380,000 years as compared to 370,000 years extrapolated by Broecker and Van Donk (1969) for the same boundary in V12–122.

An independent check on the age of the U/V boundary is provided by a recent paper by Ericson and Wollin (1968). They present a sequence of 5 long cores from the Atlantic which extend through the Brunhes Normal Epoch back to the Plio-Pleistocene boundary. If the Brunhes-Matuyama boundary is taken as 700,000 years as based on potassium-argon dates, the interpolated U/V boundary of these cores averages 375,000 years. All these numbers which range from 370,000 to 380,000 years are to be compared with the age of 290,000 years for the same boundary, if one uses the extrapolated chronology from Rona and Emiliani (1969). Consequently, the chronology of Broecker and van Donk (1969) is preferred and is used here.

Beach Terrace Dates

The uranium-series method of dating has been applied to corals, marine fossils, oölites and limestone found on ancient raised beach terraces around the world. The results from these investigations are plotted in Figure 9 against the absolute chronology and lithology of Core RC9–181. The Barbados beach-terrace dates are from Broecker *et al.* (1968); the Mediterranean dates are from Stearns and Thurber (1965); the Bahama and Florida Key results were reported by Broecker and Thurber (1965); the dating of the Pleistocene corals from the Mid-Pacific atolls

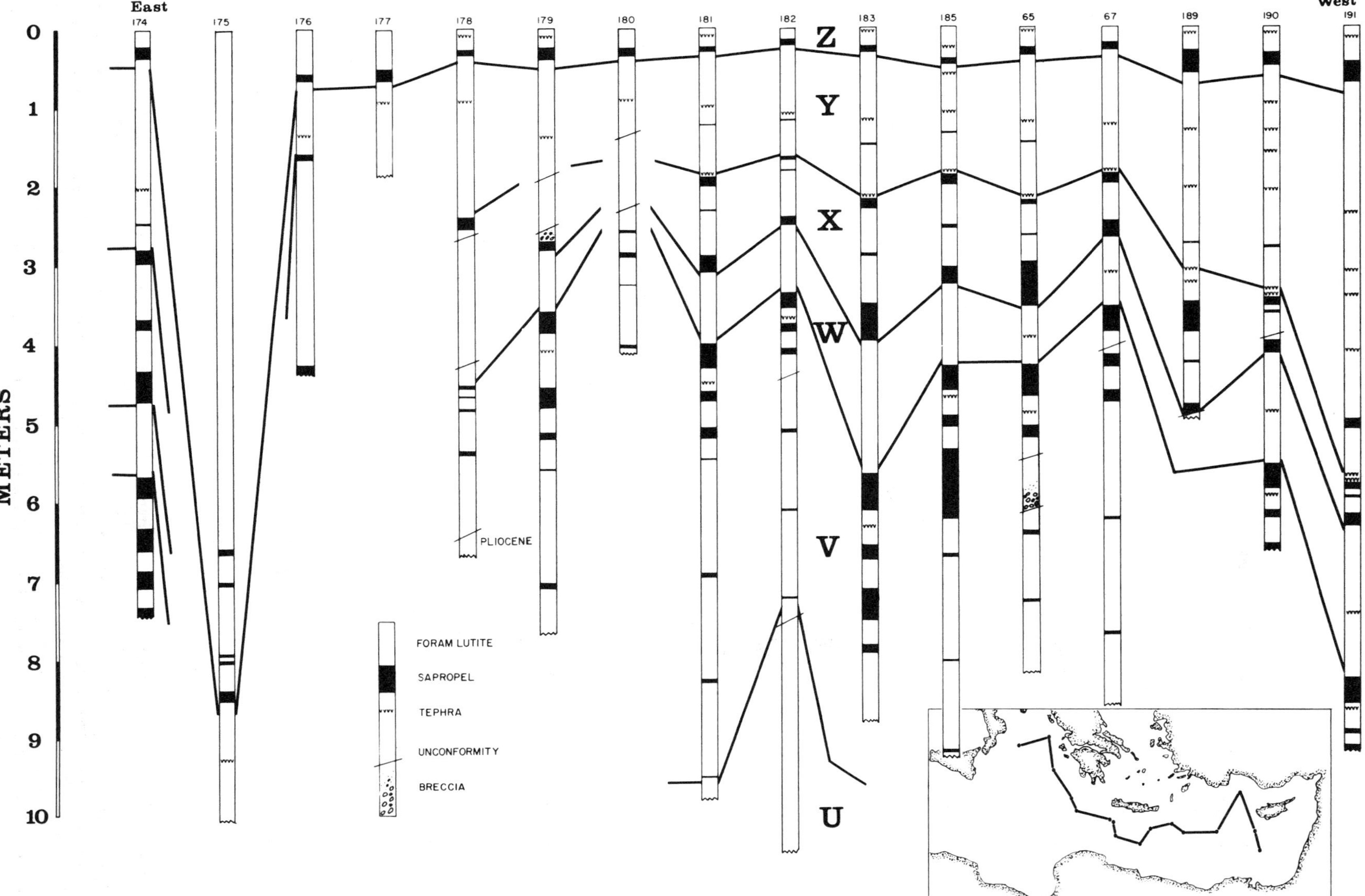

Figure 8. Correlation of faunal zones, sapropelic mud layers and tephra layers in a suite of eastern Mediterranean cores. The profile of RC9 cores extends from the eastern Nile Cone to the northern Ionian Basin, south of Calabria, Italy. Sedimentation rates in the cold W and Y zones are higher in the western part of this profile than in the eastern part, and the deposition rates in the warm X zone are anomalously low. The surfaces of unconformities shown are actual lithologic and faunal discontinuities.

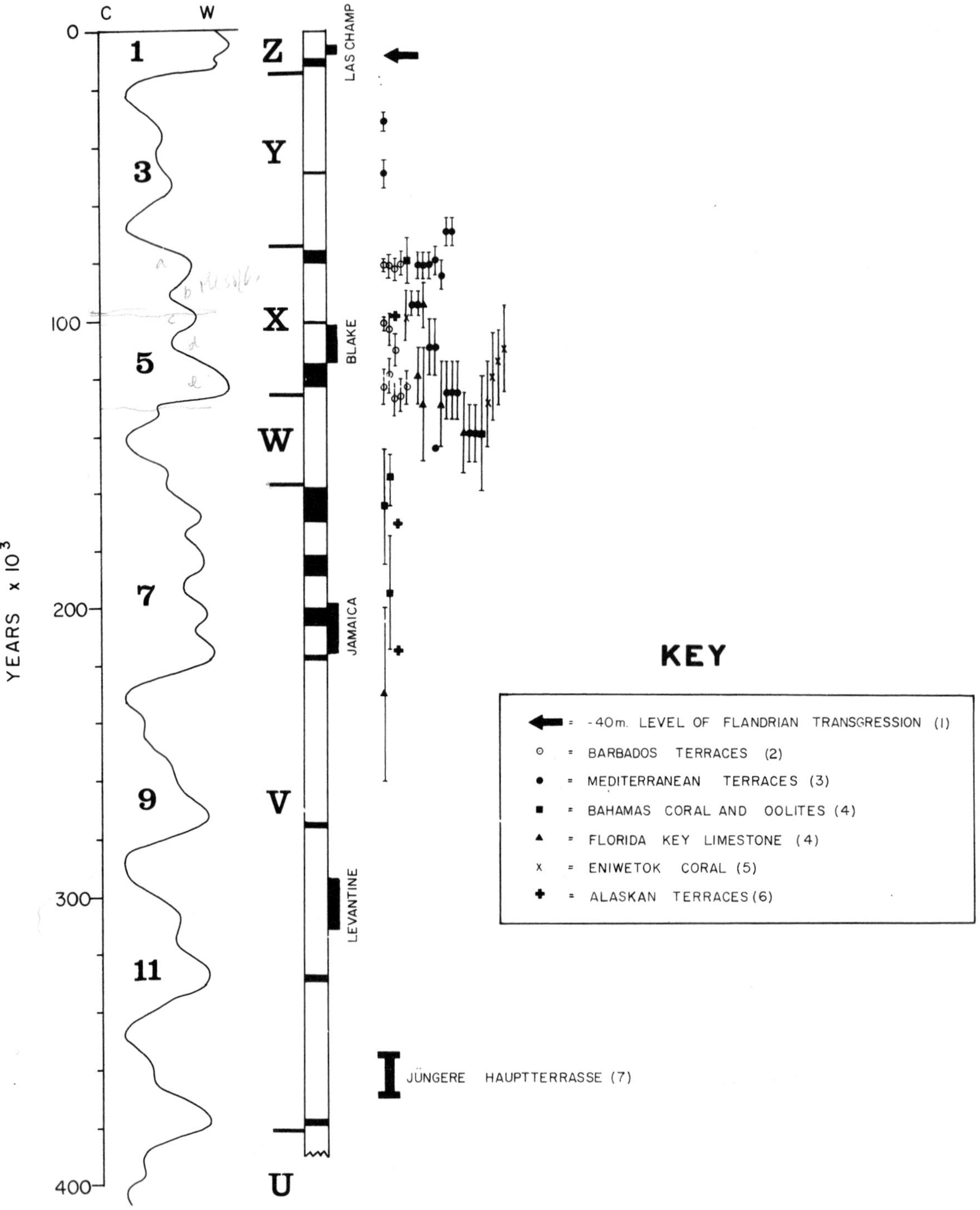

Figure 9. The generalized climate curve for the eastern Mediterranean and the lithologic sequence in RC9–181. The absolute time scale is taken from Broecker and van Donk (1969) by correlation of the base of the X zone to core V12–122 where the age of this boundary has been dated isotopically as 126,000 ± 6000 years. The short magnetic events in stages 1, 5, 7, and 11 are identified. To the right, the ages of elevated fossil beach terraces are indicated with their ranges of uncertainty. These dates have been obtained using the uranium-series method. The Jungere Hauptterrasse are volcanic tuffs found associated with glacial conditions in the Rhine Valley, and these rocks were dated by potassium-argon measurements. The references are: (1) Milliman and Emery (1968); (2) Broecker *et al.* (1968); (3) Stearns and Thurber (1965); (4) Broecker and Thurber (1965); (5) Thurber *et al.* (1965); (6) Blanchard (1963); and (7) Evernden *et al.* (1964); Lippolt (1961); and Evernden and Curtis (1965).

are from Thurber *et al.* (1965), and the Alaskan terrace dates are from Blanchard (1963). The − 40 m height of the Flandrian (Holocene) Transgression is taken at 9000 years B.P. from Milliman and Emery (1968).

The grouping of beach terrace dates in the X zone and in the upper part of the V zone, the occurrence of a date near 50,000 years, and the known Flandrian Transgression support an association of high sea level stands with the deposition of sapropelic muds.

CAUSES OF THE MEDITERRANEAN STAGNATIONS

Contrary to the deductions of Olausson (1961) the Mediterranean stagnations present a rather precise relationship to the climate fluctuations. In fact, Figure 9 shows that the stagnant phase commences on the *warming* trend of all the major temperature oscillations near the temperature maximum. The position of the Blake Event in two Caribbean cores, relative to the cooling trend of the first secondary oscillation in stage 5, provides the absolute synchroneity necessary to confirm this correlation between onset of stagnant conditions and generalized climate warming applicable to both the Mediterranean regions as well as to the Atlantic and Caribbean. Although it is still contested whether the principal factor causing the observed variations in the O^{18}/O^{16} ratio in marine carbonates result from a real change in the water temperature or in the isotopic composition of the sea water, nevertheless, Broecker and van Donk (1969, p. 170) point out that "in either case there should be at least a first order synchroneity between the isotopic composition of the $CaCO_3$ and the extent of continental glaciation." Shackleton (1967, p. 15) implies that "it is simply necessary that every faunal or isotopic curve be re-read taking cold to mean extensive continental glaciation and warm to mean glaciers reduced to their present level."

If this is done then the climate curve of Figure 9 indicates that the stagnant phases develop during the period of rapid retreat of the glaciers (*i.e.*, at times of extensive run-off and rapid rise of sea-level). Furthermore, the actual sapropelic deposition occurs not in synchroneity with the onset of the melting, although this period is accompanied in the cores by generally gray to grayish-blue reduced muds, but near the end of the melting.

This observation coupled with (1) the strong evidence for high stands of sea level on Barbados at 82,000, 103,000 and 122,000 years (presented by Broecker *et al.*, 1968); (2) the abrupt occurrence of euxinic conditions (see Figure 3); (3) anomalously low O^{18}/O^{16} values in the sapropel muds indicating a large meltwater contamination; (4) the very large abundance of *Globigerina eggeri* in these muds, and (5) the onset of the latest phase of stagnation at precisely 9,000 years B.P., when sea level rose to the −40 m level; all lend credence to Olausson's hypothesis that the principal cause of stagnation is the sudden and large contribution of additional meltwater to the eastern Mediterranean through a flushing of the Black Sea.

The volume of meltwater that would have entered directly into the eastern Mediterranean from rivers and streams during the waning of glaciation is very small compared to the paleo-drainage of rivers into the Black Sea. Olausson (1965) has shown that the geographic distribution of the European and Scandinavian ice sheets is such that during a major deglaciation rivers would contribute a volume of meltwater equal to possibly 15 times the volume of the Black Sea basin.

It is, therefore, not surprising that the Black Sea basin could become highly diluted with light, cold semi-brackish water. When sea level rises to the critical −40 meter level of the sill in the Bosporus (Caspers, 1957), the access of Mediterranean water to the Black Sea triggers eventual and rapid overturn of the latter and marks the onset of stagnant conditions in the Mediterranean.

If this is, indeed, the case for the Mediterranean euxinic phases and their accompanying sapropelic muds, then Figure 9 can be used to date precisely the sea-level curve of the eastern Mediterranean during the last 400,000 years. In particular, the occurrence of a thin layer of sapropelic mud in the middle of the Y zone indicates that sea level rose to at least −40 meters at that time. This is correlated with an Ouljian beach-terrace date from Morocco at 50,000 years and the evidence cited in Milliman and Emery (1968) that 30,000 to 35,000 years ago sea level was near its present level.

The testing of this hypothesis should provide an excellent method for correlation of the deep-sea Mediterranean climate records with the continental records preserved in terraces, marshes, lakes, and moraines. One confirmation already exists; potassium-argon dates on volcanic tuff minerals from the Jüngere Haupterrasse of the Rhine Valley (Evernden *et al.*, 1964; Lippolt, 1961; Evernden and Curtis, 1965), ranging from 370,000 to 354,000 years, coincide with the cool climatic interval between core stages 13 and 11 (see Figure 8). These volcanic materials have been correlated with the Günz or Mindel Glaciation.

PALEOSOLS

Climate changes on land can be inferred from chemical and geological weathering of soils, notably the "typical" red soils of the Mediterranean region which no longer develop on fresh surfaces. Because they are closely related to tropical soils these conspicuous Mediterranean paleosols deserve some attention. More intensive chemical weathering, with a fairly warm and seasonally rather moist climate, seem to be prerequisites for soil development of this type. Butzer (1964) has shown stratigraphic evidence that soils with characteristic red hues originated during interglacial periods. Holstein and Eem paleosols are found in soil profiles in Italy, in the Balearic Islands, and in Egypt and Libya.

In the eastern Mediterranean cores, nearly identical sequences of reddish and rosebrown hemipelagic lutites are found interbedded with lighter buff chalky lutites in the V faunal zone, but particularly on the cooling trends of stages 13, 11, and 9. Butzer (1963) has mentioned that in North Africa and Italy the red soils were frequently stripped by sheet flow and incorporated in the colluvial silts (*limons rouges*) that blanket the lowland plains. These heterogeneous beds are commonly detrital horizons and are found intercalated with littoral sediments. Because they record periods of slope wash, possibly by torrential rains and floods following seasonal droughts, it is equally likely that the *limons rouges* are washed into the sea. Butzer (1963) shows that stratigraphically, these soils date from glacio-eustatic emergences, because the horizons usually extend below present sea level beneath eolianites on the continental shelf off Egypt and Libya. If, as is suggested by Shackleton (1967), the generalized climate curve of Figure 9 is really a glacial volume indicator, the occurrence of the Levantine Event at the time of this kind of lutite deposition in stage 11 may become a way in which to cross-correlate the deep-sea records with the history of continental glaciation, because this event may also be found in the Holstein paleosol horizons.

ACKNOWLEDGMENTS

This study forms part of the author's doctoral dissertation, Faculty of Pure Sciences, Columbia University. The author acknowledges the generous support, guidance, and constructive criticism offered by his advisor, Professor Bruce C. Heezen. Most of the new materials examined here for the first time were collected on research vessels of the Lamont-Doherty Geological Observatory under the direction of Professor Maurice Ewing. His personal interest in geological and geophysical investigations of the Mediterranean has been a source of encouragement. The cores instrumental to this study were taken by Professor Ewing who, during the cruise on ROBERT D. CONRAD in 1965, encouraged the author to "delve into these pelagic sediments" which prior to then had always seemed so undecipherable.

A stimulating, exciting, and productive cooperation with John Foster is acknowledged. Discussions with Paul J. Fox, James Gardner, James D. Hays, Ansis Keneps, Marcus Langseth, Warren Prell, Eric D. Schneider, Jerry Smith, and Tsunemasa Saito have been helpful. In particular, the author wishes to acknowledge the valuable suggestions and contributions from Allan Bé, Wallace S. Broecker, Andrew McIntyre, David Needham, William F. Ruddiman, and Douglas S. Tolderlund pertaining to the faunal analyses and correlations.

The manuscript has been read critically by Neil Opdyke, Walter Pitman, John Dewey, Manik Talwani, John Sanders, and Frances Parker.

REFERENCES

Bé, A. W. H. and D. S. Tolderlund 1971. Distribution and ecology of living planktonic foraminifera in surface waters of the Atlantic and Indian Oceans. In: *The Micropaleontology of the Oceans*, eds. Funnel, B. M. and W. R. Riedel, Cambridge University Press, London, 105–149.

Blanchard, R. L. 1963. *Uranium Decay Series Disequilibrium in Age Determination of Marine Calcium Carbonate.* Ph.D. thesis, Washington University, St. Louis.

Bon Hommet, N. and J. Babkine 1967. Sur la presence d'aimantations inversées dans la chaîne des Puys. *Comptes Rendus de l'Académie des Sciences, Paris,* 264: 93.

Bradley, W. H. 1938. Mediterranean sediments and Pleistocene sea levels. *Science,* 88:376–379.

Broecker, W. S. and D. L. Thurber 1965. Uranium series dating of corals and oolites from Bahamian and Florida Key limestones. *Science,* 149: 58–60.

Broecker, W. S., D. L. Thurber, T. L. Ku, R. K. Matthews, and K. J. Mesolella 1968. Milankovitch hypothesis supported by precise dating of coral reefs and deep sea sediments. *Science,* 159: 297–300.

Broecker, W. S., and J. van Donk 1969. Isolation changes, ice volumes and the O^{18} record in deep-sea cores. *Reviews of Geophysics and Space Physics,* 8:169–198.

Butzer, K. W. 1963. Climate-geomorphologic interpretation of Pleistocene sediments in the Eurafrican subtropico. *Anthropology* (Viking Fundamental Publication), 36: 1–27.

Butzer, K. W. 1964. *Environment and Archaeology: An Introduction to Pleistocene Geography.* Aldine, Chicago, 528 p.

Caspers, H. 1957. Black Sea and Sea of Azov. *Geological Society of America Memoir,* 67: 801–890.

Dansgaard, W. 1961. The isotopic composition of natural waters, with special reference to the Greenland ice cap. *Meddelelser om Grønland,* 165: 120 p.

Emery, K. O., B. C. Heezen, and T. D. Allan 1966. Bathymetry of the eastern Mediterranean Sea. *Deep-Sea Research,* 13: 173–192.

Emiliani, C. 1955. Pleistocene temperature variations in the Mediterranean. *Quaternaria,* 3:87–98.

Emiliani, C. 1958. Paleotemperature analysis of Core 280 and Pleistocene correlations. *Journal of Geology,* 66: 264–275.

Emiliani, C. 1964. Paleotemperature analysis of Caribbean cores A254–BR–C and CP–28. *Geological Society of America Bulletin,* 75: 129–144.

Emiliani, C. 1966. Paleotemperature analysis of Caribbean cores P–6304–8 and P–6304–9 and a generalized temperature curve for the past 425,000 years. *Journal of Geology,* 74: 109–124.

Emiliani, C., T. Mayeda and R. Selli 1961. Paleotemperature analysis of the Plio-Pleistocene section et Le Castella, Calabria, Southern Italy. *Geological Society of America Bulletin,* 72: 679–688.

Ericson, D. B. 1961. Pleistocene climate record in some deep-sea sediment cores. *Annals of the New York Academy of Science,* 95: 537–541.

Ericson, D. B., M. Ewing and G. Wollin 1964. The Pleistocene epoch in deep-sea sediments. *Science,* 146: 723–732.

Ericson, D. B., M. Ewing, G. Wollin and B. C. Heezen 1961. Atlantic deep-sea sediment cores. *Geological Society of America Bulletin,* 72: 193–286.

Ericson, D. B. and G. Wollin 1956. Micropaleontological and isotopic determinations of Pleistocene climates. *Micropaleontology,* 2: 256–270.

Ericson, D. B., and G. Wollin 1968. Pleistocene climates and chronology in deep-sea sediments. *Science,* 162: 1227–1234.

Evernden, J. F. and G. H. Curtis 1965. The potassium-argon dating of Late Cenozoic rocks in East Africa and Italy. *Current Anthropology,* 6: 343–364.

Evernden, J. F., D. E. Savage, G. H. Curtis and G. T. James 1964. Potassium-argon dates and the Cenozoic mammalian chronology of North America. *American Journal of Science,* 262: 194–198.

Foster, J. H. 1966. A paleomagnetic spinner magnetometer using a fluxgate gradiometer. *Earth and Planetary Science Letters,* 6: 463–466.

Imbrie, J., and T. H. van Andel 1964. Vector analysis of heavy mineral data. *Geological Society of America Bulletin,* 75: 1131–1156.

Ku, T. L. and W. S. Broecker 1966. Atlantic deep-sea stratigraphy: Extension of absolute chronology to 320,000 years. *Science,* 151: 448–450.

Kullenberg, B. 1952. On the salinity of the water contained in marine sediments. *Meddelanden från Oceanografiska Institutet i Göteborg,* 21: 1–38.

Lippolt, H. J. 1961. *Altersbestimmungen nach der K-Ar -Methode bei kleinen Argon- und Kalium-konzentrationen.* Ph.D. thesis, Heidelberg University, Heidelberg, 82 p.

Manson, V. and J. Imbrie 1964. Fortran program for factor and vector analysis of geologic data, using an IBM 7090 or 7094/1401 computer system. *Kansas Geological Survey Special Distribution,* 13:47 p.

McGill, D. A. 1961. A preliminary study of the oxygen and phosphate distribution in the Mediterranean Sea. *Deep-Sea Research,* 8: 259–269.

Mellis, O. 1954. Volcanic ash-horizon in deep-sea sediments. *Deep-Sea Research,* 2: 89–92.

Menzies, R. J., J. Imbrie, and B. C. Heezen 1961: Further considerations regarding the antiquity of the abyssal fauna with evidence for a changing abyssal environment. *Deep-Sea Research,* 8: 79–94.

Milliman, J. D. and K. O. Emery 1968. Sea levels during the past 35,000 years. *Science,* 163: 121–123.

Ninkovich, D. and B. C. Heezen 1967. Physical and chemical properties of volcanic glass shards from Pozzuolana ash, Thera Island, and from upper and lower layers in eastern Mediterranean deep-sea sediments. *Nature,* 213: 582–584.

Olausson, E. 1961. Studies of deep-sea cores. *Reports of the Swedish Deep-Sea Expedition, 1947–1948,* 8 (4): 353–391.

Olausson, E. 1965. Evidence of climate changes in North Atlantic deep-sea cores. In: *Progress in Oceanography,* ed. Sears, M., Pergamon Press, Oxford, 3: 221–254.

Olsson, I. V. 1959. Uppsala natural radiocarbon measurements; I. *American Journal of Science, Radiocarbon Supplement,* 1: 87–102.

Parker, F. L. 1958. Eastern Mediterranean foraminifera. *Reports of the Swedish Deep-Sea Expedition 1947–1948,* (4):217–283.

Rona, E. and C. Emiliani 1969. Absolute dating of Caribbean cores P 6304–8 and P 6304–9. *Science,* 163: 66–68.

Rosholt, J. N., C. Emiliani, J. Geiss, F. F. Koczy, and P. J. Wangersky 1961. Absolute dating of deep-sea cores by the Pa^{231}/Th^{230} method. *Journal of Geology,* 69: 162–185.

Rubin, M. and H. E. Suess 1955. U. S. Geological Survey radiocarbon dates, II. *Science,* 121: 481–488.

Ruddiman, W. F. 1969. *Foraminifera of the Subtropical North Atlantic Gyre,* Ph.D Thesis, Columbia University, New York.

Ryan, W. B. F. and B. C. Heezen 1965. Ionian Sea submarine canyons and the 1908 Messina turbidity current. *Geological Society of America Bulletin,* 46: 915–932.

Schott, W. 1935. Die Foraminiferen in dem äquatorialen Teil des Atlantischen Ozeans. *Wissenschaftliche Ergebinisse der Deutschen Atlantischen Expedition auf dem Vermessungs- und Forschungsschiff "METEOR", 1925–1927,* 3 (3): 43–134.

Shackleton, N. 1967. Oxygen isotope analysis and Pleistocene temperatures re-assessed. *Nature,* 215: 15–17.

Smith, J. D. and J. H. Foster 1969. Geomagnetic reversal in Brunhes normal polarity epoch. *Science,* 163: 565–567.

Stearns, C. E., and D. L. Thurber 1965. Th^{230}–U^{234} dates of late Pleistocene marine fossils from the Mediterranean and Moroccan littorals. *Quaternaria,* 7:29–42.

Thurber, D L., W. S. Broecker, H. A. Potratz, and R. L. Blanchard 1965. Uranium series ages of coral from Pacific atolls. *Science,* 149:55.

Tolderlund, D. S. 1969. *Seasonal Distribution Patterns of Planktonic Foraminifera at Five Ocean Stations in the Western North Atlantic.* Ph.D. Thesis, Columbia University, New York.

PART 4. Coastal and Shallow Water Sedimentation: Eustatic and Tectonic Effects

Les Dépôts Quaternaires Immergés du Golfe de Fréjus (Var) France

Gilbert Bellaiche

Centre de Recherches Géodynamiques, Villefranche-sur-Mer, Alpes-Maritimes, France

RESUME

L'étude particulière du Golfe de Fréjus (Var) par sondage sismique à haute fréquence et par carottage a permis de retrouver sous la vase post-glaciaire des dépôts littoraux de sable et de galets que les datations au carbone 14 conduisent à rattacher à la fin de la dernière glaciation (Würm IV). La disparition brutale de ce matériel au niveau de la rupture de pente qui limite vers le large le plateau continental, permet de fixer à environ –80, –90m. le niveau de la mer à la fin du Würm.

Le même phénomène semble se retrouver en de nombreux points des côtes françaises de la Méditerranée et l'interprétation qui en est donnée semble pouvoir être généralisée à l'ensemble de ce littoral.

ABSTRACT

The detailed study of the Gulf of Fréjus (Var, France) by means of continuous seismic profiling (mud-penetrator) and coring, revealed littoral sand and beach pebble deposits under the post-glacial mud layer. Absolute dating with carbon-14 shows that these sediments were deposited at the end of the last glacial period (Würm IV). Their sudden disappearance at the shelf break indicates an estimated drop in sea level of about 80 to 90 m at the end of the Würm.

The same feature apparently occurs at numerous places off the French Mediterranean margin and the interpretation made in this study is probably applicable to the entire shelfbreak of this region.

INTRODUCTION

Les dépôts quaternaires immergés au large de la côte des Maures et de l'Estérel (et notamment dans le Golfe de Fréjus) ont fait l'objet depuis 1964 d'une étude approfondie faisant intervenir des méthodes très variées. Les principaux résultats obtenus dans le Golfe de Fréjus seront sommairement exposés et comparés avec ceux déjà acquis en d'autres endroits des côtes françaises de la Méditerranée afin d'essayer d'en dégager une interprétation d'ensemble cohérente.

ETUDE PARTICULIERE DES DEPOTS QUATERNAIRES DU GOLFE DE FREJUS

Rappel

Les campagnes de sismique haute fréquence (12 et 5 Kc) entreprises de 1964 à 1970 ont permis de mettre en évidence dans le Golfe de Fréjus (Var) plusieurs réflecteurs acoustiques bien individualisés. L'un d'eux, particulièrement net *et continu* (réflecteur R.A. in Bellaiche *et al.*, 1969a et Bellaiche *et al.*, 1969b, est enfoui sous une épaisseur de vase variable (de 10m au centre du golfe, à 10cm à la limite du plateau continental, vers 90 m de profondeur (Figures 1 et 2). La signification de ce réflecteur est connue grâce aux carottages réalisés au niveau de la rupture de pente limitant le plateau (Figures 3 et 4): il s'agit tantôt d'un conglomérat à ciment calcaire, tantôt d'une couche de sédiments meubles, en général granoclassés: sable grossier et galets à faciès très littoral à la base; sables de plus en plus fins vers le sommet, à faciès plus profond. Dans certains cas, les carottages ont atteint, sous cette couche meuble, une couche de vase grise beaucoup plus compacte que la vase fluide superficielle de couleur beige, et à caractère plus littoral. Parfois, c'est le simple

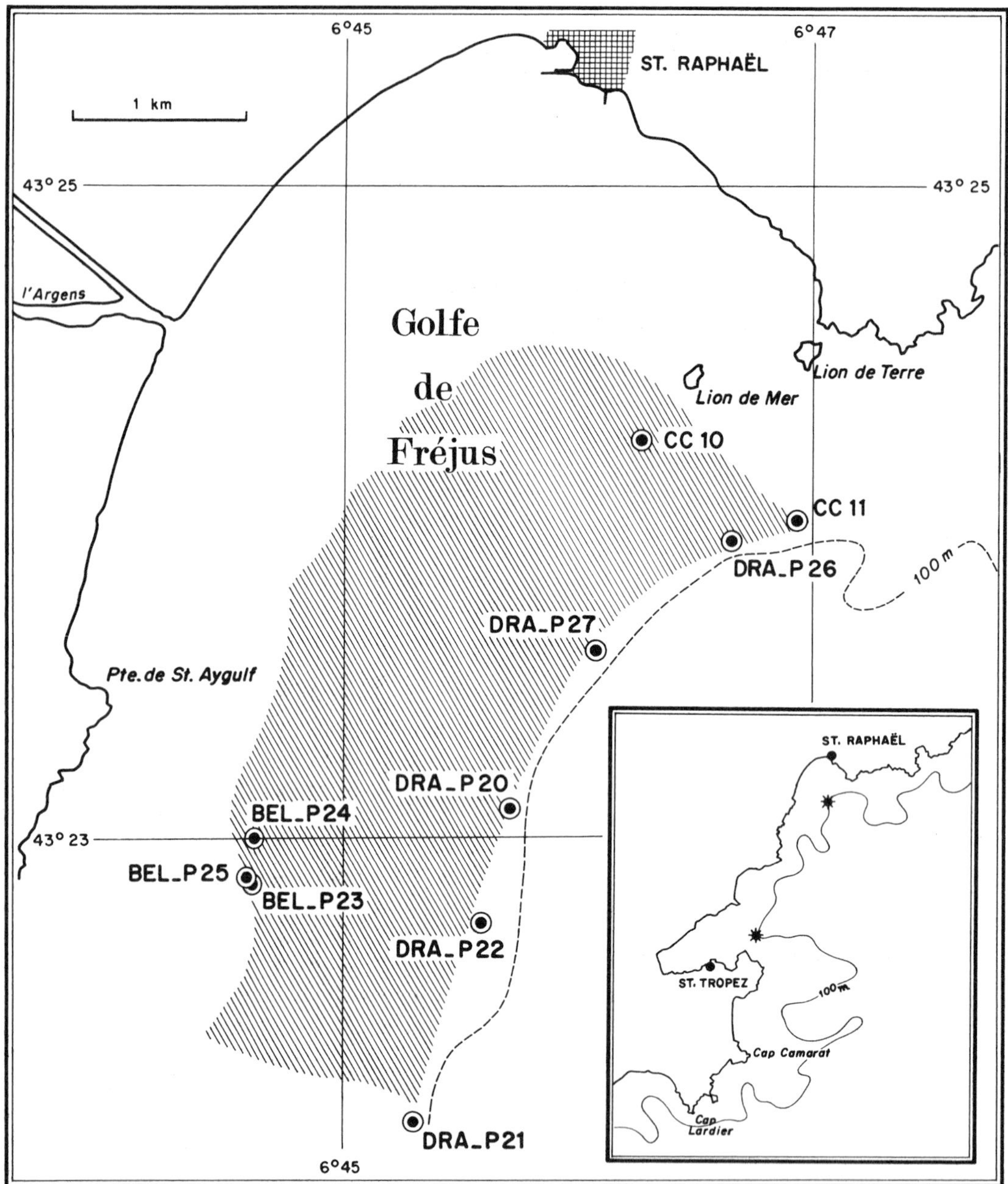

Figure 1. Cartographie du réflecteur acoustique continu R.A. (en grisé) et position des différentes carottes l'ayant atteint ou traversé. Les carottes Cc11, DRA P26, DRA P27, DRA P20, DRA P22 et DRA P21 ont été prélevées entre 80 et 90 m au niveau de la rupture de pente du plateau continental. En cartouche: les astérisques représentent l'emplacement des échantillons datés par le carbone 14.

passage de la vase sous-jacente compacte à la vase superficielle fluide qui constitue le réflecteur acoustique R.A. Un ensemble de considérations d'ordres sédimentologique, faunistique et paléoclimatique (isotopes de l'oxygène, analyses polliniques) nous a conduit à émettre l'hypothèse que vase sous-jacente, niveaux à conglomérats et couche meuble, avaient été déposés pendant la dernière époque glaciaire du Quaternaire (au Würm III et au Würm IV) alors que le niveau de la mer se trouvait à une cote très inférieure au niveau actuel (environ −90m).

Paléogéographie du Golfe de Fréjus à la fin du Würm

Une datation par le carbone 14 d'une coquille de *Chlamys* spp. recueillie dans la couche meuble séparant la vase superficielle de la vase sous-jacente, fournit un âge de 11.800 ans ± 200 ans B.P., ce qui correspond au Würm IV. On a donc là, une confirmation des résultats tirés des différentes analyses citées plus haut.

La synthèse des informations acquises dans le

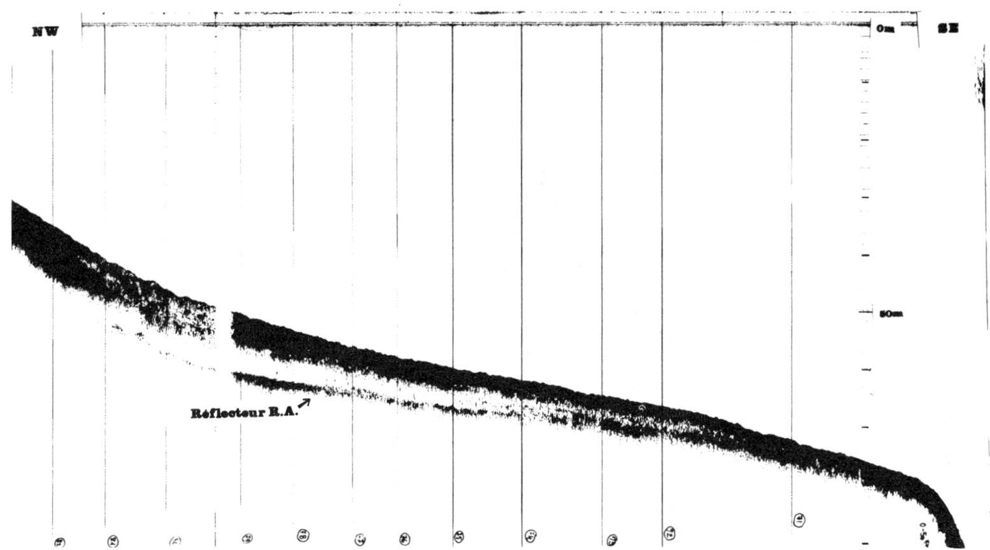

Figure 2. Coupe sismique haute fréquence (12 Kc) traversant le Golfe de Fréjus et montrant le réflecteur acoustique R.A. sous une couche de vase d'épaisseur variable.

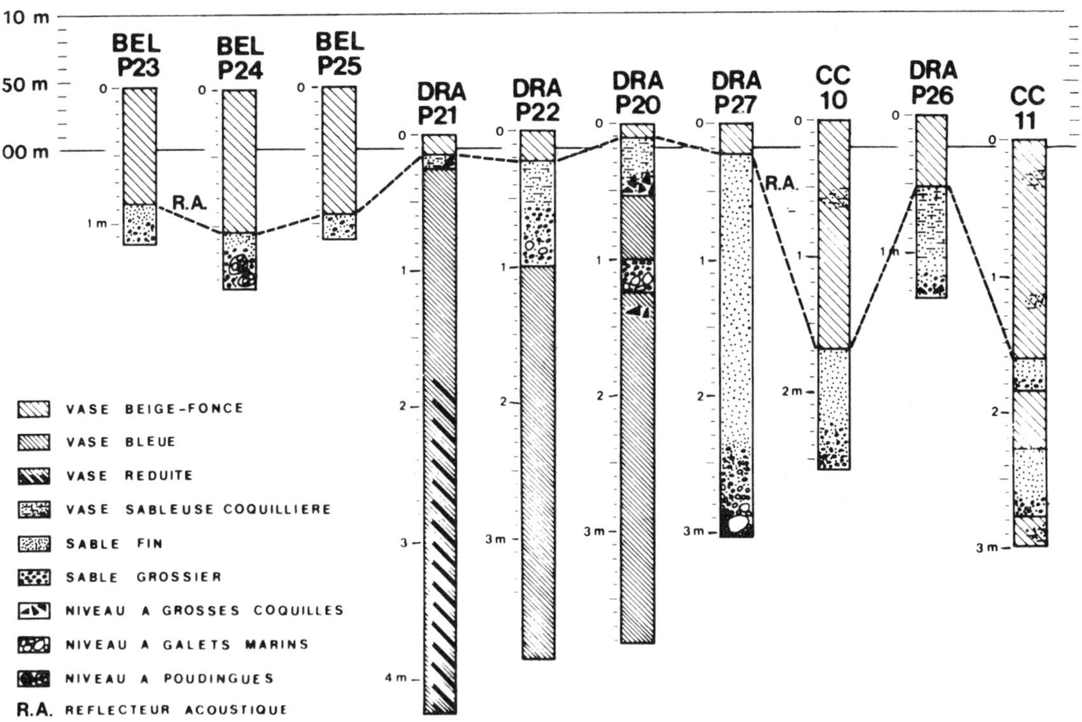

Figure 3. Description des carottes ayant atteint ou traversé le réflecteur acoustique R.A.

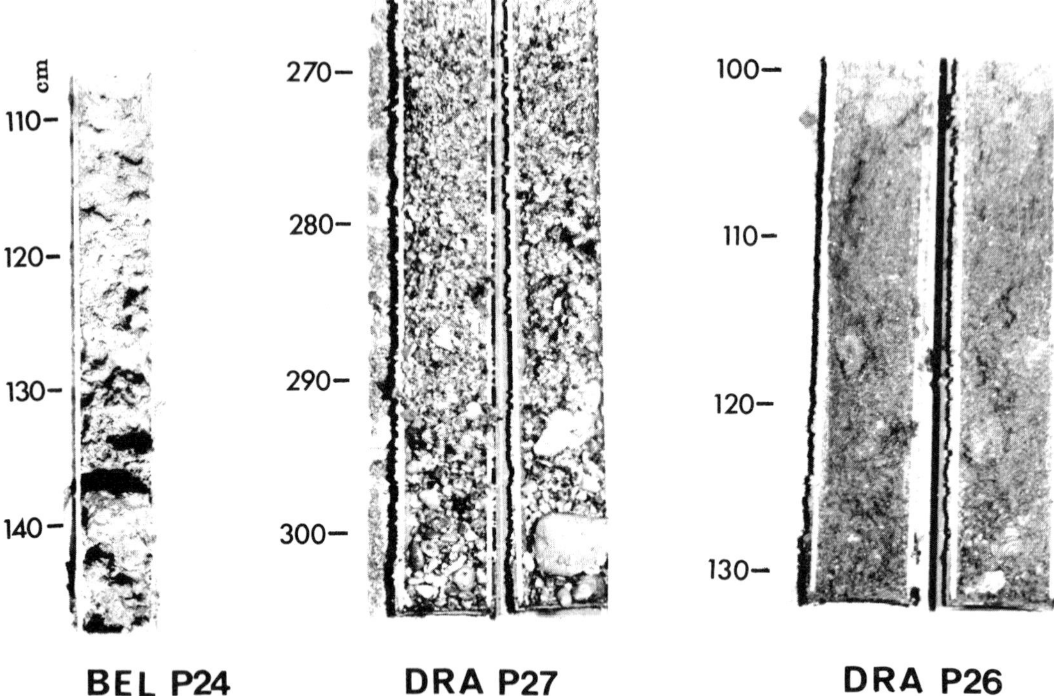

Figure 4. Photographies de la base des carottes BEL P24 (poudingues), DRA P27 (sables et galets marins à faciès très littoral) et DRA P26 (sable coquillier). Ces éléments grossiers deviennent de plus en plus fins vers le sommet où ils font place à de la vase fluide de couleur beige.

Golfe de Fréjus peut nous permettre de retracer de façon assez précise la paléogéographie de ce golfe à cette époque: la ligne de rivage atteignait alors à peu près la ligne de rupture de pente actuelle située entre 80 et 90 m. Cette cote de 90 m coïncide exactement à terre, dans la basse vallée de l'Argens avec l'amplitude de l'érosion des marnes pliocènes, qui ont été postérieurement recouvertes de formations fluvio-marines et franchement marines attribuées par Gouvernet à la transgression flandrienne. Dans la partie Sud-Est du golfe, les plages étaient tantôt sabloneuses, tantôt à galets (carottes DRA P 26, DRA P 27, Cc 11). L'Argens devait déboucher à proximité de l'endroit de prélèvement de la carotte DRAP 20 comme l'atteste la présence d'une macrofaune marquant une très nette dessalure des eaux (*Cardium glaucum*). A hauteur des carottes DRAP 21 et DRAP 22 la sédimentation était concordante et vaseuse. Vers l'Ouest, la ligne de rivage se raccordait aux reliefs rocheux de la pointe de Saint-Aygulf.

Par la suite de la remontée des eaux, consécutive au réchauffement qui a suivi cette période glaciaire, il s'est progressivement déposé dans le golfe, des sédiments de plus en plus fins et de faciès de plus en plus profond que l'on retrouve aujourd'hui par dragage dans l'ensemble du golfe au-delà d'une vingtaine de mètres.

COMPARAISON AVEC LES REGIONS VOISINES

Région de Saint-Tropez

Dans la région du Golfe de Saint-Tropez, voisine du Golfe de Fréjus (Figure 1), une datation par le carbone 14 d'une coquille de *Venus casina* récoltée à la base d'une épaisse couche de sable littoral granoclassé, prélevée par carottage à −80 m de profondeur, a fourni également un âge de 11.700 ans ±200 ans. *Venus casina* est une espèce qui accompagne souvent les thanatocoenoses quaternaires à faune froide Nord-Atlantique.

Ces thanatocoenoses sont bien connues au large de la terminaison orientale du massif des Maures. On les trouve au Sud du Golfe de Fréjus, dans le canyon de Fréjus, entre 140 et 180m. Elles renferment *Chlamys septemradiata*, *Caryophyllum clavius*, forme géante (G.B., étude en cours). Fredj (1964) a montré qu'elles accompagnaient les biocoenoses à grands brachiopodes draguées à l'intérieur d'une ceinture entourant les caps de Saint-Tropez, Camarat et Lardier entre 170 et 220 m de profondeur.

Un peu plus au Sud, Gautier et Picard (1957) ont dragué entre 95 m et 160 m sur le banc du Magaud (prolongement sous-marin oriental de l'île du Levant) des débris d'une thanatocoenose quaternaire récente accompagnant des biocoenoses à grands brachiopodes à partir de 130 m, et ennoyant une thanatocoenose quaternaire plus ancienne à *Dendrophyllia cornigera*. Cette thanatocoenose ne renferme cependant aucune coquille de faune froide.

Région de Marseille

Dans la région de Marseille (canyons de Planier, de la Cassidaigne), les thanatocoenoses à faune froide Nord-Atlantique (*Cyprina islandica*, *Modiola modiolus*, *Chlamys septemradiata*) ont été prélevées depuis 1956 par dragages et carottages entre 90 m et 250 m de profondeur. Pour Froget (communication personnelle), ces associations qui semblent correspondre à un littoral d'une centaine de mètres au-dessous du trait de côte actuel, seraient à rattacher au Würm IV.

Au large du Cap Creus (frontière franco-espagnole)

Mars (1958) avait pu établir que le rivage correspondant à la faune froide découverte par G. Pruvot au large du Cap Creus entre −70 et −200m (*Pecten islandicus, Modiola modiolus, Cyprina islandica, Chlamys islandica, Panopea norvegica* etc.) se situait entre −80 et −100m. En accord avec les idées de Bourcart (1955), il attribuait cette faune jusqu'alors considérée comme sicilienne, à la dernière période glaciaire en précisant qu'elle reflète une température beaucoup plus froide que celle qui régnait au Sicilien. Emiliani et Mayeda (1964) trouvaient pour *Cyprina islandica* des valeurs de $\delta^{18}O$ de + 4,1[1] confirmant ainsi le caractère très froid de ces thanatocoenoses. Thommeret et Thommeret (1965), grâce au carbone 14 dataient du Würm IV une coquille de *Cyprina islandica* prélevée au large Cap Creus, obtenant un âge de 13.000 ans.

Région du Cap d'Antibes (Alpes-Maritimes)

Gennesseaux et Thommeret (1968) ont daté au carbone 14 une couche de sable à faciès littoral provenant de la base d'une carotte prélevée par 100m de profondeur sur le rebord de la plate-forme littorale d'Antibes. L'âge de 14.100 ans ± 300 ans obtenu, permet de préciser que le littoral atteint à cette époque correspondant au Würm IV, coïncidait approximativement avec le rebord du plateau continental.

Au large de la Riviera italienne

Au large d'Imperia, J. P. Rehault a prélevé à faible distance du rebord du plateau continental, vers −100 m une dalle de grès calcaire à faciès littoral attribuable au Würm IV (communication personnelle).

CONCLUSIONS

Les grandes lignes de la sédimentation quaternaire depuis la fin du Würm, mises en évidence dans le Golfe de Fréjus, paraissent se retrouver d'une façon semblable à des profondeurs identiques depuis la frontière espagnole jusqu'à la frontière italienne. Cette continuité témoigne de la faible importance des phénomènes d'ordre tectonique qui ont affecté cette région depuis cette époque sans toutefois permettre d'écarter la possibilité de grands mouvements d'ensemble de cette plateforme continentale.

La disposition géographique des dépôts meubles hérités de la dernière glaciation sur le plateau continental des côtes françaises de la Méditerranée semble très proche de celle rencontrée dans d'autres régions du globe. En effet, de nombreux travaux, notamment ceux de Curray (1965, 1969) et d'Emery (1968) ont montré que d'une façon générale, les sédiments d'âge glaciaire de la plupart des plateaux continentaux du monde étaient recouverts d'une couche de vase actuelle d'épaisseur décroissante vers le large et avaient tendance à affleurer au niveau de la rupture de pente, située à des profondeurs variables selon les régions (la moyenne étant de 132 m, selon Shepard, 1963).

La cartographie et la caractérisation de ces dépôts peut se faire aisément par l'utilisation combinée des méthodes de sondage de vase et de carottages. Signalons à ce sujet l'étroite analogie dans la conduite et dans les résultats de l'étude du Golfe de Fréjus avec celle d'autres régions du globe comme le Golfe de Panama (Golik, 1968) ou le canal du Saint-Laurent (Conolly *et al.*, 1967).

On peut tenter de comparer les datations au carbone 14 obtenues pour les coquilles anciennes prélevées à la limite de la plateforme continentale des côtes françaises de la Méditerranée, avec celles réalisées sur les tourbes et coquilles littorales prélevées dans d'autres régions du globe, en particulier sur des plateformes de caractéristique voisine (notamment en ce qui concerne la position de la rupture de pente): côtes du Texas, du Mexique et de la Californie (Emery et Garrison, 1967; Milliman et Emery, 1968; Curray, 1969). Il en résulte que les datations

[1] $\delta^{18}O$: différence entre le rapport isotopique $^{18}O/^{16}O$ dans l'échantillon considéré et ce même rapport dans le standard international de référence PHB_1 Chicago exprimé selon la relation:

$$\delta^{18}O = \left[\frac{^{18}O/^{16}O \text{ échantillon}}{^{18}O/^{16}O \text{ étalon}} - 1\right] \times 1000$$

obtenues pour les coquilles méditerranéennes semblent conduire à des âges de 2.000 à 4.000 ans plus jeunes que ceux des coquilles et tourbes des régions précitées. Il serait prématuré de tenter dès à présent d'expliquer cette différence car les courbes de variation du niveau marin depuis 20.000 à 30.000 ans obtenues par les auteurs américains ont été tracées à partir d'un nuage de points très dispersés et le nombre de datations disponibles en Méditerranée nord-occidentale est encore très fragmentaire. D'autres analyses pourraient permettre de poser correctement le problème et pourront conduire éventuellement à faire la part entre les phénomènes géonomiques de nature glacio-eustatique, et les phénomènes régionaux de gauchissement à grand rayon de courbure (mouvements de surrection et de subsidence).

REMERCIEMENTS

Les datations par le carbone 14 ont été effectuées par M. et Mme. Thommeret du Laboratoire de Radioactivité appliquée, Centre Scientifique de Monaco.

BIBLIOGRAPHIE

Bellaiche, G., C. Vergnaud Grazzini, et L. Glangeaud 1969 a. Les épisodes de la transgression flandrienne dans le Golfe de Fréjus. *Comptes Rendus des Séances de l'Académie des Sciences, Paris,* 268:2765–2770.

Bellaiche, G., I. Gaudry, et C. Vergnaud Grazzini 1969 b. Paléogéographie quaternaire du golfe de Fréjus (Var). *Communication VIIIème Congrès International de l'INQUA,* Paris.

Bourcart, J. 1955. Recherches sur le plateau continental de Banyuls sur mer. *Vie et Milieu,* 6(4):492–495.

Conolly, J. R., H. D. Needham, et B. C. Heezen 1967. Late Pleistocene and Holocene sedimentation in the Laurentian Channel. *The Journal of Geology,* 75(2):131–147.

Curray, J. R. 1965. Late Quaternary history, continental shelves of the United States. In: *The Quaternary of the United States,* eds. Wright, H. E., Jr., and D. G. Frey, Princeton University Press, Princeton 723–735.

Curray, J. R. 1969. History of continental shelves. In: *The New Concepts of Continental Margin Sedimentation,* ed. Stanley, D. J., American Geological Institute, Washington, D.C. 6:1–18.

Emery, K. O. 1968. Relict sediments on continental shelves of world. *The American Association of Petroleum Geologists Bulletin,* 52(3):445–464.

Emery, K. O. et L. E. Garrison 1967. Sea levels 7,000 to 20,000 years ago. *Science,* 157:684–687.

Emiliani, C. et T. Mayeda 1964. Oxygen isotopic analysis of some molluscan shells from fossil littoral deposits of Pleistocene age. *American Journal of Science,* 262(1):107–113.

Fredj, G. 1964. Contribution à l'étude bionomique de la Méditerranée occidentale (Côtes du Var et des Alpes Maritimes—Côte occidentale de la Corse). *Fasc. 2*: La région de St. Tropez: du Cap Taillat au Cap Saint Tropez (région A_1), *Bulletin de l'Institut Océanographique de Monaco,* 63 (1311, A):55 p.

Gautier, Y. et J. Picard 1957. Bionomie du banc du Magaud. *Recueil des Travaux de la Station Marine d'Endoume,* 21(12): 28–40.

Gennesseaux, M. and Y. Thommeret 1968. Datation par le radiocarbone de quelques sédiments sous-marins de la région niçoise. *Revue de Géographie Physique et de Géologie Dynamique,* 10 (2): 375–382.

Golik, A. 1968. History of Holocene transgression in the Gulf of Panama. *The Journal of Geology,* 76(5):497–507.

Gouvernet, C. 1968. Etude géographique de la plaine du bas-Argens. Localisation des gîtes aquifères. *Annales de la Faculté des Sciences de Marseille,* 40:173–191.

Mars, P. 1958. Les faunes malacologiques quaternaires "froides" de Méditerranée. Le gisement du Cap Creux. *Vie et Milieu,* 9(3): 293–309.

Milliman, J. D. et K. O. Emery 1968. Sea levels during the past 35,000 years. *Science,* 162:1121–1123.

Shepard, F. P. 1963. *Submarine geology.* Harper and Row, New York, 555 p.

Thommeret, J. et Y. Thommeret 1965. Validité de la datation des sédiments du proche Quaternaire par le dosage du carbone 14 dans les coquilles marines. *Comité International pour l'Etude Scientifique de la Mer Méditerranée, Rapports et Procès Verbaux,* 18(3):837–843.

The Gulf of La Spezia, Italy: A Case History of Seismic-Sedimentologic Correlation

Lloyd R. Breslau and Harold E. Edgerton

U.S. Coast Guard Headquarters, Washington, D.C. and Massachusetts Institute of Technology, Cambridge, Massachusetts

ABSTRACT

The sub-bottom structure of the Gulf of La Spezia, Italy has been investigated by continuous seismic profiling and sediment dredging and coring. The seismic profiles were used to develop contour plots of subbottom layers, which were identified from the dredge and core samples.

The seismic and sedimentologic investigation revealed that while the sea floor of the present gulf is composed of featureless clay, the ancient gulf was geologically far more complex. A sand and gravel barrier-bar ran across the mouth of the gulf. Vegetation covered the upper surface of the barrier-bar and coral grew on its flanks. A lagoon formed to the landward of the bar and sediments represented by subbottom layers formed in the quiet waters of the protected lagoon.

The oldest feature observed in the gulf is an ancient drainage channel observed both landward and seaward of the barrier-bar. At some point during the rise of sea level in the Holocene, conditions were such that a cross-gulf barrier-bar was formed by longshore drift of coarse sedimentary material. With a further rise in sea level, the dynamic conditions that created the barrier-bar were removed, and the bar, vegetation, and fringing reef coral were buried by a cover of fine sedimentary material.

RESUME

La structure du substratum du Golfe de la Spezia, Italie, a été étudiée à l'aide de profils sismiques continus, de dragages et de carottages. Les profils sismiques ont révélé la topographie des couches identifiées par échantillonage.

Les études sismiques et sédimentologiques ont montré que tandis que le fond du golfe actuel est constitué d'argile sans traits bien marqués, celui du golfe ancien était, du point de vue géologique, beaucoup plus complexe. Une barre de sable et de gravier fermait l'embouchure du golfe. De la végétation couvrait la partie supérieure de cette barrière et des coraux se sont développés sur ses flancs. Un lagon a pris forme côté terre de cette barre dans les eaux protègées duquel ont pu se déposer des sédiments qui forment les réflecteurs dans le substratum.

Le relief le plus ancien observé dans le golfe est un chenal de drainage situé de chaque côté de la barrière. A un moment donné du Holocène, au cours de l'élévation du niveau de la mer, les conditions étaient telles qu'une barre a pu se former au travers du golfe sous l'action de courants de dérive longeant la côte et transportant du matériau à grain grossier. Ces conditions ont disparu lors d'une nouvelle élévation du niveau de la mer et la barre, la végétation et les coraux furent recouverts par une couche de sédiment à grain fin.

INTRODUCTION

The existence of an extensive subbottom structure in the Gulf of La Spezia was observed by Breslau and Edgerton (1965) during exploratory investigations with a continuous seismic profiler. This occurrence in shallow water close to a research center afforded an opportunity to investigate the general problem of correlation of seismic discontinuities (manifested as subbottom acoustic layering) with sedimentologic truth. Such research was therefore undertaken as part of SACLANTCEN's study of the acoustic reflectivity of the sea floor. Preliminary results of the exploratory seismic investigation were presented in the autumn of 1964 (Breslau and Edgerton, 1965) to the Comité de Morphologie et Géologie Marine of

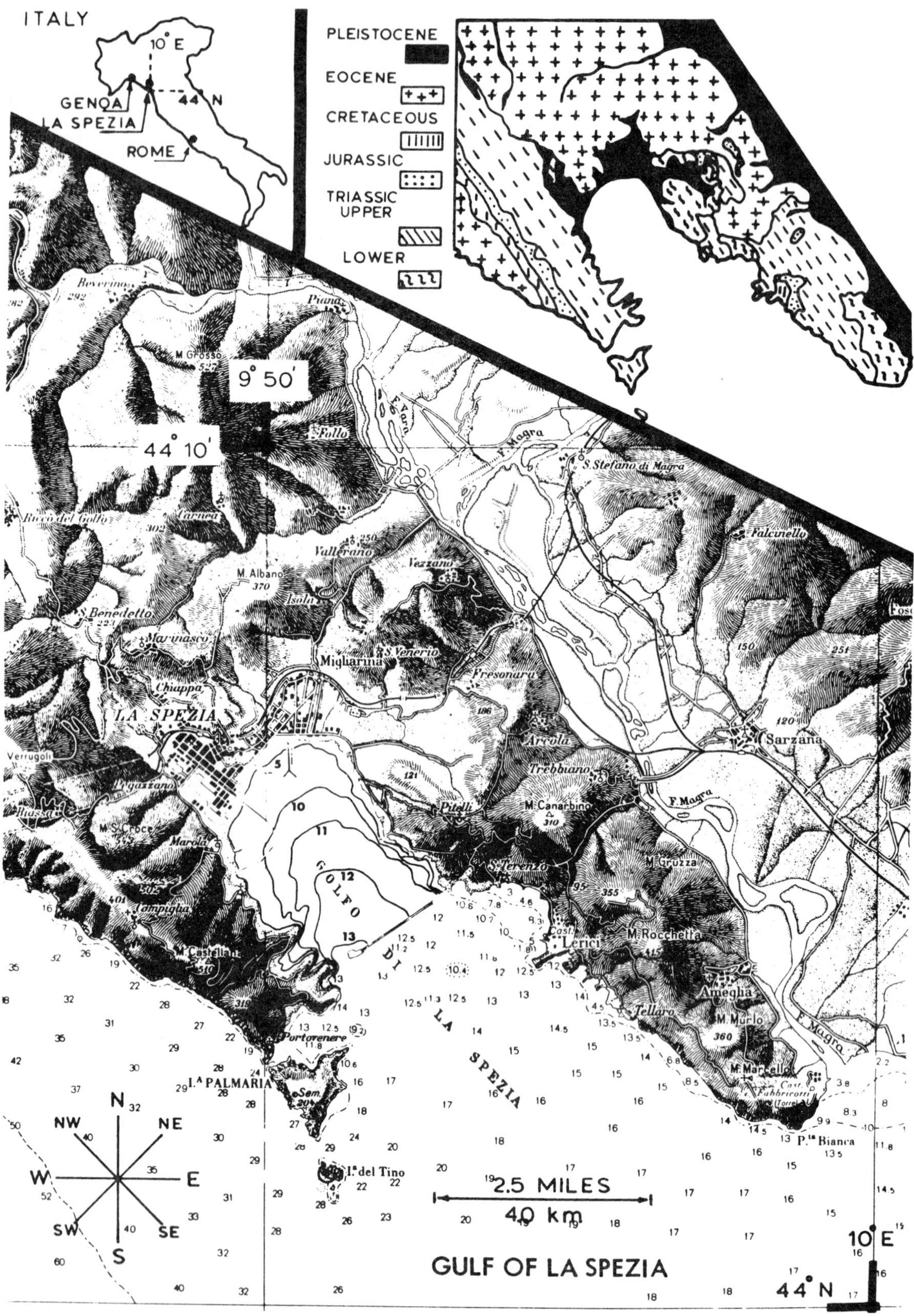

Figure 1. Location, geology, and relief of the Gulf of La Spezia region. Bathymetry and spot-heights in meters.

the Commission pour l'Exploration de la Mer Méditerranée. A detailed report of both the seismic and sedimentologic work was published by SACLANTCEN in 1968.

The Gulf of La Spezia (Figure 1) is about 75 km southeast of Genoa. It is a deep landlocked gulf with no significant river flowing into it, and has a maximum land-sheltered length of about 7 km, an average width of 4.5 km, and a minimum width of 2.8 km between Point Pezzino and San Bartolomeo. The mouth of the gulf faces southeast toward the sea. The gulf is now sheltered from the sea by a rock breakwater (constructed 1870–1880), which runs approximately SW-NE from Point Santa Maria to Point Santa Teresa.

The Gulf is ringed on its land perimeter by mountains of the Alpi Apuane (Zaccagna, 1932), a detached chain of the Tuscan Apennines. The western and eastern promontories consist of semi-parallel folds of Mesozoic and younger rocks (Capellini, 1863): an overturned anticline on the west and a normal anticline on the east. The gulf is therefore basically a tectonic feature generated in response to deep-seated thrusting forces directed from southwest to northeast. This tectonic structure was subsequently flooded to form the present gulf.

In general, the land bordering the western side of the gulf has stronger relief than the land bordering the eastern side (Da Portofino al Gombo, Istituto Idrografico, 1955). There is a small, low-relief, alluvial area on the northwest shore (where the city of La Spezia stands), and a considerably larger one on the northeast shore (Figure 1).

Depths in the gulf (with the exception of dredged areas) at present range mainly between 10 m and 15 m, with 12 m as a representative value for the inner gulf (the part sheltered by the rock breakwater). The sea floor of the inner gulf consists entirely of gray clay with virtually no bottom relief.

The earliest available bathymetric records were found on a French chart dated 1840 [Plan du Golfe de La Spezia-Côtes d'Italie (Duché de Gênes), Dépôt Général de la Marine, 1846]. Data from this chart have been used to draw the 5, 10, 11, 12 and 13 meter contours shown on Figure 1. It is interesting to note that a comparison between the 1840 chart and the present day bathymetry of undredged areas (Rada di La Spezia, Istituto Idrografico, 1962) indicates that shoaling of the inner gulf during the last 125 years on average has been less than half a meter. In addition, for most areas the sediment notations on the old French chart (*vase grise*) are the same as those in the present chart (*mud*). Apparently, the breakwater has not substantially altered either the bathymetry or the sediment type in the inner gulf.

SEISMIC OBSERVATIONS

The subbottom structure of the Gulf of La Spezia, Italy, was first investigated by continuous seismic profiling. The seismic profiler used was the Mud Penetrator (Edgerton, 1962, and Yules and Edgerton, 1964). The seismic profiles were used to develop contour plots of subbottom layers, which later were lithologically identified from measurements made on sediment cores.

The Mud Penetrator seismic profiler is a commercially available instrument that electronically generates an acoustic pulse in the water and records the arrival times of the acoustic echoes from the bottom and subbottom on an analog correlation recorder. Echoes from acoustic discontinuities are synthesized by the correlation recorder so that a graph of the bathymetry and subbottom structure along a profile is obtained directly. The principal features of the Mud Penetrator that give it the capacity to penetrate the sea floor and resolve fine structure are its high peak power output (about 105 db/1 dyne/cm^2), short pulse length (about 0.1 ms), and high repetition rate (10 or 20 pulse/sec).

The equipment is in three parts: (1) a recorder/driver console; (2) a transmitter/receiver transducer; (3) a gasoline-powered electric generator. The recorder/driver console generates the acoustic pulse and records the echo. The transducer, which has an acoustic sensitivity pattern represented by a 30° cone, is used for both transmitting and receiving the acoustic pulse. It uses a piezoelectric crystal for energy conversion, and must be submerged during operation.

This equipment was installed on the 65 ft SACLANTCEN workboat, the transducer being housed in a streamlined "fish" towed alongside the boat at a depth of approximately two meters. The survey was conducted at speeds of 2 or 3 knots.

Two separate surveys were made with the continuous seismic profiler operated from the SACLANTCEN workboat. The first exploratory investiagtion (Figure 2A) was made in 1964. Navigational position-fixing was by occasional optical bearings of land objects and steering was by compass course. Subsequent analysis of the profiles obtained during the exploratory seismic survey disclosed that the accuracy of the navigational positioning employed was inadequate to permit detailed contouring of the seismic layers, and another seismic survey (Figure 2B) was made in 1965. Navigation was by steering on a fixed landmark and position fixing was by horizontal sextant angles on land objects. This survey was used to generate the contours of subbottom seismic structure.

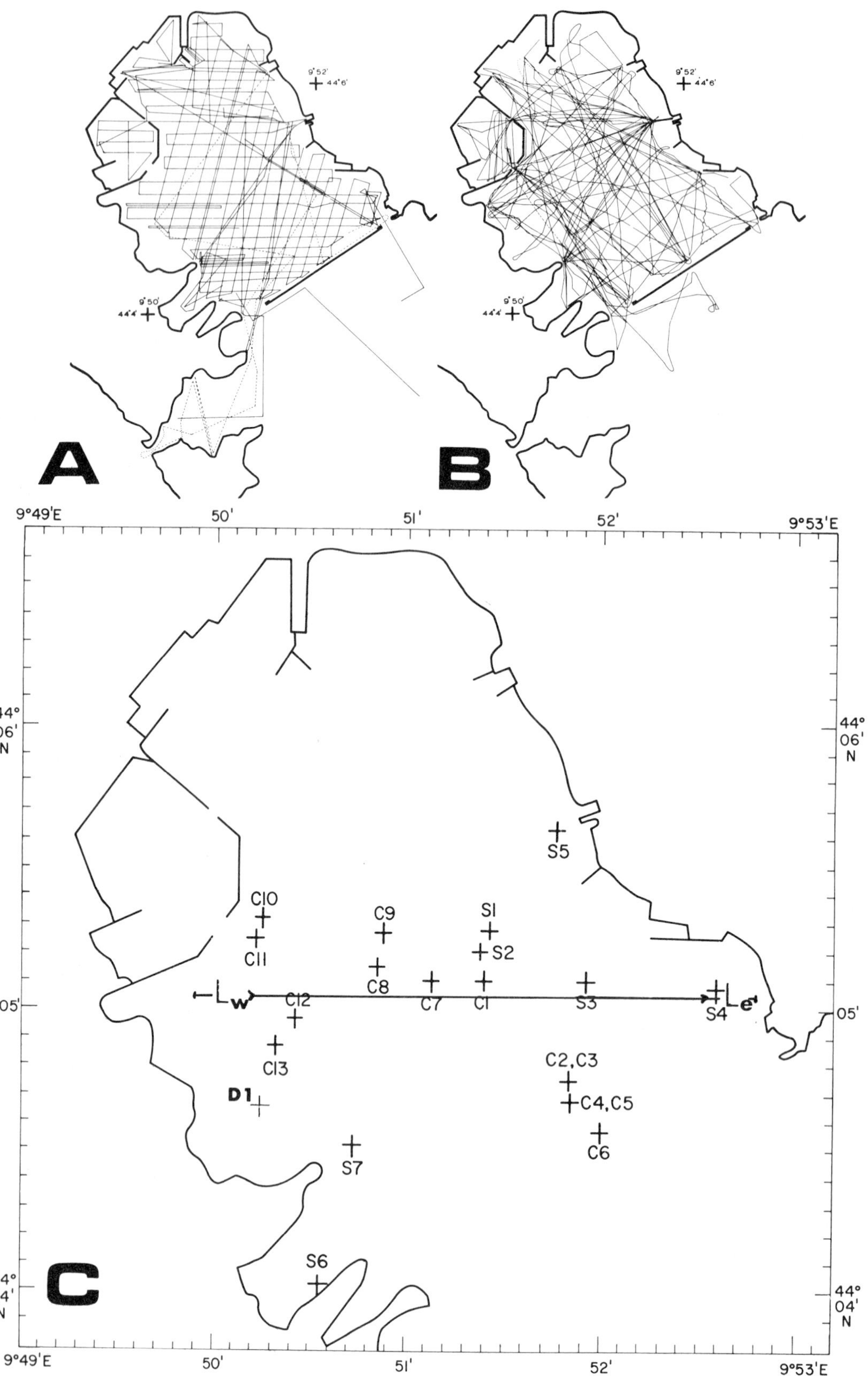

Figure 2. Ship-tracks for seismic profiling and location of sedimentologic stations in the Gulf of La Spezia: A, seismic survey-tracks, 1964; B, seismic survey-tracks, 1965; and C, surface sediment dredge stations (51–57) and wide-barrel core stations (C1–C13). Line L_W–L_E indicates position of seismic profile illustrated in Figure 3.

Some fundamentals concerning the interpretation of continuous seismic profiler records should be mentioned. Taken by itself, this type of seismic information is ambiguous. Subbottom features seen on the seismic record demonstrate the existence of subbottom structure, but the absence of such features from the seismic record does not necessarily mean that subbottom structure is absent. This is because the reception of a signal from a sediment interface depends on the magnitude of the acoustic reflection from the interface as well as on the acoustic attenuation in the overlying sediment layer. Therefore, subbottom structure might actually be present and yet not be recorded, because of masking by a highly-attenuating sediment cover.

The interpretation of the magnitude of seismic reflections must not only take account of attenuation in the overlying sediment cover, but also of the depth-dependent propagation loss in the water column, and of various gain settings on the seismic apparatus itself. In addition, the dynamic range in the recording paper used with the seismic system is only 20dB under the best laboratory conditions, and probably considerably less when used at sea. Therefore, the reflecting strengths of the seismic layers observed on the record can only be discussed in very qualitative terms.

A typical seismic profile obtained in the Gulf of La Spezia (located along line $L_W - L_E$ on Figure 2C) is presented in Figure 3. Printed annotations have been added to this otherwise unaltered seismic profile. These annotations, based on a descriptive rather than interpretative system, serve to point out the significant subbottom seismic features observed in the profiles. The following sections list and describe these features.

Deep Layer

This layer is a good reflector and poor transmitter of acoustic energy. It is the terminal reflector (deepest layer that has been reached with the particular seismic apparatus used) and appears on all the profiles. It is observed at depths ranging from 8 m to 28 m. Morphologically, it seems to represent a composite feature resulting from the superposition of a raised cross-gulf barrier-bar, on a seaward-dipping basement. The landward flank of the barrier-bar is considerably steeper than the seaward flank, and the bar's crest shoals to less than 3 m below the sea floor.

Light Dome

This feature is a good reflector and a very poor transmitter of acoustic energy. In fact, it is almost totally opaque acoustically. The light appearance on the seismic record results from the absence of any acoustic energy return after the first-arrival echo from the upper surface of the light dome. These features are commonly found on the inner and outer flanks of the cross-gulf barrier-bar, and in a zone in the northeast section of the inner gulf. The upper surface of the light domes is observed at depths ranging from 10 m to 16 m. This upper surface is usually found touching or slightly below the shallow layer, but it may also be found slightly above this layer.

Dark Patch

This feature is a partially transparent diffuse reflector. The echo return is stretched out as if the acoustic wave passed through a concentrated assemblage of small but good reflectors. Dark patches occur most prominently at depressions in the deep layer. They also occur, however, in a zone at the middle western part of the inner gulf and here they are not obviously related to a deep layer depression. The upper surface of the dark patches is found between depths of 10 m and 22.5 m. A representative depth in the major deep-layer depression landward of the cross-gulf barrier is about 20 m; in the middle western zone it is about 15 m. All these depths are approximate because it is rather difficult to specify the depths of the upper surfaces of the dark patches owing to their diffuse and irregular character.

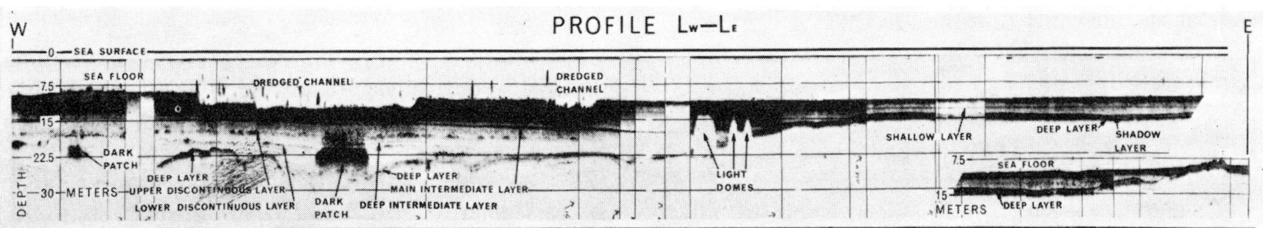

Figure 3. Typical seismic profile obtained in the Gulf of La Spezia. For location, see line $L_W - L_E$ on Figure 2C.

Main Intermediate Layer

This layer is both a good reflector and a good transmitter of acoustic energy. The echo return from this layer is strong enough to record well on the seismic record, and yet there is no evidence that the presence of this layer obscures any underlying features. It is interesting to note that there is a tendency for echo strength to increase with the depth of this layer. It is only found landwards of the inferred cross-gulf barrier-bar; its southeastern border either terminates against the barrier bar or against the light domes on the inner flank of the barrier bar. This layer is found at depths between 12 m and 17 m. Although it is not precisely concordant with the surface of the deep layer, slight depressions in this layer tend to occur over areas of depression in the deep layer.

Deep Intermediate Layer

This layer appears to be acoustically similar to the main intermediate layer. It is found within the area of development of the main intermediate layer, but is more restricted. It does not impinge against the light dome or cross-channel barrier-bar, but rather pinches out towards the southeast. It is found at depths between 16 m and 20 m. Although it is not concordant with the surface of the deep layer, there is a correlation between the depressions in both surfaces. This correlation is somewhat stronger than that observed between the main intermediate layer and the deep layer.

Shallow Layer

This layer is a weak reflector and excellent transmitter of acoustic energy. It is almost transparent acoustically and is observed on most of the profiles, appearing at about 2 to 3 m below the present-day sea floor. It probably exists even where it is not seen on the record, but is lost in the echo return from the sea floor. The weak echo strength and concordance with the sea floor topography led us to suspect that it represented an extraneous reflection from the boat. This suspicion was dispelled by an experiment in which the seismic transducer was lowered to depths of up to 10 m below the boat, during which time the shallow layer remained at a constant depth relative to the sea floor trace on the seismic record.

Shadow Layer

This layer is a good reflector and a good transmitter of acoustic energy. It is confined to a zone near the crest of the cross-gulf barrier-bar, where it is found associated with, and less than 1 m above the upper surface of, the deep layer. Its name originates from the fact that it has the appearance on the seismic record of a shadow on top of the deep layer.

Upper and Lower Discontinuous Layers

These layers are characterized principally by their disjointed or discontinuous appearance. The reflecting fragments themselves are good acoustic reflectors, which sometimes display a tendency toward crescentic shapes. These layers are observed landward of the cross-gulf barrier-bar. Where the discontinuous layers and deep intermediate layer are observed concurrently, the upper discontinuous layer occurs about 1.5 m above, and the lower discontinuous layer about 1 m below, the deep intermediate layer.

Multiple Shallow Layers

These layers are not present in Figure 3, but are seen on other profiles. In the southeastern section of the inner gulf, two layers of similar acoustic character to the shallow layer are observed within a depth zone about 1 m thick above the shallow layer.

Buried Channels

Where observed, this feature appears on the seismic record in the form of a trough-shaped depression in the deep layer. These depressions are up to 3 m deep and usually contain small-scale, semi-concordant layering within them. This feature is limited to the area landward of the cross-gulf barrier-bar. No examples of these are shown on Figure 3.

Areal Distribution of Seismic Layers

Contours and areal distributions of the deep layer, main intermediate layer, deep intermediate layer, shallow layer, multiple shallow layers, light domes, and dark patches were generated from a detailed analysis of all the seismic records obtained during the second seismic survey. A contour chart of the deep layer and an areal distribution chart of the light domes and dark patches are presented in Figures 4A and 4B respectively.

The deep layer (Figure 4A) represents the terminal layer (deepest layer reachable with the particular seismic apparatus used) in the Gulf of La Spezia. Depth contours are referenced to the sea surface. Two separate morphological features are evident: a raised cross-gulf barrier-bar (gently contoured depositional feature) and a deeply eroded older surface upon which the barrier-bar is superimposed.

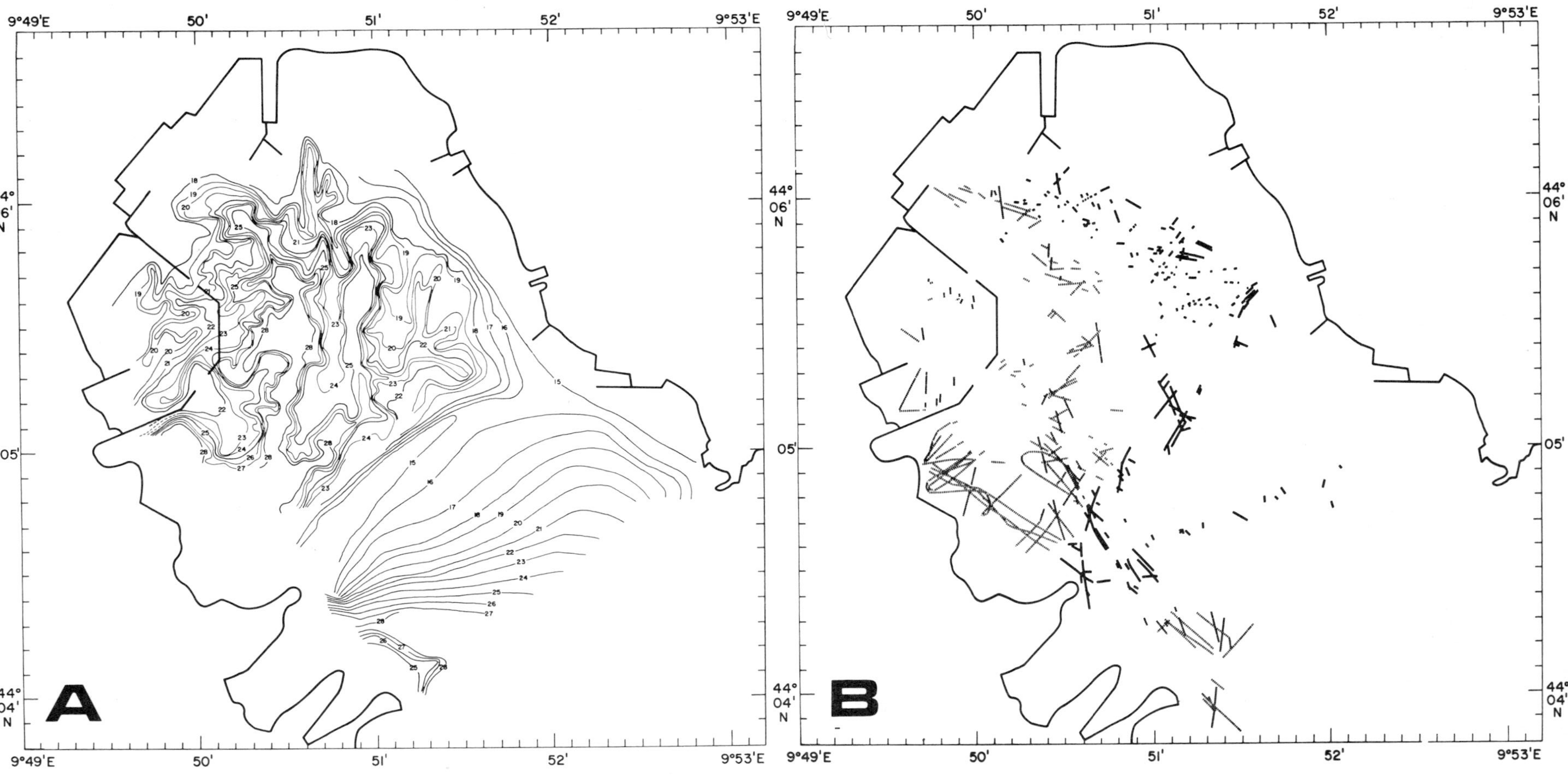

Figure 4. A, Contoured map of the terminal seismic layer in the Gulf of La Spezia; B, Areal distribution of light domes (coral: solid lines) and dark patches (? volume scatterers: dashed lines) in the Gulf of La Spezia.

There is a gap in the contouring of the landward flank of the cross-gulf barrier-bar because the presence of light domes obscured the deep layer.

The areal distribution of the light domes and dark patches is shown in Figure 4B. Clearly, both of these features have a strong tendency towards zonal rather than random distribution. The light domes occur along a narrow band on the inner and outer flanks of the cross-gulf barrier-bar, and within a broader, bank-like zone in the northeast section of the inner gulf. The dark patches occupy three general zones: a northerly-trending elongated zone landward of the cross-gulf barrier-bar, a zone in the western section just landward of the barrier-bar, and a small (possibly elongated) zone just seaward of the barrier-bar.

SEDIMENTOLOGIC OBSERVATIONS

Methodology

The continuous seismic profiling investigation provided information about the existence of subbottom layers. A sedimentologic investigation was undertaken to provide information about the actual nature of these subbottom reflectors. A standard clamshell (Van Veen) grab dredge was used to obtain undisturbed surficial sediment samples. Some exploratory coring (not indexed on Figure 2C) was performed using a sediment corer made by attaching 3 meters of 5 cm bore pipe to a standard ASW bomb filled with concrete (bomb corer). The SACLANTCEN Sphincter Corer (Kermabon *et al.*, 1963) was used to obtain undisturbed cores for detailed mass physical analysis.

The SACLANTCEN wide-barrel piston corer is a long corer with a watertight core-catcher and electrical trigger release system. It is designed specifically for obtaining undisturbed sediment samples for mass-physical analysis. When used with a 12 m coring barrel its total length is 15 m and total weight 1000 kg. A 12 cm diameter plastic liner fits inside the core barrel and retains the sediment section.

The cores were taken by gravity rather than by using the piston, because the hydrostatic head available was insufficient for piston coring. The depth that the core barrel penetrated into the sea floor was noted by sediment markings on the outside of the core barrel. This figure was recorded for later use in applying a stretch factor to the sediment sections recovered in the liner. The sediment-filled liners were analyzed in the laboratory.

Thirteen sediment cores were obtained during 1965 and 1966, using the SACLANTCEN research vessel (MARIA PAOLINA G.) when opportunities occurred during arrivals and departures from La Spezia, and during the test of coring equipment by the SACLANTCEN Oceanography Group. Seven surface sediment dredge hauls were also obtained using the SACLANTCEN work-boat (Figure 2C).

Three methods of sediment analysis were used. These, in ascending order of complexity and difficulty, are as follows: visual analysis, water content analysis and sound speed analysis. The sediments were subjected to one or more of these analyses. Detailed information on the results of these analyses is contained in Breslau and Edgerton (1968).

Seismic-Lithologic Correlation

Seismic and sedimentologic information for the coring sites are presented together in Table 1 in order to facilitate examination of the degree of correlation between them. Examination of Table 1 reveals the following information:

1. The *deep layer* (terminal layer) at the cross-gulf barrier-bar is correlated with a sand layer. Evidence is found in Cores C1, C2, C3, C4 and C5.
2. The *shadow layer,* discussed in the seismic analysis, is correlated with a peat layer. Evidence is found in Core C1.
3. The *light domes* are composed of coral. Evidence is found in Core C7.
4. The *shallow layer* cannot be related to a particular sediment type. This seismic layer was definitely penetrated by Cores C1, C2, C3, C4, C5 and C6, but no detectable variation in sediment type was observed in the cores.
5. The *dark patches* might be related to the presence of large amounts of shell material in the sediment. They are not noticeably linked to a particular sediment type. Evidence is found in Core C13.
6. The *main intermediate layer* cannot be related to any particular sediment type. This seismic layer was definitely penetrated by Cores C11 and C12, and possibly penetrated by Core C9, but no consistent variation in sediment type was detected in the cores.
7. The *deep intermediate layer* was not penetrated by any of the cores.
8. A *gravel layer* exists below the sand layer at the cross-gulf barrier-bar, but this layer was not detected by the seismic system, owing to masking by the sand (terminal) layer.

Observations 1 and 4 above are substantiated by the results obtained during exploratory sedimentologic investigations with the bomb corer. Conclusion 3 is reinforced by results obtained by the harbour dredgers, who reported: *"Nelle zone fangose si e' osservata la presenza di formazioni corallifere costituite da una specie di corallo bianco poroso di notevole*

Table 1. Seismic features and core observations.

SEISMIC INFORMATION Seismic Features and their Depth Below Seafloor					SEDIMENTOLOGIC INFORMATION		
First Seismic Feature	Second Seismic Feature	Third Seismic Feature	Fourth Seismic Feature	Sediment Sampling Site	Length of Sediment Section Recovered	Depth of Maximum Core Penetration	Sediment Type
Shallow Layer 2.4 m	Shadow Layer 3.0 m	Terminal Layer 3.4 m		C1	5.7 m	8.0 m	0–3.0 m Gray Clay 3.0–3.6 m Peat 3.6–7.7 m Sand 7.7–8.0 m Gravel
Shallow L. 2.7 m	Terminal L. 7.8 m			C2	3.65 m	5.5 m	0–5.2 m Gray Clay 5.2–5.5 m Sand
Shallow L. 2.7 m	Terminal L. 7.8 m			C3	3.45 m	6.0 m	0–5.8 m Gray Clay 5.8–6.0 m Sand
Shallow L. 2.7 m	Terminal L. 8.6 m			C4	42 m	7.0 m	0–6.9 m Gray Clay 6.9–7.0 m Sand
Shallow L. 2.7 m	Terminal L. 8.6 m			C5	4.7 m	7.1 m	0–6.6 m Gray Clay 6.6–7.1 m Sand
Shallow L. 2.7 m	Terminal L. 11.2 m			C6	4.5 m	7.2 m	0–7.2 m Gray Clay
Shallow L. 2.3 m	Light Dome 2.8 ± .5 m			C7	5.6 m	6.5 m	0–3.7 m Gray Clay 3.7–6.5 m Clay & Coral
Main. Int. L. 6.4 m	Terminal L.			C8	5.0 m	6.0 m	0–6.0 m Gray Clay
Main Int. L. 7.8 m	Terminal L. 13.8 m			C9	4.2 m	8.1 m	0–8.1 m Gray Clay
Main Int. L. 6.5 m	Deep Int. L. 10.5 m	Dark Patch 13.4 ± 1 m	Terminal L. 17.2 m	C10	4.6 m	6.0 m	0–6.0 m Gray Clay
Main Int. L. 4.2 m	Deep Int. L. 7.8 m	Terminal L. 11.2 m		C11	4.45 m	7.5 m	0–7.5 m Gray Clay
Main Int. L. 3.9m	Dark Patch 6.4 ± 1 m	Terminal L. 15.4 m		C12	5.8 m	7.5 m	0–7.5 m Gray Clay
Dark Patch 5.9 ± 1 m	Terminal L. 15.9 m			C13	7.5 m	8.0 m	0–3.8 m Lost 3.8–8.0 m Gray Clay (considerable shell material)

Note: Int. = Intermediate

durezza che veniva dragato in quantitá rilevante misto al fango", ("In the mud zones we observed coral formations made of a kind of very hard white porous coral, which was dredged in great quantity together with the mud"), (Personal communication, Agenzia Gastaldi, Naples, Italy).

Additional information bearing on the nature of the dark patches discussed in observation 5 above was obtained from a test borehole taken for civil engineering purposes in the western part of the gulf. The location of the borehole site is indicated as D1 on Figure 2C. This site is located over a dark patch whose top is 18 ± 1 m below sea level. The borehole sediment log indicates silty clay and clayey silt from the sea floor down to a depth of 28 m below sea level, where a sand layer is encountered. The annotations concerning inclusions are enlightening. At 15.5 m the log states *"Argilla limosa grigia con tracce di torba e framm. di conchiglie",* ("gray silty clay with traces of peat and shell fragments"). The next notation is at 19.5 m, where the log states *"limo argilloso con torba a grumi e diffusa e framm. di conchiglie",* ("clayey silt with peat both in patches and diffused; and with shell fragments"), (Personal communication, S.N.A.M. SA, Sondaggio 17 Seno Di Panigaglia, Studio Tecnico Geom. Celotti, Milano).

A definite conclusion has not been reached regarding the nature of the shallow and intermediate layers, except that these layers cannot be linked to any consistently occurring lithologic unit. There appear to be three ways in which the observed lack of correlation of these layers might arise: the shallow and intermediate seismic layers are artificial and are caused by spurious acoustic reflections; or, the layers are real and sedimentologic evidence concerning their existence is contained in the cores; or, the layers are real but the coring process did not adequately sample the phenomenon responsible for their existence.

The first possibility is ruled out by the fact that another investigator (personal communication, Arnie Johansen, Norwegian Defence Research Establishment, Horton, Norway), using a different seismic system (50 kHz), also noted the existence of these seismic layers. Moreover, the present authors lowered the seismic transducer almost to the sea floor to make certain that the shallow layer was not caused by a spurious reflection from the boat. The overall appearance (abrupt termination against light domes) and geographic limitation of the intermediate layers also attest their geologic credibility. The second possibility is discounted by the thorough nature of the core sampling and analysis, especially with regard to the shallow layer. At least five cores (Cl-C5) certainly penetrated this layer, because they sampled the underlying sand layer. If some lithologic discontinuity responsible for the shallow layer was present in these cores, it should at least have been detected by the high-resolution water-content analysis, since porosity is probably the most important mass physical property of sea floor sediments responsible for producing an acoustic impedance change, and the 5 cm sampling interval that was used in shorter than the wavelength of 12 kHz sound in the sediment.

There remains the third possibility—that the coring process did not adequately sample the phenomenon responsible for the seismic layering. This suggests that the seismic response creating these layers was obtained from a random distribution of acoustic scatterers located on a planar or nearly planar surface, rather than a lithologic interface. That this situation was responsible for the shallow and intermediate layers is both seismically and geologically plausible. The existence of a "layer" on the record produced by a continuous seismic profiler indicates only that moderately repetitive echoes were received during a certain period of time from some reflecting points located both within the sensitivity zone of the sonic beam and at a constant or slowly-changing distance from the ship. Since the full angle beamwidth of the transducer sensitivity cone used was 30°, it can readily be seen that the density of scatterers required to produce a "layer" is not large (one large scatterer per 5 m^2 when located on a plane 12 m from the ship). From the geological standpoint, it is perfectly reasonable to expect the random distribution of scatterers on a plane or pseudo-plane to occur in nature. The scattering agents might be biogenic material (such as the shell and peat observed in Cores C11, C13, *etc.*), sediment lenses, or primary sedimentary features such as burrows, mounds, or ripples (Shrock, 1948) located on a planar or nearly planar surface representing an ancient sea floor, and preserved in the stratigraphic column.

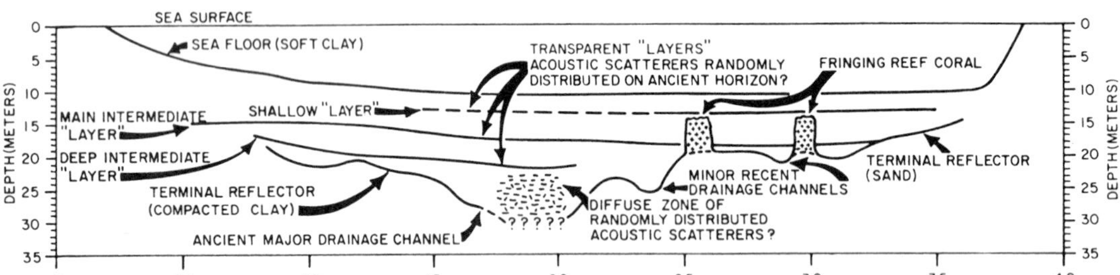

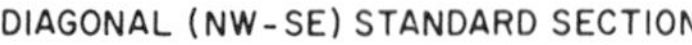

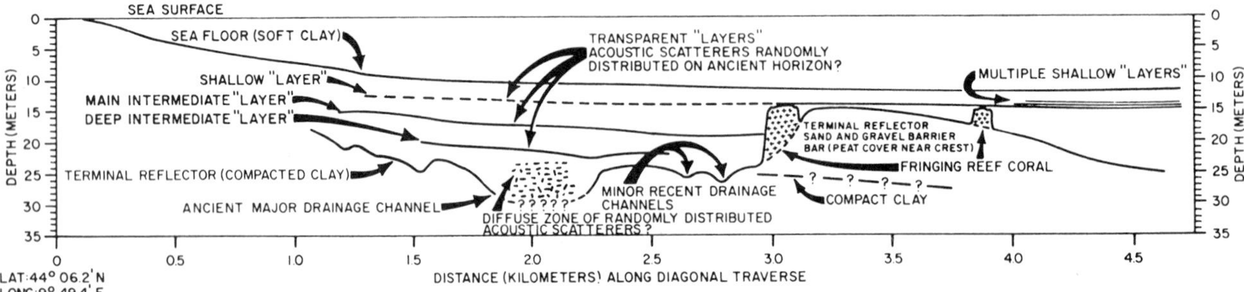

Figure 5. Profiles illustrating inferred subbottom structure and lithology of the Gulf of La Spezia. Location of profiles is indicated on Figure 6.

SUMMARY AND CONCLUSIONS

Seismic profiles indicate that a buried barrier-bar extends across the Gulf of La Spezia, and is delineated by a rise in the terminal seismic layer from more than 15 meters below the sea floor to less than 3 meters. Acoustically-opaque, dome-shaped features ("light domes") flank the barrier-bar. An acoustically-transparent layer ("shadow layer") hugs the terminal layer at the upper section of the barrier-bar. Two intermediate-depth transparent layers ("main and deep intermediate layers") terminate abruptly either against the barrier-bar itself, or against the light domes fringing the bar's landward flank. A shallow-depth transparent layer ("shallow layer") is found throughout the gulf, and apparently is independent of the bar's influence. Acoustically-scattering patches ("dark patches") are found both landward and seaward of the barrier-bar, exhibiting a definite tendency to occur along depressions in the terminal layer. The dark patches and associated depressions are interpreted as defining an ancient drainage system.

Coring has established that: (1) the terminal seismic layer in the area of the barrier-bar is related to a sand

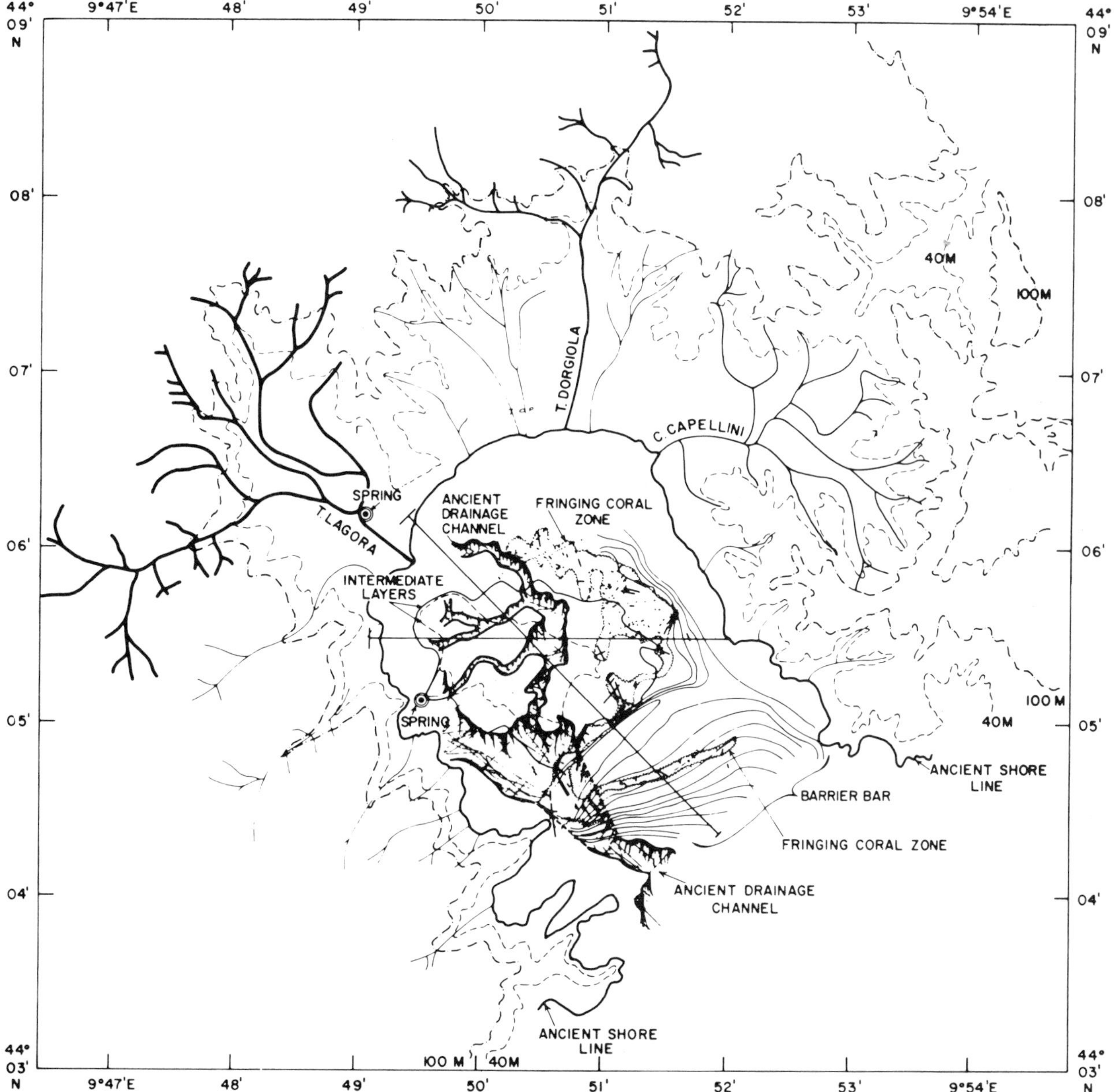

Figure 6. Map of the shallow subbottom geology of the Gulf of La Spezia. The thickness of lines indicating the river damage is relative to river size.

unit; (2) the shadow seismic layer is related to a peat stratum; (3) the light domes are composed of coral; (4) the dark patches are not clearly related to a sediment type—evidence suggests they are related to a volume distribution of acoustic scatterers; and (5) the main-intermediate and shallow seismic layers are not noticeably related to a lithologic interface. Consequently, the hypothesis is advanced that these "layers" result from a random distribution of scatterers on a plane or pseudo-plane.

Thus this study has demonstrated that most reflectors detected in high-resolution seismic profiles can be linked to conspicuous lithologic discontinuities, observable in sediment cores. However, certain well-defined seismic horizons are not correlated with major differences in sediment type but probably reflect subtle but acoustically significant sedimentologic properties, which may not be readily appreciated in the limited samples yielded by coring.

One further result of this study remains to be described—the delineation of the recent geologic history of the Gulf of La Spezia. A summary of the shallow subbottom geology of the Gulf is presented in Figures 5 and 6. On Figure 6, the area surrounding the gulf was drawn from a map compiled over a century ago (Plan du Golfe de La Spezia-Côtes d'Italie, Duché de Gênes, Dépôt Général de la Marine, 1846), and represents the area as it was before man extensively altered the shoreline and river channels.

The oldest feature observed in the gulf is the ancient, south-trending drainage channel that originated subaerially during the Pleistocene epoch. At some point during the glacio-eustatic rise of sea level in the Holocene the conditions were such that a cross-gulf barrier was formed by longshore drift of coarse sedimentary material. This material included rounded gravel. The upper surface of the barrier-bar was once at or near sea level, and was covered by vegetation, now represented by peat. Integrated into the generally accepted curve of eustatic changes in sea level (Fairbridge, 1961), this fact indicates that the bar was partially emergent probably about 8000 years ago. Coral grew on the flanks of the bar, a lagoon was formed on its landward side, and "layers", protected from the action of the sea, formed in the quiet waters of the lagoon. With a further rise in sea level, the dynamic conditions that created the barrier-bar were removed and the bar and fringing coral reef on its flanks were buried by fine sediments.

REFERENCES

Breslau, L. R. and H. E. Edgerton 1965. The sub-bottom structure of the Gulf of La Spezia, Italy (A preliminary report). *Rapports et Procès Verbaux des Réunions de la Commission Internationale pour l'Exploration de la Mer Méditerranée,* 18 (3):953–955.

Breslau, L. R. and H. E. Edgerton 1968. The sub-bottom structure of the Gulf of La Spezia, Italy. SACLANTCEN *Technical Report,* 129:82p.

Edgerton, H. E. 1962. *Instruction Manual for the "Mud Penetrator". EG&G Incorporated,* Boston, Massachusetts, 36p.

Fairbridge, R. W. 1961. Eustatic changes in sea level. In: *Physics and Chemistry of the Earth,* eds. Ahrens, L. H., F. Press, K. Rankama and S. K. Runcorn, Prentice-Hall, New York, New York, 4:99–185.

Kermabon, A., P. Blavier, and U. Cortis 1963. The SACLANTCEN sphincter corer assembly. SACLANTCEN *Technical Report,* 34:29 p.

Shrock, R. R. 1948. *Sequence in Layered Rocks.* McGraw-Hill Book Company, New York, 296–326.

Yules, J. A., and H. E. Edgerton 1964. Bottom sonar search techniques. *Undersea Technology,* 5 (2):29–32.

Zaccagna, D. 1932. *Descrizione Geologica delle Alpi Apuani, Provveditorato Generale dello Stato.* Rome, Italy, 440 p.

MAPS AND CHARTS CONSULTED

Capellini, G. 1863. *Carta Geologica dei Dintorni del Golfo della Spezia e Val di Magra Inferiore.* St. Cartografico G. Giardi, Florence, Italy.

Dépôt Général de la Marine 1846. *Plan du Golfe de la Spezia-Côtes d'Italie (Duché de Gênes).* La Marine de la République Française, Paris, France.

Istituto Idrografico 1955. *Da Portofino al Gombo,* Istituto Idrografico, Tirreno, Italy.

Istituto Idrografico 1962. *Rada di La Spezia.* Istituto Idrografico, Tirreno, Italy.

Eustatic and Tectonic Factors in the Relative Vertical Displacement of the Aegean Coast

Nicholas C. Flemming

National Institute of Oceanography, Wormley, Surrey

ABSTRACT

Sediment-filled basins and critical sill depths in the Aegean and the eastern Mediterranean are considered in the context of Pleistocene eustatic variations of sea level and tectonic movements. Seventy ancient coastal settlements in the Peloponnese and southwest Turkey furnish estimates of relative sea level changes during the last 3000 years. Most of the sites are submerged relative to present sea level, but the variation between sites of the same age is greater than the mean displacement. Multiple regression analysis separated a time-only-dependent factor equivalent to a eustatic change, and a geographically dependent factor equivalent to local tectonic variations.

From the southwest coast of Turkey a eustatic curve was derived, indicating that in the first millenium B.C. the sea level dropped from within a few centimeters of present level to 20 cm below present level at 2000 B.P. and 30 cm below at 1300 B.P. subsequently rising to present level. This curve is necessarily a smoothed representation of the actual course of events. The tectonic equations shows that the Peloponnese is doming, with subsidence of the margins at 1 to 2 m per millenium, combined with an active uplifted ridge stretching towards Crete. Southwest Turkey appears to be subsiding slowly along the main axis of the coast, with increased rates of subsidence landward and seaward. Axes of uplift extend westwards from a point south of the Cesme peninsula and west of south from the Cnidos peninsula towards Rhodes.

RESUME

Les bassins sédimentaires et la profondeur critique des seuils de la Mer Egée et de la Méditerranée orientale sont étudiés dans le contexte des variations eustatiques et des mouvements tectoniques du Pléistocène. Les observations recueillies en 70 points sur la côte du Péloponnèse et de la Turquie permettent d'en déduire des informations quant au changement relatif du niveau de la mer au cours des 3000 dernières années. La plupart de ces points (sites anciens) sont submergés et la variation entre points du même âge est plus grande que la distance moyenne de déplacement. Des analyses statistiques ont mis en évidence deux facteurs prépondérants à ces changements. L'un, ne dépendant que du temps, est semblable à une variation eustatique, l'autre, ne dépendant que de la situation géographique, rappelle des variations tectoniques.

Une courbe eustatique de la côte Sud-Ouest de la Turquie montre que pendant le premier millénaire B.C. le niveau de la mer s'est abaissé de quelques centimètres par rapport au niveau actuel puis, vers 2000 B.P., s'est abaissé de 20 cm, est encore descendu jusqu'à 30 cm en 1300 B.P. pour remonter et atteindre le niveau actuel. Cette courbe est une simplification de la succession des phénomènes. Les équations tectoniques indiquent que le Péloponnèse se soulève en forme de dome, que les marges s'abaissent à la vitesse de 1 à 2 m par millénaire alors qu'une ligne de plissement actif se forme vers la Crète. La Turquie du Sud-Ouest semble s'affaisser lentement suivant l'axe général de la côte avec vitesse plus grande côté terre et côté mer de celle-ci. Des axes de soulèvement s'étendent vers l'Ouest, partant du Sud de la péninsule de Cesme et du Sud de la péninsule de Cnide dans la direction de Rhodes.

INTRODUCTION

Hersey (1965, p. 75) states that Mediterranean sedimentary history is exemplified by formation of basins, their infilling, and subsequent folding and uplifting. However, most geophysical studies of the eastern basin have analysed horizontal movements rather than vertical (Papazachos and Delibasis, 1968; Ambraseys, 1970; and McKenzie, 1970). Understanding the sequence of events depends on quantitative estimates of rates of vertical movement, and measurement of this depends on knowledge of the eustatic changes of sea level.

The location and rates of sedimentary accumulation are affected by topographic gradients. Earth movements and sea level changes alter sill levels, reduce or increase the rate of supply of sediment from the coast, and create or destroy topographic basins. The borders of the Aegean are one of the most seismically active areas of the world, and recent earth movements have been so rapid and complex that no synthesis of the effect of these factors on sedimentary history in the Aegean can yet be attempted. The quantitative results in terms of eustatic sea level changes and rates of earth movement illustrate the speed with which the topography has changed.

Salient features of the regional structure are shown in Figure 1. Data are taken from Bogdanoff *et al.* (1964), Aubouin (1965), Ambraseys (1965), Renz *et al.* (1954), and Maley and Johnson (1971). The dominant north-northwest to south-southeast trend in the Peloponnese is reflected both in the stratigraphic boundaries and the structural zones. The west coast of the Peloponnese and the Adriatic coast of Greece are composed of Triassic and Jurassic limestones and shales with Cretaceous limestones further inland.

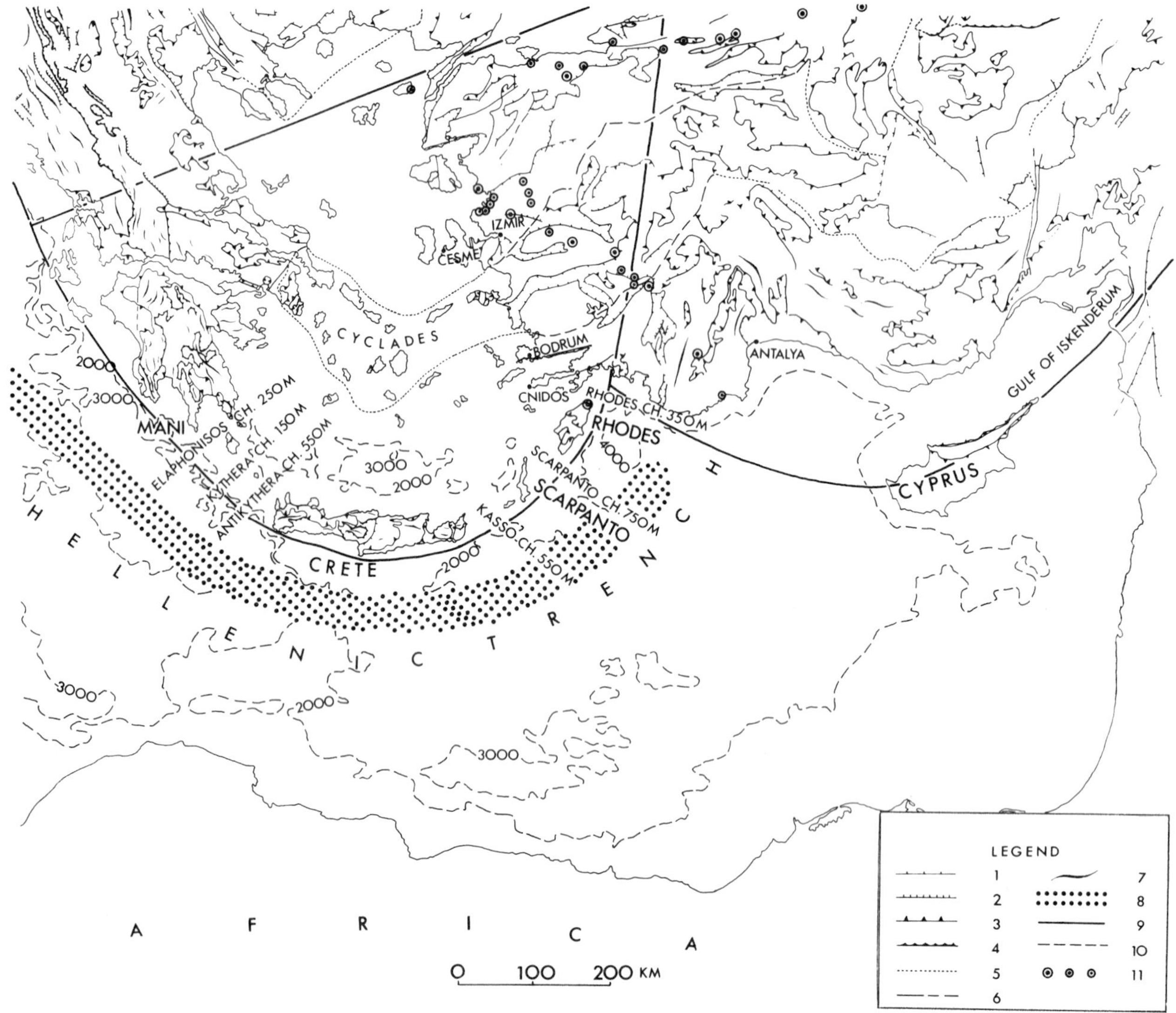

Figure 1. Structural and tectonic features of the Aegean region. Key to legend: 1, Normal faults; 2, Imbricated structure; 3, Limit of tectonic depression; 4, Limit of thrust; 5, Conventional limits of median rigid blocks; 6, Major features such as fractures or regional flexures; 7, Anticlines and anticlinoria; 8, Hellenic Trench; 9, Tectonic plate boundaries, proposed by McKenzie (1970); 10, Fractures of indeterminate type; 11, Earthquakes in Turkey which occurred during the first millenium A.D., from Ambrayses (1970). Other tectonic features from Bogdanoff *et al.* (1964).

The Mani peninsula is composed of marble of presumed Mesozoic origin, while the east coast of the Peloponnese is composed of Tertiary conglomerates, sands, clays, and marls. Aubouin (1965, p. 43–67) describes the structural geology of Greece in terms of the Alpine orogeny and the evolution of a geosyncline composed of two pairs of ridges and furrows possessing north-northwest to south-southeast axes.

Bogdanoff *et al.* (1964, sheet 15) show that the structural axes of Greece are continued in arcuate form across the Aegean in a number of approximately concentric zones. The faulting on the Turkish coast is extremely confused, with a number of north-south faults cutting across the peninsulas which are elongated east-west. Cursory investigation of the cliffs in southwest Turkey shows numerous vertical faults with slickensides. In northern Turkey the major fault system is the Anatolian fault (Ambraseys, 1965). Ambraseys (1965 and 1970) also demonstrates a zone of seismicity in an arc from Izmir to Antalya, cutting orthogonally across the trend of the Aegean arc.

Aubouin (1965) states that the eastern zones of the geosyncline system subsided in the Miocene and Pliocene, forming the Aegean basin. The Cyclades islands are the residual emerged portions of what Aubouin terms the Pelagonian zone, which, is shown by Bogdanoff *et al.* (1964, plate 15) to continue in arcuate form to the mainland of Turkey between the Cesme peninsula and the Bodrum peninsula. Aubouin (1965) states that this zone is still subsiding. Both authors indicate anticlinorial axes running north-northwest to south-southeast throughout the mainland of Greece, and along the main axis of Crete, turning northeast at the eastern end of Crete. The continuous nature of the Cretan arc is reinforced by the submarine topography taken from the Admiralty charts and from Giermann (1960).

Archaeological sites on the coast bear a specific relation to the sea level at the time of their construction (Flemming, 1969, p. 6–10), which can be measured by identifying the function of buildings which were built on the original waterline. Since the tidal amplitude in the Mediterranean is minimal, the error involved in this method is principally due to the uncertainty of establishing this relationship, which varies from 0.25 to 1.0 m from site to site. Negris (1904) made similar observations at a number of Aegean sites, and concluded that there had been widespread earth movements. Hafemann (1960) studied 37 sites in the eastern Mediterranean, of which 21 were in the Aegean. Hafemann (1960, p. 195) concluded that over the last 1600 to 1800 years there had been a eustatic rise of sea level of 2 m ± 30 cm, and that in the last 2500 years there has been a eustatic rise of sea level of 2.5 to 2.8 m.

Given the seismicity of the area and the relatively small sample of sites it is not justifiable to correlate age with average depth of sites without taking into account the great variation in depth which is found between sites of the same age. This variation can only be accounted for by earth movements, and since earth movements may on average have been more up than down, or vice versa, the average relative displacement of sites can no longer be attributed reliably to eustatic causes alone. Although several eustatic curves for the last 3000 years have been published (Jelgersma, 1961; Fairbridge, 1961; Mörner, 1969), all such curves are necessarily based on limited data from restricted areas, and it would be unjustifiable to translate such curves to the Aegean in an attempt to correct observed vertical displacements and obtain the factor due to earth movements. It is necessary, therefore, to derive a statistical method which can separate the factors without reference to data from so-called stable regions.

The purpose of the present program is to locate every ancient site on the coast of the Mediterranean and to measure its vertical displacement and age. The results of the survey of the western Mediterranean basin have already been published (Flemming, 1969). In the eastern Mediterranean the areas studied so far are the Peloponnese, exclusive of the Gulf of Corinth (Flemming, 1968 a and b), and the coast of Turkey from Izmir to Kas. The survey of the Aegean has produced data from which 70 independent estimates of relative sea level change have been made. The survey is incomplete, but significant results are already emerging.

METHODS

Field methods and analytical methods have evolved during the survey. The basic derivation of an original sea level from a group of structures is described by Flemming (1969, p. 6–10). In summary, ancient historical works, archaeological reports, original charts and navigational surveys are collated to provide a list of coastal sites, together with sketch maps, and dates of foundation or abandonment where possible. During field work the coast is searched further to locate previously unknown sites, using the topography of the coast and likely harbour sites as the primary indicator of possible settlement. At each site, buildings are measured and plotted on sketch maps using simple survey techniques with tapes, ranging poles, and pacing.

Structures are classed as follows: (1) those constructed *on land*, including houses, tombs, roads, city walls, wells, drains, gutters, storage tanks, water tanks, olive presses, religious and public buildings, steps, quarries, passages and tunnels, floors and mosaics; (2) those constructed with foundations necessarily *in the water,* including slipways, breakwaters, moles, jetties, quays, docks, fish tanks, ship channels; (3) structures which could have their foundations *on land or in the sea,* including solid towers and parts of city walls, roads on causeways, lighthouse towers, villas and baths built on solid foundations close to or in the sea. Identification of a few structures in each class allows most accurate determination of relative sea level at the time of occupation of the site, while the presence of structures in one class only provides approximate estimates or sets one limit to a possible range of values. The varying reliability of estimates of this kind can be quantified by allocating different probability weights and estimates of different magnitudes as described in the section on analysis of data.

OBSERVATIONS

In 1967 and 1968 surveys were carried out on the east and west coast of the Peloponnese respectively. The location of sites is shown in Figure 2, and the age and displacement of sites is listed in Table 1. Preliminary analysis of data from the east coast is given by Flemming (1968a). In 1969 the field survey was extended to southwest Turkey, and the site locations are shown in Figure 4, and the observations tabulated in Table 2. The archaeological sketch maps and full details of the constructions from which sea levels were derived are published in the report of the Colston Symposium on Marine Archaeology (Flemming, still in press, ed. Blackman, D.)

Table 1. Ancient sites in the Peloponnese, with age in millenia, relative displacement in meters, average rate of displacement and Bouger gravity anomaly.

Name	Age (millennia)	Depth (m)	Depth/age (m/millennium)	Bouger (mgal)
Lechaeum	2.0	0.7	0.3	04
Kenchreai	2.5	2.0	0.8	15
P. Epidaurus	2.0	2.7	1.3	35
Methane	2.0	1.0	0.5	60
Lorenzon	1.0	2.0	2.0	60
Halieis	2.4	2.7	1.1	60
Asine	3.0	2.0	0.7	20
Xarax	2.5	3.0	1.2	60
Monemvasia	2.0	1.0	0.5	60
Neapolis	1.5	0.3	0.2	55
Elaphonisos	3.5	3.0	0.9	40
Arkangelos	2.0	0.2	0.1	05
Plitra (I)	3.0	3.0	1.0	−05
Plitra (II)	1.5	2.0	1.3	−05
Kythera	4.0	0.0	0.0	10
Antikythera	5.0	−3.0	−0.6	−30
Elea	3.0	2.0	0.7	00.00
Trinisi (I)	1.0	1.0	1.0	−70
Trinisi (II)	1.0	1.0	1.0	−70
Gythion	1.5	2.5	1.5	−70
Skoutari	1.5	3.5	2.2	−70
Kotronas	1.0	1.0	1.0	−25
Tenaeron	2.2	1.0	0.4	−30
Tigani	2.0	0.5	0.3	−40
Selenitsa	1.0	1.0	1.0	−55
Stoupa	2.0	2.5	1.3	−62
Kardarayli	1.0	1.0	1.0	−65
Koroni	0.6	1.2	2.0	−90
Methoni	1.0	1.5	1.5	−85
Pylos	2.0	1.5	0.8	−90
Pheia	2.0	1.0	0.5	−130

Figure 2. Locality map for the Peloponnese region of Greece. The tips of the arrows indicate the location of ancient ruins on the coast where relative changes of sea level were measured.

ANALYSIS OF DATA

For the data from the east coast of the Peloponnese no simple correlation between age and depth of site was apparent, but a third degree surface fitted to the rate of displacement in terms of arbitrary x-y geographic co-ordinates produced low residuals. This first stage of the analysis revealed an active uplifted structure stretching slightly east of south through Kythera and Antikythera towards Crete, while there was a slight doming of the mainland of the Peloponnese. The data for the west coast were treated similarly, and again a third degree surface in terms of rate of displacement in relation to geographic coordinates produced the best fit. At the boundary of the two independent survey zones the zero displacement contour and the 2 m per millenium contour matched well, but there was a confusing pattern over the Peninsula of Mani. An attempt was made to fit a fourth or fifth degree surface to the combined data, but the irregular distribution of control points resulted in trivial solutions. To smooth the junction between the two third degree surfaces the raw data from the Mani area, together with points from the zero and 2 m per millenium contours, were taken as input for a fourth degree surface fitting program. The composite pattern thus derived is shown in Figure 3.

In analyzing the data from Turkey an attempt was made to compensate for the variation in accuracy of observations from site to site. For example, in some cases a fish tank, slipway or bollard may give an accuracy of 20 cm, while in another case a submerged breakwater may give only a rough indication of change of levels. Also, there are cases where a

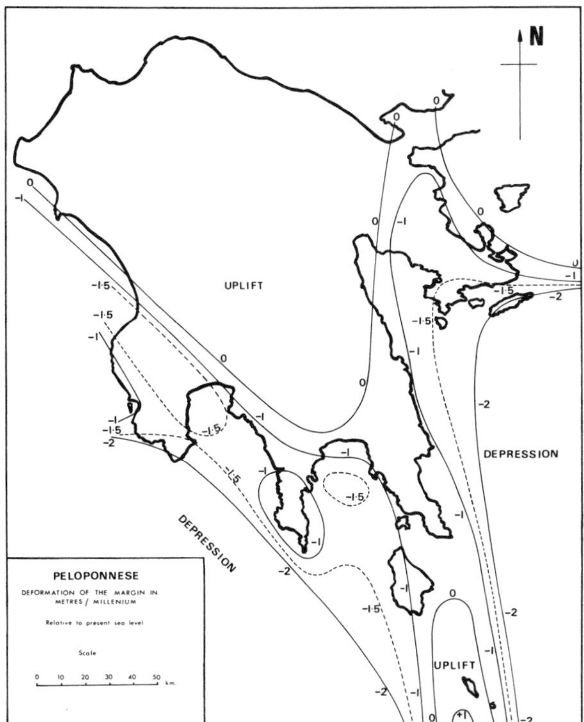

Figure 3. Computed deformation of the Peloponnese margin. The contours indicate rate of movement of the land relative to present sea level in meters/thousand years. Contour interval 1.0 m/ millenium with addition of the 1.5 m/millenium contour for detail.

submerged structure indicates a minimum necessary change of relative level, but sets no maximum. To allow for this variation, each site was allocated a best estimate of relative change, and a set of weighted probabilities for changes greater or less than the best estimate. A typical histogram illustrating the range of estimates for a single site is shown in Figure 5. The total weighting for each site is 10, with a few exceptions where the overall reliability of the site was low. The distribution of weightings to estimates, and the total weight, are listed for each site in Table 2.

The best estimates and weighted estimates were plotted against age to check for a eustatic trend or general coastal depression. The data, without the weight distributions, are plotted in Figure 6. Curve A is the best fit fourth degree curve to the total data including the weighted estimates. Clearly the older sites are on average deeper, though not by as much as suggested by Hafemann (1960). However, the scatter is extremely large, and the average trend can be explained equally by a eustatic or seismic hypothesis, or a sum of the two. The magnitude of displacement and rate of displacement were then correlated with geographic location on an arbitrary x-y grid, using third and fourth degree equations. The best fit was found for the fourth degree equation in terms of the rate of displacement. In all cases weighted estimates were used as separate data entries. The resultant contour pattern is shown in Figure 7. The contours indicate relative depression to landward and seaward of an axis curving along the main trend of the coast, with uplifted structures south of the Cesme Peninsula, and stretching south from the Cnidos Peninsula towards Rhodes.

Although no consistent eustatic trends have been detected up to this point by these methods it is certain that there has been some eustatic change in the last 3000 years, and it is desirable to identify this rather than treat it as background noise. The following analysis is an attempt to detect a eustatic component from the data for southwest Turkey, and the result should be regarded as a preliminary test of method rather than a rigorous estimate of a eustatic curve.

It is proposed that the total of relative displacement at any site can be expressed in the form: $Z = f(T) + g(x,y)T$ where Z = relative vertical displacements, T = age of site, and x,y, = geographical co-ordinates. For the area and period concerned f is put as a third degree general polynomial, and g a fourth degree polynomial. The equation is based on the assumption that tectonic movements have a constant average rate at any one site over the period concerned. Gutenberg and Richter (1954) state that seismicity has not changed significantly in historic time, while Ambraseys (1970) shows that the seis-

Table 2. Ancient sites in southwest Turkey, with age in millenia, arbitrary geographical coordinates, best estimate of relative displacement in meters, probability weighting of estimates at intervals of 25 cm, and total weight allocated to each site.

Name	Age × 10³ yrs	Geographical Coordinates		Depth Best Estimate	Weighting in 25 cm intervals							Total Weight
		X	Y	= E	−75	−50	−25	E	+25	+50	+75	
Antiphellus I	2.2	23.60	0.85	2.2	—	—	—	8	1	1	—	10
Antiphellus II	1.0	23.60	0.85	1.0	—	—	1	7	1	1	—	10
Andriake	1.0	27.50	1.60	1.0	—	1	2	4	2	1	—	10
Bargylia	2.0	5.10	11.60	0.2	—	—	1	7	2	—	—	10
Bozburun	1.0	9.20	5.80	1.0	—	3	3	4	—	—	—	10
Caryanda I	0.3	5.05	10.90	0.0	—	—	—	8	2	—	—	10
Caryanda II	1.5	5.05	10.90	0.3	—	—	—	7	3	—	—	10
Cedreae	1.5	10.70	9.45	0.3	—	—	3	6	1	—	—	10
Cesme	0.1	−6.10	23.85	0.2	—	—	—	5	—	—	—	5
Ciftlik	0.1	−6.30	23.55	0.1	—	—	—	5	5	—	—	10
Clazomenae	2.3	−1.85	24.50	1.0	—	1	1	4	2	2	—	10
Cnidos	2.3	3.25	5.90	0.0	—	1	1	7	1	—	—	10
Cyme	2.4	−0.45	28.85	1.0	—	1	2	4	2	1	—	10
Datoja	2.5	6.15	6.20	0.0	—	—	1	8	1	—	—	10
Elaea	2.3	0.50	30.85	0.0	—	—	—	3	3	2	—	8
Erithrae I	0.05	−4.55	24.65	0.25	—	—	5	5	—	—	—	10
Erithrae II	0.2	−4.55	24.65	0.5	—	—	5	5	—	—	—	10
Erithrae III	2.4	−4.55	24.65	1.40	—	1	2	6	1	—	—	10
Halicarnessos	2.4	3.75	9.75	1.0	—	—	1	6	2	1	—	10
Heraklea Latmus	2.5	4.55	14.95	1.5	—	1	1	1	1	1	—	5
Iasus I	1.0	5.25	12.35	0.0	—	—	—	7	3	—	—	10
Iasus II	2.0	5.25	12.35	0.5	—	1	3	6	—	—	—	10
Ilica	2.4	−6.70	23.9	0.5	—	2	2	2	2	2	—	10
Karatoprak	1.0	2.25	9.40	0.5	—	—	2	5	2	1	—	10
Loryma	1.5	9.05	4.65	1.0	—	1	3	6	—	—	—	10
Miletus I	2.0	2.45	15.30	1.5	—	2	3	5	—	—	—	10
Miletus II	3.2	2.45	15.30	1.50	—	—	2	5	2	1	—	10
Myndos	2.0	2.05	10.0	1.2	—	—	1	8	1	—	—	10
Myonessus	2.6	−1.15	20.85	1.0	—	1	2	4	2	1	—	10
Orhaniye	1.0	10.05	6.80	0.0	—	—	—	5	5	—	—	10
Panormus	2.6	1.95	13.90	0.0	—	1	2	4	2	1	—	10
Saranda	1.0	9.60	5.70	0.5	—	—	5	5	—	—	—	10
Sigacik I	2.3	−1.85	22.50	0.8	—	1	2	7	—	—	—	10
Sigacik II	0.1	−1.85	22.50	0.0	—	—	—	5	5	—	—	10
Smyrna I	2.5	1.50	25.65	1.0	—	—	—	6	3	1	—	10
Old Smyrna	4.0	1.50	25.65	3.0	1	2	3	3	—	—	—	10
Teos	2.3	−1.75	22.30	0.8	—	—	2	7	1	—	—	10
Urla Beach	0.3	−2.65	23.9	0.5	—	1	3	6	—	—	—	10
Yali	1.0	−4.95	24.10	0.5	—	—	—	6	3	1	—	10

micity of Asia Minor in the first millenium A.D. had a similar nature to the modern seismicity. The degree of the functions was selected on the basis that a fourth degree had already shown useful correlation for the geological component, and that a first or second degree would be inadequate to allow for probable eustatic fluctuations, while a higher degree was probably not justified by the quantity of data.

The values of the products of the appropriate powers of T, x, and y, in the terms of the general polynomial, were evaluated for each site to give an expanded expression for displacement in terms of the eighteen coefficients of the polynomial. Weighted estimates for displacement at each site were taken into account, and the resulting 378 equations were treated with a step-wise multiple regression analysis to give the coefficients. The terms in T only, and the terms in x and y, were grouped to give two separate equations, the first representing the eustatic component and the second the geological component of relative displacement.

Rounded to three figures, the eustatic equation has the form:

$$Z = 0.432 \cdot 10^{-1} \cdot T^3 - 0.317 \cdot T^2 + 0.584T$$

where Z is vertical relative displacement in meters, and T is the age of the site in thousands of years. This equation is plotted as curve B in Figure 6, and will be discussed later. The spatial equation repre-

Figure 4. Location map for southwest Turkey. The arrows indicate the location of ancient ruins on and near the coast where relative changes of sea level were measured. The grid is arbitrary.

senting the tectonic factor, also rounded to three figures, has the following form:

$$
\begin{aligned}
Z/T = & -0.483\cdot10^{-5}x^4 + 0.224\cdot10^{-3}x^3y \\
& -0.191\cdot10^{-3}x^2y^2 \\
& -0.428\cdot10^{-4}x\cdot y^3 - 0.895\cdot10^{-5}y^4 \\
& -0.736\cdot10^{-4}x^3 \\
& -0.587\cdot10^{-2}x^2y + 0.897\cdot10^{-3}\cdot xy^2 \\
& +0.257\cdot10^{-3}\cdot y^3 \\
& +0.269\cdot10^{-2}\cdot y^2 + 0.151\cdot x - 0.661\cdot10^{-1}\cdot y \\
& +0.106
\end{aligned}
$$

where x and y are arbitrary geographical co-ordinates as shown on the grid in Figure 8, where the equation is plotted in contour form.

Comparison of Figures 7 and 8 shows that the general trend of the contours is little altered by removal of the supposed eustatic component. Strictly speaking, contours falling outside the zone of the data points shown by the tips of the arrows should not be plotted, but to assist an appreciation of the form of the contoured surface additional contours have been shown in those figures. In Figure 7 the added contours reinforce the impression that there is a general de-

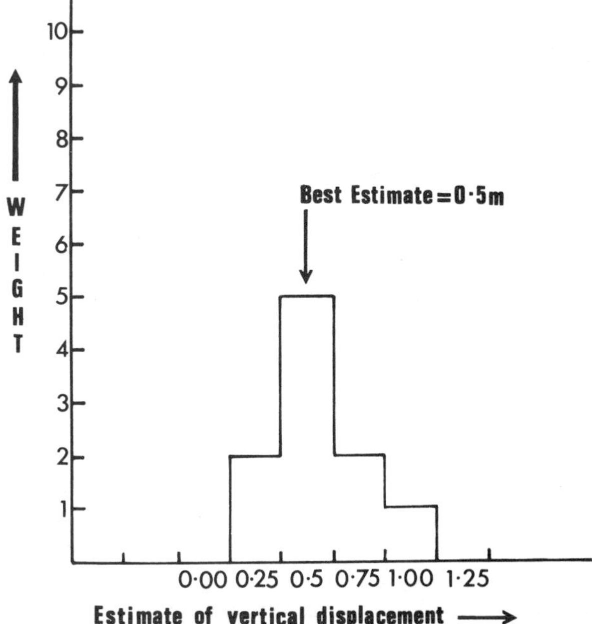

Figure 5. Histogram of the weighting of estimates of relative vertical displacement for a typical site on the Turkish coast, Karatoprak.

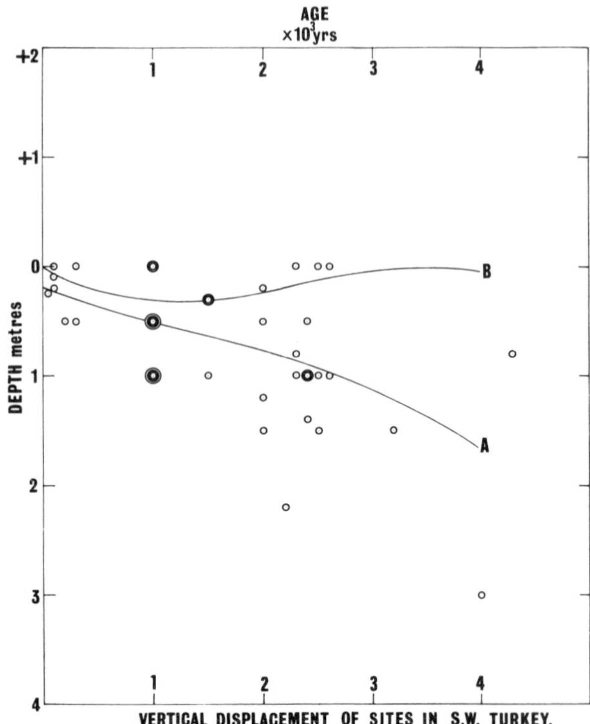

Figure 6. Graph relating the best estimates for displacement and age in thousands of years for each site on the Turkish coast. Heavy dots indicate double and triple values. Curve A is the best fit fourth degree curve for the data, including weighting of estimates. Curve B is the third degree eustatic curve derived from the composite equation for earth movement and sea level change.

pression of the land relative to sea level with a gradient to the southwest. In Figure 8 the mathematical surface outside the control of the data points has indicated uplift to the southwest. This is unfortunate in that the immediate impression no longer supports the hypothesis that there is a depression of the floor of the Aegean, but reliability should only be attributed to the contours within the data area. If the data outside the zero displacement contour are ignored the trend of uplift over the Cesme Peninsula is preserved as a monoclinal structure, as is the trend west of south over the Cnidos Peninsula.

DISCUSSION

In a study of complex natural phenomena where mathematical optimisation of possible solutions is adopted, it is essential to determine which aspects of the solution are mathematical artefacts and which have physical meaning, either necessary or possible.

In the case of the surface fitting programs without separation of the eustatic component the printout showed decreasing residuals with increasing degree of equation, up to the third or fourth degree. That is, a mathematical surface of this form departed least from the observed field data. Nevertheless, apart from errors of observation, the field data, and not the mathematical surface, represents the true distribution of vertical movements in the area. The surface provides a convenient way of averaging and smoothing the various displacements in an area, so that trends of common magnitude can be easily detected. The actual physical causes of the displacements and their magnitudes relative to each other may be broad warping, closely spaced faulting, or block faulting or formation of horsts. In this sense the impression of anticlinal structures given by the contours is a mathematical artefact. Nevertheless, whatever the specific causes and mechanisms of movement of one area relative to another, the contours illustrate the average relative movement between areas over a period of several thousand years, and thus portray the general deformation of the crust.

In the separation of eustatic and tectonic factors, a further point must be taken into account. The two physical components are presumed initially to be represented by the terms which involve time only, and those which involve space and time. In the subsequent regression analysis the only restraint on the system is that the sum square residuals should be minimised. It is possible that on mathematical grounds alone this condition may be met by adopting values of the two sets of terms which are absurd when given their physical significance, but which, when summed mathematically, achieve the effect of minimising residuals. I can see no logical method for overcoming this objection except the test of absurdity in the results. If the results are not absurd, then it is highly probable that the mathematically optimum solution is also a good physical solution.

The attempt to estimate eustatic and tectonic factors in an area where the tectonic component is large produces results which are at least reasonable. Since the area is so seismic it is possible that there may have been a trend of earth movement which had the same rate function through time at all sites in the area, that is to say, the area may from time to time have moved as a solid block. This type of movement would in the present terms be indistinguishable from a eustatic change, and thus would invalidate the proposed eustatic curve. It is important to note that the net average displacement of the coast resulting from the varying displacements of sites, each of which is being displaced at its own average rate, cannot produce such an anomaly. The only tectonic effect which can be confused with a eustatic one is the uniform synchronous movement of the whole block. Mörner (1969) detects isostatic movement of the coast in response to water loading, and this effect is also discussed by Higgins (1969). Hydro-isostatic loading of the coast may have produced a uniform response of the kind under discussion. The probability that such a fragmented area should move as a block is difficult to assess, and this illustrates the limitation of the method. However, the larger the area studied, the less error is likely to arise from this cause.

Alternative methods for separating eustatic and earth movement components have been used by Scholl and Stuiver (1967) and Mörner (1969). Scholl and Stuiver (1967) illustrate a number of relative sea level changes curves for points on the U.S. Atlantic coast between Boston and southern Florida. The curve for Florida is least displaced from present sea level, and from this it is concluded that Florida is stable, so that the curve is purely eustatic (*op. cit.*, p. 447). Mörner (1969) brought new chronological material and advanced analysis to well documented strand lines in Southern Scandinavia. From changes of gradient along a strand line of a given date the isostatic deformation is calculated, and this is subtracted from the relative shoreline displacement over the same period to give the eustatic change. The nature of isostatic recovery, with a more or less progressive change both in time and space, combined with the enormous number of dated levels, makes this study less liable to sampling error than the present work.

In interpreting the observed earth movements we must consider the structural geology of the region, gravity anomalies, seismicity, and plate tectonic boundaries.

The uplifted ridge stretching towards Crete from the Peloponnese correlates with the anticlinorium axis through Greece and Crete as shown by Aubouin (1965) and Bogdanoff *et al.* (1964, sheet 15). The complex contours around the base of the Mani coincide with an area of maximum faulting shown by Renz *et al.* (1954). The Pelagonian-Cyclades subsidence zone described by Aubouin (1965) intersects the Turkish coast in an east-west direction between Cesme and Bodrum, while the present study suggests that there are zones of lesser depression or relative uplift north and south of this region. Thus on the western margin of the Aegean the movements found in the present study correlate broadly with the axis of the anticlinorium, but on the east side the complex pattern of faulting means that the contours can only be taken as an average of many discrete vertical movements across fault boundaries.

The contours of vertical movement have been compared with the gravity data described by Woodside and Bowin (1970) and Morelli (Personal communication). Though there is no immediate numerical correlation between the magnitude of the Bouger anomaly and the direction of vertical movement up or down, there is a marked parallelism between the contours of Bouger anomaly and the contours of rate of vertical displacement. While this may derive partially from the control of both by the general trend of the coastline, two points suggest a closer correlation. Firstly the continuity of both sets of contours in the area between the Peloponnese and Crete, and secondly, and less predictably, the north-south trend towards the island of Rhodes. The anticlinal or monoclinal zone of uplift in the region of the Cesme peninsula corresponds with a general reduction in the positive gravity anomaly at the narrowest part of the Aegean between Athens and Izmir.

Ambraseys (1970) illustrates the distribution of earthquakes in the Middle East in the period 10 to 1070 A.D. (see Figure 1). Most relevant to the present study is an arc of seismicity from Smyrna, the modern Izmir, in the north-west to Andreacae in the south east. No fault zone has yet been detected along this line, but it seems possible that the area between the line and the Aegean coast may act differentially from the bulk of Asia Minor. In particular this may account for the apparent landward depression shown by the contours in Figures 7 and 8. The seismicity of the coast itself is low.

McKenzie (1970) analysed the seismicity of the Mediterranean in the broad context of plate tectonics. A compression line is assumed to traverse the Mediterranean longitudinally, with a number of subplates complicating the picture in the area of the Adriatic and Aegean (Figure 1). Net compression from north to south is of the order of 2 cm per year, with 2.6 cm per year in the eastern Mediterranean (Le Pichon, 1968). The southern Aegean is surrounded by the highest concentration of shallow and deep foci in the Mediterranean (McKenzie, 1970), and it is proposed (*op. cit.*, figure 4) that the Southern Aegean with the Peloponnese and part of Turkey comprises a distinct plate which is overriding the sea floor south of Crete. The same conclusion concerning the relationship of Crete to the sea floor south of it is reached by Caputo *et al.* (1970, their figure 2).

McKenzie (1970) shows the landward eastern boundary of the plate almost due north-south from Istanbul to Fethiye, somewhat displaced from the seismic belt detected by Ambraseys (1970). McKenzie (1970, his figure 4) shows a further plate boundary curving southeast from Fethiye through Cyprus and back into the Gulf of Iskenderun. This correlates exactly with the zone of maximum seismic energy flux computed by Ambraseys (1965). The

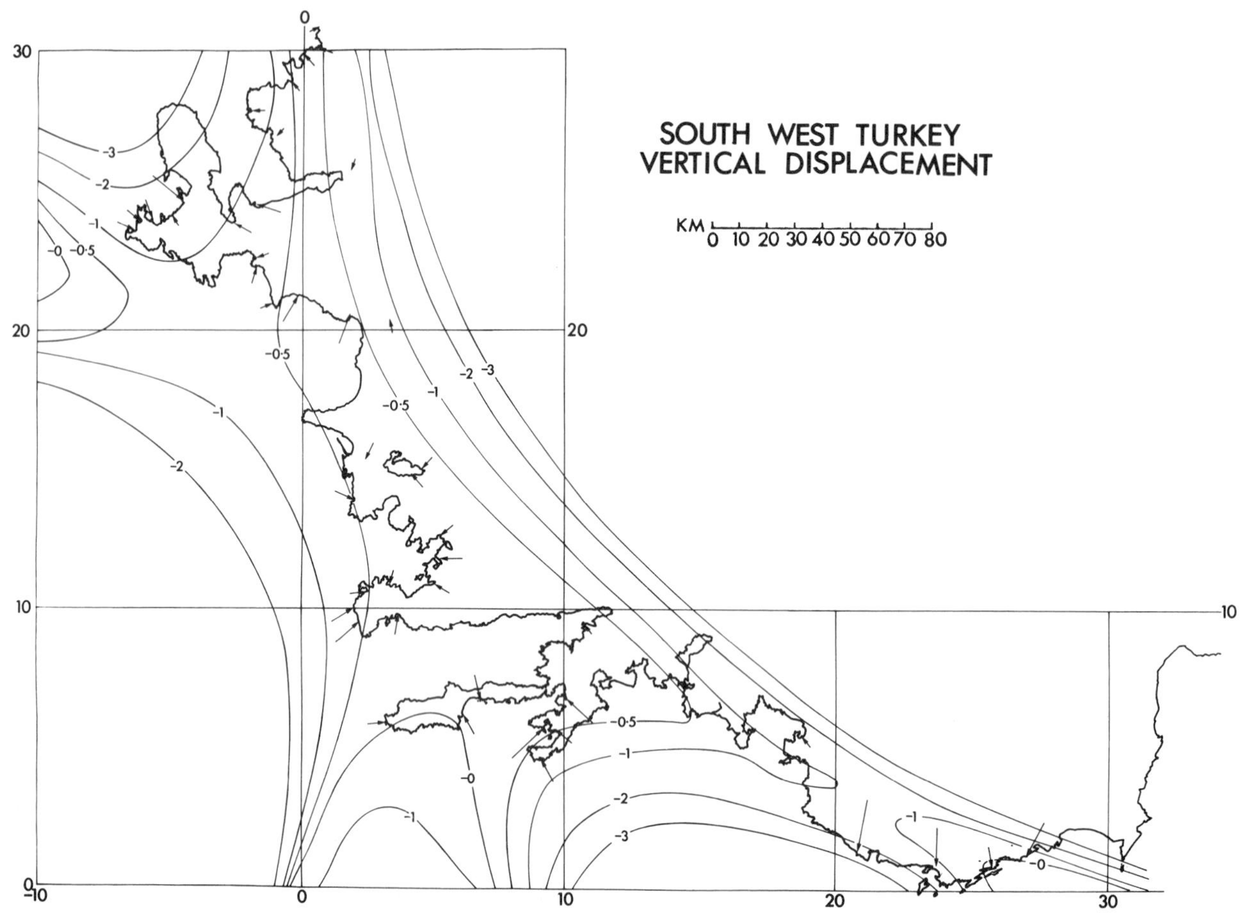

Figure 7. Contours of relative displacement of the coast of Turkey in meters/millenium. Contour interval is 1.0 m/millenium, with addition of the 0.5 m/millenium contour.

intersection of plate boundaries in the region of Rhodes – Fethiye – Andreacae may also account for the variations in vertical displacement along the coast from Fethiye to Andreacae. Woodside and Bowin (1970) state that the crust beneath the eastern Mediterranean thickens markedly from south to north, and that this is due to the overriding by the European block from the north. The discovery of Cretaceous dolomite overlying Pliocene ooze at site 127 of DSSP Leg 13 (Scientific Staff, 1970) near the northeast wall of the Hellenic trough, west of Crete, certainly suggests the presence of an overthrust or nappe structure.

In general the vertical movements detected in the present study correlate regionally with the movements which might be expected from structural, gravitational, and seismic observations, and provide the first quantitative estimates of these displacements. Where horizontal movements are of the order of 2 cm per year, vertical movements are of the order of 2 mm per year.

Figure 6, curve B, indicates a eustatic fall of sea level at 1300 B.P. of some 30 cm, while the level was within a few centimeters of present sea level at 3000 B.P. Mörner (1969) indicates a eustatic level at about −50 cm at 2000 B.P., with a number of small fluctuations about present sea level between 2000 and 3000 B.P. Fairbridge (1961) also indicates that the eustatic level fluctuated about present sea level in the first millenium B.C. and dropped in the first millenium A.D. before undergoing a number of oscillations and attaining present level.

Flemming (1969) concluded that archaeological evidence in the western Mediterranean did not indicate a net change of sea level over the last 2000 years greater than 0.5 m. Out of 179 sites only 24 were found submerged, and these were all in areas of volcanicity, recent seismic activity, or deltaic sedimentation. In that study there were two anomalous sites, Ile Ste. Marguerite (*op. cit.*, p. 27) which was submerged by less than 1 m, and Monastir (*op. cit.*, p. 63) submerged by 20 to 40 cm. Ile Ste. Marguerite, near Cannes, is not in a tectonic area, while Monastir is bordered by several sites which appear to be undisplaced. At both sites the archaeological evidence for

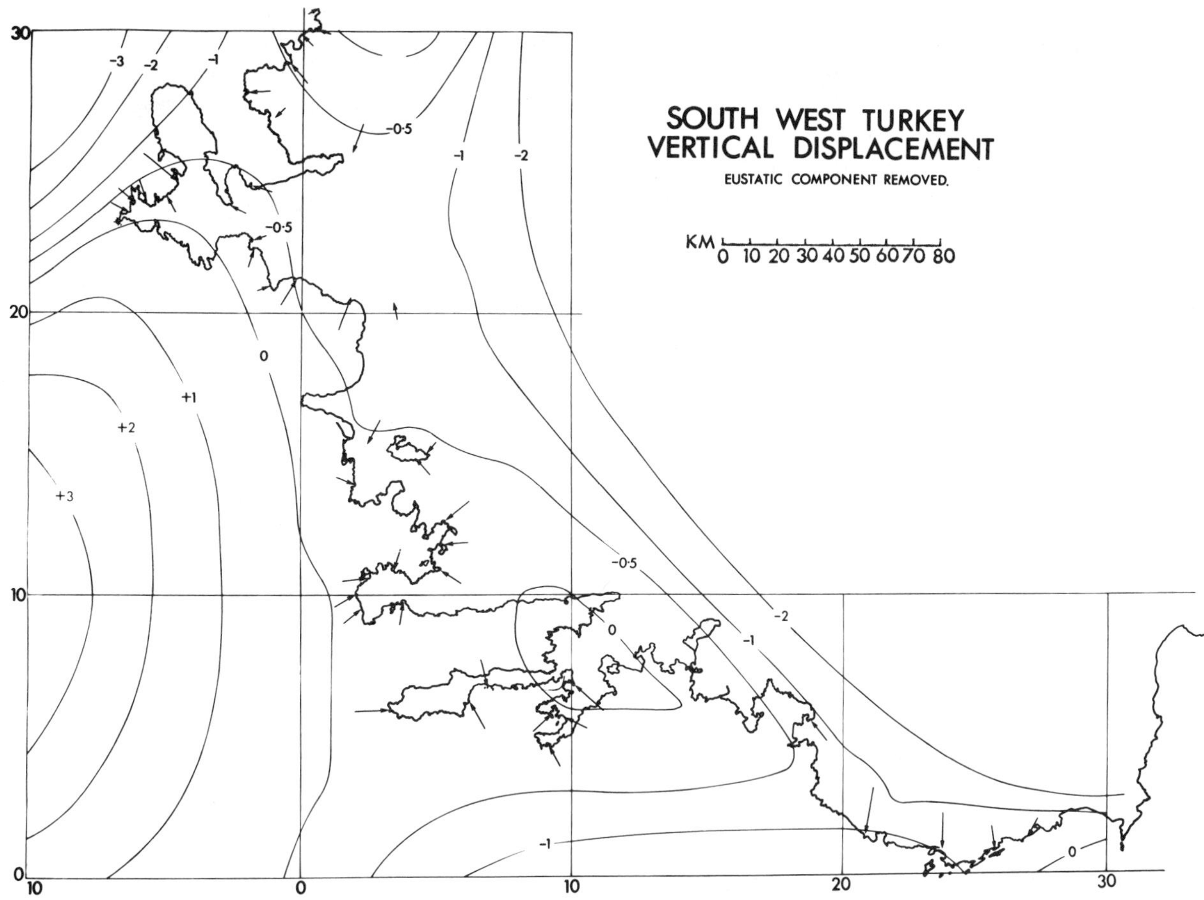

Figure 8. Contours of vertical displacement of the coast of Turkey attributed to earth movements, after removal of the eustatic factor, in m/millenium. Contour interval is 1.0 m/millenium, with the addition of the 0.5 m/millenium contour.

change was unusually detailed and accurate, and it is thus possible that their submergence could be accounted for by a eustatic change of the order of 20 cm in 2000 years, which was not detected elsewhere in the Western Basin. Thus the archaeological evidence concerning a eustatic change of level from the Western Basin is compatible with the results from southwest Turkey, and the method of separating tectonic and eustatic factors under these conditions can be regarded as reasonably reliable.

SILL DEPTHS AND SEDIMENT MOVEMENT

From Umbgrove (1947), Aubouin (1965) and McKenzie (1970) it appears probable that the earth movements detected in the present study are the short term manifestations of processes which have been operating at least since the Pliocene. In such a complex area it is possible that the direction and magnitude of vertical movements may fluctuate over short periods, but the rates of movement of 0.5 to 2.0 m per millenium are of the same order as those needed to form the basin of the south Aegean, namely 3000 m in a few million years. Thus it is a reasonable assumption that these movements have been more or less continuous throughout the Pleistocene.

Distribution of sediments in the area in the past will have been affected by relative sill levels and basin depths, while Pleistocene lowering of eustatic sea level may have dried out some sills and caused radically different circulation patterns in the Eastern Mediterranean. Wong and Zarudzki (1969) state that the principal source of sediments in the Eastern Mediterranean is the Nile River, and that the sedimentary material fails to reach the Aegean because of the Hellenic trench, and the sills between Crete, Scarpanto, and Rhodes. The same authors note that deposits in the Aegean are derived from the Greek and Turkish mainland, and from the Aegean islands.

Ninkovich and Heezen (1965) and Emery *et al.* (1966) have described the ash layers resulting from eruptions of the volcano on Santorini 25,000 and 35,000 years ago, a further link between tectonics and the sedimentary deposits. The sapropelic layers indicating periods of stagnation, described by Ninkovich and Heezen (1965), suggest that the circulation of the eastern Mediterranean basin has been altered at times by climatic variations and the blocking of various straits such as the Straits of Sicily, Messina, and those of the southern Aegean.

Critical sill depths are as follows: Straits of Sicily 400 m, Straits of Messina 100 m, Elaphonisos Channel 250 m (see Figure 1), Kythera Channel 150 m, Antikythera Channel 550 m, Kasso Channel 550 m, Scarpanto Channel 750 m, Rhodes Channel 350 m. These figures are estimated from Admiralty charts, supported by the bathymetric charts prepared by Giermann (1960). The deep basin north of Crete has several pockets with a depth of 2000 m, and one of 3000 m (Giermann, 1960), while south of Crete is the Hellenic trench with a depth of 3500 m. In the southern Aegean area the sediment thickness are greatest in the deeps north and south of Crete (Wong and Zarudzki, 1969).

Pleistocene low sea levels descended to at least 100 m and possibly 150 m (Fairbridge, 1961). Jongsma (1970) presents evidence based on radiometric dating of reef corals suggesting that the eustatic drop of sea level in the last glaciation, the Würm, was about 170 m while the drop in the Riss glaciation may have been as much as 200 m. Since these observations were made more than 100 miles offshore and there is no internal check for possible deformation of the continental margin, they must be regarded as provisional until further data are published. It is unlikely that the eustatic drop produced by earlier glaciations was as great (Flemming, 1968a, p. 281).

The vertical movements detected at either end of the Cretan arc suggest the axis of the arc is being uplifted, that is, the sills would have been deeper in the past. The broad anticlinal nature of the active folding along the arc seems certain, but whether the central axis is uplifting as a whole, whether it is buckling and fragmenting, or whether it is slowly subsiding, is not so clear. Observations on Antikythera indicate that it is rising, while the proposed uplift for the eastern end of the arc is only an extrapolation from the true data area. The high relief along the axis of the arc alternating between islands and straits suggests that relief may in fact be increasing with development of the arc structure, and that the islands are being uplifted while the channels are being depressed.

Further evidence will be required to determine the true direction of movement of the sill levels, but, on the supposition that the rate of movement is of the order of 1 m per millenium, and that it could be downwards, the maximum alteration of sill levels since the penultimate glacial maximum, say 200,000 years, would be 200 to 400 m. If the sea level were lowered 200 m the water depth over various sills would be as follows: Sicily 200 m; Messina dry to 100 m; Elaphonisos 50 m; Kythera dry to 50 m; Antikythera 350 m; Kasso 350 m; Scarpanto 550 m; Rhodes 150 m.

CONCLUSIONS

Vertical movements on the margins of the southern Aegean are of the order of 1 to 2 m per millenium, predominantly downwards. Such rates of movement have significantly altered the topography of basin structures and sill depths during the last 200,000 years, although there is insufficient evidence as yet to reach firm conclusions about most sills. The island arc from the Peloponnese through Crete to Turkey is shown to be an active anticlinal structure at its western end, and in general the vertical movements detected are in accordance with the concepts of the European block overriding the floor of the eastern Mediterranean. The rapid subsidence of the southern Aegean basin is confirmed. Alterations of sill depth will have altered water circulation in the past, and may have caused the eastern Mediterranean to be more liable to stagnation at depth under meteorologic conditions of decreased storminess and wind speeds, reduced mixing, and increased stability of thermoclines.

REFERENCES

Ambraseys, N. N. 1965. A note on the seismicity of the eastern Mediterranean: *Studia Geofisica et Geodaetica*, 9:405–410.

Ambraseys, N. N. 1970. Some characteristic features of the Anatolian fault zone. *Tectonophysics*, 9:143–166.

Aubouin, J. 1965. *Geosynclines: Developments in Geotectonics 1*. Elsevier, New York, 335 p.

Bogdanoff, A. A., M. V. Mouratov, and N. S. Schatsky 1964. Tectonics of Europe. *International Geological Congress, Subcommission for the Tectonic Map of the World*. Nauka, Moscow, 359 p.

Caputo, M., G. F. Panza, and D. Postpischl 1970. Deep structure of the Mediterranean basin. *Journal of Geophysical Research*, 75: 4919–4924.

Emery, K. O., B. C. Heezen, and T. D. Allan 1966. Bathymetry of the eastern Mediterranean Sea. *Deep Sea Research*, 13: 173–192.

Fairbridge, R. W. 1961. Eustatic changes in sea level. In: *Physics and Chemistry of the Earth*, eds. Ahrens, L. H., F. Press, K. Rankama and S. K. Runcorn, Pergamon Press, London, 4: 99–185.

Flemming, N. C. 1968a. Holocene earth movements and eustatic sea level change in the Peloponnese. *Nature*, 217: 1031–1032.

Flemming, N. C. 1968b. Derivation of Pleistocene marine chronology from morphometry of erosion profiles. *Journal of Geology*, 76: 280–296.

Flemming, N. C. 1969. Archaeological evidence for eustatic change of sea level and earth movements in the Western Mediterranean in the last 2000 years. *Geological Society of America, Special Paper*, 109: 125 p.

Giermann, G. 1960. *Topographic Chart of the Aegean.* Musée Océanographique de Monaco.

Gutenberg, B., and C. F. Richter 1954. *Seismicity of the Earth.* Princeton University Press, Princeton, 310 p.

Hafemann, D. 1960. Ansteig des Meeresspiegels in geschichtlicher Zeit. *Umschau,* 7: 193–196.

Hersey, J. B. 1965. Sedimentary basins of the Mediterranean Sea. In: *Submarine Geology and Geophysics,* eds. Whittard, W. F. and R. Bradshaw, Butherworths, London, 75–92.

Higgins, C. G., 1969. Isostatic effects of sea-level changes. *Quaternary Geology and Climate.* National Academy of Sciences. 1701: 141–145.

Jelgersma, S. 1961. Holocene sea level changes in the Netherlands. *Mededelingen van de Geologische Stichting,* Series C–IV: 9–76.

Jongsma, D. 1970. Eustatic sea level changes in the Arafura Sea. *Nature,* 228: 150–151.

Le Pichon, X. 1968. Sea-floor spreading and continental drift. *Journal of Geophysical Research,* 73: 3661–3697.

Maley, T. S. and G. L. Johnson 1971. Morphology and structure of the Aegean Sea. *Deep-Sea Research,* 18: 109–122.

McKenzie, D. P. 1970. Plate tectonics of the Mediterranean region. *Nature,* 226: 239–243.

Mörner, N. A. 1969. Eustatic and climatic changes during the last 15,000 years. *Geologie en Mijnbouw,* 48: 389–399.

Negris, P. 1904. Vestiges antiques submergés. *Athenischer Mitteilungen,* 29: 340–363.

Ninkovich, D., and B. C. Heezen 1965. Santorini Tephra. In: *Submarine Geology and Geophysics,* eds. Whittard W. F. and R. Bradshaw. Butterworth, London, 413–454.

Papazachos, B. C., and N. D. Delibasis 1968. Tectonic stress field and seismic faulting in the area of Greece. *Tectonophysics,* 7: 231–255.

Renz, C., N. Liatsikas, and I. Paraskevaidis 1954. *Geological Map of Greece.* Ministry of Coordination, Athens.

Scholl, D. W., and M. Stuiver 1967. Recent submergence of Southern Florida: A comparison with adjacent coasts and other eustatic data. *Geological Society of American Bulletin,* 78: 437–454.

Scientific Staff 1970. Deep Sea Drilling Project: Leg 13. *Geotimes,* 15: 12–15.

Umbgrove, J. H. F. 1947. *The Pulse of the Earth.* Martinus Nijhof, The Hague, 358 p. (2nd ed.).

Wong, H. K., and E. F. K. Zarudzki 1969. Thickness of unconsolidated sediments in the Eastern Mediterranean Sea. *Geological Society of America Bulletin,* 80: 2611–2614.

Woodside, J., and C. Bowin 1970. Gravity anomalies and inferred crustal structure in the Eastern Mediterranean Sea. *Geological Society of America Bulletin,* 81: 1107–1122.

PART 5. Coastal and Shallow Water Sedimentation: Carbonate Sediments

Mediterranean Beachrock Cementation: Marine Precipitation of Mg-Calcite

Torbjörn Alexandersson

University of Uppsala, Uppsala

ABSTRACT

Intertidal beachrock is found on degrading beaches on the Mediterranean coasts of Morocco, Spain, Italy, Greece and Cyprus. It appears that initial cementation takes place within the beach deposit, and that the rock becomes exposed when the beach is eroded. Mechanical erosion cuts into the seaward parts, occasionally in the form of ridge-furrow systems. Loose slabs of beachrock are common in the sublittoral zone. Pottery and other artifacts occur cemented into the rock, and lithification is apparently active at the present time.

According to X-ray diffractometry and electron microprobe analysis the cement is calcite with 13 to 16 mole % $MgCO_3$ in solid solution; this corresponds to a partition coefficient of 0.03 for Mg if sea-water is the source for the cement. The Mg-calcite is a direct precipitate, not a recrystallization product. The fabric is *hard micrite, friable micrite* and *fringe cement*, each form the result of a specific growth mechanism. In material from Rhodes, one fringe is dolomite, with the composition $(Ca_{.61}Mg_{.36}Fe_{.03})CO_3$.

The cement of Mediterranean beachrock thus represents a major occurrence of marine-precipitated Mg-calcite. In this respect the region is different from the Gulf of Mexico and the Caribbean where the corresponding cement is aragonite. It appears that precipitation of aragonite is a subordinate process in the Mediterranean at the present time.

RESUME

Sur les côtes du Maroc, de l'Espagne, de l'Italie, de la Grèce et de Chypre il y a, dans la zone de marée des plages en dégradation, du grès de plage. Il semble que la cimentation prenne place parmi les dépôts de plage; l'érosion met ensuite la roche à nu. L'érosion mécanique agit sur les parties proches de la mer, formant çà et là des systèmes de sillons et de crêtes; les plaques détachées de grès de plage sont nombreuses dans la zone sublittorale. Des poteries et autres artéfacts se rencontrent, cimentés dans les roches, et on suppose que les sédiments sont en voie de solidification.

Selon les résultats des mesures de la diffraction des rayons X et du micro-analyseur à sonde électronique, le ciment est constitué de calcite avec 13–16 mole % de $MgCO_3$ en solution solide, ce qui correspond à un *partition coefficient* de 0,03 pour Mg si l'eau de mer est la source du ciment. La Mg-calcite provient directement d'une précipitation et non d'une recristallisation. Elle présente trois textures: *micrite dure, micrite friable* et *ciment des franges*, chacune résultant d'un mécanisme spécial de croissance. Dans des matériaux provenant de Rhodes, une frange est formée de dolomite dont la composition est la suivante: $(Ca_{.61}Mg_{.36}Fe_{.03})CO_3$.

En Méditerranée, la cimentation du grès de plage est produite principalement par précipitation de Mg-calcite inorganique et marine. Sur ce point, la région diffère du Golfe du Mexique et de la Mer des Antilles où le ciment équivalent est de l'aragonite. Il semble que la précipitation d'aragonite soit à l'heure actuelle, un processus secondaire dans la Méditerranée.

Mg-CALCITE IN SEDIMENTS

The presence of the isomorphous solid solution series calcite-magnesite in sedimentary carbonates was originally demonstrated in marine biogenic calcites (Chave, 1952), and for several years all magnesian calcites described from the sedimentary environment were also of organic origin (*e.g.*, Chave, 1954, Goldsmith *et al.*, 1955, Goldsmith and Graf, 1958, and Chave, 1962). It became widely accepted that under earth-surface conditions magnesium is incorporated in calcites by metabolic processes only, an opinion which was supported by numerous laboratory experiments on precipitation of carbonates from seawater. Such experiments indicate that Mg ions in the solution inhibit the growth of calcite nuclei and cause precipitating calcium carbonate to crystallize as aragonite (Monaghan and Lytle, 1956, Lippman, 1960, Ingerson, 1962, Kitano *et al.*, 1962, Usdowski, 1963, Simkiss, 1964, de Groot and Duyvis, 1966, McCauley and Roy, 1966, and Taft, 1967). Since the Mg:Ca ionic ratio in seawater is about 5, direct inorganic precipitation of Mg-calcites should not occur under marine conditions: aragonite should be the preferred polymorph.

However, as pointed out previously by Zeller and Wray (1956), that picture is over-simplified since many factors (temperature, pH, concentration of other ions, solubility, crystal size, rate of processes) affect the precipitation of carbonates and the properties of the minerals formed. Glover and Sippel (1967) synthesized highly metastable Mg-calcites with as much as 60 mole % $MgCO_3$ in solid solution by reactions carried out at atmospheric pressure and in the temperature range 0° to 40°C. Mainly on account of field observations the importance of inorganic Mg-calcite formation is now gradually being recognized, and the number of reported field occurrences is rapidly increasing.

FIELD OCCURRENCES OF INORGANIC Mg-CALCITE

In the Coorong lagoon and associated saline lakes in South Australia Mg-calcite is formed as small, free particles in the water mass during periods of luxuriant growth of *Ruppia maritima,* and aragonite-Mg-calcite-dolomite sequences occur as modern and Pleistocene sediments in the area (Alderman and Skinner, 1957; Alderman, 1959; Alderman and von der Borch, 1963; Skinner, 1963; and von der Borch, 1965).

In Lake Balaton, Hungary, Mg-calcite with 6 to 8.5 mole % $MgC0_3$ is precipitated as particles smaller than 6.3μm because of dense and extensive growth of phytoplankton during the warmer seasons. Authigenic protodolomite occurs in the clay fraction of most of the bottom muds (Müller, 1970).

Friedman (1964, 1968) described as micritic Mg-calcite the cement in lithified carbonate sediments from 275 m water depth on top of Atlantis Seamount west of the Azores, and such cement is now well-known from deep-sea sediments (Gevirtz and Friedman, 1966; Milliman, 1966; Russel *et al.*, 1967; Fischer and Garrison, 1967; Milliman *et al.*, 1969).

Recent Mg-calcite cement with 12 to 18 mole % $MgCO_3$ also occurs under a variety of shallow–intertidal–supratidal marine conditions; for instance in shallow-water carbonate sediments in the Persian Gulf (Shinn, 1969), in hypersaline lagoons on South Bonaire, Netherlands Antilles (Lucia, 1968), in beachrock from the Red Sea and the Canary Islands (Friedman, 1968), in beachrock and in sandy tidal flats in the Persian Gulf (Taylor and Illing, 1969) and in beachrock in the Mediterranean (Alexandersson, 1969).

In some cases Mg-calcite cementation appears to be restricted to the micro-environment within various marine organism communities; for instance, in algal-*Millepora* reefs in Bermuda (Ginsburg *et al.*, 1967), in carbonate sands in a sponge mat on top of the submerged barrier reef off the west coast of Barbados (Macintyre *et al.*, 1968), within certain algal frameworks in the Mediterranean (Pérès, 1967 and Alexandersson, 1969) and in coral reefs off Jamaica (Land and Goreau, 1970). Mg-calcite also forms as micritic envelopes and microscopic infillings in biogenic carbonate grains in Florida Bay and to some extent that process may be related to the activities of boring algae (Winland, 1968).

Some field observations indicate that in the constant or recurrent presence of seawater Mg-calcite may form as a recrystallization product of aragonite (Milliman *et al.*, 1969; Taylor and Illing, 1969). However, according to the calculated stability fields for $CaCO_3$ polymorphs in seawater, a stabilizing reaction should go in the direction from Mg-calcite to aragonite, which is regarded as the thermodynamically stable form under marine conditions (Winland, 1969).

Inorganic Mg-calcite is also known to form under subaerial diagenetic conditions, and the Mg-content may even exceed the values common in marine calcites. Hay and Iijima (1968) report coarse rice-shaped calcite aggregates from the palagonite tuffs on Oahu, Hawaii, with as much as 28 mole % $MgCO_3$ in solid solution.

MEDITERRANEAN BEACHROCK CEMENTATION

General

No current theory concerning marine physico-chemical precipitation and diagenesis of carbonates is applicable to all field observations, and the Mg-calcites are particularly difficult to account for. Their importance for dolomitization is recognized (Schlanger, 1957, Gross, 1965, and Buchbinder and Friedman, 1970), but their relation to natural marine conditions is not well understood. When it was found that beachrock in some regions of the Mediterranean was cemented by Mg-calcite (Alexandersson, 1969) it seemed worthwhile to continue the investigation in other areas, and to give special attention to field relationships and to the mineralogy and fabric of the

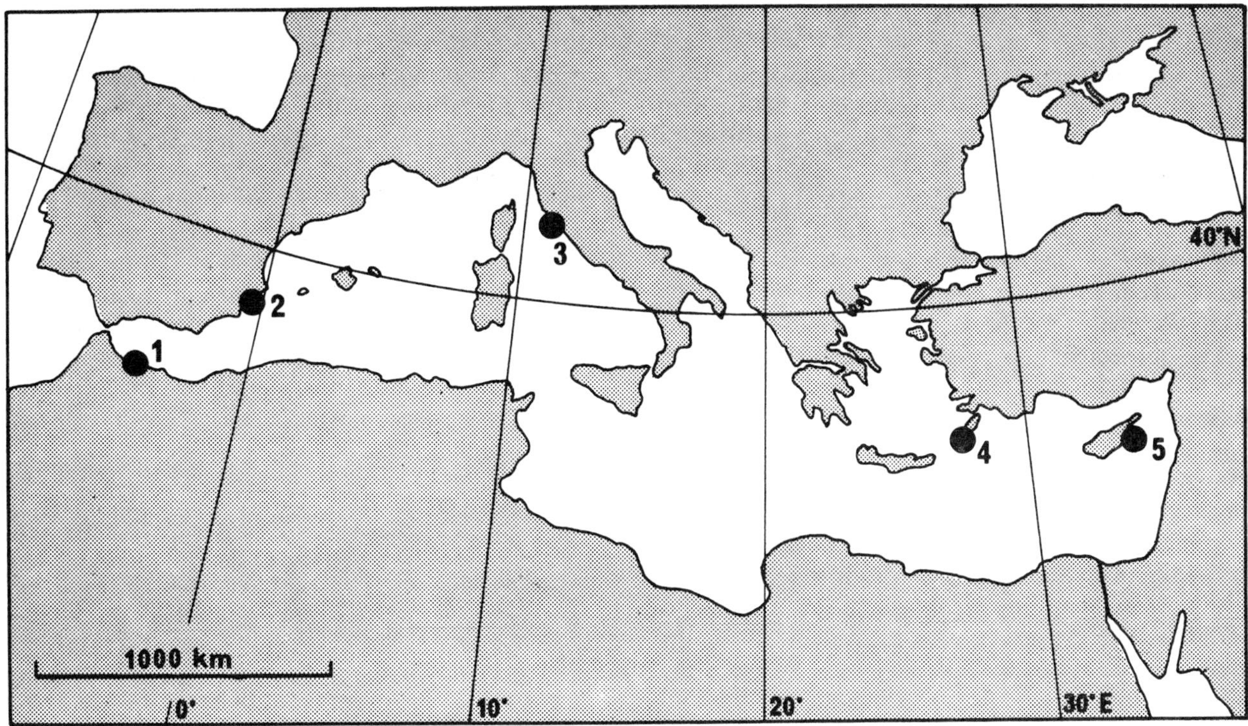

Figure 1. Areas with observed Mg-calcite cementation of beachrock. 1, The Mediterranean coast of Morocco; 2, the Spanish coast south of Alicante; 3, the Tuscan coast of Italy; 4, the Greek islands of Rhodes and Karpathos; and 5, the coast of Cyprus. The distance between area 1 and 5 is about 3,500 km.

beachrock cement. Data from that extended study are reported in the present paper.

The investigated areas are: (1) the Mediterranean coast of Morocco near the Strait of Gibraltar, (2) the Spanish coast south of Alicante, (3) the Tuscan coast of Italy, (4) the Greek islands of Rhodes and Karpathos, and (5) the coast of Cyprus. Surveys were also made on the island of Ibiza in the Balearics, along the Gulf of Hammamet in Tunisia, and along the Dalmatian coast south of Dubrovnik in Yugoslavia, but no beachrock was found. The distance covered in a west-east direction is about 3,500 km (map in Figure 1).

Previous Work

Beachrock has been described previously from many localities in the Mediterranean; for instance, from Israel (Emery and Neev, 1960; Schattner, 1967; Gavish and Friedman, 1969), from Lebanon (Fevret and Sanlanville, 1965), from Turkey (Taillefer, 1964), from the islands of Rhodes and Karpathos (Alexandersson, 1969), from the island of Crete (Boekschoten, 1962 and 1963), from the Greek mainland (Mistardis, 1963 and 1964), from northern Italy (Bloch and Trichet, 1966; and Stefanon 1969), and from the southeastern coast of Spain (Russell, 1962; Mabesoone, 1963). Goudie (1969) discusses the age and distribution of Mediterranean beachrock in general and gives a short account of 19th century observations.

Mechanisms of Beachrock Formation

Although beachrock has been extensively studied by many workers in different regions it is still not possible to attribute this kind of cementation to a specific process, for instance evaporation, mixing of fresh and saline waters, or metabolic activity by algae or bacteria; nor is it known if one and the same mechanism is responsible in all cases. From a geological point of view the processes are rapid, and lithification may be completed in less than 10 to 15 years (Schmalz, 1969). However, it is not known for what length of time precipitation really takes place; the cement may form by slow continuous growth at levels of moderate supersaturation, or grow only occasionally under extremely favorable conditions. This is a complication in field studies of the pertinent physico-chemical parameters, since it is difficult to determine whether cementation is active or not during the period of investigation.

METHODOLOGY

For the present study, shore investigations were carried out, together with observations of the sublittoral zone by means of SCUBA diving. Oriented rock samples and samples of adjacent unconsolidated sediments were taken along traverses from the supralittoral zone and out into the sublittoral. Fossil beachrock horizons in subaerial positions above the present beaches were not included in the study.

In the laboratory the general distribution of carbonate minerals in rock samples and grain-mount thin sections was studied on polished and etched surfaces with staining methods as described by Friedman (1959) and Warne (1962). The samples were treated mainly with alizarin red-S in acid solution for establishing the presence of calcium carbonate, Feigl's solution for the distinction between calcite and aragonite, and finally alizarin red-S in basic solution for separation of high- and low-Mg calcite.

The preliminary determinations made with staining methods were checked and continued with X-ray powder diffractometry, using Ni-filtered Cu/Kα radiation. As a rule scans were made in one direction over the interval 25° to 34° 2θ with a scanning speed of $\frac{1}{2}$° per minute, and in both directions over the interval 29.0° to 30.5° 2θ with a scanning speed of $\frac{1}{4}$° per minute; the paper speed was 1 cm per minute. Silicon metal was used as an internal standard. The amount of $MgCO_3$ in solid solution in calcite was calculated from the reduction of the $d_{10\bar{1}4}$ interplanar spacing using the methods and diagrams of Chave (1952), Goldsmith *et al.* (1955) and Goldsmith and Graf (1958). As demonstrated by Runnells (1970), the $10\bar{1}4$ reflection of calcite is not well suited for determination of the proportions of individual minerals in mixtures; the intensity of this reflection being highly sensitive to isomorphous substitution of the cations.

Elemental composition was studied with electron microprobe analysis on aluminum-coated petrographic thin sections without cover glasses, and with atomic absorption spectrophotometry on dissolved material. Total carbonate content of sediments was calculated from the weight loss after treatment with dilute hydrochloric acid. Scanning electron micrographs were made from etched or freshly broken gold-coated specimens.

BEACHROCK AND SHORE CONDITIONS

Water Data

The salinity of the surface waters in the Mediterranean is 36.5 ‰ near Gibraltar and increases in an easterly direction to more than 39.5‰ around Cyprus. The mean annual temperature range of the surface waters is 15° to 21°C near Gibraltar, and 18° to 26°C around Cyprus (Bruns, 1958). Variations of water level due to tides are small, usually less than 30 cm (British Admiralty Publication No. 200).

Shore Conditions

Most commonly beachrock is found in intertidal positions but it also occurs in the sublittoral zone. At some localities, for instance in the shallow bay east

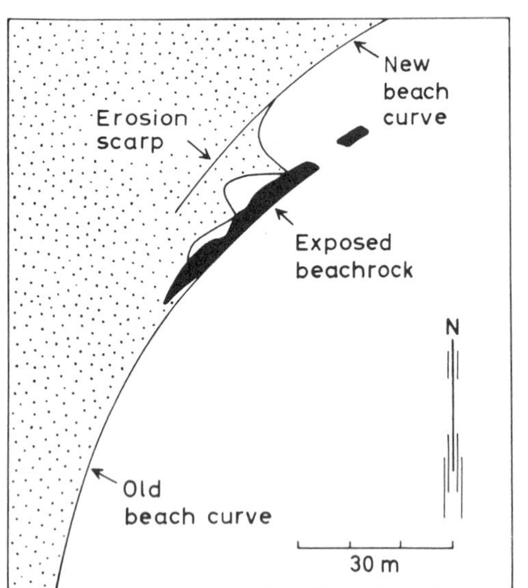

Figure 2. Beachrock north of Larnaca, Cyprus. Erosion of the beach leads to gradual exposure of the rock. When the unconsolidated sand beneath the beachrock is eventually removed by wave action, the rock breaks into flat slabs which come to rest in the sublittoral zone. The photograph was taken in a southwest direction in July 1970; hammer indicates the position of the shore-line in August 1969.

of Stazousa Point on the north coast of Cyprus, stray patches of beachrock lie in series along the coast, from several hundred meters offshore and up on to the present beach. Such series have been explained as a product of repeated beachrock cementation along a transgressing shoreline; by mapping and dating submerged beachrock outcrops, Stefanon (1969) attempted to reconstruct changes of coastline in the north Adriatic Sea.

Rocks at the shore-line often display a seaward dip, about equal to the slope of the foreshore (Figure 2); a feature which has been described from many localities and commonly is regarded as a diagnostic characteristic of beachrock (*e.g.*, Ginsburg, 1957; Krauss and Galloway, 1960; Russell, 1962).

In many cases a morphology of that kind simply reflects the inclined bedding of the sediments in the foreshore. Cementation begins at grain contacts, and as the number of grain contacts per unit volume of sediment increases with decreasing grain size, a fine-grained bed will initially form a more coherent rock than a coarser layer. Even if cementation is uniform, vertical variations of grain size and sorting in the foreshore sediments will result in beds of differing resistance; the well-known morphology of a beachrock outcrop is then eventually a product of selective erosion.

However, under some conditions, variations in cementation result in a series of "beds" which are independent of the original stratification. The lithifying processes are active only in a narrow vertical interval, rarely exceeding 1 m, and usually not more than 20 to 40 cm. The cementation varies in intensity within the interval, and this often results in firmly cemented zones with almost occluded interstices, alternating with weakly cemented zones with very little cement and remaining high porosity (Figure 6). At numerous localities on the north coast of Cyprus, such zones of differential beachrock cementation pass with a seaward dip through horizontal coastal deposits. Erosion, controlled by the cement zonation, has created a surface morphology, which gives the impression of a series of inclined foreshore beds (Figure 3).

In general, beachrock exposures occur on beaches affected by erosion, and the initial cementation certainly takes place at some depth within the beach deposit, and not at the sediment surface. On many beaches the exposed rock represents only a small fraction of all material lithified by beachrock cementation. Thus, on the curved beach north of Larnaca, Cyprus (which is more than 10 km long) beachrock is exposed only at the intersection between the arcuate old shoreline, and a new curved surface which is formed by the erosion proceeding at present (Figure 2).

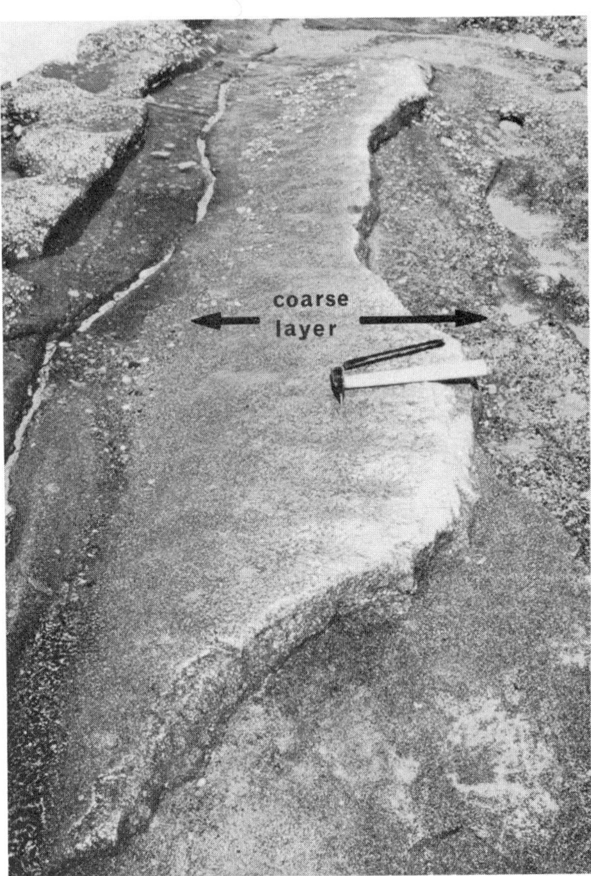

Fihure 3. Horizontal bedding in inclined beachrock, west of Stazousa Point, Cyprus. As seen from the coarse layer exposed to the left and the right, the bedding in the lithified sediment is horizontal. The apparent seaward dip of the rock is due to bands of Mg-calcite cement, which cut obliquely through the deposit. The shoreline is visible in the upper left corner. Hammer handle 0.3 m.

By gradual erosion of the old beach, more beachrock becomes exposed, while at the same time the distal, northern end of the exposed patch breaks up into flat slabs as the underlying sand is removed. Numerous such slabs are scattered in the sublittoral zone along the new beach.

This relationship between degrading beaches and exposure of beachrock is in agreement with the observations of Russell (1962) and Cooray (1968).

Ridge-Furrow Systems

Where the gradient of a beach is steep and backwash is strong, corrasion by wave action often cuts a rock surface into a system of nearly equi-spaced, sub-parallel ridges and furrows, perpendicular to the shoreline. Such systems are frequently associated with pot-holes, although the two forms can occur separately. The origin of the initial spacing of furrows is not clear, nor those factors that decide size and proportions, but once a pattern is established it is

Figure 4. Underwater photograph of beachrock front, west of Lapithos, Cyprus. The Mg-calcite cemented sediment is a coarse fluviatile deposit, different from the present beach sand. No seaward dip of the rock is apparent. Mechanical erosion works mainly during periods of strong turbulence when the corrading material is carried in suspension; there is no tendency for formation of ridge-furrow systems. Depth 0.5 m, hammer handle 0.3 m.

accentuated by its own channeling effect on water and sediment movement.

Ridge-furrow systems are common in beachrock outcrops (McLean, 1967); they also form in cohesive materials of other origin, and may develop in offshore positions within depths where wave action can move sediment as bedload (Alexandersson, in press). Barnes (1965) described a Middle Ordovician ridge-furrow system in limestone; Neev and Emery (1967) found similar systems, which they termed pseudo-ripple marks, in the sublittoral zone in the Dead Sea.

Ridges and furrows only develop where erosion is related to bedload transport of material. In environments where the abrading effect depends on the fine-grained sediments usually carried in suspension, rocks may be rounded and irregularly sculptured but no furrowing takes place (Figure 4).

Composition of Beach Sediments and Beachrock Grain Component

Normally the unconsolidated beach material at beachrock localities corresponds closely to the grain components in the beachrock. In both unconsolidated deposits and beachrock the grain size varies between medium sand and gravel, and as a rule the sorting is good. To a great extent the material on the beaches is land-derived, and in mineralogical composition it reflects the local geology. Quartz, feldspars, detrital dolomite and low-Mg calcite are the dominating minerals, and the amount of metastable Mg-calcite and aragonite is relatively small. Of the metastable carbonates 70 to 90% is Mg-calcite, mainly supplied by coralline algae and porcellaneous benthonic foraminifera (X-ray diffractograms of bulk sand samples, Figure 5*a*). Observed values of total carbonate content (= soluble in dilute HCl) in beach sands vary between less than 25% by weight and more than 99% by weight, with a mode around 50%. The high values are found in limestone areas where most of the carbonate is recycled limestone of diverse ages, and the mineral is mainly stable low-Mg calcite. The implications of a high carbonate content under such circumstances are different from those in a tropical area, where high production of Recent carbonate material from corals, algae, *etc.* leads to a large proportion of metastable phases in the sediment.

In Figure 5 the carbonate composition of Mediterranean high-carbonate beach sands from Ibiza (98 wt. % carbonate) and Rhodes (96 wt. %) is compared with the composition of beach sands from Barbados and Jamaica in the West Indies. A Mediterranean low-carbonate sand from Cyprus (25 wt %) is also included. The Mediterranean sands consist of low-Mg calcite, high-Mg calcite, some detrital dolomite, and small amounts of aragonite; the non-carbonate minerals are quartz and feldspar (Figure 5*a*). The sediments from Barbados and Jamaica are pure carbonate sands from beaches in the lee of living coral reefs; the minerals are predominantly aragonite and Mg-calcite, together with some low-Mg calcite (Figure 5*b*).

Fragments of pottery appear as a component in certain rocks, especially along the coasts of Rhodes and Cyprus. In a few cases, iron and glass artefacts were found cemented into the rocks, and the maximum age for cementation under such circumstances can be no more than a few hundred years.

BEACHROCK CEMENT

General Appearance

The color of the cement is usually uniform within a single bed, and varies on fresh surfaces from pure white to yellow, or occasionally almost brown. The darker colors are associated with the crypto-crystalline forms of cement.

The amount of cement in the pore space shows considerable variation over vertical intervals of only a few centimeters. This phenomenon is associated with variations in the amount of cement actually formed and cannot be ascribed to variations in porosity

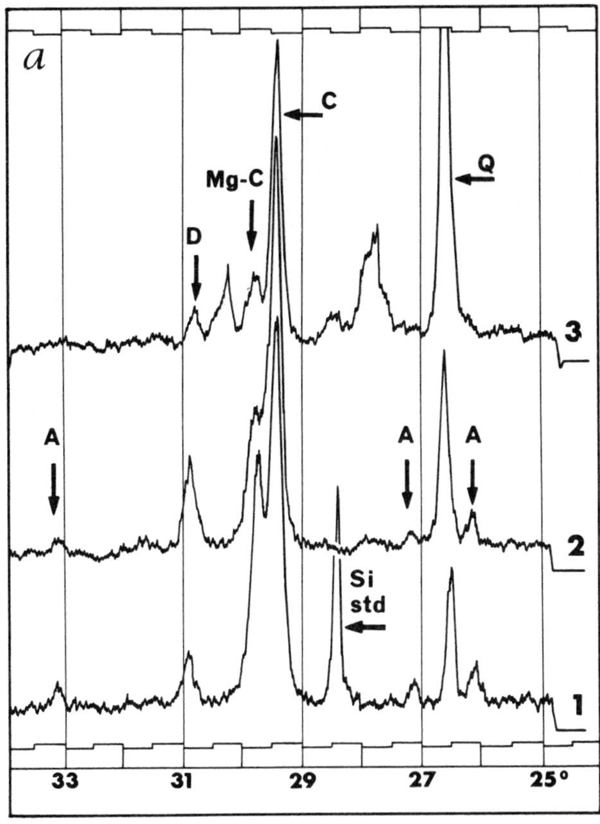

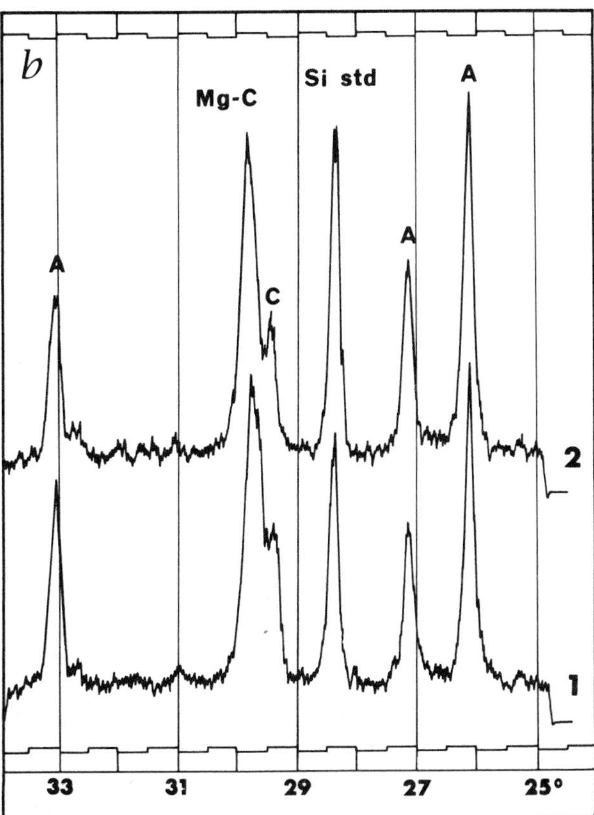

and permeability caused by changes of the primary sedimentary characteristics, such as grain size and sorting. Thus, zones of advanced cementation and occluded pores alternate with zones where grains are cemented at grain contacts only, and the boundaries between the zones are distinct. Moreover, the cement zonation does not always follow the bedding in the lithified sediment (Figures 3 and 6).

Mineralogy and Composition

X-ray diffraction analyses of cement regularly show a main phase of calcite with a *d* $10\bar{1}4$ spacing of about 2.99Å, which corresponds to Mg-calcite with 13 to 16 mole % $MgCO_3$ in solid solution (Figure 7). According to electron microprobe analyses, the Mg-content may vary within these limits in a single thin section; no regional variation between the investigated areas could be detected.

Small foreign particles are dispersed as inclusions in the cement (see the following discussion on micrite), and the amount needed for powder X-ray diffraction analysis is usually not quite pure. The inclusions correspond mineralogically to the grain component of the beachrock, and to the unconsolidated deposits on the beach in question (Figure 7, diffractometer patterns 1, 4, 5).

Micritic Mg-calcite is the cementing agent in widespread sublittoral lithification of algal frameworks in the Mediterranean (Alexandersson, 1969). Cement of that kind has the same composition and mineralogy as the beachrock cement (Figure 7).

Encrusting, grayish-white carbonate scales are occasionally found on beachrock surfaces in the spray-zone, out of reach of abrasion by waves. They are best developed on surfaces facing the spray, where their observed maximum thickness is about 0.2 mm. According to X-ray diffraction analysis their mineral form is aragonite. The scales, which postdate the Mg-

Figure 5. X-ray powder diffraction patterns of Mediterranean and West Indian carbonate sediments. Ni-filtered Cu/Kα radiation, interval 25° to 34°2θ. A = aragonite, C = calcite, Mg-C = Mg-calcite, D = dolomite, Q = quartz, Si std = internal silicon standard. *a*, Mediterranean beach sands: 1, Cala Conta, Ibiza, W. Mediterranean, 98 wt. % carbonate; 2, Ladhiko, Rhodes, E. Mediterranean, 96 wt. % carbonate; 3, North of Larnaca, Cyprus, E. Mediterranean, 25 wt. % carbonate. Stable low-Mg calcite is the dominant carbonate mineral in all cases; the most important metastable phase is Mg-calcite, while there is very little aragonite. *b*, West Indian beach sands: 1, Discovery Bay, Jamaica, pure carbonate sand; 2, Freshwater Bay, Barbados, pure carbonate sand. Both localities are situated on shores fringed by living coral reefs; the sands are composed mainly of metastable aragonite and Mg-calcite.

Figure 6. Hard micrite and friable micrite cement. *a*, Beachrock cemented by hard micrite. Vertical section, cut and polished surface. Most of the rock is cemented at grain contacts only and porosity is high. Two bands of more advanced cementation are indicated by arrows; here interstices are almost filled and porosity is low. The boundaries between the different zones are distinct. Resistance to erosion varies with the degree of cementation and to a great extent the morphology of a beachrock outcrop is controlled by this factor (*cf.* Figure 3). Scale bar 2 cm. *b*, Beachrock cemented by friable micrite. Vertical section, broken surface. In the lower part of the specimen interstices are filled by the white Mg-calcite micrite, while in the upper part some pore space remains empty. Compared to hard micrite, the banding is less distinct. A scanning electron micrograph of this micrite is shown as Figure 10*b*. Scale bar 3 cm.

calcite cement, are the only form of non-skeletal aragonite observed in connection with the beachrock material.

It is surprising that a cement, as uniform in composition as the Mg-calcite, is found in the littoral zone where the potential for variation is great. The mixing of natural waters presents many possibilities for chemical reactions (Runnells, 1969), and it is hard to think of a beach where no mixing takes place. Nearshore subsurface fluids show considerable variations of CO_2-pressure, pH, carbonate saturation level, salinity, *etc.* (de Groot, 1969; Schmalz, 1969) and such variations must be further accentuated by equilibration with different mineral assemblages in different beaches.

On the basis of the data collected in the present study it is not possible to identify the primary cause of beachrock cementation, or to determine whether the triggering mechanism is, for instance, mixing of natural waters, evaporation, or biologic activity. However, from the Mg-content of the cement it is possible to draw some conclusions concerning the Mg/Ca ratio in the precipitating fluids and the probable origin of these fluids.

Considering the budget of the material needed for cementation there are two alternatives: (1) supply of additional dissolved carbonate from sea-water, and (2) solution–reprecipitation of carbonate minerals already present in the beach. In the first process the properties of the cement should be related to the composition of seawater, and they should also be similar to those observed in indisputably marine cements. In the second process, the properties of the cement should be controlled ultimately by the mineral composition of the beach sediments, and variations between beaches might be expected.

Mg-Content and Relation to Seawater

The incorporation of foreign ions in solid solution in a growing crystal is an equilibrium phenomenon; as long as the amount of substituting ions is reasonably small the process is governed by the laws of chemical partition of elements (a review of the principles and their applications to geology is given by McIntire, 1963). Those ions which most easily fit in the normal lattice of the crystal are accepted in largest quantities, for instance Sr in aragonite and Mg in calcite. The equilibrium constant for partition of a microcomponent between two co-existing phases is called the distribution constant or partition coefficient, and is commonly defined by the equation:

$$\left(\frac{\text{Tr}}{\text{Cr}}\right)_{\text{solid}} = K\left(\frac{\text{Tr}}{\text{Cr}}\right)_{\text{liquid}}$$

Tr is the microcomponent or "tracer", and Cr is the macrocomponent or "carrier"; for values of K greater than unity the crystals growing from a liquid are enriched in microcomponent with respect to the liquid, while for K values less than unity the crystals are impoverished in microcomponent with respect to the liquid.

The distribution coefficient for partitioning Sr between aragonite and an aqueous phase at 1 atmosphere pressure is known over a range of temperatures; at 25°C it is about 1 (Oxburgh *et al.*, 1959;

Holland *et al.*, 1964; Kinsman, 1965). As an approximation K is 0.02 to 0.06 for partitioning Mg between calcite and an aqueous phase at earth surface conditions, and the value is supposed to increase with temperature (Winland, 1969). With the Mg/Ca ionic ratio in seawater taken as 5.2, the Mg-content in the beachrock cement, and also in the sublittoral cement from the algal frameworks, corresponds to a partition coefficient of about 0.03 for Mg in calcite under Mediterranean conditions.

Within reasonable limits the partition coefficient does not depend on actual concentrations, only on the ratio between the ions in question. The mixing of seawater with meteoric water, or with fresh subsurface waters in the beach, will therefore be of little consequence for the composition of the Mg-calcite cement.

Inorganic marine calcites reported from the Gulf of Mexico and the Caribbean have a slightly higher Mg-content than the Mediterranean material; 18 mole % $MgCO_3$ in solid solution in recrystallized algae from British Honduras (Purdy, 1968), and 18.5 mole % in internal cement in Jamaican reefs (Land and Goreau, 1970). If seawater is the source for the calcites, these values correspond to a partition coefficient for Mg of about 0.04. Because of the scarcity of data it is not known if the cited figures are representative for the whole West Indian region, nor is it possible to attribute the increase to any particular factor; however, it is in agreement with the temperature dependence found by Glover and Sippel (1967).

Laboratory precipitation of calcium carbonate from natural seawater by physico-chemical methods (Glover and Sippel, 1967) and by microbial growth (Malone and Towe, 1970) yielded Mg-calcite with approximately 14 mole % $MgCO_3$ in solid solution.

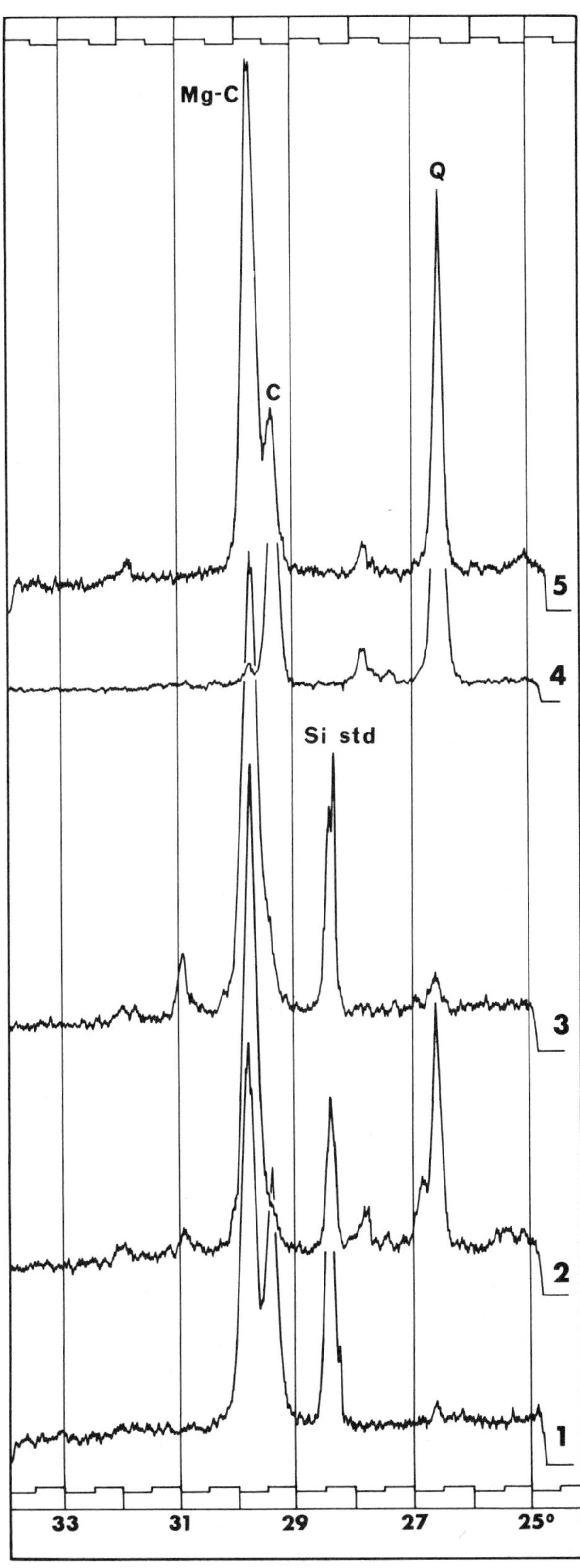

Figure 7. X-ray powder diffraction patterns of Mediterranean Mg-calcite cement. Legend as in Figure 5.1, Intertidal beachrock cement, Vurgundha, Karpathos. The grain component of the beachrock consists of detrital low-Mg calcite; a fine fraction is dispersed as small inclusions in the micrite cement. The same micrite is also shown in Figures 6*b*, and 10*b*. 2, Beachrock cement from the Ligurian coast, Italy. Courtesy by J. P. Bloch, Paris. The locality is described by Bloch and Trichet (1966). The grain components and the corresponding small inclusions are mainly terrigenous particles. 3, Sublittoral Mg-calcite cement from lithified algal framework 14 m water depth, Lindos, Rhodes. This cement doubtless formed in a submarine position; the process is described by Alexandersson (1969). 4, Beach sand, south of Alicante, Spain. Carbonate content 33 wt. % . The mineral composition is mainly quartz and detrital calcite. 5, Beachrock cement, lithifying the sand shown as 4. A fine fraction of the beach sand is dispersed as small inclusions in the cementing Mg-calcite micrite.

Mg/Ca Ratio in Fresh Water Solutions

Russell (1962 and 1963) suggested that circulating fresh water is competent to dissolve calcareous beach sands, and that the initial cementation of beachrock takes place at the groundwater table where dissolved calcium carbonate is precipitated, mainly in the form of calcite. In the present study no field observations indicate a process of that kind; furthermore, it appears that dissolution of beach material in fresh water gives a Mg/Ca ratio that is too low for formation of the observed Mg-calcite.

A laboratory simulation of the process was made. Sediments from three beaches in Italy, Rhodes, and Cyprus were thoroughly washed to remove all traces of seawater. After drying, the material was placed in beakers and distilled water, rendered acid by bubbling through CO_2, was added to the sediment surfaces. The samples were then stored in a CO_2-atmosphere at room temperature (23°C) for three months to allow the pore solutions to approach a state of equilibrium. After that time the fluids were withdrawn and immediately analysed for Mg, Ca and Fe by atomic absorption spectrophotometry (Table 1).

The Mg/Ca ionic ratio in the solutions is 2.3 to 2.5; if the partition coefficient for Mg is assumed to be 0.03, the product of solution-reprecipitation by fresh groundwater in beach sediments of that kind should be calcite with not more than 7 mole % $MgCO_3$ in solid solution.

CEMENT FABRIC AND GROWTH CONDITIONS

Terms

There is as yet no generally accepted terminology for descriptions of fabric (= size and mutual relations of crystals) of cement. The term *micrite* was originally introduced as a contraction of *microcrystalline calcite* to designate limestones with a recrystallization fabric composed of crystals less than 4 μm in size (Folk, 1959, 1962 and 1965), but the word is now widely used as a purely descriptive term for a microcrystalline fabric and even aragonite micrite is recognized (Friedman, 1968). The term is used with a similar broad meaning in this paper and the size limit of 4 μm is not strictly observed.

Table 1. Magnesium, calcium and iron in fresh-water pore solutions from beach sediments. Concentrations in ppm.

Sediment sample locality	Mg	Ca	Fe	Mg/Ca ionic ratio
Beach surface, Castiglione, Italy.	178	130	<0.1	2.3
Water depth 1.5 m Golden Beach, Rhodes.	196	130	<0.1	2.5
At water table, N. Larnaca, Cyprus.	165	118	<0.1	2.3

Micrite envelope is a specific term for a feature which "differs from an oolitic coat in that it is not an encrusting addition to the shell, but a replacement and cuts across the original shell fabric where this is preserved" (Bathurst, 1964, p. 365). The phenomenon has its own importance in diagenesis (Bathurst, 1966; Winland, 1968), and therefore the term is not applied to coatings of micrite cement even if these completely surround particles.

As pointed out by Dunham (1969), usage of the term *drusy* has been very conflicting and pertaining to the fabric of cement the word no longer has any definite meaning. It is therefore avoided in this paper; thin coatings of oriented crystals are referred to as *fringes*.

Fabric

The fabric of the cement depends on the growth conditions, and primarily on those factors which control the crystallographic form of the precipitate. Thus, aragonite cement in beachrock from the West Indies commonly occurs as discrete laths or needles, 80–100 μm in length and with a preferred orientation normal to the detrital grain surface (Figure 8). Micritic Mg-calcite cement in the Mediterranean material consists of equidimensional crystals with a random orientation, while coating fringes comprise slightly elongated, oriented crystals. Both kinds of fabric are quite different from the aragonite cement.

Micrite

Two genetically different forms are distinguished: (1) hard micrite, in which new crystals form in firm contact with the pre-existing rigid fabric, and (2) friable micrite, which is formed by weakly interconnected crystals. The difference is readily recognizable in the field. Thus friable micrite disintegrates and gives the water a milky appearance when a sample is taken, and it is easily crushed by a needle. The hard micrite, on the other hand, is rigid and coherent from its inception as small annulae at grain contacts, or as minute lumps growing into the pore space, and it resists the point of a needle. Hard micrite is the most common form, found in 80 to 90 % of all samples collected.

The differences in growth habit are best studied with the scanning electron microscope. In hard micrite new crystals nucleate on the surfaces of the pre-

Figure 8. Aragonite cement in beachrock from Barbados, West Indies. Scanning electron micrographs of broken surfaces. *a*, Orientation of aragonite needles is approximately normal to the detrital grain surface. Scale bar 100 μm. *b*, Intergrown, discrete aragonite laths with pointed terminations. Compared to the compact calcite cement, the aragonite fabric is an open structure where much void space remains. Scale bar 10 μm.

existing fabric and grow in firm contact with their substrate. Numerous rosette-shaped growth centers occur on globular or columnar crystal aggregates which have a diameter of 25 to 40 μm; crystal size ranges from the detection limit of 0.1 to 0.2 μm, up to 6 to 8 μm (Figure 9). The gradually extending fabric fills the intergranular pore space in the form of pellet-like globulae or anastomosing branches, which create a framework in the void. On continued growth these structures disappear, as the space becomes filled by a fabric of randomly oriented crystals in close contact with each other (Figure 10*a*).

Friable micrite consists mainly of platy crystals, less than 2 μm in size. They occur as discrete particles or in weakly bonded aggregates of poorly defined morphology. Specific growth centers have not been observed (Figure 10*b*).

In thin section the growth aggregates of hard micrite are seen as rounded or branching structures in many voids; they grade into uniform micrite where no record of such forms is preserved (Figures 9*a* and 11). The interior of all micrite appears as a uniform paste, with a brownish color in transmitted light. Laminations or growth layers do not occur, and electron microprobe scanning for Ca, Mg, Fe and Sr shows only minor variations in the elemental composition, without any regular pattern. This agrees with the observations from the scanning electron microscope; growth appears to take place simultaneously at a multitude of sites in the fabric, and not at a distinct surface of accretion.

The typical micrite contains a large amount of extraneous particles as inclusions. This extraneous material consists mainly of the trapped fine fractions of the lithified sediment together with sponge spicules, foraminifers and fragments of shell and calcareous algae. Particles are evenly dispersed in the cement and must have been supplied by interstitial transport many times during the growth of the cement; micrite growing downwards in a void has about the same amount of inclusions as its counterpart growing upwards. The extraneous material does not floor or fill interstices in the manner of an internal sediment (*cf.* Dunham, 1969); the absence of internal sedimentation suggests a rapid interstitial circulation.

Successive incorporation of foreign particles in the cement is a process compatible with the assumed growth mechanism for hard micrite. Particles in interstitial transport are trapped in the irregular framework which is formed by the branching growth structures; continued growth eventually leads to a firm cementation in a fabric of uniform micrite.

Micrite and Organic Influence

Under various conditions micrite occurs in close association with organisms and organic matter; for example, in micrite envelopes (Bathurst, 1966; Winland, 1968), in indurated fecal pellets (Illing, 1954; Ginsburg, 1957), and in altered benthic foraminifers and other skeletal grains (Kendall and Skipwith, 1969). Although the relations between micrite and

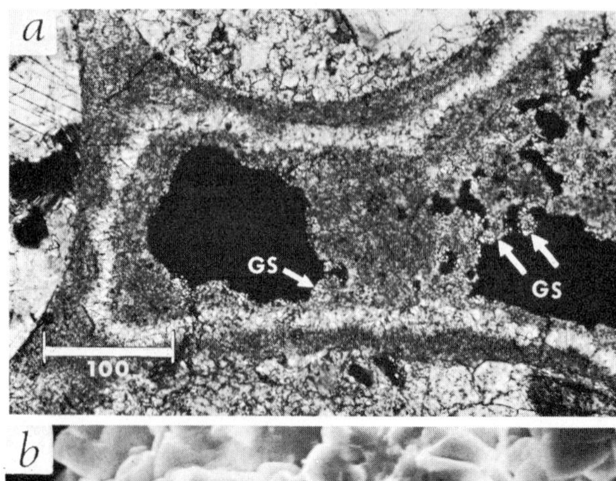

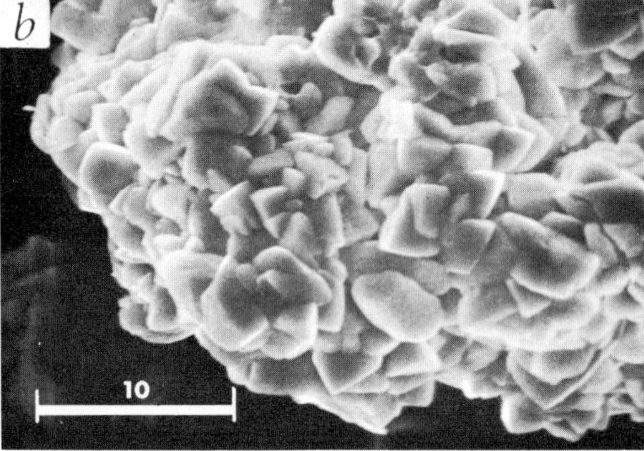

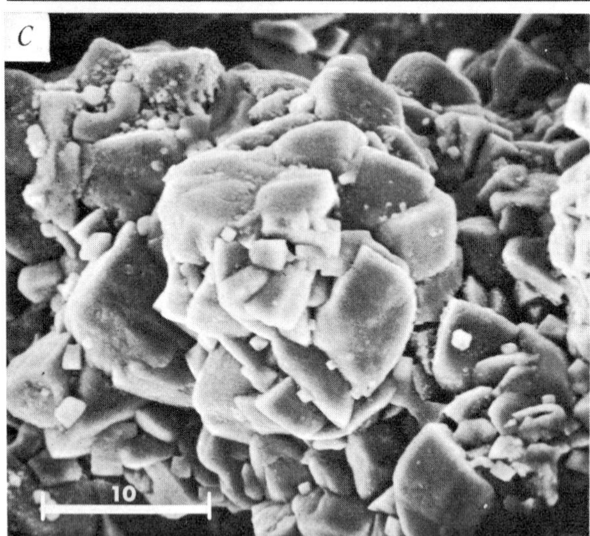

Figure 9. Growth structures in hard micrite. Beachrock from south of Alicante, Spain. *a*, Thin section, crossed polarizers. The light band is a calcite fringe, gray cement is micrite. The cement appeared in the order micrite-fringe-micrite. The latest micrite shows a framework of typical growth structures (GS) which extend into the void, but the record of such structures in the final cement is limited. Scale bar 100 μm. *b*, Scanning electron micrograph. Part of a branch-like growth structure which spans an interparticle void. Crystals are arranged in rosette-shaped growth centers; in thin sections these appear as radiating clusters. Scale bar 10μm. *c*, Scanning electron micrograph. Surface with a spectrum of crystal sizes. The upper size limit is approximately 8 μm; numerous small crystals, mostly nucleated at the junction between pre-existing crystal faces, indicate growth in progress. Scale bar 10μm.

associated organisms usually are obscure, several authors have suggested that the organisms play a vital part in the carbonate-generating processes, and that an organic influence is probable in most occurrences of shallow-water micrite.

Organic influence of this kind may depend on one or several of a series of mechanisms, and the extent of organic control may vary considerably from case to case. For example: (1) *Direct metabolic formation* of poorly organized skeletal material, such as protective or supporting tubes, capsules, *etc.* The organic control is complete; the organisms trigger precipitation, and they usually determine mineral polymorphic form, fabric and structure of the precipitate. (2) *Indirect triggering of precipitation* by metabolic activity, for instance, by algal photosynthesis or by microbial decomposition of organic matter. The organism activity is decisive for precipitation, but inorganic factors in the environment may control mineral polymorph, fabric and structure of the precipitate. (3) *Influence of organic substances* on the crystallization of carbonates. Organic substances effective in this connexion are the proteinaceous matrices of shells and other skeletal parts (Wilbur and Watabe, 1963), and soluble organic compounds such as amino acids, peptides, glycoprotein, *etc.* (Kitano and Hood, 1965). The substances are not the cause of precipitation, but in a sedimentary environment where crystal growth is in progress they may affect the polymorphic crystal form and the growth habit of crystals, and thereby also the fabric of a precipitate.

There is no obvious connexion between organisms and micrite in the Mediterranean beachrock. The organic content of the cement is very low; staining with Methylene blue on slightly etched surfaces affects only the scarce organisms, but there is no general staining of the uniform micrite, nor any preferential staining of the dark centers in the pelletoid micrite. The central parts of the pellet-like globules are dark in transmitted light only, and not when viewed in incident light on fractured or polished surfaces; they are resolved into a micromosaic of clear crystals by successive grinding of an ultra-thin section. The dark tint seen in ordinary thin sections is assumed to be an effect of the stacking of many crystallites.

From the scanning electron micrographs it is evident that there is no organic mucus in the pore space; no specimens figured in this paper were treated with, for example, sodium hypochlorite, hydrogen peroxide, or similar solutions in order to remove organic substances.

Both the internal ultra-structure of the pelletoid micrite, and the transition to uniform micrite cement are shown in scanning electron micrographs in Figure 11. A vertical section was cut from a beachrock sample

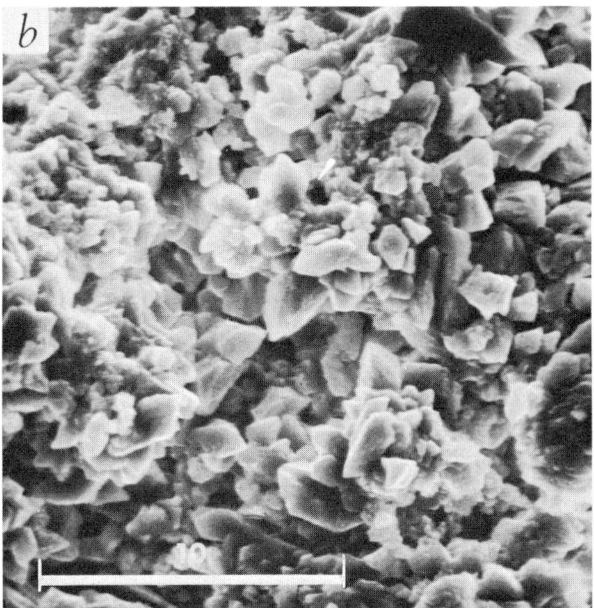

Figure 10. Scanning electron micrographs of hard micrite and friable micrite. *a*, Hard Mg-calcite micrite in intergranular position. A late growth stage; an indistinct framework is still visible although most of the pore space is filled by the rigid fabric. All crystals are firmly based on their substrate; apparently no material was precipitated in suspension and settled out of the liquid. Scale bar 10 μm. *b*, Friable Mg-calcite micrite from the specimen shown in Figure 6 *b*. The crystal size is generally less than 2 μm and the shape varies. Irregular radiating aggregates are common. The interparticle bonding is weak; material tends to spall off under the electron beam in the microscope. Scale bar 10 μm.

with well-developed pellet-like globules, and the surface was polished and then cleaned in ultra-sound. Small pieces were etched in buffered EDTA-solution at pH 7 for 2 minutes to 2 hours, gently rinsed in distilled water and alcohol, air-dried and mounted on specimen holders. The activity of EDTA is very gentle, and where the crystalline material is removed any organic tissues, algal filaments, *etc.* are left in relief above the surface; any water-soluble compounds, however, go into solution.

Neither the pelletoid micrite, nor the uniform micrite shows any internal organization that suggests an organic origin or a biological function, and the crystalline fabric contains no organic matrix. Algal filaments, small calcareous tubes, and empty borings occur here and there in the calcite, but they are not related to the growth structures of the cement. The transition from pellet-like globules to uniform micrite is evidently a result of continued nucleation and growth of crystals on exposed surfaces in the open framework, and there is virtually no variation in crystal morphology during this process.

Thus the field and laboratory data of this study do not indicate that direct metabolic calcification, and organism-triggering of precipitation are important processes for the cementation of beachrock. However, this material is characterised by a correspondence in mineralogy between the dominating skeletal carbonate in the sands, and the cement (Figures 5*a* and 7), and this correspondence may reflect an indirect organic influence on the mineral form of the cement. The skeletal grains contain an organic matrix, principally a protein-polysaccharide complex, which is decisive for the mineralogy and growth of crystals during the metabolic calcification (Wilbur and Watabe, 1963). The organic matrices of the biogenic carbonate assemblage in a beach sand might influence the mineral form of the cement, provided that (1) organic compounds from the skeletal matrices migrate through the deposit in the pore solutions, and (2) the migrating substances retain the ability of the matrices to induce formation of a particular mineral polymorph during precipitation. A biogenic Mg-calcite environment should then favor inorganic precipitation of Mg-calcite, and a biogenic aragonite environment should favor precipitation of aragonite.

The natural relationships are doubtless less simple than those outlined above but it is interesting to note that aragonite dominates in the West Indian sands described here, and it is also the form of the associated beachrock cement (Figures 5*b* and 8); there is also a correspondence in mineralogy between organic framework and cement in the cemented algal cup reefs of Bermuda (Ginsburg *et al.*, 1967), and in the cemented sublittoral algal crust in the Mediterranean (Alexandersson, 1969). However, there seems to be no similar correspondence in the lithified coral reefs off Jamaica, which are cemented by Mg-calcite although the frame is mainly aragonite (Land and Goreau, 1970).

Fringes

In thin sections, fringes appear as bands of clear, birefringent crystals which rim present and former voids. The material is a post-depositional addition and has the same Mg-content as micrite, but it contains no extraneous particles which correspond to the inclusions in the micrite. Some rocks are cemented by a fringe only (Figure 12), others by micrite only, but associations of the two forms of cement are common. Repeated changes between fringe- and micrite-cementation are recorded in many sequences (Figure 9*a*). Where fringes occur completely embedded in micrite, there are no signs to indicate that they formed through recrystallization of micrite cement.

Individual crystals are rod- or wedge-shaped, typically 8 to 12 μm in length and with a length/width ratio of 3 to 4. Where they terminate in an open pore space, crystal terminations are acute and fresh, and faces show no indication of leaching or dissolution (Figures 13 and 14).

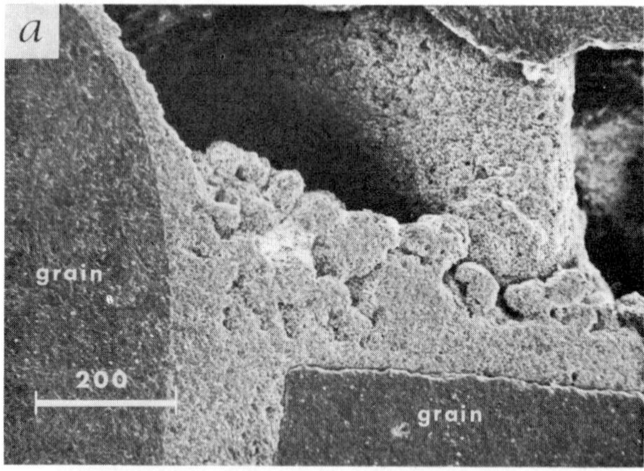

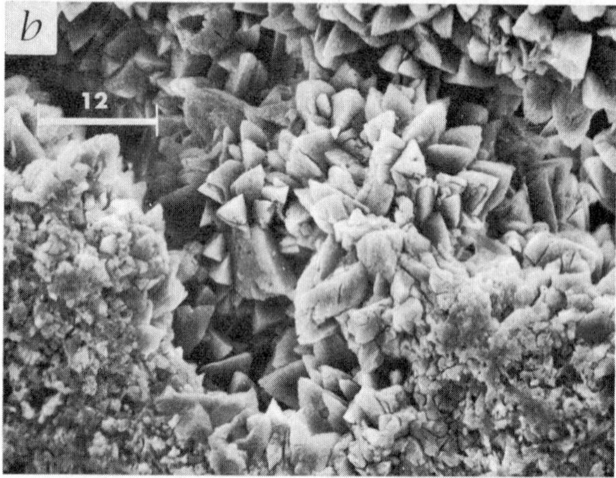

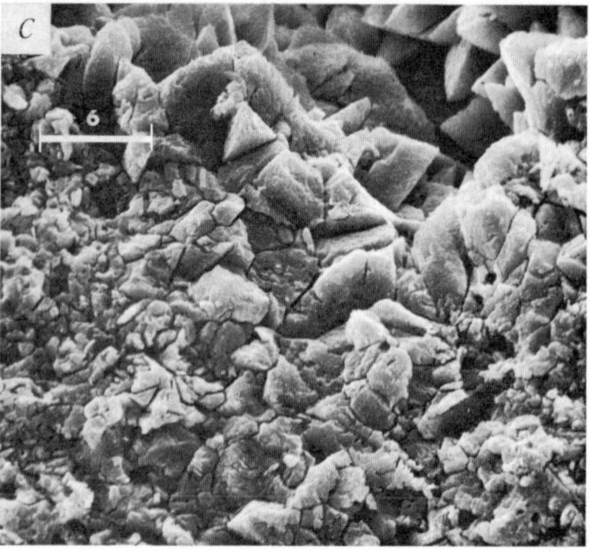

Figure 11. Internal ultrastructure of pelletoid and uniform hard micrite. Scanning electron micrographs; polished sections, 2 minutes etching in EDTA. Beachrock from Rhodes, Greece. *a*, The section includes three sand grains; in the background a fourth grain is coated by micrite. Pellet-like globules of variable size grade into uniform micrite without any real change of fabric. Organisms, internal structures, etc. do not occur. Scale bar = 200 μm. *b*, Pellet-like globules, partly sectioned. The interior of the globules consists of a dense micro-fabric of irregular crystallites. Scale bar = 12 μm. *c*, Pellet-like globule, mainly sectioned. The crystals are divided into subunits by the etching process; there is no water-insoluble organic matrix left after removal of the dissolved calcite. Scale bar = 6 μm.

Growth Characteristics of Fringes

(1) *Distribution is not affected by substrate properties.* All surfaces exposed in the pore system of a sediment are not equally suitable as substrates for crystal nucleation. The substrate influence depends on such factors as surface adsorption and lattice matching between substrate and precipitate (Walton, 1967); parameters which vary considerably from grain to grain in most sediments. Nevertheless, substrate effects on the distribution of fringes are virtually absent.

(2) *There is no lattice continuity between crystals in fringe and substrate.* From energy considerations, continued growth of pre-existing crystals corresponding to the precipitate is a more favorable process than nucleation of new crystals, and optically continuous overgrowths are well known from lithified sediments. Such continuity is absent in the present material, and it appears that this is the rule in marine lithification (*cf.* Milliman, 1966; Fischer and Garrison, 1967; Friedman, 1968).

(3) *Crystals are oriented.* Oriented crystal growth

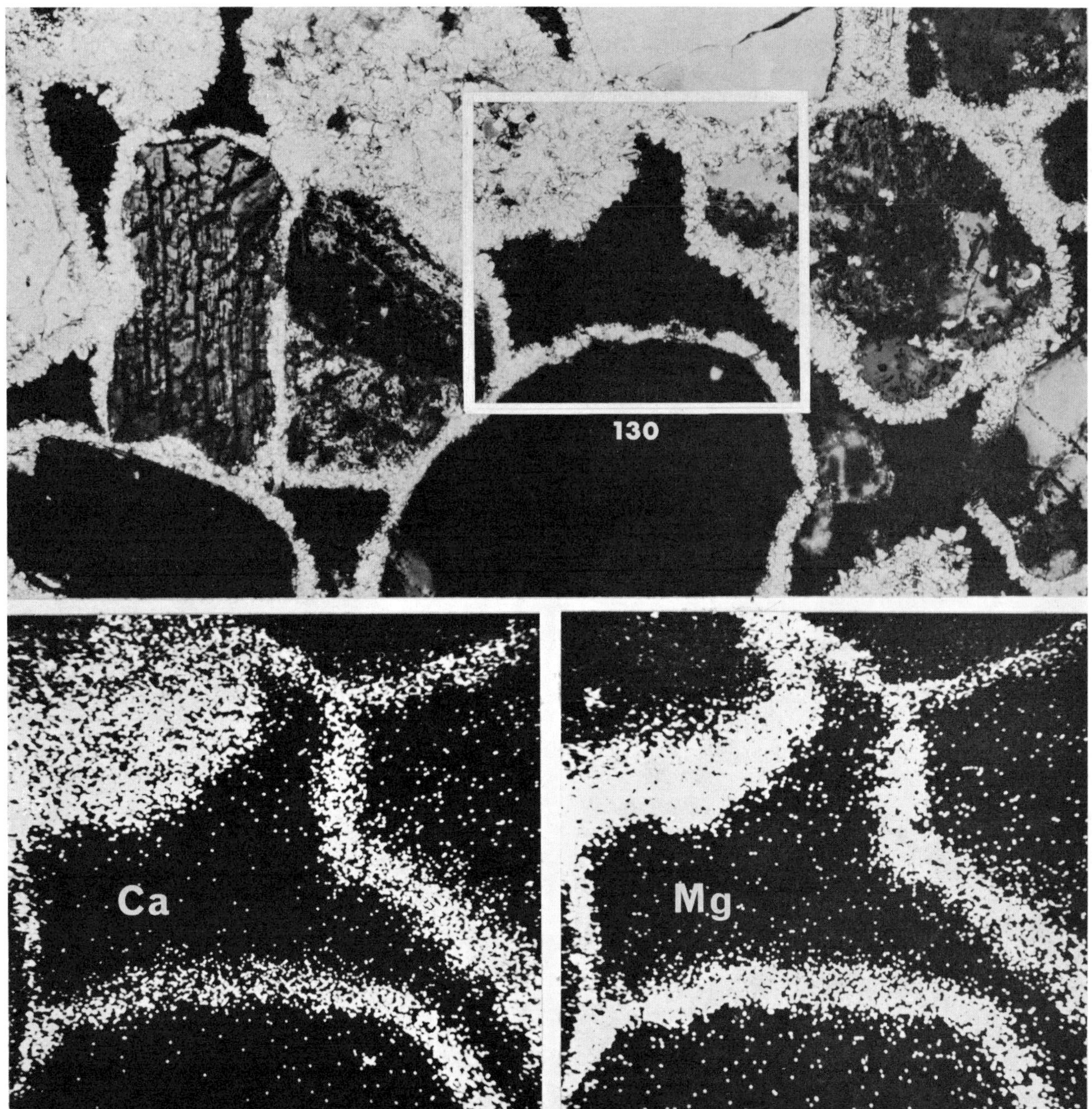

Figure 12. Distribution of Mg and Ca in fringe-cemented beachrock, Cyprus. Micrograph of thin section, under crossed polarizers, together with electron beam-scanning images of the same field showing distribution of Ca and Mg. The square indicates the area scanned; side of square is 130 μm. The grain in the upper left part of the scanned area is low-Mg calcite, and Mg is present in the encrusting fringe only. No micrite occurs in this rock (field conditions are indicated in Figure 2).

may be due to a saturation gradient in the pore fluid, to rapid growth of favorably oriented seeds in a random arrangement, or to epitaxy, the two-dimensional lattice accord between substrate and overgrowth (Walton, 1967). Presumably epitaxial control is active in the case of parallel growth which leads to fibrous crystals, normal to the growth surface (Figure 14*a*). Commonly, however, the orientation is less distinct, comparable to the growth of aragonite needles (*cf.* Figures 8, 13 and 14*b*); in this case more than one mechanism may be active.

(4) *Fringes are isopachous.* The uniform thickness is a property stressed by Land and Goreau (1970). Simultaneous crystal nucleation throughout the entire fringe-cemented deposit and subsequent slow precipitation from a homogeneous solution should lead to this kind of fabric (see Walton, 1967; Krauskopf, 1967). A spectrum of crystal sizes should appear if

nucleation took place successively during the growth period (*cf.* Figure 9*c*). However, the crystals in a fringe are almost identical in size.

(5) *Fringe thickness is limited.* The usual thickness of single-layered Mg-calcite fringes in the present material is 10 to 20 μm, exceptionally 30 to 40 μm. Thicker fringes consist of more than one layer of crystals. The Mg-calcite fringe figured by Land and Goreau (1970, page 459) is about 20 μm thick. It seems possible that there exists an upper size limit for Mg-calcite crystals under these conditions; to the author's knowledge a coarse Mg-calcite fabric has not been reported from the marine environment.

Dolomite Fringe

In a beachrock sample from a water depth of 2 m near Kamiros on the western coast of Rhodes, the fringe consists of calcian dolomite with the composition $(Ca_{61}Mg_{36}Fe_{03})CO_3$. The rest of the pore space is filled by Mg-calcite micrite with 13 mole % $MgCO_3$ in solid solution (Figure 15). The sediment grains consist mostly of low-Mg calcite and various silicate minerals, a deposit which closely corresponds to the sand on the present beach. Beachrock cement from the beach has the same fabric of fringes and micrite, but this cement is the usual Mg-calcite and no dolomite occurs.

The association indicates selective dolomitization, preferentially affecting the calcite fringes and leaving unaffected the grain component of the rock as well as the micrite cement. Dolomitization by exsolution of Mg-calcites (Land, 1967) is obviously not the mechanism in the present case since the Mg-content of the surrounding calcite is unchanged; nor is the fabric in accordance with that process which results in dolomite crystals scattered in a recrystallized matrix (Schlanger, 1957). The Mg source must be outside the sediment. However, the retention of the lattice orientation of the host is in agreement with the observations by Land (1967). He also found that a coarse-crystalline Mg-calcite from echinoderms was more prone to dolomitization than a micro-crystalline fabric from skeletons of red algae, in spite of the higher Mg-content in the algae.

Selective dolomitization of micritic envelopes on skeletal carbonates in the Miocene Ziqlag reef in Israel has been described by Buchbinder and Friedman (1970).

DISCUSSION

Distribution, Age and Origin

The occurrence of beachrock at the present sea level over great parts of the Mediterranean suggests that the lithifying process is active today. Diving observa-

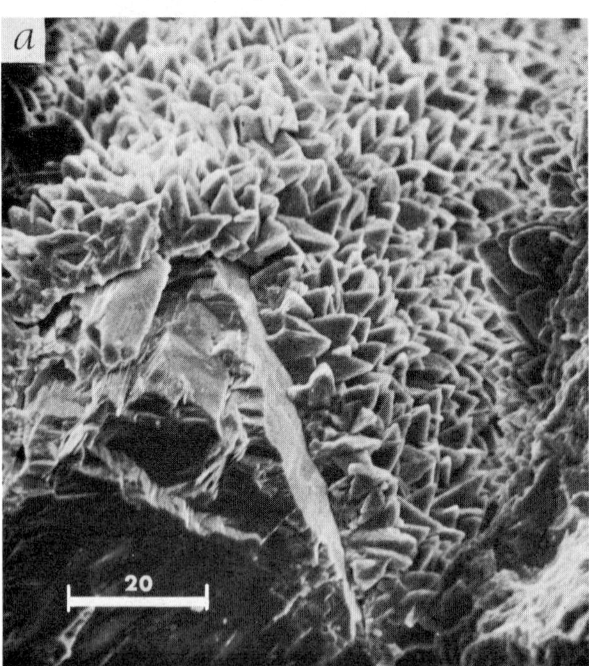

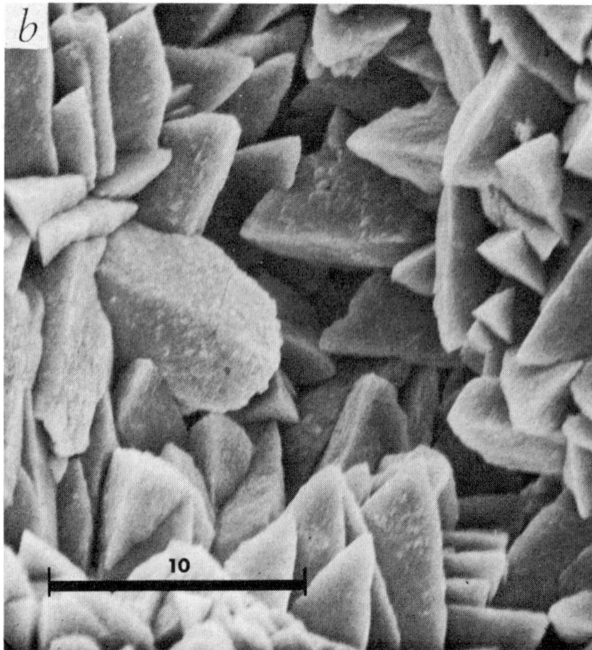

Figure 13. Scanning electron micrographs of calcite fringe. *a*, Sediment grains are covered by a continuous coat of Mg-calcite crystals. Under natural conditions the pore fluid has virtually no contact with the original grain component. Scale bar 20 μm. *b*, Microrelief of fringe surface. The crystals in this picture are based on three detrital grain surfaces, to the left, at the bottom, and to the right. Note junction of crystals in upper centre. Scale bar 10 μm.

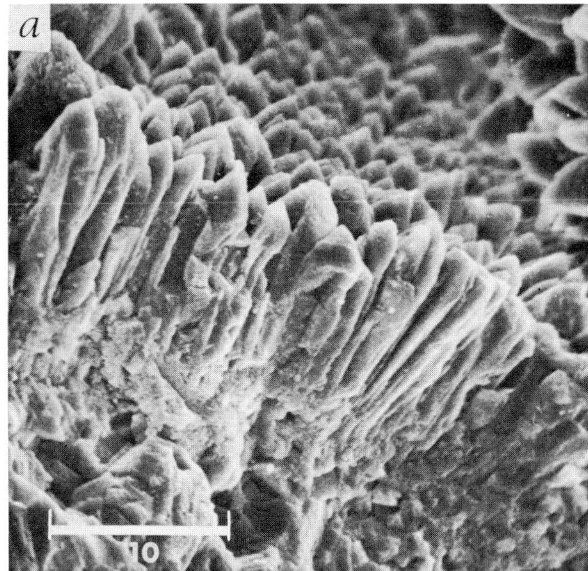

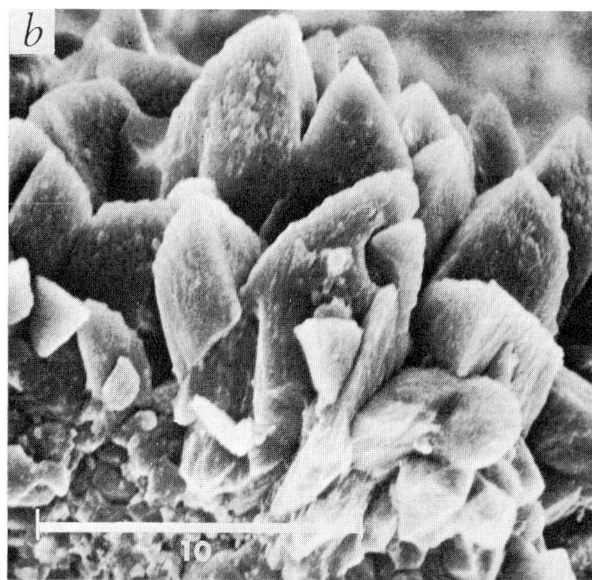

Figure 14. Orientation of crystals in Mg-calcite fringe. Scanning electron micrographs of broken surfaces. *a*, Parallel, elongated crystals with uniform orientation normal to the covered surface. In all observed cases fibrous crystals have this perpendicular orientation. Scale bar 10 μm. *b*, Inclined growth of fringing crystals. In the Mediterranean material this is the most common orientation of fringe cement. Note cavity caused by high-energy face in crystal junction at right centre (*cf.* Figure 13*b*). Scale bar 10 μm.

tions show that beachrock also occurs in offshore positions, in places forming a series of layers which probably represent repeated beachrock cementation during a rise in sea level.

The close connection between beachrock exposures and degrading beaches implies that at least the initial cementation takes place within the deposit and not at the beach surface. Scanning electron microscopy supports this opinion since the crystals in cement from recently exposed rock-ledges are clean and fresh, and have not been attacked by algal and animal borings. Microscopic borings are common in material from outcrops where field conditions indicate a long period of exposure on the shore.

Beachrock at the shoreline shows no evidence of physico-chemical erosion; on the contrary, sand trapped in pockets in the surface usually becomes cemented by secondary calcite cement. In the littoral zone abrasion by wave action is the important erosive process; in the sublittoral zone various boring and destructive organisms, such as sponges, polychaete worms, echinoids, and limpets, contribute to the erosion.

The mineral composition of the lithified deposits varies from predominantly terrigenous sands with 20 to 25 wt. % total carbonate, to almost pure carbonate sediments where detrital low-Mg calcite is the main mineral. The interstitial physico-chemical conditions may vary considerably in a single beach, both spatially and through time (Schmalz, 1969), and there must be still greater potential for variation between a number of beaches where the mineral assemblages, the organism content, and to some extent also the climatic factors are different. Nevertheless, the composition of the Mg-calcite cement is almost uniform over a distance of 3,500 km, and the Mg-content varies only a few mole % around a mode of 14 mole % $MgCO_3$ in solid solution.

Approximately the same Mg-content occurs in a number of marine calcites from various areas and depths (Fischer and Garrison, 1967; Purdy, 1968; Winland, 1968; Taylor and Illing, 1969; de Groot, 1969; Land and Goreau 1970). Moreover the Mg-calcite cement which lithifies Mediterranean algal frameworks in the sublittoral zone down to a water depth of at least 20 m has about 14 mole % $MgCO_3$ in solid solution (Alexandersson, 1969).

The uniform composition of the beachrock cement, and the close similarity between this kind of cement and various calcites from the marine environment, indicates that the beachrock cement is of marine origin. However, this does not necessarily mean that the fluids at the locus of cementation correspond exactly to normal seawater. On the contrary, variations in salinity, temperature, pH, carbonate saturation level, etc. almost certainly occur in connection with the interstitial transport.

The mechanism behind the lithifying process is not known. Neither the petrographic microscope nor the scanning electron microscope reveals any organic structures, and in the general sense of the word the process probably is inorganic. The reason for the

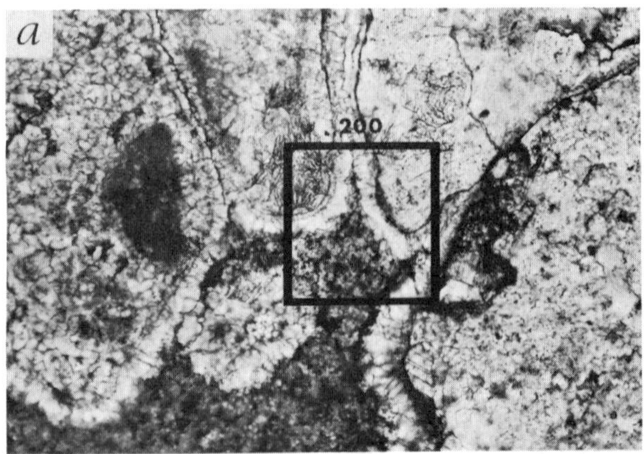

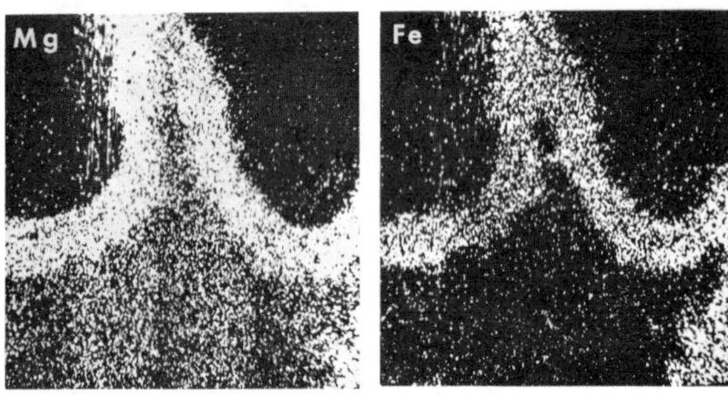

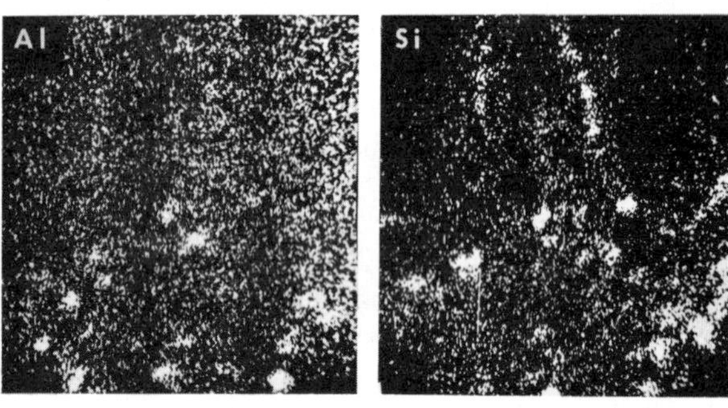

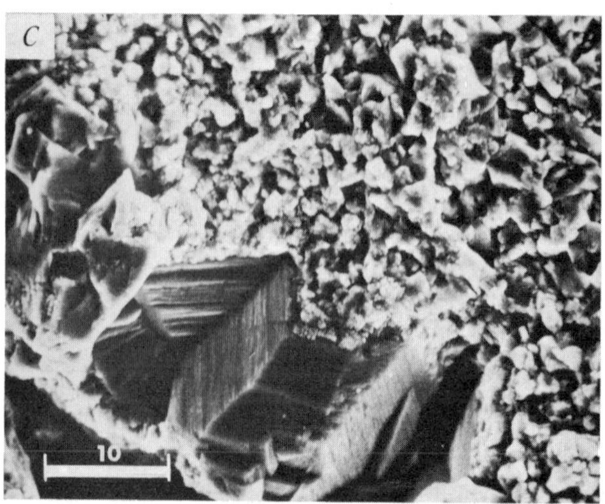

main variation in beachrock cement property, namely the variation between aragonite and Mg-calcite, also remains unknown.

By variations in the growth mechanism at least three kinds of Mg-calcite fabric are formed: hard micrite, friable micrite, and fringe cement. The small foreign particles which occur as inclusions in most micrite cements indicate rapid interstitial circulation during micrite growth. Fringe characteristics suggest slow precipitation from a homogeneous solution.

Mg-calcite micrite is the fabric generally found in connection with marine precipitation of calcite; it is supposed to form only under marine conditions (Friedman, 1968).

By analogy with laboratory results (Glover and Sippel, 1967) and field observations (Purdy, 1968; Land and Goreau, 1970), a temperature dependence leading to higher Mg-content in calcite from warmer waters might be expected. The difference in temperature between Gibraltar and Cyprus waters is 3° to 5°C and should therefore lead to an increase in Mg-content in an easterly direction. However, local variations in the present material are 2 to 3 mole% $MgCO_3$ and the number of analyses is not sufficient to reveal any regional pattern.

Aragonite does not occur in the cement, only in thin encrusting scales occasionally found in the spray-zone. Such scales postdate the cement; there are no features indicating that Mg-calcite cement was formed by recrystallization of aragonite at any stage in the lithifying process.

Mediterranean Aragonite and Mg-Calcite

Those factors which control the mineralogical form of marine carbonate precipitates are not well understood, but it seems probable that both regional and local variations occur. Gavish and Friedman (1969) report aragonite and Mg-calcite cement in beachrock from the Mediterranean coast of Israel; a combination also found in the Red Sea (Friedman, 1968) and in the Persian Gulf (Taylor and Illing, 1969). In general, however, precipitation of aragonite appears to be

Figure 15. Dolomite fringe in sublittoral beachrock, Rhodes. *a*, Micrograph of aluminium-covered thin section, crossed polarizers. The dark intergranular cement is Mg-calcite micrite with 13 mole % $MgCO_3$; the fringe is calcian dolomite of the composition $(Ca_{.61}Mg_{.36}Fe_{.03})CO_3$. Square in centre scanned for distribution of elements; side of square is 200 μm. *b*, Distribution of respectively Mg, Fe, Al and Si. High concentrations of Al and Si in the micrite are trapped terrigenous particles. *c*, Scanning electron micrograph, cut and polished surface treated 5 minutes in ultrasound. A single row of dolomite crystals is based on a grain surface in the lower left corner; the fine-grained material is micritic Mg-calcite. Scale bar 10 μm.

a subordinate process in the Mediterranean at the present time. The aragonite oöids in the Gulf of Gabes, Tunisia, are reworked material, derived from coastal Holocene oölites; on the basis of radiocarbon dates it is assumed that oöid growth ceased about 5,000 years ago (Fabricius *et al.*, 1970).

Variation through time of the mineral form of beachrock cement cannot be excluded. In regions such as the Bahaman–West Indian area, where calcium carbonate is precipitated at the present time in the form of aragonite oöids and/or aragonite mud, beachrock is usually cemented by aragonite (Ginsburg 1957; Stoddart and Cann, 1965; this paper, Figure 8). Under the conditions which obtained when aragonite oöids formed in the Mediterranean, aragonite also may have been the main cementing mineral in Mediterranean beachrock.

In submerged beachrock, a few thousand years old, from the North Adriatic Sea the cement "is beautifully crystallized in long, thin needles grown perpendicular to the grain surface and elongated towards the intergranular space" (Stefanon, 1969, p. 82); a fabric which strongly suggests aragonite. On the opposite side of the Italian peninsula, along the Tuscan and Ligurian coast, modern intertidal beachrock is cemented by Mg-calcite (Bloch and Trichet, 1966). Apparent regional variations of that kind most probably reflect variations through time.

Diagenesis

The Mg-calcites are metastable because of their content of randomly scattered Mg ions. Stabilizing reactions in the diagenetic course of events most commonly affect the Mg-content, and lead to formation of stable low-Mg calcite and/or dolomite. (1) In subaerial positions fresh water alterations result in depletion in Mg, and Mg-calcites change rapidly to low-Mg calcite without any accompanying textural changes (Friedman, 1964; Gavish and Friedman, 1969). Detailed knowledge of these processes is lacking. (2) Under certain conditions, the change proceeds as an exsolution of Mg-calcite, resulting in a mixture of stable low-Mg calcite and dolomite (Schlanger, 1957; Land, 1967). No external source for Mg is required; initially the ions occur in solid solution in the calcite. (3) An increase in total Mg-content is found where Mg ions are supplied from an external source. The process leads to dolomitization by replacement, and eventually the primary carbonate fabric may be completely replaced by dolomite. Under such conditions Mg-calcites seem to be more susceptible to dolomitization than low-Mg calcite and aragonite (Land, 1967; Buchbinder and Friedman, 1970).

In the beachrock material described here, evidence of post-cementation diagenetic changes is rare. The material does not include samples from subaerial positions, and therefore alterations due to fresh water are not found. Stabilization by exsolution is also lacking.

As indicated by the selective dolomitization of fringe cement in a beachrock sequence from Rhodes, dolomitization by replacement may take place in the sublittoral zone.

SUMMARY

Mg-calcite with 13 to 16 mole% $MgCO_3$ in solid solution forms the cement in modern Mediterranean beachrock. The uniform composition of the beachrock cement, and the similarity between this cement and various marine calcites, suggests that seawater is the source for the cement. The Mg-calcite is a direct precipitate; recrystallization of aragonite is not involved in the lithifying process. Variations in growth mechanism cause differences in fabric, but the mineral form and the elemental composition is not affected. The distribution constant for partitioning Mg between calcite and seawater is about 0.03 under the reported conditions. It appears that shallow submarine diagenesis of the resultant rock may lead to further enrichment of Mg in the cement, and eventually to the formation of some dolomites.

ACKNOWLEDGEMENTS

For discussion, encouragement and help at various stages during this study I thank Dr. Valdar Jaanusson, Stockholm, Prof. Richard Reyment, Uppsala, and Prof. Ivar Hessland, Stockholm. I am most grateful to the staff at the Department of Historical Geology, Uppsala, for competent technical assistance; special thanks go to Mr. Ulf Sturesson who made the atomic absorption analyses, and to Mr. Gustav Andersson and Mrs. Dagmar Engström who prepared the illustrations. Mrs. Karin Landgren made the French translation; the electron microprobe work was done by Mrs. Marianne Dahl at the Department of Mineralogy, Uppsala. The Scandinavian branch of JEOLCO kindly gave me access to a scanning electron microscope. I will always remember the kind and friendly people I met in the field, and particularly Mr. Yiannis Cleanthous in Kyrenia.

Part of the program was sponsored by the Swedish Natural Science Research Council under NFR contract 3045–2. Travel was supported by generous grants from the Anna Maria Lundin Foundation.

REFERENCES

Alderman, A. R. 1959. Aspects of carbonate sedimentation. *Journal of the Geological Society of Australia*, 6:1–10.

Alderman, A. R. and H. C. Skinner 1957. Dolomite sedimentation in the southeast of South Australia. *American Journal of Science*, 255:561–567.

Alderman, A. R. and C. C. von der Borch 1963. A dolomite reaction series. *Nature*, 198:465–466.

Alexandersson, E. T. 1969. Recent littoral and sublittoral high-Mg calcite lithification in the Mediterranean. *Sedimentology*, 12: 47–61.

Alexandersson, E. T. in press. Diving observations of sedimentological processes. *Confédération Mondiale des Activités Subaquatiques: Premier Symposium Comité Scientifique.*

Barnes, C. R. 1965. Probable spur-and-groove structures in Middle Ordovician limestone, near Ottawa, Canada. *Journal of Sedimentary Petrology*, 35:257–261.

Bathurst, R. G. C. 1964. The replacement of aragonite by calcite in the molluscan shell wall. In: *Approaches to Paleoecology*, eds. Imbrie, J. and N. Newell, John Wiley and Sons, New York, 357–376.

Bathurst, R. G. C. 1966. Boring algae, micrite envelopes, and lithification of molluscan biosparites. *Liverpool and Manchester Geological Journal*, 5:15–32.

Bloch, J. P. and J. Trichet 1966. Un example de grès de plage (Côte Ligure italienne). *Marine Geology*, 4:373–377.

Boekschoten, G. J. 1962. Beachrock at Limani Chersonisos, Crete. *Geologie en Mijnbouw*, 41:3–7.

Boekschoten, G. J. 1963. Some geological observations on the coasts of Crete. *Geologie en Mijnbouw*, 42:241–247.

Bruns, E. 1958. *Ozeanologie*. Deutscher Verlag der Wissenschaften, Berlin, 365 p.

Buchbinder, B. and G. M. Friedman 1970. Selective dolomitization of micrite envelopes: a possible clue to original mineralogy. *Journal of Sedimentary Petrology*, 40:514–517.

Chave, K. E. 1952. A solid solution between calcite and dolomite. *Journal of Geology*, 60:190–192.

Chave, K. E. 1954. Aspects of the biogeochemistry of magnesium. *Journal of Geology*, 62: 266–283. and 587–599.

Chave, K. E. 1962. Factors influencing the mineralogy of carbonate sediments. *Limnology and Oceanography*, 7:218–223.

Cooray, P. G. 1968. A note on the occurrence of beachrock along the west coast of Ceylon. *Journal of Sedimentary Petrology*, 38: 650–654.

Dunham, R. J. 1969. Early vadose silt in Townsend Mound (Reef), New Mexico. In: *Depositional Environments in Carbonate Rocks*, ed, Friedman, G. M., Society of Economic Paleontologists and Mineralogists, *Special Publication*, 14:139–181.

Emery, K. O. and D. Neev 1960. Mediterranean beaches of Israel. *Geological Survey of Israel Bulletin*, 26:1–24.

Fabricius, F. H., D. Berdau and K. O. Münnich 1970. Early Holocene oöids in modern littoral sands reworked from a coastal terrace, southern Tunisia. *Science*, 169:757–760.

Fevret, M. and P. Sanlanville 1965. Contribution à l'étude du littoral libanais. *Méditerranée*, 6:113.

Fischer, A. G. and R. E. Garrison 1967. Carbonate lithification on the sea-floor. *Journal of Geology*, 75:488–496.

Folk, R. L. 1959. Practical petrographic classification of limestones. *Bulletin of the American Association of Petroleum Geologists*, 43:1–38.

Folk, R. L. 1962. Spectral subdivision of limestone types. In: *Classification of Carbonate Rocks*, ed. Ham, W. E., American Association of Petroleum Geologists, Tulsa, Oklahoma, 1: 62–84.

Folk, R. L. 1965. Some aspects of recrystallization in ancient limestones. In: *Dolomitization and Limestone Diagenesis*, eds. Pray, L. C. and R. C. Murray, Society of Economic Paleontologists and Mineralogists Special Publication, 13:14–48.

Friedman, G. M. 1959. Identification of carbonate minerals by staining methods. *Journal of Sedimentary Petrology*, 29:87–97.

Friedman, G. M. 1964. Early diagenesis and lithification in carbonate sediments. *Journal of Sedimentary Petrology*, 34:777–813.

Friedman, G. M. 1968. The fabric of carbonate cement and matrix and its dependence on the salinity of water. In: *Recent Developments in Carbonate Sedimentology in Central Europe*, eds. Müller, G. and G. M. Friedman, Springer Verlag, Berlin, 11–20.

Gavish, E. and G. M. Friedman 1969. Progressive diagenesis in Quaternary to Late Tertiary carbonate sediments: sequence and time scale. *Journal of Sedimentary Petrology*, 39: 980–1006.

Gevirtz, J. and G. M. Friedman 1966. Deep-sea carbonate sediments in the Red Sea and their implications on marine lithification. *Journal of Sedimentary Petrology*, 36:143–151.

Ginsburg, R. N. 1957. Early diagenesis and lithification of shallow-water carbonate sediments in South Florida. In: *Regional Aspects of Carbonate Deposition*, eds. Le Blanc, R. J. and J. G. Breeding, Society of Economic Paleontologists and Mineralogists Special Publication 5:80–100.

Ginsburg, R. N., E. A. Shinn and J. Schroeder 1967. Submarine cementation and internal sedimentation within Bermuda reefs. *Abstracts for 1967: Geological Society of America, Special Paper*, 115:78.

Glover, E. D. and R. F. Sippel 1967. Synthesis of magnesium calcites. *Geochimica et Cosmochimica Acta*, 31:603–613.

Goldsmith, J. R. and D. L. Graf 1958. Relation between lattice constants and composition of the Ca-Mg carbonates. *American Mineralogist*, 43:84–101.

Goldsmith, J. R., D. L. Graf and O. I. Joensuu 1955. The occurrence of magnesium calcites in nature. *Geochimica et Cosmochimica Acta*, 7: 212–230.

Goudie, A. 1969. A note on Mediterranean beachrock: its history. *Atoll Research Bulletin*, 126:11–14.

de Groot, K. 1969. The chemistry of submarine cement formation at Dohat Hussain in the Persian Gulf. *Sedimentology*, 12:63–68.

de Groot, K. and E. M. Duyvis 1966. Crystal form of precipitated calcium carbonate as influenced by adsorbed magnesium ions. *Nature*, 212: 183–184.

Gross, M. G. 1965. Carbonate deposits on Plantagenet Bank near Bermuda. *Geological Society of America, Bulletin*, 76:1283–1290.

Hay, R. L. and A. Iijima 1968. Nature and origin of palagonite tuffs of the Honolulu Group on Oahu, Hawaii. *Geological Society of America, Memoir*, 116:331–376.

Holland, H. D., T. V. Kirsipu, J. S. Huebner and U. M. Oxburgh 1964. On some aspects of the chemical evolution of cave waters. *Journal of Geology*, 72:36–37.

Hydrographer of the Navy 1969. Admiralty tide tables I. *British Admiralty Publication No. 200.*

Illing, L. V. 1954. Bahaman calcareous sands. *Bulletin of the American Association of Petroleum Geologists*, 38: 1–95.

Ingerson, E. 1962. Problems of the geochemistry of sedimentary carbonate rocks. *Geochimica et Cosmochimica Acta*, 26:815–847.

Kendall, C. G. St. C. and P. A. d'E. Skipwith 1969. Holocene shallow-water carbonate and evaporite sediments of Khor al Bazam, Abu Dhabi, Southwest Persian Gulf. *Bulletin of the American Association of Petroleum Geologists*, 53: 841–869.

Kinsman, D. J. J. 1965. Coprecipitation of Sr^{2+} with aragonite from sea water at 15–95°C. (abstract). *Geological Society of America, Special Paper*, 87:88 pp.

Kitano, Y. and D. W. Hood 1965. The influence of organic material on the polymorphic crystallization of calcium carbonate. *Geochimica et Cosmochimica Acta*, 29:29–41.

Kitano, Y., K. Park and D. W. Hood 1962. Pure aragonite synthesis. *Journal of Geophysical Research*, 67:4873–4874.

Krauskopf, K. B. 1967. *Introduction to Geochemistry.* McGraw-Hill, New York, 721 p.

Krauss, R. W. and R. A. Galloway 1960. The role of algae in the formation of beachrock in certain islands of the Caribbean. *Coastal Studies Institute, Baton Rouge, Technical Report No. 11 E.*

Land, L. S. 1967. Diagenesis of skeletal carbonates. *Journal of Sedimentary Petrology,* 37:914–930.

Land, L. S. and T. F. Goreau 1970. Submarine lithification of Jamaican reefs. *Journal of Sedimentary Petrology,* 40:457–462.

Lippman, F. 1960. Versuche zur Aufklärung der Bildungsbedingungen von Kalzit und Aragonit. *Fortschritte der Mineralogie,* 38:156–160.

Lucia, F.J. 1968. Recent sediments and diagenesis of South Bonaire, Netherlands Antilles. *Journal of Sedimentary Petrology,* 38: 845–858.

Mabesoone, J. M. 1963. Coastal sediments and coastal development near Cadiz (Spain). *Geologie en Mijnbouw,* 42:29–43.

Macintyre, I. G., E. W. Mountjoy and B. F. D'Anglejan 1968. An occurrence of submarine cementation of carbonate sediments off the west coast of Barbados, W. I. *Journal of Sedimentary Petrology,* 38:660–663.

Malone, PH. G. and K. M. Towe 1970. Microbial carbonate and phosphate precipitates from sea water cultures. *Marine Geology,* 9:301–309.

McCauley, J. W. and R. Roy 1966. Evidence for epitaxial control of $CaCO_3$ phase formation as the mechanism of the influence of impurity ions. *Transactions of the American Geophysical Union,* 47: 202–203.

McIntire, W. L. 1963. Trace element partition coefficients—a review of theory and applications to geology. *Geochimica et Cosmochimica Acta,* 27:1209–1264.

McLean, R. F. 1967. Origin and development of ridge-furrow systems in beachrock in Barbados, West Indies. *Marine Geology,* 5:181–193.

Milliman, J. D. 1966. Submarine lithification of deep water carbonate sediments. *Science,* 153: 994.

Milliman, J. D., D. A. Ross and T.-L. Ku 1969. Precipitation and lithification of deep-sea carbonates in the Red Sea. *Journal of Sedimentary Petrology,* 39:724–736.

Mistardis, G. 1963. On the beachrock of southern Greece. *Deltion, Ellenikes Geologikes Etairas, Athens.*

Mistardis, G. 1964. Shoreline displacements and sea level changes during the Middle-Upper Quaternary. *20th International Geographical Congress, Abstract of Papers Supplement,* 21–22.

Monaghan, P. H. and M. L. Lytle 1956. The origin of calcareous oöliths. *Journal of Sedimentary Petrology,* 26:111–118.

Müller, G. 1970. High-magnesian calcite and protodolomite in Lake Balaton (Hungary) sediments. *Nature,* 226:749–750.

Neev, D. and K. O. Emery 1967. The Dead Sea. Depositional processes and environments of evaporites. *Israel Geological Survey Bulletin,* 41, 147 pp.

Oxburgh, U. M., R. E. Segnit and H. D. Holland 1959. Coprecipitation of strontium with calcium carbonate from aqueous solutions (Abstract). *Geological Society of America Bulletin,* 70:1653–1654.

Pérès, J. M. 1967. The Mediterranean benthos. *Oceanography and Marine Biology. An Annual Review,* Allen & Unwin, London. 5: 449–533.

Purdy, E. G. 1968. Carbonate diagenesis: an environmental survey. *Geologica Romana,* 7:183–228.

Runnells, D. R. 1969. Diagenesis, chemical sediments, and the mixing of natural waters. *Journal of Sedimentary Petrology,* 39:1188–1201.

Runnells, D. R. 1970. Errors in X-ray analysis of carbonates due to solid-solution variation in composition of component minerals. *Journal of Sedimentary Petrology,* 40:1158–1166.

Russell, K. L., K. S. Deffeyes, G. A. Fowler and R. M. Lloyd 1967. Marine dolomite of unusual isotopic composition. *Science,* 155:189–191.

Russell, R. J. 1962. Origin of beachrock. *Zeitschrift für Geomorphologie,* 6:1–16.

Russell, R. J. 1963. Beach rock. *Journal of Tropical Geography,* 17:24–27.

Schattner, J. 1967. Geomorphology of the north coast of Israel. *Geografiska Annaler (A),* 49:310–320.

Schlanger, S. O. 1957. Dolomite growth in coralline algae. *Journal of Sedimentary Petrology,* 27:181–186.

Schmalz, R. F. 1969. Beachrock formation on Eniwetok Atoll. *Bermuda Biological Station Publication No. 3, Appendix, International Carbonate Cementation Seminar.*

Shinn, E. A. 1969. Submarine lithification of Holocene carbonate sediments in the Persian Gulf. *Sedimentology,* 12:109–144.

Simkiss, K. 1964. Variations in the crystalline form of calcium carbonate precipitated from artificial sea water. *Nature,* 201: 492–493.

Skinner, H. C. 1963. Precipitation of calcian dolomites and magnesian calcites in the southeast of South Australia. *American Journal of Science,* 261:449–472.

Stefanon, A. 1969. The role of beachrock in the study of the evolution of the North Adriatic Sea. *Memorie di Biogeografica Adriatica,* 8:79–87.

Stoddart, D. R. and J. R. Cann 1965. Nature and origin of beachrock. *Journal of Sedimentary Petrology,* 35: 243–247.

Taft, W. H. 1967. Physical chemistry of formation of carbonates. In: *Carbonate Rocks: Physical and Chemical Aspects,* eds. Chilingar, G. V., H. J. Bissell and R. W. Fairbridge, Elsevier, Amsterdam, 151–167.

Taillefer, F. 1964. Le grès de plage de Viransehir. *Revue Géographique de l'Est,* 4:393–398.

Taylor, J. C. M. and L. V. Illing 1969. Holocene intertidal calcium carbonate cementation, Qatar, Persian Gulf. *Sedimentology,* 12:69–108.

Usdowski, H. E. 1963. Der Rogenstein des norddeutschen Unteren Buntsandsteins, ein Kalkoolith des marinen Faziesbereichs. *Fortschritte der Geologie Rheinland und Westfalen,* 10:337–342.

von der Borch, C. C. 1965. The distribution and preliminary geochemistry of modern carbonate sediments of the Coorong Area, South Australia. *Geochimica et Cosmochimica Acta,* 29:781–800.

Walton, A. G. 1967. *The Formation and Properties of Precipitates.* Interscience, New York, 232 p.

Warne, S. 1962. A quick field or laboratory staining scheme for the differentiation of the major carbonate minerals. *Journal of Sedimentary Petrology,* 32:29–38.

Wilbur, K. M. and N. Watabe 1963. Experimental studies on calcification in molluscs and the alga *Coccolithus huxleyi. Annals of the New York Academy of Science,* 109 (1): 82–112.

Winland, H. D. 1968. The role of high-Mg calcite in the preservation of micrite envelopes and textural features of aragonite sediments. *Journal of Sedimentary Petrology,* 38:1320–1325.

Winland, H. D. 1969. Stability of calcium carbonate polymorphs in warm, shallow seawater. *Journal of Sedimentary Petrology,* 39:1579–1587.

Zeller, E. J. and J. L. Wray 1956. Factors influencing precipitation of calcium carbonate. *Bulletin of the American Association of Petroleum Geologists,* 40:140–152.

Observations sur la Sédimentation Bioclastique en Quelques Points de la Marge Continentale de la Méditerranée

Jean J. Blanc

Centre Universitaire de Marseille, Luminy

RESUME

On expose le résultat de recherches effectuées en Méditerranée par dragages, carottages, plongées et cartographie des fonds. On définit la notion d'aires sédimentaires avec des centres de production et des zones d'accumulation du matériel bio-détritique.

Les biocoenoses des herbiers et les concrétionnements coralligènes fournissent un matériel qui évolue en relation avec les épandages terrigènes, les stocks fossiles et les facteurs hydrodynamiques. Les phénomènes eustatiques et tectoniques du Quaternaire modifieront les distributions initiales.

On souligne l'importance des formations mixtes et accumulations polygéniques résultant de mélanges et remaniements. Cette dynamique traduit une évolution pratiquement continue. Elle apparaît plus conforme aux faits que les notions basées sur une zonation bathymétrique ou granulométrique.

ABSTRACT

The result of dredging, coring, diving and sea-floor mapping in selected regions of the Mediterranean Sea are summarized, and different sedimentary zones in which biodetrial materials are formed and in which they accumulate are recognized. The biocoenose of marine flora and coralline concretions provides materials whose regional importance is affected by the supply of terrigenous sediment and fossils, and to hydrodynamic factors. Quaternary eustatic and tectonic changes have modified the original distribution of biodetrital material.

The importance of mixed and polygenic deposits resulting from mixing and reworking is noted, and a continuum of sediment types is recognized. The composition of bioclastic sediments is more closely related to parameters outlined in this paper than simply to factors of bathymetry and grain size alone.

NOTION D'AIRES SEDIMENTAIRES

Introduction

Le cadre géographique des recherches présentées concerne principalement, en Méditerranée, les régions suivantes: Provence, Corse, Golfe de Gênes, Détroit siculo-tunisien, golfes de Gabès et de Syrtes, Péloponnèse, Crète, Cyclades, Sporades, Thrace et Dodécanèse (Figure 1). Les observations ont été effectuées dans les zones littorales, par dragages, carottages et plongées. La description des fonds d'origine volcanique ne sera point abordée dans ce travail.

Les sédiments issus de terres émergées (éolianites, pyroclastites, épandages turbides), sont mêlés et juxtaposés aux sédiments proprement "thalassogènes" où la fraction bioclastique acquiert une grande importance. A l'exception de certains exemples mal définis ou peu connus, la situation des divers types sédimentaires correspond, en Méditerranée, à des conditions précises déterminées par un nombre parfois restreint de variables. On aboutit à la définition de faciès homogènes et relativement constants pour la zone étudiée. Il en résulte que les *aires sédimentaires* correspondant par exemple aux *sables isométriques*

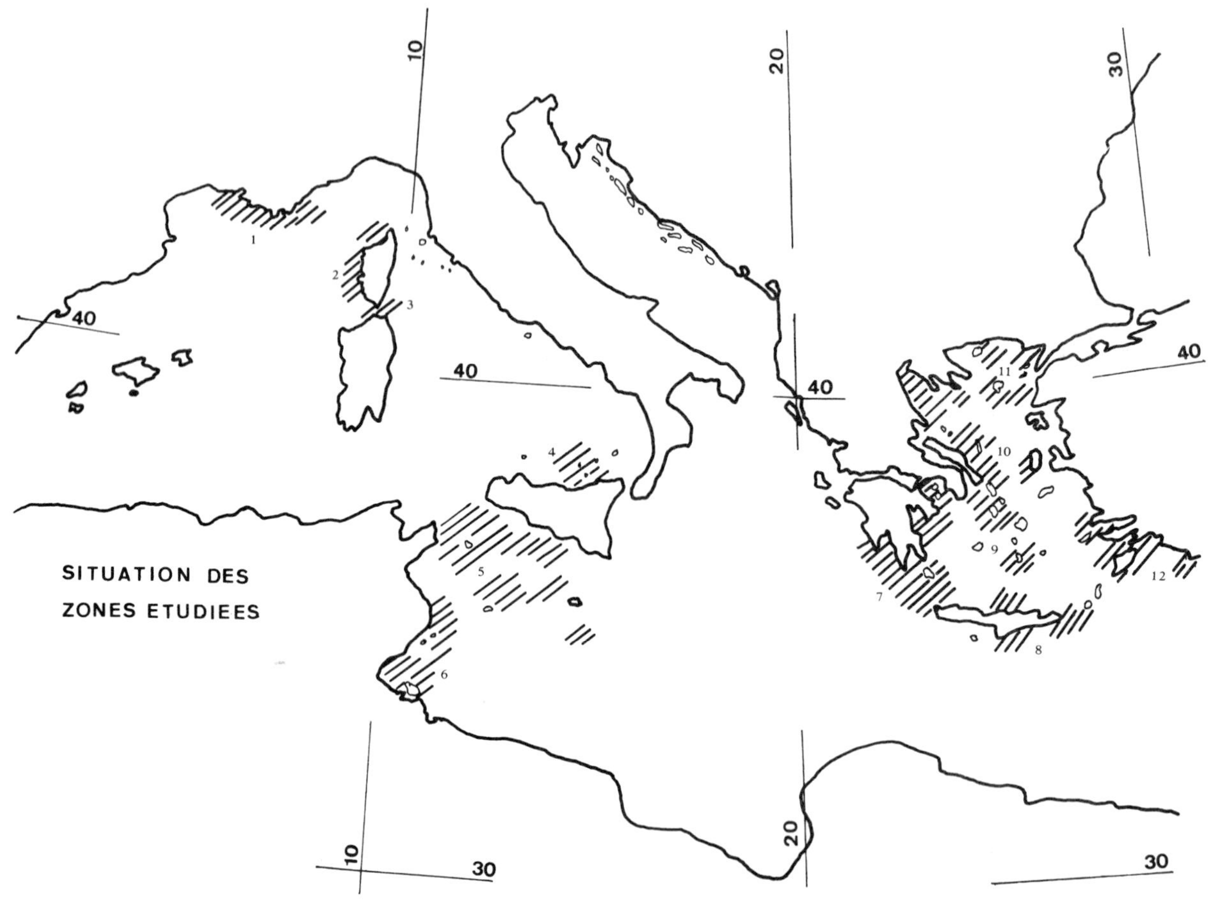

Figure 1. Situation des zones étudiées. 1–Provence, 2–Corse, 3–Détroit de Bonifacio, 4–Isles de Lipari, 5–Détroit Siculo-Tunisien, 6–Golfe de Gabès, 7–Péloponnèse, 8–Crète, 9–Cyclades, 10–Sporades, 11–Thrace, 12–Dodécanèse.

lessivés sont à la fois sous la dépendance des facteurs biologiques, hydrodynamiques et bathymétriques qu'il est possible de préciser.

Ces domaines particuliers, limités dans l'espace (et dans le temps), ou aires sédimentaires, sont généralement assez nets pour les formations bio-détritiques observées dans les milieux actuels (Picard, 1965; Blanc, 1959, 1966 et 1968). L'examen des multiples faciès nous amène ainsi à définir des zones de production et des zones d'accumulation.

Aires Génératrices de Sédiments

Les *aires génératrices de sédiments* se ramènent à deux cas essentiels: (1) *Herbiers sous-marins*, notamment herbiers à Posidonies, se développant à faibles profondeur, au fond des baies, au pied des talus, *etc.* Dans ce cas, le développement vertical des rhizomes consolide et enserre le sédiment *piégé* dans les herbiers au cours de leur croissance. Il en résulte la formation de banquettes présentant parfois une épaisseur de plusieurs mètres, nommées *mattes* (Molinier et Picard, 1952). Ces accumulations singulières aboutissent à un colmatage des criques et des fonds de baies. Le sédiment est généralement un sable hétérométrique, souvent grossier, d'origine essentiellement bio-clastique issu des peuplements autochtones ou para-autochtones en relation avec les herbiers. Les débris sont anguleux et correspondent à des éléments de rudites ou calcarénites grossières: tests brisés de Lamellibranches et Gastéropodes, Foraminifères (dominance des Miliolidae et des Elphidiidae), Ostracodes, Algues, Bryozoaires, Echinodermes, *etc.* L'étude granulométrique traduit des fluctuations irrégulières en relation avec un apport biogène aléatoire. (2) *Concrétionnements coralligènes,* formations bioconstruites, poreuses et caverneuses édifiant des placages, revêtements, cimentant les eboulis au pied des falaises et durcissant les fonds. Ces calcaires construits s'établissent sur des substrats durs: affleurements, parois de grottes ou de falaises, blocs, voire épaves. Ils peuvent encroûter les fonds meubles et constituer des radiers rigides. Le taux de sédimentation terrigène est nul ou très faible. Les organismes responsables de cette cimentation sont principalement les Algues: Rhodophycées calcaires (*Pseudolithophyllum*), Mélobésiées (*Lithothamnium*), placages encroûtants à *Neogoniolithon* et *Mesophyllum*. Interviennent encore les Bryozoaires, Polychètes, Coelentérés, Spongiaires.

Dans la structure poreuse construite s'accumulent des éléments isolés, souvent allochtones: Foraminifères, tests, *extraclasts* (micronodules remaniés et redéposés. Par convention, leur taille est supérieure à 177 microns, Folk, 1959), minéraux, *etc.* La surface de ces calcaires construits (dits "calcaires coralligènes") est irrégulière, déchiquetée. Ce sont les fonds "vifs" des pêcheurs où ancres, dragues et filets s'acrochent. La puissance des calcaires coralligènes peut atteindre plusieurs mètres par endroits, notamment au pied des falaises et talus sous-marins (Cap Sicié, Riou, Cap Caveau, près Marseille) (Blanc, 1953a; Laborel, 1960). Depuis le deuxième siècle avant l'ère chrétienne, il a pu se développer par endroits une épaisseur de 3 m de concrétions (épave et céramiques antiques concrétionnées; Grand Congloué, Marseille).

Les zones de production sédimentaire, telles que celles de l'herbier ou des fonds coralligènes, sont en relation plus ou moins étroite avec les apports détritiques issus directement du littoral. Au voisinage du talus d'érosion lié aux déferlements et au choc des vagues se disposent des formations détritiques constituant ce que Bourcart (1949) appelait le "prisme littoral". Ce matériel d'origine ébouleuse, torrentielle ou fluviatile va se disposer suivant plusieurs modalités (Figure 2):

Mode régulier: à partir du rivage et suivant une profondeur croissante, on observe la séquence suivante: prisme littoral (PL) ⟶ herbier à Posidonies (HP) ⟶ fonds "coralligènes (C).

Mode imbriqué: les aires sédimentaires correspondant à (PL), (HP) et (C) s'inbriquent sans se mêler à des distances variables du rivage, en fonction de la bathymétrie, des conditions hydrodynamiques et de l'éclairement.

Mode dispersé: il n'y a plus de séquence continue ou inbriquée. Les aires de (PL), (HP) et (C) sont dispersées à la surface du précontinent et constituent des *ensembles isolés* parmi les zones d'accumulation.

Zones d'Accumulation

Les zones d'accumulation dérivent de l'érosion et de l'épandage du matériel des aires génératrices précitées avec la participation des éléments détritiques littoraux (PL). A l'issue de l'érosion sousmarine, d'origine mécanique ou biologique, les phénomènes revêtent une certaine variété: chutes, remaniements, lévigations, dispersion, *etc.* A la diversité des peuplements animaux et végétaux s'ajoutent d'autres facteurs souvent bien moins connus: vitesse des courants sur le fond, chocs et cavitations, courants de décharge, *etc.*

Il va de soi qu'une classification méthodique, ordonnée en fonction de la seule profondeur ne saurait être retenue. Les aires d'épandage occupent de très vastes surfaces sur la marge continentale. Leur évolution complexe demeure sous la dépendance des facteurs biologiques, hydrodynamiques et géologiques.

SITUATION DES ENSEMBLES SEDIMENTAIRES

Les divers types de fond se rapportent généralement à des faciès granulométriques déterminés. Cependant, d'importantes fluctuations de caractère aléatoire sont notées dans le domaine des éboulis sousmarins, brèches et sables grossiers. Les variations granulométriques des sédiments du précontinent et du rebord sont *indépendantes de la profondeur.*

Il n'existe pas de variations sédimentaires continues tout au long du profil du précontinent. Une représentation graphique de la médiane granulométrique en fonction de la bathymétrie montre des aires groupant des familles de points. En Méditerranée, il est possible de caractériser ainsi les sables des fonds de baies, les herbiers, les sables coralligènes, les sables vaseux, les vases terrigènes côtières, les sables coquilliers du large, *etc.* (Figure 3).

Parmi les facteurs responsables de cette répartition, en plus des agents hydrodynamiques, de la turbidité et des thanatocoenoses quaternaires, nous devons mentionner les zonations de la matière organique au large des continents comme le montrent les travaux de Szekielda (1969).

A la suite de la glaciation würmienne, la remontée du niveau de la mer depuis le Würm II (phase maximale de la régression vers -110 m) a déterminé une suite de ruptures d'équilibre. Les étapes de cette transgression flandrienne (ou "versilienne") ont été précisées par l'examen des sondages dans les plaines côtières et les récents travaux portuaires (Bonifay, 1967; Gouvernet, 1965). Des turbidites à éléments infralittoraux et circalittoraux marquent l'essentiel de la régression glaciaire, puis, au cours des variations positives du niveau de la mer, on notera la présence de zonations bionomiques et de petits cycles sédimentaires de plus en plus décalés vers le haut, c'est-à-dire vers le niveau actuel. En marge des épandages terrigènes, de nombreux secteurs, en l'absence d'érosion et de sédimentation notables, laisseront subsister des biocoenoses ou thanatocoenoses fossiles, formant des calcarénites oxydées et parfois silicifiées.

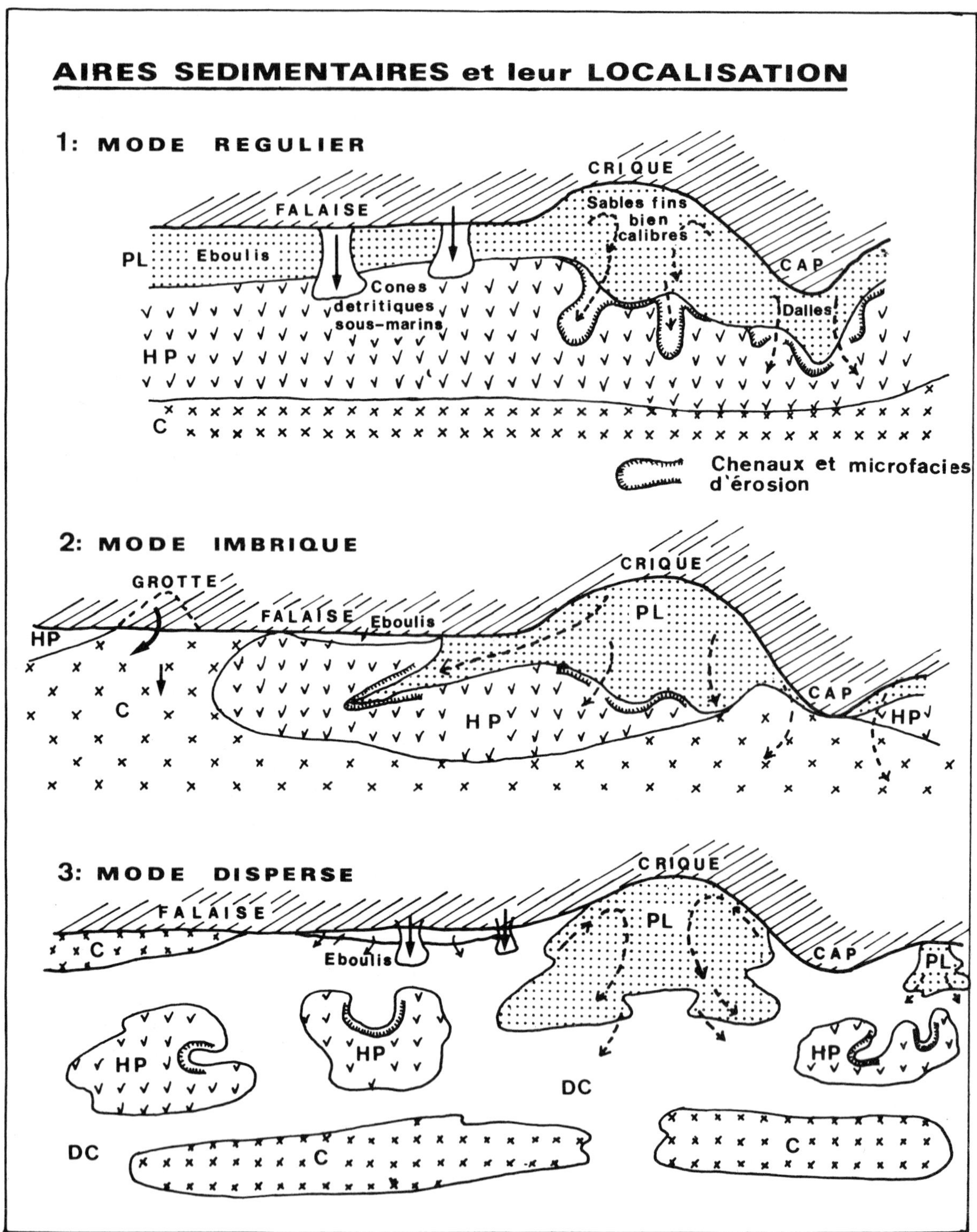

Figure 2. Aires sédimentaires et leurs localisation. PL = Prisme détritique littoral; HP = Herbiers à Posidonies; C = Concrétions et fonds coralligènes; et DC = Détritique côtier.

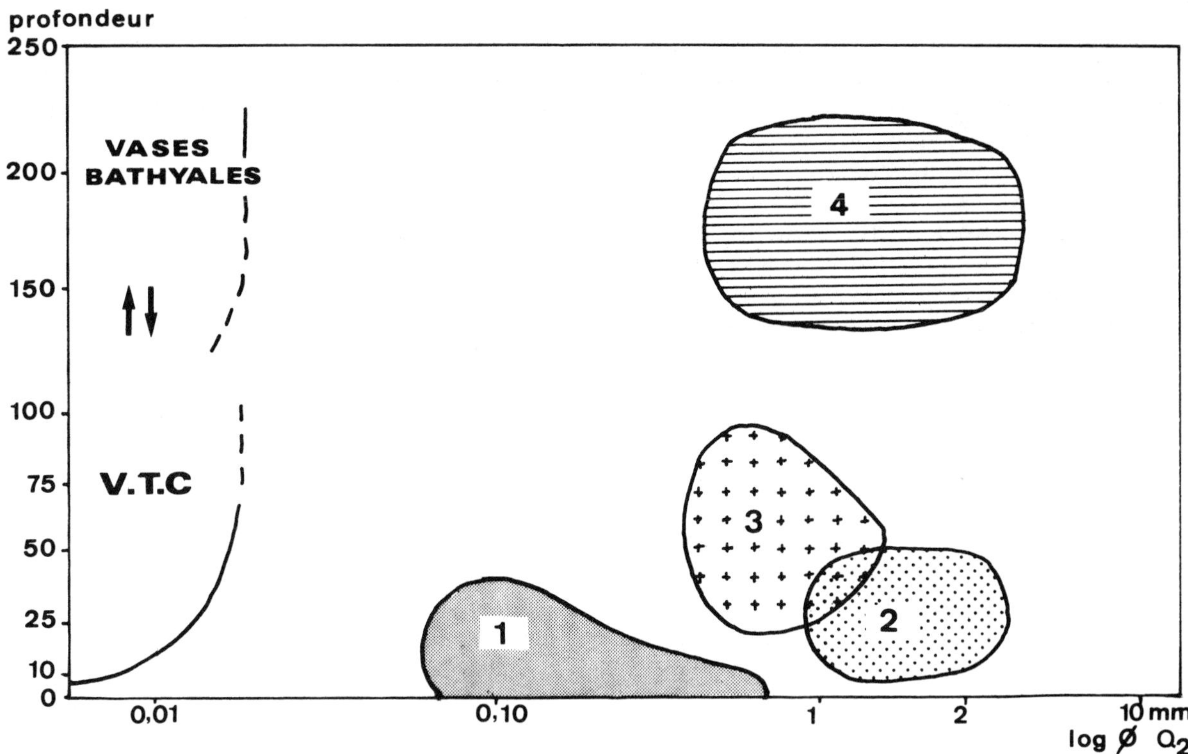

Figure 3. Aires sédimentaires représentées en fonction de la profondeur et de la granulométrie. 1 = Sables fins des plages et fonds de baies; 2 = Sables biodétritiques d'herbiers; 3 = Sables coralligènes biodétritiques; 4 = Sables biodétritiques du large (S.D.L), à thanatocoenoses quaternaires; et V.T.C. = Vases terrigènes côtières passant graduellement aux vases bathyales lorsqu'il n'y a point de "détritique du large".

DYNAMIQUE DES MILIEUX SEDIMENTAIRES NERITIQUES

Les aires génératrices de sédiments et les zones d'accumulation présentent des modes évolutifs concomitants, indépendants ou liés, rattachés aux ensembles littoraux, infralittoraux ou circalittoraux.

Talus Détritique Littoral, Zone Infralittorale

L'absence de marées notables en Méditerranée, à l'exception de secteurs localisés, souligne l'importance du choc des vagues actions de déferlements, transferts latéraux, "rip-currents" (courants sagittaux), *etc.*

A la suite des recherches de Bourcart (1949), Glangeaud (1949) et Rivière (1951), il a été confirmé que la dispersion vers le large s'avère limitée. En fait, les observations et expériences (sables et galets "marqués, *etc.*) montrent que cet étalement ne s'étend guère en deçà de quelques mètres de profondeur; le matériel revient plusieurs fois au rivage. En certains cas, la dispersion vers le large, sans retour, a pu être observée mais des recherches détaillées doivent être entreprises afin de préciser le cheminement exact des stocks détritiques d'origine littorale.

Une usure intense aboutit rapidement à un sédiment bien classé par les triages résultant des lévigations. L'évolution est pratiquement stoppée pour des indices de classement (indice de Trask) de 1,20 à 1,15. En ces cas, quelle que soit la classe granulométrique, le matériau ne s'use pratiquement plus et demeure en équilibre avec le milieu (Berthois, 1964; Blanc, 1954; Froget, 1963).

Les sédiments de mode battu ou très battu nourrissent des éboulis et talus détritiques sous-marin. Malgré la convergence des lignes de forces de la houle et des vagues, l'érosion marine demeure relativement faible et localisée sur les littoraux rocheux (Bourcart, 1952; Blanc, 1954 et Froget, 1963). Quelle est alors l'origine du matériel?

L'apport de matériel grossier est principalement issu de l'éboulis littoral sub-aérien, d'éboulements localisés parfois liés à des séismes et actions tectoniques anciennes ou récentes (rejeu de failles au Quaternaire, *etc.*). Nous citerons les exemples d'éboulis sous-marins du Puget-Devenson (Marseille), du Soubeyran (Cassis-La Ciotat), d'Istrie, de Dalmatie, des littoraux abrupts de Crète, de Cythère et d'Eubée.

En revanche, on observe le déblaiement rapide des stocks quaternaires et l'erosion des remplissages karstiques issus de cycles antérieurs (Marseilleveyre, Archipel de Riou, Péloponnèse).

Les très grocs blocs, impossibles à déplacer, sont rapidement cimentés par des concrétions biologiques au pied des falaises. Le reste du matériel (blocs anguleux) pourra se coincer et s'imbriquer en formant des éboulis vifs à très forte pente. A une certaine profondeur, à l'exception d'une érosion biologique non négligeable (Pérès, 1961; Picard, 1954), le matériel demeurera anguleux, souvent intact ou recouvert d'une concrétion superficielle, en deçà de la zone d'action des vagues. Ces talus de blocs présenteront une inclination supérieure à celle que montreraient, à l'air libre, des matériaux de nature analogue.

Bien entendu, le niveau supérieur oú se manifestent les mouvements des vagues et les "coups de bélier" dûs à l'air comprimé, sera caractérisé par une usure intense du matériel sédimentaire. Une ceinture détritique jalonnera la partie sous-marine des falaises avec, parfois, formation de petits talus "deltaïques" (Cap Sicié, près Toulon).

Lorsque la turbulence est élevée, la pente et la profondeur faibles, une partie du sédiment hétérogène sera acheminée en direction des herbiers ou vers les formations du détritique côtier. Les travaux de J. Picard et de ses collaborateurs ont souligné depuis quinze années d'observation, la mobilité de ces types de fond dans la baie de Marseille.

Un Ensemble Original: l'Herbier à Posidonies

La "matte" de l'herbier à Posidonies se développe en Méditerranée, de quelques mètres à −40 m. Son accumulation rapide est précisée par la découverte d'amphores et d'épaves antiques. Cependant les taux de sédimentation ne présentent aucune régularité: de 1 m par siècle (Moliner et Picard, 1952) à 4 m depuis l'époque romaine. Quoiqu'il en soit, on notera le dépôt d'un sable grossier, hétérométrique, presque exclusivement formé de débris biogènes autochtones et non usés. Une cimentation éventuelle du sédiment constituant ces "mattes" aboutirait à la formation de *calcirudites*: terme général désignant la catégorie des roches calcaires contenant plus de 50% d'éléments accumulés dont le plus petit diamètre se trouve supérieur à 2 mm (*in* Grabau, 1904).

En Méditerranée, ce "volant" de matériaux peut s'avérer important. Toute érosion anormale ou emprunt massif de sédiments se traduira par une rupture d'équilibre de l'environnement sédimentaire et biologique.

La croissance verticale de l'herbier amène l'exhaussement de la "matte" et une diminution de la profondeur. Il en résulte une reprise d'érosion et une dégradation rapide du fond—jusqu'à une zone plus profonde où les conditions d'installation d'un nouvel herbier et d'une nouvelle "matte", en milieu calme, seront derechef réalisées. Ainsi, l'évolution *sur place* de ces herbiers correspond à une série de cycles: croissance verticale (colmatage)—érosion, généralement à courtes périodes, laissant subsister des chenaux avec des témoins de "mattes" emboîtés.

Cette évolution cyclique paraît osciller entre deux "pôles":

1. Colmatage de certains fonds de baies, très abrités, en quelques siècles ou quelques millénaires. Moliner et Picard, (1952) ont montré, en arrière du front d'émersion de l'herbier à Posidonies, l'installation d'une zone rapidement comblée où se développent des herbiers à *Cymodocea*, *Ruppia* accompagnés d'une très riche microfaune à *Sorites* et *Peneroplis* (Blanc-Vernet, 1969). On aboutit à un sédiment hétérométrique de type *mixte* où une fraction bioclastique importante est mêlée à une matrice détritique décantée de nature sablo-argileuse (Provence, Corse, Sicile).

2. Erosion des fonds, creusement de chenaux: à très faible profondeur, l'action des vagues et les déferlements amènent le sédiment bio-détritique dans la zone de "balancement" du "prisme littoral". Ces observations ont été confirmées par le repérage de petits galets et l'examen de certaines thanatocoenoses de Foraminifères.

A partir de quelques mètres de profondeur, sous l'action de courants de décharge et de dégradations biologiques sous-marines, le sédiment de la "matte" détruite est dispersé vers la zone complexe des "sables du détritique côtier" longeant le littoral à plus ou moins grande distance du rivage. Des fronts d'érosion dans les herbiers ont ainsi été observés de −10 m (baie de Marseille) jusqu'à 15 m de profondeur. Au débouché des calanques, vers le large, un certain classement accompagné d'une faible usure se manifeste au niveau du fond. Par suite d'une diminution brusque de la compétence due à l'augmentation de la profondeur, pour les zones très exposées aux vents de large, se dépose un sédiment lessivé et classé, formant un amas aux "débouchés" de calanques (Blanc, 1968; Picard, 1965; Poizat, 1968). Le tableau 1 résume l'évolution des types mentionnés.

Le Domaine Circalittoral

Nous ne reviendrons pas sur les travaux classiques de Pérès et Picard (1958) concernant les biocoenoses *coralligènes* définies en Méditerranée et assimilables à un véritable *climax* au sens où l'entendent les écolo-

Tableau 1. Milieux sédimentaires littoraux et infralittoraux.

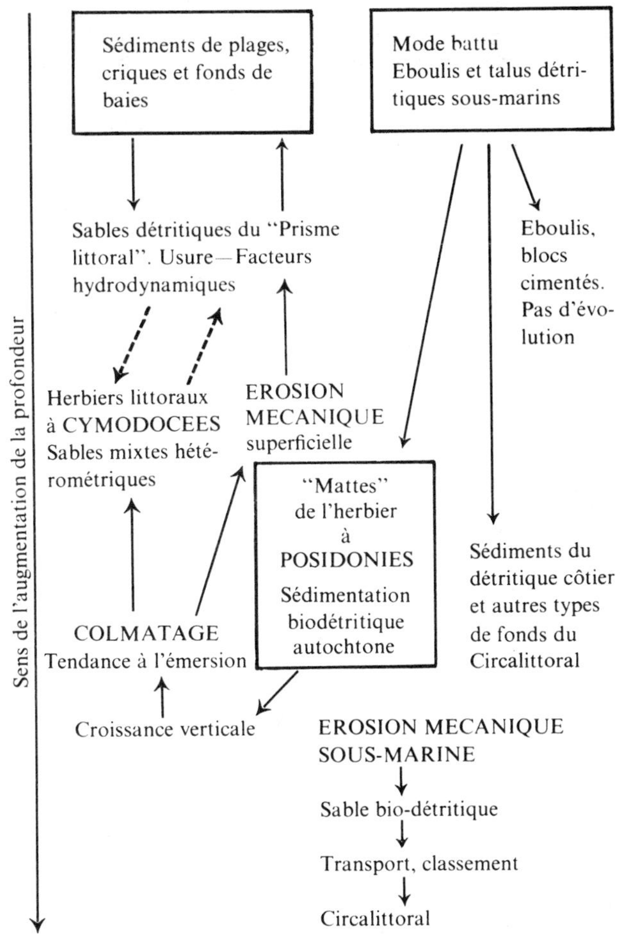

gistes. Comme pour les herbiers, les aires coralligènes sont génératrices en matériel sédimentaire, alimenté fréquemment par les débris ou concrétionnements de plusieurs biocoenoses. Sur le précontinent, elles constitueront le deuxième "pôle" parmi les zones de production.

Malgré certaines analogies avec les milieux récifaux, les environnements coralligènes montrent une répartition bathymétrique plus large: de quelques mètres à près de 100 mètres en certains cas.

Des concrétions "pré-coralligènes" peuvent s'établir sur la roche en place ou sur les matériaux grossiers et hétérométriques de l'herbier et du détritique côtier.

Puis, viennent les concrétions coralligènes proprement dites. Ce sont des calcaires construits, poreux et caverneux, enrobant des épaves anciennes, jas d'ancres et amphores, garnissant des hauts-fonds ou empâtant la base des caps rocheux ou des talus abrupts, blocs, éboulement sous-marins, les éléments constructeurs n'aboutissent jamais à des spectaculaires édifications mais à de petits massifs et placages à Rhodophycées calcaires (*Pseudolithophyllum*), Coelentérés, Bryozoaires. On notera encore l'influence des Polychètes et des Spongiaires. Pérès et Picard (1958) ont montré qu'en Méditerranée orientale le "coralligène de plateaux" peut atteindre 120 m de profondeur alors qu'il ne dépasse généralement pas 60 m dans le Bassin occidental. Il apparaît que les profondeurs importantes sont atteintes lorsque la transparence de l'eau est parfaite et le taux de sédimentation très faible. Ce cas est particulièrement réalisé en Mer Egée, dans les Cyclades et au Sud de la Crète. Sur les parois, Laborel (1960) a montré que l'extension maximum des formations concrétionnées s'effectue perpendiculairement à la direction des rayons lumineux (porches de grottes et pieds de falaises sous-marines).

On notera encore les curieuses biocoenoses coralligènes *du large*, établies près de la rupture de pente du précontinent, parfois jusqu'à 180 m de profondeur. Etudiées en "soucoupe plongeante", ces concrétionnements relativement profonds montrent la dominance des Spongiaires. S'y adjoignent: Cnidaires, Bryozoaires (*Retepora*, *Porella*), Brachiopodes (*Terebratula vitraea*), Ascidies, Polychètes, Serpulides, Echinodermes (*Echinaster sepositus*, *Cidaris cidaris*, *Antedon mediterranea*).

Les concrétions coralligènes constituent un ensemble stable, sans évolution notable. Cette dernière ne sera amorcée que si interviennent des ruptures d'équilibre par envasement et enfouissement, et érosion biologique sous-marine, beaucoup plus importante que l'érosion mécanique. On peut obtenir tous les termes de passage d'une séquence lithologique:

Calcaire construit	⟶	Brèches et calcarénites coralligènes	⟶	Calcarénites coralligènes isométriques, à thanatocoenoses de Foraminifères

Cette séquence bio-détritique, issue de l'érosion biologique sous-marine, peut se compliquer par l'introduction d'éléments également coralligènes mais issus de milieux plus ou moins éloignés, formant des *extraclasts*. En outre il arrivera que certaines formations seront plus anciennes, correspondant à des niveaux inférieurs au plan d'eau actuel [modifications bathymétriques positives (transgression flandrienne, ou versilienne) ou négatives (régression Würmienne)].

Les régressions du Würm I, et surtout du Würm II ont amené l'érosion des stocks infralittoraux et leur déplacement, et dispersion vers le large. On retrouve ces éléments remaniés dans la partie supérieure de certaines carottes et l'on sait encore qu'une partie de ces dernières peut s'incorporer aux turbidites découvertes dans les canyons (Bourcart, 1952 et 1955; Bourcart et Ottman, 1957, *etc.*).

Inversement, la remontée du niveau marin au Würm III et au Würm IV (Dryas) a laissé sur place une partie des fonds coralligènes et de leurs annexes qui se trouvent, de ce fait, situés plus au large. Il a été montré tout ce que l'ensemble des "sables détritiques du large" doit à ce stock autochtone initial (Pérès et Picard, 1958; Pérès, 1961; Blanc, 1964 et 1968; Blanc-Vernet, 1969). Le sédiment actuel est un sable bio-clastique *oxydé*, parfois silicifié, à glauconie très altérée, enrichi en éléments d'origine plus profonde (Echinides). On notera que l'augmentation de la profondeur ne se traduit point, en général, par une augmentation de la teneur en pré-colloïdes ou colloïdes. Il arrive que les ensembles construits *fossiles*, encroûtés par les oxydes de fer et de manganèse, subsistent sur d'anciens plateaux rocheux (calcaires à Bryozoaires du banc de Magaud, Iles d'Hyères, formations coralligènes du banc Jonhson). L'érosion et la dégradation de ces calcaires coralligènes, peut aboutir, au large du banc Jonhson (Méditerranée orientale) à des calcarénites et vases crayeuses riches en Bryozoaires et Foraminifères triturés, pulvérisés, passant en profondeur aux sédiments de l'étage bathyal. Une séquence analogue a été mise en évidence dans le Détroit de Caso.

Les traces des anciens littoraux würmiens peuvent, malgré la dispersion due à l'action de la mer transgressive, être jalonnées par des accumulations de galets parfois "éolisés" (−80 à −90 m, baie de Cassis) (Blanc, 1966; Froget, 1966; Bourcier, 1968; Poizat, 1968).

BIOTOPES ET AIRES D'ACCUMULATION

Ces types de fonds seront en relation avec les deux aires principales de production sédimentaire "thalassogènes": l'herbier à Posidonies et les ensembles coralligènes.

Herbiers

La croissance des herbiers à Posidonies participe au colmatage accéléré des fonds de baies depuis l'époque antique en Méditerranée (Blanc, 1959; Vita-Finzi, 1969). La "matte" arrive fréquemment à un niveau sub-émergeant déterminant un plan d'eau relativement abrité vers le littoral (Molinier et Picard, 1952). En ces lieux et dans les secteurs particulièrement protégés des fonds de baies se développe un milieu réducteur où un herbier à Cymodocées établit une "pelouse" tapissant le fond de la lagune (Provence: Port-Cros, le Brusc; Corse, Sicile, Golfe de Volo, *etc.*).

Les fonds à Cymodocées ne présentent point les caractères de rétention et de croissance des "mattes" à Posidonies. Le sédiment correspondant est également très différent. Il s'agit d'un ensemble hétérométrique à deux phases:

1. Sable biogène autochtone ou para-autochtone issu des débris d'organismes vivant dans l'herbier à Cymodocées: Foraminifères (*Miliolidae*, *Peneroplidae*, *etc.*), Echinodermes, Pelecypodes.
2. Phase sablon-poudre-précolloides aboutissant à un sédiment noirâtre, parfois très riche en eau (jusqu'à 300%), gluant, quelquefois thixotropique et riche en matières organiques. Il s'agit des produits *décantés* à l'abri issu de l'herbier émergeant ou d'une situation topographique exceptionnelle. De tels "pièges à sédiments" où se forment l'hydrotroïlite et la pyrite sont fréquents en Méditerranée.

Les courbes granulométriques sont irrégulières et montrent plusieurs stocks échelonnés. Il peut y avoir plus de 25% d'éléments fins ou ultra-fins de nature terrigène.

De petits carottages précisent les conditions de dépôt d'un sable assez argileux présentant fréquemment des lits psammitiques et des passées à Foraminifères (taphocoenoses et mortalités liées aux variations saisonnières).

En effet, la faible profondeur de ces milieux se traduit par une grande sensibilité aux écarts thermiques (de l'englacement partiel jusqu'à 28°), sursalures et dessalures locales, dérives liées aux vents accompagnées de la filtration des particules micacées au travers de l'herbier émergeant (Molinier et Picard, 1952; Blanc, 1953a, marées et seiches découvrant parfois de notables étendues (Venise, Chioggia, Caorle, le Brusc, Porto-Vecchio). Localement, il s'ensuit une sédimentation rapide et parfois, une mortalité massive de Foraminifères (Blanc-Vernet, 1969).

L'influence de la température est prépondérante au fond des baies abritées de la Méditerranée, conditionnant la répartition des biotopes et, par voie de conséquence, la sédimentation bio-détritique. Ainsi, les secteurs les plus chauds sont en étroite relation avec le développement d'herbiers à Caulerpes, notamment en Méditerranée orientale (Rhodes, Crète, Golfe de Paros, Golfe de Kos, Péloponnèse). En ces lieux, les températures de juillet-août sont comprises entre 25° et 30°. Or, durant certaines périodes de canicule, pour des eaux à 22° ou 23°, les Caulerpes peuvent apparaître temporairement en Sicile, Péloponnèse, Port-Cros et Port-d'Alon (Provence), associès aux *Peneroplidae*.

Au Sud du Péloponnèse, les fonds à Caulerpes sont associés à des herbiers à *Halophila*. Tous occupent un niveau "infralittoral", jusqu'à 40 m de profondeur. Le sédiment est un sable exclusivement bio-détritique, riche en *Miliolidae*, *Peneroplidae* et Ostracodes. L'usure est faible ou nulle. Localement, l'ensemble peut s'enrichir en éléments terrigènes fins et décantés (Paros, Volo, Talante). Les Miliolidae sont accompagnées de *Discorbis* et *Cibicides*. Les secteurs les plus chauds (Crète, Péloponnèse, Salamine, Mithylène, *etc.*) présentent une forte proportion d'Amphistegines modifiant les caractères granulométriques du sédiment.

Fonds Spongifères

Les fonds spongifères du Golfe de Gabès se rattachent à des biotopes infralittoraux (herbiers à Cymodocées, Caulerpes et Posidonies) ou circalittoraux (graviers corraligènes) (Pérès, 1961). Nous disposons de très peu de renseignements sur les sédiments correspondants. On notera la présence de courant notables et constants assurant le renouvellement des particules alimentaires en suspension et par filtration, amenant de curieuses concentrations en produits radio-actifs (cerium et praseodyme) (Boclet *et al.*, 1962). Une partie du matériel bioclastique, vraisemblablement déportée par les courants, va s'accumuler en des zones plus profondes, en milieu bathyal. Il s'agit de dépressions où se décante la matière organique (Canal de Corse). Ces observations très incomplètes montrent que les calcaires à spicules de Spongiaires peuvent se former en des conditions océanographiques et bathymétriques très variées.

Faciès Sédimentaires du Détritique Côtier

L'érosion des fonds infralittoraux et des aires coralligènes se traduit par l'accumulation d'une zone détritique meuble, ceinturant le littoral et constituant le "détritique" côtier. Il apparaît que la sédimentation oscille entre deux tendances opposées: concrétions et envasements, aboutissant à des termes calcaires ou marneux.

Concrétions: Les algues calcaires libres (*Neogonolithon*) élaborent des concrétions et pseudo-pisolites assez analogues aux "*algal balls*". Ces pisolites, encore nommées "pralines" garnissent les plateaux sous-marins balayés par les courants: détroit siculo-tunisien, banc Graham, banc de Magaud, haut-fond de Centuri, banc de Pantellaria, banc de Caso, banc Kolumbos, *etc.* Ces pisolites peuvent atteindre des tailles allant de quelques millimètres à plusieurs centimètres. Malgré certaines analogies, la participation des Mélobésiées aux concrétions exclut l'assimilation de ces nodules à des oncolites.

Ces "pralines", signalées par Dangeard (1929) en Méditerranée, sont parfois déplacées sur le fond par des profondeurs de 80 à 90 m, à la limite inférieure du développement de la vie algale.

L'accroissement de ces concrétions pisolitiques est relativement lent ainsi qu'en témoignent certains fragments de poteries. En secteurs volcaniques (banc de Pantellaria, banc Kolumbos, Santorin), les thalles concrétionnés sont interrompus par de microlaminations cendreuses correspondant aux stades d'eruptions (Blanc, 1954).

Les faciès à "pralines" peuvent être remplacés par une "gravelle" coralligène très grossière à Mélobésiées libres (*Lithothamnium calcareum*, *L. solutum*) formant jusqu'à 80% de la totalité du sédiment. Dans ces faciès assimilés au *maërl*, un "feutrage" de Rhodophycées (*Jania*) assure une cimentation cohérente des thalles de Lithothamniées (Jacquotte-Hippeau, 1962). Une lithification totalement réalisée se produit parfois dans les fonds de passes et aboutit à une biolithite ou à une biomirudite.

Contrairement à la Manche où le "maërl" peut remonter jusqu'à la zone intertidale, en Méditerranée le "maërl" se situe à des profondeurs allant de 20 à 40 m, parfois à 70 m, lorsque la transparence de l'eau est parfaite. Exceptionnellement, Picard a pu observer de telles formations jusqu'à 100 m en Mer Egée et au Sud de la Crète.

Les "gravelles" coralligènes de types "maërl" ont été observées en de nombreux secteurs de la Méditerranée (Dieuzeide, 1940; Picard, 1954; Blanc, 1958; Nesteroff, 1960).

On remarquera leur situation liée à des fonds relativement plats ou à faible pente, au-dessous de la zone d'action des vagues mais en relation avec des courants. On les localise en effet au voisinage des passes et détroits: Bonifaccio, archipel de Riou, Cyclades, détroit siculo-tunisien, *etc.* Récemment, dans des formations sous-marines attribuables au Quaternaire ancien ou moyen, un faciès proche du "maërl" a été découvert sous la forme d'une biomirudite très oxydée, au large de la Provence (Dangeard, *et al.*, 1968; Froget, 1967a). Il est probable que ces concrétions ont joué un rôle important dans la sédimentation ancienne de la marge continentale méditerranéenne au Quaternaire.

Envasements: Ce dernier montre une double origine; il peut relever des facteurs biologiques ou d'influences terrigènes. La zone du prisme littoral et les éboulis côtiers sont le siège d'une alimentation et d'un brassage continuel. L'usure du matériel est très poussée, beaucoup plus importante qu'en milieu fluviatile comme en témoignent les recherches de Cailleux (1947) et Berthois (1964).

L'usure des débris organiques est très rapide pour les premières heures du transport (transfert littoral).

Les Algues et fragments de tests présentent un "lissage" des aspérités. Puis, dès un certain état d'aplatissement et d'émoussé, l'usure des débris se stabilise ou ne progresse plus que très lentement. Le résultat est une pulvérisation du calcaire qui se trouve mis en suspension, formant des eaux troubles et blanchies. Ce phénomène des "eaux blanches" n'a rien à voir, dans le cas présent, avec le "lait de corail" ou tout autre phénomène de sursaturation à base de calcite ou d'aragonite comme cela se produit en milieu récifal. Il s'agit d'une fine poudre calcitique d'origine essentiellement *détritique*, apparaissant en nuages épisodiques après une période de gros temps et se trouvant rapidement déportée vers le large durant les heures qui suivent (littoral de Marseille à Cassis). Cette suspension aboutit au niveau du détritique côtier où elle constitue partiellement la matrice du sédiment hétérogène.

Il en résulte une *lutite* très calcaire (80% à 90% de CO_3Ca) contribuant à l'envasement des fonds de baies à la faveur des circuits décrits par les dérives et courants. Les fonds situés au large du littoral de Marseilleveyre-Puget et de la baie de Cassis montrent une zone envasée, d'abord très calcaire, puis, progressivement, vers la haute mer, le sédiment passe à une vase décantée de plus en plus argileuse, enrichie en fibres végétales et particules très fines (Poizat, 1968). La position de la *mud line* conditionne ici un *gradient d'envasement* accompagné par un appauvrissement du benthos au fur et à mesure que l'on s'éloigne du littoral et des centres de production sédimentaire (talus détritiques, herbiers, coralligène). Le ciment tend à devenir argileux et les lutites carbonatées font place à une phase très fine essentiellement illitique (Poizat, 1968; Bourcier, 1968).

Cet envasement particulier est favorisé par l'absence de courants de marées entretenant un drainage au niveau du fond. Les études granulométriques systématiques confirment la dominance des phénomènes de décantations à partir de la *mud-line*. Pour la zone non envasée du détritique côtier, la granulométrie du sédiment dépend des *allochems* dominants plutôt que de l'importance des facteurs hydrodynamiques.

L'HERITAGE WÜRMIEN ET LE FACIES DES "SABLES DETRITIQUES DU LARGE"

Ce faciès des sables détritiques du large comporte de très nombreux éléments *oxydés*. Les recherches du Centre d'Océanographie de Marseille montrent son extension au large de la Provence et de la Corse. Il a été retrouvé en Mer Egée (canal Pélago, Thaso), Détroit Siculo-Tunisien et au large de la côte algérienne.

Quoique non encore consolidés, on peut considérer ces "sables détritiques du large" comme des éléments constitutifs de calcarénites. La rubéfaction observée apparaît très fraiche et certaines espèces de la "faune celtique" (*C. islandica*) ont encore conservé leur couleur. A l'origine, le sédiment apparaît déposé en milieu réducteur. La présence de pyrite et de matière organique a été accompagnée par la formation locale de glauconie authigène moulant les cavités des Foraminifères, Bryozoaires, *etc.* (*lobe-casts*). On observe fréquemment la silicification des débris biogènes en opale et calcédonite (Foraminifères, Bryozoaires, Algues) (Bourcart et Ottman, 1957; Blanc, 1964). Puis, en dernier lieu, l'oxydation des débris s'est effectuée aux dépens des sulfures et de la glauconie. Il en résulte la teinte "rouille" des "sables détritiques du large". Cette oxydation ne correspond point aux conditions actuelles; elle paraît s'être effectuée sous une tranche d'eau plus faible ou durant une nette période de *non-sédimentation* lors d'une phase récente de la remontée du niveau marin, au Dryas ou même, peut-être, au post-glacaire (pré-Boréal ou Boréal). La situation bathymétrique *décalée* dans le temps et dans l'espace des deux thanatocoenoses (−180, −250 et −100, −150 m) confirmerait ce point de vue.

Au large de Marseille, nous avons constaté que la fraction minérale était issue d'apports fluviatiles et de stocks turbides d'origine continentale et largement déportés par les dérives. Les trajets peuvent dépasser 45 km, du Rhône au Sud de Riou. Les recherches de Picard-Tarbouriech (1969), basées sur la dispersion de minéraux d'origine alpine tels que l'épidote et le glaucophane montrent une extension encore plus grande de ces apports terrigènes. En outre, on sait que de petits galets de variolite et de quartzites, d'origine durancienne et de façonnement fluviatile, souvent éolises (*dreikanters*) ont été trouvés dispersés sur le précontinent (Marion, 1883; Pérès et Picard, 1958; Blanc, 1959) et notamment mêlés à la thanatocoenose würmienne supérieure (−100 à −150 m). On soulignera enfin, qu'une fraction d'origine détritique a contribué au colmatage des canyons où on la retrouve dans les carottages sous la forme de passées turbides.

Les "sables détritiques" du large présentent donc une composition polygénique déterminée par l'évolution sédimentaire du domaine marin à la fin des temps quaternaires. Une succession de phénomènes pourra être interprétée dans les secteurs où le taux de sédimentation est très faible. Tel est le cas de la zone profonde du précontinent à l'Est de Marseille où n'aboutit aucune rivière importante: les carottages montrent les formations du Würm IV à très faible profondeur, voire en surface comme cela a été observé en plongée (soucoupe plongeante J. Y. Cousteau) par Laborel *et al.* (1961) (thanatocoenose à "faune froide"). Dans ces conditions, l'évolution sédimentaire est réduite à l'apport bio-clastique et pélagodétritique actuel (phyllites, coccolithes, Foraminifères). Le faciès est essentiellement conditionné par l'héritage géologique et les stocks bio-détritiques fossiles et actuels. On pourra schématiser la séquence (voir Tableau 2).

Dans le cas où le taux de sédimentation est important ("pièges à sédiments", vasières, épandages fluviatiles), les formations des "sables détritiques du large", probablement enfouies—ou remplacées par le faciès des "vases terrigènes côtières", n'ont pas été observées (secteur rhôdanien, Golfe de Gênes, Golfe de Thessalonique).

Nous avons particulièrement insisté sur l'importance des formations liées à la régression würmienne, la seule relativement connue en Méditerranée. Il est probable que les autres régressions correspondant aux cycles antérieurs (Riss, Mindel, Günz) ont eu des effets relativement analogues, du même ordre que les récurrences froides à *Cyprina islandica*, observées depuis le Calabrien.

Tableau 2. Evolution sédimentaire des "sables détritique" du large.

de −120 m → Niveau actuel
Peuplements actuels
Apports pélago-détritiques
↓
3 : Etat *actuel* : SABLES DETRITIQUES DU LARGE

Apports détritiques
↓ Sables ↓ Petits galets
Sables → Canyons
Sables, Petits galets → Thanatocoenose I −100, −150 m Détritique côtier fossile Würm IV

2 : OXYDATION : Altération de (1) *Non sédimentation* Oxydes de Fe^{+++} (Dryas ou post-glaciaire)

- - - - - - - - 17.000 à 10.500 A.B.P.

Thanatocoenose II −180, −250 m Würm sup.

1 : ENVASEMENT Milieu réducteur { Silicification, Glauconie authigène, Pyrite, Réduction matière organique }

LES VASES TERRIGENES COTIERES

Les secteurs en voie de colmatage et à taux de sédimentation élevé, à proximité des estuaires et des deltas, au fond de golfes très abrités, sont caractérisés par des faciès de vases plus ou moins fines ou argileuses, désignés sous l'appellation de "*vases terrigènes cotières*" (V.T.C.) (Vatova, 1949; Gautier et Picard, 1957; Pérès et Picard, 1958).

Ces zones d'envasement présentent une composition minéralogique, une teneur en calcaire, des caractères granulométriques et souvent, une microfaune, *indépendants de la profondeur*, passant graduellement, vers −200 m, souvent sans aucune discontinuité, aux vases profondes de l'étage bathyal. Le taux de sédimentation d'origine terrigène paraît commander l'essentiel des caractéristiques du sédiment. Ce dernier peut présenter un aspect général, et parfois une microfaune à tendance profonde, en un domaine référable au circalittoral, voire à l'infralittoral (Blanc-Vernet, 1969; Blanc, 1968). On mesure les risques d'erreurs dans l'interprétation des faciès marneux appartenant aux séries anciennes et cela d'autant plus que ces lutites peuvent être assimilées, de par leur composition, aux marnes, *shales* ou argilites (*claystone*).

Au point de vue sédimentologique, ce sont des marnes ou argilites en formation oú domine l'illite généralement associée à la chlorite. Les phyllites rencontrées sont d'origine exclusivement détritique: mélange illite + quartz, dominant, puis association: chlorite, micas et interstratifiés divers. D'autres minéraux s'y adjoignent en fonction de l'environnement géologique: montmorillonite en secteurs volcaniques (Santorin, Lipari) ou ultra-basiques, ophiolites, *etc.* (Rhodes, Eubée, canal de Talante), Kaolinite au voisinage des secteurs karstiques érodés (Péloponnèse).

La granulométrie de tels sédiments présente des fluctuations importantes *indépendantes* de la profondeur, de la teneur en eau et des types de peuplements. On notera des faciès granulométriques de *décantation* et de *lévigation;* les teneurs en pré-colloïdes et en matière organique sont irrégulières, sous la dépendance des biocoenoses et de l'hydrodynamisme. En général, les médianes varient entre 4 et 10 microns. Un sablon détritique quartzeux et micacé peut représenter jusqu'à 15% et même 35% du poids total du sédiment. Il résulte d'un transport turbide par suspension ou flottation ou, en certains cas, d'un apport d'origine éolienne (Golfes de Gabès et de Hammamet).

Enfin, les zones séismiques de l'Egéide montrent un enrichissement en micas, tendant vers un faciès psammitique, pour les vases des *grabens* individualisés au Pléistocène.

La teneur en calcaire montre des fluctuations multiples en fonction du substratum, des biocoenoses et thanatocoenoses, des facteurs hydrodynamiques. On pourra distinguer plusieurs groupes:

1. Lutites calcaires: plus de 35 % de CO_3Ca. La "poudre" carbonatée constituant la majeure partie du sédiment est d'origine pélagodétritique (coccolithes) ou résulte de suspensions très fines issues de l'abrasion mécanique ou biologique des zones littorales, herbier ou coralligène. Contrairement aux sédiments polygéniques du détritique côtier, plus ou moins envasés, nous avons affaire ici à un épandage floconneux, souvent pollué, aux limites nettes (Golfes de Gabès, Thrace, Languedoc).

2. "Marnes" renfermant 25 à 35% de calcaire: tel est le cas des V. T. C. à l'ouest de Marseille, Golfe du Lion, Golfe de Gênes, Baie de Thessalonique, Dodécannèse, *etc.*). Les fractions les plus riches en calcaire renferment une abondante microfaune benthique et pélagique. Blanc-Vernet (1969) a insisté sur les analogies de cette dernière avec celle des milieux bathyaux. En fait, on observe une étroite relation entre les assemblages de Foraminifères et la nature granulométrique du sédiment, indépendamment de la profondeur. On citera l'exemple du débouché du Grand Rhône et du Golfe de Volo où des Foraminifères de la haute mer, souvent considérés comme "profonds" se rencontrent en milieu turbide circalittoral.

3. "Argilites" présentant 10 à 12% de calcaire, en bordure des littoraux cristallins, éruptifs ou métamorphiques (archipel de Santorin, Ouest de la Corse, *etc.*).

Il est intéressant de considérer les milieux à taux de sédimentation *rapide*. Dans les "vasières" au Sud de la Thrace ou au large de Gênes-Portofino, la teneur moyenne en calcaire demeure assez constante, ainsi que la granulométrie, malgré une augmentation continue de la profondeur. La proportion en carbonates est réglée par l'apport biogene: Coccolithes, Foraminiféres, Echinides. Les dragages, prélèvements à la benne, plongées et photographies sous-marines, carottages à large section, *etc.* montrent une microstructure complexe à l'intérieur du sédiment, de petits remaniements intraformationnels répétés (terriers, bioturbation), taphoceonoses à Turritelles (Golfe de Fos, Baie de Kalamata, Péloponnèse). Ces zones d'hypersédimentation présentent une multitude de traces conservées plus ou moins durablement sur le fond: pistes, "souilles", griffures de Crustacés, orifices de terriers, "pellets" et excréments. La variété ichnologique de ces fonds peut être altérée par des tassements différentiels et de petits *slumpings* affectant le sédiment gorgé d'eau.

En Méditerranée orientale, des zones isolées de la haute mer par le jeu d'actions tectoniques récentes au Pléistocène, forment de véritables "pièges à sédiments". Il se dépose dans ces "grabens" une lutite plus ou moins argileuse très voisine du faciès des V.T.C. Le sédiment est sombre, enrichi par la pyrite et l'hydrotroïlite. La couleur varie du bleu au noir. Une vase noirâtre fétide, s'accumule ainsi au fond des golfes de Volo, de Thrace et des "ombilics" du canal de Talante, présentant localement un faciès "euxinique". Là encore, il peut y avoir une convergence lithologique avec les milieux sapropéliques plus profonds.

Pour conclure sur de tels sédiments qui paraissent correspondre à des faciès marneux en formation, on notera d'abord leur répartition ubiquiste en milieu généralement circalittoral: de quelques mètres de profondeur jusqu'à 150, voire 200 m. Le sédiment très fin, réduit, riche en eau, se situe en des zones relativement abritées sans que cela constitue réellement un facteur dirimant (Golfe de Fos): les agglomérats et flocons, collés par les forces électrostatiques, peuvent résister à des mouvements notables de la tranche d'eau. Par ailleurs, la floculation doit être rapide et la sédimentation presque immédiate, du moins à faible profondeur. Mais le matériel détritique, le plus souvent très fin (sablon de quartz, poudre micacée, phyllites), élutrié puis décanté, sera fréquemment déposé assez loin des sources d'apports. En haute mer peut intervenir une sédimentation différentielle aboutissant à un certain triage des sédiments.

EROSION ET TRANSPORT; IMPORTANCE DES FACTEURS HYDRODYNAMIQUES

Parmi les zones observées en Méditerranée, nous ne considérons ici que deux types de modèles: les circuits et courants de golfes et le régime des détroits.

Circuit et Courants de Golfes

Une sédimentation bioclastique complexe a été étudiée en détail par Poizat (1970). Ici encore, les aires de production sont essentiellement représentées par l'herbier à Posidonies; la dispersion du matériel biogène s'y affectue par traction contre le fond, ensuite la sédimentation est réalisée par délestage et décantation.

Les aires hydrodynamiques sont très variées dans

le Golfe de Gabès: ce sont les réfractions des vagues vis-à-vis des écrans naturels, barrières d'îles et zones d'herbiers fonctionnant comme des filtres. Ces phénomènes sont complétés par l'action de la marée présentant en ces lieux un marnage exceptionnel de 0,80 m. Il en résulte des courants formant des circuits, amenant l'érosion du fond, creusant des chenaux dans les "mattes" de l'herbier à Posidonies et dispersant vers le large le sédiment. Ces zones, dans les herbiers, correspondent à de faibles profondeurs et renferment une riche microfaune (Castany et Lucas, 1955; Glacon, 1963).

Poizat a montré que la forme des débris bioclastiques détermine leur degré de flottabilité ainsi que leur mode de transport, notamment pour les fragments d'Algues calcaires (*Halimeda, Peyssonelia*). Un travail cartographique détaillé a montré la zonation suivante: (1) herbier à Posidonie, aire de production; (2) en bordure immédiate se déposent les débris à faible flottabilité (Gastéropodes, gros Foraminifères: *Elphidium*, Miliolidae). Le déplacement s'effectue par *traction*, et (3) plus au large, les courants drainent les allochems les plus aplatis, les débris à *flottabilité croissante (Peyssonelia, Marginopora, Nubecularia)*. Les fragments attribués aux *Halimeda* seront transportés sur les plus grandes distances.

Cette zonation biodétritique réalise un épandage non turbide colmatant en grande partie le Golfe de Gabès, recouvrant d'anciennes structures, lentilles et "oncoïdes" à *Cladocora, etc.*

En d'autres lieux, des circuits et dérives analogues s'observent mais l'absence de marées et de courants violents amènent à des résultats moins spectaculaires. On notera cependant la dégradation assez rapide des herbiers, leur ravinement formant des chenaux anastomosés, l'épandage du matériel bioclastique vers les fonds du détritique côtier (Baie de Sanary-Le Brusc, rades de Giens et d'Hyères, *etc.*).

Le Régime des Détroits

Les détroits et les passes entre les îles ont été étudiés en Méditerranée dans les secteurs suivants: Détroit Siculo-Tunisien, Canal de Corse, Canal de Rhodes, détroits de Karpathos, Caso, Cérigho et Cerighotto, Canal Pélago, archipels de Riou et des Iles d'Hyères (Provence). Ces milieux présentent les caractères généraux suivants:

1. *Importance des facteurs hydrodynamiques*: Les courants de surface peuvent atteindre 1,5 à 2 noeuds aux détroits de l'arc égéen. Au niveau du fond, le mouvement peut-être suffisant pour déterminer le remaniement de "pralines" et même certains ravinements. Le déplacement des masses d'eau sera encore accentué par de véritables courants de marées entre des bassins présentant des oscillations à caractères différents (détroit siculo-tunisien, canal de Messine), des courants de seiches (canal de Talante, détroit de L'Euripe) et même, éventuellement, la présence d'ondes internes (arc Egéen).

2. *Lévigations et triages du sédiment*: la sélection granulométrique du matériel peut-être très poussée (Blanc, 1968).

3. Eventuellement, *non-sédimentation* et *oxydation* du fond pour les secteurs particulièrement balayés par les courants, à des profondeurs variables allant jusqu'à plusieurs centaines de mètres (Canal de Malte, Détroit de Cérigho.) Comme pour les *hard-grounds* observés dans les séries anciennes, on note des phénomènes d'induration, rubéfaction et concrétionnements, enduits, *etc.* de nature calcitique ou ferrugineuse. Bien entendu, ces *hard-grounds* traduisent une discontinuité sédimentaire et non une émersion. Cependant, des actions très différentes pourront se traduire par des résultats *comparables*:

– émersion: amenant induration, lithification phréatique, rubéfaction et altération du substrat.

– parois de grottes marines battues montrant d'anciennes concrétions (stalactites et stalagmites) recouvertes d'un enduit oxydé (limonite, wad, *etc.*).

– rubéfactions profondes sur le précontinent et en bordure du talus (*knee-line*) concernant les Coraux ahermatypiques (Blanc *et al.*, 1959) ou des affleurements calcaires quaternaires. Une telle rubéfaction pourra concerner les formations meubles du détritique côtier ou du détritique du large. Nous pensons que les rubéfactions profondes mentionnées sur le précontinent, en dehors des zones à violents courants, peuvent résulter d'*oxydation* relevant d'un double mécanisme:

– émersions lors de phases régressives du Quaternaire.

– concrétionnements et formations d'enduits ferrugineux à la suite de phases rhexistasiques (altérations des aires continentales, formation des *terra rossa* fossiles du Quaternaire, *ferretto* de l'interglaciaire Mindel-Riss, *etc.*), suivies d'une érosion et d'un transport en milieu marin sous la forme de solutions.

4. *Assemblages biologiques hétérogènes*: les biocoenoses et thanatocoenoses actuelles, sub-actuelles et fossiles réalisent un mélange complexe, largement étalé dans le temps *alors que le taux de sédimentation demeure faible ou nul* (archipel de Riou, Cyclades, Sporades, Dodécanèse, Détroit Siculo-Tunisien, *etc.*). Ces dispositions aboutissent à une *condensation* de faciès entrecoupée de discontinuités.

La dérive contre le fond provoque l'érosion des mattes de l'herbier, creusant des chenaux parfois profonds de 4 à 6 m (Santorin, Anti-Paros, Riou).

Au niveau du "coralligène de plateau" l'érosion est moins spectaculaire mais elle aboutit à une véritable lévigation du sédiment qui se trouve balayé, tracté contre le fond, puis étalé dans les zones du "détritique côtier". On observe une sélection des *allochems* en fonction de leurs caractères physiques: poids, dimension, sphéricité relative, flottabilité, *etc.* Cette sélection peut accompagner un gradient granulométrique et l'on aboutit aux "sables fins bien calibrés" et "sables à Amphioxus" dont le faciès évoque certaines calcarénites bio-détritiques des milieux anciens.

L'examen des Foraminifères, "extraclasts", débris de mollusques, *etc.* traduit le *remaniement d'assemblages* avec beaucoup plus de précision que l'analyse granulométrique (Blanc-Vernet, 1969). La zone axiale des passes est souvent colmatée par un sablon ou une lutite calcaire, véritable matrice du sédiment, issue de l'évolution mécanique poussée d'un sable bio-détritique.

Des bancs à "pralines" et "*algal ball*" occupent des profondeurs généralement modérées, de 45 à 100 m, associés à de grandes Laminaires fixées sur les nodules calcaires. Au Canal de Caso, un ancien cône volcanique arrondi culmine à −80 m; il se trouve recouvert par cette "gravelle" concrétionnée, à Mélobésiées. Sur les flancs du cône, des remaniements, glissements et probablement séismes, amènent cette gravelle et ces "pralines" jusqu'à des profondeurs de 250 m, au niveau des vases bathyales à *Cidaris* et *Térebratule vitraea* (Canal Caso, banc Kolumbos, Santorin, Banc de Pantellaria).

Dans les "grabens" des détroits de Cythère (Canal Cérigho) et d'Anti-Cythère (Canal Cerighotto), les gravelles, pisolites et *algal ball* sont remaniées, en milieu très oxydant, jusqu'à des profondeurs de 570 m. La vitesse du courant de fond élimine le dépôt de la phase argileuse tandis que le substrat calcaire est altéré et rubéfié (hématite, wad) sur une épaisseur de 10 cm.

La sédimentation des plateaux sous-marins parsemant le Détroit Siculo-Tunisien, en plus des facteurs hydrodynamiques prépondérants, demeure fortement influencée par les *actions tectoniques récentes* (séismicité du Banc Graham, "Ile Julia"), le *volcanisme* (Pantellaria, Linosa) et, la *présence d'affleurements quaternaires* formant des entablements étendus au banc Talbot (Tyrrhénien) (Blanc, 1958). D'autres hauts-fonds présentent des affleurements gréseux ou marneux où l'on retrouve, jusqu'à des profondeurs importantes (−150 à −370 m) les faunes "froides" de Ficarazzi (Sicilien), Bancs des Esquerquis, Banc Hécate, Banc Médina, Banc de la Galite. On découvre encore, vers −165 m, les thanatocoenoses des "sables détritiques du large" (bancs d'El-Houaria et de Pantellaria-Vecchia).

Le taux de sédimentation demeure toujours nul ou très faible. Seuls les apports biogènes ou pyroclastiques constituent, en ces lieux, les constituants majeurs du lithofaciès. Des observations détaillées restent à entreprendre sur les détroits malgré des conditions de prélévements et d'observations difficiles. L'intérêt que ces zones présentent en Méditerranée paraît justifier la continuation des recherches.

CONCLUSIONS

1. Les sédiments biodétritiques de la marge continentale de la Méditerranée sont principalement issus des *aires de production* réalisées par les herbiers et les biocoenoses du "Coralligène". Les facteurs hydrodynamiques, la turbidité des eaux, la nature du substrat et la bathymétrie en régissent la répartition.

2. Les *zones d'accumulation* du matériel sédimentaire dérivent de l'érosion des aires de production (matériel bio-détritique allochtone) et de l'apport clastique d'origine continentale: suspension turbides, sables du "prisme littoral", éboulis, éolianites et pyroclastites, *etc.* L'érosion biologique et mécanique amène la formation d'une lutite essentiellement calcaire, mêlée à la matière organique, se déposant à une certaine distance du rivage. A ces niveaux on notera un apport pélagodétritique, d'origine minérale et biogène, devenant prépondérant au fur et à mesure que l'on s'éloigne de la zone littorale. Ainsi, les sables "détritiques côtiers", les "vases terrigènes côtières", *etc.* correspondent à des milieux sédimentaires polygéniques où les paramètres granulométriques sont indépendants de la profondeur.

3. La dynamique des milieux sédimentaires néritiques montre une dispersion du matériel sédimentaire à partir de trois pôles: apports détritiques d'origine continentale *senso latu*, herbier à Posidonies, coralligène. En fonction de la bathymétrie, des conditions hydrodynamiques et biologiques, le sédiment enrichit les "ceintures" du détritique littoral, les lutites du "terrigène côtier" et les formations bio-clastiques plus ou moins concrétionnées (Mélobésiées, "algal ball", "pralines" *etc.*). Les remaniements y sont fréquents. Néanmoins on notera des stades de rétentions et de stockages au niveau des herbiers à Posidonies entrecoupés de *cycles courts* d'érosion et de sédimentation. Les actions de colmatage l'emportent pour les zones abritées dans la mesure où se manifeste une certaine stabilité eustatique et tectonique.

4. Le faciès hétérogène des "sables détritiques du large" résulte de l'héritage würmien récent. Les éléments actuels y sont mêlés à des minéraux et thanotocoenoses se rapportant à un détritique côtier fossile (Blanc-Vernet, 1969) ou à d'autres types de

fonds. Le décalage des biocoenoses et des "ceintures" bioclastiques lors de la régression würmienne a été accompagné d'un *envasement* (milieu réducteur, silicification, pyrite et glauconie authigène) *suivi d'une phase d'oxydation et de non-sédimentation.*

5. Les lutites des "vases terrigènes côtières" correspondent à des faciès de marnes et calcaires marneux en formation. Il s'agit d'une sédimentation essentiellement réglée par l'importance de l'apport turbide indépendamment de la profondeur. L'apport biogène d'origine pélagodétritique (coccolites, Foraminifères, *etc.*) module la teneur en calcaire du sédiment qui présente de ce fait une assez grande variabilité. Les assemblages de Foraminifères, les caractères minéralogiques et granulométriques, *etc.* sont sous la dépendance de la turbidité et du taux de sédimentation qui assure, en ces lieux, la conservation des traces organiques et la présence souvent systématique de taphocoenoses.

6. Fortement influencés par le contexte géologique local et les indidences tectoniques, les facteurs hydrodynamiques réalisent le transit des sédiments. Les modalités habituelles concernent les phases très fines en suspension, étudiées par l'analyse des sestons, et les débris bioclastiques répartis et tractés en fonction de leur forme et de leur flottabilité. On peut arriver à définir une zonation des *allochems* dans les secteurs où règnent des courants de marées, circuits et dérives tourbillonnants (Golfe de Gabès, par exemple). Un autre modèle hydrodynamique est réalisé par le régime des détroits, assez spectaculaire en Méditerranée (Messine, Détroit Siculo-Tunisien, arc égéen, *etc.*) La vélocité des courants, au niveau du fond, détermine des remaniements, rubéfactions et altérations. L'absence de sédimentation et la formation de *hard-grounds* y sont fréquemment observées (Canal de Cerigho). Il en résultera la présence constante d'assemblages biologiques et de thanatocoenoses hétérogènes, témoins de niveaux quaternaires würmiens, tyrrhéniens ou plus anciens, reportés à des profondeurs variables sous l'effet des actions tectoniques récentes.

REMERCIEMENTS

L'auteur exprime sa gratitude à la Direction du Centre National de la Recherche Scientifique, au Centre National pour l'Exploitation des Océans et à Mr. J.M. Pérès, Directeur du Centre d'Océanologie d'Endoume (Marseille) qui ont permis la réalisation d'une grande partie des recherches exposées. Il convient encore de remercier tous mes collaborateurs du Laboratoire de Géologie marine et Sédimentologie appliquée de Marseille ainsi que les équipages des navires de recherches ANTEDON, CALYPSO et JEAN CHARCOT.

BIBLIOGRAPHIE

Berthois, L. 1964. Recherches sur les modalités d'usure des débris organogènes calcaires. *Bulletin de la Société Géologique de France,* 7:461–466.

Blanc, J. J. 1953a. Hydrodynamique et sédimentation des fonds de calanques. *Comptes Rendus de l'Académie des Sciences,* 237: 1173–1175.

Blanc, J. J. 1953b. Premiers résultats des recherches sédimentologiques de la "Calypso" et de la Station Marine d'Endoume à l'archipel de Riou, Marseille. *Bulletin de la Société Géologique de France,* (3) 1–2:133–146.

Blanc, J. J. 1954. Petits galets du plateau continental provençal. *Comptes Rendus de l'Académie des Sciences,* 238:1334–1336.

Blanc, J. J. 1958. Sédimentologie sous-marine du détroit siculo-tunisiens. Campagne de la "Calypso" (Août–Sept.). *Résultats Scientifiques Campagne de la "CALYPSO" (III), Institut Oceanographique,* (1):158 p.

Blanc, J. J. 1959. *Recherches sédimentologiques littorales et sous-marines en Provence occidentale,* Thèse, Paris, Annales Institut Océanographique Paris, Masson, 35:140 p.

Blanc, J. J. 1964a. Recherches géologiques et sédimentologiques. Campagne de la "Calypso" en Méditerranée nord-orientale (1960). *Résultats Scientifiques Campagne de la CALYPSO",* (6):219–270.

Blanc, J. J. 1964b. Vases bathyales et sables détritiques au large de Marseille. *Recueil des Travaux de la Station Marine Endoume,* (47)31:203.

Blanc, J. J. 1966. Le Quaternaire marin de la Provence et ses apports avec la géologie sous-marine. *Bulletin du Musée d'Anthropologie Préhistorique de Monaco,* 13:5–27.

Blanc, J. J. 1968. Sedimentary geology of Mediterranean Sea. *Oceanography and Marine Biology; Annales Revue,* 6:377–454.

Blanc, J. J., J. M. Pérès et J. Picard 1959. Coraux profonds et thanatocoenoses quaternaires en Méditerranée. *Colloque du Centre National de la Recherche Scientifique, Nice.* Topographie et Géologie des profondeurs Océaniques, 83:185–192.

Blanc-Vernet, L. 1969. Contribution à l'étude des Foraminifères de Méditerranée. Relations entre la microfaune et le sédiment, biocoenoses actuelles, thanatocoenoses pliocènes et quaternaires. *Thèse, Recueil des Travaux de la Station Marine d'Endoume,* (64) 48:315p.

Boclet, D., M. L. Drugy, G. Lambert, J. Ross, J. Thommeret, et J. Labeyrie 1962. Essai d'analyse de la radioactivité de l'éponge *"Hercinia variabilis". Colloque du Centre National de la Recherche Scientifique, Villefranche,* Océanographie Géologique et Géophysique de la Méditerranée Occidentale, 211–220.

Bonifay, E. 1967. La tectonique récente du bassin de Marseille dans le cadre de l'évolution post-miocène du littoral méditerranéen français. *Bulletin de la Société Géologique de France,* 9: 549.

Bourcart, J. 1949. *Géographie du fond des mers. Etude du relief des océans.* Payot, Paris, 420 p.

Bourcart, J. 1952. *Les frontières de l'océan.* Albin-Michel, Paris, 318 p.

Bourcart, J. 1955. Les sables profonds de la Méditerranée. *Archives Scientifiques Genève,* 8:5–13.

Bourcart, J., et F. Ottmann 1957. Recherches de géologie marine dans la région du Cap Corse. *Revue de Géographie Physique et Géologie Dynamique,* 1 (2): 66–78.

Bourcier, M. 1968. Etude du benthos du plateau continental de la baie de Cassis. *Recueil des Travaux de la Station Marine d'Endoume*, 60 (44): 65 p.

Cailleux A. 1947. L'indice d'émoussé des galets. Définition et première application. *Bulletin de la Société Géologique de France*, (5), 17: 250.

Castany, G., et G. Lucas 1955. Sur l'existence d'oolites calcaires actuelles au large de l'île de Djerba (Sud-Tunisien). *Comptes Rendus Sommaire de la Société Géologique de France*, 229–232.

Dangeard, L. 1929. Observations de géologie sous-marine. Thèse, Paris, *Annales Institut Océanographique Paris*, (1): 296 p.

Dangeard, L., M. Rioult, J. J. Blanc, et L. Blanc-Vernet 1968. Résultats de la plongée en soucoupe n°421 dans la vallée sous-marine de Planier, au large de Marseille. *Bulletin Institut Océanographique de Monaco*, 67, (1384) 21 p.

Dieuzeide, R. 1940. Etude d'un fond de pêche d'Algérie: la Gravelle de Castiglione. *Station Aquiculture Pêche de Castiglione*, Nouvelle série 1, (4):33–57.

Folk, R. L. 1959. Practical petrographic classification of limestones. *The American Association of Petroleum Geologists Bulletin*, 43:1–39.

Froget, C. 1963. La morphologie et les mécanismes d'érosion du littoral rocheux de la Provence occidentale. *Recueil des Travaux de la Station Marine d'Endoume*, 30 (45): 165–243.

Froget, C. 1966. Découverte de formations quaternaires sous-marines au banc du Veyron (Baie de Marseille). *Comptes Rendus de l'Académie des Sciences*, 263: 1352–1354.

Froget, C. 1967. Découverte d'affleurements quaternaires anciens sur le précontinent provençal au large de l'île de Riou. *Comptes Rendus de l'Académie des Sciences*, 264 (2): 212–214.

Gautier, Y., et J. Picard, 1957. Bionomie du banc du Magaud (E. des îles d'Hyères). *Recueil des Travaux de la Station Marine d'Endoume*, (12)21: 28–40.

Glaçon, G. 1963. *Foraminifères des dépôts actuels des côtes de Tunisie sud-orientale. Thèse*, Montpellier, 270 p.

Glangeaud, L. 1949. Observations sur le triage granulométrique des sédiments le long des plages sableuses à marées. *Sédimentation et Quaternaire*, 95.

Gouvernet, C. 1965. Le comblement alluvial de la basse vallée du Gapeau à Hyères (Var). *Traveau du Laboratoire de Géologie de la Faculté des Sciences, Marseille*, 8: 195–210.

Grabau, A. W. 1904. On the classification of sedimentary rocks. *American Geologists*, 33: 228–247.

Jacquotte-Hippeau, R. 1962. Etude des fonds de maërl en Méditerranée. *Recueil des Travaux de la Station Marine d'Endoume*, 26(41).

Laborel, J. 1960. Contribution à l'étude directe des peuplements benthiques sciaphiles sur substrats rocheux en Méditerranée. *Recueil des Travaux de la Station Marine d'Endoume*, 33(20).

Laborel, J., J. M. Pérès, J. Picard et J. Vacelet 1961. Etude directe des fonds des parages de Marseille de 30 à 300 m avec la soucoupe plongeante Cousteau. *Bulletin Institut Océanographique de Monaco*, 1206:1–16.

Marion, A. F. 1883a. Considérations sur les faunes profondes de la Méditerranée. *Annales du Muséum d'Histoire Naturelle de Marseille*, 1 (2):34–35.

Marion, A. F. 1883b. Esquisse d'une topographie zoologique du golfe de Marseille. *Annales du Muséum d'Histoire Naturelle de Marseille*, 1 (1):1–108.

Molinier, R., et J. Picard 1952. Recherches sur les herbiers à Phanérogames marines du littoral méditerranéen français. *Annales de l'Institut Océanographique de Monaco*, 27, (3): 157–234.

Nesteroff, W. D., 1960. Les sédiments marins entre l'Esterel et l'embouchure du Var. *Revue de Géographie Physique et de Géologie Dynamique*, 1: 17–28.

Pérès, J. M. 1961. *Océanographie Biologique et Biologie marine*. Presses Universitaires de France, 1:541 p.

Pérès, J. M. et J. Picard 1958. Manuel de bionomie benthique de la Méditerranée. *Recueil des Travaux de la Station Marine d'Endoume*, (18) 11: (nouvelle édition en 1964).

Picard, J. 1954. Les formations organogènes benthiques méditerranéennes et leur importance géomorphologique. *Recueil des Travaux de la Station Marine d'Endoume*, 18 (13):55–76.

Picard, J. 1965. *Recherches Qualitatives sur les Biocoenoses Marines des Substrats meubles Dragables de la Région Marseillaise*. Thèse, Marseille, 160 p.

Picard-Tarbouriech, F. 1969. Contribution à l'étude des minéraux lourds dans les sables littoraux de Toulon au Cap Lardier (Provence). *Tethys*, 1: 539–560.

Poizat, C. 1968. Répartition des sédiments aux débouchés des calanques du littoral des massifs de Marseilleveyre et Puget (Marseille-Cassis). *Comptes Rendus de l'Académie des Sciences*, 267 (8): 831–834.

Poizat, C. 1970. Les modalités de la sédimentation bioclastique dans le golfe de Gabès (Tunisie). *Comptes Rendus de l'Académie des Sciences*, 270 (5): 676–678.

Rivière, A. 1951. Etudes de sédimentologie littorale du Laboratoire de Géologie du S.P.C.N. à Paris. *Proceedings of the Third International Congress of Sedimentology, Groningen-Wageningen*, 211 p.

Szekielda, K. A. 1969. La répartition du matériel organique devant les côtes. *Comptes Rendus de l'Académie des Sciences*, 268: 2323–2326.

Vita-Finzi, C. 1969. *The Mediterranean Valleys, Geological Changes in Historical Times*. Cambridge University Press, Cambridge, 140 p.

Vatova, A. 1947. Caratteri della fauna bentonica dell'Alto e Medio Adriatico e Zoocenosi cui dà origine. *Publicazione della Stazione Zoologica di Napoli*, 21 (1):51–67.

Morphology and Carbonate Sedimentation on Shallow Banks in the Alboran Sea*

John D. Milliman[1], Yehezkiel Weiler[2], and Daniel J. Stanley[3]

ABSTRACT

Alboran Ridge, Western Alboran Ridge and Xauen Bank are a part of a 180 km-long linear complex of shallow banks in the Alboran Sea. The banks are volcanic in origin and capped with Pliocene-Quaternary sediments. The uniformly shallow flat tops at 100 to 120 m and prominent terraces at somewhat greater depths may reflect the influences of shallow-water erosion, perhaps during lower stands of sea level.

The surficial sediments on these banks are composed almost entirely of biogenic carbonates, due to the separation of the banks from the north African shelf. Bryozoans and coralline algae are the dominant components at shallower depths, with molluscan fragments increasing with increasing depth. Below 270 m, planktonic foraminifera dominate the carbonate suite. Magnesian calcite (contributed by bryozoans and coralline algae) dominate all the shallow-water sediments, with much smaller amounts of aragonite and calcite. Deeper sediments contain much more calcite, but magnesian calcite, probably present as secondary void fillings, is also important.

Below a depth of about 80 m, a large fraction of the biogenic sediment is relic; carbon-14 dating shows that much of the material was deposited during the early Holocene, when sea level was lower. Deposition probably occurred at depths between 70 and 100 m, although the accumulation of specific components varied with time and environmental conditions.

RESUME

Le seuil d'Alboran et le banc du Xauen font parti d'un complexe rectiligne de plateformes peu profondes situé dans la Mer d'Alboran. Ces plateformes, d'origine volcanique, sont recouvertes de sédiments d'origine Pliocène-Quaternaire. Les sommets (100–120 m) aplatis de façon uniforme et, plus bas, les terraces principales sont vraisemblablement les témoins d'une érosion en eau peu profonde ayant eu lieu quand le niveau de la mer était plus bas.

Les sédiments qui recouvrent ces plateformes sont composés en majeure partie de carbonates biogéniques ce qui est une conséquence de la séparation des plateformes avec la plateforme africaine. A de faibles profondeurs les algues bryozoaires et corallines sont prédominantes avec des traces de fragments de mollusques qui deviennent plus abondantes avec la profondeur. Au-dessous de 270 m les foraminifères planctoniques prédominent dans la série des carbonates. Les sédiments d'eau peu profonde sont constitués principalement de calcite de magnésium (provenant des algues bryozoaires et corallines) et en moindre quantité d'aragonite et de calcite. A de plus grandes profondeurs, bien que les sédiments contiennent d'avantage de calcite on trouve également en quantités notables la calcite de magnésium.

Au-dessous d'une profondeur de 80 m la majeure partie du sédiment est d'âge antérieur au Quaternaire; la datation au carbone 14 montre que la plupart du matériau a été déposée au début du Holocène lorsque le niveau de la mer était plus bas. Le dépôt s'est vraisemblablement formé à des profondeurs comprises entre 70 et 100 m à l'exception de certaines composantes dont l'accumulation a été fonction du temps et des conditions d'environnement.

*Contribution Number 2808 of the Woods Hole Oceanographic Institution.

[1]Woods Hole Oceanographic Institution, Woods Hole, Massachusetts; [2]The Hebrew University, Jerusalem; and [3]Smithsonian Institution, Washington, D.C.

INTRODUCTION

A northeast-southwest trending ridge, called the "Seuil d'Alboran" (Giermann *et al.*, 1968) divides the narrow Alboran Sea into two separate basins (Figure 1). This marked topographic high, which rises abruptly more than 400 m above the adjacent deep-sea floor, consists of three major morphologic units; from the northeast to the southwest, the Alboran Ridge, the Western Alboran Ridge (including its extension to the west, Tofiño Bank) and Xauen Bank (Figure 1). These ridges combine to form a 180 km long physiographic complex (named hereafter the *Alboran Line* to avoid confusion with the Alboran Ridge proper). This complex extends from near the mid-point of the boundary between the western Balearic Basin and the Alboran Sea (36°00′N, 2°45′W) to a site 15 km north of the Moroccan coast (35°22′N, 4°28′W).

South of the Alboran Line lies the Eastern Alboran Basin, with depths exceeding 1000 m. The Western Alboran Basin (Stanley *et al.*, 1970), located north of Xauen Bank, is about 1500 m deep. The narrow Alboran Strait separates Djibouti Bank to the north from the Western Alboran Ridge and Alboran Ridge and it also connects the Western Alboran Basin to the Balearic Sea.

A north-south seismic profile across the Alboran Ridge at 3°10′W longitude shows approximately 400 m of stratified Pliocene-Quaternary sediments covering a presumably volcanic extrusive basement (Glangeaud *et al.*, 1967). A second profile crossing the Western Alboran Bank at about 35°45′N latitude and 3°35′W longitude also shows sediment draping over basement (Huang and Stanley, this volume, their figure 5). Much of what is known about the geography and geology of this area, however, is based on studies of the volcanic Isla del Alboran by Salvator (1898) and subsequent workers (see Giermann *et al.*, 1968 and references therein). The island, composed of volcanic-sedimentary sequences, presumably of late

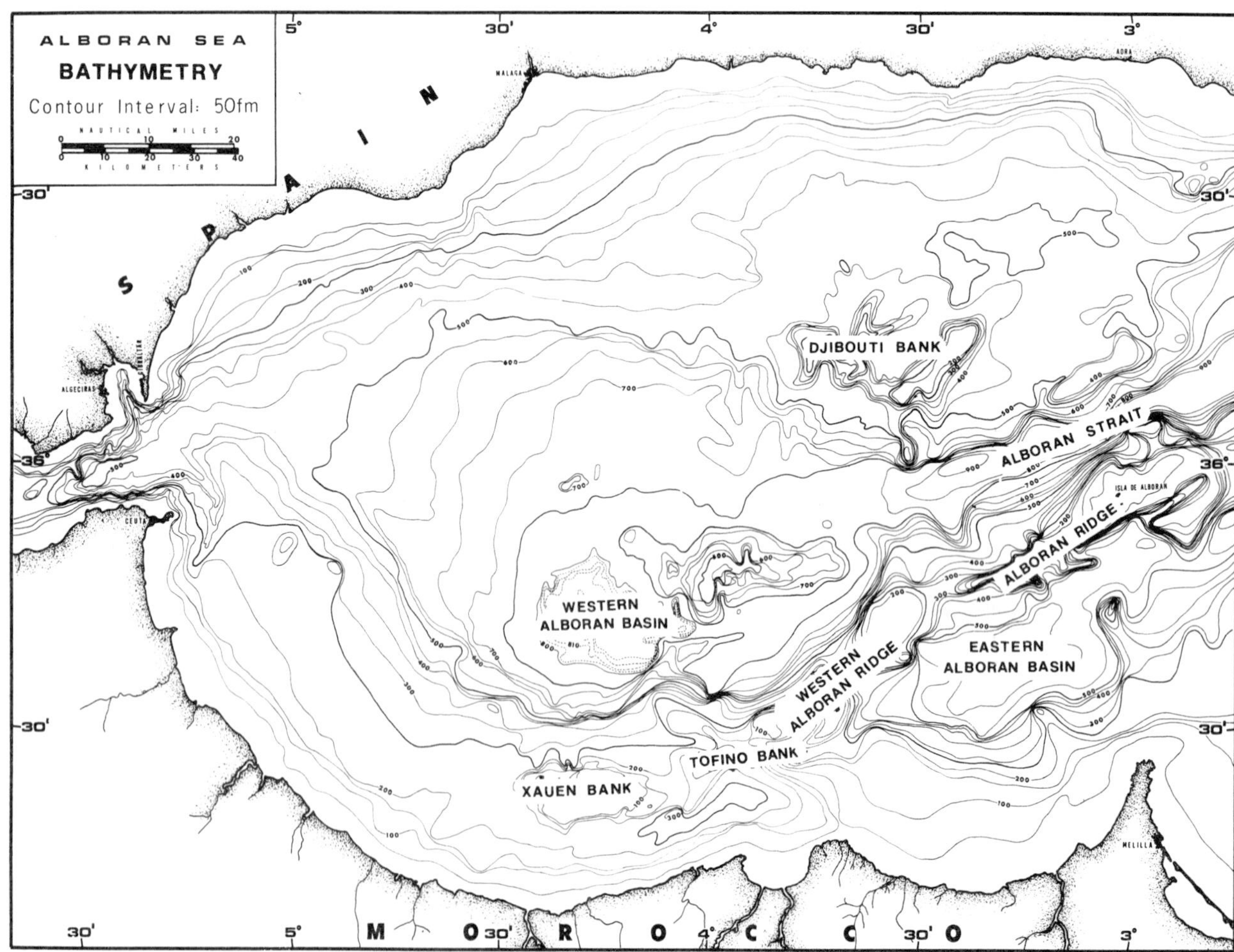

Figure 1. Reference map of the Alboran Sea, showing locality names used in text. Modified from base charts of SACLANT ASW Research Centre and Huang and Stanley (this volume).

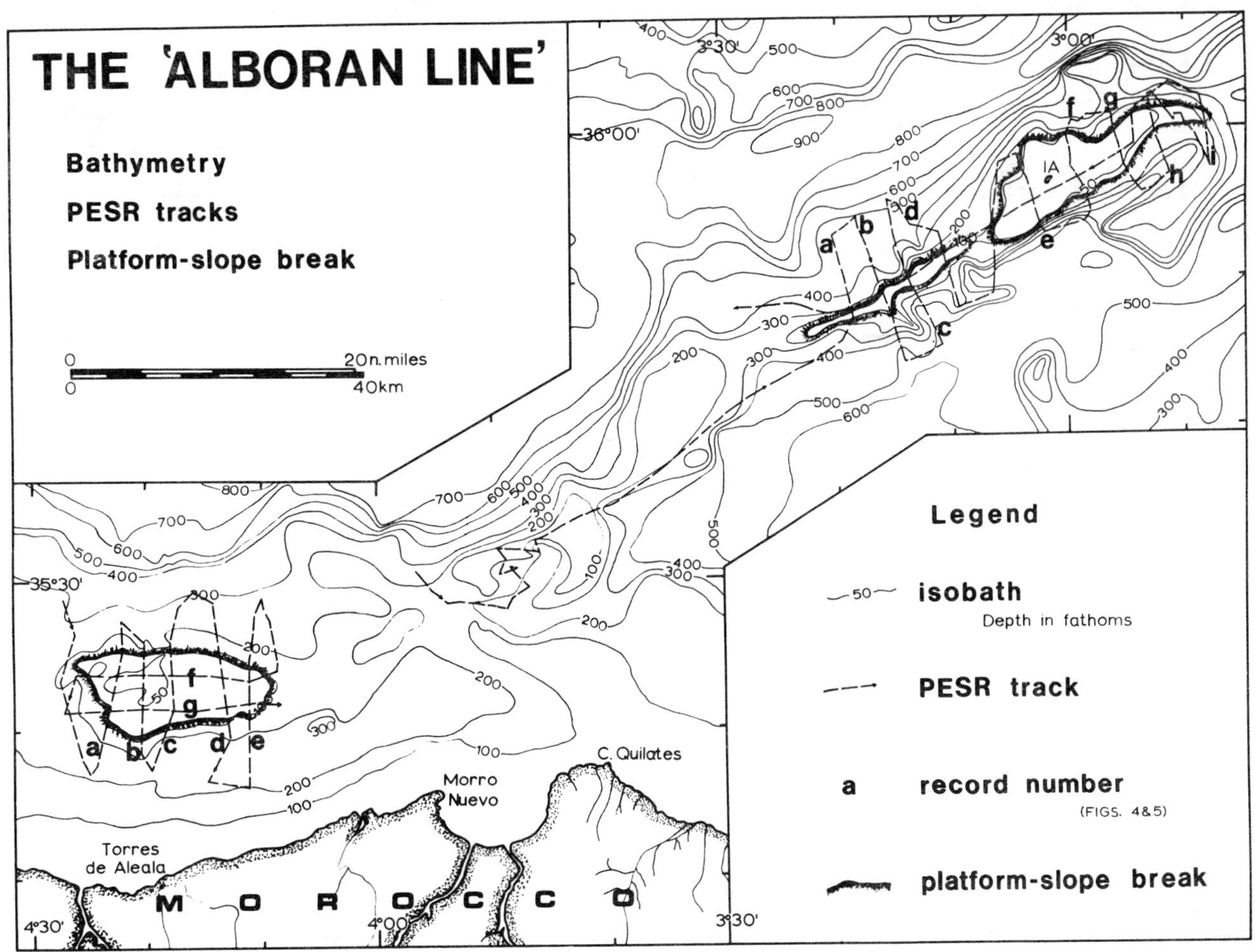

Figure 2. Chart of the bank and ridge (*Alboran Line*) province in Alboran Sea north of Moroccan coast, showing bathymetry and location of PESR track lines (see Figures 4 and 5).

Miocene age, has been re-examined in recent years by Eriksson (1961) and Gaibar-Puertas (1969).

This paper reports the results obtained aboard the USCGC ROCKAWAY cruise $RoSm_1$ which visited the Alboran Line in August, 1970. The study basically consisted of two parts: (1) a study of the topography and microtopography, using PESR (Precision Echo Sounding Record) surveys together with oblique angle deep-sea photography, and (2) a study of the bottom sediments sampled from the area.

Approximately 450 km of PESR survey lines were obtained during the cruise (Figures 2, 4 and 5). Xauen Bank was traversed by both north-south and east-west lines, whereas the narrower Alboran Ridge was traversed only by north-south parallel lines (Figure 2). Western Alboran Bank was traversed by 2 north-south and east-west lines. In total, 15 traverses were made normal to the long axis of the ridge complex, and 4 parallel to it. Navigation was achieved by both radar and land fixes.

A total of 15 successful camera and 16 Shipek bottom grab sample lowerings were made. In addition, 13 dredge hauls were recovered from six stations (Table 1; Figure 3). Where possible, camera stations were occupied adjacent to PESR survey lines. The camera (a 35 mm Alpine Geophysical Model 311) was triggered by bottom contact and photographs were collected at 15 to 20 second intervals as the ship drifted. The foreground in the photographs represents a distance of approximately 150 cm, and the area ranges from 2 to 3 m^2. A total of 221 frames were available for study.

Petrographic analyses of the sediment samples included quantitative identification of carbonate components, determination of carbonate mineralogy by x-ray diffraction, and analysis of carbonate content by acid leaching. After carbonate was removed, the noncarbonate fraction was studied petrographically. In addition, gravel, sand, silt and clay contents were calculated by standard sieve and pipette analyses. A total of 10 samples and sample splits were submitted for carbon-14 age dating.

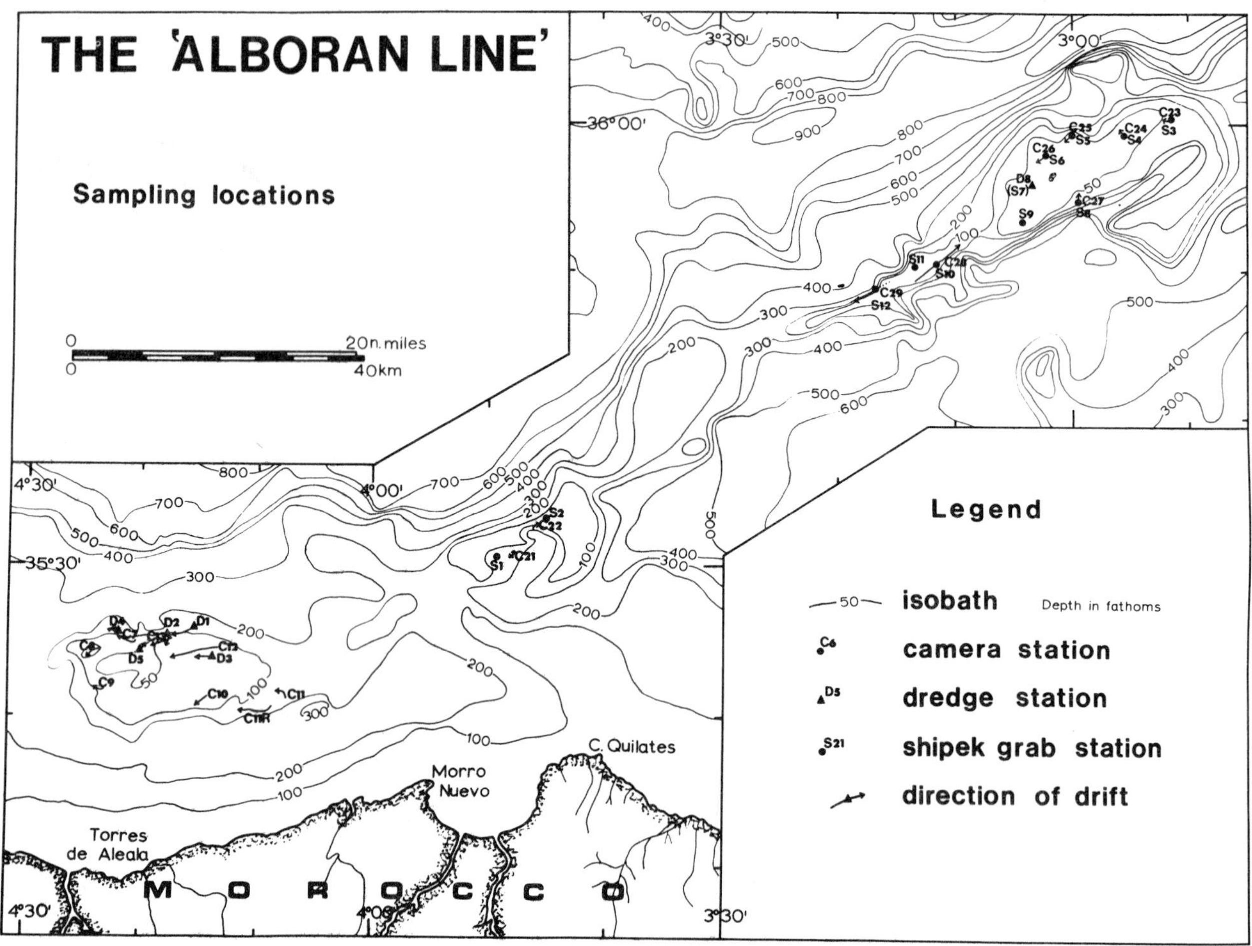

Figure 3. Chart showing sample and camera stations. Arrows delineate drift of ship during occupation of each station.

GENERAL TOPOGRAPHY

All the banks on the Alboran Line possess flat summits at depths between 100 and 120 m (55 to 65 fms); saddles 350 to 500 m deep separate the different banks (Figures 4 and 5). Evidence of some Quaternary tectonic activity in the Alboran Sea has been provided by Ryan *et al.* (1970), but it is probable that the topography of the upper bank surfaces has been affected more directly by major eustatic oscillations. The depths of the upper bank and ridge surfaces coincide closely with the probable low stand of sea level during Würm II (approximately–120 m at 15 to 19 thousand years ago, according to Curray, 1965 and Milliman and Emery, 1968), suggesting a possible planation of this surface during lowered sea level.

Local relief on the bank tops generally does not exceed 20 m. An exception occurs on the Alboran Ridge in the vicinity of Isla del Alboran, which rises from –120 m to 12 m above sea level. Locally the flat bottom is interrupted by shallow depressions, low bank-like ridges and a few channels (Figure 4). A ridge extending a few fathoms above the leveled plane at the bank edge is observed on Xauen Bank and on the Alboran Ridge. Since the platform is capped mainly by biogenic debris (some of which may be reef builders) it is possible that these bank edge ridges may be relict reefs, similar to those seen on outer shelves in other parts of the world (Macintyre and Milliman, 1970, and references therein). The dearth of present sedimentological data, however, does not allow any definite conclusions. Saddles which cut through these ridges in traverses **e** and **h** at the Alboran Ridge (Figure 5) may have developed by erosive current activity. A few channels occur on the edge of the platform (Figures 4 and 5); when present they are prominent on echo sounder records and in bottom photographs. One of these channels, several meters deep and a few tens of meters wide, occurs in the southeast corner of Xauen Bank, close to camera station C11.

Table 1. Location of sampling (dredge hauls and ship grabs) and submarine camera stations occupied on Xauen Bank, Western Alboran Ridge and Alboran Ridge during $RoSm_1$ cruise (August, 1970). Additional data includes depths, number of camera frames at each station, and ship drift direction while operating the sampling devices or camera (see also Figure 2).

Station	Bank	Location		Ship	Depth		Submarine Camera Data			Depth		Number
		Latitude	Longitude	drift	fm	m	Station	Latitude	Longitude	fm	m	of frames
D1A	Xauen Bank	35°26.0′ N	04°15.7′ W	260°	164	300						
D1B	Xauen Bank	35°26.0′ N	04°15.7′ W	260°	↓	↓						
D1C	Xauen Bank	35°26.0′ N	04°15.7′ W	260°	116	212						
D2A	Xauen Bank	35°25.3′ N	04°17.3′ W	190°	78	143	C13	35°24.4′ N	04°18.2′ W	50	92	23
D2B	Xauen Bank	35°25.3′ N	04°17.3′ W	190°	62	113						
D3A	Xauen Bank	35°24.0′ N	04°14.0′ W	270°	65	118	C10	35°21.4′ N	04°21.4′ W	53	97	16
D3B	Xauen Bank	35°24.0′ N	04°14.0′ W	270°	↓	↓	C11R, C11*	35°21′ N	04°10′ W	115	210	10
D3C	Xauen Bank	35°24.0′ N	04°14.0′ W	270°	72	131	C12	35°24.6′ N	04°13.5′ W	58	106	16
D4A	Xauen Bank	35°25.5′ N	04°22.2′ W		126	231	C7*	35°25.0′ N	04°25.5′ W	52	95	—
D4B	Xauen Bank	35°25.5′ N	04°22.2′ W				C8*	35°25.0′ N	04°22.3′ W	51	93	—
D5A	Xauen Bank	35°24.5′ N	04°20.1′ W	50°	50	92	C9	35°21.8′ N	04°23.3′ W	50	92	15
D5B	Xauen Bank	35°24.5′ N	04°20.1′ W	50°								
D8 (S7)	Alboran Ridge	35°26.2′ N	03°03.1′ W		10	18						
S1	W.A.R.	35°30.5′ N	03°49.2′ W		138	252	C21	35°30.9′ N	03°47.9′ W	130	228	17
S2A	W.A.R.	35°33′ N	03°45′ W		51	91	C22	35°32.6′ N	03°46.6′ W	57	104	18
S2B	W.A.R.	35°33′ N	03°45′ W									
S3	Alboran Ridge	36°00.8′ N	02°51.8′ W	280°	52	95	C23A–B	36°00.8′ N	02°51.8′ W	52	95	A–14, B–9
S4	Alboran Ridge	35°59.6′ N	02°55.3′ W	330°	45	82	C24A–B	35°59.6′ N	02°55.3′ W	45	82	A–6, B–13
S5A–B	Alboran Ridge	35°59.2′ N	03°00.5′ W	240°	45	82	C25*	35°59.2′ N	03°00.5′ W	45	82	—
S6A	Alboran Ridge	35°58.6′ N	03°01.2′ W	240°	48	88	C26	35°58.6′ N	03°01.2′ N	48	88	13
S6B	Alboran Ridge	35°58.6′ N	03°01.2′ W	240°								
S6C	Alboran Ridge	35°58.6′ N	03°01.2′ W	240°								
S8	Alboran Ridge	35°55.3′ N	02°59.6′ W	360°	55	101	C27	35°55.3′ N	02°59.6′ W	55	101	16
S9A	Alboran Ridge	35°53.8′ N	03°03.1′ W		40	73						
S9B	Alboran Ridge	35°53.8′ N	03°03.1′ W									
S9C	Alboran Ridge	35°53.8′ N	03°03.1′ W									
S10	Alboran Ridge	35°49.2′ N	03°13.8′ W		99	181	C28	35°49.2′ W	03°13.8′ W	99	181	18
S11	Alboran Ridge	35°51′ N	03°13′ W		89	162						
S12	Alboran Ridge	35°49′ N	03°17′ W	240°	58	106	C29	35°49′ N	03°17′ W	58	106	17

W.A.R.—Western Alboran Ridge *No readable results

Some wave-like patterns were observed on Xauen Bank platform (mainly on the west side of the bank). These are best developed on traverses **g** and **f**, but other patterns also occur on the northern part of the bank (traverses **b** and **e**). The steep slopes of these asymmetric ridges dip towards the center of the bank, forming flat, saucer-like depressions. Although it is possible that these ridges may be sand waves, the textural parameters of the bottom sediments and measured current velocities and direction (see below) suggest that they are not. Giermann *et al.* (1968) concluded that the "waves" are inclined strata, a few decimeters thick. It is possible that the strata were inclined as a result of small faults, related to the formation of the main bank escarpment.

The slopes surrounding the Alboran ridges are generally quite steep. Measured gradients are as high as 23°, although they average about 10°. Several prominent nick points occur on the slope at Xauen Bank, at 145 and 200 m (80 and 110 fms), and a major break is observed near 440 m (240 fm). The slopes off the Alboran Ridge are generally steeper (usually 5° to 15°), and the depths of breaks in slope are less uniform than off Xauen Bank. Nevertheless, a few nick points are observed at about the 440 m isobath, and in some localities (such as the northern slope of the Alboran Ridge) a well-defined terrace is indicated.

MICROTOPOGRAPHY

Oblique deep-sea photography has supplemented the bathymetric records and also has facilitated the interpretation of small topographic features on the sea floor (Figures 6 to 8). A flat bottom is common over most of the area (Figure 6); most of the irregularities result from biogenic bioturbation. Several pictures, however, indicate channels several decimeters deep and a meter or so wide, located on the northern edge of Xauen Bank. Another feature, presumably a channel filled with some fine sediment, is observed on the northern edge of the Alboran Ridge, close to

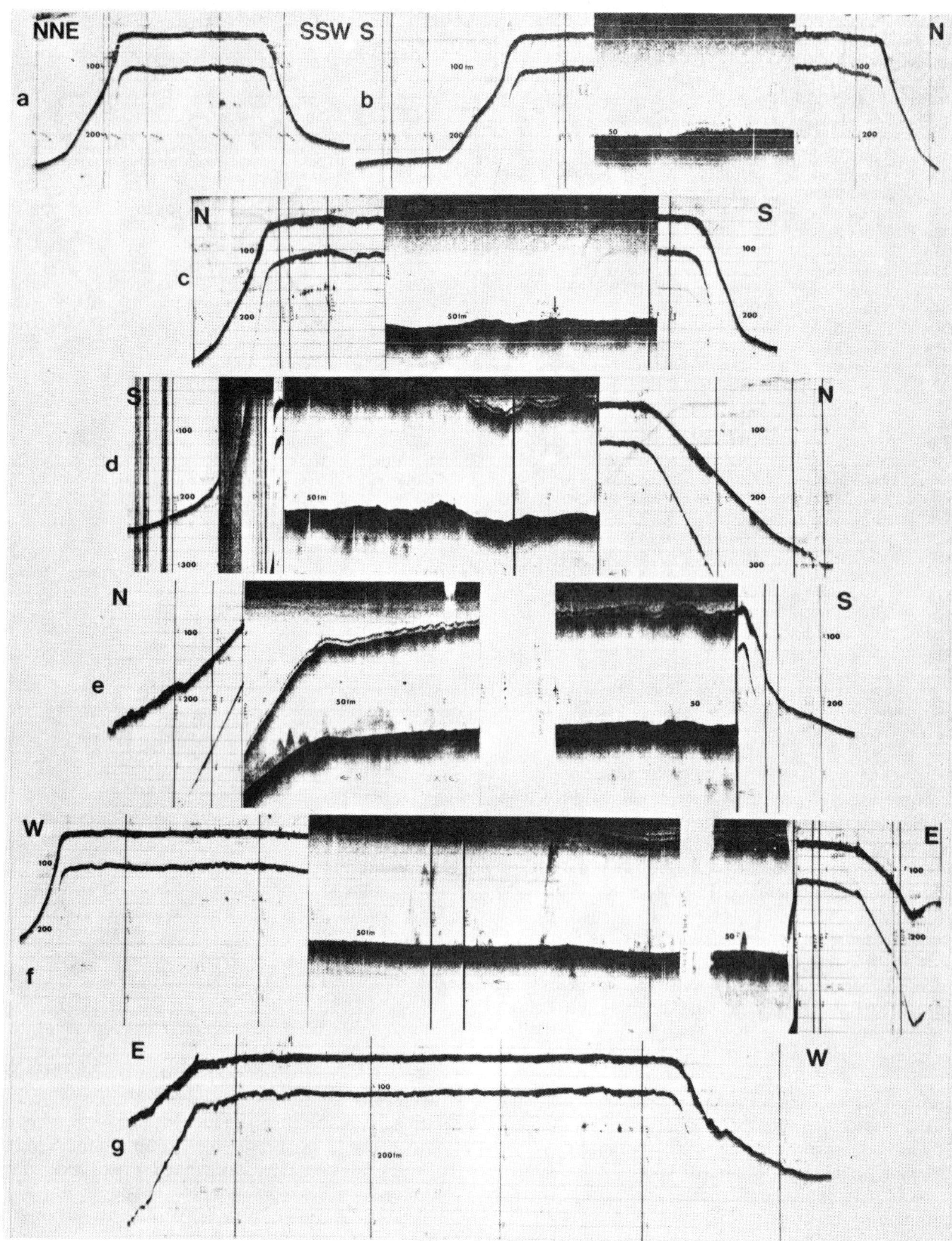

Figure 4. Echosounder (PESR) records across Xauen Bank (track lines shown in Figure 2).

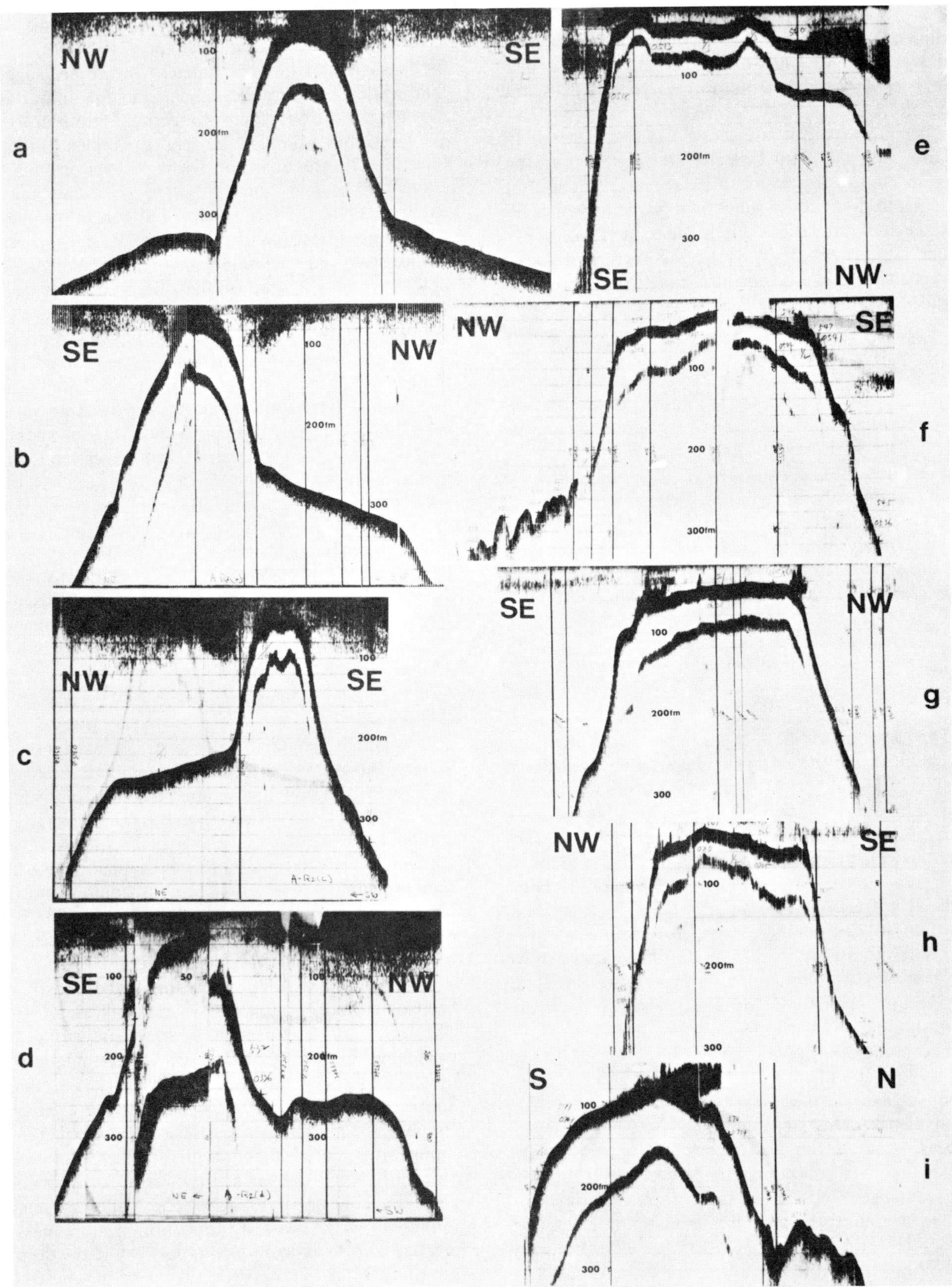

Figure 5. Echosounder (PESR) records across the Alboran Ridge (track lines shown in Figure 2).

traverse **g**. At this site an elongate tongue of fine sediment extends across an irregular bedrock, probably a biogenic limestone. The fine material also partly covers and surrounds pelecypod valves and algal nodules.

The most typical surface type is one covered by coarse sediment with little or no relief (Figure 6e, Figure 7c). In a few localities (Figures 7a and b), fragments (10 to 20 cm long) of volcanic rock, heavily encrusted with various biota have been noted. At other stations pebbles consist of aggregates of coarse to very coarse sand, cemented together by coralline algae.

Shallow depressions, circular to oval in shape and 5 to 20 cm in diameter, are very common (Figure 6a–c). These depressions probably were produced by organic activity, perhaps the resting sites of fish. Small, dome-like mounds, also attributed to biogenic activity, are present, but more rare than the depressions. A few markings (Figure 7b) may have been produced artificially by the camera. No depressions observed can be attributed to the effects of currents. Although large rounded pebbles do lie on the sand bottom, there is no evidence of pebble movements (*i.e.*, grooves or shadow scours) (Figure 7c).

SEDIMENTS

Texture

The size distribution of samples examined by mechanical analysis shows a remarkable uniformity (Table 2, Figures 9, 11, and 13). Most samples exhibit a modal size between 1 and 4 mm; a few samples also have a secondary mode between 250 and 500 microns. Only four samples contain more than 10 percent silt and clay (D1, S1, S3, and S6), but in none does this fraction exceed 20 percent. Another indication of the dearth of fine size sediment is the fact that only about 5 percent of the photographs show any evidence of clouds of mud brought into suspension by the impact of the compass or camera rig on the bottom.

Except for the portions coarser than 4 mm and finer than about 100 microns, the size distribution in most sediment samples behaves lognormally (Figure 13). Such distributions suggest sorting by some physical process, and currents do appear to be potentially important to the bottom. Shallow depths probably are influenced by the Atlantic water entering the Strait of Gibraltar and circulating in a clockwise direction in the Alboran Sea (Lucayo, 1968; Huang and Stanley, this volume, their figure 4). Bank surfaces below 100 m probably are swept by the westward flowing, denser intermediate (Levantine Water) Mediterranean water mass (Wüst, 1961; Lacombe and Tchernia, this volume, their figure 3).

Measured textural parameters, however, suggest that while these circulation patterns are important, they are not the only factors affecting bottom processes. First, the observed current directions (Figures 10 and 12) are able to orient several groups of epibenthic animals into preferred positions (Figure 8 c and d), but the lack of clear-cut current produced bed forms is noted. Either the currents are too weak to transport surficial sediment or at least they were not active at the time of the observations (August, 1970). Second, most of the sorting coefficients lie between 1.0 and 1.8, which are indicative of poorly-sorted sediments. If the current system were affecting the entire sediment, better sorting coefficients should be expected. Third, although the larger grains (coarser than about 0.7 mm) show evidence of mechanical abrasion (rounded and smoothed edges), the finer material is comparatively fresh in appearance. Large scale sediment movement should tend to round these grains. The general impression, therefore, is that although some local sediment movement may occur, large scale displacement of sand and gravel size sediments is not likely. The absence of fine-grained sediment, however, does indicate the winnowing action by the ambient current regime.

Composition

All samples contain predominantly calcium carbonate; the average carbonate content is greater than 95 percent, and only one sample (S1) contains as little as 90 percent. Noncarbonate fraction generally is composed of glauconite, although S3 and S11 both contain appreciable amounts of angular quartz and rock (volcanic?) fragments. The glauconite is present mostly as internal molds within foraminifera and other porous carbonate components, such as bryozoans and echinoid spines.

The carbonate fraction is dominated mostly by bryozoans and coralline algae at shallower depths. The bryozoans are present both as encrusting and branching forms. On the basis of examination of three samples (D4, S6 and S11), Alan H. Cheetham, Smithsonian Institution, has reported at least 49 species to be present. Most species (32) have membraniporiform zoarial types, with a lesser number of vinculariiform (6), cellariiform (4), celleporiform (3) and adeoniform (3) species. The shallowest sample (S6) contains the greatest proportion of more massive erect specimens (adeoniform types), while the deepest sample (D4) is composed primarily of encrusting membraniporiform types. The general species composition of the bryozoans is similar to those assem-

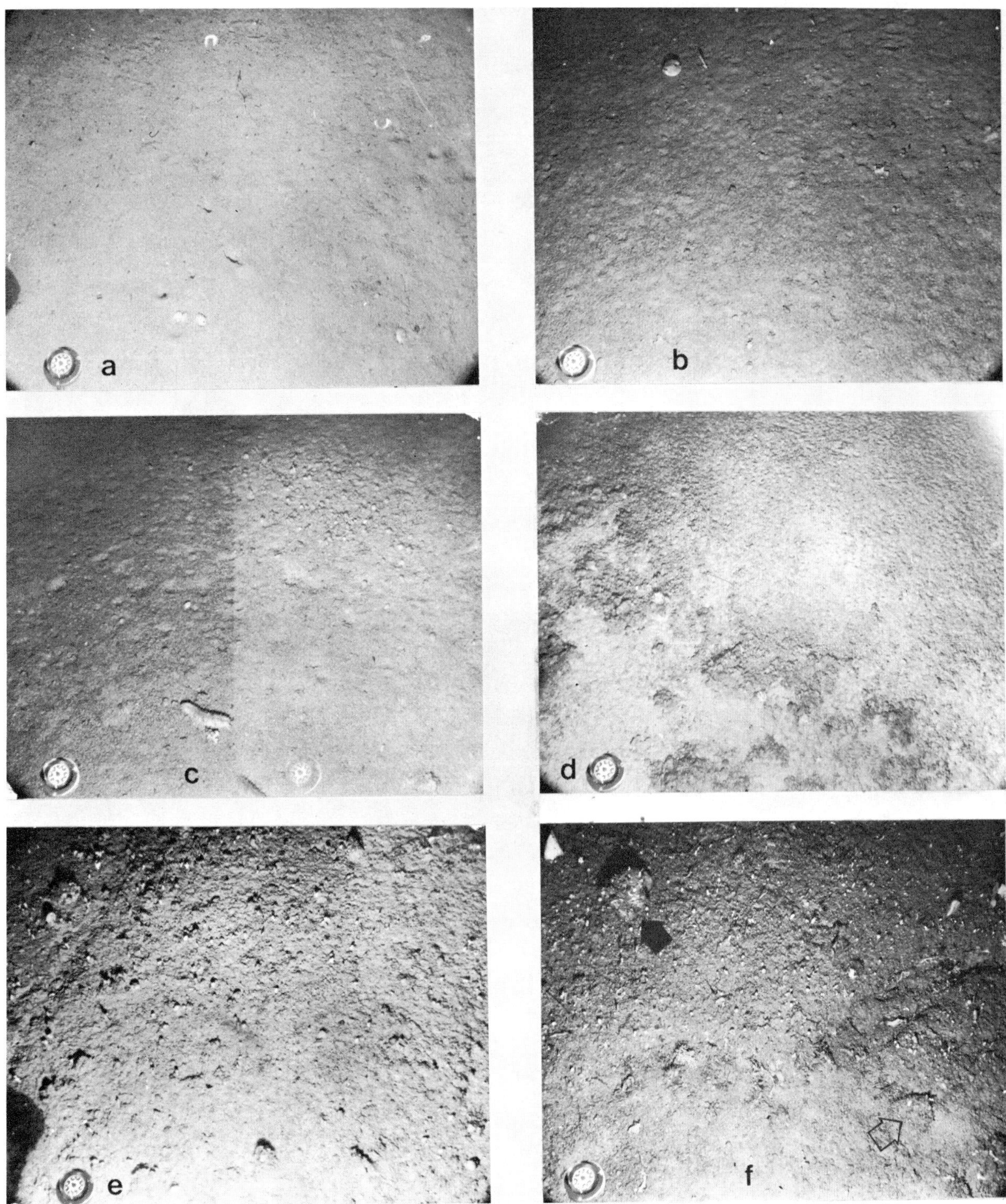

Figure 6. Bottom photographs showing different bottom types on the bank and ridge province of the Alboran Sea. Diameter of the compass is about 65 mm. (a) Flat smooth bottom with occasional depressions made by pelecypods; also some trails. Pelecypod valves, some hinged, are buried in fine to medium sand. C11R, southeast margin of Xauen Bank, traverse e. (b) Medium grained, flat bottom modified by small depressions (made by echinoderms or burrowing pelecypods). Note scattered fragments and sessile fauna. Southern edge of Alboran Ridge, traverse d, sampling station S10. (c) Holothurians lying on a flat, coarse to very coarse sand bottom. Note numerous shallow depressions of organic origin, and pelecypod assemblage at upper right margin. C28, southern edge of Alboran Ridge, traverse d, sampling station S10. (d) Outcrop or crust of biogenic limestone which covers much of entire bank platform. C13, northern part of the Xauen Bank, traverse c, sampling station D2. (e) Sessile forms and scattered faunal fragments on flat, very coarse sand bottom. C13, northern part of Xauen Bank, traverse c, sampling station D2. (f) Algal balls (arrows), stems and bryozoans on flat, very coarse bottom. C29, western part of Alboran Ridge, traverse b, sampling station S12.

Table 2. Textural and compositional properties of samples collected on Xauen Bank, Western Alboran Ridge, and Alboran Ridge (additional information given in Table 1).

		Granulometry (statistical parameters after Inman, 1952)											Percent Minerals			
		Percent														
Station	Depth m	gravel	sand	silt + clay (mud)	$Mo\phi$	$Md\phi$	$M\phi$	$\sigma\phi$	$\alpha\phi$	$\beta\phi$	Percent $CaCO_3$	Carbonate Assemblage	Calcite	Mg-Calcite	Ara-gonite	Non-carbonate components
D1A	300	11	75	14	−1.5	0.8	0.9	1.5	0.07	1.8	94	planktonic foraminifera, mollusks	15	65	20	glauconite
D1B		39	52	9	−0.7	−0.5	−0.15	1.55	−0.42	2.06		deep-water coral				
D1C	212											deep-water coral				
D2A	143	16	83	1	−1.5	−0.7	−0.35	1.65	0.21	0.45	99	bryozoan sand	15	60	25	glauconite
D2B	113	16	84	tr	−1.5	1	0.45	1.45	−0.38	0.52						
D3A	118	8	86	6	−1.5	1.2	1.05	1.35	−0.11	1.70	99	bryozoan sand	15	70	15	glauconite
D3B		17	76	7	−0.7	0.2	0.3	1.4	0.07	1.75		bryozoan sand				
D3C	131	20	79	1	−0.2	0.1	0.1	1.2	0	0.71						
D4A	231	5	89	6	1.3	1.1	0.9	1.0	−0.2	2.45	99	bryozoan sand	5	75	20	glauconite
D4B	231											algal balls				
D5A	92	14	84	2	0.7	0.2	0.3	1.1	0.09	0.64	99	bryozoan sand	0	75	25	glauconite
D5B	92											encrusted mollusks				
D8 (S7)	18											mollusks and algal balls				
S1	252	tr	80	20	2.3	2.3	3.8	2.2	0.68	0.59	90	planktonic foraminifera	45	35	20	glauconite
S2A	91											bryozoan sand	5	75	20	glauconite
S3	95	8	79	13	0.3	0.6	0.85	1.45	0.17	1.86	99	mollusks	20	65	15	angular quartz
S4	82	29	70	1	0.3	−0.1	−0.15	1.85	−0.03	0.78		algal balls, bryozoans	10	65	25	
S5A	82	15	84	1	−0.3	1.1	0.3	1.3	0.62	0.73						
S5B	82	16	83	tr	−1.5	1	−0.15	1.35	0.85	0.52						
S6A	88	27	53	20	−1.5	0	2.3	3.7	0.62	2.6	97	bryozoans, algae	15	60	25	mostly glauconite, some quartz
S6C	88	10	81	9	−0.3	0.4	1.2	1.5	0.53	1.23						
S8	101	1	93	6	1.7	1.6	1.5	1.1	−0.09	1.82	92	benthonic foraminifera	20	65	15	glauconite
S9	73										99	algal balls, bryozoans				glauconite
S10	181	29	69	2	−1.5	−0.2	0.1	1.5	0.2	0.57	95	bryozoans, mollusks	25	55	15	glauconite
S11	162										92	bryozoans	15	70	15	quartz and rock fragments
S12	106	45	52	3	−1.5	−0.9	−0.6	1	0.3	0.9						

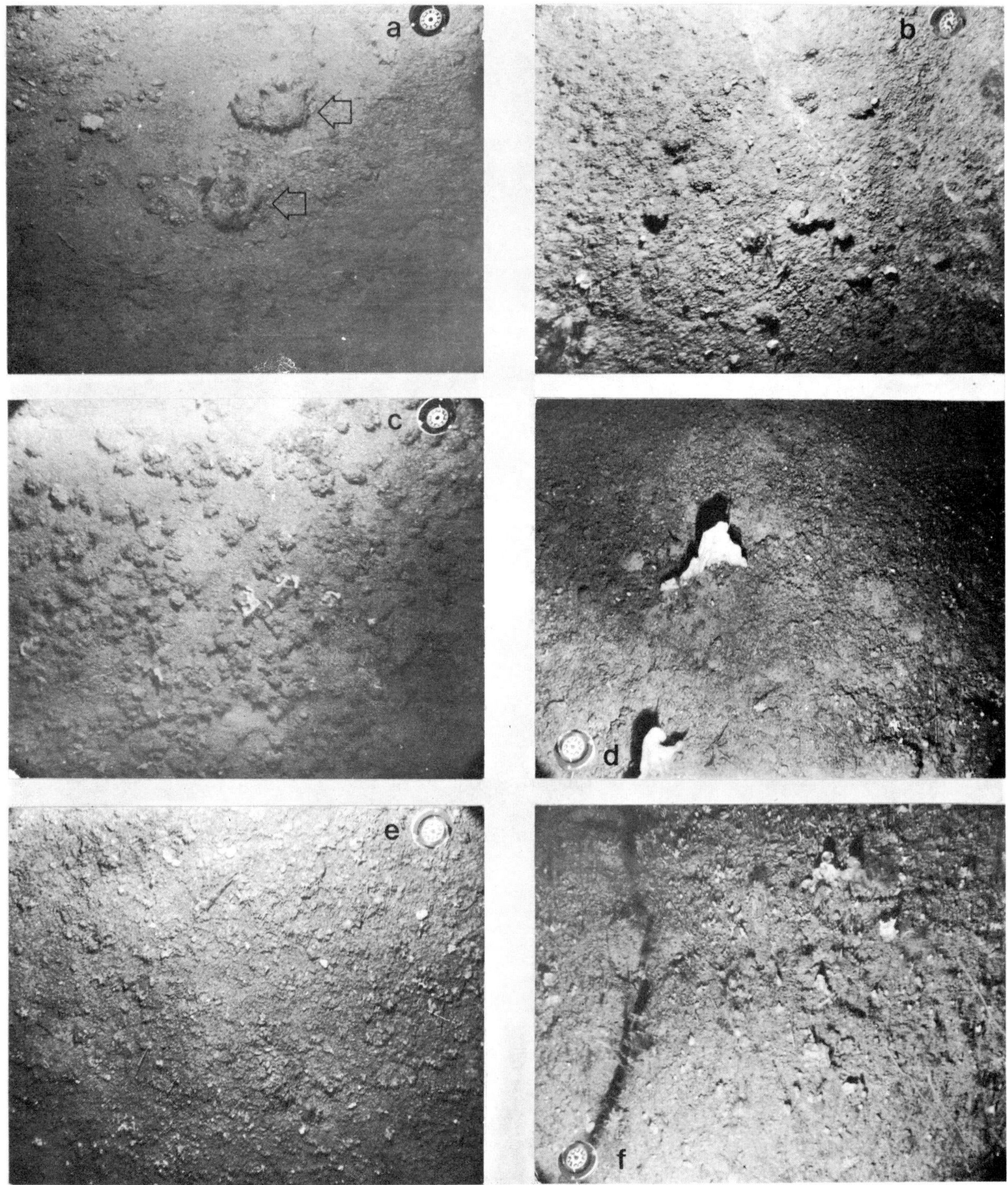

Figure 7. Bottom photographs showing different bottom types. Diameter of compass is about 65 mm. (a) Algal balls (arrows) on flat, pebbly bottom. Coarse biogenic debris is concentrated in the upper left corner. C9, southwest margin of Xauen Bank, traverse a. (b) Algal balls composed of coarse sand coated with coralline algae. Tracks may have been made by the camera frame dragged along the bottom. C13, northern part of Xauen Bank, traverse c, sampling station D2. (c) Concentrations of algal balls on flat, medium sand bottom. C24A, northern edge of Alboran Ridge, traverse g, sampling station S4. (d) Calcareous algal plates (?) partly buried in a very coarse organic debris. C29, western part of Alboran Ridge, traverse b, sampling station S12. (e) A flat, coarse sediment, covered by living algae, bryozoans and crinoids. C22, Western Alboran Ridge. (f) Rich biota attached to a coarsely grained, pebble-covered bottom. A linear depression, appearing as long shadow on left side, may be a joint or a small fault. C22, Western Alboran Ridge.

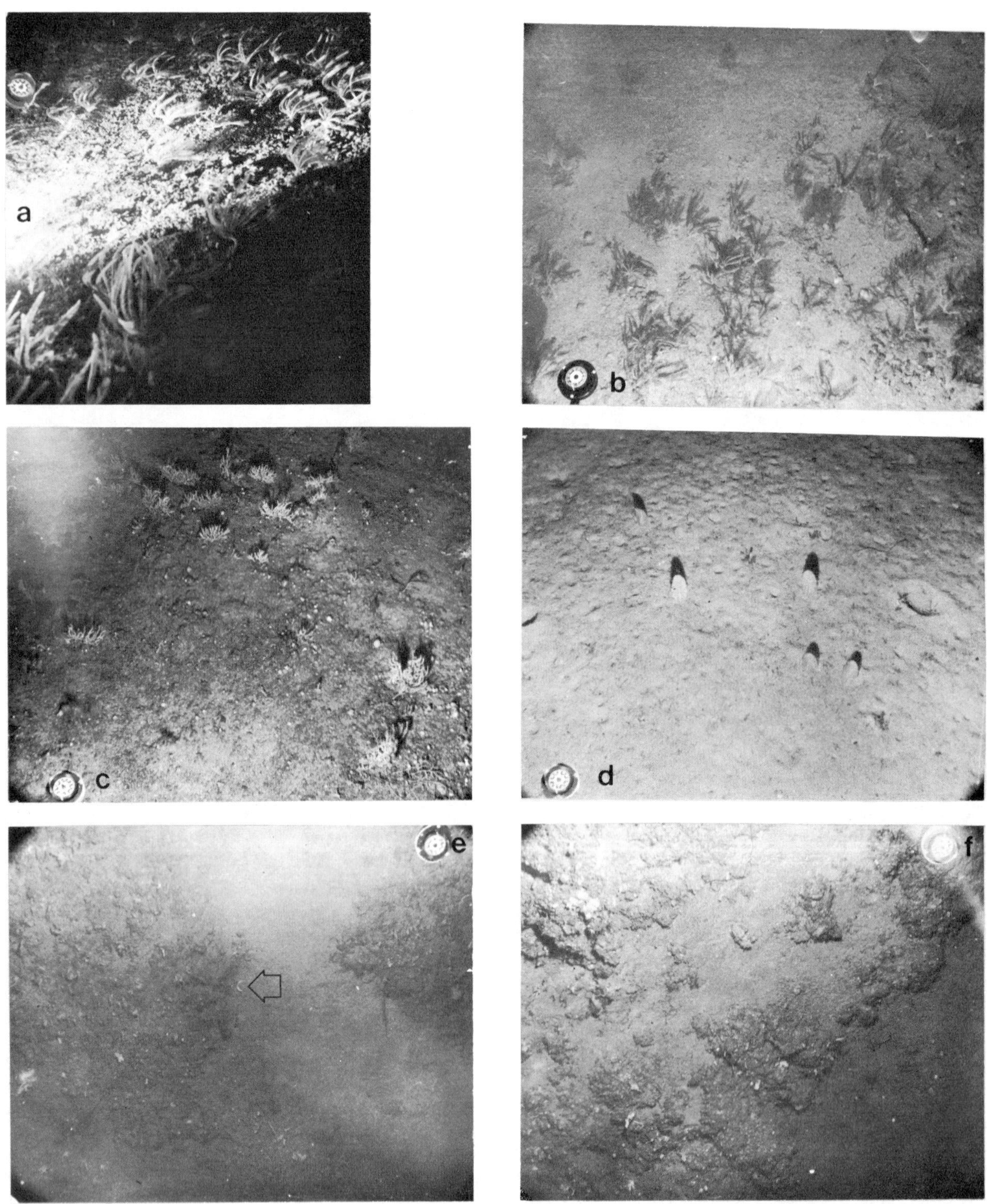

Figure 8. Bottom photographs showing bottom current activity and channeling. Diameter of compass about 65 mm. (a) Crinoids heeling in a strong, southerly current. However, no signs of current ripples are seen in the coarse bottom sands. C10, southern margin of Xauen Bank, traverse g and d. (b) Crinoids attached to fine-grained sediments and leaning slightly to the south. Small ridges in the center and right sides may be old ripple marks. C9, southwest margin of Xauen Bank, traverse a. (c) Gorgonian corals grow in preferred orientation due to prevailing current regime. Pebbles of fragmented biogenic debris are scattered throughout the coarsely grained bottom. C22, Western Alboran Ridge. (d) Sea pen grow in preferred orientation, perhaps due to the prevailing current regime. Flat, medium grained bottom is pitted with small depressions. Note the large oval depression on the right of the photo. (e) A 50 cm wide channel filled with fine to medium sand. Photos e and f were taken in sequence and the channels in both photos may be the same. Bedrock may be a reefoid (biostromal) limestone. The pelecypod valve (arrow) is buried in the channel fill. C24, northern edge of Alboran Ridge, traverse g, sampling station S4. (f) A channel cut in a very coarse grained biogenic limestone is filled with fine to medium sand. The irregular wall of the channel is well defined. C24, northern edge of Alboran Ridge, traverse g, sampling station S4.

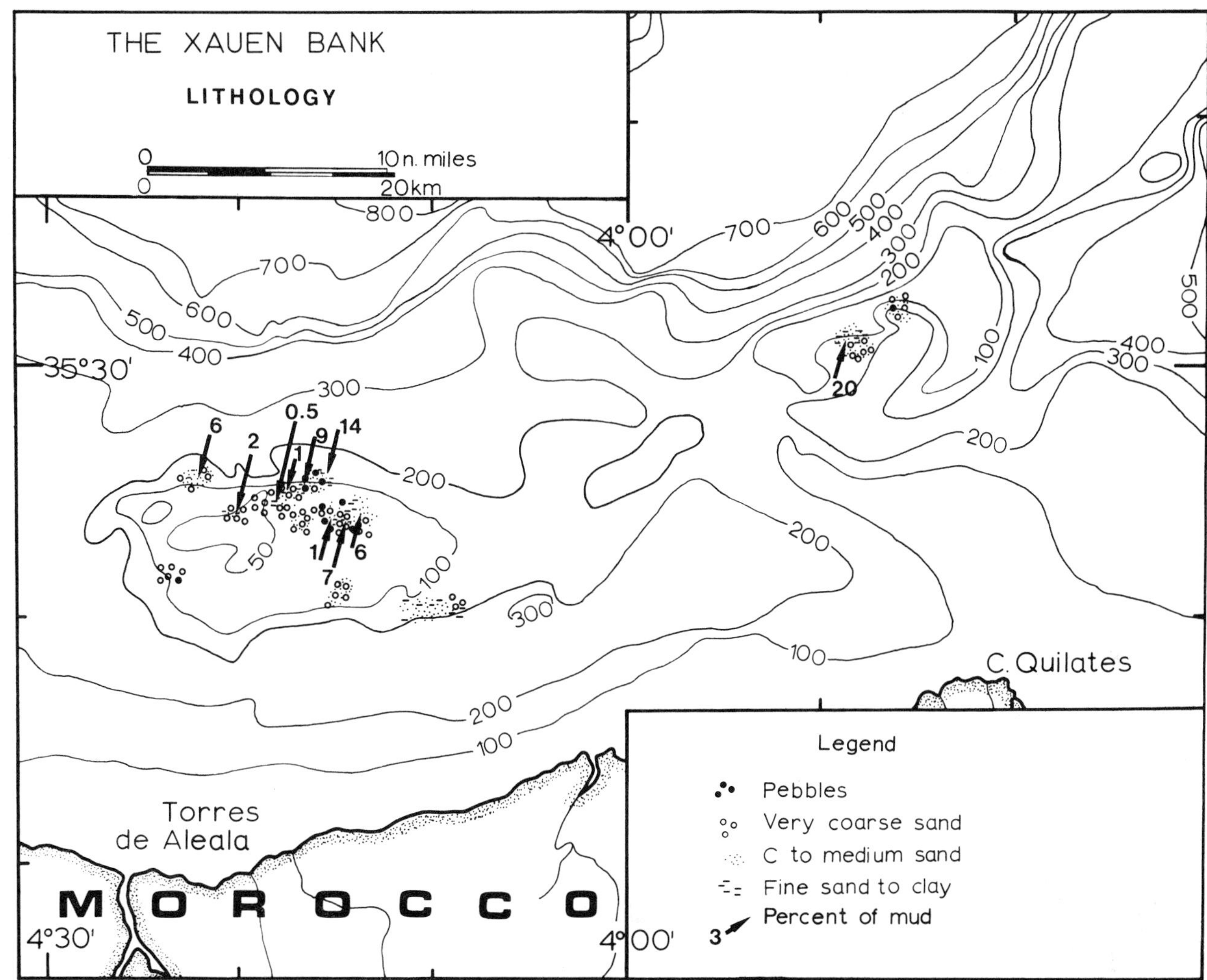

Figure 9. Texture (based on sample analysis and photographs) on Xauen Bank and Western Alboran Ridge.

blages found off the Rhône Delta (Lagaaij and Gautier, 1965) and Algeria (Caulet, 1970).

Coralline algae are present both as massive algal nodules, some greater than 10 cm in diameter, and as branching fragments. Large concentrations of this latter component are contained in *maërl* sands (French term for biogenic sands containing large quantities of the coralline algae *Lithothamnium calcareum* and *L. corallinoides*).

These sediments, rich in bryozoan-coralline algae, are remarkably similar to sediments that occur throughout the northern African Shelf, from Algeria (Caulet, 1970, this volume; Leclaire, 1970) to western Morocco and Spanish Sahara (Summerhayes, 1970: Milliman, MS in preparation) and the Canary Islands (Müller, 1969). The sediments also resemble those from other parts of the Mediterranean (Lagaaij and Gautier, 1965; Blanc, 1968) as well as other parts of the world (such as Australia, Wass *et al.*, 1970). Apparently the lack of terrigenous material, together with the subtropical water temperatures is conducive to such biogenic accumulations.

Mollusks, especially bivalves, are important components in nearly all the sediments from the Alboran Ridge and banks. Other organic remains, such as barnacles, echinoid spines and benthonic foraminifera also are important in the shallower water sediments, although usually in smaller concentrations. With increasing water depth, mollusks and benthonic foraminifera increase in abundance (Figure 14). Below about 270 m most of the carbonate is composed of planktonic foraminifera, with some deep-sea corals present locally.

The specific distribution of the various biogenic components and the sediments are interrelated. Gravel and coarse sand tend to be composed of coralline algae (both nodules that are larger than 4 mm and algal branches that are somewhat smaller), encrusting

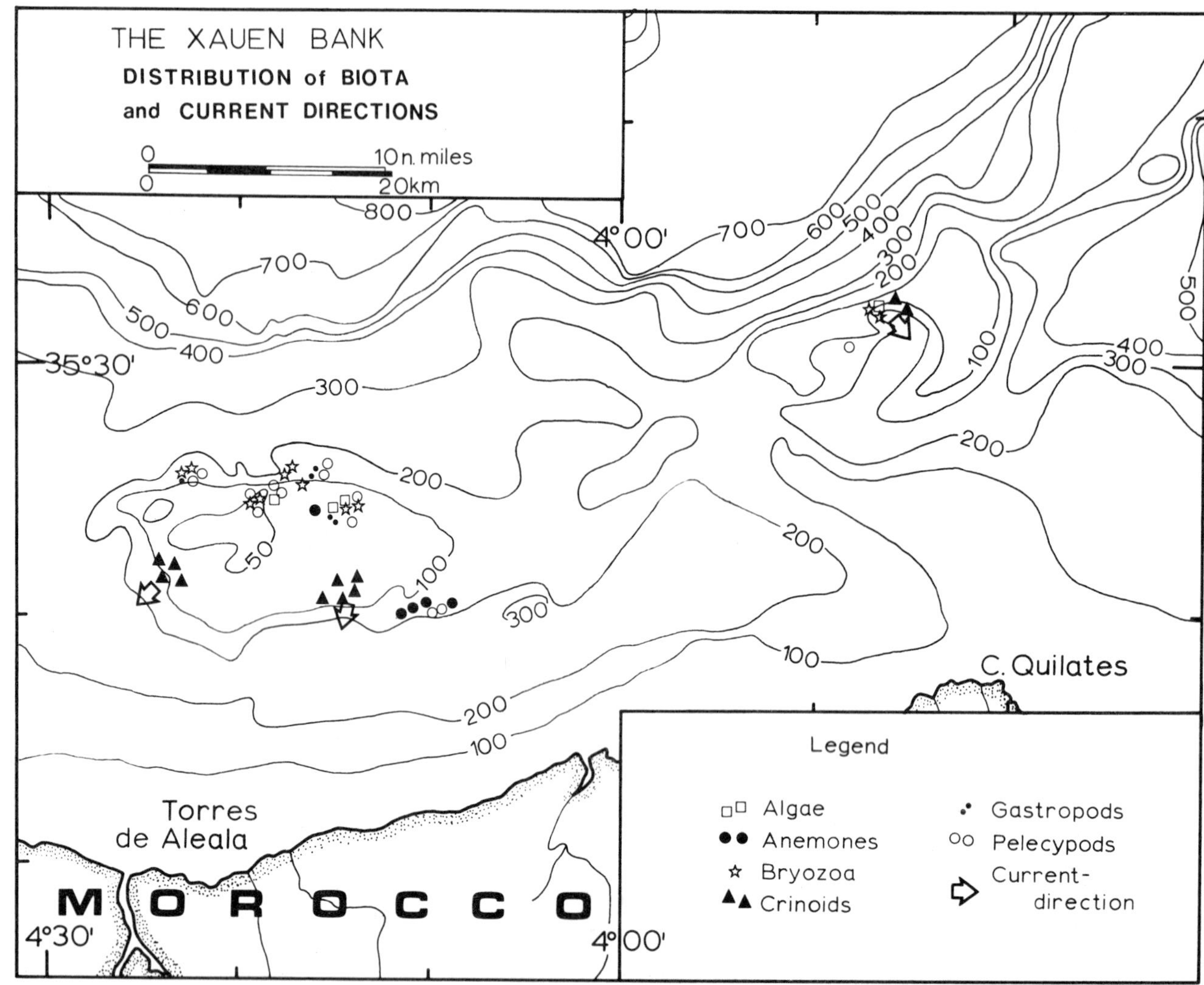

Figure 10. Biota on Xauen Bank and Western Alboran Ridge as observed in bottom photographs. Arrows indicate probable current direction.

and branching bryozoans, and mollusk shells. Medium to coarse sands contain less coralline algae and higher concentrations of bryozoans and mollusk shell fragments. Tests of large foraminifera also are present in this size fraction. The fine sand fraction contains large quantities of foraminifera (both benthonic and planktonic), mollusk fragments and echinoid spines; fragments of more delicate branching bryozoans are also common.

The dominant carbonate mineral is magnesian calcite, usually with 8 to 12 mole percent $MgCO_3$ in solid solution within the calcite lattice. Most samples contain between 65 and 75 percent magnesian calcite, with distinctly lesser amounts of aragonite and calcite. The source of the magnesian calcite in most samples is mostly from the coralline algae and bryozoans, but to a lesser extent it is also derived from minor constituents, such as benthonic foraminifera and echinoid spines. Sample S1 is an exception, in that it contains 45 percent calcite; but even this planktonic foraminifera sediment contains a surprising amount of magnesian calcite (35 percent), suggesting that secondary void fillings may be an important source of magnesian calcite. Aragonite comes from mollusks and, to some degree, from bryozoans.

Age of Sediments

The age of the carbonate-rich sediments on the Alboran Ridge and banks is difficult to estimate. Many workers (such as Blanc, 1968; Wass *et al*, 1970) have suggested that much of the coralline algae and bryozoan debris found on modern continental shelves represent Holocene accumulations. Wass *et al.* (1970) concluded that the bryozoan-rich sediments on the southern Australian Continental Shelf are younger

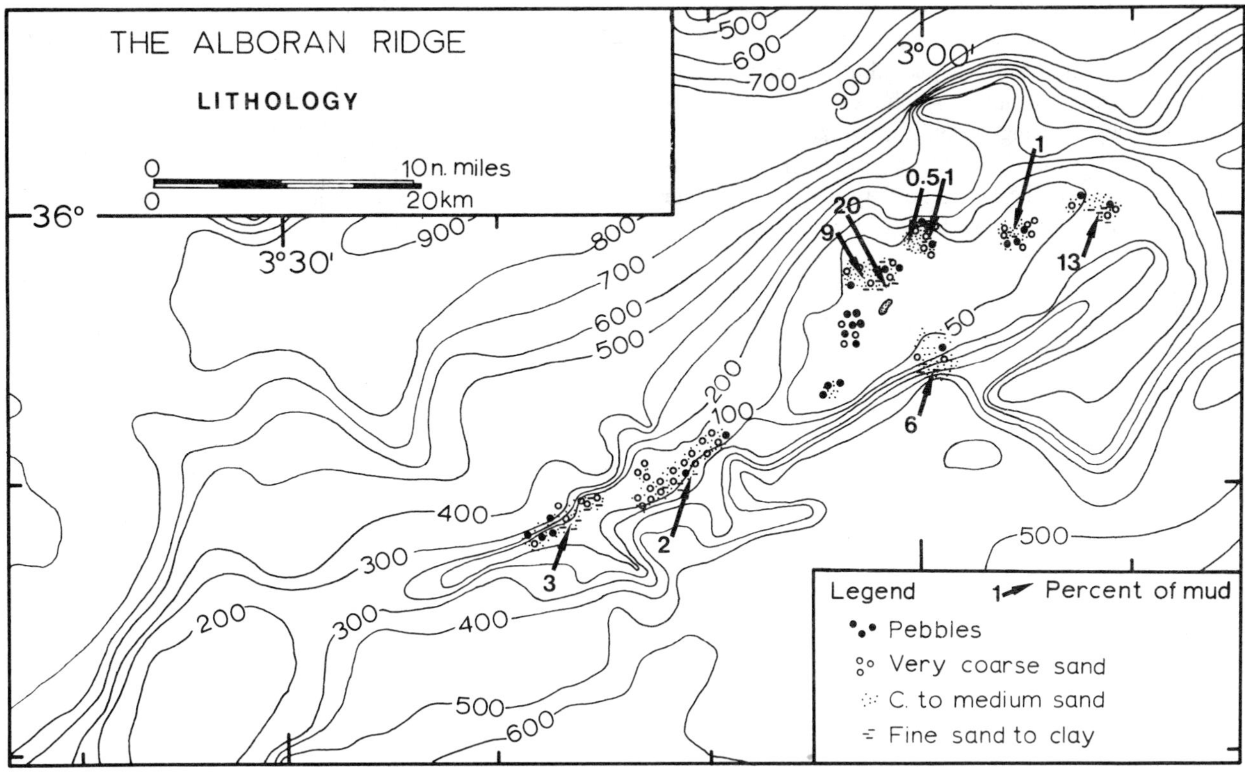

Figure 11. Texture on Alboran Ridge as observed in bottom photographs.

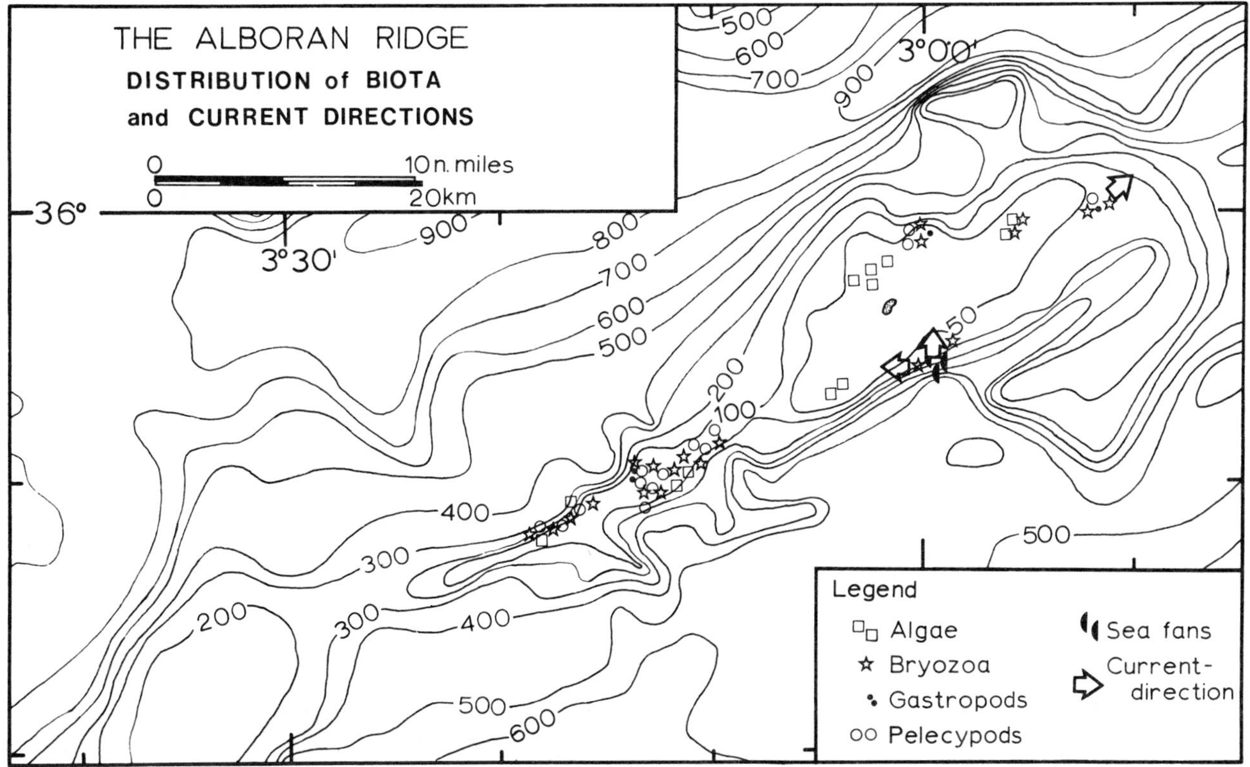

Figure 12. Biota on Alboran Ridge as observed in bottom photographs. Arrows indicate probable current direction.

than 4000 years. While some of the components on the Alboran Ridge and banks may be contemporary, most of the sediments are suspected to be older. Nearly all the biogenic components have been reworked extensively by boring sponges, algae and fungi. Very few grains look fresh. The presence of glauconite fillings in many of the biogenic fragments is another indication of the reworked nature of these grains. Furthermore, the organisms that are presently living at ambient depths (as viewed by deep-sea photographs) are not the primary sedimentary contributors, but rather include such forms as crinoids, anemones, sea fans, sponges, sea grass and algae. Other benthonic organisms include fish, echinoids, pelecypods, gastropods, starfish, chitons and holothurians (Figures 6–8). Similarly, living biogenic components are notably lacking from most of the sediments.

In order to delineate more exactly the age of the sediments and the sedimentary components, ten carbon-14 analyses were run (Table 3). Other than two algal nodules from 54 and 81 meters that have contemporary ages, all samples exhibit ages older than 3 thousand years. Generally the deeper the water depth of the sediment, the older the age of the carbonate (Figure 15), suggesting that many of the components lived during the previous transgression of sea level. Comparison of the age dates with the Holocene sea level (Figure 15), however, indicates that most of the sediments were not deposited near sea level, but rather in depths between 70 and 100 meters.

Table 3. Carbon-14 age dates of selected samples from Alboran Sea banks.

Sample Number	Material Dated	Age (in years B.P.)
DIA	Total Sample	13,430 ± 360
D4A	Total Sample	11,605 ± 235
D4B	Algal Ball	11,385 ± 145
S2A	Total Sample	3,840 ± 120
S3	Total Sample	6,605 ± 180
S4	Algal Ball	295 ± 45
S6	Algal Ball	modern
S11	Bryozoans	3,085 ± 160
S11	Gastropods	6,590 ± 280
S11	Coralline algal *maerl*	6,930 ± 240

The relic ages of these sediments does not mean that all the components were deposited contem-

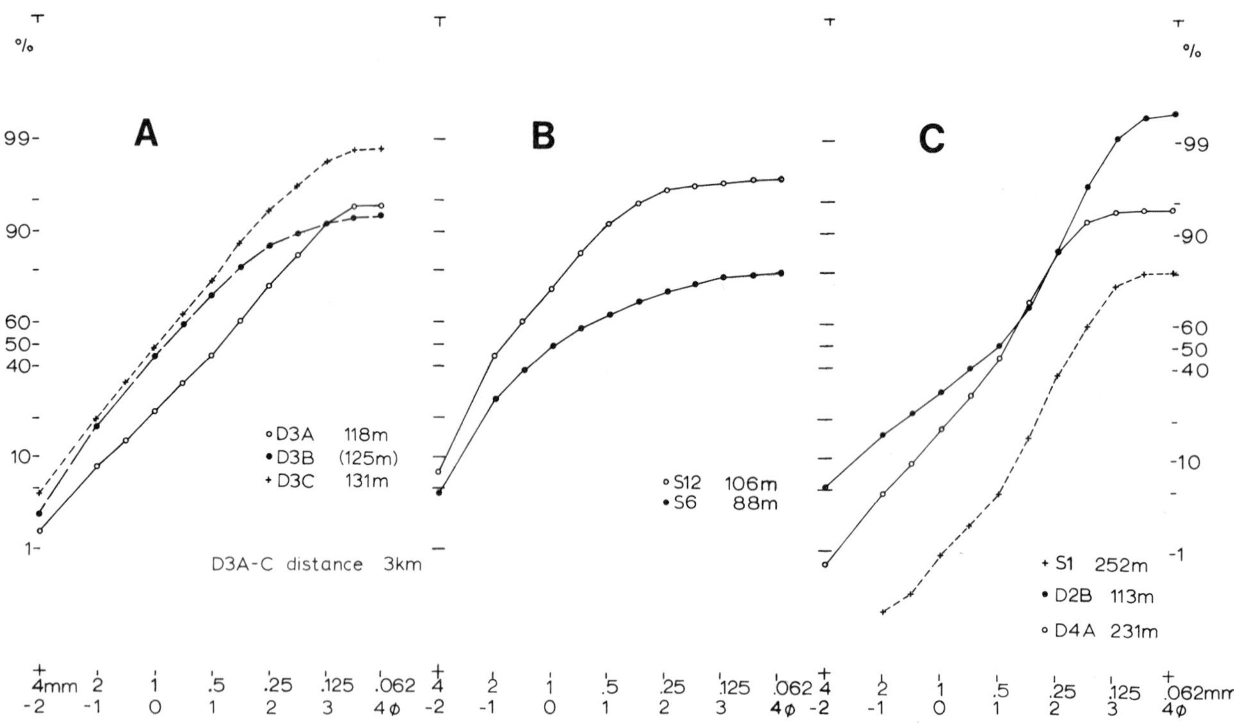

Figure 13. Selected cumulative size curves of eight samples, showing the three principal types of sediment on the shallow banks in the Alboran Sea. In type A the lognormality of the sand-size fractions is well pronounced, and the amounts of gravel and silt are conspicuously small. These samples were taken in three consecutive dredge hauls 1 to 2 km apart on Xauen Bank. In type B the coarse fraction comprises about 50 percent of the total sample. The cumulative curves in type C have an inflection between 250 and 350 microns. This break is probably due to the addition of planktonic foraminifera to a relic sand that originally had a lognormal distribution.

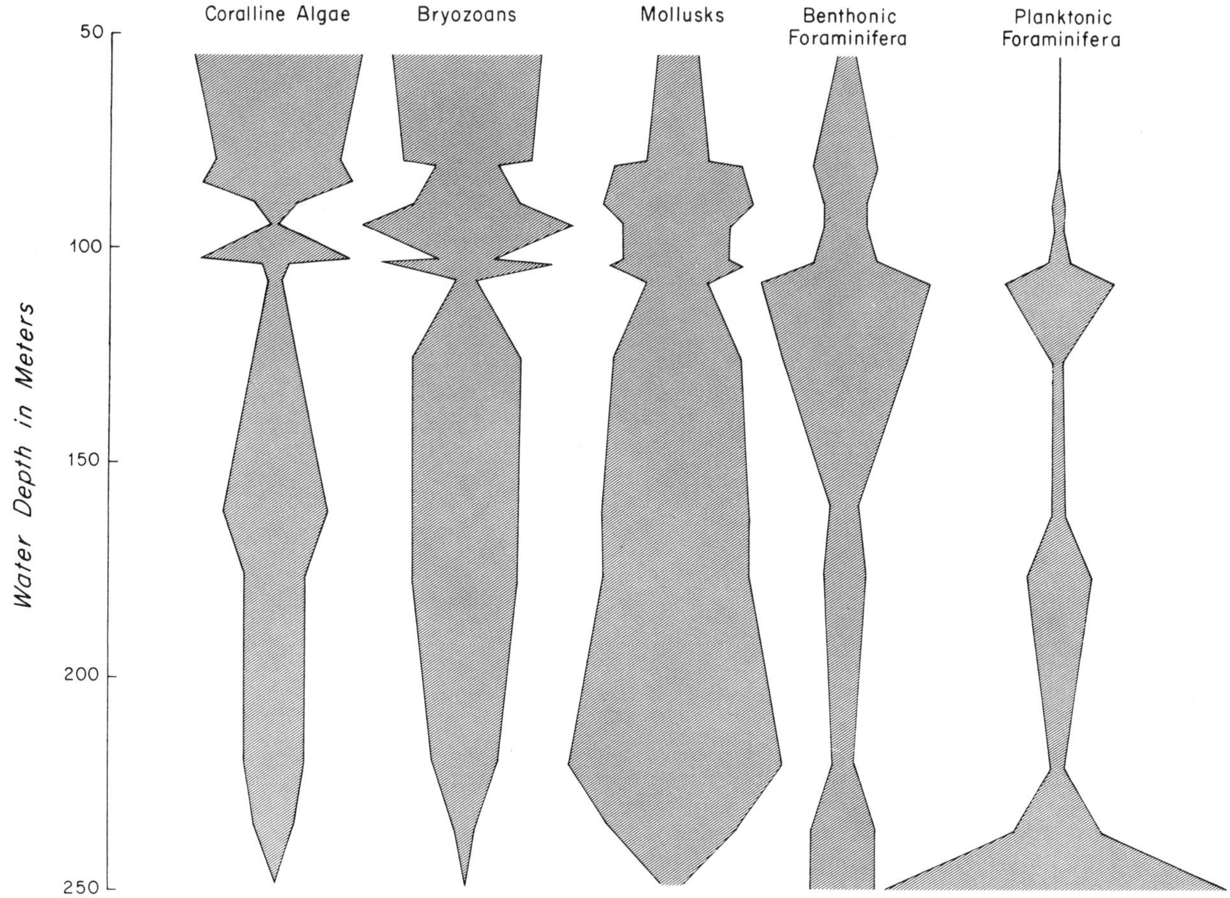

Figure 14. The relative variation of coralline algae, bryozoans, mollusks, benthonic foraminifera and planktonic foraminifera with water depth on the Alboran Ridge bank sediments.

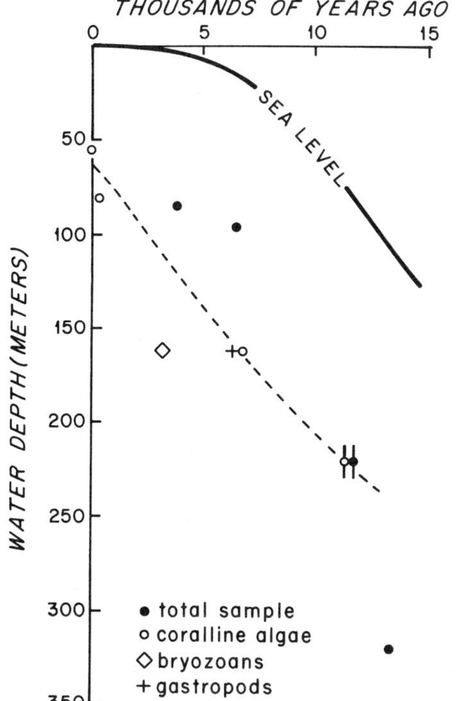

Figure 15. Water depth and age of various Alboran Ridge bank sediments and sedimentary components compared to the Holocene sea level curve of Milliman and Emery (1968). Most sediments seem to have formed in water depths between 70 and 100 meters.

poraneously, since environmental factors controlling the distribution of the various components are not the same for each organism. For example, light undoubtedly plays an important role in the distribution of coralline algae, but the zoarial type and species association of bryozoans may depend more upon substrate morphology (Caulet, 1970). As environmental parameters change during the transgression of sea level, one might expect a succession within the carbonate components, each succeeding component contributing to the final carbonate assemblage within the sediment. An obvious example of such a succession would be the deposition and mixing of modern planktonic foraminifera into a relic shelf sediment (Milliman *et al.*, 1968). Another and more subtle example is seen in sample S11, in which three individual components were dated by carbon-14 analysis. The bryozoans within the sample are distinctly younger (3085 years B.P.) than either the gastropods (6590 years) or the branching coralline algae (6930 years) (Table 3). Sea level 6500 to 6900 years ago was some 15 meters lower than it was 3200 years ago (Figure 15); apparently this slight shift in sea level was sufficient to kill or severely limit the coralline algae and gastropods, while not affecting the bryozoans.

SUMMARY

The carbonate-rich sands and gravels that lie atop the three Alboran Sea banks are the result of isolation from the North African mainland which has resulted in little influx of terrigenous sediment, as well as the current regime that has prevented the accumulation of fine material. The shallow-water carbonates are composed of large quantities of coralline algae and bryozoans, while the deeper sediments are dominated by mollusks and planktonic foraminifera. Carbonate mineralogy generally reflects the biogenic components within the sediment, with magnesian calcite being the dominant mineral. Deeper sediments contain more calcite, but the large amounts of magnesian calcite within these planktonic foraminifera-rich sediments infer void filling by internal cements. In most sediments the noncarbonate fraction is composed exclusively of glauconitic molds, but sediments near Isla del Alboran also contain volcanic debris.

Most of the carbonate components forming the Alboran bank sediments are relic; they are composed of reworked biogenic debris that does not reflect the present-day living populations of benthonic organisms. The carbon-14 age of the sediments increases greatly with water depth, suggesting that the sediments mostly accumulated during the last transgression of sea level. The depth of deposition appears to have been relatively deep, probably between 70 and 100 meters. The carbonate components within any one sample, however, reflect a succession of carbonate facies, each having been deposited during different periods and environmental conditions. In this respect the sediments on the Alboran banks reflect a thanatocoenosis accumulation rather than a biocoenosis.

ACKNOWLEDGMENTS

The authors thank the Captain, officers and men of the U.S.C.G.C. ROCKAWAY for their help in the work at sea. Dr. G. Kelling, Miss M. A. Delserre and Miss J. N. Valette aided in the collection of samples and photography. We are indebted to Dr. Robert Stuckenrath, Radiation Biology Laboratory, Smithsonian Institution for providing carbon-14 age determinations. Drs. W. H. Adey, A. H. Cheetham, and G. A. Cooper, Department of Paleobiology, Smithsonian Institution identified fauna and provided valuable advice on their ecology. Drs. K. O. Emery and Jens Müller helped with their reviews of this paper.

Support for this study was provided by the Smithsonian Institution Research Award No. 427235 (D.J.S.) and the Smithsonian Foreign Currency Program for Travel Funds (Y.W.), and from contract N00014–66–C–0241 with the Office of Naval Research, and contract GX28193, National Science Foundation, from the International Decade of Ocean Exploration (IDOE) (J.D.M.)

REFERENCES

Blanc, J. J. 1968. Sedimentary geology of the Mediterranean Sea. In: *Oceanography and Marine Biology, Annual Reviews,* George Allen and Unwin, London, 6:377–454.

Caulet, J. 1970. *Les sédiments Organogènes du Précontinent Algérien.* Thèse de Doctorat d'Etat, Université de Paris, C.N.R.S. AO 4491:503 p.

Caulet, J. 1972. Recent biogenic calcareous sedimentation on the Algerian continental shelf. In: *The Mediterranean Sea: A Natural Sedimentation Laboratory,* ed. Stanley, D. J., Dowden, Hutchinson and Ross, Inc., Stroudsburg, Pennsylvania, 261–277.

Curray, J. R. 1965. Late Quaternary history, continental shelves of the United States: In: *The Quaternary of the United States,* eds. Wright Jr. H. E. and D. G. Frey, Princeton University Press, Princeton, 723–735.

Eriksson, G. K. 1961. Granulométrie des sédiments de l'Ile d'Alboran, Méditerranée occidentale. *Geological Institute Bulletin, University of Uppsala,* 40:269–284.

Gaibar-Puertas, C. 1969. Estudio geológico de la isla de Alborán (Almeria) I: las rocas eruptivas. *Acta Geológica Hispánica,* 4: 72–80.

Giermann, G., M. Pfannenstiel and W. Wimmenauer 1968. Relation entre morphologie, tectonique et volcanisme en Mer d'Alboran (Méditerranée occidentale). Résultats préliminaires de la campagne JEAN CHARCOT (1967). *Comptes Rendus des Séances de la Société Géologique de France,* 4:116–117.

Glangeaud, L., C. Bobier and G. Bellaiche 1967. Evolution néotectonique de la Mer d'Alboran et ses conséquences paléogéographiques. *Comptes Rendus de l'Académie des Sciences de Paris,* 265:1672–1675.

Huang, T. C. and D. J. Stanley 1972. Western Alboran Sea: sediment dispersal, ponding and reversal of currents. In: *The Mediterranean Sea: A Natural Sedimentation Laboratory,* ed. Stanley, D. J. Dowden, Hutchinson and Ross, Inc., Stroudsburg, Pennsylvania, 521–559.

Inman, D. L. 1952. Measures for describing the size distribution of sediments. *Journal of Sedimentary Petrology,* 22:125–145.

Lacombe, H. and P. Tchernia 1972. Caractères hydrologiques et circulation des eaux en Méditerranée. In: *The Mediterranean Sea: A Natural Sedimentation Laboratory,* ed. Stanley, D. J. Dowden, Hutchinson and Ross, Inc., Stroudsburg, Pennsylvania, 25–36

Lagaaij, R. and Y. V. Gautier 1965. Bryozoan assemblages from marine sediments of the Rhone Delta, France. *Micropaleontology,* 11:39–58.

Leclaire, L. 1970. *La sédimentation Holocène sur le Versant Méridional du Bassin Algéro-Baléares (Précontinent Algérien)*. Thèse, Université de Paris, C.N.R.S. AO 4492:551 p.

Lucayo, N. C. 1968. Contribucion al conocimento del Mar de Alboran, I. Superficie de Referencia. *Boletin Instituto Español de Oceanografia*, 135:28 p.

Macintyre, I. G. and J. D. Milliman 1970. Physiographic features on the outer shelf and upper slope, Atlantic continental margin, southeastern United States. *Geological Society of America Bulletin*, 81:2577–2598.

Milliman, J. D. and K. O. Emery 1968. Sea levels during the past 35,000 years. *Science*, 162:1121–1123.

Milliman, J. D., O. H. Pilkey and B. W. Blackwelder 1968. Carbonate sedimentation on the continental shelf. Cape Hatteras to Cape Romain. *Southeastern Geology*, 9:245–267.

Müller, J. 1969. *Mineralogisch-Sedimentpetrographische Untersuchungen an Karbonatsedimenten aus dem Schelfbereich um Fuerteventura und Lanzarote (Kanarische Inseln)*. Unpublished Ph.D. thesis, Heidelberg University (W. Germany), 99 p.

Ryan, W. B. F., D. J. Stanley, J. B. Hersey, D. A. Fahlquist and T. D. Allan 1970. The tectonics and geology of the Mediterranean Sea. In: *The Sea*, ed. Maxwell, A. E., John Wiley and Sons, New York, 4(2):387–492.

Salvator, L. 1898. *Alboran*. Mercy Sohn, Prague, 89 p.

Stanley, D. J., C. Gehin and C. Bartolini 1970. Flysch-type sedimentation in the Alboran Sea, Western Mediterranean. *Nature*, 228:979–983.

Summerhayes, C. P. 1970. *Phosphate Deposits on the Northwest African Continental Shelf and Slope*. Unpublished Ph.D. thesis. University of London, 282p.

Wass, R. E., J. R. Conolly, and R. J. Macintyre 1970. Bryozoan carbonate sand continuous along southern Australia. *Marine Geology*, 9:63–73.

Wüst, G. 1961. On the vertical circulation of the Mediterranean Sea. *Journal of Geophysical Research*, 66:3261–3271.

Recent Biogenic Calcareous Sedimentation on the Algerian Continental Shelf

Jean Pierre Caulet

Muséum National d'Histoire Naturelle, Paris

ABSTRACT

Quantitative analysis of the texture and the composition of 282 samples from the Algerian continental shelf shows that Algerian biogenic calcareous sediments are sands and silts composed primarily of bioclastic fragments, including calcareous algae, bryozoa, pelecypods and gastropods. For the most part, less than 10 percent of terrigenous material occurs in these sediments.

From ecologic analysis of their thanatocenoses, these sediments can be divided into three groups: (1) modern biogenic sediments resulting from the accumulation of skeletal debris of existing benthic organisms, (2) relict biogenic sediments dredged between a depth range of 80 m to 150 m and composed of organic remains deposited during the Late Würm, and (3) residual biogenic sediments derived from submarine outcrops and composed of intensely reworked calcareous red algae and bryozoa.

The bathymetry and the rate of deposition of terrigenous materials are the chief factors influencing modern biogenic sedimentation, while transportation induced by wave-currents is only effective between sea level and the 30 m contour. All the observations show how the rise in sea level and the increased terrigenous sedimentation resulting from climatic changes that followed the last glacial epoch have dominated the Holocene evolution of this specialized type of sedimentation.

RESUME

L'analyse quantitative de la texture et de la composition de 282 échantillons, dragués entre 0 et –200 m sur le précontinent algérien, montre que les sédiments organogènes de cette zone sont des sables et des vases renfermant généralement moins de 10% de particules terrigènes et riches en débris bioclastiques, algues calcaires, bryozoaires, bivalves et gastéropodes.

D'après l'analyse écologique des thanatocénoses qu'ils renferment, ces sédiments peuvent être répartis en: (1) sédiments organogènes actuels, résultant de l'accumulation des organismes vivant dans les biocénoses benthiques du plateau continental, (2) des sédiments organogènes reliques, installés entre –80 m et –150 m et formés de débris organogènes (maërl, bivalves) s'étant déposés lors de la dernière grande régression ou de la transgression flandrienne, (3) des sédiments organogènes résiduels, peu nombreux et comportant des restes d'algues calcaires ou de bryozoaires d'origine anté-tardiglaciaire, remaniés pendant les grandes régressions pléistocènes.

Tous les résultats indiquent que les principaux facteurs de la sédimentation organogène locale sont le taux des apports terrigènes fins et la profondeur, les courants dus à la houle n'intervenant qu'entre 0 et –30 m. Ils montrent aussi comment la remontée du niveau marin et l'accroissement généralisé des apports terrigènes fins provoqué par les changements climatiques survenus après la dernière phase glaciaire sont intervenus, au cours de la période holocène, pour modeler l'évolution de cette sédimentation si particulière.

INTRODUCTION

Occurrences of biogenic calcareous sediments have frequently been reported from various continental shelves of the western Mediterranean Sea. Among the earliest examples reported in the geological literature are bioclastic sands on the French continental shelf (Dangeard, 1923 and Berthois, 1939) and biogenic calcareous sands in the Straits of Sicily (Blanc, 1958). More recently, coquina zones on the Rhône Delta have been described (van Straaten, 1960). Most of these bioclastic sediments are inferred to be relict, as shown by studies of numerous relict faunas (Mars and Picard, 1958).

Little published work is yet available on the biogenic calcareous sediments from the Algerian continental shelf known as *sables coralligènes*. Those off Habibas Islands (western Algeria) were dredged by Dangeard in 1929, but the first detailed studies were published subsequently by Leclaire *et al.* (1965) and Caulet (1963, 1967, 1968, 1969a, 1969b, 1970). The area of investigation comprises the widest parts of the Algerian continental shelf between the shore and the shelf-edge (near 100–150 m contours). Off the coast of Oranie (1°10′ W to 0°20′ E) 386 bottom sediment samples were collected during two one-month cruises of the LOUIS BOUTAN (1960–1963). One hundred and eighty-three samples were dredged on the eastern continental shelf north of the Bône—La Calle coast (7°40′E to 8°50′E) in August 1960 and early summer 1963. Two hundred samples were collected between Algiers and Philippeville on very narrow parts of the continental borderland. The samples were chosen to represent the extreme and average depths as well as the different coastal topographic and bathymetric conditions.

This paper, which is modified from material in Caulet (1972), presents data from the analytical studies (texture, constituent particles, species, *etc.*) of 282 samples of biogenic calcareous sediments and tries to explain the provenance and depositional history of these sediments. Moreover, special attention is given to the distribution of macro-organic detritus with the aim of using the distribution of these sediment components to decipher paleoenvironments.

GENERAL DESCRIPTION OF THE ALGERIAN CONTINENTAL MARGIN

The principal physiographic features of this area are described in papers by Rosfelder (1955) and Leclaire (1968). A set of 11 topographic charts compiled by Leclaire (1968) from all available soundings of the French Hydrographic Office is a source of information about geological processes and events.

The continental shelf of Algeria off the irregular coast and capes is generally quite narrow, with an average width of 1 or 2 km. In the embayments, however, the shelf widens (10–20 km) with a seaward limit of 120–150 m north of western Algeria (the Gulf of Oran: 0°30′W—1°10′W and the Gulf of Arzew: 0°20′W to 0°30′W) and 150 m off the Gulf of Bône (eastern Algeria, 7°40′E to 8°00′E). Beyond the Gulf of Bône, the continental slope is less steep and there are a considerable number of rocky banks (Banc Le Sec, Archipel de la Galite) along the outer margin as far as the Straits of Sicily.

The widest parts of the Algerian continental shelf have not yet approached equilibrium in either morphology or sediment cover. Well-defined relict topographies with numerous terrace levels are visible off Oran, Arzew and Bône at various depths (50 m to 90 m). Relict bioclastic sands mantle these topographic zones and form an almost continuous belt along the shelf edge. Equilibrium shelf facies muds occur in the inner to middle depressions of these areas (*les vasières*). Distribution of terrigenous sands is restricted almost entirely to the shore zone out to depths of from 10 to a maximum of 30 m (see Leclaire, this volume).

Currents, Salinity and Waves

Lacombe and Tchernia (1960 and 1965), Furnestin, (1960) and Furnstein and Allain (1962a and 1962b) have assembled a large amount of data concerning the general physical oceanography of the western Mediterranean Sea; thus only a very brief résumé of current patterns and salinity distributions off Algeria is given here.

Current systems, in this area, owe their distribution to the effect of the surface North Atlantic water which flows into the Mediterranean through the Strait of Gibraltar. As the water enters through the Strait it follows the coast of North Africa. At the longitude of Oran, this Atlantic current extends to a depth of 100–200 m with an average velocity of 0.6 knots and a salinity of 36.6‰. Near Algiers the velocity decreases to about 0.5 m/sec. (Leclaire, unpublished data). Beyond Bône, current speeds are unknown but the salinity may exceed 37.0‰ in winter. During the summer the range of temperature is from 16°–17° in the Strait of Gibraltar and increases to above 22° beyond Algiers. In winter the surface temperatures are everywhere about 14°.

Off the large embayments the Atlantic current induces clockwise circulations. The speeds of these surface currents are very low (0.3 km/h in the Bay of Algiers) and quickly decrease below the surface (1 cm/sec at 15 m contour near the bottom at Arzew, in Caulet, unpublished data).

There are no tidal currents, as the average tidal range is below 10 cm. Wave-induced currents can only transport sand particles, and classic nearshore circulation systems have been described (Caulet, 1963 and Leclaire, 1963) near the surf zone, which extend usually to the 6–9 m contours.

Sediment Supply

As quantitative investigations of living Algerian benthonic fauna have not been completed, it is difficult to estimate the bulk of modern biogenic materials. Only Sparck (1931) has tried to count and weigh the macrobenthos of the muds off Algiers, which were collected with a Petersen grab. He found only a few pelecypod shells in about 1 m^3 of mud but no estimation of the macrobenthos living on other substrates and particularly in *coralligène* environments has been made up to now.

Kruger (1950) made quantitative studies of nanophytoplankton off Algiers. She found in one liter about 30,000 to 100,000 coccoliths, 125,000 to 300,000 dinoflagellae, 6,000 to 40,000 diatoms and 600,000 to 1,000,000 flagellates. Bernard and Lecal-Schlauder (1953) estimated that there are 0.1 grams of coccoliths in 1 m^3 of water.

Most of the silica shell destruction apparently takes place in the upper few hundred meters of the sea (Berger, 1968) and only a minor part of the planktonic calcareous skeletons originally delivered is preserved in the sediments of the aerated and shallow areas (Berger and Soutar, 1970). Therefore, current investigations with the scanning electron microscope show that only a very minor part of these plankton constituents remain in the biogenic sediments off Algeria, where they always constitute less than 1 percent (Caulet, 1972).

To date, only limited information is available concerning the character and the quantity of terrigenous supplies carried out to sea off Algeria. A few estimates have been made recently by Leclaire (1972) of the total quantity of suspended sediments annually supplied to the sea by the area; they range from about 40 to 60 million tons.

ANALYTICAL TECHNIQUES

For size-analysis, samples were dried at room temperature and then divided into representative portions of 100 grams each with an Otto splitter. The fraction coarser than 62 microns was sieved through 15 Tyler sieves, and the finer portion was then cleaned in an ultrasonic tank and washed through a nest of two sieves (50 microns and 25 microns). Cumulative weight percentages were plotted on probability paper on the phi scale to produce a precise cumulative curve. The phi median diameter and the phi deviation were obtained by percentile measures (Rosfelder, 1961).

Constituents (greater than 100 microns) were sorted out and estimated in 25 gram portions under a binocular microscope. In the present study, the following constituents have been selected as most significant: numerous organic remains such as bioclastic debris of unidentifiable origin, pelecypod shells, gastropods, bryozoa, calcareous red algae, foraminifera and terrigenous debris (land-derived minerals such as quartz, feldspar, zircon and small fragments of sandstones, limestones, *etc.*). Echinoid, brachiopod, pteropod and coral fragments occur as minor constituents in these sediments.

The quantitative evaluation of the composition of each coarse-fraction (greater than 100 microns) of each sample was accomplished by identifying and counting 300 particles. From the collections made on the Algerian continental shelf, the following numbers of species were found: 10 calcareous red algae, 70 bryozoa, 75 pelecypods, 86 gastropods, 8 brachiopods and about 10 hexacorallia.

The distribution, the ecologic characteristics and the rate of accumulation of all these species are discussed elsewhere in detail (see Caulet, 1972).

MODERN BIOGENIC SEDIMENTS OF THE INNER SHELF

Modern biogenic sediments consist chiefly of skeletal debris derived from pre-existing organisms. The skeletal composition of sediments on the inner and middle sectors of the Algerian continental shelf indicates that they were deposited in equilibrium with modern environments.

On the inner parts of the shelf (10 to 60 m isobaths), three major types of modern biogenic sand can be recognized: *maerl*[1] sands, bioclastic sands and shelly fine sands. Three groups of organisms contribute to the bulk of skeletal material in the sediments: calcareous algae, gastropods and pelecypods. Detritus from bryozoa, pteropods and brachiopods are insignificant throughout the area. All identifiable species are derived from infralittoral and circalittoral benthonic communities.

The *modern maerl sands* contain more than 50 percent calcareous red algal concretions. Mud makes up little more than one percent of the material and gravel populations are highest in these biogenic sands (never less than 60%).

The bulk of the gravel is of maerl origin and is derived from indigenous fauna. Numerous red to purplish-blue living pralines and branches of the two species of *Lithothamnium* are commonly dredged with it. Except for calcareous red algal concretions,

[1]The term *maerl* is used for accumulations of debris of *Lithothamnium calcareum* and *Lithothamnium corallioides* (calcareous red algae).

Table 1. Composition of modern biogenic sediments of the inner shelf (western region).

Parameter examined	Sample Number		
	A.71	O.406	O.407
Depth/m	32	15	40
Median diameter (mm)	2.00	1.50	1.00
Phi deviation	2.5	0.6	0.4
Bioclastic fraction	14%	91%	58%
Calcareous algae	73%	3%	3%
Bryozoa	4%	1%	4%
Pelecypods	2%	+	14%
Gastropods	1%	1%	5%
Echinoids	+	+	+
Brachiopods			
Hexacorallia			
Foraminifera	+		2%
Terrigenous fraction		+	2%
Pelitic fraction	5%	2%	9%
	Maerl sand	Bioclastic sand	Shelly sand

+ = trace, in all Tables

other consituents form only a negligible part of the sediment (Table 1). The few identifiable gastropods (*Bittium reticulatum, Gibberula miliaria, Rissoa cimex, Chauvetia minima*) and pelecypods (*Cardita calyculata, Glanz trapezia, Venus casina*) are typical of current-dominated infralittoral communities and particularly of the Posidonia mats biocenosis (Pérès and Picard, 1964). Terrigenous particles (generally coarse to fine sand size quartz) have been transported from the adjacent beach swash-zone environments. They never constitute more than 10 percent of the sediment because an increasing sand supply stops the growth of calcareous red algae.

Algerian maerl sands were dredged mostly off Oranie (Figures 1 and 2) and in the Castiglione Bay (west of Algiers) near Posidonia mats on rocky beds. The general location of these areas is indicated in Leclaire (this volume, his figures 1 and 2). Their depositional environment is like those of the maerl sands described by Jacquotte (1962) on the Provençal inner shelf. East of Algiers, maerl sands have only been reported at one site, on the Aïn Barbar Shoal between Cap de Fer and Bône (Caulet, 1972), although few living specimens of the two species of red algae are recognized in this area (Lemoine, 1957). The factors which control the distribution of maerl sands are certainly ecologic in nature but they remain, as yet, largely unknown.

Reworked and worn calcareous debris from the infralittoral benthonic communities form the major part (more than 90 percent) of the *bioclastic sands*. These sands are found near maerls and off Posidonia

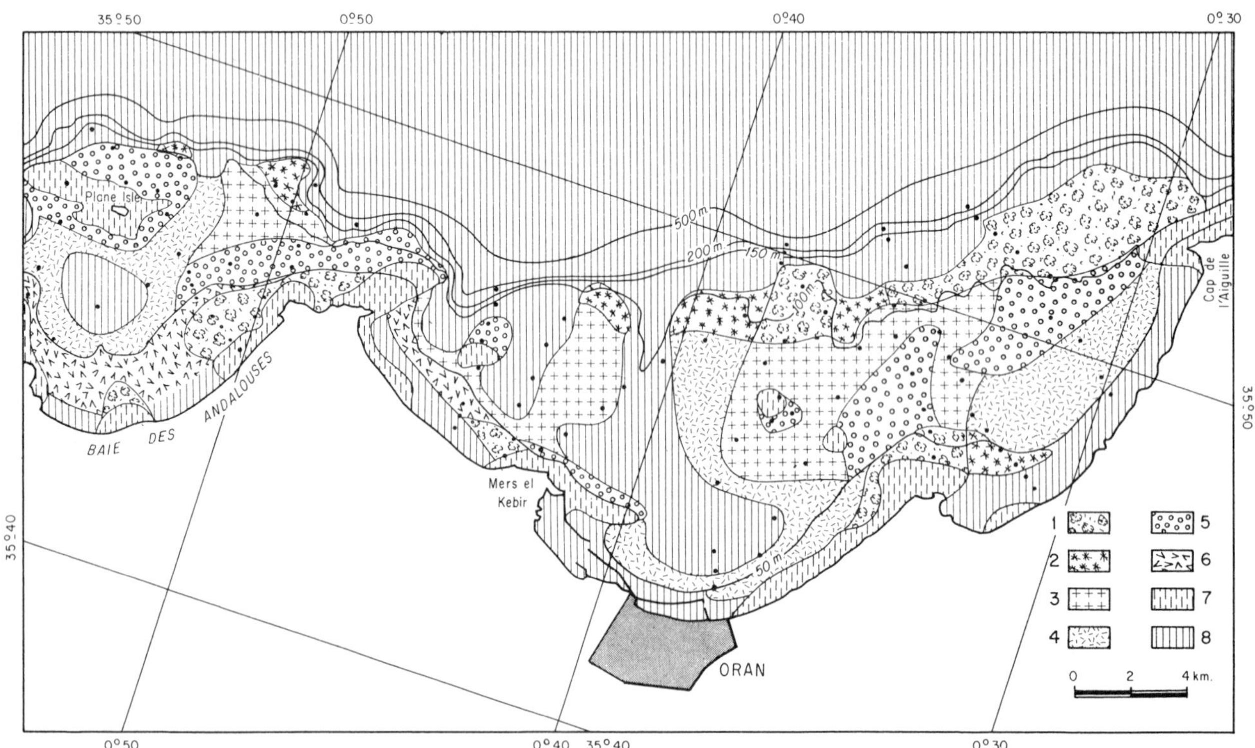

Figure 1. Map showing the distribution of biogenic sediments in the Gulf of Oran. (1) maerl sands, (2) muddy maerl sands, (3) shelly sands, (4) muddy shell sands, (5) bryozoan sands, (6) bioclastic sands, (7) rock outcrops, and (8) terrigenous sediments.

beds (Figures 1 and 2), from the 20 m to 40 m isobaths. Size-distributions and low percentages of fines show these sands occur generally in areas where wave-currents are strong enough to winnow and stir sand-size particles. Terrigenous debris percentages are highest near the beach swash-zone where samples have polymodal frequency curves.

The *shelly fine sands* are confined to deeper inner shelf portions off west Algeria (Figures 1 and 2) between beaches and mud-floored depressions (*les vasières*). The fact that these biogenic sands are medium sands deficient in fines indicates that they have been deposited under conditions of high agitation. The identifiable valves of pelecypods which constitute the greatest part of the shelly coarse-fraction (Table 1) are derived from two main sources, Many specimens are derived from infralittoral communities: *Jagonia reticulata* from the beach fine sands biocenosis and *Loripes lacteus* from the beach muddy fine sands biocenosis. Others come from the nearshore terrigenous sands biocenosis *i.e., Angulus donacinus*, or the middle shelf muddy sands communities *i.e., Timoclea ovata, Parvicardium nodosum, Nuculana deltoidea.* Moreover, two characteristic species of the transition environments biocenosis *i.e., Angulus distortus* and *Lembulus pella* (Pérès and Picard, 1964) can be found in these sediments. Thus, shelly sands of the inner shelf appear to be transitory sediments deposited in transition environments.

MODERN BIOGENIC SEDIMENTS OF THE MIDDLE SHELF

On the middle part of the Algerian continental shelf (between 50 m to 80 m, on average), another three major types of modern biogenic sand can be found: encrusting calcareous red algal sands, bryozoan sands and muddy shell sands. The bulk of the skeletal material is derived from a circalittoral environment and, as shown by the types of species, much of the debris-forming biota has grown in or near its place of deposition. Therefore, grain-size analysis of these sediments appears to be of no special value. Generally, they are coarse to medium sands with a low percentage of mud (less than 5 percent on average).

Both *encrusting calcareous red algal* and *bryozoan sands* are commonly dredged on rocky substrates where the *"biocénoses coralligènes"* (Pérès and Picard, 1964) find cohesive surfaces for growth. Thus, *coral-*

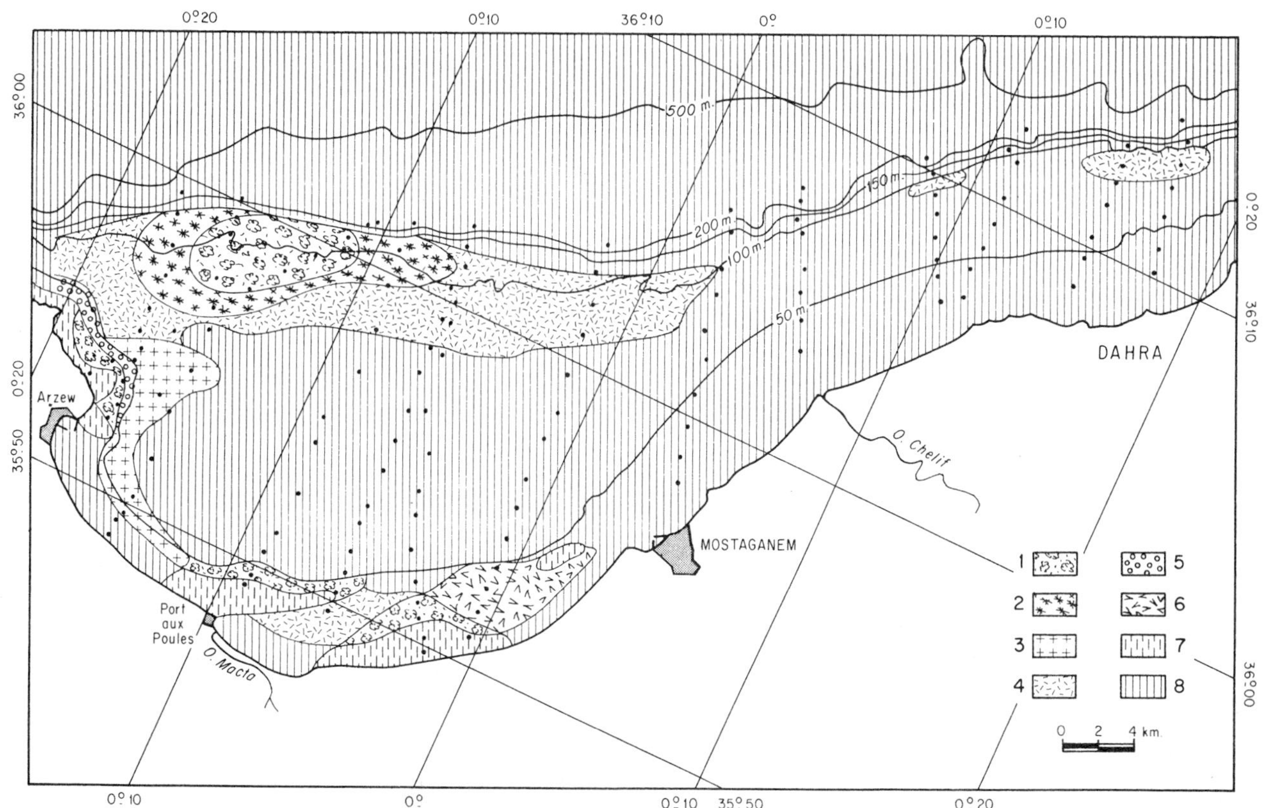

Figure 2. Map showing the distribution of biogenic sediments in the Gulf of Arzew (legend as in Figure 1).

ligène sands mantle: (1) the hard tops of shoals such as the Banc de Corail (Figure 1) north of Oran, the Seiche de Djidjelli (Caulet, 1972), the Banc le Sec (Figure 4), (2) the rocky sides of islands (see around Iles Habibas and Ile Plane, Figure 1), (3) and some relict features such as fossil dunes or terraces north of Oran (Figure 1), Bône (Figure 3) and Cap Rosa (Figure 4).

In these sediments, the percentage of pelecypod valves is largely reduced whereas the gastropod content increases (Table 2). Two species contribute to the bulk of encrusting calcareous red algae fractions: *Pseudolithophyllum expansum* and *Lithothamnium fruticulosum*. Bryozoa belong to 70 different forms; of these only 9 occur in more than 10 of the 282 samples that yielded bryozoa and 3 in more than 50 percent of the same samples, *i.e., Porella cervicornis, Adeonella calveti* and *Cellaria salicornioides*.

Off the coasts of Oranie, encrusting calcareous algal sands are strictly limited to the area around the islands and shoals, and red algal detritus is not encountered in bryozoan sands which occur generally between the 50 to 90 m contours (Figure 1). Beyond Algiers, the composition of the two sediment types is much more similar as both contain nearly equivalent percentages of bryozoa and algae (Table 2). Nevertheless, each type of sediment represents a distinct bathymetric range because bryozoa are accustomed to greater depths. Thus, fragments of numerous vinculariiform zoaria (*Palmicellaria elegans, Palmicellaria aviculifera, Diporula verrucosa, Buchneria fayalensis, Tervia irregularis*), well adapted for life in deep or sheltered waters (Lagaaij and Gautier, 1965), constitute a great part of the deeper bryozoan sands (70 to 120 m).

Table 2. Composition of modern biogenic sediments of the outer shelf.

	Bryozoan sands		Calcareous algal sands		Muddy shell sands
	W. region	E. region	W. region	E. region	W. region
Sample Number	O.141	B.402	O.379	B.339	O.118
Depth/m	75	80	70	70	75
Median diameter (mm)	1.70	1.81	1.00	2.51	0.75
Phi deviation	1.26	1.30	1.27	1.30	1.63
Bioclastic fraction	32%	24%	66%	30%	50%
Calcareous algae		12%	11%	36%	
Bryozoa	51%	55%	3%	14%	+
Pelecypods	6%	3%	3%	3%	19%
Gastropods	5%	1%	1%	6%	3%
Echinoids	+	+	+	+	
Brachiopods		+		+	
Hexacorallia	4%	1%	+	5%	
Foraminifera	+	2%	4%	1%	6%
Terrigenous fraction			2%	3%	4%
Pelitic fraction	1%		8%		20%

Muddy shell sands have been dredged on flat-bottom areas near vasières off the western coasts of Algeria (Figures 1 and 2). The proportion of fines increases from 5 percent to 20 percent near the central parts of the vasières. The macro-organism assemblages in these sands are very different from the inner areas and attest a change of environment. All the identifiable debris originated from the circalittoral muddy sands biocenosis. Among these, the valves of pelecypods (especially of *Timoclea ovata, Nuculana deltoidea, Parvicardium nodosum* and *Parvicardium papillosum*) are very numerous. Other biogenic populations are present in minor proportions (Table 2). The bryozoa group is here represented by less than 10 forms, of which the chief types (*Cupuladria canariensis, Cupuladria doma, Cellaria sinuosa, Crisia* sp.) are species well adapted to circalittoral clay substrates. Near vasières, there is an increase in the proportion of biogenic skeletal remains derived from mud communities.

RELICT BIOGENIC SEDIMENTS OF THE OUTER SHELF

On the Algerian continental shelf, relict biogenic sediments derived from original calcareous biogenic sediments include numerous remains of benthic fauna and flora that are characteristic of shallower depths. The well-known *faunes froides* of the western Mediterranean continental margins, and particularly of the French continental shelf (summary in Mars and Picard, 1958) never occur here, probably because Algerian relict sediments are relatively coarse sands.

On the outer parts of the Algerian shelf (80 m to 120 m isobaths), relict biogenic sediments constitute most of the coarse sediments. Three groups can be recognized.

The *maerl relict sediments* are the most typical and constitute the most widespread group. They are confined to the western region, off the coasts of Oranie between the 90 m to 200 m contours (Figure 1). They contain a high proportion of maerl detritus and bioclastic particles all stained by iron oxides. The amount of other constituents is insignificant (Table 3). Some stained molluscan valves of infralittoral origin (*Posidonia* mats biocenosis) such as *Cardita calyculata, Glanz trapezia, Venus casina, Gibberula miliaria, Rissoa cimex, Rissoina bruguieri,* occur in a great number in maerl relict sediments. However, no typical Pleistocene or pre-Pleistocene fauna can be

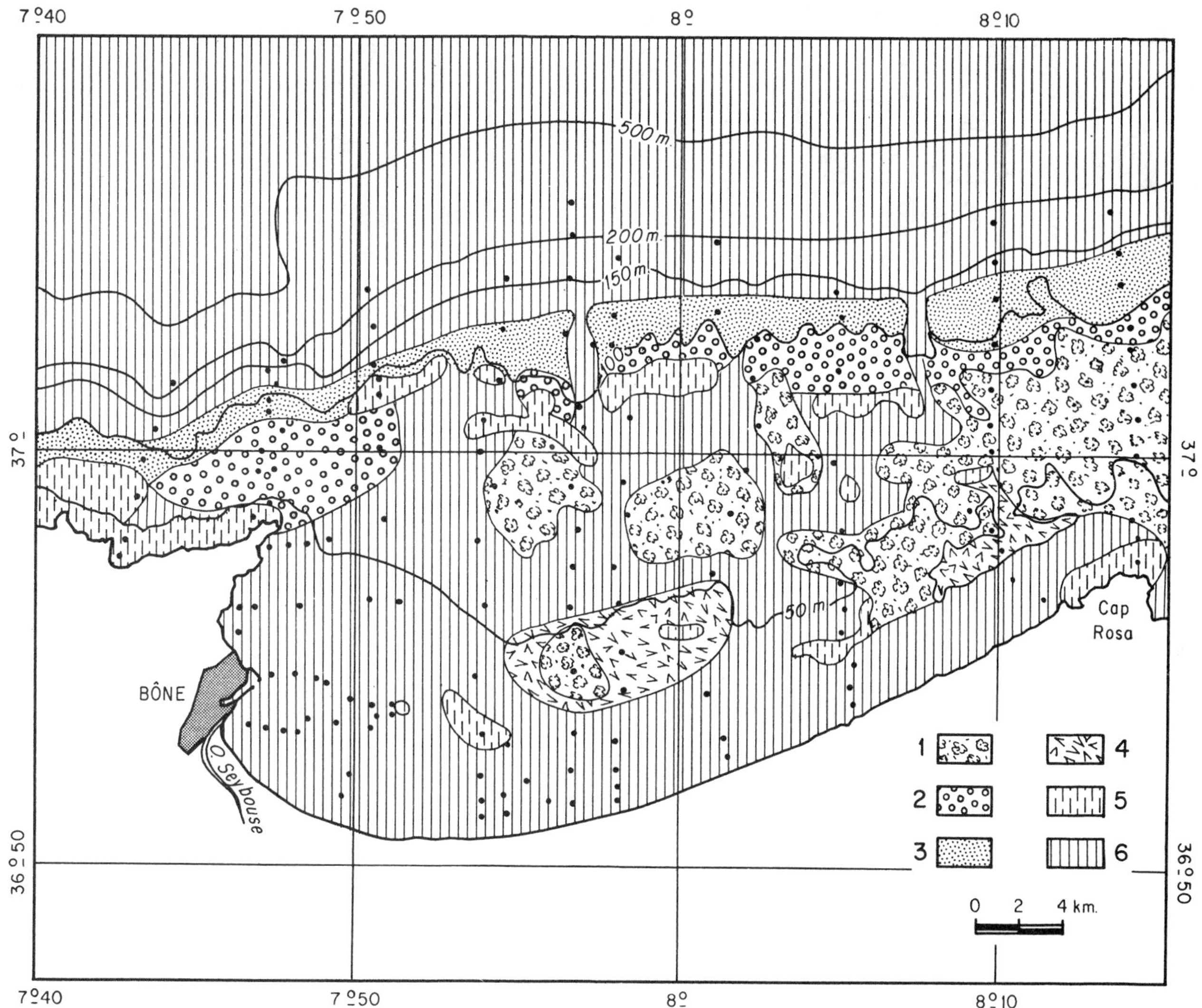

Figure 3. Map showing the distribution of biogenic sediments in the Gulf of Bône. (1) encrusting calcareous algal sands, (2) bryozoan sands, (3) muddy bryozoan sands, (4) muddy bioclastic sands, (5) rock outcrops, and (6) terrigenous sediments.

Table 3. Composition of relict biogenic sediments of the outer shelf.

	Maerl relict sand	Relict muddy shell sand	Relict muddy bryozoan sand
Sample Number	O.132	B.437	B.7
Depth/m	200	100	110
Median diameter (mm)	0.75	0.56	0.43
Phi deviation	1.70	1.80	1.65
Bioclastic fraction	46%	47%	29%
Calcareous algae	23%	1%	
Bryozoa	+	7%	15%
Pelecypods	2%	11%	1%
Gastropods	+	1%	2%
Echinoids	+	+	+
Brachiopods			
Hexacorallia		1%	
Foraminifera	3%	7%	11%
Terrigenous fraction	2%		11%
Pelitic fraction	21%	23%	28%

found in the biogenic fractions. Much of the fine fraction consists of clay material. These clay particles constitute from 7 percent to 25 percent of the sediment. The amount of terrigenous sand ranges from about 1 percent to 5 percent and probably is of residual origin. North of Oran, these particles are reworked, wind-rounded quartz, generated from the wave-erosion of ancient beach dunes or shore-terraces, such as those seen at present along the Bay of Andalouses, west of Oran. Off Arzew, the terrigenes are rounded calcschist detritus originating from the Dj. Orousse' *horsts* (Gourinard, 1958) and deposited by subaerial agents during a low stand of sea level.

Since the isopleth maps of maerl detritus (Figure 5) show that the distribution of deep maerl fractions in the outer shelf closely resembles that of the modern maerl sands of the inner shelf (they both decrease seawards), it appears that the deep relict maerl

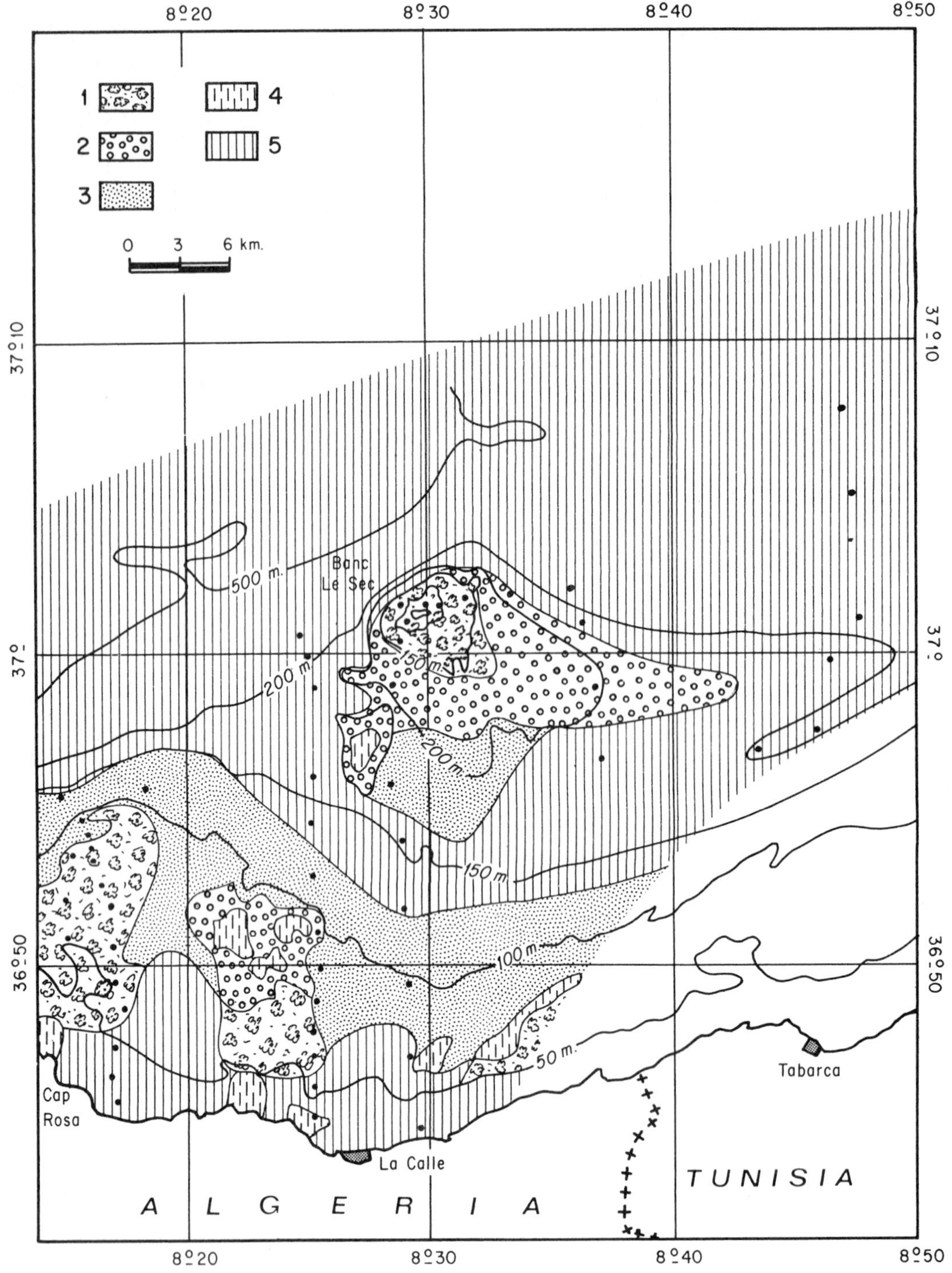

Figure 4. Map showing the distribution of biogenic sediments in the La Calle-Tabarca area (legend as in Figure 3).

deposits have remained undisturbed since their deposition. Since it is generally believed that the rise of sea level was rapid from the outset of the Holocene transgression (Shepard and Curray, 1967), it may be deduced that the deep maerls were deposited, for the most part, during the Last Glacial epoch, before the Holocene transgression. After the rise of sea level, maerl sands off Oran remained uncovered until modern muds began to mantle the middle and outer parts of the shelf. North of Arzew, however, coralligene biocenoses have already covered this facies.

Close examination of the distribution of relict

maerl fraction shows that the average low stand of the sea level during the last Pleistocene glaciation was about 90 m north of Oran. Published data indicate that the last worldwide low stand generally exceeded 100 m (Guilcher, 1969), and van Straaten (1965) has concluded that during the Late Würm, lowering of the Adriatic sea level ranged from–110 m to–120 m. It thus appears that the Oranian continental shelf has been uplifted during the past 19,000 years. This confirms observations made in the Oranian hinterland by Gourinard (1958) who demonstrated that all of the topographic features on the Oranian coast (Dj. Santon, Montagne des Lions, Dj. Orousse) have been rising since the Late Cenozoic. Laffitte (1942 and 1950) also has described warped Pleistocene strata near Mostaganem.

In the Gulf of Arzew, the distribution of relict maerl coarse fractions is somewhat different (Figure 6) and shows that in this region (1) uplift of the outer shelf nearer the great *horst* of Dj. Orousse was greater and deposition of maerls persisted in this area during the beginning of the Holocene transgression, and (2) the vasière area of the middle shelf, marking the submarine extension of the subsiding system of the Cheliff basins (Perrodon, 1957) was either a subsidence zone or a stable region during the early Holocene.

East of Mostaganem, relict maerl occurs only off Chenoua Mountain, in the Bay of Castiglione, west of Algiers (Caulet, 1972). This facies has never been recognized in the eastern region and it appears that factors controlling maerl distribution along the Algerian coast have remained unchanged from the Late Würm.

Relict muddy shell sands can be found, for the most part, in a few areas, *i.e.*, the Bay of Castiglione, the Gulf of Bône (north of the Seiche de Aïn Barbar) (Caulet, 1972) and northwest of Cap Rosa. Just five samples were dredged, which were confined to very small depressions below the relict terraces where a source of mud is most important. More than 60 percent of these samples consist of calcareous bioclastic particles and mud makes up more than 25 percent of the material (Table 3). Among the relict detritus are several valves of infralittoral molluscs such as *Glanz trapezia* and *Rissoa auriscalpium*, all stained by iron oxides and bored by *Clione* sponges. No typical pre-Pleistocene or Pleistocene fauna occurs in these sediments. But, as in the relict maerl sands, numerous eolian-rounded quartz grains, of residual origin, and probably reworked by the sea from neighbouring Pleistocene terraces, can be recognized (Table 3), particularly off Cap Rosa.

Relict muddy bryozoan sands occur only on the outer part of the eastern Algerian continental shelf between the 100 m and 150 m contours, near the shelf-edge, off Bône to the La Calle coast (Figures 3 and 4). These muddy sands contain modern biogenic materials with numerous living forms and two kinds of relict debris. The living fauna, *i.e. Adeonellopsis distoma, Cupuladria canariensis, Cellaria sinuosa, Bathyarca pectunculoides* and their debris (Group I, Table 5) are representative of circalittoral and bathyal communities well adapted to muddy substrates in quiet and deep environments where infralittoral or coralligene biocenoses do not exist. It appears that the stained debris of coralligene bryozoa (*Adeonella calveti, Myriapora truncata*) and coralligene pelecypods (*Acar pulchella, Pteromeris corbis*) (Group II, Table 5) are relict; these were probably deposited during the Holocene transgression when the sea was approaching its present level and before the great influx of mud took place. Thus the valves of infralittoral molluscs (*Gibberula miliaria, Cardita calyculata*) of Group III (Table 5) were deposited, before the debris of Group II, in nearshore environments off beaches and coasts with dunes during the last low stand of the sea. There is, as in the maerl relict sediments, a significant amount (more than 10 percent) of eolian quartz (*rond-mats à émoussé-luisants*), probably reworked from Pleistocene terraces during the Late Würm. The slope, which is now mantled by the relict muddy bryozoan sands, is an ancient nearshore talus.

Table 4. Composition of relict biogenic sediments of the middle and inner shelf.

	Muddy bryozoan sands	Muddy maerl sands off Kristel	Shelly muds off La Macta
Sample Number	B.352	O.373	O.213
Depth/m	60	40	30
Median diameter (mm)	1.00	1.00	0.50
Phi deviation	1.5	0.4	0.8
Bioclastic fraction	42%	36%	51%
Calcareous algae		28%	3%
Bryozoa	17%	+	1%
Pelecypods	2%	3%	10%
Gastropods	3%	2%	+
Echinoids		+	+
Brachiopods			
Hexacorallia	4%		
Foraminifera		2%	1%
Terrigenous fraction	7%	3%	1%
Pelitic fraction	24%	24%	30%

RELICT BIOGENIC SEDIMENTS OF THE BORDERLAND AND MIDDLE SHELF

Other kinds of relict biogenic sediments were dredged on the narrow borderlands between Algiers and Bône and on the middle portions of the widest area of the eastern continental shelf (Caulet, 1972). Generally, they are muddy sands with a high proportion of clay (more than 20 percent). The minor shelly fraction is of modern origin and several living pelecypods and gastropods (*Turritella communis*) from local *mixticoles* biocenoses can be found. However, the bulk of coarse sand, of biogenic origin, is derived from coralligene fauna (bryozoa) and flora (incrusting calcareous algae) which do not live at present in these areas. All this coralligene detritus has been reworked or stained and is regarded as relict material because it is not in equilibrium with the present environments (Emery, 1968). These *muddy bryozoan sands* and *muddy calcareous algal sands* extend to a depth of 60 m on the Kabyle borderland (Baie des Jeunes Filles, between Bougie and Philippeville) and from 60 m to 70 m off La Calle (Figure 4), and it follows that the bryozoans and calcareous algae were deposited after the last rise of sea level and shortly after the formation of the now-relict coralligène populations of the outer parts of the shelf.

This observation confirms the conclusion that the supply of mud to the Algerian shelf began to increase relatively recently, probably after sea level had nearly achieved its near-present height, about 5,000 or 6,000 years B.P., according to the general consensus of opinion (see Curray, 1969).

RELICT BIOGENIC SEDIMENTS OF THE INNER SHELF

On the inner parts of the continental shelf of Oranie (40 m to 50 m isobaths) other anomalous biogenic sands can be seen. Among them, *muddy maerl sands* off Kristel (between Oran and the Cap de l'Aiguille) (Figure 1) are the most typical. They resemble modern maerl sands of the Oranie region but they do not contain any living maerl components and clay grades constitute more than 10 percent of the sediment (Table 4). The living pelecypod species (*Angulus distortus, Lembulus pella*) which were found in these muddy maerl sands are very characteristic of the biocenoses of unstable (transition) environments (Pérès and Picard, 1964). The presence of these species shows that the rate of mud accumulation is increasing in this area at present. Moreover, these muddy maerl sands appear to represent transitory sediments between the true maerl sands and the modern muds of the nearshore prism.

The *shelly muds* dredged from the nearshore talus off

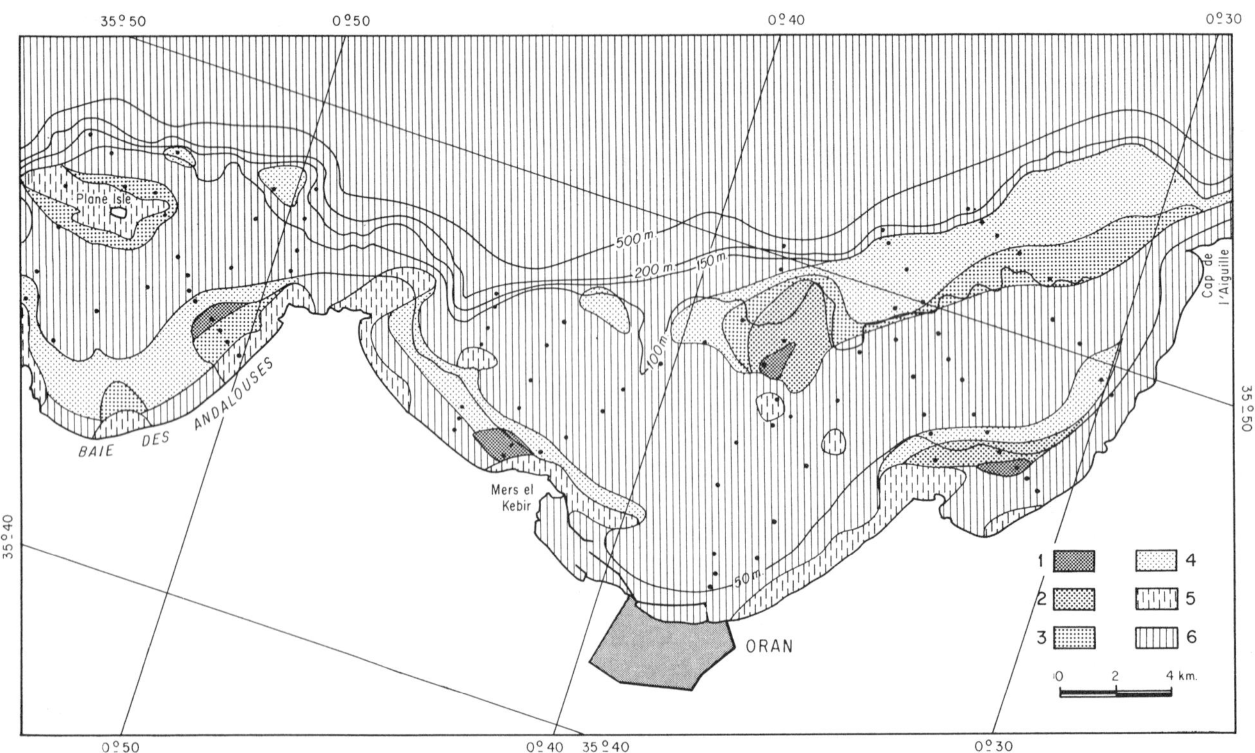

Figure 5. Isopleth map of maerl detritus in the Gulf of Oran. (1), 20%, or more, of maerl detritus, (2) 10% to 20%, (3) 5% to 10%, (4) less than 5%, (5) rock outcrops, and (6) terrigenous sediments.

La Macta river, near Port aux Poules (Figure 2) are shelly sands gradually being buried by clay that is episodically supplied by the La Macta river. They contain scattered dead maerl debris with numerous valves of *Spisula truncata* and *Mactra corallina* from the nearshore muddy sand biocenosis, some valves of *Angulus distortus* (transition environment) and rare debris of the brackish species *Cardium edule*.

From these data, it may be concluded, as Curray (1965) suggested, that muds only began to spread across the Algerian shelf floor during the recent reduction in rate of sea level rise, and they still must be actively extending seaward.

RESIDUAL BIOGENIC SEDIMENTS

The term "residual sediment", one of the original genetic categories, was defined by Emery (1952) as sediment generated from the weathering of underlying rocks. These residual sediments are difficult to distinguish from detrital sediments when both of them are of terrigenous origin (Emery, 1968). Swift (1969) states that—"in the broad dichotomy of relict versus modern sediments, these residual sediments belong to the relict group, yet their faunal and petrographic characteristics may be quite distinctive". But, on the Algerian continental shelf, well-defined residual biogenic materials can be recognized in modern sediments.

The most obvious residual materials occur in *muddy shell sands* off Dahra. These muddy shell sands mantle a few small outer parts of the narrow shelf, east of Mostaganem (Figure 2). They are completely surrounded by modern muds and contain more than 20 percent clay. The bulk of the coarse particles is of shelly origin (Table 6) and numerous valves of living mixticoles pelecypods (*Timoclea ovata, Nuculana fragilis*) can be found in the pelecypod fraction. But a detailed examination of macro-assemblages shows several reworked zoaria of a Plio-Pleistocene bryozoan characteristic of warmer waters: *Biflustra savartii* forma *delicatula* (Caulet and Redier, 1968). As these zoaria are stained by hydroxides of iron, it is probable that these residual biogenic fragments are generated from the reworking of Plio-Pleistocene materials by subaerial agencies during the last glacial recession. Moreover, a few valves of relict infralittoral pelecypods deposited during the Holocene transgression are mixed with the modern and residual biogenic detritus.

The only genuine residual sediments which were found are the *muddy bioclastic sands* dredged from about 40 m and extending to a depth of 50 m in the

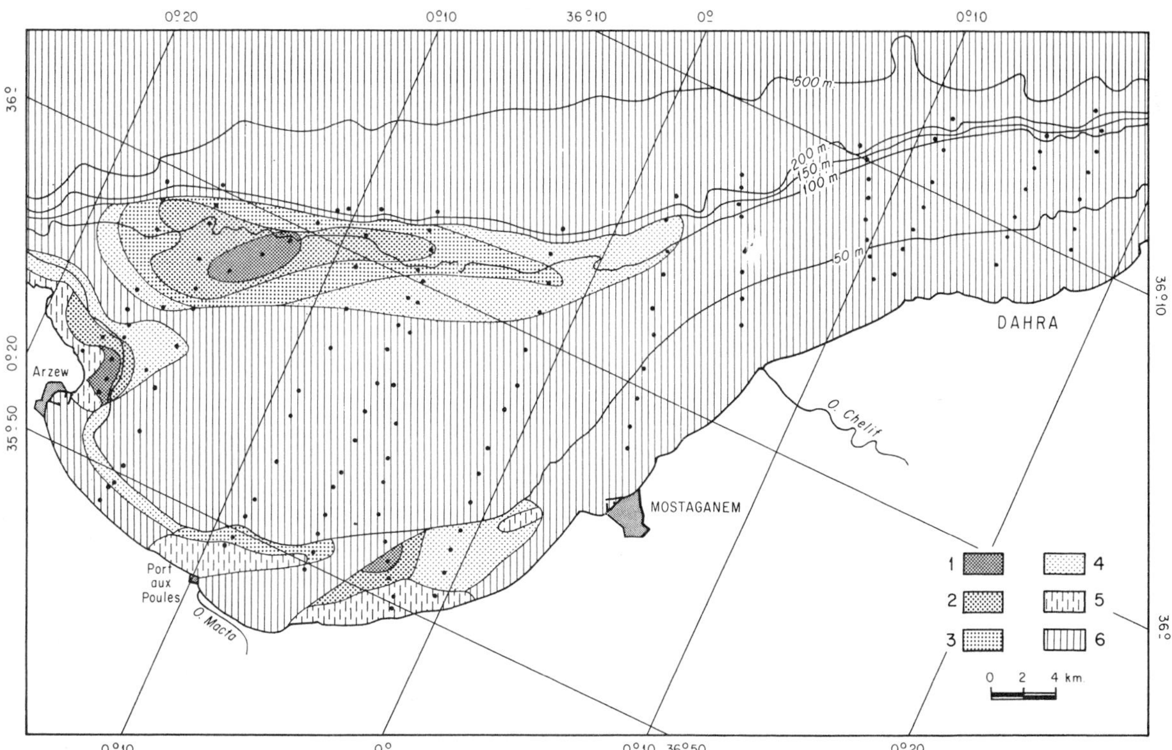

Figure 6. Isopleth map of maerl detritus in the Gulf of Arzew (legend as in Figure 5).

Table 5. Composition of relict muddy bryozoan sands (abundance of each species is expressed as percentage of specimens for each fraction).

Sample Number	2	451	458	488	489	508
GROUP I:						
Adeonellopsis distoma				1	3	
Cupuladria canariensis	1			1	1	
Cellaria sinuosa	5	5	9		12	1
Cellaria salicornioides	7		4	7	21	2
Palmicellaria elegans	2			1	7	
Bathyarca pectunculoides	11	3	5	3	12	3
Digitaria digitaria		2	1	5		
Nuculana deltoidea	6	1	3	3	3	
Timoclea ovata	3	3	3	2		3
Similipecten similis	3	7	1	1	9	2
Parvicardium nodosum	3	1	2	2	1	3
Retusa semisulcata		2	2			2
Ringicula conformis	1					1
Philine catena	1					1
GROUP II:						
Porella cervicornis	6	7	4		1	9
Adeonella calveti	7	13	13	3	5	5
Myriapora truncata	1	1				
Acar pulchella	1	3	3		1	2
Pteromeris corbis	3				2	2
Striarca lactea		2	2		3	1
GROUP III:						
Margaretta cereoides		1				
Schismopora tubigera	4	1	10			
Glanz trapezia	2		1	1	1	
Cardita calyculata			1			
Venus casina	1		1			1
Alvania cimex	2	4				
Gibberula miliaria	1	1	1		4	3
Chauvetia minima	1					2
Bittium reticulatum	3	6	6			3
Astraea rugosa		1	1			

Table 6. Composition of selected residual biogenic sediments.

	Muddy shell sands off Dahra	Bioclastic sands off eastern region
Sample Number	O.277	B.462
Depth/m	88	45
Median diameter (mm)	0.56	0.65
Phi deviation	1.2	1.5
Bioclastic fraction	28%	66%
Calcareous algae		+
Bryozoa	3%	4%
Pelecypods	12%	1%
Gastropods	+	1%
Echinoids	+	
Brachiopods		+
Hexacorallia		+
Foraminifera	4%	2%
Terrigenous fraction	11%	2%
Pelitic fraction	50%	22%

Gulf of Bône (Figure 3). Generally, the residual debris comprises strongly reworked biogenic particles of coralligene origin. Numerous Adeoniform zoaria detritus and rounded pieces of encrusting calcareous algae, which have been modified by subaerial weathering, compose the greater part of the sediment, together with a great number of red-stained bioclastic particles (Table 6). These residual sediments were dredged near fossil terraces (*les plattiers*) which are probably built with Plio-Pleistocene materials (Leclaire, 1972); it is inferred, therefore, that they were generated from the weathering of these terraces during the last stages of the Holocene transgression. Unfortunately the carbon-14 method does not provide exact determinations with small pieces of reworked biogenic debris of this type (Thommeret and Thommeret, 1965), and if the residual nature of this coralligene debris is correctly interpreted, it may only be concluded at present that this material was initially formed and deposited long before the Late Würm.

FACTORS GOVERNING DEPOSITION OF MODERN BIOGENIC SEDIMENTS

Modern biogenic sediments can roughly be defined by their size distribution characteristics, coarse fraction percentages and skeletal composition. Since size distribution, coarse fraction and skeletal components are subordinate to the environment, it may be possible to determine the chief agencies responsible for the deposition of modern biogenic sediments by studying their relations with the major environmental factors such as depth, sediment supply, and salinity. However, particle size analysis is not significant in this context because grain-size parameters relate mainly to transportation of material of relatively uniform shape, density and genetic type. Therefore, on the Algerian continental shelf, only the coarse fraction studies, using isopleth maps, and the *ecologic characteristics* of species show significant relationships.

Influence of Mud Supply

This is the most important single factor in modern biogenic sedimentation on the Algerian continental shelf. Terrigenous coarse sands are confined to the nearshore area (*prisme littoral* of Bourcart, 1960, and *nearshore modern prism* of Swift, 1969). Thus, only silt and clay which have by-passed the coastal hydraulic traps as suspended load can spread across the shelf and blanket the relict morphology and existing biotal assemblages. This influx of mud has a twofold influence on biocoenosis growth. It buries the solid substrate which the larva of many invertebrate organisms need for growth. Mud also smothers benthonic groups such as bryozoa, calcareous algae, brachiopods, and gastropods. Thus, the coralligène and maerl assemblages, which produce the major part of the modern biogenic contribution, are the most seriously affected by the advent of mud. The relative abundance of bryozoan zoaria and calcareous algae concretions varies inversely with the rate of clay sedimentation (Caulet, 1972). The isopleth maps of bryozoan debris and calcereous algal concretions show an inverse relationship between the abundance of these constituents and the mud percentage in the sediments near vasières. Moreover, these benthonic organisms cease to live and accumulate on substrates where the percentage of clay exceeds 10%.

On the other hand, pelecypods are better adapted to relatively high turbid environments and to occasional rapid sedimentation. Thus, their remains are most abundant in biogenic sediments where the proportion of fines ranges from 30 to 40 percent. An isopleth map of pelecypod detritus (Figure 7) shows that the abundance of such material increases near the mud-dominated regions of the central area of the continental shelf. Nevertheless, if the clay proportion exceeds 30 percent, the dilution of the living specimens by inorganic material increases and most species lose their tolerance against the supply of fines. Thus, the remains of pelecypods become rarer in such sediments (Caulet, 1972).

Mud supply not only influences the composition of the biocenoses and consequently the composition of biogenic sediments, but also eventually stops all biogenic sedimentation. It appears to be the chief factor controlling modern and probably recent biogenic sedimentation.

Influence of Depth

In general, there is no obvious relationship between depth and the rate of deposition of many constituents on the Algerian continental shelf. On the other hand, depth plays an important role in the distribution of some components. Highest populations of bryozoan detritus are confined between the 60 m and 80 m isobaths; the highest amounts of gastropod shells are found in shallower environments (10 m to 40 m isobaths). Moreover, the distribution of many species is related to water depth: typical infralittoral species such as *Margaretta cereoides, Glanz trapezia, Homalopoma sanguinea* do not extend to a depth of more than 50 m, and most of the near bathyal species only occur below the 50 m isobath: *Bathyarca pectuncu-*

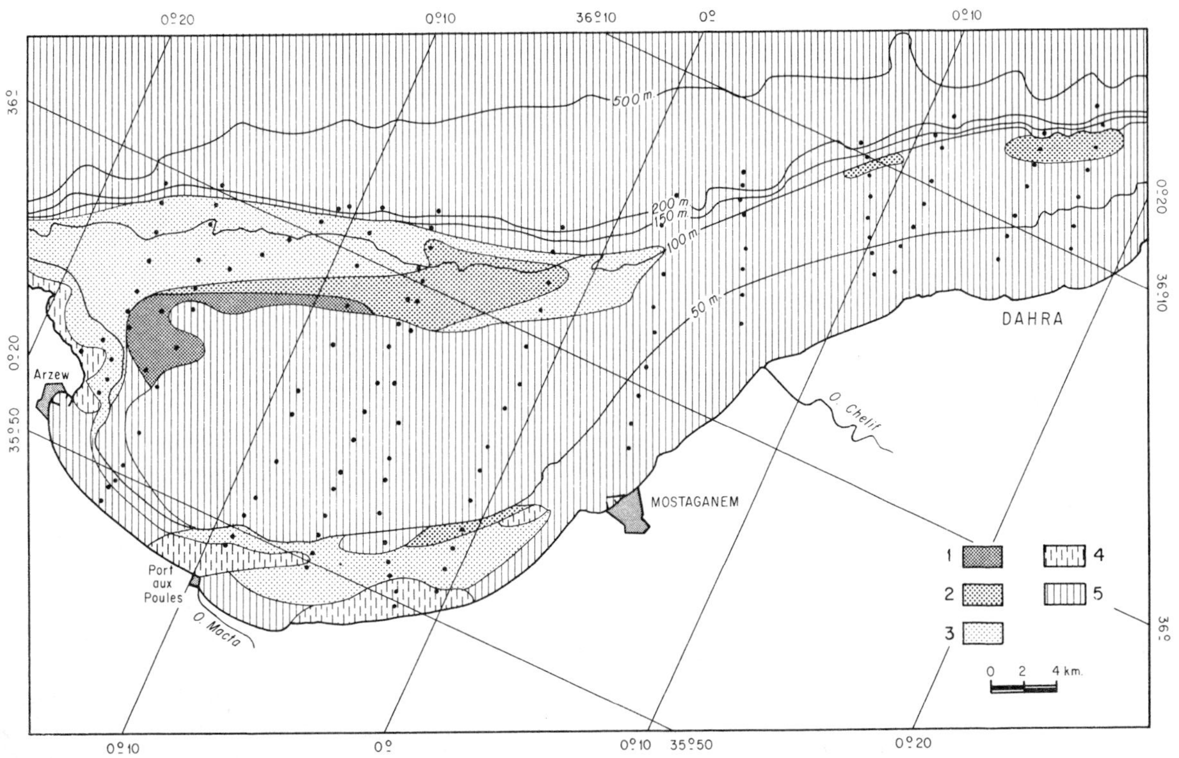

Figure 7. Isopleth map of pelecypod detritus in the Gulf of Arzew. (1) more than 10% of shell detritus, (2) 5% to 10%, (3) less than 5%, (4) rock outcrops, and (5) terrigenous sediments.

loides, Buchneria fayalensis, Adeonellopsis distoma. Therefore, the relative supply of various types of skeletal material varies with depth, and the highest rates of deposition of many species are related to depth. Finally, depth appears to influence the composition more than the abundance of biogenic fractions. Thus, accumulations of calcareous algal debris are encountered at depths ranging from 10 m to 70 m, but maerl deposits occur only from 10 m to 50 m depth, while encrusting algal debris (*Pseudolithophyllum expansum* and *Lithothamnium fruticulosum*) accumulates between 50 m and 70 m.

Since destruction of skeletal material is related to biological agents as well as mechanical processes, there is no obvious relationship between the total abundance of bioclastic material and depth.

Influence of Salinity

Salinity is of little importance because surface waters of uniform salinity extend down to a depth of 200 m all along the coast of Algeria.

Nevertheless, there are a few species of Atlantic origin (*Cupuladria canariensis, Cupuladria doma*) that principally inhabit the communities of the western region (from Oran to Mostaganem). Although they have favorable substrates east of Bône, these species are most abundant in the muddy sands off the coasts of Oranie (Caulet, 1969b). Conversely some species such as *Acar pulchella*, an eastern Mediterranean species which requires typical coralligene environments (Caulet, 1969b), occur only between Algiers and La Calle and are never found in the western region. Since organisms related to salinity are not important sediment contributors, it appears that salinity also influences the composition of debris populations more than their total abundance.

Influence of Biotic Factors

Biotic factors inherent in the communities can modify the abiotic factors and change the biocoenosis equilibrium (Pérès and Picard, 1964). They can also affect the composition of biogenic sediments. The regional variation in the composition of coralligene biogenic sands provides a good example: there is no bryozoan debris mixed with calcareous algal concretions in the western coralligene sands, but in the eastern coralligene sands the bryozoan fraction is often as important as the total amount of calcareous algae (Table 2). However, closer examination reveals that the calcareous algae are significantly different in nature.

In the eastern region, the most common species are encrusting forms which provide a favorable substrate for bryozoan growth, while in the western region, the branch-like maerl debris do not furnish good substrates for the young bryozoan larvae.

As modern biogenic sediments are constituted by no more than 10% of total living species, there is no full homology between thanatocoenosis and biocoenosis on the Algerian continental shelf. However, both the rates of deposition of the major modern groups of organisms and the accumulation of their species are highest in areas where these groups and their species can find their best environmental conditions for development. It appears that the *ecologic significance* of the most common living calcareous organisms is not altered by present processes of sedimentation.

CONCLUSIONS

This reconnaissance study of Algerian carbonate sediments indicates that they are, for the most part, biogenic sands with an insignificant proportion of terrigenous material. The particle size distribution of these sands in the Oranian gulfs and in the eastern region shows no progressive change with increasing distance from the shore or with increasing depth, as might be expected from their highly diversified composition and their mixed provenance. Instead, the particle size distribution is related chiefly to the biological composition of these sediments rather than to physical processes.

The coarsest sands, which were dredged mainly from the middle shelf, are composed of coralline algae (pralines) and bryozoan debris. Bryozoa attain local importance, especially near rocky banks and on hard bottoms. Wass *et al.* (1970) describe a similar distribution of bryozoa in sediments of the southern Australian shelf where bryozoan skeletal remains are generally the dominant constituents. Bryozoan sands were also dredged in the western English channel but they are composed of *Cellaria*, which occurs in areas where the currents are stronger (Boillot, 1964). Coralline algae generally predominate in the Algerian coarse biogenic sands, as in the bioclastic sands of the northeastern Gulf of Mexico, where algal concretions of *Lithothamnium* are numerous (Gould and Stewart, 1955). However, benthonic foraminifera which are common in the Florida and Barbados carbonate sediments (Macintyre, 1970) are insignificant in the bryozoan-algal sands of Algeria, which are always associated with shelf-banks. These sands are thought to be "actively developing", like the algal-foraminiferal sands of the Sahul shelf (van Andel and Veevers, 1967), whereas algal sands of the northeastern Gulf of Mexico are considered relict-residual in origin. As Williams (1963) suggested for the foraminiferal-algal sands of the Yucatan shelf, it seems that

Table 7. Distribution of the principal biogenic coarse fraction components in different environments of the Algerian margin.

Environments	Maximum rate of deposition	Minimum rate of deposition
Nearshore Environments (0 to 50m)	Maerls Bioclastic fraction Gastropods	Pelecypods Bryozoa
Middle Shelf: Flat Bottom without Bed Rock	Pelecypods	Bryozoa Encrusting calcareous algae Gastropods
Middle Shelf: Rocky Flat Bottom	Encrusting calcareous algae	
Middle Shelf: Rocky Topography	Bryozoa	
Banks and Shoals	Gastropods	Pelecypods
Shelf-edge	Foraminifera	Encrusting calcareous algae
Top of the Slope	Foraminifera	Calcareous algae Bryozoa Pelecypods Gastropods

the sediments off western Florida are mixed, with relict, residual and modern components.

The deep maerl sands and deep bryozoan sands, which mantle the Algerian shelf-edge areas between depths of 90 m to 150 m, also appear to be mixed sediments. The maerl sands appear to have been formed in shallow water (10 to 40 m) during the Late Würm, when the shore line coincided approximately with the present outer shelf margin. They contain numerous infralittoral skeletal remains and debris such as maerl, pelecypods and gastropods, forms which were probably living in a Posidonia mat biocenosis. These maerl sands do not occur in the eastern region where there are shelly muds with abundant relict shallow water molluscs. But the relict shelly muds here do not mantle areas as broad as those on the U.S. South Atlantic shelf (Pilkey, 1964; Merrill *et al.*, 1965; Milliman *et al.*, 1968) probably because there were no estuarine or shallow muddy environments on the narrow Algerian borderland and because shelf modern muds have blanketed most of the relict depressions. The deep bryozoan sands of the eastern region are also mixed sediments. They contain shallow water assemblages deposited during the Late Würm, together with coralligène bryozoan debris which accumulated during stillstands of the Holocene sea level rise and before the shelf floor muds began to spread across the shelf. So it appears, as Curray (1965) has suggested, that shelf floor muds began to spread across the shelf during the recent reduction in the rate of sea level rise. Careful study of the patterns of relict sediments on the Algerian shelf demonstrates that the relict bryozoan sands lie on rough bottoms of the outer shelf (outer belt), while the muddy sands occur in the transition zones near central vasières. This would confirm that the seaward flow from the surf zone, which conveys fine particles outward, is compensated on the middle shelf by the effects of landward currents induced by great storm waves (see the theoretical and experimental investigations of Lhermitte, 1958). The bottom current becomes more turbulent landward where it is in contact with the relict morphologic features; this results in the winnowing of muds from these areas.

If relict is "used in sense of not being related to present environmental conditions or, in the case of sediments, not being related to a source on the continent by existing paths of transport" (Curray, 1969, p. JC-VI-11), it then appears that some biogenic muddy sands of the middle or the inner shelf are relict. The bryozoan or coralline algal muddy sands of the eastern region of the Algerian shelf contain skeletal remains which were deposited during the Holocene transgression. The muddy maerl sands of the Oranian inner shelf are accumulations of the very recent past, and show that the mud blanket is still actively extending seaward.

Genuine residual sediments were not dredged on the Algerian shelf, but residual materials (Pleistocene eolian quartz and Pleistocene corroded debris of coralline algae and bryozoa) were found in numerous relict muddy sands. Like most of the carbonate sediments which occur on modern shelves, the Algerian carbonate sediments probably range in age from Late Pleistocene to Present.

The normal environmental range of most recent species of the Mediterranean is reasonably well known (Pérès and Picard, 1964) and close attention therefore has been paid to the composition of biogenic sediments and to the "ecologic meaning" of their relict fauna and flora, in order to define more accurately the Recent history of the Algerian continental shelf. It appears that, during the Late Würm, the greater part of this shelf was exposed to subaerial weathering under subarid climatic conditions (Brunnacker, 1969). However, the remains of terrestrial plants or continental faunas have not been found to date, as they have on the U.S. Atlantic continental shelf (Whitmore *et al.*, 1967). Erosional processes and eolian transport during the early Pleistocene gave rise to great dune systems on the widest parts of the Oranian shelf and eastern region. Around the coastal *horsts* (Dj. Orousse,

Dj. Chenoua) successive ramparts of coarse detrital material were created, principally in the western region (Perrodon, 1957). At the same time, maerls and bioclastic sands were deposited in shallow water off these slopes and off the shore terraces, on the present shelf-edge. Shelly sands accumulated between Posidonia mats and along the beaches, and nearshore talus deposits prograded seaward.

As the sea level rose, infralittoral biocenoses began to shift across the shelf to their present location. Precoralligene biocenoses and then coralligene biocenoses progressively settled on rocky substrates such as Pleistocene shore terraces, consolidated dunes and slopes. However, during the initial rapid rise in sea level (Shepard and Curray, 1967) little biogenic material was deposited, so that Late Würm deposits remained unburied. Infralittoral mollusc shells found in the relict materials of the middle shelf are not abundant, probably because contemporaneous mud supplies were not important. At this time, the Algerian hinterland was mantled by a thick herbaceous cover (Quezel, 1963), while swamps inundated the flat coastal plains (Perrodon, 1957). After the climatic change to subarid conditions, which occurred around 6,000 or 5,000 yr B.P. (Flint and Brandtner, 1961), the herbaceous cover of the Algerian hinterlands was destroyed by increased erosion. Terrigenous materials began to fill the shelf depressions and progressively blanketed the adjacent areas where they buried coralligene or infralittoral assemblages.

This rapid erosion of rocks and soils of Algeria has been progressively accelerated by human deforestation (post-Roman period), resulting in an increase of muddy material carried out to sea. New transition environmental conditions are introduced which involve the progressive destruction of a great number of sub-present marine biocenoses. The present mode of biogenic sedimentation is steadily trending towards extinction beneath a mud blanket. Carbonate sediments off Algeria thus appear to form "transitory facies", characteristic of areas of non-equilibrium morphology.

ACKNOWLEDGMENTS

The writer is grateful to Professors G. Kelling and D.J. Stanley for helpful criticism and suggestions. Many thanks go to Professors R. Laffitte and G. Lucas for their interest and encouragement. Dr. S. Freinex and Dr. L. Redier assisted in the identification of Pelecypods and Bryozoa.

REFERENCES

Berger, W. H. 1968. Radiolarian skeletons: solution at depth. *Science*, 159:1237–1238.

Berger, W. H. and A. Soutar, 1970. Preservation of plankton shells in an anaerobic basin off California. *Geological Society of America Bulletin*, 81:275–282.

Bernard, F. and J. Lecal-Schlauder, 1953. Role des flagellés calcaires dans la sédimentation actuelle en Méditerranée. XIXe *Congrès Géologique International, Comptes Rendus de la 19e Session, Section IV*, 4:11–23.

Berthois, L. 1939. Contribution à l'étude des sédiments de la Méditerranée occidentale. *Annales de l'Institut Océanographique de Monaco*, 20 (1):50.

Blanc, J. J. 1958. Sédimentologie sous-marine du Détroit de Sicile. *Annales de l'Institut Océanographique de Monaco*, 34:91–126.

Boillot. J. 1964. Géologie de la Manche occidentale. *Annales de l'Institut Océanographique de Monaco*, 62:1–120.

Boulaine, J. 1957. *Etude des sols des plaines de Chélif.* Thèse Faculté des Sciences, Alger. 1:575p.

Bourcart, J. 1960. Les divers modes de sédimentation observés en Méditerranée occidentale. *International Geological Congress 21st*, Copenhagen, Report of Sessions, 23:7–18.

Brunnacker, K. 1969. Affleurements de loess dans les régions nord-méditerranéennes. *Revue de Géographie Physique et de Géologie Dynamique*, série 2, II (3):125–334.

Caulet, J. 1963. Etude des plages entre Arzew et Port aux Poules. *Cahiers Océanographiques*, 15 (9):617–637.

Caulet, J. 1967. Les sédiments meubles à brachiopodes de la marge continentale algérienne. *Bulletin du Muséum National d'Histoire Naturelle, Paris*, 39 (4):779–792.

Caulet, J. 1968. Sur les accumulations de bryozoaires dans les sables organogènes grossiers du précontinent algérien. *Comptes Rendus de l'Académie des Sciences, Paris*, 266:449–451.

Caulet, J. 1969 a. Les sédiments organogènes du Golfe d'Arzew (Algérie). Présentation d'une esquisse au 1/150,000° de la répartition des principaux faciès. *Comptes Rendus Sommaires de la Société Géologique de France*, 5:160–161.

Caulet, J. 1969 b. Contribution à l'analyse des sédiments organogènes du précontinent algérien. Variations régionales de composition des accumulations de bryozoaires et de lamellibranches. *Bulletin du Muséum National d'Histoire Naturelle, Paris*, 41 (3): 801–816.

Caulet, J. 1972. *Les sédiments organogènes du précontinent algérien.* Thèse de Doctorat d'Etat, Université de Paris. Mémoires Muséum, Paris, 25:1–289.

Caulet, J. and L. Redier, 1968. Sur la présence de matériel plio-quaternaire dans quelques sédiments de la plate-forme continentale mostaganémoise (Algérie). *Comptes Rendus Sommaires de la Société Géologique de France*, 9:325–326.

Curray, J. R. 1965. Late Quaternary history, continental shelves of the United States. In: *The Quaternary of the United States*, ed. Wright, H.E. and D.G. Frey, Princeton Press, Princeton 723–735.

Curray, J. R. 1969. History of continental shelves. In: *The New Concepts of Continental Margin Sedimentation*, ed. Stanley, D.J., The American Geological Institute, Washington, D.C., 6:1–18.

Dangeard, L. 1923. Recherches de géologie sous-marine en Méditerranée (croisière du Pourquoi-Pas?). *Comptes Rendus de l'Académie des Sciences, Paris*, 177:1048–1050.

Dangeard, L. 1929. Observations de géologie sous-marine. *Annales de l'institut d'Océanographie de Monaco*, 6 (1):1–293.

Emery, K. O. 1952. Continental shelf sediments of southern California. *Geological Society of America Bulletin*, 63:1105–1108.

Emery, K. O. 1968. Relict sediments on the continental shelves of

World. *American Association of Petroleum Geologists Bulletin,* 52 (3):445–465.

Flint, R. F. and F. Brandtner, 1961. Climatic changes since the last interglacial. *American Journal of Science,* 259:321–328.

Furnestin, J. 1960. Hydrologie de la Méditerranée occidentale (Golfe du Lion, Mer d'Alboran, Mer Catalane, Corse Orientale). *Revue des Travaux de l'Institut Scientifique et Technique des Pêches Maritimes, Paris,* 24 (1):5–121.

Furnestin, J. and C. Allain 1962a. L'hydrologie algérienne en hiver. *Revue et Travaux de l'Institut Scientifique et Technique des Pêches Maritimes, Paris,* 26, (3)277–309.

Furnestin, J. and C. Allain, 1962b. Nouvelles observations sur l'hydrologie de la Méditerranée Occidentale (entre Alger et le 40e parallèle). *Revue et Travaux de l'Institut Scientifique et Technique des Pêches Maritimes, Paris,* 26 (3):309–317.

Gould, H. R. and R. H. Stewart, 1955. Continental terraces sediments in the northeastern Gulf of Mexico. In: *Finding Ancient Shorelines,* ed. Hough, J. L., *Society of Economic Paleontologists and Mineralogists,* Special Publication, 3:2–19.

Gourinard, Y. 1958. Recherches sur la géologie du littoral oranais. Epirogénèse et nivellement. *Publications du Service de la Carte Géologique de l'Algérie, Nouvelle Série,* 6:1–195.

Guilcher, A. 1969. Pleistocene and Holocene sea level changes. *Earth Science Reviews,* 5:69–97.

Jacquotte, R. 1962. Etude des fonds de maërl en Méditerranée. *Recueil des Travaux de la Station Marine d'Endoume, Marseille,* 26 (41).

Kruger, D. 1950. Variations quantitatives des protistes marins au voisinage du Port d'Alger durant l'hiver 1949–1950. *Bulletin de l'Institut Océanographique de Monaco,* 978:2–20.

Lacombe, H. and P. Tchernia, 1960. Quelques traits généraux de l'hydrologie méditerranéenne d'après diverses campagnes hydrologiques récentes en Méditerranée et dans le proche Atlantique. *Cahiers Océanographiques,* 12 (8):527–547.

Lacombe, H. and P. Tchernia, 1965. Océanographie physique méditerranéenne. *Rapports et Procès-verbaux de la Commission Internationale pour l'Exploration Scientifique de la Mer Méditerranée,* 18 (3):1065–1066.

Laffitte, R. 1942. Plissements post-pliocènes et mouvements quaternaires dans l'Algérie occidentale. *Comptes Rendus de l'Académie des Sciences, Paris,* 215:372–374.

Laffitte, R. 1950. Sur l'existence du calabrien dans la région oranaise. *Comptes Rendus de l'Académie des Sciences, Paris,* 230 (2):217–219.

Lagaaij, R. and Y. V. Gautier, 1965. Bryozoan assemblages from marine sediments of the Rhône Delta. *Micropaleontology,* 11(1): 39–58.

Leclaire, L. 1963. Facteurs d'évolution d'une côte sablonneuse rectiligne très ouverte. *Cahiers Océanographiques,* 15 (8):540–556.

Leclaire, L. 1968. Contribution à l'étude géomorphologique de la marge continentale algérienne. *Cahiers Océanographiques,* 20 (6):451–521.

Leclaire, L. 1972a. *La sédimentation holocène sur le versant méridional du bassin algéro-baléares (Précontinent algérien).* Thèse Université de Paris, Mémoires Muséum, Paris, 24:391p.

Leclaire, L. 1972b. Aspects of late Quaternary sedimentation on the Algerian precontinent and in the adjacent Algiers-Balearic Basin. In: *The Mediterranean Sea: A Natural Sedimentation Laboratory,* ed. Stanley, D.J., Dowden, Hutchinson and Ross, Inc., Stroudsburg, Pennsylvania, 561–582.

Leclaire, L., J. P. Caulet, and P. Bouysse, 1965. Prospection sédimentologique de la marge continentale nord-africaine. *Cahiers Océanographiques,* 17 (7):467–479.

Lemoine, M. 1957. Les algues calcaires des fonds coralligènes du Cap Carbon. *Vie et Mileu,* supplément, 6:235–236.

Lhermitte, P. 1958. Contribution à l'étude de la couche limite des houles progressives. *Comité d'océanographie et d'études des Côtes, Paris,* 136:171p.

Macintyre, I. G. 1970. Sediments off the west coast of Barbados. *Marine Geology,* 9:5–23.

Mars, P. and J. Picard, 1958. Notes sur les gisements sous-marins à faune celtique en Méditerranée. *Rapports et Procès-verbaux de la Commission Internationale pour l'Exploration Scientifique de la Mer Méditerranée,* 15 (3):325–330.

Merrill, A. S., K. O. Emery and H. Rubin, 1965. Oyster shells on the Atlantic continental shelf. *Science,* 147:395–400.

Milliman, J. D., O. H. Pilkey and B. W. Blackwelder, 1968. Carbonate sediments on the continental shelf, Cape Hatteras to Cape Romain. *Southeastern Geology,* 9:245–267.

Pérès, J. M. and J. Picard, 1964. Nouveau manuel de bionomie benthique de la Mer Méditerranée. *Bulletin et Recueil des Travaux de la Station Marine d'Endoume, Marseille,* 31 (47):5–137.

Perrodon, A. 1957. Etude géologique des bassins néogènes sublittoraux de l'Algérie occidentale. *Publications du Service de la Carte Géologique de l'Algérie. Nouvelle Série,* 12:1–328.

Pilkey, O. H. 1964. The size distribution and mineralogy of the carbonate fraction of United States south Atlantic shelf and upper slope sediments. *Marine Geology,* 2:121–136.

Quézel, P. 1963. Paléoclimatologie du Quaternaire récent au Sahara. Changes of Climates. *Proceedings of the Rome Symposium organized by the UNESCO,* Paris.

Rosfelder, A. 1955. Carte provisoire au 1/500,000° de la marge continentale algérienne. *Publications du Service de la Carte Géologique de l'Algérie. Travaux des Collaborateurs,* 5:57–106.

Rosfelder, A. 1961. *Contribution à l'analyse texturale des sédiments.* Thèse, Université d'Alger, 11:310p.

Shepard, F. P. and J. R. Curray, 1967. Carbon 14 determination of sea level changes in stable area. *Progress of Oceanography, INQUA 1965,* 4:283–293.

Sparck, R. 1931. Some quantitative investigations on the bottom fauna at the west coast of Italy, in the Bay of Algiers. *Reports of Danish Oceanographical Expedition in Mediterranean Sea,* 3 (7):1–11.

Swift, D. J. P. 1969. Outer shelf sedimentation: processes and products. In: *The New Concepts of Continental Margin Sedimentation,* ed. Stanley, D.J., The American Geological Institute, Washington, D.C., 5:1–26.

Thommeret., J and Y. Thommeret, 1965. Validité de la datation des sédiments du proche quaternaire par le dosage du C 14 dans les coquilles marines. *Rapports et Procès Verbaux de la Commission Internationale pour l'Exploration Scientifique de la Méditerranée,* 18:837.

van Andel, Tj.H. and J.J. Veevers, 1967. Morphology and sediments of the Timor Sea. *Australia, Bureau of Mineral Resources, Geological and Geophysical Bulletin,* 83:1–127.

van Straaten, L. M. J. U. 1960. Marine mollusc shell assemblages of the Rhône Delta. *Geologie en Mijnbouw,* 39(4):105–129.

van Straaten, L. M. J. U. 1965. Sedimentation in the north-western part of the Adriatic Sea. In: *Submarine Geology and Geophysics,* eds. Whittard, W.F. and R. Bradshaw, Butterworths, London: 143–162.

Wass, R. E., J. R. Conolly and R. J. Macintyre, 1970. Bryozoan carbonate sand continuous along southern Australia. *Marine Geology,* 9:63–73.

Whitmore, F. C. Jr., K. O. Emery, M. B. S. Cooke and D. J. P. Swift 1967. Elephant teeth from the Atlantic continental shelf. *Science,* 156:1481.

Williams, J. D. 1963. The petrology and petrography of sediments from the Sigsbee Blanket, Yucatan shelf, Mexico. *Texas A and M collections, Reports of Department of Oceanography and Meteorology,* 63–12T:1–60.

Recent Sedimentation and Oolite Formation in the Ras Matarma Lagoon, Gulf of Suez

Eytan Sass, Yehezkiel Weiler and Amitai Katz

The Hebrew University, Jerusalem

ABSTRACT

Oolites are accumulating at present in the small, shallow, marine Ras Matarma Lagoon (Gulf of Suez). The triangular lagoon is divided by a N-S trending peninsula into two embayments each of which shows a marked asymmetric disposition of environments: the western part is controlled by active depositional processes (eolian supply of oolitic sands from the west and northwest), whereas the eastern beaches are characterized by erosion (caused by the locally generated waves). This study sheds light on the environmental conditions needed for oolite deposition.

The oolites accumulating in the lagoon are of both high and low energy levels (in addition to allochthonous, wind-blown ones). It is suggested that smaller ooids (modal size about 0.1 mm) are formed in suspension, whereas the larger ones (0.35 mm) are formed by rolling and saltation on the bottom at higher energy levels. A sample of oolitic sediment (without using the peeling procedure, *i.e.*, including the nuclei) yielded a carbon-14 age of 1740 ± 60 years B.P.

RESUME

Un dépôt d'oolites est en train de se former dans le petit lagon marin peu profond de Ras Matarma (Golfe de Suez). Ce lagon triangulaire est divisé par une péninsule Nord-Sud formant ainsi deux baies qui présentent toutes deux une assymétrie dans leur environnement. La partie Ouest de chacune de ces baies est le siège de processus actifs de dépôt (apports éoliens de sables oolitiques venant de l'Ouest et du Nord-Ouest) alors que les plages de la partie Est sont soumises à l'action de l'érosion des vagues locales. Cette étude révèle les conditions de milieu nécessaires au dépôt d'oolites.

L'accumulation des oolites dans le lagon (en plus de celles apportées par le vent) est le résultat de processus mécaniques dans lesquels interviennent des quantités variables d'énergie. Les petites oolites (taille moyenne 0,1 mm) sont probablement déposées par suspension alors que les plus grosses (0,35 mm) sont formées par roulage et culbutage sur le fond. Le datage au carbone 14 d'un échantillon de sédiment oolitique non dépouillé a révélé un âge de 1740 ± 60 ans B.P.

INTRODUCTION

General Setting

Oolites are a major constituent of Recent sediments along the Sinai beaches of the Gulf of Suez. The Ras Matarma Lagoon, in the northern part of the Gulf, was selected for a detailed study of these sediments. The triangular lagoon is located along the western shore of the Sinai Peninsula (32°44′E; 29°27′N) about 60 km south of the entrance to the Suez Canal (Figure 1). Our study is concerned mainly with the environments of deposition in the shallow body of water in which Recent oolitic and other sediments are accumulating.

At its northern part, at Ras Muhammad, southernmost tip of Sinai, the Red Sea bifurcates into two 250 to 300 km long embayments, namely the Gulf of Suez in the west and that of Elat (Aqaba) in the east. The Gulf of Suez is much shallower than the Gulf of Elat. The geological histories and evolution of these two gulfs also differ significantly: the Gulf of Suez already existed as a continuously subsiding basin of

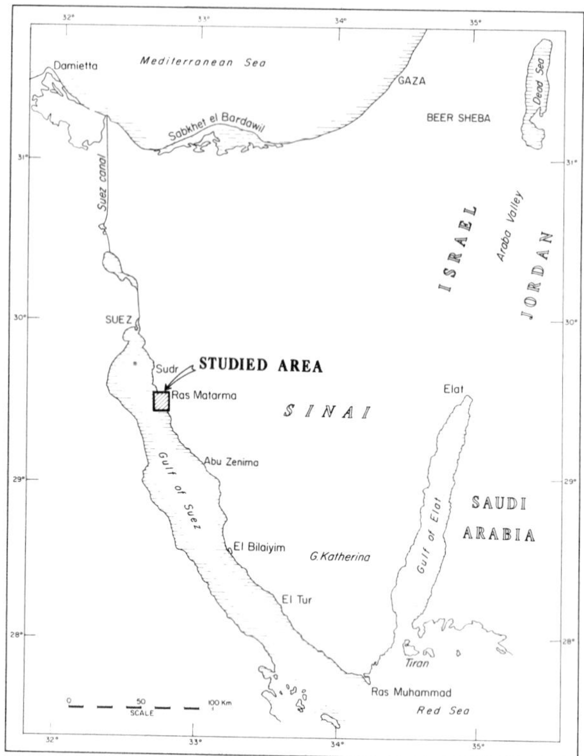

Figure 1. Location map of the Ras Matarma Lagoon in the Gulf of Suez.

sedimentation at least from the Cambrian (Said, 1962), whereas the Gulf of Elat came into being as a marine trough not earlier than the Late Neogene. Since its formation, several thousands of meters of sediments have accumulated in the Gulf of Suez; rates of sedimentation have been higher during the Neogene. This basin, consisting of thick wedges of mainly clastic sediments, has been referred to as a "taphrogeosyncline" (*cf.*, Said, 1962, comprehensive list of references on this region).

The main contribution of material (alluvial and wind-blown sand) to the Recent or Subrecent Gulf of Suez originates from the arid Eastern Egyptian desert and from northern Sinai. Moving sand dunes and ephemeral water courses forming large coarse-grained flood plains and deltas are the principal sources of clastics. This basin is a shallow (not deeper than 100 m) marine, continuously subsiding embayment, the bottom of which lies well above that of the main body of the Red Sea (about 1000 m deep, but some "holes" are deeper than 2300 m).

Coral reefs occur locally in this region, and the fringing reefs of the Sinai beaches were first described by Walther (1888). This author also called attention to, and described, the oolites in Wadi Dehese (the precise position of which is not known, but from the text it appears to be located in the vicinity of Ras Sudr). Certain aspects of the Ras Matarma region, including water and carbonate chemistry, mineralogy, and biogenic constituents will be detailed in another study.

Physiography

The Ras Matarma Lagoon (Figures 2 and 3) covers an area of about 5 km^2. It is bordered in the east and north by a flat lying sabkha which passes gradually eastwards into the fan-like, wide floodplain of the almost always dry Wadi Waradan. A higher terrain (50 to 100 m above sea level) occurs further to the east, at a distance of 3 to 5 km from the coastline. This area consists of Upper Cretaceous limestones that contain occasional chert nodules. Chert fragments form the dominant constituent of coarse clastics paving the flood plain. Limestone particles are usually much finer-grained.

An *L*-shaped bar a few tens of meters wide forms the west and southwest margins of the lagoon. This bar is connected to the mainland to the north. The main inlet, 1000 m wide, is situated at the southern end of the lagoon. Another opening, in the northern part of the western bar, is only a few meters wide and permits interchange of water with the sea only during high tide.

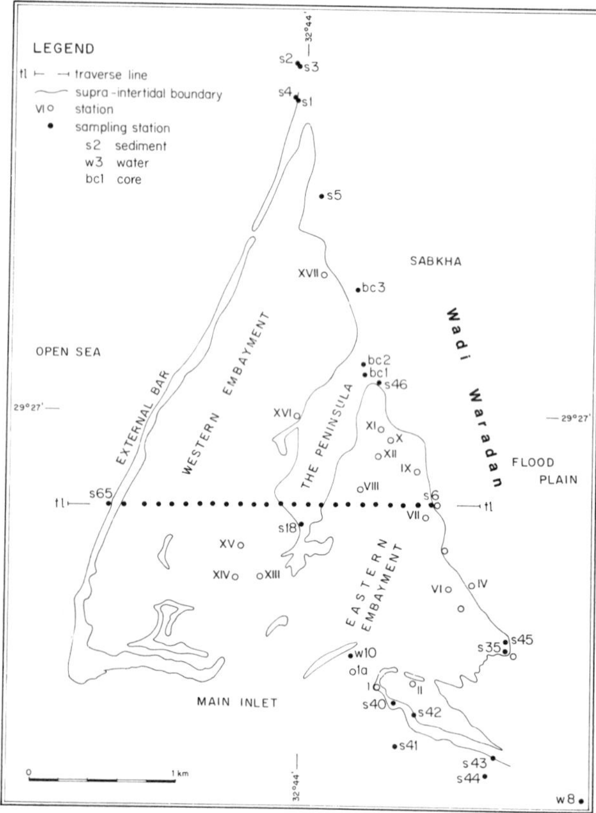

Figure 2. Sampling stations in the Ras Matarma Lagoon.

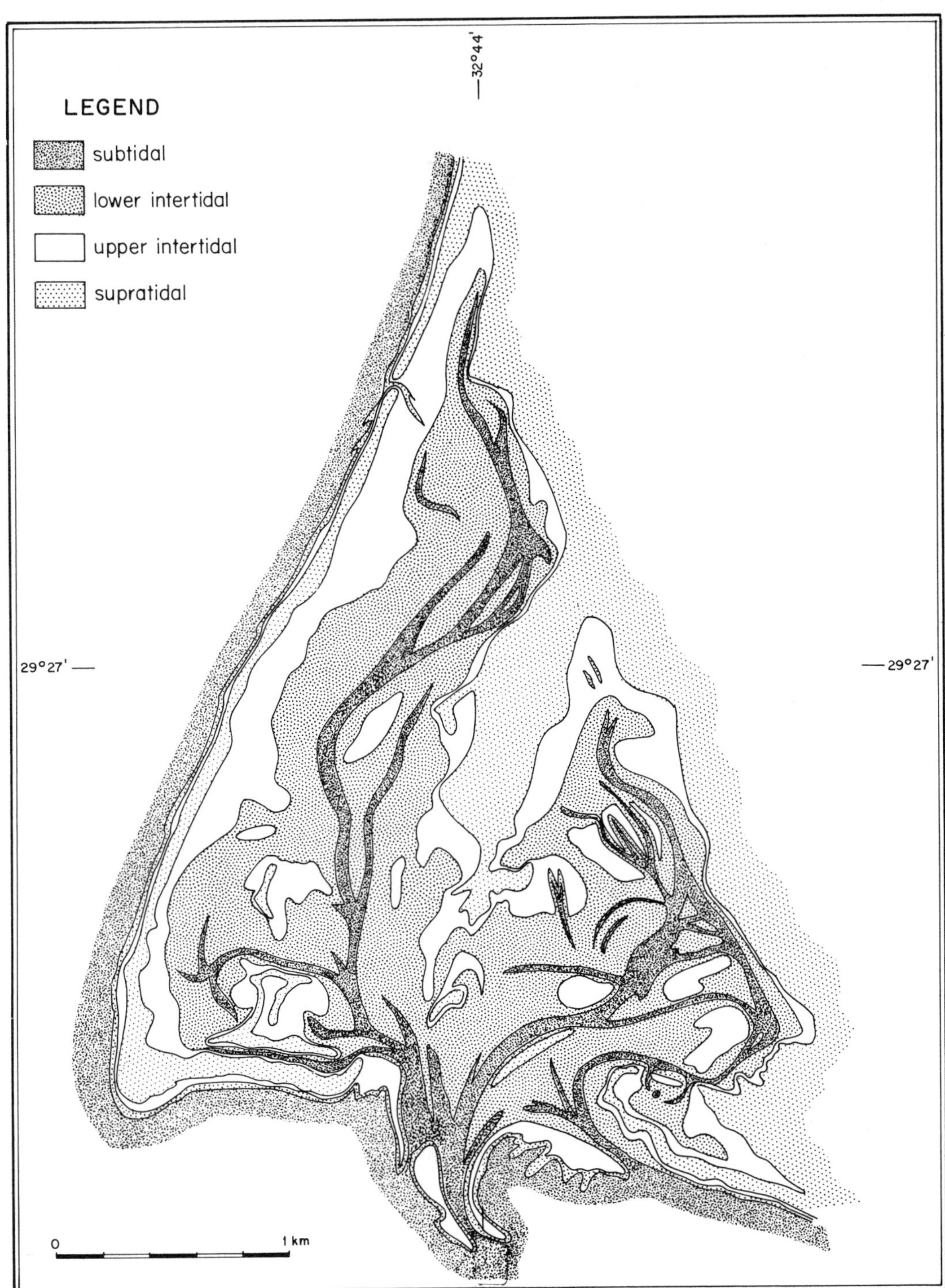

Figure 3. Physiographic features as related to tide levels in the Ras Matarma Lagoon.

A small peninsula connected to the mainland at the northern tip of the triangle divides the lagoon into two embayments (Figures 2 and 3). These embayments are hereafter called the *eastern embayment* and the *western embayment*. Some oolite-sand shoals, unstable in form and position, are found near the main inlet within the lagoon and out of it.

Most of the lagoonal area lies within the intertidal zone, and is less than 1 m deep below the high tide level. The tidal channels draining the two embayments are deeper, but nowhere within the lagoon do they extend more than 2 m below high tide level.

Climatological Data

Climatological data are available from the few meteorological stations located in the vicinity of the Ras Matarma Lagoon (Figure 1): at Abu Zenima (60 km south of the lagoon), at Abu Rudeis (10 km south of Abu Zenima) and in the town of Suez (60 km north of the lagoon). Additional data were obtained from Rosnan (1952) and Ganor (personal communication).

The climatological conditions may be summarized as follows:

Summer: High temperatures (average maximum 36°C); low relative humidity (40–50%); no precipitation; strong winds blowing persistently from the northern sector (mainly from the NW); clear sky, rarely somewhat hazy.

Winter: Mild temperatures (average minimum 20°C); relative humidity 60 to 70%; annual average precipitation 24 mm (mainly in December). Winds are strong (up to 40 knots) and steady, blowing mainly from the northern sector (N and NW); the frequency of eastern winds might be up to 7%. Occasional moderate southerly gales may occur from December to March. Annual average evaporation is estimated to be about 200 cm, or almost twice that of the Mediterranean.

Tides and Currents

Longshore currents moving toward the southeast along the eastern coast of the Gulf are continuous throughout the year.

Difference between low and high tide in the open Gulf adjacent to the lagoon is usually somewhat less than 2 m. Tidal current velocities of 1 to 0.5 knots were measured in the Gulf. Due to the shallowness of the lagoon and the narrowness of its inlet, there is a effect between the tidal ranges in the open Gulf and the lagoon. It was observed that the reduced tidal range in lagoon never exceeds 1 m. Maximum tidal current velocities measured at the inlet were usually about 2 knots (100cm/sec).

Sampling

The lagoon was studied during two seasons, *i.e.*, in September 1968 and in February 1971. Water and sediments were sampled at 18 stations in the lagoon (Figure 2). Hydrological parameters (current velocity, temperature, Eh, pH and conductivity) were measured during these periods.

In addition, numerous samples were collected along a 2500 m long east-west traverse (S6 to S65), starting from the high lying terrace in the east, through the central peninsula and western bar to the open sea in the west.

GRANULOMETRY

Technique

Friable sediments were treated with acetone for 10 hours in order to remove binding organic matter. After washing and centrifuging 4 to 5 times, sediments were separated into two fractions by wet sieving through a 62.5 micron sieve. The coarse fraction was dried and sieved using a 0.5 ϕ sieve interval, while the fine fraction was analyzed using the pipette method.

Analysis of Data

Most of the sediment samples in the present study are polymodal. The computed or graphic statistical parameters (Friedman, 1962; Folk, 1966; Inman, 1952) are most valuable in the study of unimodal populations, but when applied to polymodal distribution some of the information on the individual populations is not apparent. The granulometric data were evaluated in terms of the composing populations (after Visher, 1969) but our approach differs from Visher's in stressing the importance of modal points instead of break points. The procedure is as follows:

1. Finding the modal and break points (maxima and minima points in the distribution curve respectively). In order to find these points, use is made of the mathematical property that the first derivative of these points equals zero. The method is described in detail by Gry (1938).

2. Finding the weight percentage of the size fractions between two adjacent minima. This is easily obtained from the cumulative curve.

The grains falling between two adjacent minima points are considered as constituting a population. Each population is characterized by a modal size point, and weight percentage of the total sample.

In case of a bimodal sample, the weight percentage of one of the populations is a measure of the dominating population as well as the overall sorting of the sample.

Most of the sediments in the Ras Matarma Lagoon are composed of two populations. The coarse population has a mode which varies in the range of 0.23–0.40 mm, while the range of the fine mode is 0.07–0.12 mm. Following Groot (1955), Tanner (1958) and Visher (1965, 1969), these modes can be regarded as the bedload and suspension load respectively.

PHYSIOGRAPHIC AND DEPOSITIONAL ENVIRONMENTS

Tidal range constitutes the main factor for distinguishing the various environments in the Ras Matarma

Lagoon. Lithologic and topographic factors further serve to define subenvironments. All zones discussed are shown on Figure 4.

Subtidal Environment

The subtidal environment encompasses two distinctly different subenvironments. *Tidal channels* are a few decimeters to a few hundreds of meters wide (Figure 5a). Their depth below the surrounding intertidal flats usually ranges from a few centimeters to about 1 m. The channel sediments are composed largely of a coarse-grained population whose mode varies between 0.35 mm and 0.25 mm. The main constituents of these sediments are whole, or partly broken, skeletal fragments (*Littorina* sp. and *Marginopora* sp.), oolites, quartz grains and rock fragments.

There is one *subtidal pool* in the Ras Matarma Lagoon (Station II, Figure 2). The pool is roughly circular, with a diameter of about 50 m; its maximum depth below the surrounding flat is 100 cm. This pool is a typical low-energy environment, as indicated by low Eh values (−40 mV) of bottom waters, and the presence of muddy sediment rich in organic matter.

Intertidal Environment

The intertidal environment is subdivided into two zones on the basis of tidal range. Their boundaries are gradational.

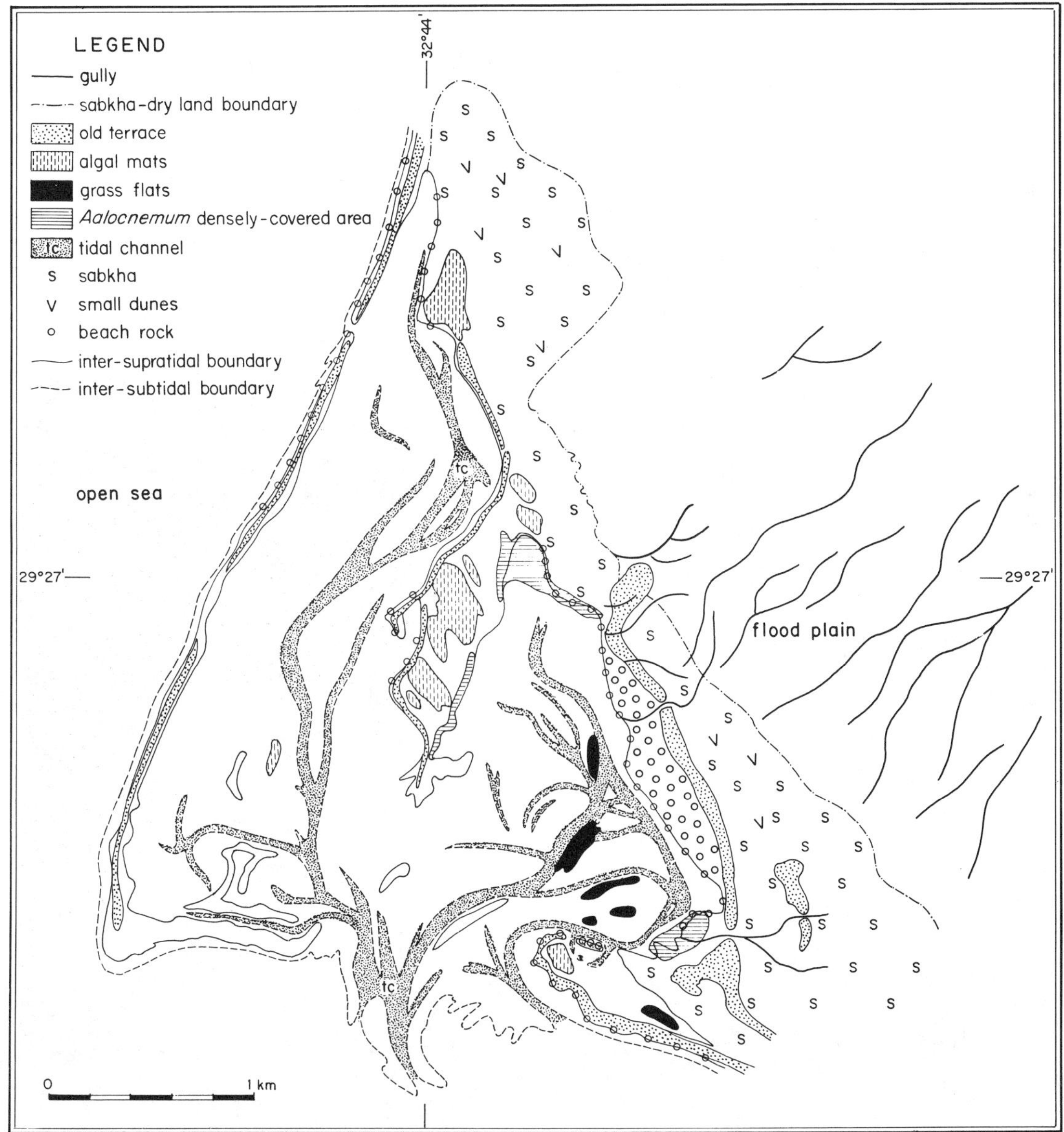

Figure 4. Environments of deposition in the Ras Matarma Lagoon.

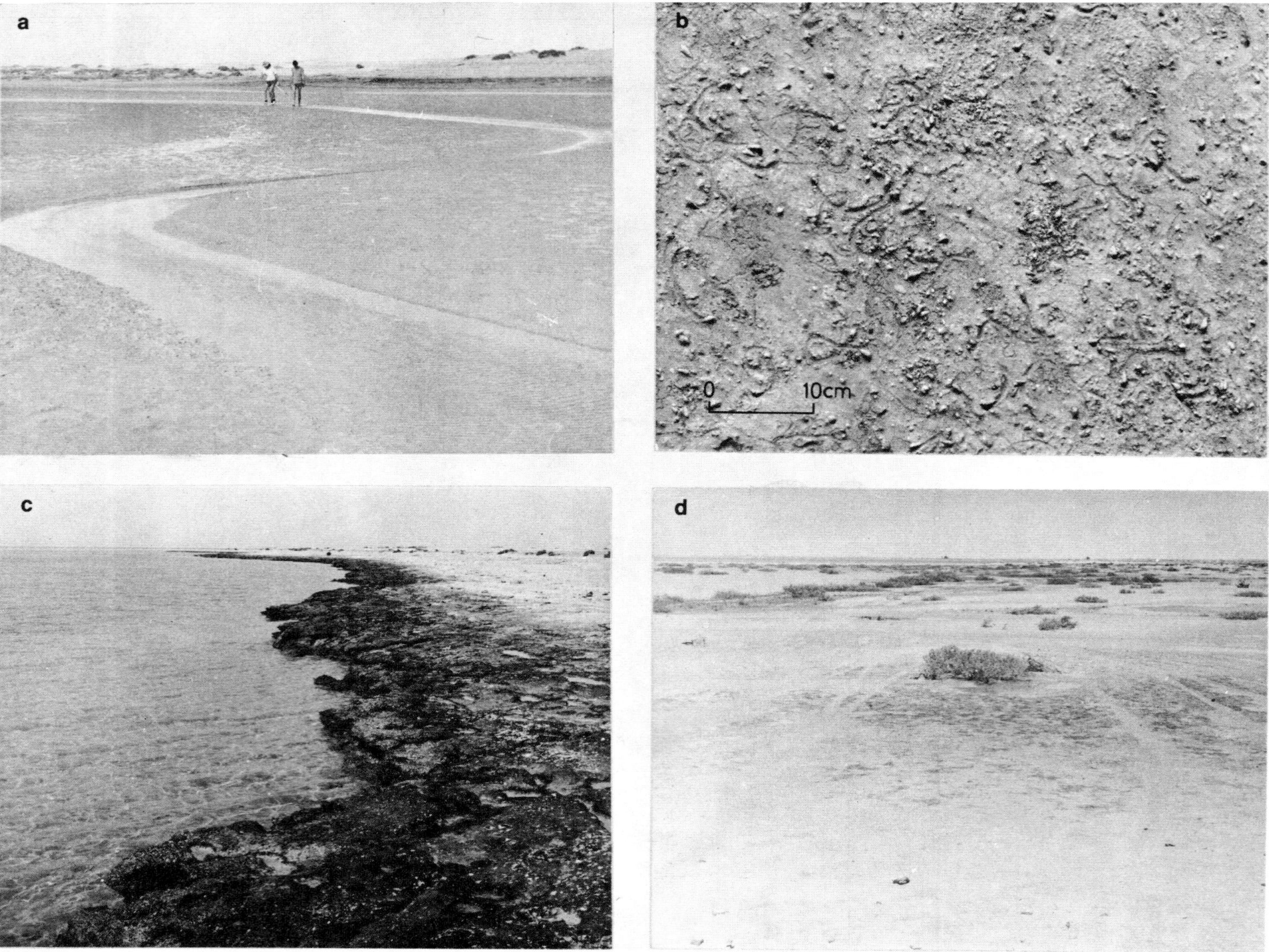

Figure 5. Photographs showing sedimentary features discussed in text. a, A tidal channel at low tide. b, *Littorina* sp. trails observed on oolitic sediment of the intertidal zone. c, Beachrock (shown at high tide) on eastern beach of the Western Embayment. d, Patches of algal mat (dark) lying on a stabilized oolite sand flat. The shrubs are *Aalocnemum* sp. Jeep tire tracks provide scale.

The *low intertidal environment* occupying most of the lagoon, is situated below the high neap tide level. It is inundated and exposed twice daily. The water salinity in summer does not exceed 45‰. The topography of this environment is flat with some rather shallow, irregular depressions which pass laterally into well-defined tidal channels.

The sediments are carbonates and are typically muddy, skeletal oolitic sands. A scatter diagram of the bedload modal size versus its abundance (Figure 6) delineates different fields for the two lagoon embayments, with only little overlap between them. In the eastern embayment the modal size is coarser than in the western one, while its abundance is lower (*i.e.*, the overall sorting is poorer in the eastern embayment compared with that of the western one).

Irregular patches of sea grasses (*Halodule uninervis* and *Halophila stipulacea*) are found locally, particularly in the shallow depressions. Burrows and mounds and reworking of sediments by crabs are common in the upper part of this environment, close to the supratidal zone.

The faunal assemblages include: (a) abundant *Littorina* which leave characteristic trails (Figure 5 b) and are ubiquitous; (b) *Marginopora*, large disc-like foraminifers concentrated locally, whose specific habitat is not clearly known; (c) *Miliolids*, small porcellaneous foraminifers, which are uncommon in the lower intertidal environment.

The *upper intertidal environment* is covered by water only during spring high tides. It occupies a relatively small part of the lagoon area due to the fact that its uppermost reaches have steeper slopes than elsewhere in the lagoon. Three zones are recognized in this environment:

1. *Aalocnemum* zone. Very dense *Aalocnemum* population covers the northern part of the eastern embayment (Figure 7). Elsewhere in the upper intertidal environment, the *Aalocnemum* shrubs are sparse or are missing altogether.

2. Salt flats. This sedimentary zone is restricted to the northeastern part of the eastern embayment. The topographic position is higher than that of the *Aalocnemum* zone, with which it shares a transitional boundary.

3. Beachrock. Beachrock is situated either at slightly higher elevation above the two former zones, or it covers the entire upper intertidal environment on relatively steep beaches. Beachrock has a typical asymmetric distribution in the lagoon, *i.e.*, in both embayments, it is found only on the eastern beaches where it forms a discontinuous belt (Figures 4 and 5c).

Algal mats are observed in the upper intertidal zone both on certain of the higher, more stable, oolite sand flats and on beachrock. The algae usually form mat-patches a few meters in diameter, or they encrust the beachrock (Figures 5d and 8a). Although most algae are alive, some older, or fossilized forms are encountered.

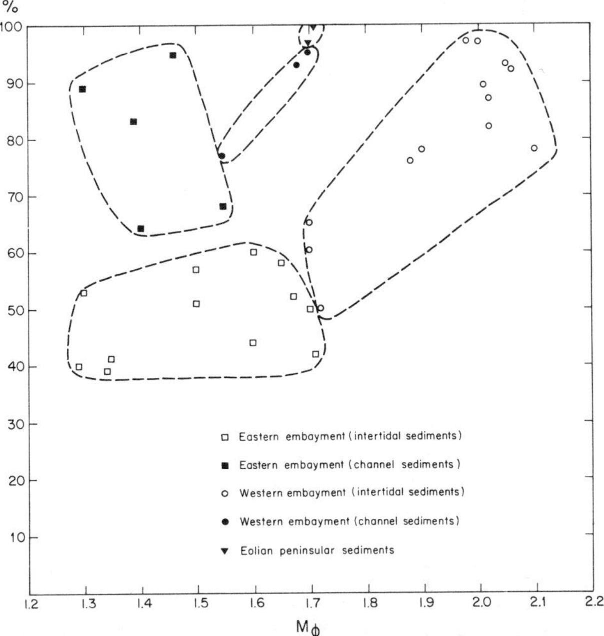

Figure 6. A scatter diagram of the bedload mode versus frequency. Each point represents a sample collected along the E–W traverse on Figure 2. Note segregation of the various environmental fields.

Supratidal Environment

Several supratidal subenvironments are recognized on the basis of rock or sediment types present and their permeabilities. A wet supratidal environment, or *sabkha*, occurs near the Pleistocene shale that crops out adjacent to the lagoon. In this area, the groundwater table is shallow (less than a meter) and saline. The capillary rise of water and its evaporation are responsible for constant wetting, and the formation of gypsum and salt in the shales. This environment, densely covered with salt-tolerant plants, is characterized by a small-mound topography (Figure 8b).

A dry supratidal environment forms over a highly permeable substrate consisting of the following lithologies:

1. Sandy, fossiliferous, pebbly rocks which are remnants of a Pleistocene terrace. Outcrops and remnants of this terrace locally crop out around the entire lagoon and appear either as disintegrated fragments or as low benches (Figure 8c).

2. Eolian sand waves and bed forms. These sand waves, typically elongate bodies, generally form behind plants and small bushes (Figure 8d). The sandy bed forms are low-lying, asymmetric features that stand out well on the otherwise flat topography. Their trend indicates prevalent wind directions.

Figure 7. Panoramic view of the northern (*Aalocnemum* densely-covered) sector of the Eastern Embayment. The intertidal zone (at low tide) is shown on the left of the photograph. Note also the sabkha, raised Pleistocene terrace and the peninsula.

Figure 8. Photographs illustrating sedimentary features discussed in the text. a, Algal crusts covering old beachrock. b, Mudcracked salt flats and mounds in the northern sabkha. Relief of largest mound in background is 2 m. c, Low "cliff" of the Pleistocene terrace, relief about 1 m. d, Eolian bed forms modified by clumps of vegetation.

PHYSIOGRAPHY AND GRANULOMETRY

An east-west trending line across the lagoon traverses most of the depositional environments in the study area (Figures 2 and 4). Detailed study of the sediments collected along this traverse reveals some trends which are indicative of the source of the sediments and processes operating in the lagoon.

Figure 3 illustrates the clear tendency of the tidal channels to develop on the eastern side of each of the lagoon embayments. A related phenomenon is the dominance of rocky beaches (beachrock and Pleistocene rocky terrace, Figure 4) on the eastern side of each embayment, and depositional beaches on the western side. Thus, the character of the western beaches is mainly determined by the accumulation of sediments transported from the west, while the eastern beaches are dominated by erosional processes.

These physiographic asymmetries are accompanied by sedimentary-granulometric asymmetries (Figure 10) and differences between the two embayments, namely:

1. On the eastern side of each of the embayments, the modal size of the bedload is coarser than on the western side.
2. The bedload modal size of most of the sediments in the western embayment is lower than in the eastern one.
3. The abundance of the bedload population on the tidal flats is lower on the average in the eastern embayment than in the western one. The western sediments are thus better sorted than the eastern ones.

Correlation between the prevalent, northwesterly wind direction and the transportation of sediments eastwards in both embayments gives a clue to the factors controlling the physiographic and sedimentary characteristics. The supply and release of eolian sediments from the west prevents the development of major tidal channels on the western side of the embayments, and drives them eastward. Furthermore, locally generated waves are weaker on the lee, or eastern side, of the external bar and of the peninsula than on their windward, or western side. As a result, the eastern sector of each embayment is one of higher wave energy.

The higher sediment supply and lower energy in the western sector of the two embayments, and lower sediment supply and higher energy in the eastern part are the determining factors controlling sedimentation.

Sediments transported by the wind to the western embayment are derived from the beach and are practically unimodal. In addition, the eroded and disintegrated beachrock and Pleistocene rock terrace constitute an important source of coarse sediment in the eastern beaches of both embayments. These sediments, by their very nature, are coarser and less sorted than the open beach oolites.

The rate of wind-transported sediment to the eastern embayment is lower, while the relative importance of coarser sediments derived from the east is higher. Thus, average modal size of the bedload population in the eastern embayment is coarser. It follows that the overall lower rate of deposition in the eastern embayment results in the deposition, locally, of muddy sediment that is less diluted by wind-driven sediment. This, in turn, results in development of more poorly sorted sediment.

OOLITES

The Ras Matarma lagoon sediments are distinguished by an association of oolites and mud. Ooids are the single most abundant textural constituent of the lagoonal sediments that also comprise eolian and open beach deposits.

This type of oolite-mud association is generally uncommon. Most of the oolitic sediments and rocks described in the literature are unimodal, and being free of a mud fraction, are believed to originate in a high energy environment (*e.g.*, Bathurst, 1968). The origin of the oolite-mud association of the type we observed may be interpreted in two ways: (1), the ooids are allochthonous, and their source is either a nearby high energy environment, or older oolitic sediments (Purdy, 1963); or (2) the ooids are autochthonous and form under low energy conditions. A brief description of the oolites and their mode of occurrence helps clarify this question.

The sand beaches of the open Gulf in the vicinity of the Ras Matarma Lagoon are composed almost exclusively of well sorted ooids whose modal size is about 0.35 mm (Figure 9a). These ooids are picked up by the winds and transported southeastward toward the lagoon, indicating that at least part of the oolites in the lagoon are allochthonous.

Microscopic examination of the lagoonal oolites reveals the existence of two size populations: a coarser one with a mode of about 0.35 mm, and a finer population with a mode of about 0.1 mm (Figure 9c). The existence of the finer population in the lagoon, and its absence in open beach sediments, points to its autochthonous, or intra-lagoonal origin. This conclusion is supported by the observation that some of the ooid nuclei in the lagoon are not known in the open beach oolites. These nuclei include miliolid fragments (Figure 9b and 9d), ostracods, pellets and aggregates of former oolites which form composite ooids (Figure 9c).

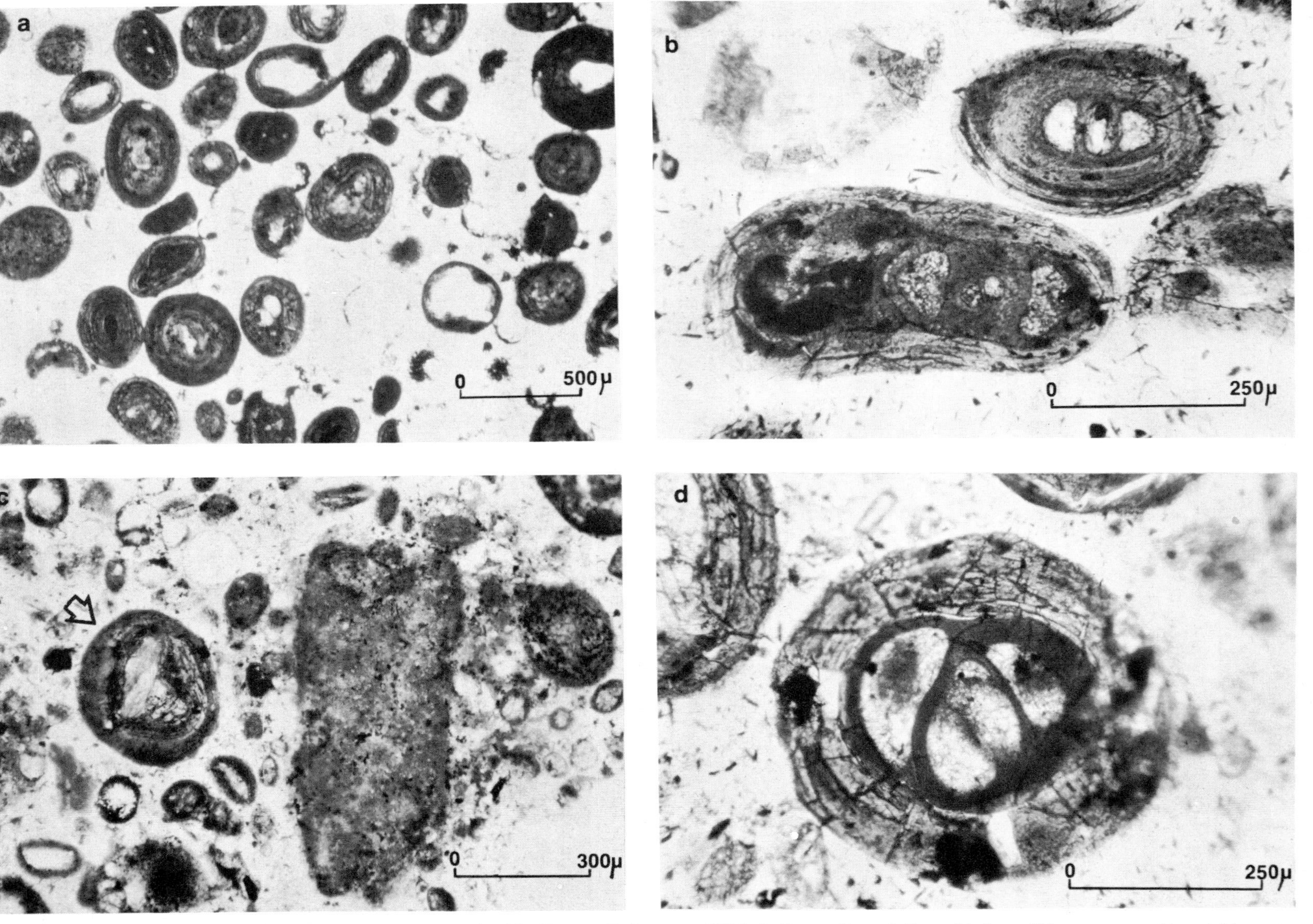

Figure 9. Photomicrographs of Ras Matarma oolites. a, Well sorted ooids (Western Embayment, BS30). b, Recent Foraminifera with drusy filling serve as ooid nuclei (open sea, east of the main inlet, BS44). c. Bimodal oolite population: some ooids (see arrow) show composite nuclei (Eastern Embayment. BS12). d, Recent Foraminifera with a drusy filling serves as nucleus of ooid (Western Embayment, BS27).

An absolute age determination of a lagoonal oolitic sediment by the C-14 method gave a date of 1740 ± 60 years B.P. The material for this analysis was taken from the size fraction consisting almost entirely of ooids. The ooids were treated as a bulk (*i.e.*, together with their nuclei), a procedure which should, evidently, provide a date somewhat older than the actual age of the oolite formation. The Recent age of the lagoonal oolites is in accord with the conclusion regarding their autochthonous formation in the lagoon.

Analysis of the grain size cumulative curves suggests that those ooids having a modal size of 0.1 mm are part of the suspension load population, while those with a coarser mode belong to the bedload population. It is probable that the development of lagoonal oolites took place in two stages:

(a) The suspension stage. The oolites start their growth over nuclei in suspension, whose minimum diameter is about 0.05 mm, and continue to grow until they reach a critical diameter of about 0.15 mm. The oolites forming during this stage may be considered as *low energy oolites.*

(b) The saltation and rolling stage. Oolitic development continues on the bottom with growth over pre-existing ooids (from the suspension stage, Figure 9c) or fragments whose diameter is larger than 0.15 mm. This type is the most common oolite described in the literature, and may be referred to as *high energy oolites*.

The observation that oolitic coatings do not develop around nuclei less than 0.05 mm in diameter is supported by X-ray and chemical data (Weiler, Katz and Sass, in preparation): samples from the mixed mud-oolite environment show a distinct minimum of the aragonite to calcite and the strontium to calcium ratios at, or about, the 16 to 62 μ size fraction of these samples. At this same grain size range, and slightly above it, a maximum of acid-insoluble detritus content (mainly quartz with small amounts of feldspar and heavy minerals) has been found.

If our "suspension stage" oolites are, in fact, *in statu nascendi*, these findings support the idea that oolites grow via mechanical attachment of pre-existing suspended aragonite needles to the oolite surface. This concept seems to be preferable to the one suggesting heterogeneous nucleation and growth of aragonite on pre-existing, extremely small (in the order of 5 unit cells) aragonite crystal debris at the ooid surface. [For a detailed discussion see the review by Bathurst (1968) and references therein]. In the first case (needle attachment) a limit, in terms of a minimum nucleus diameter, is critical as the nucleus surface curvature, relative to the needle length, increases and consequently prevents further attachment. A similar argument cannot hold for the latter mechanism (growth in place) since grain surface curvature will be negligible relative to crystal debris sizes even for extremely small nuclei.

EVOLUTION OF THE RAS MATARMA LAGOON

The Ras Matarma Lagoon is basically an erosional feature as indicated by the presence of the raised Pleistocene terrace along the external bar, as well as on the peninsula and along the eastern beaches. The sequence in evolution of the lagoon is believed to be the following:

1. Prior to the formation of the lagoon, there existed a flat, horizontal platform formed by a raised Pleistocene terrace.
2. This platform was breached and eroded, and assumed the rough form of the present day lagoon.
3. Details of the erosional processes are not known, but it is clear that they are no longer in operation and that the lagoon is now gradually filling with sediment. The fill consists of wind transported ooids from the open sea beaches west and northwest of the lagoon, locally formed ooids, skeletal fragments, aragonitic mud, and a minor contribution of alluvial material from the east.

If present processes continue, we conclude that the lagoon will be completely filled in the near geological future.

OOLITES AND PALEOGEOGRAPHIC SPECULATIONS

It is worthwhile to consider the use of oolite deposits in the Gulf of Suez and in the Mediterranean Sea in order to establish the geological relation of these two adjacent water bodies. Sites of modern oolite deposition are not known in the Mediterranean. Oolite accumulations in the recent geological past, however, are known off the coasts of Tunisia and Libya (Lucas, 1955; Fabricius *et al.*, 1970; and Fabricius and Schmidt-Thomé, this volume), about 400 km north and 1500 to 2000 km west of the Ras Matarma Lagoon. It is quite probable that oolites formed elsewhere in the Mediterranean in Quaternary time. As it now stands, the closest area of modern oolite deposition is in the Gulf of Suez, or some 150 km to the south of the Mediterranean coast.

Geological investigations have shown that the Gulf of Suez very likely was connected to the Mediterranean in Quaternary time (*cf.* Hassan and el-Dashlouty, 1970). We propose that a thorough search for, and mapping of, oolite deposits in the

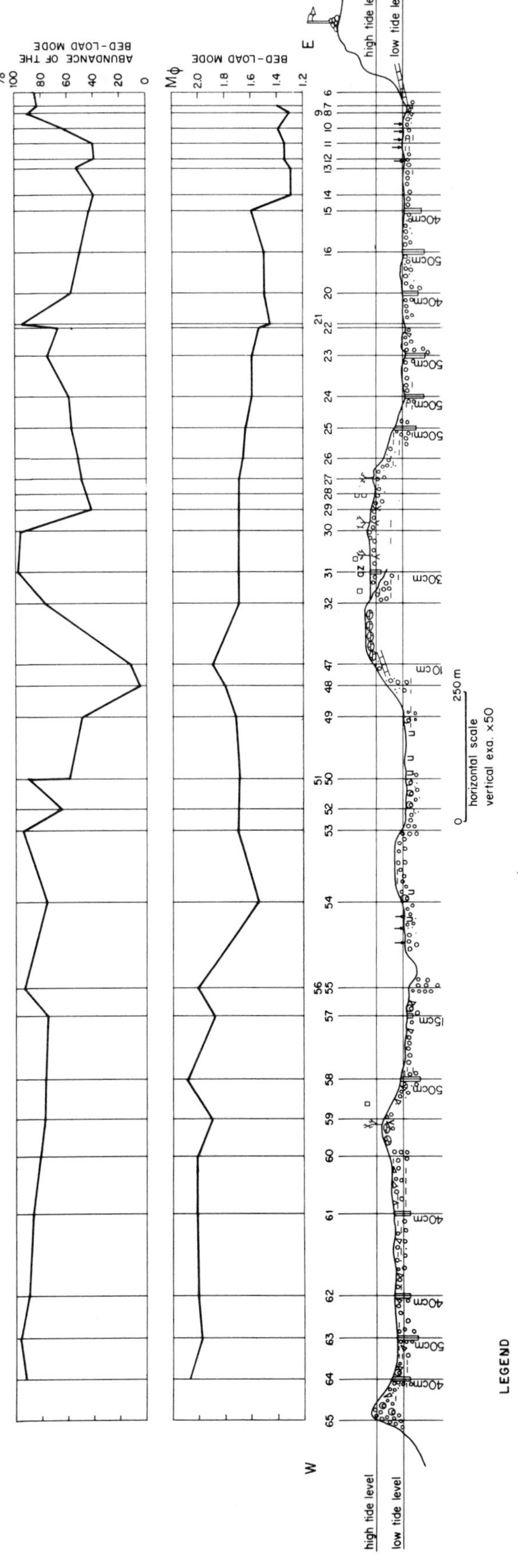

Figure 10. The E–W traverse and sample line across the Ras Matarma Lagoon (line S6–S65, Figure 2) in lower diagram. The upper diagrams show variations in the abundance of the bed-load mode (in percent of the whole sample population) and size (mode in ϕ units) along the traverse.

Mediterranean, particularly on the North Africa margin between Tunisia and Sinai would reveal information on the time of separation of these two water bodies. For instance, dating of such carbonates would provide some handle on the timing of major changes of climate (reflected in water temperature and salinity) and paleohydrology (direct connection between the Mediterranean Sea and the Red Sea). The terrigenous material carried to the sea by the Nile River has undoubtedly masked many of these earlier oolite deposits. In spite of this, it may be possible to find some oolite deposits in the land area north of Suez (the Bitter Lakes?) which are younger than those found off Tunisia. Such deposits would permit a more exact dating of the separation of the Gulf of Suez and the Mediterranean Sea in Holocene time than has been heretofore possible.

SUMMARY

The two lagoon embayments have a marked asymmetric disposition of environments: the western part of each embayment is controlled by active depositional processes, while the eastern beaches are characterized by erosion. This results from the combined effects of two factors: addition of an eolian supply of oolitic sands from the west and northwest, and modification by locally generated waves that are of higher energy at the open (eastern) part of each embayment than in the sheltered (western) part.

The oolites in the lagoon originate from an allochthonous source (*i.e.*, the nearby open-sea beaches) from which they are transported to the lagoon by the winds. Oolites in the lagoon also may have an autochthonous origin. This latter type includes low energy oolites, which form in suspension and have a modal size diameter of about 0.1 mm and high energy oolites, which form as a result of rolling and saltation on the bottom. The modal size of these oolites is about 0.35 mm.

ACKNOWLEDGMENT

We thank Dr. A. Nissenbaum, Weizmann Institute of Science, Rehovot, for carbon-14 age determination and Dr. Y. Lipkin, Department of Botany, University of Tel Aviv, for definition of the flora from the lagoon, both cited in this study. Drs. G.M. Friedman, Rensselaer Polytechnic Institute, Troy, F.H. Fabricius, Technical University, Munich, and D.J. Stanley, Smithsonian Institution, Washington, D.C., are gratefully acknowledged for critically reading the manuscript.

REFERENCES

Bathurst, R. G. C. 1968. Precipitation of ooides and other aragonite fabrics in warm seas. In: *Recent Developments in Carbonate Sedimentology in Central Europe*, eds. Müller, G. and G. M. Friedman, Springer Verlag, New York, 1–10.

Fabricius, F. H., D. Berdau, and K. O. Munnich 1970. Early Holocene ooids in modern littoral sands reworked from a coastal terrace, southern Tunisia. *Science*, 169:757–760.

Fabricius, F. H. and P. Schmidt-Thomé 1972. Contribution to Recent sedimentation on the shelves of the Southern Adriatic, the Ionian, and the Syrtis seas. In: *The Mediterranean Sea: A Natural Sedimentation Laboratory*, ed. Stanley D. J. Dowden, Hutchinson and Ross, Inc., Stroudsburg, Pennsylvania, 333–343.

Folk, R. L. 1966. A review of grain-size parameters. *Sedimentology*, 6:73–93.

Friedman, G.M. 1962. On sorting, sorting coefficients, and the lognormality of the grain size distribution of sandstones. *Journal of Geology*, 70:737–775.

Groot, J. J. 1955. Sedimentary petrology of the Cretaceous sediments of northern Delaware in relation to paleogeographic problems. *Delaware Geological Survey Bulletin*, 5:157p.

Gry, H. 1938. Eine Methode zur Charakterisierung der Kornverteilung klastischer Sedimente. *Geologische Rundschau*, 29:175–195.

Hassan, F. and S. el-Dashlouty 1970. Miocene evaporites of Gulf of Suez region and their significance. *American Association of Petroleum Geologists Bulletin*, 54:1686–1696.

Inman, D. L. 1952. Measures for describing the size distribution of sediments. *Journal of Sedimentary Petrology*, 22:125–145.

Lucas, G. 1955. Oolithes marines actuelles et calcaires oolithiques récents sur le rivage africain de la Méditerranée Orientale (Egypte et Sud Tunisien). *Bulletin Station Océanographique Salammbô* (Tunisie), 52:19–38.

Purdy, E. G. 1963. Recent calcium carbonate facies of the Great Bahama Bank, parts I and II. *Journal of Geology*, 71:334–355 and 472–497.

Rosnan, G. 1952. The Sinai Peninsula—Climate. *Meteorological Survey of Israel*, Series C:2. (In Hebrew)

Said, R. 1962. *The Geology of Egypt*. Elsevier, Amsterdam, 377p.

Tanner, W. F. 1958. The zig-zag nature of type I and type II curves. *Journal of Sedimentary Petrology*, 28:372–375.

Visher, G. S. 1965. Fluvial processes as interpreted from ancient and recent fluvial deposits. In: *Primary Sedimentary Structures and their Hydrodynamic Interpretation*, ed. Middleton, G. V., Society of Economic Paleontologists and Mineralogists Special Publication, 12:116–132.

Visher, G. S. 1969. Grain-size distribution and depositional proses. *Journal of Sedimentary Petrology*, 39:1074–1106.

Walther, J. 1888. Die Korallenriffe der Sinaihalbinsel. *Leipzig Abhandlungen der mathematisch-physischen Classe der Königlich Sächsischen Gesellschaft der Wissenschaften*, 14:437–506.

PART 6. Coastal and Shallow Water Sedimentation: Detrital Sediments

Les Minéraux Lourds de Sables de Plages et de Canyons Sous-marins de la Méditerranée Française

Solange Duplaix

Université de Paris VI, Paris

RESUME

Les minéraux lourds ont été utilisés pour caractériser les sédiments et pour rechercher leur origine et leur mode de répartition. Sur les côtes du Roussillon, le matériel sableux est pyrénéen, sur celles du Languedoc il est alpin, apporté par le Rhône, enfin, sur le rivage de la Provence une grande partie des minéraux est originaire du Massif des Maures.

Sur les plages du Golfe de La Napoule, les minéraux lourds sont transportés par le courant d'Ouest en Est, alors que dans le Golfe Juan ceux-ci s'étalent d'Est en Ouest et que les fractions légères suivent un tracé inverse jusqu'à moitié du golfe. Dans la Mer de Ligurie, les sédiments superficiels des canyons sous-marins et de la cuvette abyssale sont tributaires du Var.

ABSTRACT

Heavy minerals are useful in distinguishing the different sediment types in the Mediterranean as well as determining provenance and dispersal patterns. On the coast of France, sandy material off the Roussillon region is of Pyrenean origin, while that off the Languedoc is of Alpine origin (influence of the Rhône River). Much of the material off Provence is derived from the Maures Massif.

On beaches of the Gulf of Napoule, heavy minerals are transported by currents moving from west to east, while in the Gulf of Juan, materials move from east to west. In the Ligurian Sea, surficial sediment in submarine canyons and in the basin plain beyond were derived from the Var.

INTRODUCTION

Dans cette note seront synthétisés les divers travaux effectués depuis une vingtaine d'années, à l'aide des minéraux lourds, sur des plages et dans des canyons sous-marins de la partie française de la Méditerranée (Figure 1).

Le littoral français sera divisé en cinq régions géographiques en partant de l'Ouest vers l'Est.

DU CAP BEAR AU CAP MEJEAN

Cap Béar—Grau-du-Roi

Du cap Béar au Grau-du-Roi différentes associations minéralogiques se succèdent sur les plages. Elles sont, au Sud, formées par la hornblende et l'andalousite, auxquelles s'ajoute le grenat à Canet; puis, vers Leucate, l'andalousite devient dominante accompagnée de hornblende et d'augite (Duplaix et Lalou, 1949 et 1951; van Andel, 1955). Au Nord de Leucate et jusqu'à l'embouchure de l'Hérault, à l'augite et la hornblende se joignent la staurotide et, en faible proportion, l'hypersthène, pour constituer le spectre minéralogique de cette région (van Andel, 1955). Entre l'Hérault et Sète, la hornblende et l'augite restent abondantes, de même que l'épidote et le grenat en certains points; les proportions de staurotide et d'andalousite sont faibles, l'hypersthène est toujours présent (Duplaix et Lalou, 1949 et 1951; van Andel, 1955).

Ce schéma est celui reconnu également par Vatan (1949), mais le mode de comptage employé par cet auteur fait apparaître quelques différences, les pro-

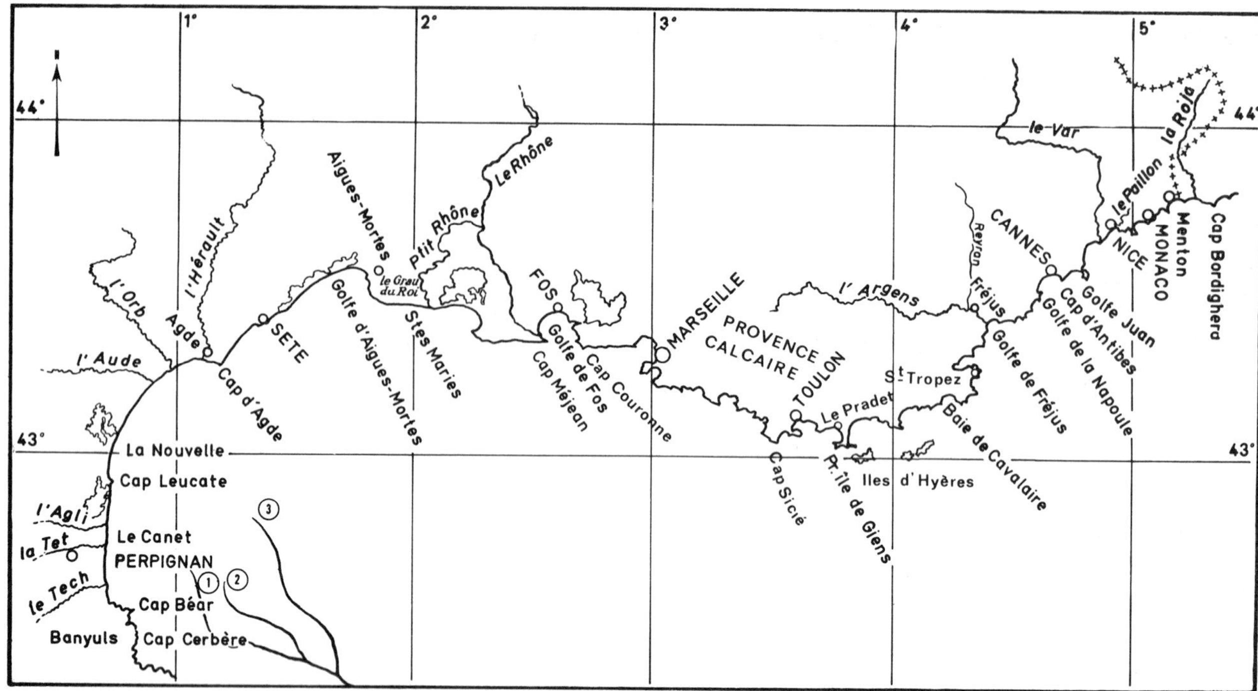

Figure 1. Carte des côtes de la Méditerranée française. Position des rechs du Roussillon: (1) rech Lacaze-Duthiers, (2) rech Pruvost, et (3) rech Bourcart.

portions d'amphibole sont, pour lui, plus fortes immédiatement après Leucate.

Plus récemment, Chassefière (1969) a confirmé que les sables des plages entre Agde et Sète sont constitués par de la hornblende et de l'augite accompagnées d'épidote et de grenat de fréquence variable, alors que les proportions de glaucophane et d'hypersthène sont constantes, ces deux derniers minéraux constituant les espèces caractéristiques.

Dans le Golfe d'Aigues-Mortes l'association minéralogique est sensiblement la même que celle trouvée près de Sète (van Andel, 1955).

Grau-du-Roi—Cap Méjean

Du Grau-du-Roi au Golfe de Fos, il n'y a pas de variations qualitatives; à mesure qu'on se déplace d'Ouest en Est ce sont toujours les mêmes espèces qui constituent l'association minéralogique, mais le minéral dominant change suivant la situation géographique du prélèvement.

Au Sud du Grau-du-Roi, la hornblende est le minéral principal associé à l'épidote, l'augite et le grenat; puis les proportions de hornblende diminuent et l'augite, le grenat ou l'épidote prennent une place prépondérante dans le spectre minéralogique: le grenat vers les Saintes-Maries-de-la-Mer, l'augite dès l'Ouest du Grand Rhône. Dans le Golfe de Fos, la hornblende redevient dominante. L'andalousite est présente en plusieurs points du littoral (van Andel, 1955).

La grande uniformité minéralogique du littoral camarguais a été reconnue également par Duboul-Razavet (1956). Cet auteur a, de même, signalé la présence d'andalousite en faible proportion. Le zircon peut être aussi relativement abondant près du Grand Rhône (Duplaix et Lalou, 1951).

A l'Est du Golfe de Fos, l'association minéralogique est nettement différente; la staurotide est l'espèce dominante, elle se groupe avec le grenat et l'épidote (van Andel, 1955).

Origine et Mode de Répartition des Sédiments

Dans la partie de la côte qui s'étend du cap Béar au Grau-du-Roi, plusieurs coupures minéralogiques sont mises en évidence par les minéraux lourds.

Tous les auteurs situent une première coupure au Nord de Leucate. Dans la partie Sud, l'andalousite est le minéral indicateur, vers le Nord-Est, c'est l'augite. Au Sud les sables sont plutôt grossiers, peu calcaires, pauvres en minéraux lourds; au Nord de Leucate ils sont plus fins, calcaires, et les minéraux lourds y sont plus abondants (Duplaix et Lalou, 1949 et 1951).

Van Andel (1955) qui a analysé un nombre important d'échantillons, place une deuxième coupure à l'Ouest de l'Hérault; dans cette région c'est la staurotide associée à l'augite qui constitue le spectre caractéristique. Plus à l'Est, les proportions d'amphibole et de grenat croissent alors que diminuent celles de l'andalousite.

Trois régions minéralogiques différentes sont ainsi définies (Figure 2). L'orgine de leurs sédiments est à rechercher à la fois dans les alluvions des fleuves côtiers, le Rhône et les apports directs des côtes.

La Têt apporte surtout de l'andalousite, mais aussi de l'amphibole, de l'épidote, du grenat, de la staurotide et un peu d'hypersthène et d'augite.

L'Agly est riche en grenat, on y trouve aussi de la hornblende, de l'épidote, du pyroxène monoclinique et de l'hypersthène.

Le Rhône charrie des minéraux issus des Alpes et du Massif Central. Certaines espèces sont communes aux deux régions, par contre, d'autres telles que glaucophane, chloritoïde, épidote verte et hornblende vert-bleuâtre sont typiques des Alpes, l'augite du Massif Central (van Andel, 1955). Toutes ces espèces sont représentatives de l'apport rhodanien dans son ensemble.

La présence sur les plages de minéraux typiques de ces différentes régions permet ainsi de retrouver l'origine et le mode de répartition des sédiments sableux.

Pour van Andel (1955), (Figure 2), l'association pyrénéenne (andalousite-hornblende commune) se retrouve sur les plages jusqu'au Nord de Leucate, les minéraux ayant été fournis par les fleuves côtiers et l'érosion de formations anciennes; cette association est remplacée, vers l'Est, par celle de la staurotide-augite. L'origine de la staurotide est assez difficile à préciser; il semble bien qu'elle doive être recherchée dans les sédiments tertiaires qu'ils soient repris par la mer ou apportés par des rivières côtières comme l'Orb (van Andel, 1955). Quant à l'augite, les projections du volcan d'Agde reprises par la suite en sont vraisemblablement la source.

A l'Ouest de l'Hérault commence l'association alpine, caractérisée par la hornblende vert-bleuâtre, l'augite, l'épidote et le grenat.

Vatan (1949) y place également la glaucophane et l'épidote verte, comme particulièrement caractéristiques de l'apport alpin.

La répartition de ces trois associations minéralogiques se fait pour Vatan (1949) d'Est en Ouest jusqu'à La Nouvelle, alors que sur la côte du Roussillon les déplacements se produisent du Sud vers le Nord.

La position des trois associations, selon van Andel, montre une répartition à peu près semblable, seule la staurotide vient, peut-être, d'apports plus locaux. Plus à l'Est, jusqu'au Golfe de Fos, les sédiments sont tributaires des apports rhodaniens.

Plateau Continental et Canyons Sous-marins

Quelques sables pris sur le plateau continental entre le Cap Leucate et l'embouchure du Tech montrent des associations minéralogiques différentes sui-

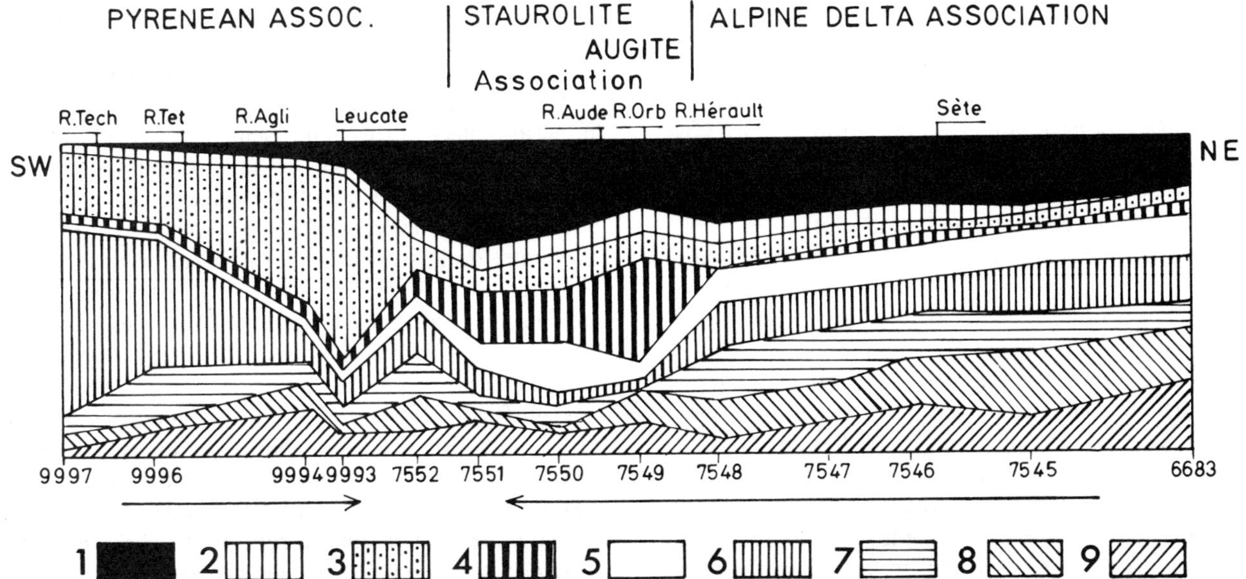

Figure 2. Variation des minéraux lourds le long des plages de la Méditerranée entre les Pyrénées et le delta du Rhône (d'après van Andel, 1955). (1) augite, (2) hypersthène, (3) andalousite, (4) staurotide, (5) autres minéraux, (6) hornblende commune, (7) hornblende vert-bleuâtre ou toutes hornblendes, (8) épidote, et (9) grenat. Les nombres à la base du diagramme représentent les numéros d'échantillons de van Andel.

vant la position géographique, la profondeur et l'âge du prélèvement (Monaco, 1970).

Les sables pris à 10 et 20 m de profondeur sont actuels. Ceux carottés de 85 à 100 m de profondeur ou dans les rechs sont quaternaires récents (post-glaciaire).

Dans la partie Nord, entre le Cap Leucate et le Sud de l'embouchure de l'Agly, à 10 m de profondeur, le minéral caractéristique est le grenat; l'amphibole y est, aussi, abondante, suivie par l'andalousite et l'épidote. Les autres minéraux sont faiblement représentés. Il n'y a pas d'évolution longitudinale nette, par contre une évolution perpendiculaire à la côte se manifeste par un enrichissement vers le large, à 20 m de profondeur, en amphibole (minéral à fort pouvoir de flottabilité) et d'épidote (petite taille). Les grenats sont encore relativement abondants (18%). Les autres minéraux se rencontrent toujours en faibles proportions.

Dans la partie au Sud de l'embouchure de la Têt, l'amphibole dominante et l'épidote forment l'association minéralogique; plus au large, vers 85 à 100 m de profondeur, ce sont l'amphibole et le grenat.

Plus au Sud, sur le banc des Canalots, au large de Banyuls, l'amphibole est, là aussi, le minéral le plus abondant; elle s'associe au grenat et à l'épidote pour constituer le spectre minéralogique. L'andalousite est peu abondante (7%) (Duplaix et Lalou, 1951). Ces associations sont très proches de celles trouvées sur les plages de cette partie de la côte, elles montrent cependant, par rapport à celles-ci, une faiblesse en andalousite.

Des carottes ont été prélevées plus au large et dans les rechs du Roussillon (Figure 1). Dans les têtes des rechs Lacaze–Duthiers et Pruvost (Monaco, 1967), la hornblende est toujours abondante, le grenat et l'épidote également. La fréquence de l'andalousite semble liée à la granulométrie des sédiments; lorsque ceux-ci sont grossiers, elle fait partie de l'association minéralogique, alors que dans les sédiments fins elle se raréfie et peut même être absente.

Tous ces résultats sont comparables avec ceux trouvés sur les plages du littoral de cette région; les sources des sédiments doivent être les mêmes.

Plus au Nord, l'analyse des sédiments de carottes prélevées dans la tête du rech Bourcart (Duplaix et Olivet, 1969) montre que les proportions d'amphibole y sont encore élevées, de même que celles de l'épidote; le grenat est moins abondant, il occupe cependant la troisième place dans l'association minéralogique. L'andalousite est peu fréquente; la présence de glaucophane et de chloritoïde, minéraux alpins (van Andel, 1955, p. 528), est à retenir. Le rech Bourcart est dans le prolongement de l'Aude qui apporte essentiellement de la hornblende et du grenat (Vatan, 1951). Les espèces trouvées dans ce canyon sous-marin sont donc originaires de sources différentes: pyrénéenne et alpine.

L'analyse des formations sédimentaires de ces rechs du Roussillon montre une évolution des sédiments avec participitation du matériel alpin dans le plus occidental.

DU CAP COURONNE AU CAP CARTAYA

Diverses études ont été effectuées entre le Cap Couronne et le Cap Cartaya, certaines sur presque tout ce littoral, d'autres sur des parties plus restreintes.

Celle de Blanc (1956b) englobe la portion du rivage comprise entre le Cap Couronne et le Cap Lardier. Cet auteur a tout d'abord défini une "association fondamentale", formée de disthène—sillimanite—staurotide—tourmaline, originaire des auréoles métamorphiques du Massif des Maures, et qu'on retrouve sur les plages par l'intermédiaire de formations d'âges variés.

Le long de la côte des Maures, premier sous-secteur métamorphique, cette association est accompagnée de minéraux plus rares ou moins caractéristiques: apatite, augite, chloritoïde, grenat, hornblende brune, hypersthène, zircon, micas.

Aux Iles d'Hyères, au même cortège des Maures, s'ajoutent des minéraux accessoires tels que: augite, chloritoïde, grenat, hornblende brune, épidote.

Dans le troisième sous-secteur, Giens—Le Pradet—Cap Sicié, le spectre minéralogique métamorphique est altéré par suite de l'éloignement du Massif des Maures et de conditions nouvelles: volcanisme, etc. La même association y est présente, mais les proportions des espèces typiques sont moins élevées. Les minéraux accessoires sont sensiblement les mêmes que dans les deux autres sous-secteurs, mais les différentes espèces sont plus localisées.

Le long de la côte de la Provence calcaire, les minéraux de l'association caractéristique des Maures sont représentés d'une manière très irrégulière avec, parfois, des concentrations très localisées, dues, en général, à la proximité de formations susceptibles de les fournir: dunes littorales le plus souvent.

Pour Blanc (1956a, p. 10) la dispersion des minéraux de l'association métamorphique se fait sensiblement de l'Est vers l'Ouest, cependant ajoute-t-il "les analyses montrent qu'à partir du cap Sicié, le courant général n'a aucune action sur la répartition des sédiments littoraux proprement dits". Le transport des minéraux de l'Est vers l'Ouest ne dépasse pas la zone métamorphique des phyllades, il y a seulement de petits transferts liés à la direction des houles.

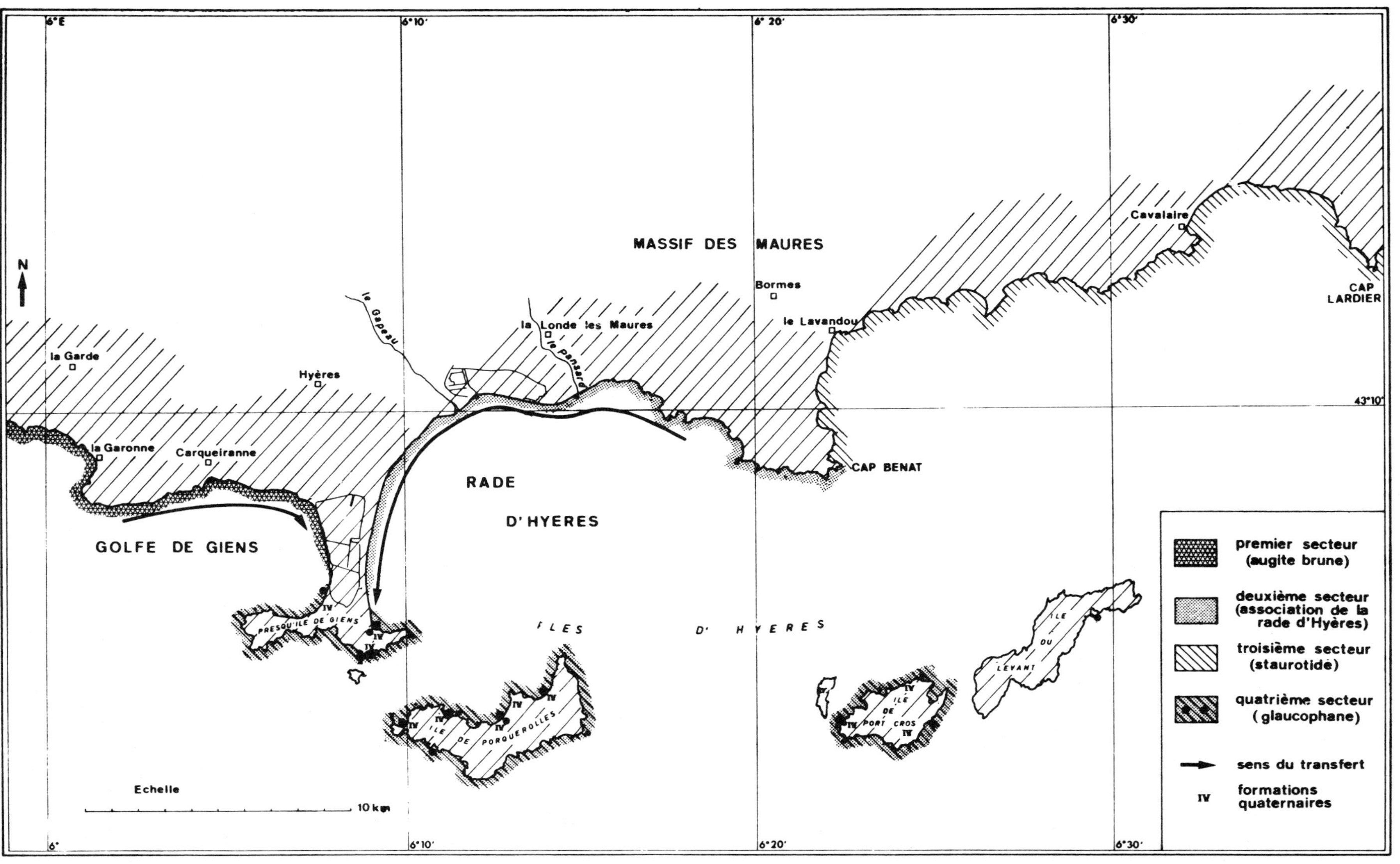

Figure 3. Répartition des minéraux lourds de Toulon au Cap Lardier (d'après Picard, 1969).

Entre Toulon et le Cap Lardier, Picard (1969) utilisant soit un minéral indicateur, soit l'association minéralogique, a caractérisé quatre secteurs distincts.

D'Ouest en Est, tout d'abord dans le Golfe de Giens (Figure 3), l'augite brune est le minéral typique, elle est issue des basaltes du Massif de Carqueiranne et est transportée d'Ouest en Est par les vagues de mistral (Picard, p. 551).

Dans la rade d'Hyères jusqu'au cap Bénat, c'est l'association minéralogique formée de chloritoïde, hornblende verte, épidote, micas, qui caractérise cette partie de côte. Ces espèces minéralogiques proviennent des sédiments du Gapeau et du Pansard, ainsi que de la côte Ouest de la presqu'ile du Cap Bénat, ils sont ensuite transportés de l'Est vers le Sud-Ouest par les vagues obliques dues au vent d'Est (Blanc, 1956a).

Dans le troisième secteur, s'étendant du Cap Bénat au Cap Lardier, la staurotide est le minéral indicateur; elle est abondante dans le Massif des Maures et se rencontre sur les plages de ce littoral.

Enfin dans le dernier secteur comprenant la presqu'île de Giens et les îles de Porquerolles et de Port-Cros, c'est la glaucophane qui est le minéral indicateur. Elle provient des formations quaternaires situées pour la plupart au voisinage des plages actuelles (Picard, 1969). Pour Froget et Picard (1968) son origine est à rechercher dans les apports d'origine durancienne, et les îles d'Hyères représenteraient la limite vers l'Est de son étalement.

La répartition de l'augite aegyrinique paraît, elle aussi, intéressante à observer. On la trouve sur les plages et le précontinent entre Marseille et Saint-Tropez (Froget et Picard, 1969). Pour ces auteurs son origine est à rechercher dans les tufs cinéritiques entièrement submergés dans ce secteur.

De la pointe du Layet, à l'Ouest du Cap Nègre, au Cap Cartaya (Figure 4), plusieurs associations minéralogiques caractérisent cette portion de côte (Burnet, 1964). Sur la partie la plus occidentale, de la plage de Cavalière au cap Cavalaire, le spectre minéralogique est dominé par la staurotide, accompagnée de grenat et de disthène; puis dans la baie de Cavalaire, et jusqu'à la plage de Gigaro, la staurotide laisse sa prédominance à la hornblende qui progresse régulièrement d'Ouest en Est, alors que baissent les proportions de staurotide, grenat, disthène. Sur la plage de la Bastide-Blanche, entre le Cap Lardier et le Cap Cartaya, l'association minéralogique est à nouveau différente, le grenat (28%) et la staurotide (24%) en sont les constituants principaux, associés à la hornblende et au zircon. Les proportions de tourmaline et d'andalousite, bien que peu élevées, sont à retenir.

Cette position des associations minéralogiques est en relation avec la situation géographique des sources:

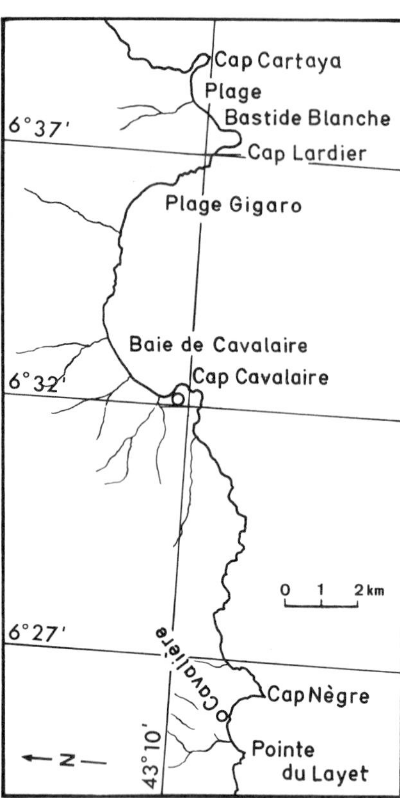

Figure 4. Carte de la Pointe du Layet au Cap Cartaya.

micaschistes à l'Ouest et amphibolites vers l'Est. Le courant d'Est étant responsable de la répartition proportionnelle des espèces principales.

GOLFE DE FREJUS

Les analyses de minéraux lourds dans la Baie de Fréjus ont été faites après la rupture du barrage de Malpasset; elles portent sur les passées sableuses superficielles (0 à 32 cm) de 5 carottes (Duplaix in Bellaiche, 1965). L'association minéralogique est formée d'amphibole très abondante, de staurotide et de grenat. Les minéraux des roches granitiques y sont aussi représentés, de même que les pyroxènes monocliniques. L'origine de ces différentes espèces est à rechercher dans l'Argens et dans le Reyran, affluent de l'Argens sur lequel avait été construit le barrage. L'Argens apporte de l'amphibole très abondante, de la staurotide et du grenat; le Reyran, de l'amphibole également, peu abondante, mais surtout du grenat et le groupement de minéraux granitiques au sens large du terme: zircon, tourmaline, monazite; du pyroxène monoclinique aussi, en aval de la Buème, petit affluent du Reyran. L'Argens et le Reyran ont, tous les deux, contribué à la formation des passées sableuses de ces carottes, avec une nette prédominance des alluvions de l'Argens. Les eaux libérées par la rupture du barrage ont repris et entraîné dans le Golfe de Fréjus les alluvions du delta de l'Argens, mais il semble bien que les sables de ce golfe devaient avoir la même composition avant cet évènement, les sables de l'Argens étant bien plus riches en minéraux lourds que ceux du Reyran.

GOLFE DE LA NAPOULE—GOLFE JUAN

Dans les sables de plages du Golfe de la Napoule et du Golfe Juan, ce ne sont pas les associations minéralogiques qui permettent de retrouver l'origine des sédiments, mais les minéraux caractéristiques: augite verte, augite brune et groupement zircon—xénotime—monazite (Duplaix et Nesteroff, 1959).

La source principale de l'augite verte est les tufs volcaniques du Cap d'Antibes; on la trouve aussi, mais en quantité nettement inférieure, dans le Massif de la Croix des Gardes, et, en proportion très faible, dans le Massif de la Maure et dans les alluvions de l'Argentière.

L'augite brune ne se rencontre que dans les alluvions de la Siagne et de l'Argentière.

La carte de répartition de l'augite verte (Figure 5) montre que celle-ci, partant du cap d'Antibes avec un chiffre très élevé de 5.000 grains, diminue progressivement à mesure qu'on suit le littoral vers l'Ouest.

L'augite brune, au contraire, s'étale d'Ouest en Est tout le long du Golfe de la Napoule, ses proportions diminuant régulièrement depuis ses sources, Argentière et Siagne; elle franchit même un peu la pointe de la Croisette. Dans le Golfe Juan on ne trouve pratiquement pas d'augite brune, sauf dans le prélèvement A 8, où les proportions de minéraux lourds sont élevées (553 grains pour un gramme de sable de la fraction considérée: 0,50–0,05 mm).

Les trois minéraux formant le groupement caractéristique se rencontrent soit ensemble, soit séparément dans toutes les sources sauf dans les tufs volcaniques. On les retrouve sur presque toutes les plages sauf le long du Cap d'Antibes et en très faible proportion dans le prélèvement A 8 (1% de zircon, < 1% de xénotime et de monazite). Le groupement des trois minéraux est pratiquement absent dans toute la partie centrale et orientale du Golfe Juan. Sa répartition dans ce golfe est intéressante à observer. Les trois espèces minéralogiques venant principalement du Bois de la Maure où elles se trouvent en inclusions dans les micas, sont entraînées en même temps que ceux-ci à quelques centaines de mètres du rivage, laissant le quartz et les feldspaths, issus de la même arène gneissique, constituer la fraction légère du littoral.

Figure 5. Répartition des minéraux lourds dans les golfes Juan et de la Napoule (d'après Duplaix et Nesteroff, 1959). (a) Augite verte, (b) augite brune, et (c) zircon, xénotime, monazite. Les chiffres représentent le nombre de grains au gramme de sable. ξ: massifs gneissiques; r, π^3: coulées de laves, conglomérats et grès permiens; t_{1-11} à J^{8-6}: plateaux calcaires (Trias et Jurassique); J_{1-11} à J^{5-1}: dolomies et calcaires du Cap d'Antibes; α: tufs volcaniques miocènes; p: Pliocène.

Ce sont alors ces quartz et ces feldspaths transitant d'Ouest en Est qui s'allient à l'augite verte venant de l'Est pour constituer les sables de plages de cette partie du Golfe Juan.

MER LIGURE

Les dépôts sous-marins de la Mer Ligure (Duplaix et Gennesseaux, 1967) proviennent soit de carottages, soit de dragages effectués dans des sédiments sous-marins quaternaires, dans les canyons du Var et du Paillon, sur la pente continentale et dans la plaine abyssale (Figures 6 et 7). Ils présentent tous une grande analogie minéralogique; l'association caractéristique y est formée d'amphibole, de grenat et de pyroxène monoclinique, avec quelques petites différences en certains points où le zircon et l'épidote peuvent être parfois relativement abondants.

Les sources de ces sables sont à rechercher dans les

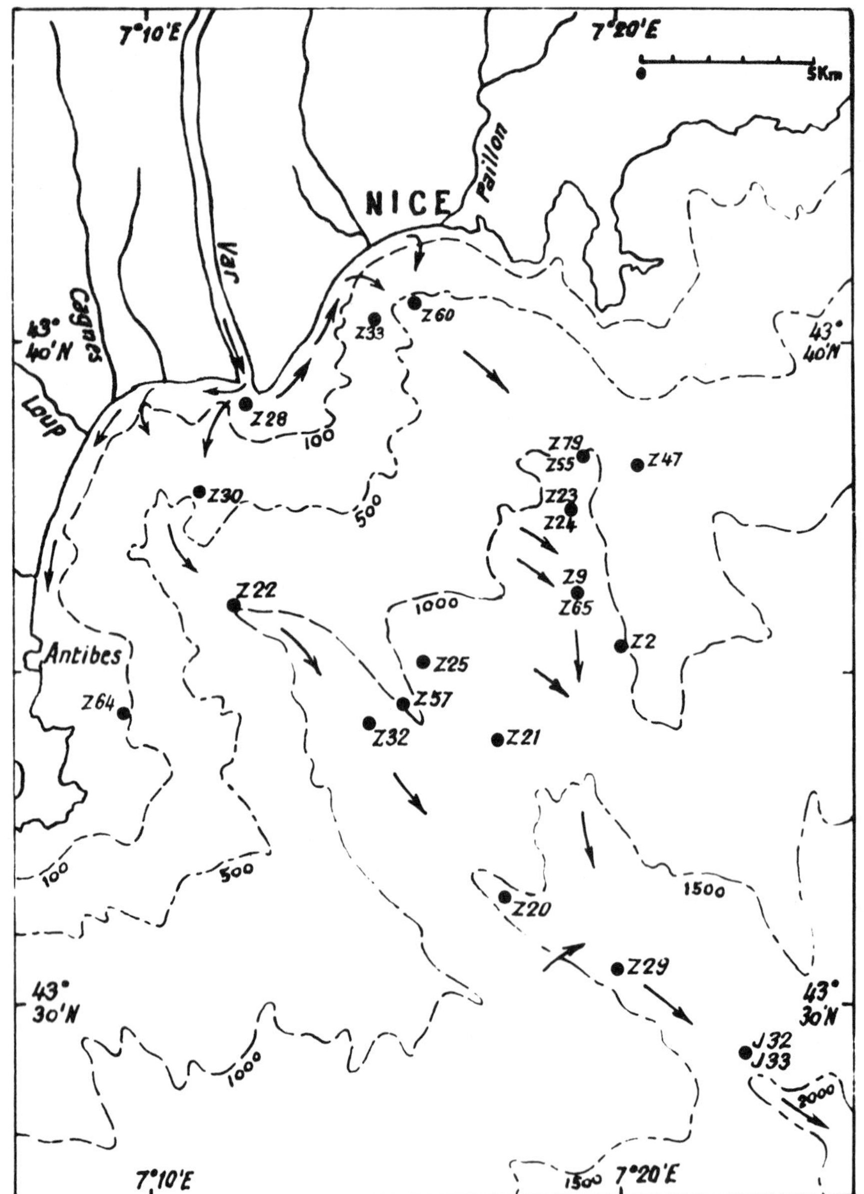

Figure 6. Carte bathymétrique de la Baie des Anges avec la position des prélèvements. Le sens du déplacement possible des sables et des galets est figuré par des flèches (d'après Duplaix et Gennesseaux, 1967).

alluvions du Var, du Paillon, de la Roya et dans les formations quaternaires et pliocènes de la région de Nice. Le Var fournit de l'amphibole abondante, du grenat, du pyroxène monoclinique; le Paillon, les mêmes espèces, mais avec une nette prédominance du grenat. L'analyse de niveaux sableux de carottes du canyon de la Roya montre que l'association minéralogique est tout à fait différente; elle est formée d'anatase abondante, de zircon, tourmaline et grenat; on doit aussi noter la présence en très faible proportion de glaucophane (Figure 8). Dans les formations quaternaires de la région de Nice, l'amphibole et le grenat sont les minéraux principaux. Dans les conglomérats pliocènes le grenat est le plus abondant, l'amphibole est peu représentée ou même absente, le pyroxène monoclinique est encore plus rare (Duplaix et Gennesseaux, 1968).

La comparaison de la minéralogie des sources et des sables sous-marins de la mer de Ligurie montre que ces derniers présentent une grande analogie avec les alluvions actuelles du Var, qu'ils aient été prélevés dans le canyon du Var, le canyon du Paillon ou la plaine abyssale (Tableau 1). Par contre les affleurements de vase ancienne de la pente continentale peuvent être rapprochés des formations quaternaires littorales. Seule la base de ces dépôts quaternaires

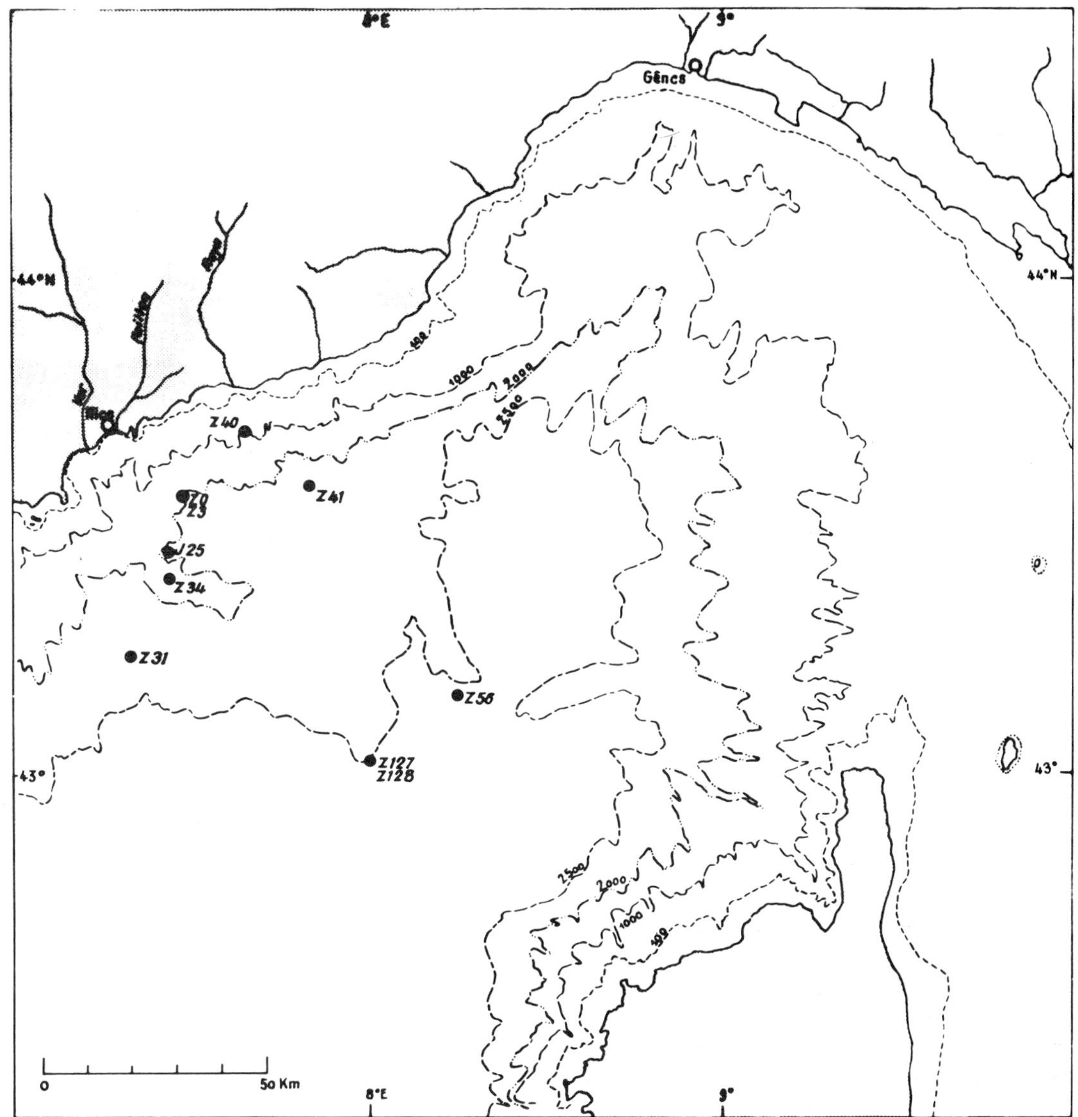

Figure 7. Carte bathymétrique de la mer de Ligurie avec la position des prélèvements (d'après Duplaix et Gennesseaux, 1967).

contient des éléments remaniés des conglomérats pliocènes. Parmi ces dépôts, les prélèvements Z O et Z 3 (Figure 7) prélevés à la base du canyon de Beaulieu, contiennent des proportions de tourmaline et d'anatase qui rappellent les sables du canyon de la Roya, de même que la présence de la glaucophane. Des apports orientaux se seraient ainsi produits à cette époque.

Tableau 1. Associations de minéraux lourds dans les différents milieux de la Mer Ligure.

	Var inférieur	Canyon du Var	Canyon du Paillon	Plaine abyssale	Vases anciennes
Tourmaline	11	4,5	3	4,5	6
Zircon	4	5,5	4	7,	5,5
Grenat	21	24	23	20,5	24
Epidote	7	10	10	12	9
Amphibole	44	30	33	28	37
Pyroxène monoclinique	5	12	18	16	1,5

Il semble bien que ces sédiments anciens ne participent pas à la composition minéralogique des sables superficiels des canyons et de la plaine abyssale qui seraient plus récents. Le Var aurait donc joué un rôle très important pour le remplissage de la cuvette ligurienne durant le Quaternaire supérieur et l'Holocène.

CONCLUSIONS

Les sources des sédiments des plages, du plateau continental et des canyons sous-marins de la Méditerranée française sont à rechercher dans les apports des

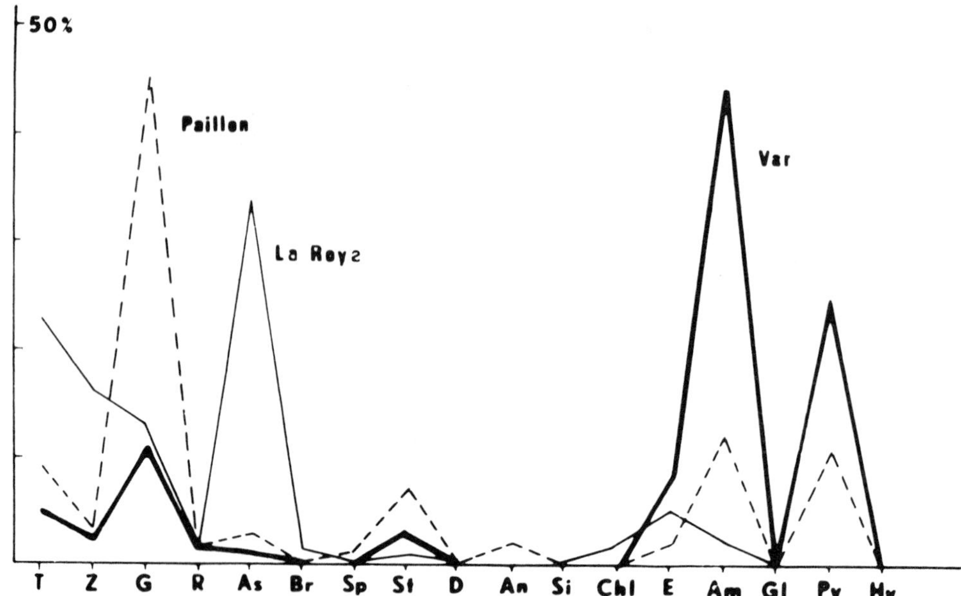

Figure 8. Composition minéralogique des sables du Var, du Paillon et de la Roya (d'après Duplaix et Gennesseaux, 1967). T = tourmaline, Z = zircon, G = grenat, R = rutile, As = anatase, Br = brookite, Sp = sphène, St = staurotide, D = disthène, An = andalousite, Si = sillimanite, Chl = chloritoïde, E = épidote, Am = amphiboles, Gl = glaucophane, Py = pyroxènes monocliniques, et Hy = hypersthène.

fleuves et les produits de l'érosion des formations bordant les côtes. Leur étalement est lié à plusieurs facteurs, sens des courants, direction des vents dominants, forme et orientation de la côte. Ceci est particulièrement sensible sur les côtes du Roussillon, du Languedoc et de la Provence.

D'autres causes interviennent également dans la répartition des sources; leurs minéraux lourds se comportent de façon différente dans un même milieu, le déplacement de ceux-ci étant fonction de la dimension, de la forme et de la densité des espèces. C'est ce qui se produit sur le plateau continental du Roussillon et dans le Golfe Juan.

D'autre part, lorsque les proportions de minéraux lourds sont différentes dans les diverses sources, on retrouve dans les sédiments à peu près la composition minéralogique de la source la plus riche, les autres peuvent alors se manifester par leurs minéraux typiques. C'est le cas pour les niveaux sableux des carottes du Golfe de Fréjus et des vases anciennes ZO et Z3.

Dans la Mer de Ligurie, les apports du Var au Quaternaire supérieur et à l'Holocène ont été si importants que ce sont les seuls qui se reconnaissent dans les sédiments superficiels des canyons du Var, du Paillon et de la plaine abyssale. Seules les vases anciennes prises sur le flanc des canyons ont une composition minéralogique différente. Ce qui laisse supposer que des sédiments, sans doute anté-würmiens, sont enfouis plus profondément sous les sédiments superficiels.

BIBLIOGRAPHIE

van Andel, T. H. 1955. Sediments of the Rhône delta, II. Sources and deposition of heavy minerals. *Koninklijk Nederlandsch Geologisch Mijnbouwkundig Genootschap,* Geologische serie XV, 3:353–556.

Blanc, J. J. 1956a. *Recherches de sédimentologie littorale et sous-marine en Provence occidentale.* Thèse, Masson et Cie, Paris, 140 p.

Blanc, J. J. 1956b. Etude minéralogique des sables littoraux du Cap Lardier au Cap Couronne (Provence). *Bulletin du Muséum d'Histoire Naturelle de Marseille,* 16:69–92.

Bellaiche, G. 1965. *Contribution à l'étude de la sédimentation actuelle dans le golfe de Fréjus.* Thèse de 3ème cycle de Géologie sous-marine, Paris, 140 p.

Burnet, M. 1964. *Etude sédimentologique des sables littoraux entre le Cap Cartaya et la pointe du Layet (Var).* Diplôme d'études supérieures, Paris, 46 p.

Chassefière, B. 1969. Etude minéralogique de la fraction lourde de sédiments de la région de Thau (Hérault). *Vie et Milieu,* série B: Océanographie, 20 (1–B):37–50.

Duboul-Razavet, C. 1956. Contribution à l'étude géologique et sédimentologique du delta du Rhône. *Mémoires de la Société Géologique de France,* 76:234 p.

Duplaix, S. et M. Gennesseaux 1967. Les minéraux lourds des sables du Var, du Paillon et de la Roya et les dépôts sous-marins de la mer de Ligurie. *Cahiers Océanographiques,* 19 (3):219–236.

Duplaix, S. et M. Gennesseaux 1968. Les minéraux lourds des formations tertiaires et quaternaires de la région niçoise (Alpes-Maritimes). *Revue de Géographie Physique et de Géologie Dynamique* (2), 10:353–374.

Duplaix, S. et C. Lalou 1949. Etude minéralogique et granulométrique des sables de plages du littoral méditerranéen. *Comptes Rendus sommaires de la Société Géologique de France,* 3:64–65.

Duplaix, S. et C. Lalou 1951. Etude pétrologique des sables du Roussillon. *Vie et Milieu,* 2 (4):501–527.

Duplaix, S. et W. D. Nesteroff 1959. Recherches sur les minéraux lourds du littoral du golfe de la Napoule et du golfe Juan. *Bulletin de la Société Géologique de France,* 7:107–111.

Duplaix, S. et J. L. Olivet 1969. Etude sédimentologique et morphologique de la tête du rech Bourcart (Golfe du Lion). *Cahiers Océanographiques,* 22 (2):127–146.

Froget, C. et F. Picard 1968. Présence de glaucophane sur les plages des îles d'Hyères; son origine probable. *Comptes Rendus de l'Académie des Sciences, Paris,* 266:2313–2315.

Froget, C. et F. Picard 1969. Présence de tufs cinéritiques à augite aegyrinique sur le précontinent méditerranéen entre Marseille et les îles d'Hyères. *Comptes Rendus de l'Académie des Sciences, Paris,* 269:1482–1485.

Monaco, A. 1967. Etude sédimentologique et minéralogique des dépôts quaternaires du plateau continental et des rechs du Rousillon. *Vie et Milieu,* série B: Océanographie, 18 (1–B):33–62.

Monaco, A. 1970. Analyse minéralogique préliminaire des sables du plateau continental du Roussillon. *Comité National d'Etude et d'Exploitation des Océans,* 68/49:22p.

Picard, F. 1969. Contribution à l'étude des minéraux lourds dans les sables littoraux de Toulon au Cap Lardier (Provence). *Téthys,* 1:539–560.

Vatan, A. 1949. Etude pétrographique des matériaux sableux côtiers du golfe du Lion entre Cap Cerbère et l'embouchure du Rhône. *Congrès Sédimentation et Quaternaire, La Rochelle,* 147–156.

Essai d'Interprétation Hydrodynamique de la Granulométrie des Sédiments Sableux, Plage de Pramousquier, Var (France)

Claude Degiovanni

Faculté des Sciences, Alger et Centre Universitaire de Marseille-Luminy, Marseille

RESUME

Les différentes zones qui constituent une plage, sont définies par les analyses granulométriques des sables. Dans la zone aérienne, les sédiments sont plus grossiers que dans le secteur marin; ils montrent de fortes concentrations en minéraux lourds, ont une asymétrie généralement positive et un bon triage; en revanche, les sables marins sont fins, pauvres en minéraux lourds, ont une asymétrie négative et un triage moyen. Les indices d'évolution (Rivière, 1952), qui conduisent à la notion de faciès granulométriques différencient eux aussi la zone aérienne du domaine marin. La première est caractérisée par des sédiments qui sont en cours d'évolution.

Les courbes en ordonnée de probabilité et l'indice de Harris, calculé pour chaque population, donnent trois faciès sédimentologiques distincts: deux sont rencontrés sur la plage aérienne et sont de faible énergie, le dernier correspond soit à un régime hydrodynamique intense, soit à des actions de moindre énergie. Dans ce dernier cas, elles traduisent une meilleure évolution du sédiment.

Les diagrammes CM (Passega, 1957) précisent les modes de transport et de dépôt des particules sur les divers secteurs de la plage. La comparaison des diagrammes CM avec les courbes en ordonnée de probabilité révèle le mode de transport des sables sur la plage: les particules se déplacent essentiellement en suspension granoclassée sur les plages aérienne et sous-marine et par traction dans la zone du ressac.

ABSTRACT

A detailed size analysis of beach sands serves to define the different zones that constitute the beach at Pramousquier (Var) France. Sands in the subaerial zone tend to be coarser than those of the foreshore and also display high heavy mineral concentrations, positive skewness and good sorting. Foreshore sands are finer, have lower heavy mineral contents and display negative skewness and fair sorting. Maturity indices (Rivière, 1952), used to define textural facies, serve to distinguish the subaerial zone from the foreshore. The former zone is characterized by less mature sediments than those in the foreshore.

The size distribution curves on probability paper and the Harris parameter computed for each population serve to distinguish three sedimentological facies: two occur on the subaerial portion of the beach influenced by a lower energy regime; the third corresponds either to a high energy regime or to a lower energy regime resulting in a more mature sediment type.

CM patterns (Passega, 1957) reflect the style of transportation involved in the different sectors of the beach. A comparison of CM patterns and curves on probability paper helps identify the type of sand transportation on the beach: grains are moved principally in graded suspensions on the subaerial and foreshore zones and by traction in the swash zone.

INTRODUCTION

Si les mécanismes de formation et d'évolution des plages des littoraux atlantiques et pacifiques du continent nord-américain sont bien connus après les travaux des auteurs américains, il n'en est pas de même pour ceux qui régissent les plages méditerranéennes. En France, les travaux antérieurs (Rivière, 1949a et b, 1952 et 1955; Vernhet, 1953a et b et 1955; Rivière et Vernhet, 1953, 1955, 1962 et 1966; Blanc, 1956a; Steinberg, 1959; Arbey, 1961a, b et c; Rivière, Arbey et Vernhet, 1961) traitent surtout des structures littorales, telles que les croissants de plage (beach-cusps) et les sinuosités de plage qui se forment sur la ligne de rivage; des rides littorales sous-marines et des transports de sable. Plus récemment, Nesteroff (1965) aborda les problèmes de la formation et de l'évolution des plages dans la région d'Antibes, à l'Ouest de Nice. Afin de compléter ces premières données, la plage de Pramousquier a été choisie du fait de son tracé rectiligne et des fortes concentrations en minéraux lourds contenues dans son sable.

Le sable étant un sédiment meuble, il ne peut être défini sédimentologiquement que par l'utilisation des méthodes statistiques. Leurs applications se font en utilisant des indices déduits de certains paramètres lus sur les courbes granulométriques cumulatives. De nombreux auteurs les avaient déjà employés pour différencier les milieux de sédimentation et les sédiments. Pour déterminer les caractères granulométriques et dynamiques des zones constituant la plage, et afin d'accroître la représentativité des résultats, les échantillons de sable ont été prélevés selon une maille relativement étroite dans les plages aérienne et sous-marine. Les relations entre les indices aident à caractériser chacune de ces zones.

A la suite des auteurs ayant déjà examiné les rapports courbes granulométriques-milieux de sédimentation, le comportement des particules a été étudié à partir des données granulométriques. L'emploi de l'indice d'évolution (Rivière, 1952), qui amène à la notion de faciès granulométriques correspondant à des faciès géologiques précis, permet de connaître le degré d'évolution des sables. Les échantillons représentés par les valeurs de leurs indices d'évolution sur des diagrammes CM (Passega, 1957) montrent les variations de faciès granulométriques dans les différents secteurs de la plage.

Le comportement dynamique des grains est analysé par comparaison entre les diagrammes CM et les courbes cumulatives en ordonnée de probabilité.

Les courbes cumulatives en ordonnée de probabilité et le calcul de l'indice de Harris (Harris, 1959) conduisent à la définition de trois types sédimentologiques.

DESCRIPTION DE LA PLAGE DE PRAMOUSQUIER ET METHODES D'ETUDE

Située à l'Est du Cap Nègre, France, la plage de Pramousquier est limitée latéralement par une côte rocheuse, escarpée, taillée dans la série des micaschistes du Cap Nègre, qui appartient au massif métamorphique des Maures. Le rivage, pratiquement rectiligne, a une longueur de 350 m, et la largeur de la bande sableuse oscille entre 5 et 30 m. Un ruisseau traverse la grève dans la partie Nord pour se jeter à la mer, mais il n'a aucune incidence sur la sédimentation sableuse, car une mare en amont de son embouchure forme un piège à sédiments (Figure 1).

La plage, faite de deux ensembles morphologiques naturels qui sont la zone aérienne et la zone sous-marine (Ottmann, 1965), présente au Nord et au Sud deux profils transversaux différents.

Au *Nord*, il s'agit du profil transversal type de la plupart des plages méditerranéennes. En partant de de l'arrière-plage, se distinguent successivement: les formations dunaires d'une dénivelée de 1 m à 1,50 m, la haute-plage étroite, la berme de la plage moyenne, la zone du ressac (swash) sur laquelle se forment parfois des croissants de plage d'une longueur d'onde de 3 à 6 m, la microfalaise immergée dont le talus est constitué par de petits galets de 1 à 3 cm de diamètre, et enfin la plage sous-marine (Figure 2). La pente générale n'est que de quelques degrés.

Au *Sud*, la présence de bâtiments à quelques mètres de la limite plage-mer a provoqué des modifications du profil transversal. On retrouve les formations dunaires, puis, de la haute-plage peu différenciée de la plage moyenne, on passe presque immédiatement à la zone du ressac; le profil sous-marin est semblable à celui décrit dans la partie Nord. Suivant les saisons et les houles, la pente peut atteindre sur la plage aérienne une dizaine de degrés, mais elle reste toujours très douce dans le domaine marin.

Par son orientation, NNE-SSO, la plage subit principalement les influences des houles de secteurs Est et Sud-Est, alors que la houle de Mistral (NO; NNO) ne l'atteint qu'après diffraction sur la pointe du Cap Nègre. Dans les deux cas, les plans de houle sont parallèles au rivage. Quelle que soit l'origine des vents, l'avant-plage est parcourue par un courant littoral (longshore current) de direction SSO-NNE.

Les échantillons de sable ont été prélevés dans les formations dunaires, sur la haute-plage, dans la berme, sur la zone du ressac et dans l'avant-plage entre la ligne de rivage et la première ligne de déferlement des vagues vers le large, suivant neuf transversales perpendiculaires au front de mer, distantes les unes des autres de 25 à 50 m (Figures 1 et 2). L'épaisseur des sédiments soumis aux effets hydrodynamiques de fond étant faible, seuls les deux ou trois millimètres superficiels ont été récoltés. Sur les prélèvements, lavés plusieurs fois à l'eau douce et séchés à l'étuve à 60°C, les minéraux lourds sont séparés des minéraux légers par densimétrie, dans du bromoforme R.P (densité 2,9), en vue de leur étude minéralogique. La séparation achevée, les deux fractions sont

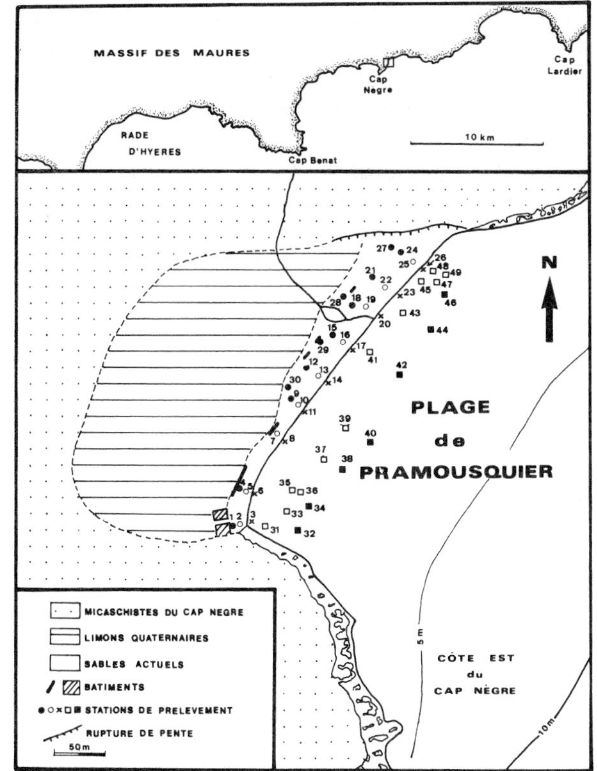

Figure 1. Situation et environnement géologique de la plage de Pramousquier, France.

pesées pour établir le pourcentage de minéraux lourds contenus dans le sédiment total, puis, elles sont analysées mécaniquement sur une colonne de tamis dont les mailles ont été mesurées au micromètre sous la loupe binoculaire. Les courbes granulométriques cumulatives sont tracées dans un système de coordonnées semi-logarithmiques à ordonnée gaussienne; le sens du cumul s'effectuant toujours des particules les plus grossières vers les particules les plus fines. Pour réduire les erreurs sur la taille des éléments, la valeur calculée de la diagonale $(a\sqrt{2})$ des mailles est portée en abscisse, car les grains anguleux possédant une dimension comprise entre (a) (côté de la maille) et $(a\sqrt{2})$ traversent les tamis selon la diagonale des mailles. L'emploi d'une ordonnée gaussienne facilite l'individualisation des populations de grains (Visher, 1969).

De la courbe granulométrique expérimentale sont tirés les histogrammes de fréquence. Les limites des classes granulométriques sont définies par les valeurs en progression géométrique de raison $\sqrt[10]{10}$. En utilisant une échelle logarithmique, des intervalles de classe égaux sont ainsi obtenus, et les histogrammes sont alors directement comparables entre eux.

L'indice de Harris (Harris, 1959) est calculé pour chaque ensemble de particules; sa valeur est d'autant plus élevée que le stock est bien trié. $\rho = \Delta\%/(\Delta D \text{ en } \phi)$, où ΔD est la différence des diamètres, lus en unités phi, des grains limitant la population, et $\Delta\%$ le pourcentage de grains contenus dans cette population.

La conversion des millimètres en unités phi ($\phi = -\log_2$, diamètre en mm), les calculs des indices granulométriques et les histogrammes sont obtenus par traitement des données sur ordinateur. Les indices calculés sont le triage σ_I (Inclusive Graphic Standard Deviation), l'acuité du mode KG (Graphic Kurtosis), l'asymétrie Sk_1 (Inclusive Graphic Skewness), la taille moyenne Mz (Mean Size), utilsés par Folk et Ward (1957); la deviation standard σ_ϕ (Phi Deviation Measure) de Inman (1952), l'indice de classement G (Rivière, 1952b) et la médiane M.

L'indice d'évolution *n* qui conduit à la notion de faciès granulométriques (Rivière, 1952b) est tiré des courbes canoniques tracées à partir des courbes

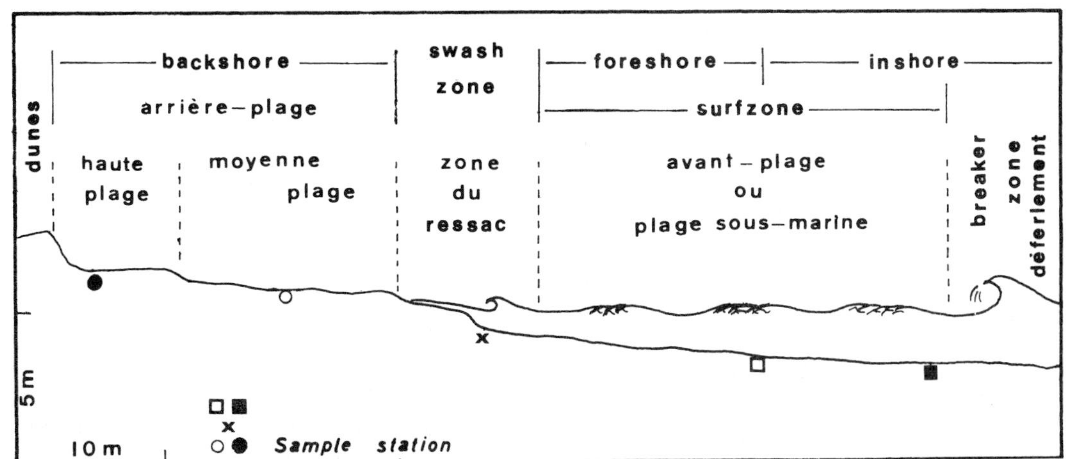

Figure 2. Profil transversal Nord de la plage de Pramousquier et correspondance des termes utilisés avec les termes anglais.

cumulatives par les déformations géométriques décrites par Weydert (1968). Selon Rivière (1952b, p. 166) "l'indice *n* permet de définir quatre faciès granulométriques correspondant à de véritables faciès géologiques de sédimentation définis d'après les phénomènes actuels", mais, (p. 161) "entre lesquels il ne semble pas y avoir de discontinuité brusque". Les quatre faciès déterminés sont: le faciès linéaire représentant des sédiments non évolués; le faciès parabolique (sédiments ayant déjà subi un début d'évolution); le faciès logarithmique caractérisant des sédiments dont l'évolution par transport est très avancée; le faciès hyperbolique qui paraît résulter d'actions naturelles de décantation. Les échantillons différenciés par les valeurs de l'indice d'évolution sont portés sur des diagrammes CM analogues à ceux de Passega (1957), mais dans lesquels le paramètre C ne représente pas le premier percentile mais le 5%, M étant la médiane. La construction des diagrammes, en employant successivement le 1% et le 5%, montre que l'utilisation du 5% en ordonnée ne fait pas varier l'enveloppe de la population, mais provoque seulement un rapprochement des points vers la droite C = M. En conséquence, le 5% a été choisi afin de réduire les erreurs de détermination graphique sur ce paramètre. Pour employer des coordonnées arithmétiques les valeurs de C et de M sont données en unités phi.

RESULTATS GRANULOMETRIQUES

La Composition Minéralogique des Sables

Les études minéralogiques des sables ont été faites sur les classes granulométriques 0,31–0,25 mm, 0,25–0,16 mm et 0,16–0,05 mm.

Les minéraux légers sont constitués par des grains de quartz (environ 90% du sédiment "léger"). L'étude morphoscopique montre 72% de quartz non usés, 25% de quartz "coins arrondis" et 3% d'arrondis. Les autres éléments sont les feldspaths potassiques et les plagioclases.

Les minéraux lourds sont plus variés et par ordre d'importance on trouve: la staurotide (62%); le disthène (10%); le grenat et la tourmaline (7% chacun); la biotite, la sillimanite, les minéraux opaques, la muscovite, l'andalousite, le rutile et le zircon constituent 14% du stock lourd. Il s'agit de "l'association métamorphique des Maures" définie par Blanc (1956b).

Les pourcentages de minéraux lourds contenus dans le sable, calculés après les séparations par densimétrie révèlent deux domaines:

1. la *plage aérienne*, sur laquelle, suivant les lieux de prélèvements, les échantillons renferment de 7 à 79% d'éléments lourds;
2. la *plage sous-marine*, relativement pauvre en minéraux lourds, n'a que des teneurs de l'ordre de 1 à 2%.

Les concentrations de minéraux lourds, très repérables sur la grève, se situent aux endroits où la pente du talus limitant la plage moyenne de la zone du ressac tend à s'adoucir, et d'une manière presque symétrique par rapport à l'axe de la baie.

A Mustang Island, Texas, Bradley (1957) note que les sédiments provenant du premier talus de la plage aérienne sont plus riches en minéraux lourds que les sables marins. La différence de pourcentage entre ces deux milieux atteignant 0,30%, il attribue cette concentration plus élevée dans le domaine aérien à un transport sélectif du vent. Or, sur la plage de Pramousquier, les écarts étant de 5% au minimum et de 78% au maximum, une action éolienne ne peut seule être envisagée, d'autres facteurs ont dû agir. Ces concentrations très importantes seront expliquées dans la deuxième partie de ce travail.

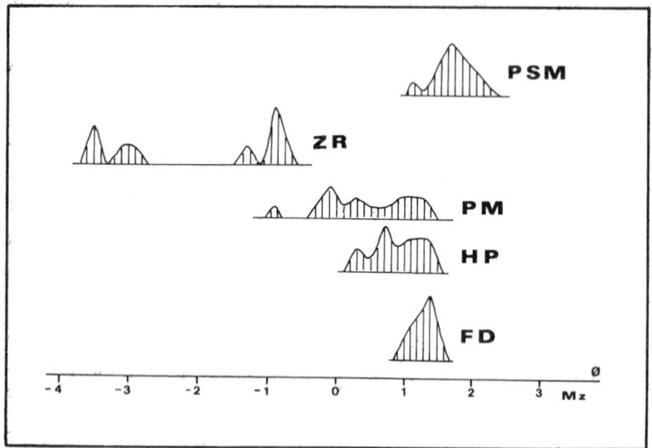

Figure 3. Histogrammes de fréquence des tailles moyennes, Mz, sur les formations dunaires (FD), de la haute-plage (HP), de la plage moyenne (PM), de la zone du ressac (ZR) et de la plage sous-marine (PSM).

Les Indices Granulométriques

Les analyses granulométriques des échantillons prélevés dans les divers secteurs de la plage montrent que malgré une certaine unité, chaque sédiment provenant d'une zone déterminée possède des caractères distinctifs. La zone du ressac est la plus grossière: les particules ont en général un diamètre supérieur au millimètre. De part et d'autre de la zone du ressac, la dimension des grains diminue; de plus, l'arrière-plage est composée de sables dont les tailles sont supérieures à celles des sables marins (Figure 3). Nesteroff (1965) et Rivière (1952a) avaient observé ces mêmes phénomènes respectivement sur les plages de la région d'Antibes et sur la plage de Saint-Aygulf, toutes deux à l'Est de Pramousquier. Les sédiments

Tableau 1. Caractéristiques granulométriques des différentes zones de la plage.

	Triage	Asymétrie	Acuité du Mode
Formations Dunaires	bien trié	très peu négative	moyennement accusée
Haute-Plage	bien trié	tendance négative	moyennement accusée
Plage Moyenne	bien trié	tendance positive	moyennement accusée
Zone du Ressac	moyennement trié	très négative à très positive	moyennement accusée
Plage Sous-Marine	moyennement trié	très peu négative	moyennement accusée

sont dans l'ensemble bien triés, sauf dans la zone du ressac et sur la plage sous-marine où ils apparaissent moyennement triés. L'acuité du mode est moyennement accusée sur la totalité de la plage; quant aux courbes de fréquence, elles sont rarement gaussiennes, tous les intermédiaires existent entre une asymétrie très négative et une asymétrie très positive: tendance négative sur les formations dunaires, la haute-plage et la plage sous-marine; tendance positive sur la plage moyenne, et très hétérogène dans la zone du ressac. Le Tableau 1 résume ces principaux caractères.

La Figure 4 représente la variation de la taille moyenne Mz en fonction de la déviation standard σ_ϕ. Les zones aérienne et sous-marine, ainsi que la zone du ressac sont formées de sédiments qui évoluent dans le même sens: les valeurs de la déviation standard diminuent lorsque la taille moyenne des particules augmente. Les sables des formations dunaires sont intermédiaires entre ceux de la plage aérienne et ceux de la plage sous-marine.

Le graphique donnant l'indice de classement G_T ($G_T = \log m/M$, où m est le diamètre du grain le plus petit, et M le diamètre du grain le plus gros) en fonction du triage σ_I, montre une population "coudée" (Figure 5). L'indice G_T représente en fait le maximum d'étalement de la courbe cumulative de fréquence. Les sables aériens et sous-marins, constituant l'ensemble AB, ont un indice de classement qui tend à augmenter lorsque le triage devient plus mauvais. La portion BC, composée d'échantillons provenant de la zone du ressac, est parallèle à l'axe des abscisses: c'est-à-dire que pour un étalement pratiquement constant des courbes granulométriques, l'indice de triage prend des valeurs différentes. Le meilleur triage s'explique par l'enrichissement en particules de certaines classes granulométriques.

Les variations latérales de la médiane M, du Sud au Nord de la plage, sont données sur la Figure 6. Dans la zone du ressac, la médiane semble diminuer vers le Sud; ce phénomène apparaît lié aux concentrations de galets se formant au Nord de la plage. Sur les haute et moyenne plages, les médianes sont de plus en plus faibles vers le Nord. Sur la plage moyenne, un rétrécissement de la largeur de la grève au niveau de l'échantillon 7 est la cause du maximum observé; les influences du ressac s'y font sentir. Dans les dunes et l'avant-plage les valeurs de la médiane sont pratiquement constantes. Les faciès granulométriques indiquent que dans la zone du ressac, les maxima de la médiane correspondent à des faciès linéaires et les minima à des faciès à tendance hyperbolique. Sur la haute plage et la plage moyenne, les minima représentent des faciès paraboliques peu évolués. En admettant qu'un sédiment évolue au cours de son transport d'un faciès linéaire vers un faciès hyperbolique, les résultantes des circulations d'eau qui sévissent sur l'arrière-plage pendant les tempêtes seront déduites des faciès granulométriques.

Les courbes granulométriques caractérisent trois domaines principaux: La *plage aérienne,* où les mi-

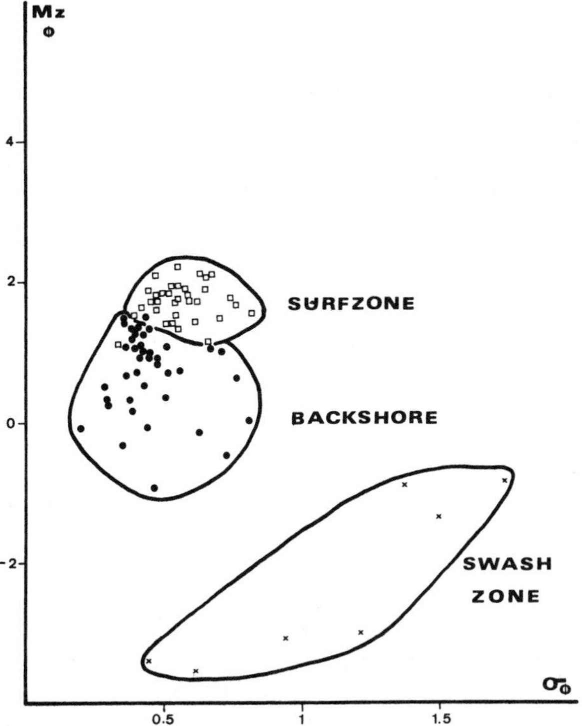

Figure 4. Variations de la taille moyenne, Mz, en fonction de la déviation standard σ_ϕ .

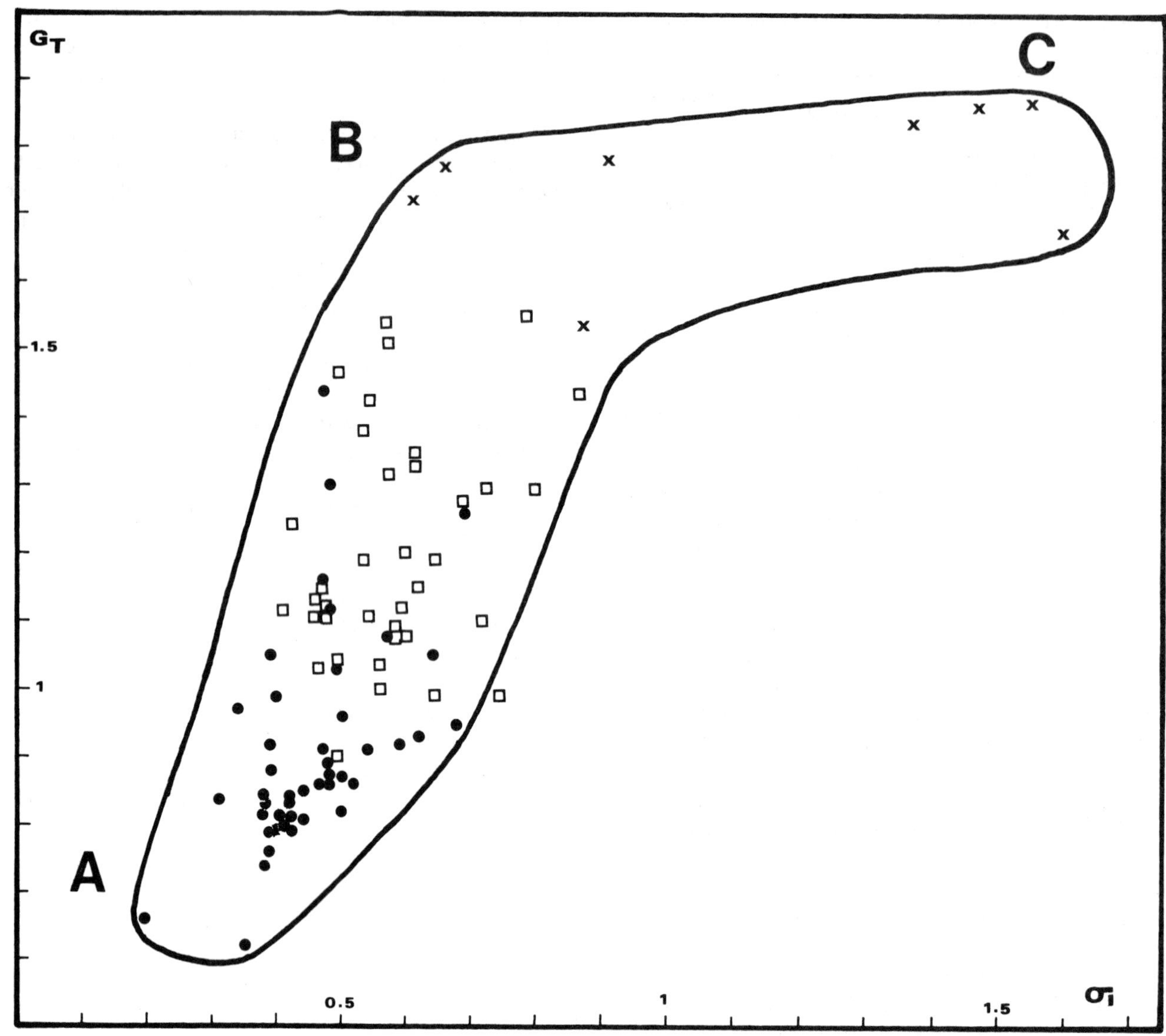

Figure 5. Variations de l'indice de classement, G_T, en fonction du triage σ_I. Les symboles sont définis sur la Figure 4.

néraux lourds sont abondants. L'asymétrie est à tendance positive et le triage est bon. La *plage sous-marine*, relativement pauvre (1 à 3%) en éléments lourds, à asymétrie négative et un triage moyen; et la *zone du ressac* avec des propriétés granulométriques variables.

Friedman (1962) avait déjà observé la variation d'asymétrie entre les sédiments marins à asymétrie négative et les sédiments de dunes à asymétrie positive: il en avait conclu que cet indice pouvait servir à différencier les deux milieux. Or, en ce qui concerne la plage de Pramousquier, les sables qui sont sous l'effet du vent à l'arrière-plage présentent une asymétrie faiblement négative. Duane (1964) note que certains sables de dune développés sur "l'Outer Banks" dans le Pamlico Sound de l'Ouest (Caroline du Nord) ont une asymétrie négative. Il attribue ce fait à une action érosive continue du vent, entraînant les particules fines; les dunes ne peuvent alors que prendre une très grande extension dans l'espace. Il faut penser que les formations dunaires peu importantes de Pramousquier correspondent au même effet dynamique que celui décrit par Duane (1964).

INTERPRETATIONS SEDIMENTOLOGIQUES

L'existence de deux types de sédiments sur les plages, l'un aérien et l'autre marin suggère que des effets dynamiques différents ont régi le dépôt et le transport des particules. Les courbes cumulatives en ordonnée de probabilité et les diagrammes CM

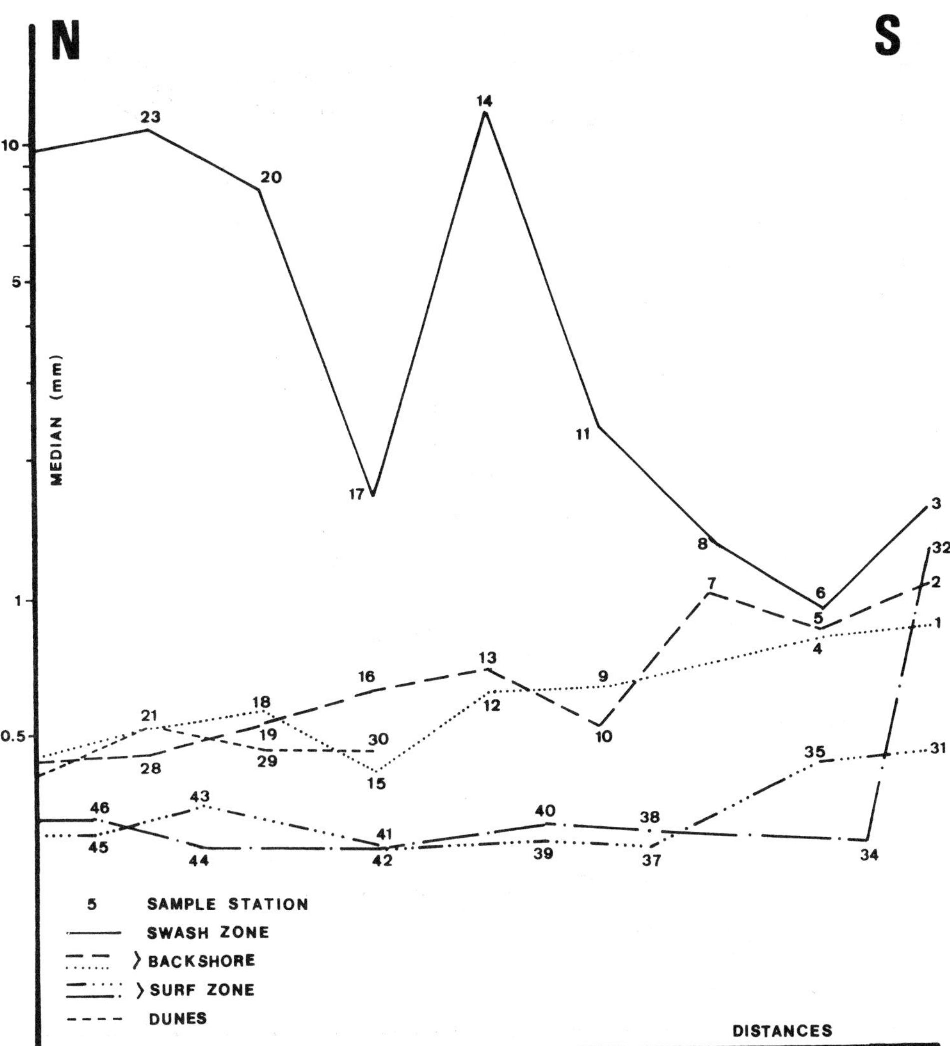

Figure 6. Variations latérales de la médiane.

(Passega, 1957) ont été tracés afin de déterminer les milieux de sédimentation et les modes de dépôt des sédiments. Les courbes en ordonnée de probabilité permettent une étude détaillée des populations élémentaires composant chaque échantillon, les diagrammes CM permettent de proposer une synthèse du transport des sables sur la totalité de la plage. L'essai d'interprétation des conditions hydrodynamiques de l'arrière-plage et de l'avant-plage est obtenu de la confrontation des résultats granulométriques tirés des courbes cumulatives.

Les Courbes en Ordonnée de Probabilité

Les sables de plage sont rarement représentés par une courbe de fréquence granulométrique d'allure parfaitement gaussienne, car si tel était le cas, les courbes cumulatives en ordonnée de probabilité seraient constituées par un seul segment de droite. Dans un milieu de sédimentation, chaque action dynamique influe, suivant sa compétence, sur des particules de certaines dimensions. Les sédiments sont ainsi composés de plusieurs populations granulométriques élémentaires. Leur individualisation peut s'établir par l'emploi d'une ordonnée gaussienne qui paraît, selon Visher (1969), le mieux rendre compte des divers phénomènes hydrodynamiques s'exerçant dans un milieu de sédimentation. Visher (1969, p. 1079) décrit une courbe type pour les sédiments de plage et postule que chaque segment de droite correspond à un ensemble de grains caractérisé par un type de transport. De son étude, il ressort que les sables de plage sont essentiellement formés de trois ou quatre populations élémentaires, mues par traction, par saltation et par suspension, respectivement désignées par les lettres C, B et A, par comparaison avec les stocks décrits par Moss (1962).

Le calcul de l'indice de Harris montre que ces

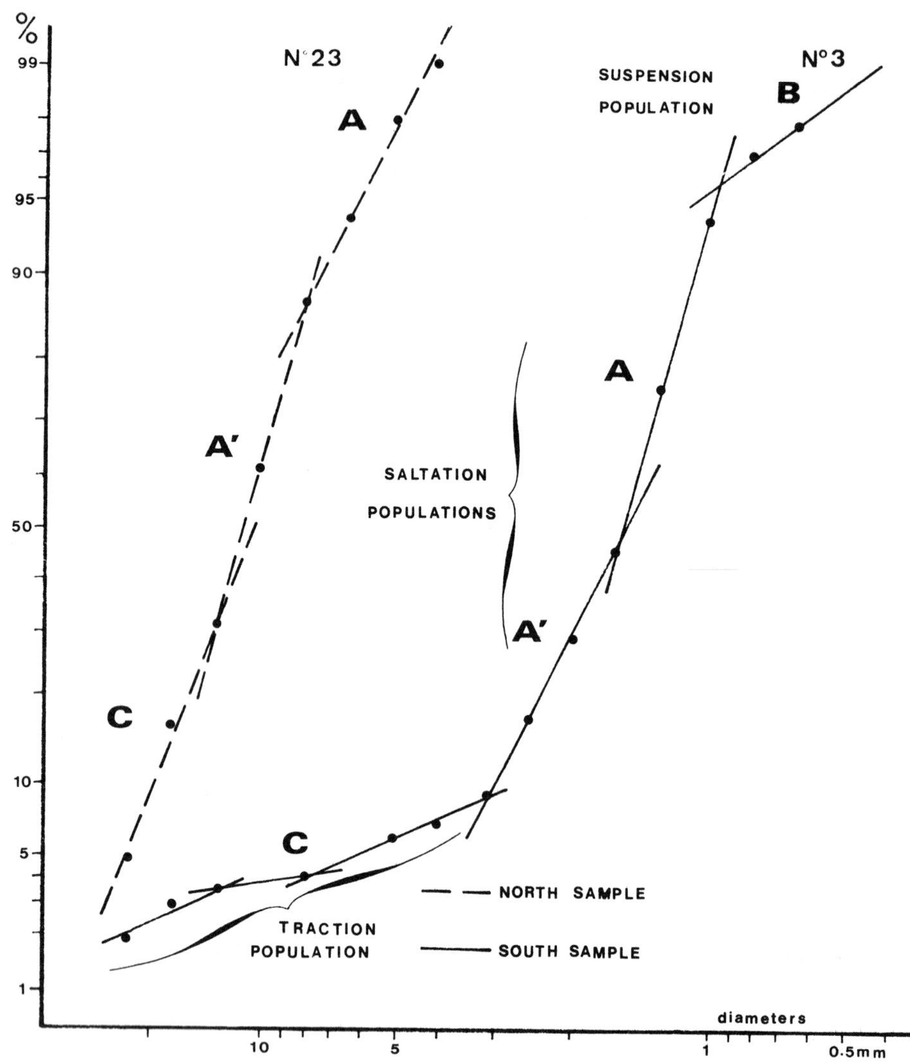

Figure 7. Courbes cumulatives de sables de la zone du ressac montrant les différences de compétence du courant de jet et de retour, au Nord et au Sud de la plage de Pramousquier.

populations ont un triage plus ou moins évolué: mauvais pour les grains déplacés par traction et suspension (ρ est compris entre 5 et 15), bon pour ceux qui sont transportés par saltation (ρ est supérieur à 30).

Les Courbes des Minéraux Légers

La Traction

Les dimensions de grains mûs par traction sur le fond, varient suivant les milieux considérés: elles sont en général plus grandes sur la plage aérienne que sur la plage sous-marine. Les valeurs minimales (lues sur l'axe des abscisses des courbes en ordonnée de probabilité) à partir desquelles les particules sont tractées se situent aux alentours de 0,80 mm sur la plage sous-marine, et entre 1,60 mm (plage moyenne) et 1 mm (dunes) sur la plage aérienne. Sous l'eau les éléments en traction ne forment que 1 à 3% du sable, alors qu'ils constituent 5 à 10% du sédiment sur la plage aérienne.

La Saltation

C'est le mode de transport le plus important, 90 à 98% des grains y sont soumis. Selon Visher (1969), cette population se divise en deux ensembles A′ et A (Figure 7): l'un représente les éléments mis en mouvement sous l'influence du jet de rive; l'autre sont les particules déplacées sous l'action du courant de retour.

Il est difficile de savoir quelle est la population qui correspond aux effets de l'un ou de l'autre de ces deux courants. Dans le domaine marin, les ensembles A′ et A montrent que les courants de jet et de retour n'ont pas la même compétence au Nord et au

Sud de l'axe de la baie de Pramousquier. Au Sud, les tailles des grains limitant A′ et A sont de 0,25 mm à 0,30 mm, tandis qu'au Nord, la séparation s'effectue au niveau de 0,40 mm. Ceci suggère qu'au Nord, les particules de diamètre compris entre 0,40 mm et 0,25 mm sont probablement soumises, alternativement, aux actions de jet et de retour.

Bien que vers le Nord, quelques échantillons donnent des courbes rappelant les sables dunaires par l'existence d'un seul stock en saltation, A′ et A sont présents sur la haute-plage.

Les formations dunaires n'ont en général qu'une population en saltation, ce qui signifie que les grains sont sous la dépendance de la seule action dynamique du vent.

La Traction et la Saltation dans la Zone du Ressac

C'est dans cette zone que l'énergie est la plus forte (Schiffman, 1965). La présence de petits galets dans ce secteur provoque l'apparition de plusieurs populations difficiles à interpréter. Le travail de Visher relatif aux sédiments de plage, n'envisage pas le comportement dynamique des sables très grossiers et des galets. Afin d'individualiser chaque type de transport, les courbes cumulatives ont été comparées avec un diagramme CM (Figure 12). Les échantillons de la zone du ressac constituent la portion NOP du diagramme CM, qui selon Passega (1967) est formée par les sédiments déplacés par traction ou roulement sur le fond. Deux sédiments de la zone du ressac sont représentés sur la Figure 7: le N°3 provient du Sud de la plage et le N°23 du Nord.

L'échantillon N°3 montre quatre populations élémentaires. La plus grossière possède un indice de Harris analogue à ceux trouvés pour les éléments tractés dans les sables fins. Il s'agit vraisemblablement des galets déplacés par traction sur le fond. Puis, viennent les deux ensembles en saltation A′ et A, enfin les particules en suspension. La caractéristique du milieu marin étant d'avoir de grandes variations d'énergie avec le temps, le segment A′ pourrait également être considéré comme un mélange d'éléments en traction et en saltation; mais dans ce cas, les points expérimentaux de cette population donneraient un alignement moins rigoureux. La courbe serait plus douce car les grains tendraient alors à s'équilibrer avec les nouvelles conditions.

L'échantillon N°23 ne présente que trois populations élémentaires. Les deux premières ont des indices de Harris comparables avec ceux des populations en saltation des sables fins. Mais, par comparaison avec le diagramme CM (Figure 12)—ce sable fait partie des quatre échantillons situés dans la portion NO—il est vraisemblable que l'ensemble le plus grossier correspond aux grains tractés, les deux autres étant les stocks en saltation. La côte rocheuse au Nord de la plage forme une barrière naturelle contre laquelle les galets viennent s'accumuler.

Ces deux échantillons tendent à montrer que l'énergie au Nord de la plage est sûrement plus élevée qu'au Sud, car les éléments tractés dans le sédiment N°3 se retrouvent en saltation dans le sable N°23.

La Suspension

Comme pour la traction, la suspension n'intéresse que 2 ou 3% du sédiment. Les tailles limites entre les population B et A s'étalent entre 0,20 mm et 0,16 mm, aussi bien sur la plage aérienne que sur la plage sous-marine. Ces valeurs diminuent pour les minéraux lourds.

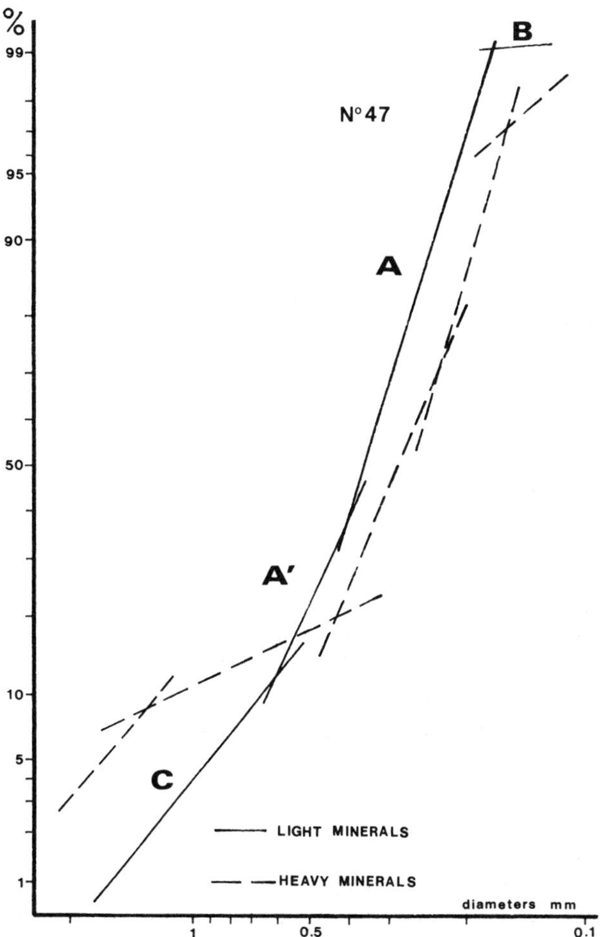

Figure 8. Courbes cumulatives d'échantillons marins riches en micas. Les minéraux lourds et légers ont la même représentation sur les Figures 9, 10, 11.

Les Courbes des Minéraux Lourds

Sur la zone aérienne, les minéraux lourds ont un comportement dynamique analogue à celui des minéraux légers, mais les tailles limites entre les différents ensembles sont plus faibles.

Les courbes cumulatives en ordonnée de probabilité des minéraux lourds contenus dans les sables marins sont plus délicates à interpréter; la présence de forts pourcentages de lamelles de micas, qui forment la fraction la plus grossière du sédiment, entraîne souvent des histogrammes bimodaux. Ces modes correspondent à deux populations de nature hydraulique distincte. Les lamelles de micas donnent le mode le plus grossier, le second représente les minéraux en grains. Cette distribution particulière des grains se traduit sur les courbes en ordonnée de probabilité par l'apparition de plusieurs segments de droite (Figure 8).

La Saltation et les Indices de Harris

Les calculs des indices de Harris pour les populations A′ et A, ont permis de définir trois faciès sédimentologiques à partir des courbes granulométriques cumulatives des minéraux légers.

Cas des Sediments dont l'Indice de Harris est plus Elevé pour le Stock A′ que pour le Stock A

Un certain nombre d'échantillons présente un ensemble A′ mieux trié que l'ensemble A (Figure 9). Ce type de sédiments se rencontre sur la haute-plage et à proximité de la première ligne de déferlement des vagues vers le large. Ce sont en général des sables dont les courbes granulométriques ont une asymétrie négative.

La présence d'éléments roulés sur le fond et l'entraînement vers le large d'une partie de la fraction fine par le courant de retour, confèrent au sable une asymétrie négative. Les grains projetés sur la haute-plage par le jet de rive ont un bon triage, car leurs dimensions sont compatibles avec l'énergie du courant, alors que ceux qui sont l'effet du courant de retour apparaissent mal triés. Ce mauvais triage est dû au fait qu'une fraction des particules susceptibles d'être ramenées vers le large reste coincée dans les espaces intergranulaires lors du dépôt instantané se produisant entre la fin du jet et la naissance du courant de retour.

Dans le secteur proche de la ligne de déferlement des vagues, la célérité du courant littoral diminue (Ingle, 1966) et la partie fine des sables ne peut plus être extraite.

Ces sables ont un faciès granulométrique essentiellement hyperbolique (zone aérienne) ou très hyperbolique (zone sous-marine). Ce dernier précise le degré d'évolution des sédiments, alors que le faciès sédimentologique rend compte de l'hydrodynamisme du milieu.

Ce type de courbe granulométrique se trouve dans des secteurs où les intensités des effets hydrodynamiques sont en général faibles: zones balayées par les courants de jet et de retour, zones proches du déferlement (breaker zone).

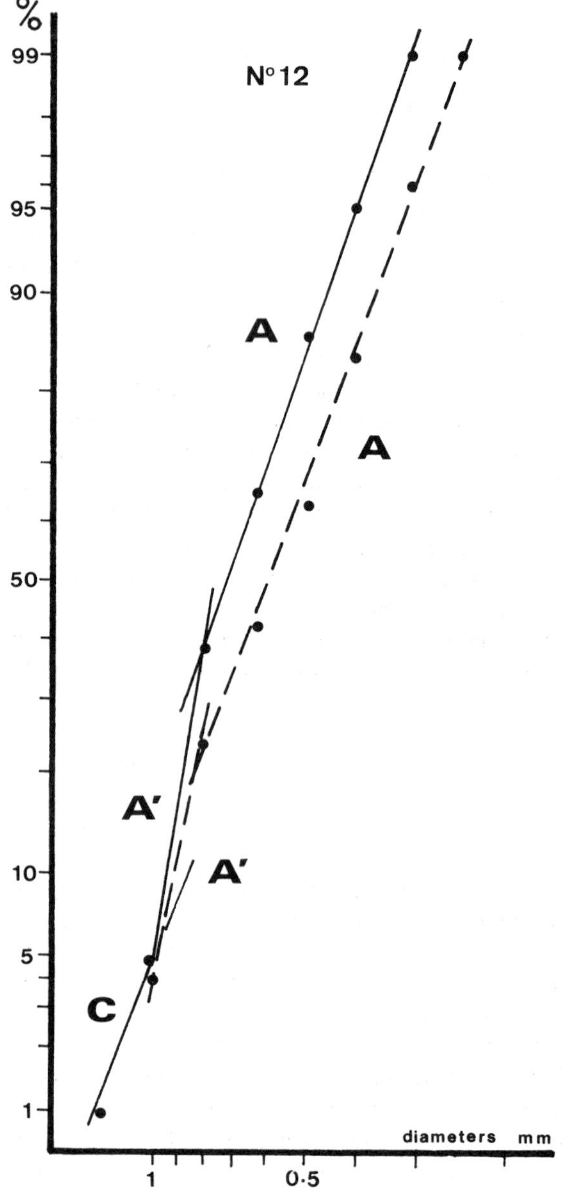

Figure 9. Courbes cumulatives d'échantillons présentant un indice de Harris plus élevé pour le stock A′ que pour le stock A.

Cas ou un Seul Stock en Saltation est Présent

Le jet de rive et le courant de retour ont dans ce cas une compétence équivalente: leurs actions agissent sur les mêmes classes granulométriques (Figure 10). Ce faciès n'est présent que dans la plage moyenne, lieu où se concentrent les minéraux lourds. Les faciès granulométriques des sables permettent de déduire le sens de déplacement des particules sur la grève: un sédiment évolue d'un faciès linéaire vers un faciès hyperbolique. Lors de la baisse du niveau de la mer après les tempêtes, les résultantes des circulations d'eau qui sévissent sur la plage aérienne, s'apparentent à celles qui opèrent dans les "croissants" de plage (Degiovanni, 1970). Elles sont de faible envergure et perdent leur énergie après de brefs parcours, les minéraux lourds se déposent par délestage au point de convergence des résultantes de ces effets dynamiques internes. Dans ces zones de concentrations, les sables ont un faciès granulométrique hyperbolique. En revanche, lorsque les sédiments sont en cours de migration, ils présentent un faciès parabolique.

Cas ou l'Indice de Harris est plus élevé pour A que pour A'

Ce faciès se rencontre dans la zone du ressac et dans le domaine marin (Figure 11). La population

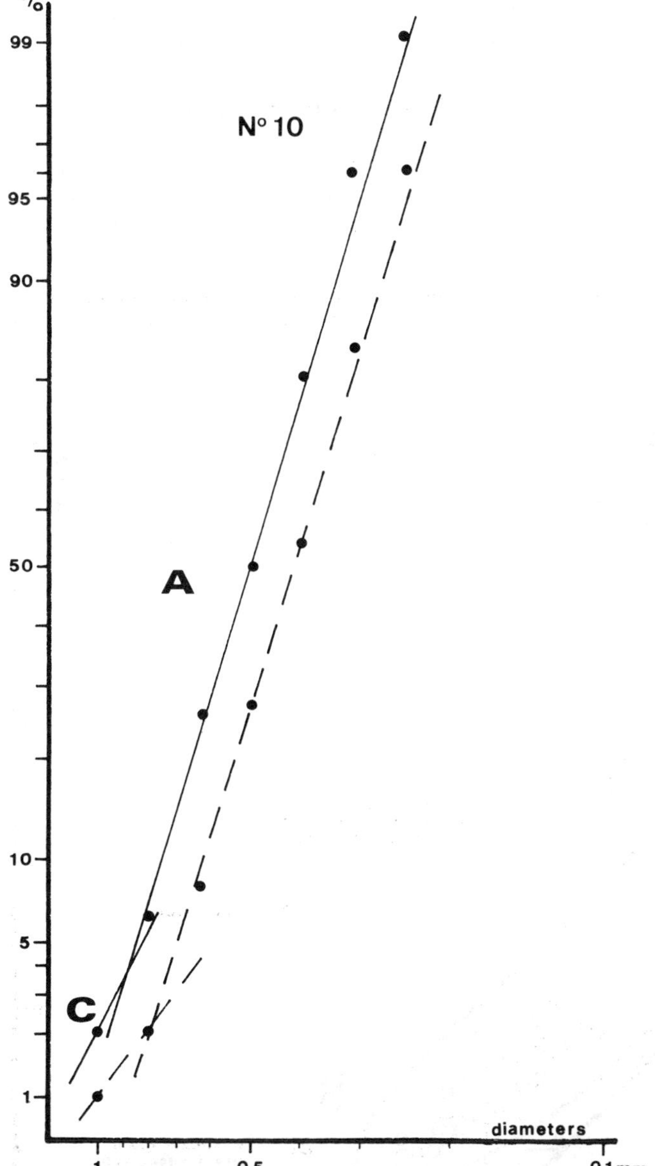

Figure 10. Courbes cumulatives d'échantillons n'ayant qu'un seul stock en saltation.

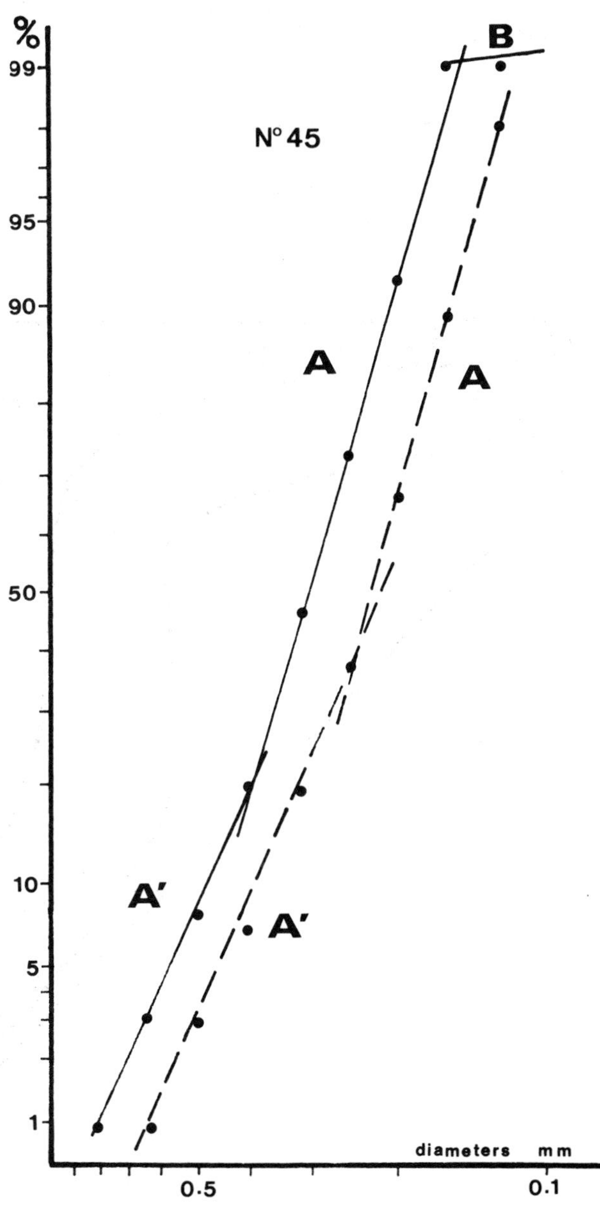

Figure 11. Courbes cumulatives d'échantillons dont le stock A est mieux trié que le stock A'.

A est mieux triée que la population A′. Sous l'eau, A′ a un indice de Harris moyen (20 à 40). Le courant littoral, efficace sur les grains déplacés initialement par le passage des vagues (Ingle, 1966), entraîne les plus petites particules de la fraction fine. L'absence des éléments les plus fins aboutit à un meilleur triage (ρ = 40–80) de l'ensemble A, et à une asymétrie négative des sables marins.

Il s'agit d'un faciès sédimentologique qui dénote des milieux dans lesquels les intensités des actions hydrodynamiques peuvent être soit faibles ou moyennes (avant-plage), soit très fortes (ressac). Ces conditions confèrent au sédiment total un triage moyen, et des faciès granulométriques très hyperboliques résultant de l'effet dû à la houle, qui conduit à une très bonne évolution du sable.

LES DIAGRAMMES CM

Les diagrammes de Passega (1957) permettent une définition plus générale du mode de dépôt des particules. Cet auteur décrit un diagramme CM complet de sédiments déposés par des courants tractifs, en indiquant que les diverses parties du diagramme représentent des particules mues par un type de transport bien déterminé (Passega, 1964, p. 831). Le diagramme est divisé par les points N, O, P, Q, R. Les portions NO et OP se composent d'échantillons tractés sur le fond par glissement ou roulement, les éléments étant épisodiquement mis en saltation. QR correspond aux sédiments transportés en suspension granoclassée.

Le graphique obtenu avec les sables de la plage de Pramousquier (Figure 12) montre qu'ils sont mis en place par des courants tractifs et que les particules de la suspension uniforme sont absentes. Trois modes de dépôt sont distingués: le roulement ou la traction (glissement et roulement sur le fond) et la suspension granoclassée.

La Traction et la Saltation

La taille Cs à partir de laquelle les grains commencent à être roulés sur le fond se situe à $-1{,}20\ \phi$. Le roulement et le glissement sont prédominants

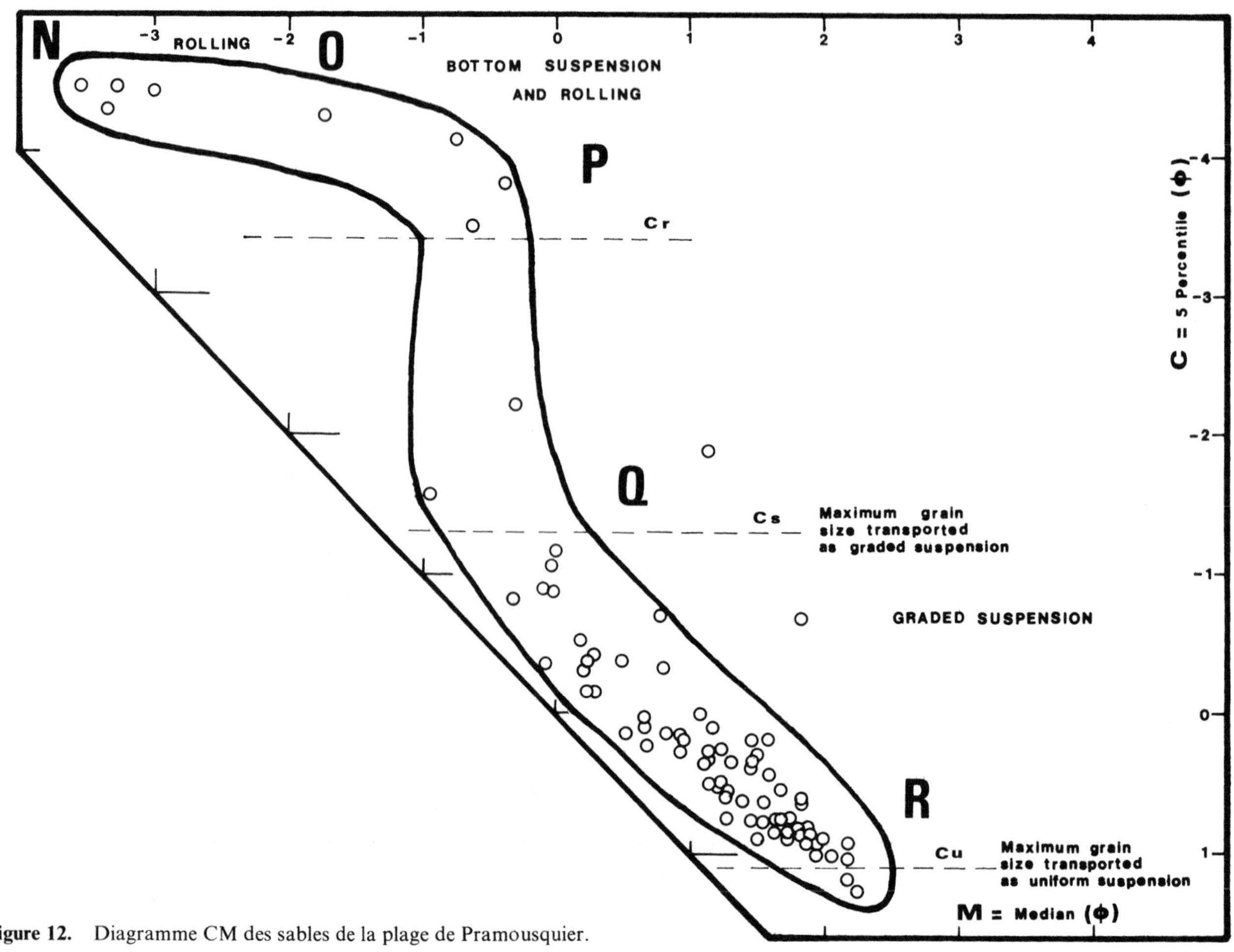

Figure 12. Diagramme CM des sables de la plage de Pramousquier.

dans la zone du ressac. Ce sont ces échantillons de sable grossier qui constituent le segment NOP. Le segment PQ contient peu de points car, selon Passega (1964), ce sont des particules dont les dimensions sont trop petites pour être roulées facilement sur le fond et trop grandes pour être transportées en suspension granoclassée. Seuls deux prélèvements tombent dans cette partie du diagramme: il s'agit d'un sable marin et d'un sable de la zone du ressac.

La Suspension Granoclassée

C'est le mode de transport principal du sédiment. Il est représenté par le segment QR qui est constitué par des sables des secteurs aériens et sous-marins (Figure 13). Comme le précise Passega (1957) une partie des grains de la suspension granoclassée peut être mue par saltation au moins temporairement. Les éléments transportés sous cette forme ont une taille comprise entre $-1{,}30\ \phi$ (Cs) et $1{,}10\ \phi$ (Cu). Les points sont concentrés vers la valeur minimale de la médiane et dispersés vers les valeurs grossières de C. Passega (1957) avait remarqué cette propriété sur les plages de l'Est de la Floride. Dans les plus faibles valeurs de la médiane, se placent les sédiments de l'avant-plage moyennement triés.

L'Indice d'Evolution n et les Diagrammes CM

Les calculs de l'indice *n* montrent que les sédiments de plage n'ont pas tous atteint le même degré d'évolution. Des diagrammes CM dans lesquels les échantillons ont été différenciés par les valeurs de *n* permet-

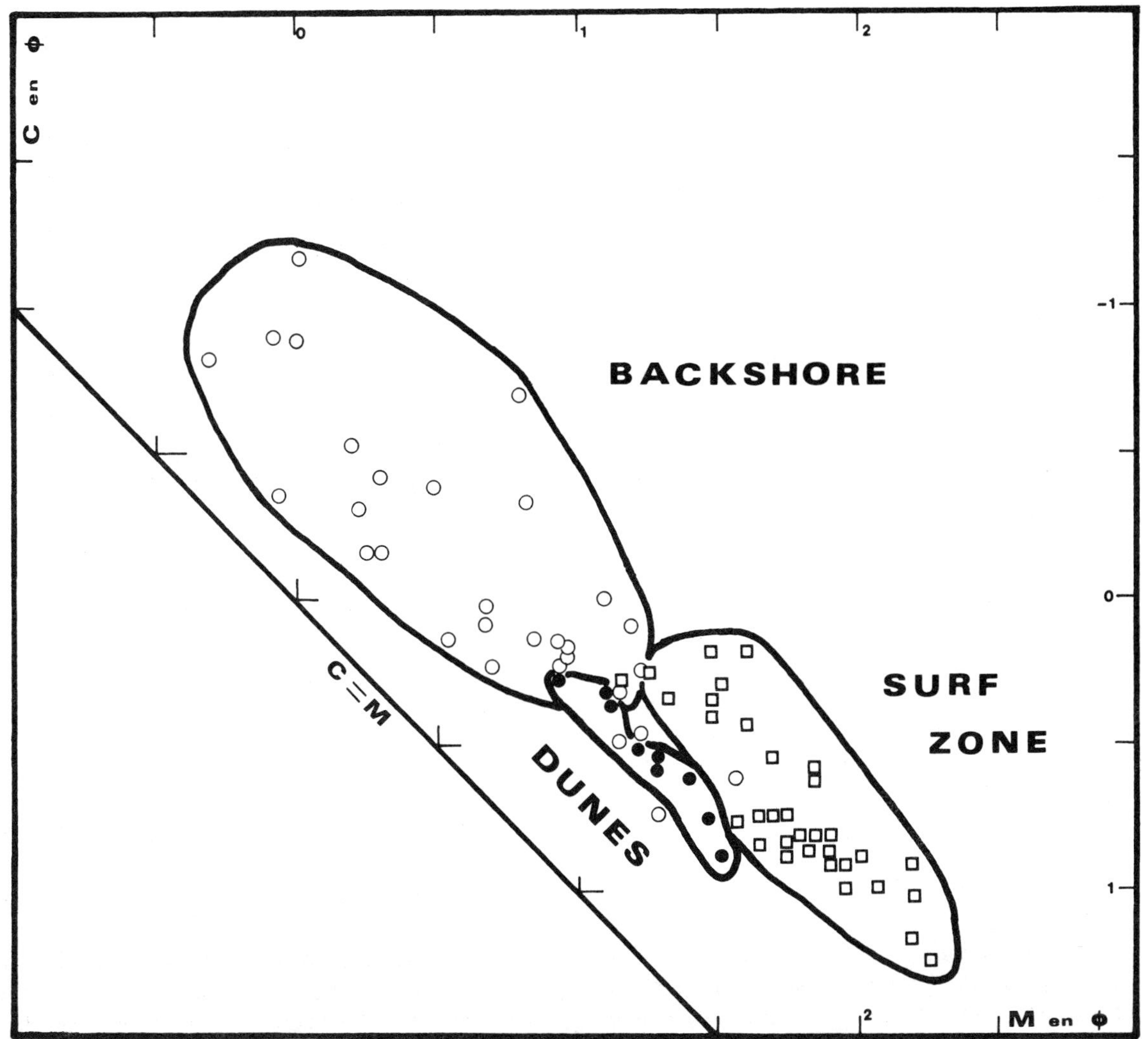

Figure 13. Répartition des échantillons des différentes zones de la plage qui composent la suspension granoclassée.

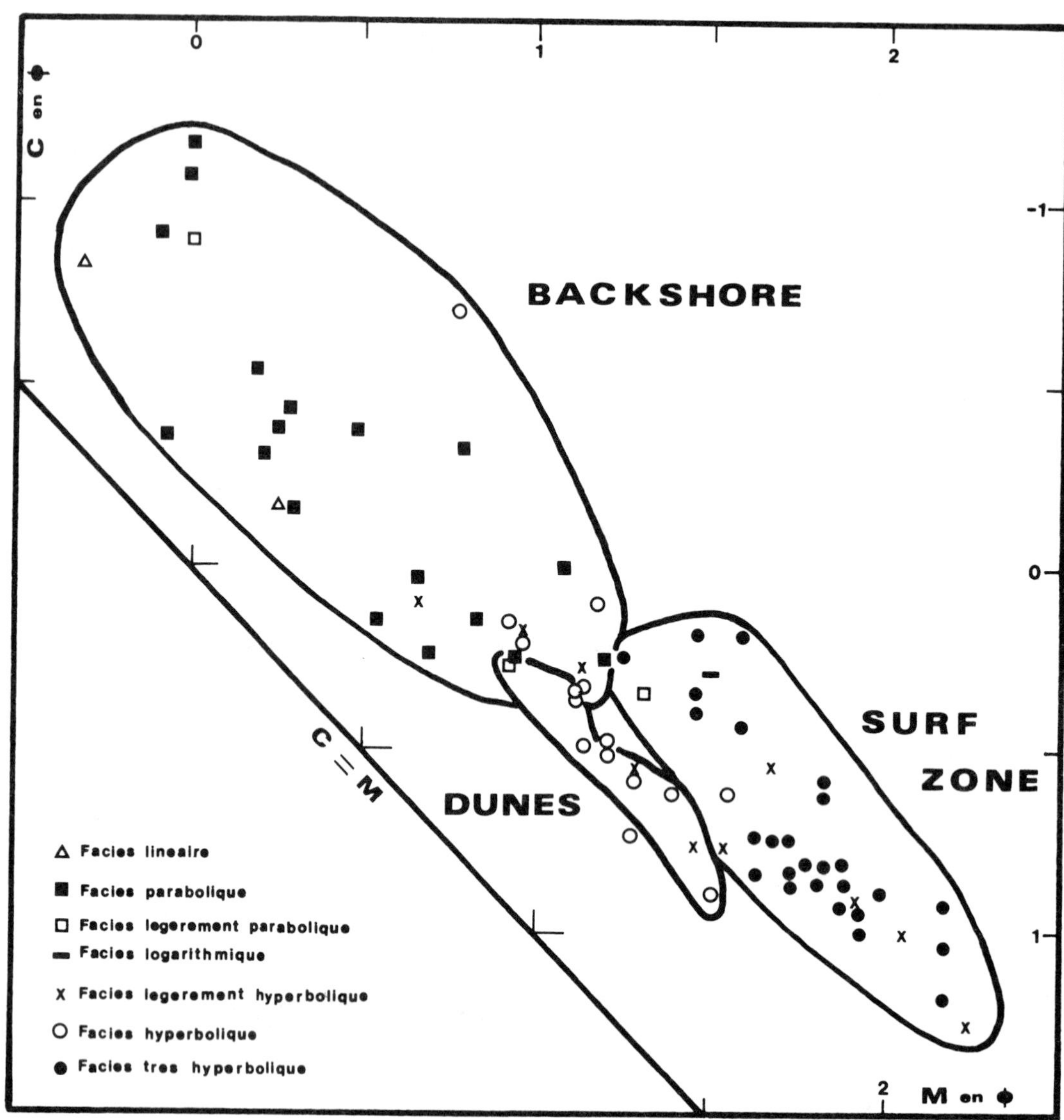

Figure 14. Evolution des faciès granulométriques dans les divers secteurs de la plage.

tent de comparer les zones de la plage (Figure 14). Ces diagrammes révèlent que l'évolution des sables s'améliore vers les petites dimensions de la médiane. La zone aérienne est constituée essentiellement de sédiments à faciès paraboliques, alors que les dunes, certains échantillons de la haute-plage et les sables sous-marins possèdent des faciès très évolués.

CONCLUSIONS

Les études granulométriques montrent que les sables de la plage de Pramousquier n'ont pas les mêmes propriétés dans les diverses zones de la plage. Ainsi, trois domaines essentiels se dégagent: la plage aérienne, la plage sous-marine et la zone du ressac.

La Plage Aérienne

Elle se compose de sables mieux triés et plus grossiers que ceux de la plage sous-marine; leur asymétrie est positive, sauf sur la haute-plage, et ils sont riches en minéraux lourds localement concentrés.

Les faciès granulométriques sont variés, mais ils indiquent des sables en cours d'évolution. Les courbes en ordonnée de probabilité permettent de mettre en évidence deux types de sédiments: l'un à éléments grossiers bien triés, l'autre avec une seule population en saltation.

C'est un secteur dans lequel les actions dynamiques sont de faible énergie. La plage aérienne est modelée par le jet de rive et le courant de retour, dont les effets sont enregistrés par les courbes granulométriques cumulatives.

La Plage Sous-Marine

Les sables sont fins et moyennement triés par un hydrodynamisme très complexe (jet de rive, courant de retour, courant littoral, mouvement périodique des vagues). L'indice de triage calculé n'est qu'un indice moyen entre les différents stocks liés aux actions dynamiques élémentaires. Les sédiments fins entraînés vers le large provoquent une asymétrie négative. L'indice d'évolution montre des faciès granulométriques très hyperboliques dénotant que les sables ont presque atteint l'état de stock sédimentaire parfait en fonction des conditions ambiantes.

Sous l'influence prépondérante du courant longitudinal, qui entraîne les particules les plus fines, la population en saltation A est bien triée dans les sables de l'avant-plage, mais près de la zone de déferlement, par suite de la diminution de la célérité du courant longitudinal l'ensemble A′ apparaît mieux trié que l'ensemble A. C'est le milieu où les conditions hydrodynamiques sont nombreuses, mais de faible énergie.

La Zone du Ressac

C'est la zone la plus hétérogène, les sédiments sont transportés par traction sur le fond, la saltation n'intervenant que dans des secteurs particuliers lorsque l'énergie des courants augmente. Les sables grossiers constituant les sédiments n'ont aucune unité granulométrique, de même que le faciès granulométrique. En revanche, le type des courbes cumulatives sont assez homogènes: la fraction fine a un meilleur triage que la fraction grossière.

Par suite des nombreuses actions hydrodynamiques, le dépôt et le transport des sables de plage sont très complexes. Les granulométries des sédiments permettent d'approcher le phénomène. Si à Pramousquier, l'influence du courant littoral n'est sensible sur les particules grossières qu'au niveau de la zone du ressac, l'effet de ce courant devient plus important sur une plage longue et rectiligne. Plusieurs auteurs, en utilisant des traceurs radioactifs ou fluorescents, ont montré le déplacement latéral des grains de sable dans l'avant-plage. Il semble que le brassage dû au passage de la houle n'ait pour rôle que la mise en suspension périodique des éléments. Ces derniers sont ensuite transportés par saltation granoclassée dans la direction des résultantes des courants tractifs agissant sur le fond. Les conditions hydrodynamiques confèrent aux sédiments, selon leur granulométrie et les zones de sédimentation, des faciès granulométriques différents qui définissent les secteurs où se concentrent les minéraux lourds.

En comparant avec d'autres études de sédimentation littorale effectuées autour de la Méditerranée, la sédimentation et la répartition des sables de plages méditerranéennes apparaissent analogues à celles des plages soumises aux courants de flux et de reflux engendrés par les marées.

REMERCIEMENTS

J'adresse mes remerciements les plus vifs au Dr. R. Passega pour avoir accepté de relire ce travail et de m'avoir fait partager son expérience à travers ses critiques et ses suggestions très constructives.

Je remercie également Mr. le Professeur J. J. Blanc et Mr. P. Weydert du Centre Universitaire de Marseille-Luminy (France), ainsi que Mr. le Professeur P. Collomb et toute l'équipe du Département des Sciences de la Terre d'Alger (Algérie) pour leurs intéressantes remarques lors de la préparation du manuscrit.

BIBLIOGRAPHIE

Arbey, F. 1961a. Etudes littorales sur la côte des Maures. I. Les croissants de plage. *Cahiers Océanographiques,* 13:381–396.

Arbey, F. 1961b. Etudes littorales sur la côte des Maures. II. Les sinuosités de plage. *Cahiers Océanographiques,* 13:569–576.

Arbey, F. 1961c. Etudes litorales sur la côte des Maures. III. Les sinuosités "rocheuses". *Cahiers Océanographiques,* 13:727 733.

Blanc, J. J. 1956a. *Sédimentologie Littorale et Sous-Marine en Provence Occidentale.* Thèse Faculté des Sciences, Paris, 140 p.

Blanc, J. J. 1956b. Etude minéralogique des sables littoraux du Cap Lardier au Cap Couronne (Provence). *Bulletin du Muséum d'Histoire Naturelle, Marseille,* 16:69–92.

Bradley, J. S. 1957. Differentiation of marine and subaerial sedimentary environments by volume percentage of heavy minerals. Mustang Island, Texas. *Journal of Sedimentary Petrology,* 27: 116–125.

Degiovanni, C. 1970. Les concentrations de minéraux lourds sur la plage de Pramousquier (Var), et leurs relations avec les indices d'évolution de A. Rivière. *Compte Rendus de l'Académie des Sciences, Paris,* 217 (D):28–30.

Duane, D. B. 1964. Significance of skewness in recent sediments, Western Pamlico Sound, North Carolina. *Journal of Sedimentary Petrology,* 34:864–874.

Folk, R. L. et W. C. Ward 1957. Brazos River Bar: a study on the significance of grain size parameters. *Journal of Sedimentary Petrology,* 27:3–26.

Friedman, G. M. 1962. On sorting, sorting coefficients, and the lognormality of the grain-size distribution of sandstones. *Journal of Geology,* 70:737–775.

Harris, S. A. 1959. The mechanical composition of some intertidal sand. *Journal of Sedimentary Petrology,* 29:412–424.

Ingle, J. C., Jr. 1966. *The Movement of Beach Sand.* Developments in Sedimentology. Elsevier Publishing Company, Amsterdam, 5:221 p.

Inman, D. L. 1952. Measures for describing the size distribution of sediments. *Journal of Sedimentary Petrology,* 22:125–145.

Moss, A. J. 1962. The physical nature of common sandy and pebbly deposits. Part I. *American Journal of Science,* 260:337–373.

Nesteroff, W. 1965. Les sédiments marins actuels de la région d'Antibes. *Annales de l'Institut Océanographique,* 43:1–136.

Ottmann, F. 1965. *Introduction à la Géologie Marine et Littorale*. Masson et Cie, Paris, 259 p.

Passega, R. 1957. Texture as characteristic of clastic deposition. *American Association of Petroleum Geologists Bulletin*, 41: 1952–1984.

Passega, R. 1964. Grain size representation by CM patterns as a geological tool. *Journal of Sedimentary Petrology*, 34:830–847.

Rivière, A. 1949a. Conditions d'existence et aspects des rides sous-marines littorales. Leur rôle dans la sédimentation côtière. *Comptes Rendus Sommaires de la Société Géologique de France*, 13:311–312.

Rivière, A. 1949b. Sur certains aspects de la morphologie littorale des plages et leurs interprétation. *Comptes Rendus de l'Académie des Sciences*, Paris, 229:940–942.

Rivière, A. 1952a. Sur la représentation graphique de la granulométrie des sédiments meubles. *Bulletin de la Société Géologique de France*, 2:145–154.

Rivière, A. 1952b. Expression analytique générale de la granulométrie des sédiments meubles. *Bulletin de la Société Géologique de France*, 2:155–167.

Rivière, A. 1955. Sur la radioactivité des sédiments actuels et récents de la côte méditerranéenne. Interprétation sédimentologique. *Bulletin de la Société Géologique de France*, 5:495–508.

Rivière, A. et S. Vernhet 1953. Sur la formation des croissants de plage et les mouvements des sédiments dans le profil. *Comptes Rendus de l'Académie des Sciences*, Paris, 237:659–661.

Rivière A. et S. Vernhet 1955. Sur l'interprétation des phénomènes d'agitation au rivage sous l'influence des vents de terre. *Comptes Rendus de l'Académie des Sciences*, Paris, 240:1451–1453.

Rivière, A. et S. Vernhet 1962. Les structures de plage à caractère périodique et leur rôle dans la morphologie littorale. In: *Colloque du Centre National de la Recherche Scientifique de Villefranche*, Mémoire: Océanographie Géologique et Géophysique de la Méditerranée Occidentale, 73–80.

Rivière, A. et S. Vernhet 1966. Etudes littorales. Contribution à l'étude des rivages du Golfe du Lion, signification sédimentologique des radioactivités naturelles. *Cahiers Océanographiques*, 18:857–899.

Rivière, A., F. Arbey et S. Vernhet 1961. Remarque sur l'évolution et l'origine des structures de plage à caractère périodique. *Comptes Rendus de l'Académie des Sciences*, Paris, 252:767–769.

Schiffman, A. 1965. Energy measurements in the swash-surf zone. *Limnology and Oceanography*, 10:255–260.

Steinberg, M. 1959. *Contribution à l'étude morphologique et sédimentologique de la côte des Maures (de Saint-Raphael à Sainte-Maxime)*. Thèse Laboratoire de Sédimentologie, Faculté des Sciences, Paris-Orsay.

Vernhet, S. 1953a. Sur un mode de cheminement littoral par migration lente de rides obliques ou perpendiculaires. *Comptes Rendus de l'Académie des Sciences*, Paris 237:1268–1270.

Vernhet, S. 1953b. Sur les transferts littoraux dans le Golfe du Lion. *Comptes Rendus de l'Académie des Sciences*, 237:1747–1748.

Vernhet, S. 1955. Influence de faibles courbures du littoral sur l'érosion des rivages sableux. Interprétation de l'allure sinueuse de caractère plus ou moins périodique du tracé des grandes plages. *Comptes Rendus de l'Académie des Sciences*, Paris, 240:336–338.

Visher, G. S. 1969. Grain size distribution and depositional processes. *Journal of Sedimentary Petrology*, 39:1074–1106.

Weydert, P. 1968. Etude sédimentologique du milieu glaciotorrentiel. Le torrent de Celse-Nièvre, Massif du Pelvoux (Hautes-Alpes). *Annales de la Faculté des Sciences, Marseille*, 40:193–213.

Répartition des Sédiments Marins près de Livourne (Mer Ligure)

Giuliano Fierro, Giovanni Battista Piacentino, et Graziella Ricciardi

Università di Genova, Gênes

RESUME

Dans cette étude nous avons considéré une zone qui appartient au plateau continental de la Toscane (Mer Ligure orientale) comprenant un haut-fond (Secche della Meloria) et dont une certaine partie reçoit les apports terrigènes de l'Arno. Les analyses ont mis en évidence une corrélation entre la radioactivité bêta, la teneur en calcaire et la composition granulométrique. Cette corrélation nous a permis de reconnaître quatre faciès de sédimentation: de haut-fond, d'embouchure, de plateau avec sédiments pélitiques et un faciès de transition. Les valeurs des paramètres qui concernent ces faciès sont comprises entre des limites assez bien définies.

ABSTRACT

Analyses pertaining to the quantity of carbonates, beta radioactivity and grain-size distribution have been made on samples collected on the continental shelf in the eastern Ligurian Sea off Tuscany. One portion of this area, a shoal (Secche della Meloria), receives some terrigenous sediments carried by the Arno River. Correlation of the lithologic parameters enables us to distinguish four sedimentary facies: shoal, river mouth, continental shelf with pelitic sediments, and a transitional facies. These facies are distinguished by well defined lithological parameters.

INTRODUCTION

Dans le cadre des études de base sur les écosystèmes des *Secche della Meloria* (Mer Ligure orientale) programmées par le Centre Universitaire de Biologie Marine et par l'Aquarium Communal de Livourne en relation avec le projet pour un Parc National Sous-Marin de la Meloria, on a entrepris des recherches sédimentologiques sur les fonds de la zone intéressée par ce projet. Il s'agit d'une zone de hauts-fonds s'étendant sur 30–40 km environ, de profondeur variable, d'un maximum de 20–25 m à un minimum de 2–3 m au sommet des hauts-fonds, qui se trouve à 6 km environ de la côte (Figure 1). Le véritable corps des hauts-fonds est caractérisé par la présence des herbières des Posidonies, succédant à des zones où les concrétions organogènes sont les plus nombreuses et à des poches de sédiments détritiques organogènes; au bord des hauts-fonds les sédiments sont surtout terrigènes avec des éléments de petite taille.

D'après les reconstructions paléogéographiques de Segre (1954), concernant le plateau continental près de Livourne, on peut penser que les Secche della Meloria représentent un témoin du paléorelief à présent submergé sur lequel se sont placé des organismes constructeurs qui l'ont recouvert. Les biogistes n'ont étudié et analysé que la surface de cette couverture, mais excluent le fait qu'il s'agisse de sédiment coralligène (Bacci *et al.*, 1969); à ce point, la nature géologique des terrains au-dessous des concrétions n'a pas encore été étudiée.

Dans une première phase des travaux, les sédiments ont été analysés du point de vue de leur texture et de leur composition en relation avec la morphologie

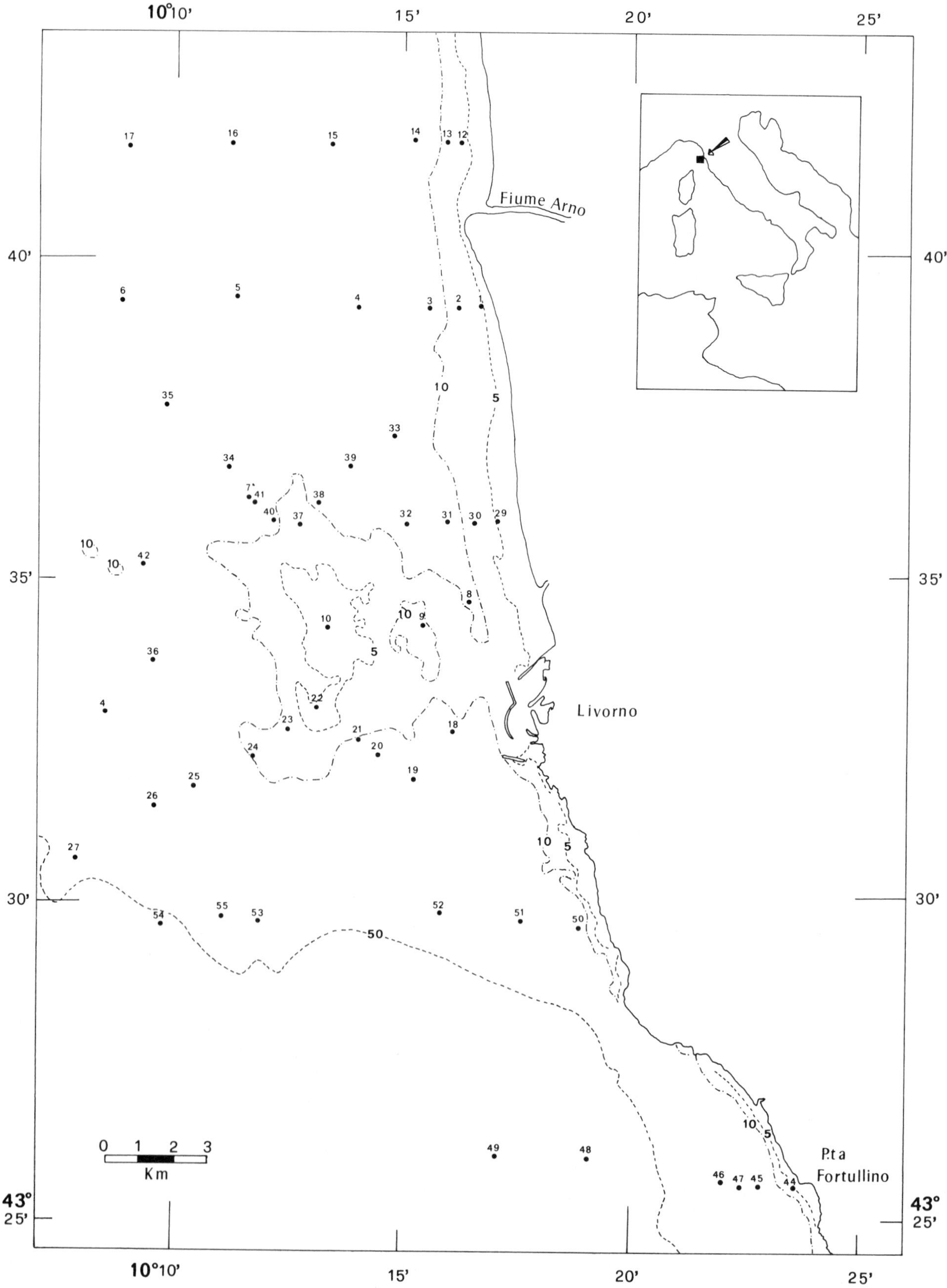

Figure 1. Position des échantillons (profondeurs en mètres).

du fond, la bathymétrie et les apports terrigènes (Fierro *et al.*, 1969). Dans une deuxième phase, sur les mêmes échantillons on a effectué des mesures de l'activité bêta totale et la détermination de la fraction inférieure à 62 microns. Ces analyses ont été ensuite comparées avec la teneur en carbonates et la valeur du diamètre moyen. Dans ce travail on a établi une corrélation entre les divers paramètres obtenus afin d'identifier plus sûrement des faciès de sédimentation dans le secteur des recherches. Le Tableau 1 donne les valeurs relatives des paramètres pris en considération.

Tableau 1. Valeurs relatives des paramètres des sédiments marins près de Livourne.

Station N°	bathy-métrie	β pc/g	carbonates %	M_z ϕ	Fraction $< 62\mu$ %
1	5	28,1	8,8	3,1	4,45
2	8	26,4	8,8	3,35	6,9
3	11	36,1	12,2	> 3,5	63,64
4	14	25,5	8,3	3,5	12,0
5	20	26,9	10,9	> 3,5	52,82
6	27	31,5	11,3	> 3,5	95,76
7	20	22,5	13,8	2,1	2,0
8	10	28,0	11,8	> 3,5	53,0
9	11	27,8	12,8	> 3,5	49,06
10	4	7,9	90,1	− 0,6	0,1
12	6	24,6	7,9	3,2	8,7
13	8	29,6	8,9	2,95	7,3
14	11	36,9	10,9	> 3,5	97,45
15	14	39,1	7,9	> 3,5	99,90
16	19	33,7	9,3	> 3,5	99,86
17	24	34,6	9,9	> 3,5	99,43
18	16	17,4	16,8	1,9	1,5
19	23	11,4	83,3	− 0,35	0,3
20	14	9,0	86,5	0,95	0,09
21	10	9,0	76,2	− 1,1	0,3
22	4	10,4	83,3	0,95	0,05
23	15	4,4	96,1	0,2	0,00
24	10	4,9	96,5	− 0,45	0,01
25	20	4,4	95,8	− 0,9	0,05
26	32	4,8	96,3	− 0,5	0,09
27	42	3,6	95,0	− 0,45	0,05
29	5	25,7	9,4	3,64	10,8
30	8	29,3	9,4	> 3,5	73,68
31	10	35,1	9,5	> 3,5	80,9
32	13	33,6	10,4	> 3,5	92,56
33	13	31,0	9,9	> 3,5	92,93
34	25	35,2	12,9	> 3,5	91,96
35	28	34,8	10,8	> 3,5	95,80
36	23	8,3	91,6	1,48	1,7
37	6	6,8	93,1	− 1,12	0,1
38	9	29,6	33,7	2,7	8,8
39	16	30,0	13,8	> 3,5	89,0
40	15	11,2	88,6	1,0	0,1
41	18	10,6	94,5	− 0,75	0,2
42	14	5,0	96,8	− 0,1	0,02
43	36	6,0	95,9	− 0,37	0,07
44	6	12,2	43,6	− 1,3	0,3
45	15	7,0	32,7	− 0,2	0,2
46	38	21,6	48,0	− 3,5	—
47	25	13,0	17,8	3,0	4,7
48	60	31,4	28,7	> 3,5	99,34
49	72	38,4	19,8	> 3,5	99,80
50	11	4,0	91,1	0,13	0,08
51	36	28,0	19,8	> 3,5	81,86
52	40	35,7	17,3	> 3,5	98,92
54	52	16,0	51,0	> 3,5	36,36
55	25	9,4	96,0	− 0,17	0,1

Dans le même secteur, en 1962, on avait déjà effectué des mesures d'activités β et γ sur les sédiments (Argiero *et al.*, 1965). Cependant il n'a pas été possible de comparer nos mesures avec celles-ci parce que les échantillons analysés par les auteurs susnommés avaient été recueillis avec une drague, tandis que les échantillons que nous avons examinés ont été recueillis par un plongeur qui a éffectué un échantillonage seulement dans la partie la plus superficielle du sédiment.

RADIOACTIVITE BETA ET TENEUR EN CARBONATES

Les valeurs de la radioactivité sont comprises entre 3,2 et 39,1 pc/g; les pourcentages des carbonates sont compris entre 7,9 et 96,8% CO_3^{--}. On peut classer les échantillons en deux groupes, suivant le graphique de la Figure 2: (a) pourcentage des carbonates compris entre 0 et 35 % avec des valeurs de radioactivité qui varient de 22 à 40 pc/g; et (b) pourcentage des carbonates compris entre 75 et 100 % avec des valeurs de radioactivité qui varient de 3 à 12 pc/g.

Six échantillons seulement s'éloignent de ces valeurs; on peut remarquer que le groupe de 44 à 47 a une distribution géographique bien définie et un faciès bien déterminé qui correspond aussi à un certain milieu.

Dans la Figure 3 on a mis en évidence la répartition latérale des valeurs de la radioactivité, groupées de la manière suivante: 0–10 pc/g; 10–20 pc/g; 20–30 pc/g; plus de 30 pc/g. On observe tout de suite que les valeurs de la radioactivité ne dépendent pas de la bathymétrie, mais seulement du type de sédiment: en effet les valeurs les plus faibles sont concentrées dans la zone du haut-fond tandis qu'au nord et au sud de celui-ci on trouve immédiatement des valeurs beaucoup plus élevées, qu'on peut retrouver aussi à proximité de la côte, lorsque celle ci n'est pas modifiée par des sédiments organogènes comme devant la Punta Fortullino. Pour la zone côtière en face de l'embouchure de l'Arno, les valeurs de la radioactivité sont de l'ordre de 25–28 pc/g. De telles valeurs sont en accord avec celles rencontrées dans les apports solides de l'Arno, recueillis près de son embouchure (25–29 pc/g.)

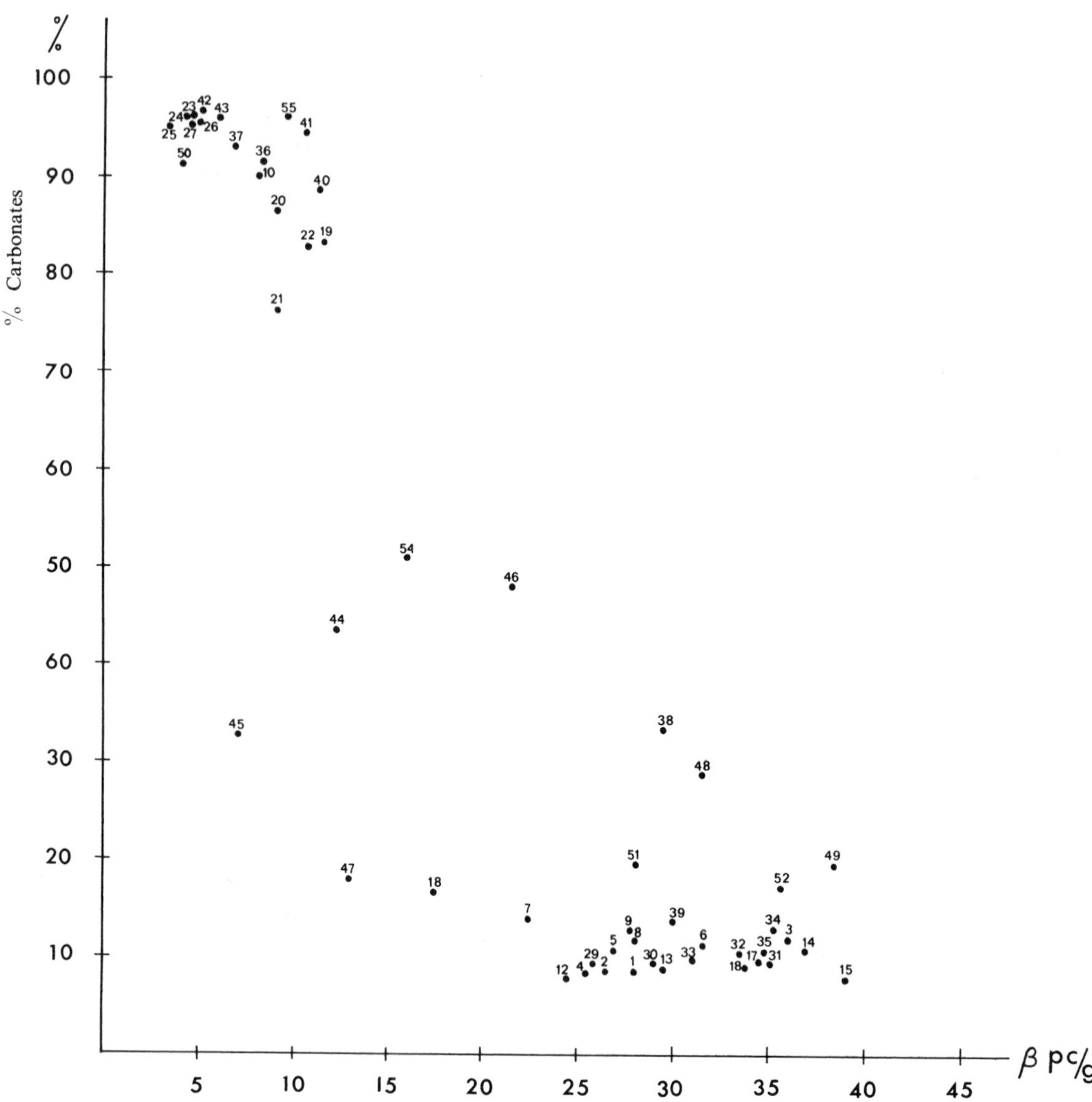

Figure 2. Rapport entre le pourcentage des carbonates et la radioactivité bêta des sédiments.

RADIOACTIVITE BETA ET DIAMETRE MOYEN

En comparant la carte de la Figure 3 à celle de la répartition granulométrique des sédiments (Figure 4) de la même zone on trouve une bonne correspondance entre les variations de la radioactivité et les variations du diamètre moyen (Bellaiche *et al.*, 1966). En particulier, les plus fortes valeurs de la radioactivité correspondent aux *vases*, les plus basses aux *sables très grossiers* ou aux *graviers très fins*. On a calculé les valeurs du diamètre moyen qui ont été considérées par rapport à la radioactivité bêta (Figure 5); dans ce cas on peut aussi distinguer deux groupements très nets: (a) $M_z < 1{,}5\phi$, radioactivité < 12 pc/g, et (b) $M_z > 1{,}5\phi$, radioactivité > 22 pc/g. En dehors de ces groupements on peut noter que les deux échantillons, no. 18 et no. 47, présentent une anomalie du rapport bêta/carbonates.

Il est évident que la radioactivité bêta est associée aux fractions les plus fines des sédiments qui, en général dans la zone que nous avons étudiée, sont pauvres en carbonates, comme on peut le voir sur la Figure 6, dans laquelle on a considéré le diamètre moyen par rapport au pourcentage des carbonates (Blanc, 1964).

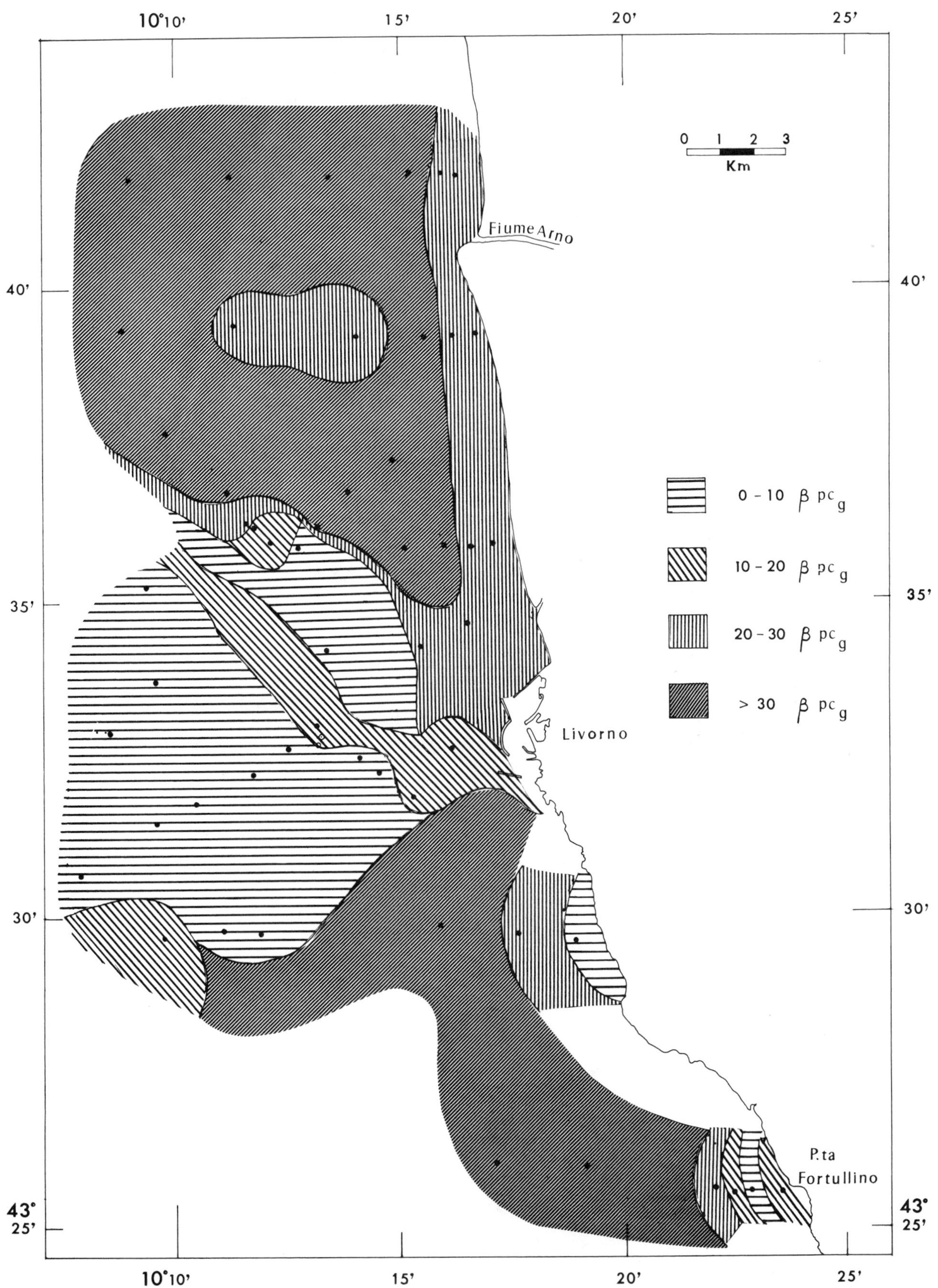

Figure 3. Carte de la répartition de la radioactivité bêta des sédiments.

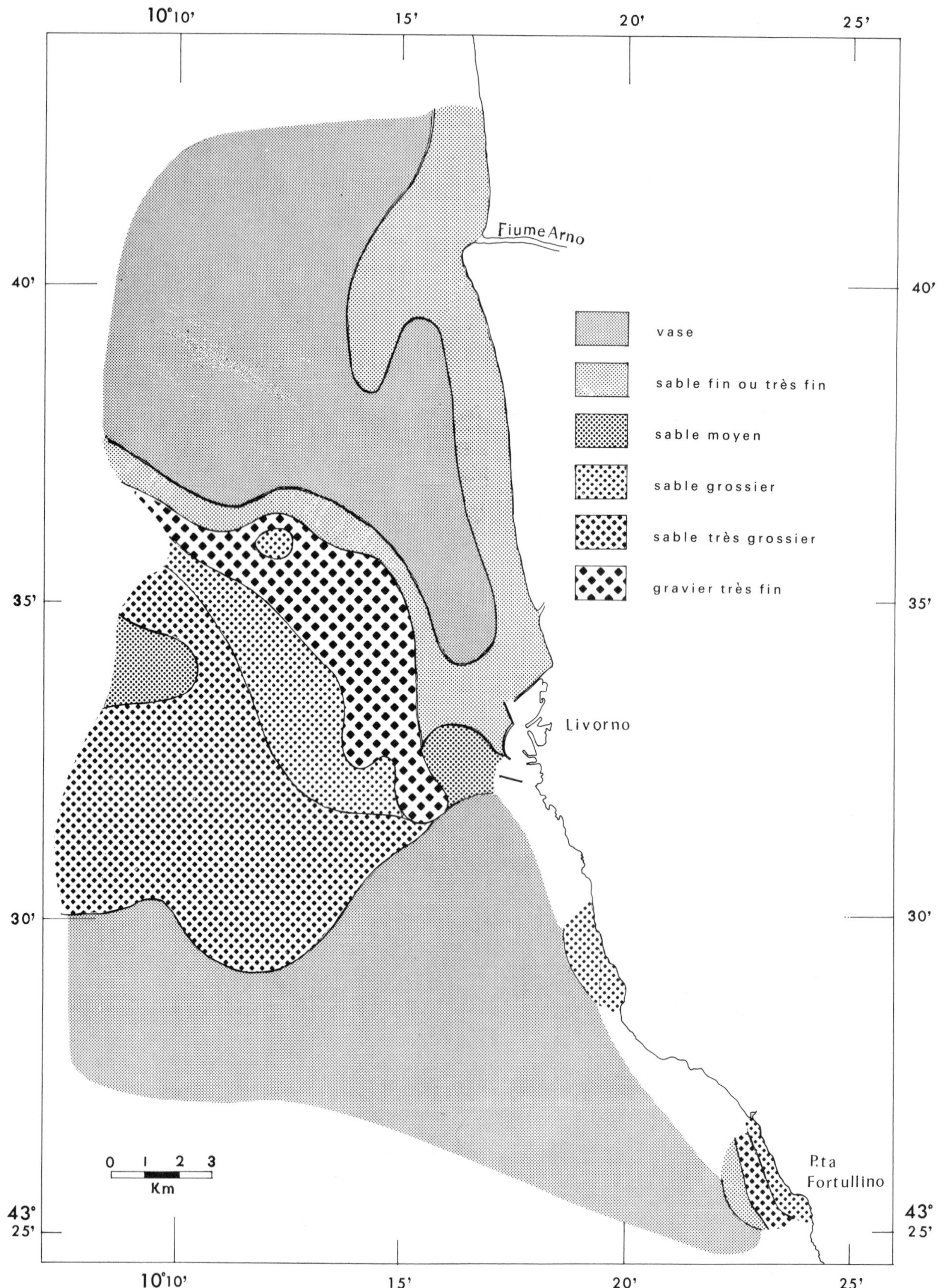

Figure 4. Carte de la répartition granulométrique des sédiments.

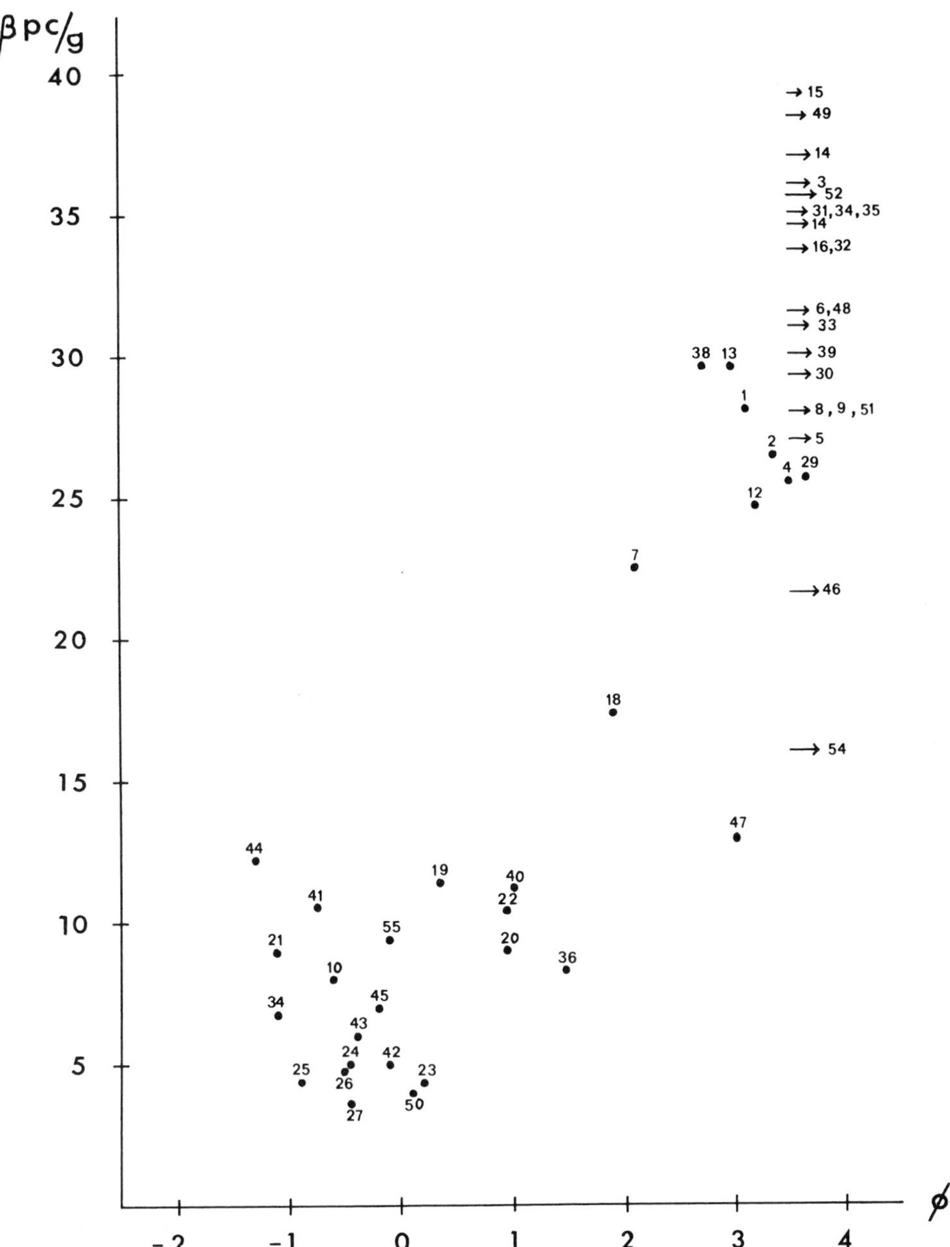

Figure 5. Rapport entre la radioactivité bêta et le diamètre moyen. (Les positions des échantillons marqués d'une flèche correspondent à des valeurs exactes de bêta et seulement à des valeurs supérieures à 3,5 ϕ pour le diamètre moyen).

RADIOACTIVITE BETA ET FRACTION INFERIEURE A 62 MICRONS

Dans la Figure 7 on a étudié les relations qui existent entre le pourcentage de la fraction inférieure à 62 microns présente dans les sédiments et la radioactivité bêta. Les sédiments organogènes restent groupés dans une zone très restreinte du graphique. En effet de tels sédiments sont caractérisés par les valeurs de bêta inférieures à 15 pc/g et leur fraction inférieure à 62 microns est plus faible que 1,5%.

Les autres sédiments dans lesquels le composant terrigène est supérieur, apparaissent suivant un alignement rectiligne. On note que la valeur de la radioactivité bêta présente une faible croissance en même temps que la fraction fine du sédiment devient plus importante.

Dans ce groupe même on peut encore retrouver trois sous-groupes: (a) bêta compris entre 15 et 30 pc/g,

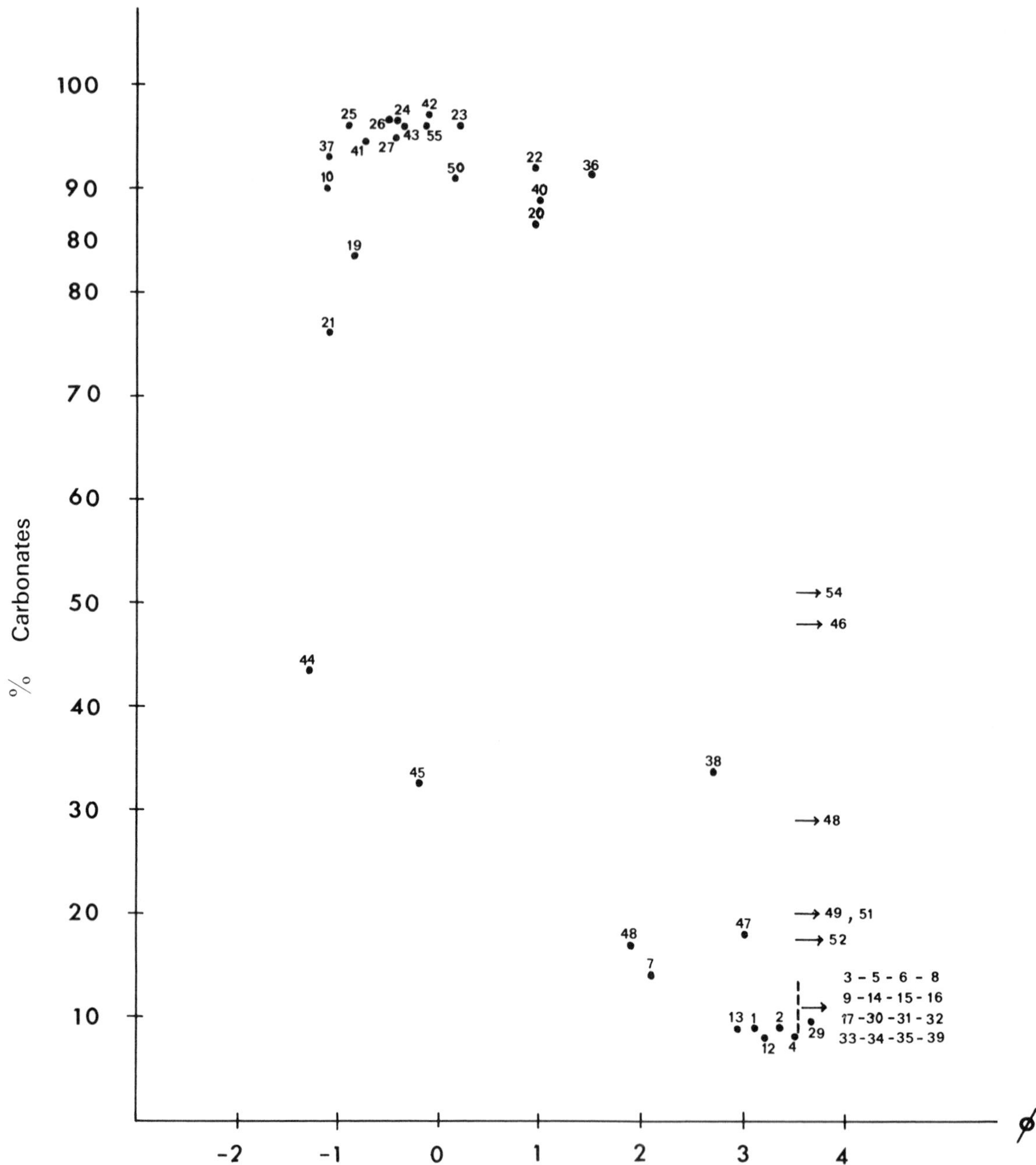

Figure 6. Rapport entre le pourcentage des carbonates et le diamètre moyen. (Pour les échantillons marqués d'une flèche les positions correspondent à des valeurs exactes des carbonates et seulement à des valeurs supérieures à 3,5 ϕ pour le diamètre moyen).

fraction < 62 microns inférieure à 12%; (b) bêta supérieur à 15 et fraction < 62 microns comprise entre 36 et 82%; et (c) bêta supérieur à 30 et fraction < 62 microns au dessus de 90%.

Comme nous le verrons par la suite, ces trois sous-groupes correspondent à des milieux de sédimentation bien définis et bien localisés dans la zone qui fait l'objet de notre étude. Afin de localiser les centres émetteurs des radiations bêta mesurées sur les échantillons, on a comparé la teneur en minéraux lourds dans les fractions 420–42 microns, les valeurs de bêta net, calculées après le titrage du potassium, et les pourcentages de fraction pélitique, selon le procédé suivi dans un travail précedent (Fierro *et al.*, 1965).

D'après ce qui ressort de cette comparaison, la radioactivité bêta net est certainement liée à la fraction fine. L'activité due aux minéraux lourds, du moins en ce qui concerne la zone que nous avons

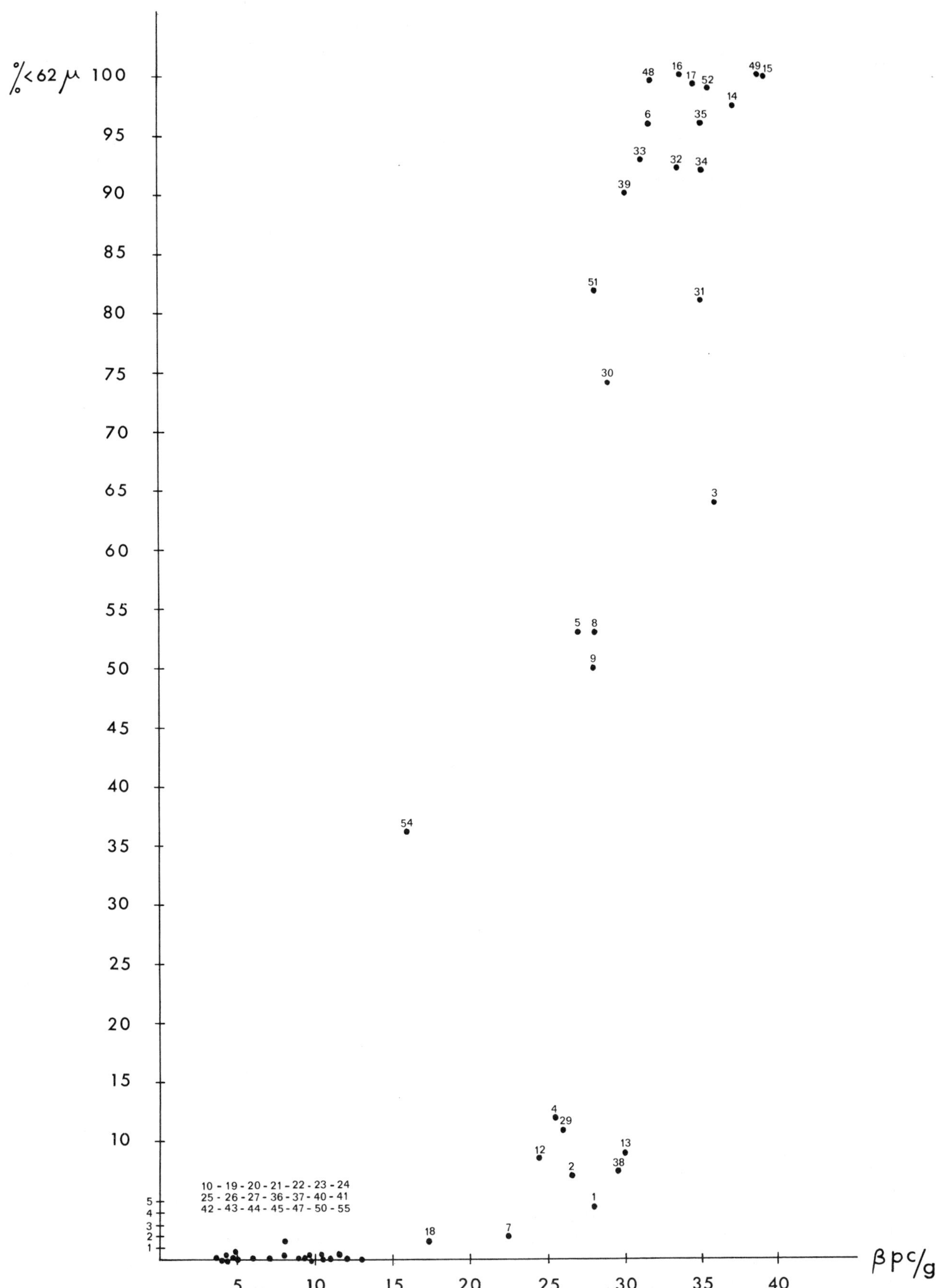

Figure 7. Rapport entre le pourcentage de la fraction de sédiment inférieure à 62 microns, et la radioactivité bêta.

étudiée, ne semble pas déterminer des variations remarquables même dans les sédiments qui présentent une faible teneur en fraction fine. Ces donnés ne s'accordent pas avec ce qui avait été trouvé par Rivière et Vernhet (1966).

IDENTIFICATION DES FACIES DE SEDIMENTATION

Les groupements des échantillons définis précédemment ont été rapportés sur une carte (Figure 8). Ces groupes sont distribués dans des zones topographiquement bien définies; il est possible de retrouver sur la carte des zones avec des faciès sédimentaires différents et avec des passages de l'une à l'autre.

Faciès des "Secche della Meloria": elle comprend les échantillons numéros 10, 19–27, 36, 37, 40–43 et 55. Il s'agit de sédiments organogènes caractérisés par une forte teneur en carbonates, par de faibles valeurs de la radioactivité et une fraction fine peu importante; le diamètre moyen est inférieur à 1,5 ϕ. A ce groupe on peut ajouter l'échantillon numéro 50 qui, tout en étant éloigné des "Secche", a les mêmes caractéristiques et provient d'un fond morphologiquement semblable.

Dans les graphiques diamètre moyen/bêta et bêta/fraction fine, avec le groupe indiqué ci-dessus nous retrouvons d'autres sédiments (numéros 44–47) qui semblent présenter les caractéristiques de la zone des "Secche"; mais un examen des autres graphiques nous conduit à reconnaître un faciès caractéristique, celui de la zone devant Punta Fortullino; elle est caractérisée par des valeurs remarquablement plus basses en ce qui concerne le pourcentage en carbonates, par des diamètres moyens supérieurs à 1,5 ϕ, tandis que les autres paramètres restent dans les limites sus-nommées.

Un autre faciès est bien défini par les valeurs des paramètres dans les échantillons numéros 1, 2, 4, 7, 12, 13 et 29 pour lesquels le pourcentage des carbonates est inférieur à 20%, la radioactivité est supérieure à 22 pc/g, la fraction fine est inférieure à 12% et les dimensions du diamètre moyen sont supérieures à 1,5 ϕ. Ces échantillons se trouvent spécialement sur la zone côtière devant l'Arno et se prolongent dans le canal entre la côte et le haut-fond avec des pointes en direction du bord septentrional de ce haut-fond.

Il n'est pas possible de replacer les échantillons numéros 18–38 et 54 qui, se trouvant sur les marges nord et sud du haut-fond, réunissent les caractéristiques des faciès voisins. Il faut aussi tenir compte du fait que l'échantillon no. 18 provient d'une zone devant le port de Livourne et qu'il pourrait renfermer des apports étrangers à la sédimentation naturelle.

Le groupe des échantillons restants limite deux zones au large du haut-fond, au nord et au sud. Il s'agit de sédiments pélitiques dont la teneur en carbonates ne dépasse pas 30% avec des valeurs de radioactivité supérieures à 22 pc/g; le diamètre moyen est supérieur à 3,5ϕ et la fraction fine est présente en grande abondance. Sur la base de la variation de la fraction inférieure à 62 microns, ce groupe de sédiments peut se subdiviser en deux sous-groupes: (a) les échantillons numéros 3, 5, 8, 9, 30, 31 et 51 dans lesquels la fraction fine varie de 35 à 85% et qui se rattachent à la zone côtière; ils sont encore influencés par les apports sablonneux de l'embouchure de l'Arno. (b) Les échantillons numéros 6, 14–17, 32–35, 39, 48, 49, et 52 dans lesquels la fraction fine dépasse 90% et qui sont distribués en deux zones au nord et au sud du haut-fond avec des sédimentations caractéristiques et semblables.

CONCLUSIONS

D'après les résultats de l'analyse des sédiments provenant de la zone limitée par l'embouchure de l'Arno et par la Punta Fortullino, qui comprend les "Secche della Meloria", on peut mettre en évidence des corrélations entre les valeurs de radioactivité bêta, du diamètre moyen, du pourcentage en carbonates et du pourcentage de la fraction inférieure à 62 microns.

Ces corrélations nous permettent d'individualiser des faciès de sédimentation pour lesquels les valeurs des paramètres sus-nommés oscillent autour de cer taines valeurs bien définies. En particulier on peut distinguer:

1. Des sédiments calcaires-organogènes à basse radioactivité contenant une faible fraction fine: ces sédiments caractérisent trois zones correspondantes aux véritables "Secche della Meloria", au littoral au Sud de la ville de Livourne et au littoral de la Punta Fortullino;
2. Des sédiments terrigènes sablonneux à haute radioactivité, peu calcaires, de diamètre moyen supérieur à 1,5 ϕ: la zone concerne l'embouchure de l'Arno et la bande côtière devant elle;
3. Des sédiments à faciès de transition; et
4. Des sédiments terrigènes à haute radioactivité contenant une abondante fraction fine, peu calcaires, distribués en deux zones au Nord et au Sud des "Secche della Meloria".

REMERCIEMENTS

Les mesures radiométriques ont été exécutées par les Laboratoires de Radioprotéction du C.A.M.E.N. (Pisa), dont nous remercions les techniciens.

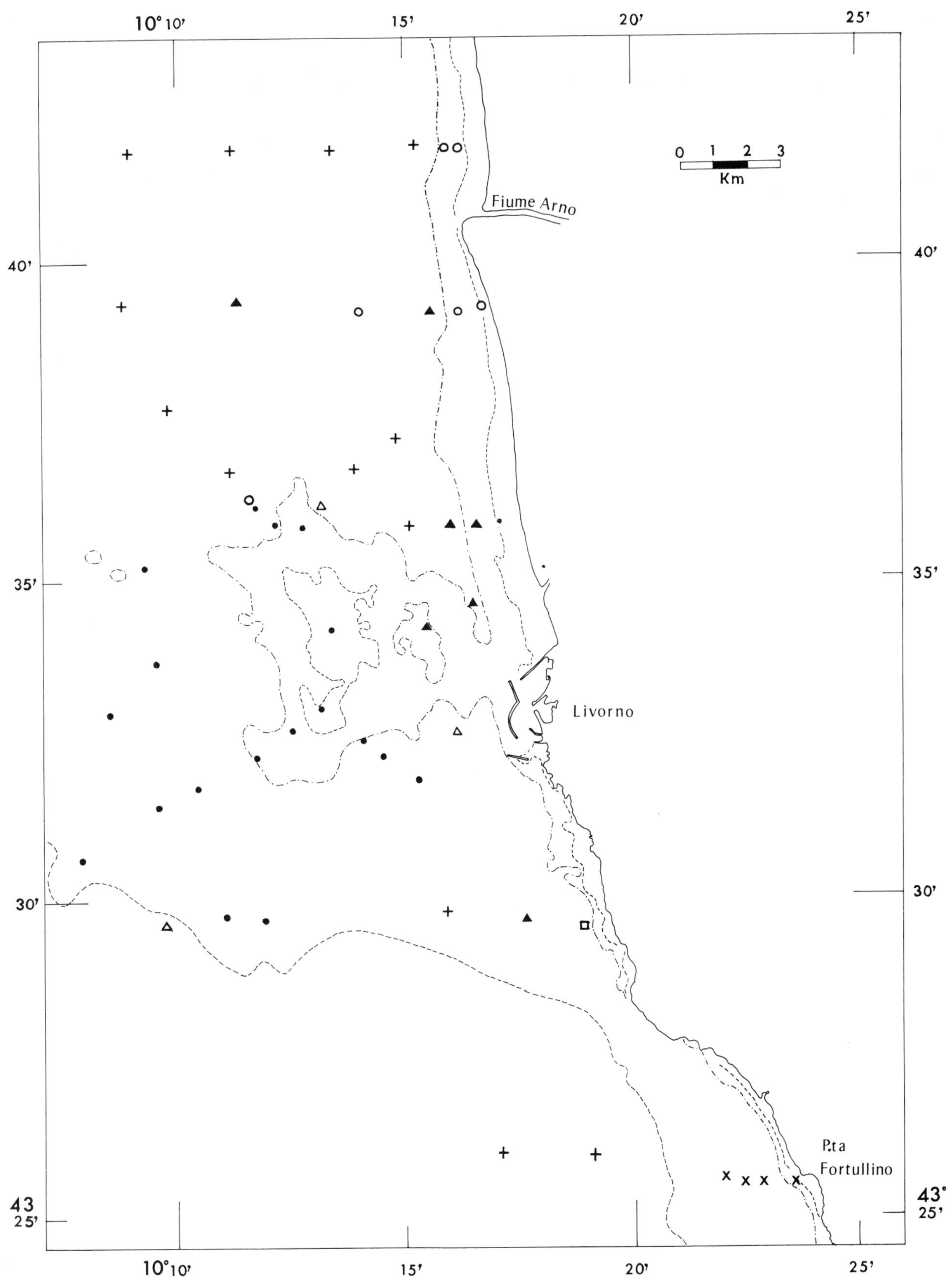

Figure 8. Carte de la répartition des sédiments d'après les rapports entre les carbonates, la radioactivité bêta, le diamètre moyen et la fraction fine. (a) Sédiments calcaires organogènes: [●] faciès des "Secche della Meloria", [□] faciès littoral au sud de Livourne, [×] faciès littoral près de Punta Fortullino; (b) sédiments terrigènes sablonneux: [○] faciès littoral près de l'embouchure de l'Arno; (c) sédiments qui renferment à la fois des apports terrigènes et organogènes: [△] faciès de transition; (d) sédiments terrigènes du large: [+] faciès des sédiments vaseux, [▲] faciès des sédiments vaseux avec une fraction sablonneuse notable.

BIBLIOGRAPHIE

Argiero L., S. Manfredini et G. Palmas 1965. Studio delle caratteristiche radioattive, biologiche ed idrografiche del Mar Tirreno. *Centro Applicazioni Militari Energia Nucleare*, S. Piero a Grado (Pisa), 17 p.

Bacci G., G. Badino, L. Lodi et L. Rossi 1969. Biologia delle Secche della Meloria. I° Prime ricerche e problemi di conservazione e di ripopolamento. *Bolletino Pesca Piscicultura e Idrobiologia*, 24 (1): 5–31.

Bellaiche G., J. L. Cheminee, J. M. Martin et G. Pautot 1966. Applications des mesures de bruits de fond gamma à l'étude de l'origine et de la dynamique des sables sous-marins. *Comptes Rendus Sommaire des Séances de la Société Géologique de France*, (10): 388–389.

Blanc J.J. 1964. Campagne de la Calypso en Méditerranée Nord-Orientale 1960. In: *Recherches Géologiques et Sédimentologiques*, Masson et Cie. Paris, 219–270.

Fierro G., F. Bedarida et G. Ricciardi 1965. Misure di radioattività nei sedimenti del Mar Tirreno, *Atti dell'Accademia Ligure di Scienze e Lettere*, 22: 222–232.

Fierro G., F. Miglietta et G.B. Piacentino 1969. Biologia delle Secche della Meloria. III° I sedimenti superficiali delle Secche e delle aree limitrofe dalla foce dell'Arno a Punta Fortullino. *Bollettino Pesca Piscicultura e Idrobiologia*, 24 (2): 115–149.

Rivière M.A. et S. Vernhet 1966. Signification sédimentologique de la radioactivité naturelle des plages du Golfe du Lion. *Comptes Rendus des Séances de l'Académie des Sciences*, Paris, 262: 440–443.

Segre, A.G. 1954. Commento alla carta submarina del Foglio 111° (Livorno) della Carta Geologica d'Italia. *Bollettino Servizio Geologico d'Italia*, 55: 827 p.

Contribution to Recent Sedimentation on the Shelves of the Southern Adriatic, Ionian, and Syrtis Seas

Frank Fabricius and Paul Schmidt-Thomé

Technische Universität, Munich

ABSTRACT

Recent sedimentologic investigations in four separate areas of the middle part of the Mediterranean Sea between the Southern Adriatic and the Syrtis Seas are described: (1) the Gulf of Manfredonia, Italy, in the Southern Adriatic Sea; (2) the shelf off Otranto, Italy, in the Strait of Otranto; (3) the island shelf between Cephalonia and Zante, Greece, in the Ionian Sea; and (4) the shoals off Djerba Island, Tunisia, in the Gulf of Gabès.

Each of these areas has its own oceanographic, climatic and geologic setting, resulting in four distinct types of sedimentation. We present four models:

1. Clastic sedimentation in a subsiding area;
2. resedimentation of relict sediments on a stable shelf;
3. an uplifted island shelf lacking clastic and biogenic sedimentation;
4. biogenic carbonate sedimentation combined with eolian clastic sedimentation.

Late Pleistocene and Early Holocene coastal depositional models of types (3) and (4) are compared. These sediment assemblages, although different in their bio-and litho-facies, are indicative of a warmer climate. The fossil terraces are very different from their present equivalents.

RESUME

On décrit des études sédimentologiques récentes effectuées dans quatre régions distinctes de la partie centrale de la Méditerranée, entre l'Adriatique du Sud et la Mer des Syrtes: (1) Golfe de Manfrédonie, Italie, Adriatique sud, (2) plateau continental au large d'Otrante, Italie, détroit d'Otrante, (3) plateau continental entre la Céphalonie et Zante, Grèce, Mer Ionienne, (4) plateforme au large de l'Ile de Djerba, Tunisie, Golfe de Gabès.

Chacune de ces régions possède ses propres caractères océanographiques, climatiques et géologiques ce qui conduit à quatre types de sédimentation:

1. Sédimentation clastique dans une zone de subsidence;
2. reprise de sédiments anciens sur une plateforme stable;
3. absence de sédimentation biogénique et clastique sur un plateau continental surélevé;
4. sédimentation calcaire biogénique combinée avec sédimentation clastique éolienne.

Les dépôts côtiers du Pléistocène supérieur et Holocène inférieur de types (3) et (4) sont comparés; bien que différents dans leurs bio- et litho-faciès, tous deux sont caractérisques d'un climat plus chaud. Les terraces anciennes sont très différentes de leurs équivalents actuels.

INTRODUCTION

Marine geologists and students of Mesozoic Tethyan sedimentation are concerned with depositional and diagenetic processes in the Mediterranean Sea. One approach to this complex problem is to study the local factors of sedimentation in selected areas. With this end in view we initiated this study in the central part of the Mediterranean Sea, between the southern Adriatic and the Syrtis Sea off Lybia. Each of the selected areas is known to have oceanographic and/or climatic and/or geologic characteristics which distinguishes it clearly from the others. Thus, it was intended to discriminate specific sedimentological situations which could be correlated with a specific local environment, in other words, to establish some models of Mediterranean sedimentation.

The areas under consideration are indicated on Figure 1. Three of them, the *Gulf of Manfredonia* (southern Adriatic Sea, Italy), the *island shelf between Cephalonia and Zante* (Ionian Islands, Greece) and the *Gulf of Gabès* (southern Tunisia, Syrtis Minor) are in a typical offshore and shallow water environment. The fourth area is a transitional one from the shelf down the slope to the deep sea of the *Strait of Otranto*.

It was our main objective to study small areas —especially shallow-water ones—and to cover them with a close-spaced grid-pattern of stations in order to distinguish all the local sedimentary facies. (Investigations of the deeper part of the Strait of Otranto are described in more detail by Hesse and von Rad, this volume).

The studies outlined here form part of the international program of *Cooperative Investigations of the Mediterranean* (C.I.M.). The present publication is concerned partly with studies recently completed (published in German), and partly with investigations in preparation.

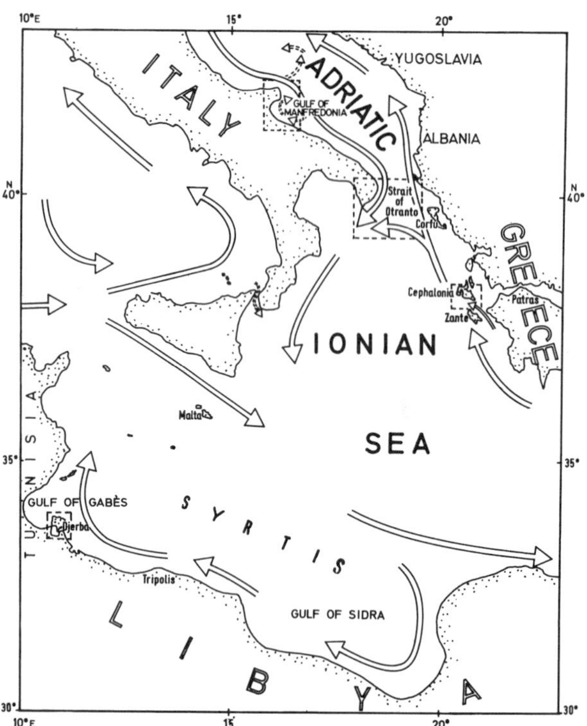

Figure 1. Generalised map of the central Mediterranean showing the southern Adriatic and the Ionian Seas: location of the four areas investigated (dashed squares). The arrows indicate the main directions (but *not* the velocity) of the surface water currents.

GENERAL ASPECTS OF MEDITERRANEAN SEDIMENTATION

The patterns of sedimentation and of sedimentary facies show a great diversity within the relatively limited area of the Adriatic, the Ionian, and the Syrtis Seas. This diversity depends mainly on the circulation of the surface waters (Figure 1), especially the surface currents sweeping the coast and parts of the shelf which are responsible for the transportation of the bulk of the clastic and river-derived material. Most of the smaller rivers of Greece, Albania and southern Italy are intermittent and therefore only of regional or seasonal importance. There are only a few perennial rivers of sedimentological importance in this region, among which the Po River surpasses all the others. The sediments derived from the Po are transported by southeasterly directed currents as far as the Strait of Otranto (Fabricius *et al.,* 1970a; Hesse *et al.,* 1971).

Clastic and dissolved matter from the rivers furnish important nutrients to the marine phytoplankton, the frequency of which can be judged from the color and transparency of the water.

Thus, in the Adriatic Sea the southeasterly-directed current crosses the mouth of the River Po and intercepts the river-derived suspended and dissolved matter. These waters are turbid and of brownish green color. Life is more abundant here than in Greek waters, which are blue and more transparent. The main reason for the generally low biocontent of the eastern Mediterranean, reflected by its deep blue and transparent water, is the extremely low content of dissolved phosphates (Ryan, 1966; Seibold 1970, p. 96).

PATTERNS OF SURFACE CURRENTS

As shown on Figure 1, the main circulation of the surface waters of the Adriatic Sea is counterclockwise. Near the peninsula of Monte Gargano the current is deflected towards the east. North of Bari it reverts to its former southeasterly trend, following the coast as far as Cape Santa Maria di Leuca. Here the current reaches maximum velocities of up to four knots (Hydrographer of the Navy, 1968). South of the Strait of Otranto the outflowing Adriatic waters are mixed with the west-flowing branch of the northwest-trending Ionian current, which emanates from the Levantine Sea, passing the 'fingers' of the Peleponnesus and the Ionian islands (Figure 1).

According to the Mediterranean Pilot (Hydrographer of the Navy, 1968), the main surface currents in the entire Mediterranean conform to a left-turning circulation pattern, whereas all the large oceans of the Northern Hemisphere have a generally clockwise surface current circulation (Dietrich and Ulrich, 1968; Emery, 1969). The left turning direction of the Mediterranean induces secondary clockwise eddies near the coasts, such as those in the Gulf of Manfredonia or the Gulf of Gabes and the Gulf of Sidra (Syrtis Sea). This general pattern can be disturbed or even reversed for some limited time by strong winds and gales, especially during the winter. Wind-produced currents are generally stronger than the normal flows and are able to pick up sediments previously deposited in shallow areas. Therefore, it is probable that movement of the sand is mainly due to the winter gales, while normal currents are able to transport only the lutum (von Rad *et al.*, 1970).

THE GULF OF MANFREDONIA (SOUTH ITALY)

The Gulf of Manfredonia has been studied by Oeltzschner and Sigl (1970). During seven expeditions (1966 to 1969), more than 300 samples were collected, including gravity—and "Kasten" (box)—cores, grab samples and beach samples. Samples from Apennine rivers north of Monte Gargano were also studied for the purpose of comparison. Other activities included SCUBA-diving for *in-situ* observations, photography, and sea floor mapping of a special area. Investigations of submarine morphology by echo-sounding and long-term metering of bottom currents in the gulf area were also conducted (von Rad *et al.*, 1970).

General Setting and Current Patterns

This gulf is the only major shallow embayment in the south Italian Adriatic coast. Three rivers, mostly intermittent ones, discharge into the Gulf; the most important is the Ofanto River, carrying the largest quantity of sediments into the sea. On its way to the coast the Ofanto crosses the volcanic area of Monte Vulture. As a result, Ofanto sediments contain a very typical heavy mineral suite, the composition of which can be used as an indicator for the direction of sediment transport within the gulf (Oeltzschner, 1971).

Because of the secondary (*i. e.*, clockwise) circulation in the Gulf of Manfredonia (see above) the sediment is transported mainly in a northerly direction. This contrasts with the general southeasterly direction of sediment transport on the Italian Adriatic coast.

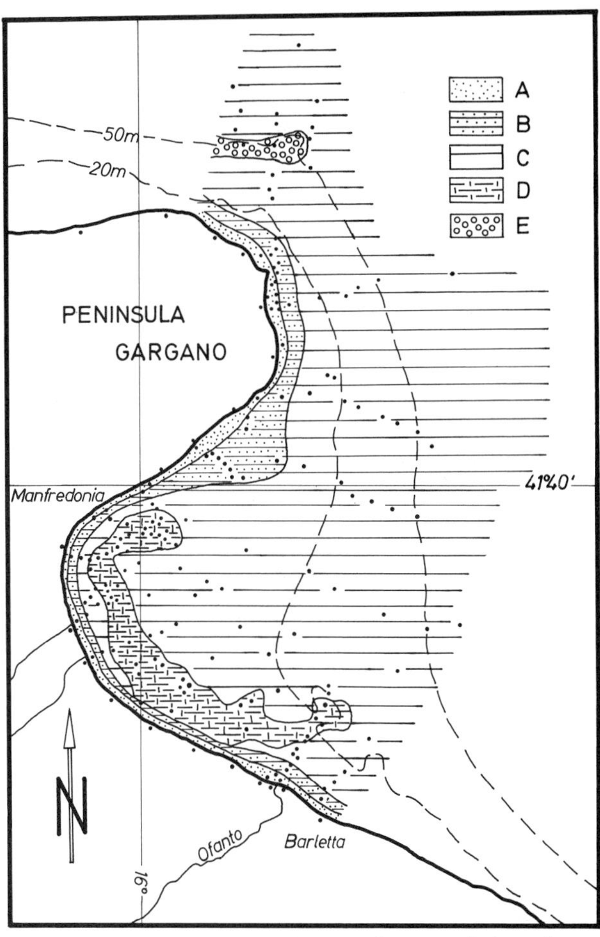

Figure 2. The Gulf of Manfredonia: sedimentary facies and sampling localities (after Sigl, 1971). Legend: A = sand; B = muddy sand; C = mud; D = *Scogliera* (shoal banks rich in benthic organisms); E = coarse relict sediments (gravel) below mud cover. Distance Manfredonia–Barletta about 25 nautical miles.

Heavy Mineral Content

The distribution of the heavy minerals, studied by Oeltzschner (1971), shows two significant trends:

1. An association attributed to an Alpine and northern Apennine source, which is brought into the Adriatic mainly by the River Po, and is dominant only in the north eastern part of the Gulf of Manfredonia, off Monte Gargano. This is indicated by the distribution of the glaucophane content (Figure 3c), one of the typical heavy minerals of the "Padanian-Apenninian Association" of Pigorini, 1968 (Oeltzschner and Sigl, 1970, p. 142). Other characteristic minerals are green hornblende, kyanite, staurolite, tourmaline, and hyperstene.

2. Pigorini (1968) discussed the "Monomineral Augitic Southern Province" of the southern Adriatic and in this region the Gulf of Manfredonia was included. However, Oeltzschner (1971) was able to show a diversity in the composition of his "Ofanto Association" which comprises augite, aegirine-augite,

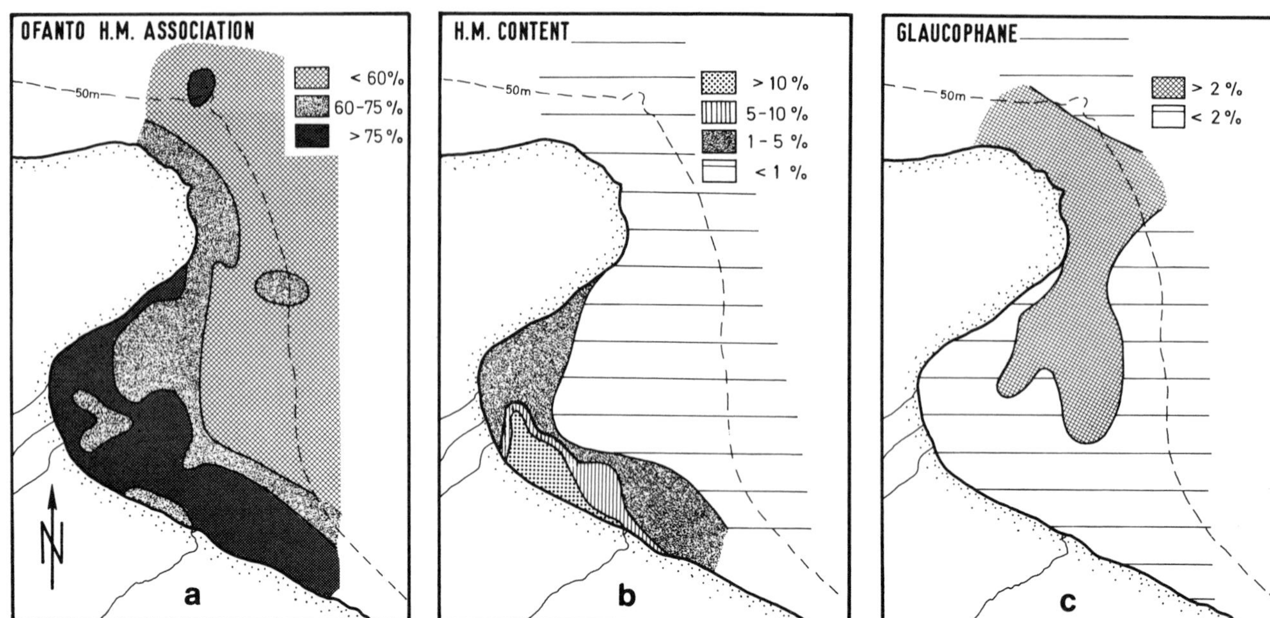

Figure 3. Distribution of the heavy mineral content of surface sediments in the Gulf of Manfredonia for the grain size fraction 0.36–0.063 mm (after Oeltzschner, 1971): a, Distribution of the "Ofanto Association", as a percentage of the total heavy mineral content (= 100%). b, Content of the heavy minerals expressed as a percentage of the total sediment. c, Distribution of glaucophane, expressed as a percentage of the total heavy mineral content.

melanite, dark hornblende and magnetite (Oeltzschner and Sigl, 1970). At the mouth of the Ofanto River, the "Ofanto Association" represents about 80 percent of the entire heavy mineral fraction. This value decreases to only 35 percent in the center and the northern part of the Gulf (Figure 3a). Parallel to the shore the heavy mineral content decreases more rapidly toward the southeast than toward the northwest and north (Figure 3b). As shown on Figure 3 the distribution of the heavy minerals of both associations display patterns consistent with the inferred clockwise current circulation.

Two irregularities contrast with the general right-turning circulation in the Gulf. A lobe of the glaucophane trends more to the west in the center of the gulf; and the south easterly extension of the Ofanto association points toward Barletta (Figure 3a). Both these anomalies may be due to a reversal of the normal Gulf circulation (right turning) by northeasterly winds and gales. These features are discussed in more detail by von Rad *et al.* (1970, p. 145).

Sedimentary Facies

With less than 20 meters of water depth near its center, the Gulf of Manfredonia (Figure 2) is an appropriate site for extensive diving operations. In this way Sigl (1971) has studied the different types of sediment covering the sea floor with respect to their morphology, composition and texture, together with the associated organisms.

The main sedimentary facies (indicated on Figure 2) are: (A, B, C) *Mobile Sediments:* (A), belt of near shore sands; (B), a transitional zone of sandy mud; (C), a flat central mud floor; (D), *Fixed sediments* ("Scogliera"); and (E), *Relict sediments.*

The muddy mobile sediments (B and C) show no stratification in the cores because of intense bioturbation. Measurements of the thickness of these muds above an acoustic reflector (sand and/or bed rock?) by Fabbri and Gallignani (in press) range from zero to five meters east of the "Scogliera" (see below) to about 30 meters east of the Gargano peninsula. Northeast of Mte. Gargano, cores penetrated older relict sediments (sand, gravel, and coquina beds) which were covered by only 20 to 70 cm of mud (U. von Rad, personal communication; *cf.*, Fabbri and Gallignani, in press).

In the southwest part of the Gulf between Barletta and Manfredonia, a broad shoal (8 to 14 meters deep) with a very irregular relief runs parallel to the coast. This shoal is called *Scogliera* ('reef') by the local fishermen.

The *Scogliera* projects above the normal sandy to muddy sea floor by one to two meters. The flanks are often very steep or even vertical, and are thus very conspicuous on the echo sounding records. The extent of the *Scogliera*, well known to fishermen but not marked precisely on the charts, was mapped using a close-spaced grid of echo sounding profiles with a total length of 540 kilometers (Sigl, 1971).

Diving investigations on several parts of the *Scogliera* revealed much diversity in the sedimentary facies and biofacies (Oeltzschner and Sigl, 1970). Parts of the *Scogliera* are formed by outcropping limestones of unknown age, which are dissected by subaerial or marine erosion. Other parts are formed by a thick incrustation of soft and hard organisms, such as calcareous red algae, sponges, scleractinians, clams, and very abundant worm tubes. Here marine life is so prolific that these places might be interpreted as "biostromes" *in statu nascendi.*

A third sub-facies of the *Scogliera* consists of stiff sandy mud containing much molluscan debris. This mud supports dense prairies of seagrass which are responsible for the accumulation of sediment by trapping and binding the sand and mud.

Large amounts of skeletal debris are concentrated at the foot of the submarine cliffs which form the flanks of the *Scogliera,* and also in is lated depressions on the flat tops of the banks. Generally, the *Scogliera* is rich in carbonate (up to 88 percent of $CaCO_3$), in contrast to the normal mud of the central part of the Gulf, which has a total carbonate content of only 20 to 30 percent.

SHELF SEDIMENTS OFF OTRANTO (SOUTH ITALY)

Relict Nature of Sediments

The Strait of Otranto, which connects the Adriatic to the Ionian Sea, was sampled by R. Hesse and U. von Rad in 1968. On three east-west profiles and one north-south profile (Figure 4) 26 box core and gravity core samples were taken. The profiles cover the Italian shelf, the shelf slope, the Corfu-Cephalonia trough, and the Apulian-Ionian ridge. The results are published in Fabricius *et al.,* 1970b (surface sediments and biofacies), in Hesse *et al.* (1971) (description of the cores, stratigraphy, and heavy mineral content), and in Hesse and von Rad (this volume) (sedimentary fabrics). In the present discussion we shall deal only with the relict sediments from the shelf, which are of special interest because of their comparison with shelf sediments from other areas.

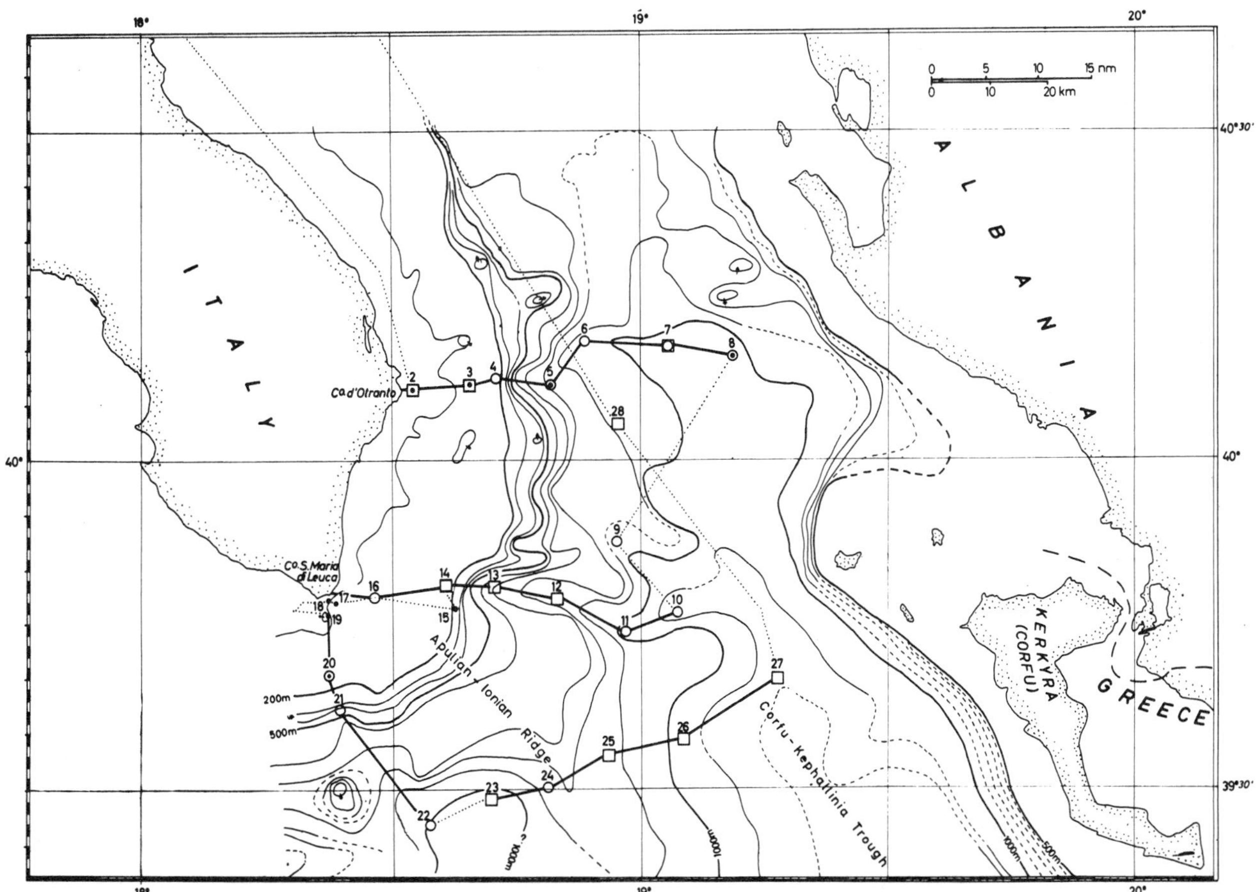

Figure 4. The Strait of Otranto: Bathymetry and location of samples. Stations: ○ = gravity cores; □ = Kögler-box cores ("Kastenlot"); ● = grab samples.

The relict character of the surface sediments in the Strait of Otranto is attested by the composition of the sand fraction, obtained from an analysis of the 20 major mineral and organic components in 26 samples from the Strait (Fabricius *et al.,* 1970b, Tables 1 to 3). Such sediments have been found mainly on their original sites of deposition, but relict material also occurs in slumped masses and in deposits resedimented by gravity currents.

Ooid Content

One of the most diagnostic criteria for the recognition of a relict sediment in this area is the content of ooids, since there is, as yet, no positive evidence for the formation of ooids anywhere in the Mediterranean at the present or in the recent geological past (Fabricius *et al.*, 1970). However, ooids with a rather thin carbonate envelope have been found in the following cores (for their location see Figure 4): in the surface layers of OT 2 (depth of 98 m), OT 3 (106 m), OT 5 (826 m); in a sand layer (turbidite?) of core OT 7 (1050 m), 160–153 cm below the top.

The texture and mineralogy of the sand fraction in core OT 5 indicate that the ooids in this area orginally were deposited on a pre-existing bluish clay in a water depth of only 150 ± 30 m, probably near or between stations OT 3 and OT 4 (Fabricius *et al.,* 1970b, p. 189). Subsequently the ooids, together with their substratum, have slumped down the slope to their present location at the foot of the continental rise. The ooids in the sand layer of core OT 7 are the only ones which show no corrosion. This suggests that, shortly after their formation on the shelf, the ooids were transported into the Corfu-Cephalonia Trough, probably by a gravity current. Since then, about 150 cm of pelagic muds, including another sand layer (about 60 cm below the core top), have accumulated above the ooid-bearing sand layer.

As yet, the exact age of the ooids cannot be stated, but they probably were formed in one of the last warm interstadial periods during late Quaternary time, *i. e.*, during the Tyrrhenian III or the Atlantic stage. To our knowledge, no friable oolitic rocks exist on the adjacent Italian coast and it therefore seems improbable that the ooids result from coastal erosion of such rocks, as is the case in the vicinity of Djerba (see below).

THE ISLAND SHELF BETWEEN CEPHALONIA AND ZANTE (GREECE)

The Ionian Islands, situated on the outer margin of the West Greek shelf are washed mainly by a northwesterly surface current which flows in from the Levantine and Aegean seas. The blue waters in the eastern Ionian Sea are deficient in suspended sediment and nutrients, and the few intermittent rivers originating from the Peleponnesus have little influence upon shelf sedimentation in this area. Therefore, the sedimentologic situation is completely different from that of the Adriatic Sea.

The shelf between the two islands of Cephalonia and Zante (Figure 5) has been studied in detail by Braune (in Braune and Fabricius, 1970 and Braune, 1971). He distinguishes two main types of sea bottom; (a) current-swept hardground and rock floor, and (b) soft sediment bottom.

Hardground

The hardgrounds and rocky bottoms (Figure 5) are most frequently encountered on the steep slopes off the west coast of Cephalonia and on the submarine connection between this island and Zante. On the top of this Pleistocene land bridge (sill depth is about 80 meters) the bottom bears only a thin veneer of shelly sand and some pockets of coarse skeletal debris. This "sediment-starved" zone extends on both sides of the 'bridge' down to depths of almost 200 m. No trace of a former beach sand was found here. It is a region representing a sedimentary discordance *in statu nascendi* and indicates more than a mere 'diastem', although the age of the rocks outcropping on the sea floor is not known. The lack of sediment in this sound may be due to a rather constant westerly current as well as to ground swells. The latter may be caused by the strong westerly winter gales which inhibit the settling of finer materials.

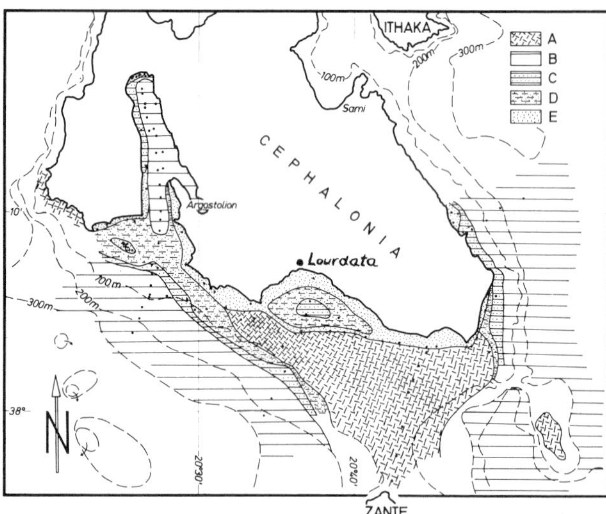

Figure 5. The island shelf off Cephalonia. Sedimentary facies, hard grounds, and sample locations (after Braune, 1971). A = rock bottom and hardground; B = mud; C = sandy mud; D = muddy sand; and E = sand.

Recent Sediments of Soft Bottom Areas

Areas with soft bottom consisting of Recent sediments (Figure 5) are found (a) in the deep parts of the sound between Cephalonia and Zante, (b) in offshore areas sheltered by the islands (east of Cephalonia), and (c) in embayments (south of Lourdata) and the Gulf of Argostolion.

The carbonate mineralogy and the content of microfossils in the Recent sediments demonstrate the very important influence of coastal and submarine erosion on sedimentation in these regions. For example the content of dolomite in the Recent sediments is highest near regions where dolomitic marls (dolomite represents 15–20% of the total carbonate content) of Miocene and Pliocene age are exposed, both on the coast and on the sea floor. Thus, knowing the composition of the source, the dolomite content of the sediment furnishes an index of the degree of local erosion. Similarly, the composition of the assemblages of foraminifera in the Recent sediments is influenced by reworking of foram-bearing marls which crop out on the sea bed.

On the other hand, in areas free from such reworking, Braune (1971) was able to erect a scheme of bathymetric zonation between 0 and 200 m of water depth using the following Foraminifera: restricted appearance of *Sphaeroidina bulloides, Trifarina sp., Uvigerina sp., Elphidium sp.,* and *Ammonia sp.*; restricted frequency of *Bulimina sp.*, and *Cassidulina sp.*

Marine Terraces of Quaternary Age

Marine coastal terraces on Cephalonia have been described recently by Braune and Fabricius (1970, p. 242). The mean values of the altitude of terraces above present sea level are given below, together with the names assigned by Woldstedt (1958) to Mediterranean marine terraces of comparable height: (a) 10–20 m (?Tyrrhenian); (b) 35–50 m (Milazzian); (c) 80–100 m (Sizilian); (d) 140–160 m (?); and (e) 180–200 m (Calabrian). Terraces (b) and (c) have numerous bored surfaces and wave-cut notches.

Recently, Braune (1971) has subdivided terrace (a) into a lower terrace (a_1) with a base ranging from 3 to 6.5 m and a top ranging from 5 to 10.8 m above present sea level, and an upper one (a_2) at 14–17.8 to 18–21.6 m. The terrace sediments contain an abundance of red calcareous algae, and are tentatively dated as Late Pleistocene (Tyrrhenian III) for the upper terrace (a_2), and perhaps Early Holocene (?Atlantic) for the lower one (a_1).

Because of the tectonic instability of the Ionian island arch (Pfannenstiel, 1960)—the last disastrous earthquake in 1953 lifted Cephaloma 0.8 m in the south and 0.2 m in the north—it is not possible to comment on Quaternary sea level changes from this local point of view. However, it seems likely that the elevated marine terraces indicate an uplift of at least 150 m during Quaternary time.

SHALLOW WATER SEDIMENTS OFF DJERBA (TUNISIA)

The littoral and sublittoral shoals off Djerba Island (southern Tunisia) are situated in the Gulf of Gabès, the western part of the Syrtis Sea. Here, a third type of sedimentation has been studied by Fabricius. The formation of the Recent littoral and sublittoral sand is governed by three factors: (1) carbonate formation by benthic organisms; (2) coastal and submarine erosion and/or mixing with relict sediments; and (3) eolian sedimentation.

Nature and Age of the Carbonate Components

The carbonate components mainly comprise green and red calcareous algae, foraminifera, bryozoans and mollusks, and less frequently corals, ostracods and echinoderms. Therefore, high-magnesium calcite and aragonite are more common than low-magnesium calcite.

The occurrence of ooids (Lucas, 1955) in the sublittoral sands is not due to formation at the present time. Radiocarbon dating of these ooids and also of coastal oolites (Fabricius *et al.*, 1970) has proved them to be of Early Holocene age. Other oolites situated on the inland side of the Quaternary oolite dunes are older, perhaps of Late Pleistocene age ("Tyrrhenian III"). It is inferred that the ooids in the Recent marine sands have been reworked from coastal and submarine outcrops. It is also possible that some of the ooids form an *in situ* relict sediment of Early Holocene age. Scanning electron microscope investigations of the aragonitic ooid envelopes by Fabricius and Klingele (1970) have revealed ultra-fine structures within these fossil ooids of the Mediterranean which are identical with those in ooids from modern tropical or subtropical marine areas. Moreover, at the present time the ooids are clearly subject to the same degree of strong mechanical and biological abrasion as other carbonate grains from the Djerba area (Figure 6). No recent formation of an oolitic coating could be detected on any kind of submarine material from this area.

Implications for Early Holocene Climate

From these observations we conclude that both in interglacial times ("Tyrrhenian") and also in the postglacial period the climatic and sedimentologic

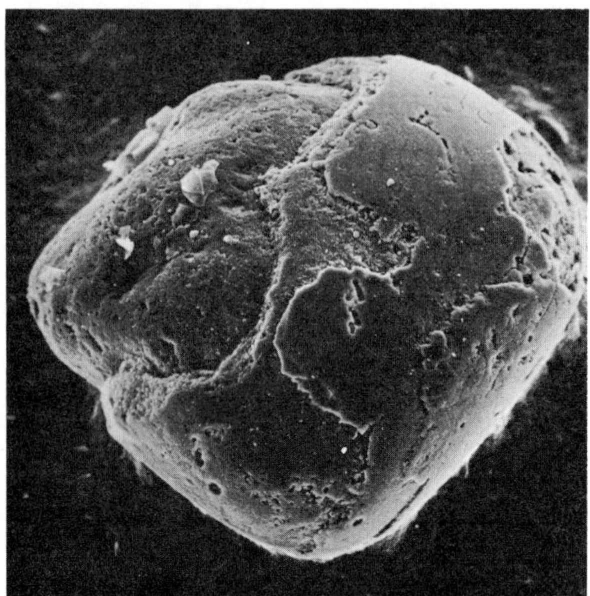

Figure 6. Ooid from marine sand off Djerba Island (southern Tunisia). Scanning Electron Micrograph showing corroded and bored oolitic coatings surrounding a quartz nucleus. Diameter of the grain: 0.5 mm (sample Fabricius KS 30).

situation of the southern parts of the Mediterranean must have been similar to that of the modern Bahama Bank, so far as the formation of ooids is concerned. From other evidences a climatic maximum is inferred during the Early Holocene, from about 7000 to 6000 years B.P., *i. e.* during the first half of the Atlantic stage. This is in accordance with the C-14 age (6010± 90 years B. P.) of the ooid coatings from the shoals off Djerba (Fabricius *et al.*, 1970a).

It must be emphasized that the comparison of the former Gulf of Gabès with the modern Bahamas is correct only in regard to the formation of ooids.

To date no real hermatypic corals or coral reefs have been found in the Mediterranean sediments of Quaternary age. This may mean that, during Quaternary time, the Mediterranean Sea did not have a constant, year-round warm climate, but instead possessed a seasonal climate where, during winter time, the water temperature dropped to levels which cannot be tolerated by reef corals. The formation of ooids and the growth of red calcareous algae would not be affected so seriously by these lower winter temperatures.

Formation of Low Mg-Calcite

Some of the marine carbonate grains become incorporated in coastal dunes. Here, the aragonite and the high Mg-calcite are exposed to the dissolving activity of the meteoric waters. Subsequently the dissolved $CaCO_3$ will be precipitated elsewhere as calcite, very often between adjacent grains, thus forming an intergranular cement. In the fossil dunes of Pleistocene and Early Holocene age this process is relatively advanced (Fabricius and Klingele, 1970, their plate 3) and these fossil coastal dunes are the main source for most of the low Mg-calcite in the marine sands, which in consequence attain their maximum content of low Mg-calcite near the shore.

Eolian Sand

The main constituents of the noncarbonate sand and silt grades are quartz and quartzite clasts, together with a small amount of feldspar. The larger grains are well rounded and many of them are frosted; the smaller ones are angular. Today the bulk of these grains are blown into the sea by offshore winds and by sand storms which transport sand from the desert hinterland. Both the frequency and the mean grain size of the quartz (including quartzite and other silicates) decrease rapidly with distance from the shore. There are two reasons for this situation. First, the larger grains, which are transported in the lower part of the atmosphere, mainly by saltation, become trapped in the rough waters of the coastal region and, once immersed, cannot be picked up again by the wind, as on land. Secondly, biogenic calcareous components become more abundant in the deeper and less agitated waters. Hence the content of quartz and other noncarbonates becomes progressively diluted.

This general situation also must have existed in the recent geologic past, with the important difference that in the agitated shallow water environment of former times all grains became coated with oolitic envelopes. Today, in the nearshore zone, these oolitic coatings are being stripped from the nuclei of quartz and quartzite fragments and these formerly wind-derived grains are thus contributing to the total amount of noncarbonate sand in this environment (Figure 6).

COMPARISON OF THE INVESTIGATED AREAS

The most important and typical discriminant features of the sediments and facies in the investigated areas are concerned with the following factors: (1) Source, availability and composition of the clastic fraction; (2) type and frequency of biogenic constituents; and (3) facies of the fossil coastal sediments of Quaternary age.

Source, Availability and Composition of Clastic Sediments

Much or most of the clastic components in the marine regions under consideration originate from:

Table 1. Synopsis of sedimentological characteristics of the study area.

REGION / SEDIMENT & FACIES	(1) GULF OF MANFREDONIA Southern Adriatic Sea (Italy)	(2) SHELF OFF OTRANTO Adriatic/Ionian Sea (Italy)	(3) OFF CEPHALONIA & ZANTE Ionian Island Shelf (Greece)	(4) SHOALS OFF DJERBA ISLAND Gulf of Gabès (Tunisia)
Grain size	Sand to clayey mud	Gravelly sand to sandy mud.	Sand and mud.	Sand to silty sand.
a) Clastic Components Composition: (Cc = calcite; Ch = chert; Q = quartz; F = feldspar)	Limestone clasts to Cc; Ch; sandstone clasts to Q, F, etc. clay minerals; volcanic material, heavy minerals.	Similar to (1); additional elements are relict Pleistocene sediments with glauconite and ooids.	Limestone clasts to Cc; Ch; marls: Cc, dolomite, clay minerals.	Eolian origin from desert hinterland: Q, quartzite clasts, and rare F. From local erosion by waves and biological erosion, transported by longshore & tidal currents: mainly ooids → AR (envelopes), → Q (nuclei), → LC (cement).
Main source:	Continental (Apennines, Mte. Gargano).	Continental and local.	Local insular and submarine erosion.	
Transported by:	Rivers (local) & longshore currents (partly Padanian-Apenninian material).	Longshore currents.	Local currents and wave action.	
Availability:	High.	Moderate.	Low.	High.
Climatic situation of source region:	Humid (mountain region) to Mediterranean (local).	Similar to (1).	Mediterranean (semihumid).	Arid.
b) Carbonate Content (% of total sediment; 24 = average)	Beach sand: 2 - 72, mud area: 15 - 24 - 30, "Scogliera": 22 - ≈50 - 76.	Mainly of biogenic origin: 33 - 39 - 50.	Gulf of Argostolion: 30 - 41 - 90; offshore muds: 31 - 64 - 98;	35 - 69 - 99: Partly of biogenic, partly of erosive origin (ooids). Mean ratio AR:HC:LC ≈ 5:4:1.
c) Biogenic (skeletal) Carbonate (organisms, relative frequency, & mineralogy; LC = (low-Mg-)calcite, HC = high-Mg-calcite, AR = aragonite.)	Mainly calcite (LC & HC not discriminated): Red algae, worm tubes, bryozoans, molluscs, echinoderms, etc. Rarely AR: "cold-water" corals (*Cladocora* sp.), green algae (scarce), molluscs.	Similar to (1) but no "Scogliera" observed.	LC: abundant Foraminifera, some molluscs; HC: rare red algae, echinoderms; AR: very rare green algae, some molluscs, etc.	AR: abundant — Halimeda; rare — red algae (indet. foliate sp.), *Cladocora* sp., molluscs, bryozoans (partly Cc). HC: frequent — other red algae, Foraminifera, echinoderms. LC: molluscs.
d) Local Settings Main direction of marine transport:	Right turning, i.e. opposite to the general direction of the main Adriatic W-coast current.	SE, veering clockwise toward SW, strong current up to 4 knots.	Westward between the islands; towards the north to the west of the islands.	Alternating tidal currents and right turning longshore currents.
Tidal influence:	None.	None.	None.	Important: max. amplitude 1.7m.
Relation between Sedimentation (S) and Erosion (E):	S > E = Accumulation.	S ⋗ E.	S < E; in some shallow areas S = O = formation of a discordance; unbalanced.	S ≧ E = accumulation.
Tectonic situation:	Slow subsidence.	More or less stable.	Episodic uplift by earthquakes.	More or less stable (the center of the gulf, probably, is sinking).
Possibility of Fossilisation:	Highly probable.	Uncertain.	Improbable.	Possible.
e) Type of Facies	Transgressive clastic sedimentation in an open coastal embayment with some "biostromal" intercalations ("Scogliera").	Current swept hardground (with sandy relict sediments) and muddy sands; more or less stagnant sedimentation.	Regressive facies in an environment of slow biogenic & clastic sedimentation. (Typical for a rising island arch under Mediterranean conditions).	Mixed sedimentation on a stable continental shelf with some coastal erosion in an arid climate: Biogenic carbonate facies, more or less diluted by eolian and clastic components.

1. river-derived material in the Gulf of Manfredonia, and, to a lesser degree, in the Strait of Otranto;
2. erosion from local coastal or submarine sources of various ages off Cephalonia, and to a certain extent, off Djerba;
3. sediments, derived from eolian abrasion and transport, off Djerba; and
4. unconsolidated relict sediments of Quaternary age off Otranto and, to a minor extent off Djerba.

In the Manfredonia area the clastic material includes the debris of carbonate rocks as well as non-carbonate detritus. In areas characterised by sand we find quartz, chert, feldspar, and heavy minerals; the mud areas are dominated by clay minerals. Off Djerba, wind-derived clastics (mainly quartz and quartzitic grains) prevail in many areas. However in other places of this area the eolian components are diluted by the products of local erosion and/or by biogenic carbonates. Between Cephalonia and Zante the supply of clastic material to the sea floor is almost restricted to local sources on the islands and submarine outcrops. The influence of the Greek mainland is not important. The clastic components are entirely recycled from limestones and marls. Thus in this area we find a large proportion of carbonates (10 percent of which is dolomite), chert and clay minerals.

Type and Frequency of Biogenic Constituents

The highest content of organically produced *carbonates* was found off Djerba. High Mg-calcite is dominant, while aragonite is rare. The main sources of the biogenic carbonate in this area are red and green algae, together with molluscs and bryozoans. Some of the sands are rich in foraminifera, such as the large tests of the *Peneroplis* group.

In the Gulf of Manfredonia, carbonate-secreting sessile organisms are especially abundant on the *Scogliera*. Common forms include algae, corals (but no genuine hermatypic types), bryozoans and worm tubes, besides many molluscs. The debris of such organisms is also of local importance elsewhere, *e. g.*, in pockets on the hardground off Otranto.

The sediments off Cephalonia are characterised by the relatively high frequency of pelagic and benthic foraminifera. In some offshore muds they form the entire sand fraction. However, the production of skeletal carbonate in the region of the Ionian island shelf is generally insignificant. This factor, in conjunction with the limited supply of clastic material, results in a generally low rate of sedimentation on the outer Ionian island shelf. A quite different situation probably existed during some epochs of the late Quaternary (see below).

Biogenic Silica produced by sponges is recorded from sediments on the *Scogliera* (Gulf of Manfredonia) but volumetrically this component is of little significance in the sediments studied.

Direct observations of organisms by diving, together with the relative frequency of skeletal remains in the sediments lead us to conclude that the ecologic conditions in the eastern Mediterranean vary conspicuously from one region to another. The main reasons for this diversity have been discussed in an earlier section. In the areas we have studied these bionomic differences are reflected in the range from the very prolific life in some parts of the Manfredonia area, compared with the restricted biotal suite found in the open sea off Cephalonia. The ecologic situation off Djerba represents a stage intermediate between the two regions mentioned above.

Facies of the Quaternary Coastal Deposits

The fossil marine shallow water deposits on the coasts of Cephalonia, studied by Braune (1971), and on Djerba (Fabricius, unpublished), are both of late Quaternary age.

Two observations are of importance. (a) There is a striking difference between the fossil terraces of both regions with regard to their bio-and lithofacies; and (b) the fossil deposits display facies which are very different from their present local equivalents.

In both regions the sediments of these marine terraces possess a high carbonate content. On Cephalonia the growth of red algae has been so prolific in the past that they have erected a framework several meters thick. The algal crusts covered rocks and other 'hard grounds' as well as a soft muddy substrate. Today, in contrast, we find only a thin veneer of red calcareous algae clinging to the rocky cliffs of Cephalonia and Zante.

This earlier situation on the Ionian Islands can be compared, perhaps, with the present appearance of calcareous algae around Djerba, although, on the sand bottoms in the Gulf of Gabès the red foliated or branched algae never form a cover or framework fixing the mobile sandy or muddy sediments. The thick deposits of calcareous algae on Cephalonia formed during at least two of the last high sea level stages of Quaternary time. These, like the oolitic deposits in the Gulf of Gabès are indicative of a warm climate.

On Djerba, the striking contrast between older and modern nearshore sand deposits is due to the total lack of Recent ooid formation, and to the paucity of organic remains in the fossil oolitic deposits (in contrast to abundant organisms and their remains in the Recent sublittoral sediments).

CONCLUSIONS

In summary, the sedimentological data (Table 1) of the four selected shelf areas of the Central Mediterranean Sea exemplify the following models:

1. The Gulf of Manfredonia is characterized by clastic river-derived sediments deposited in a subsiding area.
2. Resedimentation of relict sediments is characteristic of the shelf off Otranto.
3. The rising island shelf of Cephalonia and Zante is one without clastic and bio-carbonaceous sedimentation; local erosion and the formation of an unconformable surface is of importance here.
4. The shoals off Djerba are a model of a biogenic carbonate sedimentation pattern diluted by eolian sediments.

ACKNOWLEDGMENTS

We gratefully acknowledge the financial support of the *Deutsche Forschungsgemeinschaft* for several expeditions, and for the field and the laboratory work. We cordially thank Dr. R. Hesse (presently at Montreal, Canada) and Dr. U. von Rad (now at Hannover, Germany) who directed or assisted in the scientific and technical planning of several expeditions. We wish to thank Drs. K. Braune, H. Oeltzschner, and W. Sigl, who kindly allowed us to reproduce some of their figures and to add information from their studies to the present paper. We also extend our thanks to the many others who assisted in the field, at sea, and in the laboratory. For critical reading of the manuscript and for helpful suggestions we are indebted to Drs. J. D. Milliman, G. Kelling, and L. M. J. U. van Straaten.

REFERENCES

Braune, K. 1971. *Die rezenten und pleistozänen Sedimente des Sublitorals von Kephallinia (Ionische Inseln)—Ein Beitrag zur Meeresgeologie des Mittelmeeres.* Unpublished dissertation, Technical University Munich, 145p.

Braune, K. and F. Fabricius 1970. Geologische Beobachtungen an der Kuste und auf dem Schelf von Kephallinia (Ionische Inseln). *Geologische Rundschau,* 60:235–244.

Dietrich, G. and J. Ulrich 1968. *Atlas Zur Ozeanographie.* Meyers Grosser Physischer Weltatlas, Bibliographisches Institut, Mannheim, 7:76 p.

Emery, K. O. 1969. The continental shelves. *Scientific American,* 221 (3):107–142.

Fabbri, A. and P. Gallignani (in press). Données morphologiques et sédimentologiques sur l'Adriatique Méridionale. *Rapports et Procès Verbaux de la C.I.E.S.M. Monaco, Meeting at Rome,* 1970.

Fabricius, F., D. Berdau and K. O. Münnich 1970a. Early Holocene ooids in modern littoral sands reworked from a coastal terrace, Southern Tunisia. *Science,* 169:757–760.

Fabricius, F. and H. Klingele 1970b. Ultrastrukturen von Ooiden und Oolithen: Zur Genese und Diagenese quartärer Flachwasserkarbonate des Mittelmeeres. *Verhandlungen Geologische Bundes-Anstalt, Wien,* 1970 (4):495–617.

Fabricius, F., U.von Rad, R. Hesse, and W. Ott 1970c. Die Oberflächensedimente der Strasse von Otranto (Mittelmeer). *Geologische Rundschau,* 60 (1):164–192.

Hesse, R. and U. von Rad 1972. Undisturbed large-diameter cores from the Strait of Otranto. In: *The Mediterranean Sea: A Natural Sedimentation Laboratory,* ed. Stanley, D. J., Dowden, Hutchinson and Ross, Inc., Stroudsburg, Pennsylvania, 645–653.

Hesse, R., U. von Rad, and F. Fabricius 1971. Modern sedimentation in the Strait of Otranto between the Adriatic and the Ionian Seas (Mediterranean). *Marine Geology,* 10:293–355.

Hydrographer of The Navy 1968. *Mediterranean Pilot and Supplements.* Hydrographic Department, Admiralty, London.

Lucas, G. 1955. Oolithes marines actuelles et calcaires oolithiques récentes sur le rivage africain de la Méditerranée Orientale (Egypte et Sud Tunisien). *Bulletin Station Océanographique Salammbô (Tunisie),* 52:19–38.

Oeltzschner, H. 1971. *Eine Untersuchung über Herkunft und Verteilung der Oberflächensedimente des Golfes von Manfredonia/Süditalien mit Hilfe der Schwer- und Leitchtmineral-Analyse.* Unpublished dissertation, Technical University Munich, 179p.

Oeltzschner, H. and W. Sigl 1970. Sedimentologische Untersuchungen im Golf von Manfredonia (Südadria). *Geologische Rundschau,* 60:131–144.

Pigorini, B. 1968. Sources and dispersion of Recent sediments of the Adriatic Sea. *Marine Geology,* 6:187–229.

Pfannenstiel, M. 1960. Erläuterungen zu den bathymetrischen Karten des östlichen Mittelmeeres. *Bulletin Institut Océanographique Monaco,* 1192:60p.

von Rad, U., W. Sigl and H. Oeltzschner 1970. Bodenströmungen und Sedimenttransport im Golf von Manfredonia (Italien, Südadria). *Geologische Rundschau,* 60 (1):145–164.

Ryan, W. B. F. 1966. Mediterranean Sea. Physical oceanography. In: *The Encyclopedia of Oceanography,* ed. Fairbridge R. W., Reinhold Publishing Co., New York. 492–493.

Seibold, E. 1970. Nebenmeere im humiden und ariden Klimabereich. *Geologische Rundschau,* 60:73–105.

Sigl, W. 1971. *Die fazielle Differenzierung der Sedimente im Golf von Manfredonia der Südadria.* Unpublished dissertation, Technical University Munich, 141p.

Woldstedt, P. 1958. *Grundlinien einer Geologie des Quartärs.* Enke Verlag, Stuttgart, 2:395p.

A Study of Grain Morphology and Heavy Mineral Composition in the Sedimentary Sequence of the U.A.R.

Mahmoud M. Kholief

National Research Centre, Dokki, Cairo

ABSTRACT

Quartz sands of Paleozoic to Recent age account for much of the sediment transported to the eastern Mediterranean by the Nile River. An examination of the quartzose sediments in different localities of Egypt (U.A.R.) show these to be poorly consolidated; psammitic size grade predominates, with minor amounts of gravel, silt and clay particles.

Surface texture, roundness, transparency and crystallographic characters are the main features studied. The frequency distribution of the heavy mineral grains is also discussed.

It is concluded that wind impressions are predominant in quartz grains in the older sediments (Paleozoic to Cretaceous). In Eocene, Oligocene and Miocene sediments, there is an increase in the proportion of quartz grains showing evidence of aqueous transport. However, both wind or water surface impressions prevail in different sands during Pliocene, Pleistocene and Recent times. The distribution of the heavy mineral grains is controlled by factors of provenance and stratigraphy.

RESUME

Les sables quartzeux du Paléozoïque à nos jours forment une grande partie des sédiments transportés par le Nil dans la Méditerranée occidentale. Ces sédiments quartzeux, provenant de différentes régions d'Egypte (U.A.R.), sont peu consolidés et contiennent surtout des particules de psammite, avec de petites quantités de gravier, de silt et d'argile. L'état de surface, l'arrondi, la transparence et les caractères cristallographiques sont les principaux facteurs étudiés. La distribution de fréquence des minéraux lourds est également examinée.

Il apparaît que les impressions dues à l'action du vent dominent dans les grains de quartz provenant de sédiments plus anciens (Paléozoïque à Crétacé). Dans les sédiments éocènes, oligocènes et miocènes, il y a un accroissement de la proportion des grains de quartz transportés par l'eau. Toutefois, dans différents sables pliocènes, pléistocènes et actuels les impressions de surface qui prédominent sont dues tantôt à l'action du vent, tantôt à l'action de l'eau. La distribution des minéraux lourds dépend de l'origine et de la position stratigraphique des sables.

INTRODUCTION

The Nile is the major source of sediments in the eastern Mediterranean Basin, and much of the materials transported to the sea is of sand size. This study details characteristics of quartz sand in different localities of Egypt (U.A.R.) in order to determine provenance and pin-point source terrains.

The main factors responsible for the formation of quartz sand sediments include the nature of the source rocks, the transporting medium, the nature of the sedimentation basin, and the diagenetic history. The present work deals with petrographic characters of the sand-size quartz grains including surface texture, degree of transparency and roundness, crystallographic strain and polycrystallinity, and is also concerned with the distribution of heavy mineral grains. These features are used to deduce the conditions prevailing during the formation of the quartz sand sediments.

The sedimentary rocks in the U.A.R. overlying the Pre-Cambrian and early Paleozoic basement complex are lithologically subdivided into two clastic divisions separated by a calcareous group (Said, 1962). The older clastic division is pre-Carboniferous to Cretaceous in age and the younger one is of upper Eocene to Recent age. These clastic groups comprise arenaceous and argillaceous sediments with a few calcareous intercalations. The middle calcareous division spans the interval from Cretaceous (Cenomanian) to upper Eocene times.

The following sediments of the clastic divisions were used as a source for the samples collected: the Carboniferous and upper Paleozoic sandstones of Sinai and the area west of the Gulf of Suez, the so-called Nubian sandstones (Cretaceous?) at Aswan, the Upper Eocene sandstones at Fayum, the Oligocene sands east of Cairo, the Miocene and Pliocene sands at Wadi El Natrun and finally the Pleistocene and Recent sands of the Nile Delta proper and in the vicinity of Cairo (Figure 1). The sediments sampled are poorly consolidated or friable. The samples were chosen to represent either the whole thickness of the outcrop or an interval of more than two meters within it. In all these sediments, the sand grades predominate, and gravel, silt and clay fractions are subordinate. The geological features of the sediments discussed in the text are summarized in Table 1.

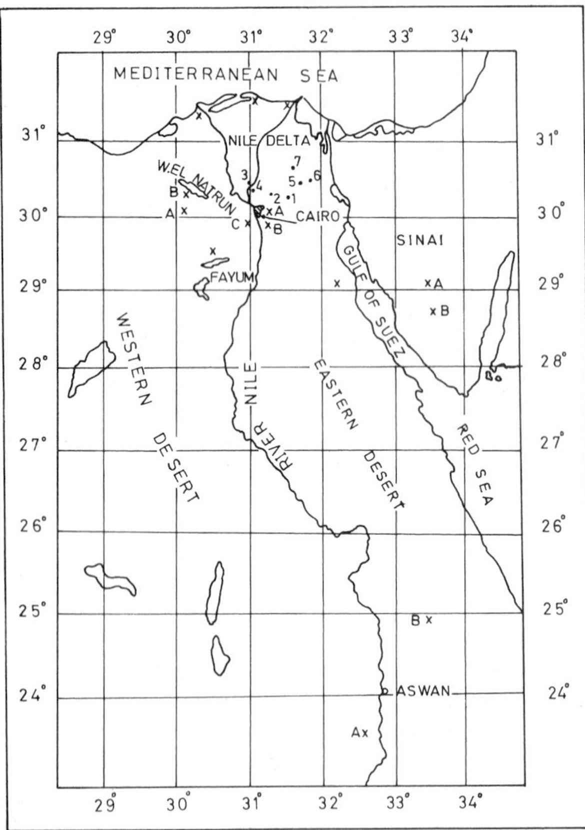

Figure 1. Map of Egypt showing general location of sand samples.

Quartz sands were prepared for analysis by washing with water to remove the fine silt and clay particles. The fine gravels, if present, were removed by sieving and grain aggregates were easily disintegrated in the poorly consolidated samples. Using an electric sieve shaker, a suitable portion of each sample was subdivided into the following size fractions (phi-scale): >0.0; 0.1 to 0.32; 0.33 to 1.0; 1.1 to 2.0 and 2.1 to 3.32. Heavy minerals were separated from the sands with bromoform.

QUARTZ GRAINS

Surface Texture

The surfaces of quartz grains carry surface impressions resulting from the action of the transporting medium (Krinsley and Donahue, 1968; Cailleux, 1968. The pitted surface of quartz grains usually results from wind action while polished surfaces occur in sand grains transported in water. Some quartz grains may show an intermediate type of surface impression which is due to the successive or repeated action of both water and wind agents during a prolonged period of transportation. The meandering ridges and graded arcs described by Krinsley and Donahue (1968) are attributed to strong wind abrasion, while the scratched and striated surface of some ancient quartz grains is attributed to diagenetic alteration. All these surface impressions are distinguished under the binocular microscope and are more easily noticeable in the coarser size grades. Two size grades were examined (0.0–0.32ϕ and 0.33–1.0ϕ) and about 100 grains in each grade were differentiated into the different textural types. The results are incorporated in Figures 2 and 3.

In the coarse grade, all the quartz grains from the older sediments from the Sinai, the west of the Gulf of Suez and from Aswan have a pitted surface texture, but in the Tertiary sediments of Fayum, Cairo A and the Wadi El Natrun A, most of the grains are either polished or show an intermediate stage and the amount of pitted grains is insignificant (Figure 2). Some of the Pliocene and the Pleistocene sediments of Wadi El Natrun B, Cairo B, C, and the Nile Delta are characterised by pitted surface texture, but in other sediments polished and/or intermediate types of surface predominate. A similar pattern of distribution was found in the finer size grade (0.33–1.0ϕ), except

Table 1. Data on sediment samples discussed in text.

Locality	Samples Collected	Age	Type of Sediments	Sedimentary Environment	References
Sinai	55	Carboniferous	Mainly sandstones with some clay bands	Probably coastal/ shallow marine	Yousef (1968); Kholief and Yousef (1969)
West of Suez Gulf	60	Carboniferous to Permo-Triassic	Sandstones with intercalated shales; some limestone and marl layers	Continental with occasional marine incursions	Nakkady (1955); Abdallah and El Adindani (1963); Yousef (1968)
Aswan	10	Early Paleozoic to upper Cretaceous	Mainly sandstones	Continental (mainly aqueous deposition)	McKee (1962); Kholief (1970a)
Fayum	4	Upper Eocene	Sandstones with shale, marl and limestone intercalations	Mixed continental and marine	Cuvillier and Blankenhorn (*in* Hume, 1965)
East of Cairo	14	Oligocene	Sandstones	Continental (mainly alluvial)	Shukri (1955); Kholief (1968)
Wadi El Natrun (A)	6	Miocene	Mainly sandstones; some intercalations of clay and limestone	Continental to Marine	Sabet *et al.* (1969); Kholief (1970a)
Wadi El Natrun (B)	4	Pliocene	Mainly sandstones	Continental (eolian)	
Nile Delta and vicinity of Cairo	18	Pleistocene	Sandstones	Continental (alluvial and eolian)	Kholief *et al.* (1969)
Mediterranean coast north of Nile Delta	12	Recent	Sandstones	Continental, deltaic to marine (beach)	Soliman (1964)

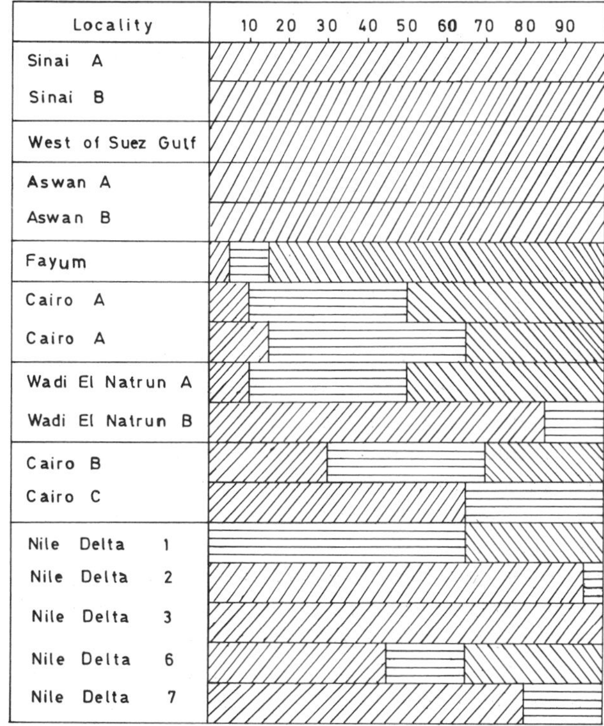

Figure 2. Relation between the average percentage of quartz grains and the type of surface texture. (Size grade: 0.0–0.32ϕ).

in the Fayum sands where the proportion of frosted grains is greater (Figure 3).

Degree of Transparency

Following the work of Shamrai (1959) and Kholief (1963) in some Miocene sediments from the USSR and that of Cailleux and Deviatkin (1969) in the Meso-Cenozoic sediments of Mongolia, the present study includes an attempt to assess the significance of the degree of transparency in the quartz grains from different sands. The degree of transparency depends essentially on the following features: (a) the type and intensity of the coating film, composed in many cases of ferruginous matter; (b) the nature of the fine sediment coating the surface of quartz grains; (c) the surface texture; and (d) the body-color of quartz grains which may be colorless, milky, smoky, etc.

Quartz grains have been categorized as transparent, semitransparent or matted and the sand fractions examined for this purpose were either washed with water or treated with 10% diluted hydrochloric acid to minimize the amount of adherent silt and clay

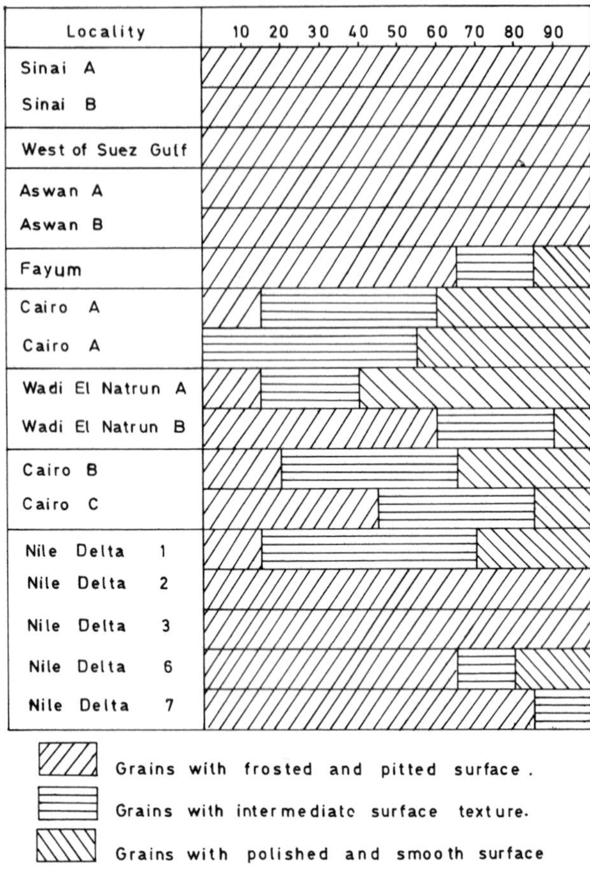

Figure 3. Relation between the average percentage of quartz grains and the type of surface texture. (Size grade: 0.32–1.0 ϕ).

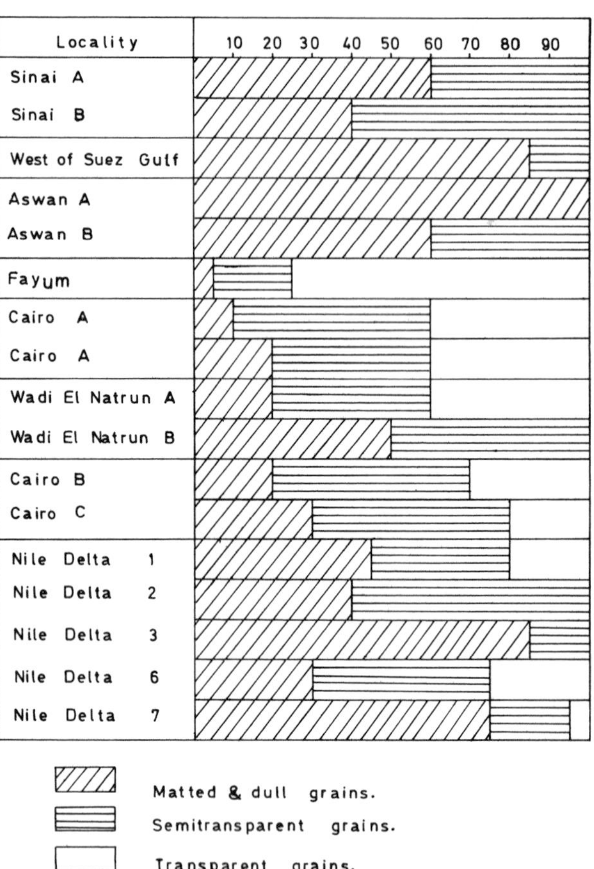

Figure 4. Relation between the average percentage of quartz grains and the degree of transparency. (Size grade: 0.0–0.32 ϕ).

and to remove most of the coating film. The results are shown on Figures 4 and 5.

In the coarse grade (0.0–0.32 ϕ), the older sediments from Sinai, the region west of the Gulf of Suez and Aswan contain a high percentage of matted grains with subordinate proportions of semitransparent grains (Figure 4). In sediments from Fayum, Cairo A and the Wadi El Natrun A, the transparent and semitransparent quartz grains predominate, while matted grains represent less than 20% of the total. Samples from Wadi El Natrun B are composed equally of matted and semitransparent quartz grains and the younger sediments of Cairo B, C and the Nile Delta contain variable proportions of all three types.

In the finer grade (0.32–1.0 ϕ) there is a relative increase in the percentage of transparent and semitransparent quartz grains in most sands. It is also apparent that an increase in the amount of pitted or polished quartz grains results in a corresponding increase in the amount of matted or transparent grains.

Degree of Roundness

Quartz exists in different forms depending on the nature of the source rocks and the stage of its formation. Usually many factors have influenced the shape of quartz grains, of which the most important is probably the nature and amount of transportation. The degree of roundness of quartz grains may be determined in several ways (Powers, 1953; Pettijohn, 1957; and Waskom, 1958). According to Crook (1968), solution may affect *in situ* the rounding of quartz grains from soils, paleosols and silcretes, to varying degrees, depending probably on the climate. However, this factor appears to be of no significance in the sands considered here because these unconsolidated sediments have suffered prolonged mechanical transportation, mainly in an arid zone. Quartz grains which have been reworked and involved in prolonged transportation usually attain a relatively high degree of roundness. At the same time the etching of grains due to solution and the formation of angular grains due to fragmentation will be of minor significance.

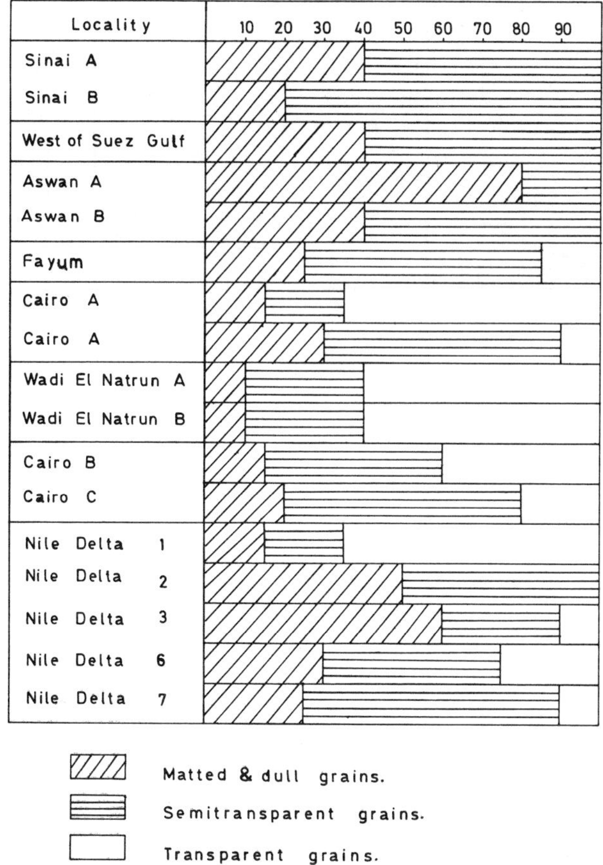

Figure 5. Relation between the average percentage of quartz grains and the degree of transparency. (Size grade: 0.32–1.0ϕ).

This study has adopted the broad classification for the external shape of quartz grains suggested by Preobrajensky and Sarkisean (1954), namely, rounded, intermediate and angular types. This study was carried out with the help of a binocular microscope and 50 to 100 grains were counted in each of three size grades (1.0–0.8 mm; 0.8–0.5 mm; 0.5–0.25 mm). These fractions were selected because they are present in all the samples and the wearing action is more effective within these size limits. The results of this study are shown in Figure 6.

Quartz grains with intermediate form predominate in all samples studied except in some sands of the Nile Delta. The percentage of rounded grains varies between 6 and 65 and is smallest in the older sands of Sinai, west of the Gulf of Suez and of Aswan, as well as in sands of Cairo and Wadi El Natrun. Rounded grains are most abundant in some sands from the Nile Delta. On the other hand, the percentage of angular grains varies between 1 and 33, and is least in some sands of the Nile Delta. Thus it appears that there is little difference in the degree of roundness of quartz grains within the samples studied, a result which may be ascribed to the complexity of the processes which affect this character.

Type of Extinction and Polycrystallinity

Detailed studies of the type of extinction and polycrystallinity in quartz grains from various rocks have been carried out by Blatt and Christie (1963), Conolly (1965) and Blatt (1967). According to these studies, the extinction in quartz may be subdivided into undulatory and non-undulatory types. The non-undulatory type of extinction is common in more mature sediments while quartz grains with undulatory extinction predominate in immature sediments. The percentage of undulatory grains also increases in the coarser size grades. The present study has investigated the type of extinction in quartz grains from two size-grades (0.8–0.5 mm and 0.5–0.25 mm).

From the study of 31 representative samples of the different sediments, it is concluded that in the coarser size grade, the percentage of quartz grains with undulatory extinction is greater than 50% in all sands except in Aswan A (Figure 7). In the finer size grade, the percentage of undulatory quartz is less than 50% in most sands (Figure 7). Sinai sands contain an abnormally high proportion of undulatory quartz and this increase is ascribed both to the influence of the source rocks and also to tectonic effects (Kholief and Yousef, 1969).

HEAVY MINERALS

The heavy mineral fraction usually represents less than 5% of the sample. Abnormal concentrations of heavy minerals, amounting to 10% or more, occur in the beach sediments of the Nile Delta and are attributed to wave-sorting effects. The heavy mineral species comprise iron ores, tourmaline, zircon, rutile, garnet, epidote, staurolite, kyanite, sillimanite, amphiboles, pyroxenes, muscovite, biotite, apatite, chlorite, anatase, brookite and andalusite (Table 2). These minerals have been described elsewhere (Kholief *et al.*, 1969; Kholief and Yousef, 1969; Kholief, 1970(a); and Morsi, 1969). The first four minerals exist in all the sands in variable proportions, but they predominate in the older sands of Sinai, west of the Gulf of Suez and Aswan. A few grains of garnet, epidote and staurolite were also recorded in these older sands. Authigenic anatase and brookite are common in the sand samples of Aswan B. In the Eocene, Oligocene and Miocene sediments of Fayum, Cairo A and Wadi El Natrun A, there is a significant

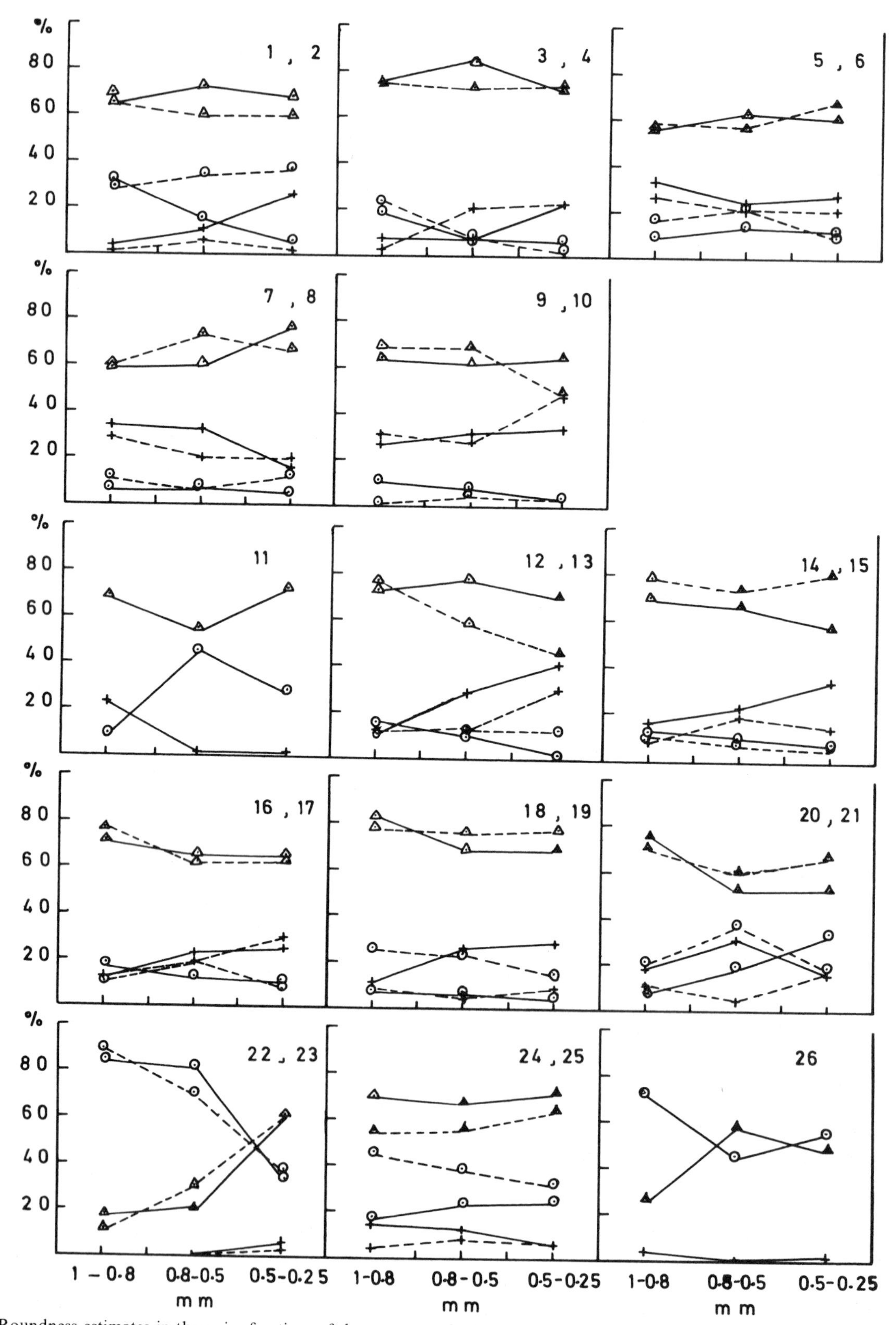

Figure 6. Roundness estimates in three size fractions of the quartz sands. Sample localities include: 1, 2- from Sinai A; 3, 4- from Sinai B; 5, 6- from region west of Suez Gulf; 7, 8- from Aswan A; 9, 10- from Aswan B; 11- from Fayium; 12–15- from CairoA; 16, 17- from Cairo B and C; 18, 19- from Wadi El Natrun A and B; 20–26- from Nile Delta (1 to 7). Signs (○,△, +) refer to rounded, subrounded to subangular (intermediate), and angular grains respectively. The solid line indicates the first sample from a given locality, while the dashed line refers to the second sample.

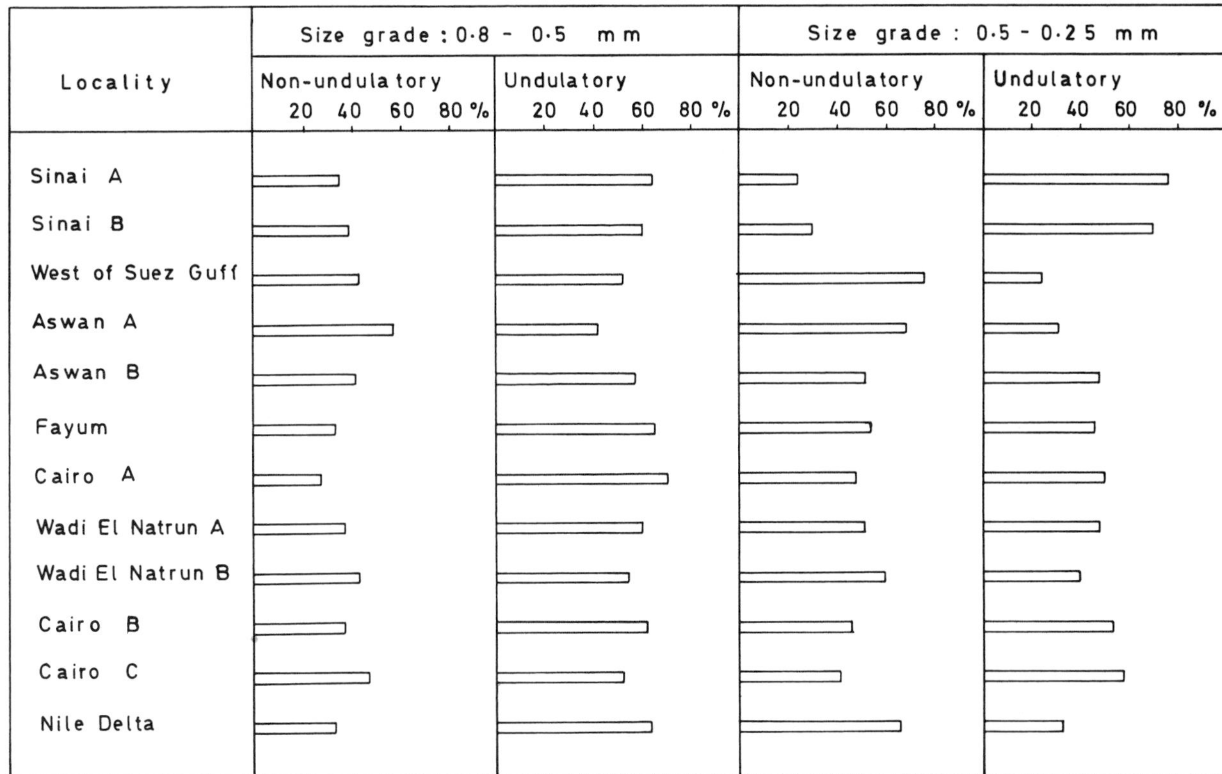

Figure 7. Average percentage of quartz grains, in two size grades, with non-undulatory and undulatory extinction.

increase in the amount of garnet, epidote, staurolite and kyanite, while sillimanite, hornblende and muscovite occur in subordinate proportions. In the Pliocene sediments of Wadi El Natrun B, epidote and amphiboles are the major non-opaque minerals while in the Pleistocene sands of Cairo (B, C) and the Nile Delta, amphiboles, pyroxenes and epidotes are the most abundant non-opaques.

Iron ore minerals consist mainly of magnetite, ilmenite, hematite, limonite and leucoxene, and occur in all sands, constituting more than 50% of the heavy fractions of Aswan, Fayum, Cairo A and Wadi El Natrun sands. Under reflected light the opaques may be subdivided into colored and black grains and counted (Table 2), and it is evident that the percentage of the black grains (magnetite and ilmenite) is relatively low in the older sediments from Sinai, the region west of the Suez Gulf and from Aswan. There is a sharp increase in the percentage of the black opaques in the succeeding sediments of Fayum and the younger formations. The predominance of the black variety in the younger sediments can be attributed partly to weathering and intrastratal solution of the older sediments and partly to the contributing source rocks.

RECENT COASTAL SANDS OF THE NILE DELTA

The samples under investigation were collected from the beach sand of the Mediterranean Sea at the two main distributaries of the Nile River (Damietta and Rosetta) and at the mouth of Lake Burullus. Additional samples were collected from coastal dunes occurring behind the Mediterranean Sea beach. Heavy minerals are concentrated to a variable degree in the beach sands, locally giving rise to placers. Most of the sands are dominated by the fine sand fraction, especially near the Nile River distributaries, but both medium and fine sands occur in the Burullus and Damietta sediments.

The morphologic and crystallographic characters of the quartz grains in these sediments have been examined and are recorded in Table 3. Little difference is apparent between the quartz-grain attributes of the beach and dune sands, although wind impressions are sometimes more conspicuous in the dune sands. Most of the sands, whether beach or dune, possess relatively low proportions of rounded grains. The few samples which contain high amounts of rounded quartz are also those which are of somewhat coarser

Table 2. Average frequency percent of the heavy mineral grains (symbol + indicates less than 0.5%).

Locality	No. of samples	Zircon	Rutile	Tourmaline	Garnet	Epidote	Staurolite	Kyanite	Sillimanite	Andalusite	Hornblende	Augite	Muscovite	Biotite	Chlorite	Apatite	Anatase	Brookite	Opaques	Black grains in 100% opaques
Sinai A	6	20.6	5.8	43.8	+	1.3	+												27.7	7
Sinai B	7	15.5	2.6	57.1		0.7											+		23.9	5
West of Suez Gulf	10	38.8	12.0	18.3															30.6	4
Aswan A	2	6.4	2.3	8.4	0.6	0.8	1.7												79.6	20
Aswan B	4	15.2	7.1	4.9		0.9	0.5						+	+	+		8.6	1.5	60.9	7
Fayum	3	6.0	4.3	3.4	3.1	8.1	2.0	2.8	+		+								68.3	65
Cairo A	12	11.0	3.3	2.2	3.0	5.2	1.8	1.4	0.6		1.0	+				+			69.2	66
Wadi El Natrun A	4	5.4	2.7	1.7	3.7	15.4	2.0	1.1	+		0.8		0.7	+	+				65.4	75
Wadi El Natrun B	4	2.2	1.3	0.8	1.6	18.8	1.8	1.6	0.6		11.7		+						59.5	65
Cairo B	1	3.9	1.0	0.8	2.6	10.6	1.6	0.7		0.8	25.1	8.7	+						41.0	74
Cairo C	1	4.1	1.6	2.2	1.9	10.8	2.4	0.7	1.3		23.6	2.5		1.3	1.6				44.6	52
Nile Delta (1)	1	3.9	+	1.7	1.7	16.4	+	+			36.4	1.2		+	+	+			37.6	79
Nile Delta (2–7)	6	2.9	5.1	2.2	1.8	15.6	1.0	+			15.4	16.5		+	+	+			42.3	

grade, suggesting a size-effect. The percentage of the undulatory quartz grains is significantly higher in the Burullus and Rosetta sands than in the Damietta samples.

SUMMARY AND CONCLUSIONS

The morphologic study of quartz grains and of heavy mineral assemblages is useful in that it serves to define attributes of terrigenous source terrains in the vicinity of the Nile. Some of these parameters, used as tracers, would be useful in provenance studies of offshore Nile Cone sediments. The following observations are made:

1. In the Paleozoic and Mesozoic sediments, wind action appears to have been generally more effective, although these sediments were deposited in a variety of environments. The Upper Eocene, Oligocene, and Miocene sediments contain a high proportion of quartz grains that were affected by aqueous transport but wind action is again prevalent in Pliocene and Pleistocene sands. All the Recent sands studied reveal the effects of aqueous transport. It is interesting to note that a similar variation in the content of the different types of quartz grains through the stratigraphic sequence of sediments in other parts of the world has been described by Cailleux and Deviatkin (1969).

2. Due to the continuous mixing of new contributions and the advent of different environments, the variation in the characters of the sand sediments is not sharply distinct. However, there is a general tendency for the amount of frosted, transparent, rounded and strained quartz grains to increase with increasing grain size. The time factor does not appear to be significant so far as the morphologic characters of the quartz grains are concerned, but it has a noticeable effect on the nature of the heavy mineral assemblage. Tectonic/diagenetic effects are indicated by the surface striations of quartz grains from the Sinai sediments, and also by the anomalous increase in the proportion of grains with undulatory extinction in the finer size grade.

3. Heavy mineral assemblages from the older sediments are dominated by the very stable minerals (tourmaline, zircon, and rutile) with a few grains of moderately stable minerals, and some authigenic minerals such as brookite and anatase. The proportion of garnet, epidote and staurolite increases in the Tertiary sands, probably reflecting the increasing significance of metamorphic source rocks. In addition to the metamorphic minerals, Pliocene sands contain a high proportion of amphiboles and in the Pleistocene and Recent sands, amphiboles, pyroxenes and epidotes are the most common non-opaque species. These minerals were derived mainly from the crystalline basement complex near the sources of the Nile in east and central Africa. The basement complex in the Eastern Desert of Egypt has also contributed additional material to these sediments since Upper

Table 3. Characters of the Recent sands of Nile Delta from Mediterranean Sea beaches and coastal dunes.

Locality	No. of samples	Surface Texture		Degree of Transparency				Degree of roundness			Undulatory quartz	Polycrystalline quartz
		Pitted	Intermediate	Polished	Matted	Semi-transparent	Transparent	Rounded	Intermediate	Angular		
Damietta												
Ras El-Bar beach	2	10	28	62	7	17	76	6	46	48	43	
Ras El-Bar beach	2	23	45	32	19	44	37	42	53	5	48	7
Gamasa beach	2	15	39	46	8	23	69	3	71	26	50	6
Gamasa coastal dune	1	21	33	46	7	22	71	5	72	23	41	8
Burullus beach	2	13	55	22	7	53	40	42	54	4	78	9
Burullus coastal dune	2	22	59	19	12	48	40	34	54	12	78	6
Rosetta beach	3	11	39	50	5	26	69	6	69	25	75	6
Rosetta coastal dune	1	14	56	30	8	61	31	2	70	28	64	5

Eocene times (Shukri, 1951). Finally the effects of intrastratal solution become more evident with increasing age of the rocks, resulting in disappearance of the unstable heavy mineral species and alteration of the iron-rich minerals.

ACKNOWLEDGMENTS

I wish to express my appreciation to Professor D. J. Stanley for his interest and encouragement and to Professor G. Kelling for useful criticism improving the text.

REFERENCES

Abdallah, A. M. and A. El Adindani 1963. Stratigraphy of Upper Paleozoic rocks on the western side of the Gulf of Suez. *Geological Survey and Mineral Resources Department*, Cairo, 25:28p.

Blatt, H. 1967. Original characteristics of clastic quartz grains. *Journal of Sedimentary Petrology*, 37:401–424.

Blatt, H., and J. M. Christie 1963. Undulatory extinction in quartz of igneous and metamorphic rocks and its significance in provenance studies of sedimentary rocks. *Journal of Sedimentary Petrology*, 33:559–579.

Cailleux, A. 1968. Etude morphoscopique de sables non marins de l'Alaska. *Compte Rendu de la Société Géologique de France*, 1:21.

Cailleux, A. and E.V. Deviatkin, 1969. Morphosculptural studies of quartz grains from the sands of Meso-Cenozoic deposits of Mongolia (in Russian). *Litologia i Poleznye Iskopayeme (Lithology and Mineral Resources)*, 5:101–109.

Conolly, J. R. 1965. The occurrence of polycrystallinity and undulatory extinction in quartz in sandstones. *Journal of Sedimentary Petrology*, 35:116–135.

Crook, K. A. 1968. Weathering and roundness of quartz sand grains. *Sedimentology*, 11:171–182.

Hume, W. F. 1962, *Geology of Egypt* (a digest of papers published in Egypt). United Arab Republic, Geological Survey and Mineral Resources Department, Cairo, 3(2):734 p.

Kholief. M. M. 1963. *Geological Structure, Material Composition and Condition of Formation of Quartz Sand Deposits at Donetsky Oblast USSR* (in Russian). Aftoreferat Ph.D. thesis, MGRI, Moscow, 31 p.

Kholief, M. M. 1968. Cross-bedding and its geological significance. *Proceedings of the Egyptian Academy of Sciences*, 21:61–65.

Kholief, M. M. 1970a. Petrographical studies on some quartz sand samples from Wadi El Natrun, Western Desert, U.A.R. *Bulletin du Institut Desert d'Egypt*, 19:105–118.

Kholief, M. M. 1970b. A note on analysis of cross-bedding measurements in the Nubian sandstones of the Eastern Desert. *Proceedings of the Egyptian Academy of Sciences*, 23 (in press).

Kholief, M. M., E. Hilmy and A. El-Shahat 1969. Geological and mineralogical studies of some sand deposits in the Nile delta, U.A.R. *Journal of Sedimentary Petrology*, 39:1520–1529.

Kholief, M. M. and M. Yousef 1969. Mineralogy of quartz sand deposits around the Gulf of Suez. *Journal of Geology Egypt*, 13:25–35.

Krinsley, D. and J. Donahue 1968. Methods to study surface texture of sand grains (a discussion). *Sedimentology*, 10:217–221.

McKee, E. D. 1962. Origin of the Nubian and similar sandstones. *Geologische Rundschau*, 52:551–587.

Morsi, M. 1969. *Geochemical Studies of Some Local Sands and their Suitability for Glass Industry*. Unpublished M.Sc. thesis, Cairo University, 140 p.

Nakkady, S. E. 1955. The stratigraphy and geology of the district between the northern and southern Galala plateaus, Gulf of Suez coast. *Bulletin Institut du Desert d'Egypt*, 33:254–268.

Pettijohn, F. J. 1957. *Sedimentary Rocks* (second edition). Harper & Brothers, New York, 718 p.

Powers, M. C. 1953. A new roundness scale for sedimentary particles. *Journal of Sedimentary Petrology*, 22:117–119.

Preobrazhensky, E. A. and S. G. Sarkisean 1954. *Minerals of Sedimentary Rocks* (in Russian). Gostoptekhizdat, Moscow, 462 p.

Sabet, A., M. Bedewi and T. Abd El Razik 1967. Beneficiation of white sands from Wadi El Natrun. *Geological Survey of Egypt*, 49:26 p.

Said, R. 1962. *The Geology of Egypt*. Elsevier, Amsterdam, 377 p.

Shamrai, E. A. 1959. *Thesis* (in Russian). Troudy Geologo-geographic Facultat, Utshenie Zapisky Rostov University, 8

Shukri, N. M. 1951. Mineral analysis tables of some Nile sediments. *Bulletin Institut du Desert d'Egypt*, 1:39–69.

Shukri, N. M. 1955. The geology of the desert east of Cairo. *Bulletin Institut du Desert d'Egypt*, 3:89–105.

Soliman, S. M. 1964. Primary structures in a part of the Nile Delta beach. In: *Deltaic and Shallow Marine Deposits* (Developments in Sedimentology), ed. Van Straaten L.M.J.U., Elsevier, Amsterdam, 1:379–387.

Waskom, J. D. 1958. Roundness as an indicator of environment along the coast of Panhandle Florida. *Journal of Sedimentary Petrology*, 28:351–360.

Yousef, M. 1968. *Geological and Mineralogical Studies of White Quartz Sands Around the Gulf of Suez*. Unpublished M.Sc. thesis, Cairo University, 149 p.

PART 7. **General Aspects of Recent Deep-sea Sedimentation**

Principal Types of Recent Bottom Sediments in the Mediterranean Sea: Their Mineralogy and Geochemistry

E. M. Emelyanov

P. P. Shirshov Institute of Oceanology, Kaliningrad

ABSTRACT

During cruises of the R/V ACADEMIK S. VAVILOV in 1959 to 1969 about 340 samples of bottom sediments were taken by grabs, gravity and piston corers. Most samples were subjected to granulometric, mineralogical, chemical and micropaleontological analyses. Maps of bottom sediments, terrigenous mineralogical provinces (scale 1:3,000,000) and distribution patterns of some minerals (including illite, kaolinite and montmorillonite) and various chemical components ($CaCO_3$, $MgCO_3$, amorphous silica, organic carbon, Fe, Mn, Ti, P, B, Au, U, Cu, Ni, Co, Mo, Cr and other trace elements) were constructed on the basis of these analyses as well as published data. Chemical composition was analyzed by normal chemical methods and by neutron activation and spectrographic methods.

The basic granulometric and genetic types of surficial sediment (layer 0–5 cm) are distinguished and described. Volcanic sediments are found in the caldera of Santorin, in the Gulf of Naples and in the area of Stromboli; ferruginous sediments of volcanic origin occur in the caldera of Santorin. Biogenic calcareous sediments, common throughout the Mediterranean area, are characterized by coarse calcareous-argillaceous concretions of irregular shape (calcite, magnesian calcite) and relatively high contents of magnesian calcite and dolomite. Calcareous oolitic sands are spread on the shelf off the arid zone of Africa.

The investigation of absolute age of the terrigenous volcanogenous minerals (80 samples) by potassium-argon method delineated age provinces. These sometime coincide with terrigenous mineralogical provinces determined earlier (Emelyanov, 1968); in some cases they are very different.

RESUME

Au cours de la croisière à bord du R/V ACADEMIK S. VAVILOV en 1959–1969, des échantillons de sédiments du fond de la Méditerranée furent rassemblés par grab et carottier. Presque tous les échantillons ont été soumis à des analyses granulométriques, minéralogiques, chimiques et micropaléontologiques. Les cartes des sédiments du fond, des provinces minéralogiques terrigènes (échelle 1:3.000000) ainsi que de la distribution de certains minéraux (comprenant illite, kaolinite et montmorillonite) et constituants chimiques divers ($CaCO_3$, $MgCO_3$, silicate amorphe, carbone organique, Fe, Mn, Ti, P, B, Au, U, Cu, Ni, Co, Mo, Cr et autres éléments rares), ont été établies suivant les résultats de ces analyses et les données publiées. La composition chimique a été déterminée selon les méthodes habituelles ainsi que par la méthode d'activation des neutron et par spectrographie.

Les principaux types granulométriques et minéralogiques de la couche supérieure (0 à 5 cm) de sédiments meubles ont été décrits. Les sédiments volcaniques se trouvent dans la caldère de Santorin, Golfe de Naples et région de Stromboli; les sédiments ferrugineux d'origine volcanique sont étalés dans la caldère de Santorin. Les sédiments calcaires biogéniques qui se développent pratiquement dans toute la zone méditerranéenne, sont caractérisés par des concrétions de calcaire argileux grossier (calcite, calcite de magnésium de forme irrégulière) et une teneur relativement haute de calcite de magnésium et dolomite. Les sables calcaires oolitiques s'étalent sur la plateforme de la zone aride d'Afrique.

L'identification de l'âge absolu des minéraux terrigènes d'origine volcanique (80 échantillons) par la méthode potassium-argon a permis de délimiter certaines zones d'âge qui coïcident parfois avec les provinces minéralogiques terrigènes délimitées auparavant par l'auteur (Emelyanov, 1968).

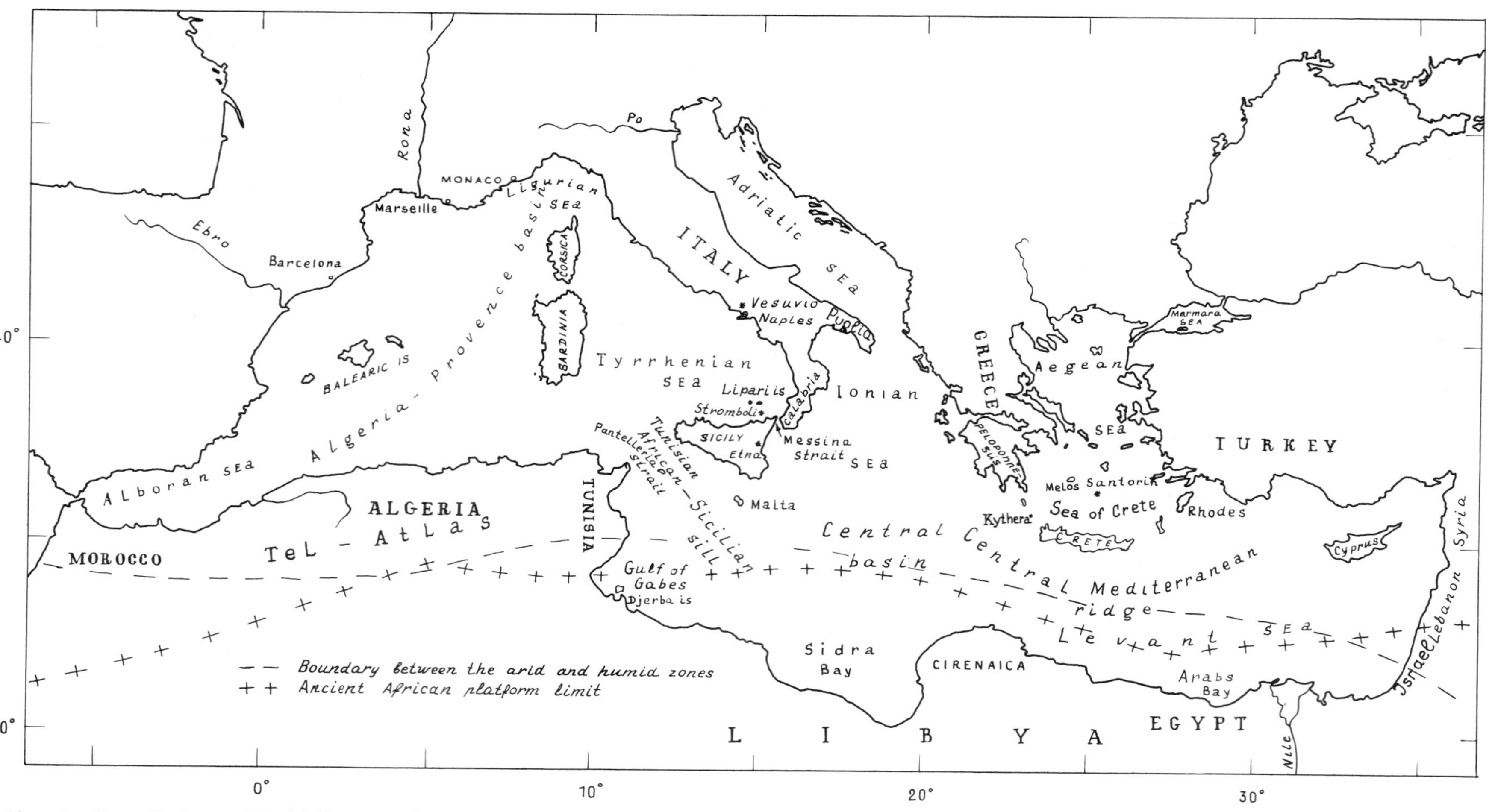

Figure 1. Generalized map of the Mediterranean Sea showing principal physiographic and climatic features.

INTRODUCTION

The Mediterranean Sea is of particular interest in studying the processes of sedimentation. The basins are located in different tectonic and climatic zones: the northern areas of the Mediterranean Sea lie in a humid climatic zone and the depositional basin is essentially geosynclinal (Figure 1); the southern areas (from Lebanon to Tunisia) are situated on a submerged platform in which lithogenesis occurs under arid conditions. The Algerian-Provençal Basin is located in an intermediate zone. The process of sedimentation shows evidence of active modern volcanism, biogenic sedimentation, chemical processes of deposition of sedimentary matter and diagenetic redistribution of elements in the upper sediment layer.

Numerous works concerning the geology of the Mediterranean Sea have been published. The principal expeditions include the Swedish deep-sea expedition in 1947–1948, numerous French expeditions (including Monaco) aboard the CALYPSO, American cruises of the VEMA, CHAIN, ATLANTIS, expeditions of Italian, Yugoslavian and Egyptian scientists, *etc.* It is also appropriate to recognize the importance of the work of those individuals, such as K. André, O. B. Boggild, A. d'Arrigo, L. Berthois and J. Bourcart, whose early studies have provided the basis for more recent investigations of sedimentation in the Mediterranean Sea. No single study, however, satisfactorily summarized the principal types of Mediterranean sediments and their general mineralogy and chemistry.

During the last decade (1959–1969) bottom sediments were collected at 340 different stations in the Mediterranean (Figure 2) by Soviet expeditions, mainly aboard the R/V AKADEMIK S. VAVILOV. These samples (214 bottom grabs and 135 cores up to 11.06 m in length) are distributed throughout most of the area of the basin, and thus the varying character of the bottom sediments and their components can be traced over the entire Mediterranean.

A total of 200 samples were subjected to a comprehensive analysis including granulometric, mineralogic, chemical and micropaleontologic investigation which was undertaken in laboratories of the Southern and Atlantic Departments of the Institute of Oceanography, Academy of Sciences USSR under the direction of the author. Standard analytical methods of the Institute of Oceanology of the USSR were used (Strakhov *et al.*, 1957). Granulometric composition was investigated by sieving and settling methods; mineralogy of the aleurite (light and heavy) fractions by immersion; clay minerals by X-ray; absolute age of terrigenous volcanic minerals by potassium-argon; $CaCO_3$, $SiO_{2amorph}$, C_{org}, Fe, Mn, Ti, P, chemical composition of carbonates, total silicate composition and uranium by wet chemical analysis; gold by neutron activation analysis; trace elements by emission spectrographic analysis (quantitative method); mineralogic composition of carbonates by diffractometric and optic methods.

The results of nearly all these analyses, including other data published before 1963, were first summarized by this author in 1964 (Emelyanov, 1964 and 1967) but a comprehensive account has not yet been published. Data published previously include the granulometric composition of sediments (uppermost stratum, 0–5 cm) and their mode of origin (Emelyanov, 1965a), carbonates (Emelyanov, 1965b), mineralogy of coarse fractions (Emelyanov, 1968), clay minerals (Rateev *et al.*, 1966), amorphous silica (Emelyanov, 1966a), organic matter and uranium (Kochenov *et al.*, 1965), titanium (Emelyanov, 1966a), gold (Anoshin *et al.*, 1969), boron (Sukhorukov and Emelyanov, 1969), absolute age of terrigenous minerals (Emelyanov *et al.*, 1973) and others. The volcanic sediments in the area of Santorin were considered in a separate work by Butuzova (1969).

The distribution of the principal sediment types, $CaCO_3$, iron and manganese and the location of geological stations of Soviet and other expeditions were shown on 13 colored charts of the Atlantic Ocean at a scale 1:20,000,000 (Emelyanov *et al.*, 1966; Joint Geophysical Committee, 1969). The present paper is a synthesis of all data published at present. It includes an account of the granulometric and genetic composition of the upper layer (0–5, or sometimes 0–10 and 0–20 cm) of bottom sediments over all the Mediterranean Sea, together with their mineralogy and chemistry, as well as deductions concerning the processes of sedimentation.

GRANULOMETRIC COMPOSITION OF SEDIMENTS

The following sediments are distinguished in the Mediterranean Sea, according to the classification adopted in the Institute of Oceanology, Academy of Sciences USSR (Bezrukov and Lisitsin, 1960): gravel, sands, coarse aleurites (coarse silt), fine-aleuritic mud (fine silty mud), aleurite-pelitic mud and pelitic mud (clay) (Figure 3). Gravelly sediments are not common; they are sometimes encountered in the northern part of the Aegean Sea and on the African-Sicilian sill (shell material), and occasionally in the canyons off Provence, France and some regions where the continental slope is steep.

Sands are widely distributed in the northern part

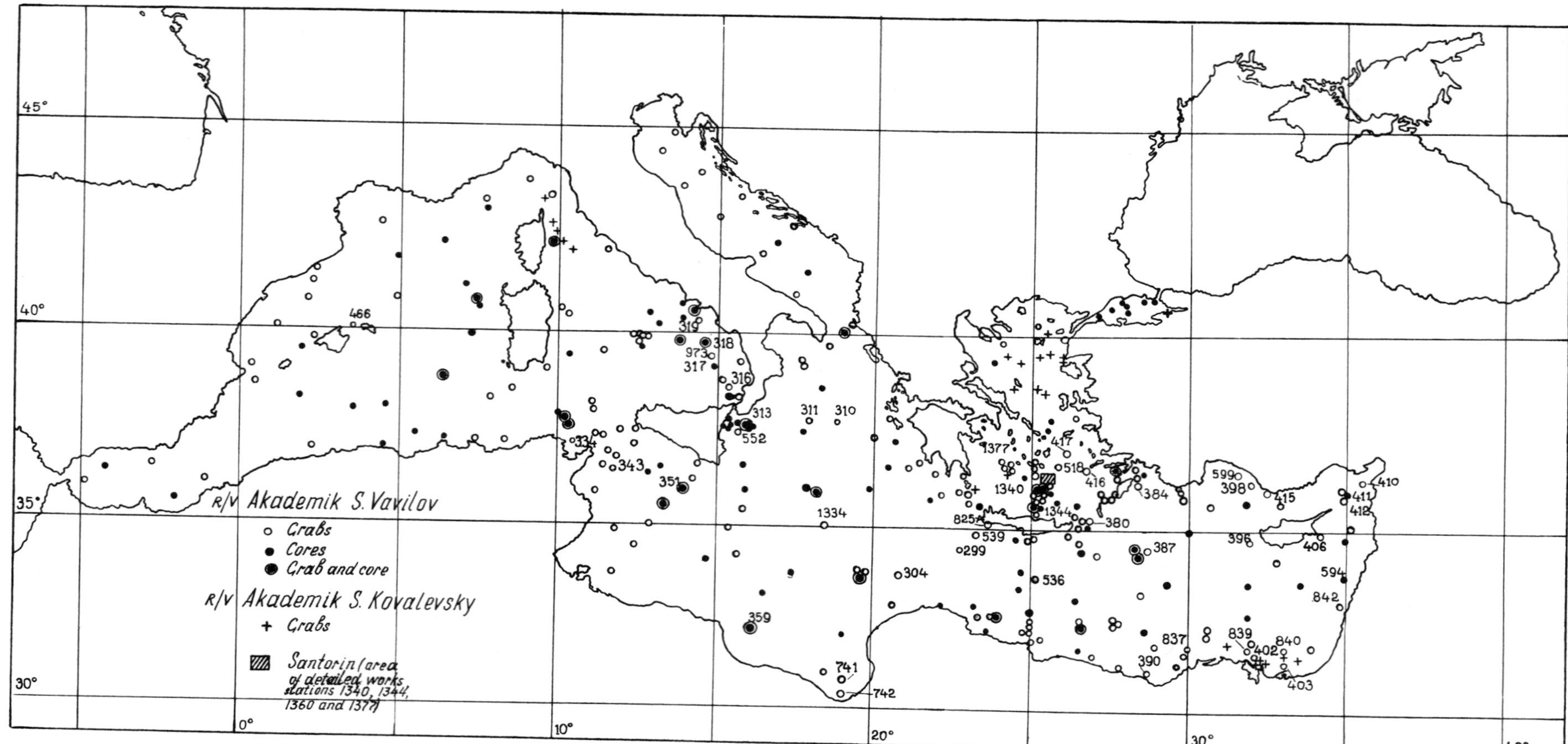

Figure 2. Location of geological stations occupied by R/V AKADEMIK S. VAVILOV in 1959–1968 and R/V AKADEMIK S. KOVALEVSKY in 1960.

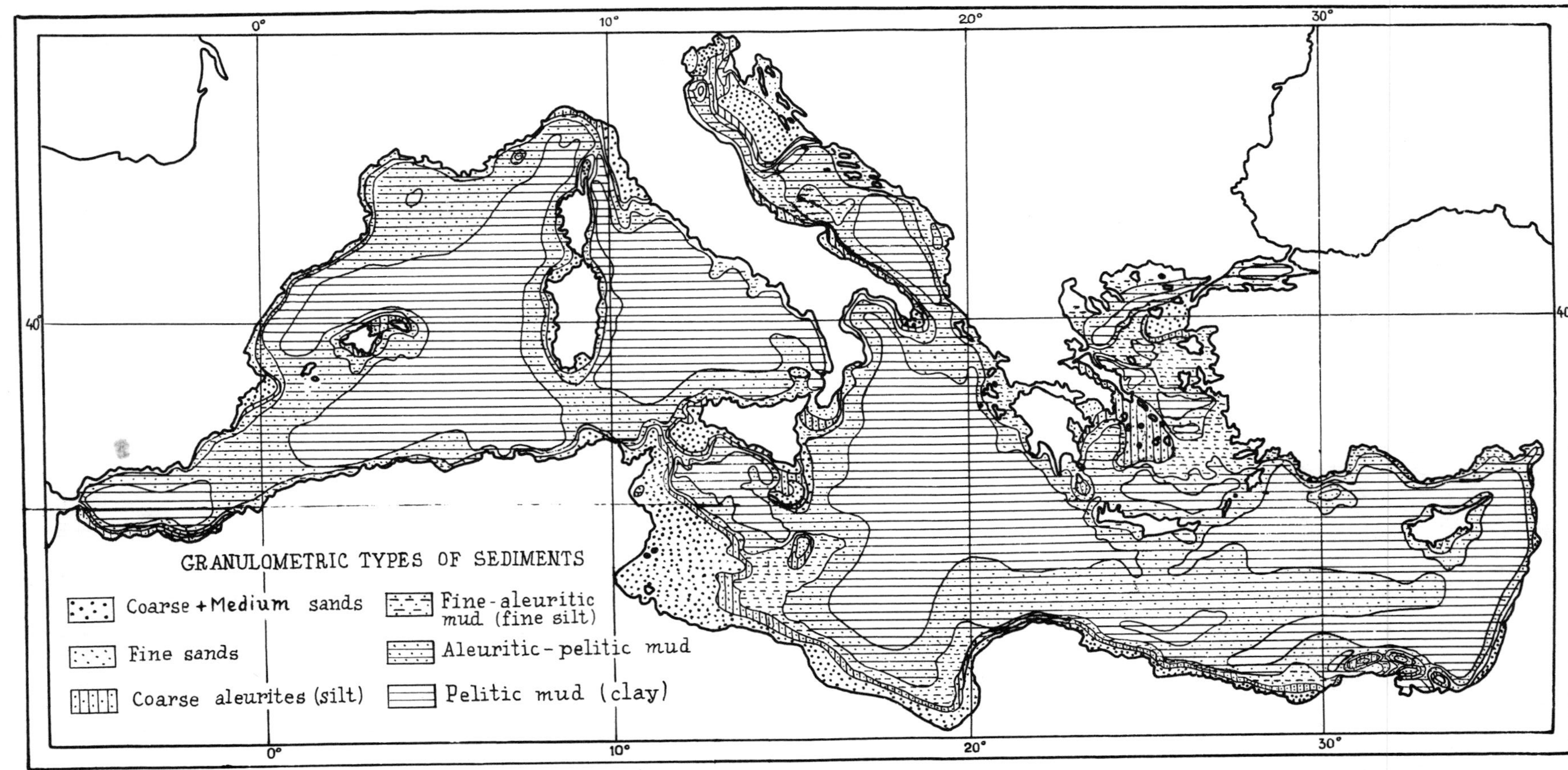

Figure 3. Distribution of the principal granulometric types of bottom sediments in the uppermost layer (0–5 cm) of the Mediterranean Sea (according to Emelyanov, 1965a with some modifications).

of the Adriatic Sea, on the African Shelf and the African-Sicilian sill and are the result of an active hydrodynamic regime on broad shelves with a level bottom. Sands in these regions are relict and occur at depths to 200 m (Pigorini, 1968; van Straaten, 1965). These sands originally accumulated at shallow depths and in coastal environments and are now found at greater level. Sands are found at even greater depths (down to 740–1830 m) in straits, near the sites of submarine slides, and in regions of turbidity current activity.

Coarse aleurites occur in the uppermost part of the continental and island slopes and separate sands from fine-aleuritic mud. However, this type of sediment is extremely rare.

Fine-aleuritic muds are transitional between typical muds and nonconsolidated sediments (coarse aleurites and sands). They occur on the upper and middle parts of the continental slopes and are also widespread in the Aegean Sea.

Aleuritic-pelitic mud is a widely distributed type of sediment which is characteristic of the Mediterranean Sea. This type of sediment occupies a narrow band in geosynclinal regions adjacent to the base of continental and island slopes. In preplatform sections, especially in arid zones where the supply of terrigenous material is low and the processes of biogenic deposition are intensive, aleuritic-pelitic muds form a continuous cover (for instance, on the central Mediterranean Ridge between Africa and Crete).

Pelitic muds occur in depressions more than 2500 m deep. At shallower depths they occur in basins and near the deltas of large rivers, as mentioned by d'Arrigo (1936).

GENETIC TYPES OF SEDIMENTS

Classification

We have distinguished four major genetic types of sediments:

I. *Terrigenous sediments* (< 30% of $CaCO_3$)
  - Terrigenous (< 10% of $CaCO_3$)
  - Terrigenous, low calcareous (10 to 30% of $CaCO_3$)
  - Terrigenous, low calcareous-low ferruginous-low manganese (10 to 30% $CaCO_3$; 5 to 10% of Fe; 0.20 to 5.0% of Mn)
  - Terrigenous, low calcareous-low ferruginous (10 to 30% of $CaCO_3$; 5 to 10% of Fe).

II. *Volcanic sediments*
  - Volcanoclastic, low calcareous (pyroclastic and fragments of volcanite make up more than 10%; 10 to 30% of $CaCO_3$)

III. *Biogenic sediments* (>30% of $CaCO_3$)
  - Foraminiferal, coccolithic foraminiferal
    - Calcareous (30 to 50% of $CaCO_3$)
    - High calcareous (> 50% of $CaCO_3$)
  - Foraminiferal, coccolithic and pteropodal
    - Calcareous (30 to 50% of $CaCO_3$)
    - High calcareous (>50% of $CaCO_3$)
  - Shelly
    - Calcareous (30 to 50% of $CaCO_3$)
    - High calcareous (> 50% of $CaCO_3$)
  - Foraminiferal and shelly
    - Calcareous (30 to 50% of $CaCO_3$)
    - High calcareous (> 50% of $CaCO_3$)
    - Calcareous, low manganese (30 to 50% of $CaCO_3$; 0.20 to 5.0% of Mn)

IV. Chemogenic and biogenic high calcareous sediments (> 50% of $CaCO_3$)
  - Oolitic
  - Oolitic and shelly
  - Biogenic sediments with high content of calcareo-argillaceous concretions, calcite and dolomite.

Terrigenous Sediments

Terrigenous sediments occur in the geosynclinal areas of the Mediterranean, in the Algerian-Provençal Basin and in the region of the Nile delta-front (Figure 4).

1. The sediments in geosynclinal areas (the Sea of Marmara, northern parts of the Aegean, Adriatic, Ionian and Levant seas, eastern parts of the Tyrrhenian Sea) are highly variable both in granulometric composition and physical properties (Emelyanov, 1965a). Usually they are of gray color, their moisture content ranging from 24 to 81 per cent. The sediments are characterized by a relatively average content of chemical components (Tables 1, 2, 3) but they differ considerably in mineralogic composition as a consequence of their differing provenance. Each mineralogic province (Figure 5) has its own distinctive mineral suite; the quartz-to-feldspar ratio is usually less than 1. For example, plagioclase, hornblende, epidote-zoisite are typical of the sediments from the Sea of Marmara. The sediments from the northern part of the Adriatic Sea include feldspar, quartz, calcite, garnet, green hornblende, fibrous amphiboles, and apatite. The light subfraction (sp. gravity less than 2.9) of the eastern part of the Tyrrhenian Sea is characterized by volcanic glass and ash as well as feldspar and quartz. The heavy subfraction is dominated by clinopyroxenes. The highest content of mica, especially muscovite, and abundant slate fragments and ore minerals are recorded in the sediments south of Messina Strait and Calabria. In the area of Crete-Rhodes-Anatolia, colored mica,

Table 1. Chemical composition (%) of terrigenous (< 10% $CaCO_3$) bottom sediments of the Mediterranean Sea.

	Sands		Coarse aleurites	Aleuritic-pelitic mud		Pelitic mud (clay)			Sediment type
	313	402	552	317	842	594	837	839	Station number
	1833 m	23 m	2136 m	3252 m	510 m	1674 m	724 m	26 m	Water depth
Composition									
$CaCO_3$	3.20	8.17	6.00	8.19	7.50	4.55	6.37	3.64	
$SiO_{2\,amorph}$	1.35	1.86	0.75	1.32	—	0.87	—	—	
C_{org}	0.40	0.20	0.40	0.40	0.80	1.00	0.90	1.30	
Fe	—	0.47	4.06	4.56	5.99	—	—	—	
Mn	—	0.02	0.07	0.08	0.10	0.11	—	—	
Ti	—	0.12	0.42	0.37	0.94	0.94	—	—	
P	—	0.01	0.08	0.09	—	0.08	—	—	

Table 2. Chemical composition (%) of some terrigenous low calcareous (10–30% $CaCO_3$) sediments of the Mediterranean Sea.

		Sands			Fine aleuritic mud	Aleuritic-pelitic mud			Pelitic mud (clay)		
	Station	403	516	825-A	518	Range of values	Mean	Number of samples examined	Range of values	Mean value	Number of samples examined
Composition	Water depth	21 m	95 m	18 m	708 m						
$CaCO_3$		14.10	25.92	12.96	11.14	13.19–29.56	22.00	14	11.83–29.11	20.86	33
$SiO_{2\,amorph}$		2.01	0.75	—	2.93	0.76– 0.99	0.85	6	0.55– 1.00	0.81	16
C_{org}		0.20	0.70	0.60	0.20	0.30– 1.20	0.58	13	0.40– 1.20	0.70	31
Fe		0.14	1.96	—	1.68	3.48– 5.60	4.30	8	3.14– 6.19	4.16	26
Mn		0.02	0.04	—	0.07	0.04– 0.19	0.10	10	0.02– 0.46	0.09	25
Ti		0.14	0.28	—	0.13	0.23– 0.62	0.40	9	0.09– 0.78	0.38	23
P		0.07	0.02	—	0.02	0.03– 0.09	0.06	6	Trace–0.10	0.06	14

especially yellow biotite, and plagioclase and garnet are dominant. The same minerals usually prevail in the suspended matter of the northern Mediterranean (Emelyanov and Shimkus, this volume).

Terrigenous minerals can be traced up to 300 km from their sources in geosynclinal regions (Emelyanov, 1968). In the deep area south of Messina Strait, coarse sediments (sand, coarse aleurites) have been deposited by turbidity currents (Ryan and Heezen, 1965). Transportation of mica in this area is observed from Sicily to Sidra Bay, a traverse of about 600 km (Figure 6). Probably this mica is transported by deep currents even further eastward. The high content of yellow and brown mica characteristic of the Sicily and East-Sicilian mineralogical province may be traced nearly as far as the Lebanon-Cyprus coasts.

Illite is the most abundant clay mineral[1] (above 50–70 per cent); the second most important is kaolinite (30–50 per cent). Montmorillonite is rare (Figure 7), although it is somewhat more abundant (and illite is correspondingly less common) in the area of Rhodes-Anatolia-Cyprus. Montmorillonite is probably transported to this area by the Nile.

The absolute ages of terrigenous and volcanic minerals in sediments of geosynclinal regions are not constant but vary within the range from 87 to 200 m.y. The youngest minerals occur in the northern part of the Aegean Sea, while the oldest occur in Sidra Bay and Levant Sea.

2. Terrigenous sediments in the Algerian-Provençal Basin are texturally pelitic and aleurite-pelitic muds. The upper layer of muds contains higher moisture (42.1 to 62.6%). It is gray and yellow in color. In pelitic muds, the subcolloidal fraction (less than 0.001 mm) is the most prevalent, whereas sandy material and coarse aleurites are quite rare (Emelyanov, 1965a). These are typical deep-sea sediments of the transitional zone between geosynclinal and platform regions on the one hand, and between arid and humid climatic conditions on the other hand. The

[1] X-ray diffraction records were taken with URS-5OYM equipment operating at 32 kv, 8 mA, with system slits of 0.25; 0.25; 0.5 mm; the rate of scanning was 4°/min. Samples were prepared in the form of a thin layer on a glass slide. The X-ray method is described by Sudo *et al.* (1961) and Rateev *et al.* (1966).

Table 3. Total chemical composition (in per cent) of the upper sediment layer in the Mediterranean Sea.

Station Number	Co-ordinates Latitude (N)	Co-ordinates Longitude (E)	Depth, in m	SiO_2	Al_2O_3	Fe_2O_3 total	TiO_2	MnO	P_2O_5	CaO	MgO	Na_2O	K_2O	Loss on ignition	Total	Corg	CO_2	$CaCO_3$
								Coarse aleurites										
359	32°11′0 N	16°11′5 E	297	31.60	4.66	3.69	1.00	0.04	0.09	26.55	2.93	1.19	1.23	26.3	98.9	0.5	19.5	44.3
								Fine aleuritic mud										
316	38°43′6 N	15°20′3 E	1707	46.16	20.58	7.17	0.90	—	0.15	10.90	0.94	2.67	2.54	9.7	100.7	0.4	—	10.6
351	36°03′2 N	14°00′3 E	567	30.60	6.29	4.46	0.80	0.04	0.11	26.50	2.29	1.19	0.99	26.1	99.3	0.4	23.1	52.5
								Aleuritic-pelitic mud										
417	37°07′1 N	26°09′5 E	306	11.20	4.60	2.30	1.10	0.05	—	37.65	5.70	2.12		36.2	100.9	0.7	30.0	68.2
304	33°47′3 N	20°53′3 E	2575	20.96	5.18	4.12	0.50	—	0.11	33.70	1.80	1.28	0.92	31.2	99.8	0.7	28.0	63.7
334	37°17′4 N	10°28′5 E	183	32.80	10.50	4.30	0.60	0.01	0.15	20.75	1.10	1.62	1.10	25.0	98.0	0.9	16.4	37.3
406	34°54′4 N	34°26′6 E	820	27.34	10.56	5.94	0.50	0.15	—	23.90	3.20	1.91	1.10	26.2	100.8	1.2	10.5	23.9
410	36°13′7 N	35°33′8 E	175	28.48	10.99	6.04	0.60	0.04	0.10	20.90	5.40	2.00	1.17	25.0	100.7	0.6	16.2	36.8
416	36°32′2 N	26°48′6 E	587	27.42	9.31	4.04	1.00	0.09	0.08	24.10	4.35	1.85	1.41	26.3	100.1	0.4	19.5	44.3
								Pelitic mud										
310	37°50′1 N	18°51′6 E	3382	29.10	12.08	4.82	0.90	0.08	—	21.60	4.16	2.67		25.2	100.6	0.7	18.4	41.9
311	37°51′4 N	17°55′7 E	2791	32.50	11.87	5.08	0.80	0.09	0.12	20.00	1.42	2.57	2.44	24.3	101.5	0.5	8.2	18.6
319	40°26′3 N	14°20′6 E	784	44.12	18.60	6.90	0.60	0.22	—	8.85	2.88	2.08	3.17	14.0	101.4	0.7	5.8	13.2
343	36°31′8 N	11°39′0 E	289	28.40	9.64	4.38	0.70	0.02	—	23.85	2.20	1.93	1.35	27.3	99.7	0.5	17.5	39.8
380	35°22′0 N	27°17′5 E	1231	30.12	10.41	5.34	0.60	0.05	0.14	23.65	1.80	1.42	1.18	25.0	99.7	0.9	15.1	34.3
384	36°10′3 N	28°20′4 E	2771	33.00	8.98	6.42	1.00	0.06	—	17.70	7.70	3.00		22.9	100.7	1.0	11.1	25.4
387	34°28′8 N	28°43′2 E	2351	18.60	7.31	3.69	trace	0.06	0.09	31.00	3.26	1.96	1.03	33.2	100.2	0.5	24.9	56.6
396	34°39′4 N	31°58′5 E	2295	26.80	9.02	5.08	1.00	—	0.09	24.20	3.45	1.87	0.98	27.1	99.4	0.4	24.0	54.6
−398	36°11′0 N	31°59′3 E	2203	36.12	10.50	6.70	0.60	0.08	—	19.00	3.90	1.99	1.79	20.6	101.2	1.2	10.5	23.9
411	35°57′0 N	35°04′8 E	1150	12.18	4.11	2.04	0.60	—	—	36.90	5.73	2.80		36.0	100.5	—	—	—
412	35°47′8 N	34°55′4 E	698	29.86	10.69	6.81	0.50	0.06	0.12	21.30	4.05	1.97	1.13	25.0	101.5	0.5	15.8	35.9
415	36°00′0 N	32°31′2 E	1063	33.60	12.14	5.86	0.80	0.04	—	19.00	4.26	2.92		22.2	101.0	0.4	19.5	44.3

Stations 311, 384, 398, 406, 411 rich in terrigenous sediment; stations 316 and 319 contain volcanoclastic sediments and sediments rich in pyroclastic material; all other samples contain biogenic sediments. Analyses were performed in the Chemical Analytical Laboratory of the Near Black Sea-Geological Exploration Expedition (Odessa).

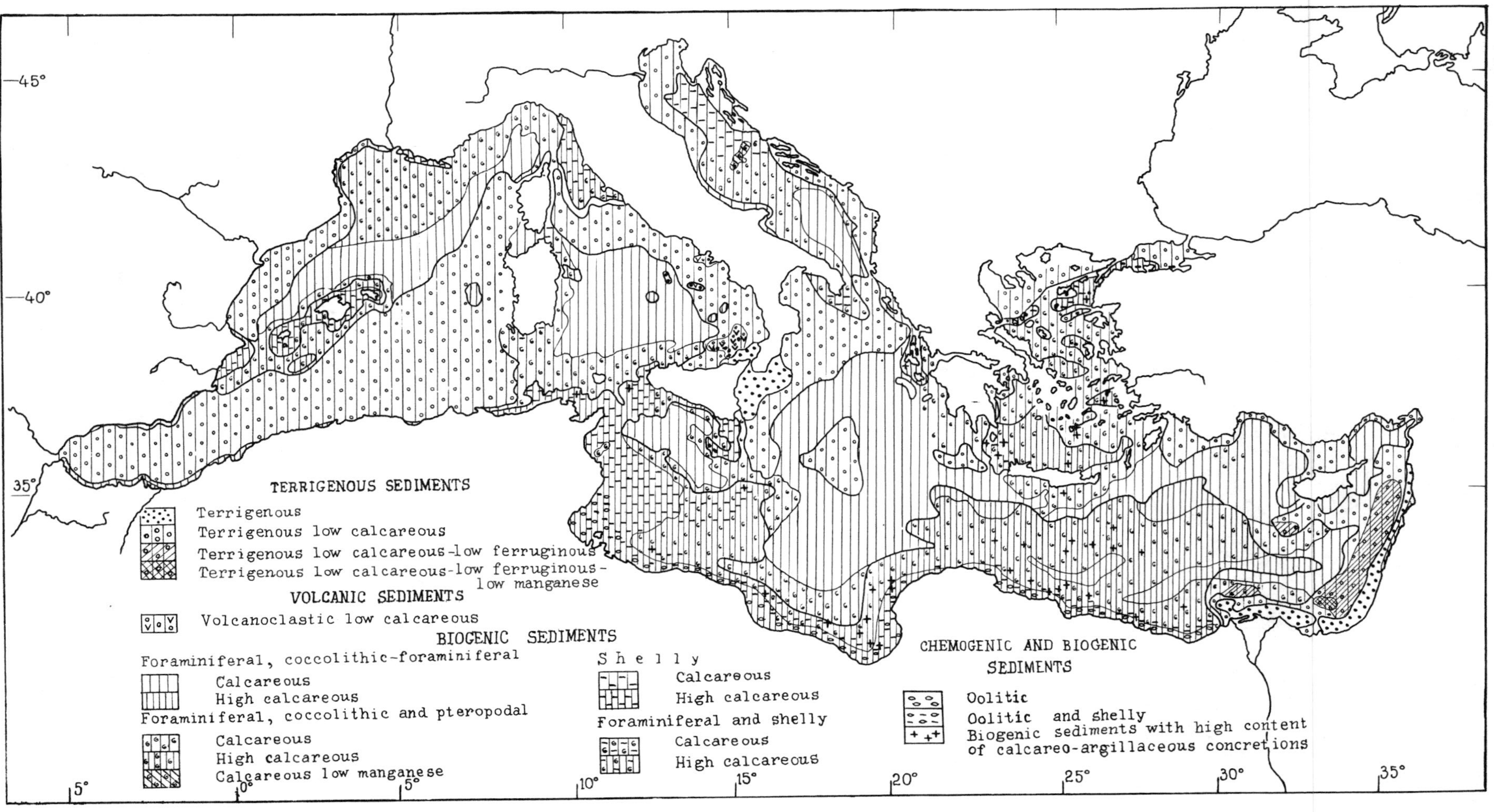

Figure 4. Distribution of bottom sediments (uppermost layer 0–5 cm; sometimes 0–10 or 0–20 cm). Genetic classification discussed in text.

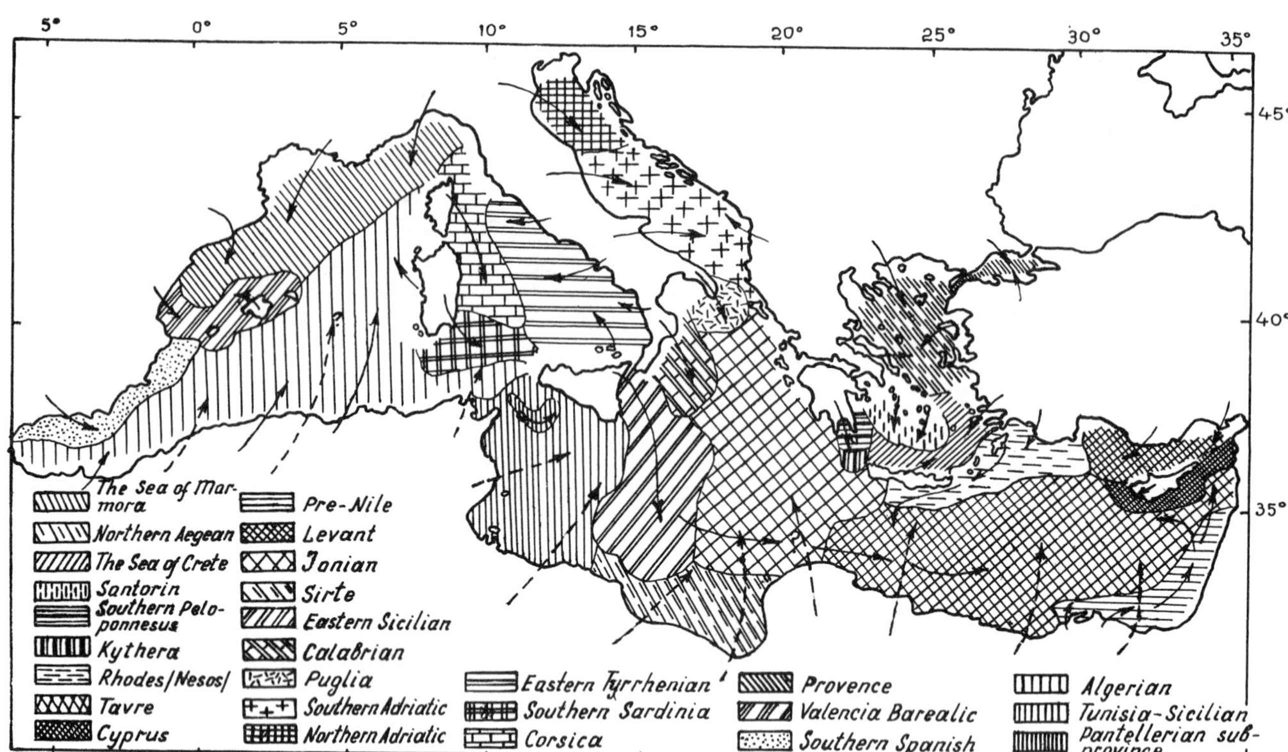

Figure 5. Mineralogical provinces of the coarse-aleuritic (0.1–0.05 mm) fraction of bottom sediments from the upper layer of the Mediterranean Sea (Emelyanov, 1968, with some modifications). The solid arrows show the main sources and supply paths of detrital material; the dotted arrows show those of eolian material.

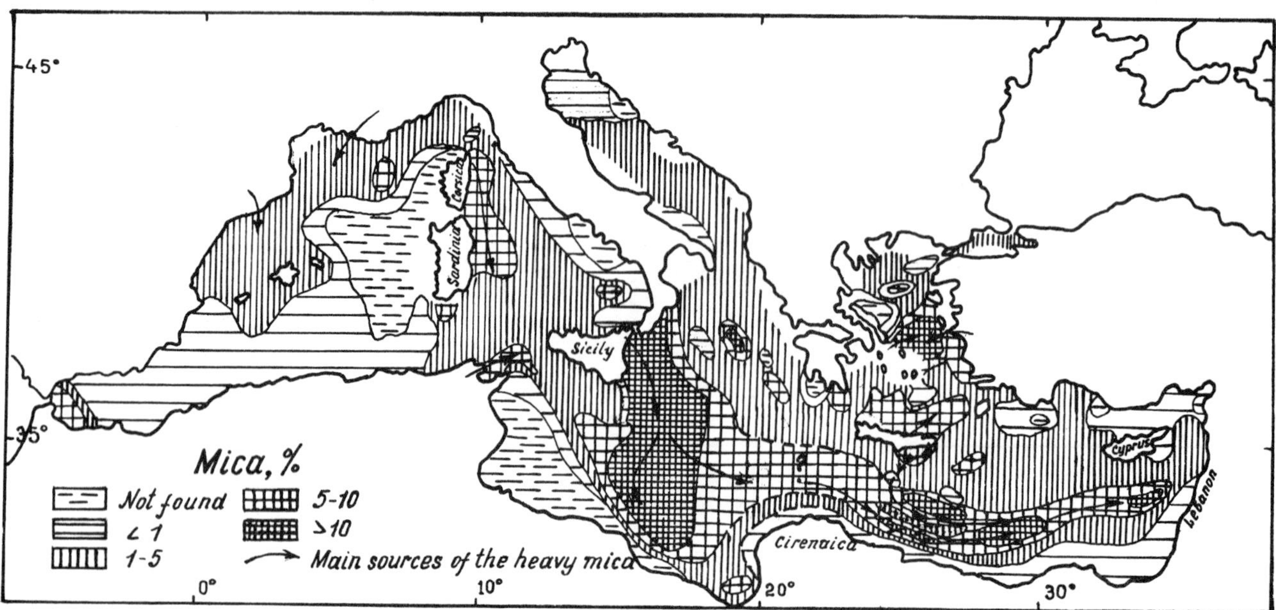

Figure 6. Distribution of yellow and brown mica in heavy (specific gravity > 2.9) coarse-aleuritic (0.1–0.05 mm) subfraction from the upper layer of Mediterranean Sea sediments (220 stations).

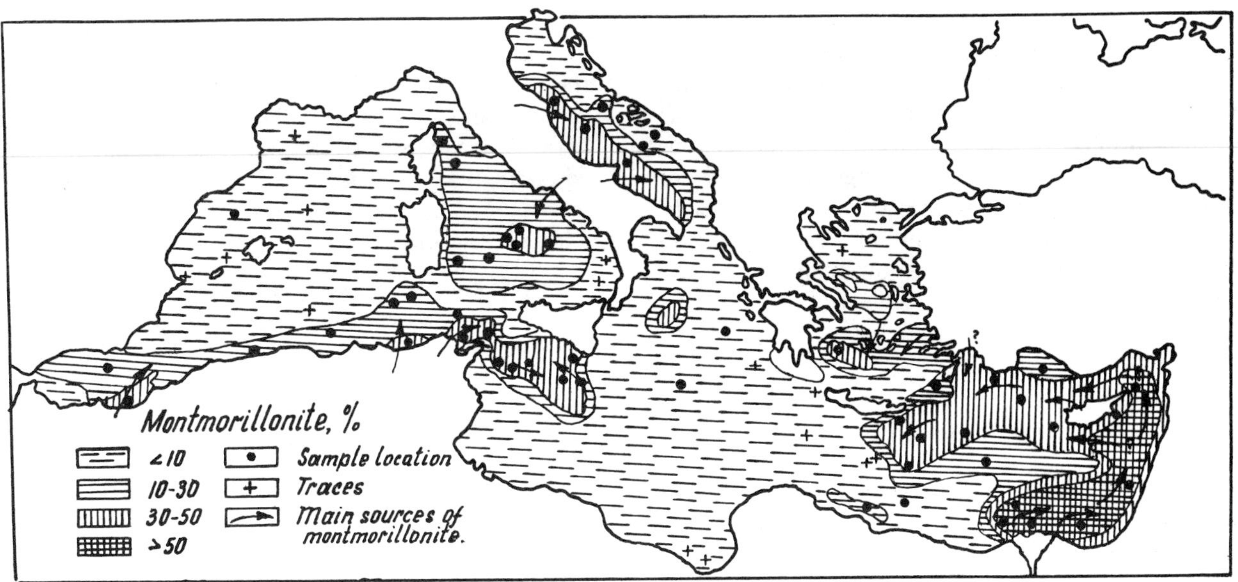

Figure 7. Distribution of montmorillonite in the upper layer of bottom sediments of the Mediterranean Sea (in % of total fraction less than 0.001 mm, devoid of carbonates). According to Rateev *et al.*, 1967.

mineralogic composition of the given climatic province is generally mixed, being affected by a great variety of source factors. The sediments are characterized by a rather low content of fragmental minerals (especially heavy ones) of coarse aleuritic grade. Quartz, stable minerals (rutile, brookite, anatase, tourmaline and zircon), limonite-hematite, and epidote-zoisite usually prevail in the detrital mineral suite. Contents of mica, amphiboles and pyroxenes are very low.

Illite is the most common clay mineral (above 70 per cent). Contents of kaolinite and montmorillonite are somewhat higher (up to 10–30 per cent), but only near the continental slope of Africa.

The absolute ages of terrigenous and volcanic minerals in southern geosynclinal regions are 100 to 150 m.y., and 150 to 200 m.y. in the central regions.

3. In the Nile submarine delta the distribution of terrigenous sediments is related to the high input of detrital material carried by this river. This transportation of material can be traced distinctly as far as Cyprus-Syria by characteristic physical properties as well as by granulometric, mineralogic and chemical analyses.

According to the distribution of montmorillonite the finest detrital particles of the Nile input actually reach the continental slope of Turkey.

The Nile terrigenous sediments are generally close to quartz sands and pelitic muds. Quartz sands lie on the shelf and grains of quartz and feldspar are often covered by a ferruginous coating, which may be indicative of their eolian origin. There are high quantities of ore minerals, clinopyroxenes, hornblende and epidote. Absolute ages of terrigenous minerals in the sands range from 335 to 430 m.y., *i.e.* they are approximately as old as the sand of the Arabian Desert (about 400 m.y.).

Muds occur in limited regions on the shelf and spread continuously on to the continental slope and basin floor between the Nile Delta, Cyprus and Syria.

The muds are characterized by high water content (from 55.3 to 68.0%) and gray color (sometimes with spots or rust color). The muds in this area belong to the Pre-Nile province, with a characteristic mineral content, similar to the mineralogic composition of muds from the Nile River (Emelyanov, 1968).

Among the light minerals quartz, colored mica and calcite are the most common. The quartz-to-feldspar ratio is much greater than 1. The percentage of heavy minerals in sediments is rather high (from 4.29 to 8.29%). Clinopyroxenes, epidote-zoisite, ore minerals, hornblende, zircon are the most common heavy minerals. Clay minerals include 60 to 80% of montmorillonite and 10 to 30% of kaolinite.

The absolute ages of terrigenous minerals in the muds range from 85 to 200 m.y. Generally, the younger sediments are also finer in grade and contain higher proportions of clay minerals. This suggests that the clay minerals are considerably younger than detrital material from African deserts and may have developed from eolian erosion of volcanic rocks which occur in the upper course of the Nile River.

Since the muddy deposits supplied by the Nile are markedly enriched in iron (7.8%) and titanium (1.38%), the concentration of these components in muds of the Pre-Nile province is greatest as compared with the other terrigenous marine sediments (Emelyanov, 1966b). Many small globules of ferric hydrate and acicular rutile are found in pelite. Acicular rutile probably results from decomposition of terrigenous minerals, especially mica. The highest content of manganese (0.46%) was found at station 840. Sediments with the highest contents of Fe and Mn still fall within the low ferruginous-low manganese category of our classification. The muds of the Pre-Nile province are also characterized by a higher content of C_{org} (0.6–1.3%).

Volcanic Sediments

Volcanic sediments are characterized by a strictly local distribution: they occur in the lagoon of Santorin, the submarine dome of Stromboli, and the Gulf of Naples (Figure 4).

The volcanoclastic sediments of the Santorin lagoon are represented by sands, coarse aleurites and fine aleuritic muds (Figure 8). They consist mainly of the weathered products of volcanite, volcanic glass, feldspar, hypersthene and magnetite (Table 4). The sediments are enriched in iron, manganese and, to a lesser degree, in phosphorus, copper and vanadium (Table 5). The titanium content of the sediments is the same as in sediments outside the caldera, *i.e.*, several times less than in the volcanoclastic deposits of the Denmark Strait (Atlantic), where we have estimated up to 1.96% of Ti and up to 0.33% of P. The content

Table 4. Principal components (%) in the coarse-aleuritic fraction (0.1–0.05 mm) of volcanoclastic sediments in and adjacent to the caldera of Santorin (according to Butuzova, 1969).

Components	Within the caldera proper			Beyond the caldera		
	minimum	maximum	mean (25 samples)	minimum	maximum	mean (16 samples)
Fragments of volcanogenous rocks	28	58	42	6	37	20
Volcanic glass	15	43	20	31	68	50
Feldspars	10	25	14	2	15	9
Pyroxenes	3	7	4	1	3	2
Ore minerals	1	4	3	1	3	2
Siliceous organisms	2	7	4	trace	5	2.5
Calcareous organisms	5	10	6	15	66	42
Chemogenic carbonates	1	3	2	1	14	10

Table 5. Distribution of Fe, Mn, Ti, P, V, Ni, Co, Cu (in%) in the volcanoclastic sediments in the lagoon of Santorin (according to Butuzova, 1969).

Elements	Type of sediments	Number of samples	Natural dry sediments			On carbonate free basis		
			min.	max.	average	min.	max.	average
	Sands	6	3.38	5.76	4.40	3.95	6.19	4.67
Fe	Coarse aleurites	10	4.08	4.79	4.52	4.24	5.17	4.93
	Fine aleuritic mud	19	4.23	17.15	5.97	4.37	17.54	6.37
	Sands	6	0.10	0.14	0.12	0.11	0.15	0.14
Mn	Coarse aleurites	10	0.12	0.18	0.15	0.13	0.19	0.17
	Fine aleuritic mud	19	0.20	0.62	0.35	0.23	0.69	0.38
	Sands	6	0.32	0.51	0.41	0.37	0.52	0.44
Ti	Coarse aleurites	10	0.34	0.42	0.39	0.38	0.43	0.40
	Fine aleuritic mud	19	0.29	0.42	0.38	0.33	0.44	0.40
	Sands	6	0.05	0.08	0.06	0.06	0.08	0.07
P	Coarse aleurites	10	0.05	0.06	0.06	0.05	0.07	0.07
	Fine aleuritic mud	19	0.04	0.07	0.06	0.04	0.08	0.07
	Sands	6	76	130	107	79	141	116
$V(10^{-4})$	Coarse aleurites	9	101	178	142	105	201	152
	Fine aleuritic mud	19	117	188	155	121	203	166
	Sands	6	20	40	27	20	47	29
$Cr(10^{-4})$	Coarse aleurites	10	16	33	26	16	38	28
	Fine aleuritic mud	19	18	40	31	19	45	34
	Sands	6	12	18	14	12	21	15
$Ni(10^{-4})$	Coarse aleurites	10	10	28	17	10	32	18
	Fine aleuritic mud	19	14	34	22	15	38	23
	Sands	6	10	18	14	10	19	15
$Co(10^{-4})$	Coarse aleurites	10	12	21	16	13	24	18
	Fine aleuritic mud	19	12	26	18	13	29	20
	Sands	6	24	63	37	25	71	41
$Cu(10^{-4})$	Coarse aleurites	10	34	66	47	35	72	51
	Fine aleuritic mud	19	32	66	48	35	70	52

of chromium, nickel and cobalt is somewhat less than in calcareous deposits beyond the caldera. The data show that volcanoclastic and hydrothermal processes favor an increase in iron and manganese concentrations and, to a lesser degree, an increase in phosphorus, vanadium and copper; but at the same time the sediments are not enriched in chromium, nickel and cobalt.

Around the submarine volcanic dome of Stromboli, volcanoclastic sediments are found at a depth of 1797 m (Station 316: 38°43′6″N, 15°20′3″W). They are fine-aleuritic muds of yellow-gray color with a high content of black opaque particles of volcanic ash (43.3% of the light 0.1–0.05 mm subfraction), colorless glass (15.2%) and sodic plagioclase (28.9%); the heavy subfractions consist of detrital rock and various opaque aggregates (38.9%) as well as clinopyroxenes (43.0%).

Volcanoclastic sediments with prevalent volcanic glass are also found in the Gulf of Naples at depths varying from 0 to 800 m (Müller, 1961 and 1964). These sediments appear to result from weathering of volcanics in the adjacent coastal areas. The chemical composition of volcanoclastic sediments and muds rich in pyroclastics is given in Table 6.

No bottom sediments with a content of pyroclastic material exceeding 10% has been encountered anywhere else in the Mediterranean. Thus, the role of volcanism in sedimentation appears to have been less over the past several thousand years than during the Pleistocene. Previous studies (Mellis, 1948; Norin, 1958; Duplaix, 1958; Ryan *et al.*, 1965) and our own data have revealed widespread deposits of Pleistocene ash which occur in layers up to 10–22 cm thick and are distributed not only near volcanoes but also in areas such as the Tyrrhenian Abyssal Plain, the Ionian Sea and the northern part of the Levant Sea (Mellis, 1954; Ninkovich and Heezen, 1965), some hundreds of kilometers from the eruptive sites.

Ferruginous deposits of volcanic origin are being formed now in the southwestern part of Neo Kamenis (Figure 8) where the thickness of a bright rusty ferruginous stratum reaches 70 cm. Beneath this layer, black ferruginous mud is found. The rust-colored

Table 6. Granulometric and chemical composition (% per dry weight of sample) of volcanoclastic and terrigenous (enriched with pyroclastics) sediments in the Tyrrhenian Sea and Sea of Crete.

Station No.	316	317	318	319	518
Depth	1797 m	3252 m	1139 m	784 m	708 m
Latitude (N)	38°43′6	39°10′7	39°30′9	40°26′3	36°39′6
Longitude (E)	15°20′3	14°54′8	14°32′1	14°20′6	25°51′1
			Grain size distribution		
> 0.1 mm	11.62	1.84	3.28	3.72	1.89
0.1–0.05 mm	20.03	13.10	3.87	0.67	2.34
0.05–0.01 mm	25.53	34.13	17.00	10.91	50.50
0.01–0.005 mm	17.29	11.63	—	11.33	17.13
0.005–0.001 mm	11.24	20.02	—	32.15	17.94
< 0.001 mm	14.29	19.28	—	41.22	10.19
0.1–0.01 mm	45.56	47.23	20.87	11.85	32.84
0.01–0.001 mm	28.53	31.65	—	43.48	35.07
< 0.01 mm	42.82	50.93	75.85	84.70	45.26
Md	0.016	0.009	—	0.002	0.012
So	3.4	4.3	—	3.1	1.8
			Chemical composition		
$CaCO_3$	10.50	9.19	25.00	13.20	11.14
$SiO_{2 amorph.}$	—	1.32	1.00	0.87	2.93
$C_{org.}$	0.40	0.40	0.80	0.70	0.20
Fe	4.80	4.56	4.00	4.36	1.68
Mn	—	0.08	—	0.17	0.07
Ti	—	0.37	—	0.36	0.13
P	—	0.09	—	—	0.02
Content of volcanic glass and ash in the fraction 0.1–0.05 mm (%)	58.5	33.0	23.3	74.5	88.9

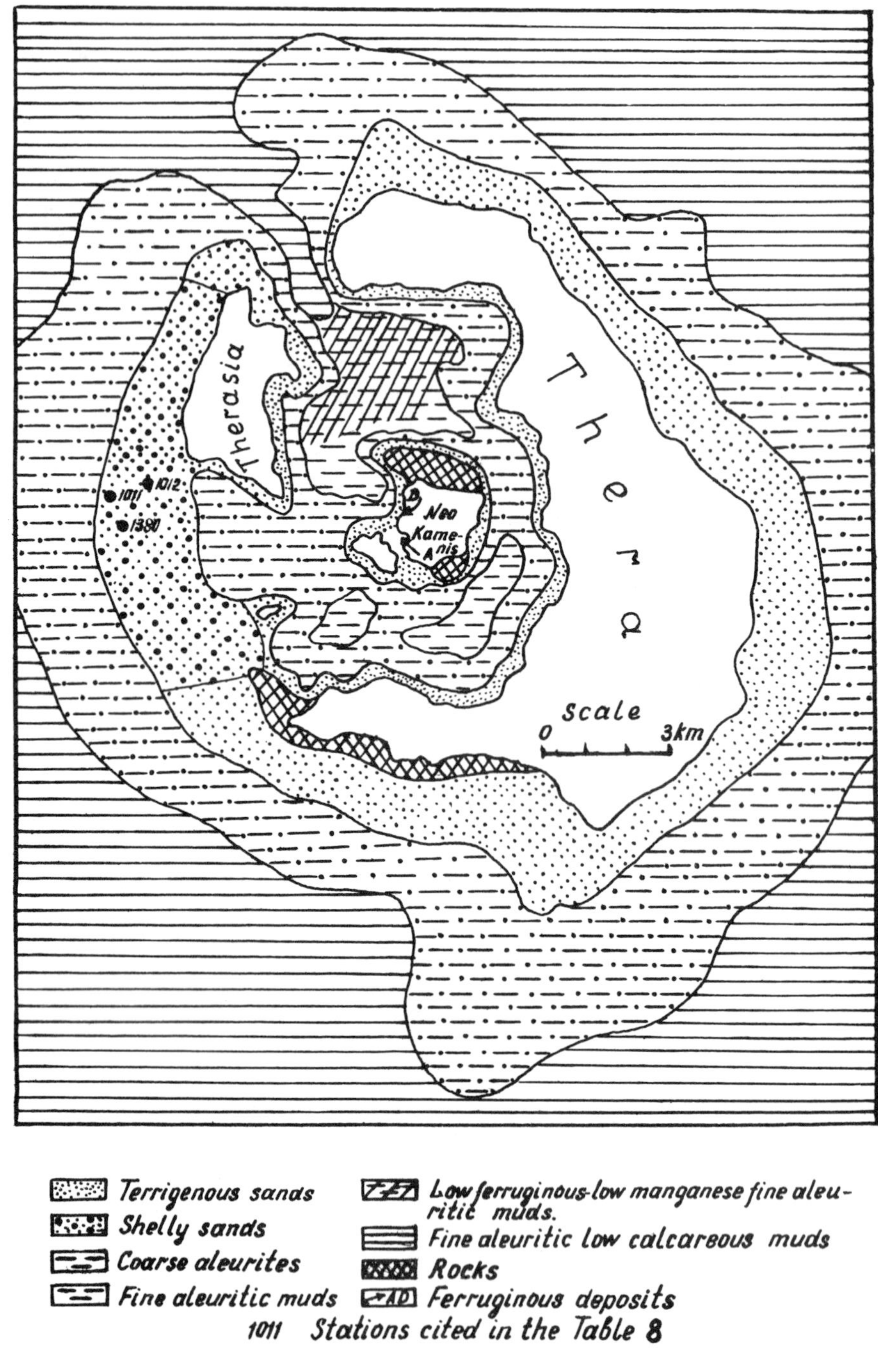

Figure 8. Distribution of sediment types in the vicinity of Santorin volcano, Aegean Sea (Butuzova, 1969).

sediment consists of amorphous ferruginous masses; sometimes particles of ash, fragments of andesite, diatomaceous skeletons, and authigenic vivianite also occur in these masses. In the black underlying stratum, hydrotroilite and amorphous masses of brownish and greenish color prevail. The ferruginous sediments are somewhat enriched in amorphous silica, P_2O_5 (up to 1.64%), Pb (0.0285%), Zn (0.00139%), and B (0.02%). The percentage of V, Ni, Cu, Cr and Co is low (Table 7). The upper brown stratum is readily distinguished from the underlying black stratum by chemical composition (Table 7).

These ferruginous sediments are ascribed to submarine volcanic activity (Butuzova, 1969). The influence of post-volcanic hydrothermal processes upon sedimentation is only evident within the lagoon of Santorin.

Table 7. Chemical composition (in%) of ferruginous sediments in various bays of Neo-Kamenis located within the caldera of Santorin (according to Butuzova, 1969)[a]

Sample No.	69–I	69–II	214	215–I	215–II	216–I	216–II	216–III	216–V
SiO_2	19.19	18.08	8.77	13.13	17.48	22.77	24.59	24.59	72.70
TiO_2	—	0.01	—	trace	—	trace	—	—	0.13
Al_2O_3	1.93	1.20	0.30	0.60	0.31	1.26	0.99	0.37	0.97
Fe_2O_3	50.53	49.05	53.34	47.43	44.87	30.57	21.82	41.48	0.12
FeO	0.95	0.18	0.43	0.43	0.57	4.14	6.50	1.22	0.22
CaO	1.56	1.52	1.53	1.28	1.53	1.65	2.34	1.05	0.85
MgO	0.18	0.70	0.99	1.05	0.90	1.40	1.40	0.48	0.05
MnO	0.23	0.26	0.04	0.01	0.06	0.02	0.06	0.06	0.02
P_2O_5	0.32	0.39	0.21	0.25	0.14	1.64	0.73	1.37	0.01
Na_2O	2.59	2.57	4.77	6.75	5.79	6.75	8.76	0.89	0.41
K_2O	0.42	0.42	0.48	0.60	0.70	0.68	1.03	0.22	0.30
H_2O	11.17	11.26	11.32	13.27	11.14	9.72	6.75	9.13	8.23
H_2O^+	7.80	9.22	10.31	5.58	8.44	5.91	9.36	13.18	5.97
CO_2	1.86	1.62	0.70	1.20	1.16	1.36	3.10	—	0.06
C_{org}	1.21	1.02	2.13	1.30	1.22	3.50	3.45	2.32	3.40
SO_3	—	—	—	0.23	—	0.80	—	—	—
Spyr.	—	0.66	0.04	0.47	0.83	0.76	0.13	0.07	—
Cl	—	1.42	3.33	7.86	4.85	7.47	6.07	0.56	0.17
$S_{element}$	—	0.14	0.08	—	0.06	1.73	1.44	1.46	3.52
V	0.0038	0.0036	0.0051	—	—	0.0099	—	—	—
Cr	—	0.0033	—	—	—	0.0003	—	—	—
Ni	0.0033	0.0005	—	—	—	0.0013	0.0014	—	—
Co	0.0003	—	—	—	—	0.0003	—	—	—
Cu	0.0005	0.0006	0.0009	0.0027	—	0.0022	0.0031	—	—
B	—	—	—	0.02	—	0.01	—	—	—
Pb	—	—	—	0.0285	—	0.0013	—	—	—
Zn	—	—	—	0.0139	—	0.0022	—	—	—

[a] Samples 69–I. 69–II—ferruginous film covering blocks of andesite and dacites near edge of water in bay D; Sample 214–red sediment from the upper stratum of bay D; samples 215–I, 215–II—red sediment from the upper stratum of various parts of bay A; samples 216–I, 216–II—black sediment from the lower stratum of various parts of bay A; Sample 216–III—black sediment (Sample 216–II) with sea water salts washed out; Sample 216–V—the same sediment treated by alcohol saturated with HCl.

Biogenic Sediments

Biogenic sediments are the most widely distributed type in the Mediterranean Sea. In the arid platform part of the basin they occur from the coast to depths of 3500 to 4000 m. In geosynclinal sections and near the mouths of large rivers the biogenic sediments are separated from the shore by a wide zone of terrigenous deposits. The northern geosynclinal regions of the sea and the Algerian-Provençal Basin are characterized by calcareous biogenic sediments with $CaCO_3$ varying from 30 to 50%. Highly calcareous sediments with $CaCO_3$ ranging from 50 to 90.7% predominate in the arid zone of the basin (from the Tunisian Strait to the Nile prodelta) and in some sections of the geosynclinal zone (the Sea of Crete, the Balearic Islands).

The sediments in shallow waters of the Aegean Sea and the Sea of Crete consist of shell sand and foraminiferal and coccolithic-foraminiferal muds. In the deeper waters, globigerinas are the most common foraminiferal group. The muds of the Sea of Crete contain abundant coccoliths, such as *Syracosphaera pulchra* (Lohman) and *Helicosphaera carteri* (Wallich) and others (Ninkovich and Heezen, 1965). Lithothamnium nodules occur in the shallow-water sediments while in the areas of Santorin and the Sea of Crete, irregular calcareo-argillaceous concretions up to 5 to 10 cm in size are also found. Chemical analysis indicates up to 6.94% of $MgCO_3$ in the sediments of the Sea of Crete but according to diffractometric and optical analyses these sediments carry much Mg-calcite (molecular per cent of $MgCO_3$ is 8.0). There are also crystals of diagenetic calcite, Fe-carbonates and dolomite.

The mineralogic composition of detrital material in the central part of the Aegean Sea is highly variable and may be assigned to five mineralogical provinces (Figure 5). One of these is characterized by volcanic glass, clinopyroxenes and orthopyroxenes, another by basaltic hornblende, and yet another province is defined by alkaline amphiboles (riebeckite, arfvedsonite), *etc.* As for clay minerals, illite (30–90%) and kaolinite (10–30%) prevail here. Montmorillonite is abundant only in the volcanic island arcs (Melos-Santorin-Nesos) where it comprises 10 to 50% of the less than 0.001 mm fraction. The age of terrigenous minerals of volcanic origin is 30 to 145 m.y.

In the northern shelf part of the Adriatic Sea, shell sands with a low content of $CaCO_3$ (30 to 32%) are abundant. Sometimes they are followed by terrigenous (quartz) sediments. These sands have been studied intensively (Morović, 1951; van Straaten, 1965; Pigorini, 1968). Their thickness is not great (usually less than 1 m) and they have formed in shallow coastal zones. The sands are predominantly composed of metamorphic minerals, such as quartz, colorless garnet, fibrous amphiboles, hornblende, and epidote. The absolute age of the clastic minerals is 135 to 140 m.y.

The sediments in the southern half of the Adriatic Sea consist of foraminiferal muds, with $CaCO_3$ content ranging from 30.0 to 32.1%. Thus, like the shell sands, they fall very close to the terrigenous deposits boundary. According to Pigorini (1968) the detrital material belongs to three mineralogical provinces. The main (augitic or central) province is characterized by augite, rock fragments, apatite, zeolite, prismatic grains of epidote, calcite, Fe-Mn (?)-calcite and dolomite. Clay minerals are composed of illite (40–90%) and montmorillonite also occurs (30–50%) along the Apennine coast of the Italian peninsula. The absolute age of detrital minerals is 130 to 140 m.y. In the Albanian province, epidote predominates, while the percentages of green hornblende, garnet and augite are lower.

The Tyrrhenian Sea is rich in foraminiferal muds with a $CaCO_3$ content reaching 45% and sometimes 55%. The clastic material of these muds belongs to three mineralogical provinces. The eastern part of the sea is characterized by a considerable admixture of ash, colorless glass and clinopyroxenes whereas in the northwestern part metamorphic minerals predominate (alkaline and fibrous amphiboles, basaltic hornblende, kyanite, mica, epidote). These minerals were probably derived from Corsica. The province southeast of Sardinia has abundant epidote and zircon and less pyroclastic material. Among clay minerals, montmorillonite (10–30%) and illite (50–70%) prevail but along the periphery of the basin kaolinite is dominant. The absolute age of the clastic minerals is in the range of 152 to 190 m.y.

Foraminiferal and coccolithic-pteropodal-foraminiferal muds with $CaCO_3$ content ranging from 33.9 to 48.2% cover the Provençal Basin between the coast of France and the Balearic Islands. The foraminifera include *Globigerina bulloides, G. glutinata, G. quinqueloba, G. inflata, G. rubra* (Parker, 1958). In addition, benthonic foraminifera, pteropods, and coccoliths are also abundant. The clastic part of the sediments is dominated by material carried by the Rhône and Ebro rivers. Mixed mineral components are characteristic of this province. There is a markedly higher content of calcite, dolomite, Fe-carbonates and garnet. Clay material is mainly illite, with some admixture of kaolinite. The absolute age of the minerals near the mouth of Rhône is very high (255 m.y.). Seaward from the mouth of the Rhône, the proportion of deposits from the ancient French shield decreases and the age of detrital material near the Ligurian Sea (of Alpine provenance) decreases to 104 m.y.

The extensive area of the African-Sicilian sill is generally covered by bioclastic (relict?) sands. They mainly comprise molluscan shell detritus and extend to a depth of 200 m. Shell material is dominated by *Arca diluvii, Pectunculus pilosus, Pecten sp., Cardium sp., Venus sp., Turritella sp., Nassa sp., Spondylus sp., Aporrhais pespelecani* and others, as well as by remains of corals, calcareous algae and sea urchins. Carbonate content fluctuates from 38.2 to 92.3 per cent and $MgCO_3$ from 2.0 to 8.0 per cent. Carbonates are principally Mg-calcite and sometimes aragonite (40.5% of sediment weight at station 345). Dolomite occurs frequently. Terrigenous components consist mainly of stable minerals: magnetite, ilmenite, zircon, rutile and eolian quartz. Detrital material has been derived from African deserts under the influence of winds. We have not found saussurite and spinel in the Gulf of Gabès although these minerals prevail in this province, according to Berthois (1939).

Biogenic sediments are widely distributed between Cyrenaica-Crete-Alexandria. They are mainly highly calcareous (50–68.9%), coccolithic-pteropodal-foraminiferal muds which are rather firm (moisture content ranging from 30.9 to 41.3%), tough and viscous and light brown in color. In the coarse fractions we can distinguish a great number of fragile, transparent shells of pteropods, *Clio pyramidata.* The foraminiferal component consists of planktonic species: *Globigerina bulloides* d'Orb., *G. dubia* d'Orb., *G. rubra, G. inflata, Orbulina universa* d'Orb., *Uvigerina mediterranea* Hofker (Parker, 1958). Numerous benthonic foraminifera are also recorded (from 10 to 2000 shells per 1 gm of sediment, Korneva, 1966) and among them *Gyroidina neosoldanii* (Brotzen), *Articulina tubulosa sequenza* and *Usbekistana charoides* (Jones and Parker) prevail (Korneva and Saidova, 1969). The abundant coccoliths include *Coccolithus huxleyi* (Lohm.) Kamptner, *C. atlanticus* (Cohen), *Braarudosphaera bigelowi* (Gran), *Coccolithites variosus* (Cohen), *C. mendicus* (Cohen), *Syracosphaera pulchra* (Lohm.) (identifications by M. G. Ushakova). *Cyclococcolithus leptoporus* (Murray, Blackman) Kamptner, *Helicosphaera carteri* (Walich) Kamptner,

Rabdosphaera claviger (Murray and Blackman), *Ceratolithus cristatus* (Kamptner) appear in minor amounts. Coccolithic-pteropodal-foraminiferal muds in this area are characterized by the general occurrence of coarse (1–10 cm in cross section) calcareo-argillaceous concretions of diagenetic origin (Emelyanov, 1965b). In the area of Station 536 (between Crete and Africa, depth 2130–2167 m) concretions cover about 40% of the floor area.

$CaCO_3$ content in the concretions varies from 65.35 to 91.5 per cent, $MgCO_3$ from 4.06 to 9.77 molecular per cent, Fe is rare, Mn represents from 0.008 to 0.90 per cent. The material of the concretions is mostly Mg-calcite.

Concretions are concentrated in the uppermost layer of sediments, *i.e.*, in the sediment-water interface where sharp changes in the physico-chemical environment occur. They are dissolved in the lower layers of the sediment where a marked increase of the alkaline-chlorine coefficient takes place (from 0.120–0.122 in the pre-bottom water layer at 0–10 cm depth, to 0.136–0.169 in pore water layers, at 21–40cm depth) (Emelyanov and Chumakov, 1962).

According to Blanc (1969), coral-algal and coral-shelly high calcareous sediments are more or less widely distributed in the shallow parts of the Mediterranean Sea, especially along the coast off Provence, in the Tunisian Strait, and in the Sea of Crete. However, typical coral sediments are not recovered.

The chemical composition of biogenous calcareous sediments is shown in Tables 3, 8, 9 and 10.

Chemogenic Sediments

Chemogenic sediments occur in limited areas of the Mediterranean. Oolitic calcareous sediments occur in the Gulf of Gabès, near Djerba Island, Sidra Bay, and near the Egyptian coast. They probably cover coastal parts of the African shelf from Alexandria to Tunisia. According to Lucas (1955), calcareous oolitic sands from a depth of 1 m and collected at a distance of 200 m from the shore in Arabs Bay (Egypt) were studied by Lafitte in 1944. Water in this region was milky because of suspended matter and the sands in the bottom layers included 80% of oolitic grains; grains with nuclei (quartz, shell detritus) as well as grains without nuclei were observed. Grains with nuclei were characterized by a thick concentric lamination. The oolitic sands from the Egyptian coast were described as pseudo-oolites (Hilmy, 1951). More detailed study of the sands carried out by Lucas (1955) and this author (station 390, Table 11) showed that they consisted of oolitic grains with some admixture of pseudo-oolites, the latter being well rounded fragments of shell detritus or calcareous grains transported from the African deserts. Some of the pseudo-oolites can settle at sea owing to abrasion of coastal deposits of ancient sedimentary rocks. Such rocks containing pseudo-oolites have been discovered in Sidra Bay (Wood, 1964). In this area (Station 742, Table 11) heavy minerals (ore minerals, hornblende) often form the nuclei of ooliths, which then have a specific gravity

Table 8. Chemical composition (in %) of calcareous sediments from stations in the area of Santorin volcano (according to Butuzova, 1969).[a]

Component	Shell sands				Calcareous pelitic mud		
	St. 1011	St. 1012	St. 1380	St. 1360	St. 1340	St. 1344	St. 1377
i.r.	33.20	34.48	16.29	4.44	29.18	22.94	26.16
Al_2O_3	(x)	—	1.68	1.61	0.07	0.58	1.26
Fe_2O_3	1.28	1.78	0.46	0.25	0.98	1.00	0.37
FeO	—	—	0.14	0.07	0.29	0.29	0.22
MnO	—	—	—	—	0.06	0.06	0.09
CaO	30.29	27.99	39.94	47.93	29.52	32.85	31.32
MgO	2.55	3.32	2.71	1.56	3.54	4.06	3.88
P_2O_5	—	—	0.05	0.05	0.09	0.09	0.09
CO_2	25.40	23.84	33.25	39.34	24.78	28.22	26.80
NaCl	—	—	2.34	1.15	2.36	—	2.34
SO_3	—	—	0.63	0.68	0.49	—	0.48
$CaSO_4$	—	—	1.07	1.16	0.83	0.85	0.83
$Ca_3(PO_4)_2$	—	—	0.10	0.11	0.20	0.20	0.20
$CaCO_3$	54.07	49.96	70.49	84.69	51.88	57.80	55.15
$MgCO_3$	5.33	6.95	4.33	3.26	5.12	5.38	4.37
MgO	1.07	1.61	0.64	—	1.09	1.49	1.62
Calcite	47.75	41.72	65.35	80.82	45.80	51.52	49.96
Dolomite	11.65	15.18	9.47	7.13	11.20	11.76	9.56

[a]Abbreviations: i.r. = insoluble residue; (x) = not determined.

Table 9. Chemical composition (%) of biogenic calcareous (30–50% $CaCO_3$) sediments of the Mediterranean Sea.

Components	Sands			Coarse aleurites	Fine aleuritic mud			Aleuritic-pelitic mud			Pelitic mud (clay)		
	Range of values	Average	Number of samples	St. 359, 297 m	Range of values	Average	Number of samples	Range of values	Average	Number of samples	Range of values	Average	Number of samples
$CaCO_3$	31.4–37.80	33.61	3	44.30	37.26–49.12	43.55	5	30.02–48.66	40.06	25	30.02–49.80	38.20	41
$SiO_{2\,amorph}$	—	0.36	1	—	0.74–0.84	0.79	2	0.48–1.80	0.90	11	0.54–1.86	0.92	24
C_{org}	—	0.30	3	0.50	0.40–0.60	0.50	4	0.30–1.10	0.58	25	0.30–1.00	0.58	41
Fe	0.87–1.39	1.13	2	1.87	0.96–3.55	2.26	2	2.32–3.83	2.98	18	2.27–4.11	3.04	32
Mn	—	0.06	1	0.03	0.03–0.06	0.05	2	0.008–0.18	0.06	20	0.02–0.17	0.07	34
Ti	—	0.19	1	0.22	0.25–0.29	0.27	2	0.17–0.42	0.26	18	0.19–0.47	0.27	34
P	—	0.08	1	—	0.04–0.05	0.04	2	0.02–0.07	0.05	9	0.04–0.07	0.05	19

Table 10. Chemical composition (%) of biogenic high calcareous (>50% $CaCO_3$) sediments of the Mediterranean Sea.

Components	Sands			Coarse aleurites			Fine aleuritic mud			Aleuritic-pelitic mud			Pelitic mud (clay)		
	Range	Average	Number of samples	St. 741 298 m	St. 466 223 m	Average	Range	Average	Number of samples	Range	Average	Number of samples	Range	Average	Number of samples
$CaCO_3$	52.98–01.90	71.42	23	68.00	81.40	74.70	50.30–77.10	62.00	11	50.30–68.91	57.04	28	52.53–67.08	56.38	12
$SiO_{2\,amorph}$	0.20–1.02	0.58	17	—	0.24	0.24	0.57–0.94	0.73	6	0.56–1.45	0.78	17	0.60–1.03	0.80	11
C_{org}	0.30–1.20	0.63	23	0.60	0.40	0.50	0.30–0.70	0.44	11	0.20–0.90	0.55	25	0.25–0.60	0.44	12
Fe	0.62–4.07	1.83	15	—	0.86	0.86	0.49–1.79	1.46	9	1.74–3.00	2.14	21	1.50–2.87	2.20	11
Mn	trace–0.062	0.023	18	—	0.008	0.008	trace–0.04	0.03	9	0.04–0.20	0.07	20	0.02–0.07	0.05	10
Ti	trace–0.20	0.08	17	—	0.05	0.05	0.14–0.57	0.25	9	0.11–0.45	0.22	19	0.14–0.28	0.20	10
P	0.04–0.09	0.05	11	—	—	—	0.03–0.06	0.05	7	0.03–0.05	0.04	11	0.03–0.05	0.04	8

above 2.9. Off Djerba Island oolitic sands have been observed to a depth of 13 m. According to Fabricius *et al.* (1970) they are not modern but are reworked from coastal sedimentary rocks.

The oolitic sands are characterized by a $CaCO_3$ content ranging from 90.05 to 91.70 per cent but they are poor in Fe, Mn, Ti, P, and $SiO_{2\ amorph}$ (Table 11).

According to Lucas (1955), the oolites appear in coastal zones where turbulence is low. At depths of 30 to 100 m, typical oolitic sands are replaced by shell sands with an admixture of oolites. It is most likely that at these and greater depths the oolitic grains are transported by winds or by currents from the beach. Some oolitic grains are found at a depth of about 190 m.

Manganese sands cover one of the peaks of a submarine volcanic mountain in the southeastern part of the Tyrrhenian Sea. They contain 9.11 % of manganese, but are poor in Fe, Ti, and P (Table 11). Some fractions of this sample are characterized by the following manganese content: 1.0–2.0 mm: 7.66%, 2.0–3.0 mm: 6.47%, more than 3.0 mm: 5.79%. Black manganese incrustations of irregular shape containing rounded grains are the most common types. Besides manganese concretions, sandy or gravelly sediments contain a great number of detrital shells of molluscs, corals and pieces of volcanic ash, all covered with a manganese sheath or coating.

Table 11. Granulometric and chemical compositions (in %) of oolitic and manganese sands in the Mediterranean Sea.[a]

Fraction (mm)	Oolitic sands		Manganese sands
	St. 742 36 m	St. 390 61 m	St. 973 130 m
	Grain size distribution		
>1.0	26.30	6.89	—
1.0–0.5	19.00	5.67	—
0.5–0.25	44.10	26.28	—
0.25–0.1	8.50	43.42	—
0.1–0.05	2.26	13.15	5.83
0.05–0.01	0.49	0.82	9.01
<0.01	traces	3.70	25.86
>0.1	97.00	82.33	59.39
Md	0.55	0.19	—
So	1.8	1.8	—
Chemical Components	*Chemical composition*		
$CaCO_3$	90.05	91.70	55.28
$SiO_{2\,amorph.}$	—	0.09	—
$C_{org.}$	0.30	1.00	0.54
Fe	0.27	0.23	1.29
Mn	traces	traces	9.11
Ti	0.02	traces	0.11
P	0.06	0.02	0.08

[a] St. 742 = 30°21′2 N, 19°06′3 E; St. 390 = 31°34′8 N, 28°42′9 E; and St. 973 = 39°28′2 N, 14°49′0 E.

MINERALOGIC COMPOSITION OF SEDIMENTS

Some eighty-eight mineral species are recorded in the upper stratum of the Mediterranean Sea (0.1–0.5 mm fraction). The most abundant are terrigenous minerals (75 species) followed by biogenic, chemical and pyroclastic minerals. Some minerals have several possible origins. Volcanogenous and terrigenous minerals of sandy fractions and those of coarse-aleuritic fractions are absent from pelagic units.

Investigation of the 0.1 to 0.25 mm fraction serves to distinguish a total of 17 mineralogic provinces (Emelyanov, 1968). They differ in mineralogic character only within 50 to 100 km of shore. Beyond the shelf the mineralogic differences between the sandy fractions disappear completely or become indistinct.

In coarse-aleuritic fractions (0.1–0.05 mm) of sediments from geosynclinal sections, the most abundant minerals are plagioclases, quartz, yellow biotite (Figure 6), colorless glass, and opaque particles of ash. The heavy fractions are frequently dominated by such minerals as riebeckite, arfvedsonite, actinolite-tremolite, basaltic hornblende and garnet (Figure 9), all of which are rare in deep-sea sediments. In some provinces considerable amounts of chloritoid, apatite, chlorite, and zeolite also occur. The sediments in geosynclinal areas are characterized by a great variability of mineralogic composition, due to the location of sediment sources, stability range of some minerals and the high content of rock fragments. Resistant minerals, such as magnetite, common hornblende, zircon, rutile-brookite-anatase, tourmaline, quartz (frequently of eolian origin), and plagioclase are most widely distributed in the preplatform and central parts of the Mediterranean Sea. The 22 mineralogic provinces recognized in these areas are characterized by a wider areal extent and a smaller degree of mineralogic variation than exist in the provinces of the geosynclinal sections of the basin.

Carbonates of fragmental, chemical and diagenetic origin are typical of the upper stratum of sediments in the Mediterranean Sea. They include: calcite, aragonite, Fe-Mn(?)-calcite, dolomite, Fe-carbonates (ankerites), and siderite (Figure 10). As mentioned above, oolites occur on the African shelf (in the arid zone). They consist of aragonite and, to a lesser degree, calcite. Calcareous-clay concretions are composed of magnesian calcite ($MgCO_3$ sometimes reaches

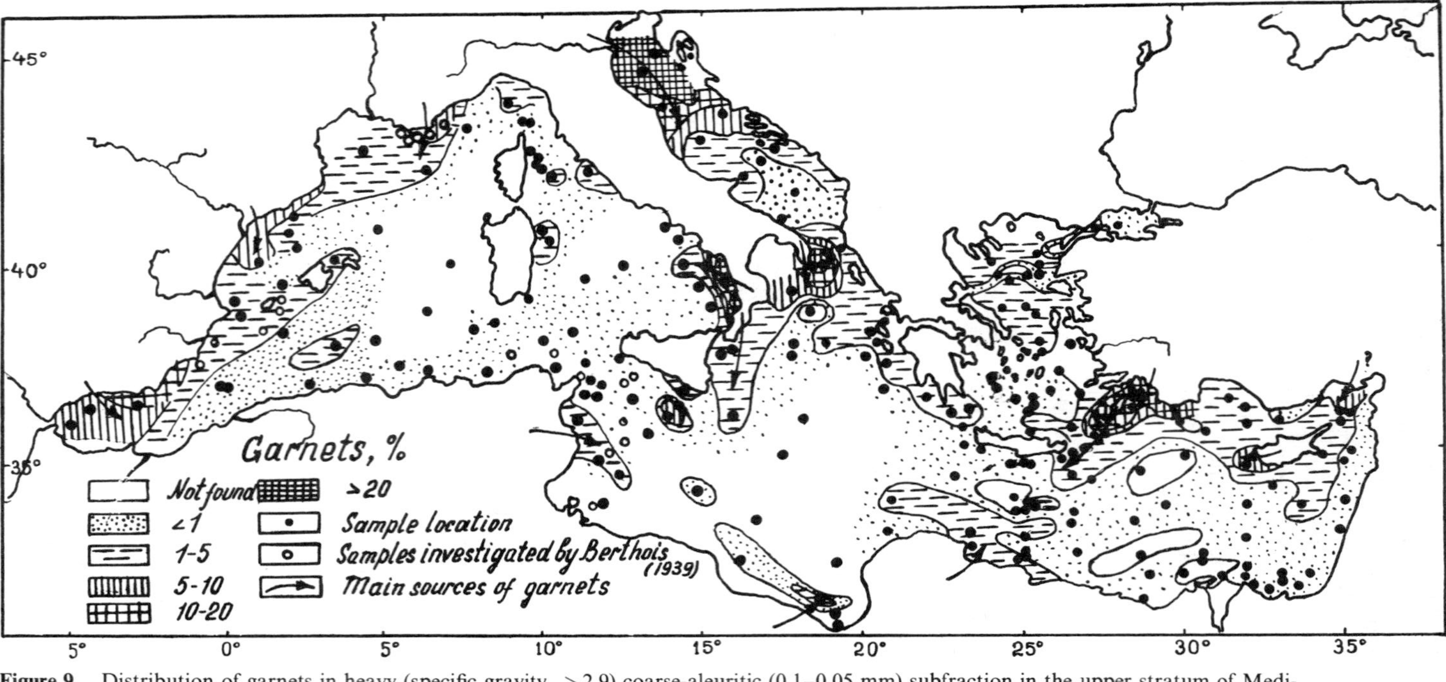

Figure 9. Distribution of garnets in heavy (specific gravity >2.9) coarse-aleuritic (0.1–0.05 mm) subfraction in the upper stratum of Mediterranean Sea sediments (220 samples).

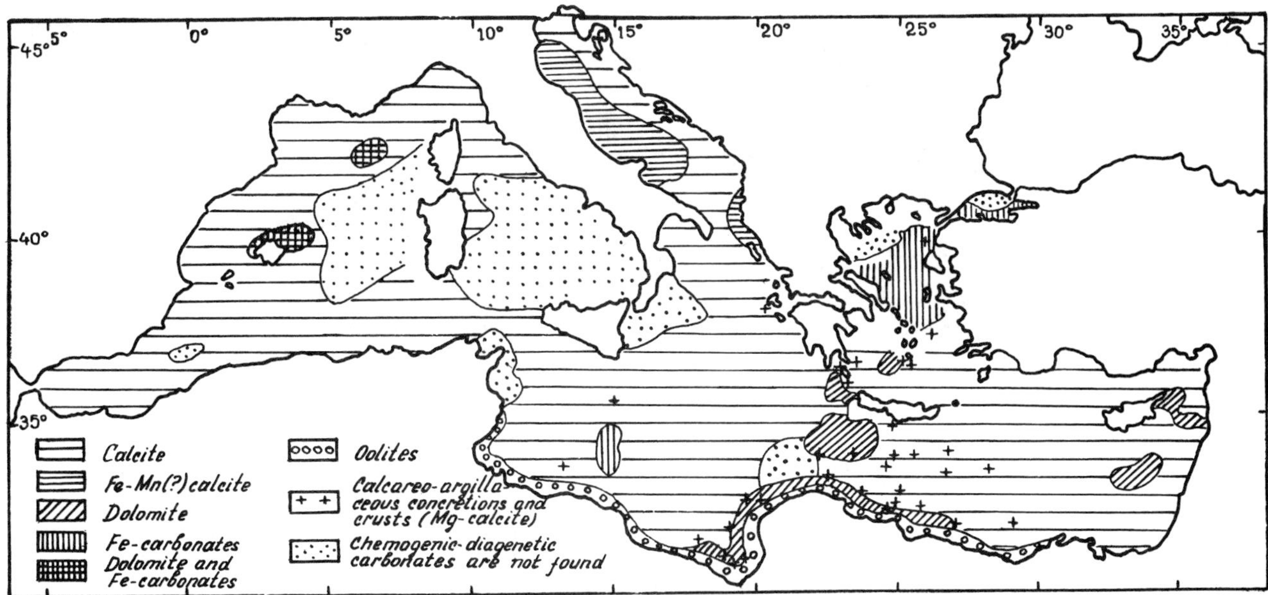

Figure 10. Sites of chemogenic-diagenetic carbonate formation in the coarse fractions of the upper layer of Mediterranean bottom sediments.

9.77 molecular per cent). Calcite crystals are transparent, wedge-shaped or needle-shaped. Sometimes the edges of crystals meet in a point forming "a starlet". Scalenohedra of calcite are also observed in pumice pores. Grains of calcite of irregular scalenohedral form are found in the sediments at 125 out of a total of 133 stations and sometimes they make up 60% of light coarse-aleuritic subfraction. Calcite crystals occur occasionally as chains. Such chains are observed not only in sediments but also in suspended matter and this is indicative of their chemical formation.

Fe-Mn(?)-calcite is relatively abundant in the sediments of the Adriatic Sea (from 3.7 to 4.4% of light coarse-aleuritic subfraction). Very often crystals of this calcite form "starlets" or chains. The refractive index (n_ω) varies in the range 1.670 to 1.678.

Negligible amounts of dolomite occur at most stations (Figure 11). The maximum content is recorded at Station 539 (17.6% of light subfraction) and in Station 299 (38.9% of heavy subfraction). This mineral is represented by rhombohedra with zonal structure. Sometimes rhombohedra form chains (station 840).

Fe-carbonates (mainly ankerite) may reach 34.9% of heavy subfraction (Station 539) and 18.0% of light subfraction (Station 599). They are rhombohedral grains of irregular form with a refractive index (n_ω) ranging from 1.682 to 1.75. The following composition of heavy subfraction was determined at Station 539: ankerite ($N_\omega = 1.692$) 45%, Fe-ankerite (n_ω from 1.718 to 1.734) 30.0%. Some grains of Fe-ankerite (n_ω is about 1.702) may be magnesite. Such grains reach 1% of the subfraction.

Siderite is extremely rare in the Mediterranean Sea. It occurs as grey spherulitic grains and sometimes as rhombohedra with $n_\omega = 1.875$ and $n_\varepsilon = 1.633$.

Calcite may be bioclastic, terrigenous, chemical or diagenetic in origin. Fe-Mn(?)-calcite, ankerite and dolomite are of terrigenous and diagenetic origin, while siderite is of terrigenous origin.

According to diffractometric analysis of the total sample of sediments, the calcareous fraction is composed of calcite, Mg-calcite (4 mole % of $MgCO_3$) aragonite and dolomite. Calcite is the prevalent mineral and contains from 0 to 9.8 mole per cent of $MgCO_3$ (Figure 12). Aragonite fluctuates from 0 to 44%, and dolomite from 0 to 13% of total sediment weight. The maximum content of aragonite is characteristic of shallow waters. Dolomite is concentrated in areas with a high percentage of ankerite in the coarse-aleuritic fraction. Mg-calcite is most abundant in the arid zone (especially between Crete and Cyrenaica, on the African-Sicilian sill, and in the Sea of Crete, see Figure 12). These regions are also characterized by a maximum content of $MgCO_3$ as determined by chemical analysis.

Fractions less than 0.001 mm are dominated by three minerals: illite, kaolinite and montmorillonite (Rateev *et al.*, 1966). A high content of montmorillonite (more than 50%) is recorded only in the Pre-Nile province and a somewhat higher value (30–50%) is obtained from some sections of the geosynclinal zone and within the African-Sicilian sill (Figure 7). Kaolinite is abundant (30–50%) in the area of Tel-Atlas-Sicilian-eastern part of the Mediterranean Sea. In the central parts of the basin, kaolinite is less than

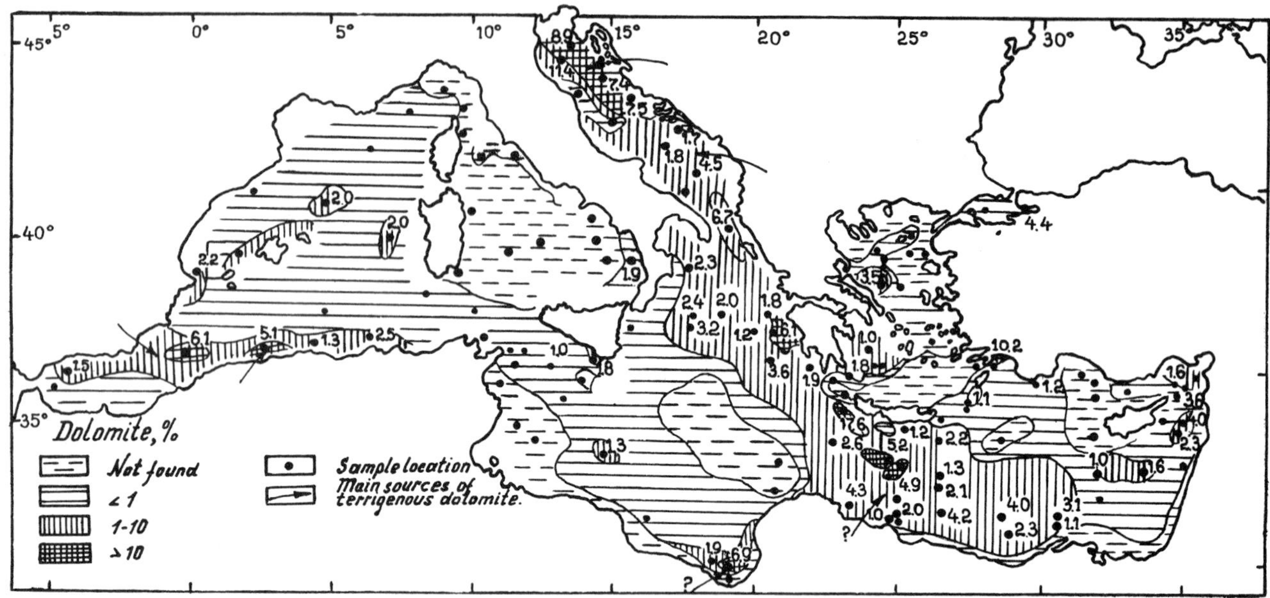

Figure 11. Distribution of dolomite in light (specific gravity < 2.9) coarse-aleuritic (0.1–5.05 mm) subfraction in the upper layer of sediments of the Mediterranean Sea (220 samples).

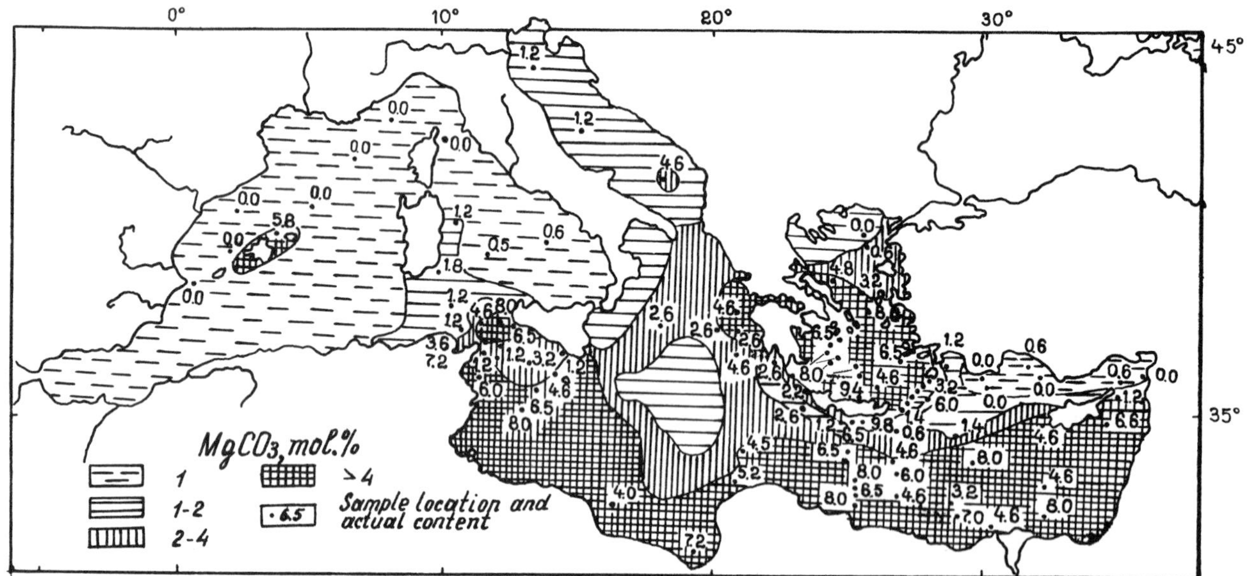

Figure 12. Distribution of $MgCO_3$ entering calcite lattice in the upper layer of bottom sediments (according to X-ray analysis, in mole per cent).

10%. Illite prevails in the Algerian-Provençal and central basins and in the geosynclinal zone.

AGE PROVINCES OF TERRIGENOUS VOLCANIC MINERALS

The absolute age of terrigenous-volcanogenic minerals in the Mediterranean sediments varies within the range of 0 to 3 m.y., up to 430 m.y. (Emelyanov *et al.*, 1973). The minimum ages are characteristic of volcanoclastic sediments (0 to 50 m.y.), while the maximum ages characterize quartz in the Nile Delta front. On the whole, the age of the minerals increases from the northern part of the Aegean Sea (50–100 m.y.) towards the African platform (350–430 m.y.; Figure 13). In geosynclinal regions the youngest material is coarse (pyroclastic), while in platform areas the pelitic material is the youngest ($<$ 100 m.y.). In the transitional zone both coarse and pelitic minerals are approximately of the same age, *i.e.*, 150 to 200 m.y.

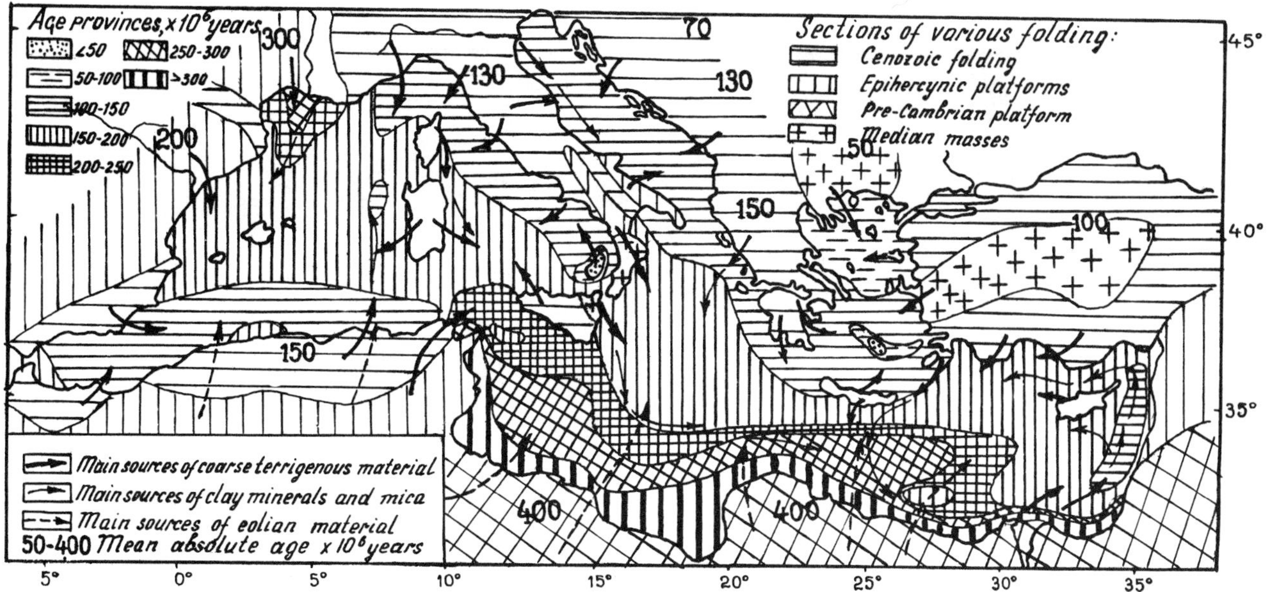

Figure 13. Age provinces of terrigenous-volcanogenous minerals in the upper layer of Mediterranean Sea sediments (Emelyanov *et al.*, 1973) related to tectonic provinces of the adjacent lands (according to Emelyanov *et al.*, 1964).

DISTRIBUTION OF PRINCIPAL CHEMICAL COMPONENTS

$CaCO_3$ (total) constitutes 3.2 to 92.3 per cent of the sediment (Emelyanov, 1965b). It attains a maximum value (more than 70%) in the sediments from shallow arid regions of the basin, and the minimum amounts (less than 10%) are found in terrigenous sediments from the Pre-Nile province and in the vicinity of volcanoclastic deposits.

$MgCO_3$ total (determined chemically in 2% HCl extraction) varies from 0 to 6.2% of the sediment weight. The distribution is shown in Figure 14, in which the maximum concentrations of $MgCO_3$ are shown to occur in carbonate-rich sediments in the arid zone and in the Sea of Crete.

Amorphous silica ($SiO_{2\,amorph}$) of probable biogenic origin is rare; the amount detected varies from traces to 2.01 per cent (on average, 0.5–1.0%). On a carbonate-free basis, the amorphous silica content increases to 4.3 per cent (on average 1–2%).

Organic Carbon (C_{org}) values vary from 0.2 to 1.6 per cent. The maximum content (more than 0.6%) is observed in the pre-estuarine parts of the sea and in coastal areas of the geosynclinal zone, while minimum values (0.2–0.4%) occur in the central part of the Levant Sea and near Sidra Bay (Figure 15). In the dark or brown-green (sapropelic) layers of cores, C_{org} concentrations increase sharply, sometimes attaining 7.35% (Station 1319, 171–180 cm level, Sevastyanov, 1968).

Total Iron (Fe) content ranges from 0.07 to 6.19 per cent (excluding the sediments from the lagoon of Santorin). The highest concentrations of this element (4.0–6.2%) were determined in muds of the Pre-Nile province, and the lowest values are found in the highly calcareous sediments and the quartz sands of coastal areas of the African platform (less than 1%). On the whole, Fe content increases with an increase in the pelitic character of the sediment and a decrease in $CaCO_3$. Fe content varies up to 0.5–17.0 per cent on a carbonate-free basis (on average, 3–5%). In the upper stratum of sediments the iron resides principally in the detrital fraction. Some 15 to 30 per cent of total Fe is mobile ferrum (Fe^{3+} and Fe^{2+}; Sevastyanov, 1968). Pyritic iron is very rare or entirely absent.

Manganese (Mn) in sediments ranges from trace to 9.11 per cent (on average, 0.05–0.10%). Higher concentrations are found in the muds of the Pre-Nile province (0.10–0.46%), in the Tyrrhenian Sea (0.10–0.17%) and some other limited regions. Ore concentrations of this element occur only at station 973. Mn content on a carbonate-free basis increases up to 0.01–0.60 per cent (excluding Station 973), averaging from 0.10 to 0.20 per cent.

Some oxidized interlayers of mud are characterized by very high Mn concentrations (up to 5.25% in Station 1334, at 22–24 cm level; Sevastyanov, 1968), which are apparently related to diagenetic redistribution of the elements. Such an abrupt enrichment is a result of migration and precipitation of insoluble compounds dispersed as very fine black grains in brown muds.

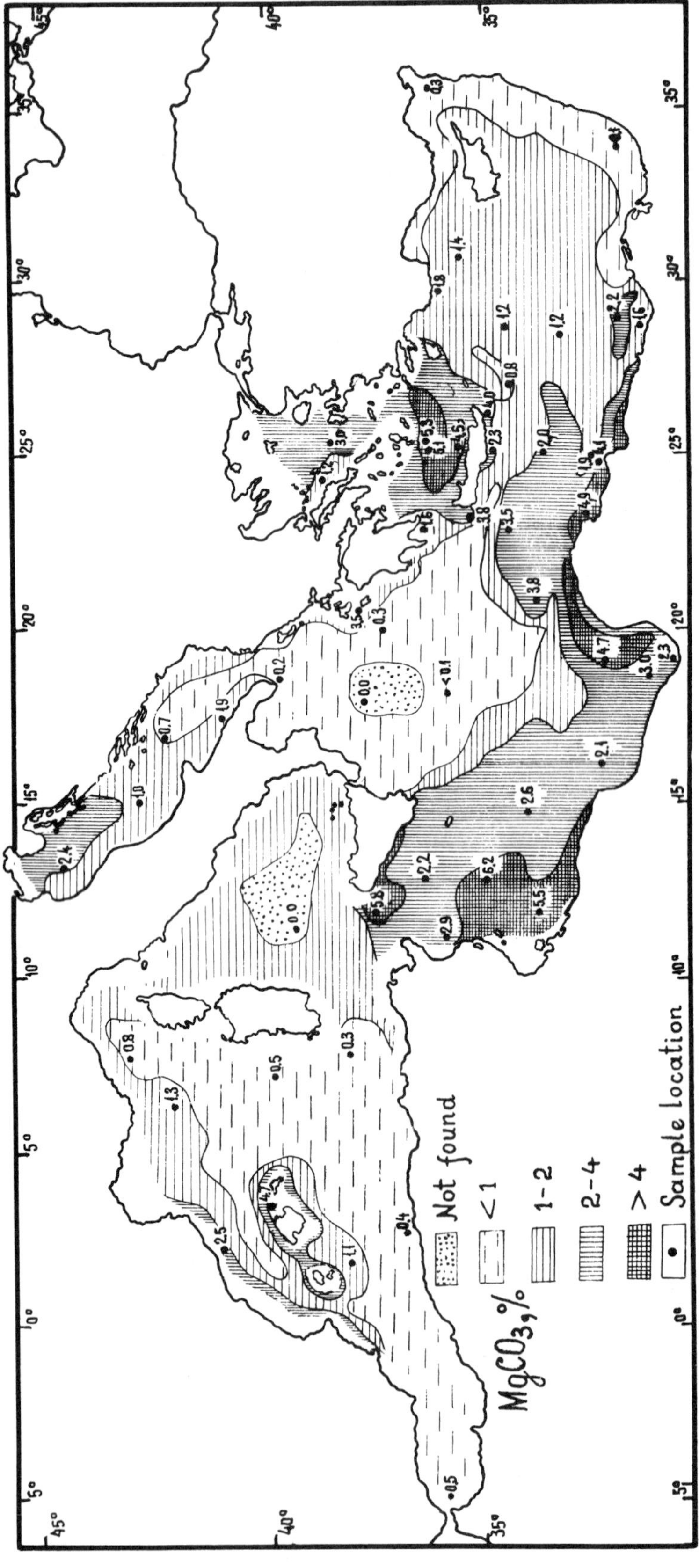

Figure 14. Distribution of $MgCO_3$ (total carbonate) in the upper layer of bottom sediments of the Mediterranean Sea (% of dry sediment weight). Determinations are made by chemical method in 2% of HCl extraction.

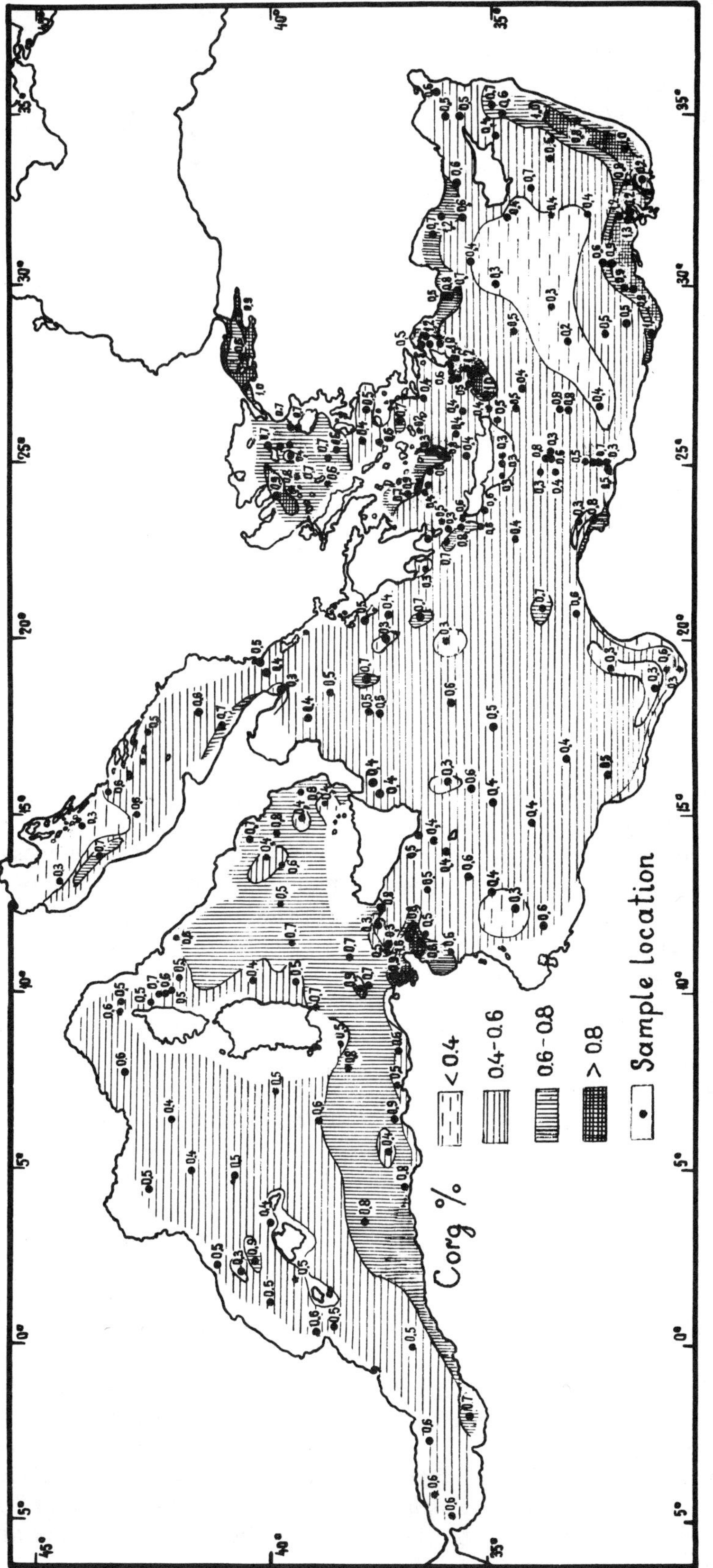

Figure 15. Distribution of per cent organic carbon ($C_{org.}$) in the upper layer of bottom sediments.

Titanium (Ti) ranges in value from traces to 0.94 per cent in natural sediments (on the average 0.3–0.4%) and from traces to 1.03% on a carbonate-free basis (on average, 0.4–0.6%). The maximum Ti content occurs in deposits supplied by the Nile. Ti content is directly proportional to pelite and inversely proportional to $CaCO_3$ (Emelyanov, 1966b).

Phosphorus (P) content varies from 0.01 to 0.21 per cent (on average, 0.04–0.06%) but reaches 0.80 per cent on a carbonate free basis (averaging 0.05–0.10%).

DISTRIBUTION OF TRACE ELEMENTS

Gold content in the upper stratum of sediments ranges from 1.1 to 9.0 p.p.b. In the terrigenous muds (20 samples) the average content is 1.3 p.p.b. (range: 0.6 to 4.2); in biogenic-calcareous sediments (46 samples) 3.1 p.p.b. (range: 1.1 to 1.65); in volcaniclastic muds from the lagoon of Santorin (1 sample), 14.0 p.p.b. On the whole, the northern parts of the sea are richer in gold (3 to 9 p.p.b.) than the southern areas (1 to 3 p.p.b.).

The sediments are rather poor in *uranium* (0.2 to 4.0 p.p.m.). The concentration increases up to 12 p.p.m. only in Station 789, near Barcelona. Uranium is closely connected with organic matter; higher percentages of uranium occur in sediments richer in organic carbon. This relationship becomes particularly obvious when comparing the sediments of the Mediterranean with those of the Black Sea (Figure 16). Uranium accumulation in the Black Sea sediments is caused by high concentrations of organic matter, the latter being absorbed by uranium from sea water. This absorption is promoted by a low oxidation-reduction potential of bottom sediments (Kochenov *et al.*, 1965). Such uranium concentrations do not take place in the Mediterranean due to the low content of organic matter and the high value of the oxidation-reduction potential. In the Mediterranean, uranium concentrations are high (reaching 35 p.p.m.) only in gray-green sapropelic layers of those cores with a high C_{org} content, a situation similar to that which exists in the Black Sea (Baturin *et al.*, 1967).

According to the data obtained at 92 stations, *boron* comprises 25 to 300 p.p.m. of the sediment (average 96 p.p.m.) and 44 to 368 p.p.m. on a carbonate-free basis (average 148 p.p.m.). Boron content increases with an increase in the pelite admixture and decreases with an increase in $CaCO_3$.

Quantitative-spectrographic analysis of 142 samples shows that concentrations of *nickel, cobalt, chromium, vanadium, copper* and *molybdenum* in most of the samples vary within the range (in p.p.m.): Ni, 10–

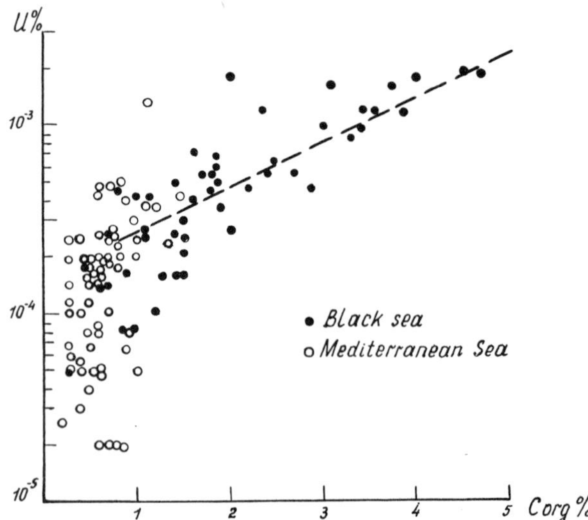

Figure 16. Correlation of uranium versus Corg. in surficial sediments in the Black Sea and Mediterranean Sea (according to Kochenov *et al.*, 1967).

300; Co, 10–30; Cr, 10–300; V, 5–120; Cu, 20–100; Mo, 1–5. Concentrations of these elements sometimes appear to exceed the above figures, but we do not consider these high values as fully reliable. The areal distribution of the elements listed above is rather complicated. We can only conclude in general terms that higher concentrations of some elements (nickel for instance) occur in the geosynclinal zones whereas minimum values are observed along the coast of the African platform and in the abyssal parts of the Algerian-Provençal Basin. For other elements the reverse situation holds. However, it is noticeable that these elements, together with uranium, tend to be concentrated in sapropelic bands rich in organic matter (Baturin *et al.*, 1967) or in distant oxidized muds with a high manganese content (Sevastyanov, 1968). The heterogenous distribution of trace elements in certain sediment layers is explained by their diagenetic redistribution, which results from two processes: migration of trace elements into the oxidized zone and formation of insoluble compounds here (mainly with MnO_2 hydrate), and secondly, fixation of the elements in the reduction layer within bands enriched in organic matter (Sevastyanov, 1968, p. 13).

Strontium and especially *barium* content is not constant in Mediterranean sediments but varies within wide limits. Strontium concentrations are usually 0.02 to 0.2 per cent, increasing with the $CaCO_3$ content. Barium concentrations are usually 0.002 to 0.04 per cent and do not appear to be correlated with the $CaCO_3$ content.

RECENT SEDIMENTATION IN THE MEDITERRANEAN SEA: A DISCUSSION

Geological, climatic and physico-chemical features of the Mediterranean Sea are reflected in some specific features of the surficial bottom sediments of this basin. In contrast to other enclosed or semi-enclosed seas (the Black, Caspian and Aral seas), the Mediterranean Sea is characterized by a very low planktonic and benthonic biomass, an unusually high temperature in the deep water layers (+ 13° C), low specific alkalinity, rather high pH and carbonate oversaturation of water. In terms of physico-chemical features, the Mediterranean Sea is more like the equatorial Atlantic than most enclosed seas.

The poor terrigenous supply to the Mediterranean Sea and the low plankton productivity result in a relatively low concentration of suspended matter in this basin, averaging in the upper water layers about 0.5 to 1.0 mg/l. This amount is somewhat less than in the Black and Caspian seas and is close to that in the Atlantic Ocean. Hence Mediterranean sedimentation rates are similar, locally, to those in the open oceans.

Sediment distribution in the Mediterranean Sea is largely controlled by mechanical factors: coarse-grained sediments occur in shallow waters and on submarine highs, *i.e.*, in areas subjected to high hydodynamic activity. Only occasionally are coarse-grained sediments found at great depths (in straits or submarine canyons and other regions with strong bottom currents). Fine-grained sediments occur in deep areas and cover depressions. In areas near river-mouths where there is an abundant supply of fine fragmental material, pelitic muds sometimes occur to depths as small as 10 to 20 m, *i.e.*, to wave-base.

Thus, the distribution of sediment types is, first of all, dependent upon the bottom morphology. In those areas with a broad shelf, coarse-grained sediments cover broad zones and are conspicuous on bottom sediment charts (Figure 3). In areas where the shelf is narrow, coarse-grained sediments cover very narrow belts and pass into zones of fine pelitic mud not far from the coast.

The distribution of sediment types is also affected by climatic conditions. In the humid zone of the Eastern Mediterranean (the Levant and Ionian Seas), coarse-grained sediments (sands) are seldom found deeper than 50 to 100 m (excluding straits affected by strong deep currents), and pelitic muds cover most of the continental slope. In the arid zone (except for the near-Nile region) sands occur down to outer shelf depths (100–250 m), fine-aleuritic muds to depths of 700 m and even 1700 m, and aleurite-pelitic ones cover almost the whole sea floor down to 2500 to 3200 m.

In these northern sectors of the Mediterranean adjacent to mountainous coasts, terrigenous and low-calcareous sediments occur because here there is a relatively high supply of detrital material to the sea.

Sediments of the arid zone are usually of the high-calcareous ($CaCO_3$ content from 50 to 93 per cent) type. A great amount of calcareous material generally has a higher median diameter than the admixed terrigenous material and there is, therefore, a rather spurious increase in the area covered by coarse-grained sediments that extend down to great depths in the arid zone. This calcareous material is seldom controlled by the laws of mechanical differentiation since it is formed and deposited independently of the distance from the coast.

The different granulometric and mineralogic composition of the sediments deposited in the humid and arid zones results in their displaying different bulk physical properties. Muds of the humid zone are characterized by a higher moisture content (40–55 per cent), a semi-fluid character, soft consistency and feeble viscosity. Those of the arid zone are less moist (30–45 per cent), more compact and strongly viscous. This helps explain the maximum development of slumping and turbid flows observed in the northern sectors, *i.e.*, the Algerian-Provençal Basin, the Tyrrhenian and Ionian seas.

Geological and climatic conditions of the Mediterranean Sea are further reflected in some specific properties of granulometric, mineralogic and chemical composition of the sediments. Thus, in the humid zone, chemical weathering results in sediments that are supplied to the sea mainly in the form of pelite. Sandy and coarse-aleuritic fractions are not abundant in sediments of this region, and the low concentration of the coarse-aleuritic fraction makes it difficult to isolate coarse aleurites as a distinct sediment type. The boundary between sand and mud seldom involves enrichment in the coarse-aleuritic fraction. Instead, increasing depth generally corresponds with an increase in the pelite content of the sands and finally the latter change into poorly sorted, fine-aleuritic and aleuritic-pelitic muds. Thus, in sediment histograms, the coarse-aleuritic fraction seldom gives rise to clearly defined maxima.

In the arid zone, the considerable amount of sandy and aleuritic materials in the sediments reflects the importance of an eolian supply.

Different genetic sediment types are distinguished not only by the differing granulometric properties and the variation in carbonate content but also by changes in the concentration of iron, manganese, titanium and other components in the sediments.

The formation of these genetic sediment types is closely related with the geological and geomorphological structure of the basin, the climatic conditions and marine chemistry. Terrigenous deposits are accumulated in the humid zone (under warm and humid climatic conditions) where an active supply of fragmental terrigenous material coincides with a deficiency of carbonates. Even where the absolute amounts of accumulated carbonates are high they are usually diluted with terrigenous material.

Highly calcareous, biogenic and oolitic sediments are principally formed in the arid zone adjoining the African platform, but the maximum carbonate values occur in belts of medium-grained sediment, that is, on the deeper part of the shelf, on the continental slope and on submarine highs.

In the arid zone (the Levant Sea, Central and Ionian Basin, Sidra Bay and the African-Sicilian sill), terrigenous content and sediment grade are closely, but inversely, related; the sands contain only 10 to 20 per cent of terrigenous components while the deep-sea muds contain 40 to 50 per cent (the carbonate content concomitantly decreasing from 90–80 per cent in sands to 60–50 per cent in muds). Hence, in this zone, the granulometric composition of the calcareous material and that of the noncalcareous (terrigenous) fraction differ greatly, and the real distribution pattern of the non-calcareous material is distorted. In the arid zone, the rate of carbonate formation exceeds the supply of terrigenous material. Thus calcareous material is here the main component of the bottom deposits, the granulometric composition of which is governed by the calcareous components.

If carbonates were removed from sediments then fine sediments would appear much more widespread than is shown on Figure 3, since they would then be shown as occurring in the arid zone as well as on submarine elevations, the continental slope and on some shelf areas. The sediments occurring on the continental slope of the northern region, where there is an increased supply of terrigenous material, would then appear much coarser than sediments of the southern continental slope, adjoining the arid African lands.

Thus, sediment character is affected by the active formation of carbonates in the Mediterranean Sea. This carbonate formation begins at the surface of the sea, continues in deep water layers (generation and development of organogenous and chemogenic carbonates) and comes to an end in the upper sediment layer (diagenetic carbonate formation). Therefore, sediments are enriched in carbonates some 5 to 10 times, as compared with the suspended matter.

The carbonate content in the sediments ranges from 2 to 93 per cent. Biogenic low-magnesian calcite (less than 4 molecule per cent of $MgCO_3$) is the predominant mineral in the humid zone and biogenic highly-magnesian calcite (from 4 to 8 molecule per cent of $MgCO_3$) is characteristic of the arid zone and the Aegean Sea.

Substantial amounts of $MgCO_3$ forming chemical and diagenetic accumulations are observed only in the arid zone, where dolomite as well as magnesian calcite occurs in calcareous-argillaceous concretions. Ferroan and manganese carbonates are found in much smaller amounts, and ankerite and siderite are encountered in only a few areas.

On the other hand, it is established (Strakhov, 1969, page 291) that in seas of the arid zone "The grand scale of accumulation of magnesium carbonate in the sediments here in the form of dolomite is striking." But in arid basins ". . . the increased intensity of Mg accumulation in arid basins is accompanied by a radical rearrangement of the very mechanism of its deposition. Purely biogenic extraction of $MgCO_3$, characteristic of the sedimentational process in humid seas, remains an essential process in arid regions, but it is negligible and becomes decreasingly significant as increasing salinity of the basins leads to rapid extermination of the fauna. Chemical precipitation of dolomite from the water quickly becomes an important process and then becomes the sole process" (Strakhov, 1969, p. 292). In the Mediterranean Sea, similar processes do not take place, as the salinity is relatively low, and as a result, extinction of faunas is not observed and the biota remain the main sources of carbonate.

Thus, as to carbonate accumulation, this basin must be considered as one located in the intermediate zone between typical humid and arid sedimentation conditions. Hence, formation of dolomite is in its initial stages. This conclusion is supported by the occurrence of dolomite crystals, probably of chemogenic origin, in the suspended matter on filters and by somewhat higher amounts of fresh and pure Fe-Mn(?)-calcite rhombohedra (in the Adriatic Sea) and dolomite in the upper sediment layer. The formation of these crystals is probably related with diagenetic transformation of the calcareous material. The role of such diagenetic processes in carbonates is of great importance in humid as well as arid environments according to Strakhov (1969). The calcareous-argillaceous concretions rich in $MgCO_3$ which are commonly found in the uppermost layer of sediment and are widespread in the deep areas of the eastern Mediterranean Sea, suggest the operation of these early diagenetic processes. The situation probably results from the following: (a) interstitial carbonate-rich waters rise to a level near the sediment-water interface where the sharp change in the alkaline-chloric coefficient results in precipitation of some carbonates; (b) some enrichment of intersitial waters in magnesium carbonate may result from partial dissolution of shell material.

Thus, the waters of the Mediterranean Sea, containing insignificant carbonate reserves, form a medium of rapid carbonate return (*i.e.*, precipitation from water) and redistribution. In this respect, the Mediterranean differs from the Black Sea, the deep waters of which are enriched in carbonates (Strakhov, 1951).

The low organic carbon content of Mediterranean sediments (on average the Black Sea sediments contain thirty six times as much) is ascribed to the paucity of phytoplankton and the rapid decay of organic matter in the warm, oxygen-saturated waters of the Mediterranean. For similar reasons, the low diatom population leads to a dearth of biogenic silica. *Thus the Mediterranean sediments are uniquely characterised by a high carbonate content and a deficiency of amorphous silica and organic carbon.*

The distribution of such elements as iron, manganese, titanium and phosphorus is affected to a lesser degree by climatic conditions. All these components occur in the fine-grained sediments, but they are much less abundant in sediments of the arid zone than in those from the humid areas because of their dilution by carbonates. Of all these elements, titanium and iron are the most terrigenous (they are diluted with carbonates to a large degree). These elements are most common in fine terrigenous and low calcareous sediments of the humid zone. Phosphorus is the most organogenous and mobile element, and phosphorus values seldom depend on the dilution with carbonates and the climatic zonation. In its distribution, manganese tends to fall between iron and phosphorus.

With respect to mineralogic composition, the sands reveal that sedimentary terrigenous-volcanogenic material is supplied only to shallow areas of the Mediterranean basin. Only restricted amounts of sand reach the deep parts, and the mineral composition of this fraction is rather monotonous. The coarse-aleuritic fraction is more enriched in heavy minerals as compared with the sandy one. Hence, a detailed mineralogic analysis of the aleurite material clearly reflects the geological and petrographical features of the bordering lands.

A mineralogic study of the sand and aleurite fractions enables the sources and transport paths of the sediment to be traced and also allows determination of the relative lateral persistence of the various components. Mineralogic provinces of sandy sediments maintain their specific features (their physiognomy) only in the shallow inner shelf while mineralogic distinctions become blurred on the outer shelf. However, specific mineralogic features of the coarse aleurite material often may be traced to outlying deep sea provinces (those of fine aleuritic fractions to central sea areas). Sometimes (in near-mouth areas and other regions subject to strong currents) these features remain distinctive all the way into the central parts of the Mediterranean depressions.

Near shore sediments most closely reflect petrographic features of the adjacent land. Increasing distance from land results in a diminution of this petrographic reflection and it ultimately disappears in the sediments of abyssal basin areas.

Lateral displacement of the mineralogic province from the associated source is observed only in zones of strong longshore currents (*e.g.*, from the mouth of the Nile in the direction of Israel and Lebanon).

Such minerals as ilmenite, magnetite, limonite-hematite, epidote, tourmaline, zircon, rutile, sphene, apatite, quartz, plagioclase, calcite and dolomite are almost evenly distributed over the Mediterranean floor. They are ubiquitous minerals, being supplied from almost all adjoining lands. Their utility for determining sources, transport paths and migration distance of sedimentary material is usually not too significant.

Hydrodynamic basin properties are best reflected in the distribution of terrigenous minerals. Strong longshore currents confine the coarse-grained material to the nearshore regions. In areas where these currents are diverted from the land (especially in deep-sea regions) this fragmental material may reach the abyssal basin areas (east of Sicily, near-Nile province, *etc.*).

Easily destroyed or locally derived minerals such as augite-diopside, basaltic hornblende, tremolite-actinolite, glaucophane-arfvedsonite, garnet, muscovite, biotite, hypersthene, volcanic colorless glass and ash, kyanite, common hornblende and quartz are the most important for tracing provenance and transport paths. Garnet, augite-diopside, hypersthene, alkaline amphibole, basaltic hornblende, volcanic ash and glass occur nearest to the source while common hornblende and quartz (due to their widespread occurrence and stability), and especially muscovite and biotite, are transported over the greatest distances. Mica is perhaps the best indicator of sediment transport paths.

Highly varied mineralogic provinces are observed in the northern geosynclinal areas, surrounded by rugged uplands drained by numerous rivers. In areas of the Mediterranean adjacent to arid regions, the mineralogic complexity of the detrital sediments is much reduced and the provinces are more extensive and monotonous.

The small amount and mixed mineralogy of the heavy coarse-aleuritic subfraction in deep-sea (abyssal) sediments indicate that an analysis of sandy and coarse-aleuritic fractions in this area is not always useful in delineating sources or supply-routes of abyssal sediments. However, the clay minerals can help in this respect. Hydromica is the main clay mineral in the Mediterranean Sea, followed by kaolinite, then montmorillonite. Chlorite, halloysite and palygorskite are rather rare.

The origin of hydromica is connected with erosion of chestnut and brown soils and terra-rossa. The bladed grains of this mineral, characteristic of authigenous formation, are rarely observed. They originate during the early diagenetic stage, as a result of the trioctahedral hydromica change (Rateev *et al.*, 1966). Small amounts of kaolinite are encountered over the whole Mediterranean area, but kaolinite is more abundant in the bordering regions than in the central zones of the sea. Kaolinite, like hydromica, originates from soil erosion.

Montmorillonite is principally of volcanogenic origin and its maximum abundance is in areas of weathered igneous rocks. As this mineral is distributed over the area unevenly it is a good indicator of source and supply paths of fine material. In this manner, for instance, the movement of Nile-derived sediment may be traced extensively.

The absolute age determination of terrigenous-volcanogenic minerals is a good additional lithological method. This method has allowed estimation of the geographical range of the younger detritus derived from the Alpine folded belt and that of the ancient material from the African platform. The young material prevails in the northern parts of the eastern Mediterranean Sea and near recent structures of the Algerian-Provençal Basin. The material of the African platform is principally observed south of Sicily and the central Mediterranean Ridge (rise). Mixed material, supplied both from the north and the south, has accumulated between these zones.

The boundaries of these age provinces sometimes coincide with those of the terrigenous-mineralogic ones. However, the boundaries are generally not coincident in areas where the mineralogic provinces adjoin a land area with different tectonic structures (*e.g.*, when they are located near the boundaries separating platforms or old median masses from regions of young folding). Moreover, in some areas the absolute age of the sandy-aleuritic material does not agree with that of the pelitic fraction. For example, sandy aleuritic material is determined to be younger than the pelitic detritus in the zone of Recent and Quaternary volcanism. In the zone of Alpine folding, the age of the sandy-aleuritic material and that of the pelitic fraction are nearly the same, but in the zone of the Epihercynic platform the pelitic material is somewhat younger, and adjacent to the Pre-Cambrian African platform the age of sandy-aleuritic minerals is 2 to 4 times that of pelitic minerals.

The relatively young age of minerals of the Tyrrhenian Sea and the Algerian-Provençal Basin indicates that the admixture of eolian material from the African deserts to sediments of this part of the Mediterranean is very small.

The role of volcanic material in Mediterranean Sea sedimentation is relatively insignificant. Charts of volcanic glass and opaque ash particle distribution in the upper sediment layer indicate that substantial amounts of pyroclastics from Vesuvius and Stromboli occur in that region of the Tyrrhenian Sea between the Lipari Islands (Stromboli, *etc.*) and the Bay of Naples. Volcanogenic sediments formed as a result of precipitation from hydrothermal solutions are found only in local parts of the lagoon of Santorin.

CONCLUSIONS

The data presented in the previous sections demonstrate the great variability in composition and distribution of the bottom sediments in the Mediterranean Sea. The formation of individual sediment types is determined by the relative influence of several factors. These factors are clearly reflected in the granulometric, mineralogic and chemical composition of the sediments and in their physical properties. Geosynclinal regions are characterized by sediments with a particularly variable composition; on the other hand, in preplatform regions the composition is more homogenous. The composition of the sediments was originally imprinted by chemical and diagenetic processes, resulting in the accumulation of different types of carbonates. Volcanic influence, despite many active volcanoes, is of minor importance in Recent sedimentation.

In the terms of the general distribution of sediment types and their composition, the Mediterranean Sea is transitional between a true basin and an open ocean; the sediments display some features characteristic of both platform areas and trench seas, and others more commonly found in the open oceans situated in analogous climatic zones.

ACKNOWLEDGMENTS

The author expresses his gratitude to Professors A. P. Lisitzyn, J. R. Curray and D. J. Stanley, and particularly to Drs. T. A. Anderson, T.-C. Huang and G. Kelling for their considerable effort in modifying the paper into English and for their useful suggestions that improved the text.

REFERENCES

Anoshin, G. N., E. M. Emelyanov and G. A. Perezhogin 1969. Gold in recent sediments of the northern part of the Atlantic Ocean Basin (in Russian). *Geokhimiya,* 9 : 1120–1129.

d'Arrigo, A. 1936. *Richerche sul regime dei litorali nel Mediterraneo.* Stabilimento Tipografico Aternum, Rome, 172 p.

Baturin, G. N., A. V. Kochenov and K. M. Shimkus 1967. Uranium and trace elements in cores of bottom sediments from the Black and Mediterranean seas (in Russian). *Geokhimiya,* 1 : 41–50.

Berthois, L. 1939. Contributions à l'étude des sédiments de la Méditerranée occidentale. *Annales de l'Institut Océanographique, Paris,* 20 : 1–47.

Bezrukov, P. L. and A. P. Lisitsin 1960. Classification of sediments of modern sea basins (in Russian). *Transactions of the Institute of Oceanology,* Academy of Sciences USSR, 32 : 3–14.

Blanc, J. J. 1969. Sedimentary geology of the Mediterranean Sea. *Oceanography and Marine Biology, Annual Review,* 6 : 377–454.

Butuzova, G. Yu. 1969. *Recent Volcano-Sedimentary Iron-Ore Process in Santorin Volcano Caldera (Aegean Sea) and its Effect on the Geochemistry of Sediments* (in Russian). Nauka, Moscow, 112 p.

Duplaix, S. 1958. Etude minéralogique des niveaux sableux des carottes prélevées sur le fond de la Méditerranée. *Reports of the Swedish Deep-Sea Exploration, 1947–1948,* 8 (2) : 137–166.

Emelyanov, E. M. 1964. *Some features of recent sedimentation in the Mediterranean Sea* (in Russian). Thesis, Shirshov Institute of Oceanology, Academy of Sciences of the USSR, Moscow, 1–17.

Emelyanov, E. M. 1965a. The granulometric composition of modern sediments and some features of their formation in the Mediterranean Sea (in Russian). In: *The Main Features of the Geological Composition, the Hydrological Regime and Biology of the Mediterranean Sea,* ed. Fomin, L., Nauka, Moscow, 42–67.

Emelyanov, E. M. 1965b. The carbonate content of modern bottom sediments of the Mediterranean Sea (in Russian). In: *The Main Features of the Geological Composition, the Hydrological Regime and the Biology of the Mediterranean Sea,* ed. Fomin, L., Nauka, Moscow, 71–84.

Emelyanov, E. M. 1966a. Distribution of authigenic silica in suspension and in sediments of the Mediterranean Sea (in Russian). *Geokhimiya Kremnezena* (*Geochemistry of Silica*), 284–294.

Emelyanov, E. M. 1966b. Titanium in the sediments of the Mediterranean Sea (in Russian). *Litologia i Poleznye Iskopayeme (Lithology and Mineral Resources),* 6 : 3–18.

Emelyanov, E. M. 1967. The investigation of the Mediterranean floor. In: *International Dictionary of Geophysics,* Pergamon Press Limited, Oxford, 1–4.

Emelyanov, E. M. 1968. Mineralogy of sandy-aleuritic fractions of modern sediments of the Mediterranean Sea (in Russian). *Litologia i Poleznye Iskopayeme (Lithology and Mineral Resources),* 2 : 3–21.

Emelyanov, E. M. and V. D. Chumakov 1962. Some data on study of muddy waters of the Sea of Marmara and the Mediterranean Sea (in Russian). *Doklady Akademiya Nauk SSSR,* 143 (3) : 701–704.

Emelyanov, E. M., O. V. Mikhailov, V. N. Moskalenko and K. M. Shimkus 1964. Main features of the tectonic structure of the floor of the Mediterranean Sea. Geology of bottom of oceans and seas. In: *Reports of Soviet Geologists, 22nd Session of Geological Congress in Delhi.* Publications of the Academy of Sciences USSR, Moscow, 97–113.

Emelyanov, E. M., L. S. Lukoshevichus, I. P. Svirenko, A. V. Soldatov, B. A. Koshelev, A. P. Lisitsin, A. V. Illyin, V. M. Litvin and Yu M. Senin 1966. Sedimentation in the Atlantic Ocean, II. In: *International Oceanographical Congress* (Abstracts), Nauka, Moscow, 157–158.

Emelyanov, E. M., A. Ya. Krylov, Yu. I. Silin, K. M. Shimkus and Ya. Ya. Tzovbun 1973. Processes of recent sedimentation in the Mediterranean Sea (according to data of absolute age of terrigenous minerals) (in Russian). *Litologia i Poleznye Iskopayeme (Lithology and Mineral Resources),* (in press).

Emelyanov, E. M. and K. M. Shimkus 1972. Suspended matter in the Mediterranean Sea. In: *The Mediterranean Sea: A Natural Sedimentation Laboratory,* ed. Stanley, D. J. Dowden, Hutchinson and Ross, Inc., Stroudsburg, Pennsylvania, 417–439.

Fabricius, F. H., D. Berdou and K. O. Münnich 1970. Early Holocene oöids in modern littoral sands reworked from a coastal terrace, Southern Tunisia. *Science,* 169 : 757–760.

Hilmy, M. E. 1951. Beach sands of the Mediterranean coast of Egypt. *Journal of Sedimentary Petrology,* 21 : 109–120.

Joint Geophysical Committee 1969. Map 3—map of investigations (Soviet expeditions), map 4—map of investigations (foreign expeditions), map 5—types of bottom sediments, map 6—distribution of calcium carbonate, map 7—distribution of iron, map 8—distribution of manganese (in Russian). *General Board of Geodesy and Cartography at Ministry Council of the USSR,* Moscow.

Kochenov, A. V., G. N. Baturin, S. A. Kovaleva, E. M. Emelyanov and K. M. Shimkus 1965. Uranium and organic matter in sediments of the Black and Mediterranean seas (in Russian). *Geokhimiya,* 3 : 302–313.

Korneva, F. R. 1966. Distribution of foraminifera in surface stratum of sediments in the eastern part of the Mediterranean Sea (in Russian). *Okeanologiya,* 6 : 817–822.

Korneva, F. R. and H. M. Saidova 1969. Stratigraphy of sediments in the eastern part of the Mediterranean Sea by benthic foraminiferas (in Russian). In: *The Main Problems of Micropaleontology and Organogenous Sedimentation in Oceans and Seas,* Nauka, Moscow, 188–192.

Lucas, G. 1955. Oolithes marines actuelles et calcaires oolithiques récents sur le rivage africain de la Méditerranée orientale (Egypte et Sud Tunisien). *Bulletin Station Océanographique Salammbo* (Tunisie), 52 : 19–38.

Mellis, O. 1948. The coarse-grained horizons in the deep-sea sediments from the Tyrrhenian Sea. *Göterborgs Kungl. Vetenskaps-Och Viterhets-* Samhälles Handlinger, 6 Följden, series B., 5 (13) : 45–72.

Mellis, O. 1954. Volcanic ash horizons in deep-sea sediments from the Eastern Mediterranean. *Deep Sea Research,* 2 : 89–92.

Morović, D. 1951. Composition mécanique des sédiments au large de l'Adriatique. *Institut za Oceanografiju i Ribarstvo, Split,* 3: 1–21.

Müller, G. 1961. Die rezenten Sedimente in Golf von Neapel. 2. Mineral-Neu-und-Um-bildungen in den rezenten Sedimenten des Golfes von Neapel. Ein Reitrag zur Unwandlund vulkanischer Gläser durch Halmyrolyse. *Beiträge zur Mineralogie und Petrologie,* 1 : 1–20.

Müller, G. 1964. Die Korngrossenverteilung in den rezenten sedimenten des Golfes von Neapel. In: *Deltaic and Shallow Marine Deposits* (Developments in Sedimentology), ed. van Straaten, L. M. J. U., Elsvier, Amsterdam, 1 : 282–292.

Ninkovich, D. and B. C. Heezen 1965. Santorini Tephra. In: *Submarine Geology and Geophysics,* eds. Whittard, W. F. and R. Bradshaw, Butterworths, London, 413–453.

Norin, E. 1958. The sediments of the central Tyrrhenian Sea. *Reports of the Swedish Deep-Sea Expedition 1947–1948,* 8 (1): 1–136.

Parker, F. L. 1958. Eastern Mediterranean foraminifera. *Reports of the Swedish Deep-Sea Expedition, 1947–1948,* 8 (2) : 217–283.

Pigorini, B. 1968. Sources and dispersion of recent sediments of the Adriatic Sea. *Marine Geology,* 6 : 187–230.

Rateev, M. A., E. M. Emelyanov and M. B. Kheirov 1966. Conditions for the formation of clay minerals in contemporaneous sediments of the Mediterranean Sea. *Litologia i Poleznye Iskopayeme (Lithology and Mineral Resources)*, 4:6–23.

Ryan, W. B. F. and B. C. Heezen 1965. Ionian Sea submarine canyons and the Messina turbidity current. *Geological Society of America Bulletin*, 76:915–932.

Ryan, W. B. F., F. Workum, Jr. and J. B. Hersey 1965. Sediments on the Tyrrhenian abyssal plain. *Geological Society of America Bulletin*, 76:1261–1282.

Sevastyanov, V. F. 1968. Redistribution of chemical elements as a consequence of oxidation-deoxidation processes in sediments of the Mediterranean Sea. *Litologia i Poleznye Iskopayeme (Lithology and Mineral Resources)*, 1:3–15.

van Straaten, L. M. J. U. 1965. Sedimentation in the northwestern part of the Adriatic Sea. In: *Submarine Geology and Geophysics*, eds. Whittard, W. F. and R. Bradshaw, Butterworths, London, 143–162.

Strakhov, N. M. 1951. Limestone-dolomitic facies of recent and old basins (in Russian). *Trudy Institut Geologicheskikh Nauk, Akademiya Nauk SSSR*. 125, geol. seriya, 45:1–331.

Strakhov, N. M. 1969. *Principles of Lithogenesis*. Consultants Bureau, New York, 2:609.

Strakhov, N. M., G. I. Buchinski, L. V. Poustovalov, A. V. Khabakov and I. V. Khovrova 1957. *Methods of Study of Sedimentary Rocks* (in Russian). Gosgeoltechyzdat, Moscow, 2:528 p.

Sudo, F., K. Oinuma and K. Kobayashi 1961. Mineralogical problems concerning rapid clay mineral analysis of sedimentary rock. *Acta Universitatis Carolinae-Geologica, Supplementum*, 1.

Sukhorukov, F. V. and E. M. Emelyanov 1969. Boron in bottom sediments in the northeastern part of the Atlantic Ocean basin. *Doklady Akademiya Nauk SSSR*, 187:1153–1156.

Wood, L. E. 1964. Pseudo-oolites of Northern Libya: their occurrence and origin. *Journal of Sedimentary Petrology*, 34:661–663.

Sur la Sédimentation Argileuse Profonde en Méditerranée

Hervé Chamley

Centre d'Océanographie, Luminy, Marseille

RESUME

La répartition des minéraux argileux a été déterminée au long de 25 carottes profondes, par diffraction des rayons X et microscopie électronique. Afin de déterminer sa signification historique, cette répartition a été confrontée avec divers aspects des sédiments, comme la teneur en sable, carbonates, l'analyse chimique et granulométrique, et la microfaune. Plusieurs cas de sédimentation argileuse sont identifiés:

1. Héritage continental strict: carottes de Méditerranée occidentale sans turbidites (ex: sommets de dômes). Les argiles issues du continent, plus ou moins altérées par les climats quaternaires, permettent, du fait de la modestie des actions diagénétiques, de reconstituer une paléoclimatologie précise.
2. Héritage perturbé: carottes marquées par des turbidites grossières (ex: axe du Rhône) ou fines (ex: Sud de Matapan), ou par des niveaux pyroclastiques (Méditerranée orientale). L'interprétation climatique est souvent compliquée, l'interprétation chronologique souvent simplifiée par ces passées remarquables.
3. Transformation: les niveaux sapropéliques de Méditerranée orientale montrent une dégradation de la montmorillonite et des minéraux mal cristallisés.
4. Néoformation en milieu chimique basique: sédiments de fosse profonde, liés à une période ancienne chaude et humide (Sud de Matapan). Les ions, issus du lessivage continental en même temps que des argiles banales, auraient été concentrés jusqu'à permettre la formation d'attapulgite.
5. Néoformation en milieu volcanique: les cendres issues de l'érosion de Santorin subissent une évolution en montmorillonite cependant que les diatomées sont dissoutes.

D'une manière générale les argiles de Méditerranée occidentale témoignent essentiellement de leur histoire continentale avant le dépôt; celles de Méditerranée orientale reflètent fréquemment le milieu de dépôt.

ABSTRACT

The distribution of clay minerals in 25 deep sea cores has been examined by X-ray diffraction and electron microscopy. In order to determine its geological significance, this distribution has been compared with various parameters, such as sand and carbonate content, chemical composition, size analysis, and microfauna. Several types of argillaceous sedimentation have been identified:

1. Undisturbed detrital deposits, *i.e.*, cores from western Mediterranean without turbidites, such as those on domes. The clays transported from continents, somewhat altered by Quaternary weathering, serve to interpret the paleoclimatology since subsequent diagenetic processes affecting the clays have been weak.
2. Disturbed detrital deposits, *i.e.*, cores with sandy (Rhône Fan) or clayey (Matapan Trench) turbidites, or with pyroclastic layers (eastern Mediterranean). Climatological interpretation is in many cases difficult, although these units serve as chronological markers.
3. Transformed detrital deposits, *i.e.*, the sapropel units in the eastern Mediterranean, display poor preservation of montmorillonite and poorly crystallized clay minerals.
4. Authigenesis in basic environments, *i.e.*, formation of attapulgite in Matapan Trench during an early warm, humid period. Such ions, weathered from continental sources along with those of other more common clay minerals, were thus concentrated so as to permit the formation of attapulgite.
5. Authigenesis in the volcanic environment, *i.e.*, fine ash from Santorini, are altered to montmorillonite while diatoms are dissolved.

The clay minerals from the western Mediterranean can generally be used as provenance indicators, while those from the eastern Mediterranean often reflect conditions in the depositional environment.

TRAVAUX ANTERIEURS

Les études publiées sur les argiles méditerranéennes ont trait, soit à la répartition et à l'origine des différents minéraux dans les bassins, soit à l'influence des actions liées à la diagenèse, soit encore à l'influence des actions continentales précédant l'érosion des minéraux et leur sédimentation.

Le premier groupe d'études montre que les argiles méditerranéennes sont essentiellement héritées des continents. Ce fait ressort des synthèses (Chamley *et al.*, 1962; Nesteroff *et al.*, 1963; Biscaye, 1965; Griffin *et al.*, 1968; Rateev *et al.*, 1966 et Rateev *et al.*, 1968), mais aussi des études plus localisées, qui sont variées (Vernhet, 1956; Blanc, 1958a, 1958b et 1964; Nesteroff et Sabatier, 1958–1959; Ngoc Cau *et al.*, 1959; Monaco, 1965; Chamley, 1967a, 1968a et b).

Le second groupe d'études montre que les actions diagénétiques sont très modestes dans les sédiments carottés (Chamley *et al.*, 1962; Monaco, 1965; Chamley, 1968a), sauf dans les secteurs volcaniques du Sud de la mer Tyrrhénienne, où s'observent quelques transformations minéralogiques (Norin, 1953; Grim et Vernet, 1961; Müller, 1961). Ces observations sont du reste applicables à l'océan mondial (Griffin *et al.*, 1968; Rateev *et al.*, 1968).

Le troisième groupe d'études, dont les conclusions sont subordonnées aux résultats des deux premiers, conduit à penser que la plupart des mélanges argileux méditerranéens est représentative des climats continentaux sous lesquels ils ont été érodés (Chamley, 1967b, 1968b, 1969a et 1969b; Peronne, 1967; Blanc-Vernet *et al.*, 1969).

BUT DU TRAVAIL ET DOMAINE D'INVESTIGATION

Des recherches récentes montrent que, si les trois ensembles de résultats qui viennent d'être évoqués

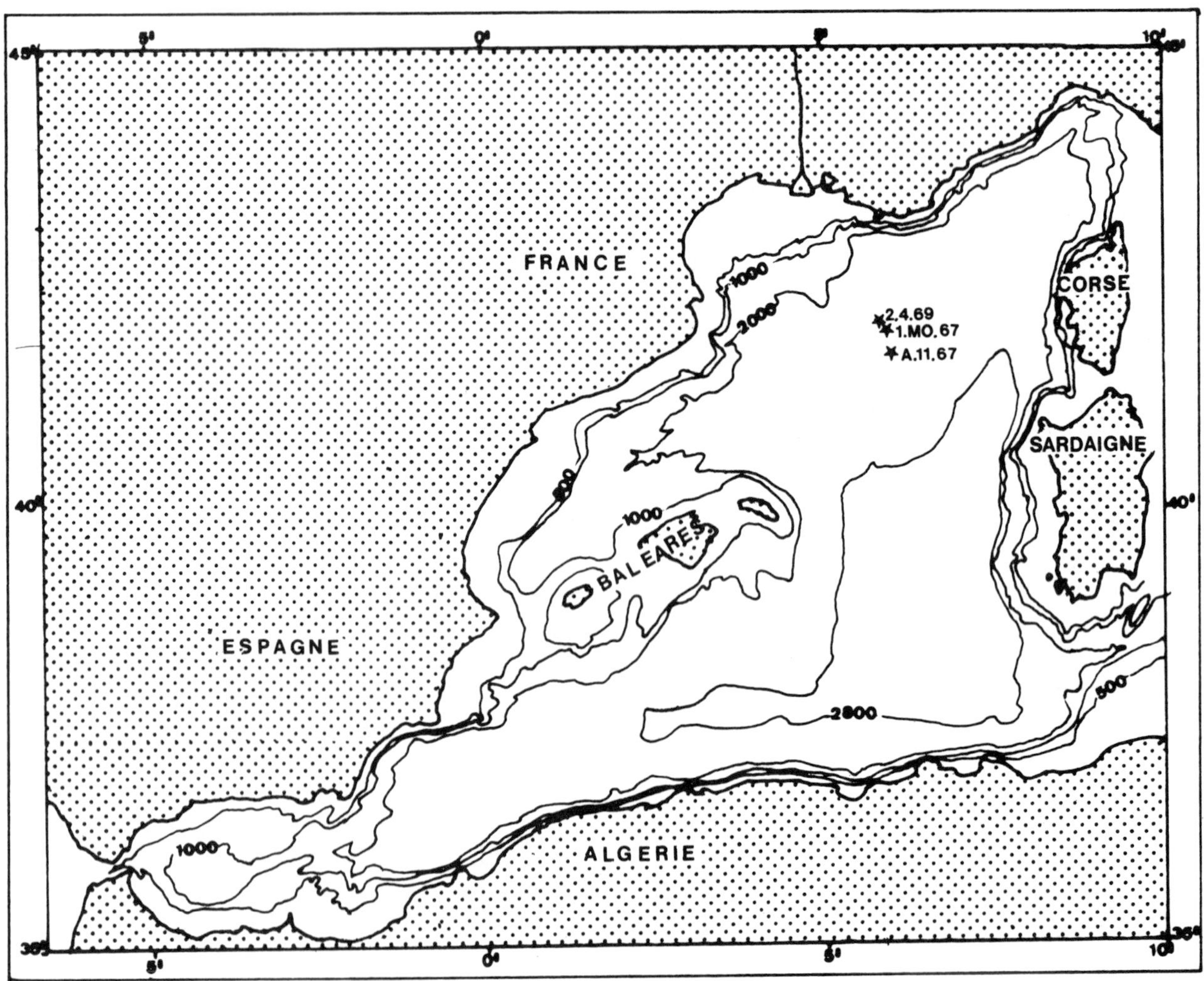

Figure 1. Méditerranée occidentale; situation des carottes citées.

sont bien établis, il existe cependant en Méditerranée des modèles de sédimentation argileuse plus variés que ceux qu'ils impliquent. Nous nous proposons ici de faire le tour d'horizon de ces différentes recherches, à partir d'exemples tirés de l'étude de 25 carottes profondes.

Les régions étudiées (Figures 1 et 2) correspondent au Bassin Méditerranéen nord-occidental, aux côtes de la Libye et aux fonds de Méditerranée orientale situés de part et d'autre de l'arc égéen externe (fosse de Matapan au Sud du Péloponnèse, bassins sud- et nord-crétois).

Les carottes proviennent des missions des navires océanographiques CALYPSO (1964 en Méditerranée orientale, 1965 au large de la Tripolitaine), JEAN CHARCOT (1966 et 1969 en Méditerranée occidentale, 1967 en Méditerranée orientale) et ORIGNY (1963 en Méditerranée occidentale).

METHODES D'ETUDE

Les carottes, longues de 300 à 800 cm, ont fait l'objet d'un prélèvement de 3 cm de longueur, tous les 10 cm en moyenne, moins lorsque les variations de texture l'exigeaient.

Les techniques utilisées pour l'étude de la fraction argileuse sont classiques: ce sont principalement la diffraction des rayons X sur des agrégats orientés de particules décarbonatées inférieures à 2 microns, et la microscopie électronique sur des évaporats de suspensions brutes (Beutelspacher et van der Marel, 1968; Brown, 1961; Lucas *et al.,* 1959). Des données complémentaires ont été apportées par les analyses thermiques différentielle et pondérale (Mackenzie, 1957).

Différentes analyses ont été pratiquées sur d'autres fractions des sédiments, dans le but de replacer la sédimentation argileuse dans un contexte plus général. Il s'agit des dosages de la fraction grossière (inférieure à 50 microns), du CO_3Ca total, des éléments chimiques majeurs de la phase argileuse, du pH, de mesures de la couleur, de la granulométrie, de l'étude optique des résidus grossiers et fins.

Dans les confrontations entre la fraction argileuse et d'autres composants des sédiments, les données de Foraminifères occupent une place privilégiée. Les comparaisons ont été principalement effectuées avec les travaux de Blanc-Vernet (1969, et ce volume).

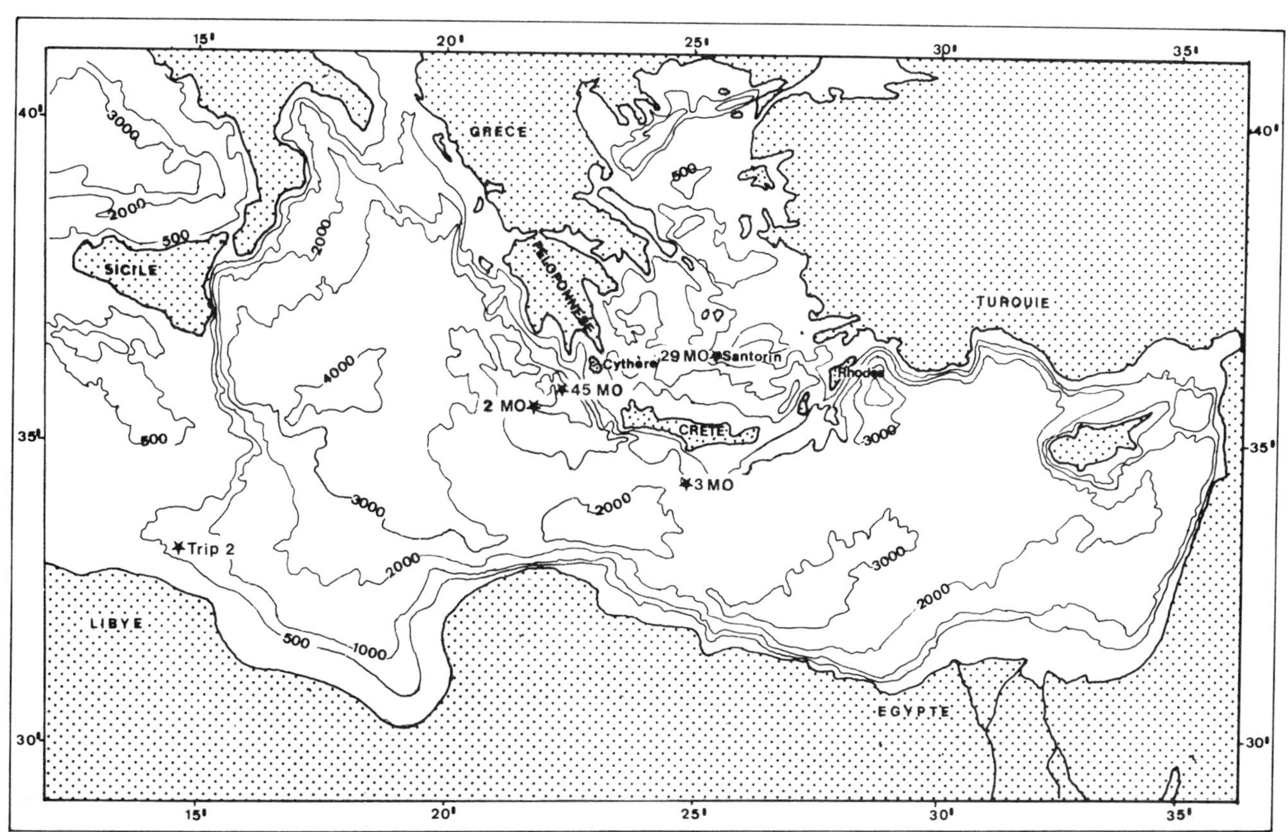

Figure 2. Méditerranée orientale; situation des carottes citées.

La manière la plus sûre de définir le type et l'intensité du climat dont peut témoigner une argile héritée consiste principalement à mesurer la largeur du pic de l'illite (10Å) à mi-hauteur, en $1/10°\theta$, sur l'essai glycolé (indice n_a): cette largeur est proportionnelle au degré d'altération (Chamley, 1969a). Des renseignements complémentaires s'obtiennent à partir d'autres mesures effectuées sur les diagrammes de rayons X (Chamley, 1969a) et des clichés de microscopie électronique (Chamley, 1969b). Elles concourent toutes à mettre en évidence et à chiffrer le degré d'hydrolyse des minéraux, désignant un climat plus ou moins froid-sec, plus ou moins chaud-humide.

RESULTATS

Chaque type de sédimentation rencontrée sera décrit à l'aide d'une carotte choisie comme exemple. Puis, dans le chapitre des discussions, on tentera une généralisation à partir de l'ensemble des résultats obtenus.

Méditerranée Nord-Occidentale: Carotte 1 MO 67

Prélevée par 2.450 m de profondeur dans la zone des dômes de Méditerranée occidentale (Menard *et al.*,

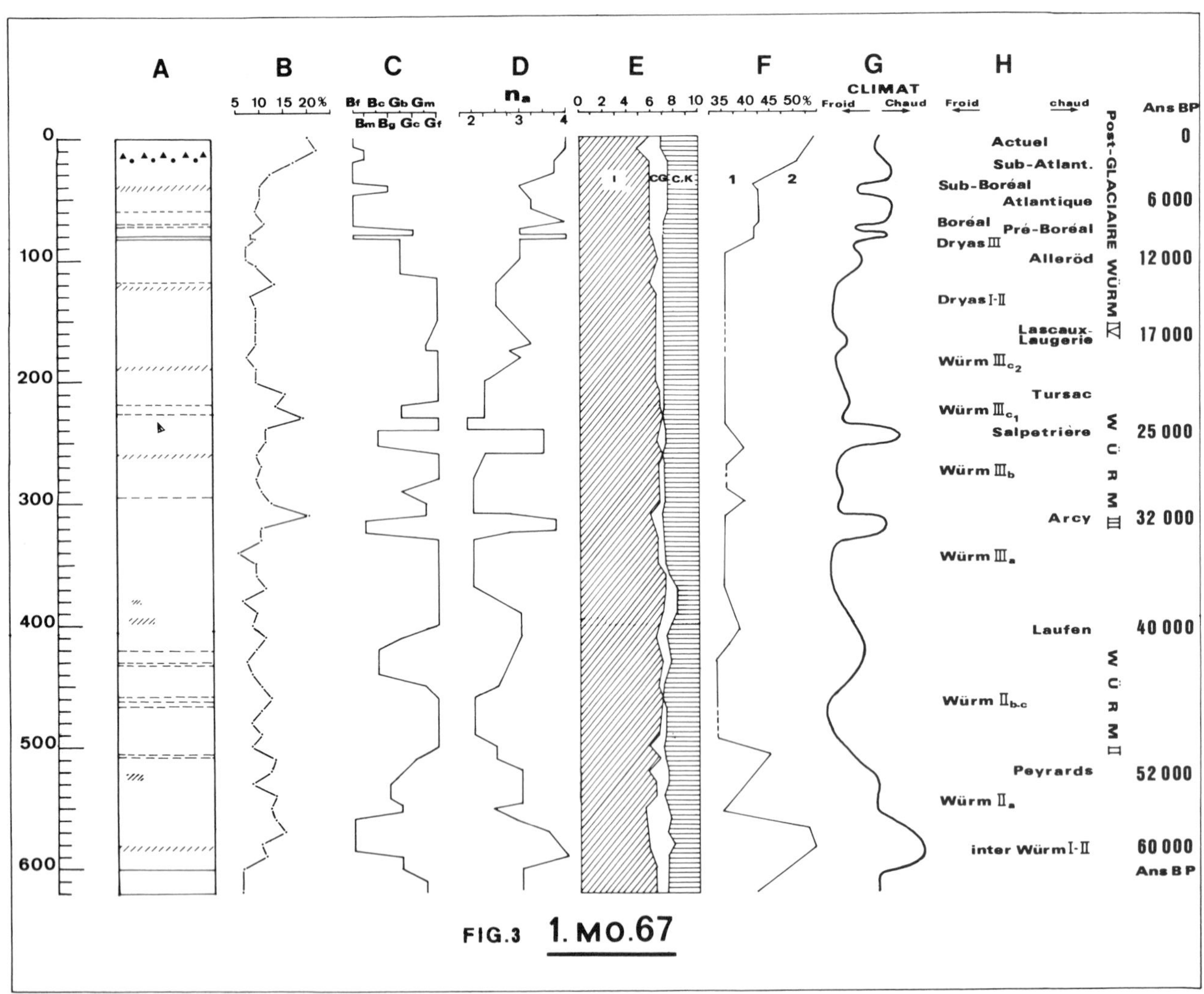

Figure 3. Légende pour les carottes 1 MO 67 et A 11 67 (Figure 5). **A,** Texture: pointillé = sable, blanc = vase, trait continu = contact tranché, tireté = taches étalées, hachures obliques = tachetures, cercles = Foraminifères, triangles = Ptéropodes, et triangle barré = Pélécypodes. **B,** Pourcentage de sable (particules supérieures à 50 microns). **C,** Couleur des suspensions argileuses décalcifiées: B = beige, G = gris, f = foncé, m = moyen, et c = clair. **D,** Etat de l'illite: largeur à mi-hauteur, en $1/10°\ \theta$, du pic à 10Å, sur l'essai glycolé (indice n_a). **E,** Proportions relatives, en 1/10, des minéraux argileux: I = illite, CG = complexe gonflant (montmorillonite surtout), et C, K = chlorite (dominante) et kaolinite. **F,** Analyse chimique, limite en % de deux groupes d'éléments: 1, CaO, perte au feu, Mn_4O_4, et 2, SiO_2, Al_2O_3, MgO, Fe_2O_3, TiO_2 (ordre d'abondance décroissant). **G,** Climat: trait plein = sédimentation normale, et tireté = turbidite. **H,** Interprétation chronologique.

1965; Glangeaud, 1966), la carotte 1 MO 67 (41°52′N, 05°52′E), longue de 620 cm, est constituée de vase fine, homogène, troublée seulement par une passée à Ptéropodes et Foraminifères à—10 cm (Figure 3, A). La fraction grossière est présente en faible quantité, 10% en moyenne, et subit des variations de faible amplitude (Figure 3, B). La couleur de la vase varie du gris foncé au beige foncé (Munsell 2,5 Y 4/0 à 7,5 YR 5/4). Ces variations sont plus ou moins rapprochées mais jamais brutales (Figure 3, C); elles correspondent fréquemment aux variations de la proportion de fraction sableuse, les niveaux les plus beiges étant les plus grossiers. Les niveaux superficiels témoignent d'une certaine brunification, due à la présence de matières organiques encore non oxydées.

L'étude par diffraction des rayons X montre des proportions relatives assez constantes entre les minéraux argileux: l'illite est, tout au long de la carotte, dominante sur la chlorite, le complexe gonflant (montmorillonite surtout, interstratifiés illite-montmorillonite et chlorite-montmorillonite) et la kaolinite; cette composition est celle des vases de Méditerranée occidentale liées principalement aux apports du Rhône (Chamley *et al.*, 1962). Cependant de modestes variations, qui se révèlent parfaitement significatives, s'observent au long de la carotte (Figure 3, E, et Chamley, 1969a); elles concernent surtout l'abondance relative du complexe gonflant, qui croît lorsque la couleur de la vase est plus beige. Ces variations correspondent par ailleurs à celles de l'indice d'ouverture de l'illite (n_a), qui augmente dans les passées plus beiges (Figure 3, D). Une étude de détail montre enfin que ces mêmes passées sont plus pauvres en chlorite, en feldspath, plus riches en kaolinite et en minéraux altérés (minéraux simples à réflexions larges, minéraux interstratifiés gonflants de type irrégulier).

L'étude morphologique (microscopie électronique) permet d'établir de nettes différences entre les niveaux gris à illite fermée et les niveaux beiges à illite ouverte (Figure 4, A et B): les premiers renferment des particules aux contours nets, les seconds des particules aux bords effrangés, souvent morcelés et flous. Cette opposition se retrouve dans une certaine mesure dans la fraction sableuse des sédiments, les niveaux beiges contenant généralement des minéraux plus altérés (phyllites surtout). Enfin on note dans les niveaux beiges une proportion plus importante de Coccolithes.

L'analyse chimique des éléments majeurs montre qu'ils sont répartis en deux groupes: les uns, liés aux argiles, varient en opposition avec l'indice n_a (cristallinité de l'illite): SiO_2, Al_2O_3, Fe_2O_3, MgO, TiO_2; les autres, liés aux carbonates, varient parallèlement à n_a : CaO, perte à 1000°C (Figure 3, F). D'autre part le rapport SiO_2/Al_2O_3 diminue dans les passées de mauvaise cristallinité.

En résumé, la carotte 1 MO 67 est caractérisée par une sédimentation fine, que confirme du reste l'analyse granulométrique par densimétrie, montrant un dépôt par décantation. Les particules héritées du continent subissent des variations au long de la carotte, qui sont sans polarité et significatives; ces variations sont parallèles ou symétriques les unes par rapport aux autres, et concernent des paramètres variés: couleur, teneur en sable, abondance, cristallinité et morphologie des minéraux, surtout argileux, composants chimiques des argiles et des carbonates.

Méditerranée Nord-Occidentale: Carotte A 11 67

Prélevée par 2.600 m de profondeur à une quinzaine de milles nautiques au Sud de la carotte 1 MO 67, la carotte A 11 67 (41°37′N, 05°52′E) en diffère fortement par sa texture (Figure 5, A): longue de 445 cm, elle comprend plus de 300 cm de sable gris, réparti en trois passées principales. Le reste de la carotte est constitué de vase, beige au sommet (Munsell 7,5 YR 5/4), gris foncé ailleurs, particulièrement entre les passées sableuses (2,5 Y 4/0) (Figure 5, C). Les contacts entre le sable et la vase sont rectilignes ou sinueux, mais toujours tranchés; des "galets" de vase se trouvent parfois à la base des passées grossières, ainsi que de nombreux ptéropodes. La mesure du pourcentage de fraction grossière illustre bien ces variations de texture (Figure 5, B).

L'étude par diffraction des rayons X montre une grande homogénéité au long de la carotte (Figure 5, E et D): les minéraux, dont les proportions sont analogues à celles de la carotte précédente (illite dominante), ne subissent des variations, modestes, que dans les 50 cm supérieurs et entre 120 et 140 cm. Ces variations sont de même nature que pour la carotte 1 MO 67. Ainsi une illite plus ouverte correspond à un complexe gonflant plus abondant, et à des minéraux plus altérés: la couleur de la vase correspondante est plus beige, cependant que les contours des particules argileuses sont plus flous.

L'étude des niveaux sableux permet de préciser leur origine:

1. *Morphoscopie, microfaune*: les grains sont parfois usés par le vent ou l'eau; par ailleurs les rares Foraminifères benthiques rencontrés (*Elphidium, Quinqueloculina*) appartiennent à des associations peu profondes (Blanc-Vernet, 1969). Ces deux arguments suggèrent une origine littorale.

2. *Minéralogie*: les espèces identifiées forment un mélange complexe, très voisin de celui présent à l'embouchure du Rhône (van Andel, 1955); par ailleurs

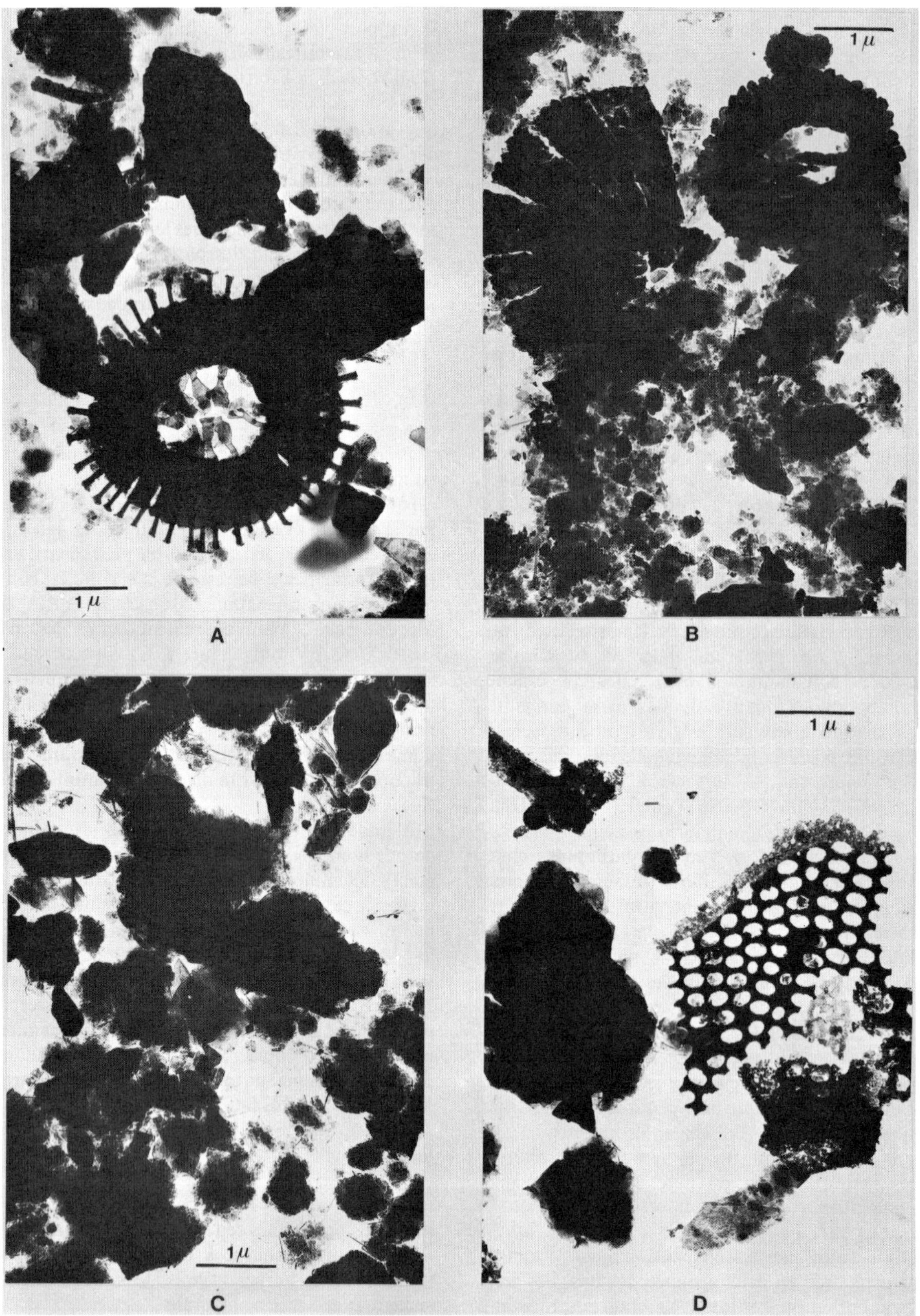

Figure 4. Micrographies électroniques. Carotte 1 MO 67, Méditerranée nord-occidentale: A, Niveau de type froid (× 13.500). B, Niveau de type chaud (× 13.500). Carotte 3 MO 67, Sud de la Crète: C, Vase banale (× 13.500). D. Sapropel (× 13.500).

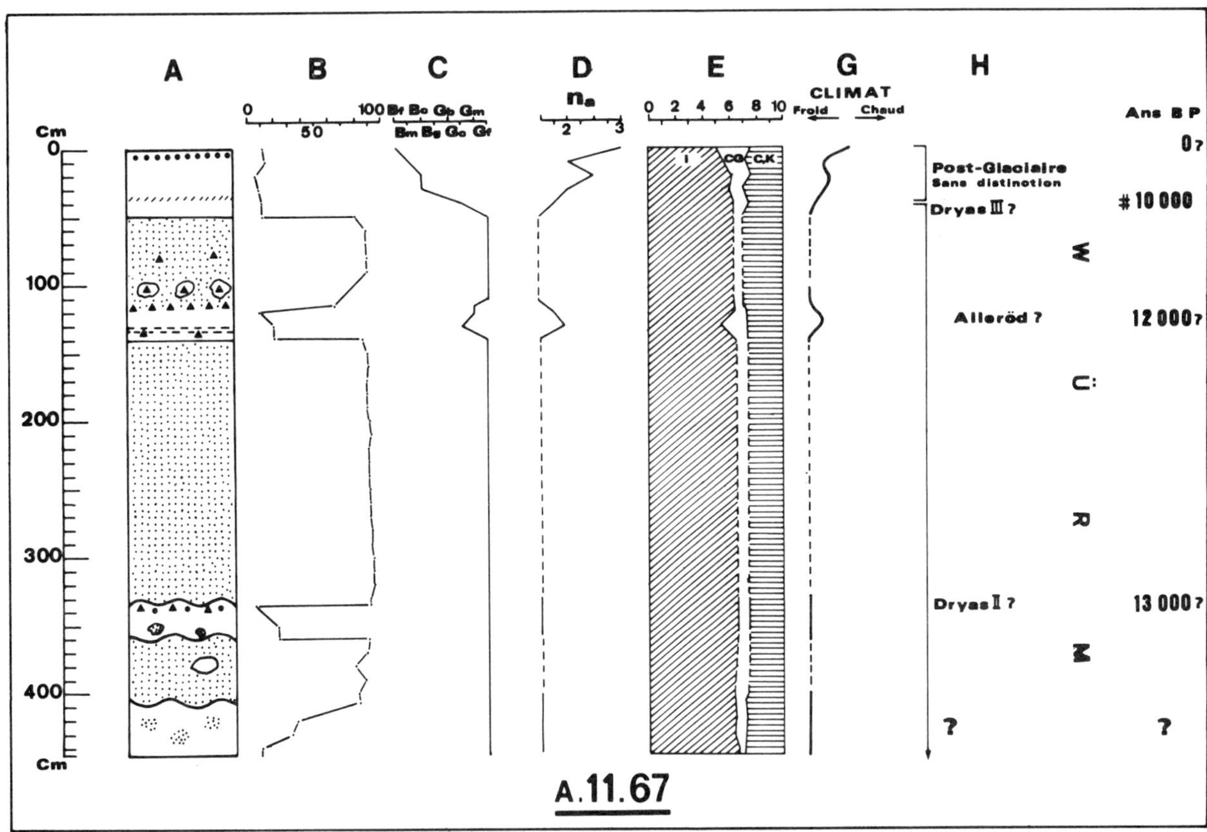

Figure 5. Résultats d'étude de la carotte A 11 67 (légende, voir Figure 3).

la carotte est située dans l'axe du canyon prolongeant le Rhône.

3. *Granulométrie*: on observe un tri très net. Ainsi dans la passée sableuse comprise entre 330 et 140 cm, le pourcentage de sable (supérieur à 50 microns) passe de 94 à 85%, et les médianes des courbes cumulatives de 0,135 à 0,084 mm).

En résumé, la carotte A 11 67 correspond à une sédimentation détritique pélagique, interrompue par d'abondantes turbidites littorales granoclassées, d'origine vraisemblablement rhodanienne. Les variations dans la composition et l'état de la fraction fine sont peu nombreuses: mais cependant, synchrones et significatives, elles sont analogues à celles observées pour la carotte précédente (1 MO 67).

Méditerranée Orientale, Bassin Sud Crétois: Carotte 3 MO 67

Prélevée par 1950 m de profondeur à une trentaine de milles au Sud du Cap Littinos (Crète), la carotte 3 MO 67 (34°25′N, 24°50′E) est située dans la zone des "montagnes de Ptolémée", au Nord de la dorsale est-méditerranéenne (Emery *et al.*, 1966) (Figure 2). Longue de 450 cm, elle est constituée de vase fine, interrompue par une passée de sable volcanique de 135 à 137 cm, à la base de laquelle se trouve un lit d'organismes planctoniques. De tels niveaux pyroclastiques sont bien connus en Méditerranée orientale, et ont été rapportés aux principales éruptions historiques de Santorin, archipel situé au Nord de la Crète (Mellis, 1954; Ninkovich et Heezen, 1965). La vase est de couleurs variées, allant du beige foncé (Munsell 7,5 YR 5/4) à différents gris, dont les nuances sont parfois mauves, parfois verdâtres. La principale originalité de la sédimentation fine réside dans la présence de 5 passées noires (5Y 3/1), de type sapropélique, épaisses de quelques cm à près de 40 cm, fréquemment riches en organismes planctoniques (Ptéropodes surtout). De tels sapropels ont été signalés en Méditerranée orientale à diverses reprises (Olausson, 1960 et 1961; van Straaten, 1966; Pastouret, 1969), et correspondent à des phases de stagnation des eaux.

Les analyses diffractométriques pratiquées sur la carotte montrent une légère prédominance de la montmorillonite sur l'illite; l'ensemble chlorite-kaolinite est d'importance secondaire. Cela correspond aux résultats connus dans le Bassin Oriental (Chamley *et al.*, 1962). Quant aux variations relevées au long de la carotte, elles sont dans le cas général de même nature que celles de la carotte 1 MO 67: le degré d'ouverture de l'illite, les proportions relatives des minéraux argileux, *etc.* varient de manière synchrone et tout à fait comparable aux données de la Figure 3.

Il n'en est pas de même pour les sapropels. Ces niveaux sont en effet originaux à divers points de vue:

1. L'analyse minéralogique montre la quasi-disparition de la montmorillonite. D'autre part le pic de l'illite apparaît systématiquement très étroit.

2. La couleur noire correspond à une forte teneur en matières organiques. Celles-ci ont été appréciées par calcination durant une heure, à 400°C: la perte de poids, rattachable à la matière organique et à l'eau (peu abondante), est de deux à quatre fois plus forte dans les sapropels que dans les vases banales. Les teneurs en matière organique mesurées par Pastouret (1969) dans des carottes au voisinage de Rhodes sont égales à 10% pour les sapropels, contre 1% pour les vases banales. On note une grande abondance de pollen, de fibres végétales diverses, de chitine, et de vertèbres de poissons.

3. Les sulfures, et particulièrement la marcassite, sont abondants, d'après les courbes d'analyse thermique différentielle.

4. La microscopie électronique montre une grande pauvreté en argile, limitée à de grosses particules bien cristallisées; la taille de ces particules est responsable de la bonne cristallinité des minéraux, et de l'illite en particulier. On remarque par ailleurs une abondance extraordinaire de Diatomées, d'espèces variées et bien conservées (Figure 4, D); ce fait, déjà pressenti par Olausson (1961) et Emery *et al.* (1966), s'oppose aux observations effectuées sur la plupart des vases méditerranéennes: en particulier les vases limitrophes des sapropels, qui contiennent des phyllites de toutes tailles aux contours variés, sont dépourvus de Diatomées (Figure 4, C). Les coccolithes sont parfois peu abondants, et presque toujours très fins; cela va de pair avec la fragilité des tests de Ptéropodes et Foraminifères, observée à la loupe.

5. Le pH, reconstitué à partir de 10 g de sédiment sec complété à 100 g avec de l'eau distillée, est dans les sapropels, inférieur de 0,8 à 0.9 unité à celui des vases banales (7,6 à 7,8 contre 8,5 à 8,6).

En résumé, la carotte 3 MO 67 apparaît constituée de deux groupes de sédiments fins: d'une part les vases grises à beiges dont les constituants sont banaux et ont un comportement relatif analogue à ceux des vases de Méditerranée occidentale; d'autre part les sédiments sapropéliques, dont les constituants minéraux et organiques présentent des particularités marquées.

Méditerranée Orientale, Fosses Sud de Matapan: Carotte 45 MO 67

Les fosses de Matapan, situées au Sud du Péloponnèse, sont les plus profondes de Méditerranée. La carotte 45 MO 67, longue de 365 cm, à été prélevée à 4.420 m de profondeur dans le groupe des fosses Sud (35°53′N, 22°21′E). Sa partie supérieure, jusqu'à 45 cm de profondeur, est classique: il s'agit d'une vase beige rougeâtre surmontant une vase grise. En-dessous apparaissent des boues calcaires blanches, qui alternent avec des vases rose-rougâtres fréquemment marquées par des figures de remaniement (Chamley et Millot, 1970).

L'analyse roentgenographique montre également deux types de sédiments:

1. Les vases grise et rouge supérieures contiennent de l'illite et de la montmorillonite en quantités voisines; la chlorite et la kaolinite sont subordonnées. Cette composition est celle de tous les fonds situés au large du Péloponnèse.

2. Les boues blanches contiennent, en plus des minéraux précédents, de l'attapulgite, dont l'abondance diminue dans les passées roses. La structure en lattes de ce minéral est nette (Figure 6, A).

L'opposition entre les vases supérieures et inférieures apparaît encore à d'autre égards: les boues blanches sont calcaires, très riches en cuivre, pauvres en quartz détritique, et caractérisées par une faune de Foraminifères et de Coccolithophoridées d'affinité intertropicale (Blanc-Vernet et Chamley, 1971). Ces caractères sont d'autant mieux marqués que les niveaux sont plus blancs.

En résumé, la carotte 45 MO 67 renferme, sous quelques dizaines de centimètres de vase banale, des boues blanches, calcaires, à attapulgite, de faciès climatique chaud, et dont le dépôt a été plus ou moins contrarié par des vases rose-rougeâtres, déterminant une sédimentation en séquences.

Méditerranée Nord-Orientale, Archipel Volcanique de Santorin: Carotte 29 MO 67

Situé au Sud des Cyclades, l'archipel de Santorin correspond à un ensemble volcanique ancien, effondré en mer, et duquel a surgi un ensemble volcanique récent; il a fait l'objet d'études variées (in Blanc, 1958b). La carotte 29 MO 67, longue de 730 cm, a été prélevée dans la partie Sud de l'ancienne caldeira, à 285 m de profondeur (36°22′,75N, 25°21′,1E). Elle est constituée de vase et de sable volcaniques, issus de l'érosion rapide des falaises cendreuses de l'archipel.

Etudié aux rayons X, le niveau de surface montre une quasi-absence de minéraux, si ce n'est des traces de montmorillonite, d'illite et de chlorite. Les niveaux sous-jacents montrent une augmentation de la montmorillonite, très bien cristallisée, cependant que les autres minéraux restent à l'état de traces. Cette augmentation est forte dans les premières dizaines de centimètres, lente au-dessous. Elle est relativement moins forte dans les niveaux plus sableux, mais la polarité observée est indépendante de la granulométrie.

La microscopie électronique confirme ces faits: le niveau de surface ne comprend pratiquement pas d'argile, mais des cendres amorphes, opaques aux électrons (Figure 6, B); les niveaux sous-jacents contiennent des phyllites, qui se développent sur les bords des cendres volcaniques (Figure 6, C); il s'agit de la montmorillonite identifiée aux rayons X. Par ailleurs le niveau de surface contient de nombreuses diatomées (Figure 6, B), cependant qu'en profondeur les frustules sont très rares et très corrodés (Figure 6, D); cette observation diffère de nombreuses observations antérieures qui, depuis Humboldt (1839), décrivent la constance de l'association cendres volcaniques—diatomées. La profondeur à laquelle les diatomées ont pratiquement disparu (180 cm) correspond au ralentissement constaté dans le développement de la montmorillonite.

Par ailleurs le sédiment est totalement dépourvu de calcaire et de matière organique. L'analyse chimique montre une constance rigoureuse du rapport SiO_2/Al_2O_3; l'analyse granulométrique confirme la finesse des boues volcaniques.

En résumé, la carotte 29 MO 67 est constituée de boue volcanique amorphe à diatomées, qui passe rapidement en profondeur à une boue volcanique à montmorillonite, sans diatomées.

DISCUSSION

Le Milieu Pélago-Détritique Non Perturbé

Les variations minéralogiques et sédimentologiques observées sur la carotte 1 MO 67 (Méditerranée occidentale) s'expliquent toutes de manière simple si on les met en parallèle avec les variations observées dans les sols actuels sous divers climats (Millot, 1964): une hydrolyse croissante, correspondant à un climat plus chaud et plus humide, et de bonnes conditions de drainage, détermine une dégradation croissante des minéraux (élargissement des réflexions de l'illite, diminution de la chlorite, développement des interstratifiés gonflants, détérioration des bords des particules), une augmentation de la kaolinite néoformée à partir des éléments résiduels, une rubéfaction liée à la libération des oxydes de fer, enfin une augmentation de fraction grossière, souvent calcaire, favorisée par la fragmentation des roches.

En période froide et sèche, c'est l'inverse qui s'observe. Ainsi les argiles héritées sont, au long des carottes, le reflet des climats successifs (Chamley 1968b, 1969b). Une confirmation en est donnée par la confrontation avec les données de la microfaune (Blanc-Vernet, 1969; Blanc-Vernet *et al.*, 1969; Rotschy *et al.*, 1969): les niveaux à illite ouverte correspondent à des faunes chaudes, les niveaux à illite fermée à des faunes nord-atlantiques.

Il est donc possible, en chiffrant les données de l'analyse minéralogique et sédimentologique (Chamley, 1969a), d'obtenir une courbe paléoclimatique à partir de la fraction sédimentaire issue du continent (Figure 3, G). Déduite principalement de la mesure de l'ouverture de l'illite, corrigée des effets éventuels de diagenèse, cette courbe conduit à une interprétation chronologique (Figure 4, H), fondée sur la comparaison avec les données acquises sur le continent (Escalon de Fonton, 1966). Par voie de retour elle permet de préciser certains épisodes devinés sur le continent (Escalon de Fonton, 1968, p. 290), épisodes dont les témoins sont parfois érodés, modifiés ou masqués (Fedoroff, 1969). Enfin elle permet d'apprécier la vitesse de la sédimentation pélago-détritique au cours des périodes climatiques successives; la valeur moyenne obtenue dans le Bassin Occidental (vase humide) est de 10 cm/1.000 ans (Chamley,1968a).

On définit ainsi un premier milieu de sédimentation argileuse, celui de l'*héritage simple*, qui se révèle d'une grande utilité pour la connaissance des paléoclimats. Un tel milieu se rencontre en Méditerranée occidentale, particulièrement sur les dômes de la plaine bathyale, qui échappent aux écoulements turbides. Au large des côtes de la Tripolitaine se rencontre une sédimentation analogue, encore que davantage soumise aux variations de l'apport détritique littoral (carotte Trip 2; 33°06′N, 14°31′E); par ailleurs on y trouve en assez forte proportion de l'attapulgite, dont l'habitus montre qu'elle est héritée du continent africain, riche en palygorskites dans le Crétacé terminal et l'Eocène (Millot, 1964). L'attapulgite, dont la réflexion basale à 10,5 Å est très proche de celle de l'illite, rend la signification paléoclimatique des minéraux difficile à lire. En Méditerranée orientale le milieu de sédimentation détritique non perturbée se trouve dans des carottes variées (ex. 3 MO 67), et particulièrement à l'extérieur de l'arc Cerigho-Crète-Rhodes; les niveaux sapropéliques, nous le verrons, constituent une exception importante.

Le Miliéu Pélago-Détritique Non Pertubé

Lorsque la sédimentation pélagique est interrompue par des venues turbides, comme dans le cas de la

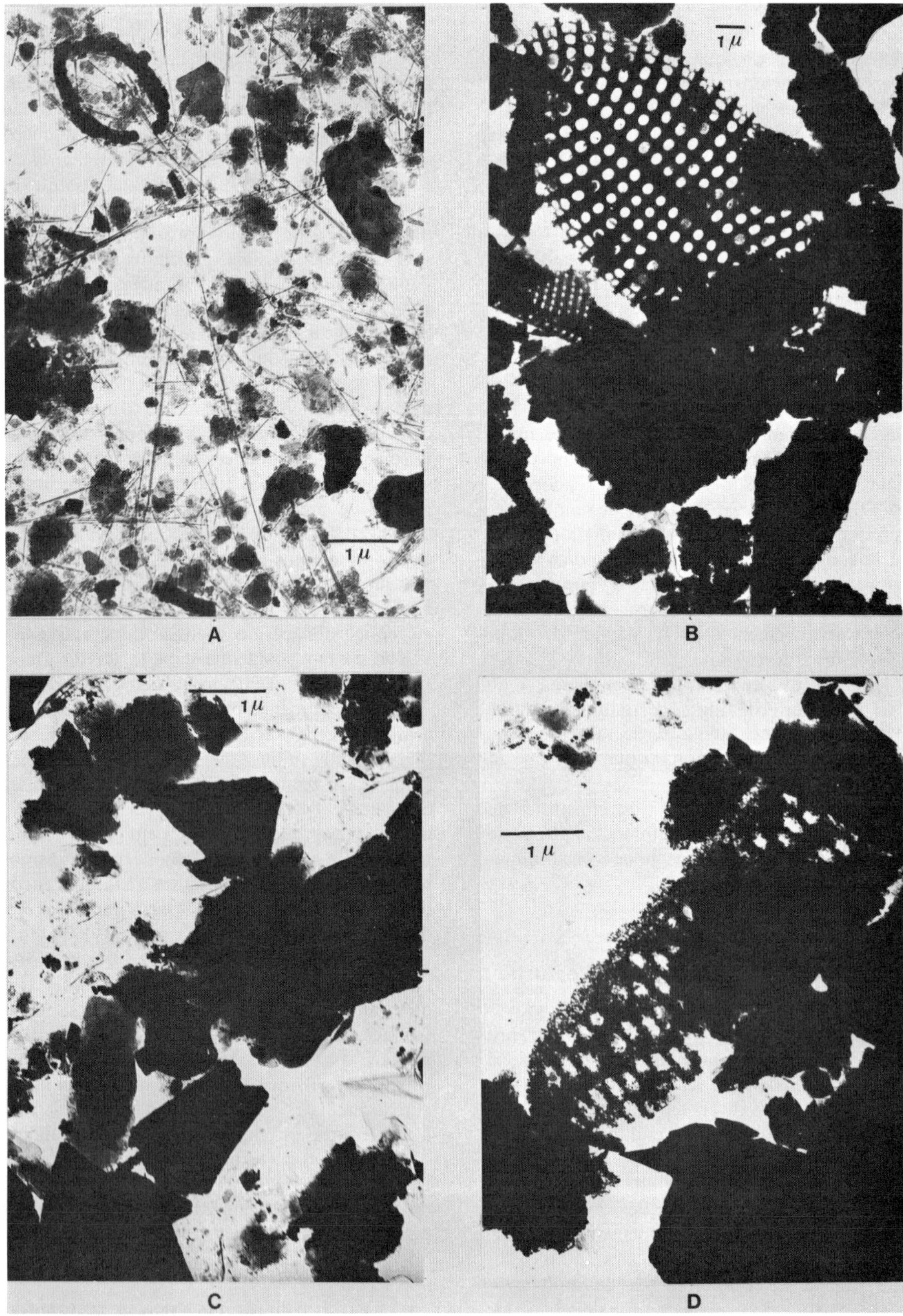

Figure 6. Micrographies électroniques. Carotte 45 MO 67, fosse de Matapan: A, Boues blanches à attupulgite (× 13.500). Carotte 29 MO 67, caldeira de Santorin: B, 0 cm: boue volcanique à Diatomées (× 4.500), C, 180 cm: boue volcanique à montmorillonite (× 13.500), D, 180 cm: diatomée fortement dissoute (× 13.500).

carotte A 11 67 (Figure 5), le message climatique, tel qu'il apparaît dans le milieu non perturbé, est respecté, mais seulement dans les passées autochtones; de ce fait, s'il est facile de dessiner une courbe climatique à partir de l'étude minéralogique (Figure 5, G), celle-ci donnera des renseignements délicats à interpréter. Il faut d'abord reconnaître avec sûreté les passées accidentelles, de manière à les éliminer dans la reconstitution historique; cela, qui nécessite une étude sédimentologique détaillée (granulométrie, morphoscopie, etc.), est facile dans une carotte tranchée comme la A 11 67 (Chamley, 1970), mais devient difficile lorsque les perturbations sont ténues. Il faut ensuite rapporter les sédiments autochtones à une période donnée (Figure 5, H), ce qui est souvent impossible à cause des lacunes et remaniements déterminés par les venues allochtones. Enfin la vitesse de sédimentation moyenne est généralement forte (plusieurs dizaines de cm/1.000 ans), et ne donne pas d'indication sur les variations de l'apport continental.

On définit ainsi un second milieu, celui de l'*héritage contrarié*, dont l'utilisation en paléoclimatologie est souvent délicate; cet inconvénient ne concerne pas seulement la méthode minéralogique, mais toutes les méthodes relatives aux sédiments marins. Par contre ce milieu peut apporter de nombreux renseignements sur la dynamique de la sédimentation profonde (origine des apports turbides, paléocourantologie, *etc.*). C'est le modèle de sédimentation le plus fréquemment rencontré dans le Bassin Méditerranéen occidental, et se trouve généralement lié à des périodes froides, ce qui aide à l'interprétation chronologique. Il est également assez bien représenté en Méditerranée orientale. Les turbidites sont parfois abondantes et dans ce cas on observe généralement un granoclassement, que les venues allochtones soient grossières (carotte A 11 67, Figure 5) ou fines (décantation de toute la fraction grossière: carotte 2 MO 67, fosses de Matapan, 35°41′N, 21°40′E). Parfois les turbidites sont ménagées; c'est le cas de nombreuses carottes du bassin algéro-provençal et de mer Egée (ex. 2 4 69, 42°07′N, 05°43′E), qui recquièrent une analyse très serrée si l'on désire séparer le message climatique continental du message dynamique marin. Enfin les niveaux de pyroclastites du Bassin Oriental représentent d'autres perturbations qui constituent de précieux repères chronologiques.

Le Milieu Sapropélique

On peut envisager d'expliquer la quasi-absence de la montmorillonite et des minéraux argileux petits et mal cristallisés dans les sapropels de deux manières:

1. Ces minéraux n'ont pas été apportés à la mer durant les phases de stagnation. Cela paraît peu vraisemblable, pour deux raisons principales: les dépôts sapropéliques n'ont pas été instantanés, mais ont duré quelques centaines, voire quelques milliers d'années (in Pastouret, 1969). Les sapropels ne peuvent donc correspondre à un arrêt des apports continentaux. Cela est du reste attesté par l'abondance des pollens dans ces niveaux.

Si les minéraux absents n'avaient pas été apportés durant les épisodes sapropéliques, ils ne l'auraient pas été davantage dans les dépôts banaux sous- et sus-jacents: l'homogénéité des apports argileux dans la sédimentation quaternaire est en effet un fait universellement reconnu. Il en résulte que la montmorillonite serait néoformée dans l'ensemble des sédiments, et que la néoformation serait contrariée seulement dans les sapropels. Cela est très improbable, malgré l'augmentation relative de la montmorillonite constatée vers la base de certaines carottes; on sait en effet que la montmorillonite de Méditerranée orientale est essentiellement détritique (Chamley *et al.*, 1962; Rateev *et al.*, 1966; Chamley, 1967a).

2. *Le milieu sapropélique s'oppose à la conservation de la montmorillonite et des argiles mal cristallisées (illite ouverte, interstratifiés,* etc.) (Figure 3, D). Cette hypothèse est vraisemblable. Un argument en sa faveur est l'opposition systématique constatée entre les données climatiques de la microfaune et des pollens d'une part, et celles des argiles d'autre part: les premières désignent des températures plutôt élevées, les secondes des températures basses. Cette opposition est unique en soi et montre que le milieu sapropélique a modifié le message continental porté par les argiles.

On identifie ainsi un troisième milieu de sédimentation argileuse, *le milieu sapropélique, dont l'empreinte sur les minéraux argileux efface celle du climat continental, par destruction* de la montmorillonite et de la plupart des particules argileuses petites et mal cristallisées. Cette destruction explique l'amélioration *apparente* de la cristallinité de l'illite, qui n'est en fait qu'une *conservation des phyllites les plus résistantes,* c'est-à-dire les plus grandes et les mieux cristallisées. Un tel phénomène peut être rapporté à une *transformation par dégradation*, analogue à celles des sols en milieu lessivé; il ne paraît pas avoir été décrit dans les sapropels anciens ou modernes.

Il reste à expliquer les causes intimes de cette dégradation, et ses liens éventuels avec les autres particularités sédimentaires (abondance des matières organiques, des diatomées, *etc.*). Il n'est pas exclu d'envisager un parallèle avec certains milieux pédologiques, riches en matières organiques acidifiantes et complexantes, qui déterminent une mise en solution préférentielle de l'alumine, et par conséquent une destruction des argiles (milieu de la podzolisation).

Le milieu sapropélique se rencontre à des profondeurs variées dans les carottes de Méditerranée orientale, depuis l'Adriatique jusqu'à la mer du Levant. Son empreinte sur les divers constituants sédimentaires est plus ou moins marquée; elle paraît plus modeste dans le bassin nord-crétois, et plus forte à l'extérieur de l'arc Cerigho-Crète-Rhodes. Mais dans tous les cas elle paraît s'exercer de manière tranchée: cela permet aux minéraux de conserver leur signification paléoclimatique à proximité immédiate de ces niveaux remarquables.

Le Milieu de la Sédimentation Chimique Basique

La présence d'attapulgite dans les fosses situées au Sud de Matapan (carotte 45 MO 67), sous une couverture de vase banale, a été attribuée à une *néoformation en milieu chimique basique* (Chamley et Millot, 1970; Blanc-Vernet et Chamley, à l'impression), selon le schéma défini par Millot (1964) dans les séries tertiaires d'Afrique. Cette hypothèse génétique s'applique mieux à ce gisement que celles émises pour les autres gisements de palygorskites marines: héritage, altération sous-marine de matériel volcanique ou néoformation en milieu hydrothermal (bibliographie in *op. cit.*). Les arguments les plus importants sont la présence d'un milieu très calcaire, d'une microfaune d'affinité tropicale, l'absence d'attapulgite dans les sédiments actuels du même secteur, l'absence d'indices minéralogiques hydrothermaux (zéolithes, palygorskites en agrégats, *etc.*).

Ainsi se trouve défini un quatrième milieu de sédimentation argileuse, le chimique basique. Il correspond à une sédimentation abyssale, interrompue par des venues détritiques glissées des flancs de la fosse de Matapan. D'âge vraisemblablement interglaciaire (Tyrrhénien ou Antetyrrhénien), ces sédiments originaux ont pu être recoupés par carottage grâce à une vitesse de sédimentation très faible dans les couches supérieures.

Le Milieu Volcanique

L'évolution minéralogique constatée à partir des cendres fines déposées dans la caldeira effondrée de Santorin (Carotte 29 MO 67) est à rapporter à l'halmyrolyse, phénomène bien connu dans les zones à sédimentation volcanique (in Millot, 1964). L'analyse chimique et la microscopie électronique permettent ici de mieux comprendre ce phénomène (Figure 6, B à D); la montmorillonite se développe rapidement grâce à la silice des diatomées, dont le potentiel de dissolution est très grand. Le support de cette croissance est constitué par les verres volcaniques, qui offrent à leur périphérie les éléments chimiques complémentaires et une matrice; ils sont d'autant plus facilement sollicités que leur taille est plus petite.

Le cinquième milieu de sédimentation argileuse identifié est donc celui de la *néoformation diagénétique de la montmorillonite dans les boues volcaniques,* accompagnée d'une dissolution des diatomées. Il est limité à Santorin. En particulier il n'a pas été reconnu au sein des passées pyroclastiques des carottes de mer ouverte, probablement du fait de différences granulométriques et peut-être chimiques.

CONCLUSIONS

La Méditerranée comprend divers types de sédimentation argileuse: héritage simple, héritage perturbé, transformation par dégradation en milieu sapropélique, néoformation chimique basique, néoformation diagénétique en milieu volcanique. Les minéraux simplement hérités, les plus répandus, contribuent à reconstituer les paléoclimats continentaux. Les minéraux formés dans les bassins contribuent à définir les milieux de sédimentation actuels et quaternaires. Ainsi les minéraux argileux, selon leur gisement, constituent le reflet soit des phénomènes produits sur les continents, soit des phénomènes propres au milieu marin.

BIBLIOGRAPHIE

van Andel, T. H. 1955. Sediments of the Rhône Delta. II. Sources and deposition of heavy minerals. *Verhandelingen van het Koninklijk Nederlandsch Geologisch Mijnbouwkundig Genootschap,* Geologische Serie, 15:515–556.

Beutelspacher, H. et H. W. van der Marel. 1968. *Atlas of Electron Microscopy of Clay Minerals and their Admixtures.* Elsevier Publishing Company, Amsterdam, 333 p.

Biscaye, P. E. 1965. Mineralogy and sedimentation of recent deep-sea clays in the Atlantic Ocean and adjacent seas and oceans. *The Geological Society of America Bulletin,* 76:803–832.

Blanc, J. J. 1958a. *Recherches de sédimentologie littorale et sous-marine en Provence occidentale.* Thèse, Masson et Cie. Editeurs, Paris, 140 p.

Blanc, J. J. 1958b. Recherches géologiques et sédimentologiques en Méditerranée nord-orientale. *Résultats Scientifiques des Campagnes de la CALYPSO,* 8(3):158–211.

Blanc, J. J. 1964. Vases bathyales et sables détritiques au large de Marseille. *Recueil des Travaux de la Station Marine d'Endoume,* 37:203–230.

Blanc-Vernet, L. 1969. Contribution à l'étude des Foraminifères de Méditerranée. Relations entre la microfaune et le sédiment. Biocoenoses actuelles, thanathocoenoses pliocènes et quaternaires. *Recueil des Travaux de la Station Marine d'Endoume,* 64: 251 p.

Blanc-Vernet, L. 1972. Données micropaléontologiques et paléoclimatiques d'après des sédiments profonds de Méditerranée. In: *The Mediterranean Sea: A Natural Sedimentation Laboratory,* ed. Stanley, D. J., Dowden, Hutchinson and Ross, Inc., Stroudsburg. Pennsylvania, 115–127.

Blanc-Vernet, L. et H. Chamley 1971. Sédimentation à attapulgite et *Globigerinoides trilobus f. dehiscens* (P. et J.) dans une carotte profonde de Méditerranée orientale. *Deep-Sea Research.*, 18:631–637.

Blanc-Vernet, L., H. Chamley et C. Froget 1969. Analyse paléoclimatique d'une carotte de Méditerranée nord-occidentale. Comparaison entre les résultats de trois études: foraminifères, ptéropodes, fraction sédimentaire issue du continent. *Palaeogeography, Palaeoclimatology, Palaeoecology*, 6:215–235.

Brown, G. (Editor) 1961. *The X-ray Identification and Crystal Structures of Clay Minerals*. Mineralogical Society (Clay Minerals Group), London, 544 p.

Chamley, H. 1967a. Quelques modalités de la sédimentation argileuse marine aux environs de l'île de Crète. *Bulletin du Musée d'Anthropologie Préhistorique de Monaco*, 14:25–48.

Chamley, H. 1967b. Possibilités d'utilisation de la cristallinité d'un minéral argileux (illite) comme témoin climatique dans les sédiments récents. *Comptes-Rendus de l'Académie des Sciences, Paris*, 265:184–187.

Chamley, H. 1968a. La sédimentation argileuse actuelle en Méditerranée nord-occidentale. Données préliminaires sur la diagenèse superficielle. *Bulletin de la Société Géologique de France*, 7:75–88.

Chamley, H. 1968b. Sur le rôle de la fraction sédimentaire issue du continent comme indicateur climatique durant le Quaternaire. *Comptes-Rendus de l'Académie des Sciences, Paris*, 267:1262–1265.

Chamley, H. 1969a. Intérêt de l'étude chiffrée des minéraux argileux par diffraction des rayons X, pour la connaissance des paléo-climats. *Colloque de la Compagnie Générale de Radiologie*: Méthodes analytiques par rayonnements X, Montpellier, 99–106.

Chamley, H. 1969b. Intérêt paléoclimatique de l'étude morphologique d'argiles méditerranéennes. *Tethys*, 1:923–926.

Chamley, H. 1970. Signification paléoclimatique des sédiments argileux quaternaires de Méditerranée occidentale. *Colloque de Stratigraphie: Tendances Actuelles de la Stratigraphie*, Orsay, 15p.

Chamley, H. et G. Millot 1970. Séquences sédimentaires à attapulgite dans une carotte profonde prélevée en Mer Ionienne (Méditerranée orientale). *Comptes-Rendus de l'Académie des Sciences, Paris*, 270:1084–1087.

Chamley, H., H. Paquet et G. Millot 1962. Minéraux argileux de vases méditerranéennes. *Bulletin du Service de la Carte Géologique d'Alsace et de Lorraine*, 15(4):161–169.

Emery, K. O., B. C. Heezen et T. D. Allan 1966. Bathymetry of the eastern Mediterranean Sea. *Deep-Sea Research*, 13:173–192.

Escalon de Fonton, M. 1966. Du Paléolithique supérieur au Mésolithique dans le Midi Méditerranéen. *Bulletin de la Société Préhistorique Francaise*, 113:66–180.

Escalon de Fonton, M. 1968. Problèmes posés par les blocs d'effondrement des stratigraphies préhistoriques du Würm à l'Holocène dans le Midi de la France. *Bulletin de l'Association Française pour l'Etude du quaternaire*, 4:289–296.

Fedoroff, N. 1969. *Les pédogenèses quaternaires en France*. 1 brochure in 4° (ronéotypé). Grignon, 20 p.

Glangeaud, L. 1966. Les grands ensembles structuraux de la Méditerranée occidentale d'après les données de Géomède L. *Comptes-Rendus de l'Académie des Sciences, Paris*, 262:2405–2408.

Griffin, J. J., H. Windom et E. D. Goldberg 1968. The distribution of clay minerals in the World Ocean. *Deep-Sea Research*, 15: 433–459.

Grim, R. E. et J. P. Vernet 1961. Etude par diffraction des minéraux argileux de vases méditerranéennes. *Bulletin Suisse de Minéralogie et de Pétrographie*, 41: 65–70.

von Humboldt, A. 1839. Geognostiche und physikalische Beobachtungen über der Vulkane der Hochebene von Quito. *Monatsbericht der Königlische Preuss Akademie des Wissenschaften, Berlin*, 245 –253.

Lucas, J., T. Camez et G. Millot 1959. Détermination pratique aux rayons X des minéraux argileux simples et interstratifiés. *Bulletin du Service de la Carte Géologique d'Alsace et de Lorraine*, 12:21–31.

Mackenzie, R. C. (Editor) 1957. *Differential Thermal Investigation of Clays*. Mineralogical Society (Clay minerals group), London, 456 p.

Mellis, O. 1954. Volcanic ash-horizons in deep-sea sediments from the eastern Mediterranean. *Deep-Sea Research*, 2:89–92.

Menard, H. W., S. M. Smith et R. M. Pratt 1965. The Rhône deep-sea fan. In: *Submarine Geology and Geophysics*, eds. Whittard, W. F. and R. Bradshaw, Butterworths, London, 271–285.

Millot, G. 1964. *Géologie des Argiles*. Masson Editeurs, Paris, 499 p.

Monaco, A. 1965. Evolution de quelques sédiments argileux de la Méditerranée occidentale (entre Carthagène et Mostaganem). *Bulletin de la Société Géologique de France*, 7:521–529.

Müller, G. 1961. Die rezenten Sedimente im Golf von Neapel. *Beiträge zur Mineralogie und Petrographie*, 8:1–20.

Nesteroff, W. D. et G. Sabatier 1958–1959. Etude minéralogique de vases bleues méditerranéennes. *Bulletin de la Société Francaise de Minéralogie et de Cristallographie*, 81:380 et 82:72.

Nesteroff, W. D., G. Sabatier et B. C. Heezen 1963. Les minéraux argileux dans les sédiments du bassin occidental de la Méditerranée. *Comptes-Rendus de la Commission Internationale pour l'Etude Scientifique de la Mer Méditerranée*, 17(3): 1005–1007.

Ngoc Cau, H., W. Donoso et G. Sabatier. 1959. Minéralogie de quelques vases marines de la région de Monaco. *Bulletin de la Société Française de Minéralogie et de Cristallographie*, 82:380.

Ninkovich, D. and B. C. Heezen 1965. Santorini Tephra. In: *Submarine Geology and Geophysics*, eds. Whittard, W. F. and R. Bradshaw, Butterworths, London, 413–453.

Norin, E. 1953. Occurrence of authigenous illitic mica in the sediments of the central Tyrrhenian Sea. *Bulletin of the Geological Institute of the University of Uppsala*, 34:279–284.

Olausson, E. 1960. Description of sediment cores from the Mediterranean and the Red Sea. *Reports of the Swedish Deep-Sea Expedition, 1947–1948*, 7(3): 285–334.

Olausson, E. 1961. Studies of deep-sea cores. *Reports of the Swedish Deep-Sea Expedition, 1947–1948*, 8(4):335–391.

Pastouret, L. 1969. *Contribution à l'Etude des Sédiments Quaternaires Récents de Méditerranée Orientale (Région de Karpathos-Rhodes)*. Thèse de troisième cycle, Marseille, 107 p.

Peronne, D. 1967. *Contribution à l'Etude Sédimentologique de Sondages Sous-Marins*. Thèse de troisième cycle, Paris, 171 p.

Rateev, M. A., E. M. Emelyanov et M. B. Kheirov. 1966. Particularités de la formation des minéraux argileux dans les sédiments actuels de la Méditerranée (en russe). *Litologia i Poleznye Iskopayme, (Lithology and Mineral Resources)*, 4:6–23.

Rateev, M. A., Z. N. Gorbunova, A. P. Lisitzyn et G. I. Nosov. 1968. Zonation climatique des minéraux argileux dans les sédiments de l'Océan mondial (en russe). *Okeanologiya*, 18: 283–311.

Rotschy, F., C. Vergnaud-Grazzini, G. Bellaiche et H. Chamley. 1969. Paléoclimatologie d'une carotte prélevée sur le dôme Alinat dans la plaine abyssale ligure. *Résumés des Communications du VIIIème Congrès de l'INQUA*, Paris, p. 64.

van Straaten, L. M. J. U. 1966. Micro-malacological investigation of cores from the southeastern Adriatic Sea. *Koninklijk Nederlandsch Akademie van Wetenschappen, Amsterdam*, 69: 429–445.

Vernhet, S. 1956. Etude chimique et minéralogique de quelques sédiments méditerranéens de moyenne et grande profondeur. *Comptes-Rendus de l'Académie des Sciences, Paris*, 242:1049–1052.

Geotechnical Properties of Submarine Sediments, Mediterranean Sea

George H. Keller and Douglas N. Lambert

National Oceanic and Atmospheric Administration, Miami, Florida

ABSTRACT

Investigation of the mass physical properties (sediment type, shear strength, water content, unit weight, and porosity) of sediment cores from 27 sites in the Mediterranean Sea provides an insight into the general distribution of these properties within the surficial deposits. Deposits of the western portion of the basin, with an abundance of turbidites, tend to possess higher shear strengths and unit weights than do those sediments blanketing the eastern Mediterranean. Much of the easternmost Mediterranean is strongly influenced by the deposition of the Nile-contributed sediments which are fine grained and display very low shear strengths (less than 50 g/cm^2) and unit weights (1.40–1.47 g/cc), but relatively high water contents (103 to 124 percent) and porosities (73 to 77 percent).

A study of the vertical variation of selected mass properties clearly reveals the unique characteristics of both the turbidites and sapropelic mud layers found in the Mediterranean. Turbidite sequences commonly possess relatively high shear strengths and unit weights as well as low water contents and porosities. Sapropelic muds exhibit just the opposite characteristics. Using plasticity as a means of delineating sediments, the easternmost Mediterranean deposits are readily distinguished by their high plasticity indices in contrast to the much lower values observed in sediments from the central and western parts of the basin. Four samples tested for degree of consolidation possess characteristics similar to those of overconsolidated material. This is attributed to some form of cementation or "ionic bonding" and not to previous overburden pressures.

RESUME

L'examen des propriétés de masse (types de sédiments, shear, teneur en eau, poids spécifique et porosité) de carottes prélevées dans des bassins de la Mer Méditerranée donnent un aperçu de la distribution générale de ces propriétés dans les dépôts supérieurs. Les dépôts du Bassin Occidental, avec un grand nombre de turbidites, ont une résistance à la rupture par cisaillement plus haute et un poid spécifique plus élevé que ceux du Bassin Oriental. Une grande partie, à l'Est de ce bassin, est fortement influencée par le dépôt des sédiments d'apport du Nil, avec grains fins, shear (moins que 50 g/cm^2) et poid spécifique (1.40 à 1.47 g/cc) faibles, mais avec teneur en eau (103 à 124 pour cent) et porosité (73 à 77 pour cent) relativement élevées.

Une étude de la variation verticale des propriétés de masse dégage nettement les caractéristiques des turbidites et couches de boue sapropélique recontrées dans la Méditerranée. Les successions des turbidites présentent généralement shear et poid spécifique relativement élevés, et une teneur en eau et porosité faibles. Les boues sapropéliques présentent des caractéristiques renversées. En utilisant la plasticité, les couches des sédiments de la partie Est de la Méditerranée se distinguent par leurs indices de plasticité élevés en comparaison des valeurs bien inférieures observées dans les sédiments des parties centrale et occidentale du bassin. Quatre échantillons dont on a étudié le degré de consolidation présentent des traits similaires à ceux des matériaux surconsolidés. On peut attribuer ce fait à un type de cimentation ou "ionic bonding," et non à la compression exercée par les terrains de couverture.

INTRODUCTION

Marine geotechnique is a relatively young field, but one which is receiving considerable attention today. It is the area of study which deals with the investigation of the engineering and geological aspects of the sediments and rocks comprising the sea floor. Specifically, it is more commonly used to denote the study of the mass chemical and physical properties of the electrolyte-gas-solid system of the sea floor and the response of this sedimentary system to applied static and dynamic loading.

One of the earliest studies in marine geotechnique was that of Arrhenius (1952) in which relative strength measurements were made on sediment cores collected from the Pacific during the Swedish Deep-Sea Expedition 1947–1948. Shortly thereafter, a number of U. S. Navy scientists commenced a research program into the mass physical properties of deep-sea sediments. These early efforts of Hamilton and Menard (1956), Hamilton (1959), Richards (1961, 1962), and Moore (1962), along with the offshore foundation studies by Fisk and McClelland (1959) for drilling platforms, have served as an impetus for the increased interest in marine geotechnique of both marine geologists and civil engineers.

Recent studies have provided an insight into the areal distribution of various mass physical properties in the North Pacific and North Atlantic basins (Keller and Bennett, 1968), the stability of submarine deposits (Morgenstern, 1967; Bryant and Wallin, 1968; and Morelock, 1969), and the consolidation characteristics of submarine sediments (Hamilton, 1964; Richards and Hamilton, 1967; and Bryant *et al.*, 1967).

Studies of the mass physical properties of deep-sea sediments in the Mediterranean basin have been limited. To our knowledge only six such studies have been published. Richards (1961, 1962), in his initial reports on the mass properties of deep-sea sediment cores, provided data from three sites in the vicinity of the Strait of Gibraltar. Einsele's (1967) study of eight sediment cores from just north of the Nile Delta made available the first detailed information on the mass properties of surficial deposits off the Nile River. Off the coast of Tel-Aviv, Israel, Almagor (1967) investigated the shear strength, water content, and consolidation characteristics of nine sediment cores collected from water depths ranging from 35 to 1100 meters. Studies attempting to correlate the acoustical and mass physical properties of Mediterranean sediments on a regional scale have been reported by Horn *et al.* (1968), and on a more local basis by Ryan *et al.* (1965) and Kermabon *et al.* (1969) in the Tyrrhenian Sea.

The present study has assembled all the available data, published and unpublished, relating to sediment distribution, shear strength, water content, unit weight, and porosity in the Mediterranean basin (Table 1). It is based on data obtained from 96 sediment cores representing 27 areas in the Mediterranean. The majority of the cores were collected and analysed by, or on contract for, the U. S. Naval Oceanographic Office and, as yet, the data are mainly unpublished. Included in the 96 cores are four collected by the OCEANOGRAPHER in 1967 and analyzed by the authors.

Because of the nature of the measurements made on the cored sediment, disturbance of the sediment fabric during sampling is of utmost concern. Even under ideal sampling conditions with the best available sampler, truly undisturbed samples cannot be obtained from the sea floor. Since the data presented here are from several sources representing different types of samplers (both piston and gravity corers) and sampling techniques, it is clear that some degree of variation in measured properties is inevitable, based solely on sampling disturbance. The degree to which these properties have been affected by sampling cannot now be ascertained and, for the purpose of this discussion, no attempt has been made to adjust these data for disturbance. Obviously disturbed samples, *e.g.*, displaying distorted bedding due to piston sucking, were discarded.

Cores used in this study vary in length from 0.18 m to 9.4 m with an average of about 3.5 m. Considering the short core lengths relative to the scale of the Mediterranean basin, it was decided to average the respective parameter values, except for sediment type, over the entire length of each core. For example, if ten shear strength measurements were made on a core, the average of these values was used to derive the areal distribution of the respective properties shown in later figures. Subsample values from each core were critically evaluated to exclude any which were obviously in error as a result of sampling or testing procedures.

GENERAL SEDIMENTATION PATTERNS IN THE MEDITERRANEAN

The Mediterranean Sea occupies a zone of considerable crustal activity, lying along one of the major earthquake belts of the world. Consequently, tectonism and volcanic activity have played a major role in determining the morphological characteristics of the present sea floor. In conjunction with tectonism, detritus from the surrounding land mass has contributed much to modify or smooth out the basin floor irregularities.

Morphologically, the Mediterranean basin is divided distinctly into western and eastern sectors (Figure 1). The western portion is by far the simpler of the two with regard to structure and topography. It consists primarily of the Balearic Abyssal Plain (largest in the Mediterranean), the Tyrrhenian Abyssal Plain and their adjacent continental margins. Another prominent feature is the Rhône Fan which covers an extensive area of the northwestern portion of the Mediterranean basin. There appears to be little doubt that the River Rhône has provided the sediments needed to build this feature (Menard *et al.*, 1965).

With its median ridge, trenches, continental border-

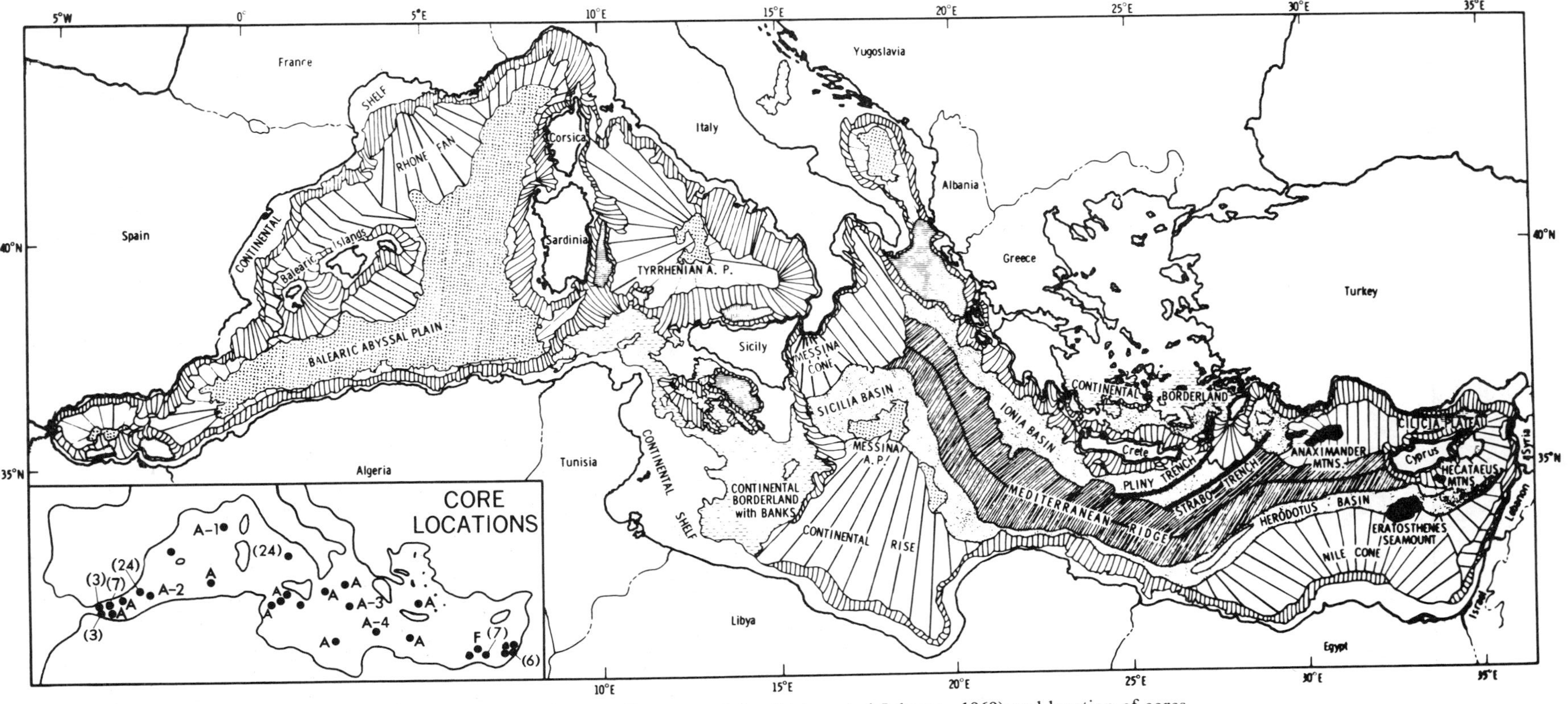

Figure 1. Diagrammatic display of physiographic provinces of the Mediterranean (after Watson and Johnson, 1969) and location of cores used in this study (insert). Numbers in parentheses indicate number of cores at each site. See also figures 7, 8 and 10.

Table 1. Sources and total number of measurements of geotechnical properties used in this study.

Source	Area	no. Cores	Ave. Core Length (cm)	Shear Strength	Unit Weight	Water Content	Porosity	Atterberg Limits	Type of Corer
Horn *et al.* (1967)	Entire Med.	14	748	954	338	338	338	—	Ewing
Kermabon *et al.* (1968)	Tyrrhenian Sea	21	807	—	3255	3255	3255	—	Sphincter
*P. Blavier, C. Gehin, & F. Kögler	Tyrrhenian Sea	3	792	166	238	238	238	—	Sphincter
Richards (1962)	Strait of Gilbraltar	3	180	30	33	33	33	23	Kullenberg
Gehin & Blavier (1969)	Alboran Sea	7	510	—	681	681	681	—	Sphincter
Einsele (1967)	Nile Cone	8	200	53	14	36	14	13	Kastenlot
Almagor (1964)	Israel Margin	9	280	170	219	219	219	93	Hydroplastic (modified)
†*NAVOCEANO	Off Palomares, Spain	24	66	81	57	59	4	—	Hydroplastic
†*NAVOCEANO	Strait of Gilbraltar	3	82	6	6	24	6	—	Kullenberg
OCEANOGRAPHER (this study)	Eastern & Western Mediterranean	4	167	14	17	48	11	14	Hydroplastic

†U.S. Naval Oceanographic Office.
*Unpublished data.

land, limited abyssal plains, and numerous volcanoes and seamounts, the eastern sector of the Mediterranean presents a relatively complex area of bottom morphology. The most prominent feature of the eastern Mediterranean is a broad median ridge (Mediterranean Ridge) extending from the continental margin of Italy to the island of Cyprus. Although the median line is not denoted by an axial rift valley, the general relief and steep local slopes indicate a tectonic origin for the ridge (Emery *et al.*, 1966). A number of basinal enclosures such as the Sicilia, Herodotus and Ionia Basins are not entirely smooth-floored and have not been designated as abyssal plains (Watson and Johnson, 1969). The Nile and Messina Cones are deposits resulting from rapid sedimentation and slumping. The source for the former is obvious, but the same cannot be said for the Messina Cone. Indications are that northerly currents concentrate sediment in the vicinity of the strait between Italy and Sicily. Numerous slumps and turbid flows resulting from periodic seismic shocks of this sediment accumulation have amassed the deposits now comprising the Messina Cone (Ryan and Heezen, 1965). For a more detailed discussion of the morphology of the Mediterranean basin, the reader is referred to the studies of Pfannenstiel (1960), Goncharov and Mikhailov (1964), Hersey (1965), Emery *et al.*, (1966), and Watson and Johnson (1969).

The Mediterranean Sea is one of relatively active and diverse sedimentation as might be anticipated in an essentially land-locked basin. Average sedimentation rates for the Quaternary have been estimated at 10 to 30 cm/1000 years by Mellis (1954) with an overall average rate of 0.5 cm/1000 years since Cretaceous time (Wong and Zarudzki, 1969).

In addition to the widespread occurrence of calcareous oozes and terrigenous lutites, there appears to have been a long history of turbidity currents, slumping, and ash falls in the Mediterranean. A unique aspect of Mediterranean deposits is the occurrence of numerous sapropelic mud layers in the eastern sector of the basin. These Pleistocene muds are indicative of an anerobic depositional environment brought on by stagnant bottom water, a condition not found presently in the Mediterranean Sea.

The Adriatic and Aegean Seas have been purposely omitted from this discussion owing to the lack of any mass physical properties data in these areas. Only in the section pertaining to surface sediment have these two seas been included.

SURFACE SEDIMENT TYPES

Sediment distribution (Figure 2) is based mainly on a compilation of published data from the U. S. Naval Oceanographic Office (1965) and the Interdepartmental Geophysical Committee of the Academy of Science, USSR (1969). Additional data have been obtained from the files of the U. S. Naval Oceano-

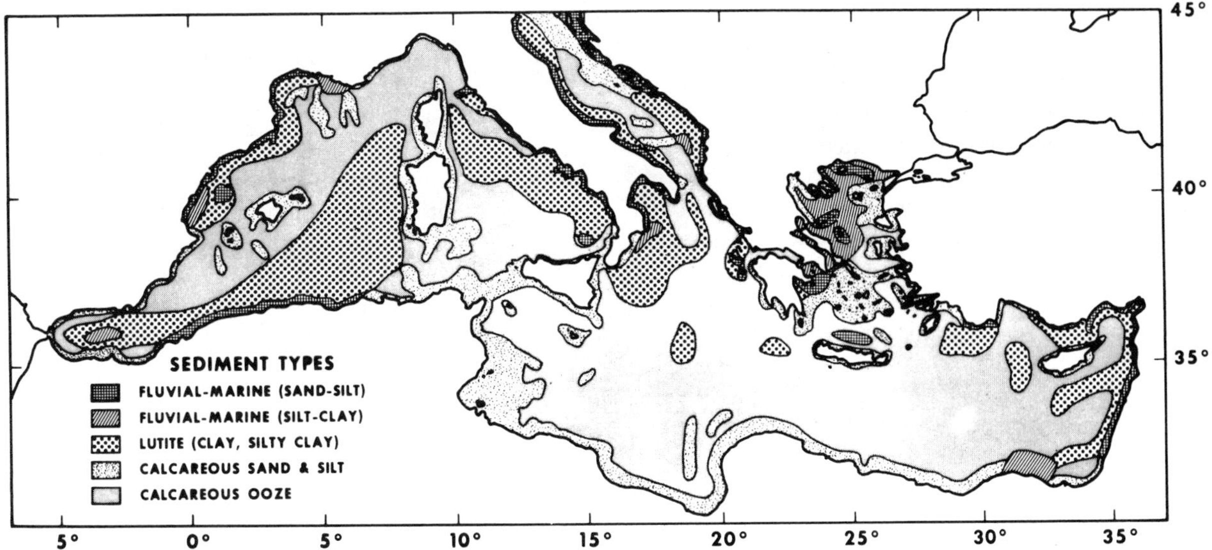

Figure 2. Surface sediment distribution.

graphic Office, U. S. Navy contract reports from Texas Instruments Incorporated (1967), and such published studies as the Swedish Deep-Sea Expedition 1947–1948 (Olausson, 1960, 1961), Ninkovich and Heezen (1965), and Emery *et al.* (1966).

Lack of a common sediment classification system among the various sources noted above has made the task of collating these data difficult and, at best, the results can only be considered as a broad generalization. An effort has been made here to use a simple classification of five sediment types which tends to unite the majority of the systems previously used: (1) fluvial marine (sand-silt), representing the coarse fraction (larger than 0.062 mm); (2) fluvial-marine (silt, clayey silt), the finer fraction (0.062 to 0.002 mm); (3) lutite (silty clay, clay), the finest fraction (smaller than 0.002 mm) of material primarily of terrigenous origin and containing less than 30 percent calcium carbonate; (4) calcareous sand and silt, consisting mainly of shell fragments and coralline debris of sand and silt-size particles; (5) calcareous ooze, used here to denote sediment composed of at least 30 percent calcium carbonate which is in the form of skeletal material from various planktonic animals and plants. *Globigerina* and pteropod oozes are those most commonly found in the Mediterranean. Volcanic ash, which is widespread in the eastern Mediterranean, does not readily fit into the classification used here. However, to maintain this system, ash is classed as fluvial-marine for the purpose of this discussion.

It is readily apparent (Figure 2) that much of the Mediterranean basin is blanketed by either lutite or calcareous oozes. The latter constitute the predominant sediment type and occur primarily in those areas not strongly influenced by high current energies, turbidity currents, major rivers, strong volcanic activity, or reefs. Lutite is present in areas influenced by terrestrial drainage such as the far eastern Mediterranean, off the coast of Spain and France, in the Adriatic Sea, and on those continental rises and abyssal plains where turbidity currents are most common. The largest lutite deposits are those covering the Balearic Abyssal Plain which receives much of its sediment from turbidity currents originating on the Rhône Fan (Menard *et al.*, 1965) and from portions of the African margin (Heezen and Ewing, 1955).

The relatively large lutite deposit in the Tyrrhenian Sea extends across both the continental rise and abyssal plain provinces. This distribution may possibly be attributed to the northwestern flow of intermediate water carrying terrigenous fines away from Italy. Another area of extensive lutite is that of the Messina Cone and Messina Abyssal Plain where Ryan and Heezen (1965) report active slumping and flow of turbidity currents.

Coarse-grained calcareous deposits are prominent along the southern margin of the Mediterranean where large shell and coralline concentrations occur. Similar occurrences are common in the vicinity of the Balearic islands, Corsica, Sardinia and Sicily as well as among many of the islands in the Aegean Sea. Surprisingly, few areas of fluvial-marine (silt, clayey silt) material are reported among the deposits of the Mediterranean.

The surface sediment distribution map (Figure 2) is deceiving in that it delimits only the surface deposits and may not adequately present the true sedimentological setting. It is particularly misleading in an area such as the Nile Cone where great quantities of sediment (Wong and Zarudzki, 1969, reported 120 million tons; Holeman, 1968, reported 57 million tons) enter the basin annually. Only after examining seismic reflection records and core samples is it obvious that Nile-contributed sediments extend over a much greater portion of the eastern Mediterranean than is revealed by the map. To a lesser extent, a similar corollary can be drawn from the Rhône River deposits in the northwestern portion of the Mediterranean.

Of particular interest to the sedimentologist, but not obvious from the map, is the influence of volcanism on the sedimentary régime, especially in the eastern Mediterranean. Numerous ash layers have been noted throughout the province (Ninkovich and Heezen, 1965), but only in a few areas, such as north of Crete, has ash been reported in the surface deposits (Horn *et al.*, 1968).

SHEAR STRENGTH

Sediment cores used in this study consisted primarily of fine-grained cohesive material with a few laminae of fine sand occurring in a number of the samples. Laboratory shear strength measurements of relatively weak, saturated sediments are commonly accomplished by one of three tests: vane shear, unconfined compression, or fall cone. A vane shear test consists of inserting a small four-bladed vane into the sample and applying an increasing torque until shear occurs (Evans and Sherratt, 1948). The unconfined compression test basically involves loading a vertically oriented cylindrical sample, unconfined in all lateral directions, at a constant rate until failure. A more detailed discussion of these two tests as applied to submarine sediments has been presented by Richards (1961).

The fall cone test was developed in Sweden and has received wide usage in the study of various Scandinavian clays. This test is quite simple and consists of relating the penetration of a falling metal cone into a sediment to the strength of the material. A complete description of this test has been presented by Hansbo (1957).

Depending on the type and strength of the sediment, it is sometimes difficult to compare shear strength measurements made by these three techniques. There does, however, seem to be a better correlation between the vane shear and unconfined compression tests than between either of these two and the fall cone measurements, particularly at lower shear strength values (Flaate, 1965, and Einsele, 1967).

Shear strength measurements on 14 of the 96 cores used in this study were made with the fall cone and can be identified by the prefix "A" in Figure 1. These values do appear to be slightly higher in some cases than shear strengths determined by the other testing techniques. Caution is in order when referring to the shear strength distribution pattern (Figure 3).

Even with the difficulty of comparing these various test data, some generalizations can be drawn from the available shear strength values. In viewing the overall shear strength distribution, a distinct difference is noticed between the eastern and western sectors of the Mediterranean basin. In the western Mediterranean, the Balearic Abyssal Plain appears to be blanketed with sediments of relatively high shear strength (>200 g/cm^2). These are the highest average values yet observed in the Mediterranean. As pointed out by others (Heezen and Ewing, 1955; and Menard *et al.*, 1965), the Rhône Fan and the Balearic Abyssal Plain are largely composed of turbidites. Turbidite layers within any core display distinctly different values (*e.g.*, higher shear strength and unit weight) than are found associated with the matrix of the core as a whole thus, when averaged into the entire core, generally resulting in higher shear strengths for the overall abyssal plain and outer fan deposits. This contrast is further discussed in a later section dealing with vertical variations of mass properties at a select number of sites. Westward towards the Strait of Gibraltar, few if any turbidites are reported and shear strengths are commonly lower than those observed for the Balearic Abyssal Plain.

The influence of the Nile in the eastern Mediterranean is clearly apparent from the distribution pattern of the various mass physical properties. Our study indicates that the Nile-contributed sediments possess shear strengths lower than most others yet reported for the Mediterranean basin. A marked contrast in shear strength values is noted between the offshore areas of the rivers Nile and Rhône. Although turbidites are common to both areas, they appear to occur in greater abundance in the vicinity of the Rhône and may be responsible for the associated higher shear strengths found there. Calcium carbonate and organic carbon analyses from both these areas indicated a notable contrast between the respective deposits. Nile sediments are relatively high in organic carbon, as high as 1.194 percent, in comparison to values of the order of 0.180 percent for the general area of the Rhône Fan. Calcium carbonate content is slightly higher in the Rhône Fan deposits than was observed on the Nile Cone. The combination of lower calcium carbonate and higher organic carbon may well have contributed to the lower shear strengths occurring in the eastern Mediterranean. Variations in mineralogy and grain size between the two areas undoubtedly are also a contributing factor to the observed strength differences. Relatively low shear strengths are also found in the Messina Cone and Messina Abyssal

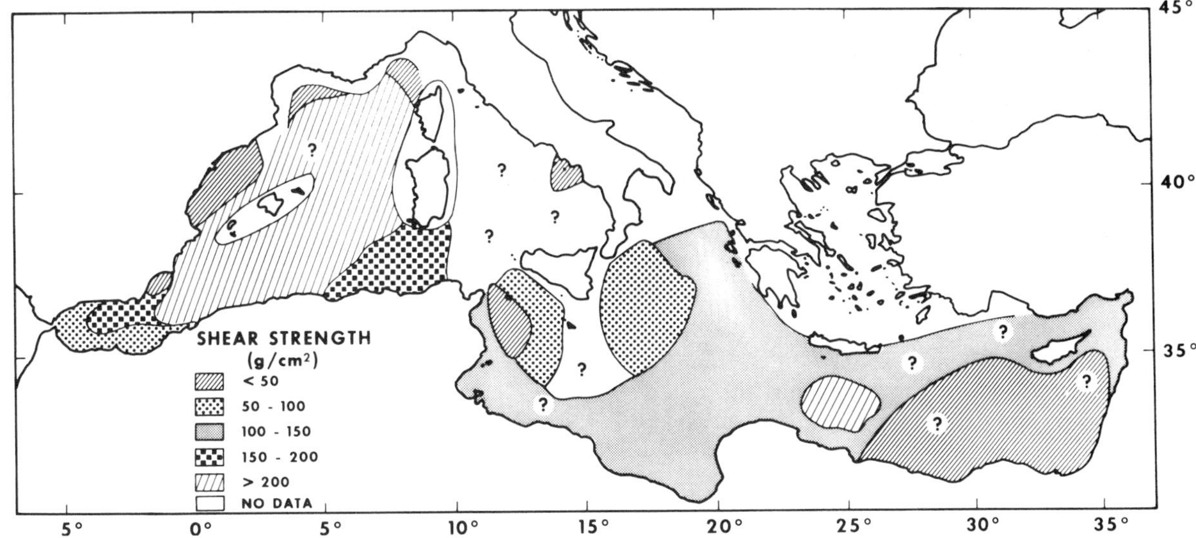

Figure 3. Average shear strength distribution.

Plain sediments which are areas of slumping and sediment disturbance (Ryan and Heezen, 1965).

WATER CONTENT

Water content is used here as the ratio, in percent, of the weight of water to the weight of oven dried (110°C) solids in a given sediment mass. It is determined by weighing a representative fraction of the sample, oven-drying at least over night, cooling in a desiccator and reweighing.

Overall water content variation in the Mediterranean basin appears to be rather moderate (Figure 4). The major portion of the abyssal plain and rise sediments have water contents ranging from 50 to 100 percent, much the same as has been reported for Atlantic sediments (Keller and Bennett, 1968). This relatively moderate amount of variability compared to Atlantic and Pacific deposits, may possibly reflect the strong influence turbidity currents have over much of the basin and thereby lead to some degree of lateral uniformity among the submarine sediments. Marginal deposits in the western and central Mediterranean Sea are characterized by relatively low water contents (< 50 percent) as are commonly found in association with coarser material. North of the island of Crete, the low water content reflects the unusally high volcanic ash content of these deposits.

Distribution of the Nile lutites, with their characteristic high water content, is clearly evident from Figure 4. The highest values yet observed in the

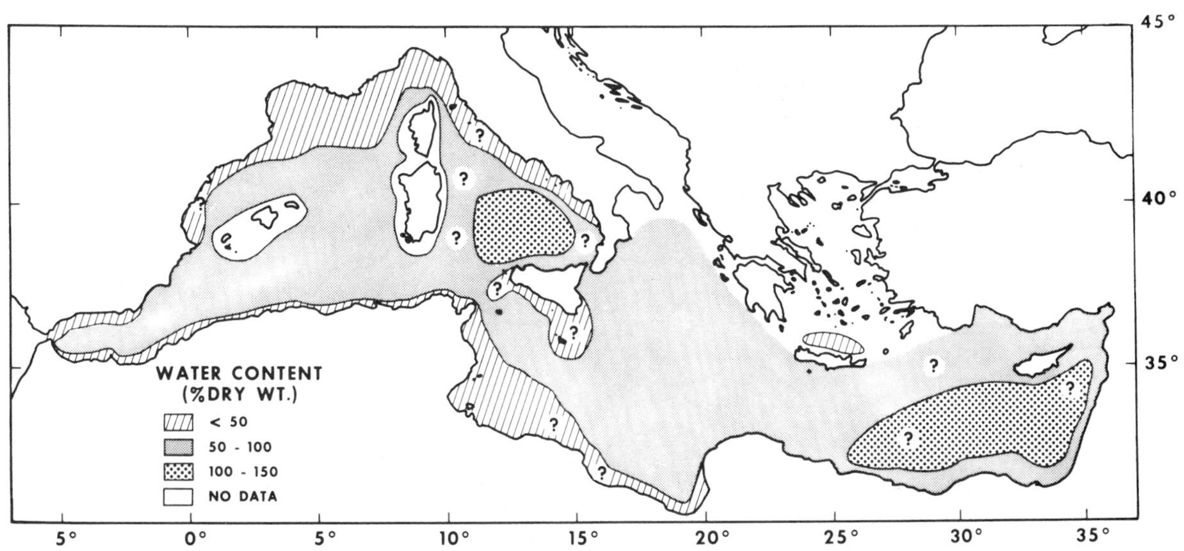

Figure 4. Average water content distribution.

Mediterranean are those associated with the Nile sediments. Influence of these deposits on the normally lower water content sediments of the coastal margins is plainly evident off Israel where higher than expected values are found. The Tyrrhenian Abyssal Plain is the only other locale in the Mediterranean reported to have average water contents greater than 100 percent. In contrast to the Nile Cone, the outer Rhône Fan deposits possess rather low water contents. This may be possibly ascribed to a greater concentration of coarse-grained turbidites in the Rhône-contributed sediments. The relatively low water contents of the Balearic Abyssal Plain are not surprising when it is recalled that a prominent portion of these deposits is attributed to turbidity currents from both the Rhône Fan and Algerian areas.

UNIT WEIGHT

Unit weight, or wet bulk density as referred to by some investigators, is the weight per unit of total volume of a sediment mass. Submarine sediments are usually 100 percent saturated or are sufficiently close to 100 percent to allow the term saturated unit weight (the in-place unit weight) to be substituted for unit weight. It is that value which is presented here.

Unit weights ranging from 1.50 to 1.75 g/cc tend to characterize a large portion of the eastern Mediterranean basin deposits, whereas values between 1.75 to 2.00 g/cc predominate over much of the western portion of the basin (Figure 5). Nile-derived lutites possess some of the lowest unit weights found in the Mediterranean. These unit weights, as do the other mass physical properties, clearly identify these Nile sediments.

Low unit weight (1.35 g/cc) in the Tyrrhenian basin are attributed to the concentration of lutite found there. Unit weights from the Rhône Fan and Balearic Abyssal Plain are the highest yet reported from the Mediterranean. These high values undoubtedly are related to the great abundance of turbidites and the mineralogy associated with these sequences. High unit weights also occur along the southern margin of Spain and in the Strait of Gibraltar. The relatively low unit weights found off the northern margin of Morocco may possibly reflect the characteristics of material entering the basin from Morocco and northwestern Algeria. Prevailing westerly surface currents may have contributed to the observed distribution pattern.

POROSITY

Porosity is the ratio, expressed as a percent, of the volume of voids in a given mass to the total volume of the sediment mass. It was determined from calculations based on the measured water content, unit weight, and grain specific gravity values.

The lutites of the Nile-contributed sediments as well as those of the Tyrrhenian Abyssal Plain clearly stand out as those deposits possessing the highest average porosities yet found in the Mediterranean basin (Figure 6). Porosities of 60 to 70 percent tend to predominate throughout most of the basin. A relatively low average porosity appears to characterize the Rhône Fan as might be anticipated in this depositional environment of numerous coarse-grained turbidites.

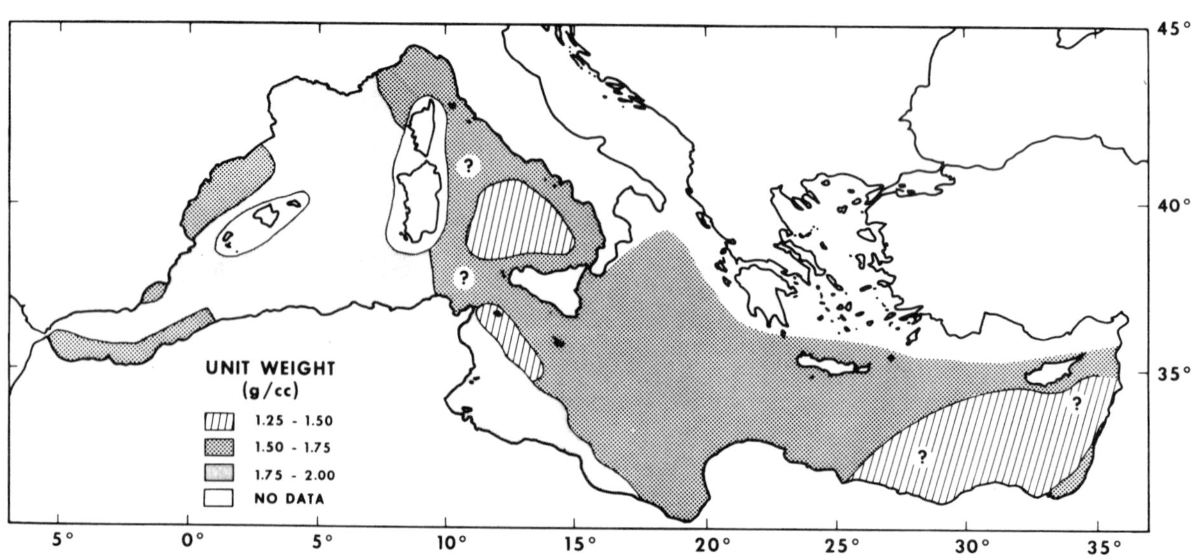

Figure 5. Average unit weight distribution.

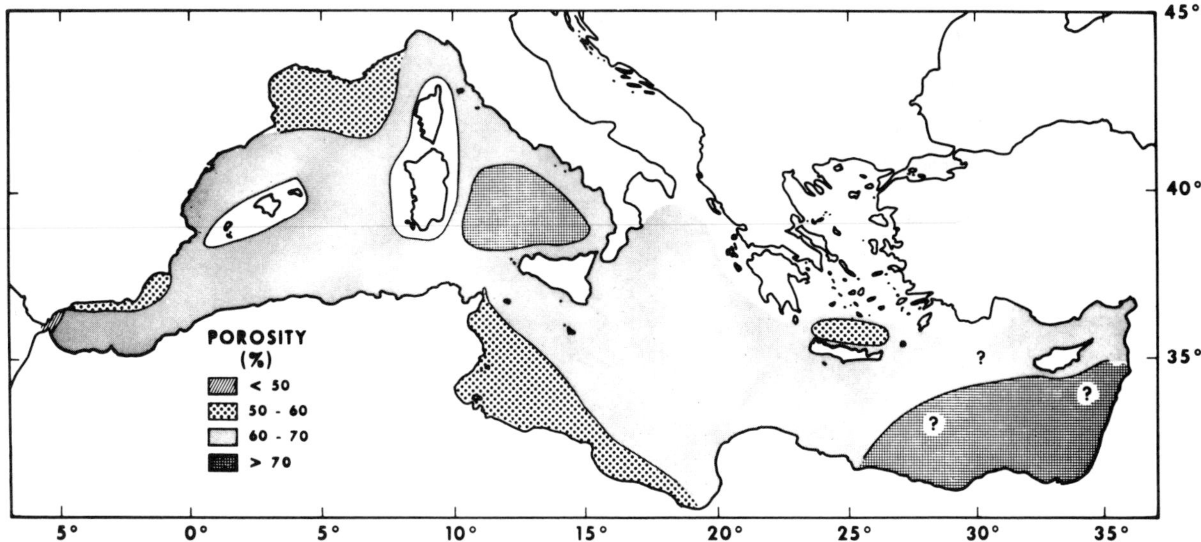

Figure 6. Average porosity distribution.

Similar low porosities are displayed by the fine grained ash and pumice deposits occurring north of the island of Crete. Average porosities of less than 50 percent are found in association with relatively coarse-grained sediments such as in the Strait of Gibraltar, and off the east coast of Tunisia.

VERTICAL VARIATION OF MASS PROPERTIES

The diverse sediments and depositional conditions in the Mediterranean are strongly reflected in the mass properties of the respective deposits. A series of profiles from various parts of the Mediterranean is presented in Figures 7 and 8 to show the variation of selected mass properties with depth as well as the interrelationship of these properties. Two cores from Horn *et al.* (1967) were selected from the Balearic Abyssal Plain, one just off the Rhône Fan (Core A-1) and the other from the southwest sector of the abyssal plain (Core A-2). As noted earlier, the Balearic Abyssal Plain deposits are largely composed of turbidites originating from the Rhône Fan and the continental margin of northern Africa. This depositional environment is clearly shown in cores A-1 and A-2, each of which display a number of sandy layers (turbidites) within a matrix of calcareous ooze. Although not distinct in the profile, core A-1 possesses 14 turbidite sequences and core A-2 has 5 such layers (Horn *et al.*, 1967). The influence of these layers on the overall properties of the deposits is obvious from the profiles. In contrast to the calcareous ooze, the turbidites exhibit higher percentages of sand, high shear strength, low water content and porosity, and high unit weight. Calcium carbonate content is often higher in the turbidite zone, but not in all cases, owing to probable differences in the source areas for the respective turbidites.

Core A-3 (Figure 8) was collected from the Messina Abyssal Plain, approximately 275 km southeast of Italy (Horn *et al.*, 1967). There is little doubt, based on the study by Ryan and Heezen (1965), that much of the sediment reaching this plain results from turbidity currents and slumping on the Messina Cone. Variation of the mass properties with depth appears to be strongly influenced by the silt concentration. The presence of a turbidite between 100 cm and 110 cm is noted by distinct changes in the mass properties. An overall increase in silt content from 110 cm to 680 cm is associated only with a gradual decrease in water content and porosity and a slight increase in unit weight. In the presence of relatively large silt percentages, only shear strength appears to be significantly affected by minor changes in sand content.

The lower 200 cm of the core presents an interesting number of relationships. Increased sand contents at intervals 770 cm, 850 cm, and 885 cm indicates possible turbidites although none were reported (Horn *et al.*, 1967). The unusual aspect of these three zones is the relative increase in water content and porosity with a decrease in unit weight. This is in contrast to that observed in turbidites from the Balearic Abyssal Plain. Although a reason for this relationship is not clear, the increased percentage of calcium carbonate found in these same intervals may indicate a concentration of Foraminifera in the sand layers. This in turn could explain the lower unit weights (forams being filled with water) and possibly the higher water contents.

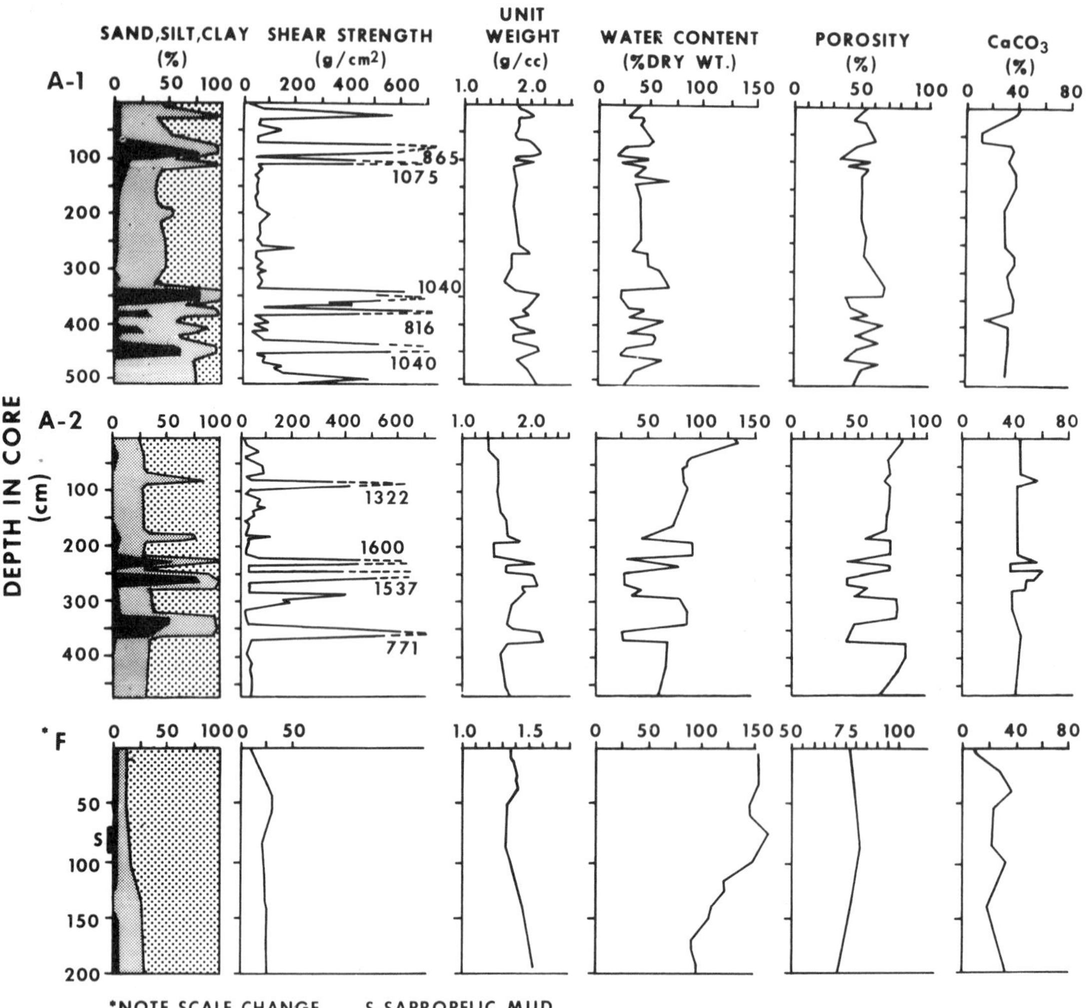

Figure 7. Vertical variation of selected mass properties from cored turbidite sequences (A-1, A-2) and the Nile Cone (F). See Figure 1 for core locations.

Core A-4 from the Mediterranean Ridge, southwest of the island of Crete (Horn *et al.*, 1967) represents a calcareous ooze sequence with minor variation in silt and sand content throughout the sampled interval. Of particular interest are the three sapropelic mud layers that occur at 145 cm, 190 cm, and 225 cm below the top of the core. These sediments are characterized by significant increases in water content and porosity, along with a notable decrease in unit weight. Porosities as high as 83 percent are found in association with the sapropelic muds. Shear strength of these muds appears to be slightly less than that found in the adjacent sediment. The marked increase of shear strength at the 325 cm depth reflects the notable increase in calcium carbonate content.

In contrast to the other areas of the Mediterranean basin, sediments comprising the upper portion of the Nile Cone display only a slight variation in their mass properties to a depth of at least 200 cm, as seen in a core collected by the OCEANOGRAPHER (Core F). Within this interval, the sediments consist primarily of clay-size material (73 to 89 percent), minor amounts of silt (9 to 25 percent), and traces of sand (1 to 3 percent). Water content and porosity are considerably higher than those observed elsewhere in the Mediterranean, having values of 150 to 162 percent and 77 to 82 percent respectively in the upper 100 cm of the sampled interval. Below 100 cm a general decrease in these values takes place. Except for the weaker surfical layers, shear strength is almost uniform with depth. Examination of the unit weight and calcium carbonate content profiles from core F indicates a somewhat parallel pattern of distribution with depth. The low calcium carbonate and unit weight noted at a depth of approximately 80 cm corresponds to the occurrence of a layer of sapropelic mud. The influence of this layer is also clearly noted in the water content profile.

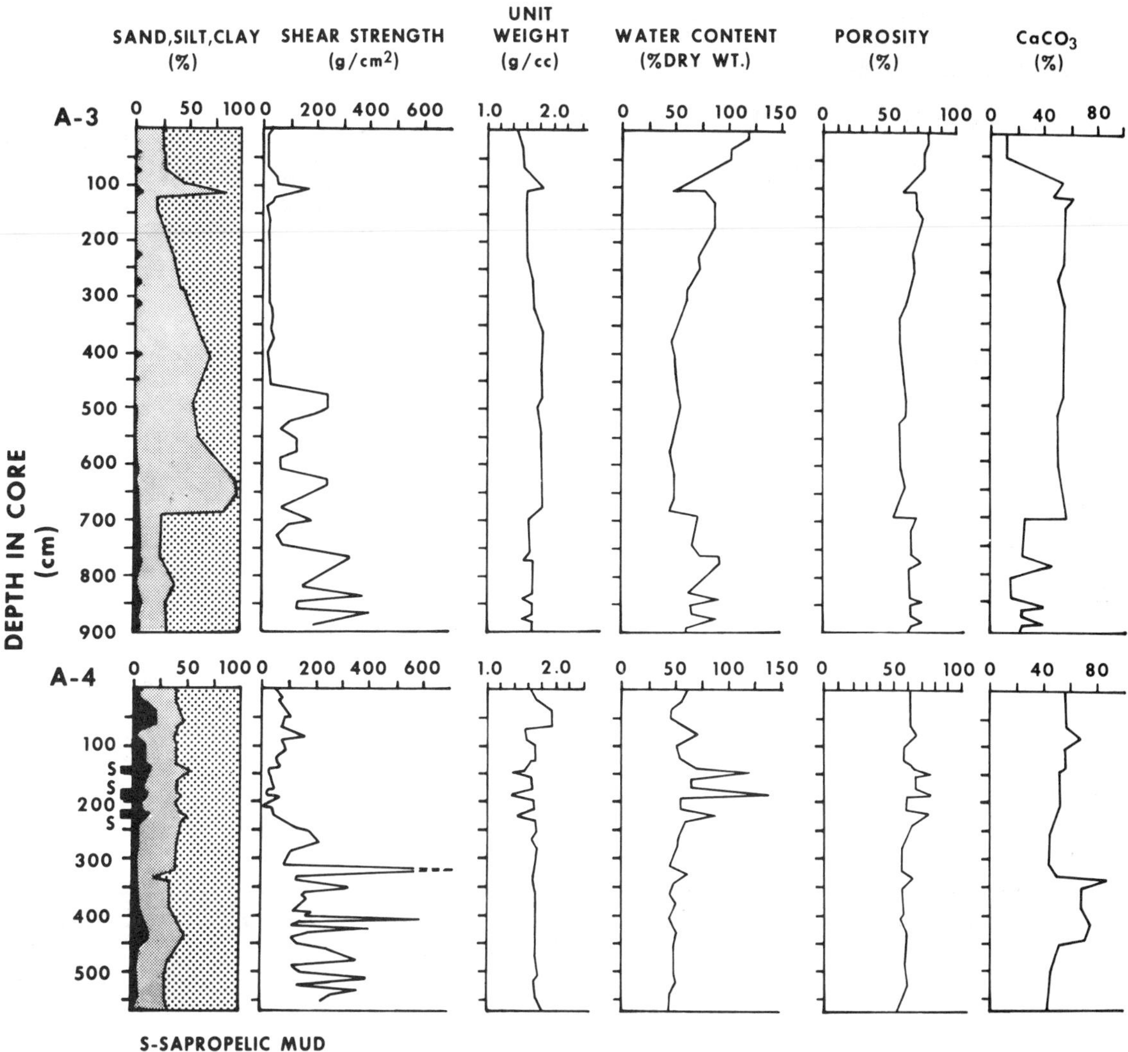

Figure 8. Vertical variation of selected mass properties in cores from the Messina Cone (A-3) and the Mediterranean Ridge (A-4). See Figure 1 for core locations.

PLASTICITY CHARACTERISTICS

Based on a sediment's plasticity, Casagrande (1948) devised a method for classifying fine-grained deposits. This technique employs a plasticity chart, the ordinate being the plasticity index and the abscissa the corresponding liquid limit. On the chart an "A"-line is drawn to represent an empirical boundary between inorganic clays, commonly above the line, and organic clays along with inorganic silts below it. Although this is a relatively simple classification and easily determined, it serves to quickly characterize sedimentary deposits not only as to their plastic properties, but as to their source as well. A detailed discussion of this classification can be found in most soil mechanics text books.

In an attempt to classify the Mediterranean sediments, a comparison of subsamples from a number of cores was made using the method proposed by Casgrande. A distinction is readily made between those samples collected in the eastern Mediterranean and those from the central (south of Sicily) and western portions of the basin (Figure 9). Sediments influenced by the Nile River are either highly organic or highly plastic and generally plot above the "A"-line or very close to it. Although primarily derived from the Nile River, sediments from off Israel possess a distinct characteristic of their own and are easily identified from the chart. Plasticity is generally higher and organic carbon content frequently lower than is found in the Nile Cone deposits. In contrast, sediments from the central and western Mediterranean display lower plasticity indices which agree reasonably well with those reported for open Atlantic deposits (Richards, 1962).

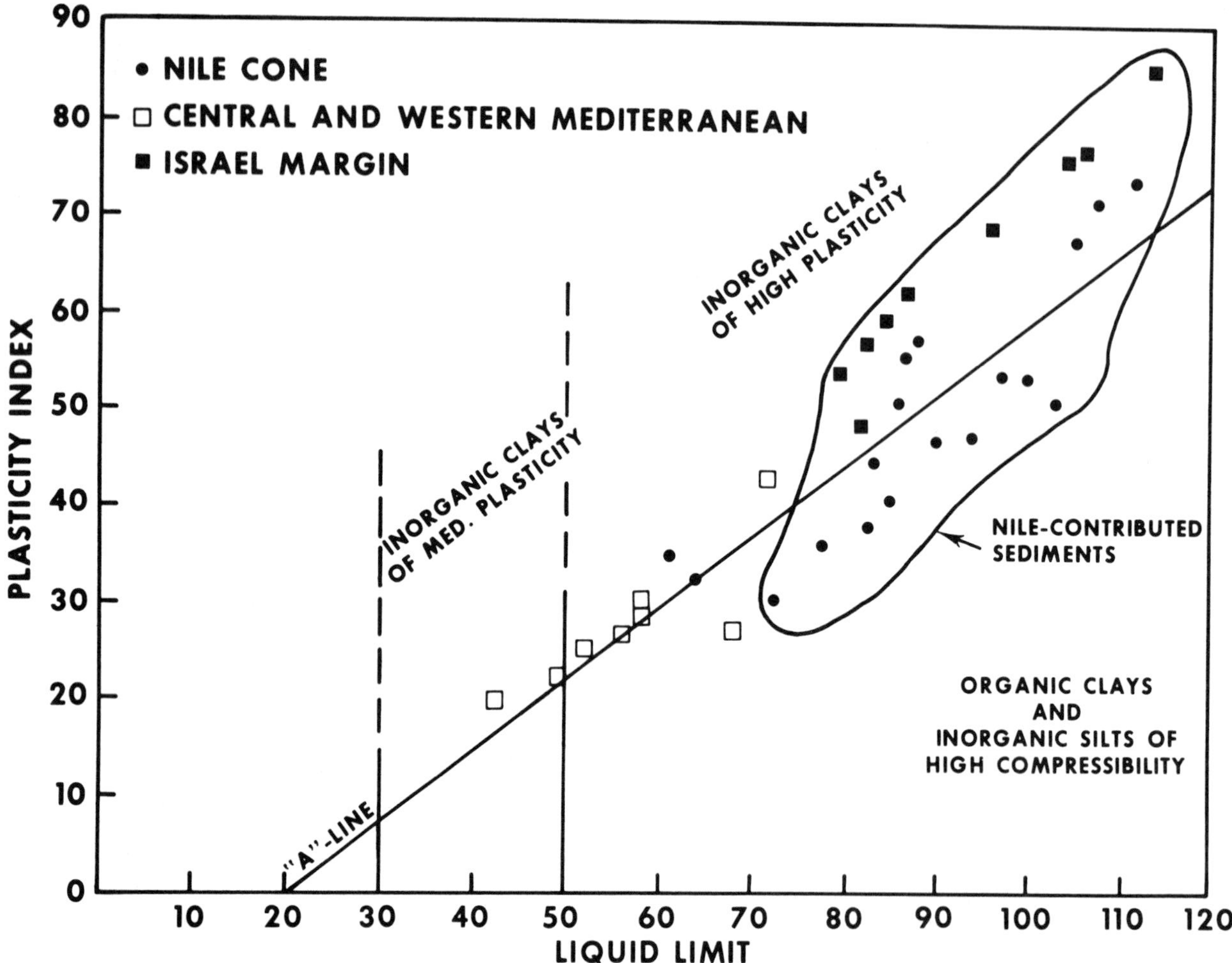

Figure 9. Plasticity chart and the classification of Mediterranean sediments.

CONSOLIDATION CHARACTERISTICS

In an effort to investigate the consolidation or compaction (as used by geologists) characteristics of the Mediterranean sediments, four samples were selected for consolidation tests. Consolidation, as used here, refers to the reduction of the volume of a sediment under an imposed load. In the case of a saturated sediment, this can only result when there is a loss of pore water. This load can be man-made (*e.g.*, installation of a structure) or result from pressure exerted by the overlying sediment. The consolidation test is commonly used in the field of soil mechanics to determine the amount of settlement from any given load. It also is used to provide an insight into the depositional history of the particular deposit by indicating the loading conditions that have affected the deposit. For a detailed discussion of this test, the reader is referred to a basic text of soil mechanics as well as to the study by Richards and Hamilton (1967) dealing with consolidation characteristics of submarine sediments.

Each of the four samples tested revealed that they were overconsolidated. Normally (in the soil mechanics sense) this would indicate that the sediment has been subjected to a greater load than is presently exerted by the overlying material. On land this is usually explained by erosion of previously deposited material. Erosion is known to take place on the sea floor, but it is highly unlikely in the case of the samples studied here. Core F, for example, was collected from the Nile Cone where deposition rather than erosion is currently taking place. A subsample of this core from a depth of one meter displayed a preconsolidation pressure, P_c, (pressure under which sediment has been previously loaded as defined in soil mechanics) of 0.072 kg/cm^2 and an overburden pressure, P_o, (present field load on the sampled interval) of 0.035 kg/cm^2 clearly indicating a state of overconsolidation. The void ratio (ratio of the volume of voids to the volume of solids) versus log of pressure curve shown in Figure 10 is basically similar to those of the other tested samples. Similar studies of submarine sediments from the Atlantic and Pacific (Richards and Hamilton, 1967) and the Gulf of Mexico (Bryant *et al.*, 1967) have also reported "over-consolidated" sediments. Since it is highly improbable that these deposits are "overconsolidated" (as the term is used in soil mechanics) there must be some unique aspect or property of submarine sediments that results in this increased strength. It has been suggested that cementation or unusually strong interparticle forces are responsible for this increased strength (Rittenberg *et al.*, 1963; Richards and Hamilton, 1967). As yet, little is known about this phenomenon, but there seems to be little doubt that such factors as ionic bonding and/or solution and redeposition within the sediment of various cations, (*e.g.*, silicon, iron, calcium or phosphorus) tend to alter the structural strength of the mass.

The finding of "overconsolidated" sediments in the Mediterranean basin is not surprising in light of similar observations reported elsewhere. It was somewhat surprising, however, to find that core F also displayed "overconsolidated" characteristics. This core was collected from the upper portion of the Nile Cone which is an area of relatively rapid deposition. In areas of rapid sedimentation, such as off the Mississippi River, which is admittedly an extreme case, sediments are commonly found to be underconsolidated owing to insufficient time for consolidation (dissipation of pore water pressure) to take place (Fisk and McClellan, 1959). For the type of sediment involved, it is apparent that the deposition rate on the upper Nile Cone is not so rapid that the pore water pressure cannot dissipate.

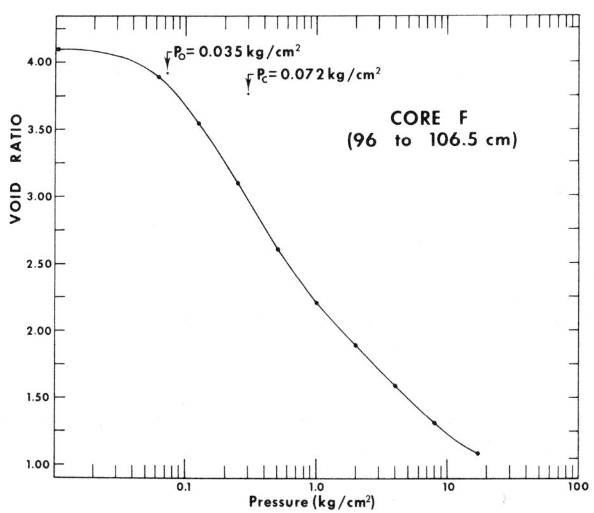

Figure 10. Consolidation test, void ratio-pressure curve.

SUMMARY

A total of 96 sediment cores from 27 different sites in the Mediterranean Sea were examined for their mass physical properties. Although these relatively few cores cannot possibly represent all the unconsolidated deposits occurring in the basin, some generalizations can be made regarding the distribution of selected mass physical properties and the range of these values can be anticipated for the surficial layers of the sea floor.

Morphologically, the Mediterranean basin can be divided into eastern and western sectors. The western Mediterranean consists largely of the Balearic and Tyrrhenian Abyssal Plains and the Rhône Fan. To the east, the basin becomes complex with a median ridge, continental borderland, few abyssal plains, and the Nile and Messina Cones. The contrast between the eastern and western portions of the Mediterranean is also evident in the sediments associated with each of the two provinces.

1. Submarine sediments west of Italy are mainly composed of terrigenous lutites derived from the River Rhône and the margins of Africa and Italy. Turbidity currents appear to be the major mechanism by which these lutites as well as fine sands are deposited in this portion of the Mediterranean. *Globigerina* and pteropod oozes are the predominant surfical deposits in the eastern Mediterranean. Although the Nile River is a major source of lutite, its influence on the sediment distribution in the eastern part of the basin is deceiving when only the surface deposits are considered. It is obvious from an examination of various mass properties with depth, even in relatively short sediment cores, that the unique characteristics of Nile-contributed sediment are rather extensive. Also, not revealed in the surface deposits, but of major importance to the overall sedimentary properties in the eastern Mediterranean, are the numerous tephra and sapropelic mud layers.

2. Average shear strengths vary from 16 to 227 g/cm^2; the lower values commonly being associated with the Nile-contributed lutites and higher values with the turbidite deposits off the Rhône Fan. The western Mediterranean, with its greater abundance of turbidites, displays higher average shear strengths overall than do those sediments in the eastern sector of the basin.

3. Water content varies from 30 to 124 percent, but commonly occurs in the range of 50 to 100 percent throughout the most of the Mediterranean basin. Only in the Tyrrhenian basin and in those areas influenced by the Nile do water contents greater than 100 percent occur. Continental margin deposits tend to be coarser-grained and possess relatively low water contents. Fine-

grained ash and pumice deposits in the vicinity of the island of Crete also display low water contents.

4. Unit weight values are found to range from 1.35 to 1.90 g/cc throughout the Mediterranean. The western Mediterranean is blanketed primarily by sediments displaying unit weights of 1.75 to 1.90 g/cc. This same property in the eastern Mediterranean commonly varies from 1.65 to 1.70 g/cc except in the easternmost areas where Nile-contributed sediments display unit weights of 1.40 to 1.47 g/cc.

5. Average porosities vary only slightly (60 to 68 percent) throughout most of the Mediterranean basin. The highest observed values occur in the Tyrrhenian basin and in the extreme eastern Mediterranean where Nile-derived lutites predominate. Low porosities (50 percent) are found to characterize sediments from the outer Rhône Fan, the fine-grained ash and pumice occurring north of Crete, and those deposits along the eastern margin of Tunisia which are composed largely of relatively coarse-grained calcareous material.

Variation of the selected mass physical properties within the upper 3 to 4 m of the sea floor is mainly controlled by changes in sediment texture. Turbidites in the western Mediterranean are clearly distinguished from the surrounding sediments by their high shear strengths and unit weights as well as low water contents and porosities. Just the reverse is observed in the case of the numerous sapropelic mud layers occurring in the eastern Mediterranean. Nile-contributed lutites, which cover much of the eastern Mediterranean display decidedly different plasticity characteristics than those sediments from the central and western portions of the basin. The lutites are highly plastic whereas those from other parts of the Mediterranean possess distinctly lower plasticity indices.

Examination of the consolidation or compaction (as commonly used in geology) characteristics of sediments from both the eastern and western Mediterranean indicates that they are similar to those of "overconsolidated" deposits. There is little doubt that these sediments are not "overconsolidated" (as defined in soil mechanics), but have increased in strength by means of ionic bonding or some form of "cementation" so as to behave in a manner indicating an "overconsolidated" state.

ACKNOWLEDGMENTS

Thanks are due to David R. Horn of the Lamont-Doherty Geological Observatory and John Bethell of the SACLANT ASW Research Centre, La Spezia, Italy, for providing us with data on sediment mass properties. We are indebted to P. Blavier and C. Gehin of the SACLANT Centre and F. Kögler of Kiel University for allowing us to use some of their unpublished data in this study. We are most grateful for the assistance provided us by the officers and crew of the NOAA OCEANOGRAPHER during the collection of sediment cores for this study. We acknowledge with many thanks the critical review of this paper by Louis Butler, Harris B. Stewart, Jr., Adrian F. Richards, and Richard Bennett.

REFERENCES

Almagor, G. 1964. Studies of sediments in core samples collected from the shelf and slope off Tel-Aviv–Palmakhim coast. *Geological Survey of Israel Report.* QGR/2/64:80p. (unpublished).

Almagor, G. 1967. Interpretation of strength and consolidation data from some bottom cores off Tel Aviv–Palmakhim coast, Israel. In: *Marine Geotechnique,* ed. Richards, A. F., University of Illinois Press, Urbana, 131–153.

Arrhenius, G. 1952. Sediment cores from the East Pacific. *Reports of the Swedish Deep-Sea Expedition 1947–1948,* 5(1): 227 p.

Bryant, W. R., P. Cernock and J. Morelock, Jr. 1967. Shear strength and consolidation characteristics of marine sediments from the Western Gulf of Mexico. In: *Marine Geotechnique,* ed. Richards, A. F., University of Illinois Press, Urbana: 41–62.

Bryant, W. R., and C. S. Wallin 1968. Stability and geotechnical characteristics of marine sediments, Gulf of Mexico. *Transactions, Gulf Coast Association of Geological Societies,* 18:334–356.

Casagrande, A. 1948. Classification and identification of soils. *American Society of Civil Engineers, Transactions,* 113:901–931.

Einsele, G. 1967. Sedimentary processes and physical properties of cores from the Red Sea, Gulf of Aden and off the Nile Delta. In: *Marine Geotechnique,* ed. Richards, A. F., University of Illinois Press, Urbana, 154–169.

Emery, K. O., B. C. Heezen and T. P. Allan 1966. Bathymetry of the Eastern Mediterranean Sea. *Deep-Sea Research,* 13:173–192.

Evans, I. and G. G. Sherratt 1948. A simple and convenient instrument for measuring the shear resistance of clay soils. *Journal of Scientific Instruments and Physics in Industry,* 25:411–414.

Fisk, H. N. and B. McClelland 1959. Geology of continental shelf off Louisiana: its influence on offshore foundation design. *Geological Society of America Bulletin,* 70:1369–1394.

Flaate, K. 1965. A statistical analysis of some methods of shear strength determination in soil mechanics. *Norwegian Geotechnical Institute Publication,* 62:8p.

Gehin, C. and P. Blavier 1969. Numerical results of the analysis of sea-bottom cores, Vol. 2 Alboran Sea. *NATO SACLANT ASW Research Centre, La Spezia, Special Report* (unpublished), M-52:30p.

Goncharov, V. P. and O. V. Mikhailov 1964. New data on the bottom relief of the Mediterranean. *Deep-Sea Research,* 11:625–628.

Hamilton, E. L. 1959. Thickness and consolidation of deep-sea sediments. *Geological Society of America Bulletin,* 70:1399–1424.

Hamilton, E. L. 1964. Consolidation characteristics and related properties of sediments from experimental MOHOLE (Guadalupe site). *Journal of Geophysical Research,* 69:4257–4269.

Hamilton, E. L., and H. W. Menard 1956. Density and porosity of sea-floor surface sediments off San Diego, California. *Bulletin of the American Association of Petroleum Geologists,* 40:754–761.

Hansbo, S. 1957. A new approach to the determination of the shear strength of clay by the fall-cone test. *Royal Swedish Geotechnical Institute Proceedings,* 14:47p.

Heezen, B. C. and M. Ewing 1955. Orléansville Earthquake and turbidity currents. *Bulletin of the American Association of Petroleum Geologists,* 39:2505–2514.

Hersey, J. B. 1965. Sedimentary basins of the Mediterranean Sea. In: *Submarine Geology and Geophysics,* eds. Whittard, W. F. and R. Bradshaw, Butterworths, London, 75–89.

Holeman, J. N. 1968. The sediment yield of major rivers of the world. *Water Resources Research,* 4(4):737–741.

Horn, D. R., B. M. Horn and M. N. Delach 1967. Correlation between acoustical and other physical properties of Mediterranean deep-sea cores. *Texas Instruments, Inc., Technical Report,* 2; 152p.

Horn, D. R., B. M. Horn and M. N. Delach 1968. Correlation between acoustical and other physical properties of deep-sea cores. *Journal of Geophysical Research,* 73:1939–1957.

Interdepartmental Geophysical Committee of the Academy of Sciences, USSR. 1969. *Bottom Sediment Chart.* Main Administration for Geodesy and Cartography, Moscow, USSR.

Keller, G. H., and R. H. Bennett 1968. Mass physical properties of submarine sediments in the Atlantic and Pacific basins. *Proceedings, 23rd International Geological Congress,* 8:33–50.

Kermabon, A., C. Gehin and P. Blavier 1968. Numerical results of the analysis of sea-bottom cores, Vol. I, Naples and Ajaccio zones. *NATO SACLANT ASW Research Centre, La Spezia, Special Report* (unpublished), M-46:14p.

Kermabon, A., C. Gehin, P. Blavier and B. Tonarelli 1969. Acoustic and other physical properties of deep-sea sediments in the Tyrrhenian Abyssal Plain. *Marine Geology,* 7:129–145.

Mellis, O. 1954. Volcanic ash-horizons in deep-sea sediments from the Eastern Mediterranean. *Deep-Sea Research,* 2:89–92.

Menard, H. W., S. M. Smith and R. M. Pratt 1965. The Rhône deep-sea fan. In: *Submarine Geology and Geophysics,* eds. Whittard, W. F. and R. Bradshaw, Butterworths, London, 271–284.

Moore, D. G. 1962. Bear strength and other physical properties of some shallow and deep-sea sediments from the North Pacific. *Geological Society of America Bulletin,* 73:1163–1166.

Morelock, Jr., J. 1969. Shear strength and stability of continental slope deposits, Western Gulf of Mexico. *Journal of Geophysical Research,* 74:465–482.

Morgenstern, N. R. 1967. Submarine slumping and the initiation of turbidity currents. In: *Marine Geotechnique,* ed. Richards, A. F., University of Illinois Press, Urbana, 189–220.

Ninkovich, D., and B. C. Heezen 1965. Santorini Tephra. In: *Submarine Geology and Geophysics,* eds. Whittard W. F. and R. Bradshaw, Butterworths, London, 413–452.

Olausson, E. 1960. Description of sediment cores from the Mediterranean and Red Sea. *Reports of the Swedish Deep-Sea Expedition 1947–1948,* 8(3):287–334.

Olausson, E. 1961. Studies of deep-sea cores. *Reports of the Swedish Deep-Sea Expedition 1947–1948,* 8(4):337–391.

Pfannenstiel, M. 1960. Erläuterungen zu den Bathymetrischen Karten des Östlichen Mittelmeeres. *Bulletin de l'Institut Océanographique,* Monaco, 1192:60p.

Richards, A. F. 1961. Investigations of deep-sea sediment cores, I. Shear strength, bearing capacity, and consolidation. *U. S. Navy Hydrographic Office, Technical Report,* 63:70p.

Richards, A. F. 1962. Investigations of deep-sea sediment cores, II. Mass physical properties. *U. S. Navy Hydrographic Office, Technical Report,* 106:146p.

Richards, A. F., and E. L. Hamilton 1967. Investigations of deep sea sediment cores, III. Consolidation. In: *Marine Geotechnique,* ed. Richards, A. F., University of Illinois Press, Urbana, 93–112.

Rittenberg, S. C., K. O. Emery, J. Hulseman, E. T. Degens, R. C. Fay, J. H. Reuter, J. R. Grady, S. H. Richardson and E. E. Bray 1963. Biogeochemistry of sediments in experimental MOHOLE. *Journal of Sedimentary Petrology,* 33:140–172.

Ryan, W. B. F. and B. C. Heezen 1965. Ionian Sea submarine canyons and the 1908 Messina Turbidity Current. *Geological Society of America Bulletin,* 76:915–932.

Ryan, W. B. F., F. Workum, Jr. and J. B. Hersey 1965. Sediments of the Tyrrhenian Abyssal Plain. *Geological Society of America Bulletin,* 76:1261–1282.

Texas Instruments Incorporated 1967. 1965–67 North Atlantic Ocean, Norwegian Sea and Mediterranean Sea Area 6. *U. S. Naval Oceanographic Office Report,* Contract No. N62306–1687, 48p.

U. S. Naval Oceanographic Office 1965. *Oceanographic Atlas of the North Atlantic Ocean. Section V. Marine Geology.* Washington, D. C. 700; 71p.

Watson, J. A. and G. L. Johnson 1969. The marine geophysical survey in the Mediterranean. *International Hydrographic Review,* 4 6:81–107.

Wong, H. K. and E. F. K. Zarudzki 1969. Thickness of unconsolidated sediments in the Eastern Mediterranean Sea. *Geological Society of America Bulletin,* 80:2611–2614.

PART 8. Suspended Sediments and Regional Dispersal Patterns

Suspended Matter in the Mediterranean Sea

E. M. Emelyanov and K. M. Shimkus

P. P. Shirshov Institute of Oceanology, Kaliningrad and Gelendgik

ABSTRACT

About 800 samples of suspended matter were collected by filtration at 180 stations during the Soviet expeditions on R/V AKEDEMIK S. VAVILOV from 1959 to 1968. Minimum suspension concentrations of 0.2 to 0.5 mg/l from the upper 1 m surface layer were recorded in the Levant Sea southwest of the Nile Delta. Maximum concentrations were found in the Aegean Sea (2.3 mg/l), Sidra Bay (1.7 mg/l), the Algerian-Provencal basin (1.9 mg/l) and the Adriatic Sea (1.65 mg/l). In the 100 to 3000 m depth layer, concentrations varied within the range of 0.4 to 1.0 mg/l.

Microscopic examination of about 150 samples revealed that the amount of suspended particles per liter of water fluctuates from 0.2×10^6 to 5×10^6. Organic matter distinctly dominates in suspensions, measuring 40 to 95 per cent by weight. Terrigenous material makes up 15 per cent, carbonates 5 per cent and siliceous particles 3 to 10 per cent. The following components are distinguished in the suspended matter: coccoliths, foraminifera, chemogenic calcite crystals, dolomite, radiolarians, silicoflagellates, clay aggregates, mica, quartz, feldspar, green and brown hornblendes, actinolite-tremolite, clinopyroxene, epidote, zoisite, garnet and other minerals.

Chemical, colorimetric and kinetic methods have revealed the following composition of suspended matter per 1 liter of water: $CaCO_3$ to 9 μg (5.1 per cent of suspended matter by weight); total silica, to 207 μg (47.5 per cent); amorphous silica, to 156 μg (39.0 per cent); Fe, to 43 μg (14.2 per cent); cent); Mn, to 0.46 μg (0.162 per cent); Ti, to 1.10 μg (0.30 per cent); Ni, to 0.024 μg (0.006 per cent); and Co, to 0.0105 μg (0.0016 per cent).

RESUME

Au cours des expéditions soviétiques, entre 1959 et 1968, à bord du R/V AKEDEMIK S. VAVILOV quelque 800 échantillons ont été recueillis par filtration. A un mètre de profondeur de la couche de surface, des concentrations de 0,2 à 0,5 mg/l (suspension minimum) ont été enregistrées dans la Mer du Levant, au Sud-Ouest du delta du Nil. Des concentration maxima ont été enregistrées dans la Mer Egée (2,3 mg/l), la Baie de Sidra (1,7 mg/l), le Bassin Algéro-Provençal (1,9 mg/l) et l'Adriatique (1,65 mg/l). Dans la couche située entre 100 et 3.000 mètres de profondeur, les concentrations varient entre 0,4 et 1,0 mg/l.

L'étude microscopique de quelque 150 échantillons a montré que la proportion de matière en suspension par litre d'eau varie de $0{,}2 \times 10^6$ à 5×10^6. Les matières organiques dominent nettement dans les suspensions à raison de 40 à 95% en poids. Les matériaux terrigènes comptent pour 15%, les carbonates 5% et les particules de silice de 3 à 10%. Dans les matières en suspension on distingue: coccolites, foraminifères, cristaux de calcite chémogénique, dolomite, radiolaires, silico flagélés, agglomérats d'argile, mica, quartz, feldspath, hornblende vertes et marrons, actinolite-trémolite, clinopyroxène, épidote, zoïsite, grenat et autres minéraux.

Les méthodes chimiques, colorimétriques et cinétiques ont mis en évidence la composition de la matière en suspension dont les valeurs maxima par litre sont indiquées: $CaCO_3$, 9 μg (5,1% en poids de matière en suspension); silice totale, 207 μg (47,5%); silice amorphe, 156 μg (39,0%); Fe, 43 μg (14,2%); Mn, 0,46 μg (0,162%); Ti, 1,10 μg (0,30%); Ni, 0,024 μg (0,006%) et Co, 0,0105 μg (0,0016%).

INTRODUCTION

Studies of concentration, granulometry and chemical composition of suspended matter throw light upon a number of unresolved problems such as: (a) the genesis of sedimentary material in seas and oceans; (b) the mode of transport of such material into a basin; (c) the relative importance of various factors such as river sediment load, the supply of material from abrasion, volcanism, biogenic activity, eolian and chemogenic factors; (d) changes in material during transport; and (e) the accumulation rate of sediments. The granulometry, mineralogy, and chemical composition of suspended sediments ultimately determine the composition of bottom sediments. For these reasons, in recent years the study of suspended matter has become an integral feature of studies of modern sedimentation processes.

Only a few papers have dealt with the nature and importance of suspended matter in the Mediterranean Sea. Earlier works by Jerlov (1953 and 1958), Rakestraw (1958) and Ochakovsky (1965) on optical properties of water in connection with suspension concentrations provide some information concerning the turbidity of water in the sea. Again, in 1957 Lisitsin (1961a and b; 1964), on board the research vessel OB, used the method of separation to collect 4 samples of suspension from the upper 7 m thick layer of water and determined the granulometric and chemical compositions of these samples.

METHODOLOGY

From 1960 to 1966, scientists aboard the research vessel R/V AKADEMIK S. VAVILOV collected about 400 samples from 185 stations. Water samples were taken from different levels by plastic bottles of 5 and 7 liter capacities. The suspensions were filtered (Lisitsin, 1961b) on membrane biological filters 35 mm or sometimes 110 mm in diameter, with 0.7μ pores (for chemical purposes, with 0.5μ and 0.3μ pores).

Salts were removed from the particulate matter by washing with 10 to 20 ml of distilled water. The filters were dried in dessicators. Total suspension concentrations on filters were determined by weighing. Subsequently, portions of each filtrate were subjected to granulometric and mineralogical analysis and some samples were chemically analysed.

Granulometry and mineralogy of suspensions were determined under the polarizing microscope ($\times 160$–$\times 720$). The following procedure was adopted to facilitate counting (Bogdanov, 1965). One fourth of each filter was stained with a solution containing 4 per cent of yellow potassium cyanide and 5 per cent hydrochloric acid (Kuznetsov, 1949). Thus, carbonates were dissolved and all particles containing iron were colored blue. After washing in water, the preparation was treated with a 3 per cent erythrosine solution in 5 per cent carbolic water, staining organic detritus bright red and leaving nonbiogenic, "non-ferruginous" particles unstained. This method permitted discrimination and counting of genetically heterogenous particles. Counting of particles was carried out separately for each type of constituent. Stained and unstained parts of each filter were then mounted on slides in Canada balsam, dried, and analysed under the microscope. Carbonates, siliceous and detrital particles were studied in unstained filtrates and organic matter and ferruginous material were analyzed in the stained ones.

Under low magnification a field with a moderate number of particles was chosen, and about 1000 to 2000 particles on 7 to 10 fields were counted. The equivalent number for the whole area of the filter and the total volume of filtered water was calculated on the assumption that all the particles were distributed evenly on the filter.

Calculation of particle abundance was carried out separately for each of the following principal types of material: terrigenous (non-carbonate), ferruginous (mainly hydrous ferric oxide), carbonates (terrigenous, biogenic and chemogenic), organic detritus and biogenic silica.

In the calculations, all particles were divided into 6 size fractions: >0.05 mm; 0.05 to 0.25 mm; 0.025 to 0.010 mm; 0.010 to 0.005 mm; 0.005 to 0.0025 mm; and 0.0025 to 0.0010 mm. The volume of particles in each fraction was measured, assuming a spherical shape for all particles. Finally the obtained volume was recalculated in terms of weight. The following densities were assumed: 2.6 for terrigenous particles; 4.0 for ferruginous particles; 2.6 for carbonate particles; 1.0 for organic detritus; and 2.2 for siliceous particles. A total of 91 samples from 41 stations were analyzed in this manner. The calculated summary weight of all types of particle as a rule did not coincide with the total weight concentrations, the discrepancy being $\pm$ 20–30 per cent in most cases, and even reached $\pm$ 100 per cent. To correct this discrepancy, the difference between calculated and actual weights was redistributed proportionally among all the suspension types and within the particle types the difference was distributed proportionately across all six size fractions. The complete calculations were carried out for only 19 stations (Figure 1).

In addition to granulometric and mineralogical determination of particulate composition, the concentrations of silica and of 5 trace elements (Fe, Mn, Ti, Ni, Co) were determined. Total and amorphous silica were determined by colorimetry on a separate filter or on one quarter of a 110 mm diameter filter. The weight of the sample was 1 to 3 mg. Iron and titanium were determined with violet pyrocatechol for iron and tyron (1,2-dioxibensol-3,5-disulfoacid disodium salt) for titanium (Emelyanov, 1968; Emelyanov *et al.*, 1971). Manganese was determined by the kinetic method, using the reaction of oxidation of diethylaniline by potassium periodate in a phosphate-citrate buffer (Yatsimirsky *et al.*, 1971). Iron, titanium and manganese were determined in a 1–3 mg sample collected on one 35 mm diameter filter.

Nickel and cobalt were studied by the kinetic method in a 1–3 mg sample collected on a separate

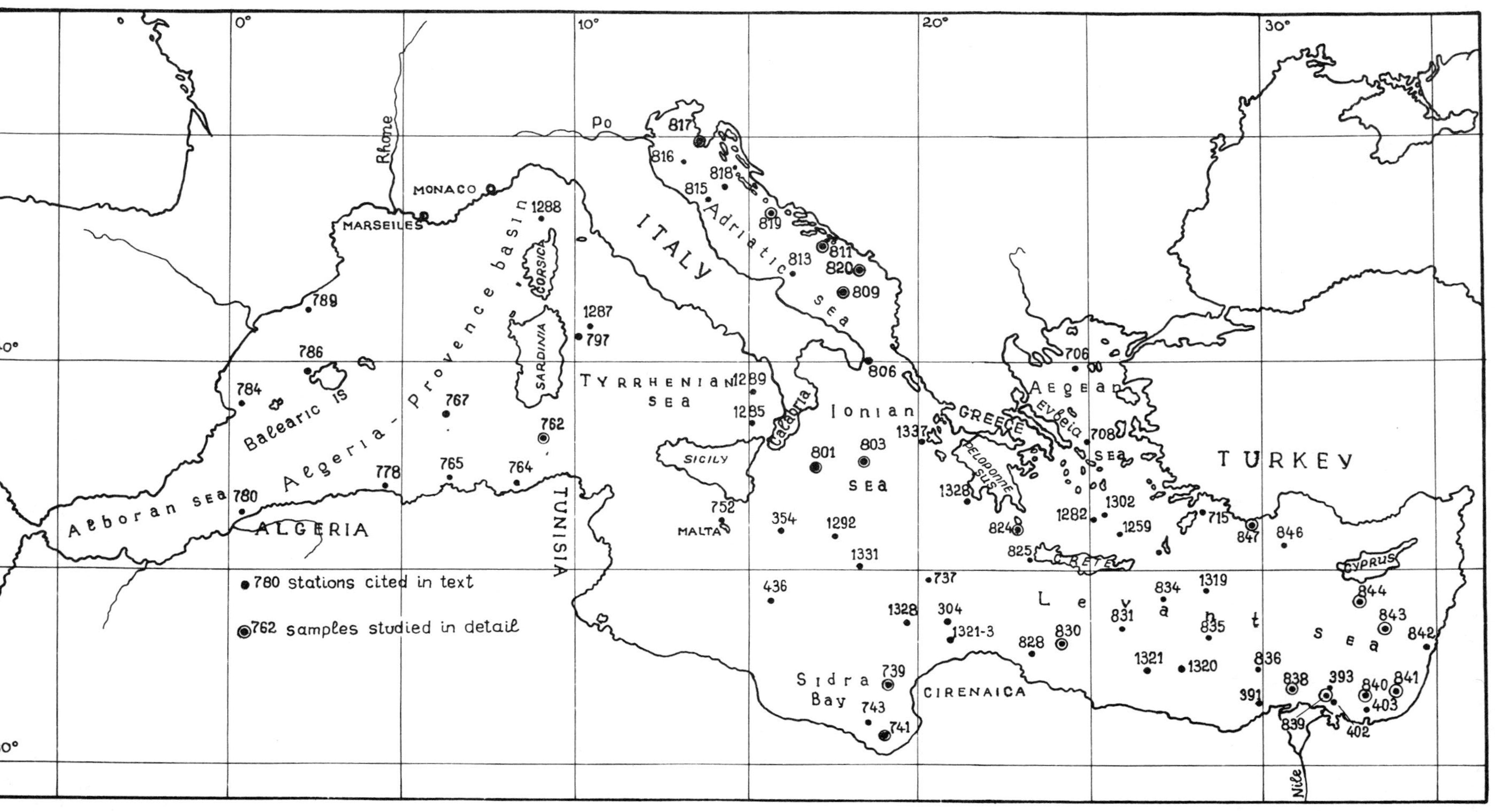

Figure 1. Generalised map of the Mediterranean Sea, showing location of stations sampled.

filter. For nickel the catalytic reaction of tyron-diphenilcarbazone–hydrogen peroxide was used. Quantitative determination of cobalt was based on the oxidation of alizarin by hydrogen peroxide at pH 12.4 (Yatsimirsky *et al.*, 1970).

PARTICULATE CONCENTRATION

In the upper layers in different areas the average particulate concentration is 0.5 to 1.5 mg/l. The concentration has a considerable range in some areas (Figure 2). Along the Tyrrhenian shores of Sardinia and near Evbeia in the Aegean Sea the average suspension concentrations are 2.90 and 2.15 mg/l respectively. These are the highest concentrations in the surface layer of the Mediterranean Sea, not counting coastal regions near the mouths of rivers, and are attributed mainly to high phytoplankton production in these areas (because organic detritus predominates) but partly also to abundant terrigenous supplies.

In coastal areas and especially near river-mouths the particulate concentration varies in accordance with the intensity of sediment delivery from the coast. Thus, near the mouth of the Nile during the autumn flood of October 3, 1959 at Station 402, 20 kilometers offshore (Figures 1 and 2), the maximum concentration of suspension was 4.6 mg/l. At the same time, at stations 403, 393, and 391, located about the same distance from the mouth of the river but outside the current pattern that carries turbid Nile waters in a northeast direction, the concentration was only 0.4 mg/l, comparable in turbidity to the waters of the open sea (Emelyanov, 1962). Analogous rapid decreases in suspension concentration with increasing distance from the mouth of a river or from the coast were noted in other regions. Jerlov (1953) points out the same phenomenon for the area near the mouth of the Nile.

The amount of suspended material tends to decrease toward the surface, the main decrease in concentration occurring in the upper 100 m of water. This decrease varies in different parts of the sea. For instance, the decrease in concentration in the Levant Sea is very marked; even at depths of 100 m, the concentration is only 0.30 to 0.55 mg per liter (Figures 3 and 4). Only 3 stations out of 15 show higher values. In contrast, rather high concentrations are found in the 100 to 200 m deep layers of the Aegean Sea. Elsewhere in the Mediterranean Sea the average particulate concentrations at depths greater than 100 m are moderate, ranging between 0.5 and 0.8 mg/l. In waters deeper than 100 to 200 m, particulate concentration tends to be almost constant, from 0.3 to 0.6 mg/l. and only at Station 739 in Sidra Bay, at depths ranging from 800 to 1000 m, were high concentrations encountered (to 2.15 mg/l).

In the near-bottom layers (5 m to 50 m off the bottom) particulate concentration tends to increase again as compared with values at intermediate levels. This apparently results from scouring of the upper layer of sediment by near-bottom currents, especially on the shallow and rugged bottom topography of the Aegean Sea. This scoured material is augmented by sediments precipitated from higher levels, as is clearly seen in the distribution of particulate matter near the Nile mouth (Figures 2 and 5). In this area, particulate concentration decreases off-shore from the river mouth only in the upper layer. Near the bottom and at intermediate depths, high particulate concentrations persist for great distances from the delta along the axes of sea currents, confirming that the bulk of material quickly settles to some depth and then is dispersed by currents within a large area, including Cyprus and the coasts of Turkey. This is in good agreement with the distribution data for some minerals and principal chemical components in bottom sediments (see Emelyanov, this volume). This mode of sediment distribution explains the low particulate concentrations in surface waters within the near-mouth area, both outside the current mainstreams and in the regions of intensive water-mass displacement.

It should be noted that for the early autumn period of September, 1959 we obtained somewhat greater particulate contents of 0.4 to 0.8 mg/l in the upper layers of the Levant Sea and slightly lower values of 0.2 to 0.5 mg/l in deeper levels. In contrast, for the late autumn and early winter period of November to December 1960, very low values of 0.28 to 0.54 mg/l were obtained in the upper levels whereas in deeper levels (100–200 m) suspension concentration values were 0.5 to 0.8 mg/l, close to those for the entire Mediterranean Sea. During the winter-to-spring period from January to May 1962, maximum suspension contents were found at almost all levels (Figures 2 and 4), particularly in the upper layer of Sidra Bay and in the Algerian-Provençal Basin.

These variations of particulate concentration in the upper layers, as well as in deeper waters, are attributed mainly to seasonal changes in the nutrient content of the water. During the winter, mixing of waters delivers more nutrients from deeper levels to the surface, resulting in plankton blooming at the end of the winter and in the spring. In the summer and autumn, conditions for plankton development are less favorable and hence suspension concentrations are lower.

Particulate concentrations obtained by the separation method (with centrifuge) are generally lower in value than ours. Separators probably do not catch a

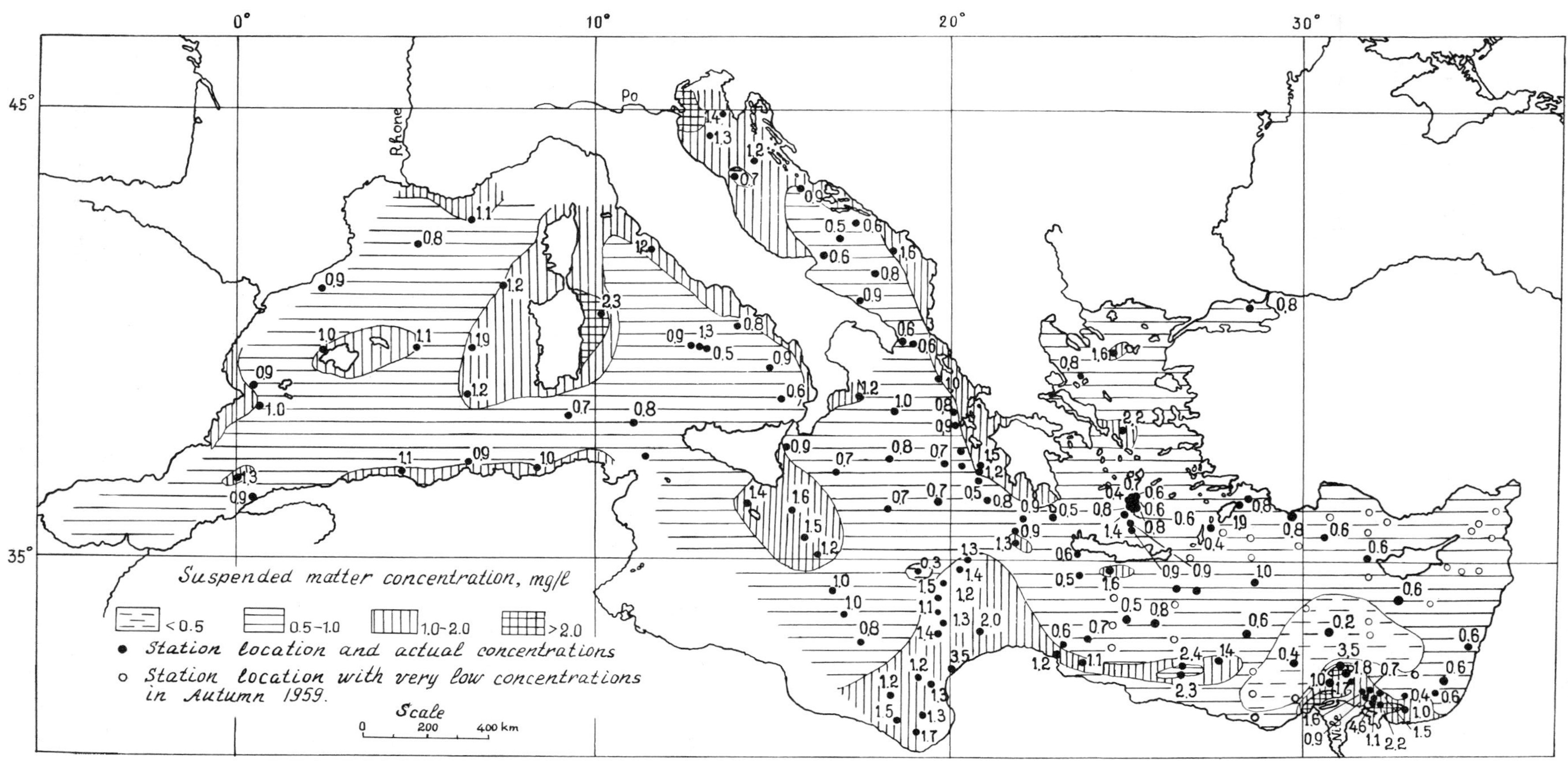

Figure 2. The quantitative distribution of suspended matter (in mg/l) in the upper water layer of the Mediterranean Sea.

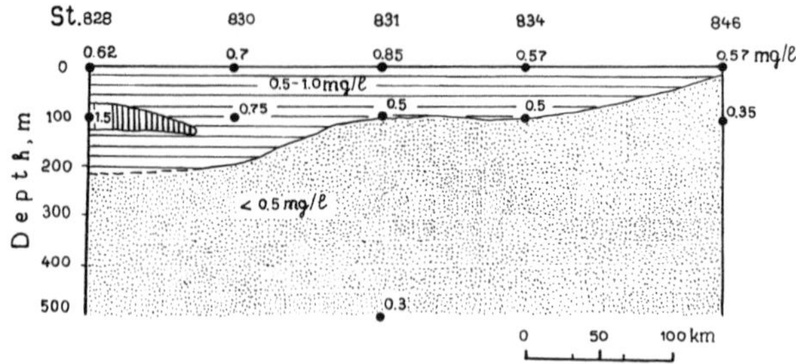

Figure 3. Suspended matter concentration (in mg/l) for a profile from Cyrenaica (at left) to Turkey (at right).

considerable portion of the particulate matter, especially the fine clay.

The total amount of suspended material in the Mediterranean Sea is less than in the Black Sea and is about the same as in the Atlantic.

NUMBER OF PARTICLES

The number of suspended particles in 1 liter of open ocean water varies widely, from 0.2×10^6 to 1.5×10^6, but averages about 0.7×10^6 (Table 1). Near rivers

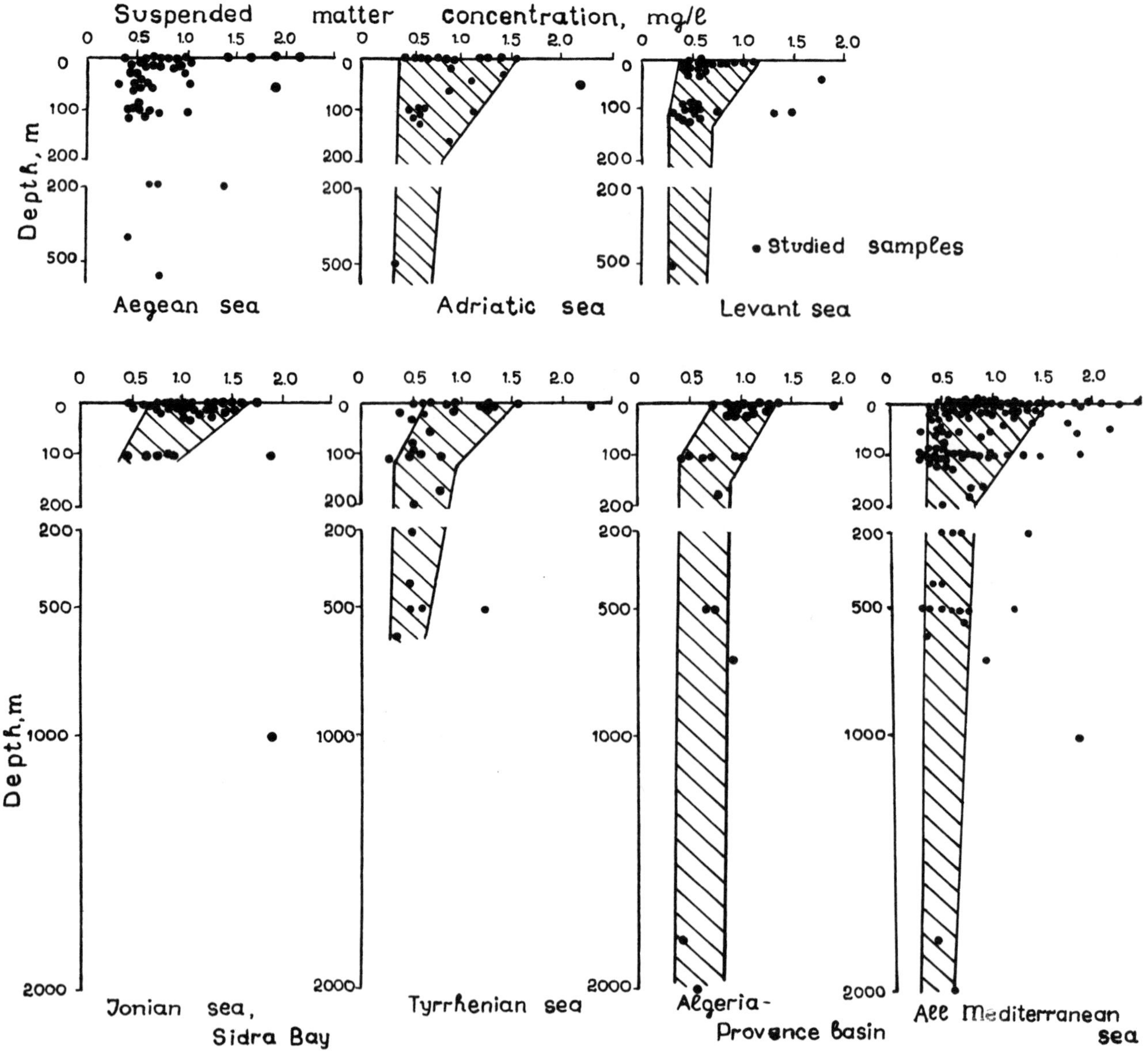

Figure 4. The vertical distribution of suspended matter (in mg/l) in different parts of the Mediterranean Sea. Areas of high sample density are shaded.

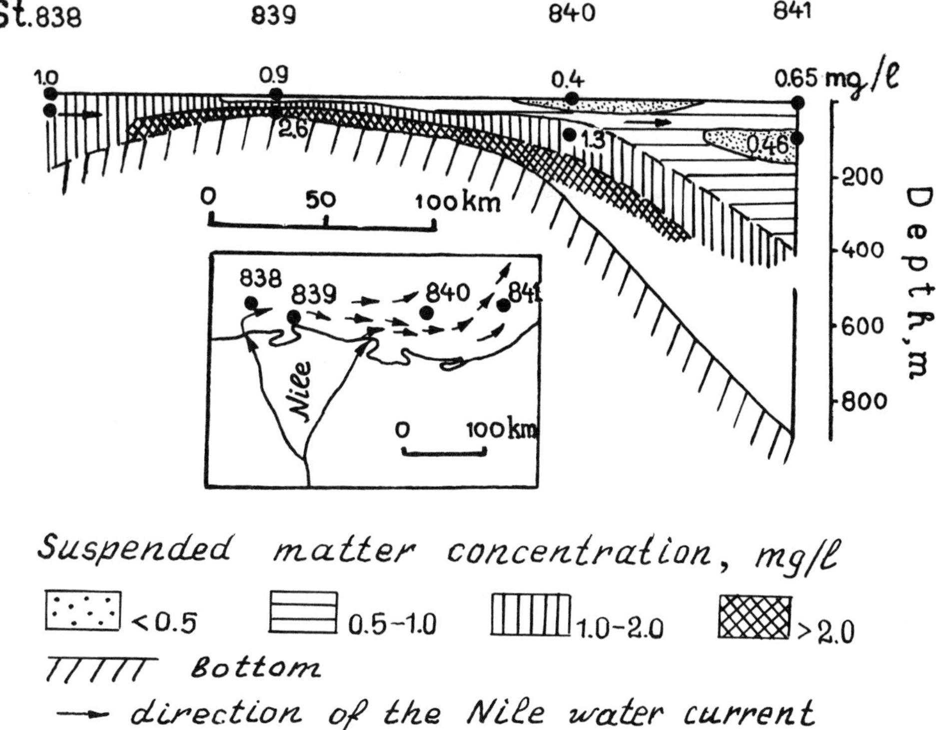

Figure 5. The suspended matter distribution (in mg/l) in the region of the Nile Delta (May, 1962). Optical data by Jerlov (1953) have been incorporated.

such as the Nile and Po, the number of particles rises to between 2×10^6 and 5×10^6. The mimimum numbers (0.2×10^6 to 0.4×10^6 particles per liter), were observed in southern regions of the Ionian Sea and in the Levant Sea.

The same stations show rather high variations in the number of particles with depth. Both at the surface and at depth, maximum amounts of suspended material were observed inshore and near river-mouths, while minimum amounts occur in the open parts of the basin, particularly in its eastern portion.

Terrigenous particulate matter is more abundant than all other particulate types in the surface layer (1 m thick), averaging about 368,400 particles per liter (Table 1). Individual stations 764, 789 and samples over the submarine Nile Delta show amounts up to 1.5×10^6 particles per liter. During high water, near river-mouth areas the number of particles are even greater. Minimum numbers of abiogenic particles (60,000 to 300,000) were observed in southern areas of the Ionian Sea-Sidra Bay and the Levant Sea, the basin areas which do not receive runoff.

At depth, the amount of terrigenous material decreases somewhat. Very high contents of terrigenous material at 40 and 50 meter depths were found only in the Adriatic Sea. Analogous increases are likely near river mouths. A sharp increase in the number of terrigenous particles accompanies a decrease in their grain size. The smallest pelitic particles (2.5–1 μ) at all depths comprise 90 to 95 per cent of the total suspension.

The maximum number of small pelitic particles occurs in the hydrodynamically active surface layer. At depths greater than 300 m, the number of small pelitic particles is less by 10 to 30 per cent. Especially high amounts of very fine material were observed over the delta-front region of the Nile.

The number of particles of ferruginous material in the Mediterranean Sea is comparatively small, varying from several hundred to 32,600 per liter. These materials are most abundant near river mouths and in the Adriatic.

Carbonate suspended particles are second in abundance to terrigenous material in the upper layer of the sea. The average amount is about 200,000 per liter, or about 32 per cent of the total suspension. The highest values of suspended carbonate were observed in the Aegean Sea, in the area of Sidra Bay-Malta, and in the inshore areas of the Provençal basin. The lowest contents occur in the open basin. The number of particles increases sharply with depth in most cases, particularly in the 50 to 100 m layer. In waters deeper

Table 1. The mean number of suspended particles (coarser than 1 micron) of different particulate types in the Mediterranean Sea.

Particulate type		0–2 m (19 stations)		40–50 m (stations 815, 816)		100 m (12 stations)		500 m (stations 765, 767, 806, 831)		900 m (station 806)		2000 m (station 765)	
		pc/l	%	pc/l	%	× 1000 pc/l	%	× 1000 pc/l	%	× 1000 pc/l	%	x 1000 pc/l	%
		Total number of suspended particles											
All types	Min.	326.1	100	4390.2	100	240.4	100	526.4	100	—	100	97.6	100
	Max.	1261.9	100	5367.9	100	1823.5	100	929.7	100	—	—	—	—
	Mean	621.1	100	4879.0	100	803.6	100	722.2	100	1436.3	100	976.9	100
		Number of particles in different particulate types											
Terrigenous material	Min.	58.5	10.7	888.9	20.2	141.9	29.2	37.8	5.5	—	—	—	—
	Max.	1103.3	87.4	1207.7	22.5	532.1	73.9	463.2	49.8	—	—	—	—
	Mean	368.4	58.6	1048.3	21.4	316.3	39.7	207.3	25.8	357.1	24.9	283.2	29.0
Ferruginous material	Min.	0.3	0.05	18.5	0.3	8.2	0.5	4.4	0.8	—	—	—	—
	Max.	32.6	4.0	59.8	1.4	100.6	16.9	21.8	3.2	—	—	—	—
	Mean	7.8	1.4	39.2	0.8	26.4	4.0	10.4	1.5	6.5	0.4	17.8	1.8
Carbonate material	Min.	30.5	5.6	2015.3	45.8	22.3	3.8	408.2	46.7	—	—	—	—
	Max.	519.0	84.4	2458.8	45.9	1255.3	86.0	577.2	84.0	—	—	—	—
	Mean	197.2	31.6	2237.0	45.8	416.5	50.1	473.1	68.4	1051.7	73.2	639.6	65.5
Organic material (detritus)	Min.	25.6	2.8	182.6	4.2	18.5	1.2	12.5	2.3	—	—	—	—
	Max.	103.3	14.5	265.5	4.9	76.1	14.1	49.6	7.2	—	—	—	—
	Mean	45.2	7.8	224.0	4.6	42.1	5.9	30.3	4.1	18.8	1.3	35.7	3.7
Siliceous material	Min.	0.3	0.1	1243.5[a]	26.4	0.3	0.04	0.5	0.1	—	—	—	—
	Max.	20.7	1.4	1417.4[a]	28.3	12.0	1.1	1.6	0.2	—	—	—	—
	Mean	2.5	0.4	1330.5[a]	27.4	2.3	0.2	1.1	0.1	2.2	0.2	0.5	0.1

[a]Spines 2.5–1μ prevail.

than 50 m they comprise up to 73.2 per cent of the whole suspension.

Organic suspended matter, usually colorless transparent particles of vegetable detritus, brown and black plant fiber and scraps of fiber, ranks third in number of particles and amounts to about 1.3 to 7.8 per cent of the whole suspension at different levels (Table 1). Organic detritus maxima of 26,000 to 103,000 particles per liter were observed in the upper layers, and the mean value for 19 stations was 45,200 particles per liter or 7.8 per cent of suspended particles larger than 1 μ. In the Adriatic Sea the organic content is higher, about 224,000 particles per liter. In the open sea the number of particles decreases with depth.

Amorphous silica particles comprise very little of the suspended material, their mean content being 0.1 to 0.4 per cent of the whole suspension, or 300 to 20,700 particles per liter. Only at stations 815 and 816 in the North Adriatic were higher contents of diatoms, radiolarians, shell spines, and spicules observed. Siliceous material percentages are less in deep suspensions than in upper layers of the bottom sediments.

WEIGHT OF VARIOUS SUSPENSION TYPES

Organic detritus forms the bulk of the suspended mass constituting from 0.33 to 1.32 mg/l (Table 2), some 37 to 96 per cent of the whole suspended material. The maximum concentration of this material is observed in the uppermost 50 m of water. Ferruginous particles are next in abundance by weight, averaging 0.08 to 0.09 mg/l (5.7 to 15.2 per cent of the total suspension) in the uppermost 100 m of water.

The third most abundant suspension type by weight is represented by the carbonates, which vary from traces to 0.22 mg/l in the surface to 100 m layer, the

Table 2. The mean weight of different particulate types in the Mediterranean Sea.

Particulate type		Depths, m											
		0–2 m (19 stations)		40–50 m (stations 815, 816)		100 m (12 stations)		500 m (4 stations)		900 m (station 806)		2000 m (station 765)	
		Overall content of different types											
		mg/l	%	mg/l	%	mg/l	%	mg/l	%	mg/l	%	mg/l	%
All Types	Min.	0.4	100	1.4	100	0.3	100	0.3	100	—	—	—	—
	Max.	1.7	100	2.2	100	0.9	100	0.9	100	—	—	—	—
	Mean	0.9	100	1.8	100	0.6	100	0.7	100	0.9	100	0.6	100
		The content of different suspension types											
Terrigenous material	Min.	0.004	0.4	0.02	1.4	0.00	0.0	0.01	1.8	—	—	—	—
	Max.	0.24	15.9	0.04	1.8	0.21	23.4	0.12	41.7	—	—	—	—
	Mean	0.04	3.7	0.03	1.6	0.03	4.3	0.05	12.6	0.04	4.5	0.02	3.1
Ferruginous material	Min.	0.002	0.3	0.04	1.8	0.02	2.9	0.03	4.3	—	—	—	—
	Max.	0.59	59.1	0.14	10.0	0.25	35.7	0.36	40.4	—	—	—	—
	Mean	0.08	10.1	0.09	5.9	0.09	15.2	0.12	16.6	0.04	4.8	0.14	21.9
Carbonate material	Min.	0.003	0.4	0.05	2.3	0.0	0.0	0.02	4.3	—	—	—	—
	Max.	0.22	30.3	0.10	7.1	0.20	40.0	0.24	34.8	—	—	—	—
	Mean	0.05	5.1	0.08	4.7	0.06	8.6	0.07	12.7	0.40	45.9	0.02	3.1
Organic material (detritus)	Min.	0.33	37.0	0.65	29.5	0.22	27.7	0.10	30.4	—	—	—	—
	Max.	1.32	96.2	0.65	46.4	0.77	91.2	0.48	85.4	—	—	—	—
	Mean	0.70	78.8	0.65	37.9	0.40	67.7	0.27	50.6	0.38	43.2	0.39	60.0
Siliceous biogenic material	Min.	0.00	0.00	0.49*	35.0	0.00	0.0	0.00	0.0	—	—	—	—
	Max.	0.27	19.0	1.42*	64.5	0.28	35.2	0.20	29.0	—	—	—	—
	Mean	0.03	2.2	0.95	49.6	0.03	4.2	0.05	7.4	0.01	1.6	0.07	10.9

*There are some large spicules in these samples.

mean value being 0.05 to 0.08 mg/l (4.9 to 8.6 per cent of the total suspension). The weight abundance of calcareous suspended matter increases considerably with depth, down to 900 m. At depths of 500 m, the mean weight of calcareous suspensions is 0.07 mg/l, or 12.7 per cent of the bulk suspension. At individual stations, carbonate content values substantially exceed these mean values, and the relative abundance of carbonates in water layers some 100 to 1000 m deep reaches 40 to 46 per cent. In such cases the carbonates form the most abundant type by weight.

The fourth component is terrigenous, noncarbonate matter, averaging 0.03 to 0.04 mg/l (1.6 to 4.3 per cent of the bulk suspension), both in the near-surface layer and in deeper waters.

Siliceous biogenic remains form the least abundant suspended material by weight. They average about 0.02 to 0.03 mg/l (2.2 to 4.2 per cent of the total suspension) in the surface-to-100 m layer. As stated above, samples from stations 815 and 816, containing great amounts of siliceous material, are exceptions to this rule.

GRANULOMETRIC AND MINERALOGIC COMPOSITION OF TERRIGENOUS (NON-CARBONATE) SUSPENSIONS

In the upper layers of the sea, terrigenous particles in the size fractions 0.05 to 0.025 mm, and less than 0.0025 mm, are most abundant by weight. At depths of 100 to 500 m the dominant particles are larger than 0.05 mm (Table 3). On the whole terrigenous particles are rather unevenly distributed across the size spectrum.

For mineralogic purposes we have considered only particles larger than $3\,\mu$. Finer particles are too small to determine.

The Aegean Sea

The most abundant mineral particles here are clay aggregates and non-determinable grains. Their content varies from 1 to 13 per cent of the total suspended particles and averages 5 per cent. Mica plates, usually colorless, yellow and, rarely, green together comprise about 5 per cent, although at Station 708 they amount

Table 3. Vertical size-frequency distribution of terrigenous particles in Mediterranean Sea suspensions.

	The overall content of suspension mg/l	Terrigenous Material		Distribution of particle weight (mg/l) according to fractions					
		mg/l	% of total suspension weight	>0.05 mm	0.05–0.025 mm	0.025–0.01 mm	0.01–0.005 mm	0.005–0.0025 mm	0.0025–0.001 mm
				Depth: 0–2 m, 19 stations					
Min.	0.4	0.004	0.4	0.000	0.000	0.000	0.000	0.000	0.001
Max.	1.7	0.236	15.9	0.000	0.222	0.015	0.016	0.015	0.114
Mean	0.8	0.060	5.9	0.000	0.022	0.003	0.002	0.003	0.010
				Depth: 40–50 m, stations 815, 816					
Min.	1.4	0.02	1.4	0.000	0.000	0.000	0.009	0.003	0.008
Max.	<2.2	0.04	1.8	0.000	0.000	0.016	0.010	0.007	0.008
Mean	1.8	0.03	1.6	0.000	0.000	0.008	0.010	0.005	0.008
				*Depth: 100 m, 12 stations**					
Min.	0.3	0.00	0.0	0.000	0.000	0.000	0.000	0.000	0.000
Max.	0.9	0.21	23.4	0.196	0.000	0.026	0.016	0.017	0.008
Mean	0.6	0.03	4.3	0.016	0.000	0.007	0.003	0.003	0.003
				Depth: 500 m, 3 stations					
Min.	0.3	0.01	1.3	0.000	0.000	0.000	0.000	0.001	0.000
Max.	0.9	0.12	41.7	0.045	0.075	0.003	0.003	0.004	0.003
Mean	0.7	0.05	12.6	0.015	0.025	0.002	0.001	0.002	0.001
				Depth: 900 m, station 806					
	0.9	0.04	4.5	0.000	0.000	0.026	0.004	0.005	0.003
				Depth: 2000 m, station 765					
	0.6	0.02	3.1	0.000	0.011	0.004	0.001	0.003	0.002

*Two stations at 125 m and 165 m in the Adriatic Sea included.

to 15 per cent. The various kinds of mica and their aggregates are often larger than any other particles. In some mica plates we managed to measure the 2V angle (colorless mica in station 715 had 2V = 5°). As separate grains we observed plagioclase, quartz, hornblende with extinction angle of about 5° (Station 706), colorless volcanic glass (Station 715), and various unidentifiable minerals.

The Levant Sea

Suspensions from this area differ from those of the Aegean Sea in having less clay aggregates, usually about 1 to 2 per cent. Different unidentifiable grains and mica minerals, represented by colorless and yellow flakes, predominate. They sometimes amount to 11 per cent (Station 836). Among the other mineral grains quartz, plagioclase, hornblende and monoclinic pyroxene (Station 835) were observed.

Suspensions over the submarine delta of the Nile (stations 838-840) are characterized by the abundance of fine-pelitic (1–3 μ) material, represented by isotropic and, rarely, anisotropic flakes. Stations 841, 842 and 843 show smaller amounts of this pelitic dust, due apparently to its deposition by settling into deeper waters. In the area between Cyprus and Turkey (stations 844 and 847), the content of suspended matter again increases, together with particle size, so that the suspensions do not differ much from those of the Aegean Sea.

Bernard (1959) has found high concentrations of corroded sandy particles of 2 to 30 μ in the southern part of the East Mediterranean Sea (3×10^6 to 175×10^6 particles per liter). These values exceed those along the coasts of Monaco by 100 times and those of Algeria by 3,000 times. Maximum sand-grain concentrations were observed in the 0 to 800 m layer north of Cyrenaica. According to Bernard (1959), the sand grains are eolian and are delivered to the sea from the African deserts.

The Ionian Sea-Sidra Bay

The northern part of the Ionian Sea contains very small quantities of mineral particles. Mica flakes prevail but individual grains of epidote, zoisite, plagioclase, brown hornblende (Station 737), clay aggregate, quartz and various unidentified grains are also found.

In the northern part of the sea and particularly near Calabria the suspension concentration is very low. Mineral particles occur as individual grains and the background of the filter is completely clean. Clayey flakes appear only in the area west of the Peloponnesus. In the Sidra Bay area, increased concentrations of quartz and plagioclase were observed, amounting to 4 per cent of the total suspension (Station 743).

The Adriatic Sea

Particulate content is very low in the southern part of the sea. Pelitic particles are nearly absent and the background of the filter is clean. In the northern part of the sea, pelite appears, but the content is considerably less than in the area off the Nile mouth. These data indicate that a considerable amount of the Po sediment discharge is not transported for great distances within the surface waters. The highest content of suspended matter is observed at Station 816, nearest to the mouth of the Po. Mica is most abundant among the large mineral grains in Adriatic suspensions. Thus, at Station 813 colorless mica plates constituted 8.5 per cent. Occasionally, clayey materials and aggregates occur in large quantities (10 per cent in Station 818). Grains of quartz, plagioclases, colorless garnet (Station 819) and colorless fibrous amphiboles (Station 819), possessing a birefringence of about 0.030 and an extinction angle of about 5°, also occur.

The Tyrrhenian Sea

Colorless and yellow mica plates are most abundant, averaging 4 to 6 per cent of the total suspension. Some grains of epidote and glauconite (Station 762), clay aggregates and indeterminable grains were observed. The amount of mineral suspension in this sea is small.

Algerian-Provençal Basin

In the northern part, the composition of suspended mineral matter is similar to that of the Tyrrhenian Sea. Colorless, yellow and the rare green (Station 784) mica and various unidentified particles (which amount to 24 per cent of the whole suspension) are the dominant constituents in this area. In Station 789 the filter indicates high pelite concentrations.

In the southern part of the basin, especially at inshore stations, the content of mineral suspension was considerably greater than that of the northern part. At stations 764, 765, 778 and 780 the following material was observed: many clay aggregates (up to 4 per cent in Station 764), some grains of plagioclase, quartz and mica flakes and a great number of unidentifiable mineral grains. Very little terrigenous suspension was observed at stations situated far from the shore.

In general, the Mediterranean Sea contains fewer mineral (mainly terrigenous) particles than was believed earlier (Emelyanov, 1962), when organogenic particles were included with mineral particles. Only after using our staining method could we clearly distinguish between these two types of particle.

The very rare occurrence of pyroclastic material in suspension in the Mediterranean Sea should be noted. Volcanic ash was never observed and volcanic glass was found in suspension only once.

ORGANIC MATTER: CONCENTRATION AND COMPOSITION

In 4 separate samples of suspensate from the surface-to-7 m layer, between 7.77 and 17.04 per cent organic carbon (0.010 to 0.015 mg/l, Table 4) was measured. The total amount of organic matter (seston) suspended in the upper 50 m layer was measured by the oxidation method (Sushchenya, 1961a and b). Dry weight amounts of seston in the Mediterranean Sea vary from 206 to 299 mg/m^3 or 0.21 to 0.30 mg/l (Table 5). These values are almost half those of the Black Sea.

The organic matter (ashless) portion amounts to 0.13–0.19 mg/l (Table 5) or 60 to 67 per cent of the total seston, averaging 64 per cent. Assuming mean suspensate content to be 1 mg/l we estimate organic matter to constitute some 13 to 19 percent of the total material in the upper water layer, which is in agreement with data which Lisitsin (1964), obtained by the chemical method in centrifuged samples of suspensions from the Mediterranean Sea (Tables 4, 9). It should be noted, however, that absolute organic carbon concentrations for these samples (0.007–0.020 mg/l, Table 4) are 10 times lower than values obtained by membrane ultra-filtration. Centrifuges probably fail to retain a considerable portion of suspended particles, especially pelitic ones. Thus the data of Sushchenya (1961a and b) are more reliable. However, in Sushchenya's opinion even those data are somewhat

Table 4. Mean content of C_{org} and chlorophyll in Mediterranean Sea suspensions (the layer thickness is 0–7 m) (according to Lisitsin, 1964; Kutyurin and Lisitsin, 1961).

Region	Mean suspension concentration according to the separation method, mg/l	C_{org} % of suspension concentration	C_{org} mg/l	Sum of suspension, biogenic components %	Sum of suspension, biogenic components mg/l	Chlorophyll content mg/l	Chlorophyll content dry weight mg/l	Chlorophyll content % of organic matter	Mean suspension concentration (mg/l) in the same regions according to filtration method (our data)
Levant Sea	0.20	7.77	0.015	12.97	0.026	—	—	—	0.6
Ionian Sea	0.09	11.75	0.011	17.29	0.020	—	0.30	0.15	1.2
Tunisian coast	0.07	14.85	0.010	18.63	0.010	0.03	0.36	0.14	0.90
Alboran Sea	0.09	17.04	0.015	23.09	0.020	0.06	0.71	0.25	1.1
Mediterranean, Mean	0.11	12.90	0.013	12.59	0.019	0.04	0.45	0.18	1.0

underestimated due to the incomplete oxidation of organic matter during analysis. This point is corroborated by subsequent descriptions of the organic detritus.

Suspended organic matter is represented by (a) phytoplankton, (b) zooplankton, and (c) organic detritus. The relative significance of phyto- and zooplankton in the seston content for the surface-to-50 m layer and apparently for the surface-to-200 m layer is comparatively small (Table 6). Phytoplankton masses in the Aegean Sea and in the Ionian Sea, and probably in other parts of the Mediterranean Sea, amount to 4 to 8 per cent of the seston, and zooplankton masses amount to 4 per cent. Thus 92 to 88 per cent of the seston is represented by organic detritus and by plankton undetected by net hauls.

The major specific and biomass group in phytoplankton is represented by dinoflagellates, the second group is the diatoms. Next in importance are the coccolithophorids, silicoflagellates, small flagellates and other forms.

It is known that phytoplankton and zooplankton are concentrated in the uppermost 100 or 200 m of water, the maximum phytoplankton content in the Mediterranean Sea being observed in the surface-to-25 m and in the 75-to-100 m layers. From 200 m downwards, phytoplankton concentration decreases, whereas zooplankton occurs in marked quantities down to depths of 500 to 1000 m.

The distribution of organic detritus is of great importance for determining the mechanism which delivers organic matter into deep layers and to the bottom of the sea. Microscope analyses reveal that detritus composes about 98 to 100 per cent of the

Table 5. Seston and chlorophyll amount and phytoplankton biomass in the upper layer of the Black Sea and the Mediterranean Sea (according to Sushchenya, 1961a and b).

Region	Depth m	Number of samples	Seston amount in 1.0 m³: in mg O_2	Seston: dry weight mg	Seston: ashless matter mg	Chlorophyll amount: in mg/l	Chlorophyll: in g under 1 m²	Chlorophyll: in % of seston	Phytoplankton biomass, mg/m³: dry weight	Phytoplankton biomass, mg/m³: wet weight
Black Sea (pre-Bosporus region)*	0	10	363(221–552)	377	243	1.29(0.60–2.90)	0.065	0.53	43(20–97)	215(100–485)
Sea of Marmora	0	3	225(208–243)	235	151	0.63(0.55–0.70)	0.063	0.39	21(18–23)	105 (90–115)
Aegean Sea	0	13	198(161–296)	206	133	0.44(0.31–0.49)	0.047	0.33	15(10–16)	75 (50– 80)
	50	6	200(149–254)	208	134	0.49(0.31–0.87)		0.33	16(10–29)	80 (50–149)
Levant Sea (south	0	2	198(181–215)	206	133	0.31(0.25–0.36)	0.035	0.23	10(8–12)	50(40–60)
of island of Rhodes)	50	2	223(216–229)	231	149	0.38(0.34–0.42)	—	0.25	13(11–14)	65(55–70)
Ionian Sea	0	8	246(130–319)	256	172	0.25(0.17–0.36)	0.030	0.15	8 (6–12)	40(30–60)
	50	6	226(170–304)	233	151	0.35(0.28–0.42)	—	0.23	12 (9–14)	60(45–70)
Adriatic Sea	0	14	285(150–510)	299	194	0.44(0.25–0.84)	—	0.23	15 (8–28)	75 (40–140)
	50	11	260(175–436)	271	174	0.37(0.25–0.67)	—	0.21	13 (8–22)	65 (40–110)

*The depth of the photosynthetic layer in the Black Sea is accepted at 75 m, and in the Mediterranean Sea, 100 m. Ranges are given in brackets.

seston, although the data of Sushchenya (1961a) indicate somewhat lower values of about 92 per cent.

Organic detritus is represented by various noncarbonate, gray and colorless fibers and grains of irregular and oval shapes. In suspensions containing particles larger than 0.01 mm, colorless and gray, oval and irregular grains are usually predominant. In suspensions of particles 1 to 5 μ in size, various fibers and small oval particles predominate, which are completely invisible in unstained filters.

Table 6. Relative significance of zooplankton and phytoplankton in the seston component for 0–50 m layer (Sushchenya, 1961a).

Region	Seston mg/m^3	Zooplankton		Phytoplankton	
		mg/m^3	% of seston	mg/m^3	% of seston
Pre-Bosporus area of the Black Sea	377	16.8	4.5	43.0	11.4
Aegean Sea	207	8.7	4.0	15.5	7.5
Ionian Sea	245	8.7	3.5	10.5	4.3

Maximum amounts of organic detritus were found at the sea surface (layer 0–2 m., Table 7; Figure 6). Particles coarser than 1 μ in each liter of water numbered 26,000 to 103,000, averaging 45,000, or 8 per cent of total particle numbers at the 19 stations shown in Table 7. Other stations, especially those in the Aegean Sea, showed particle numbers of 200,000 to 300,000. We observed maximum organic detritus in the inshore area of the Mediterranean Sea and in Sidra Bay (two stations which yielded 54,000 and 103,000 particles per liter). Minimum values of about 30,000 were determined in the Levant and in some stations in the northern part of the Ionian and the Adriatic Seas. It is obvious that detritus and plankton are distributed in almost the same manner. Organic detritus in deep layers usually does not exceed 5 to 20 per cent of the total particle number (Tables 1 and 7).

Maximum weights of organic detritus, about 1 mg/l,

Table 7. Number distribution of organic detritus in Mediterranean Sea suspensions (according to particle calculation by microscope).

	Total number of all particles in suspension (× 1000 pc/l)	Number of organic detritus particles		Number distribution of organic detritus according to fraction (× 1000 pc/l)					
		(× 1000 pc/l)	% of total number	> 0.05 mm	0.05–0.025 mm	0.025–0.01 mm	0.01–0.005 mm	0.005–0.0025 mm	0.0025–0.001 mm
Depth: 0–2 m, 19 stations									
Min.	326.1	25.6	2.8	0.0	1.1	3.8	2.2	0.0	0.0
Max.	1261.9	103.3	14.5	4.4	38.6	29.4	39.1	15.5	0.0
Mean	620.7	45.2	7.8	1.2	12.3	14.7	15.5	1.4	0.0
Depth: 40–50 m, stations 815, 816									
Min.	4390.2	182.6	4.2	0.0	15.2	14.1	14.1	0.0	0.0
Max.	5367.9	265.5	4.9	2.2	15.2	16.3	37.0	114.2	217.6
Mean	4879.0	224.0	4.6	1.9	15.2	15.2	25.6	57.1	108.8
Depth: 100 m, 13 stations									
Min.	240.4	18.5	1.2	0.0	2.2	3.3	5.1	0.0	0.0
Max.	1823.5	76.1	14.1	4.4	26.1	21.4	35.9	21.7	0.0
Mean	801.2	42.1	5.9	1.4	8.7	10.7	17.4	4.0	0.0
Depth: 500 m, 4 stations									
Min.	526.4	12.5	2.3	0.0	3.3	3.8	4.0	0.0	0.0
Max.	929.7	49.6	7.2	1.5	7.3	10.1	11.8	23.8	0.0
Mean	714.5	30.3	4.1	0.6	5.1	7.2	7.0	10.4	0.0
Depth: 900 m, station 806									
	1436.3	18.8	1.3	1.6	4.9	4.6	7.6	0.0	0.0
Depth: 2000 m, station 765									
	976.9	35.7	3.7	0.7	6.8	12.6	6.9	0.0	0.0

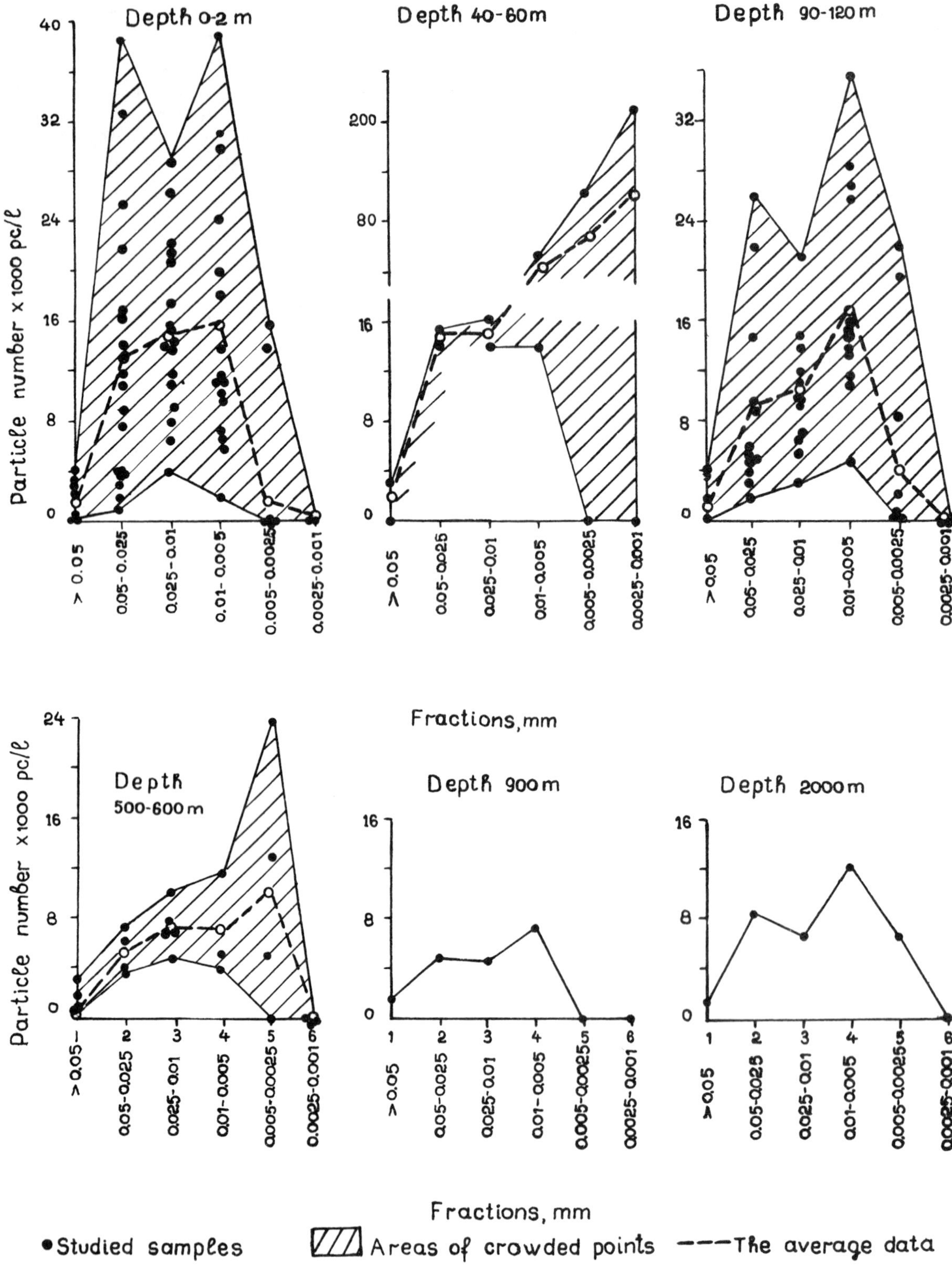

Figure 6. The size-frequency distribution of organic detritus particles in Mediterranean Sea suspensions, according to water depth.

were found in Sidra Bay, in the inshore areas of the Ionian Sea near the Balkan peninsula, in the Messina Strait, and in the northern part of the Adriatic (Table 8). In the central parts of the Mediterranean Sea, the mass of organic detritus is small. In the Levant Sea, comparatively small amounts of organic detritus were observed, usually not exceeding 0.71 mg/l even near the submarine delta of the Nile.

Organic detritus at the sea surface ranges from 37 to 96 per cent of the suspension by dry weight. Maximum percentages were observed in Sidra Bay and in the Adriatic (75 to 96 per cent of the total suspension, mean values being 86 and 83 per cent respectively). In the surface waters of the Levant Sea the percentage of detritus ranges from 37 to 95 per cent of the total suspension with a mean value of about 76 per cent.

In Adriatic waters 40 to 50 m deep, the content of detritus is rather high: 0.65 and 1.22 mg/l (Table 8). With depth, the detritus content tends to decrease and at 100 m depth in the Mediterranean Sea it ranges from 0.22 to 0.63 mg/l, or 46 to 90 per cent of the total mass, mean values being 0.37 mg/l and 68 per cent (Figure 7).

At 500 m depths the content of detritus is still lower: 0.10–0.48 mg/l. Minimum contents of 0.1 and 0.2 mg/l or 32 and 31 per cent of the total suspension occur in the central parts of the Levant Sea and the Algerian-Provençal Basin.

Two determinations in waters deeper than 900 m revealed very small quantities of organic detritus, especially in the open parts of the sea. Inshore along the coasts of Tunisia and in waters 2000 m deep, organic detritus amounts to 0.39 mg/l or 60 per cent of total suspension.

Our calculated values of organic detritus content are twice as large as the data on seston (Tables 5 and 8). It seems that the results of Sushchenya (1961a and b)

Table 8. Weight distribution of organic detritus in Mediterranean Sea suspensions (according to particle calculations by microscope).

	Total content of suspension, (mg/l)	Organic detritus content		Organic detritus distribution according to fractions (mg/l)					
		(mg/l)	% of the total suspension weight	>0.05 mm	0.05–0.025 mm	0.025–0.01 mm	0.01–0.005 mm	0.005–0.0025 mm	0.0025–0.001 mm
Depth: 0–2 m, 19 stations									
Min.	0.4	0.33	37.0*	0.00	0.17	0.02	0.00	0.00	0.00
Max.	1.7	1.32	96.2	0.42	0.90	0.21	0.08	0.001	0.00
Mean	0.9	0.70	78.8	0.11	0.50	0.08	0.01	0.001	0.00
Depth: 40–50 m, stations 815, 816									
Min.	1.4	0.65	29.5	0.00	0.59	0.05	0.01	0.00	0.00
Max.	2.2	0.65	46.4	0.16	0.45	0.05	0.01	0.00	0.00
Mean	1.8	0.65	37.9	0.08	0.52	0.05	0.01	0.00	0.00
Depth: 100 m, 11 stations									
Min.	0.3	0.22	27.7	0.00	0.08	0.01	0.001	0.00	0.00
Max.	0.9	0.77	91.2	0.25	0.45	0.09	0.01	0.001	0.00
Mean	0.6	0.40	67.7	0.09	0.23	0.04	0.004	0.001	0.00
Depth: 500 m, 4 stations									
Min.	0.3	0.10	30.4	0.00	0.09	0.01	0.001	0.00	0.00
Max.	0.9	0.48	85.4	0.15	0.30	0.04	0.003	0.001	0.00
Mean	0.7	0.27	50.6	0.06	0.18	0.03	0.002	0.001	0.00
Depth: 900 m, station 806									
	0.9	0.38	43.2	0.16	0.20	0.02	0.002	0.001	0.00
Depth: 2000 m, station 765									
	0.6	0.39	60.9	0.06	0.30	0.02	0.003	0.001	0.00

*Per cent of the total suspension at this station

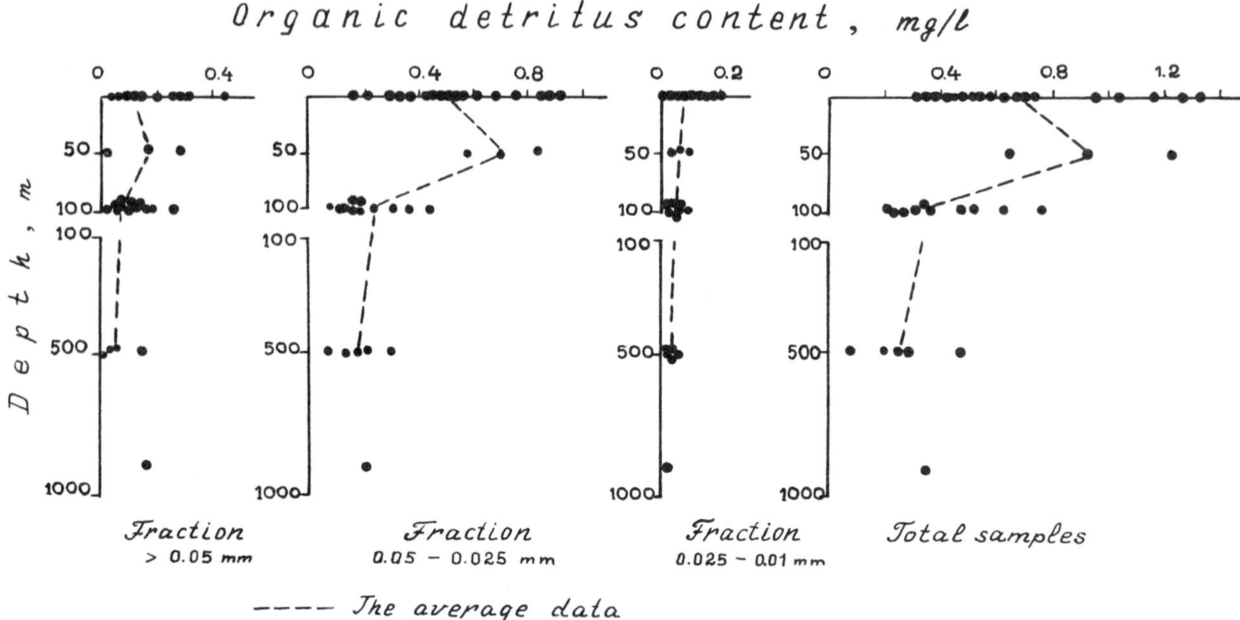

Figure 7. The vertical distribution of organic detritus weight (in mg/l) in Mediterranean Sea suspensions, according to size-fraction.

are actually underestimated as stated earlier. The portion of suspended organic carbon in the upper layers averages 0.70 mg/l.

CARBONATES

According to the data of 4 representative samples (Table 9) obtained by the separation method of Lisitsin (1964) the carbonate content in surface suspensions of the Mediterranean Sea ranges from 2.5 to 5.1 per cent of the total suspensate weight, or 0.002 to 0.009 mg/l. Our calculated data show that carbonate particles at the sea surface constitute from 0.003 to 0.22 mg/l or from 0.48 to 32.2 per cent of the suspension weights (Table 10). The weight of calcareous particles increases with depth, averaging as much as 0.076 mg/l at 40 to 50 m depths, 0.060 mg/l at 100 m and up to 0.097 mg/l at 500 m. This last value tends to remain constant down to the very bottom in the eastern Mediterranean, according to our data. Particles larger than 0.02 mm prevail in weight, except at the deepest station.

Calcareous organogenous material at the surface of the Mediterranean forms a small portion of the total carbonate material, about 20–30 per cent, based on our data. The calcareous material is represented by single 5 to 50 μ grains of shelly detritus, probably plates of pteropod shells and foraminifera. Material composed exclusively of foraminifera is rare. Coccoliths, in contrast, are very numerous and sometimes number several million per liter of water. Coccolith material is most abundant in the Algerian-Provençal Basin.

Calcareous material, presumably biogenic, is often represented by small particles only (less than 3 to 5 μ). Unfortunately this pelitomorphic carbonate was not distinguished from pelite-grade terrigenous and chemogenic calcite.

In the Aegean Sea, abiogenic carbonates are widely distributed. Almost all the filters from this area are covered with calcareous dust which amounts to as much as 53 per cent of the total suspension, but averages about 30 per cent. The dust includes different irregular-shaped grains, flakes, and, in the central part

Table 9. Chemical content of centrifuge-separated suspensions from the upper water layer (0–7 m) of the Mediterranean Sea (according to Lisitsin, 1964).

Chemical components	Levant Sea		Ionian Sea		Tunisian coast		Alboran Sea	
	mg/l	%	mg/l	%	mg/l	%	mg/l	%
$CaCO_3$	0.009	4.3	0.004	5.1	0.002	2.5	0.004	4.4
C_{org}	0.0150	7.8	0.011	11.8	0.010	14.9	0.015	17.0
SiO_2 amorph	0.002	0.9	0.0004	0.5	0.0009	1.3	0.0015	1.7
Fe	0.020	8.6	—	510	—	—	—	—
Mn	0.0001	0.5	—	—	—	—	—	—
TiO_2	0.0018	0.9	—	—	—	—	—	—

Table 10. Distribution of carbonate particles in Mediterranean Sea suspensions.

	Total content of suspension, (mg/l)	Calcareous matter		Distribution of particle weight according to fractions, mg/l					
		mg/l	% of total suspension weight	>0.05 mm	0.05–0.025 mm	0.025–0.01 mm	0.01–0.005 mm	0.05–0.0025 mm	0.0025–0.001 mm
				Depth 0–2 m, 19 stations					
Min.	0.4	0.003	0.4	0.000	0.000	0.000	0.000	0.001	0.000
Max.	1.7	0.22	32.3	0.153	0.039	0.043	0.024	0.031	0.015
Mean	0.9	0.05	5.1	0.008	0.004	0.014	0.005	0.006	0.004
				Depth: 40–50 m, stations 815, 816					
Min.	1.4	0.05	2.3	0.000	0.000	0.016	0.000	0.003	0.016
Max.	2.2	0.10	7.1	0.000	0.000	0.040	0.009	0.020	0.016
Mean	1.8	0.08	4.7	0.000	0.000	0.028	0.004	0.012	0.016
				Depth: 100 m, 12 stations					
Min.	0.3	0.000	0.00	0.000	0.000	0.000	0.000	0.000	0.000
Max.	0.9	0.219	39.79	0.123	0.202	0.017	0.010	0.017	0.012
Mean	0.6	0.060	8.55	0.010	0.033	0.006	0.004	0.004	0.004
				Depth: 500 m, 3 stations					
Min.	0.3	0.015	4.33	0.000	0.000	0.000	0.003	0.002	0.002
Max.	0.9	0.240	34.80	0.070	0.144	0.028	0.011	0.011	0.004
Mean	0.7	0.097	17.09	0.023	0.048	0.009	0.007	0.006	0.003
				Depth: 900 m, station 806					
	0.9	0.399	45.88	0.137	0.226	0.000	0.000	0.016	0.009
				Depth: 2000 m, station 765					
	0.6	0.02	3.83	0.000	0.000	0.007	0.009	0.000	0.004

of the Aegean sea, rare rhombohedra and their various aggregates and chains. Large (aleurite) grains of calcite and dolomite are very rare, forming irregular grains and, very rarely, rhombohedra.

Carbonates are rare in the Levant Sea, as compared with the Aegean Sea. Considerable quantities of carbonate (up to 16 or 18 per cent of the total number of suspended particles) are found only in the area between Turkey and Cyprus. They are represented by various irregular grains and fragments. At Station 847 grains of calcite form aggregates and chains in the shape of stars.

In the Ionian Sea-Sidra Bay area, suspended carbonate occurs in large quantities (9 to 21 per cent at stations 304 and 739). In Sidra Bay calcareous dust covers almost the entire background of the filters. Large crystals and irregular calcite grains and rounded rhombohedral grains of dolomite often occur in this area.

At stations 436, 354, and 752, east of Malta, calcareous dust also covers almost the whole filter background. Regular rhombohedra of calcite and dolomite, forming stars and chain aggregates, are often seen among large grains. Calcareous material in this area averages 35 per cent of the total particulate content.

Pelitomorphic calcite in the Adriatic is virtually absent. Calcite, in the form of irregular fragments or sometimes 3 to 5 μ rhombohedra, comprises 5 to 11 per cent of the suspensate in the central part of the Adriatic (stations 819, 820, 813). In the Tyrrhenian Sea, suspended carbonates forming 1 to 5 μ grains occur in small amounts. Single grains of calcite, dolomite and siderite were distinguished.

Great quantities of pelitomorphic material occur in the northern and southern parts of the Algerian-Provençal Basin, mainly in inshore areas. South of Marseilles the content attains 17 per cent, and near the Algeria-Tunisia coast, about 12 to 14 per cent (stations

764, 765, 778, 780). In every inshore area we observed great concentrations of large calcite fragments, amounting to 8 per cent at Station 765. In the central part of the Algerian-Provençal Basin, suspended carbonate is negligible.

Microscopic studies show that carbonates at the surface of the Mediterranean Sea occur in quantities greater than expected in view of the data determined from separation samples (Table 9). Previous values can probably be attributed to the fact that many small particles of calcareous dust are not retained by centrifuges.

Calcareous dust particles (1–3 μ) and larger grains of calcite originate from two sources: terrigenous or chemogenic. Hexagonal or rounded shapes of some fragments indicate a terrigenous origin. Such hexagonal and rounded grains are characteristic of the Algerian-Provençal Basin, the Adriatic and Sidra Bay. Ideally shaped crystals and their star and chain aggregates point to a chemogenic origin. Such crystals are most abundant in the southern part of the Ionian Sea-Sidra Bay, near Malta and in the Aegean Sea.

Some carbonate dust is probably formed by the destruction of shelly material. Unfortunately, it is still impossible to distinguish such particles from terrigenous or chemogenous pelitomorphic calcite.

High concentrations of carbonate also were observed in the deep layers of the eastern Mediterranean Sea. These concentrations are best represented by the calcareous dust in the Sidra Bay area where carbonates constitute 30 to 50 per cent of total suspended matter.

Our data indicate enrichment of suspensions in carbonate formed by organic and chemical processes operating at depth. This enrichment continues in the upper layer of bottom sediments due to diagenesis (Emelyanov, 1965). Thus, bottom sediments in the Mediterranean Sea generally contain from 30 to 92 per cent calcium carbonate.

SILICA

The concentration of total suspended silica indicated by our data varies from trace amounts to 207μ g/l, or up to 47.5 per cent of total particulate weight (Table 11). The proportion of amorphous silica in suspension is smaller, from traces to 156μg/l, or up to 39.0 per cent of the total suspensate. The highest concentrations of amorphous and total silica were observed at stations 1282 and 1288 in spring time. This phenomenon seems to be caused by the intensive development of plankton, especially of diatoms, in these regions. This is confirmed by the fact that in samples taken in spring, most of the total silica is represented by the amorphous form. In contrast, samples taken in autumn are relatively deficient both in amorphous and in total silica. Microscopic data indicate a total content of amorphous silica in surface waters ranging from 0 to 19.0 per cent (Table 12). Silica particles usually are coarser than 0.05 mm. Smaller particles prevail only at the surface.

Amorphous silica is represented by shells and fragments of diatoms, radiolarians and flagellates. Material composed entirely of shell is rarely observed on the filters but where present the shell consists mostly of radiolarian tests. Among skeletal remains the most abundant are siliceous spines. Radiolaria and spines are widely distributed in the Aegean Sea and in the Adriatic, that is, in the northern areas of the basin. Here the spines constitute 5 to 15 per cent of the total suspended matter. The length of the spines reaches 80μ and they may be as much as 6μ thick. Sometimes fragments of thick siliceous spicules and other siliceous organic fragments are observed on the filters.

In the deeper layers (more than 500–1000 m) of the Mediterranean Sea, the enrichment of suspensions by precipitation of silica was not observed; both here and near the surface of the sea crystalline silica occurs very rarely and in single specimens.

The paucity of suspended authigenic silica is fully confirmed by microscopic studies. The small numbers of siliceous remains in suspension account for the small quantities ($<$ 1 per cent) of amorphous silica in the bottom sediments (Emelyanov, 1966).

IRON

Total iron content, based on analyses of one separation sample, was 8.7 per cent of the total suspension, or 20 μg/l (Table 9). Colorimetric studies of filtration samples show that iron is only 0.7 to 14.2 per cent of the total suspensate or 1.5 to 43.0 μg/l (Table 11). The average content of iron in most of the analyzed samples varies from 1 to 5 per cent or from 5 to 10 μg/l.

In the Mediterranean Sea, as in the oceans and other seas, suspended iron occurs in two basic forms: (a) absorbed by organogenous material (organic detritus), and (b) mineral particles.

Ferruginous mineral particles amount to about 90 to 95 per cent of the total suspended iron in the samples studied. These particles occur predominantly in the 0.05 to 0.005 mm fraction (Table 13). Most pelite particles contain iron as small globules of hydrous ferric oxide. With an increase in the diameter of the particles the admixture of iron in them falls sharply and the element is almost entirely lacking in size-fractions larger than 0.05 mm. In these fractions, ferruginous particles occur only as single grains.

The content of suspended ferruginous particles

Table 11. Chemical composition of suspended matter in the Mediterranean Sea.

Station	Coordinates	Depth m	Suspension concentration (mg/l)	SiO_2 total µg/l	SiO_2 total %	SiO_2 amorph µg/l	SiO_2 amorph %	Fe µg/l	Fe %	Mn µg/l	Mn %	Ti µg/l	Ti %	Ni × 10^{-2} µg/l	Ni × 10^{-2} %	Co × 10^{-3} µg/l	Co × 10^{-3} %
Suspended matter collected on a membrane filter with 0.5 µ pores (October—November, 1964)																	
1319	34°24′2 N	0	1.7	4	0.2	1	0.1	—	—	—	—	—	—	—	—	—	—
	28°25′1 E																
1320	32°27′2 N	0	0.2	18	7.2	9	3.6	16.6	6.6	0.08	0.032	0.10	0.04	—	—	3.4	1.4
	27°39′8 E	40	0.3	35	11.3	3	1.0	8.8	2.8	0.01	0.003	0.33	0.11	—	—	—	—
		65	0.2	27	12.3	10	4.5	1.5	0.7	0.06	0.027	0.02	0.01	—	—	8.3	3.8
		100	0.9	70	7.7	6	0.7	8.2	0.9	0.13	0.014	0.20	0.02	—	—	3.1	0.3
		1000	0.7	17	2.3	14	1.9	4.8	0.7	0.02	0.003	0.31	0.04	2.0	0.3	5.9	0.8
		3000	0.2	traces		traces		6.1	2.4	0.05	0.020	0.05	0.02	—	—	5.1	2.0
1321	32°25′7 N	0	2.2	10	3.3	traces		10.2	0.5	0.11	0.005	0.50	0.02	—	—	—	—
	26°38′8 E																
1321–1	33°11′0 N	0	0.1	3	5.0	3	5.0	8.5	14.2	0.025	0.042	0.15	0.25	—	—	—	—
	20°56′7 E																
1328	33°34′9 N	0	0.9	14	1.6	1	0.1	27.3	3.1	0.13	0.015	1.10	0.12	1.3	0.2	6.8	0.8
	20°46′6 E	100	0.3	traces		traces		4.7	1.4	0.01	0.003	0.30	0.09	—	—	—	—
		1000	0.6	11	1.9	traces		4.4	0.7	0.04	0.007	0.06	0.01	—	—	—	—
		2500	0.4	16	3.8	11	2.6	7.2	1.7	0.09	0.021	0.73	0.17	1.6	0.4	7.0	1.7
1331	35°05′2 N	0	0.4	—	—			43.0	11.6	0.10	0.027	0.55	0.15	2.4	0.6	3.5	0.9
	18°17′5 E																
1337	38°06′6 N	0	0.6	14	2.2	14	2.2	35.8	5.6	0.46	0.072	0.50	0.08	1.0	0.2	10.5	1.6
	20°20′0 E																
1338	35°55′6 N	0	1.4	60	4.3	60	4.3	10.8	0.8	0.09	0.006	0.55	0.04	—	—	—	—
	22°15′2 E																
Suspended matter collected on a membrane filter with 0.3 µ pores (March-May, 1964)																	
1259	Aegean	0	0.3	—	—	—	—	10.0	3.3	0.01	0.004	0.55	0.18	—	—	—	—
	Sea	100	0.4	—	—	—	—	16.8	4.2	0.12	0.031	0.15	0.04	—	—	—	—
		500	0.3	—	—	—	—	7.0	2.3	0.12	0.042	<0.02	<0.01	—	—	—	—
		1000	0.1	—	—	—	—	5.5	5.5	0.10	0.100	<0.02	<0.02	—	—	—	—
		2000	0.2	—	—	—	—	5.5	2.8	0.11	0.056	<0.02	<0.02	—	—	—	—
1282	35°58′0 N	0	0.2	207	41.4	120	24.0	10.5	5.2	0.02	0.012	0.55	0.28	—	—	—	—
	25°03′7 E																
1285	38°38′4 N	0	0.4	101	15.5	30	4.6	5.0	1.1	0.02	0.005	1.05	0.23	—	—	—	—
	15°05′0 E																
1287	40°52′2 N	0	0.4	—	—	—	—	10.0	2.5	0.01	0.003	0.80	0.20	—	—	—	—
	10°34′7 E	100	0.5	—	—	—	—	6.0	1.2	0.08	0.016	0.75	0.15	—	—	—	—
		500	0.2	—	—	—	—	5.8	2.9	0.01	0.006	0.55	0.28	—	—	—	—
		1000	0.3	—	—	—	—	9.0	3.0	0.01	0.004	0.80	0.27	—	—	—	—
		2000	0.1	—	—	—	—	5.8	5.8	0.01	0.012	0.30	0.30	—	—	—	—
1288	43°24′6 N	100	0.4	190	47.5	156	39.0	—	—	—	—	—	—	—	—	—	—
	08°59′5 E	500	0.6	154	25.3	139	23.2	—	—	—	—	—	—	—	—	—	—
		1000	0.3	137	45.7	—	—	—	—	—	—	—	—	—	—	—	—
		1500	0.2	37	18.5	26	13.0	—	—	—	—	—	—	—	—	—	—
1289	39°16′7 N	10	0.2	—	—	—	—	7.9	4.5	0.04	0.025	0.02	<0.01	—	—	—	—
	15°02′7 E																
1292	35°56′1 N	0	0.2	—	—	—	—	8.5	4.2	0.12	0.062	0.05	0.02	—	—	—	—
	17°26′1 E	100	0.1	—	—	—	—	10.0	10.0	0.15	0.150	—	—	—	—	—	—
		500	0.1	—	—	—	—	6.8	6.8	0.16	0.162	0.12	0.12	—	—	—	—
		1000	0.1	—	—	—	—	4.2	4.2	0.16	0.162	<0.02	<0.02	—	—	—	—
1302	36°10′2 N	0	0.3	82	27.3	43	14.3	—	—	—	—	—	—	—	—	—	—
	25°17′3 E	100	0.2	73	36.5	34	17.0	—	—	—	—	—	—	—	—	—	—
		500	0.1	35	35.0	9	9.0	—	—	—	—	—	—	—	—	—	—

Table 12. Distribution of siliceous particles in Mediterranean Sea suspensions.

	Total content of suspension, (mg/l)	Content of siliceous material		Distribution of particle weight according to fractions, (mg/l)					
		mg/l	% of total suspension weight	>0.05 mm	0.05–0.025 mm	0.025–0.01 mm	0.01–0.005 mm	0.005–0.0025 mm	0.0025 0.001 mm
				Depth: 0–2 m, 19 stations					
Min.	0.4	0.00	0.0	0.000	0.000	0.000	0.000	0.000	0.000
Max.	1.7	0.27	19.0	0.000	0.120	0.020	0.027	0.185	0.054
Mean	0.9	0.03	2.2	0.000	0.006	0.006	0.002	0.010	0.003
				Depth: 40–50 m, stations 815, 816					
Min.	1.4	0.49	35.0	0.450*	0.000	0.018	0.006	0.003	0.008
Max.	2.2	1.42	64.5	1.371*	0.000	0.027	0.006	0.005	0.009
Mean	1.8	0.95	49.6	0.910	0.000	0.022	0.006	0.004	0.009
				Depth: 100 m, 12 stations					
Min.	0.3	0.00	0.0	0.000	0.000	0.000	0.000	0.000	0.000
Max.	0.9	0.28	35.2	0.262	0.000	0.050	0.001	0.000	0.000
Mean	0.6	0.03	4.2	0.022	0.000	0.010	0.000	0.000	0.000
				Depth: 500 m, 3 stations					
Min.	0.3	0.00	0.0	0.000	0.000	0.000	0.000	0.000	0.000
Max.	0.9	0.20	29.0	0.176	0.024	0.002	0.000	0.000	0.000
Mean	0.7	0.05	7.4	0.059	0.008	0.001	0.000	0.000	0.000
				Depth: 900 m, station 806					
	0.9	0.01	1.6	0.000	0.000	0.014	0.000	0.000	0.000
				Depth: 2000 m, station 765					
	0.6	0.07	10.9	0.050	0.020	0.000	0.000	0.000	0.000

*There are some large spicules in these samples

larger than 1μ varies from 1 to 591 μg/l (on average, about 33 to 90 μg/l) in the Mediterranean Sea, as determined by microscopic calculations (Table 13). The percentage of these particles in the whole suspensate is about 5 to 15 per cent. However, if iron absorbed by organogenous material is taken into account, ferruginous contents are slightly higher, with values of about 60 μg/l. Mineral ferruginous particles are represented by two types: (a) translucent brown, red and brown-yellow forms of hydrous ferric oxide, and (b) black opaque particles. Hydrous ferric oxides are widely distributed in the Mediterranean Sea and represent a considerable part of the suspensions. They are observed throughout the whole suspended grain size spectrum, but their number increases progressively with decrease in particle dimensions. However, sometimes the dimensions of hydrous oxide particles reach 0.15 mm, as at Station 825. Black ore minerals also occur widely, usually forming single grains, but this type is never abundant. In the Aegean Sea they amount to 1 or 2 per cent, and their dimensions reach 0.17 mm (Station 708).

MANGANESE, TITANIUM, NICKEL AND COBALT

Data on these minor elements are shown in Table 11. As can be seen, the Mediterranean Sea suspensions are poor in manganese, titanium, nickel and cobalt. Relative amounts of these elements are usually smaller than in the bottom sediments. Data for the Baltic Sea (Emelyanov, 1968 and 1969) and the Indian Ocean (Lisitsin, 1964) show that manganese, titanium

Table 13. Distribution of ferruginous particles in Mediterranean Sea suspensions.

	Total content of suspension (mg/l)	Content of ferruginous material		Distribution of particle weight according to fractions (mg/l)					
		mg/l	% of total suspension weight	>0.05 mm	0.05–0.025 mm	0.025–0.01 mm	0.01–0.005 mm	0.005–0.0025 mm	0.0025–0.001 mm
				Depth: 0–2 m, 19 stations					
Min.	0.4	0.001	0.3	0.000	0.000	0.000	0.000	0.000	0.000
Max.	1.7	0.59	59.1	0.153	0.137	0.571	0.027	0.001	0.000
Mean	0.9	0.08	10.1	0.008	0.013	0.053	0.008	0.000	0.000
				Depth: 40–50 m, stations 815, 816					
Min.	1.4	0.04	1.8	0.000	0.000	0.024	0.013	0.000	0.000
Max.	2.2	0.14	10.0	0.000	0.000	0.081	0.056	0.000	0.000
Mean	1.8	0.09	5.9	0.000	0.000	0.052	0.034	0.000	0.000
				Depth: 100 m, 12 stations					
Min.	0.3	0.02	2.9	0.000	0.000	0.000	0.004	0.000	0.000
Max.	0.9	0.25	35.7	0.000	0.126	0.125	0.134	0.003	0.000
Mean	0.6	0.09	15.2	0.000	0.018	0.048	0.023	0.000	0.000
				Depth: 500 m, 3 stations					
Min.	0.3	0.03	4.3	0.000	0.000	0.011	0.002	0.000	0.000
Max.	0.9	0.36	40.4	0.087	0.195	0.025	0.011	0.004	0.000
Mean	0.7	0.12	16.6	0.022	0.056	0.031	0.005	0.001	0.000
				Depth: 900 m, station 806					
	0.9	0.04	4.8	0.000	0.000	0.038	0.004	0.000	0.000
				Depth: 2000 m, station 765					
	0.6	0.14	21.9	0.000	0.114	0.018	0.006	0.001	0.000

and nickel particles are most common in the terrigenous type of suspension; cobalt in suspensions rich in organogenous detritus. These conclusions are confirmed in general by the Mediterranean Sea data.

CONCLUSIONS

The above observations are summarized as follows:

1. Average suspension amounts in the upper layers of the Mediterranean Sea vary from 0.2 to 4.6 mg/l and in deeper levels from 0.1 to 2.5 mg/l. Minimum concentrations were observed in the open areas of the basin, especially in the arid zone, with a maximum near the mouth of the Nile. In near-bottom waters some increase in suspension concentration was observed as compared with intermediate depths.

2. Total numbers of suspended particles in each liter of surface water range from 326,100 to 1,261,900 particles and sometimes reach 5×10^6. In descending order, the highest proportion of particles belong to terrigenous, noncarbonate suspensates (386,400 particles per liter on average), the second to carbonate material (197,200 pc/l on average), next organic detritus (45,200 pc/l on average), and biogenic siliceous remains are least abundant (2,500 pc/l on average). At deeper levels, the calcareous suspension content increases sharply and becomes predominant, while there is a decrease in organogenous matter and terrigenous noncarbonate material.

3. By weight distribution, organogenous matter is predominant, ranging from 0.30 to 1.32 mg/l (0.70 mg/l on average) or about 37 to 96 per cent of the total suspension near the surface of the sea. From 88 to 92 per cent of the suspended organic matter is repre-

sented by organic detritus (grains from 0.05 to 0.005 mm). Organogenous matter content decreases with depth: in the 100 m layer it averages 0.37 mg/l or 68 per cent; in the 500 m layer, 0.27 mg/l or 43 per cent.

4. The quota of carbonate material in the surface water is 0.48 to 32.2 per cent of the total suspension (0.003 to 0.22 mg/l). The weight of calcareous material increases with depth. Carbonates are represented by coccoliths and shelly detritus (grains from 5 to 50 μ), terrigenous irregular-shaped grains, chemogenic calcite crystals and regular rhombohedra of dolomite. Carbonates of chemogenic origin prevail in the Sicily-Malta-Sidra Bay area and in the Aegean Sea. Terrigenous carbonates are most abundant in the Adriatic and Aegean Seas and in Sidra Bay.

5. The amount of terrigenous non-carbonate particles in the surface waters of the Mediterranean Sea ranges from 58,500 to 1,103,300 pc/l or from 0.004 to 0.24 mg/l. In the deeper layers, suspended terrigenous matter is somewhat less abundant. Terrigenous particle size ranges from 1 μ to 0.17 mm. Indeterminable grains, clay aggregates, mica and quartz are most common. Plagioclase, hornblende, epidote, zoisite, brown hornblende, colorless fibrous amphiboles, garnet and monoclinic pyroxene occur in smaller amounts. These minerals are also the most common in bottom sediments. The great amount of mica in suspension and in the bottom sediments is one of the characteristics of the Mediterranean Sea.

6. Suspended volcanic products are rather rare in the Mediterranean Sea.

7. The concentration of total suspended silica varies from traces to 207 μg/l or up to 47.5 per cent of total particulate weight, and suspended amorphous silica reaches 156 μg/l or 39.5 per cent. The highest concentrations of SiO_2 were determined in spring, the lowest in autumn. Amorphous silica is represented by shells and fragments of diatoms, radiolarians, flagellates, and spines.

8. Total iron content varies from 1.5 to 43.0 μg/l or from 0.7 to 14.2 per cent. Iron occurs in two forms: (a) mineral particles (about 90 to 95 per cent of total iron) and (b) absorbed by organic detritus. Mineral particles are represented by small globules of hydrous ferric oxide and black opaque particles.

9. Suspended manganese content varies from 0.01 to 0.46 μg/l or 0.003 to 0.162 per cent, titanium from 0.02 to 1.10 μg/l or <0.01 to 0.30 per cent, nickel from 0.010 to 0.024 μg/l or 0.002 to 0.006 per cent, cobalt from 0.0031 to 0.0105 g/l or 0.0003 to 0.0038 per cent. Manganese, titanium and nickel are most common in the terrigenous type of suspended matter, and cobalt in the organogenous material.

ACKNOWLEDGMENTS

The authors express their gratitude to Drs. A. P. Lisitsin, G. Kelling, J. W. Pierce, K. S. Rodolfo and D. J. Stanley for their critical review of the manuscript and for offering useful suggestions improving the text.

REFERENCES

Bernard, F. 1959. Données sur l'abondance de sable corrodé dans les eaux de la Méditerranée orientale. *Revue de Géographie Physique et Géologie Dynamique*, 2:113–119.

Bogdanov, Yu, A. 1965. *The formation and distribution of the biogenic sedimentary material in the Pacific Ocean* (in Russian). Unpublished Thesis, Institute of Oceanology, Moscow: 25 p.

Emelyanov, E. M. 1962. Some data on the Black and Mediterranean seas suspension. *Okeanologiya*, 2:664–672.

Emelyanov, E. M. 1965. The carbonate content of modern bottom sediments of the Mediterranean Sea (in Russian). In: *The Main Features of the Geological Composition, the Hydrological Regime and the Biology of the Mediterranean Sea*, ed. Fomin, L., Nauka, Moscow, 71–84.

Emelyanov, E. M. 1966. Distribution of authigenous silica in suspension and in sediments of the Mediterranean Sea (in Russian). *Geokhimiya Kremnezena (Geochemistry of Silica)*: 284–294.

Emelyanov, E. M. 1968. Study trace amounts of iron, manganese and titanium in the suspension of the Baltic Sea (in Russian). *Litologia i Poleznye Iskopayeme (Lithology and Mineral Resources)* 6:43–52.

Emelyanov, E. M. 1969. Microquantities of nickel and cobalt in suspension in the Baltic Sea and the Atlantic Ocean (in Russian). *Geokhimiya*, 10:1269–1273.

Emelyanov, E. M. 1972. Principal types of recent bottom sediments in the Mediterranean Sea: their mineralogy and geochemistry. In: *The Mediterranean Sea: A Natural Sedimentation Laboratory*, ed. Stanley, D. J., Dowden, Hutchinson and Ross, Inc., Stroudsburg, Pennsylvania, 355–386.

Emelyanov, E. M., I. K. Blazhis, R. I. Yuryavichyus, R. J. Paeda, Ch. A. Valyukyavichus, and I. I. Yankauskas 1971. The minor elements determination in sea water and suspension. The determination of iron, cobalt and titanium in suspension (in Russian). *Okeanologiya*, 11:1116–1125.

Jerlov, N. G. 1953. Particle distribution in the ocean. *Reports of the Swedish Deep-Sea Expedition, 1947–1948*, 3 (3):73–125.

Jerlov, N. G. 1958. Distribution of suspended material in the Adriatic Sea. *Archivio di Oceanografia e Limnologia, Venezia*, 11 (2):227–250.

Kutyurin, V. M. and A. P. Lisitsin 1961. Vegetable pigments in the suspended material and bottom sediments of the Indian ocean (in Russian). *Oceanological Research, X Section of IGY Program*, Publications of the Academy of Sciences USSR, Moscow, 3.

Kuznetsov, S. I. 1949. The application of microbiological methods to the study of organic matter in basins (in Russian). *Mikrobiologia*, 18:43–52.

Lisitsin, A. P. 1961a. The distribution and composition of suspension in the Indian Ocean water (in Russian). *Okeanologicheckiye Issledovaniya*, 3:59–89.

Lisitsin, A. P. 1961b. The distribution and composition of suspended matter in seas and oceans (in Russian). In: *The Recent Sediments of Seas and Oceans*. Publications of Academy of Sciences USSR, Moscow, 175:231.

Lisitsin, A. P. 1964. Distribution and chemical composition of suspended matter in the waters of the Indian Ocean (in Russian).

In: Oceanological Research. X Section of IGY Program, Publications of the Academy of Sciences USSR, Moscow, 10: 1–136.

Ochakovsky, Yu. E. 1965. On dependence of the release index on suspensions contained in the sea (in Russian). *Trudy Instituta Okeanologii AN SSSR*, 77:35–70.

Rakestraw, N. W. 1958. Particulate matter in the oxygen-minimum layer. *Journal of Marine Research*, 17:429–431.

Sushchenya, L. M. 1961a. Some data about seston amount in waters of the Aegean, Ionian and Adriatic seas (in Russian). *Okeanologiya*, 1:664–669.

Suschchenya, L. M. 1961b. The chlorophyll content in plankton of the Aegean, Ionian and Adriatic Seas (in Russian). *Okeanologiya*, 1:1039–1045.

Yatsimirsky, K. B., E. M. Emelyanov, V. K. Pavlova and Ya. S. Savichenko 1970. Determination of microquantities of nickel and cobalt in small weighed portions of marine suspended matter (in Russian). *Okeanologiya*, 10:1111–1117.

Yatsimirsky, K. B., E. M. Emelyanov, V. K. Pavlova and Ya. S. Savichenko 1971. The determination of minor elements of copper and manganese in small samples of sea suspension (judging by the Baltic Sea and Atlantic Ocean) (in Russian). *Okeanologiya*, 11:730–735.

Mineralogical Differentiation of Sediments Dispersed from the Po Delta

Bruce W. Nelson

University of South Carolina, Columbia, South Carolina

ABSTRACT

The Po River carries sediment that consists of montmorillonite, muscovite, chlorite, serpentine, quartz, feldspar, calcite, and dolomite. After dispersal in the Adriatic Sea, fine-grained suspensions are rich in muscovite, chlorite, and serpentine, while montmorillonite is deficient. Montmorillonite disappears from the saline suspensions very close to the delta and must be deposited at the mouths of distributaries. The rates of change in sediment concentration in front of the delta conform to those expected by simple gravitational settling. Therefore, the mineralogical differentiation that occurs in the sediment suspensions apparently is not caused by flocculation effects that might be expected to occur when fresh-water suspensions mix with sea water. Bottom sediments in the Adriatic Sea become enriched in montmorillonite along vectors away from the Po Delta beyond the pro-delta slope. The differentiation that occurs in bottom sediments is not related to that occurring in the sediment suspensions as they move away from the delta. Bottom currents must re-work the pro-delta slope sediments and increase the proportion of fine-grained minerals—principally montmorillonite—at greater distances from the delta.

High magnesian calcite occurs in bottom sediments of the Adriatic Sea below a depth of 150 meters. It must be derived from benthic organisms that live in the deeper waters of the basin.

RESUME

Le Pô transporte des sédiments qui sont constitués de montmorillonite, muscovite, chlorite, serpentine, quartz, feldspath, calcite et dolomite. Après dispersion dans l'Adriatique, les suspensions à grains fins sont riches en muscovite, chlorite, et serpentine et ne contiennent pas de montmorillonite. Celle-ci disparaît des suspensions salines très près du delta et est déposée aux embouchures des affluents. Les variations dans les concentrations de sédiments en avant du delta sont conformes à celles rencontrées dans les dépôts formés sous l'action de la force de pesanteur. En conséquence, la variation minéralogique au sein des suspensions de sédiments n'est pas due à l'effet de floculation contrairement à ce qu'on peut attendre lorsque les suspensions en eau douce se mèlent à l'eau de mer. Les sédiments du fond de l'Adriatique s'enrichissent en montmorillonite au long de vecteurs ayant pour origine le delta du Pô et s'étendant au-delà de la pente pro-delta. Les variations qui ont eu lieu dans les sédiments de fond n'ont aucun rapport avec celles des suspensions qui s'éloignent du delta. Les courants de fond agissent sur les sédiments de la pente pro-deltaique accroissant ainsi la proportion des minéraux à grains fins, surtout montmorillonite, à une plus grande distance du delta.

La calcite à haute teneur en magnésium des sédiments de fond apparaît à une profondeur de 159 mètres. Elle doit provenir d'organismes vivant dans les eaux plus profondes du bassin.

INTRODUCTION

The deltas of the Mediterranean Sea provide classical examples of precipitation of river-borne sediment by sea water. The observation that large Mediterranean rivers deposit their sediment loads almost immediately upon entering the sea has led to the conclusion that the sediment carried in suspension by the river is deposited after flocculation with sea salt. Study of the mineralogy of suspended and bottom sediments shows, however, that in the case of the Po Delta flocculation in the sea seems not to be important in determining the size range and composition of sediments distributed into the Adriatic.

Instead, normal gravity mechanisms and circulation patterns explain the essential features of both horizontal and vertical suspended sediment concentrations and compositions near the delta. The variations in bottom sediment mineralogy similarly depend on bottom current movements and size sorting.

DELTA FRONT HYDROGRAPHY AND THE PATTERNS OF SEDIMENT DISPERSAL

The relation between sediment dispersal and hydrography at the mouths of the deltaic distributaries of the Po has been described earlier (Nelson, 1970). A highly stratified salt wedge occupies the mouths of the distributaries at high tide where it is arrested by fresh-water river flow. At low tide the river extrudes the salt wedge to a position seaward of the river mouth bar. In contrast to the small tidal range in much of the Adriatic, the mean tidal range at the delta front is about 60 centimeters. The tidal energy causes the river and sea water to mix together in a characteristic pattern close to the front of the delta. At low tide both the suspended and saltation sediment load of the river are transported directly into the sea; the coarsest particles are dropped on the seaward slope of the delta front platform. Even at high tide, the river generally is able to transport all of its load directly into the sea, because the terminal flow velocities allow the river to move the coarsest particles over the salt wedge and eject them into the sea.

A study of the distribution of temperature, salinity, and suspended sediment during the spring discharge period of 1963 led to the picture of inferred water movements in front of the delta illustrated by Figure 1. At the surface a thin layer of low salinity water

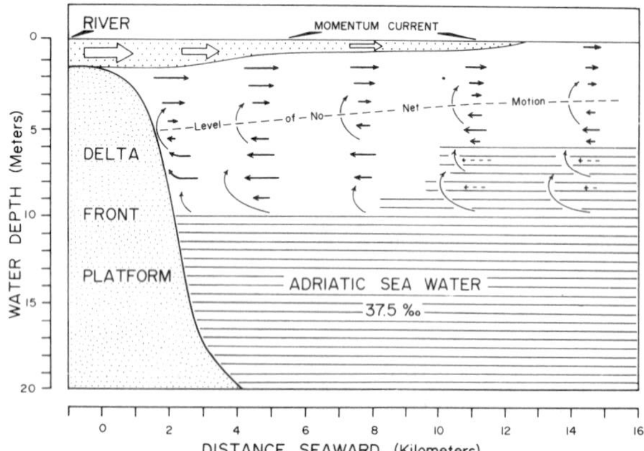

Figure 1. Circulation in front of the Po Delta at high fresh-water discharge. Magnitude and direction of arrows indicate inferred net water movements. Open arrows indicate significantly higher velocities. From field data collected in April and May, 1963.

(momentum current) issues from the distributary with velocities initially of 100 centimeters/second. At peak fresh-water discharges they may be higher. The momentum current may extend as far as 25 km beyond the delta, but as it moves seaward it gradually loses momentum, drops suspended sediment, and gains salinity (by means of advective mixing with the water below). At intermediate depths close to the delta, but extending to the surface beyond the limits of the momentum currents, lies a water layer of intermediate salinity (25 to 30‰). This layer shows a density-driven, two-layer, circulation pattern. Above the level of no net (horizontal) motion a current moves away from the delta with velocities that are modest compared to the surface momentum currents but competent to transport significant quantities of suspended sediment; below the level of no net motion a slow current moves towards the delta supplying salt water to the upper layer by vertical advective mixing. This two-layer density current is the primary agent that disperses suspended sediment and freshwater away from the distributaries. Below a depth of 10 m is a reservoir of high salinity Adriatic Sea water which does not participate directly in the dispersal of suspended sediment, although tidal and longshore currents may redistribute sediment that initially is deposited upon the sea bed.

Suspended sediment concentrations in the surface layer immediately in front of the delta at peak freshwater discharges reach 350 mg/l, but normally concentrations of 100 to 150 mg/l occur. Within a momentum current sediment concentration increases from the surface to the base, where a thin zone of turbulent eddies occupies the transition from surface to intermediate salinity water. This suggests that the normal sediment concentration gradient of the freshwater stream has been transferred more or less intact to the momentum current beyond the discontinuity of the delta front. As momentum decreases seaward the eddies dissipate, the ability of the surface current to transport coarse particles decreases, and ultimately its competence is lost.

Within the upper part of the intermediate salinity water, suspended sediment concentrations decrease from 10 to 20 mg/l near the surface to 1 to 5 milligrams/liter near the level of no net motion. The competence of this layer probably decreases with depth and with distance seaward of the delta.

Within the high salinity deep water, sediment concentrations usually are 1 mg/l or less, but close to the delta front concentrations up to 5 mg/l are common. These concentrations probably result from particles that have dropped out of the surface and intermediate layers and are settling to the bottom where they will accumulate on the pro-delta slope.

Suspended sediments discharged by the Po are caught up by the prevailing surface drift currents, which carry them down the Adriatic along the Italian coast. These drift current have velocities of about 50 cm/sec and seem competent to transport a certain fraction of the suspended load until the open water of the Mediterranean Sea is reached. The region south of the Po Delta thus lies in a "dispersion shadow" of suspended particles derived by surface drift from the delta front dispersal system. Samples collected in this dispersion shadow from the surface to 5 m in depth about 40 km south of the main distributary contain 4 mg/l of sediment in water of 35‰ salinity. A series of hydrographic sections along a north to south transect through the dispersion shadow typically shows a decrease in sediment concentration with distance from the delta (Figure 2) produced by (1) sediment dropping from suspension and (2) dilution caused by advective addition of salt water as the sediment suspensions move southward. The sediment composition of water in the dispersion shadow typically follows the trend of the lower dashed line in Figure 2. The composition of the intermediate water layer from west to east in front of the delta typically follows the trend of the upper dashed line.

The distribution of suspended sediment shows that a large fraction of the total sediment load is lost to the delta front platform and to the pro-delta slope. The largest fraction of the total load is transported by saltation and consists of relatively coarse particles which the momentum currents drop within a very short distance of the mouth of the distributary. Figure 2 shows that most of the sediment in suspension does not precipitate immediatly in front of the delta. It is dropped progressively from suspensions that are increasing in salinity, and some is carried for great distances away from the distributaries by the prevailing currents.

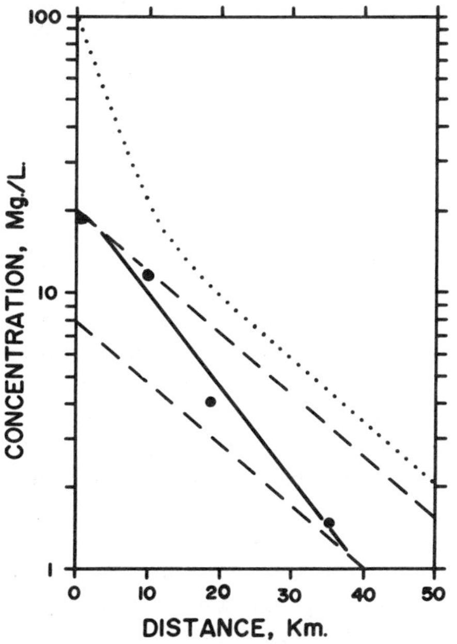

Figure 2. Suspended sediment distribution in front of the Po Delta. Dotted curve shows maximum concentrations in surface momentum currents. Area bounded by dashed lines is the general distribution of turbidity in the intermediate water below the surface currents. Solid line is fitted to a specific series of observations.

MINERALOGY OF THE SEDIMENTS

In order to determine if flocculation affects the suspended sediments, the mineral composition in waters of different salinity was analyzed. The solids from more than two hundred water samples were obtained on Millipore filters, and the mineralogical constituents were identified by X-ray diffraction methods, even when initial concentrations were 2 to 3 mg/l (Figure 3). Montmorillonite, chlorite, muscovite, serpentine, quartz, alkali feldspar, calcite, and dolomite occur in the suspended sediments. A few notes on individual mineral identifications follow.

Mineral Identification

Montmorillonite gives a 14 Å spacing in untreated samples that after glycerol solvation expands to 17.7 Å. All diffraction traces in Figure 3 are of glycerol solvated samples. The montmorillonite in suspended sediments beyond the mouth of the Po expands with some difficulty. Figure 3B was exposed to glycerol for 100 hours before expansion took place. Clay sized montmorillonite should expand within a few hours, as it does in the bottom sediments.

Muscovite gives the strong diagnostic peak at 10 Å and an integral sequence of lower orders.

Chlorite gives a 14 Å first order that may be observed in some glycerol solvated samples. It is enhanced after heat treatment at 600° C. The second and fourth order peaks at about 12° and 25° are resolved from serpentine, which gives peaks at slightly larger angles.

Serpentine gives diagnostic peaks at 7.25 Å and 3.62 Å. Gallitelli (1965) observed similar peaks in samples from the *argille scagliose* in the Apennines, which he attributed to kaolinite. Similarly, Quakernaat (1968, p. 66) thought he identified kaolinite in the sediments of rivers that enter the Adriatic from the Italian coast. However, the spacing is too large for kaolinite, and the peaks persist beyond heat treatment at 500°C, the usual upper stability limit for kaolinite. In fact, the 7.25 Å peak remains after heat treatment at 600°C, when the chlorite reflection near it disappears. The mineral is properly identified as serpentine, and its occurrence in these suspensions is notable.

The relative intensity of the quartz peaks at 22° and 26.6° is a very sensitive indicator of the quartz content of the samples. In Figure 3D the peak intensity at 26.6° is due mostly to the third order muscovite, which occurs at the same angle, while in Figure 3A the very strong peak reflects abundant, probably coarse, quartz in the suspension.

The peaks at 3.18 Å and less frequently at 3.23 Å identify an alkali variety of feldspar (Na-K rich).

The calcite peak at 29.5° and the dolomite at 31° also are sensitive to variations in composition of these minerals, when coarser particles are abundant.

Compositional Variations Observed

Significant changes in mineral composition occur when Po River water mixes with sea water and the suspended sediment disperses (Figure 3). Among the non-phyllitic minerals a rapid loss in relative intensity of the diagnositic peaks shows that particles consisting of these minerals settle out rapidly as the velocity of the transporting currents decreases. Quartz, calcite, and to a somewhat lesser degree feldspar, occur more abundantly near the bottom of the fresh-water stream than near the surface and drop from suspension first in front of the delta. The relative changes may be gauged by reference to an internal standard provided by the 5.0 Å muscovite reflection. In Figure 3A the reference peak for quartz (4.25 Å) is nearly twice as intense as the standard. A continuous decrease in intensity takes place between Figures 3B, 3C and 3D until the quartz peak is less than half as strong as the reference muscovite peak. Beyond this point a certain amount of fine-grained quartz remains in suspension. Feldspar and calcite concentrations show a similar decline. Feldspar and dolomite persist longest in suspension as the sediment is dispersed in the sea, which suggests that the suspended sediment in the river contained abundant silt and clay sized particles of these minerals. All of the changes in composition among the non-phyllitic minerals seem to be explained by the progressive failure of the transporting currents after disemboguing to compensate for the settling velocities of the individual particles.

Among the phyllitic minerals, the most obvious change is the progressive decrease in montmorillonite. The initial fresh-water suspended sediment is rather rich in montmorillonite (Figure 3A). (The peak height for montmorillonite is rather modest, but one must remember that diffraction effects from this mineral are less efficient than from muscovite). A significant loss of montmorillonite occurs very quickly at very low salinity (Figure 3B). The relative decrease is even more pronounced than that of calcite and quartz and suggests that the decrease in concentration of suspended sediment between the river and the delta front (Figures 3A and 3B) results from loss of montmorillonite-rich sediment. Beyond the front of the delta, montmorillonite is lost progressively with distance (Figures 3C and 3D). Montmorillonite was detected in all suspensions more concentrated than 10 mg/l, and all of the data indicate that a real and progressive loss in montmorillonite occurs as Po River water is discharged into the Adriatic Sea.

The data also show an apparent increase in relative intensity of the 10 Å muscovite peak as suspended sediment is transported farther from the delta. Within the river, the 10 Å peak is about twice the intensity of the 5 Å reference peak. In the dispersion shadow, the difference is three times or more, and the change seems to be progressive. It seems likely that this change is due to the changing particle size distribution of the muscovite, for finer grained particles should give a relatively more intense 10 Å reflection.

Chlorite and serpentine intensities bear an almost constant relation to the reference muscovite peak. Thus, little differentiation occurs during transport, probably because these minerals are concentrated in the silt and clay size range.

Particle size analysis of the suspensions reveals decreases in the proportion of larger particles, even within the range 2 to 16 microns. Thus, the clay and silt composition is very sensitive to current velocity and distance from the fresh-water source. Montmorillonite that drops from suspension must be associated with relatively coarser particles, although usually one thinks of this mineral as being finely dispersed. Perhaps the composition and salinity of the river water cause the montmorillonite to be transported as aggregates along with larger particles in the suspended load. In any case, the most long lived sediment suspensions under the salinity and current conditions that exist in the surface waters of the Adriatic Sea are rich in muscovite, chlorite, and serpentine. They contain smaller amounts of quartz, calcite, dolomite, and feldspar. They are impoverished, in comparison to the source sediment suspensions, in montmorillonite.

Vertical variations in mineral composition also are important in analyzing sedimentation processes in front of the Po Delta. Although X-ray diffraction traces are not illustrated, they exhibit the following features. Within the fresh-water portion of the river channel little variation in composition of the phyllitic minerals occurs from the top to the bottom of the water column at high flows, in spite of the increase

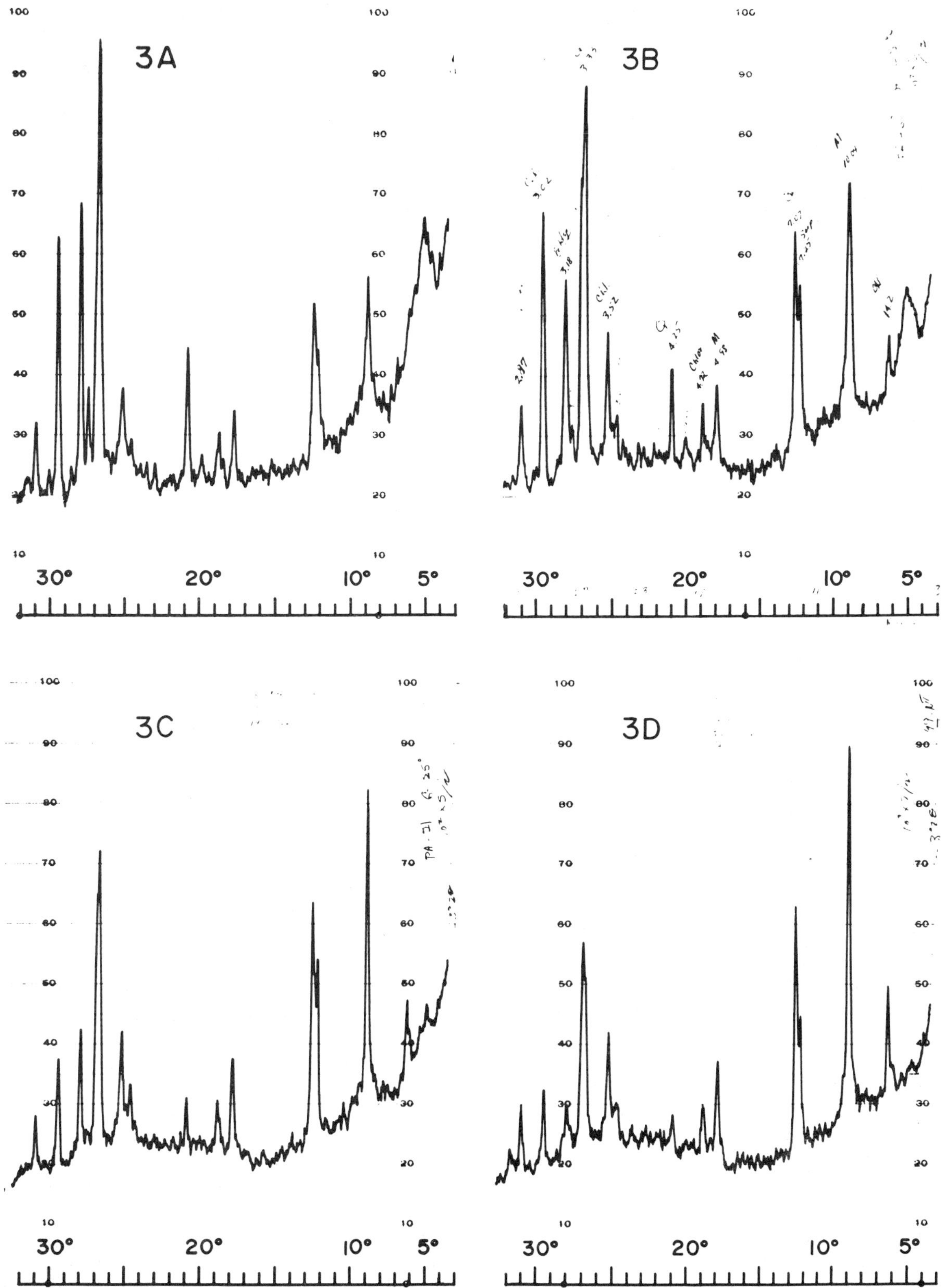

Figure 3. X-ray diffraction traces of suspended sediments. A, Po River at Pila, 6 kilometers above the mouth, surface, 92 milligrams/liter, 0.03%; B, Station no. 2, in front of the bar, surface, 47 milligrams/liter, 5.6‰; C, Station no. 4, limits of momentum flow 5 kilometers seaward of the bar, surface. 19 milligrams/liter 14.1‰; D, Station no. 9, dispersion shadow 8 kilometers downdrift from momentum current, surface, 14 mg/l, 26.3‰. (See Nelson, 1970, for details).

in sediment concentration towards the bottom. At moderate flows an increase in the relative amount of montmorillonite takes place towards the bottom along with the increase in sediment concentration. At low flows the suspended sediments tend to be poor in montmorillonite.

Table 1 summarizes the progressive changes that take place at moderate and higher fresh-water discharges when fresh-water is mixed with sea water in front of the delta. The slight increase in montmorillonite content between the top and the bottom of the momentum current parallels that observed in fresh-water at moderate flow and is related to the increased quantities of larger and non-phyllitic mineral particles transported near the bottom. The abrupt decrease in montmorillonite content below the momentum current corresponds to similarly rapid decreases in the non-phyllitic minerals. Thus, montmorillonite follows the larger and heavier detrital particles along the gradient of increasing velocity and seems to be independent of the changes in salinity. It appears reasonable to attribute the change in concentration of both montmorillonite and non-phyllitic minerals simply to gravitational effects.

Table 1. Compositional changes in transitional water masses.

	Depth, meters	Salinity ‰	Suspended Sediment milligrams/liter	Montmorillonite (*a*)
1. Po River at Pila				
Surface (*b*)	0.3	0.0	92	170
Mid-depth	2.5	0.0	98	172
Bottom	5.0	0.0	125	176
2. Momentum Current				
Surface (*c*)	0.3	5.6	47	78
Mid-depth	0.6	14.2	78	92
Bottom	1.3	20.5	101	98
3. Intermediate Water				
Surface	3.0	26.9	19	26
Bottom	5.0	28.1	17	26

[a]Montmorillonite peak intensity compared to 5 Å muscovite = 100; [b]corresponds to Figure 3A; and [c]corresponds to Figure 3B. The momentum and intermediate currents are from station 2 (Nelson, 1970) in front of the distributary mouth bar.

MINERALOGY OF THE BOTTOM SEDIMENTS

A group of 23 bottom sediment samples obtained from van Straaten (1970) was analyzed by X-ray diffraction methods (Figure 4), and typical diffraction traces show that with one exception the minerals identified are the same as those in the suspended sediments.

Occurence of High Magnesian Calcite in Adriatic Sea Sediments

The X-ray diffraction peak resolved at 3.005 Å on the high angle side of the calcite peak (Figure 4D) is produced by high magnesian calcite, a mineral unique in the Adriatic to the deep water facies. Following Goldsmith and Graf (1958) this spacing estimates a composition containing about 12 per cent $MgCO_3$ in the calcite. This mineral never is detected in the source sediments of the Po River, nor in any of the bottom sediments found along the Italian coast, nor on the shallow shelf north of Ortona (Figure 5), although it is ubiquitous in the sediments east and southeast of Ortona at depths exceeding 150 meters. The mineral appears not to be detrital nor derived from shallow water sediments, because it cannot be detected in these sources. High magnesian calcite is probably biogenic and derived from the shells or skeletons of benthic organisms, a variety of which contain magnesian calcite components and probably occur in the deep water environments. Friedman (1965) found the mineral in the deep water sediments of the Gulf of Aqaba and the Red Sea, although he was unsure whether or not it had formed in place in the basin. The Adriatic Sea occurrence appears to be the first clear demonstration that high magnesian calcite forms within a deep basin facies. Conditions that lead to formation of magnesian calcite in the basin facies at moderate depths merit further study.

Compositional Variation Between Different Facies

Compositional variations involving the fine-grained minerals in the bottom sediments are rather subtle, but three mineralogical associations or facies seem to exist. Their distribution is shown in Figure 5. Bottom sediments near the Po Delta contain relatively less montmorillonite, calcite, and quartz than those from the shallow shelf near the eastern Italian coast. Figure 4C shows the strong montmorillonite peak that is observed frequently in the bottom sediments from this latter area. Statistical tests (Table 2) reveal the greater calcite and quartz contents of the Italian coastal shelf sediments. Even though the samples from the shelf between Ravenna and Ortona lie at an average water depth of 43 m (those from

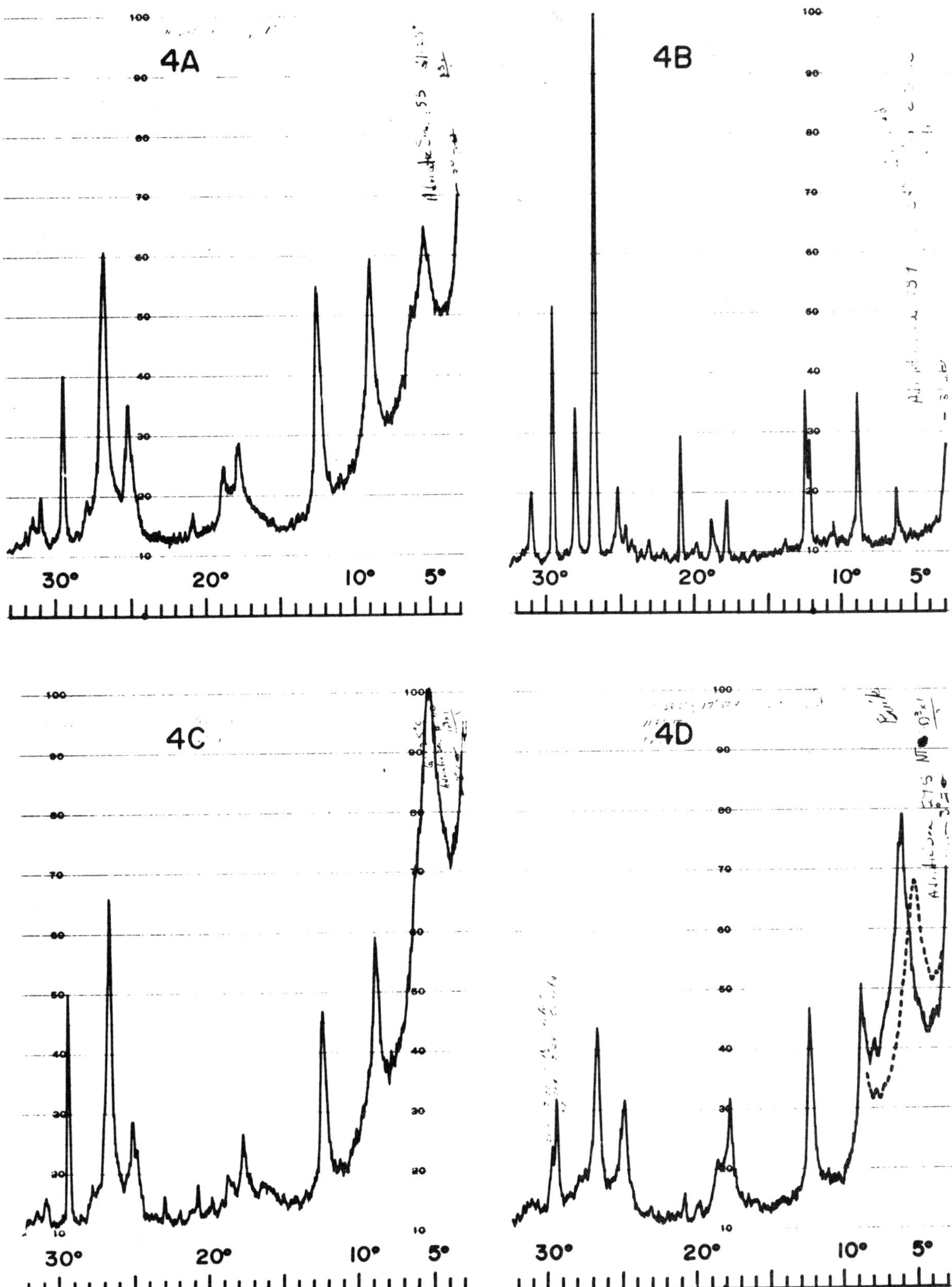

Figure 4. X-ray diffraction traces of bottom sediments. A, mouth of Pila, 13 meters depth, van Straaten no. 155, typical of samples from north of Ravenna; B, mouth of Bastimento, 9 meters depth, van Straaten no. 159, coarse fraction only. C, San Benedetto, 25 meters depth, van Straaten no. 70, typical of samples from Ravenna to Ortona; D, Brindisi, 1161 meters depth, van Straaten no. 295, typical of samples from depths greater than 150 meters southeast of Ortona. Solid line, untreated; dashed, solvated.

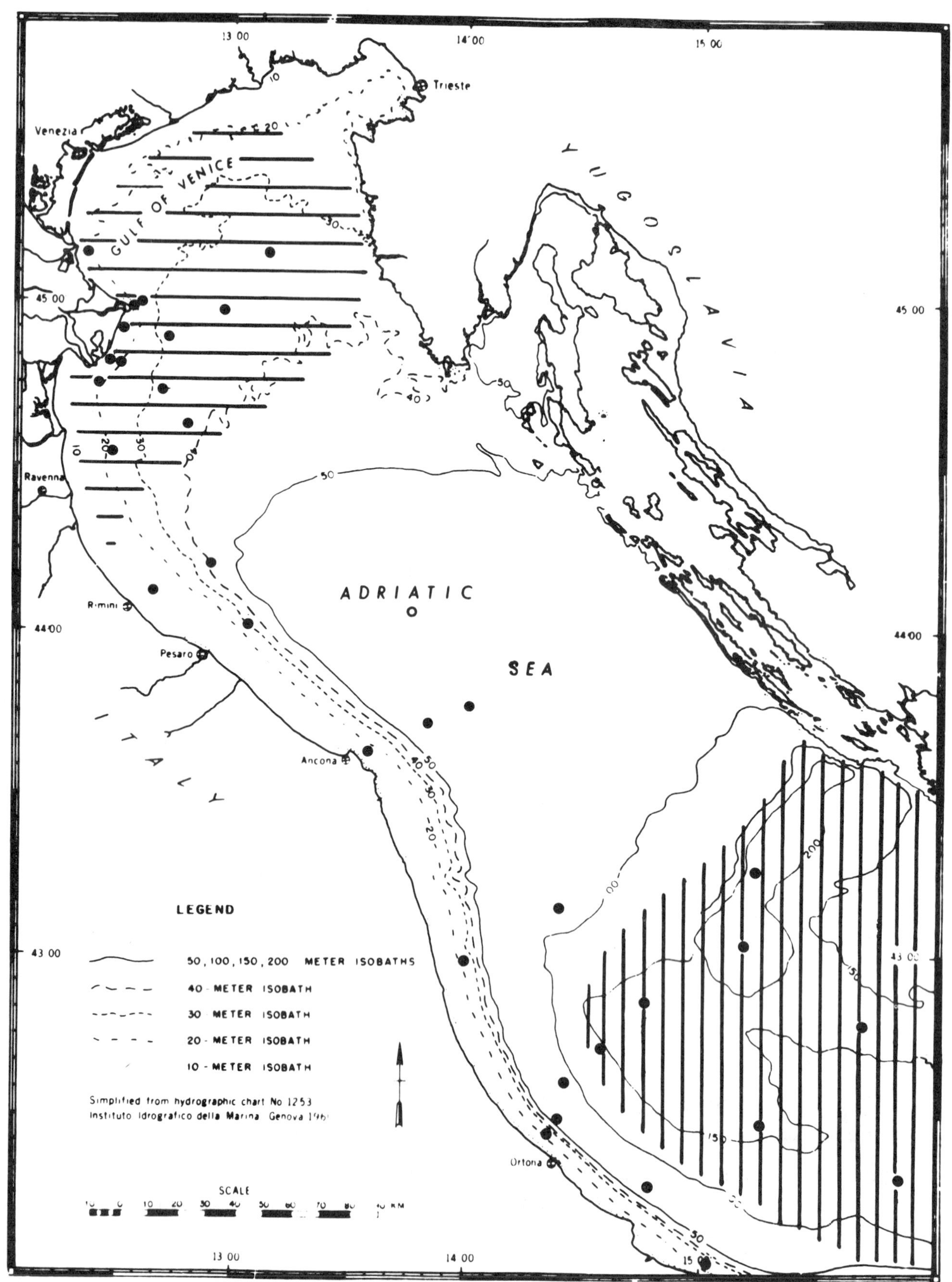

Figures 5. Distribution of bottom sediment mineral associations. Horizontal pattern is the pro-delta slope facies; clear pattern in open shelf facies; vertical pattern is deep basin facies.

north of Ravenna lie at an average depth of 23 m), the higher calcite and quartz contents probably result from nearshore currents that transport coarse sediment southward along the Italian shoreline. These sediments are also richer in quartz, calcite, and feldspar than those from the deep-water basin east of Ortona (Table 2).

The montmorillonite content of the deep-water basin sediments is not significantly different from that of the Ravenna-Ortona coastal shelf sediments. On the other hand, the basin sediments are richer in montmorillonite than the bottom sediments close to the Po Delta. Table 2 demonstrates that montmorillonite is enriched in the shelf and basin sediments that lie beyond the pro-delta slope of the Po Delta.

Even after the above differences have been noted, it must be said that the bottom sediment mineralogies exhibit a great deal of sameness throughout the area that has been studied. Variations in montmorillonite composition are relatively subtle. The change in fine-grained quartz and calcite contents of the shelf sediments is not great. The amount of magnesian calcite in the basin sediments is relatively small.

Comparison of the Bottom Sediments and Suspended Sediments

Figures 3 and 4 show that the suspended sediments appear rather different from the bottom sediments. Serpentine occurs in all of the suspended sediments, but it is not detected in the fine fraction of any bottom sediment. On average, the suspended sediments show consistently higher 10 Å muscovite peak intensities, more feldspar, more dolomite, less calcite, and much less montmorillonite (Table 2).

Diffraction traces from the coarse fraction, which consists mostly of silt, from samples near the delta and on the shallow shelf along the Italian coast frequently have the appearance of Figure 4B. The resemblance to the mineralogy of the residual sediment suspensions is clear, particularly the characteristic serpentine peaks. This similarity suggests that the residual sediment suspensions from the Po ultimately are incorporated in the bottom sediments. The predominant fraction of the bottom sediments, however, probably is not formed from the stable residual suspensions dispersed in the surface waters, in view of the general dissimilarity between Figure 3D and Figures 4A, C, and D.

Figure 3A does bear a strong resemblance to Figure 4A, however. If the relatively coarse particles of feldspar, calcite, and quartz are dropped as soon as the river leaves its mouth, the resulting mixture resembles that deposited close to the delta in 13 m of water; the relative proportions of montmorillonite, muscovite, and chlorite are about the same. Although the residual sediment suspensions contrast with Figure 4A, the pro-delta slope sediments

Table 2. Mineral composition of Adriatic Sea sediments.[a]

	Statistic	Muscovite	Chlorite	Montmorillonite	Serpentine	Quartz	Feldspar	Calcite	Mg-Calcite	Dolomite
All Suspended Sediments	$\bar{X}_1$	280	210	66	D	50	135	156	ND	103
	s_1	46.1	27.6	47.4		62.4	102.9	92.4		36.7
All Bottom Sediments	$\bar{X}_2$	201	225	210	ND	47	60	229		58
	s_2	23.8	29.2	64.8		20.3	15.4	109.8		33.6
$\bar{X}_1 - \bar{X}_2$	t	7.4	1.9	13.7		0.2	3.1	2.6		4.6
		***	*	***	***		***	***		***
Bottom Sediments Near Po Delta	$\bar{X}_3$	198	215	143	ND[b]	31	40	162	ND	63
	s_3	19.0	23.6	20.8		13.5	25.5	33.0		38.3
Bottom Sediments Ravenna to Pescara	$\bar{X}_4$	214	229	253	ND[b]	62	67	340	ND	72
	s_4	20.3	35.0	56.2		17.7	17.0	101.6		34.7
$\bar{X}_3 - \bar{X}_4$	t	1.6	0.9	4.8		3.8	2.4	4.5		0.4
				***		***	*	***		
Bottom Sediments Below Pescara	$\bar{X}_5$	188	229	227	ND	27	34	155	D	39
	s_5	28.2	27.8	51.8		22.4	23.4	62.2		16.1
$\bar{X}_3 - \bar{X}_5$	t	0.7	1.0	4.0		0.4	0.5	0.2		1.4
				***					***	
$\bar{X}_4 - \bar{X}_5$	t	2.1	0	0.9		3.4	3.1	41.7		2.3
		*				***	***	***	***	*

[a] Composition in terms of diagnostic peak intensity referred to 5.0 Å muscovite intensity (= 100).
[b] Detected in the course fraction of some samples. D=Detected. ND=Not detected.
* Significant at the 95% confidence level.
*** Significant at the 99.9% confidence level.

contain material of similar composition and consist apparently of the two fractions recombined, *i.e.*, the residual suspensions and the coarse load dropped at the delta front.

The enigmatic enrichment of montmorillonite, and possibly fine-grained calcite and quartz, in the bottom sediments on the coastal shelf and in the basin south of Ravenna is an unresolved problem. While it may be assumed that sediments in these areas receive contributions from the residual surface suspensions, the silt fraction is only a minor part of the bulk sediments, a large proportion of which is of clay size (van Straaten, 1970). A source must be found for the montmorillonite-rich fraction, and it is not an obvious one, unless it is inferred that the prodelta slope sediments are re-worked by bottom currents and that this leads to concentration of materials richer in montmorillonite.

DISCUSSION

Clay Distribution and Dispersal

A number of previous investigations have shown montmorillonite to be enriched in the bottom sediments that lie seaward of a detrital source. For example, van Andel and Postma (1954) observed regional changes in clay mineral abundance in front of the Orinoco Delta not unlike those in the Adriatic, and Porrenga (1966) found enrichment of montmorillonite in the deep-water sediments in front of the Niger Delta. They attributed the distributions to differential transport of clay minerals, saying that clay poor in montmorillonite deposits in fresh and brackish water near the delta while clay rich in montmorillonite deposits in marine areas. Their interpretations were based on the differences in settling velocities of various clay minerals as observed by Whitehouse and co-workers. Supplementing earlier work, Whitehouse *et al.* (1960) showed that kaolinite, illite, muscovite, and chlorite flocculate at low salities and settle at rates of 11 to 15 meters/day; in contrast, montmorillonite settles at a rate of 0.03 meter/day at low salinities and does not achieve its maximum settling velocity of 1.3 meters/day until nearly full marine salinities are reached. Thus, it was expected that montmorillonite would be enriched and transported farther as fresh-water suspensions mix with ocean water and disperse.

The progressive changes observed in sediment suspensions dispersed from the Po are different from those predicted by the settling velocities of the flocculated constituent minerals, since montmorillonite becomes deficient and muscovite, chlorite, *etc.* are enriched progressively as salinity increases. Since montmorillonite is more difficult to detect the farther seaward the suspensions travel, montmorillonite must be eliminated along with the coarsest particles as they are dropped from suspension. Therefore, differential transportation in the surface currents of the kind described by Whitehouse *et al.* (1960) cannot explain the distribution pattern of bottom sediment mineral assemblages in the Adriatic Sea.

Evaluation of Flocculation as a Depositing Mechanism

Po River water has an initial salinity of 0.03 ‰, it carries calcite and dolomite in suspension, and it contains dissolved Ca and Mg ions. Gherardelli and Canali (1960) determined concentrations of about 50 ppm of calcium, about 10 ppm of magnesium, and around 200 ppm of total dissolved solids. Under these conditions the clay minerals are likely to be saturated with Ca and Mg and carried in suspension as aggregates from the very beginning. In fact, clay floccules are found in the concentrated sediment carried by the river near its bed. Since abrupt decreases in sediment concentration occur between the river and the surface momentum currents before the suspensions contain much salt (compare the concentrations in Table 1), the aggregated clay minerals must follow the larger non-clay particles to the bottom before there is an opportunity for flocculation by sea salt to affect them. By the time advected sea water increases the salinity of the suspensions issuing from the river to a significant level, they already contain the residual mineral assemblage that represents but a minor fraction of the initial suspended sediment load. Thus the major part of the river load deposited near the Po Delta must not be influenced significantly by flocculation effects.

The fact that during relatively low discharge the suspended sediment within the distributaries consists of montmorillonite deficient sediment similar to Figure 3D (plus calcite, feldspar, and quartz) suggests that the residual mineralogy is as much the stable product of suspension in fresh-water as it is in saline. Consequently, the mineralogy of the residual suspensions is a characteristic of stream sediments in general, rather than a floccular differentiate.

Gravitational Settling of Clayey Sediment from a Saline Current

Further understanding of the behavior of fine-grained sediment in saline flowing water was obtained by Krone (1962) who conducted flume studies in which the bed material was free to interact with the particles in suspension. Deposition is sensitive to sediment concentration, particle size distribution, and the internal shearing stresses of the moving fluid. At high velocities the shearing stresses tend to break up floccules, which have low shearing strengths, until they settle with velocities comparable to or less than those of individual clay and silt particles. Above a critical level of internal shearing, flocculation does not enhance depositional rates. The critical level is achieved in a channel of 10 m depth when the water flows with an average velocity of 70 cm/sec. At shallower depths, such as in the Po distributaries, the critical value should be reached at lower velocities, below those found in the distributaries. Krone found that at suspended concentrations less than 300 mg/l aggregation is too slow to cause deposition from a moving fluid. For example, with particles of 0.3μ diameter at a concentration of 300 mg/l the suspension requires about 100 seconds to form a floccule 2μ in diameter. Such a suspension loses particles very slowly, even at very low flow velocites. At 6 cm/sec only 3 per cent of the sediment concentration is lost per hour; at 15 cm/sec the loss is 0.5 per cent per hour. The flow velocities of the drift currents in the Adriatic Sea exceed these critical velocities, and at their relatively low suspended sediment concentrations it should be expected that sedimentation will not be particularly enhanced by flocculation. Also, since the observed clearing rates for these suspensions are higher than those in Krone's experiments, flocculation does not seem a likely mechanism to control sedimentation from them.

The fundamental law governing the suspended sediment concentration of a water mass moving out of contact with the bottom has been provided by Sundborg (1956, p. 221), as follows.

$$S_t = S_0 \cdot e^{-[c \cdot \Phi(c)/d]t}$$

Here S_0 is the average initial suspended sediment concentration; S_t is the concentration after a time, t; e represents the natural logarithm; c is the settling velocity of a particular grain size; d is the depth of the current; and $\Phi(c)$ is an unknown function of the grain size and flow parameters. This relationship predicts that the sediment concentration will decrease logarithmically with time or distance of transport. A similar relation describes the suspended sediment distributions in parts of some estuaries and in front of deltas. The decrease in sediment thickness with distance seaward of deltaic distributaries is also related to this function. The sedimentation rate, or slope of the function, is sensitive to particle size, because the settling velocity increases with the square of the particle diameter. Some qualitative sediment deposition functions are given in Figure 6. The important parameter $\Phi(c)$ has been evaluated from Sundborg's curves for the distribution of suspended material of various sizes with depth in rivers. For a sea current flowing without contact with a bed, the flow parameters would be somewhat similar to those in the upper tenth of a stream bed 400 cm in depth, and Sundborg's distributions of particle sizes at this level has been employed to estimate $\Phi(c)$.

The results show that material of clay size will be carried virtually indefinitely, even at current velocities as low as 25 cm/sec. Particles in the fine silt range will be transported great distances, as indicated for the 16 micron particles (Figure 6), but the coarse silt is very sensitive to current velocity and will be lost from the current at a relatively high rate. Particles of very fine sand (125μ) have a very low stability in suspension, even when the current velocity is 100 cm/sec. Such particles will drop very close to the delta front, and, therefore, the deposits only at a relatively short distance from the delta already will be rather fine-grained.

If the flow parameters needed to evaluate $\Phi(c)$ in currents flowing free of a bed were available, the horizontal distribution of suspended sediment in front of a delta could be related to simple gravitational settling in a quantitative way. In the absence of such data, we can only compare the observed rate of change of particle concentration in the residual

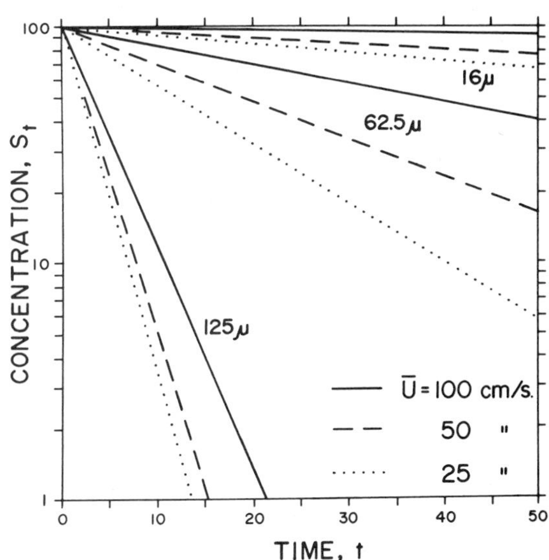

Figure 6. Calculated sedimentation function for free-flowing currents. Curves show expected concentrations after given times for currents of 25, 50, and 100 cm/sec mean velocity and for particles of 16, 62.5, and 125 microns average diameter. The curve for 4 microns lies parallel to the abscissa along the upper boundary of the figure. For details of calculation, see text.

suspensions that move outwards from the Po Delta with the calculated curves. The distribution of suspended particles observed (Figure 2) is similar to what might be expected from polydisperse suspensions settling according to the equation presented above. The similarity between Figure 2 and the curves in Figure 6 suggests that the saline suspensions in front of the Po Delta are clearing by gravitational settling. It would, of course, be desirable to study this relationship quantitatively.

DIFFERENTIATION OF BOTTOM SEDIMENT MINERAL ASSEMBLAGES AND CONCLUSIONS

Montmorillonite-rich sediment deposits immediately in front of the distributaries, and, except for the separation of sand-sized particles on the upper part of the delta front platform, this sediment is not differentiated mineralogically from the main suspended sediment load of the Po. A differentiation does occur between the pro-delta slope sediments and those facies that are far removed from the delta. The differentiation of clay minerals in the bottom sediments probably occurs by means of bottom currents that rework the deposits of the pro-delta slope. Sediments of the deep basin facies may be derived from the pro-delta slope sediments simply by enrichment in montmorillonite, while those along the Ravenna to Ortona shelf require selective addition of fine-grained quartz and calcite as well.

The surface circulation in the Adriatic (Nelson, 1970) brings high salinity water northward along the Yugoslavian coast until it is deflected by the fresh-water outflow of the northern Italian rivers. Lower salinity water moves southward along the Italian side of the Adriatic until it discharges into the Ionian Sea. This surface pattern resembles the lateral circulation of relatively broad and shallow estuaries, where saline water moves upstream towards the fresh-water source on one side and lower salinity water moves downstream on the other side. In such estuaries the flow is in the same direction from the top to the bottom of the water column and changes in direction occur from one side of the estuary to another. In view of the breadth to depth ratio of the Adriatic it seems possible that the lateral circulation of the surface layers might also drive the bottom water, at least in the northern half of the sea where the water is shallow. In any case, we know that at certain seasons of the year the net circulation pattern of the water near the bottom is directed down the Adriatic towards the Ionian Sea. Zore-Armanda (1969) observed that in winter, because of density contrasts, the Adriatic bottom water flows into the Ionian Sea (Figure 2 in the cited work shows the Adriatic - Ionian winter exchange relationship schematically). Thus, an appropriate pattern of net water movement near the bottom exists to move resuspended bottom sediment from the pro-delta slope near the head of the Adriatic to the deeper water of the basins in the southern part of that sea. It is only necessary to show the possibility that the bottom sediments once deposited can be resuspended and reworked. Evidence for resuspension has been given by Jerlov (1958, p. 241–242). He observed a generally high particle content in the water near the bottom throughout the Adriatic, and a detailed study of two stations showed a logarithmic increase in sediment concentration from about 20 m in depth to the bottom at 30 to 35 m. This is below the surface circulation. The suspended sediment distribution conforms to that expected if a tidal current resuspends and transports the bottom sediment. Thus the instantaneous tidal velocities resuspend, and the net exchange of water transports.

This type of bottom circulation would lead to the reworking of pro-delta slope sediments and their transport both southward along the Italian coast and into the deep water basins. If particles of montmorillonite are on average smaller than those of the other components of the bottom sediments, they would become concentrated downstream. In addition, the colloidal characteristics of montmorillonite also would lead to downstream augmentation. Downstream enrichment in fine-grained quartz and calcite might also be expected. The distribution of sandy facies in the central and western Adriatic (van Straaten, 1970) reflects the lack of significant deposition of fines in these parts and may confirm that elutriation is active in this area.

The small streams that drain into the Adriatic from the eastern slope of the Appenines may also contribute important quantities of the required mineralogical constituents to the nearshore bottom sediments south of Ravenna. The bottom sediments close to the coastline often are richest in montmorillonite; the possibility of supplemental detrital sources should be investigated. Since the surface waters in this part of the Adriatic are always very clear, montmorillonite must be transported along the bottom and in order to enrich the sediments in the basin, montmorillonite would have to be transported across the shelf rather than directly downstream with the presumed bottom water circulation. Consequently it seems unlikely that this supplementary source of montmorillonitic sediment can adequately account for the facies differentiation.

ACKNOWLEDGMENTS

This work was supported by National Science Foundation Grant GP–525.

REFERENCES

van Andel, T. H. and H. Postma 1954. Recent sediments of the Gulf of Paria. Reports of the Orinoco Shelf Expedition, I. *Verhandelingen der Koninklijke Nederlandse Akademie van Wetenschappen, Afd. Natuurkunde,* Eerste Reeks, 20:245 p.

Friedman, G. M. 1965. Occurrence and stability relationships of aragonite, high-magnesium calcite, and low-magnesium calcite under deep sea conditions. *Geological Society of America Bulletin,* 76:1191–1196.

Gallitelli, P. 1965. Ricerche su alcune argille della formazione argilloscistosa appenninica. *Atti della Societa Toscana di Scienze Naturali, Pisa,* 117 (A):88–124.

Gherardelli, L. and L. Canali 1960. Enquète sur les caractéristiques chimiques et physicochimiques des eaux du Pô. *International Association of Scientific Hydrology Bulletin,* 19:24–36.

Goldsmith, J. and D. Graf 1958. Relation between lattice constants and composition of the Ca-Mg carbonates. *American Mineralogist,* 43:84–101.

Jerlov, G. 1958. Distribution of suspended material in the Adriatic Sea. *Venice Archivio di Oceanografia e Limnologia,* 11:227–250.

Krone, R. B. 1962. *Flume studies of the transport of sediment in Sedimentation—Modern and Ancient,* ed. Morgan, J. P., Society University of California, Berkeley, California, 110 p.

Nelson, B. W. 1970. Hydrography, sediment dispersal, and recent historical development of the Po River delta, Italy. In: *Deltaic Sedimentation—Modern and Ancient,* ed. Morgan, J. P., Society of Economic Paleontologists and Mineralogists Special Paper, 15:142–184.

Porrenga, D. H. 1966. Clay minerals in recent sediments of the Niger delta. In: *Clays and Clay Minerals,* Proceedings of the 14th National Conference, Pergamon Press, Oxford, 221–233.

Quakernaat, J. 1968. *X-ray analyses of clay minerals in some recent fluviatile sediments along the coasts of central Italy.* Dissertation, Amsterdam, 105 p.

van Straaten, L. M. J. U. 1970. Holocene and late Pleistocene sedimentation in the Adriatic Sea. *Geologische Rundschau,* 60:106–131.

Sundborg, A. 1956. The River Klaralven, a study of fluvial processes. *Geografisca Annaler,* 38:127–316.

Whitehouse, U. G., L. M. Jefferey, and J. D. Debbrecht 1960. Differential settling tendencies of clay minerals in saline waters. In: *Clays and Clay Minerals,* Proceedings 7th National Conference, Pergamon Press, Oxford, 1–79.

Zore-Armanda, M. 1969. Water exchange between the Adriatic and the eastern Mediterranean. *Deep Sea Research,* 16:171–178.

Origin and Dispersal of Holocene Sediments in the Eastern Mediterranean Sea*

Kolla Venkatarathnam, Pierre E. Biscaye, and William B. F. Ryan

Lamont-Doherty Geological Observatory of Columbia University, Palisades, New York

ABSTRACT

Holocene sediments from the eastern Mediterranean Sea have been analyzed for calcium carbonate concentration, fine-fraction mineralogy, Rb/Sr ratio and Rb-Sr "apparent age" and for concentrations of four transition metals: Cu, Mn, Zn and Fe. Because of possible ambiguities in resolving sediment sources with similar mineralogies, the clay mineral characteristics have been subjected to Q-Mode factor analysis. The major sources of aluminosilicate detritus to the sediments of the eastern Mediterranean are suggested by distinctive clay mineral assemblages. Q-Mode factor analysis resolved three factors which account for 98.6% of the variability in the data and which are consistent with the sediment sources suggested by the mineral assemblages. Rb/Sr ratios and Rb-Sr "apparent ages" also confirm these sources.

The principal sources are: Nile River detritus which is carried northeast along the coast of Israel, Lebanon and Syria; sediment from the southeastern Aegean Sea (and possibly the coast of Turkey) which is carried by the Levantine Intermediate Water into the Hellenic Trough; sediment from the south-western Aegean which is carried by bottom currents westward between Crete and Peloponnesus into the Hellenic Trough; sediments from the Adriatic Sea transported by cold bottom currents into the Ionian Basin; and detritus from northern Africa transported by winds and which, because of the lack of dilution by current-transported detritus, constitutes a major detrital source for sediments of the Mediterranean Ridge and southwestern Levantine Basin.

The concentration of calcium carbonate is highest in the same areas as those influenced by eolian sediments and for the same reason—lack of dilution by current-transported aluminosilicate

RESUME

Les concentrations en carbonate de calcium, la minéralogie de la fraction fine, le rapport Rb/Sr et "l'âge apparent Rb-Sr" ainsi que les concentrations de quatre métaux de transition: Cu, Mn, Zn et Fe, ont été analysés dans les sédiments Holocène de l'Est de la Méditerranée. Les caractéristiques minérales des argiles ont été soumises à l'analyse statistique de "Q-Mode factor" afin d'éviter les ambiguités qui pourraient apparaître quand on cherche à distinguer les sources des sédiments qui ont une minéralogie semblable. Les sources principales de détritus alumino-silicates des sédiments de l'Est Méditerranéen sont suggérées par la présence de différents assemblages minéraux des argiles. L'analyse par le "Q-Mode factor" permet de résoudre trois facteurs qui interviennent pour 98,6% dans les variations qui apparaissent dans les données minéralogiques et les résultats sont en accord avec les sources sédimentaires suggérées par les assemblages minéraux. Les rapports Rb/Sr et les "âges apparents Rb-Sr" confirment également ces sources.

Les sources principales sont: (1) les détritus du Nil qui sont charriés vers le Nord-Est, le long des côtes d'Israël, du Liban et de la Syrie; (2) les sédiments provenant de la Mer Egée (et peut-être de la côte turque) qui sont charriés par les eaux intermédiaires du Levant dans la fosse hellénique; (3) les sédiments provenant du Sud-Ouest de la Mer Egée qui sont charriés par les courants de fond vers l'Ouest, entre la Crête et le Péloponèse, dans la fosse hellénique; (4) les sédiments de la Mer Adriatique qui sont transportés par les courants froids de fond dans la partie Nord du bassin Ionique; (5) les sédiments qui sont transportés par les vents de l'Afrique du Nord et qui, parce qu'ils ne sont pas dilués par la matière détritique

*Lamont-Doherty Geological Observatory Contribution No. 1895.

detritus. The concentration of iron appears to be entirely related to the several sources of aluminosilicate detritus with calcium carbonate acting only as a diluent. It is not possible from our data to distinguish between biogenic and detrital control of the distribution of copper, manganese and zinc.

transportée par les courant de fond, constituent une source détritique tres importante pour les sédiments de la dorsale méditerranéenne et le bassin Levantin du Sud-Ouest.

Les teneurs en carbonate de calcium sont les plus élevées dans les régions mêmes qui sont influencées par les sédiments éoliens et elles le sont pour la même raison: les détritus alumino-silicatés transportés par les courants de fond ne diluent pas le carbonate de calcium. La teneur en fer semble être entièrement dépendante des diverses sources de détritus alumino-silicatés, le carbonate de calcium agissant seulement comme un diluant. A partir de nos données il n'est pas possible de distinguer quel est le rôle joué par les processus biogènes et détritiques dans la distribution de Cu, de Mn et de Zn.

INTRODUCTION

The geologic and oceanographic history of an ocean basin is written in its sedimentary record. Deciphering this record requires first an understanding of the factors governing the distribution of sediments. Resolving the bulk sediment into its several components and their characterization by several parameters facilitates distinction of their diverse origins. These data permit inferences to be made concerning mechanisms of sediment transport to and dispersal paths and mechanisms within the basin. Our study of the eastern Mediterranean Sea has these immediate and ultimate goals.

Previous work on sediments of the eastern Mediterranean has been described by Emelyanov 1965a and b) who studied both granulometry and carbonate distributions. Rateev *et al.* (1966) made a general survey of the clay minerals of the entire Mediterranean, while Chamley (1962) and Biscaye (1964) analyzed a few samples from this region. Chamley (1971, thesis, University D'Aix-Marseille; this volume), reports additional analyses in the eastern Mediterranean and summarizes the scattered results of several workers. The clay mineral analyses discussed in this paper have been in part reported by Venkatarathnam and Ryan (1971). The oceanic circulation patterns in the Mediterranean have been more extensively studied (Nielson, 1912; Pollack, 1951; McGill, 1961; Wüst, 1961; Moskalenko and Ovchinnikov, 1965; Ovchinnikov and Plakhin, 1965).

In this paper we discuss the distribution of calcium carbonate and the major sources of nonbiogenic sediment to the eastern Mediterranean as distinguished by the mineralogy of the fine fraction. Ideally, sediment sources should be distinguished by mineral phases confined to a single source. Because this situation seldom exists we have applied Q-mode factor analysis to aid in the distinction of sediment sources which differ only slightly in fine-fraction mineral composition. We report analyses of the Rb, Sr and Sr-isotope characteristics of the sediment and integrate these results with the conclusions drawn from the mineralogy and factor analysis. The dispersal paths of sediments from these sources are analyzed in conjunction with present knowledge of the surface and abyssal circulation of the eastern Mediterranean. The concentrations of four transition metals are also reported and their relation to the distribution of both the biogenic and non-biogenic sediments is discussed.

SAMPLES AND ANALYTICAL METHODS

Holocene sediments were obtained by sampling the tops of gravity (trigger) cores from the Lamont-Doherty core library (the 10th and 14th cruises of VEMA, the 9th cruise of ROBERT D. CONRAD, and cores from Task Area 6, U.N.S.O.O. taken from the ATLANTIC SEAL) and from the 61st cruise of the Woods Hole oceanographic vessel CHAIN, through the courtesy of E. F. K. Zarudski. The locations of these samples are shown in Figure 1.

Calcium carbonate was analyzed on a gasometric apparatus described by Hülsemann (1966). Our overall precision using this apparatus was $\pm$ 3%. Techniques of preparation and x-ray diffraction analysis of the $<2\mu$ nonbiogenic fraction have been described in Venkatarathnam and Ryan (1971). Rubidium and strontium concentrations were determined by x-ray fluorescence (described in Biscaye and Dasch, 1971) in most cases on the same $< 2\mu$ fraction. The entire carbonate-free fraction of the six samples chosen for strontium isotope analysis was also analyzed for rubidium and strontium concentrations by x-ray fluorescence. Carbonate was destroyed by buffered acetic acid (pH = 5). The precision of these analyses is approximately five percent (1σ).

Strontium separations and isotope analyses by mass spectrometry were made by the methods described in Biscaye and Dasch (1971). Precision of the Sr^{87}/Sr^{86} ratio is approximately 0.0005 (1σ) as determined by replicate analyses of a single deep-sea

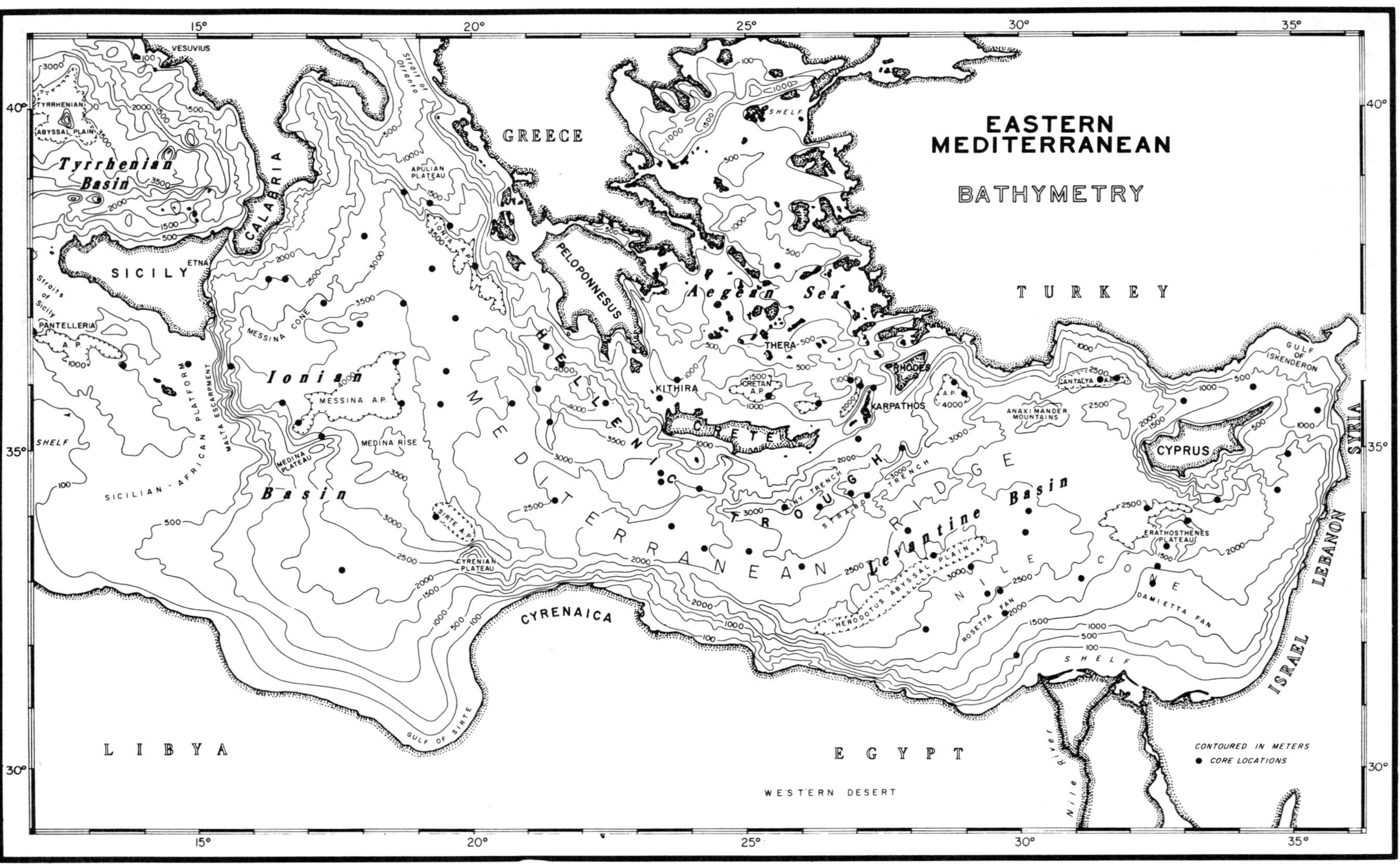

Figure 1. Core locations and bathymetry of the eastern Mediterranean Sea.

sediment sample. Replicate analyses of the Eimer and Amend standard strontium carbonate (Lot No. 492327) give a Sr^{87}/Sr^{86} value (normalized to $Sr^{86}/Sr^{88} = 0.1194$) of 0.7077 ± 0.0003. Iron, nickel, copper and manganese were analyzed by means of a Perkin Elmer Model 303 atomic absorption spectrophotometer. These analyses were made on separate aliquots of raw core containing both carbonate and clays, but the results are also expressed on a carbonate-free basis. These samples, taken from trigger weight cores, are not subject to the contamination discussed by Bender and Schultz (1969). LDGO trigger weight cores have always been stored in plastic liners. These analyses were made at the same time, on the same spectrophotometer and using the same standards as those of Bender and Schultz (1969) and we estimate the same analytical precision.

RESULTS AND DISCUSSION

Calcium Carbonate

The major features of the distribution of the biogenic and nonbiogenic sediments of the eastern Mediterranean are shown in the map of calcium carbonate concentration (Figure 2). With the exception of some of the $> 70\%$ $CaCO_3$ samples along the North African coast, which represent physico-chemically precipitated oolitic carbonate (Emelyanov, 1965b), the distribution shown in Figure 2 represents accumulated carbonate plant and animal tests from pelagic organisms. This biogenic debris is diluted to varying degrees by alumino-silicate detritus. The major lithologic features of the region are therefore the central high-carbonate zone coinciding with the Mediterranean Ridge and which grades south into areas of even higher carbonate concentration along the African coast and the areas of low carbonate—the Nile cone and the coastal region to the northeast, the southern Aegean region, the northern Ionian Basin—coinciding with major terrigenous sediment sources.

Transition Metals

Table 1 lists the concentrations of calcium carbonate, and of copper, manganese, iron and zinc determined on the bulk sample from thirty of the core tops shown in Figure 1. The data are listed both as determined on the whole sample and, following the usual practice, as calculated on a carbonate-free basis. The tacit assumption in calculating data on a

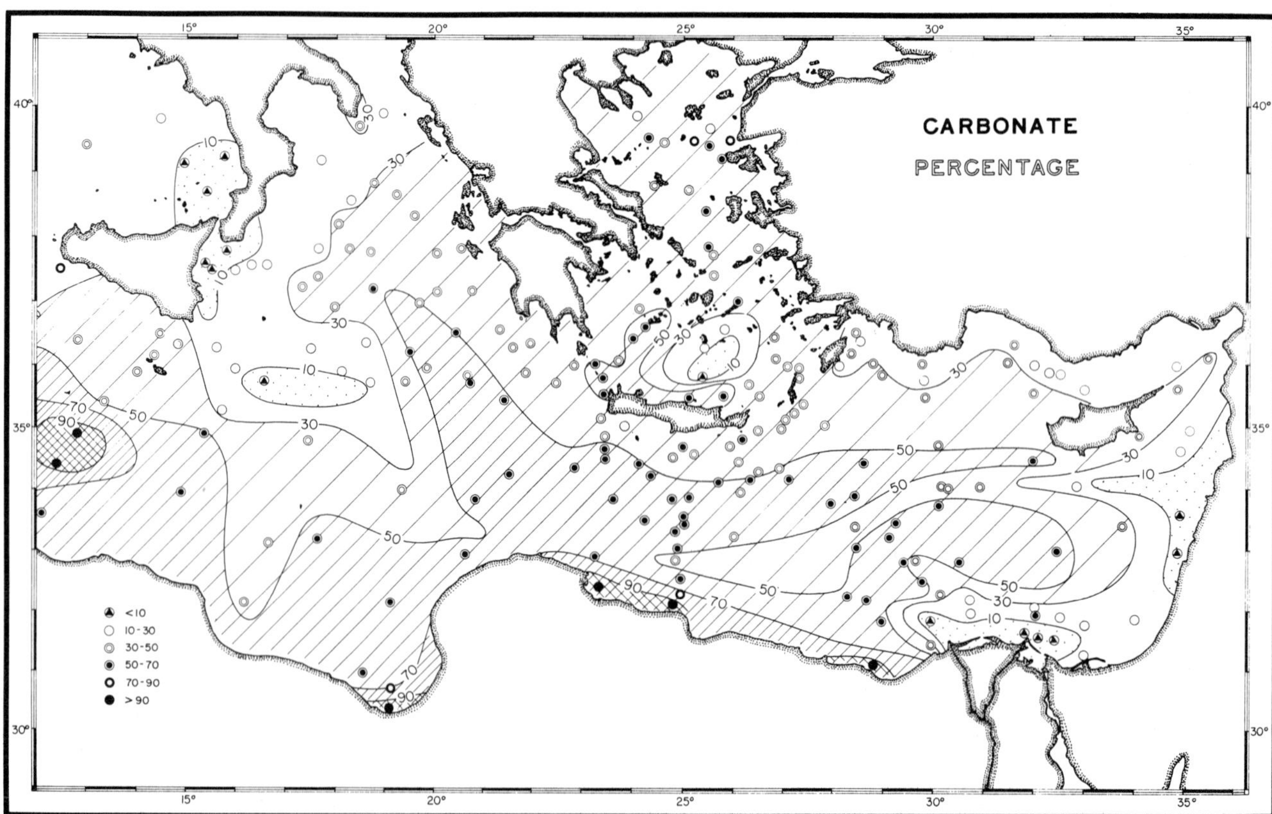

Figure 2. Distribution of calcium carbonate, eastern Mediterranean Sea (includes data of Emelyanov, 1965b).

Table 1. Transition metal and calcium carbonate concentrates in Holocene samples from the eastern Mediterranean Sea. Transition metal concentrations are given both as analyzed, on the whole sample, and as calculated on a carbonate-free basis.

Sample	$CaCO_3$ %	ppm Cu		ppm Mn		ppm Fe($\times 10^3$)		ppm Zn	
		Whole Sample	$CaCO_3$ free	Whole Sample	$CaCO_3$ free	Whole Sample	$CaCO_3$ free	Whole Sample	$CaCO_3$ free
RC9–174	49.8	70	139	1130	2250	89	177	60	120
RC9–175	16.6	71	85	1700	2040	220	264	98	118
RC9–177	54.4	65	143	825	1810	61	134	46	100
RC9–180	52.5	62	131	850	1790	72	151	54	113
RC9–181	56.0	62	142	832	1890	20	45	51	115
RC9–182	64.6	51	144	715	2020	45	127	47	134
RC9–183	52.6	57	121	850	1790	59	124	52	110
V10–22	50.1	42	85	1000	2010	70	141	62	124
V10–24	46.2	44	92	1000	1860	122	227	—	—
V10–25	40.4	64	108	1000	1690	74	124	63	105
V10–26	45.9	86	159	1160	2150	81	150	60	111
V10–27	35.3	78	120	1170	1820	46	71	82	126
V10–48	45.9	40	74	1160	2150	72	133	51	95
V10–51	46.1	42	78	1010	1880	130	241	—	—
V10–52	37.5	58	92	887	1420	70	112	53	84
V10–53	51.2	63	129	975	2000	68	139	53	108
V10–54	40.9	60	101	926	1570	63	106	45	77
V10–56	43.7	60	107	850	1520	69	123	46	81
V10–58	32.4	51	75	1110	1640	79	117	60	89
V10–60	31.0	60	87	910	1320	92	133	62	90
V10–61	32.1	60	88	1110	1630	100	147	69	101
V10–67	58.1	57	137	1010	2420	67	160	375	895*
V14–128	53.9	55	119	737	1600	54	117	53	114
V14–130	49.7	55	110	840	1670	56	112	376	747*
V14–131	49.7	62	120	734	1410	58	111	54	103
V14–132	50.2	47	95	850	1720	44	90	50	102
V14–133	47.0	45	84	832	1570	156	287	55	103
V14–135	11.1	50	57	951	1070	97	237	200	225
V14–136	8.9	77	85	735	807	130	143	110	126
V14–138	31.8	26	38	650	953	91	132	72	105

*Not included in Figure 3.

carbonate-free basis is that calcium carbonate itself contributes none of the trace metals to the sediment and acts only as a diluent. These data are presented in Figure 3 in which both the whole sample and carbonate-free values of each transition metal are plotted along with regression lines for all data except zinc.

Factors and processes affecting the trace element concentrations in marine sediments have been discussed by many workers (*e.g.*, Krauskopf, 1956; Turekian and Schutz, 1965; Boström *et al.*, 1969; and others). Sevast'yanov (1968) reported on Ti, Cr, No, Ni, Co and Cu contents of a few cores from the eastern Mediterranean Sea. We note in our data that for both copper and manganese the correlation coefficient indicates some significance to the correlation between percent calcium carbonate and transition metal concentration calculated on a carbonate-free basis. The same analytical results, however, calculated on a whole sample basis show no correlation between trace metal and carbonate which means that one cannot distinguish between the carbonate and alumino-silicate phases as the source of these metals. It is therefore probable that the positive correlation between ppm copper and manganese (carbonate-free) and percent carbonate is an artifact of the exclusion of carbonate in the calculation. We recognize the possibility that the copper and manganese concentrations may actually be controlled by another phase that has not been distinguished in our analyses, but whether or not this is so, we do not feel that the source of these metals is discernable in our data.

In the case of the iron analyses presented in Figure 3

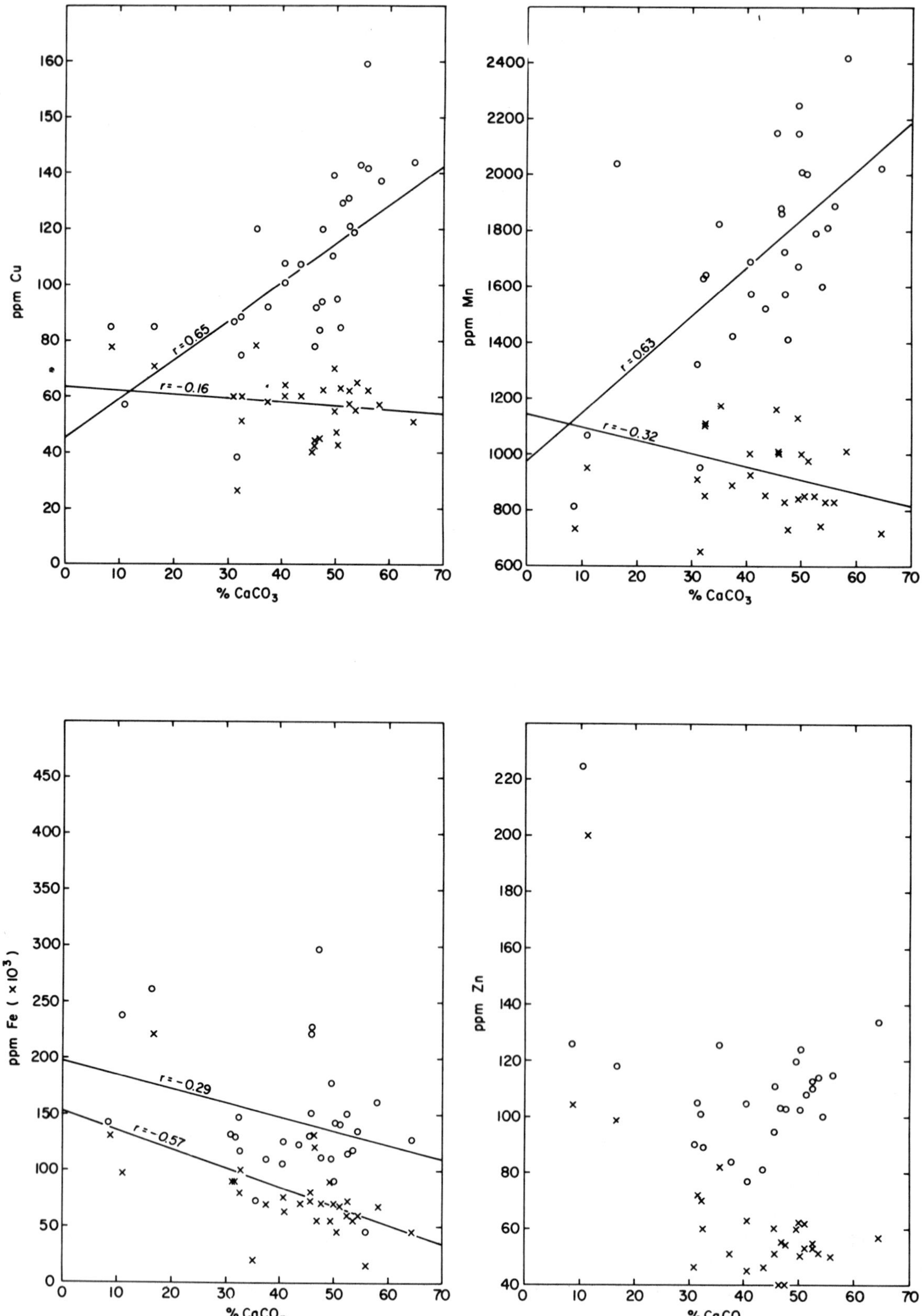

Figure 3. Concentrations of copper, manganese, iron and zinc versus percent calcium carbonate in thirty Holocene samples from the eastern Mediterranean Sea. X's = transition metal concentration in whole sample (as analyzed). O's = transition metal concentration calculated on a carbonate-free basis.

the data calculated on the whole sample basis (the x's) have a regression line whose correlation coefficient has some significance. The line suggests that sediment with no carbonate, *i.e.*, entirely alumino-silicate detritus, would have an iron concentration around 150,000 ppm and 100 percent carbonate sediment would have no iron. When this diluting effect of the carbonate is eliminated by calculating on a carbonate-free basis (the o's) the data are more scattered and the significance of the correlation drops from −0.57 to −0.29. This corroborates our conclusion that the iron is carried in the alumino-silicate phase. The increased scattering probably just reflects sediments from different detrital sources with varying iron concentrations.

The data for zinc show such a degree of scatter that we do not feel that meaningful conclusions can be drawn from them.

Clay Minerals: Distribution and Factor Analysis

The noncarbonate fine-fraction of the sediments in the eastern Mediterranean, as in the world oceans, consists primarily of four clay minerals—illite, smectite, kaolinite and chlorite. These mineral names, as used here and in most other studies of deep-sea sediments, represent major clay mineral groups distinguished principally on the basis of basal x-ray diffraction peaks. (The term smectite follows current usage for the group of expandable clay minerals in the same family as montmorillonite.) Maps showing the distribution of these minerals and a parameter related to smectite crystallinity in the $< 2\mu$ fraction have been given in Venkatarathnam and Ryan (1971). Six clay mineral assemblages based on these distributions, have been outlined, five of which were named after appropriate geographic areas of the eastern Mediterranean (the Nile, southeast Aegean, Kithira, Messina and Sicilian) while the sixth was termed the Kaolinite-rich assemblage.

The clay mineral data from about ninety core tops have been subjected to a Q-mode factor analysis. The purpose of this treatment is to overcome the possible built-in interdependences inherent in clay mineral analyses (because they are calculated on an assumed 100% basis) and to refine our understanding of the sources of sediments in the eastern Mediterranean.

Factor Analysis

A factor analysis similar to that of Imbrie and van Andel (1964) was run on the percentage abundances of four clay minerals and v/p ("crystallinity" index) of smectite of 90 samples analyzed. Each component was first percent-range-transformed. In the Q-mode factor analysis, attention is focused on the samples on the basis of all the variables (clay minerals) with the objective of finding the minimum number of end members (as identified by certain actual samples) in terms of which the sediment variability can be explained. Factor loadings are then calculated for each sample as the degree to which its mineral composition can be explained in terms of each end member.

In the present study, the Q-mode analysis has selected three factors which can explain 98.6% of the variability in the samples. The three factors are designated here as A, B, and C. The extremal (reference) sample of Factor A is characterized principally by high amounts of smectite with high v/p ratios; that of Factor B by high kaolinite; and Factor C by high chlorite and illite (Table 2). The distributions of the different factor loadings are shown in Figures 4 to 6.

Very high values of Factor A loadings are confined to the eastern Levantine Basin and the eastern Nile cone and moderately high values occur in the westernmost part of the Ionian Sea south of Sicily (Figure 4). Examined in detail, differences within the Levantine Basin and between this basin and the Aegean Sea become apparent. Within the Levantine Basin, the Factor A loadings decrease from east to west and from the Nile River towards the Mediterranean Ridge. This is seen most dramatically in the decrease of v/p ratio from the Nile towards the Mediterranean Ridge (Figure 7). An area of very high crystalline (0.5 to > 0.6) smectite in the southern Aegean Sea extending to the Mediterranean Ridge is also not apparent in the Factor A loadings. Also the sediments of the southern Aegean Sea have lower kaolinite/chlorite ratios compared to the Nile sediments (Figure 8).

The sediments with high Factor A loadings in the Nile cone and the easternmost Levantine Basin are clearly derived from the Nile source. The dispersal of the Nile sediments is affected by the easterly directed surface currents which are part of the main Mediter-

Table 2. Clay mineral percentages and v/p ratios of extremal samples of factors A, B, and C.

Factor	Core No.	Smectite	v/p	Kaolinite	Chlorite	Illite
A	CH61–66	69	0.65	20	4	7
B	RC9–182	28	0.30	32	6	4
C	RC9–186	27	0.42	10	14	49

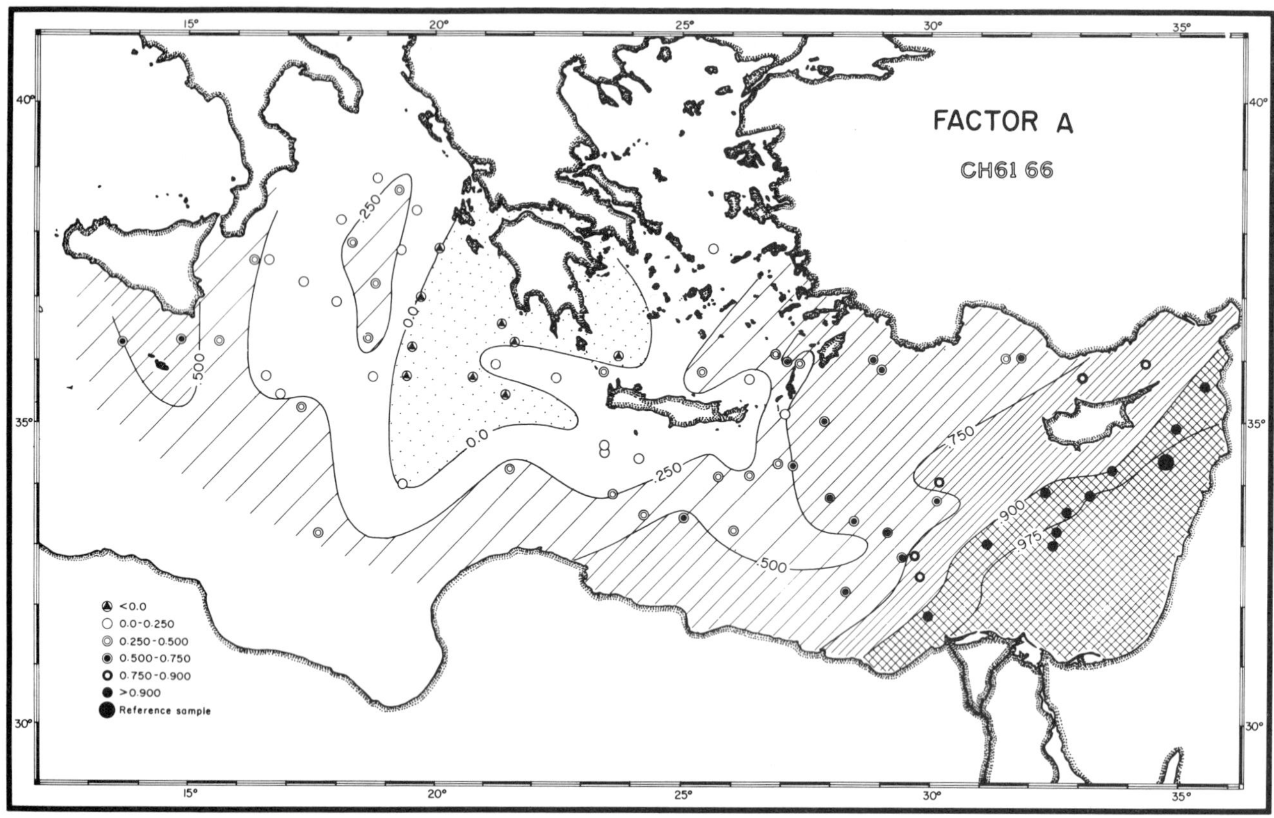

Figure 4. Distribution of Factor A loadings. CH61–66 (largest black dot) is extremal, or reference sample.

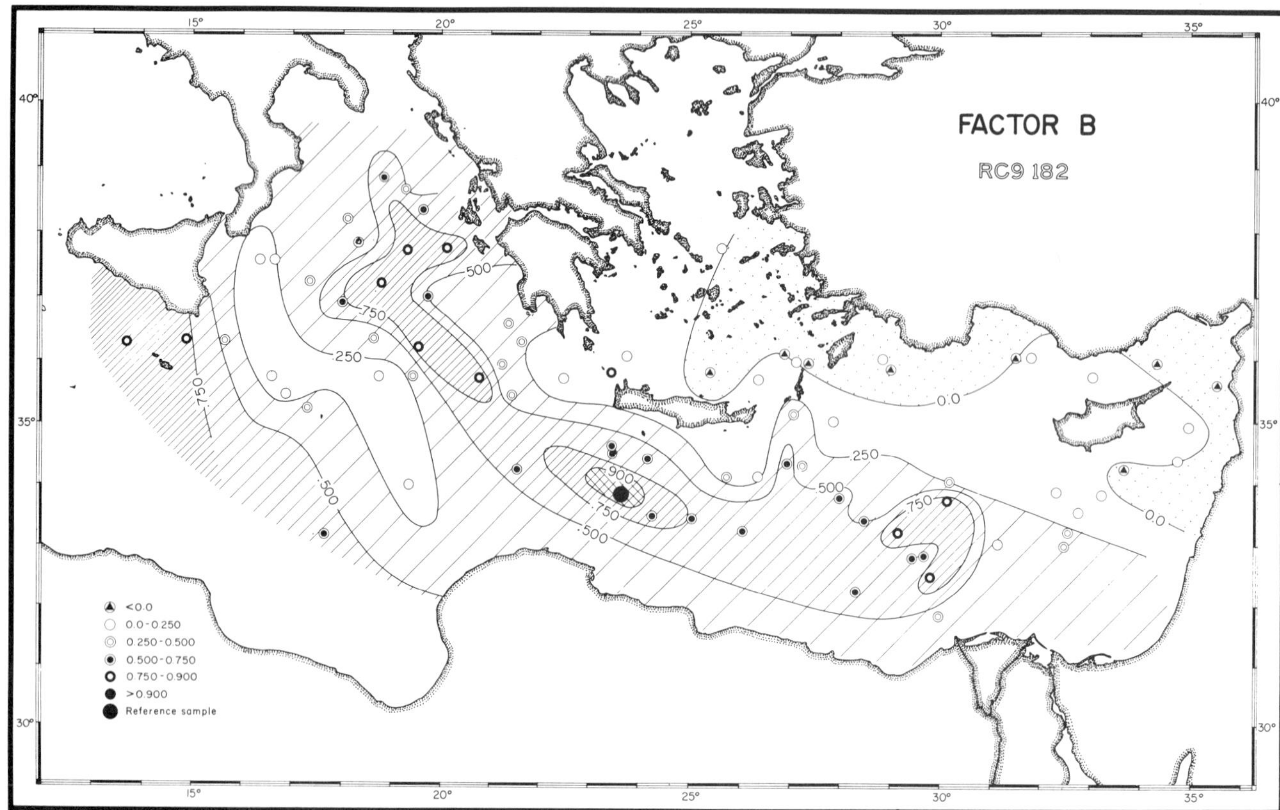

Figure 5. Distribution of Factor B loadings. RC9-182 (largest black dot) is extremal, or reference sample.

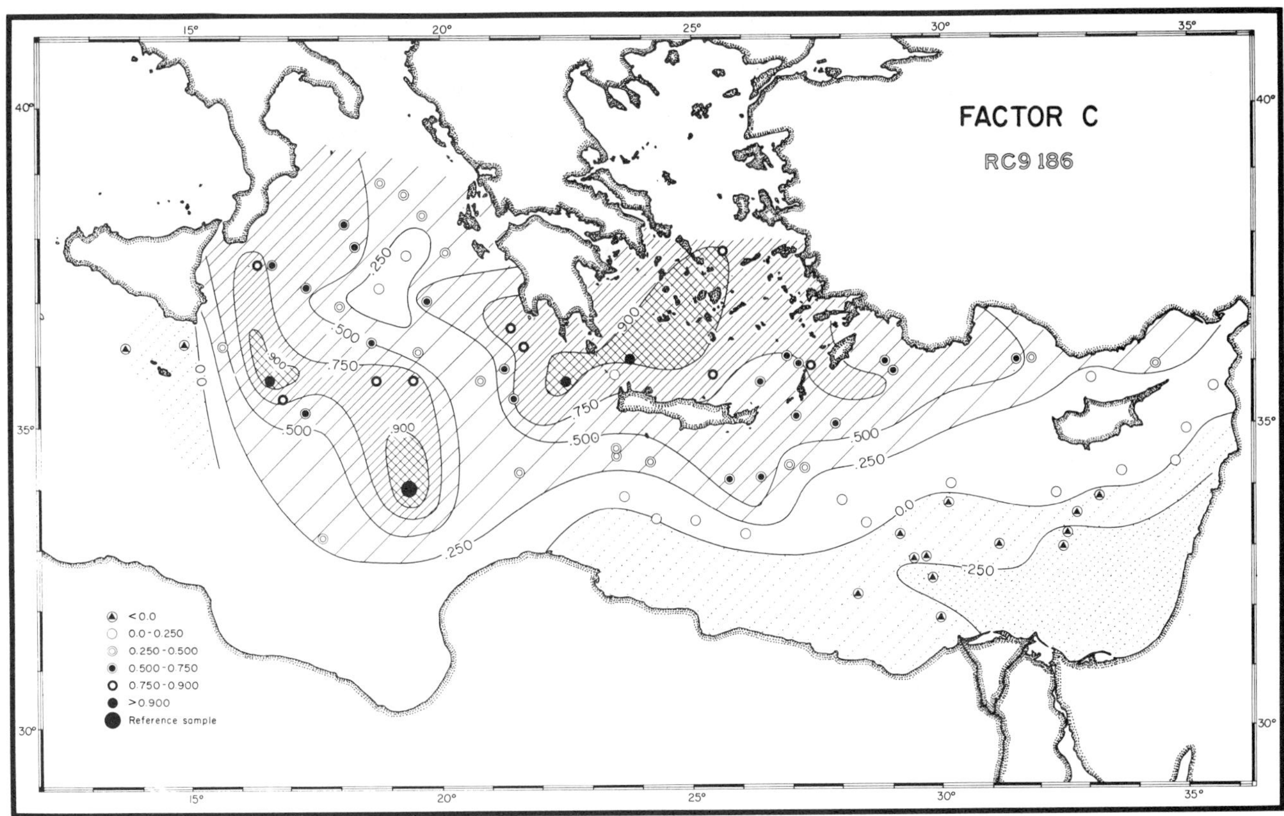

Figure 6. Distribution of Factor C loadings. RC9–186 (largest black dot) is extremal, or reference sample.

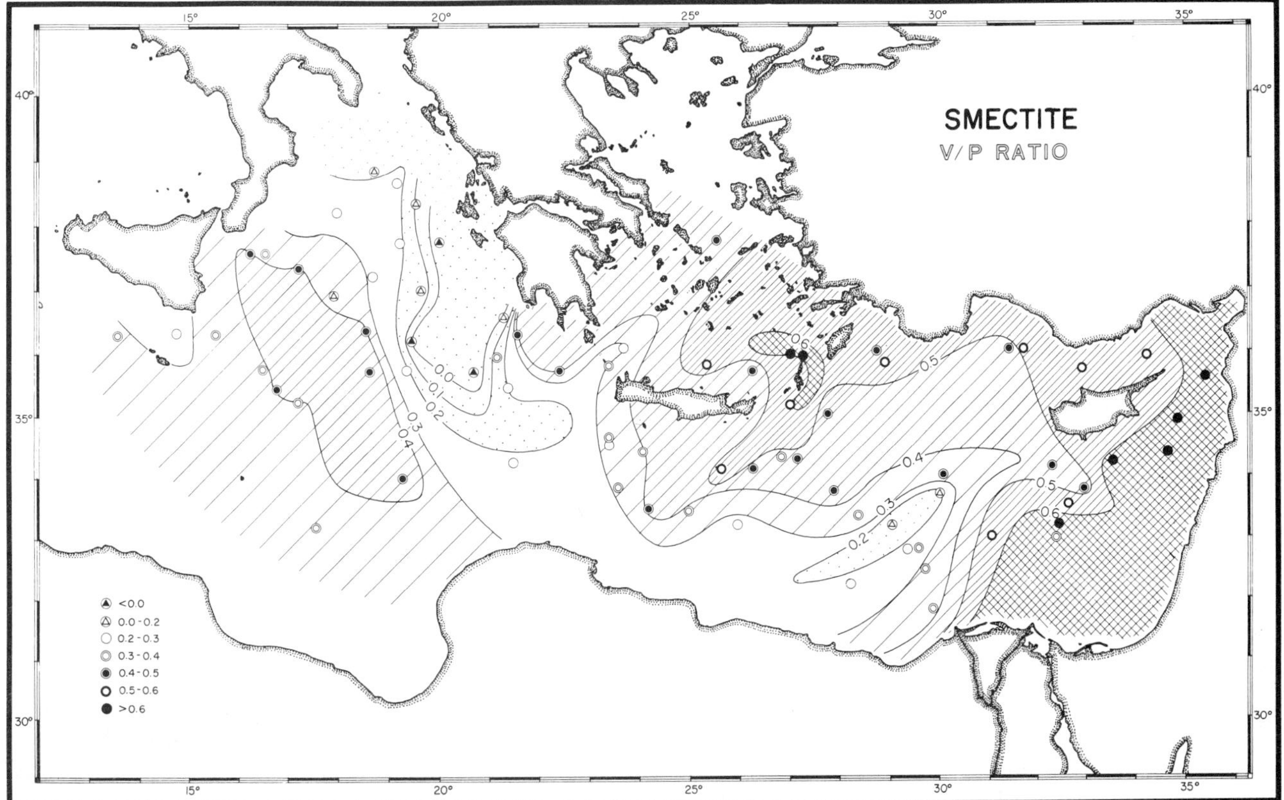

Figure 7. Distribution of v/p ("crystallinity") index for smectite.

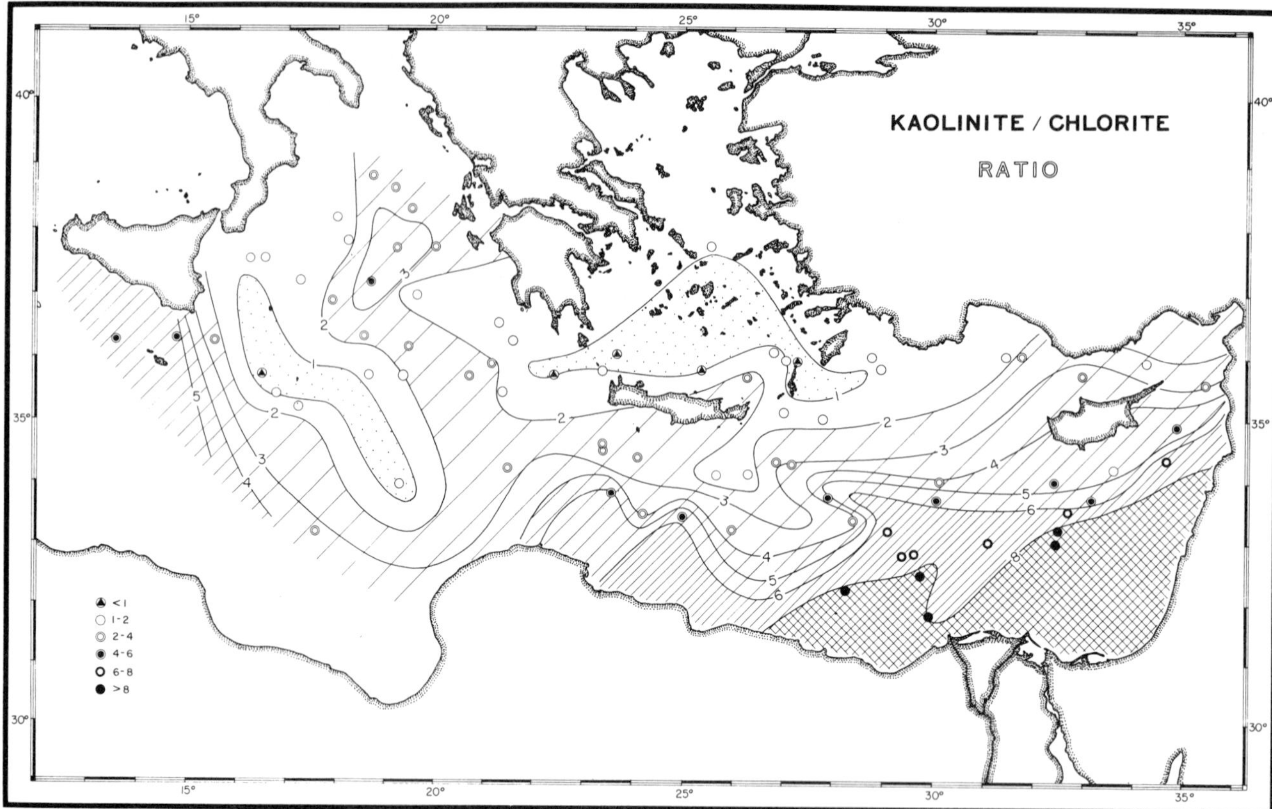

Figure 8. Distribution of Kaolinite/Chlorite ratio.

ranean counter-clockwise gyre (Figure 9A). This inference is supported by the observation of Zarudzki *et al.* (1969) based on seismic profiling that the thickness of unconsolidated sediments increases eastward across the Nile cone. The presence of low amounts of $CaCO_3$ in the eastern Nile cone and the eastern Levantine Basin provides further support for this view as does the transport of even coarse-sized Nile sand as far as the shores of Lebanon (Emery and George, 1963; Emery and Neev, 1960). Differentiation of Nile sediments seems to have occurred as the finer-sized smectite is transported farther eastward relative to other clay minerals. The distribution of Factor A and particularly the v/p of smectite indicate that little sediment is directly transported across the Ridge into the northern Levantine Sea. The turbidite sands present at depth in ten cores taken in the western Nile cone and Herodotus Abyssal Plain contain considerable amounts of clinopyroxene (order of 35%), hornblende (55%) and epidote (10%) and clearly have been derived from the Nile and transported into the area by turbidity currents. This contrasts with the Holocene situation when transport of Nile sediments into the western cone and Herodotus Abyssal Plain has been less important. The extension of well crystallized smectite from the southeast Aegean Sea onto the Mediterranean Ridge southeast of Crete is the result of transport by intermediate water from the Aegean Sea (Figure 9B).

The high Factor B loadings are confined for the most part to the Mediterranean Ridge, to the Herodotus Abyssal Plain and western Nile cone, and to the westernmost Ionian Sea south of Sicily. The high Factor B loadings of the Ridge generally coincide with the zone of high $CaCO_3$. The desert soils west of the Nile River contain dominant amounts of kaolinite (Elgabaly and Khadar, 1962). We think that the high Factor B loadings on the Ridge reflect the eolian transport of desert soils, not diluted by the addition of material carried by bottom currents. The increased haze as one approaches Africa is well known to mariners.

Factor C loadings are high in the Messina cone-Sirte Abyssal Plain, and in the Hellenic Trough and the Aegean Sea near Peloponnesus. These two high Factor C areas are separated from each other by the zone of high Factor B loadings of the Mediterranean Ridge. The high Factor C sediments of the Messina cone-Abyssal Plain may have been derived from the Adriatic Sea and parts of southern Calabria (Duplaix, 1958). The distribution of these sediments follows the known paths of the Ionian deep water from its

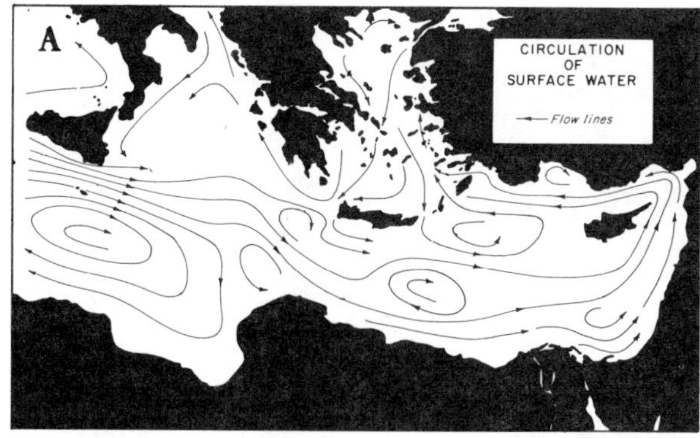

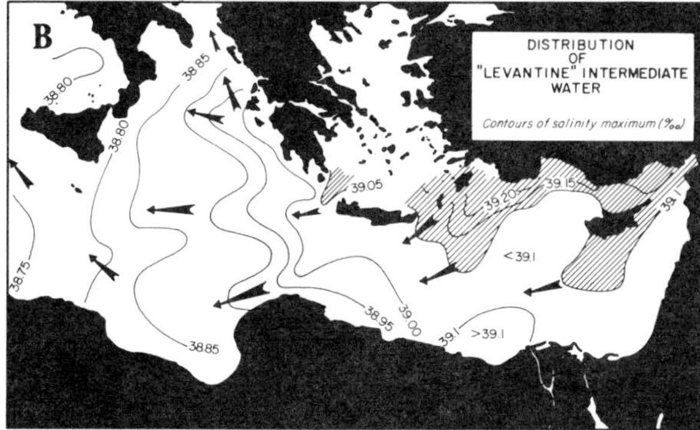

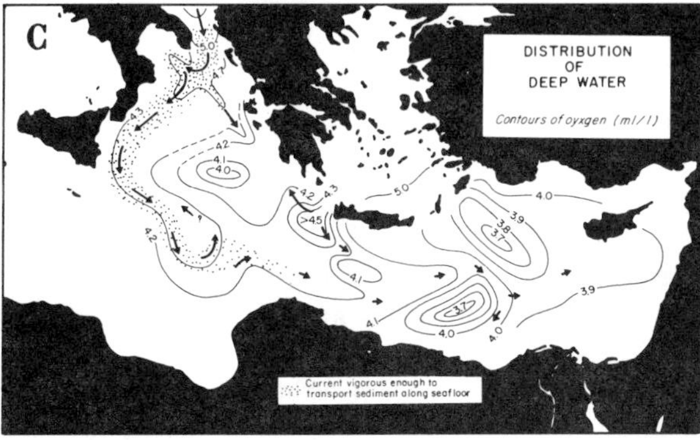

Figure 9 A, Surface water circulation; B, distribution of the "Levantine" intermediate water mass as shown by contours of high salinity; and C, distribution of deep water as shown by contours of dissolved oxygen (from Moskalenko and Ovchinnikov, 1965).

origin in the southern Adriatic Sea (Figure 9C).

The high Factor C clays of the Hellenic Trough and the nearby Aegean Sea, have probably been derived from the soils of the Peloponnesus and nearby land. The greater restriction of Factor C sediments to the deep areas of the Hellenic Trough suggests that the deep cold water mass spilling out of the Aegean Sea is responsible for such dispersal (Figure 9C).

Rb/Sr Ratio and "Apparent Age"

The distinction of the various sediment sources contributing to a given area is facilitated in proportion to the number of independent variables which may be measured in the accumulated sediments. As pointed out by several workers (Biscaye, 1965; Rateev *et al.*, 1969; Griffin *et al.*, 1968) the mineralogy of deep-sea fine fraction detritus primarily reflects the type and intensity of weathering in the continental source areas. Studies of the rubidium, strontium, strontium-isotope system in deep-sea sediments (Dasch, 1969; Biscaye and Dasch, 1971; and Biscaye and Dasch, in preparation) have shown that these parameters reflect certain geological characteristics of the continental source areas, *i.e.*, lithology and geologic age of the parent rocks from which the soil sediments formed. We have made measurements of these parameters on a limited number of samples as a test of the validity of the sediment sources and dispersal patterns inferred above.

We measured the Rb/Sr ratio on the same $< 2\mu$ (carbonate-free) size fractions from which the x-ray mineral analyses had been obtained. The results of these measurements are presented in Figure 10. Although the sample distribution is not uniform, a good correlation with some of the distributions based on mineralogy may be seen. In the South Atlantic Ocean, where oceanic bottom currents are important in sediment dispersal, Biscaye and Dasch (1971) have shown that the distribution pattern of this parameter may reflect the action of these currents as much as it does the source of the sediments. That is, the Rb/Sr ratio of sediment from a given source may change during transport by currents because of processes of differential sedimentation or winnowing. While this may introduce ambiguities in distinguishing sediment sources, it may also be helpful in accentuating current activity.

Along the eastern coast of the Mediterranean the Rb/Sr ratio around 0.7 is consistent with the mineralogical distinction of the high-crystallinity, montmorillonite-rich sediment source associated with the Nile and transported northeastward by the major counterclockwise current gyre. The sharp transition to Rb/Sr values around 1.1 along the south coast of Turkey is consistent with the dominance of another sediment source. This is probably more closely related mineralogically to the southeastern Aegean source as suggested by all three factor maps (Figures 4–6). It may, however, represent a sediment source not distinguishable in terms of mineralogy.

To the west the high Rb/Sr ratios (1.3 to 1.8) in the Ionian Basin confirm the distinction of the illite- and chlorite-rich Messina assemblage distinguishable also by high Factor C loadings (Figure 6). The other illite- and chlorite-rich sediment source (the Kithira, coming from the Aegean Sea between the Peloponnesus and Crete) may also have a high Rb/Sr ratio as suggested by the two samples analyzed from that region (1.6 in

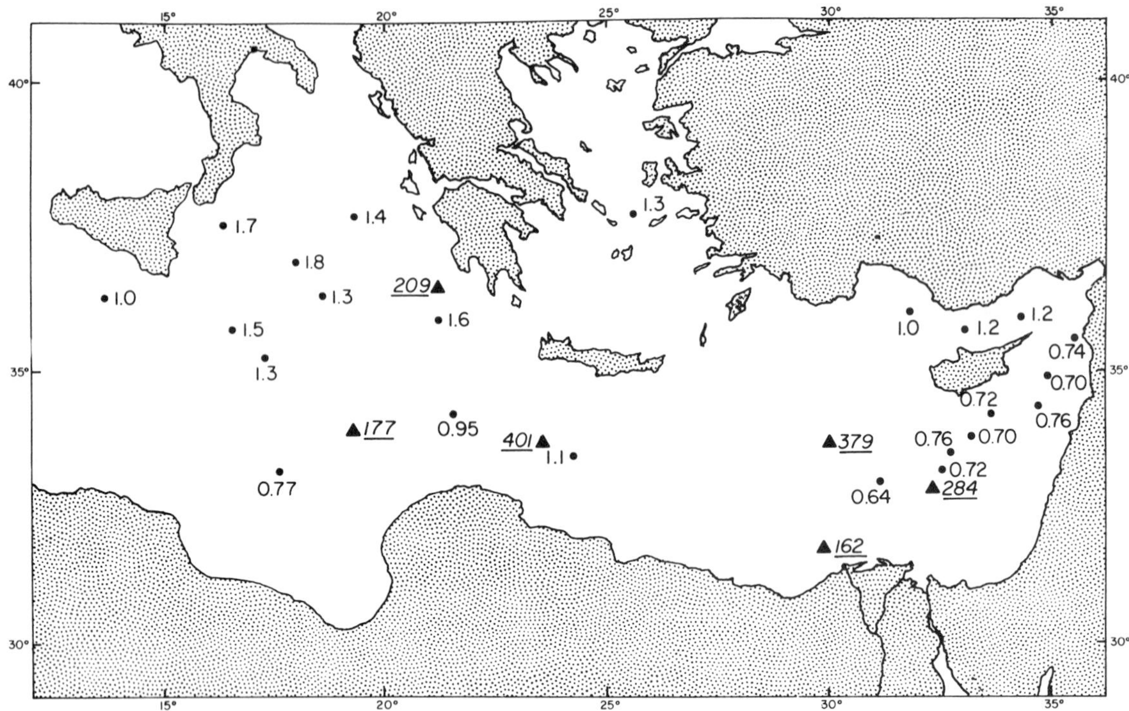

Figure 10. Distribution of Rb/Sr ratio in carbonate-free, $< 2\mu$ fraction (circles) and Sr-isotope "apparent ages" of carbonate-free, bulk samples (triangles). Apparent age values (underlined) are in "million years".

the Hellenic Trough and 1.3 in the Aegean; both within the 0.900 Factor C loading contours of Figure 6). The samples to the south along the Mediterranean Ridge (Rb/Sr around 1.0) again have significantly lower ratios than either the Messina or Kithira assemblages and strengthen the distinction of the principally eolian, kaolinite-rich source represented in high loadings of Factor B in Figure 5.

In addition to the Rb/Sr analyses of the $< 2\mu$ fraction, six samples representing extremal or near-extremal values of the three Q-mode factors were chosen for measurement of Rb/Sr and Sr^{87}/Sr^{86} on the whole, noncarbonate fraction. Biscaye and Dasch (1971, and in preparation) have shown that the "apparent age" of the sediments calculated from Rb/Sr and Sr^{87}/Sr^{86}, is related to the geologic age of the rocks underlying the continental provenance area of the sediments. The apparent age is also less affected than either Rb/Sr or Sr^{87}/Sr^{86} values themselves by current transport and other sedimentation processes discussed above. The analytical results and the factor loadings of the six samples are shown in Table 3. The apparent age of these samples is also plotted on Figure 10. (We emphasize that, although this parameter is given in units of millions of years, this represents neither the age of sedimentation—in this case Holocene—nor the exact geologic age of the source area.)

Based only on apparent age, the two samples representing the A factor (principally the Nile source) cannot be distinguished from the two representing the C factor (principally the Messina and Kithira assemblages, *i.e.,* the Adriatic and southwestern Aegean sources). (They are, however, different in Rb/Sr ratios.) The apparent age of the two samples representing the B factor, however, are distinctly "older" than the other four samples. The high apparent age (401 my) for RC9-182 supports, in our view, the conclusion based on mineralogy that the important source of Mediterranean Ridge sediment is wind-borne detritus from northern Africa. The Paleozoic

Table 3. Rb, Sr-isotope characteristics of near-extremal samples from Q-mode factor analysis.

Core	Rb/Sr($\pm\sigma$)	Sr^{87}/Sr^{86}	"Apparent Age" (million years)	Q-mode factor and percentage loading
RC9–174	0.661(0.031)	0.7122	284	A–0.980
V14–129	0.495(0.013)	0.7076	162	A–0.926
RC9–182	0.836(0.013)	0.7185	401	B–1.000
RC9–177	0.734(0.037)	0.7161	379	B–0.821
RC9–186	1.50 (0.020)	0.7155	177	C–1.000
V10–27	1.57 (0.015)	0.7182	209	C–0.846

and Precambrian rocks of the northern Sahara represent the geologically oldest terrain around the eastern Mediterranean.

A similarly strong eolian influence for the other kaolinite-rich sample representing Factor-B (RC9-177) is more difficult to believe. This sample is from the Herodotus Abyssal Plain northwest of the Nile cone where one would expect a strong Nile influence. However, all of the mineral distribution maps (Venkatarathnam and Ryan, 1971), the Q-mode factor loading maps (Figures 4–6) and the calcium carbonate distribution (Figure 2) all indicate that the Nile detritus is transported dominantly to the northeast, away from the Herodotus Abyssal Plain. Thus both the mineralogy and the "old" apparent age suggest a common source for the Holocene Herodotus Abyssal Plain sediments and for those of the Mediterranean Ridge.

SUMMARY

Holocene sediments from the eastern Mediterranean Sea have been analyzed to determine calcium carbonate concentration, fine-fraction mineralogy, Rb/Sr ratio and Rb/Sr "apparent age" and for the concentrations of four transition metals. The mineralogical characteristics have been analyzed by Q-mode factor analysis. Based on these results we draw the following conclusions concerning sedimentary sources and processes in the eastern Mediterranean.

1. Calcium carbonate, principally from pelagic plant and animal tests, is relatively abundant along the Mediterranean Ridge probably because here it is relatively free from dilution by the alumino-silicate detritus found in adjacent topographic lows. Sediments of the southwestern Levantine Basin (including the Herodotus Abyssal Plain) also contain high carbonate concentrations which again are attributed to reduced accumulation of alumino-silicate detritus despite proximity to the Nile Delta, probably because sediment from the Nile River is at present being diverted by currents to the northeast.

2. Six distinctive clay mineral assemblages identify source areas of alumino-silicate detritus in the eastern Mediterranean. The assemblages are named after the regions where they are most prominently developed: Nile, southeastern Aegean, Kithira, Messina, Sicilian, in addition to a Kaolinite-rich assemblage which characterizes the Mediterranean Ridge and the southwestern Levantine Basin (Venkata-

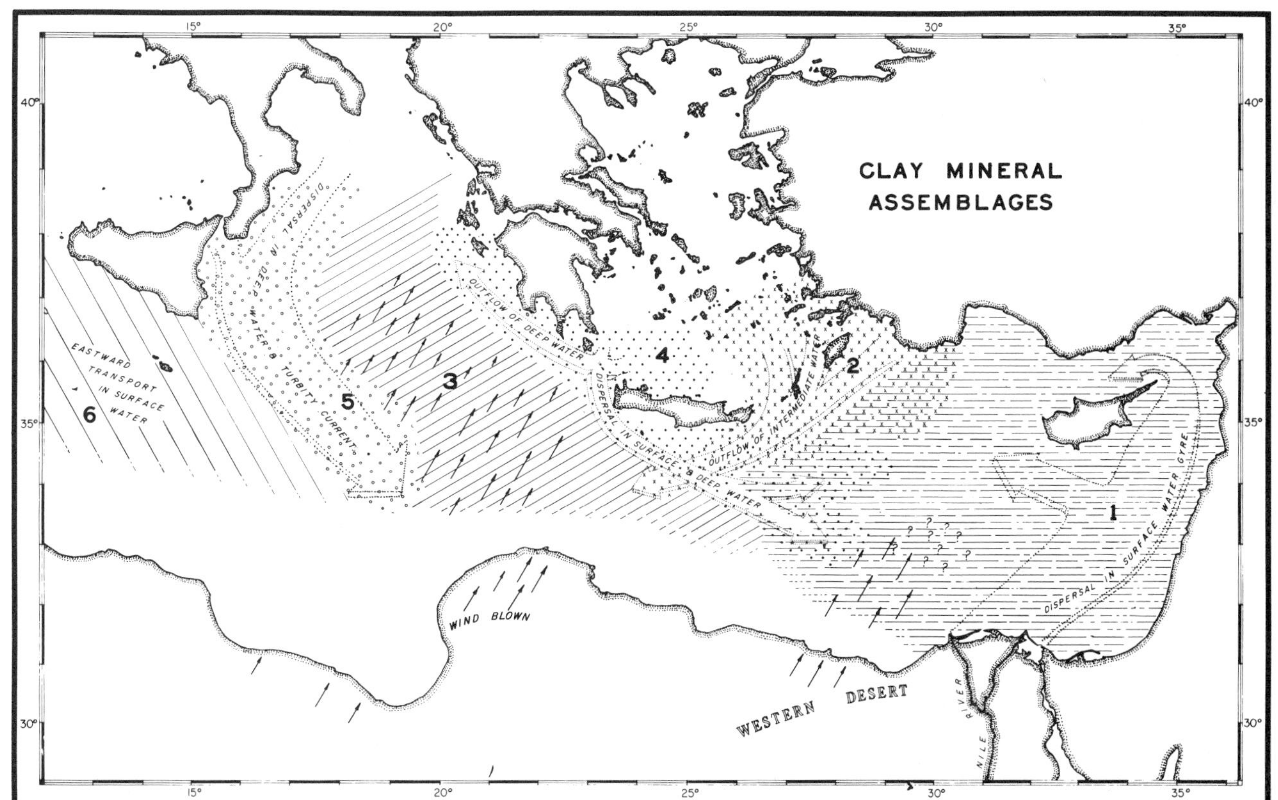

Figure 11. Detrital sources and dispersed paths of clay mineral assemblages in Holocene sediment in the eastern Mediterranean Sea. 1, Nile assemblage; 2, Southeast Aegean assemblage; 3, Kaolinite-rich assemblage; 4, Kithira assemblage; 5, Messina assemblage; 6, Sicilian assemblage.

rathnam and Ryan, 1971). These sources and their dispersal paths are summarized in terms of these assemblages in Figure 11.

3. Q-mode factor analysis of the clay mineral data confirms most of the features of the mineral distributions and has delineated three source factors which account for 98.6% of the variability in the data.

4. Rb/Sr ratios of the $< 2\mu$, carbonate-free sediment are consistent with the mineral assemblages and sources indicated by factor analysis.

5. Rb/Sr ratios and Sr-isotope "apparent ages" of six samples representing extremal or near-extremal Q-mode factors are consistent with the mineralogically determined sediment source end members.

6. The concentration of Fe is controlled entirely by the several sources of alumino-silicate detritus with calcium carbonate acting only as a diluent. The correlation between calcium carbonate percentage and concentration of Cu and Mn calculated on a carbonate-free basis is an artifact of the assumption inherent in that mode of calculation. Our data do not permit a distinction between the several possible controls on the distribution of Cu, Mn or Zn.

ACKNOWLEDGMENTS

The authors thank E. F. K. Zarudzki of Woods Hole Oceanographic Institution for samples from the CHAIN-61 cruise. We gratefully acknowledge the technical assistance of Adele Hanley and the critical comments of Michael Bender and P. J. Fox of Lamont-Doherty Geological Observatory, E. Julius Dasch of Oregon State University and Gilbert Kelling of the University of Wales. Some of the laboratory work was carried out when the senior author was a UNESCO Fellow; some was done while the second author was a Visiting Researcher at the SNPA laboratories, Pau, France for which thanks are due to Georges Kulbicki. This study was also in part supported by National Science Foundation Grant No. GA-580, Atomic Energy Commission Contract No. AT(11-1) 3132, and by Office of Naval Research Contract No. N-00014-67A-0108-0004. The latter contract, along with National Science Foundation Grant No. GA-29460, support the operation of LDGO ships which were used to raise most of the cores used in this study.

REFERENCES

Bender, M. L. and C. Schultz 1969. The distribution of trace metals in cores from a traverse across the Indian Ocean. *Geochimica et Cosmochimica Acta*, 33:292–297.

Biscaye, P. E. 1964. *Mineralogy and Sedimentation of Deep-Sea Sediment fine fraction in the Atlantic Ocean and Adjacent Seas and Oceans*. Ph. D. thesis, Yale University, 86p.

Biscaye, P. E. 1965. Mineralogy and sedimentation of recent deep-sea clay in the Atlantic Ocean and adjacent seas and oceans. *Geological Society of America Bulletin*, 76:803–832.

Biscaye, P. E. and E. J. Dasch 1971. The rubidium, strontium, strontium-isotope system in the deep-sea sediments; Argentine Basin. *Journal of Geophysical Research*, 76:5087–5096.

Boström, K., M. N. A. Peterson, O. Joensuu and D. E. Fisher 1969. Aluminium-poor ferromanganoan sediments on active oceanic ridges. *Journal of Geophysical Research*, 74:3261–3270.

Boström K. and D. E. Fisher 1971. Volcanogenic uranium, vanadium and iron in Indian Ocean sediments. *Earth and Planetary Science Letters*, 11:95–98.

Chamley, H. 1962. Minéraux argileux de vases Méditerranéennes. *Bulletin du Service de la Carte Géologique, Alsace Lorraine*, 15 (4): 161–169.

Chamley, H. 1972. Sur la sédimentation argileuse profonde en Méditerranée. In: *The Mediterranean Sea: A Natural Sedimentation Laboratory*, ed. Stanley, D. J., Dowden, Hutchinson and Ross, Inc., Stroudsburg, Pennsylvania, 387–399.

Dasch, E. J. 1969. Strontium isotopes in weathering profiles, deep-sea sediments and sedimentary rocks. *Geochimica et Cosmochimica Acta*, 33:1521–1552.

Duplaix, M. N. 1958. Etude minéralogique des niveau sableux des carrottes prélevées sur fond de la Méditerranée. *Reports of the Swedish Deep-Sea Expedition, 1947–1948*, 8:137–166.

Elgabaly, M. M. and M. Khadar 1962. Clay mineral studies of some Egyptian desert and Nile alluvial soils. *Journal of Soil Science*, 13:333–342.

Emelyanov, E. M. 1965a. The granulometric composition of modern sediments and some features of their formation in the Mediterranean Sea (in Russian). In: *The Main Features of the Geological Composition, the Hydrological Regime and Biology of the Mediterranean Sea (Osnovnye Cherty Geologisheskogo Stroenia, Gidrologicheskogo Rezhima i Biologii Stredizemnogomona)*, ed. Fomin, L., Nauka, Moscow, 42–67.

Emelyanov, E. M. 1965b. The carbonate content of modern bottom sediments of the Mediterranean Sea (in Russian). In: *The Main Features of the Geological Composition, the Hydrological Regime and Biology of the Mediterranean Sea (Osnovnye Cherty Geologisheskogo Stroenia, Gidrologicheskogo Rhezima i Biologii Stredizemnogomona)*, ed. Fomin, L., Nauka, Moscow, 71–84.

Emery, K. O., and C. J. George 1963. The shores of Lebanon. *The American University of Beirut, Miscellaneous Papers in the Natural Sciences*, 1:1–10.

Emery, K. O., and D. Neev 1960. Mediterranean beaches of Israel. *Israel Geological Survey Bulletin*, 26:1–23.

Goldberg, E. D. and G. O. S. Arrhenius 1958. Chemistry of Pacific pelagic sediments. *Geochimica et Cosmochimica Acta*, 26:525–544.

Griffin, J. J., H. Windom and E. D. Goldberg 1968. The distribution of clay minerals in the world ocean. *Deep-Sea Research*, 15: 433–459.

Hülsemann, J. 1966. On the routine analysis of carbonate in unconsolidated sediments. *Journal of Sedimentary Petrology*, 36:622–625.

Imbrie, J. and T. H. van Andel 1964. Vector analysis of heavy mineral data. *Geological Society of America Bulletin*, 75:1131–1156.

Krauskopf, K. B. 1956. Factors controlling the concentrations of thirteen rare metals in the sea water. *Geochimica et Cosmochimica Acta*, 9:1–32.

McGill, D. A. 1961. A preliminary study of the oxygen and phosphate distribution in the Mediterranean Sea. *Deep-sea Research*, 8:259–269.

Moskalenko, L. V. and I. M. Ovchinnikov 1965. The water masses of the Mediterranean Sea (in Russian). In: *The Main Features of the Geological Composition, the Hydrological Regime and Biology of the Mediterranean Sea (Osnovnye Cherty Geologisheskogo

Stroenia, Gidrologicheskogo Rhezima i Biologii Stredizemnogomona), ed. Fomin, L., Nauka, Moscow, 119–131.

Nielson, J. N. 1912. Hydrography of the Mediterranean and adjacent waters. *Reports of the Danish Oceanographic Expedition, 1908–1910*, 1:77–191.

Ovchinnikov, I. M. and E. A. Plakhin 1965. The historical observations of the hydrological researches in the Mediterranean Sea (in Russian). In: *The Main Features of the Geological Composition, the Hydrological Regime and Biology of the Mediterranean Sea (Osnovnye Cherty Geologisheskogo Stroenia, Gidrologicheskogo Rhezima i Biologii Stredizemnogomona)*, ed. Fomin, L., Nauka, Moscow, 84–107.

Pollack, M. J. 1951. The sources of deep water of the eastern Mediterranean Sea. *Journal of Marine Research*, 10:128–152.

Rateev, M. A., E. M. Emelyanov and M. B. Kheirov 1966. Conditions for the formation of clay minerals in contemporaneous sediments of the Mediterranean Sea. *Lithology and Mineral Resources (Litologia i Poleznye Iskopayeme)*, 5:418–431.

Rateev, M. A., Z. N. Gorbunova, A. P. Lisitzyn and G. L. Nosov 1969. The distribution of clay minerals in the oceans. *Sedimentology*, 13:21–43.

Revelle, R. R., M. Bramlette, G. Arrhenius and E. D. Goldberg 1955. Pelagic sediments of the Pacific. *Geological Society of America Special Paper*, 62:221–236.

Sevast'yanov, V. F. 1968. Redistribution of chemical elements as a consequence of oxidation-deoxidation processes in sediments of the Mediterranean Sea. *Lithology and Mineral Resources (Litologia i poleznye Iskopayeme)*, 1:3–15.

Turekian, K. K. and D. F. Schutz 1965. Trace element economy in the oceans. *Narragansett Marine Laboratory, University of Rhode Island, Occasional Publication*, 3:41–89.

Turekian, K. K. and J. Imbrie 1966. The distribution of trace elements in deep-sea sediments of the Atlantic Ocean. *Earth and Planetary Science Letters*, 1:161–168.

Venkatarathnam, K., and W. B. F. Ryan 1971. Dispersal patterns of clay minerals in the sediments of the eastern Mediterranean Sea. *Marine Geology*, 11:261–282.

Wüst, G. 1961. On the vertical circulation of the Mediterranean Sea. *Journal of Geophysical Research*, 66: 3261–3271.

Zarudzki, E. F. K., H. K. Wong and J. D. Phillips 1969. Structure of the eastern Mediterranean (abstract). *Transactions of the American Geophysical Union*, 50:208.

Données du Sondage Sismique Continu Concernant la Sédimentation Plio-Quaternaire en Méditerranée Nord-Occidentale

Geneviève Alla, Daniel Dessolin, Olivier Leenhardt et Serge Pierrot

Musée Océanographique de Monaco, Monaco

RESUME

Une grande partie de la plaine abyssale et de la pente continentale du Nord de la Méditerranée occidentale est couverte d'un quadrillage sismique (sparker et Flexotir). On analyse les différents réflecteurs caractéristiques, leur distribution géographique et on propose une identification stratigraphique de ces réflecteurs.

On distingue ainsi 4 zones paléogéographiques présentant un "faciès sismique" propre: Mer Ligure, zone Sud des Maures, Golfe du Lion, et Nord-Baléares. Trois ensembles stratigraphiques sont décrits: un Miocène supérieur horizontal, non identifié (D), un Miocène terminal évaporitique et salifère (B et C), un Plio-Quaternaire (A) hémipélagique à la base, à sédimentation plus variée et plus grossière au sommet. La néotectonique et la tectonique salifère ont contribué au modelé actuel du Plio-Quaternaire de cette région. Ont en donne quelques exemples.

ABSTRACT

Continuous seismic profiles, using a 9000 Joules sparker, a Chesapeake M16 hydrophone and a graphic recorder (Alden and EG&G), have been obtained over a large portion of the abyssal plain and continental slope and rise in the western Mediterranean Sea. The various major reflectors and their geographical distribution are described, and stratigraphic interpretations are made.

Four paleogeographic zones are recognized on the basis of the sub-bottom surveys: Ligurian Sea, region south of the Maures Massif, Gulf of Lions, and the Northern Balearic region. Three stratigraphic horizons are noted: Upper Miocene (undifferentiated, D), Uppermost Miocene (evaporitic and salt sequences, B and C), and Plio-Quarternary (hemipelagic at the base becoming coarser and lithologically more varied at the top of the unit, A). Three different types of tectonic processes are believed to have determined the present shape of the basin in this part of the Mediterranean Sea.

HISTORIQUE

En 1957, Alinat et Cousteau (1962) découvrent, sur le fond plat de la plaine abyssale ligure, de petites collines de 60 m de haut, de 1 à 2 km de diamètre. Bourcart (1960) pensait que ces collines étaient peut-être volcaniques.

Le premier sondage sismique continu est réalisé sur cette plaine par Hersey (*cf.* Fahlquist et Hersey, 1969) en 1961, à bord du CHAIN. La pénétration est alors insuffisante pour que, du profil réalisé qui passait sur une série de dômes non perçants, on puisse déduire autre chose que l'existence de mouvements tectoniques sous le fond.

En 1962, Menard découvre d'autres collines, plus importantes, au Sud de Toulon (Menard *et al.*, 1965, taches noires sur la Figure 2), qu'ils lèvent en bathymétrie. Menard *et al.* (1965) passent en revue les hypothèses possibles pour expliquer ces collines et suggèrent qu'il s'agit de dômes de sel.

En 1964, nous avons pu utiliser la CALYPSO pour un premier sondage sur la colline A (42°47′N, 07°41′E) avec le boomer 13.000 Joules. Nos données sont en en nombre trop limité pour que nous puissions publier l'hypothèse que suggère l'enregistrement, mais le CHAIN, alors en Méditerranée, peut reconnaître la même zone, et Hersey (1965) avance l'hypothèse de dômes de sel. Alinat *et al.* (1966) la confirment à la suite d'une reconnaissance plus complète. Puis, L. Glangeaud, dirigeant la campagne GEOMEDE I, donne une théorie de l'ensemble de la Méditerranée occidentale, proposant un sel d'âge triasique: il définit la zone A (où se trouvent les dômes) comme le prolongement du subbriançonnais alpin, thèse que défend ensuite Guillaume (1967). La CALYPSO précise en 1966 la limite Nord de la zone des dômes (Glangeaud *et al.*, 1966); alors que le JEAN-CHARCOT avait reconnu le Dôme T (Sud de Toulon) en 1967, la WINNARETTA-SINGER étudie en détail, par sismique, le dôme A, après une étude bathymétrique et magnétique du BANNOCK (Leenhardt, 1968).

La CATHERINE-LAURENCE reconnaît la zone Sud d'Imperia, complétant des profils de la CALYPSO et de la WINNARETTA-SINGER, et effectue divers profils au large des Maures et de l'Esterel. Les publications du Centre de recherches géodynamiques de Villefranche-sur-Mer (Bellaiche, 1969; Mascle, 1968; Mauffret, 1969; Pautot, 1969; Rehault, 1968) utilisent les profils du Musée océanographique complétés, dans plusieurs travaux, par ceux de la CATHERINE-LAURENCE.

En 1968 et 1969, les croisières du FRANÇOIS-BLANC (Alla *et al.*, 1969) et de la MURIEL (Alla, 1970) poursuivent ces études vers l'Ouest. L'exploration du bassin occidental de la Méditerranée s'est effectuée en 1967 à bord du JEAN-CHARCOT (croisière GIBRALTAR II), du TEREBEL, puis en 1970, par le projet ANNA et les campagnes GEOMEDE III et POLYMEDE du JEAN-CHARCOT.

Les principaux travaux américains sont ceux de Watson et Johnson (1969). Un profil du METEOR (Alinat *et al.*, 1970) coupe longitudinalement la plaine abyssale au Nord de l'Algérie. Ces derniers travaux ne sont pas intégrés à la présente étude.

Par contre, nous avons pu utiliser le profil AUGUSTA, tiré en Flexotir (Montadert *et al.*, 1970), qui apporte la preuve directe de l'existence du sel, confirmée par les forages ultérieurs du GLOMAR-CHALLENGER, qui permettent de dater le sel.

Nous nous limitons ici à l'étude de 5.000 milles environ de sondages sismiques continus (Figure 1). Dans certaines zones, des quadrillages à maille serrée (1 à 5 milles) complètent les profils de grande reconnaissance.

DONNEES TECHNIQUES

Les appareils utilisés pour obtenir les résultats discutés ici sont:

1. un sparker E.G.&G. 9.000 Joules à 3 électrodes, traîné à 25 m environ derrière le navire, immergé à 4 ou 5 m de profondeur.
2. une flûte Chesapeake M 16, traînée à 250 ou 270 m derrière le sparker, selon la taille du navire utilisé.
3. deux enregistreurs graphiques PGR Alden 419 (enregistrant sur 3,75 s d'échelle) ou E.G.&G. 254 (enregistrant sur 2,5 ou 3,3 s d'échelle).

A la vitesse de 6 ou 7 noeuds, l'appareillage permet une pénétration de l'ordre de 1,5 s (temps-double) sous le fond. Cette profondeur correspond, nous le verrons, au moins à l'ensemble du Plio-Quaternaire. La résolution, de l'ordre de 60 ms, favorise l'étude détaillée de ces séries.

L'exagération verticale des enregistrements dépend de la vitesse du navire, du type d'enregistreur utilisé ainsi que de l'échelle et du programme d'enregistrement. Le rapport de l'échelle verticale à l'échelle horizontale varie dans les illustrations présentées de 1/3 à 1/9, soit nettement moins que la majeure partie des enregistrements présentés dans la littérature où l'on voit des rapports d'échelle allant de 1/10 à 1/40. L'effet de la déformation est donc moins considérable qu'à l'accoutumée.

ANALYSE SISMIQUE

On remarque immédiatement une certaine homogénéité dans l'aspect des enregistrements sismiques en plaine abyssale. Ainsi, il est possible de définir une coupe théorique type des séries. Une analyse plus détaillée permet, en outre, de mettre en évidence les principaux réflecteurs caractéristiques. Comme ces réflecteurs présentent des variations de caractère selon de grandes régions paléogéographiques, on distinguera plusieurs "faciès sismiques".

Coupe Théorique Type

En étudiant les profils depuis le fond marin jusqu'à la limite de pénétration obtenue (1,5 s environ avec le sparker, 3 s avec le Flexotir), on distingue une succession très constante de quatre grandes unités (Figure 2): ensemble supérieur A, ensemble B, ensemble C, et ensemble inférieur D.

Ensemble A

Cet ensemble situé sous le fond, est plus ou moins épais. Il se subdivise dans certaines régions en deux séquences caractéristiques:

1. Une *séquence supérieure* litée, riche en hautes fréquences, limitée à sa base par un réflecteur bien net, à 350 ms en moyenne, sous le fond: *Horizon G*, dont

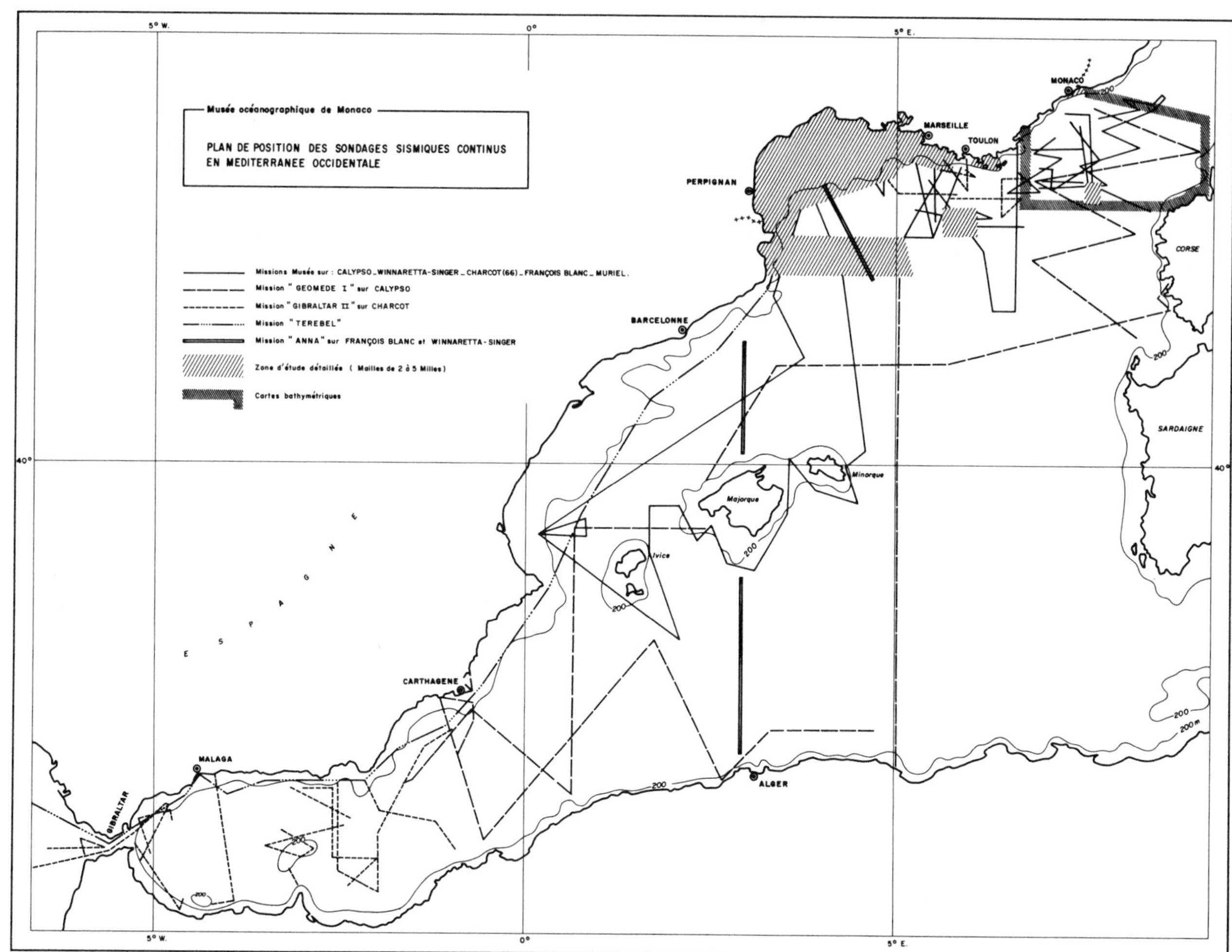

Figure 1. Plan de position des sondages sismiques continus en Méditerranée occidentale.

le caractère est défini comme suit (Leenhardt, 1970): "forte amplitude, fréquence de l'ordre de 65 Hz; trois phases régulières; l'amplitude de la troisième phase s'estompe à l'approche des accidents".

2. Une *séquence inférieure* claire, contenant des réflecteurs très clairs et souvent discontinus dont le caractère ne peut être décrit mais qui sont conformes, dans l'ensemble, au réflecteur G.

Ensemble B

Cet ensemble est formé de réflecteurs bien nets et toujours conformes. Son épaisseur varie de 120 à 270 ms. Il se présente sous deux aspects différents:

1. A l'Est de la zone étudiée et dans la zone A, définie par Glangeaud *et al.* (1966), il est formé d'une série de phases sur 180 à 260 ms, de forte amplitude, de fréquence 60 Hz, d'aspect très cassé et souvent courbé en sommets d'hyperboles. C'est ce que nous appelons *Horizon H* (Alinat *et al.,* 1966).

2. A l'Ouest de la zone étudiée, dans le Golfe du Lion et le bassin Nord Baléares, cette séquence se subdivise en deux horizons bien distincts. A la base, l'*Horizon K* (Leenhardt, 1970): "puissant, de fréquence 65 Hz, trois phases toujours visibles, la première est d'aspect tireté, les suivantes sont, de plus, ondulées. A la base de la réflexion, on observe de petites diffractions qui forment la signature". L'horizon K est à environ 1,2 s sous le fond. Il est surmonté en parfaite conformité par l'*Horizon J* qui correspond soit à deux phases, soit à une série assez épaisse dont la base se trouve à 200 ms du réflecteur K.

Ensemble C

Un ensemble transparent se trouve sous l'ensemble B. Il est rarement traversé par le sparker. Il s'ondule fréquemment jusqu'à donner des dômes qui percent les séries supérieures. Par endroits, près des dômes, il peut disparaître.

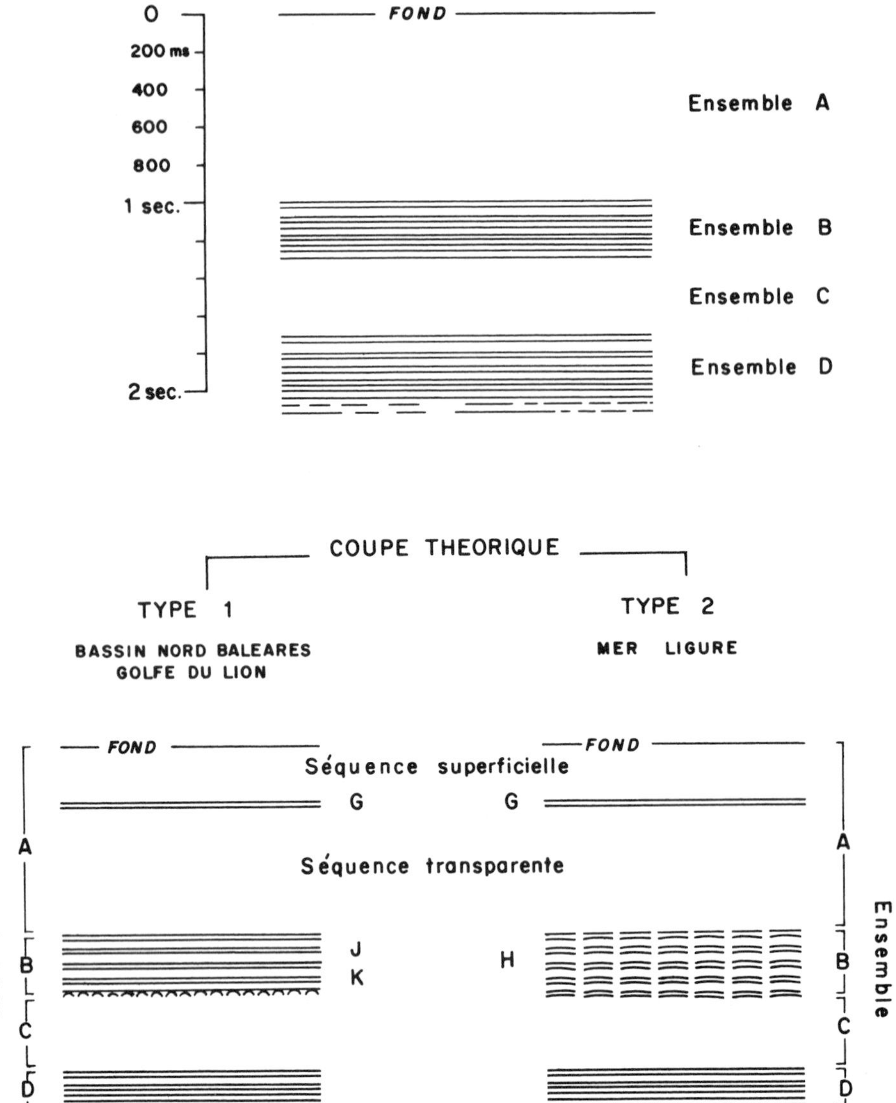

Figure 2. Coupe théorique montrant les quatre grands ensembles et les différentes séquences.

Ensemble D

C'est l'ensemble le plus profond reconnu jusqu'ici. Il débute par le *réflecteur L* (Montadert *et al.*, 1970), formé de trois phases nettes, régulières, horizontales, à près de 2 s sous le fond et se poursuit par une série de réflecteurs conformes. Le réflecteur L n'a été enregistré en sparker qu'uniquement sur les bordures. Il est défini par les enregistrements Flexotir, qui atteignent par endroits le substratum (Golfe de Valence).

De nombreux auteurs ont décrit les principaux horizons de cette coupe dans diverses régions de la Méditerranée occidentale mais en utilisant parfois un même nom pour des réflecteurs différents ou un nom différent pour des réflecteurs identiques, sans se référer aux définitions pré-existantes. La Figure 3 donne un tableau de concordance.

Discussion de l'Individualité de la Couche Transparente

La distinction entre la séquence litée et la séquence transparente saute à l'oeil. Il est cependant permis de s'interroger sur sa signification. Dans le Golfe de Cadix, l'allure de l'ensemble superficiel est la même qu'en Méditerranée. A partir d'enregistrements magnétiques, Dessolin (communication orale) montre que les amplitudes de réflexion se répartissent suivant la courbe type de l'amortissement d'un système oscillant (décroissance *grosso-modo* logarithmique). Il est donc permis de se demander si l'énergie de l'impulsion

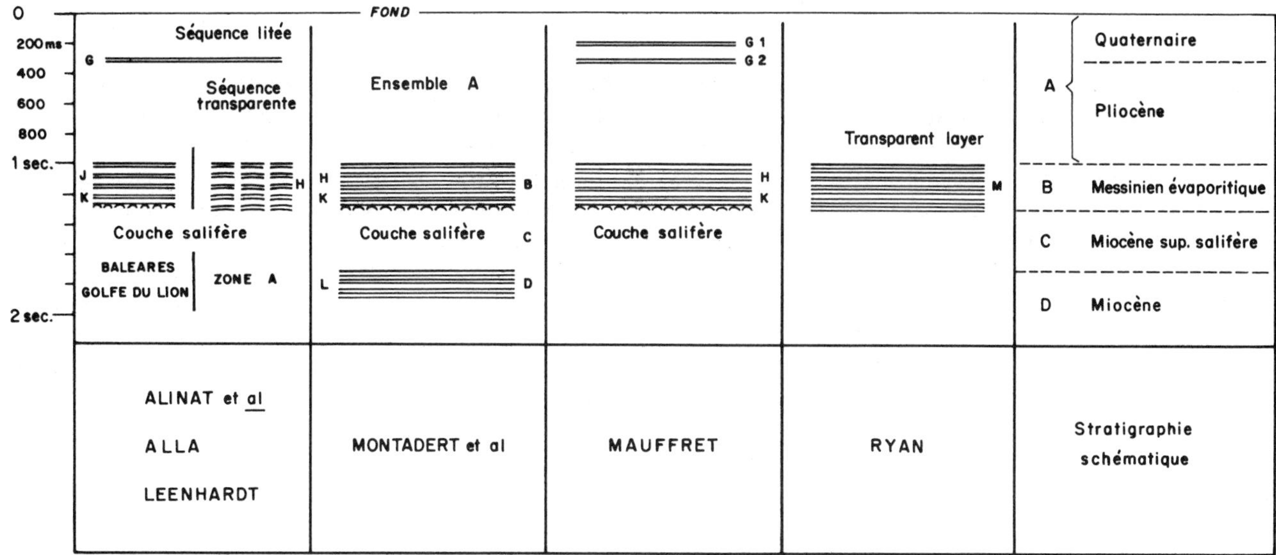

Figure 3. Tableau détaillant les séquences et les ensembles et montrant la corrélation entre les dénominations des différents auteurs.

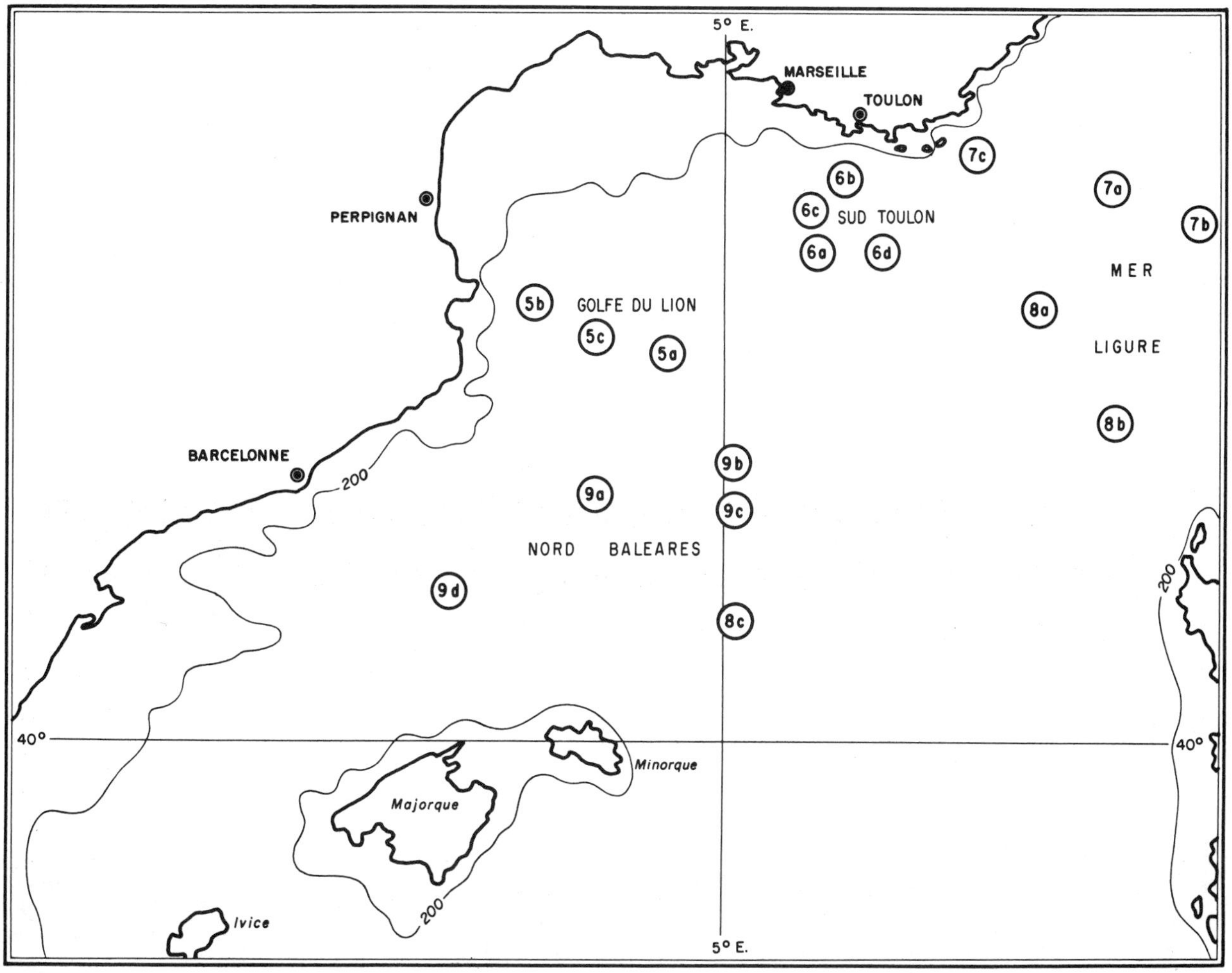

Figure 4. Positions des illustrations présentées (Figures 5–9).

sismique n'est pas progressivement consommée par les réflexions successives jusqu'à n'en laisser qu'une faible quantité au-delà d'une faible profondeur. Dans cette hypothèse, la séquence transparente serait analogue à la séquence litée et seule la diminution progressive de l'énergie vers le bas serait responsable de l'allure de la zone transparente: les réflecteurs observés dans cette zone sont conformes, même s'ils sont faibles et discontinus, à ceux observés dans la zone litée.

Nous préférons penser que la séquence transparente correspond à un processus régulier de sédimentation et que la séquence litée traduit un dépôt irrégulier. En effet, l'épaisseur de la séquence litée n'est pas constante. Elle peut s'étendre sur près d'une seconde de pénétration. Elle peut aussi être très mince ou absente. Nous développerons ultérieurement cette discussion.

Variations Régionales et Passages Latéraux

Les unités et les réflecteurs décrits dans cette coupe type subissent des variations suivant de grandes zones.

On peut ainsi distinguer quatre grandes régions paléogéographiques présentant un "faciès sismique" propre (Figure 4): Golfe du Lion, région Sud Maures, Mer Ligure, et région Nord Baléares.

Golfe du Lion (Figure 5)

Sur la pente, nous trouvons une épaisse couverture sédimentaire (ensemble A) qui diminue d'importance vers le large. Elle est affectée de phénomènes de glissement, de creusement, d'érosion et les biseaux sont nombreux. Sa forte puissance est en relation avec les apports sédimentaires du Rhône, d'une part, et des rivières pyrénéennes, d'autre part. Le sparker ne permet pas de voir les séries sous-jacentes. En plaine abyssale, l'ensemble A présente des réflecteurs subhorizontaux. La séquence superficielle montre encore de nombreux biseaux. Le réflecteur G est peu visible. La séquence transparente ne se distingue pas nettement de l'ensemble, les réflecteurs étant nombreux, bien que discontinus.

L'ensemble B sous-jacent, peu épais (150 ms), se compose d'un réflecteur K atypique surmonté de l'horizon J, souvent réduit au triplet inférieur. Cet ensemble plonge sous la couverture meuble au Nord-Ouest et à l'Ouest. Vers l'Est et le Sud, l'horizon J s'épaissit et devient plus net et la série est affectée de larges ondulations. Des failles NE-SO affectent toute la couverture (Alla *et al.,* 1969).

Région Sud Maures (Figure 6)

Vers l'Est du Golfe du Lion, dans la zone Sud des Maures, le schéma devient différent: l'aspect des séries évoluant lentement mais avec continuité de passage d'une zone à l'autre, bien que toute la région pyrénéo-provençale soit affectée de failles importantes.

Dans la plaine abyssale (Alla, 1970), l'ensemble B, peu épais, présente un horizon K typique de forte amplitude, surmonté d'un horizon J discontinu et souvent réduit au réflecteur inférieur (doublet ou triplet) (Figure 6a). L'ensemble supérieur A, subhorizontal, bien divisé en séquence litée limitée par un horizon G typique net et en séquence transparente, s'épaissit sur la pente où apparaissent des phénomènes de glissements et des canyons plus ou moins remblayés (Figure 6c).

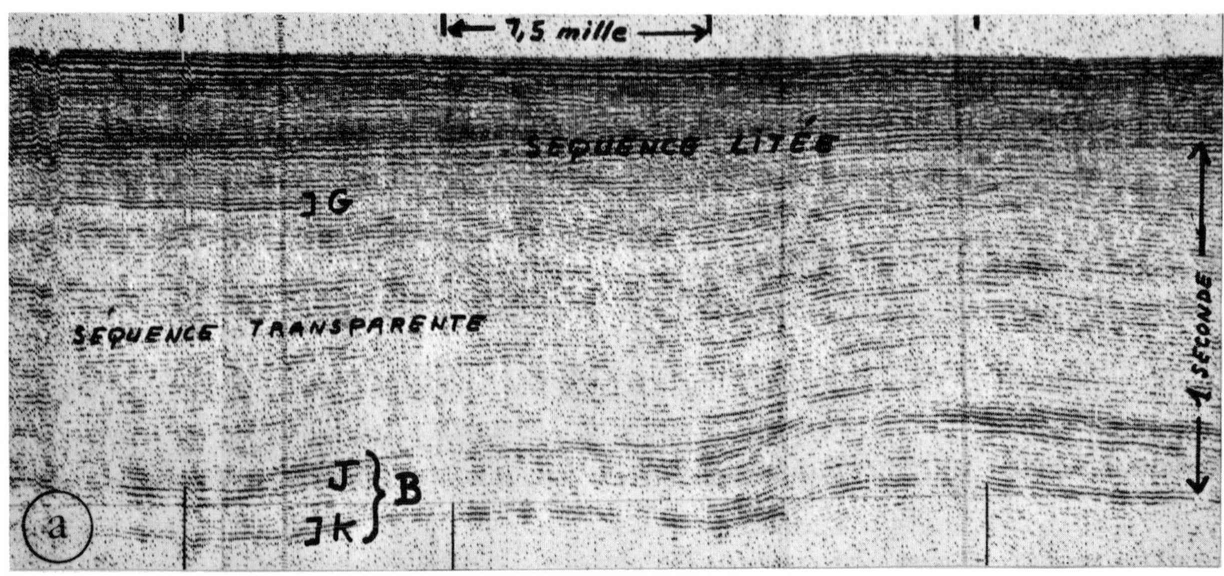

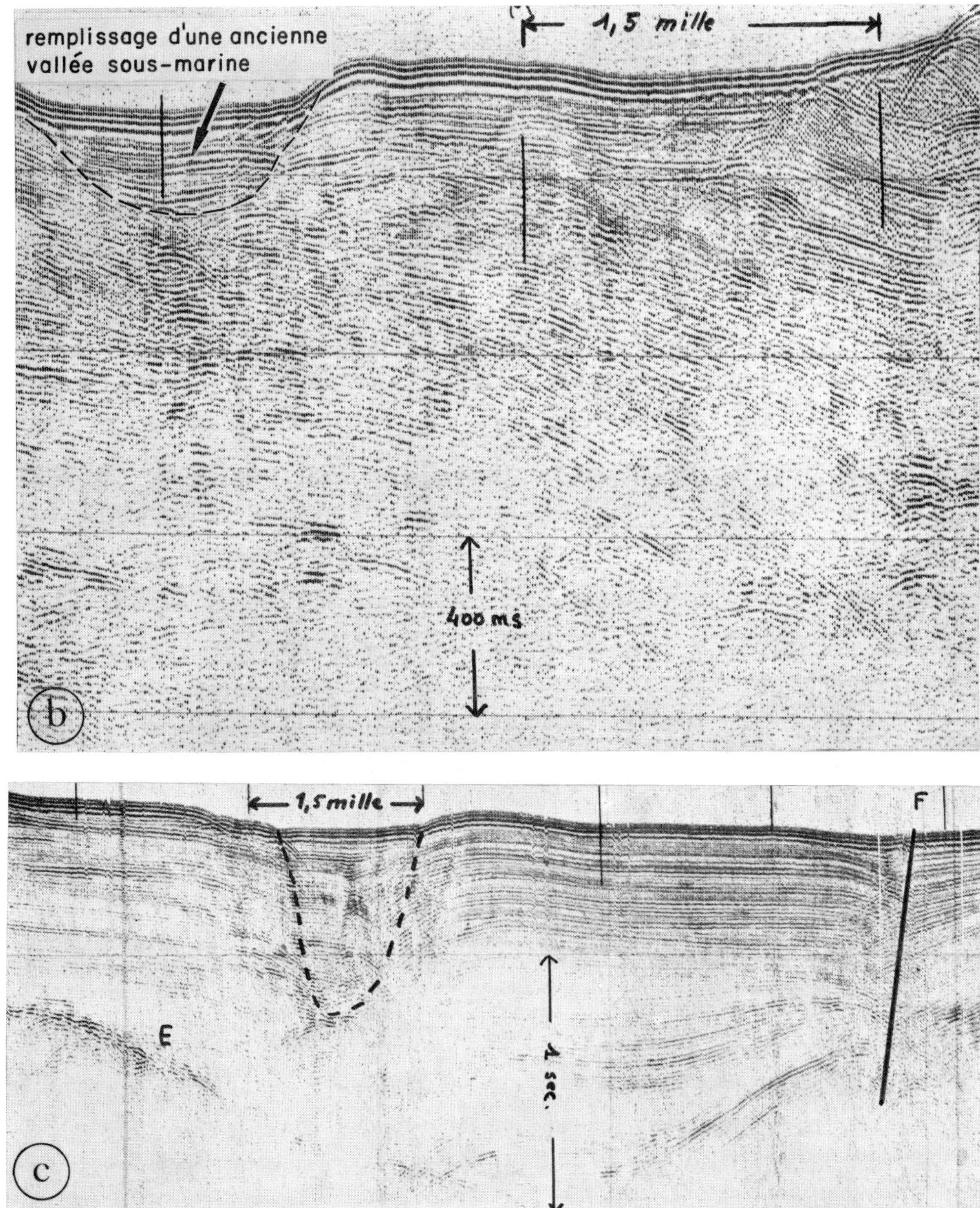

Figure 5. Golfe du Lion (sparker 9.000 Joules).

a, Coupe sismique en plaine abyssale. Profil 8F41. La séquence superficielle se biseaute du Nord vers le Sud. Réflecteur G peu net. Nombreux réflecteurs dans la séquence transparente. L'ensemble B est affecté de larges ondulations (rapport d'échelle 1/3). (Coupe S-N de gauche à droite).

b, Coupe sur la pente. Profil 8F30. Coupe meuble épaisse perturbée par des glissements et des creusements (rapport d'échelle 1/7). (Coupe E-O de gauche à droite).

c, Coupe au pied du glacis. Profil 8F34. Surface d'érosion (E). Canyon remblayé. Faille, F (rapport d'échelle 1/5). (Coupe O-E de gauche à droite).

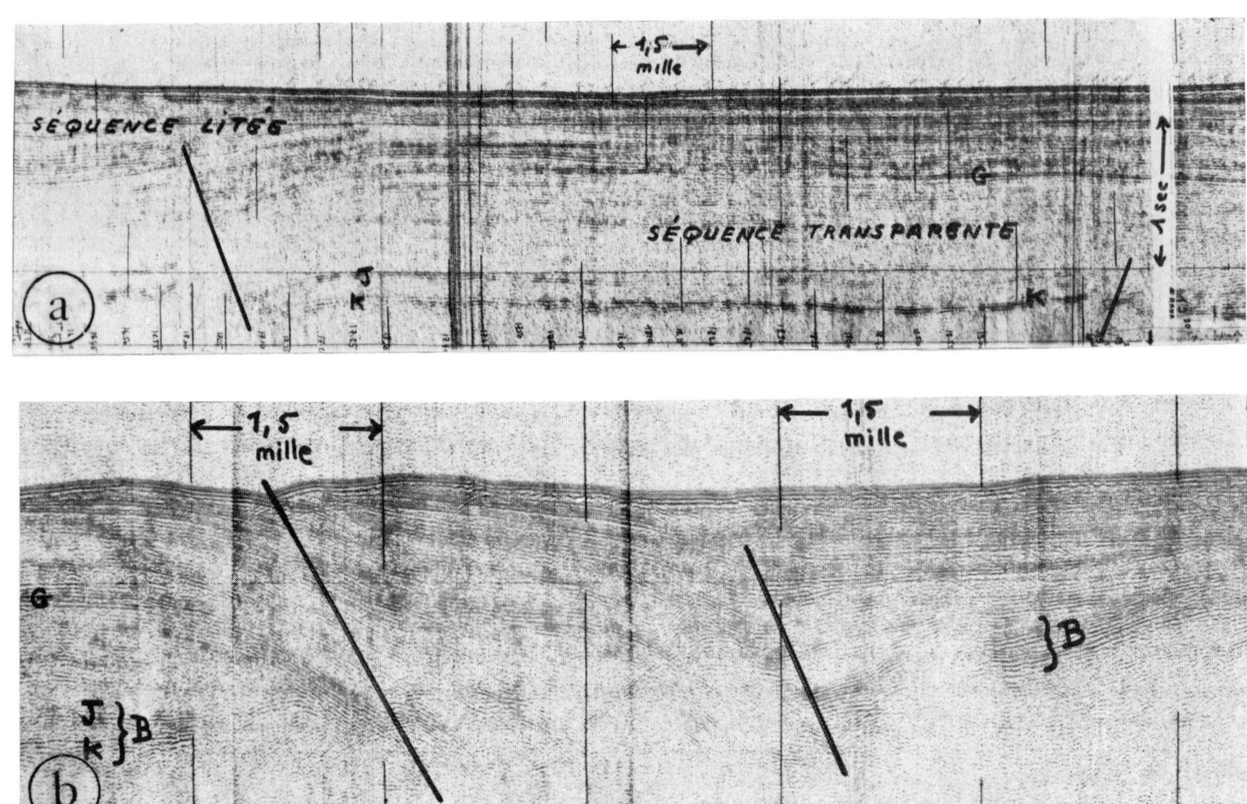
1,5 mille
SÉQUENCE LITÉE
G
SÉQUENCE TRANSPARENTE
J
K
1 sec
a
1,5 mille
1,5 mille
G
B
J
K
B
b

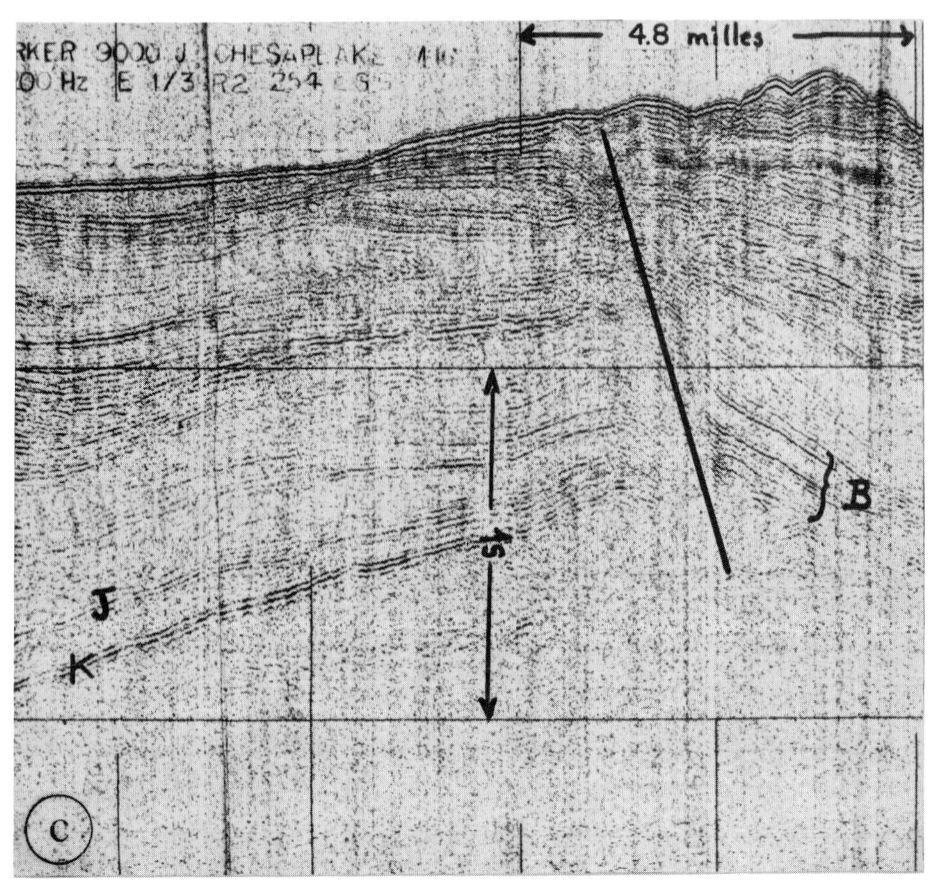
4.8 milles
B
J
K
c

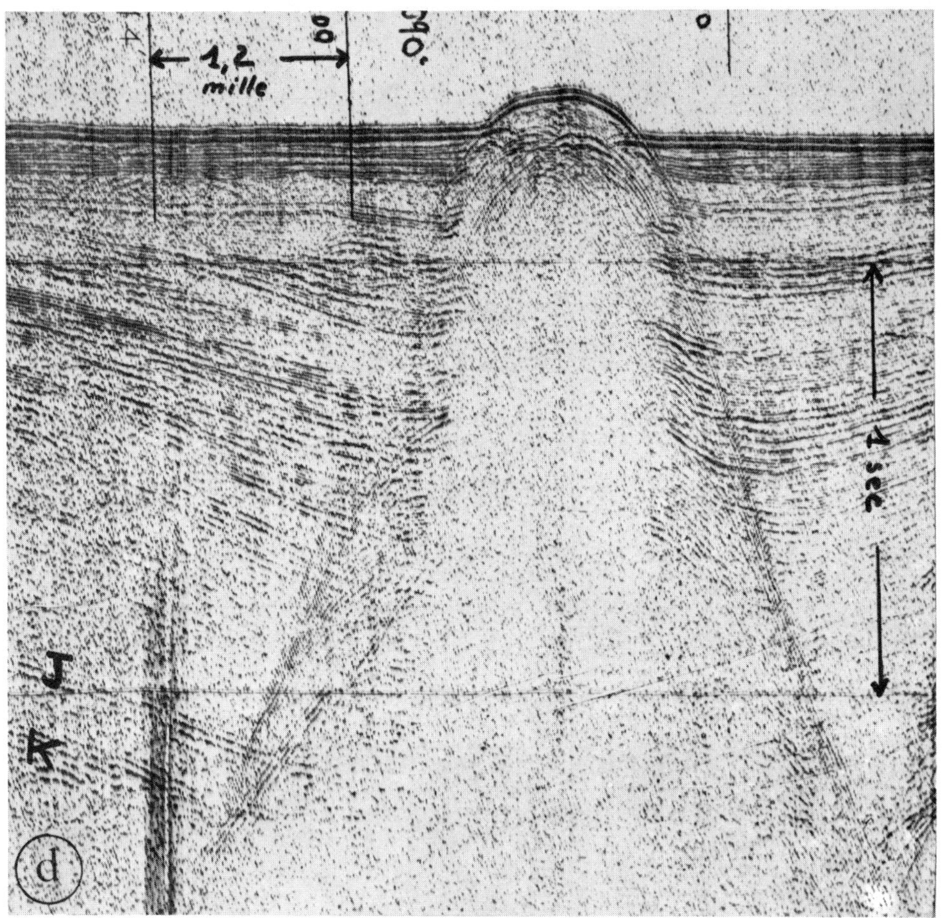

Figure 6. Zone Sud-Toulon (sparker 9.000 Joules).
a, Coupe en plaine abyssale. Profil 9M10. Séquence superficielle épaisse. Horizon G net. Horizon K typique. Horizon J réduit au triplet inférieur. Faille. Biseaux (rapport d'échelle 1/4). (Coupe N-S de gauche à droite).
b, Coupe sur pente provençale. Profil 8F5. Changement de caractère de l'ensemble B au passage d'un accident (rapport d'échelle 1/3). (Coupe S-N de gauche à droite).
c, Coupe sur la pente provençale. Profil 6J3. Remontée des horizons sur la pente. Biseaux. Failles. Niveau K typique (rapport d'échelle 1/7). (Coupe SE-NO de gauche à droite)
d, "Dôme franc". Profil 6B3. Dôme massif, perçant. Dépressions périphériques primaires (rapport d'échelle 1/4). (Coupe O-E de gauche à droite).

A l'approche de la marge continentale, la couverture sédimentaire est fortement affectée par des failles Est-Ouest d'âge récent, abaissant le substratum des Maures en marche d'escalier vers le Sud. L'ensemble B change de caractère au passage de la plus importante de ces failles avant de se biseauter sur la pente (Figure, 6b).

L'ensemble C, à peine ondulé dans le Golfe du Lion, donne ici naissance à des "dômes francs", isolés, massifs, plus ou moins alignés suivant une direction NO-SE. Ces dômes percent les séries sus-jacentes et sont entourés de dépressions périphériques importantes (Figure 6d). Leur présence a influencé d'une manière certaine la sédimentation locale, ainsi que le montrent les profils.

Mer Ligure (Figures 7 et 8)

Vers la Mer Ligure (Figure 7), l'aspect change brutalement et il n'y a continuité de passage que dans les réflecteurs supérieurs.

– L'*ensemble A* est moins épais. On peut pointer (Mauffret, 1968) un réflecteur G et dans la séquence superficielle, à 130 ms sous le fond, un réflecteur G_1 "amplitude moyenne, deux phases de fréquence 65 Hz, précédées d'une petite phase (à 125 Hz) constituant la signature" (Leenhardt, 1970). Cet horizon est constant et régulier. La séquence transparente est très nette.

– L'horizon H forme l'*ensemble B.* Cet horizon est haché, peu visible entre les dômes et présente des

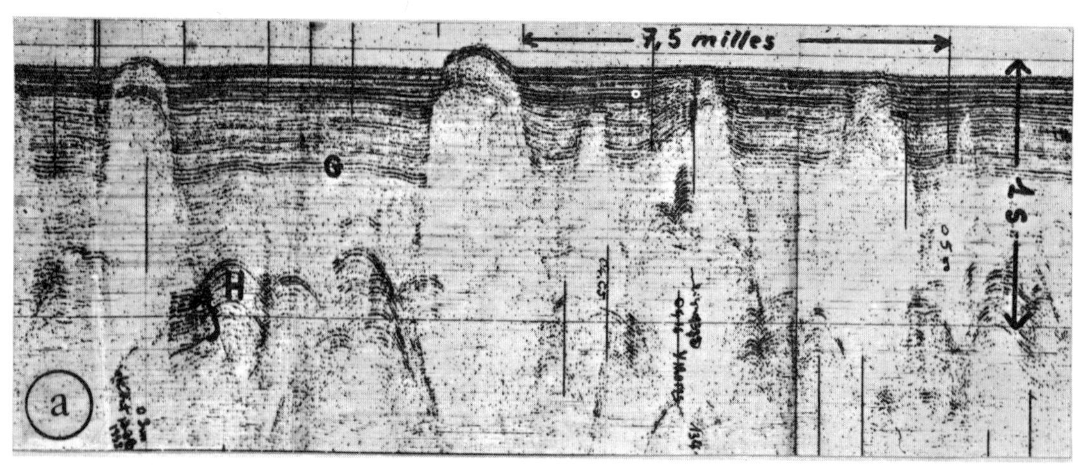
7,5 milles
G
H
1 s.
a

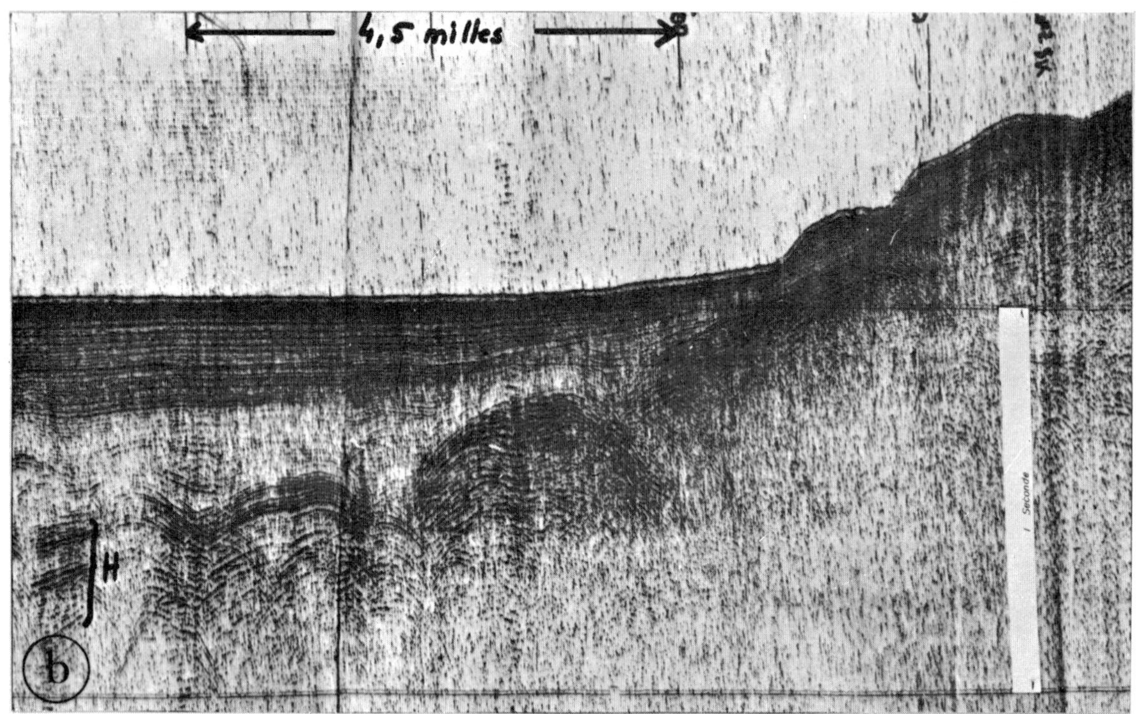
4,5 milles
H
b

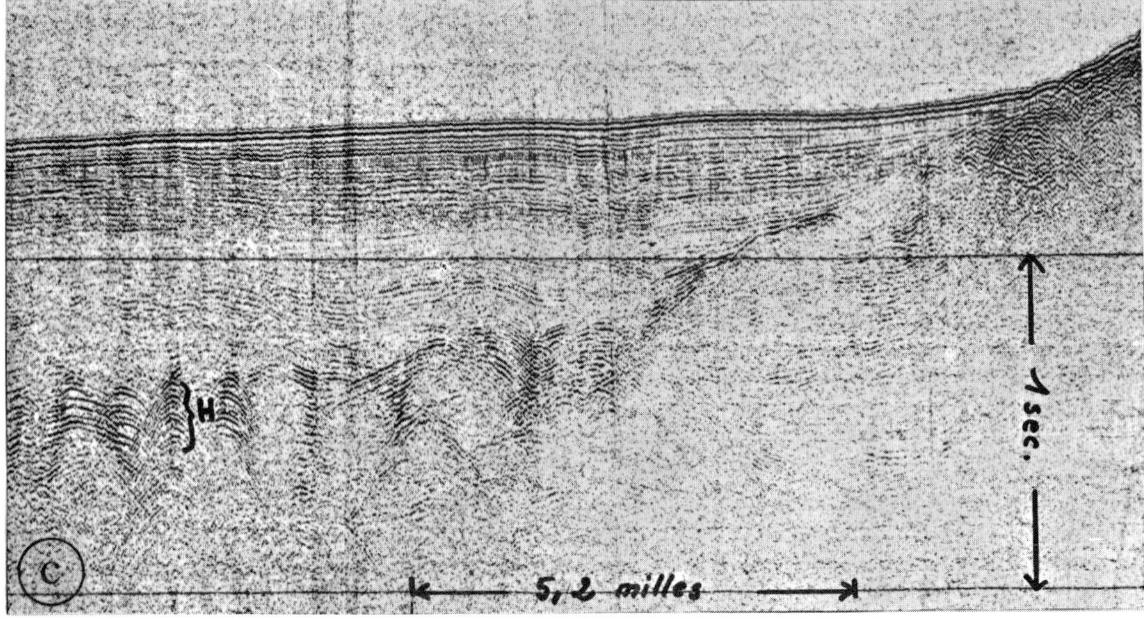
H
1 sec.
5,2 milles
c

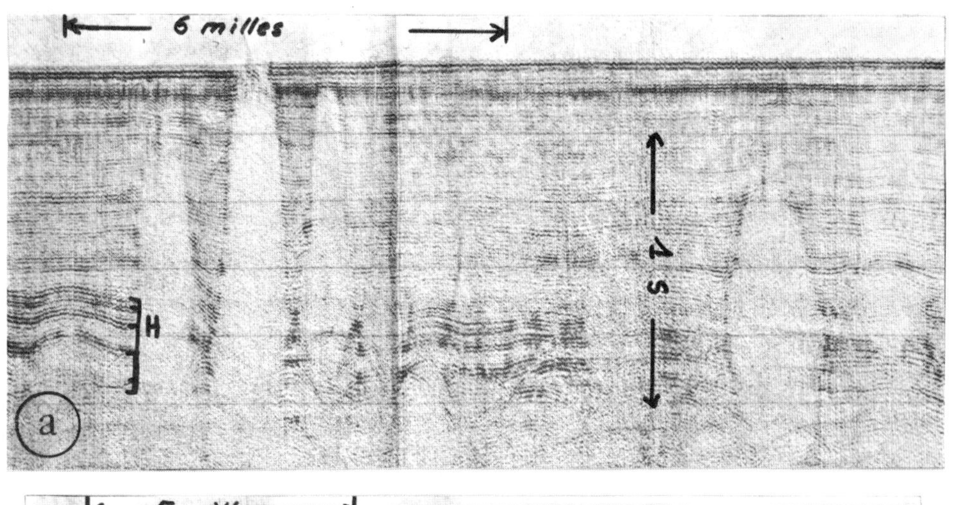

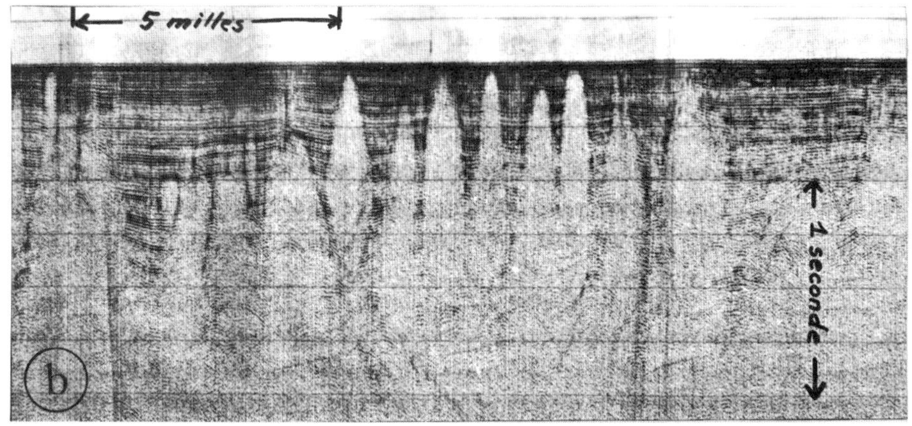

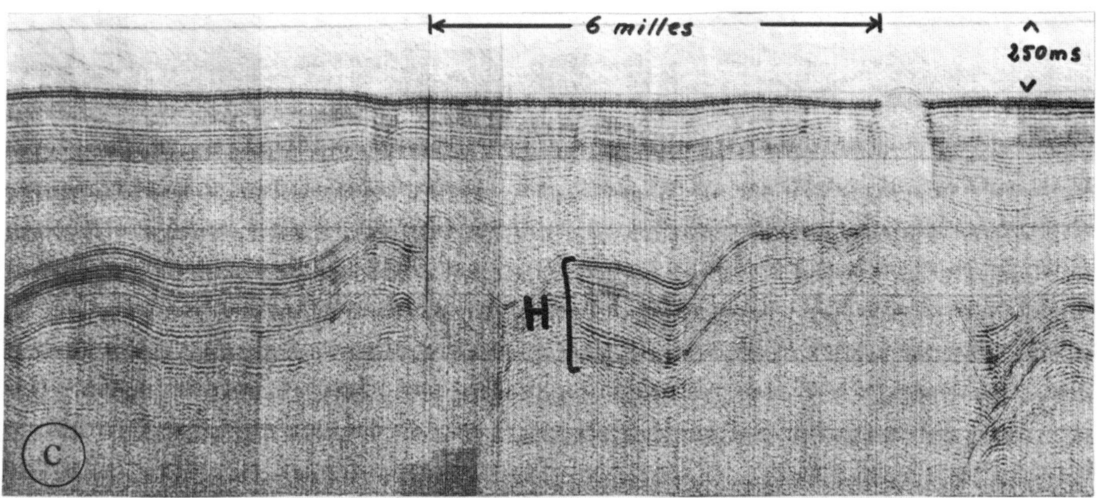

Figure 8. Sud-mer Legure (zone A), sparker 9.000 Joules.
a, Coupe en plaine abyssale. Profil CG4. Séquence supérieure peu marquée. Réflecteurs discontinus. Horizon H plus net. Dômes moins nombreux (rapport d'échelle 1/7). (Coupe E-O de gauche à droite).
b, Bordure corse. Profil CG4. Remontés des réflecteurs. Dômes filiformes (rapport d'échelle 1/7). (Coupe NO-SE de gauche à droite).
c, Limite du bassin Nord-Baléares. Profil CG16. Ensemble A peu épais. Horizon H devenant moins perturbé (rapport d'échelle 1/7). (Coupe N-S de gauche à droite).

Figure 7. Mer Ligure (sparker 9.000 Joules).
a, Coupe en plaine abyssale. Profil 7C3. Séquence litée et horizon G net. Séquence transparente. Horizon H haché et peu net. Nombreux dômes (rapport d'échelle 1/9). (Coupe SE-NO de gauche à droite).
b, Coupe sur la bordure corse. Profil 6B2 (rapport d'échelle 1/7). (Coupe NO-SE de gauche à droite).
c, Coupe sur bordure provençale. Profil 6J1. Biseautage des réflecteurs sur le substratum, avec traces d'érosion (rapport d'échelle 1/7). (Coupe SE-NO de gauche à droite).

variations de caractère de l'Ouest vers l'Est (Leenhardt *et al.*, 1970).

Sur la bordure provençale (Figure 7c), les horizons se biseautent sur la remontée du substratum avec très peu de perturbations. Sur la bordure corse (Figure 7b), les phénomènes sont très semblables. Toute la série est hachée de nombreux dômes provenant de l'ensemble C.

Vers le Sud (Figure 8), l'aspect des séries est peu différent du schéma ci-dessus. L'ensemble A devient moins épais et moins bien divisé en deux séquences. A l'approche de la Corse, les dômes deviennent de plus en plus filiformes (Figure 8b). A l'approche des Baléares, l'ensemble B (horizon H) devient moins haché (Figure 8c) avant de passer brutalement à un autre faciès.

Nord Baléares (Figure 9)

Au Sud de l'Espagne, dans le bassin Nord Baléares, l'ensemble A est relativement homogène, la séquence B, épaisse, présente un horizon K atypique et un horizon J à nombreuses phases. Vers l'Est, cette séquence est marquée par de larges anticlinaux influençant les séries supérieures. Vers la côte espagnole, les réflecteurs se biseautent progressivement sur le substratum (Montadert *et al.*, 1970).

Le passage de ce bassin Nord Baléares se fait en continuité par lente modification du caractère des réflecteurs vers le Nord (Golfe du Lion) et le Nord-Est (zone Sud Maures); par contre, vers l'Est, on entre dans le "faciès ligurien" par des modifications brutales. Il n'est pas encore possible de préciser si ce contact est dû à un accident.

INTERPRETATION STRATIGRAPHIQUE: IDENTIFICATION DES SERIES

En s'appuyant sur les forages de la Compagnie Française des Pétroles-Métropole dans le Golfe du Lion, du TEREBEL au pied de la pente continentale provençale, et ceux, plus récents, du Leg XIII du GLOMAR CHALLENGER on peut ratacher les horizons sismiques décrits aux séries stratigraphiques.

L'*ensemble A,* d'une épaisseur moyenne de 800 à 1200 m correspond au Plio-Quaternaire. Les forages donnent une première série de sédiments lités formés de sables grano-classés, de vases fines, localement de couches plus grossières à galets. Mais la résolution de notre méthode ne permet pas de distinguer des couches aussi fines que des turbidites. Nous préférons, sans pouvoir le démontrer, penser que les miroirs de la séquence litée correspondent à des lacunes ou à des accidents de sédimentation dont les dates et les modes d'action restent à préciser. Cette irrégularité se traduirait sur les enregistrements par l'apparence litée de la première série de l'ensemble A.

Par contre, la séquence transparente correspondrait à une sédimentation homogène pélagique à hémi-

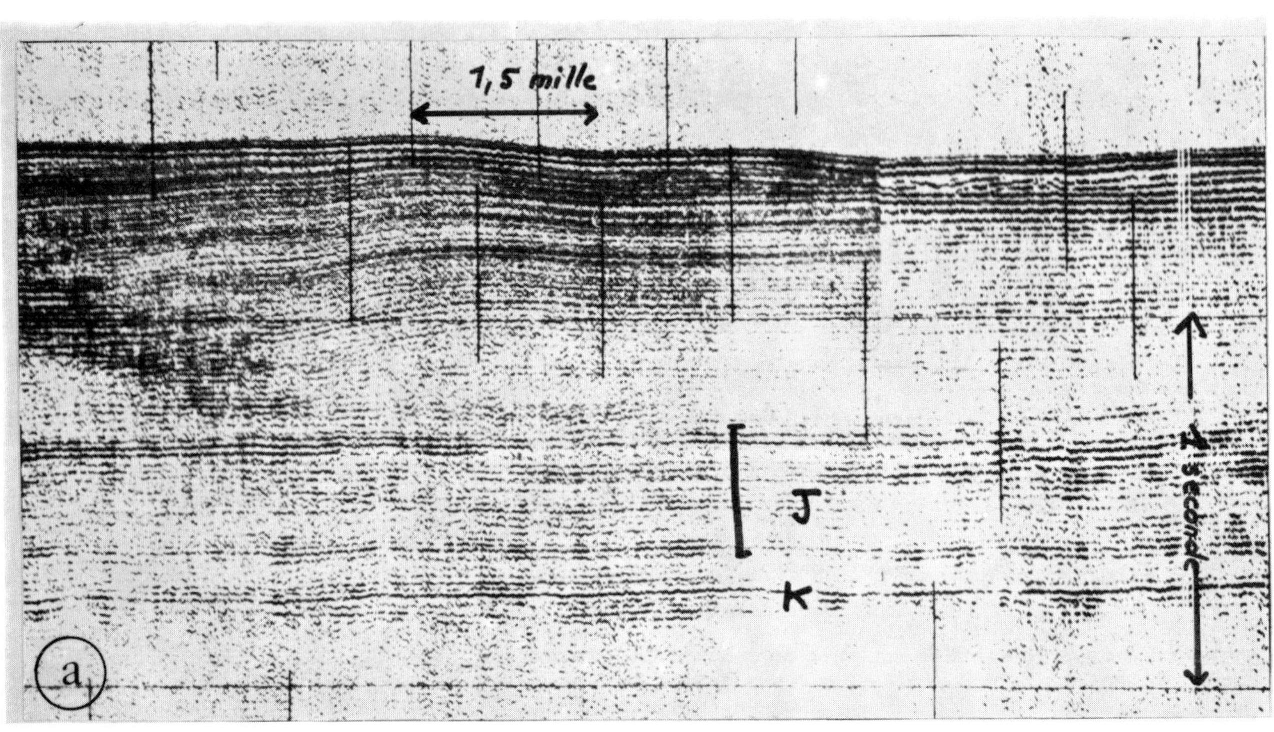

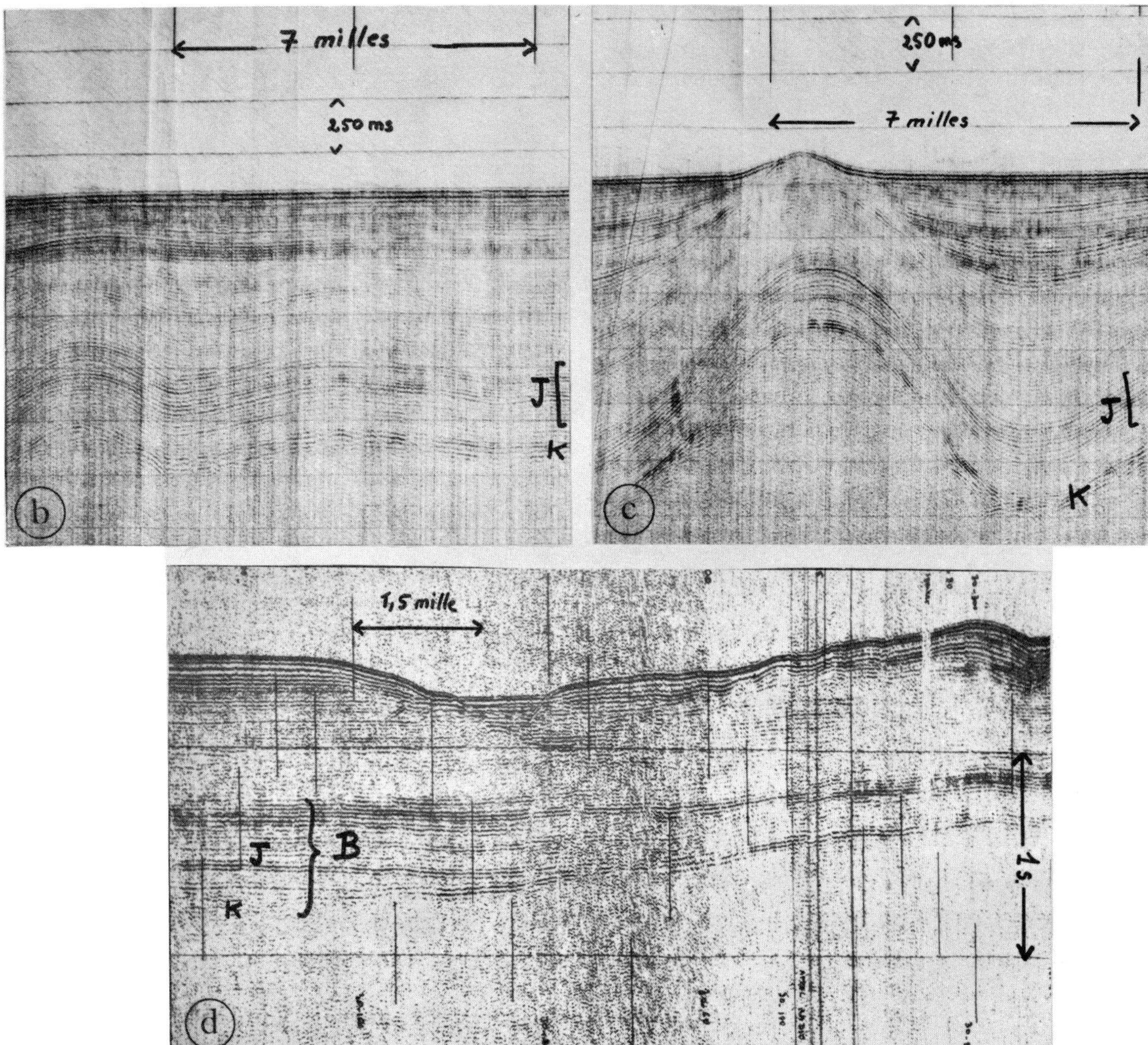

Figure 9. Zone Nord-Baléares (sparker 9.000 Joules).
a, Coupe dans la partie Ouest. Profil 9S1. Séquence litée nette. Séquence transparente mince. Ensemble B épais avec un horizon K et un horizon J (rapport d'échelle 1/3). (Coupe NO-SE de gauche à droite).
b, Coupe dans la partie Est. Profil CG16. Séquence transparente plus épaisse. Ensemble B ondulé (rapport d'échelle 1/8). (Coupe NO-SE de gauche à droite).
c, Anticlinal salifère. Profil CG16 (rapport d'échelle 1/8). (Coupe N-S de gauche à droite).
d, Coupe sur la bordure espagnole. Profil 9S1. Biseautage des réflecteurs de l'ensemble B sur la remontée du substratum (rapport d'échelle 1/4). (Coupe N-S de gauche à droite).

pélagique, plus ancienne et éventuellement reprise par des courants de fond.

La limite Pliocène-Quaternaire est mal précisée pour l'instant. Elle pourrait correspondre à l'horizon G au-dessous duquel se produit un changement net dans le type de sédimentation.

L'*ensemble B,* 400 à 800 m d'épaisseur, n'a été atteint que par les forages du GLOMAR-CHALLENGER et se révèle constitué d'une série évaporitique datée du Messinien et comprenant des marnes, du gypse, de l'anhydrite, des dolomies, du sapropel, et localement des lits de diatomites. Ces dépôts s'étendent de façon très uniforme sur toute la Méditerranée occidentale.

Le passage latéral de l'horizon H aux horizons J-K que, seule, l'étude du caractère met en évidence, pourra s'expliquer par les modifications dans la composition de cette suite évaporitique.

Ce changement de faciès correspond à une réalité

géophysique soulignée en sismique réfraction par Fahlquist et Hersey (1969). Ces auteurs mettent en évidence des vitesses différentes à l'Est et à l'Ouest du 6e méridien pour une couche ayant la profondeur et l'épaisseur de l'ensemble B.

La Figure 10 donne une répartition géographique des différents faciès sismiques de l'ensemble B.

L'*ensemble C:* deux démonstrations géophysiques de la nature salifère de la couche C ont été données. L'une procède par élimination (Leenhardt, 1968); l'autre constitue une preuve directe de la nature de la couche et rapporte de façon certaine le sel au Miocène (Montadert *et al.,* 1970).

Les forages du Leg XIII du GLOMAR-CHALLENGER confirment l'existence du sel et permettent de le dater.

Ce sel, d'épaisseur variable, en moyenne 400 m, a donné naissance à de nombreux dômes. On distingue plusieurs types de dômes. Près des Baléares, il s'agit seulement de bombements, appelés anticlinaux salifères; au Sud de Toulon, les dômes sont isolés et de grande taille, ce sont des "dômes francs"; dans la zone A, les dômes sont nombreux et de diamètre plus modeste; vers la Corse, ils deviennent très fins, "corsides". La Figure 11 résume cette classification des diverses structures en dômes que Mauffret (1969) avait suggérée.

L'*ensemble D:* cet ensemble n'a pas été atteint par forage. On ne peut donc lui attribuer une nature et un âge certain. Dans le bassin Nord Baléares, ces réflecteurs horizontaux surmontent le substratum oligocène plissé (Montadert *et al.,* 1970).

LES TYPES TECTONIQUES

Etablir un schéma structural complet de la Méditerranée suppose de pouvoir intégrer les unes aux autres

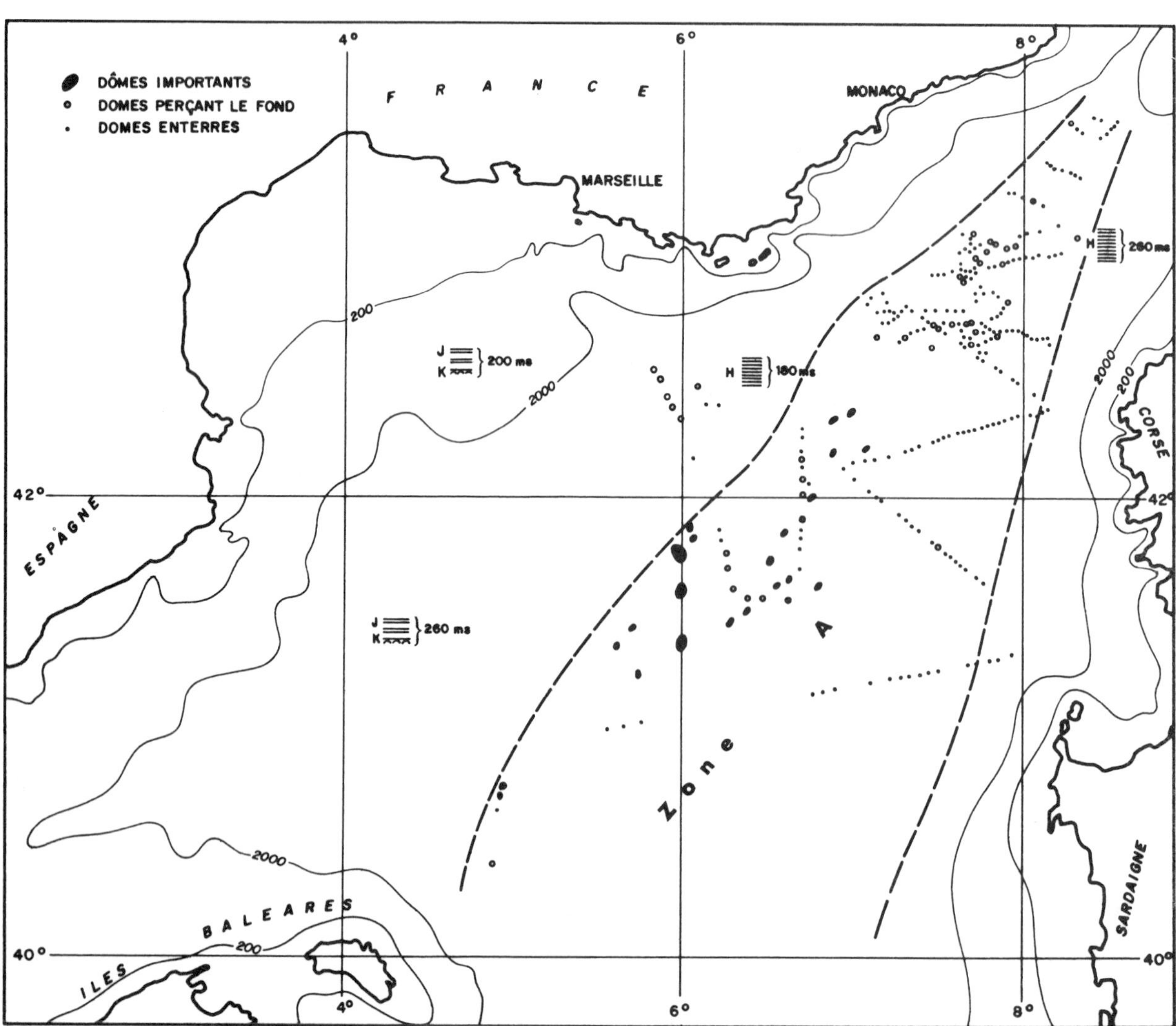

Figure 10. Position des différents dômes de sel du Nord du bassin signalés dans la littérature.

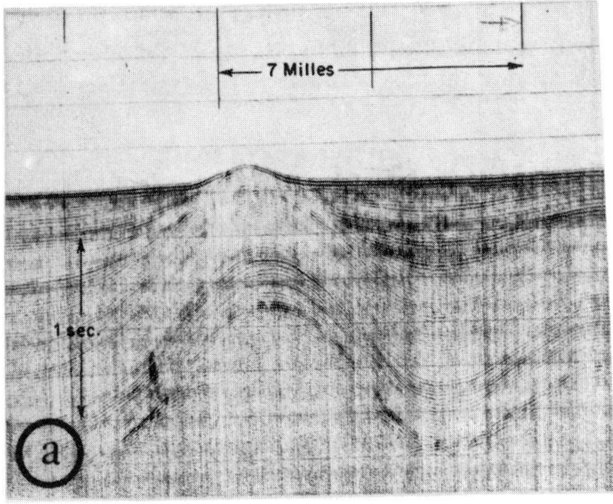

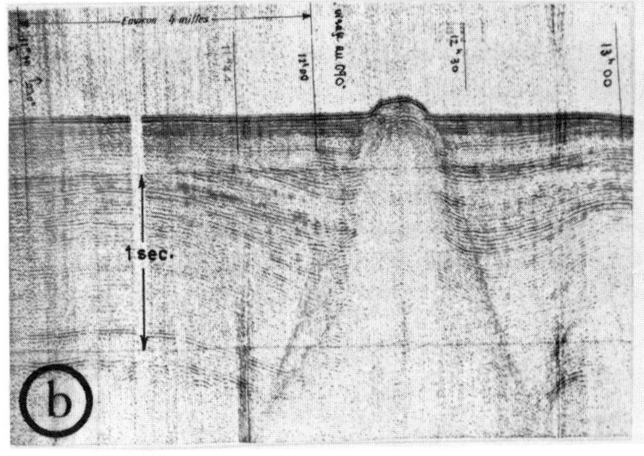

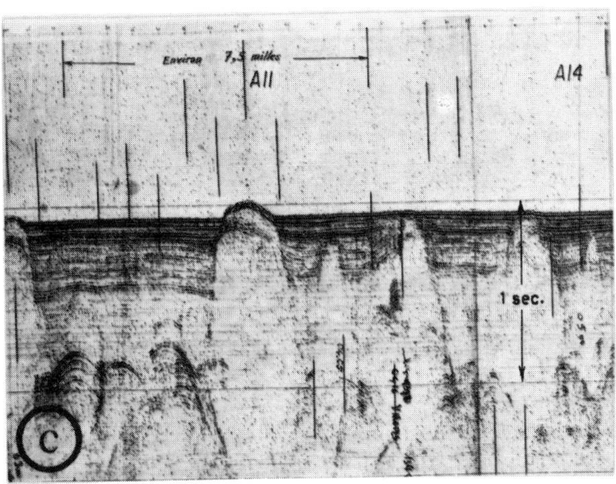

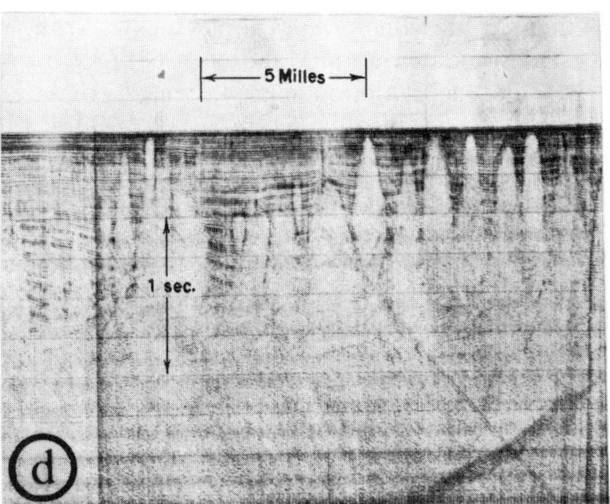

Figure 11. Classification des structures en dôme de la Méditerranée occidentale (Sparker 9.000 J).
a, Anticlinal salifère. Profil CG16 (rapport d'échelle 1/6). (Coupe N-S de gauche à droite).
b, "Dôme franc". Zone Sud Toulon. Profil 6B3 (rapport d'échelle 1/4). (Coupe O-E de gauche à droite).
c, Zone A. Structure A (rapport d'échelle 1/8). (Coupe SE-NO de gauche à droite).
d, "Corsides", Profil CG4 (rapport d'échelle 1/3). (Coupe NO-SE de gauche à droite).

des données variées. La séismologie d'observatoire, la sismique réfraction (pour une part), la gravimétrie et le magnétisme montrent la Méditerranée occidentale "déchirure" de la croûte (J. Rothé, communication orale). Le sondage sismique continu et quelques résultats de réfraction ne s'appliquent qu'au remplissage récent. Glangeaud (1968) est l'un des seuls à avoir tenté cette synthèse, en y incorporant les résultats des innombrables observations géologiques effectuées à terre. Mais on n'est pas toujours sûr de la nature des terrains sous-jacents au Miocène terminal mis en évidence par Montadert *et al.* (1970).

Nous avons signalé plus haut les accidents tectoniques propres à chaque zone. Nous ne pouvons ici qu'en donner un sommaire: les analyses sont encore trop incomplètes pour les présenter en un ensemble logique.

Sur les marges, une tectonique d'effondrement est apparente. Elle date du Ponto-Plio-Quaternaire (Glangeaud *et al.*, 1967).

Certains profils présentés (Figure 6 a, b et c) montrent un réseau de failles d'effondrement mis en évidence au Sud de Toulon (Alla, 1970) et qu'il reste à relier avec précision aux structures pyrénéo-provençales auxquelles elles sont sûrement rattachées.

Des traces d'érosion sont nettes sur les plateaux (Leenhardt, 1963) et sur les pentes. Nous les identifions comme des traces de la régression fin-Miocène.

Sur les bordures, encore, nous observons les phénomènes pelliculaires décrits à petite échelle par Glangeaud *et al.* (1968) sur le plateau continental du Golfe du Lion et en Ligurie par Rehault (1968). Nous en avons donné des exemples (Alla *et al.*, 1969). Les Figures 5b et 5c en montrent deux illustrations.

Mais le modelé essentiel des couches sédimentaires récentes du bassin abyssal est causé par la tectonique salifère. Montadert *et al.* (1970) ont montré que le sel repose sur un ensemble (D) plat et horizontal. Seules de grandes failles coupent par places cet ensemble. Le phénomène de compression que Glangeaud *et al.* (1966) ont cru voir dans un dôme du Nord-Est de la Mer Ligure semble inexistant après le profil AUGUSTA: ce gros dôme a été pris pour un anticlinal déjeté.

La tectonique salifère (voir par exemple, Figure 9) se manifeste soit par des anticlinaux et des synclinaux à grand rayon de courbure (domaine du réflecteur K, soit par une fracturation du réflecteur H. Plus haut dans la série, les sédiments épousent l'ensemble B et sont plus ou moins rebroussés au contact des dômes. L'étude détaillée des strates au contact des dômes montre souvent les dépressions périphériques classiques et indique que la montée des dômes est synchrone des dépôts plio-quaternaires. Mauffret (1969) propose de considérer une évolution dans le temps comme dans l'espace: la montée du sel aurait commencé à l'Est et les anticlinaux de la région Ouest n'auraient pas encore évolué en dômes. La sédimentation vaseuse récente empâte la majeure partie des reliefs. Cependant, nous avons pu observer (Alla et Leenhardt, 1971) des affleurements sur un dôme T_1 au Sud de Toulon lors d'une plongée bathyscaphe.

CONCLUSIONS

Le domaine abyssal du Nord de la Méditerranée occidentale présente une formation supérieure homogène, subdivisée en quatre unités principales allant du Miocène supérieur au Quaternaire.

Le Miocène supérieur est salifère et évaporitique. Le Plio-Quaternaire présente une sédimentation abondante, hémipélagique à sablo-vaseuse avec des épisodes franchement grossiers. La tectonique salifère et la néotectonique récente contribuent à modeler ces sédiments.

Le modelé des sédiments résulte de grandes failles liées aux traits tectoniques majeurs de la région, d'actions plus locales résultant des phénomènes néotectoniques et de la tectonique salifère.

REMERCIEMENTS

Les travaux présentés ici ont été effectués avec le concours du Centre national pour l'exploitation des océans (CNEXO), du Centre national de la recherche scientifique (CNRS), de la Délégation générale à la recherche scientifique (Comité exploitation des océans), de la Société Nationale des Pétroles d'Aquitaine, de la Compagnie Française des Pétroles-Métropole, de Elf-Re et du Research Committee of the National Geographic Society, ainsi que des campagnes océanographiques françaises.

BIBLIOGRAPHIE

Alinat, J. et J. Y. Cousteau 1962. Accidents de terrain en mer de Ligurie. In: *Océanographie géologique et Géophysique de la Méditerranée Occidentale.* Colloque National du CNRS, Villefranche, avril 1961, 121–123.

Alinat, J., G. Giermann et O. Leenhardt 1966. Reconnaissance sismique des accidents de terrain en mer Ligure. *Comptes Rendus de l'Académie des Sciences, Paris,* A et B, 262:1311–1314.

Alinat, J., K. Hinz et O. Leenhardt 1970. Quelques profils en sondage sismique continu en Méditerranée occidentale. *Revue de l'Institut Français du Pétrole,* 25:305–326.

Alla, G. 1970. Etude sismique de la plaine abyssale au sud de Toulon. *Revue de l'Institut Français du Pétrole,* 25:291–304.

Alla, G., D. Dessolin, H. Got, O. Leenhardt, A. Rebuffatti et R. Sabatier 1969. Résultats préliminaires de la mission FRANÇOIS BLANC en sondage sismique continu. *Vie et Milieu,* 20 (2-B):211–220.

Alla, G. et O. Leenhardt 1971. Découverte d'un affleurement de *caprock* sur le sommet d'un dôme de sel (Dôme T_1 Sud-Toulon) avec le Bathyscaphe. *Comptes Rendus de l'Académie des Sciences, Paris,* D 272:1347–1349.

Bellaiche, G. 1969. *Etude Géodynamique de la Marge Continentale au Large du Massif des Maures (Var) et de la Plaine Abyssale Ligure.* Thèse d'Etat, Paris, 221p.

Bourcart, J. 1960. Carte topographique du fond de la Méditerranée occidentale. *Bulletin de l'Institut Océanographique de Monaco,* 57 (1163):20 p.

Dessolin, D. *Contribution à l'Etude du Golfe de Cadix en Sondage Sismique Continu.* Diplôme d'ingénieur, 40 p. (communication orale).

d'Erceville, I. and G. Kunetz 1963. Sur l'influence d'un empilement de couches minces en sismique. *Geophysical Prospecting,* 11: 115–121.

Fahlquist, D. A. and J. B. Hersey 1969. Seismic refraction measurements in the western Mediterranean Sea. *Bulletin de l'Institut Océanographique Monaco,* 67 (1386):52 p.

Glangeaud, L. 1966. Les grands ensembles structuraux de la Méditerranée occidentale d'après les données de *GÉOMÈDE I. Comptes Rendus de l'Académie des Sciences, Paris,* D 262:2405–2408.

Glangeaud, L. 1968. L'évolution de la Méditerranée et la dérive continentale. Communication orale au *Colloque International du CNRS,* Villefranche, septembre 1968.

Glangeaud, L., J. Alinat, C. Agarate, O. Leenhardt and G. Pautot 1967. Les phénomènes ponto-plio-quaternaires dans la Méditerranée occidentale d'après les données de GÉOMÈDE I. *Comptes Rendus de l'Académie des Sciences, Paris,* D 264:208–211.

Glangeaud, L., J. Alinat, J. Polvèche, A. Guillaume and O. Leenhardt 1966. Grandes structures de la Mer Ligure: leur évolution et leurs relations avec les chaînes continentales. *Bulletin de la Société Géologique de France,* 8:921–937.

Glangeaud, L., G. Bellaiche, M. Gennesseaux et G. Pautot 1968. Phénomènes pelliculaires et épidermiques du rech Bourcart (Golf du Lion) et de la Mer Hespérienne. *Comptes Rendus de l'Académie des Sciences, Paris,* D 267:1079–1083.

Guillaume, A. 1967. *Contribution à l'Etude Géologique des Alpes Liguro Piémontaises.* Thèse de 3e Cycle, Paris. 190 p.

Hersey, J. B. 1965. Sedimentary basins of the Mediterranean Sea. In: *Submarine Geology and Geophysics,* eds. Whittard, W. F. and R. Bradshaw, Butterworths, London, 75–91.

Leenhardt, O. 1963. Un sondage sismique continu sur le plateau continental près du Planier (Marseille). *Comptes Rendus de l'Académie des Sciences, Paris,* D 257 : 1541–1544.

Leenhardt, O. 1968. Le problème des dômes de la Méditerranée occidentale. Etude géophysique d'une colline abyssale, la Structure A. *Bulletin de la Société Géologique de France,* 10:497–509.

Leenhardt, O. 1970. Sondages sismiques continus en Méditerranée occidentale. Enregistrement, analyse, interprétation. *Mémoires de l'Institut Océanographique de Monaco,* 1 : 120 p.

Leenhardt, O., S. Pierrot, A. Rebuffatti et R. Sabatier 1970. Sub-sea floor structure South of France. *Nature,* 226:930–932.

Mascle, J. 1968. *Contribution à l'Etude de la Marge Continentale et de la plaine abyssale au large de Toulon.* Thèse de 3e Cycle, Paris, 98 p.

Mauffret, A. 1968. *Etude des Profils Obtenus au Cours de la Campagne GÈOMÉDE I au large des Baléares et en Mer Ligure,* Thèse de 3e Cycle, Paris, 95 p.

Mauffret, A. 1969. Les dômes et les structures "anticlinales" de la Méditerranée occidentale au nord-est des Baléares. *Revue de l'Institut Français du Pétrole,* 24 : 953–960.

Menard, H. W., S. M. Smith et R. D. Pratt 1965. The Rhône deep-sea fan. In: *Submarine Geology and Geophysics,* eds. Whittard, W. F. and R. Bradshaw, Butterworths, London, 271–285.

Montadert, L., J. Sancho, J. P. Fail, J. Debyser et E. Winnock 1970. De l'âge tertiaire de la série salifère responsable des structures diapiriques en Méditerranée occidentale au nord-est des Baléares. *Comptes Rendus de l'Académie des Sciences, Paris,* D 271 :812–815.

Pautot, G. 1969. *Etude Géodynamique des Terminaisons Sous-Marine de l'Esterel.* Thèse d'Etat, Paris, 100 p.

Rehault, J. P. 1968. *Contribution à l'Etude de la Marge Continentale au Large d'Imperia et de la Plaine Abyssale Ligure.* Thèse de 3e Cycle, Paris, 95 p.

Ryan, W. B. F. 1969. *The Floor of the Mediterranean.* Ph. D. Thesis, Columbia University, New York, 236 p.

Ryan, W. B. F., D. J. Stanley, J. B. Hersey, D. A. Fahlquist et T. D. Allan 1970. The tectonics and geology of the Mediterranean Sea. In: *The Sea,* 4, ed. Maxwell, A. E., Wiley-Interscience, New York, 387–492.

Watson, J. A. et G. L. Johnson 1969. The marine geophysical survey in the Mediterranean. *International Hydrographical Review,* 46: 81–107.

Sedimentation in the Vicinity of the Strait of Gibraltar

Gilbert Kelling and Daniel J. Stanley

University of Wales, Swansea and Smithsonian Institution, Washington, D. C.

ABSTRACT

Three major linear depressions: the bifurcating Strait Channel, the Gibraltar (Algeciras) Canyon and the Ceuta Canyon, meet in a deep, flat basin at the eastern end of the Strait of Gibraltar. A programme of precision echo-sounding, bottom photography, grab and short-core sampling has yielded the first integrated analysis of recent sedimentation in this critical region. The walls and floor of the Strait Channel are either rocky or 'ledged' (current-swept) with a veneer of bluff-coloured moderately sorted gravel and sand, largely of biogenic origin. Rock-outcrops are rare in the walls of Gibraltar and Ceuta Canyons, where there is a thick sediment cover of olive terrigenous sandy silt. Textural and shape parameters distinguish these canyon sands from those of the Strait Channel and indicate significant differences in the nature of the transportation-deposition process in these two regions.

The orientation of a prolific suite of erosional and depositional sedimentary features observed in bottom photographs indicates consistently westward-flowing bottom currents in the Strait Channel, and axial, down-canyon flow in the Gibraltar and Ceuta depressions. However, the divergent orientations of scours and ripples suggest a more complex regime in the deep zone of confluence. Terrigenous fractions from the three major depressions may be differentiated petrographically, on the basis of both light and heavy mineral assemblages, and are ascribed to similar but distinguishable suites of source-rocks, comparable to the northern Rif and the Malagan Betics, together with contribution from local, probably submarine, sources. Thus coarse sediment supplied via Ceuta and Gibraltar canyons to the basin at the eastern end of the Strait is largely reworked westward, but finer grades probably spill east into the Alboran Basin; a process aided, if not controlled, by the tidally controlled reversals of water-movement in the region south of Gibraltar.

RESUME

Un bassin profond à fond plat et situé à l'extrémité Est du Détroit de Gibraltar est le site du point de rencontre de trois importantes dépressions linéaires. Ces dépressions sont: le chenal proprement dit du détroit qui se divise en deux branches, le cañon de Gibraltar (ou d'Algeciras) et le cañon de Ceuta.

La première synthèse des analyses de la sédimentation moderne de cette zone critique a pu être réalisée grâce à un programme comprenant mesures précises d'échos sonores, photographies, et échantillonage du fond. Les parois et le fond du chenal proprement dit sont soit rocheuses soit terracées (érodées par le courant) et sont couvertes d'une fine couche rouge jaunâtre composée de gravier et sable biogéniques à grain fin. Les affleurements de la roche sont rares dans les parois des cañons de Gibraltar et de Ceuta qui sont couvertes d'une épaisse couche de sable silteux terrigène de couleur olive. Les sables de ces cañons se distinguent de ceux du chenal proprement dit par leurs textures et leurs formes qui indiquent des différences significatives quant à la nature des mécanismes de transport et de dépôt dans ces deux régions.

L'orientation d'une séquence sédimentaire révélée par les photos du fond indique que les courants de fond coulent dans la direction de l'Ouest dans le chenal du détroit et suivant l'axe dans les dépressions de Gibraltar et de Ceuta. L'orientation divergente des scours et ripples suggère néamoins un régime plus complexe dans la zone profonde du confluent. Les fractions terrigènes des trois dépressions sont distinguables du point de vue pétrographique aussi bien dans les associations de minéraux lourds que dans celles des minéraux légers, et il est possible de leur associer des sources, similaires mais distinctes, semblables à la partie Nord du Rif et à la Cordillère Bétique de Malaga, complémentées par des sources locales et probablement sous-marines. En conséquence, le sédiment grossier apporté à l'Est du détroit par les canyons de Gibraltar et de Ceuta sont remaniés dans la direction de l'Ouest alors que les sédiments plus fins se déversent à l'Est dans le Bassin d'Alboran sous l'influence du phénomène d'inversion des mouvements de l'eau cause par les marées dans la zone au Sud de Gibraltar.

INTRODUCTION

The Strait of Gibraltar, the Pillars of Hercules of ancient navigators, delineates the westernmost sector of the Mediterranean Sea and serves as its only connection with the eastern North Atlantic Ocean. This narrow, relatively shallow east-west trending passage is about 14 km in minimum width, and 50 km in length, and occupies an area of approximately 1,200 km^2. It is bordered on the north by the southernmost tip of the Iberian peninsula (Tarifa to Algeciras, Spain, and Gibraltar) and is bounded to the south by Tangier and Cap Spartel to Ceuta on the northernmost tip of Morocco.

The Strait has long received attention because of its obvious maritime importance. Military considerations during and following World War II accentuated the number of studies of its morphological and physical oceanographic characteristics (*cf.* summary in Lacombe *et al.*, 1964) by countries interested in its obvious strategic aspects. Surprisingly little is still known, however, of sedimentation in this critical zone of current interchange although reports of numerous cable breaks (Heezen and Johnson, 1969), shipwrecks and other phenomena indicate that the Strait is a region of intense erosion and bottom sediment transport.

This study focuses on sediment transport and dispersal patterns both in the Strait proper and in the adjacent areas. Although the predominant water mass flow patterns are reasonably well established (less dense Atlantic water inflow above the denser Mediterranean water outflow), there are indications that the bottom sediments are not uniformly transported westward along the entire length of the Strait. Studies of the Alboran Sea to the east (Stanley *et al.*, 1970; Huang and Stanley, this volume) indicate that some sediment at least, derived from the eastern portion of the Strait, has been transported eastward into the Mediterranean in the recent geological past. This study sheds light on this problem, as well as on the provenance and dispersal paths of sediment transport in submarine canyons associated with the Strait area.

METHODOLOGY

A cruise on the USCGC ROCKAWAY in August 1970 had as one of its missions the definition of bathymetry and bottom conditions in the Strait and adjacent

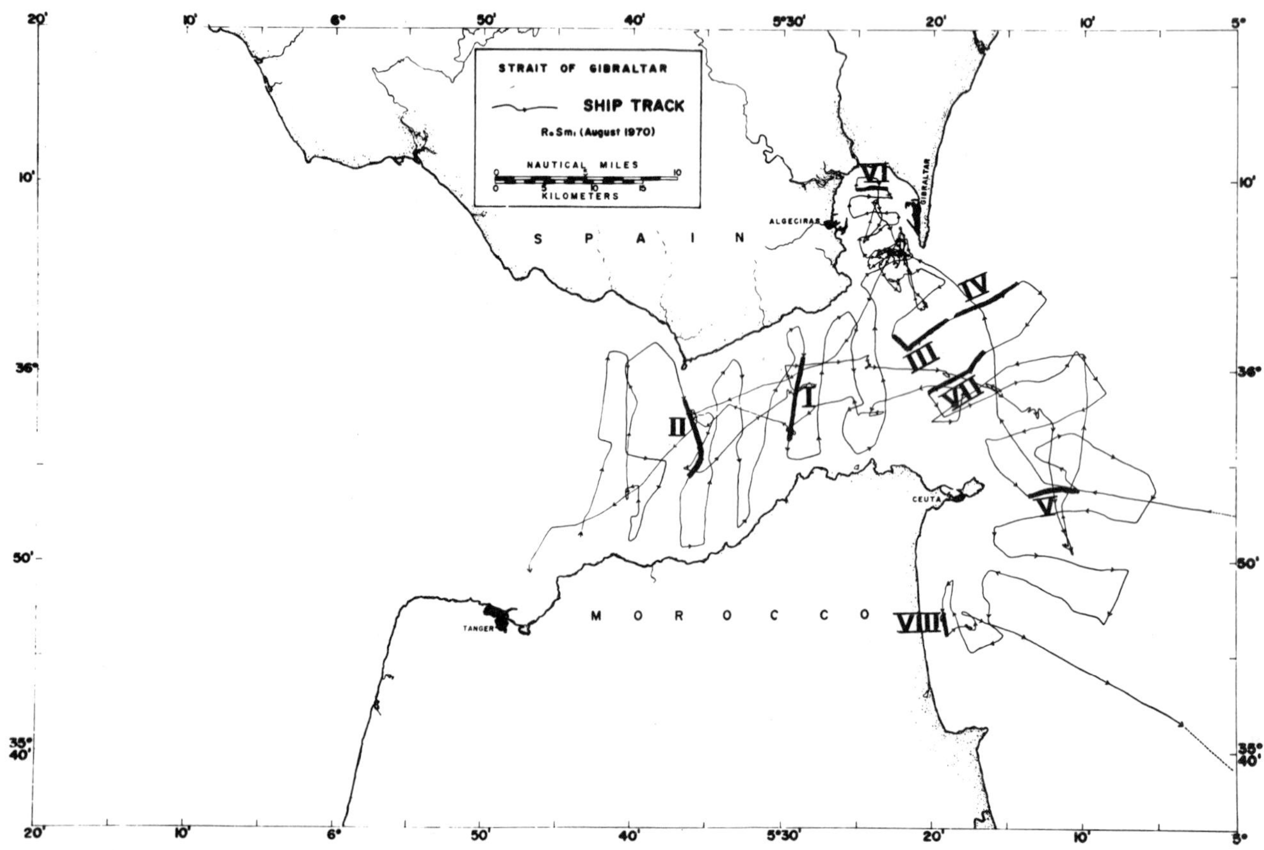

Figure 1. Ship's track for research cruise RoSM$_1$ of U.S.C.G.C. ROCKAWAY. Numbers (I, II, *etc.*) refer to precision echo-sounding records reproduced in Figure 4.

regions. Over 800 km of PESR (Precision Echo Sounding Record) lines were obtained (Figure 1). Navigation was achieved both by radar and land fixes obtained at 10 to 15 minute intervals and the quality of positioning is considered very good to excellent. The first part of the cruise was reserved for bathymetric profiling of the Strait, Algeciras Bay and adjacent canyons (Figure 1), and the second part was reserved for observing conditions on the bottom at 18 stations (Figure 5, Table 1). At seventeen of these, a bottom camera was lowered. At 8 stations, a Shipek bottom grab was used to recover surficial material from the sea floor.

The underwater photographic rig included a 2 m vertical frame on which was mounted, from top to bottom, an Alpine Geophysical camera (Model 311), a transponder (Model 295 S) to indicate bottom contact, and a strobe (Model 311 L) for illumination. Oblique photography emphasized the relief of bottom features (the camera was oriented at 38° from the vertical while the strobe was oriented at 50°). A 6.5 cm diameter compass, serving to orient bottom features at each station, was mounted on a long (64 cm) arm at the base of the camera frame, so as to appear on the lower left of each photo frame.

The camera was triggered by bottom activation in order to obtain a reasonably uniform photo-to-photo coverage. Photos were collected at 15 to 20 second intervals and the station was occupied sufficient long to obtain about 16 to 18 photographs.

The camera lens was about 85 inches (216 cm) above the bottom at contact. The trigger consisted of a lead-weighted micro gravity corer (liner about 3.8 cm in diameter and 15 cm in length). The small core samples aid in determining bottom lithology, and are most useful in interpreting bottom photographs. Bottom configuration, lithology, current activity and fauna are the principal parameters observed and measured in photographs (method described in Stanley and Kelling, 1968).

BATHYMETRY AND MORPHOLOGY OF THE STRAIT AREA

General

The Strait of Gibraltar has been the subject of numerous bathymetric studies and the main morphological features of this area are well known (see Giermann, 1961, pp. 4–6, for a review of recent

Table 1. Location of camera and grab stations.

Station no.	Instrumentation	(Beginning of Station) Lat. N.	Long. W.	Depth Fms.	(m)	Drift Distance N. mls	(km)	Drift Direction
1	Camera/short corer	36°06.2′	05°22.6′	210	(384)	0.4	(0.73)	SE
2A	Camera/short corer	36°06.7′	05°24.7′	25	(46)	0.3	(0.55)	NE
2B	Camera/short corer	36°08.9′	05°23.9′	120	(220)	0.5	(0.82)	SSE
3	Camera/short corer	36°05.6′	05°24.2′	38–42	(70–77)	0.4	(0.73)	N
4	Camera/short corer	36°05.0′	05°21.8′	294–265	(538–485)	0.4	(0.73)	ESE
6A	Camera/short corer	35°46.0′	05°18.8′	49–140	(90–256)	0.5	(0.92)	NE
6B	Camera.	35°46.5′	05°18.1′	140– 65	(256–119)	0.7	(1.28)	E
30	Camera/short corer	36°03.3′	05°20.3′	375	(686)	0.3	(0.55)	SSE
31	Camera/short corer	35°57.7′	05°36.3′	350	(640)	0.6	(1.10)	E
	Shipek Grab	35°57.7′	05°35.1′	342	(626)	—	—	—
32	Shipek Grab	35°54.6′	05°35.7′	182	(333)	—	—	—
33	Camera/short corer	35°57.0′	05°29.3′	445	(814)	0.3	(0.60)	SE
	Shipek Grab	35°56.8′	05°29.7′	448	(820)	—	—	—
34	Camera/short corer	36°00.0′	05°29.6′	285–250	(522–458)	0.4	(0.73)	N
	Shipek Grab	35°59.0′	05°28.8′	415	(760)	—	—	—
35	Camera/short corer	36°00.7′	05°24.3′	315–350	(576–641)	0.5	(0.82)	SSE
	Shipek Grab	36°00.5′	05°24.0′	345	(631)	—	—	—
36	Camera/short corer	36°00.0′	05°19.0′	480–485	(878–889)	0.6	(1.10)	E
	Shipek Grab	35°59.8′	05°19.1′	480	(878)	—	—	—
37	Camera/short corer	35°58.1′	05°13.0′	425–440	(778–805)	0.8	(1.46)	S,NNE
	Shipek Grab	35°57.9′	05°13.3′	447	(818)	—	—	—
38	Camera/short corer	35°50.9′	05°11.0′	240–225	(439–412)	0.5	(0.82)	S
39	Camera/short corer	35°57.3′	05°20.0′	370–375	(677–686)	0.4	(0.64)	E
40	Camera/short corer	35°57.8′	05°23.4′	450–460	(824–842)	0.6	(1.10)	W
	Shipek Grab	35°57.9′	05°24.1′	460	(842)	—	—	—

publications). However, less information is available on the bathymetry of Gibraltar (Algeciras) Canyon and Ceuta Canyon. Bathymetric data from the 800 km of PESR lines have enabled slight modification of existing charts of the Strait of Gibraltar and have established in more detail the course and morphology of both the Gibraltar and the Ceuta canyon-systems (Figure 2).

The western part of the Strait Channel originates near the prominent northwest-trending feature known as the Ridge, north of Tangier. The Channel may be traced as a complex, single feature into a deep hollow south of Tarifa, but eastwards from this point the depression divides into two branches, separated by an elongate ridge shoaling to 250 fm (458 m) at its western end (Figure 3, profile G-H, J-K). In general the southern branch is slightly deeper but the profiles of both branches are U-shaped or sub-acute. The marginal slopes bordering both sides of this part of the Strait are dissected by numerous small ravines and canyons, most of which can be traced upwards to the 50 fm (92 m) isobath, that is, to just below the local shelf-break. Both flanks of the central ridge are comparably dissected. The two branches of the Strait Channel reunite in a further trough, south of Punta del Carnero, which marks the deepest part of the Strait (Figure 2). Eastwards from this confluence the bottom topography is more muted and the axis of the broad, flat-floored Channel runs closer to the north flank of the Strait (profile L-W on Figure 3). However, the narrowest portion of the shelf and the steepest, most dissected marginal slope occurs off Punta Leona, on the south side of this part of the Strait. Here, slopes exceeding 15° are common.

In contrast to the Strait Channel the two other major depressions in this region behave as normal submarine canyons, originating on the bordering shelf and gradually enlarging with increasing depth. Although shorter, Gibraltar Canyon is more deeply incised than Ceuta Canyon, with rougher, more steeply inclined walls (Figure 3). The maximum relief and the most acute profiles in Gibraltar Canyon occur at the entrance to Algeciras Bay. The varying asymmetry of the profiles shown in Figure 3 indicates the meandering character of the axis of Gibraltar Canyon.

Ceuta Canyon is bounded to the east by the smooth topography of the north-trending Ceuta Ridge. The west flank of this depression is formed by a more steeply inclined, moderately dissected slope which merges into the broad shelf between Ceuta and Cabo

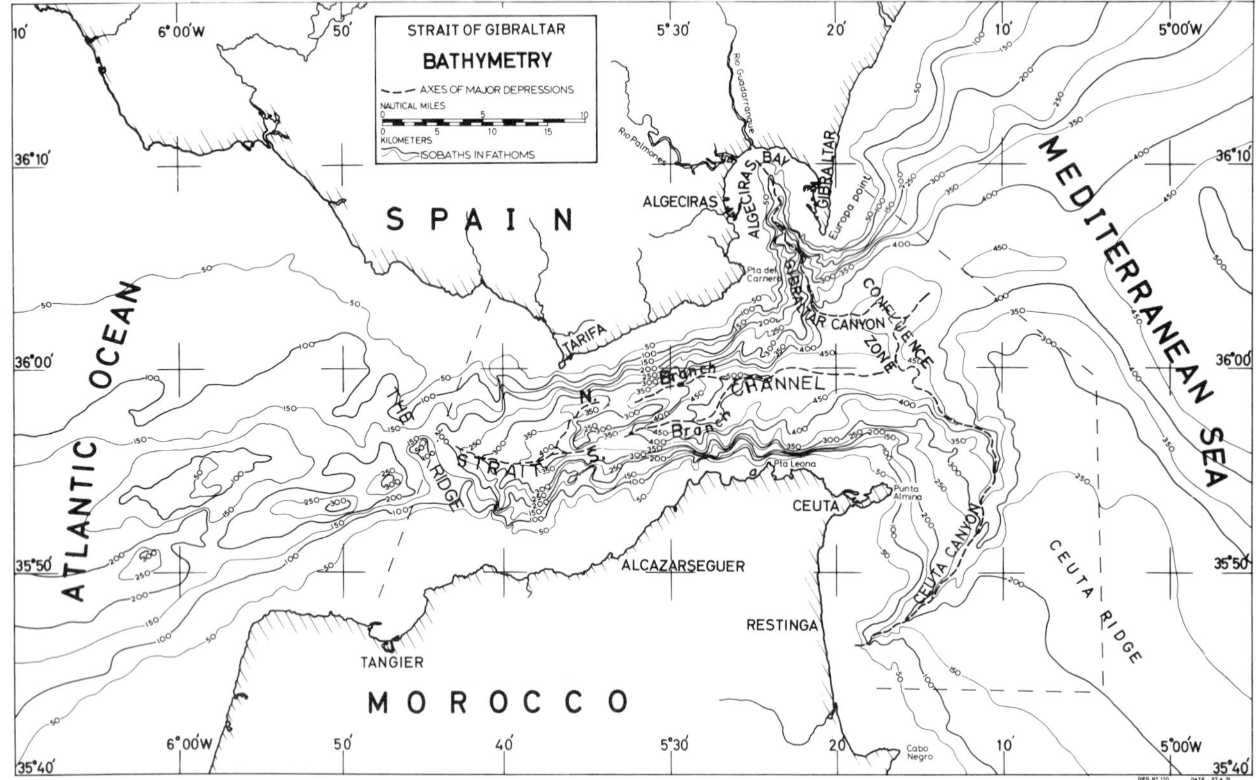

Figure 2. Bathymetry of the Strait of Gibraltar area, illustrating the location of the three major depressions. Isobaths at intervals of 50 fathoms. Dashed lines indicate outer limits of area of present survey.

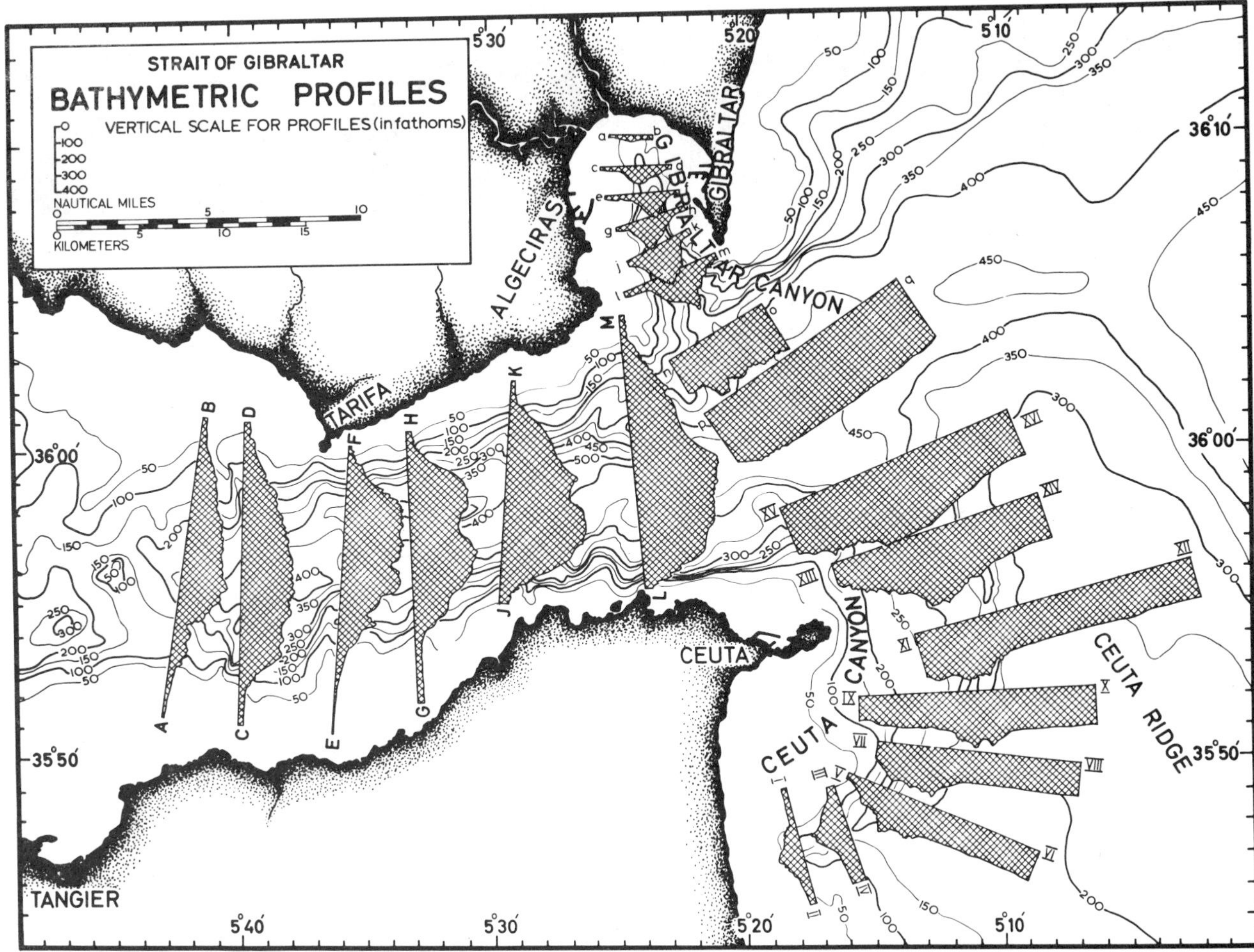

Figure 3. Selected bathymetric profiles, constructed from precision echo-sounding records.

Negro (Figure 2). Canyons profiles (Figure 3) are generally acute and terraced in the shallow part of this depression but beyond an axial depth of about 357 fm (686 m) the profiles indicate a broader, less corrugated axial zone, which becomes essentially smooth beyond a depth of 440 fm (805 m). The only major tributary to join Ceuta Canyon along its 21 mile (38 km) length enters tangentially from the east at an axial depth of 310 fm (567 m).

The zone of confluence of the Strait Channel, Gibraltar Canyon and Ceuta Canyon is represented by a broad and deep, smooth or minutely terraced region south of Gibraltar (Figure 3, profile p-q). The junction of the Strait Channel and Ceuta Canyon occurs on the 36° latitude line at a depth of about 480 fm (878 m), while Gibraltar Canyon may be traced to a slightly shallower ponded area (460 fm or 842 m) a few miles to the north. An interesting feature of the latter area is a pair of minor ridges, about 10 fm (18 m) high and 0.5 miles (0.9 km) apart, resembling channel-levees. These were encountered on two bathymetric profiles and their position is indicated on Figure 5. A broad, gently inclined hollow connects the Gibraltar 'pond' with the Strait-Ceuta confluence and a wide col, skirting the north end of Ceuta Ridge at a maximum depth of 445 fm (841 m), separates the ponded area from the western flank of the Alboran Basin (Figure 2). A linear depression, the Gibraltar Submarine Valley, extends downslope from the junction of the Ceuta and Gibraltar Canyons to the Western Alboran Basin, approximately 90 km to the southeast (Stanley *et al.*, 1970; Huang and Stanley, this volume).

In addition to determining bathymetry, the character of the precision echo sounding record furnishes evidence of bottom morphology (Heezen and Johnson, 1969). Six major physiographic types are recognised (Figure 4) and their general distribution is indicated on Figure 5. The terminology adopted is partly descriptive and partly interpretative, Interpretation is based on experience acquired elsewhere (Kelling and Stanley, 1970; Heezen and Johnson, 1969) combined with direct evidence from bottom photographs.

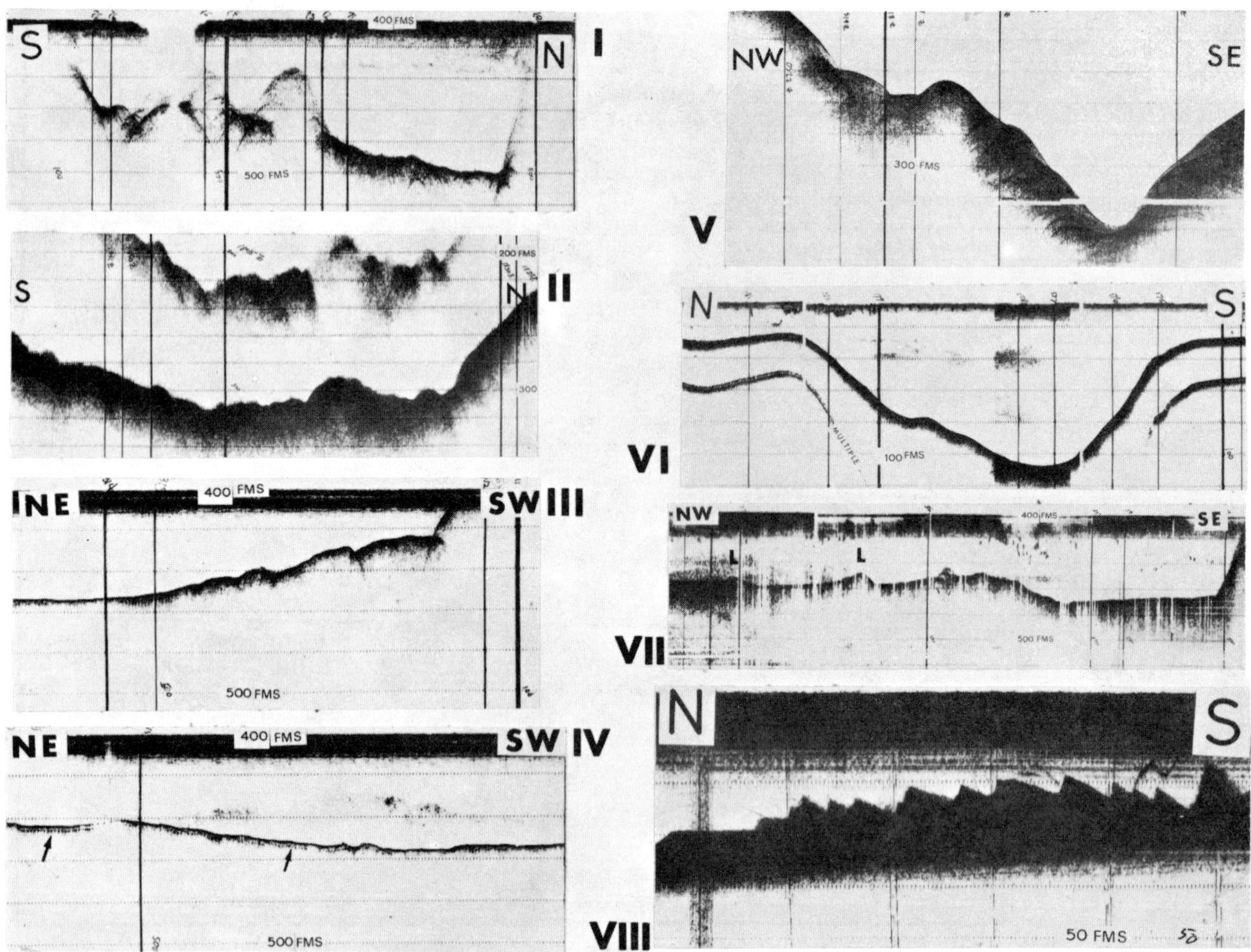

Figure 4. Photographs of PESR records illustrating the six main types of bottom physiography. Location of profiles indicated in Figure 1. I, Rocky bottom trace on left; probable lodged topography of Strait Channel floor on right. II, Rocky bottom trace. III, Ledged sloping topography with smooth floor to left. IV, Smooth, flat basin-floor with indications of sub-bottom at left (arrows); minor ledging near profile centre. V, Gently undulating topography. VI, Profile of Gibraltar Canyon head; smooth, sloping topography with smooth flat shelf physiography at left and right. VII, Zone of confluence profile: smooth, flat Strait Channel at right, slightly ledged and smooth floor of Gibraltar Canyon mouth at left. L-probable levees. VIII, Sand-wave physiography on shelf (Ceuta Canyon head).

Rocky Physiography

The PESR trace of this type of bottom is characterised by prolonged, overlapping parabolic echoes which indicate abrupt and frequent changes in depth (Figure 4, I and II). In shallow water the parabolic nature of the echoes may be masked but the irregularity of the trace is still diagnostic. This type of bottom morphology is encountered frequently in the Strait of Gibraltar but is less common in the two associated canyons. Within the Strait, rocky topography is typically developed on both marginal slopes and on the median ridge and is more sporadically encountered on the adjacent shelves (Figure 5). Rocky slopes characterise the deeply incised part of Gibraltar Canyon, just south of Europa Point but are generally lacking in the shallow part of the canyon and the neighbouring shelf. Similarly, rocky topography in Ceuta Canyon is confined to small areas near the head and near the confluence with the Strait Channel.

Ledged Physiography

This type of bottom generally merges with the rocky areas and it is sometimes difficult to distinguish these two related physiographic forms on PESR profiles. Ledged topography is typically indicated on the echo-sounding record by a strong, relatively smooth solid echo with frequent minor 'steps' of a few fathoms relief (Figure 4, I and III). Traces of this type are often interrupted by occasional hyperbolic echoes, indicating rock outcrops, and smaller, regularly spaced undulatory features [5 to 10 fms (9–18 m) high, 0.1 to 0.2 miles (180–366 m) in wave

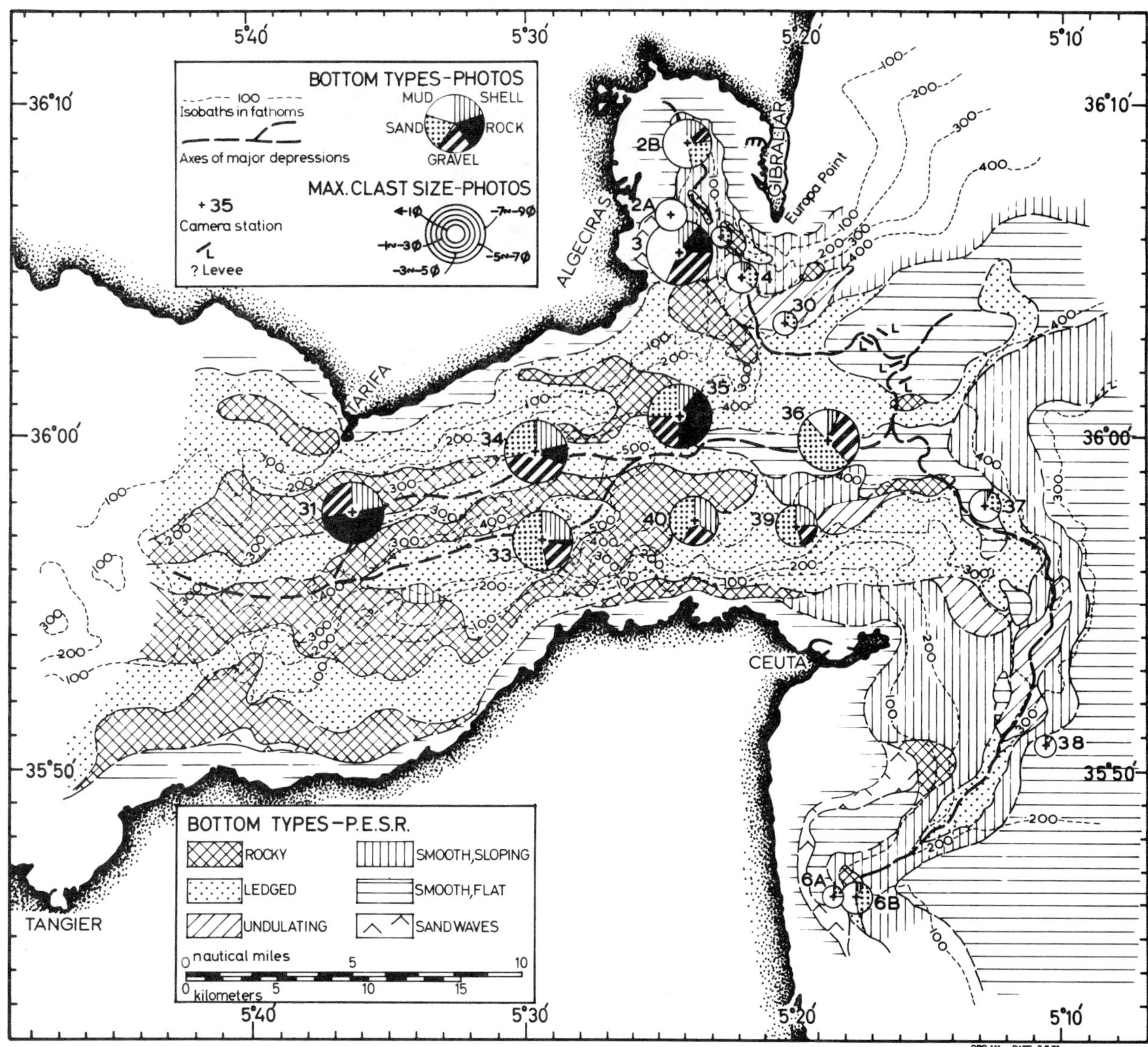

Figure 5. Distribution of bottom types as inferred from PESR profiles and from bottom photographs. The pie-diagrams represent the relative areas occupied by the bottom types indicated, as aggregated from all frames obtained at each photographic station. The diameter of the circle is proportional to the diameter of the maximum observed clast and is expressed in phi-units.

length], which are probably small sand waves. Photographs of the sea floor from areas of ledged topography generally reveal evidence of strong current activity and the presence of sand, gravel and rock. This topography probably corresponds to the Type III, current-swept physiography of Heezen and Johnson (1969), and is interpreted as an essentially rocky floor blanketed by a variable thickness of mobile sand and gravel. Ledged topography is typical of the outer shelf and upper slope on either side of the Strait and also characterises the floor of the bifurcating sections of the Strait Channel (Figure 5). The same physiographic type is encountered on the slopes bordering the zone of confluence at the eastern end of the Strait, and a narrow tongue of ledged bottom appears on the col separating the zone of confluence from the Western Alboran Basin to the east. Scattered patches of ledged or current-swept morphology also occur on the floor of both Gibraltar and Ceuta canyons.

Undulating Physiography

In some respects this type of bottom is the most difficult to distinguish and interpret. The characteristic PESR trace (Figure 4, V) is a sloping surface surmounted by a series of gently arcuate, irregularly spaced undulations which may or may not overlap one another. The undulations are usually between

10 and 50 fm (18–91 m) high and 0.3 to 0.6 n. miles (0.5–1.1 km) across and parabolic echoes are either absent or poorly developed. The only photographs obtained on this type of bottom (Station 30) reveal a relatively smooth surface of sandy silt, with few sedimentary structures.

This physiographic variety usually intervenes between a rocky or ledged bottom and a smooth topography and this fact suggests that the undulating physiography may be ascribed to a rocky surface thickly mantled by relatively fine sediment. Significantly, this type of bottom morphology is absent from the Strait but is patchily developed on the walls of the middle section of Ceuta Canyon and also occurs on the flanks of Gibraltar Canyon, where that depression emerges from Algeciras Bay (Figure 5).

Smooth, Flat Physiography

Several profiles of both deep and shallow areas are characterised by smooth, near-horizontal traces on which the bottom relief seldom exceeds 2 fm (3.7 m). In deep water, the returns from this type of surface are generally diffuse but may include sub-bottom reflectors (Figure 4, IV) while the echo from a shallow smooth bottom is usually solid and prolonged (Figure 4, VI). Photographs from areas of smooth topography reveal surfaces of variable aspect. Deep stations in Ceuta Canyon (37, 38) display a relatively featureless topography of muddy or silty sediment. However, the smooth floor of the eastern end of the Strait Channel (Station 36; *cf.* Allan, 1966, his Station 5) is composed of sand and gravel, strongly scoured and with some ripples. The shallow, smooth areas of Algeciras Bay and the shelf south of Ceuta (stations 2A, 3, 6A) are characterised by featureless or biologically disturbed surfaces of silty aspect, with rare patches of rock or gravel. However, there is evidence (Allan, 1966, his Figure 3c) that the smooth inner shelf of the Strait area is characterised by rippled and scoured sand and gravel. These observations suggest that this type of bottom morphology may arise wherever a thick mantle of sediment acquires a smooth surface, either through strong and prolonged current activity (as in the Strait Channel) or because of uniform settling of fine sediment (as on the shelf areas).

Smooth flat topography is encountered on the inner shelf throughout most of the areas studied and is also widespread on the Ceuta Ridge, east of Ceuta Canyon (Figure 5). The basins marking the confluence of the Strait Channel with Ceuta Canyon and with Gibraltar Canyon are also characterised by this type of bottom.

Smooth, Sloping Physiography

The PESR traces across some inclined areas of the sea-bed display characteristics similar to those which distinguish the previous physiographic type, although the smooth surface may be modified by occasional gentle undulations, a few fathoms in height (Figure 4, VI). Photographs of such regions (stations 2B, 4, 6B; *cf.* Allan, 1966, his figures 4b, 4c) generally reveal a silty or sandy bottom with evidence of biological reworking and occasional small ripples.

The upper slopes bordering both Gibraltar Canyon and Ceuta Canyon are of this type and a similar morphology is encountered on both flanks of the broad trough north of Ceuta Ridge (Figure 5).

Sand Wave Physiography

Although small sand waves are probably present in other areas, including the eastern part of the Strait Channel, the term is here reserved for relatively large, regularly spaced features encountered on some shelf-areas. These bed-forms have amplitudes varying from 2 to 8 fm (3.7–14.6 m) and wavelengths of 0.1 to 0.2 n. miles (180–360 m) and they commonly appear to have strongly asymmetric profiles (Figure 4, VIII), although this may be a function of the ship's heading. Sand-wave topography is developed in the southwest part of Algeciras Bay and more extensively around the head of Ceuta Canyon (Figure 5) where the apparent steep faces point downslope. A small area of sand-waves, apparently facing west, occurs at the western end of the Strait, some 8 miles (14 km) northeast of Tangier.

SURFICIAL SEDIMENT DISTRIBUTION

As indicated in the preceding section, sea-floor photographs from a total of 17 stations provide evidence of the distribution of sediment types in the vicinity of the Strait of Gibraltar. Strong surface currents and (especially in the Strait) contrary movement of bottom water account for on-station drift varying from 0.25 to 0.8 miles (0.5–1.5 km), and the average of 14 analysable frames per station is thus representative of a significant area of the sea-bed at each location (Table 1).

Additional and confirmatory data relating to the surficial sediment distribution was obtained by a program of sampling of the sea-bed by means of the Shipek grab and the small coring device attached to the trigger-weight of the camera. Although complementary, these two methods of determining the

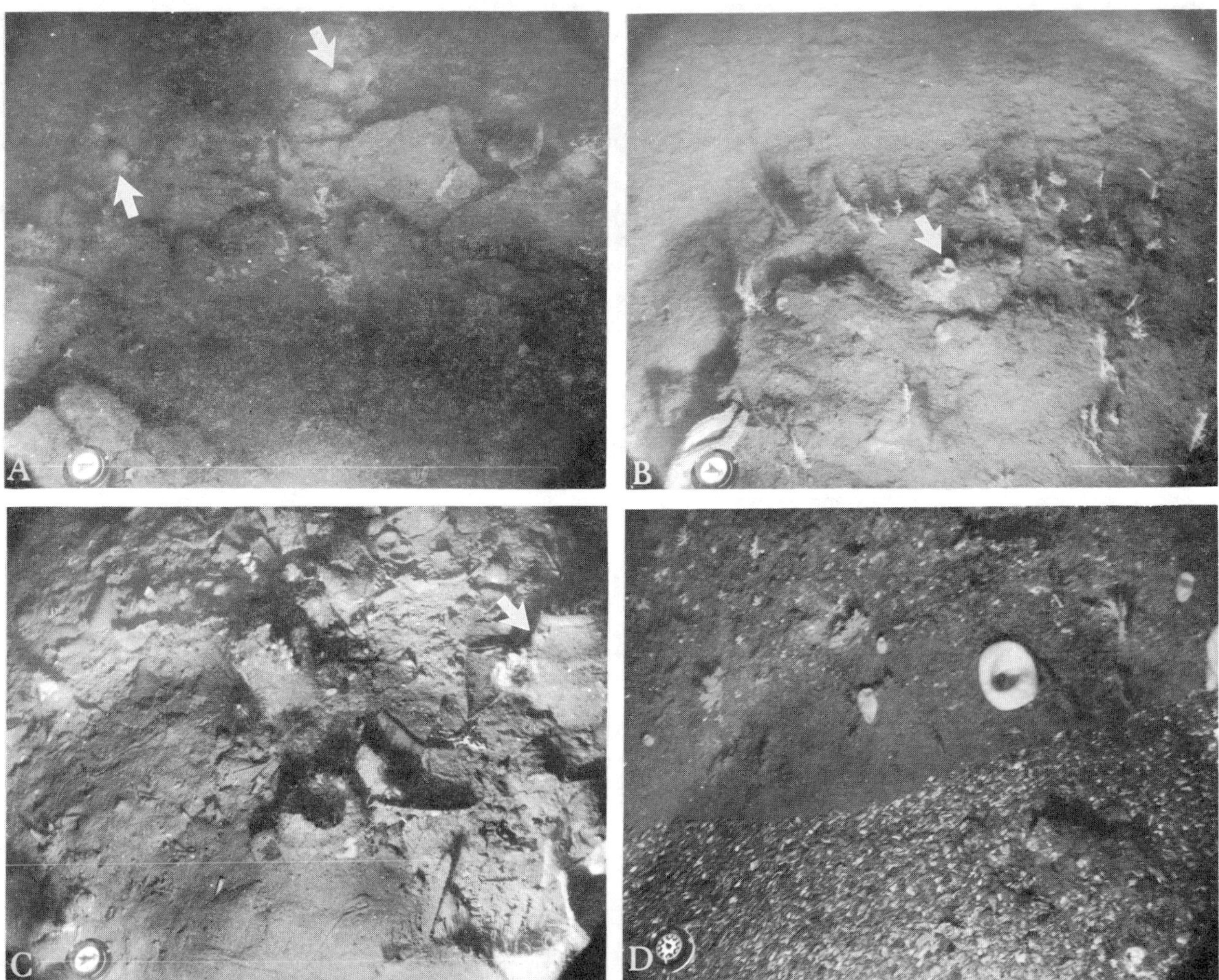

Figure 6. Bottom photographs (cruise RoSM$_1$). A, Station 3, frame 5 (depth: 75 m). Outcrop of probable dipping bedded sediment, with blocks and mantle of sandy silt. Biotal elements comprise mainly sea-urchins (arrowed) and gorgonian corals. B, Station 3, frame 4 (depth: 75 m). Bottom composed of bioturbated sandy silt, mantling talus blocks. A cup-sponge (centre, arrowed), gorgonians and ?holothurians (lower left) are also present. C, Station 3, frame 7 (depth: 75 m). Sandy silt mantling broken pottery (probable mediaeval oil- or wine-jars). Note undamaged jar on the right margin (arrowed); sea-anemones and sea-urchins utilising the pottery, and various bottom trails and tracks. D, Station 31, frame 7 (depth: 640m). Outcrop of dark rock, furnishing firm base for sponges, bryozoa and various coelenterates. Coarse pebble-gravel in foreground.

nature and distribution of bottom sediments are qualitatively different (the traverse versus the point approach) and therefore are treated separately in the following account.

Bottom Photography

As indicated on Figure 5, all stations in the Strait Channel reveal photographic evidence of gravel, sand and appreciable quantities of 'shell' (probably including coralline and bryozoan debris). Significant outcrops of rock were observed only in the three stations (31, 34, 35) in the north branch of the Channel. At Station 31 the outcropping rock appears to be a dark bedded sediment, dipping at about 20° to the northwest and is mantled by coarse, shelly gravel showing evidence of current-scour (Figure 6D). Solitary sponges and bryozoan thickets are the most prominant biotic elements here. At stations 34 and 35 the rock-surfaces are almost horizontal and thickly encrusted with bryozoa and other lime-secreting organisms. Evidence of current scour is conspicuous at station 35 where the thin veneer of shelly sand and gravel is often arranged in patches and streamers. The efficacy of current winnowing and scouring at this station is further attested by the observation of man-made objects which lie, free of any sediment

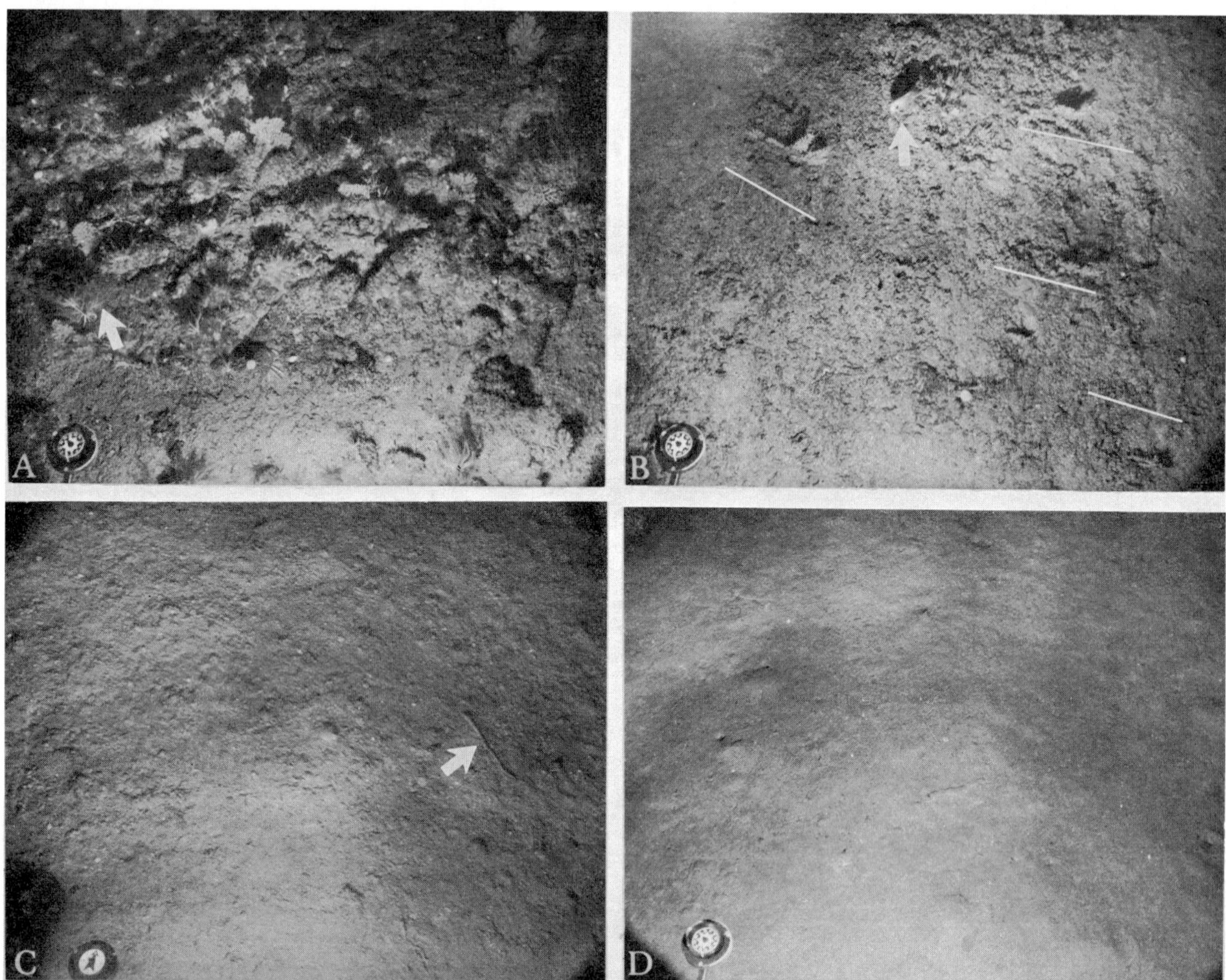

Figure 7. Bottom photographs (cruise $RoSM_1$). A, Station 34, frame 13 (depth: 510 m). Probable rock and coarse gravel bottom, thickly encrusted with bryozoa and various coelenterates, including crinoids (one is arrowed) and gorgonian corals. Some coarse sand in the foreground. B, Station 35, frame 4 (depth: 590 m). Rock-surface with thin veneer of sand and encrusting calcareous organisms, probably bryozoa and gorgonians. Note tendency for parallel arrangement of coral-thickets (trend indicated), probably perpendicular to local current-flow. Small fish at upper centre of photograph (arrowed). C, Station 6A, frame 16 (depth: 250 m). Bioturbated silty mud surface with numerous mounds and trails. Note worm at centre right (arrowed). D, Station 30, frame 5 (depth: 685 m). Bioturbated sandy silt, with burrows and mounds; partly buried ripple-crests trend from left to right across the photograph.

cover, on the hard sea-floor. The eight photographs obtained at Station 31 and the twelve recovered from Station 35 reveal a relatively homogenous type of bottom. However, the sea-bed at station 34 is apparently more diverse in character, displaying areas of rippled sand in addition to the scoured and encrusted gravel and the rock typical of the other two stations. The ripples include both asymmetrical linguoid (Figure 8A) and straightcrested, apparently symmetrical, types (Figure 8C). Angular talus-blocks up to 60 cm in diameter occur at all three North Branch stations (Figure 5), furnishing additional proof of the importance of erosional processes in this area of the Strait.

By contrast, stations from the south side of the Strait (33, 39, 40) reveal little evidence of outcropping rock and few large blocks (Figure 5). Instead this area is characterized by rippled and scoured shell-sand with occasional patches of fine gravel and rare, subangular cobbles which achieve their maximum observed size (25 cm) at Station 33. Ripples of variable character were observed in all but three of the forty-nine frames obtained from these three stations and concentration of coarser shell-debris in ripple-troughs is a common feature. Many of the ripples observed at Station 40 display rounded crests and traces of the internal lamination are sometimes discernible on the flanks of the ripples. This morphology suggests an old, inactive, or stabilised ripple field, probably undergoing partial degradation by scouring. Benthic organisms are relatively rare in this area but include a few echinoids and clumps of a dark weed-like organism.

Station 36 falls within the bathymetric limits of the Strait Channel but it displays a unique combination of features for this region. The twenty-five photographs obtained at this station represent a drift distance of 0.6 miles (1.1 km) and imply an average spacing between successive frames of about 145 feet (44 meters). Almost all frames reveal a relatively smooth sea-floor, consisting of silty sand at the beginning (western part) of the station but passing gradually eastwards into a zone of mixed sediment, with dominant sand accompanied by fine gravel and occasional sub-rounded blocks up to 25 cm across (Figures 5 and 8B). A prominent feature of the easternmost frames is the abundance of dark, rounded pebbles and granules. A few linguoid ripples occur in the silty sand of the early frames but gravel-streaks, sand-shadows and other scour-features are typical of the coarse sediment at the eastern end of this station. The intense current activity and associated sediment transport which is indicated by these bottom features is confirmed by the frequent tilting of the camera and the drifting clouds of disturbed sediment which obscure several of the frames. The relative lack of

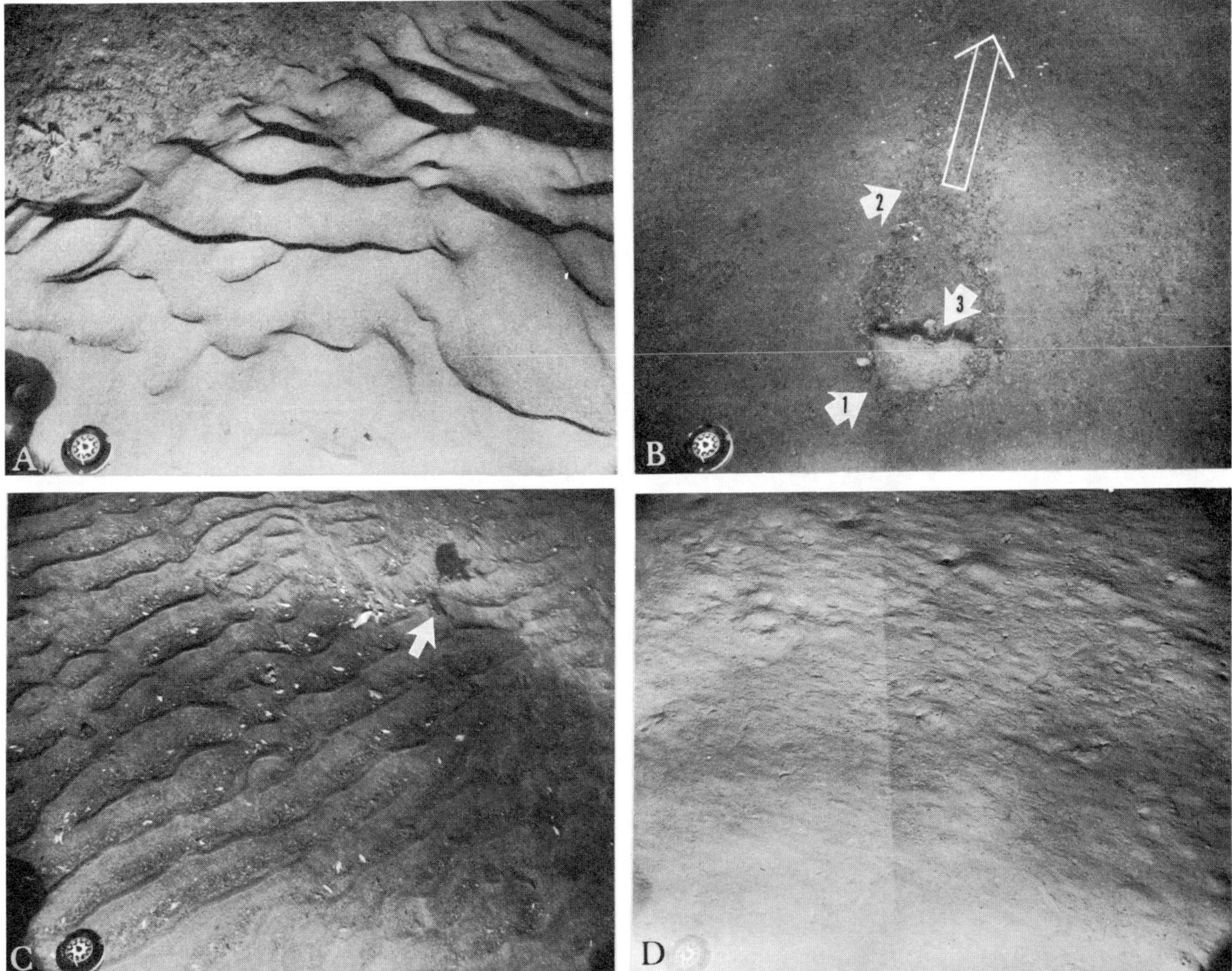

Figure 8. Bottom photographs (cruise $RoSM_1$). A, Station 34, frame 12 (depth: 490 m). Strongly linguoid ripple marks formed in carbonate sand, with sandy gravel in upper left corner. Ripples indicate powerful westerly bottom current flow. B, Station 36, frame 15 (depth: 880 m). Fine gravelly sand bottom with large isolated block surrounded by shallow moat (1) and gravel-train on lee or down-current side (2). Note also sea-anemones growing in shelter of the block (3). Large arrow indicates west south westerly bottom flow.

C, Station 39, frame 7 (depth: 680 m). Almost straight-crested near-symmetrical ripple-marks formed in carbonate sand. Note concentration of coarse shell-depris in ripple-troughs and slight scouring perpendicular to ripple-crests at top right. Small fish (with larger shadow) is arrowed.

D, Station 37, frame 1. (depth: 800 m). Bioturbated silt, with pronounced scours trending from upper left to lower right, modifying burrowed and mounded surface. Note "crag-and-tail" effect around hollows at upper right of picture.

benthic organisms also may be ascribed to a high rate of sediment transport.

Photographic stations within Algeciras Bay present a marked contrast to the Strait in the prevalence of fine surficial sediment and the paucity of shell-debris. Poor visibility is a feature common to most of the Algeciras Bay photographs and is attributed to the high concentration of suspended fine sediment in the near-bottom waters. Two shelf-stations (2A and 3) on the west side of the Bay display a sea-floor dominated by silt or fine sand and extensively modified by the mounds, trails and burrows indicative of a prolific and varied in-fauna. However, Station 3 also reveals evidence (in 6 out of 15 frames) of large, angular talus-blocks and possible outcrops of bedded rock, dipping at about 40° to the south or southeast (Figure 6A), all embedded in the mantle of fine sediment. One frame also reveals a profusion of broken pottery, including wine or oil-jars (Figure 6C). These fragments and the blocks of rock support an abundance of sessile organisms, including sea-anemones, sponges and sea-fans. Evidence of current-activity at Station 3 is confined to a few, partly buried ripples, some 'moating' around the talus-blocks and oriented, uprooted and flat-lying tubes of the polychaete worm *Hyalinoecia*. More conspicuous scour-structures were observed at Station 2A, together with a few *Hyalinoecia* tubes, which here are protruding out of the sea-bed and inclined parallel to the direction of scouring.

Photographs taken at stations within Gibraltar Canyon proper (1, 2B, 4, 30) generally exhibit the rather monotonous aspect of a bottom composed of silt and fine sand, with mounds, trails and other biogenic traces. Subrounded small cobbles and pebbles occur sporadically at station 2B, near the head of the canyon. Partly buried low-amplitude ripples, faint scour-structures and oriented polychaete worm-tubes furnish evidence for relatively weak bottom currents in the axial zone of the canyon.

The Ceuta Canyon stations (6, 37, 38) are also characterised by fine sediment and a general lack of shell-debris (Figure 5). Maximum clast-size for these stations never exceeds 5 cm. Station 6 extends from the shelf across the shallow canyon head to an axial depth of 140 fm (255 m) and, because of the excessive length of the traverse (1.2 miles, 2.2 km) two sub-stations (6A and 6B) have been erected at this locality. Photographs at Station 6A reveal a relatively smooth and muddy sea-floor, often obscured by clouds of suspended sediment, presumably created by the impact of the camera-frame or trigger-weight. Biogenic trails, mounds and hollows occur, together with a few low-amplitude ripples. The initial frames of Station 6B, taken near the canyon axis, show a more varied bottom of sandy aspect, with minor scours and obscure ripples. Followed eastwards up the flank of the canyon, the sea-floor gradually reverts to a bioturbated silty mud, similar to station 6A.

A biogenically reworked mud-floor is also characteristic of Station 38, on the western edge of Ceuta Ridge, although scour-structures are common in photographs from the north end of the traverse. Station 37 lies close to the axial zone of Ceuta Canyon, near its junction with the Strait Channel. Here, almost all of the 14 frames display a strongly scoured sandy silt bottom with numerous biogenic mounds which frequently exhibit elongation due to 'silt-streaking' and moating (Figure 8D).

Grab and Short Core Samples

Data from samples collected by grab or short core broadly confirm conclusions derived from the photographic study (Figure 9), although it should be emphasized that in some cases the grab-sample was obtained from points up to 1.2 miles (2.2 km) distant from the photographic station. The sediment texture is expressed in terms of the relative proportions of gravel, sand and mud (silt and clay grades), as defined on the Wentworth scale, obtained from the bulk grab or core sample. Supplementary data were provided by incorporating textural analyses published by Frassetto (1966). The resulting distribution map (Figure 9) involves substantial extrapolation, especially in the Strait and on Ceuta Ridge, and does not provide a precise indication of the nature of the sea-floor throughout the region. This is illustrated by the fact that *rock* has been recorded only at Station 35 (where three successful triggerings of the grab failed to secure a sediment sample), but the photographic evidence of other rock-outcrops is not incorporated in Figure 9 .

Samples for the Strait Channel are exclusively composed of sand and gravel, in varying proportions, and are light in color (buff to pale brown). As pointed out by Frassetto (1966) a tongue or fan of this coarse sediment extends eastwards from the zone of confluence of the Strait Channel with Gibraltar Canyon (Figure 9).

In contrast, samples from Gibraltar and Ceuta canyons are generally admixtures of sand and mud, olive green in color, and short cores indicate interlamination of these sediments in layers a few centimeters thick. The proportion of sand appears to be higher in the Gibraltar Canyon samples than in those from the Ceuta region. Intermixed gravel, sand and mud occur in restricted areas on the west shelf of Algeciras Bay and in the lower reach of Ceuta Canyon, near the junction with the Strait Channel (Figure 6).

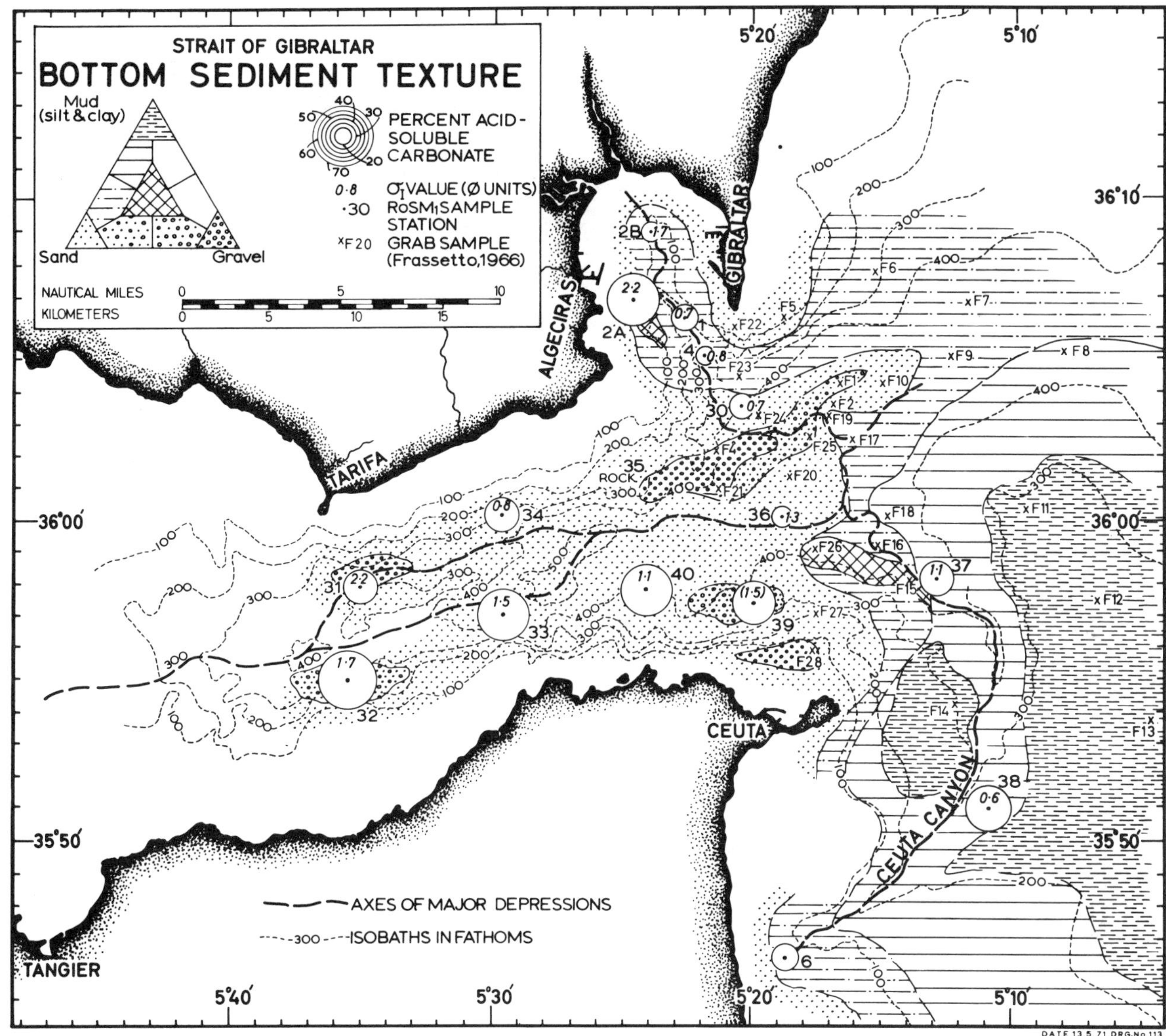

Figure 9. Distribution of surficial sediment texture as determined from size analysis of grab and core samples. Additional data is recalculated from analyses published by Frassetto (1965).

Determination of the relative carbonate content of the gravel and sand fractions in these samples was achieved by treating the washed fractions with cold 10% HCl. Carbonate-rich samples are characteristic of the Strait, and especially the south side. Here there is some tendency for a westward increase in carbonate content (Figures 5 and 9), mainly represented by coral, algal and bryozoan debris and abraded bivalve shells. Sample 2A, from the west shelf of Algeciras Bay, is also exceptionally rich in carbonate, mainly finely comminuted bivalves. However, samples from both Gibraltar and Ceuta canyons are carbonate-deficient, although an abundance of foraminifera inflates the carbonate fraction in samples 37 and 38.

TEXTURE AND PETROLOGY OF THE SEDIMENTS

Texture

Grain-size analyses of sand and gravel samples from the Strait area and the associated canyons were carried out by means of a modified Woods Hole sedimentation column. Short cores from the canyons generally reveal inter-laminated sand and mud but marked disturbance in the core generally has resulted in some mixing of these layers. Moreover, most grab-samples also represent an adventitious mixture of fine and coarse sediment. Accordingly all samples were washed free of fine silt and clay [less than 5 ϕ (31 microns) fraction] before grain-size analysis.

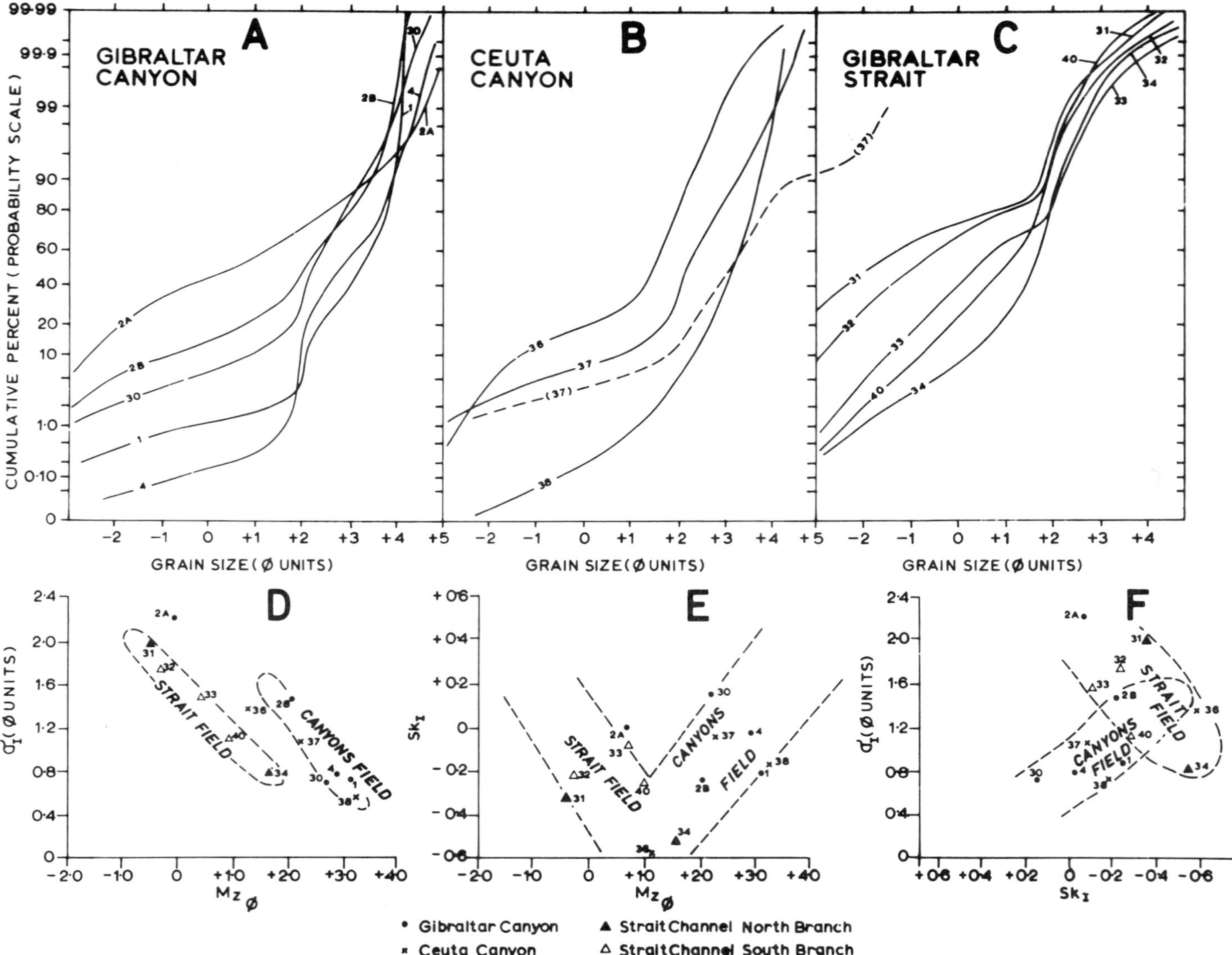

Figure 10. Textural attributes of sand and gravel fractions from grab and short core samples.

Impregnated thin-sections suggest the original presence of a fine silt and clay matrix in some of the sands, particularly those from Gibraltar Canyon and Ceuta Canyon. To assess the effect of prior washing on the size-distribution obtained from such sediments, a full granulometric analysis was undertaken of an unwashed sand forming an apparently undisturbed layer in the short core from Station 37. The cumulative curve representing this analysis is indicated by a dashed line in Figure 10B and the relevant size-parameters are indicated by figures in parentheses in Table 2. If this analysis is a valid guide it appears that originally the canyon sands may contain ten percent or more of lutite grade material and consequently will be more poorly sorted and more positively skewed than the tabulated results indicate.

Almost all the samples analyzed reveal a bimodal or polymodal grain-size distribution (Figure 10, A-C), with a significant truncation-point at about 2.0ϕ. This point is generally regarded as marking the break between traction and saltation grain populations (Fuller, 1961; Visher, 1969). The only samples approaching unimodality are 2A, the sole analyzed representative of the shallow shelf zone, and 38, from the upper east wall of Ceuta Canyon. However the cumulative curves for samples from both canyons are readily distinguishable from those samples taken from the Strait Channel, the canyon curves displaying slightly steeper (better sorted) fine tails and flatter, (poorly sorted) coarse modes (Figure 10, A and B). The coarse grades (coarse sand, granules and small pebbles) of canyon samples include some abraded shell together with highly rounded rock-fragments which display a wide but 'normal' range in size. This compositional diversity accounts for the poor size-sorting of these grades in the canyon sediments. However the coarse components of the Strait samples are almost entirely of biogenic origin—small abraded bivalve shells, coral and bryozoan stems—and probably represent a local contribution with an organically limited spectrum of available sizes, accounting for the better coarse-mode sorting of these samples. Moreover, removal of this local biogenic contribution would greatly improve the sorting of the remaining terrigenous fraction, and reveal the essential similarity of the size-distribution of sand-grade sediments from the canyons and the Strait. The correlation between sorting and carbonate content which is inferred above is demonstrated in Figure 11 and it is significant that on this diagram the locale appears to have little influence on the relationship, some points representing Strait samples occuring in close proximity to others from the canyon regions.

The grain-size parameters listed in Table 2 broadly conform to the pattern suggested by the cumulative curves. All the coarse sediments of the Strait are poorly sorted (verbal limits of Folk, 1968) and display marked negative skewness. There is some indication of a westerly decrease in sorting (Figure 9). Samples from the shallow part of Algeciras Bay (2A, 2B) exhibit granulometric characteristics similar to the Strait sands (Table 2). However, sands from Gibraltar and Ceuta canyons are finer, better sorted and display only slight negative skewness or, in the case of sample 30, slight positive skewness. Sample 36, from the zone of confluence, is better sorted than most Strait samples (Figure 10) but exhibits strong negative skewness (Table 2), and thus appears intermediate between the two main groups in terms of size-attributes.

Table 2. Grain size parameters; gravel and sand fractions. Data expressed in ϕ units.

Sample	Md	Mz	σ_1	Sk_1
1	3.33	3.17	0.71	−0.24
2A	0.55	0.52	2.22	−0.01
2B	2.13	1.81	1.44	−0.18
4	2.95	2.93	0.77	−0.03
30	2.45	2.23	0.70	+0.15
31	−1.50	−0.50	1.98	−0.35
32	−0.70	−0.33	1.71	−0.21
33	0.55	0.63	1.51	−0.09
34	1.85	1.62	0.79	−0.54
36	1.70	1.17	1.35	−0.59
37	2.33	2.31	1.07	−0.09
(37)	(3.10)	(3.13)	(1.29)	(+0.03)
38	3.35	3.27	0.55	−0.18
40	1.10	0.95	1.08	−0.27

Plotted against one another the size-parameters of the analysed samples reveal a number of separable trends (Figure 10, D, E, F). In aggregate, samples from the entire region display a broadly antipathetic relationship between mean grain size and inclusive graphic standard deviation (Figure 10D), an observation which accords with the well-established sinusoidal trend between these two parameters for material of this grade (Folk, 1968, p. 5–6). This antipathetic relationship is considerably strengthened if data from the Strait Channel samples are separated from the canyon parameters, which appear to form a distinct field, of consistently finer grade. Significantly, sample 2A (from the shelf-zone) and sample 36 (from the deep zone of confluence) again plot in an intermediate position.

The scatter diagram for mean size versus inclusive skewness (Figure 10E) is more complex but the distribution is essentially V-shaped and may represent the mid-portion of another sinusoidal curve typical of sands from continental (Folk and Ward, 1957), shallow marine (Inman, 1949) and deep marine

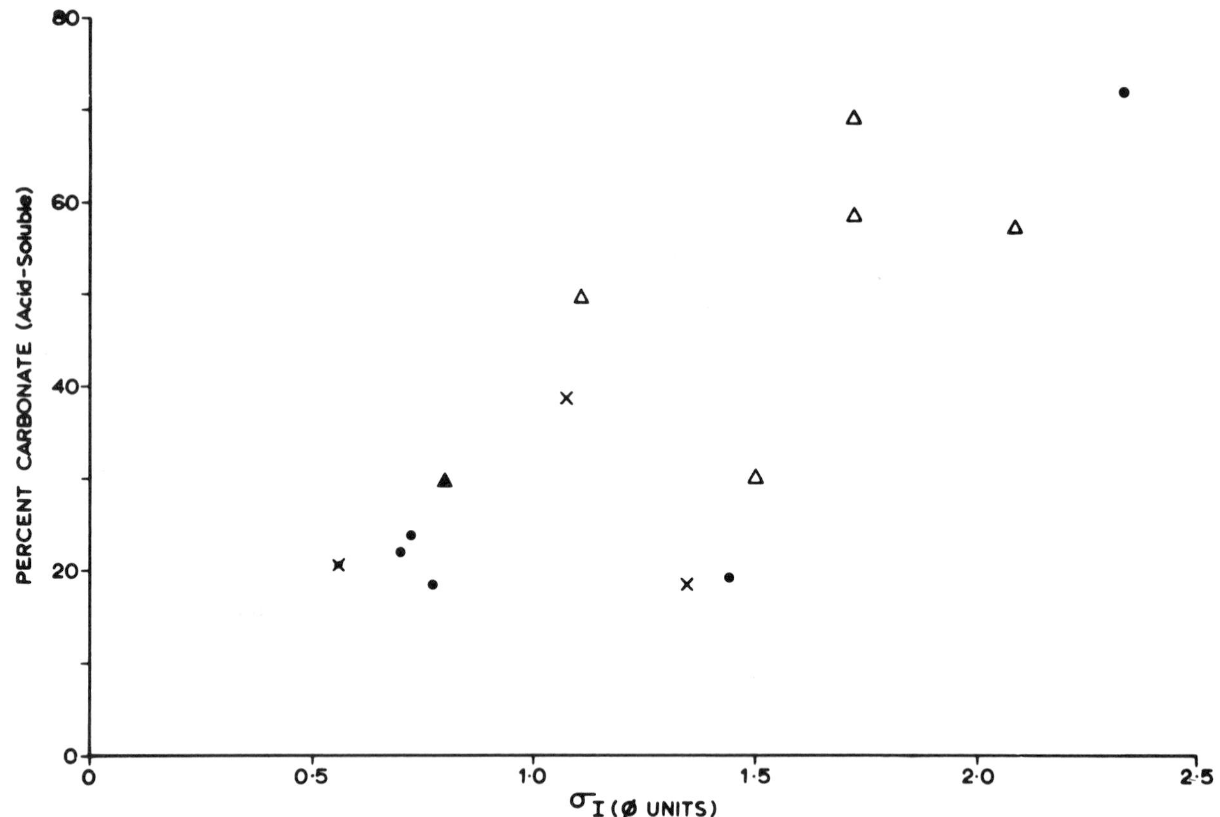

Figure 11. Relationship between bulk sediment sorting value (σ_1) and acid-soluble total carbonate content. KEY: Dots—Gibraltar Canyon; crosses—Ceuta Canyon; open triangles—Strait Channel, south side; solid triangles—Strait Channel, north side.

(Hubert, 1964) environments. As in the cases quoted above, the Gibraltar samples reveal maximum coarse-skewness near the coarse-medium sand boundary (1.0 ϕ) and again the Strait Channel field is distinct from the finer canyons field (Figure 10E).

These results confirm the well-known grain-size dependence of sorting and skewness measures in natural sediment populations. However, plotting skewness against sorting enables this grain-size effect to be minimised and may provide a more accurate assessment of the hydraulic forces responsible for creation of the different grain-size distributions. At first sight the relationship between sorting and skewness in the Gibraltar sands appears random (Figure 10F) but further inspection suggests assignment of samples from the canyons and from the Strait to two overlapping fields of contrasting trend. The canyon samples apparently become more positively skewed with a decrease in sorting value while improved sorting results in increasingly negative skewness in samples from the Strait Channel. The trend displayed by the canyons sediments is comparable to that found in some littoral environments (Hails and Hoyt, 1969, their figure 9) and probably results from progressive suppression of the coarse shell-fraction, the principal component of the tractional load (coarser than 2 ϕ). Conversely, the trend exhibited by samples from the Strait is ascribed to the relative degree of elimination of the dominantly terrigenous saltation load (finer than 2 ϕ). Impregnated thin-sections suggest an inherent lack of fine material in the Strait samples and the modification of the size-distribution resulting from sampling and analytical techniques is probably negligible in these sediments.

Thus sands from the canyons are predominantly composed of saltation and suspension load materials, with subordinate tractional components, whereas the Strait Channel sediments are composed largely of tractional elements with minor contributions from the saltation load. This conclusion is clearly important with respect to the hydraulic processes responsible for deposition of these two groups of sands. The canyon sediments exhibit the tendency towards positive skewness which appears to be a common feature of turbidite deposits (Ericson, *et al.*, 1961; Middleton, 1962; Horn *et al.*, 1971) whereas the strong negative skewness of the Strait samples indicates the importance of powerful, persistent tractional bottom flow in this region. The characteristics of the shelf sample 2A and of sample 36 from the deep confluence zone are significant in this discussion. Sample 2A possesses attributes closely similar to the Strait sediments and evidently was formed by comparable tractional processes. Sample 36 generally plots in the area of overlap of the canyon and Strait fields and apparently represents an almost equal mixture of traction and saltation elements, probably produced by modification of an original canyon sediment through tractional reworking in the zone of confluence.

Light Minerals and Clasts

Examination of impregnated thin-sections of sands from the Gibraltar area reveals that *biogenic components* constitute from 12.0% to 63.6% of the sands and are the dominant element of samples from the Strait and from the Algeciras Bay shelf (Table 3), confirming the chemical evidence cited earlier (Figure 9). Calcareous and aragonitic megabenthos is the most abundant biotal category and includes bivalves, brachiopods, gastropods, echinoid plates, corals, bryozoa and abundant algal encrustations. These components are generally fragmental and abraded and a few are polished and manganese-coated, suggesting long exposure on the sea-floor. On the other hand, some Strait samples (notably 31 and 32) contain abundant, freshly broken stems of coral. Microfaunal elements, principally foraminiferal tests, occur sparingly in most samples and are more common in the Ceuta Canyon sediments (Table 3). A profusion of species is represented, with both planktonic and benthonic forms. Most tests are empty but partial infilling by poorly organised glauconite or by hydrotroilite and pyrite (often framboidal) has been observed in some sands from the canyons. Fragments of phosphatic material (probably bone) ranging up to 15 mm in diameter, are frequently encountered, together with irregular and twisted pieces of carbonaceous debris, which includes both normal vegetable detritus and (especially in samples from Algeciras Bay) numerous pieces of coal and coke!

Quartz is the principal inorganic component and generally constitutes between 50 and 75 percent of the total terrigenous fraction. The simple (unstrained) variety forms an average of 72 percent of the total quartz (range 63–78), compared with means of 19 and 9 percent for the strained (undulatory) and composite varieties respectively. No systematic areal variation in the relative proportions of these quartz types has been discerned.

The roundness of quartz may be used as a measure of morphologic maturity in arenaceous sediments. For this purpose 300 grains from each available sample were assigned to the scale devised by Powers (1953) and the results are expressed in Figure 12. The grains were taken from the principal sand-grade mode (2.0–2.5 ϕ) to avoid possible size-bias in the roundness estimates. The roundness distributions of all the analysed samples are similar with subangular grains dominant in all but one sand (33). Very angular and well rounded grains are relatively

Table 3. Composition of light mineral assemblages.

Mineral	Sample Number Gibraltar Canyon						Ceuta Canyon					N. Strait			S. Strait			
	1	2A	2B	4	30	Average	6	36	37	38	Average	31	34	Average	32	33	40	Average
Quartz	42.9	18.4	49.1	47.1	43.3	40.2	28.7	43.3	38.1	35.8	36.5	35.2	56.5	45.8	21.8	22.6	32.1	25.3
K Feldspar	—	0.4	0.6	1.0	1.5	0.7	1.7	0.9	2.0	1.0	1.4	2.6	3.0	2.8	1.9	2.8	1.9	2.2
Plagioclase	7.1	3.1	3.5	8.8	4.4	5.4	4.6	1.8	2.5	3.4	3.1	2.6	2.0	2.3	0.6	0.9	0.5	0.7
Sedimentary Clasts	2.1	3.1	2.3	6.4	5.9	4.0	17.8	15.2	9.4	2.5	11.2	13.5	3.5	8.5	14.7	7.1	15.3	12.4
Igneous Clasts	1.4	0.4	1.2	2.9	1.0	1.4	1.1	2.3	—	0.5	1.0	1.3	1.5	1.4	—	—	0.9	0.3
Metam. Clasts	6.4	2.2	6.9	2.9	7.9	5.3	—	6.5	2.5	2.0	2.8	4.5	9.0	6.7	1.9	0.9	2.3	1.7
Micas	6.4	3.1	2.9	9.8	2.5	4.9	6.9	1.4	2.0	4.9	3.8	0.6	1.0	0.8	1.3	0.9	1.9	1.4
Megabenthos ('Shell')	7.9	55.7	12.7	5.4	9.9	18.3	17.8	10.1	7.4	20.6	14.0	32.7	8.5	21.1	47.4	53.8	27.9	43.0
Microfauna	5.7	0.9	5.2	2.0	9.4	4.6	5.2	7.8	18.3	8.8	10.0	3.2	4.0	3.6	5.1	4.7	10.2	6.7
Phosphatic/carbonaceous material	5.7	7.0	3.5	4.4	4.9	5.1	5.2	1.4	2.5	2.9	3.0	—	2.0	1.0	2.6	0.9	2.8	2.1
Glauconite	3.6	1.8	2.3	1.5	3.4	2.5	2.9	—	5.4	4.4	3.2	—	—	—	1.3	1.4	0.5	1.1
Others	10.7	3.9	9.8	7.8	5.9	7.6	8.0	9.2	9.9	13.2	10.1	3.8	9.0	6.4	1.3	3.8	3.7	2.9

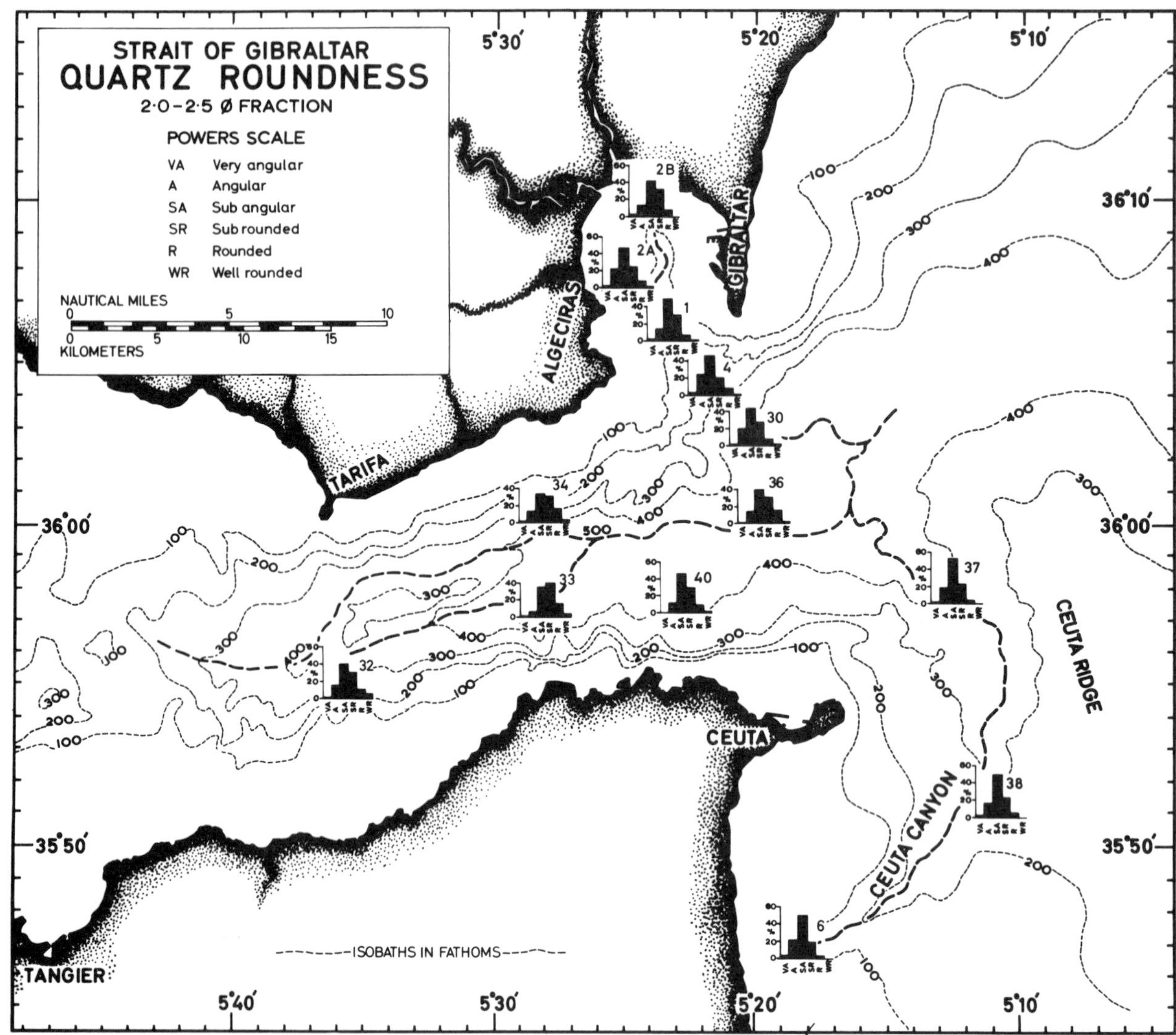

Figure 12. Percentage distribution of roundness classes in quartz grains from grab and core samples.

scarce, although the latter category is more prominent in the Strait samples. The Ceuta Canyon histograms are almost symmetrical, with a strong subangular mode (Figure 12) while the Gibraltar Canyon sands also possess prominent subangular modes in distributions slightly skewed towards the rounded classes. A general down-canyon diminution of the subangular mode is apparent from the Gibraltar Canyon analyses (Figure 12) and an interesting related feature is the occurrence in sample 30 from the mouth of this canyon of a conspicuous proportion of well-rounded quartz grains in the 0.0 ϕ to -0.5 ϕ grade. Grains of this type recur in all the Strait samples and probably represent contributions from eolian sources. Round-skewed distributions characterise the Strait sands (including 36) but there the subangular mode is muted (Figure 12).

Thus the Ceuta and Gibraltar depressions apparently furnish distinctive quartz-roundness populations to the zone of confluence. The enhanced roundness of the Strait sands may be ascribed either to an independent source of more mature grains (perhaps by down-slope funnelling of sand from the narrow, scoured shelves on either side of the Strait) or to shape-sorting and shape-modification of a mixed population of sand from the canyons which is being re-worked westward from the zone of confluence. A final decision between these two alternative propositions is not possible on the available data but the evidence from heavy mineral assemblages and from bottom current data, considered below, tends to favor the reworking hypothesis.

Rock-fragments constitute the second major terrigenous component (range: 8 to 38 percent of total terrigenous fraction) and comprise variable proportions of a number of rock-types. Carbonate clasts dominate the sedimentary rock-types, and rounded granules and grains of a buff-colored, coarsely crystalline dolomite are particularly common in samples from Ceuta Canyon and the south side of the Strait. Subrounded clasts of an iron-stained dolomicrite occur in most Strait sands, accompanied by rare fragments of silicified limestone. Fragments of limonite-stained or calcite-cemented protoquartzite with well-rounded grains, and of welded and sutured orthoquartzite also characterise the Strait samples. Shiny dark granules of chert, some highly cleaved and showing indistinct traces of radiolaria, are conspicuous in the coarse fractions of samples 36, 37 and 39 near the mouth of Ceuta Canyon, and smaller chert-fragments occur widely in the Ceuta and Strait sands. Subangular pieces of mudstone and fine siltstone are the principal sedimentary clasts in the Gibraltar Canyon samples.

Fragments of igneous origin are not abundant in these sediments, but a few grains of highly altered basaltic and/or rhyolitic rocks occur in most sands. Small clasts of a picotite-bearing serpentinite were observed in two Gibraltar Canyon samples (1 and 4). Acid plutonic or hypabyssal rocks are unrepresented.

Metamorphic clasts are more abundant in aggregate than the igneous types but are somewhat restricted in variety. Schistose and granulitic rock-types are the most common, with quartz-mica-schists and quartz-chlorite-schists predominant, followed by foliated quartz-granulites and metaquartzites, occasional hornfelsic types and rare epidosites.

Feldspar is a widespread but seldom an important element in the samples examined (Table 3), and represents between 3 and 11 percent of the terrigenous components (average is 7.3 percent). Both plagioclase and potash feldspar are represented, and the former is more abundant in aggregate. Plagioclase composition ranges from albite to labradorite, the sodic varieties being most common. All plagioclase grains are altered to some degree, usually by kaolinisation. Orthoclase is the main variety of potash feldspar and generally occurs as relatively rounded or equant grains, cloudy through alteration, which is nevertheless less marked than in the plagioclase group. Microcline and grains with perthitic texture are rare.

The *mica* family is well represented in sediments from Gibraltar Canyon and from the shallow portion of Ceuta Canyon but the Strait sands are deficient in mica. Brown and greenish-brown biotite is the dominant variety and flakes of this mineral display varying degrees of alteration, usually to chlorite. Partially altered flakes of muscovite are uncommon.

The composition of the terrigenous fraction of the sands has been determined by point-counting an average of 400 points from each impregnated thin-section (Table 3). Plotted on a Q-F-R diagram (Figure 13A) all the sands fall within the subgreywacke field of Pettijohn (1958) but there is no distinction discernible between the different depressions. However, the relative proportions of different types of rock-fragment are evidently controlled by geographic factors (Figure 13C). Thus sands from the south side of the Strait and from Ceuta Canyon are characterised by high proportions of sedimentary clasts (mainly dolomitic limestones) whereas Gibraltar Canyon and the north side of the Strait are characterised by a preponderance of metamorphic fragments, schists in the case of the canyon, metaquartzites in the sands from the north Strait.

The relative abundance of plagioclase, potash feldspar and quartz also appears to be geographically controlled (Figure 13B). Sands from Gibraltar Canyon consistently contain low relative amounts of potash feldspar (less than 2 percent relative) while the Strait samples carry relative amounts in excess of 5 percent. Ceuta Canyon sands contain intermediate quantities of potash feldspar. Moreover, most Gibraltar Canyon sands are proportionately richer in plagioclase than samples from elsewhere (Figure 13B).

The significance of these compositional differences will be considered more fully in a later section.

Heavy Minerals

Heavy minerals form a significant proportion (1.0 to 13.8 per cent) of the terrigenous sand fractions and include a wide variety of species. The aggregate suite is dominated by relatively unstable minerals (Table 4; Figure 14) and the stable species such as zircon, tourmaline and rutile are less prominent. Volumetrically significant nonopaque species and varieties which were identified comprise, in order of decreasing aggregate abundance: pyroxenes, principally orthopyroxene (pale green, prismatic schillerized bronzite-enstatite; with much rarer hypersthene, pleochroic in pink and green), and subordinate clinopyroxene (mainly pale green-pale brown augite and traces of gray diopside); garnet, mainly colorless with numerous small inclusions but some pinkish grains also occur; micas, dominated by green-brown biotite, with yellow-green clinochlore, rare penninite and partly chloritised muscovite; kyanite, ice-blue fractured prisms; andalusite, usually crammed with carbonaceous inclusions and gradational into chiastolite; amphiboles, mainly green or pale brown common hornblende, with rare acicular actinolite and flakes of glaucophane; dolomite, distinctive pale

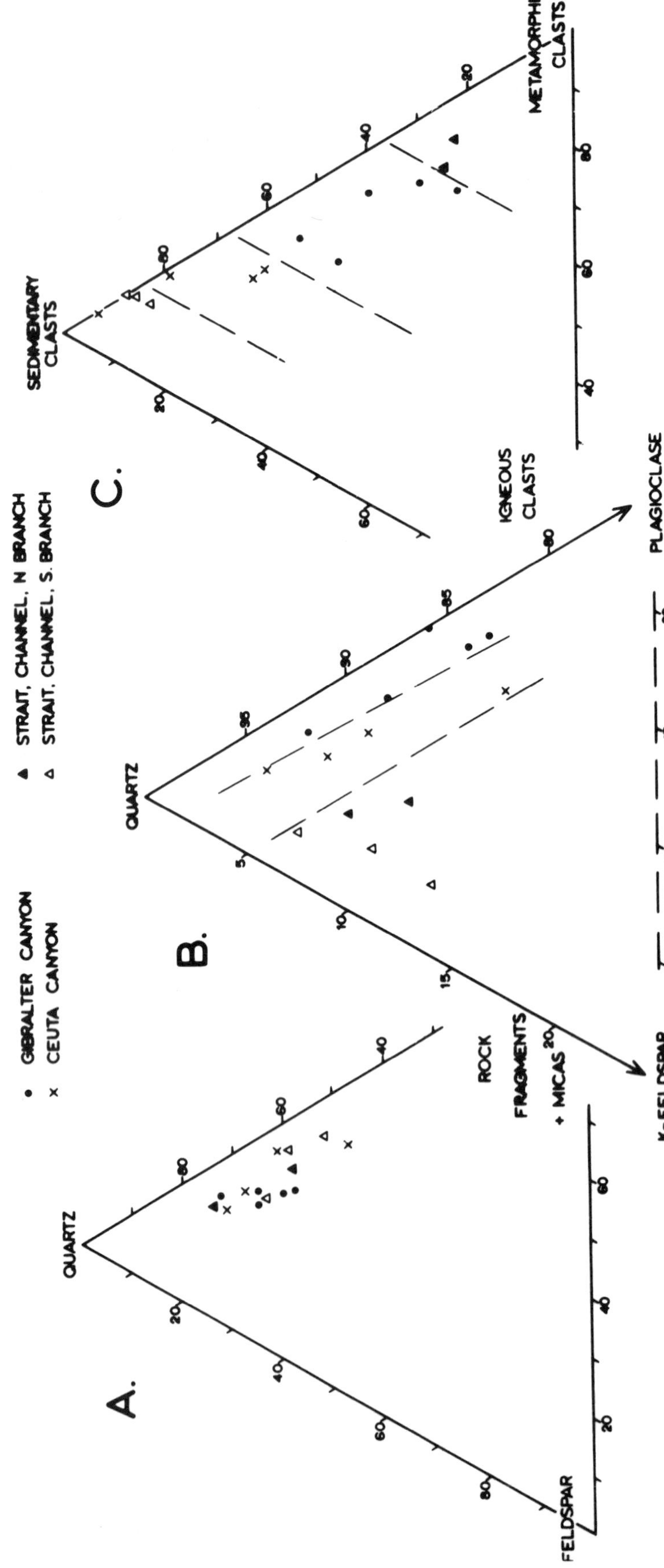

Figure 13. Petrologic attributes of sand fractions from grab and core samples.

Table 4. Percentage composition of heavy mineral assemblages.

Mineral	Sample Number Gibraltar Canyon						Ceuta Canyon				S. Strait					N. Strait
	1	2A	2B	4	30	Average	36	37	38	Average	32	33	39	40	Average	34
Zircon	0.9	2.6	1.7	1.8	1.2	1.6	2.5	—	1.4	1.3	2.3	3.1	2.5	1.0	2.2	0.6
Tourmaline	4.3	1.0	—	1.8	1.2	1.7	—	2.2	3.3	1.9	2.9	3.7	2.6	1.5	2.7	0.5
Rutile	1.0	0.5	4.0	3.0	1.7	2.1	2.5	2.9	3.3	2.9	2.8	2.5	—	1.5	1.7	0.6
Opaques	7.3	10.4	10.9	9.5	5.8	8.8	17.5	13.8	22.2	17.8	30.8	35.0	20.7	15.0	25.4	9.3
Garnets	5.9	14.6	14.9	11.8	17.5	12.9	28.4	20.4	10.4	19.7	3.4	5.6	15.9	17.6	10.6	18.6
Andalusite	3.4	7.8	1.7	1.8	—	2.9	2.0	3.3	—	1.8	3.4	1.2	3.5	2.9	2.9	7.6
Chiastolite	8.3	5.7	8.6	7.7	8.2	7.7	1.0	2.2	1.3	1.5	1.7	0.6	3.4	4.5	2.7	9.3
Kyanite	6.3	5.2	6.3	13.0	8.8	7.9	3.9	8.3	4.2	5.5	3.4	3.1	2.4	2.1	2.7	2.3
Staurolite	2.9	2.1	1.7	2.4	1.7	2.2	3.9	2.2	0.9	2.3	1.1	1.2	2.5	3.4	2.0	2.3
Sillimanite	1.5	2.6	—	1.2	1.2	1.3	2.5	2.2	0.9	1.9	1.7	1.9	1.0	1.6	1.6	—
Micas	17.8	6.8	5.7	8.9	8.8	9.6	1.0	3.3	11.8	5.4	3.0	3.7	3.9	4.0	3.6	4.1
Orthopyroxenes	22.5	18.1	22.0	19.5	27.5	21.9	20.6	14.3	11.0	15.5	8.0	8.6	12.1	28.6	14.5	29.8
Clinopyroxenes	2.4	5.2	2.3	1.8	1.2	2.6	3.9	8.8	4.7	5.8	14.4	18.5	9.7	2.0	11.2	6.9
Amphiboles	3.9	4.7	7.5	4.7	5.3	5.2	3.4	4.4	4.7	4.2	8.0	4.9	4.9	2.4	5.1	2.3
Dolomite	4.8	6.8	5.7	5.9	5.3	5.7	1.0	2.7	9.0	4.2	1.7	1.9	2.4	6.0	3.0	1.3
Sphene	1.5	1.6	2.3	0.6	—	1.2	—	1.7	2.4	1.4	2.3	0.5	2.7	1.4	1.7	—
Epidotes	1.9	1.6	0.6	3.0	1.2	1.7	1.0	3.3	5.2	3.2	6.3	4.3	3.7	2.1	4.1	1.2
Spinels	0.9	1.6	2.3	0.6	2.9	1.7	4.4	2.8	0.9	2.7	1.7	1.2	2.3	1.0	1.6	1.7
Apatite	—	—	—	1.2	—	0.2	—	—	0.9	0.3	—	0.7	—	—	0.2	0.6
Olivine	1.0	0.8	1.2	0.8	—	0.8	0.4	1.2	0.7	0.8	—	0.5	0.8	—	0.3	—
Others	0.5	0.4	0.4	0.5	0.6	0.5	0.2	—	0.7	0.3	0.8	1.5	2.9	1.4	1.7	1.0

buff rhombs, often somewhat abraded, soluble in concentrated HCl; epidote group, including hackly, bottle green grains of genuine epidote and rare gray-green prisms of zoisite; staurolite; rutiles, foxy-red and much rarer golden yellow varieties, both generally well-rounded; tourmaline, mainly ovoid brown grains; spinels, including a coffee-brown isotropic variety (?picotite) and rare dark red octahedra; zircon, colorless to pale green or brownish, rounded prisms, often zoned; sillimanite, acicular and yellowish fibrous grains; sphene; olivine, egg-shaped or lobate grains, partly or entirely serpentinised; apatite. Other minerals which occur in trace amounts include anatase, corundum (?sapphire), monazite and chloritoid.

Although most ferromagnesian grains exhibit some degree of alteration, the proportion of highly altered grains is generally low (between 3 and 7 percent). This suggests either that processes of chemical weathering have not been effective in the source-rocks or more probably, that altered grains have been selectively extracted from the heavy mineral assemblage during transport. Physical abrasion of these less resistant grains in regimes of high hydraulic energy furnishes a likely cause of removal.

The opaque minerals are dominated by magnetite and less abundant leucoxene-coated ilmenite but irregular masses of hydrotroilite or pyrite, probably authigenic and sometimes representing foraminifera-fillings, are not uncommon in the canyon assemblages. Occasional subrounded, equant grains of dark brown chromite occur and are particularly common in sands from the Strait. One interesting feature of the opaque minerals is the occurrence of limonite and haematite coatings, which mask many of the more rounded ferrous grains in all samples, even in those canyon sands which carry abundant pyrite. The degree of rounding of the coated grains indicates an origin in a more abrasive, oxidising environment (?desert, coastal dunes) and the anomalous association with sulphides may reflect the short-lived nature of sand deposition within the reducing environment of the canyons, allowing survival of the oxides.

Further evidence concerning the transport and abrasion history of the sands is provided by the shape of the heavy mineral grains. For any given species, grains of comparable size are appreciably more rounded in sands from the Strait area than in the canyon sediments. This relationship is most readily established in relatively unstable minerals, such as the pyroxenes, which are represented by angular to subangular grains in the sands of Algeciras Bay, Gibraltar Canyon and Ceuta Canyon but occur predominantly as subrounded to rounded prisms in the Strait assemblages. Such an observation, common to all the heavy mineral species, provides clear evidence for a substantial degree of shape-modification of the sand grains during transport through the canyons and in the Strait.

Qualitatively, the aggregate suite of heavy minerals remains constant throughout the investigated area. However the quantitative composition of the heavy mineral assemblage (as determined by point-counting 300 grains from each sample) varies appreciably with location and three main suites may be recognised and linked to the three major depressions (Figure 14). For this purpose the relative proportions of six groups of heavy minerals have been determined. The groups comprise: zircon, tourmaline and rutile (ZTR—the stable group, probably reflecting a sedimentary source); the opaques (part sedimentary, part igneous in origin); garnets (mainly metamorphics, including regional and contact metamorphism); the metamorphic group (kyanite, staurolite, sillimanite, andalusite/chiastolite and biotite mica); orthopyroxenes (from basic or ultrabasic igneous sources); amphiboles plus clinopyroxenes (mainly basic igneous but possibly including metamorphic contributions).

The Gibraltar Canyon-Algeciras Bay suite is dominated by the metamorphic group and orthopyroxenes, which together constitute more than 50 per cent of each assemblage (Figure 4). This suite is relatively deficient in the ZTR group and the opaques. Another feature of the Gibraltar suite (not indicated on Figure 14) is the relative abundance of detrital dolomite rhombs (Table 4). The sole analysed sample from the north side of the Strait (34) yields a very similar suite and for this purpose must be included with the Gibraltar samples. The restricted quantity of sand recovered from Station 31, south of Tarifa on the north branch of the Strait Channel, yielded a number of heavy mineral grains which was considered insufficient for quantitative analysis. However, subjective assessment indicates a strong affinity with Sample 34.

The Ceuta Canyon samples (including Sample 36) provide a suite in which the metamorphic group and orthopyroxene are less conspicuous but garnet, opaques and the ZTR group together constitute approximately half the suite (Figure 14).

Samples from the south side of the Strait generally yield an assemblage in which opaques and the igneous group (amphiboles plus clinopyroxenes) are dominant and the ZTR group is more prominent. Orthopyroxene and the metamorphic group are correspondingly deficient. However the suite from Sample 40 is unusual in displaying some attributes of the Ceuta Canyon association (low metamorphic group, moderate proportion of opaques) and some features of the Gibraltar suite (abundant orthopyroxenes, low proportion of igneous group minerals).

SEDIMENT PROVENANCE AND MODIFICATION

Sediments from Gilbraltar Canyon are characterised by subangular quartz, plagioclase, biotite, and metamorphic clasts with subordinate shale-fragments, orthopyroxenes and metamorphic group minerals (Figures 13 and 14). This mineral assemblage attests the predominance of igneous (basic and ultrabasic) and metamorphic sources for these sands and the relatively unaltered and angular condition of the ferromagnesian and metamorphic minerals indicates their primary, first-cycle character. The abundance of andalusite/chiastolite in these sediments (Table 4) may reflect the importance of contact-metamorphism in the source-rocks, in addition to the regional or stress metamorphism responsible for the kyanite, staurolite and sillimanite grains. Sedimentary rocks probably contributed some of the more rounded quartz grains (although morphological evidence of recycling is lacking) together with the shale-clasts, but the impoverished ZTR group indicates that sedimentary sources were relatively insignificant. The dolomite rhombs which form a characteristic element of the Gibraltar heavy mineral suite may have originated either in a sequence of carbonate sediments[1] or in a metamorphic terrain. In more general terms, the importance of basic or ultrabasic (probably gabbro and peridotite) sources is indicated not only by the abundance of schillerized orthopyroxene but also by the widespread occurrence of the spinel minerals especially picotite and chromite.

The Ceuta Canyon sands carry a high proportion of angular quartz, and modest amounts of plagioclase and potash feldspar. Metamorphic clasts are less common but fragments of limestone and sandstone are abundant (Figure 13). These features suggest the importance of sedimentary source-rocks for the Ceuta sediments, and the increased proportion of opaques and the ZTR group heavy minerals provides corroboration for this view. Nevertheless, the diminished amounts of orthopyroxene and metamorphic group minerals still indicate important contributions from basic/ultrabasic and metamorphic sources. Significantly, in view of its equivocal textural status, the confluence zone Sample (36) is petrologically identical with the Ceuta Canyon sediments.

The light and heavy mineral fractions of sands from the Strait are marked by an apparent dichotomy in character but the terrigenous portions reveal a generally enhanced maturity when compared with

[1]Dolomite rhombs are common in (?) Jurassic Gibraltar Limestone (Bailey, 1952, p. 162).

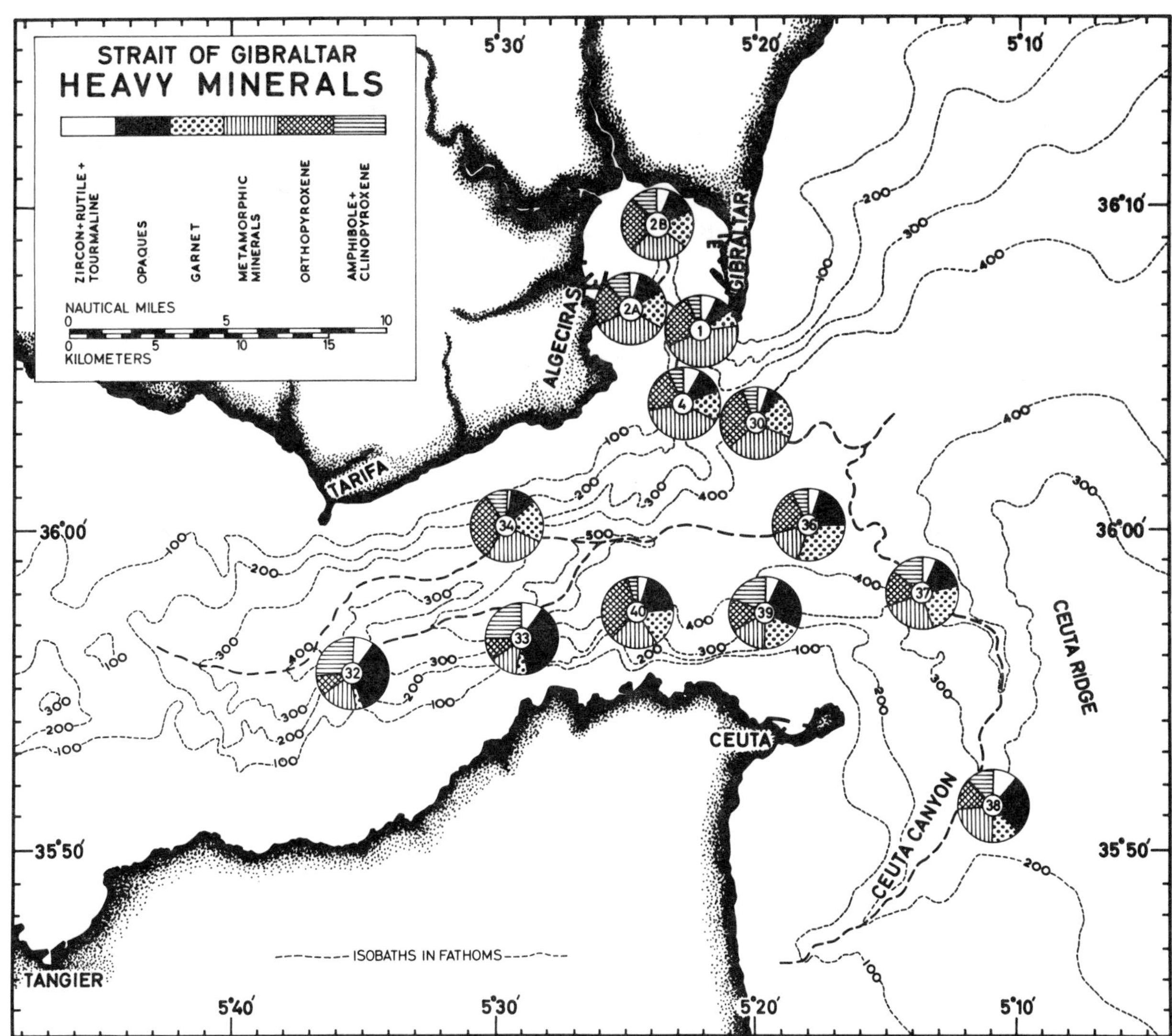

Figure 14. Percentage distribution of six principal groups of heavy minerals in grab and core samples.

the canyon sediments. The north and south sides of the Strait are clearly separable on the basis of feldspar and clast proportions (Figure 13) and heavy mineral suites (Figure 14). Sands from the south side of the Strait are characterised by an abundance of sedimentary fragments (limestones, sandstones, cherts), rounded quartz grains, dominant potash feldspar, and enhanced proportions of opaques and stable (ZTR group) heavy minerals—all features indicative of a sedimentary source. On the other hand, significant contributions from metamorphic and basic igneous rocks are indicated by the remainder of the heavy mineral suite in which the dominance of the amphibole-clinopyroxene group over the orthopyroxenes suggest some change in the nature of the contributory basic igneous rocks (perhaps gabbros replacing the peridotites?). Compositionally, Sample 40 is unique for this area in that the light minerals indicate a strong affinity with the Ceuta Canyon sediments, whereas the heavy mineral suite is very similar to that of the Gibraltar Canyon sands.

Sedimentary source-rocks appear to be less important in the North Strait sands where equal proportions of the two feldspar types and an increased percentage of metamorphic and igneous clasts (Table 3) again suggest predominance of basic igneous and metamorphic contributions—a view reinforced by the nature of the heavy mineral suite, virtually identical with that of the Gibraltar Canyon sediments.

Not all the compositional and morphologic features of the Strait sands are explicable in terms of provenance. The diminished total content of feldspar and micas, the survival of only the resistant siliceous rock-fragments and the enhanced roundness of the quartz grains all point to a degree of compositional maturity probably achieved by a process of modification and elimination of the original terrigenes by hydraulic processes operating within the regime of sedimentation. Confirmation of this hypothesis is provided by the heavy minerals, since grains of the same species (for example, orthopyroxenes) invariably exhibit better rounding in the Strait samples than in those from the canyons.

Thus it is possible to account for the petrologic features of the Gibraltar area sediments in terms of two principal associations—one, represented by the Gibraltar-Algeciras Bay samples, derived predominantly from ultrabasic and metamorphic sources, the other, represented by the Ceuta Canyon sands, derived mainly from sedimentary rocks of mixed carbonate-clastic facies, with subordinate basic igneous and metamorphic contributions. Partial elimination of less stable minerals within the Strait area has resulted in considerable modification of these associations and the production of the two discrete subassociations which characterize the north and south parts of the Strait respectively. Moreover the possibility of some independent contribution of minerals such as the amphibole-clinopyroxene group cannot be discounted. Nevertheless, the North Strait sands are genetically linked with the Gibraltar-Algeciras sources whereas the South Strait samples betray their affinity with the Ceuta Canyon provenance. However, the apparent exclusiveness of this dichotomy is broken by Sample 40, whose unique petrology integrates the characteristics of both associations and suggests the possibility of at least some cross-Strait mixing of terrigenous sediment.

BOTTOM CURRENT FLOW AND SEDIMENT TRANSPORT

Because of its critical role in the interchange of Mediterranean and Atlantic water the Strait of Gibraltar displays circulatory attributes which are highly unusual, if not unique. In consequence the water-mass properties of this area have been studied in considerable detail (see Lacombe *et al.*, 1964, and Heezen and Johnson, 1969, for reviews of existing knowledge). The prime factor in the local hydraulic regime is the net eastward inflow at the surface of less saline Atlantic water and the net westward discharge of a bottom layer of salty Mediterranean water (Kuenen, 1950, pp. 42–43). The constricted nature of the Strait obviously exerts a profound influence on the rate of exchange of these two water-masses and high surface and bottom current velocities may be inferred from observed salinity-temperature profiles (see, for example, Wüst, 1961). The upper boundary of the bottom water appears to lie at a depth of about 75 to 100 fm (140–185 m) throughout the Strait proper, but forms a westward-descending wedge ("The Mediterranean Undercurrent" of Heezen and Johnson, 1969), west of the prominent sill or transverse ridge which crosses the Strait northeast of Tangier. However, both long-term (seasonal) and short-term (diurnal or tidal) fluctuations in the boundary between the surface and bottom water-masses have been determined in the region of the Strait (Wüst, 1961; Lacombe *et al.*, 1964; Frassetto, 1965; Heezen and Johnson, 1969).

Metered measurements of current direction and velocity, usually by means of profiles, are available for much of the Strait area (Lacombe *et al.*, 1964; Frassetto, 1965; Heezen and Johnson, 1969) and provide verification of many of the conclusions derived from the salinity-temperature studies. Westward-directed near-bottom flows exceeding 100 cm/sec have been recorded from the western end of the Strait (Allain, 1964) but the near-bottom vectors from stations within and to the east of the Strait proper vary both in direction and magnitude. However, the majority indicate general westerly flow at velocities ranging from 10 to 80 cm/sec (Lacombe *et al.*, 1964; Frassetto, 1965; see Figure 15). Moreover the current-profiles are complex, with evidence of tidal reversals, in some instances extending to the deep sea-bed, and evidence for internal waves and severe internal turbulence is cited by Lacombe *et al.* (1964). Maximum velocities and maximum variability in speed and direction are encountered at stations on the narrow shelves and upper slopes around the Strait.

The observations summarised above relate to the structure and dynamics of the water-mass. These may leave their impress on the underlying sediments in the form of sedimentary structures which, when observed in bottom photographs, indicate the mode and direction of sediment transport and thus record the nature of the flow which effectively impinged on the sea-floor for some finite period of time. Both depositional and erosional structures provide valuable indirect evidence of the character of such flows, while direct evidence of present current activity is gained from the orientation of certain marine organisms.

The most abundant depositional features are ripple-marks, sand and gravel streamers and lineations and alignment of shell-debris and pebbles.

Ripple-marks exhibit considerable diversity in form and dimensions. Most appear to be asymmetrical in profile, although the oblique illumination employed in the photographs sometimes renders it difficult or even impossible to detect the steeper face of the ripples. This problem is most acute where, as is frequently

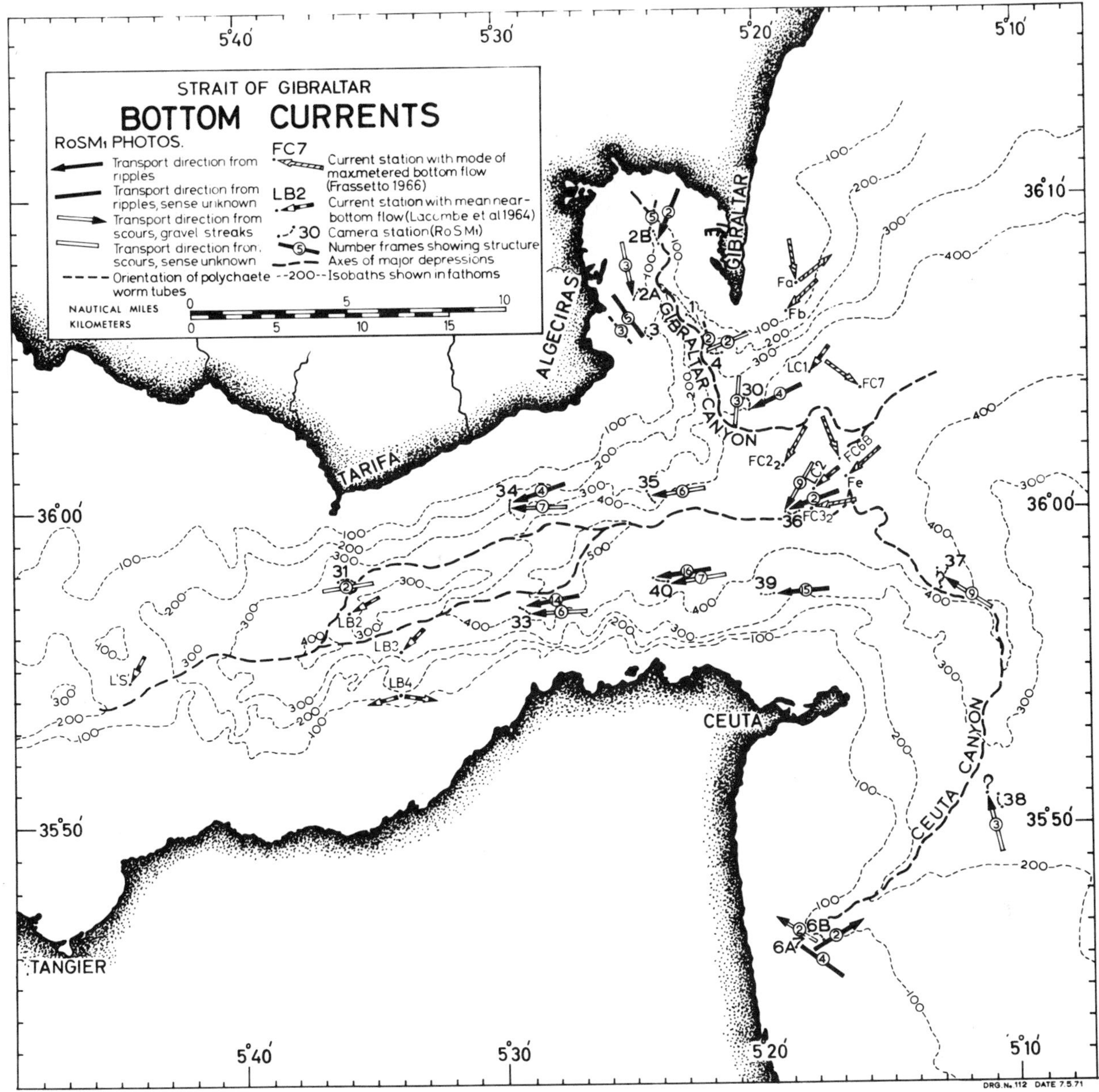

Figure 15. Mean sediment transport vectors as determined from the orientation of different types of current-formed structure present in bottom photographs. The number of frames displaying a given structure at one station is indicated. Directly metered near-bottom flow directions (published by Lacombe *et al.*, 1964, and Frassetto, 1966) are included for comparison.

the case, the ripple-crests are relatively straight and parallel and some of these forms appear to be genuinely symmetrical. Short-crested, linguoid ripples are also common in the Gibraltar area and asymmetry is readily detectable in these forms because of the pronounced 'wings'. True cell-like interference patterns have not been observed and cross-cutting sets of ripples were found at only one station (33). Within the Strait most of the ripples are sharp-crested and the troughs are often marked by concentrations of fine terrigenous gravel and shell debris. Such ripples are regarded as recently formed features but there are other forms which display rounded crests, a thin but uniform veneer of fine sand or silt and, occasionally, excavation of the internal lamination which is then exposed as traces on the flanks of the ripples. Ripple-fields exhibiting these attributes probably indicate some degree of stabilisation and a general, if temporary, halt to tractional processes in that area. More complete burial of ripples beneath a substantial silt-mantle was detected at some stations near the head of both canyons (2B, 3, 6A). Ripple wave-lengths observed range from 8 to 35 cm and amplitudes vary between 1 and 10 cm although, as indicated earlier, the suspected existence of dunes and small sand-waves, of dimensions too large to be appreciable in the few square meters of the photograph frame, minimizes the quantitative value of the dimensional data from perceptible ripple-forms. Within the smaller ripple category, longer wave-length appears to be linked to increasing coarseness of the sand.

Particularly in the Strait area, ripple-fields often pass laterally into relatively smooth areas dominated by *lineated bands or streamers* of sand with scattered fine gravel, including shell. Such bands are usually 10 to 100 cm wide and apparent blurring of the detail in photographs of such areas probably indicates active bed-load transport, suggesting that an upper flow regime condition exists here. Partial to strong *alignment of inequidimensional shells* and *pebbles* is often associated with the lineated areas and this orientation is almost invariably parallel to that inferred as the current path from any associated ripples or scour-features. Another depositional feature is provided by arrays of roughly *parallel tubes* of a polychaete worm, probably *Hyalocoelia*. These stalk-like tubes, about 5 to 10 mm in diameter and up to 15 cm long, are composed of agglutinated sand grains and in the life position they project obliquely out of the sea-floor. However the death of the organism or an increase in current energy resulting in erosion of the substrate promotes collapse of the tubes which then lie flat on the sea-bed and may become aligned into subparallelism by rolling under current influence. In this area such tubes are confined to silty bottoms and are generally aligned perpendicular to the prevailing transport vectors, as deduced from associated current-structures, although a crudely rectilinear bimodal distribution was observed at Station 2B.

Erosional structures observed in bottom photographs of this region include a variety of linear or torose scour-marks, and current-crescents with associated sand—or mud shadows. *Scour-marks* are best preserved on firm surfaces of mud or fine sand where they generally display strong parallelism. Most are delicate structures, only a few millimeters deep and a few centimeters wide, usually overlapping, but occasionally isolated. Another form of scour mark, exhibiting more marked relief, is associated with erosional modification of the biogenic mounds and hollows which characterise the muddier areas (Figure 8D). This process gives rise to pronounced elongation or stream-lining of the mounds in the direction of scouring (*cf.* Heezen and Hollister, 1964, their figure 6). Deep subcircular hollows of biogenic origin often occur at one end of the mound (Figure 8D). This relationship may enable determination of the actual sense of current movement since sediment ejected by the animal in the course of excavation will presumably accumulate on the down-current side of the resulting hollow, forming a mound which tapers down stream. A similar situation will result from mechanical enlargement of an initially small biogenic by a subsequent scouring flow—a process analogous to that envisaged for the formation of flute-marks in ancient sediments. However, since the sense of current flow inferred from these features is conjectural the mean current-vectors for stations exhibiting such structures, but lacking corroborative evidence from other sedimentary features, are qualified by a question-mark (Figure 15, stations 37, 38).

In contrast to the scour features just described, *current crescents* are most abundant on sandy or gravelly surfaces, where they are invariably associated with obstacles such as pebbles or cobbles. The most striking examples of such structures were obtained at Station 36 (zone of confluence) where large blocks of local bed-rock display the characteristic 'moated' depression around the obstacle and the tail of finer sediment formed on the protected lee or down current side (Figure 8D). Smaller 'crag-and-tail' features occur around pebbles or shells isolated on a rippled surface of sand.

Organisms conveying information with respect to the prevailing current flow include tube-building polychaete worms, certain types of coelenterate and sponges. In life the polychaete worm-tubes project obliquely out of the sea-bed and occasionally display a parallel alignment and a dominant sense of inclination which is inferred to be in a down-current direction. At stations 2B and 3 (Algeciras Bay) independent evidence from ripples confirms this conclusion (Figure 15). Within the strongly scoured and rocky north branch of the Strait Channel encrusting bryozoans and gorgonian corals represent virtually the only benthonic forms, and thickets of the latter group of organisms appear to be preferentially aligned perpendicular to the main direction of flow deduced from associated scour-features (Plate 2B). Finally, small sponges, crinoids and sea-anemones occasionally may be observed to heel over in a preferred, down-current direction.

Measurement of the orientation and geometry of current-formed sedimentary structures[1] in compass-

[1] Current-activity in the form of scours and aligned worm-tubes is indicated in photographs from Station 1 but unfortunately the compass is not visible in the frames and therefore no orientation can be assigned to the current flow here.

oriented photographs from each of 16 camera-stations in the Gibraltar area enables assessment of the principal directions and modes of sediment transport in this region (Figure 15). A distinction has been maintained between the flow directions deduced from depositional structures (mainly ripple-marks), and those inferred from erosional features (scours and current-crescents). Aligned polychaete worm-tubes (either flat-lying or inclined) represent a third category. Where doubt exists as to the correct sense of current motion (for example, where only apparently symmetrical ripple-marks are present), only a two-directional trend is indicated on Figure 15. For each station, current-data derived from each of the three categories of structure listed above were aggregated separately and vector means calculated. The degree of reliance which may be placed upon the mean estimate is indicated to some extent by the number of frames from the station in which the relevant structures occur (Figure 15). The range of current orientations within any one station is generally small, usually less than 35° for ripples, and less than 20° for scours and aligned worm-tubes, indicating general uniformity of the current-regime over areas of the sea-bed as much as 0.8 miles (1.5 km) in width.

The most striking feature of the regional sediment transport pattern, as revealed in Figure 15, is the remarkably consistent westerly flow indicated within both the north and south branches of the Strait Channel. Here there is good agreement between directions obtained from depositional and erosional features. Westerly or west-southwesterly flow is also prevalent in the broad, relatively flat zone of confluence of the three major depressions (stations 36 and 30) but here the erosional scour-marks display considerable divergence from the more consistently oriented depositional structures (Figure 15). Station 30, on the gentle slope north of the ponded mouth of Gibraltar Canyon, exhibits a partly buried rippled morphology, in places modified by later, slope-parallel scouring.

The shallow stations (2A and 3) on the west side of Algeciras Bay apparently reveal uniform southerly transport (Figure 15), although partial burial of the ripples at Station 3 introduces an element of uncertainty as to the sense of movement here. However, some corroborative evidence is provided by inclined polychaete worm-tubes at this station which are consistently heeled over in a south-southeasterly direction (Figure 15).

Down-canyon flow is indicated by ripples and inclined worm-tubes at Station 2B in the axial zone of the head of Gibraltar Canyon but Station 4, on the east wall of this canyon at the entrance to Algeciras Bay, again reveals slope-parallel scours of uncertain sense, which are perpendicular to flat-lying, rolled polychaete worm-tubes. The westerly trend of the scours at this station suggests affinity with the main Strait flow-system, rather than any down-canyon transport.

As described earlier, photographs from the long traverse of Station 6, at the head of Ceuta Canyon, have been assigned to two sub-stations. The more shallow one (6A) reveals scour-marks indicating northwesterly, cross-head transport, accompanied by ripples of corresponding trend but uncertain asymmetry. The deeper substation (6B) provides evidence of down-canyon movement of rippled fine sand (Figure 15). The two remaining Ceuta Canyon stations (37 and 38) exhibit pronounced scour-effects which also appear to indicate down-canyon flow but the uncertain current-attitude of the scour-modified mounds and burrows introduces some doubt into this interpretation.

A comparison of these inferred sediment transport directions with observed current-meter data (expressed as the mode or mean of the flow within 50 m of the sea-floor, and derived from Lacombe *et al.*, 1964 and Frassetto, 1966) reveals a high degree of correspondence for appropriate localities, including a considerable diversity of current orientation in the zone of confluence (Figure 15). The virtual absence from the sediments of any evidence of the important tidal reversals of near-bottom flow recorded by Frassetto (1965) and Allan (1966) may be ascribed to fortuitous sampling of the sea-bed at equivalent periods of the tide. More plausibly, the apparent anomaly may indicate that the tidally reversed easterly flow is generally inadequate in power to effectively rework the coarse sediment of the Strait area. The structures of the sea-floor thus tend to preserve only the impress of the more powerful westerly components. Support for this interpretation is forthcoming from a consideration of the size characteristics of sediment samples from stations on the Gibraltar-Ceuta traverse. Allan (1966, p. 6) reports that in this area the strength of the near-bottom current varies from 0.3 knots (15 cm/sec) easterly to 0.6 knots (30.5 cm/sec) westerly. The maximum size of terrigenous material in the grab-sample from Station 36, near the middle of this traverse, is 5 mm. Graphs published by Hjulstrom (1939), Nevin (1946), Sundborg (1956) and Heezen and Hollister (1964) all indicate that flow-velocities of 30 to 50 cm/sec are necessary before particles of this diameter can be entrained and transported. Thus the recorded easterly flow (15 cm/sec) would be inadequate to move or substantially modify sediment of this grade but the westerly flow, at 30.5 cm/sec, is just capable of transporting such material.

DISCUSSION

The information summarised above demonstrates that the Strait of Gibraltar region comprises three major depressions, each characterised by unique morphologic and sedimentologic features and merging with one another in a wide and deep zone at the eastern end of the Strait which is marked by low relief and a current-swept topography. The faulted nature of the Strait appears beyond dispute (Giermann, 1961; *cf.* le Pichon *et al.*, 1971), but subsequently the area has undergone significant modification by processes of marine erosion to create the bifurcating Strait Channel. Simultaneously, deposition athwart the eastern end of the Strait has resulted in the northward growth of the Ceuta Ridge and may have caused the conspicuous westward diversion of Ceuta Canyon, which occurs at the 420 fm (770 m) isobath.

An extensive programme of bottom photography and precision echo-sounding has confirmed previous accounts of the rocky, gravelly nature of the marginal slopes and bordering shelves in the Strait. However, the deep floor of the Strait Channel is more extensively blanketed by sediment (sand and gravel) than earlier studies suggested. The distribution of bottom types and of maximum clast-size (Figure 5) and the nature and dimensions of sedimentary structures, indicate that the north branch of the Strait Channel is at present subject to more intense current activity than the south branch.

Algeciras Bay and Gibraltar Canyon, unlike the Strait, possess an almost continuous blanket of fine sand and silt, but this forms a relatively thin veneer above the bed-rock, which frequently emerges through the sediment mantle in the canyon walls and on the adjacent shelf. In conjuction with the short length and high axial gradient, these features indicate that Gibraltar Canyon is a comparatively youthful and presumably active feature. Its connection with a river valley (Rio Guadarrangue) cannot be regarded as coincidental. On the other hand, Ceuta Canyon, with a thick blanket of silt, a low axial gradient, and no immediate connection with a permanent river, displays somewhat less evidence of present-day activity.

Textural analyses have revealed the dual influence of terrigenous and biogenic sources, the former being quantitatively more important in sands from the canyons, the latter source contributing more to the Strait sediments. Size-distributions of canyon sands are dominated by saltation-suspension elements, suggesting turbidity current origin, whereas sands and gravels from the Strait including the confluence zone, are essentially tractional deposits, from which most finer materials have been winnowed by the powerful bottom currents prevalent in this area.

Slight but important differences in sediment petrology demonstrate essentially dual provenance for the canyon sediments. The Gibraltar Canyon association is predominantly derived from basic/ultrabasic and metamorphic sources and this provenance poses a problem with respect to the local geology, since the surrounding land is composed entirely of Mesozoic and Tertiary sediments (Figure 16). Moreover the Rio Guadarrangue does not drain the appropriate type of terrain. The nearest direct source of the relevant minerals and clasts on the Spanish side of the Strait appears to be in the Sierra Bermeja, some 50 km to the northeast (Figure 16), within the main Betic zone of southeastern Spain. Here extensive bodies of partly serpentinised peridotite are intruded into a Palaeozoic clastic/carbonate sequence which exhibits varying degrees of regional and contact metamorphism. Garnet, kyanite, staurolite, andalusite, albite and biotite are common in these rocks and dolomitic marbles are widely developed (Blumenthal, 1933; Hoeppner *et al.*, 1964). Assuming this area to be the ultimate source of much of the Gibraltar sediment, there remains the problem of transporting this material to the Gibraltar area and into Algeciras Bay.

Two possibilities suggest themselves:

1. A number of short streams drain the south flanks of the Sierra Bermeja and presumably convey the requisite detritus to the coast between Estepona and Marbella. Longshore-drift may then transport this material to the southwest, eventually rounding Europa Point to enter Algeciras Bay (Figure 16). Some support for this inferred longshore movement is provided by current-meter data at stations 'a' and 'b', just east of Europa Point (Frassetto, 1966), where tidally reversing flows were encountered but the strongest currents, exceeding, 1.5 knots (76 cm/sec), were directed to the south or southwest (Figure 15).

2. Alternatively, minerals and clasts from the Sierra Bermeja may be conveyed to the coast between Gibraltar and Estepona by the Rio Genal, an important tributary of the Rio Guadiaro (Figure 15) and then transported by longshore drift over the remaining 20 km to Gibraltar. The short coastwise transport involved in this explanation accords more fully with the relative compositional immaturity of the sediments, since prolonged exposure to littoral and sublittoral processes induces a high degree of compositional and morphologic maturity. Additional support for this view is derived from the slightly decreased orthopyroxene-metamorphic group content of the Algeciras Bay samples (2A and 2B) and the corresponding enhancement of the ZTR-opaques proportion, suggesting dilution from sedimentary

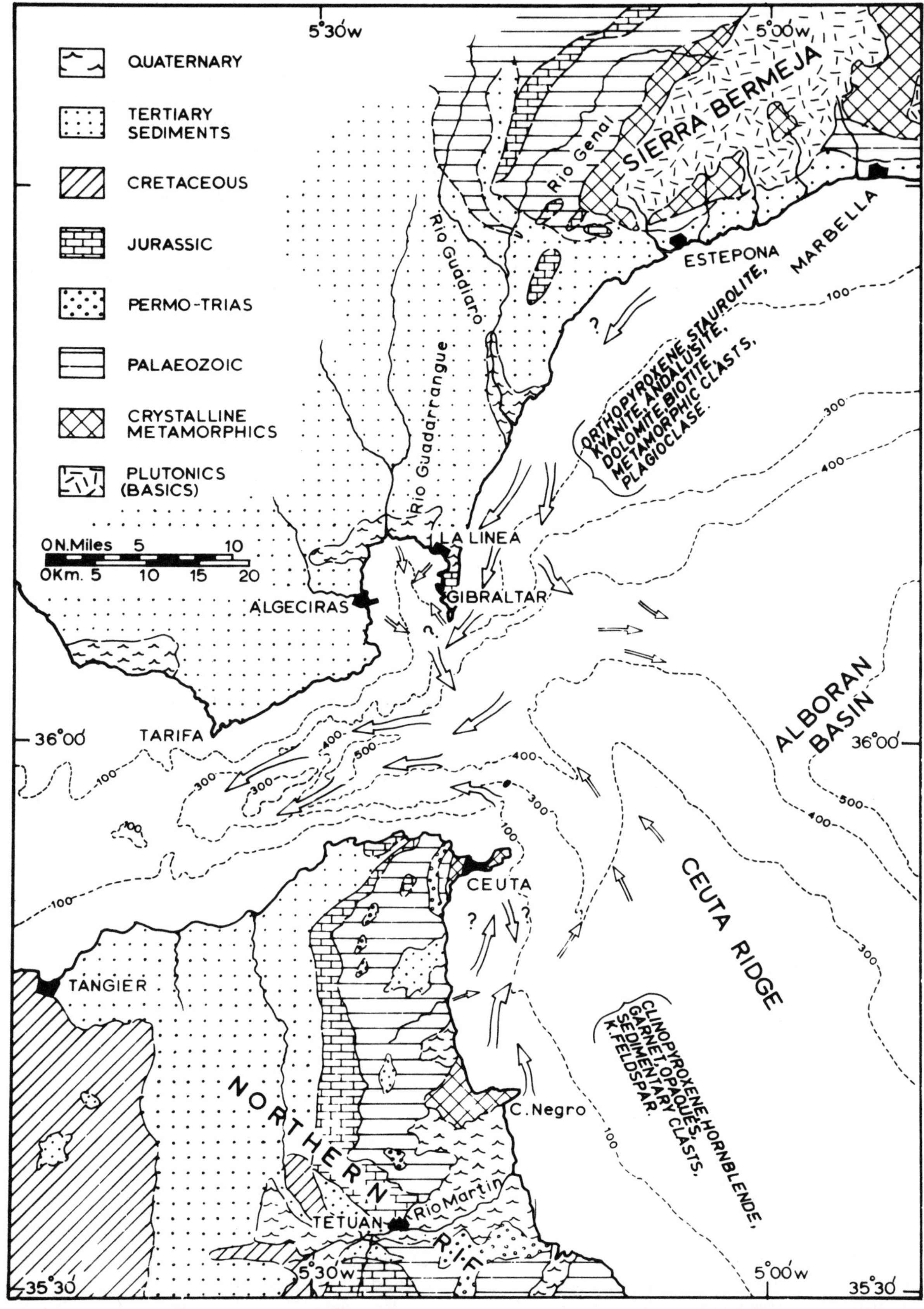

Figure 16. General geology of the western Betics and the northern Rif with principal paths of sediment transport as inferred from sediment petrology and direct or indirect measurements of bottom current flow. Size of arrows is proportional to relative importance of transport path.

sources, possibly supplied by the Rio Guadarrangue and Rio Palmones, or by coastal erosion. Moreover some portion of the sediment within Algeciras Bay probably was conveyed directly to the inner part of the bay through the La Linea-Gibraltar gap which, with an average elevation of about 10 m, remained submerged until very late glacial (Mousterian) time (Giermann, 1962).

One important consequence of the inferred longshore drift of sediment along the Estepona-Gibraltar coast (and the strong bottom currents observed near Europa Point) is that the large gullies and canyons which indent the shelf-margin here may intercept drifting sediments and funnel it downslope, probably into the Alboran Basin (Figure 16), as suggested elsewhere (Stanley *et al.*, 1970; Huang and Stanley, this volume).

The Ceuta Canyon mineral association is also an unstable one but the influence of ultrabasic and metamorphic source-rocks is less pronounced and clasts and minerals derived from sedimentary rocks correspondingly more common. The adjacent terrain provides a ready supply of sedimentary detritus but the source of the igneous and metamorphic minerals is not directly apparent. Crystalline metamorphics (including quartz-rich hornblende-gneisses and garnet-mica-schists) form the promontories of Ceuta and Cabo Negro (Figure 16) and are intruded by relatively small bodies of peridotite (Fallot, 1937, p. 294). These ultrabasic rocks, while petrographically identical with the Sierra Bermeja suite, occupy much smaller areas than their Betic counterparts (Fallot, 1937, p. 303) and hence may be expected to be less prominent in any suite of Rif-derived sediments.

While it again appears necessary to invoke coastwise movement of detritus and interception by the headward region of Ceuta Canyon (Figure 16), it is not clear whether the principal source of the ultrabasic and metamorphic minerals is the Ceuta peninsula or Cabo Negro. Both alternatives are portrayed in Figure 16, although the localised northwest transport vectors recorded near the Ceuta Canyon head (Station 6A, Figure 15) tend to favor the Cabo Negro source.

The distinctive petrologic features of terrigenous fractions from the floor of the Strait are attributed to the effects of increasing compositional and morphologic maturity resulting from prolonged reworking and eventual westward transport of canyon-supplied sediment. Sediments from both sides of the Strait display diagnostic petrologic characters which can be linked with the appropriate canyon-association. Such a distinction presumably reflects the rapidity and consistency of the reworking process and this view is reinforced by the remarkable uniformity of the sediment-affecting flow-directions observed in the Strait (Figure 15). Further modification and dilution of the sediments flooring the Strait may be achieved by lateral, down-slope addition of predominantly biogenic detritus, conveyed through the numerous gullies and ravines cut in the marginal slopes.

The zone of confluence at the eastern end of the Strait has a critical role in the sedimentation pattern of this region. Frassetto (1965) interpreted this region as a fan, built out eastwards by sediment supplied through the canyons by turbidity currents and constrained within the Strait. The essentially flat, depositional topography of this area, the presence of channel-levees and the lenticular distribution of surficial sediment types (Figure 9) all attest to the essential validity of the fan interpretation and indicate the predominant role of Gibraltar Canyon in the supply of sediment. However the prevalence of scouring and the diversity in current flow observed in this area (Figure 15) indicate a complex regime from which westward transfer of sediment ultimately results (Figure 16). The presence of the broad, silt-draped swell or col north of Ceuta Ridge (Figure 9) precludes the possibility of continuous eastward tractional movement of coarse sediment by this path. However, it is probable that finer materials, winnowed from the canyon sediments by the powerful westerly bottom currents and transferred by internal waves and turbulence (Lacombe *et al.*, 1964) to the east-flowing near-surface layer, are conveyed to the Alboran Basin, together with suspensates (mainly clay minerals) derived from Atlantic sources (see Huang and Stanley, this volume).

ACKNOWLEDGMENTS

We are indebted, first of all, to the U. S. Coast Guard and its Oceanographic Unit, Washington, for providing ship time (Cruise $RoSM_1$) necessary for conducting this study. The Captain, officers and men of the USCGC ROCKAWAY are thanked for their support in our work at sea. Mr. A. Church, Alpine Geophysical Association, provided assistance in all phases of underwater photography during the cruise. Mesdemoiselles M. A. Delssere, Monaco, and J. N. Valette, Perpignan, also assisted in the collection of data at sea. The technical assistance in processing sediment samples and drafting was provided by Mr. J. Rogers, Mr. S. Osborn and Mr. H. Sheng. The paper was critically reviewed by Dr. T. C. Huang, University of Rhode Island, and Dr. Y. Weiler, Hebrew University, Jerusalem. Financial support for this study was provided to G. K. by the United Kingdom NERC research grant GR/3/899 and to D. J. S. by Smithsonian Research Foundation grants 234230 and 436330.

REFERENCES

Allan, T. D. 1966. Underwater photographs in the Strait of Gibraltar. *NATO Technical Memorandum SACLANT ASW Research Centre, La Spezia,* 116: 19 p.

Allain, C. 1964. L'hydrologie et les courants du détroit de Gibraltar pendant l'été de 1959. *Revues des Travaux de l'Institut Technique Maritime,* 28 (1): 3–102.

Bailey, E. B. 1952. Notes on Gibraltar and the Northern Rif. *Quarterly Journal of the Geological Society of London,* 108: 157–176.

Blumenthal, M. 1933. Der Paleozoikum von Malaga als tectonische Leitzone im Alpidischen Andalusien, *Geologische Rundschau,* 24: 104–149.

Ericson, D. B., M. Ewing, G. Wollin, and B. C. Heezen 1961. Atlantic deep sea sediment cores. *Geological Society of America Bulletin,* 72: 193–286.

Fallot, P. 1937. Essai sur la géologie du Rif septentrional. *Notes et Memoires du Service des Mines et de la Carte Géologique du Maroc,* 40: 558 p.

Folk, R. L. 1968. *Petrology of Sedimentary Rocks.* 2nd Edition, Hemphills, Austin, 161 p.

Folk, R. L. and W. C. Ward 1957. Brazos River bar: a study in the significance of grain size parameters. *Journal of Sedimentary Petrology,* 27: 3–26.

Frassetto, R. 1965. Report on the oceanographic cruise GIB V, 7 April–8 May 1965 and some preliminary results on temperature and sound velocity measurements. *NATO Technical Report, SACLANT ASW Research Centre, La Spezia,* 45: 30 p.

Frassetto, R. 1966. Discussion on the distribution of sea floor types at the eastern entrance to the Strait of Gibraltar, as revealed by grabs. *NATO Technical Memorandum, SACLANT ASW Research Centre, La Spezia,* 117: 18 p.

Fuller, A. O. 1961. Size characteristics of shallow marine sands from Cape of Good Hope, South Africa. *Journal of Sedimentary Petrology,* 31: 256–261.

Giermann, G. 1961. Erläuterungen zur bathymetrischen Karte der Strasse von Gibraltar. *Bulletin de l'Institut Océanographique Monaco,* 1218: 1–28.

Giermann, G. 1962. Meeresterrassen am nordufer der Strasse von Gibraltar. *Bericht der Naturforschenden Gesellschaft zu Freiburg, i. Br.,* 52: 111–118.

Hails, J. R. and J. H. Hoyt 1969. The significance and limitations of statistical parameters for distinguishing ancient and modern sedimentary environments of the lower Georgia coastal plain. *Journal of Sedimentary Petrology,* 39: 559–580.

Heezen, B. C. and C. D. Hollister 1964. Deep-sea current evidence from abyssal sediments. *Marine Geology,* 1: 142–174.

Heezen, B. C. and G. L. Johnson 1969. Mediterranean undercurrent and microphysiography west of Gibraltar. *Bulletin de l'Institut Océanographique Monaco,* 69: 1–95.

Hjulström, F. 1939. Transportation of detritus by moving water. In: *Recent Marine Sediments,* ed. Trask, P. D., Murby & Co., London, 5–31.

Hoeppner, R., P. Hoppe, St. Durr and H. Mollat 1964. Ein Querschnitt durch die Betischen Cordilleren bei Ronda (SW Spanien). *Geologie en Mijnbouw,* 43: 282–284.

Horn, D. R., M. Ewing, M. N. Delach and B. M. Horne 1971. Turbidites of the northeast Pacific. *Sedimentology,* 16: 55–69.

Huang, T. C. and D. J. Stanley 1972. Western Alboran Sea: Sediment dispersal, ponding and reversal of currents. In *The Mediterranean Sea: A Natural Sedimentation Laboratory,* ed. Stanley, D. J., Dowden, Hutchinson and Ross, Inc., Stroudsburg, Pennsylvania, 521–559.

Hubert, J. H. 1964. Textural evidence for deposition of many western North Atlantic deep sea sands by ocean bottom currents rather than turbidity currents. *Journal of Geology,* 72: 757–785.

Inman, D. L. 1949. Sorting of sediments in the light of fluid mechanics. *Journal of Sedimentary Petrology,* 19: 51–70.

Kelling, G. and D. J. Stanley 1970. Morphology and structure of Wilmington and Baltimore submarine canyons, eastern United States. *Journal of Geology,* 78: 637–660.

Kuenen, P. H. 1950. *Marine Geology.* Wiley, New York, 568 p.

Lacombe, H., P. Tchernia, C. Richez and L. Gamberoni 1964. Deuxième contribution à l'étude du détroit de Gibraltar. *Cahiers Océanographiques,* 16: 283–314.

le Pichon, X., J. M. Auzende, G. Pautot, S. Monti and J. Franchetau 1971. Deep sea photographs of an active seismic fault near Gibraltar Straits. *Nature,* 230: 110–111.

Middleton, G. V. 1962. Size and sphericity of quartz grains in two turbidite formations. *Journal of Sedimentary Petrology,* 32: 725–742.

Nevin, C. 1946. Competency of moving water to transport debris. *Geological Society of America Bulletin,* 74: 1057–1062.

Pettijohn, F. J. 1958. *Sedimentary Rocks,* 2nd Edition, Harper, New York, 718 p.

Power, M. C. 1953. A new roundness scale for sedimentary particles. *Journal of Sedimentary Petrology,* 23: 117–119.

Stanley, D. J. and G. Kelling 1968. Photographic investigation of sediment texture, bottom current activity and benthonic organisms in the Wilmington submarine canyon. *United States Coast Guard Oceanographic Report,* 22: 95 p.

Stanley, D. J., C. E. Gehin and C. Bartolini 1970. Flysch-type sedimentation in the Alboran Sea, western Mediterranean. *Nature,* 228: 979–983.

Sundborg, A. 1956. The river Klaralven—a study of fluvial processes. *Geografiska Annaler,* 38: 127–316.

Visher, G. L. 1969. Grain size distributions and depositional processes. *Journal of Sedimentary Petrology,* 39: 1074–1106.

Wüst, G. 1961. On the vertical circulation of the Mediterranean Sea. *Journal of Geophysical Research,* 66: 3261–3271.

Western Alboran Sea: Sediment Dispersal, Ponding and Reversal of Currents

Ter-Chien Huang[1] and Daniel J. Stanley

Smithsonian Institution, Washington, D.C.

ABSTRACT

Two contrasting hypotheses have been proposed to explain Mediterranean oceanographic circulation in Würm time: one suggests circulation in a silled basin where runoff and precipitation exceed evaporation; the other, a circulation where evaporation exceeds precipitation and runoff (a pattern not drastically different from the present one, but with a lower exchange of water mass at the Strait of Gibraltar). A petrologic examination of cores from the Western Alboran Sea between Morocco and Spain allows these hypotheses to be tested. Cores consist almost entirely of hemipelagic clay and silt with only minor thin sand layers. The stratigraphy of these sections based on 21 C^{14} determinations shows an almost uninterrupted record of deposition during the past 20,000 years.

A major petrological change occurs at about 10,000 years B. P. when the rate of sedimentation also decreased rapidly from about 60 to 130 cm/1000 years to about 30 to 40 cm/1000 years in the Western Alboran Basin plain. Rates on the slope during the late Quaternary averaged about 20 cm/1000 years. These observations, coupled with more stratification of the mud in the lower portions of cores, indicate stronger bottom and turbidity current activity of the sea floor in Würm time. The high frequency of bioturbation and current-produced lamination of silt and clay and the lack of sapropel layers in core sections older than 10,000 years show that the Western Alboran Basin plain was not stagnant as in Mediterranean basins to the east, although there is a suggestion of somewhat increased reducing conditions.

The increased proportion of pelagic tests relative to terrigenous material in the Holocene mud section indicates the more important role of suspension transport and hemipelagic deposition in recent time. Wind intensity also decreased in Holocene time as shown by lesser amounts of wind-blown sand in upper mud sections. The increase of the kaolinite/chlorite ratio and of planktonic foraminifers, and concurrent decrease of the montmorillonite/illite ratio,

RESUME

Deux hypothèses contradictoires ont étés offertes pour expliquer la circulation des eaux océaniques dans la Méditerranée, au Würmien. L'une suggère une circulation dans un bassin dans lequel précipitation et écoulement sont supérieurs à l'évaporation. L'autre suggère une circulation avec évaporation supérieure à l'écoulement et précipitation (situation très semblable à l'actualité mais comportant un moindre échange d'eau au niveau du Détroit de Gibraltar). Ces hypothèses sont mises à l'épreuve grâce à l'examen d'échantillons prélevés en Mer d'Alboran occidentale entre le Maroc et l'Espagne. Les échantillons sont presque entièrement constitués d'argile hémipélagique et de silt avec quelques couches de sable fin. L'étude stratigraphique des carottes, basée sur le datage au carbone 14, indique un dépôt continu durant les 20.000 dernières années.

Un changement capital a pris place aux environs de 10.000 ans B. P. alors que la vitesse de sédimentation a décrue de 60 à 130 cm/1000 ans à 30 ou 40 cm/1000 ans dans la plaine du Bassin d' Alboran occidental. A la fin du quaternaire, la vitesse moyenne sur la pente était à peu près 20 cm/1000 ans. Ces observations, et le fait que la stratification soit plus marquée dans la partie inférieure des carottes, indiquent un courant de fond plus fort et une plus grande activité de courant de turbidité au fond de la mer au Würmien. La haute fréquence de bioturbation et lamination du silt et de l'argile due au courrant, ainsi que l'absence de couches de sapropel parmi les couches plus anciennes que 10.000 ans, indiquent que le Bassin d'Alboran n'était pas stagnant au contraire des bassins méditerranéens orientaux, bien qu'il y ait une indication d'accroissement de conditions réductrices.

L'accroissement du nombre de foraminifères et matières pélagiques dans les strates d'âge Holocène révèle l'importance du rôle de transport par suspension et de dépôt hémipélagique pendant les époques récentes. Ainsi que le montre la faible quantité de sable

[1] Present address: Graduate School of Oceanography, University of Rhode Island, Kingston, Rhode Island 02881 U.S.A.

of carbonate debris and of wood fibers also indicate a change in sedimentation regime at about 10,000 years B. P.

The Alboran Basin sediment is mainly derived from two intrabasinal areas: one is situated in the northwest quadrant of the Alboran Sea, near Gibraltar, and this area contributed mainly terrigenous material; the second source region comprises the banks and ridges south of the Western Alboran Basin, which supplied carbonate debris.

The lower rate of Holocene ponding in the Western Alboran Basin plain reflects a rapidly reduced rate of erosion of sediments on slopes of the Alboran Sea. Changes in sedimentation appear correlative with a notable alteration in geography and hydrologic conditions in the western Mediterranean, including a reversal of current regime that probably occurred at the Strait of Gibraltar at the beginning of Holocene time. Before 10,000 years B. P., less dense Mediterranean water outflowed above denser Atlantic water entering the Strait and then flowing into the Alboran Sea. A thin sand layer consisting almost entirely of foraminifera and carbonate debris, organisms influenced by this change of circulation, was emplaced at this time.

transporté par le vent, l'intensité de ce dernier a décru pendant le Holocène. L'accroissement du rapport kaolinite/chlorite et des foraminifères planktoniques d'une part, et le décroissement du rapport montmorillonite/illite ainsi que des débris de carbonate et des fibres végétales d'autre part, indiquent un changement du régime de sédimentation aux environs de 10.000 ans B. P.

Le sédiment du Bassin d'Alboran provient largement de deux régions propres au bassin, l'une, située dans le quadrant N. O. de la Mer d'Alboran près de Gibraltar fournissant des matériaux terrigènes, l'autre, située dans une zone englobant les rives et crêtes au Sud de la partie occidentale du Bassin d'Alboran, fournissant les débris de carbonates.

Le remplissage (ponding) holocène plus faible dans la plaine du bassin reflète une érosion décroissante des sédiments sur les pentes de la Mer d'Alboran. Les changements de la sédimentation semblent être en corrélation avec une altération notable des facteurs géographiques et hydrologiques dans la Méditerranée occidentale, entrainant une inversion du régime présent; celle-ci a sans doute pris place au début du Holocène dans le détroit de Gibraltar. Avant 10.000 ans B. P., les eaux méditerranéennes, moins denses que celles de l'Atlantique, ont débordé au-dessus de ces dernières à leur entrée dans le détroit, les forçant à se déverser dans la Mer d'Alboran. Une fine couche de sable a été déposée à cette époque; elle est presque entièrement constituée de foraminifères et de débris d'organismes carbonatés probablement accumulés lors de l'inversion de la circulation.

INTRODUCTION

The Alboran Sea, a partially land-locked depression in the western Mediterranean Sea enclosed between the Spanish and Moroccan coasts, is connected through the Strait of Gibraltar to the Atlantic on the west, and via the Alboran Trough to the Balearic Basin on the east (Figure 1). This study describes the regional sediment distribution and dominant dispersal patterns that accompanied climatic and hydrographic changes in this critical region during Late Pleistocene to Holocene time.

Earlier studies have shown that marked eustatic changes of sea-level, as well as climate and vegetation (Eriksson, 1965) occurred at the end of the Pleistocene. These in turn are believed to have affected the hydrography of the Mediterranean, including the Alboran Sea, by altering the exchange pattern of Atlantic and Mediterranean waters through the Strait of Gibraltar. Two contrasting hypotheses explain the paleo-oceanic circulation in the Mediterranean Sea during Würm time. One suggests that the Mediterranean was an "estuary-type" basin with deep Atlantic water inflow below less dense Mediterranean outflow through the Strait of Gibraltar (Bradley, 1938; Kullenberg, 1952; Mars, 1963; and Leclaire, 1970). The other theory maintains that the Mediterranean was a "lagoonal-type" basin with net outflow of a deep layer of more saline Mediterranean water, *i.e.*, a circulation system not drastically different from the present one, but with a somewhat lower exchange of water masses at the Strait of Gibraltar (Anderson, 1965). The present study tests these hypotheses by comparing sedimentary patterns in Würm time with those existing at present.

Of particular interest is determination of the origin of the mud that accounts for 95% of the total core section collected in the Alboran Sea. The mode of sediment dispersal from the adjacent slopes into the Western Alboran Basin also requires consideration. Processes responsible for deposition of sediments in bathymetric lows in enclosed basins [called *ponding* (Hersey, 1965a)], include turbidity current transport and slumping. Sedimentation in this deep-sea basin is also influenced by pelagic settling, bottom currents, organisms and wind transport (Stanley *et al.*, 1970). The relative importance of each of these processes is evaluated in the context of the sub-Recent history of this region.

PHYSIOGRAPHY AND HYDROGRAPHY

Bathymetric charts of the Strait of Gibraltar and the western Mediterranean Sea have been compiled by Giermann (1961, 1962). Bathymetry of this area has been subsequently revised and charts modified from Stanley *et al.* (1970, his figure 3) and Gehin *et al.* (1971, their figure 1) with additional data collected by the present authors are illustrated in Figures 1 and 2. The Western Alboran Basin is roughly bounded by the 500 fm contour line with a flat basin plain (called the Western Alboran Basin plain) in the center portion. The plain is defined by the 800 fm (1463 m) isobath. The diamond-shaped configuration of the basin is

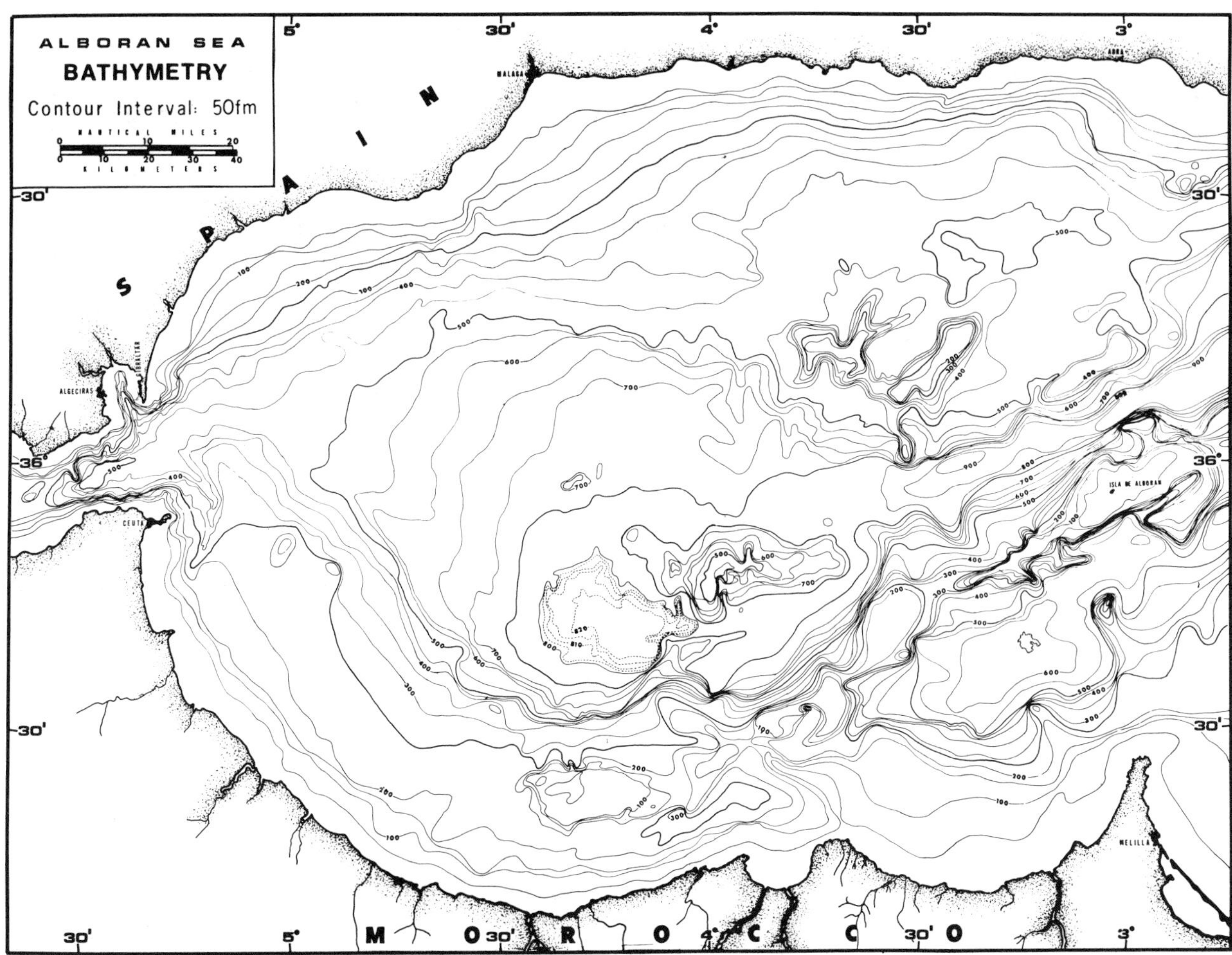

Figure 1. Bathymetry of the Alboran Sea. Note the submarine valleys off Gibraltar and Spain that cut across the slope and descend to the flat floor of the Western Alboran Basin plain. The almost continuous complex of banks and ridges off Morocco trends NE-SW. This chart is modified from one prepared by SACLANT ASW Research Centre, La Spezia with supplemental data from USCGC ROCKAWAY cruise $RoSm_1$. Depth in fathoms. Navigational control by Loran and radar.

probably controlled by the dominant NE-SW and NW-SE tectonic trends of this region (Giermann *et al.*, 1968). The Gibraltar Submarine Valley, west of the basin, links the basin plain with the Strait of Gibraltar and the Ceuta and Gibraltar Canyons (Figures 1 to 3). This valley is about 80 km long, 5 to 10 km wide, and compound *i.e.*, characterized by several channels near the head (Figure 3, profile G-H). The Alboran Knolls (the Western Alboran Seamounts of Stanley *et al.*, 1970) display high relief and irregular topography and lie NE of the basin plain (Figure 3, profiles C-D, I-J). The Alboran Trough (Figure 3, profile C-D), is a deep flat depression which connects the Western Alboran Basin with the Balearic Basin; this NE-SW trending trough lies between the Alboran Knolls and the Western Alboran Ridge. The Malaga Low, a small basin bounded by 700 fm (1280 m) isobath is situated between the Western Alboran Basin and the submarine slope off Spain. Several submarine valleys extend downslope from the Spanish margin toward the Malaga Low and the Western Alboran Basin. One of these, the Malaga Valley, passing through the Malaga Low, is also compound (Figure 3, profile E-F).

Locally, slopes from Africa, ranging to 11 degrees, are somewhat steeper than those off Spain. A series of banks and ridges, such as the Xauen and Tofiño Banks and the Western Alboran Ridge, form a linear SW-NE trending submarine chain. The Eastern Alboran Basin is bounded by the 600 fm (1097 m) contour line, and is characterised by a very flat floor (Figure 3, profile A-B).

Hydrographic investigations of this portion of the Mediterranean Sea have been conducted by among others, Sverdrup *et al.* (1942, p. 644), Lacombe and Tchernia (1960), Wüst (1960, 1961), Anderson (1965), Lucayo (1968) and Heezen and Johnson (1969). The present general circulation pattern of Mediter-

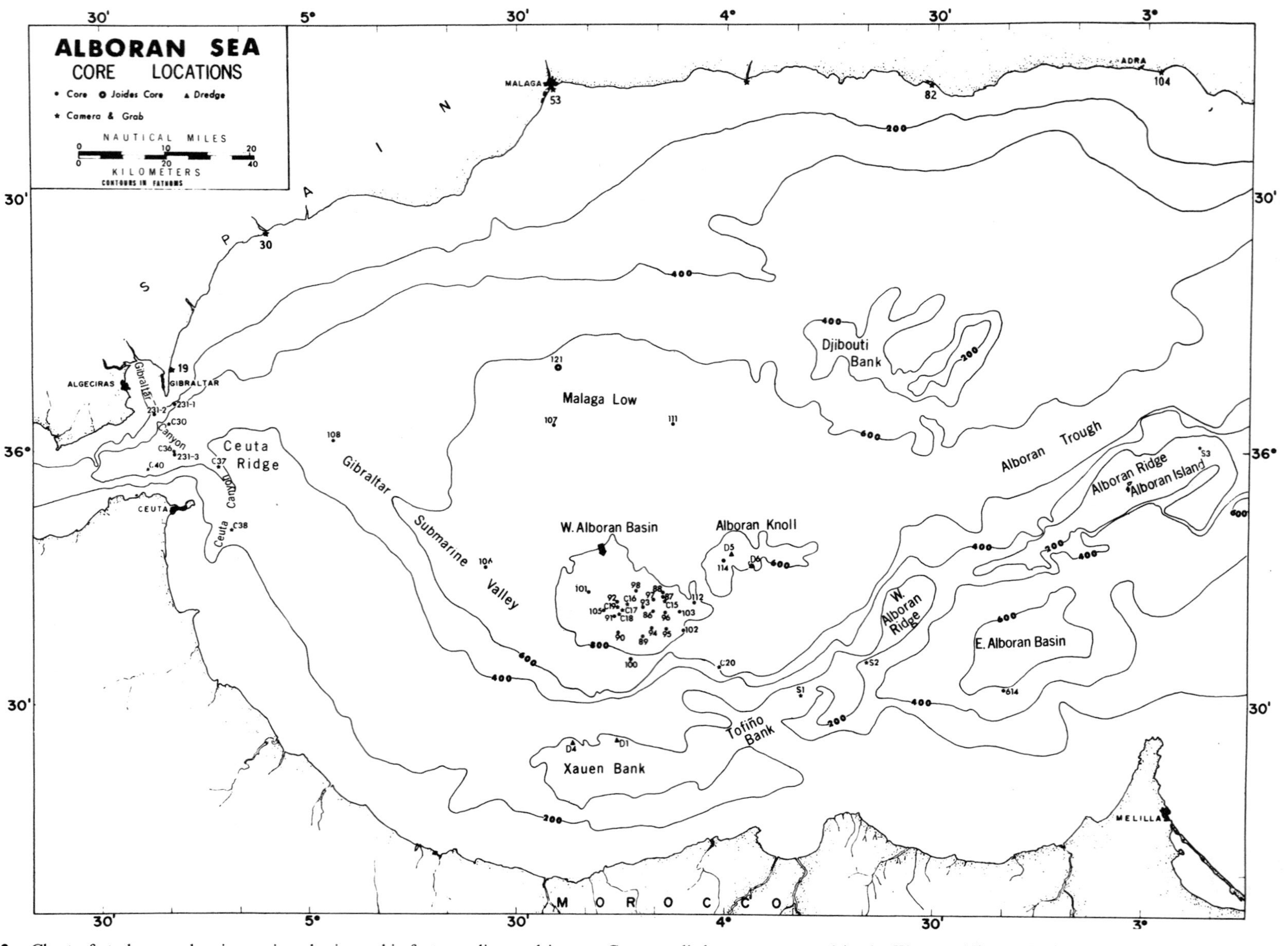

Figure 2. Chart of study area showing major physiographic features discussed in text. Cores studied are concentrated in the Western Alboran Basin plain. Note position of JOIDES-DSDP Leg 13 Core 121 in the Malaga Low. Depth in fathoms.

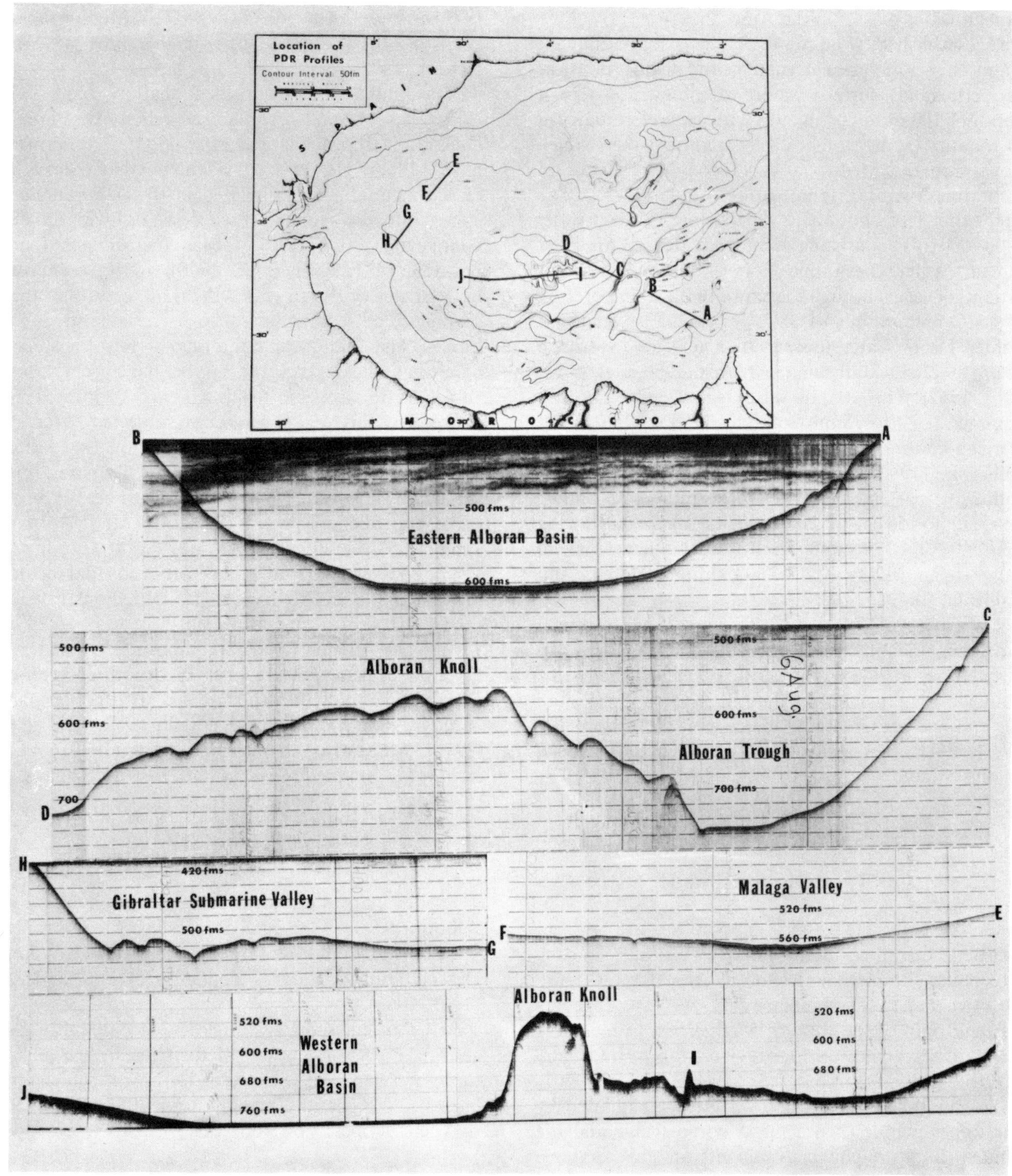

Figure 3. Selected bottom profiles in the study area showing major bathymetric patterns. Profiles A through H (PDR tracks from CONRAD cruise 9, 1965; courtesy of W. B. F. Ryan); Profile I-J (a PEST track from USCGC ROCKAWAY Cruise $RoSm_1$, 1970). Sub-bottom reflectors in PDR profiles show ponding in the Eastern Alboran Basin and Alboran Trough and the sediment fill in the Gibraltar and Malaga Valleys.

ranean water masses in the Alboran Sea is reasonably well established (Figure 4). Three semi-permanent water masses are present, each having distinct physical properties: (1) surface water, extending down to a depth of 100 m to 200 m, and with a water core depth of about 75 m, consists of Atlantic water which flows into the Mediterranean through the Strait of Gibraltar (with a temperature range of 15–25°C and salinity of about 37.0‰; (2) intermediate water (200–600 m), consisting mainly of denser Mediterranean water (Levantine Water) outflowing to the Atlantic Ocean through the Strait (with a temperature of 15°C and salinity of 38.5‰); and (3) somewhat colder Deep Water (deeper than 600 m) (Sverdrup *et al.*, 1942; Lacombe and Tchernia, 1960, and Wüst, 1961, 1962). The Atlantic water is best defined during summer at 20 to 75 m depth chiefly along the North African continental slope, as shown by Lacombe and Tchernia (1960, their figure 3). The velocities of inflowing currents through the Strait of Gibraltar range up to 150 cm/sec, and velocities of the outflowing denser water encountered at depths near 200 m range from 100 cm/sec to 300 cm/sec (Heezen and Johnson, 1969).

Lucayo (1968) working with actual and empirical data concluded that there is a surface cyclic (clockwise) circulation in the Alboran Sea. The cyclic paths coincide roughly with the bottom topography (Figure 4), and the center of the major gyre lies above the Western Alboran Basin and Alboran Knolls discussed in this study.

THE GEOLOGICAL FRAMEWORK

Tectonic Setting

Geologically, the western Alboran Sea lies between the Betic Cordillera on the Iberian Peninsula and the Moroccan Rif Mountains of North Africa (Figure 1). Vogt *et al.* (1971) summarize the two basic concepts of the origin of the western Mediterranean including this area. One proposes that positive land areas (nuclei) foundered (perhaps in early Mesozoic time) and became vertically assimilated by the upper mantle (hence the Alboran Ridge and the Xauen Bank might be remnants of this craton). Hersey (1965 b), van Bemmelen (1969), Glangeaud (1970) and Giermann (1961, 1968) among others, support this hypothesis. A more recent view, however, implies some form of axial accretion and sea-floor spreading. Interpretation of the JOIDES core in the Alboran Sea (Hsü, 1971; Nesteroff *et al.*, this volume), structure of the deep-stratum in Mediterranean basins (Caputo *et al.*, 1970), seismicity and fault planes (McKenzie, 1970), magnetic properties (Vogt *et al.*, 1971) and the Ostracoda of the western Mediterranean (Benson, this volume) support this type of plate-tectonics.

The knolls and steep-sided shallow banks and ridges have been interpreted as relict Mio-Pliocene volcanic features (Glangeaud *et al.*, 1967; Giermann *et al.*, 1968). However, it is known that over 1 km of sediments has accumulated in the Alboran Basin since Pliocene time[1] (Ewing and Ewing, 1959; Glangeaud *et al.*, 1967; Texas Instruments Incorporated, 1967; Ryan *et al.*, 1970), and a recent sub-bottom survey (Ryan *et al.*, 1970) indicates that thick sequences of sediment also cover certain ridges and knolls. The folding and faulting of Plio-Pleistocene deposits indicate that the region has recently been subjected to orogenic displacement. Additional evidence that this region has been subjected to crustal movement in recent time is shown by the vertical displacement of Quaternary terraces (Giermann, 1961; Butzer and Cuerda, 1962) and a relative high frequency of earthquake epicenters (Barazangi and Dorman, 1969).

The concurrent nature of sedimentation and tectonic movement in the Mediterranean is demonstrated in numerous studies including those of Carey (1958), Ewing and Ewing (1959), Fahlquist (1963), Hersey (1965b), Texas Instruments Incorporated (1967), Glangeaud *et al.* (1967), Glangeaud (1970), Giermann *et al.* (1968), van Bemmelen (1969), Caputo *et al.* (1970), McKenzie (1970), Ryan *et al.* (1970), Wong *et al.* (1970), Auzende *et al.* (1971), and Vogt *et al.* (1971).

Sedimentation in the Alboran Sea

Earlier marine studies bearing on sediments in the Alboran Sea and the adjacent western Mediterranean sea floor include those of Kullenberg (1947), Weibull (1947), Duplaix (1958), Todd (1958) and Eriksson (1961, 1965).

More recent sedimentological and paleontological studies of this part of the Mediterranean Sea include those by Giermann *et al.*, 1968 (sediments on shallow banks); Vergnaud-Grazzini and Bartolini, 1970 (oxygen istopes); Benson, this volume (ostracods);

[1] JOIDES-DSDP Core 121, north of the Alboran Basin plain, confirms earlier interpretations of seismic profiles, *i.e.*, revealing a sequence of sediments about 1 km thick, above a basement of granodiorite rock which was recovered at a core depth of 868 m. This basement is covered by some 200 meters of marls and oozes of Upper Miocene age (Ryan and Hsü, 1971; Nesteroff *et al.*, this volume). Interbedded mud and sand strata occur between about 300 m and 680 m; the sands, some of them graded, were deposited during the Pliocene. The uppermost 300 m of olive gray mud and marly ooze represent the Quaternary sequence.

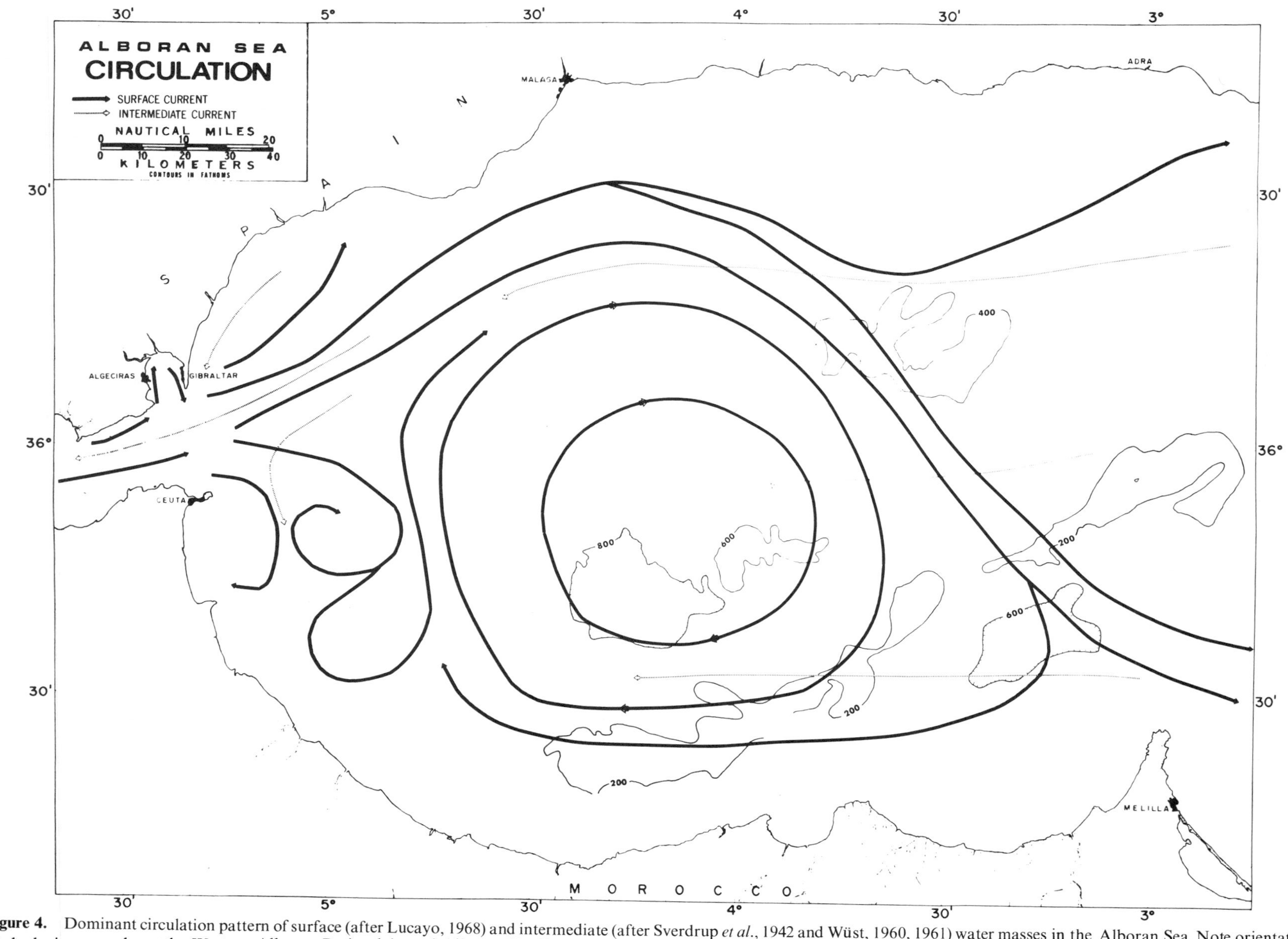

Figure 4. Dominant circulation pattern of surface (after Lucayo, 1968) and intermediate (after Sverdrup *et al.*, 1942 and Wüst, 1960, 1961) water masses in the Alboran Sea. Note orientation of clockwise gyre above the Western Alboran Basin plain and Alboran Knolls.

Emelyanov, this volume (principal sediment types); Emelyanov and Shimkus, this volume (suspensates); Kelling and Stanley, this volume (sediments of the Gibraltar Strait), and Milliman *et al.*, this volume (carbonate sediment of the shallow banks). It will be shown in later sections that Eriksson's (1965) detailed investigation of Core 210 in the Western Balearic Basin about 180 miles ENE of the Western Alboran Basin plain is pertinent to the present study.

A continuous seismic profile made by R/V CONRAD cruise 9 in 1965 and extending from the Eastern Alboran Basin, across the Alboran Knolls and the Malaga Low to the Spanish slope south of Malaga is shown in Figure 5 (profile A-B). Sedimentary strata are generally concordant, particularly in lows such as the Eastern Alboran Basin. Deposits on the northern margin of the Alboran Trough have been uplifted and offset on the eastern portion of the Alboran Knolls. An acoustically transparent "basement" is seen below the stratified sediments in much of the area and the sedimentary cover lies discordantly on this irregular basement. Seismic reflectors underlying the Eastern Alboran Basin, Alboran Trough, Malaga Low and Spanish slope suggest alternations of acoustically transparent and opaque layers. Reflectors clearly show the cyclic nature of sedimentary sequences as well as the sediment ponding in all topographic lows (Figure 5, profile A-B).

Particularly pertinent to the present study is the analysis of even higher resolution sub-bottom (PGR) profiles made by Gehin *et al.* (1971). These authors emphasize turbidity and normal bottom currents as predominant ponding processes. The flysch-like nature of the late Quaternary deposits in the western Alboran Sea has been discussed by Stanley *et al.* (1970).

PROCEDURES

Materials Examined in the Study

Materials examined during the course of this study include: (a) forty-nine large diameter (13 cm) sphincter cores (see Figure 2 and Table 1) collected during R/V MARIA PAOLINA G. (SACLANT) cruises in the Western Alboran Basin and adjacent areas [a tight network of some 2200 km of PGR profiles was also obtained on these cruises (Stanley *et al.*, 1970; Gehin *et al.*, 1971)]; (b) fifteen Shipek grab and dredge samples and photographs obtained at forty-four camera stations in this area during USCGS ROCKAWAY cruise $RoSm_1$ in August 1970; (c) sample cuts from three short cores collected from the Gibraltar Canyon area by the R/V ZELIAN (NAVOCEANO) cruise in 1964 and one Kullenberg piston core collected from the Eastern Alboran Basin

Table 1. Core, grab, camera and dredge locations in the Alboran Sea (see Figure 2).

Station	Latitude	Longitude	Water Depth in fms (m)	Length (cm)
		I. Core Locations		
86	35°41.3′	4°10.0′	823 (1504)	670
87	35°43.0′	4°08.9′	820 (1499)	630
88	35°43.4′	4°08.9′	820 (1498)	590
89	35°38.3′	4°11.6′	822 (1503)	590
90	35°38.7′	4°15.2′	821 (1500)	530
91	35°40.6′	4°15.8′	825 (1508)	635
92	35°42.3′	4°15.3′	823 (1504)	730
93	35°41.8′	4°11.6′	823 (1505)	710
94	35°39.3′	4°10.3′	825 (1507)	575
95	35°39.2′	4°08.2′	827 (1511)	700
96	35°41.1′	4°08.5′	823 (1504)	570
97	35°42.6′	4°10.0′	823 (1504)	700
98	35°43.7′	4°12.6′	823 (1504)	720
100	35°36.0′	4°13.7′	775 (1417)	440
101	35°43.5′	4°19.5′	802 (1466)	600
102	35°39.0′	4°05.8′	812 (1484)	620
103	35°41.2′	4°06.4′	821 (1500)	590
105	35°41.4′	4°17.4′	818 (1495)	490
106	35°46.5′	4°34.4′	714 (1305)	500
107	36°03.3′	4°24.5′	689 (1260)	490
108	36°01.6′	4°56.5′	490 (895)	300
111	36°03.4′	4°07.4′	708 (1293)	540
112	35°42.2′	4°04.4′	805 (1472)	410
114	35°47.1′	4°00.0′	443 (809)	400
231–1	36°05.2′	5°19.1′	382 (699)	63
231–2	36°04.3′	5°22.1′	339 (619)	74
231–3	35°59.2′	5°19.1′	492 (899)	100
614	35°38.0′	3°05.0′	538 (983)	910
121	36°10.0′	4°23.0′	611 (1116)	*
		II. Grab and Camera Locations		
C14	35°44.5′	3°55.7′	660 (1206)	
C15	35°43.6′	4°06.9′	820 (1499)	
C16	35°40.8′	4°44.0′	822 (1503)	
C17	35°39.6′	4°44.0′	822 (1503)	
C18	35°39.2′	4°15.0′	825 (1508)	
C19	35°42.2′	4°15.6′	823 (1504)	
C20	35°29.7′	4°55.1′	770 (1407)	
S1	35°31.7′	3°49.0′	138 (252)	
S2	35°32.6′	3°46.6′	47 (86)	
S3	36°00.5′	2°51.8′	53 (97)	
C30	36°03.5′	5°20.3′	370 (676)	
C36	35°59.9′	5°19.2′	478 (876)	
C37	35°58.1′	5°13.0′	440 (804)	
C38	35°50.9′	5°11.2′	240 (439)	
C40	35°57.8′	5°23.4′	456 (834)	
		III. Dredge Locations		
D1	35°26.0′	4°15.7′	164 (300)	
D4	35°25.5′	4°22.2′	128 (234)	
D6	35°46.0′	3°56.0′	653 (1194)	
D7	35°48.0′	3°59.0′	468 (855)	

*Joides Core 121 penetrated 867 meters

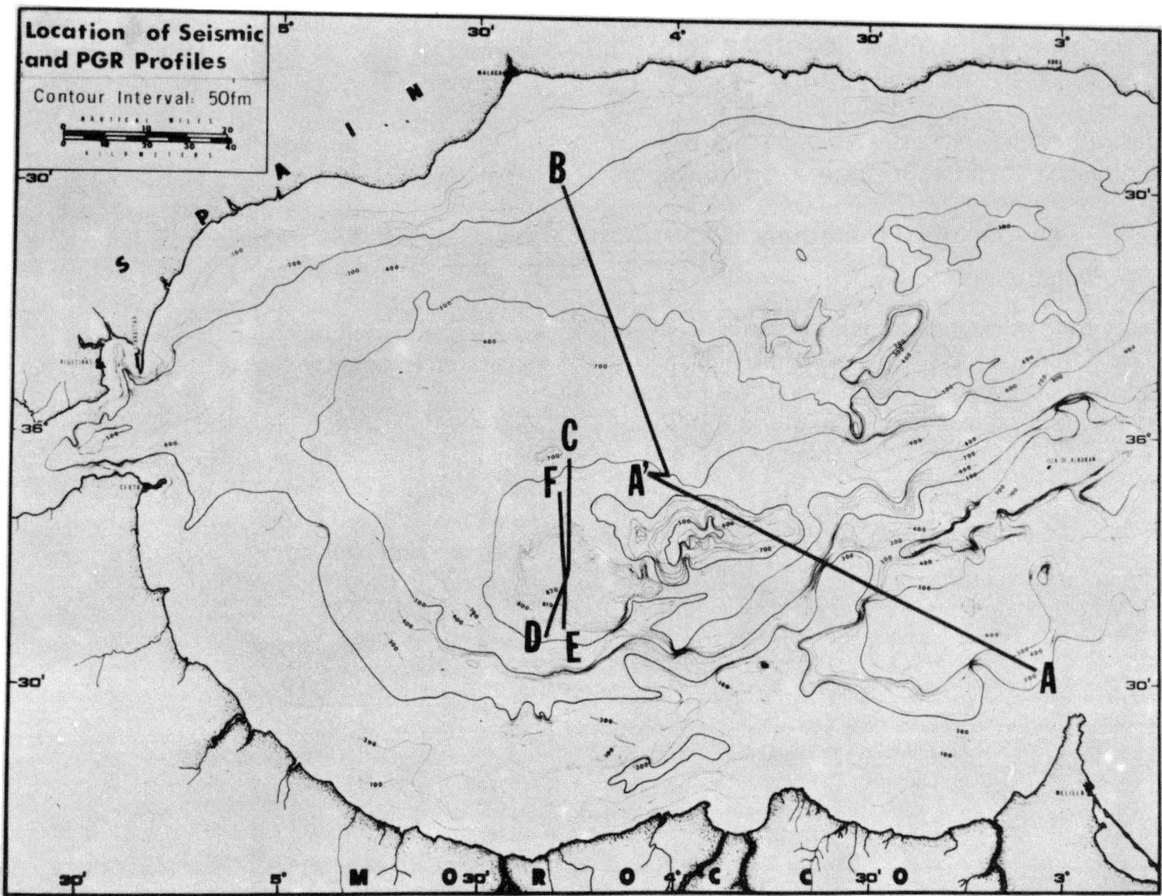

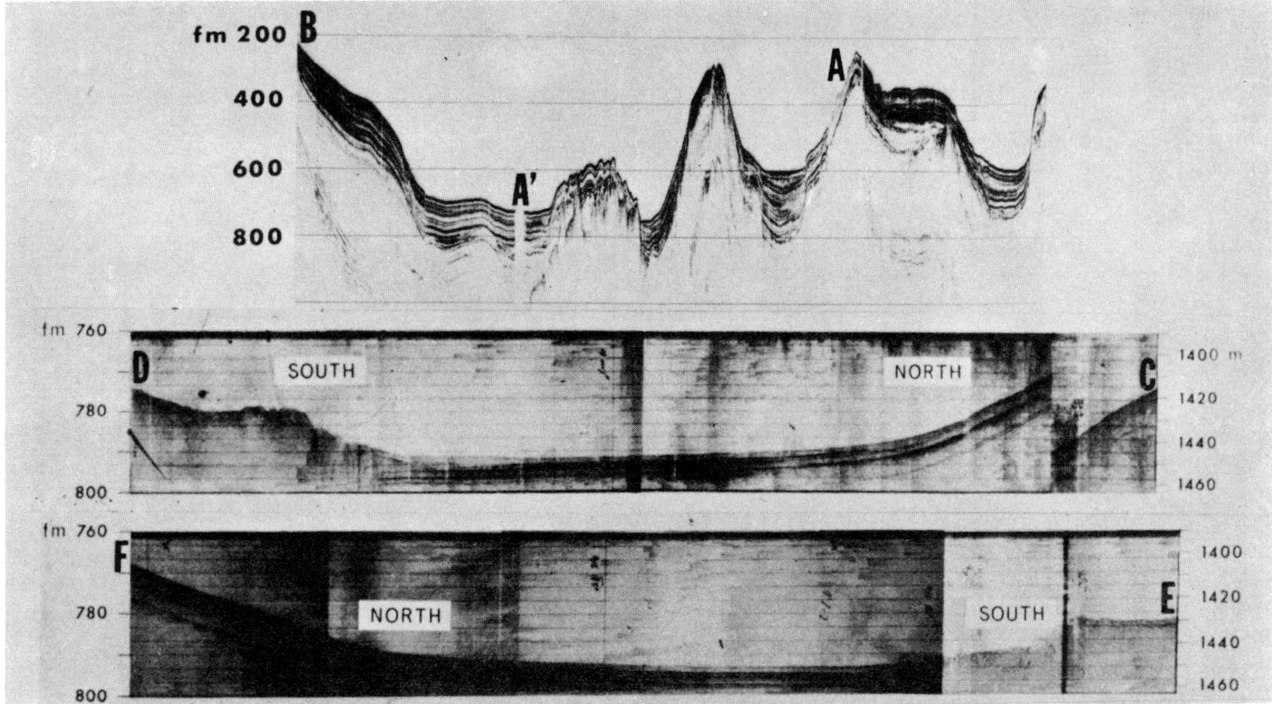

Figure 5. Seismic sub-bottom reflection and PGR profiles across sectors of the Alboran Sea. Profile A-B, seismic profile made during cruise R/V CONRAD 9 (1965) showing cyclic nature of sedimentary sequences and ponding, particularly in the Eastern Alboran Basin (profile courtesy of W. B. F. Ryan). PGR profiles C-D and E-F made during R/V MARIA PAOLINA G. cruise, January 1969, across the Western Alboran Basin plain show the continuous nature of reflectors and ponding (profiles courtesy of C. E. Gehin, SACLANT ASW Research Centre, La Spezia).

by the R/V ATLANTIC SEAL cruise in 1967; (d) shipboard notations and analyses of JOIDES core 121 were provided by the Leg scientific party; and (e) approximately one hundred samples were collected along the southern coast of Spain between Gibraltar and Adra for provenance determination.

Laboratory Techniques

X-ray Radiography

Twenty-four undisturbed sphincter cores were selected for X-radiography. X-rays were made of split 13 cm-wide cores using an industrial Norelco X-ray camera, and a modified standard technique was followed (Hamblin, 1962; Bouma, 1964; Huang and Goodell, 1970).

Textural and Compositional Analyses

Sedimentary structures, as revealed in radiographs, are related closely to texture and compostion. Samples were selected in those portions of the core where prominent changes in the structures were noted. Five cores (Table 2) were selected for detailed analysis and samples were taken from these at 60 cm intervals where the core appears homogeneous.

All samples were washed to remove salt content, and passed through a 62μ sieve with a Calgon dispersive agent. The samples from sand layers were subjected to size analysis by a modified Woods Hole Rapid Sediment Analyzer settling tube (for the sand fraction) and pipetting the fine fraction. Textural data (after Inman, 1952) were obtained using computer analyses.

Mineralogical data were obtained by X-ray diffraction and optical microscopy. Clay mineral analysis and identification (after Warshaw and Roy, 1961) was based on X-ray diffraction of the less than 2μ oriented clay aggregates and their response to glycolation. The semi-quantitative analysis was based on the product of the basal peak height and the

Table 2. Composition of sediment components in the coarse fractions of mud layers.

Sample No.[1]	Depth from top of core (cm)	Planktonic Foraminifera with Hydrotroilite	Planktonic Foraminifera without Hydrotroilite	Benthonic Foraminifera	Carbonate Debris	Plant Remains	Quartz	Heavy Minerals	Others
107–1	20	90.7	0	0.7	1.7	0.7	4.0	1.3	1.0
107–2	60	96.7	0	0.3	1.0	0.3	0.7	0.3	0.7
107–3	120	91.3	2.0	0	2.7	0.3	2.3	0.7	0.7
107–4	180	87.0	2.7	0.7	2.3	0.7	3.3	1.7	2.0
107–5	240	67.0	8.7	1.0	11.0*	1.3	7.3	2.3	1.7
107–3p	257	34.0	9.7	1.3	19.0*	2.7	20.3	3.3	2.0
107–3t	274	29.3	6.0	0.7	23.7*	2.3	30.3	4.7	2.3
107–3b	296	56.7	11.3	0.7	10.7	1.3	16.0	2.3	1.0
107–6	340	53.7	18.0	1.3	14.0	1.0	10.7	1.3	0.7
107–7	400	29.3	15.0	1.3	19.3*	2.3	28.0	3.7	1.0
107–1p	464	22.7	9.3	1.7	22.7*	1.3	34.0	5.0	5.3†
107–1s**	20	60.3	0	0	6.0	1.0	27.7	4.0	1.0
107–3s**	120	60.7	0	0	13.0	0.7	24.3	3.7	1.0
107–5s**	180	17.3	3.7	0	27.3	0.7	41.7	8.3	1.0
107–7s**	400	7.3	3.3	0	31.0	0.7	49.3	10.7	1.0
106–1	0	91.3	0	0.7	4.0	0.7	2.7	0.7	0.7
106–2	60	96.0	0	0.3	1.0	0.7	1.0	0.3	0.7
106–3	120	96.3	0	0.3	0.7	0.7	1.0	0.3	0.7
106–4	180	90.0	0	0.3	2.7	1.0	4.0	1.3	0.7
106–5	240	69.0	7.7	0.3	4.0	1.3	12.7	4.0	0.7
106–6	300	65.7	9.3	0.3	11.7*	1.0	8.7	2.7	0.7
106–7	360	69.3	12.3	0.7	10.7*	1.7	4.0	1.7	0.7
106–8	420	40.0	40.0	1.0	14.7*	1.7	2.7	1.0	0.7
106–9	498	43.7	14.7	0.7	22.3*	1.0	12.0	4.7	1.0
106–1s**	0	52.7	0	0	14.0	1.0	24.0	4.0	1.0
106–3s**	120	55.7	0	0	21.3	0.7	18.0	3.3	1.0
106–5s**	240	30.7	5.7	0	27.3	2.0	26.0	7.0	1.3
106–7s**	360	10.0	5.0	0	33.3	2.7	38.3	9.3	1.3
106–9s**	498	5.3	3.0	0	36.7	2.3	41.3	10.0	1.3

Table 2. (Cont.)

Sample No.[1]	Depth from top of core (cm)	Planktonic Foraminifera with Hydrotoilite	Planktonic Foraminifera without Hydrotoilite	Benthonic Foraminifera	Carbonate Debris	Plant Remains	Quartz	Heavy Minerals	Others
111–1	0	93.3	0	0.7	2.0	1.0	1.3	0.7	1.0
111–2	60	94.3	0	0.7	2.3	0.7	0.7	0.3	1.0
111–3	120	90.3	0	1.0	4.0	1.0	2.0	0.7	1.0
111–4	180	81.0	4.0	0.3	8.0	1.3	4.0	1.0	1.0
111–5	240	67.0	12.7	0.7	10.7	1.0	6.0	1.0	1.0
111–6	300	58.7	11.3	0.7	16.0	1.0	9.7	1.3	1.0
111–7	360	50.0	22.3	1.0	18.0	1.3	6.7	1.0	0.7
111–8	420	40.0	16.0	1.3	24.0*	2.0	14.0	1.3	0.7
111–9	480	45.0	15.0	0.7	27.7*	1.0	9.0	0.7	0.7
111–10	537	47.0	16.0	0.7	30.0*	0.7	4.3	0.3	1.3
101–1	0	88.0	0.7	0.7	8.3	1.0	0.7	0	0.7
101–2	60	90.7	0	0.3	6.7	0.7	1.3	0	0.3
101–3	120	90.7	0	0.3	7.7	0.3	0	0	1.0
101–4	180	71.3	23.7	1.0	3.3	0	0	0	0.7
101–5	240	53.0	18.0	1.3	24.0*	0.7	2.7	0	0.7
101–6	300	59.7	19.7	1.0	9.0	0.7	8.0	1.3	0.7
101–4p	320	55.7	14.0	1.7	7.3	2.0	19.3	2.3	1.0
101–4b	335	63.0	14.7	0.7	13.3	0.7	6.0	1.0	0.7
101–7	360	60.0	29.7	0.7	8.3	0	1.7	0	0.7
101–8	420	65.0	21.0	1.0	11.0	0	1.0	0	0.3
101–9	480	33.7	30.0	1.3	19.3*	1.3	12.7	0.7	1.0
101–10	540	48.0	12.0	1.0	30.7*	1.0	5.3	0.7	0.7
101–11	598	37.3	29.3	1.3	17.7*	1.3	11.7	0.7	0.7
95–1	0	96.0	0	0.3	2.3	0.7	0	0	0.7
95–2	60	90.0	0	0.7	3.3	1.0	2.3	0.7	2.0
95–3	120	94.3	1.7	0	2.7	0.3	0.7	0.3	0
95–4	180	94.7	2.0	0	3.7	0.3	0.7	0.3	1.7
95–5	240	70.0	22.0	0	4.0	0	0.7	0	0
95–6	300	57.0	18.0	1.0	17.3	0	5.0	0.7	1.0
95–6p	306	53.7	19.0	0.3	18.3	0.7	7.0	0.3	0.7
95–6b	318	58.0	18.0	0.3	14.3	0.3	8.7	0.7	0.7
95–7	360	42.0	21.3	0.3	24.0*	1.0	9.7	0.7	1.0
95–8	420	38.7	34.0	0.7	20.3*	0.3	5.3	0	0.7
95–9	480	39.0	31.0	0.3	24.0*	0.7	4.0	0	0.7
95–10	540	40.0	38.3	0.7	17.3*	0.7	2.3	0	0.7
95–11	600	53.7	18.0	0.3	23.7*	0.3	1.0	0	0.7
95–12	698	52.0	10.7	1.3	21.3*	2.7	12.3	3.3	0.7
C14		93.3	0	0.3	4.0	0.7	0.7	0.3	0.7
C15		95.7	0	0	2.7	0.3	1.0	0	0.3
C16		96.3	0	0	2.0	0.7	0.7	0	0.3
C17		93.7	0	0	3.7	0.3	1.3	0.3	0.7
C18		92.7	0	0.7	3.3	1.0	1.7	0	0.3
C19		92.7	0	0	4.7	1.3	0.7	0	0.7
C20		94.3	0	0.3	2.7	0.7	1.3	0.3	0.3

* Abundant molluscan shell hash
† Volcanic glass shard
** Coarse silt fraction
[1] First part of sample number refers to the core number shown in Figure 2; second part of sample designation refers to the core section sampled.

In addition, sand samples were pulverized for carbonate content determination. The magnesium content of calcite was estimated by the shift of the (101) calcite line (Chave, 1952). The intensity ratios, or relative amounts, of aragonite, low-Mg calcite and high-Mg calcite were computed using the method developed by Huang and Pierce (1971).

The major mineralogical components of the sand grades and some coarse silt fractions were determined by 300 grain counts. The relative percentages of heavy minerals in sand layers (62–250μ) were identified using 200 grain counts. The heavy minerals of the silt fractions were identified from powder diffractograms.

C^{14} Dating

A total of 21 core samples were dated by standard carbon-14 radiometric methods in order to determine sedimentation rates. Samples were directly hydrolyzed with acid to produce CO_2 prior to dating. The coarse fraction ($>62\mu$) of the mud sections just above and below the sand and silt key stratigraphic beds were dated. It was necessary to sieve about a 10 to 20 cm core section of mud in order to obtain enough carbonate shell for dating.

BOTTOM FEATURES NOTED ON PHOTOGRAPHS

Photographs made of the flat sea floor in the Western Alboran Basin plain and adjacent area (Cruise USCGC ROCKAWAY $RoSm_1$) show that organisms and their burrows, tracks, and mounds are the most prominent features on the muddy bottom (Figure 6). The presence of organisms was recorded at five stations on the basin plain (C-15 to C-19 on Figure 2) at a depth of about 1500 m, and at one station (C-14)

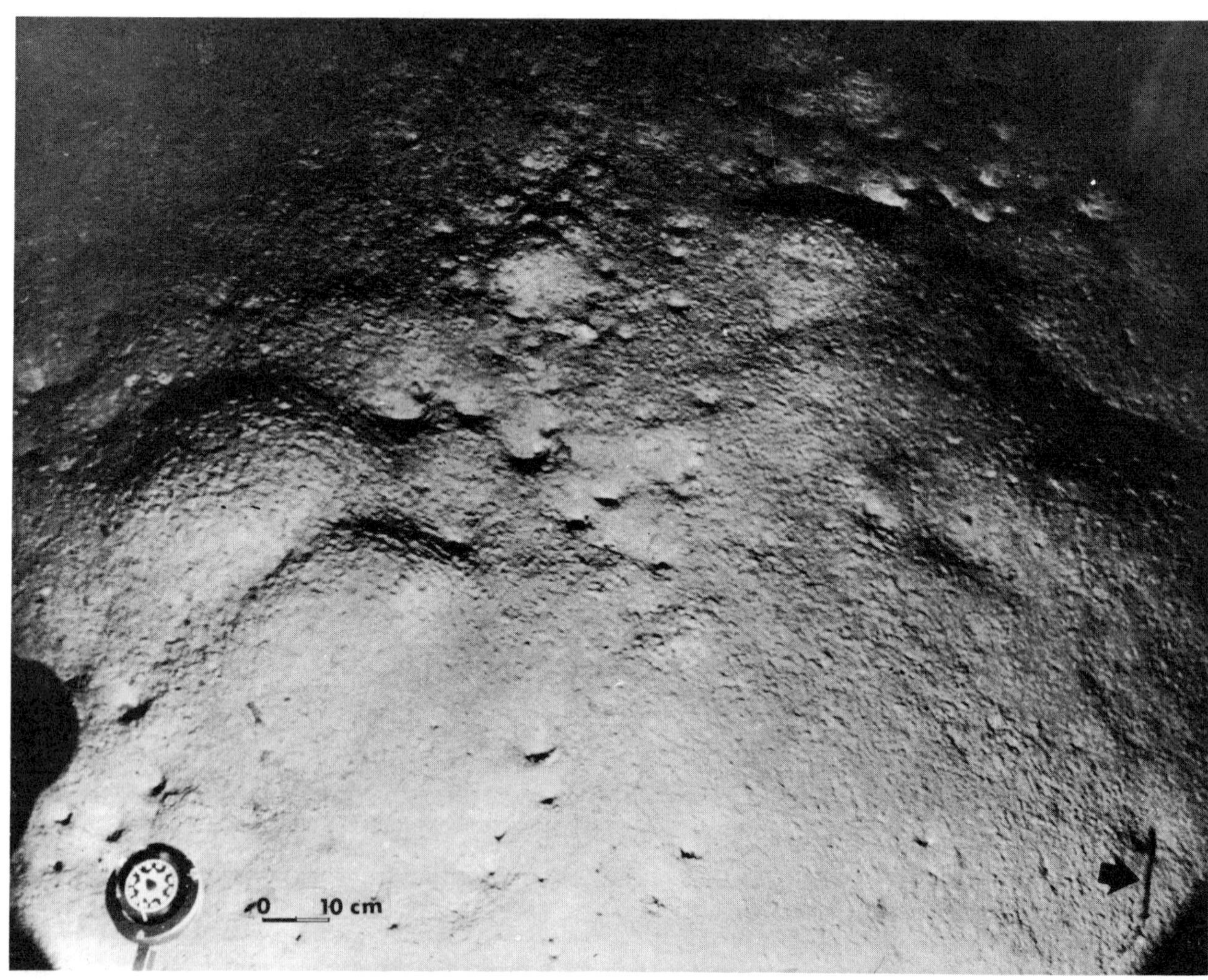

Figure 6. Bottom photograph (Station C-17, approximately 1500 m, area covered: 8.7 m²) showing typical bioturbated mud bottom in the Western Alboran Basin. Large and small mounds and depressions produced by organisms such as deep-stars cover the sea floor. Note worm track (arrow) in lower right part of photo. Entire surface is pitted and presents a granular texture (possible faecal pellets).

on the Alboran Knolls (Table 3). These bathyal and abyssal forms include brittle stars and starfish, *Galathea* (a deep-sea crustacean) and fishes. However, sessile organisms are dominant. No bottom current markings were noted, and evidence of present current activity, if any, was not recorded. This is in sharp contrast with the current swept surfaces of Xauen Bank and Alboran Ridges (Milliman *et al.*, this volume) to the south, and the Spanish margin near the Strait of Gibraltar (Kelling and Stanley, this volume).

The photographic survey shows that the entire muddy floor of the basin plain is apparently reworked by organisms. The main bottom features include large and small mounds, depressions, worm tubes, and deep star imprints (Figure 6). A comparison of these features noted on 75 photographs at the 5 basin plain stations indicates no significant regional difference in the frequency of markings within the basin plain proper (Table 3). The distribution pattern of organisms that produced these features is thus believed to be reasonably uniform throughout the basin. Small depressions, however, were more abundant on the Alboran Knoll. Photographs at Station C-20 (base-of-slope) at a shallower depth (1100 m), about 5 n mi southeast of the basin plain, show a lower density of bioturbation features.

Table 3. Distribution of the bottom features on photographs collected at six bottom camera stations (see Figure 2).[a]

	Alboran Knolls	Western Alboran Basin plain				
	C14	C15	C16	C17	C18	C19
No. of Photographs	17	17	18	18	10	12
Large mound:						
with crater	1.00	0.59	0.67	0.33	0.30	0.83
without crater	0.23	1.29	2.28	2.78	0.90	0.75
Small mound:						
with crater	1.06	1.12	1.33	1.28	1.90	1.58
without crater	2.29	2.47	3.00	1.78	3.20	0.08
Depression:						
large	5.06	4.65	2.39	3.55	4.80	3.25
small	14.12	4.12	9.72	7.05	4.70	4.75
elongate	0.06	0.18	0.61	0.28	0.30	0.16
Worm tube	0.06	0.18	0.05	0.17	0.30	0
Deep-star imprint	0.18	0.12	0.11	0.22	0.20	0.08
Galathea sp.	0	0.06	0	0	0.10	0
Benthic fish	0	0	0.05	0.05	0	0

[a] Numbers refer to the average number of bottom features noted per photograph.

STRATIGRAPHY AND LITHOLOGY OF THE ALBORAN CORES

A PGR survey in the Western Alboran Basin (Stanley *et al.*, 1970; Gehin *et al.*, 1971) shows the remarkable lateral continuity of sub-bottom reflectors in the upper 10 fm (18 m) of section. Used as isochrons, the sub-bottom reflectors show progressive thickening of the strata from the base-of-slope to the Western Alboran Basin plain (Figure 5, profiles C-D, E-F). There are at least 4 mappable reflectors at depths of

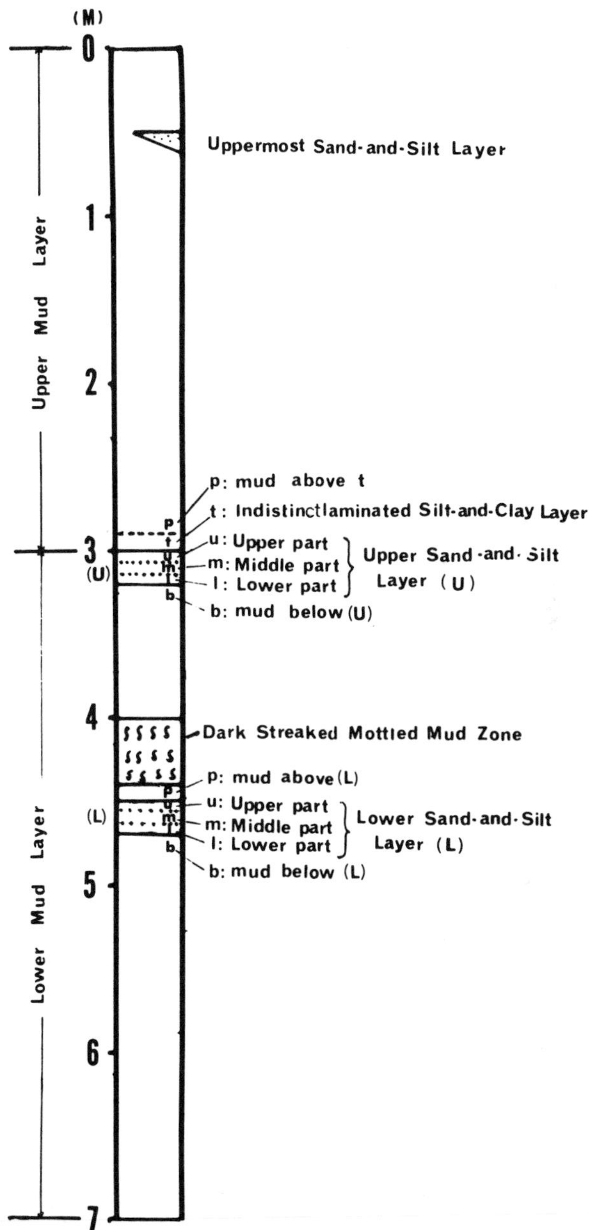

Figure 7. Generalized stratigraphic sequence of a typical 7 m long core in the Western Alboran Basin plain (this sequence is based on analyses of 29 cores ranging from 4 to 7 m in length).

about 0.4, 1.5, 2.5 and 4.5 fm (0.74, 2,74, 4.57 and 8.23 m) respectively in the sub-bottom of the Basin plain. Correlation of these reflectors is detailed in Gehin *et al.* (1971) who refer to the upper three reflectors as turbidites "a", "b", and "c". More than 49 sphincter cores (coring method detailed in Kermabon *et al.*, 1966) and other cores collected in the western Alboran Sea serve to establish the lithology of the uppermost 7 m of sedimentary section seen in the PGR profiles.

Generally at least 95 percent of the core-length consists of hemipelagic mud (mixture of terrigenous and pelagic components) with only a few thin sand and silt layers in addition. The three upper continuous reflectors noted in PGR profiles are correlated with three thin sand and silt layers retrieved in cores. Both PGR records and cores show that these layers extend continuously across most of the Western Alboran Basin plain. Correlation of reflectors is limited to the basin plain: it is difficult to correlate the thin sand layers of the basin plain with the sands found occasionally in slope cores. The basin plain sequence is as follows (Figure 7): (1) The *Lower-sand-and-silt-layer* (L) occurs at about 450 cm depth; it consists of both terrigenous and bioclastic components. (2) The *Upper sand-and-silt layer* (U) is found at about 300 cm depth, and is composed mainly of whole planktonic foraminiferal tests. (3) The *Uppermost sand-and-silt layer* occurs only sporadically at a depth of about 50 cm; it also contains mostly foraminiferal tests. Samples were collected from upper and lower portions of each sand-and-silt unit. Such samples are designated with symbol u and 1 (Figure 7 and Table 4). Where a sand layer is

Table 4. Percentages of sediment components in the coarse fractions of sand and silt layers.

Sample No.[1]	Depth from Top of Core (cm)	Planktonic Foramin-ifera	Benthonic Foramin-ifera	Carbonate Debris	Plant Remains	Eolian Quartz	Non-Eolian Quartz	Feldspar	Heavy Minerals	Volcanic Glass	Others
I. Upper sand-and-silt layer											
86–5u	328	92.5	2.0	3.0	1.0	2.0	0	0	0.5	0	1.0
86–51	331	70.5	2.5	4.0	6.0	9.0	0	0.5	0.5	0	2.0
87–5u	292	76.2	2.3	11.6	7.0	4.6	0	0	0	0	0
87–5m	294	88.5	0.5	8.5	2.0	3.5	0	0	1.0	0	1.0
87–51	299	90.7	0.7	6.7	0	0	0	0	0	0	0
88–5u	269	56.3	2.0	22.3	6.7	12.0	0	0	0.7	0	0
88–5m	272	70.0	0.7	13.0	8.3	6.3	0	0	0.7	0	1.0
88–51	275	73.7	1.0	17.3	2.3	5.0	0	0	0	0	0.7
89–5u	259	61.0	0.7	30.0	3.0	1.3	0	0	0	0	0
89–51	262	36.0	1.3	21.7	1.0	1.3	0	0	0	0	0.7
91–6u	219	75.0	1.5	20.5	0	0	0	0	0	1.0	1.0
91–61	220	65.0	2.0	27.7	0	1.7	0	0	0	0.6	0.7
92–6u	346	51.0	1.0	38.0	3.7	3.7	0	0.3	1.3	1.0	0
92–61	349	72.0	3.3	20.7	1.7	1.3	0	0	0	0	0.3
93–6u	323	58.0	0.7	35.0	2.7	3.3	0	0.3	0	0	0.3
93–61	326	73.0	1.0	19.7	2.3	3.7	0	0	0	0	0
94–4u	312	58.7	2.0	29.0	4.3	5.0	0	0	0.3	0	0.7
94–4m	315	64.7	1.7	31.3	0.3	0.7	0	0	0.3	0	0.3
94–41	318	91.3	8.7	0	0	0	0	0	0	0	0
95–6u	312	73.3	0	13.3	5.7	7.0	0	0	0.7	0	0
95–61	315	76.7	0.3	10.7	1.3	9.0	0	0	1.3	0	0.7
96–5u	312	56.0	1.0	35.0	1.0	5.3	0	0	0	0	1.0
96–51	315	76.7	0.3	30.7	0	1.0	0	0	0	0	0.7
97–6u	308	63.3	0	27.0	3.0	5.0	0	0.3	0.7	0	0.7
97–6m	310	41.0	0	32.7	1.3	15.3	1.3	0.3	7.3	0	0
97–61	312	82.3	0.3	15.3	0.3	1.0	0	0	0	0	0.7
98–6u	298	6.7	0	28.3	3.3	51.7	3.3	2.3	3.3	0	1.0
98–61	301	65.3	1.0	29.3	0.7	2.7	0	0	0.3	0	0.7
102–5u	271	64.7	0.7	7.7	2.7	19.7	0.3	0.3	2.7	0	1.0
102–51	273	57.3	1.0	25.3	4.0	7.7	0	0	2.0	0	1.7

Table 4 (Cont.)

Sample No.[1]	Depth from Top of Core (cm)	Planktonic Foramin-ifera	Benthonic Foramin-ifera	Carbonate Debris	Plant Remains	Eolian Quartz	Non-Eolian Quartz	Feldspar	Heavy Minerals	Volcanic Glass	Others
103–5 u	265	45.3	1.7	27.3	4.0	16.0	0	0	3.3	0	0.7
103–5 m	268	60.0	0.3	19.0	4.0	7.3	0	0	4.0	0	1.7
103–5 l	272	59.0	0.3	25.3	1.3	10.0	0	0	1.3	0	0.3
105–4 u	238	65.3	0.3	19.3	3.0	8.3	0	0	2.7	0	0.3
105–4 l	240	53.3	3.0	36.7	0.3	4.7	0	0	1.3	0	1.0
II. Lower sand-and-silt layer											
90–1 u	466	15.0	11.0	74.3	0	0	0	0	0	0	0
90–1 l	468	20.0	10.3	65.3	0	0	0	0	0	0	0
91–4 u	384	14.7	0	33.3	46.7	0.3	0	0	2.0	0	1.7
91–4 l	386	31.0	0	26.7	37.3	1.3	0	0	1.0	0	1.0
91–3 u	547	18.3	0.7	26.3	13.0	25.3	1.0	1.7	14.0	0	0
91–3 l	540	33.0	0.7	24.7	27.3	8.3	1.0	0	4.7	0	0.3
92–3 u	532	3.3	3.0	12.3	4.3	46.0	17.0	2.0	14.0	0	1.0
92–3 m	540	6.0	0	39.3	15.3	20.7	13.3	0.7	4.3	0	0.3
92–3 l	547	2.0	0	15.3	2.3	49.3	10.3	3.3	16.0	0	0
93–3 u	509	24.3	0.3	30.0	11.0	29.0	1.3	0	4.0	0	0
93–3 l	510	16.7	0.3	34.0	1.3	38.7	1.0	1.0	6.0	0	0.7
97–4 u	477	19.3	0	17.3	36.3	17.7	4.0	0	5.0	0	0.3
97–4 l	480	24.7	0.3	31.3	10.7	23.7	4.3	0	2.7	0	0.3
98–4 u	455	28.3	0.3	14.7	37.7	16.7	0.3	0	1.3	0	0.3
98–4 l	458	20.3	0.7	18.0	4.0	42.7	3.0	3.7	4.0	0	0.7
101–4 u	323	23.7	0	15.0	28.3	26.7	2.3	1.0	2.3	0	0.3
101–4 m	326	19.3	0.7	14.3	1.3	46.7	4.0	3.3	6.7	0	0.3
101–4 l	329	73.7	0.7	16.0	0.7	6.7	0	0	1.3	0	1.0
107–3 u	280	26.3	0.3	13.3	3.3	41.7	3.3	2.7	7.3	0	1.7
107–3 l	285	26.7	1.0	27.0	1.0	35.0	2.7	1.0	5.0	0	0.7
III. Graded sand layer											
107–1 u	474	3.3	1.3	33.3	1.3	36.3	12.7	2.0	7.0	3.0	0.3
107–1 m	476	0.3	1.3	20.7	0.7	34.7	10.0	3.3	11.3	4.3	0.3
107–1 l	479	1.0	0.3	30.0	3.3	32.0	11.7	2.3	13.3	4.7	0.3
108–1 u	265	9.3	0.7	34.0	5.3	36.3	5.0	3.0	5.0	0.7	1.7
108–1 m	272	0.7	2.3	26.0	1.1	40.7	16.0	3.0	7.7	1.7	0.3
108–1 l	280	0.3	0.7	25.0	0.7	36.7	18.3	0.7	14.7	2.0	0.3
IV. Sediments in shallower areas											
231–1		8.3	0.7	23.7	1.0	47.3	7.0	2.7	3.3	3.3	2.7
231–2		3.7	1.0	20.7	3.7	52.3	7.7	1.7	4.7	2.7	2.0
231–3		6.7	0.7	17.7	2.3	59.3	10.7	3.3	6.0	4.0	1.0
C30		6.0	1.3	19.3	2.0	46.0	7.3	3.7	10.7	2.3	1.0
C36		7.0	1.0	27.3	1.0	34.0	9.3	4.0	12.7	5.3	0.7
C37		5.0	1.3	31.0	1.3	30.7	11.7	3.0	9.7	5.7	0.7
C38		8.0	0.7	27.0	2.0	33.3	10.0	1.7	13.3	3.3	0.3
D1		16.0	4.3	70.3	0	4.0	0	0	1.3	0	0.7
D4		4.0	1.3	93.3	0	0.7	0	0	0	0	0.7
D6		95.0	0.7	1.0	0.7	2.0	0	0	0	0	0.7
D7		95.3	0.3	2.0	0.7	1.0	0	0	0	0	0.7
S1		55.7	2.7	37.3	2.7	1.3	0	0	0	0	0.3
S2		7.3	6.0	85.7	1.0	0.7	0	0	0	0	0.3

[1]First part of sample number refers to the core number shown in Figure 2; second part of sample designation refers to the core section sampled.

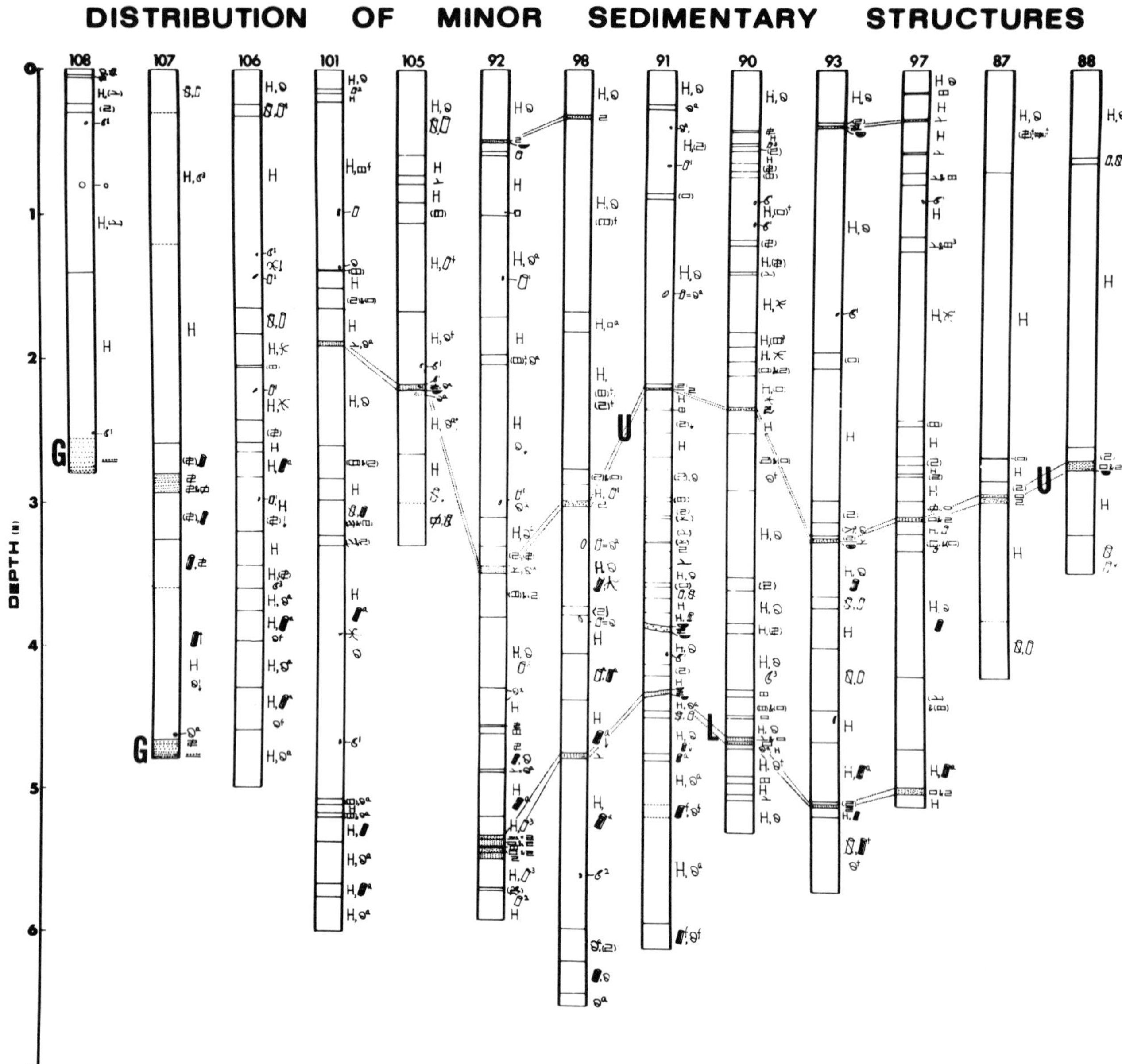

Figure 8. Distribution of minor sedimentary structures in Alboran Sea cores. U: Upper sand-and-silt layer; L: Lower sand-and-silt layer; G: Graded Sand layer. Position of cores shown on Figure 2.

thicker than 2 cm, the middle portion of the sand layer was sampled (symbol m is used in such cases). All sand samples taken were about 0.5 cm thick. In most cores the Upper sand-and-silt layer lies below an indistinct laminated silt and clay layer; a sample of this layer was also collected (identified with the symbol t). Samples were also collected from the overlying (symbol p) and the underlying (symbol b) mud layers for comparison.

Based on the mineralogy of the coarser fractions ($>62\mu$), the mud sections can be divided into *Lower* and *Upper mud layers* with a boundary at about 300 cm core-depth; this boundary coincides closely with the Upper sand-and-silt layer (U). The coarse fraction of the Upper mud layer consists almost entirely of foraminiferal tests and that of the lower mud layer is mainly composed of terrigenous and bioclastic components. A dark-streaked, mottled layer, about 50 to 70 cm thick, occurs at about 400 to 500 cm within the Lower mud layer and consists of hydrotroilite and organic matter.

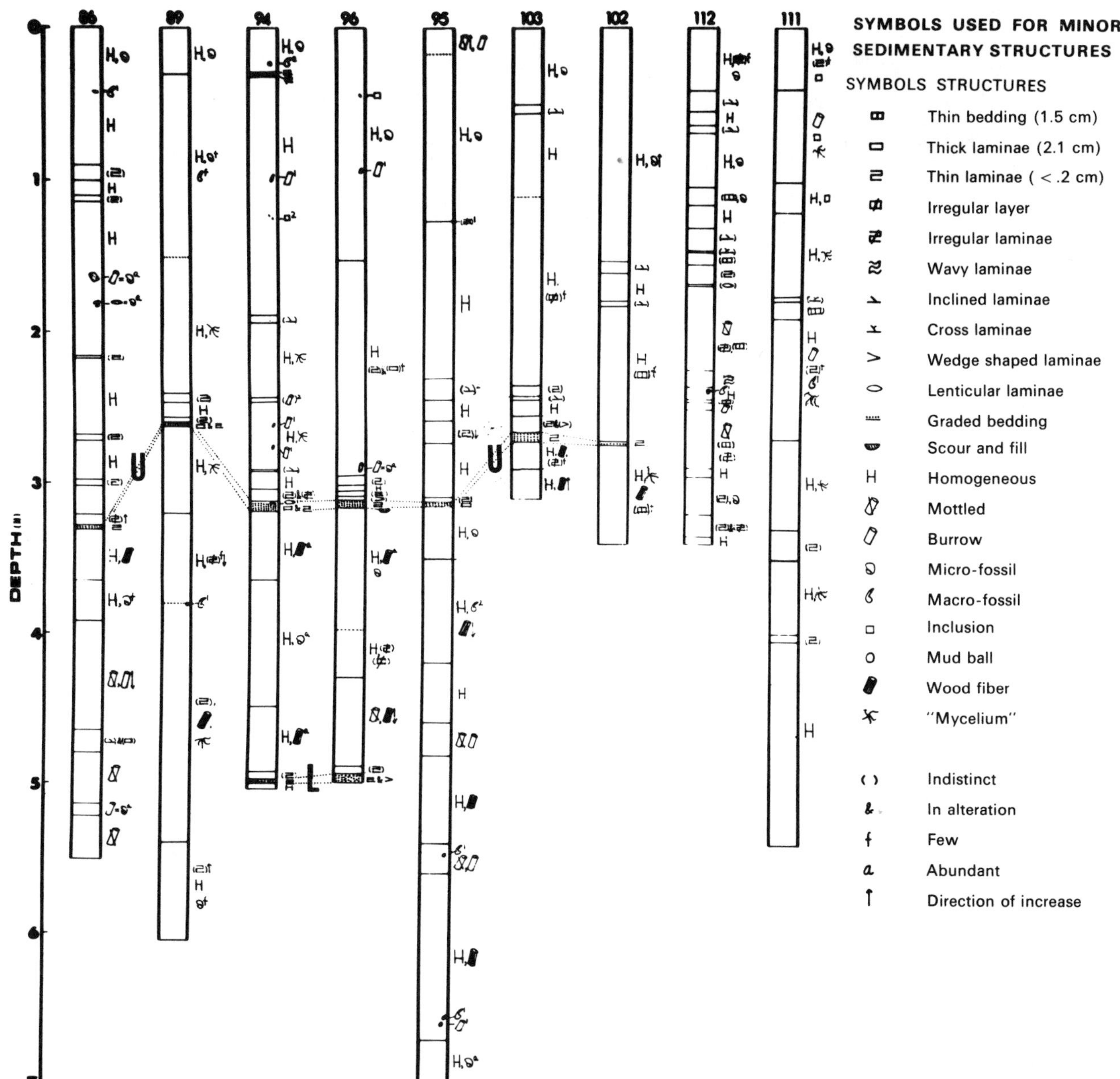

Figure 8. (cont.)

MINOR SEDIMENTARY STRUCTURES

Nature and Distribution

The distribution of internal sedimentary structures in 22 of the 29 spincter cores as revealed by the X-radiographs is presented in Figure 8. A general description of these structures and the possible processes that produced them is outlined by Huang and Goodell (1970). Actual examples of the more important structures as noted in X-radiographs are shown in Figure 9.

The most important sedimentary type in the basin is a mud that appears homogeneous ("structureless") even in X-radiographs. Of secondary importance are bioturbated (Figure 9, I) and laminated (Figure 9, A, E) muds. These three types account for more than 85% of all basin sediments. Sand-and-silt layers that extend across the lower slope and basin plain (Figure 5) are characterized by distinct horizontal laminae (Figure 9, A, B), cross-laminae (Figure 9, F) and scour-and-fill (Figure 9, B). In some cases, the sand-and-silt layers are bioturbated. The *Uppermost* sand-and-silt layer, at about 0.5 m, is also easily distinguished. Graded sand layers (Figure 9, C) are found in submarine valley and slope (Cores 107, 108) areas (Figure 8, G). Some of the mud sections in the cores appear more structureless than others, for example the Upper mud layer above the Upper sand-and-

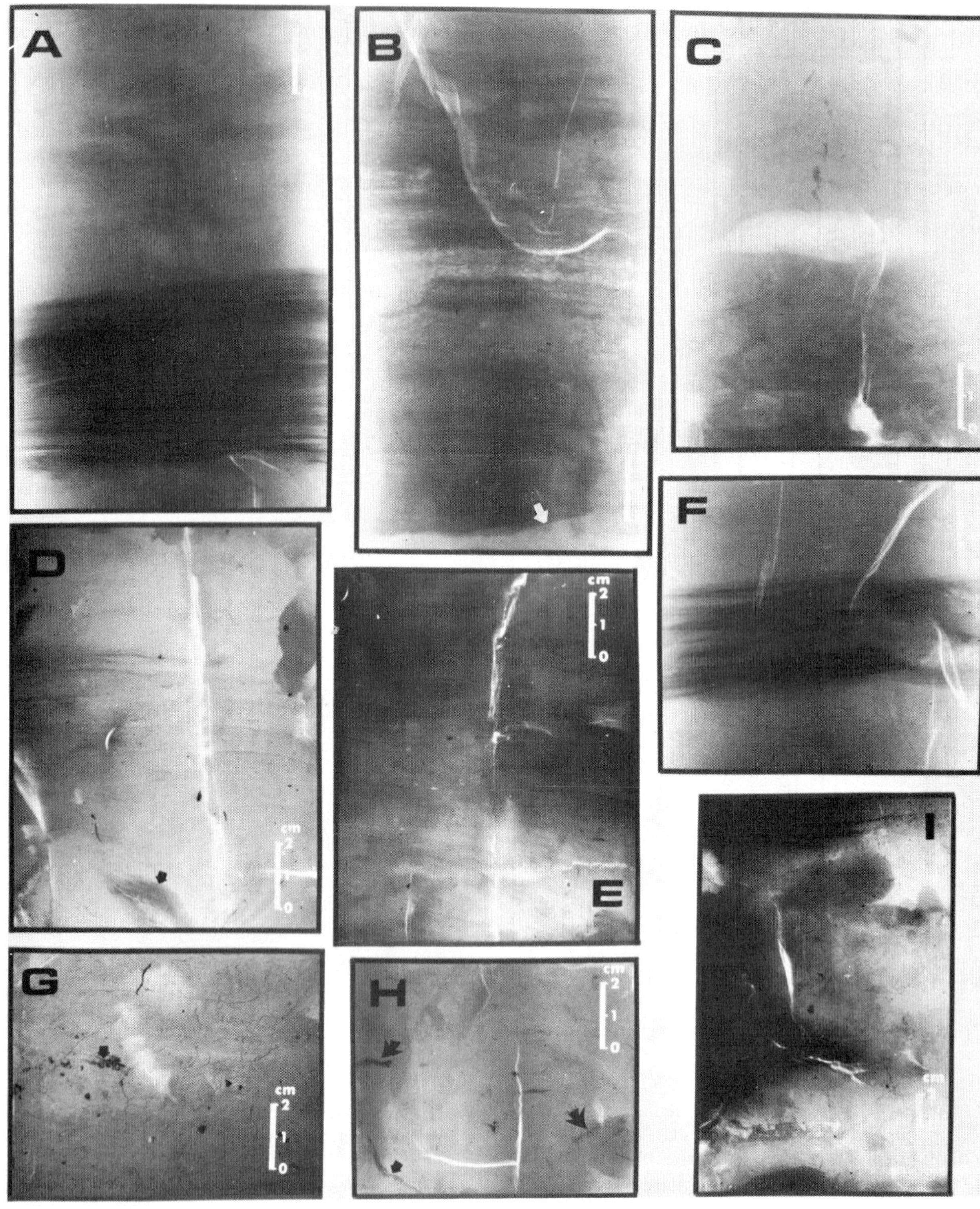

Figure 9. Representative minor sedimentary structures noted on radiograph positives of Alboran Sea cores. A, Distinct thin laminae (sand and silt) covered by an indistinctly laminated mud (silt and clay). B, Well-defined 1 cm thick carbonate sand bed, and alternating thick and thin laminae above erosional (scour-and-fill) surface (arrow). C, Graded sand bed; grading shown by progressive darkening toward base of sand unit. D, Repeated graded bedding in mud (middle part of photo) and burrows (arrow). E, Inclined and wedge-shaped laminae in mud. F, Cross laminated sand-and-silt layer (U) in homogeneous mud. G, *Mycelium* (arrow, dark filaments extending from a pocket of dissolved foraminiferal tests). H, Wood fibers (arrows) scattered in homogeneous mud. I, Mottles of organic origin and burrows in mud. Dessication cracks appear white in photographs.

silt layer (*e.g.*, Cores 87, 88, 102, 103). In general, the Lower mud layer (below the Upper sand-and-silt layer, U) is more varied and displays laminae, plant fibers and bioturbation.

Texture and Composition Interpreted from Radiographs

In all radiograph positives, coarser grained materials appear darker than finer sediments (Figure 9), so that relative grain size can be determined. Sections displaying indistinct laminae or homogeneous "structureless" units consist mostly of mud.

Radiographs also provide information on composition. For example, concentrations of foraminiferal tests in mud produces a distinct "micro-fossil structure" (Figure 9, G). The regular arrangement of quartz grains, heavy minerals, carbonate debris and carbonaceous fragments (plant remains) in the sand-and-silt layers produces distinct laminae (Figure 9, B). Thin laminations in mud in some cases are produced by the regular arrangment of eolian quartz grains. Somewhat larger dark inclusions may consist of iron nodules, fish teeth or ear bones. Filamentous hair lines, called *mycelium,* may be reprecipitated carbonate produced after dissolution (Figure 9, G).

TEXTURE AND COMPOSITION OF THE MUDS

The hemipelagic mud sections consist mainly of terrigenous clay and silt size material (about 90%) with minor amounts of organic fragments, such as foraminifera and ostracoda. The uppermost 70 to 140 cm section in cores is generally a soft, yellowish, olive gray (5Y 6/4 to 5Y 5/2) oxidized mud. The lower core sections are stiffer, olive greenish gray (5Y 5/1) silty clay. Visible concentrations of carbonate debris, mostly shell material and foraminiferal tests, are common in almost all cores. A darker section of mud some 50 to 70 cm thick occurs at about 4 to 5 m in some cores, mostly those from the Western Alboran Basin plain.

Coarse Fraction

The mud includes approximately 5% of coarse fraction (>62μ) of which 30 to 90% are planktonic foraminifera (Table 2). The foraminifera (mostly of the fine sand grade) thus control the size distribution of the coarse fraction. The amount of foraminifera in two cores (106 and 107) on the slope increases from 1.5% (by weight of the total sample) at the top of the core to slightly over 2% at a depth of about 1 m, and then decreases rapidly down the cores to less than 1% (Figure 10). The actual amount of foraminifera is somewhat higher than shown because fragmented, partially dissolved foraminifera and those smaller than 62μ were not included in the calculations.

The three major components of the coarse fraction are planktonic foraminifera, carbonate (shell fragments, *etc.*) debris and eolian quartz (Figure 10). The accessory components include mostly plant

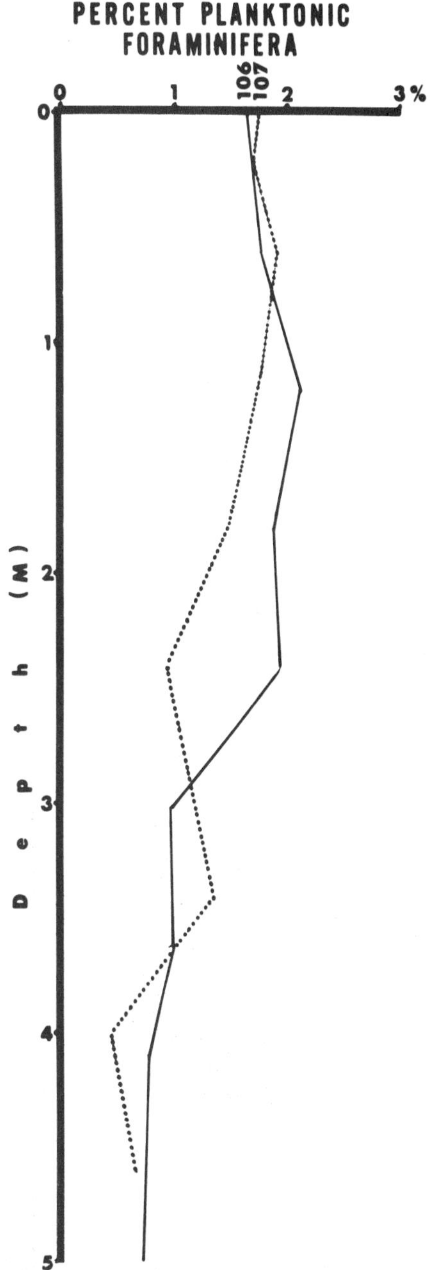

Figure 10. Percentage (by weight) of planktonic foraminiferal tests in the >62μ fraction in two cores (106, 107) collected on the slope northwest of the Western Alboran Basin plain.

remains, benthonic foraminifera, and heavy minerals (Table 2). The relative percentage of planktonic foraminifers generally decreases downward in cores at about 2 to 3 m where there is a sharp reduction of tests. Below this depth carbonate debris and eolian quartz generally increase toward the base of cores.

The assemblage of planktonic foraminifera in the cores is similar to that found by Todd (1958) in the Swedish R/V SKAGERACK Core 210 some 180 nautical miles ENE of the Western Alboran Basin. In five cores examined, two of which are shown in Figure 10, *Globigerina bulloides* and *Globigerinoides ruber* together exceed 50% of the total foraminifera. In all of these cores the number of foraminifera filled with opaque hydrotroilite increases at a depth of about 4 m to 5 m (Figure 11). This depth roughly coincides with the dark, streaked and mottled mud layer (Figure 5) consisting of hydrotroilite and organic matter.

The carbonate debris includes calcareous skeletal

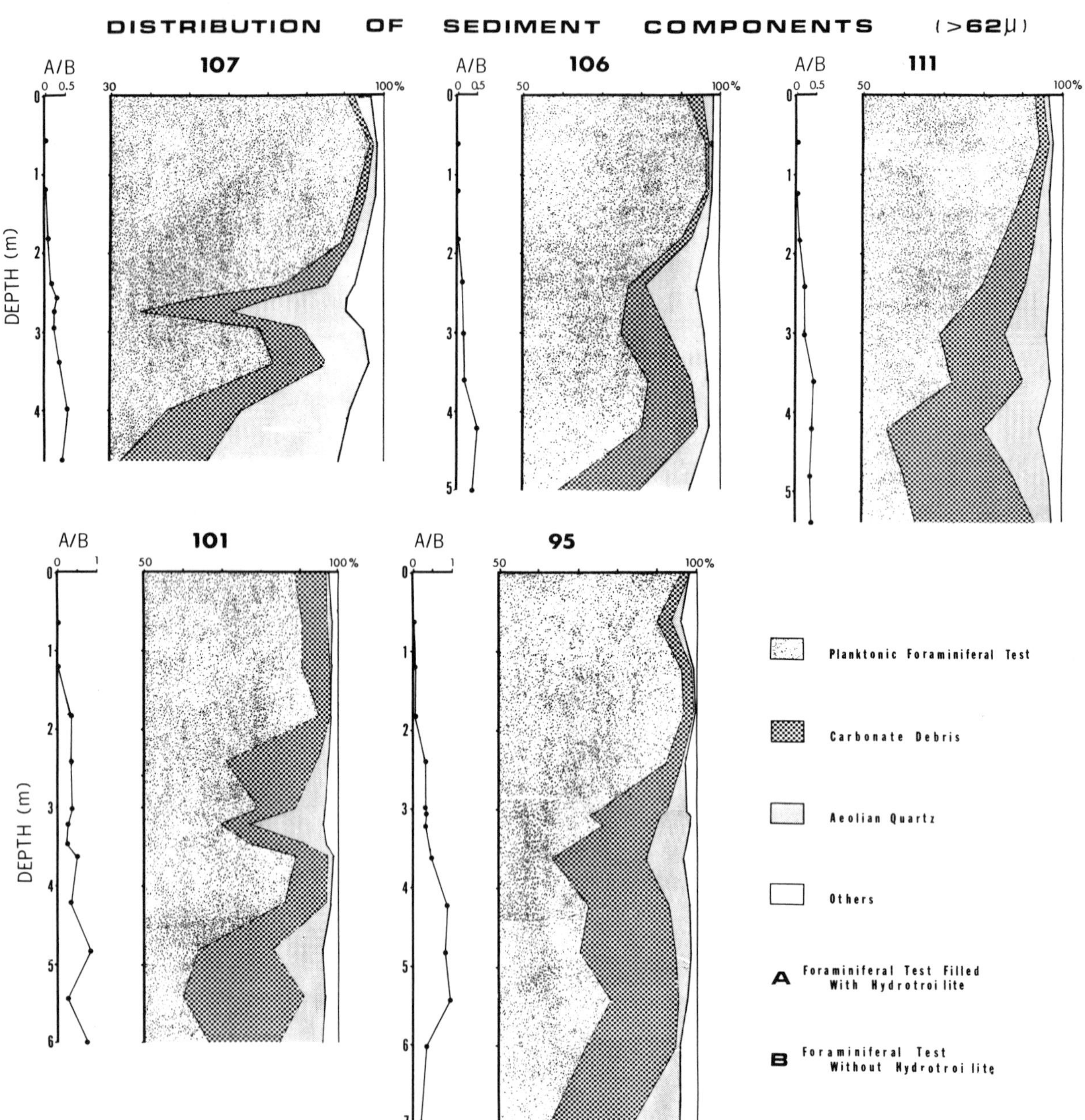

Figure 11. Distribution of major components in the coarse fraction (> 62μ) in Upper and Lower mud layers. Slope cores (106, 107 and 111); Western Alboran Basin plain cores (95, 101). Note the decrease of relative percentages of planktonic foraminifera at a depth of about 2.5 m in slope Core 107 and at 3.5 m in basin plain Core 95.

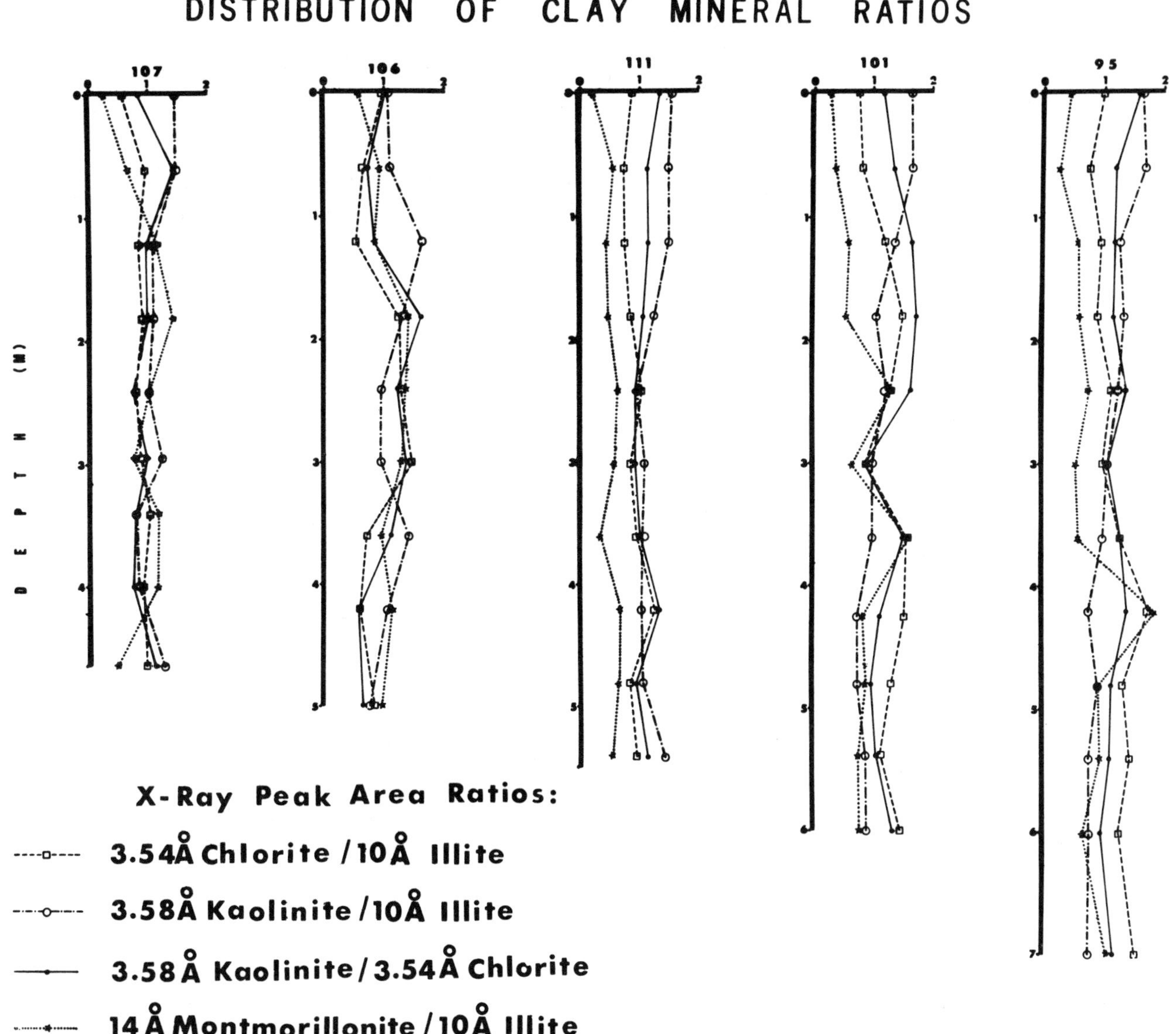

Figure 12. Distribution of clay mineral ratios. Slope cores (106, 107 and 111); Western Alboran Basin plain cores (95, 101). Note that all clay mineral ratios diverge upward from a core depth ranging from below 1 m in the slope Core 107 to about 3 m in the basin plain Core 95.

remains, planktonic and benthonic organisms. The skeletal bioclasts include, in order of importance, fragments of mollusc, bryozoa, coral, algae and ostracods, echinoid spine and sponge spicules. These forms (not identified as to species) are similar to those found on Xauen Bank and adjacent ridges in the Alboran Sea (Milliman *et al.*, this volume). Evidence that they are derived from shallower water is provided by the abraded (reworked) nature and the high content of high-Mg calcite.

Quartz of wind-blown origin in Alboran Sea sediments is usually coated with a reddish brown or orange yellow film of iron oxides, is rounded to subrounded in shape, and displays wind attrition surfaces. A similar type of quartz is described by Eriksson (1961, 1965). In contrast, "non-eolian" quartz is transparent, non-coated and of varied, angular to subrounded shape. This latter type of quartz, subordinate in the mud layers of the basin plain cores, is abundant in slope sediments, especially in graded sand layers (Figure 8, Cores 107 and 108; also Table 4).

Heavy mineral suites in mud are similar to those found in the sand-and-silt layers (Table 5), and these are discussed in a later section. Plant remains (Figure 9, H), mainly carbonaceous detritus, increase down-

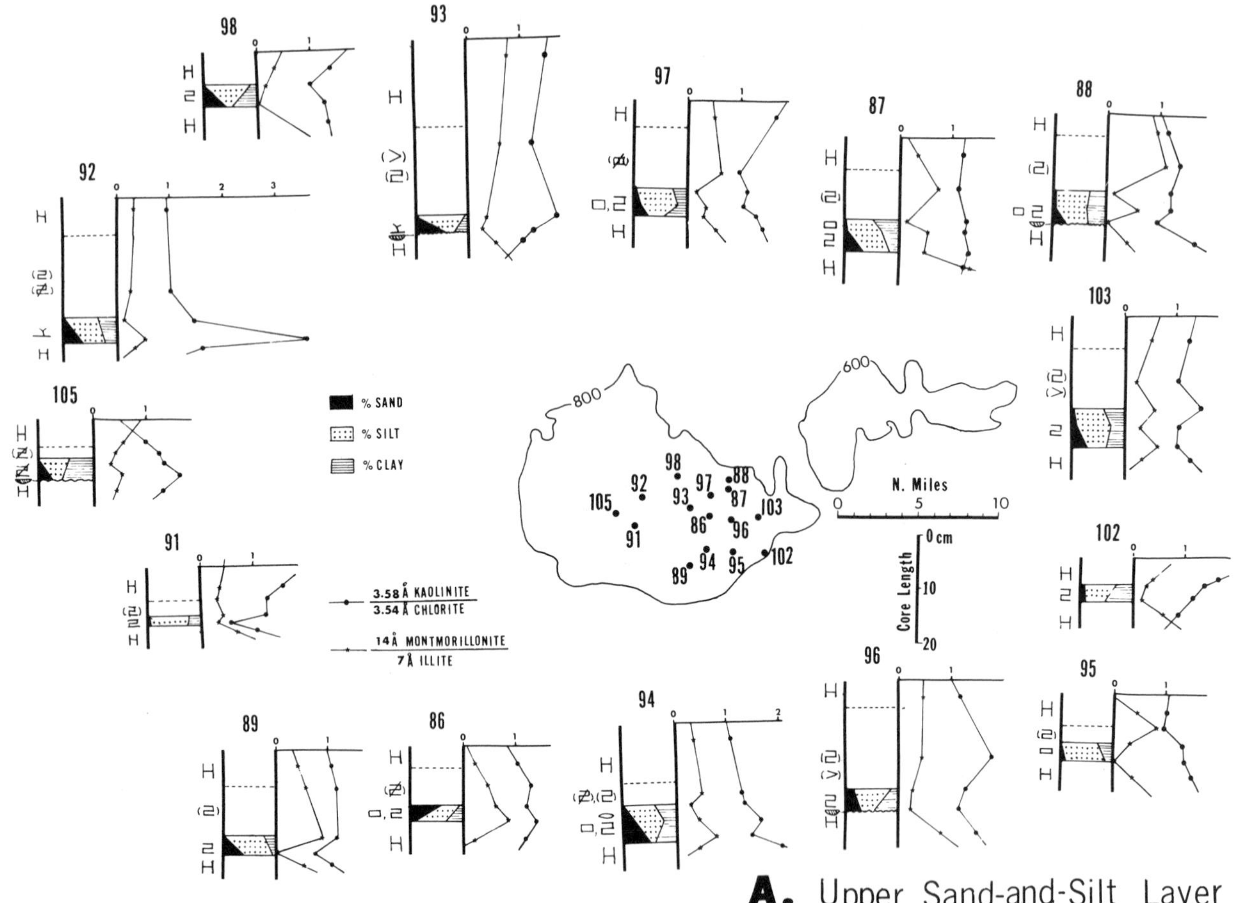

Figure 13. Distribution of percent sand, silt, clay, minor sedimentary structures, and clay mineral ratios of the Upper (**A**) and Lower (**B**) sand-and-silt layers and Graded Sand (**C**) Layer. Clay ratios of the mud above and below the sand layers are also shown.

ward in amount, together with quartz content (Table 2, Figure 8). The plants are usually wood fibers of reed, cotton or grass, as identified by Eriksson (1965).

The coarse fraction in the mud layers just above (p) and below (b) of the Upper (U) sand-and-silt layer is similar in composition to the coarse layer (U). The same is true for the Lower (L) sand-and-silt layer and adjacent mud (Figure 11 and Tables 2 and 4). However, in the case of the Upper sand-and-silt layer, the underlying mud (b) contains more planktonic foraminifera than the overlying mud (p).

Fine Fraction

In the fine grades ($< 62\mu$), 3 to 11% by weight is composed of calcareous nannoplankton, mostly coccoliths and other organic fragments and the remainder of the mud is mainly terrigenous clay with minor amounts of silt-grade quartz, heavy minerals and an unidentified opaque substance.

A comparison has been made of the composition of coarse silt (44μ to 62μ) and of sand ($>62\mu$) in the same samples in Cores 106 and 107 (Table 2). There are relatively fewer foraminifera, more eolian quartz and heavy mineral grains in the silt fraction than in the sand of the same sample.

Kaolinite, chlorite, illite and montmorillonite and mixed-layered clay minerals occur throughout all of the cores. The variation of peak area ratios of these clay minerals in five selected cores is illustrated in Figure 12. Between 0 and about 2 to 3 m in the cores: (1) chlorite increases downwards while kaolinite increases upwards; (2) montmorillonite increases downwards; and (3) all ratios tend to diverge from one another upwards. A marked change in the ratios occurs between 1.0 and 2.5 m in slope cores, and at about 3 m in the basin plain cores (these depths correspond with the depth of the Upper (U) sand-and-silt layer, traced across the basin plain) and there is less variation of the ratios below this level.

In summary, two mud layers (Upper mud and Lower mud layers) are distinguished on the basis of miner-

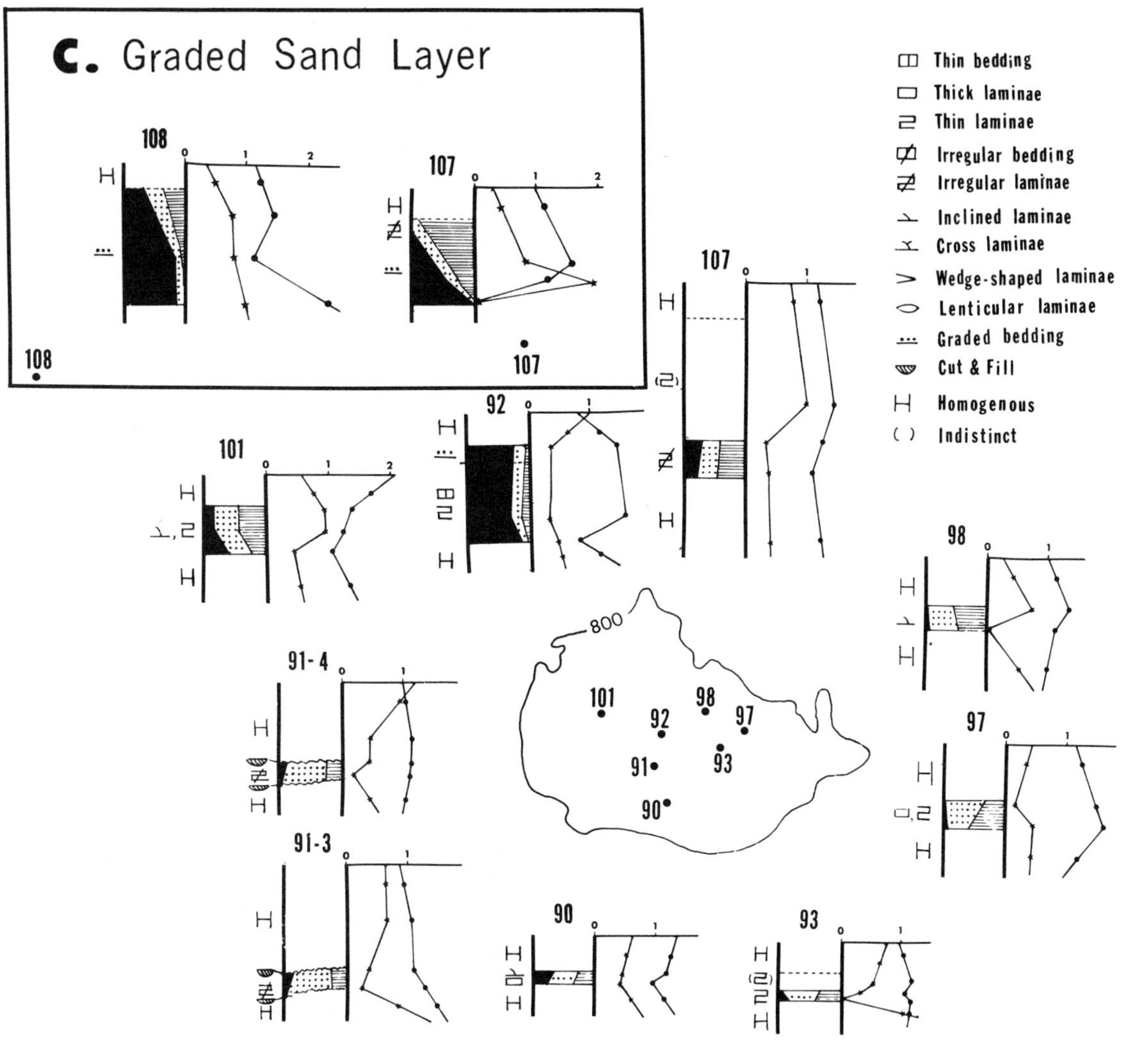

Figure 13. (cont.)

alogy. *The boundary between these two layers is the well-defined Upper sand-and-silt layer*. The sedimentary structures as well as mineralogy of both coarse and fine grades are different in the Upper and the Lower mud layers.

TEXTURE AND COMPOSITION OF THE SAND-AND-SILT LAYERS

Texture

Upper Sand-and-Silt Layer (U)

This layer in most of the cores, except Core 86, shows graded bedding, *i.e.*, the percentage of sand decreases upwards with a concurrent increase of silt and clay (Figure 13, A). In general, the following vertical sequence of minor sedimentary structures can be recognized: sharp base, showing in some instances cut-and-fill; distinct thin laminae becoming fine upwards; a relatively sharp sand-to-mud boundary; indistinct laminated mud; and, finally, homogenous structureless mud (Figure 13 A). Mean grain size varies regionally: the grade of the lower part of the layer increases toward the southern part of the basin plain, and the grade of the upper part of the layer increases toward the center (Figure 14). The distribution of sorting (standard deviation) indicates that sediments having a larger mean grain size are also better sorted.

The mean grain size and standard deviation exhibit

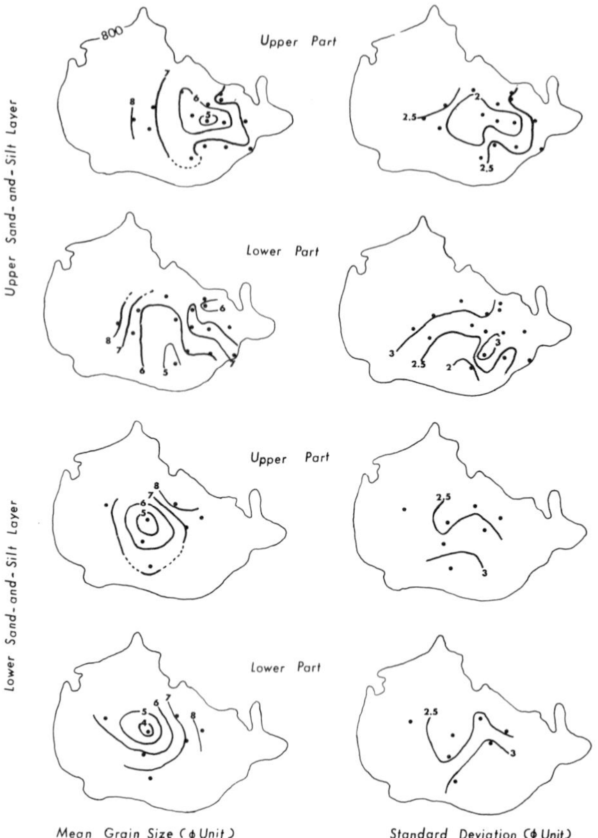

Figure 14. Areal distribution of mean grain size and standard deviation of the base and top of the Upper (U) and Lower (L) sand-and-silt layers in the Western Alboran Basin plain.

a positive linear relation (Figure 15A). A linear, negative trend is shown on the mean grain size versus skewness plot (Figure 15B). Most of the skewness values are positive, indicating an excess of finer grain sizes.

The thickness of this unit varies regionally, increasing from 1.8 cm in Core 91 to about 7 cm in Core 103 (Figure 13A). There is a general thickening of the layer in the eastern and the northeastern sectors of the basin plain. Where it is thickest, this sand layer also tends to be coarser. The total volume of sand, based on an average of 4 cm thickness and an area of 310 km^2 (delineation based on PGR coverage) is about 12,000 m^3.

Lower Sand-and-Silt Layer (L)

Three out of eight cores examined reveal an upwards increase in the percentage of sand and a concomitant decrease in silt and clay in this layer (Figure 13B). Unlike the Upper sand-and-silt layer, only one core (91) shows a cut-and-fill structure in the lower layer, and only Core 93 shows indistinct lamination in the overlying mud. Thus layer (L) lies within a section of homogenous mud, the Lower mud layer.

The areal distribution of texture shows that there is an increase of mean grain size and of sorting values toward the center of the basin plain (Figure 14). A mean grain size versus standard deviation plot is somewhat more complex than that of the Upper sand-and-silt layer: there is a suggestion of an inversion at about 5 ϕ unit. Grain size is negatively correlated with skewness (Figure 15D).

The thickness of this unit also varies regionally, increasing from 1.5 cm in Core 93 to about 16 cm in Core 92 (Figure 13, B). Like layer U, there is a general thickening of layer (L) toward the center of the basin plain; this thicker sand layer is almost always coarser grained. The total volume of the layer, based on an average of 4 cm thickness and an area of 250 km^2, is about 10,000 m^3.

Graded Sand Layers

Of all of the cores collected on the Western Alboran Basin slope and submarine valleys, only two cores (Core 108, depth: 280 cm; Core 107, depth 476 cm) display graded sand layers, composed mainly of coarse terrigenous components. In the Eastern Alboran Basin, a 10 cm thick graded sand layer composed of coarse sand to granule size bioclastic components (mainly of molluscan shell hash) was also found in Core 614 (core depth: 575 cm). This core was collected at a depth of 535 fm (984 m).

The graded sands show a decrease in the amount of sand and silt from the base to the top of the layer (Figure 13 C). All three have a sharp base and an indistinct upper boundary. The relation of mean grain size to standard deviation and to skewness is similar to that of the Lower sand-and-silt layer (Figures 15 C and 15 D).

Composition

There are obvious differences in the composition of the Upper and the Lower sand-and-silt layers (Table 4). The Upper sand-and-silt layer consists predominantly of planktonic foraminifera whereas the Lower layer and also the Graded Sand Layer contain more terrigenous detritus, primarily quartz, feldspar, heavy minerals, plant remains and carbonate debris.

Non-eolian quartz is concentrated primarily in the Lower sand-and-silt and Graded Sand layers. Little or no volcanic glass was found in either of the sand-and-silt layers; however, a few glass shards occur in the Graded Sand (Cores 107 and 108, Table 4).

The intensity ratio, or relative amounts, of aragonite and high-Mg calcite is higher in the Lower sand-and-silt layer than in the Upper one. The Graded Sand contains more aragonite and high-Mg calcite than either Upper or Lower sand-and-silt layers (Table 6). Sediments from shallower areas surrounding the Western Alboran Basin plain contain higher amounts of aragonite and high-Mg calcite than those in the deeper water, including the basin plain (Table 6). Banks and ridges in the surrounding area also consist largely of carbonate debris and planktonic foraminifera, with minor amounts of terrigenous material including some eolian quartz and volcanic debris (Table 4). Clays within the basal and upper parts of the sand-and-silt layers and the Graded Sand were also examined (Figure 13). Generally, the morillonite/illite ratio and the kaolinite/chlorite ratio both increase upward as the grain size of the sediment decreases.

The general composition of the heavy minerals in the Lower sand-and-silt and the Graded Sand layers are similar (Table 5). These suites consist mainly of pyroxene, biotite, epidote and opaque minerals. There is, however, a somewhat higher percentage of metamorphic species (staurolite, kyanite, sillimanite and andalusite) in the Graded Sand layers. The assemblage in the Graded Sand Layer is similar to that found in the Strait of Gibraltar area (Kelling and Stanley, this volume) and samples (Nos. 19, 30) collected along the Spanish coast (see Figure 2). Coastal sediments west of Malaga contain more orthopyroxene and metamorphic minerals, and less epidote and opaques than samples east of Malaga (Table 5). There is a slight difference in the proportions of certain major minerals between the Strait and Alboran Basin sediments. For example, some accesory minerals such as rutile and apatite are not found in the Alboran Basin plain sands.

SEDIMENTATION RATES

The geographic distribution of cores and the core isochrons inferred from the 21 carbon-14 dates are

Table 5. Heavy mineral percentages.

Sample No.[1]	Amphi-bole	Pyrox-ene	Tour-maline	Epi-dote	Zircon	Stau-rolite	Garnet	Sphene	Biotite	Kyanite	Silli-manite	Anda-lusite	Opaque	Others
A. Alboran Sea sediment														
91–3u	2	25	1	17	2	1	9	1	21	1	1	0	16	3
91–3l	2	32	1	16	4	0	6	0	20	3	1	1	12	2
92–3u	3	23	2	8	1	1	11	1	25	1	1	1	18	3
92–3m	2	25	3	7	2	2	15	1	21	2	0	3	17	2
92–3l	3	30	2	7	2	1	10	1	24	1	0	1	15	3
93–3u	2	38	2	12	2	0	6	0	22	3	1	0	20	0
93–3l	2	28	0	14	2	0	12	0	14	2	0	2	14	4
97–4u	1	32	2	8	0	0	2	0	28	0	0	1	22	4
97–4l	2	26	2	10	0	0	4	0	32	2	0	2	16	4
98–4u	2	34	0	10	2	2	4	0	20	2	0	2	20	2
98–4l	5	37	1	7	0	0	4	0	16	1	0	1	26	2
101–4u	2	32	0	12	1	1	3	0	25	1	0	1	20	2
101–4m	2	26	2	16	0	0	2	0	22	0	0	2	24	2
101–4l	2	32	2	8	0	0	4	0	24	0	0	0	22	2
107–3u	5	30	2	11	2	0	6	0	24	0	0	1	17	2
107–3l	6	27	2	15	1	0	9	0	23	0	1	1	13	2
107–1u	6	30	2	8	8	2	11	0	18	1	1	3	8	2
107–1m	7	38	2	10	4	2	10	0	14	0	0	2	9	2
107–1l	6	30	0	11	2	2	12	0	20	0	0	1	15	1
108–1u	5	32	1	6	2	1	8	1	23	1	1	2	15	2
108–1m	3	38	1	15	1	2	5	1	21	2	2	1	7	2
108–1l	3	32	2	10	6	4	10	2	12	2	3	3	9	4
B. Spanish coastal sediment														
19	6	33	2	10	1	1	10	1	13	1	1	7	9	4
30	4	53	1	5	1	1	10	1	4	2	2	8	5	3
53	7	17	2	14	2	1	9	0	20	4	1	6	14	3
67	6	12	0	16	1	0	9	0	24	0	0	5	19	5
82	8	13	0	13	1	1	28	0	13	0	0	7	22	4
104	7	10	1	12	2	1	7	0	23	0	0	4	26	6

[1] First part of sample number refers to the core number shown in Figure 2; second part of sample designation refers to the core section sampled.

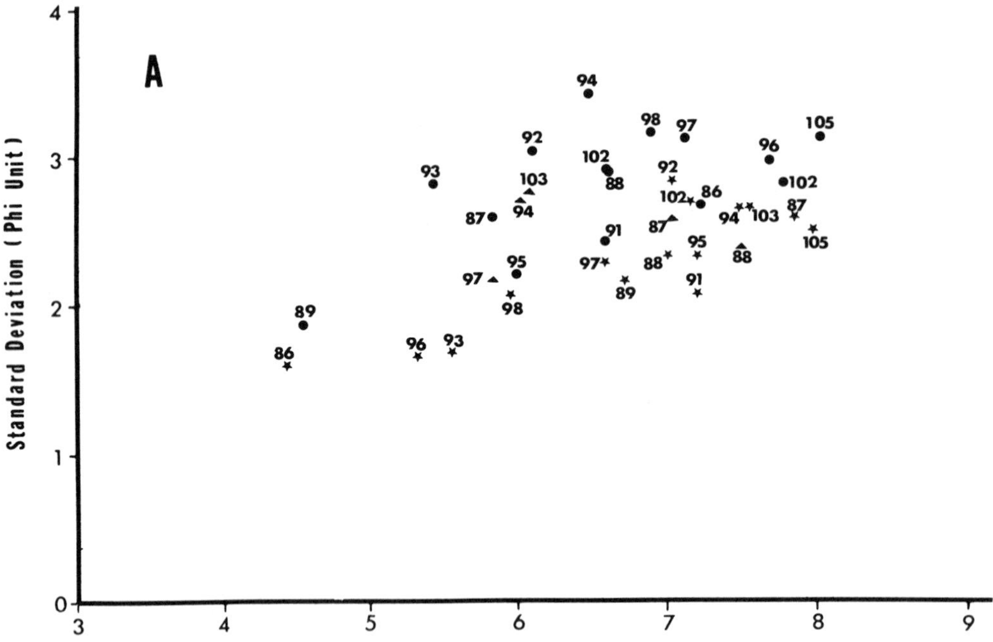

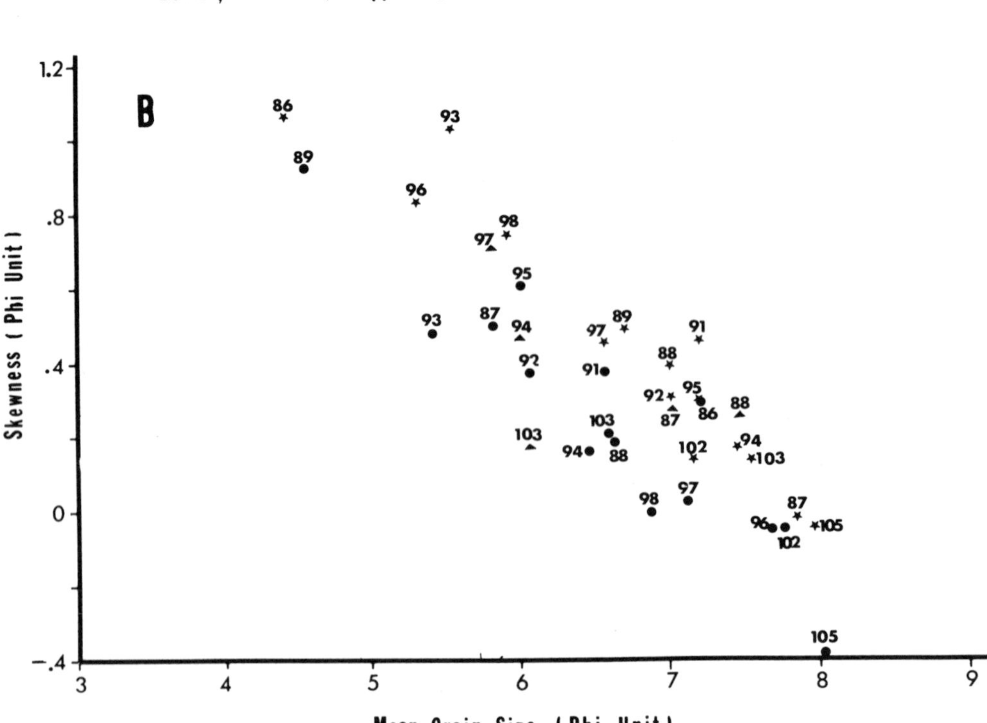

Figure 15. Mean grain size versus standard deviation and skewness of sand samples in the Western Alboran Sea. A and B: Upper sand-and-silt layer (U); C and D: Lower sand-and-silt layer (L) and Graded Sand layer.

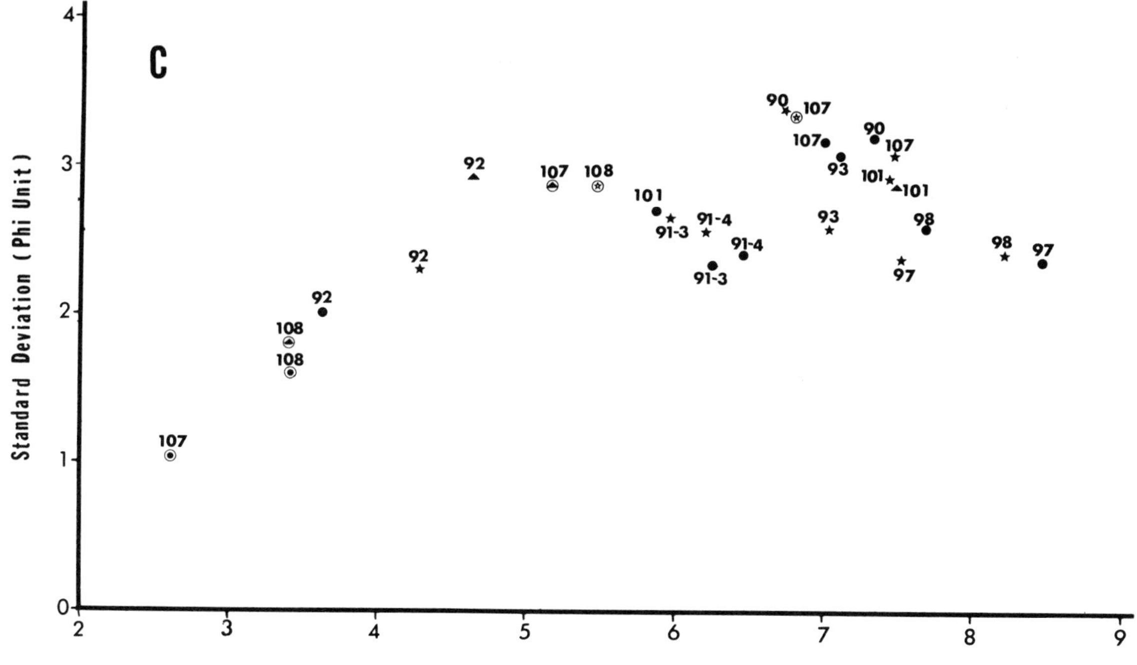

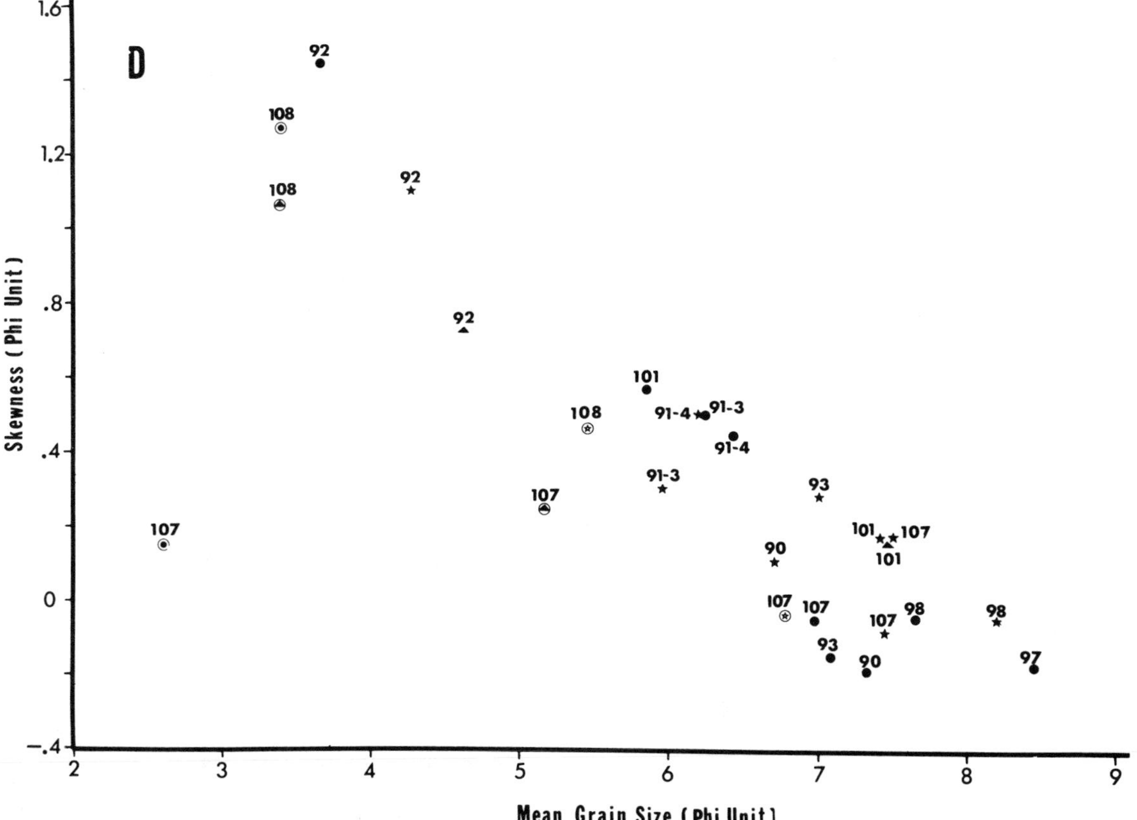

Figure 15. (cont.)

Table 6. Intensity ratios of the carbonate minerals in sand and silt layers.

Sample No.	Aragonite	Low-Mg Calcite	High-Mg Calcite
I. Upper sand-and-silt layer			
86–5u	2.1	84.5	13.4
86–5l	3.7	81.6	14.7
87–5u	4.8	86.8	8.4
87–5m	2.5	91.0	6.5
87–5l	3.3	93.9	2.8
	4.9	90.7	4.4
88–5m	2.9	92.6	4.6
88–5l	4.5	90.0	5.5
89–5u	5.2	80.6	14.2
89–5l	4.6	86.5	8.4
91–6u	5.7	88.0	6.3
91–6l	5.0	87.9	7.1
92–6u	1.9	96.7	1.4
92–6l	2.7	92.8	5.5
93–6u	2.7	92.6	4.7
93–6l	3.1	92.2	4.7
94–4u	3.3	93.4	3.3
94–4m	3.4	93.2	3.4
94–4l	2.1	91.3	6.6
95–6u	4.3	89.6	6.1
95–6l	7.1	92.2	7.8
96–5u	3.6	92.8	3.6
96–5l	5.3	88.6	6.1
97–6u	5.5	90.8	3.7
97–6m	4.2	92.5	3.3
97–6l	2.0	93.3	4.7
98–6u	3.5	92.9	3.6
98–6l	2.8	94.3	2.9
102–5u	2.4	89.4	8.2
102–5l	0	97.5	2.5
103–5u	6.0	86.9	7.1
103–5m	5.9	91.4	3.2
103–5l	5.0	83.6	7.4
105–4u	5.3	90.5	4.2
105–4l	6.9	89.7	3.4
II. Lower sand-and-silt layer			
90–1u	9.9	55.3	34.8
90–1l	4.3	85.1	10.6
91–4u	5.2	90.1	4.7
91–4l	7.5	85.7	6.8
91–3u	7.0	86.6	7.4
91–3l	6.8	85.5	7.7
92–3u	4.3	92.4	3.3
92–3m	3.5	86.2	10.3
92–3l	4.4	91.2	4.4
93–3u	4.2	86.4	9.4
93–3l	3.8	93.0	3.2
97–4u	3.3	93.3	3.4
97–4l	3.4	93.2	3.4
98–4u	3.2	90.3	6.4
98–4l	3.0	91.1	5.9
101–4u	0	95.9	4.1
101–4m	0	96.0	4.0
101–4l	0	97.7	2.3
107–3u	5.2	87.1	7.7
107–3l	5.2	87.1	7.7

Table 6. (Cont.)

Sample No.	Aragonite	Low-Mg Calcite	High-Mg Calcite
III. Graded sand layers			
107–1u	5.1	81.6	13.3
107–1m	7.5	83.6	8.9
107–1l	22.1	51.6	26.3
108–1u	18.2	75.3	6.5
108–1m	8.4	70.5	21.1
108–1l	9.0	74.2	22.8
614–29	11.0	48.7	40.3
614–30	29.0	23.3	47.7
IV. Sediments in shallower areas			
D1	7.5	40.0	52.6
D4	9.9	32.2	57.9
D7	1.9	94.9	3.2
S1	8.4	57.9	33.7
C20	1.9	94.2	3.9
C30	9.3	68.0	22.7
C38	7.3	65.8	26.8

illustrated in Figure 16. Progressive increase of the thickness of the strata between isochrons from cores on the slopes toward those in the basin plain clearly indicates that the sedimentation rate has been higher in the basin plain during the Holocene (or since about 10,000 years B. P.). The sedimentation rate during the past 10,000 years in the basin plain is 30 to 40 cm/1000 years [values closer to 40 cm/1000 years are measured (Figure 17)]. This is about twice as fast as the rate on the slope areas which ranges from 12 to 28 cm/1000 years (an average of about 20 cm/1000 years)[1].

Furthermore, there is a very marked increase of sedimentation in the basin plain below a core depth of 2.5 to 3 m, a depth which corresponds to a period of time from 8,000 to 11,000 years B.P. This sudden increase in rate, is also discernible in slope Core 111, but at a shallower core depth (< 2 m). However, in some slope areas no corresponding increase in sedimentation rate is apparent, and some cores (100, 108, 107) actually reveal a slight decrease in the rate in the Lower mud layer. One C^{14} date shows that the foraminifera of the Upper sand-and-silt layer (U) were deposited at about 9125 ± 410 years B. P. The mud section above it (Upper mud layer) is thus entirely of Holocene age, while the mud below the Upper sand-and-silt layer is of Pleistocene

[1] Comparable values were obtained independently by Bartolini and Gehin (1970) and Vergnaud-Grazzini and Bartolini (1970) who used a different method. They report, for instance, a rate of 27 cm/1000 years in the Core 80 and 26 to 30 cm/1000 years in Core 68.

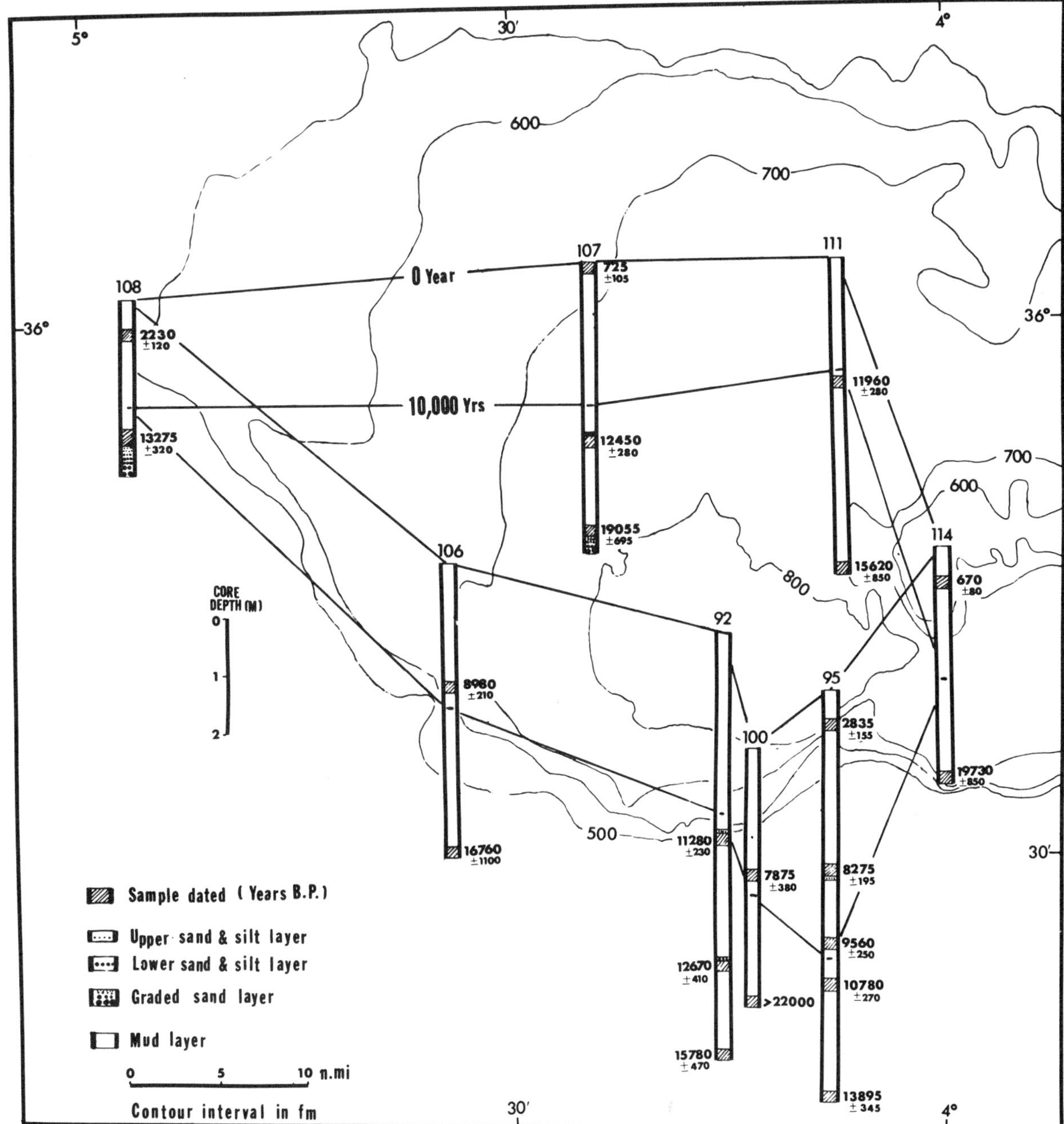

Figure 16. Core isochrons, based on 21 carbon-14 age determinations in eight Alboran Sea cores. Note progressive thickening of Holocene section (between 0 and about 10,000 B. P.) from the slopes toward the Western Alboran Basin.

age. The Lower sand-and-silt layer was emplaced subsequently to about 12,500 years B. P. The coarse Graded Sand layers in the two much shallower slope cores appear to be different in age and are older than the sand-and-silt layers of the basin plain. One was deposited before 13,275 years B. P. (Core 108), and the other (Core 107) before 19,055 years B. P.

The ages of carbonate-rich sediments from the shallow bank and ridge surfaces to the south of the basin plain are discussed in Milliman *et al.* (this volume).

DISCUSSION

Sediment Provenance and Dispersal

Two major sources for the Western Alboran Sea sediments have been proposed (Stanley *et al.*, 1970): an intrabasinal source area (from adjacent slopes, banks, and the coasts of Morocco and Spain), and an extrabasinal source (from the Atlantic and Balearic regions beyond the Alboran Sea).

Weathered detritus furnished to the Moroccan and Spanish coasts by streams draining the Rif and Betic

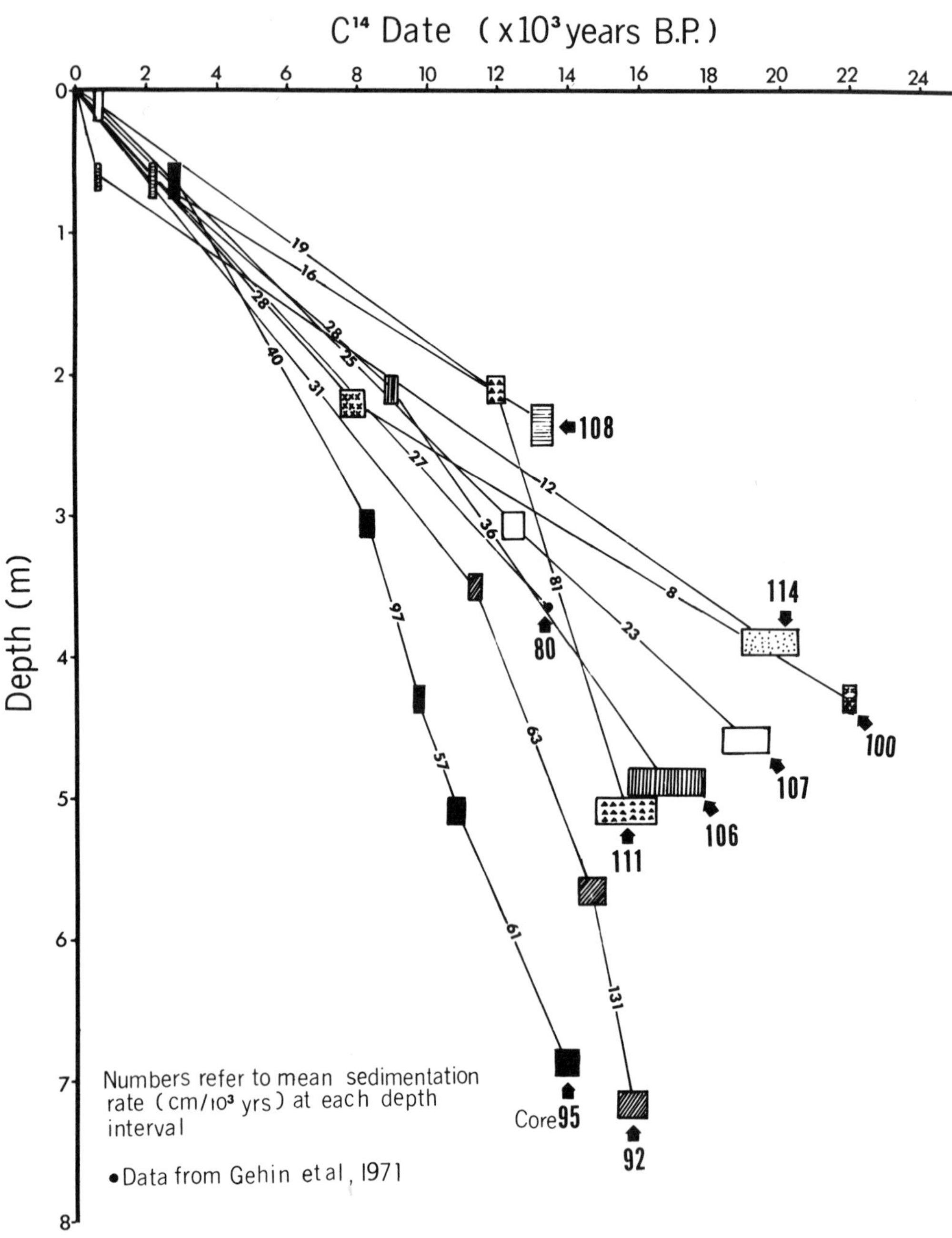

Figure 17. Sedimentation rates in the Alboran Sea, based on twenty-two carbon-14 age determinations. Data on Core 80 reported in Gehin *et al.* (1971). Slope cores: 80, 100, 106, 107, 108, 111, 114; Western Alboran Basin plain cores: 92, 95.

chains is compositionally complex since source terrains include Paleozoic to Quaternary sediments and plutonic and metamorphic rocks. The sediments sampled along the coast and in the Alboran Sea are mixtures of these varied rock suites.

Analysis of grab samples confirms the existence of two distinct intrabasinal source areas for sediments in the Western Alboran Basin. One source area lies near the eastern portion of the Strait of Gibraltar and the adjacent Spanish slope. Sediments from this region are distinguished by higher amounts of terrigenous components such as quartz, feldspar, mica, heavy minerals and carbonaceous (plant) matter. The other source area comprises the ridges, banks and knolls located to the south and east of the Western Alboran Basin, where sediment is composed largely of biogenic carbonate debris.

Kelling and Stanley (this volume) report that the terrigenous fraction of the Gibraltar Strait sediments may be assigned petrographically to two main distinguishable suites of source rocks, comparable to the northern Rif and the Betics, together with contributions reworked from local, probably submarine, sources. Heavy mineral assemblages from the Spanish coast described in this paper indicate two distinct source terrains with a boundary in the region of Malaga. Coastal sediments west of Malaga are characterized by abundant orthopyroxenes, suggesting a predominant basic igneous origin, while the abundant andalusite and garnet are probably derived from associated contact metamorphic rocks. The coastal sediments east of Malaga, however, include abundant opaque heavy minerals derived, in part, from sedimentary and igneous rocks. Regional or stress metamorphic minerals are less common here. All sediments on the Spanish coast include 'stable' heavy minerals such as zircon and tourmaline.

The heavy minerals of the Western Alboran Basin indicate derivation from several of the above source areas during late glacial to Holocene time. In slope sediments, the heavy minerals consist of mixed assemblages showing transport from both coastal areas west of Malaga and the Strait of Gibraltar region. The Western Alboran Basin plain sediments, however, include a mixture of mineral suites from the Strait of Gibraltar as well as the coastal areas east and west of Malaga.

It is noteworthy that the Alboran Sea sediment distribution varies regionally. There is a relatively high concentration of terrigenous components in cores on the northern part of the basin. On the other hand, cores collected in the southern part of the basin reveal a higher proportion of carbonate debris. The composition of the sediment in this area is strongly influenced by the adjacent banks and ridge system. These features not only shed substantial quantities of carbonate material northward into the deep basin but also act as a barrier, restricting the supply of terrigenous detritus, derived from the Moroccan mainland, to the Western Alboran Basin. The sediment distribution is thus, to a large degree, due to difference in the amount and composition of the sediment contributions from the Moroccan and Spanish source areas.

Although little data is available on the discharge rates of the short streams draining the mountainous area surrounding the Alboran Sea, the recorded annual rainfall on the adjacent Spanish coast clearly is appreciably higher than that on the corresponding part of Morocco (Mikesell, 1961; Way, 1962). This suggests that the Spanish mainland, at least at present, provides more sediment to the Western Alboran Basin, and would account for the observed distribution and nature of the terrigenes. The importance of the seasonal nature of snow melt and river flow on sediment transport is discussed in a later section.

The clay mineralogy serves to emphasize the importance of extrabasinal sources. The types and relative amounts of clay minerals are similar in all cores penetrating recent Alboran sediments, probably because homogenization of fine minerals has occurred as sediments entered the Alboran Sea. Clay minerals include illite, kaolinite, chlorite, montmorillonite and mixed layered clay, all of which also occur in modern sediments of the Atlantic and the Balearic Sea (Biscaye, 1965; Emelyanov and Shimkus, this volume) as well as in Tertiary sections as found in Mediterranean JOIDES cores (Nesteroff *et al.*, this volume). Biscaye (1965, *his* figures 4 to 7) has shown that kaolinite and chlorite are equally abundant in sediments of the Alboran Sea and of the Atlantic Ocean west of the Strait of Gibraltar, but are less abundant in sediments from the Balearic Sea. Moreover, the amount of montmorillonite in the Alboran Sea is lower than in basins to the east. Illite content remains about the same throughout the Atlantic Ocean and the western Mediterranean Sea and this mineral can be used as a stable index to compare the proportions of the other clay minerals. On this basis, observed ratios of different clay minerals in the Recent Alboran sediments (Figure 12) suggest that at present the Atlantic surface current may contribute a somewhat greater proportion of the chlorite and kaolinite-rich clay to the Alboran Sea from the Atlantic than is derived from the Balearic Basin to the east. However, during Würm time, when the relative amount of chlorite was increased with respect to kaolinite (Figure 12), clay provenance was different Montmorillonite is generally derived from alteration of volcanic rocks and is therefore a more sensitive indicator of source than most of the other clay minerals. At present, little montmorillonite is being supplied to the Alboran area, either by the Atlantic surface currents or the Mediterranean Undercurrent. However, the enhanced proportions of this mineral in the Würm sediments (Figure 12) indicates the influence of volcanic sources to the east (possibly as far as the Tyrrhenian Sea) where montmorillonite is still abundant in recent clays (Nesteroff *et al.*, 1965). This observation has important consequences with respect to the Würm hydraulic regime, which are discussed later. It is also possible that the presence of montmorillonite reflects erosion of volcanic rocks such as the Xauen and Alboran Ridge within the Alboran Sea area during the low eustatic stands of sea level.

Major processes responsible for sedimentation in the Alboran Basin during late Würm and Holocene time include: eolian transport, pelagic differential settling, both high and low velocity turbidity currents, bottom currents, and perhaps, turbid layer flow. Stanley *et al.* (1970) also cite evidence for slumping on the slope southwest of the Western Alboran Basin plain.

Eolian Transport

Locally, wind-blown grains are an important component of bottom sediments. An example is the coastal sediments of southern Mallorca to the east where Butzer and Cuerda (1962) record 20 to 40 percent "windworn" quartz. The most probable sources of wind-blown material in the western Mediterranean Sea are (1) the Sahara Desert proper and adjacent areas, (2) the tablelands of Spain, Morocco and Algeria, and (3) the coastal areas (Eriksson, 1965). At present, eolian materials are derived mainly from mountain areas and coastal regions surrounding the Alboran Sea, and not directly from the Sahara Desert.

Coastal dune fields around the Alboran Sea are fairly limited in size today but were probably extensive during the lower sea-level stages of the Pleistocene (Eriksson, 1965; Butzer, 1964). This conclusion is corroborated by the higher proportions of eolian quartz in the late Pleistocene lower mud Layer of the Western Alboran Basin plain.

It is difficult to evaluate how much of the eolian material has been directly wind-blown into the Alboran Sea and how much has been transported by fluvial processes to the sea. It is probable that at least some of the eolian quartz found in the mud and sand of the bank, ridge and knoll areas (Table 4) was transported by wind directly into the sea. Eriksson (1961) has reported wind-blown quartz of fine sand and silt grade on Alboran Island, requiring wind-velocities of storm force (Bagnold, 1942) to achieve transport across more than 30 kilometers of open sea.

The abundance of eolian quartz in thin laminae in the cores indicates that some wind-blown quartz was subsequently affected by pelagic differential settling and then may have been concentrated by bottom currents.

Pelagic Settling

This process of sedimentation is represented in the cores by indistinct lamination of silt and clay and by homogenous, or structureless, mud (Figure 8). The relatively rapid currents at present flowing eastward at the surface and westward at depth are competent to transport nearly all the silt and clay-size sediment found in the cores. Lacombe and Tchernia (1960, this volume), Allan (1966), Frassetto and Crose (1965) and Kelling and Stanley (this volume) all provide evidence of a particularly strong westward flowing bottom current (Mediterranean Undercurrent) in the Strait of Gibraltar area. This westerly current is capable of transporting coarse sediment up to 5 mm in diameter in the Strait; the shallower easterly currents are inadequate to move such coarse sediment but are capable of moving some fines to the Alboran Sea (Kelling and Stanley, this volume). In addition, fines are seasonally provided to the Alboran Sea from the Spanish and Moroccan coasts by rivers draining the adjacent highlands.

Once the sediments are carried into the Alboran Sea, semipermanent Atlantic water surface currents and Mediterranean and deep water undercurrents modify the settling paths of suspensates. It appears that the present stratified circulation effectively concentrates somewhat different suspensates in the different water masses. Thus, surface currents entering from the Atlantic Ocean through the Strait of Gibraltar carry a suspended load rich in kaolinite and chlorite and disperse these rapidly along their clockwise cyclic paths, while the colder, more saline Mediterranean Undercurrent transports materials of both intra- and extra-basinal source (including much of the montmorillonite) towards the Strait. The dense but slow-moving Deep Water mass also concentrates suspensates which settle from layers nearer the surface of the sea. The settling time of particles through these three water masses is likely to vary considerably, and the effectiveness of differential settling becomes more important when suspensates reach the cold, saline deep water masses which migrate slowly over the Western Alboran Basin plain.

It has been postulated that during the Würm, current speeds at the Strait were considerably higher than at present (Anderson, 1965) and it is probable that the stratification of the western Mediterranean water mass was somewhat different to that existing today. However, pelagic transport would have been enhanced during Würm time due to the greater stratification of denser water masses and the higher concentrations of suspensates.

Bottom Currents

Analysis of cores and bottom photographs suggests that bottom currents are not significantly active at present in the Western Alboran Basin plain. However, the enhanced activity of bottom currents during the Würm is indicated by evidence which includes: (1) the concentration of sand-size terrigenes, foraminiferal tests and carbonaceous matter within laminated mud layers; (2) minor scour-and-fill structures in some cores; and (3) cross-, inclined- and wedge-shaped laminae of silt and clay that result from migration of ripple-foresets. Similar types of lamination occur in the mud layers of the Adriatic Basin (van Straaten, 1970). The occurrence of cross-stratification is variable from core to core (Figure 8) indicating that ripple-mark formation was either a local phenomenon or that ripple marks commonly were destroyed by organic activity, soon after formation. All available sedimentary structures and bed-form data indicate that these Würm bottom currents were of low-flow regime character.

High Velocity Turbidity Currents

The relatively coarse graded sands in cores collected on the slope and channel areas (Cores 108 and 107) and the graded carbonate sand in the Eastern Alboran Basin (Core 614) are believed to result from this type of mechanism. These sands comprise a predominant graded unit with little or no lamination. Since some burrows occur in the overlying mud layer, the laminated units may have been reworked subsequently by bottom-living organisms or they may have been destroyed by bottom currents. The graded sands were clearly derived from shallow water environments, as indicated by their mixed terrigenous and carbonate components, benthic organisms and plant matter. The high content of aragonite and high Mg-calcite of this deposit (Table 6) also indicates derivation from shallower sources (Huang and Pierce, 1971).

Low Velocity Turbidity Currents

This type of current flow is genetically difficult to distinguish from gravity-assisted bottom currents (Bartolini and Gehin, 1971) but flows of this type probably deposited the Upper and Lower sand-and-silt layers. The Upper sand-and-silt layer is remarkably rich in planktonic tests, and the basal portion consists of alternating, very thin laminae of test-rich and test-poor silty sand. Similar carbonate turbidites have been recorded by Nesteroff (1961, 1962), Kuenen (1964), Eriksson (1965) and van Straaten (1967, 1970).

A number of questions arise with regard to this graded and laminated foram-rich deposit, mainly pertaining to the source of the carbonates, the triggering agent or agents and the depositional mechanisms. The intervening mud usually contains only 2 to 3 percent foraminifera. In what manner, therefore, could foraminiferal tests be concentrated so as to make up 95% or more of sand-sized components in a sand layer that is 1 to 3 cm thick over an area of about 310 km^2 (representing a volume of about 62,000 m^3)? We conclude the following:

1. The source is intra-basinal. The concentration of foraminifera which can account for 95% of the sand-sized components resulted from sudden productivity or a catastrophic decimation of planktonic forms with subsequent winnowing. In addition, erosion by the Mediterranean Undercurrent, of the adjacent slopes, knolls, banks and ridges that surround the Western Alboran Basin plain probably also contributed to the deposition of this layer. This is substantiated by the presence of sand-sized terrigenous sediment components in the Upper sand-and-silt layer that are similar to those in the underlying mud layer.

2. The mechanism or mechanisms that deposited the layer on the base of slope and in the basin plain is not an obvious one. Evidence favoring turbidity current and gravity-controlled bottom flow is present in the form of grading, the concentration of the coarsest grains in the central, deepest portion of the basin plain, the near uniform thickness of the layer and its continuous nature in the base-of-slope region and across the basin plain. On the other hand, evidence favoring the influence of normal bottom currents is found in the laminated nature of the sands, thin sharp (truncated) tops as well as bases, and the presence of bioturbation suggesting a relatively slow accumulation of the layer. The resulting deposit reflects a relatively rapid input of material into the lower part of the basin, and this deposit was almost immediately modified by existing bottom currents, active in these deeper environments.

3. Thus, deposition was probably not instantaneous nor resulting from a single gravity-controlled episode, but accumulated from a continued grain-by-grain settling and traction transport. If a turbidity current was indeed involved, it must have been of low density and low velocity (Moore, 1969). Sedimentary structures and textures indicate a waning of current energy throughout the time of deposition. Because of the multiple mechanisms involved, the resulting deposit is difficult to differentiate from sediment deposited by a bottom current containing large amounts of suspensate, and a similar mechanism has been described by Ewing and Thorndike (1965) as a gravity-assisted bottom current.

The Lower sand-and-silt layer (L) displays many of the same features and perhaps can be attributed to the same processes. The high terrigenous content of this layer is related to the difference of the source materials. It should again be emphasized that the mineralogy of this sand and that of the coarse fraction in the underlying mud is similar. The source area was probably topographic highs (including carbonate banks) surrounding the Western Alboran Basin plain. Topographic highs were eroded by rapidly flowing bottom currents which concentrated sands. How were the sands transferred from the slope to the basin plain? The importance of downslope gravity flow is again demonstrated by the general continuity and uniformity of the layer and the concentric, centripetal arrangement of the mean grain size and sorting isopleths (Figure 14). The absence of this layer was noted in some of the cores suggesting that, locally, sediment is trapped and channelized in small depressions towards the very center of the basin plain.

The nature of the 18 cm thick Lower sand-and-silt layer (L) in Core 92, near the center of the basin plain, is of particular interest because the maximum grain size is observed in the middle part of the unit and there is a form of symmetrical grading both upwards and downwards from this horizon. Moreover, the middle coarse portion consists of alternating laminae, some rich in carbonaceous and carbonate debris, the others rich in terrigenous minerals. It is not certain that a single high velocity, high density turbidity current flow could produce this type of complex deposit. An alternative explanation is that strong bottom currents may have modified the turbidite shortly after its deposition in the vicinity of Core 92.

Additional evidence for reworking is derived from Core 91 where layer (L) consists of two identical sand layers containing laminae of segregated plant remains and other terrigenous components separated by a thin mud layer. Both laminated layers show erosional basal surfaces and truncated tops, suggesting that in this case, the Lower sand-and-silt layer was reworked after initial deposition, probably by bottom currents or by subsequent turbidity currents.

Turbid Layer Flows

Repetitions of a few thin (less than 1 cm thick) indistinct mud layers, some of them graded, have been noted in core radiographs and these units are commonly associated with thin laminae, believed to result from bottom current activity. The graded mud layers may have originated from periodic, gravity-assisted bottom currents having a high suspensate load. Another mechanism has also been invoked: mud-rich turbidity currents depositing *mud turbidites*. Such currents would be related to turbid underflows induced by seasonal flood waters (Moore, 1969; Stanley *et al.*, 1970). There is still some question as to whether flood water can actually produce a turbid layer flow. Measurement of suspended load introduced by flood water in the Santa Barbara Channel area has led Drake and Gorsline (1971) to conclude that thermal stratification is a major obstacle to the suggested process of turbid layer flow. Instead, the suspensates are dissipated by oceanic currents at the shelf edge and reconcentrated in the deep water through differential settling. However, during the Würm, flood waters may have been dense enough to penetrate this barrier of dense water to initiate a turbid layer flow, and it is noteworthy that graded mud horizons are, in fact, found only in the lower mud layer.

SEDIMENT PONDING

Two important features of sub-Recent sedimentation in the Western Alboran Basin have emerged from the stratigraphic correlation of the cores and from C-14 age dating of cored sequences. First, the thickness of sediment separating the Upper and Lower sand-and-silt layers increases progressively from the surrounding slope areas toward the Western Alboran Basin plain. Secondly, Holocene sedimentation rates are not constant throughout the basin but increase from about 20 cm/1000 years in slope areas to 40 cm/1000 years in the basin plain. Furthermore, we have found that the rate increases rapidly to an average of 80 cm/1000 years in the western Alboran Basin plain during late Würm time. On the other hand, sedimentation rates near the banks, ridges and adjacent slope areas have been maintained at a lower but almost constant value during the past 20,000 years.

A marked change in rate (Figure 17) is noted between 11,000 to 8,000 years B.P. (at about the time of deposition of the Upper sand-and-silt layer). These observations lead us to conclude that the basin plain has acted as an enclosed sediment trap during the late Pleistocene and Holocene, and that ponding as described elsewhere in the Mediterranean Sea by Hersey (1965a) has been a major factor in sediment accumulation in this area.

This ponding has resulted from several processes:

1. The long-term effect of the cyclic motion of the surface water masses in the Alboran Sea has concentrated suspensates above the Western Alboran Basin plain.
2. Deep water masses, including the Mediterranean Undercurrent, have probably eroded sediment from topographic highs, and redeposited these materials in adjacent lows, thus concentrating both coarse and fine particles in such areas as the Western Alboran Basin.
3. Occasional high-velocity turbidity currents transported some coarse material downslope from shelf and upper slope areas, via submarine channels. More frequent low-velocity turbidity currents (and/or gravity-assisted bottom currents) eroded and transported slope sediments downslope.

Erosion of the topographic highs and subsequent downslope transport thus has resulted in a progressive shift of material toward the basin plain, and the rates of erosion and deposition must have varied from area to area during the late Quaternary due to climatic and and eustatic factors. For example, the reduced thickness of the Lower mud layer (late Pleistocene) in Cores 100 and 114 suggests a higher rate of erosion of the slope than in the region near Core 111 during the same period (Figure 17).

EVIDENCE OF CURRENT REVERSALS IN THE ALBORAN SEA

One of the most striking features of the post-glacial history of the Western Alboran Sea is the marked change in the rates and nature of the sedimentation which may be discerned in the basin plain core sequences. This change coincides with the end of the Würm period, at about 11,000 years B. P. (Broecker *et al.*, 1960; Ericson *et al.*, 1961; Ericson and Wollin, 1968), and may be attributed to changes affecting the sedimentary processes operating in this region. In the Lower mud layer, the high frequency of bioturbation and the lack of sapropel layers show that in Würm time the Western Alboran Basin plain was not stagnant, as were some Mediterranean basins to the east (*cf.* van Straaten, 1970, this volume; Ryan *et al.*, 1970, this volume) The increased occurrence of hydrotroilite, however, suggests that more reducing conditions may have prevailed during the Würm. Moreover, the higher frequency of current-produced lamination in the Lower mud layer clearly indicates stronger bottom current activity at this time, a conclusion which is confirmed by the relatively high late Würm sedimentation rate on the basin plain (60 to 130 cm/1000 years).

However at about 11,000 years B. P. the rate of sedimentation in the Western Alboran Basin plain dramatically declined to about 30 to 40 cm/1000 years (Figure 17) and simultaneously the proportion of coarse terrigenous components in the Alboran sediments became greatly reduced with a concomitant increase in the amount of planktonic carbonates (Figure 11). Specifically there is a lower proportion of eolian quartz in the Upper mud layer (possibly pointing to a decrease in wind intensity during the Holocene) and a concomitant decrease in coarse carbonate debris and wood fibers. The kaolinite/chlorite ratio tends to increase in the Holocene core sections while the proportion of montmorillonite diminishes (Figure 12). It is noteworthy that these compositional changes have also been observed by Eriksson (1965) and Leclaire (1970) in cores collected off Algeria and in the western Balearic basins. Moreover, a decrease in sedimentation rate similar to that in the Alboran basin (from 67 cm/1000 years in the period 13,000 to 11,000 B. P. to 27 cm/1000 years in the Holocene) has been recorded in the western Balearic Basin by Eriksson (1965).

The Alboran core analyses, therefore, show that as Holocene time began, sediment dispersal by wind and by bottom and turbidity currents became relatively less important than hemipelagic deposition. In essence, the rate of erosion of the slopes bordering the Alboran Sea decreased, with a concomitant decrease in the rate of deposition, or ponding, in the Western Alboran Basin plain.

The remarkable change in hydraulic regime which is revealed by the cored sediments may be attributed partly to the effects of change in the contemporary climate on the weathering and drainage of the surrounding lands and partly to changes in sea-level and the marine circulation system.

For example, the enhanced proportions of chlorite, eolian quartz and plant debris in the late Würm Lower mud layer probably reflect the more widespread development of vegetation and the colder, drier climate prevailing at that time (Butzer, 1964; Eriksson, 1965). Reliable evidence of Pleistocene permafrost is lacking in the Mediterranean region during Würm glacial time. However, cold climate phenomena such as solifluction, colluviation and valley alluviation are common in the Betic and the Rif chains. Accordingly, during the early Würm, flash floods probably were more frequent and may have been capable of carrying a larger load for greater distances into the Mediterranean than at present (Butzer, 1964; Vita-Finzi, 1969). Moreover, during the Würm glacial maximum, sea-level was at least one hundred meters lower than at present (Zeuner, 1952; Guilcher, 1958; Fairbridge, 1961) and rivers thus transported sediment directly onto the subaerially exposed narrow shelf margins. A Würm paleogeographic and paleo-hydrographic reconstruction (Figure 18) suggests that large amounts of sediment, derived from the emerged margins and islands, were transported directly into the Western Alboran Basin.

Compositional variations cannot be attributed directly to climatic or eustatic factors (sea level was only 20 m below the present stand 10,000 years ago) alone. More probably they reflect important changes in the circulatory system of the Alboran Sea. The presence in the Würm lower mud layer of coarse terrigenous material derived mainly from the Strait of Gibraltar and the Spanish coast indicates the operation of relatively powerful east-flowing bottom currents ("Atlantic Undercurrent"). Again, the relative enrichment of the Würm sediments in the clay mineral montmorillonite, which is probably derived from volcanic areas east of the Alboran

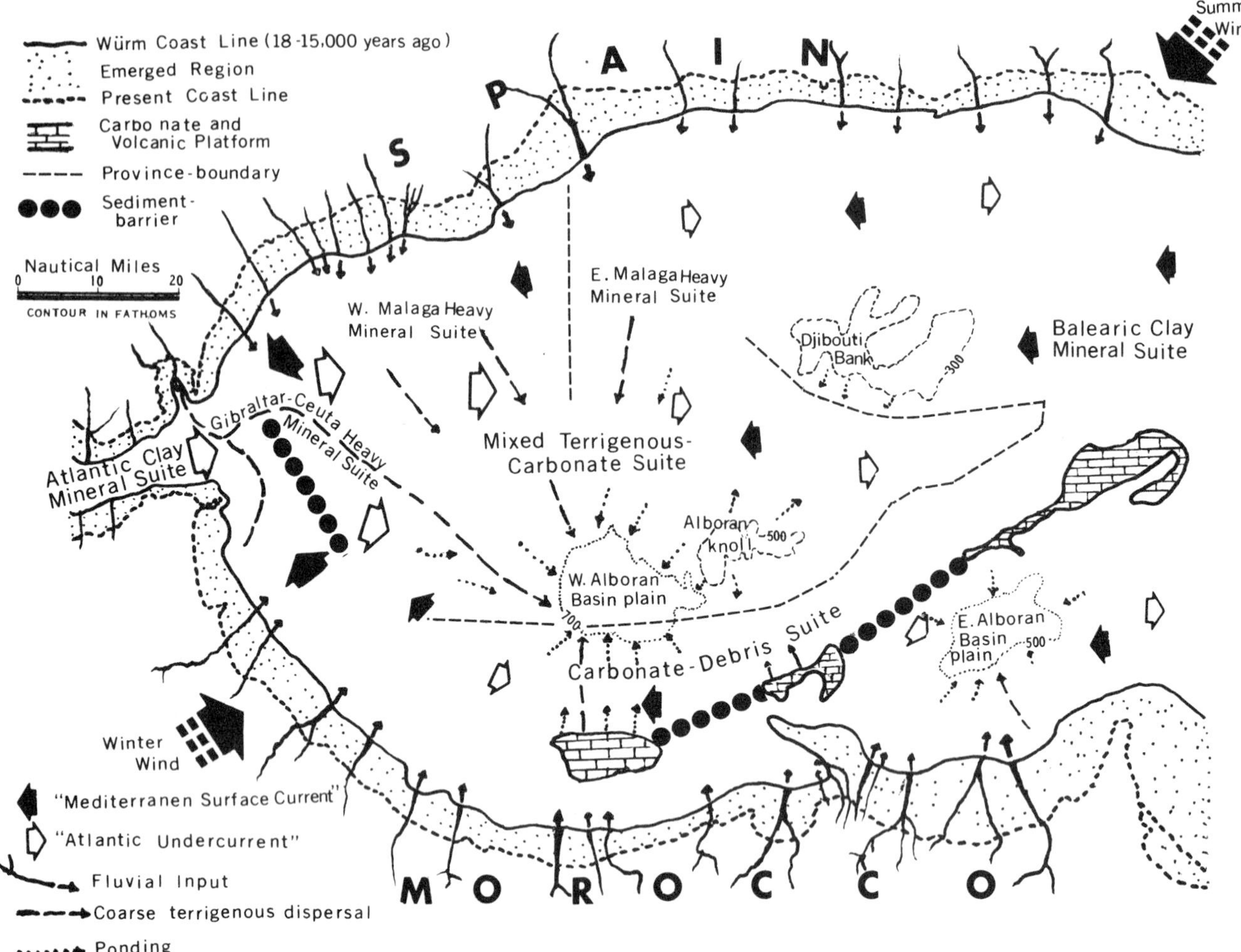

Figure 18. A schema showing suggested provenance, paleogeographic and paleohydrographic patterns in the Alboran Sea during the Würm. Note that a reversal of current regime is indicated, with a denser "Atlantic undercurrent" flowing into the Alboran Sea. The shelves were exposed, as were the upper surfaces of the banks. Note the consequently higher influx of fluvial sediments onto the slopes. The Balearic clay mineral suite consists of more montmorillonite, less kaolinite and chlorite than that in the Atlantic clay mineral suite.

Sea, suggests the operation of a west-moving surface current system ("Mediterranean Surface current") (Figure 18).

These conclusions imply a late glacial circulatory system in the western Mediterranean which was the reverse of that now existing, and accords with the theories of Kullenberg (1952), Mars (1968), Leclaire (1970 and 1972), Vergnaud-Grazzini and Bartolini (1970) and Ryan (1972) who have suggested that a reversal of water mass flow through the Strait of Gibraltar occurred at the end of Pleistocene time. Prior to this reversal, an "estuary-type" circulation existed in the Mediterranean, with less dense Mediterranean water outflowing above denser Atlantic water which flowed into the Alboran Sea through the Strait of Gibraltar. The lower density of the Würm Mediterranean is attributed to the lower evaporation rates and the higher run-off from the surrounding lands, consequent upon the more temperate climatic conditions obtaining in this region during late glacial time.

Evidence from the Alboran Sea cores, already cited, suggests that this reversal probably occurred during the period between 11,000 and 9,000 years B. P., a time which coincides with marked changes in temperature which have been observed in cores from the Atlantic Ocean (Ericson *et al.*, 1961). Changes of the magnitude envisaged here clearly must have had profound biological consequences, and such changes have been noted in a study of oxygen isotope analysis on foraminiferal tests (Vergnaud-Grazzini and Bartolini, 1970). Therefore, it is particularly significant that the changes in foraminiferal productivity, which occurred at some time between 10,000 and 9,000 years B.P. and is recorded by the Upper sand-and-silt layer of the Alboran area, ushers in the new circulatory pattern which became established in the Holocene.

SUMMARY

1. The Late Pleistocene to Holocene sediments in the Western Alboran Basin and on adjacent slopes and knolls consist almost entirely of hemipelagic mud with only a few thin interbedded sandy layers.

2. Sedimentation rates average about 20 cm per 1000 years for the late Pleistocene-Holocene section in slope and knoll areas, but 40 cm per 1000 years during the Holocene and 80 cm per 1000 years during the Würm in the Western Alboran Basin plain. The change in rate recorded from the basin plain reflects changes in climate and in sea level during late Quaternary time which, in turn, produced a different circulation pattern in the western Mediterranean.

3. Alboran Basin sediments are derived mostly from intra-basinal source areas. The two major intra-basinal source areas are the region near the Strait of Gibraltar and the southeast coast of Spain in the northwest quadrant of the Alboran Sea, together with the banks and ridges south of the Western Alboran Basin plain. The rate of sediment supply from both extra- and intra-basinal provinces was considerably higher in the Würm than in the Holocene.

4. The core isochrons and high-resolution sub-bottom records show a progressive thickening of strata from slope areas toward the Western Alboran Basin plain. Textural parameters from the sand-and-silt layers are centripetally distributed, with maximum grain size and best sorting in the central part of the basin plain. These observations demonstrate that sediment ponding has been significant in this area. The ponding processes are multiple and include: wind transport, differential pelagic settling, high and low velocity turbidity currents and bottom currents. The latter two mechanisms have displaced sediment downslope onto the basin plain.

5. Petrological changes suggest that as Holocene time began, sediment dispersal by wind and by bottom and turbidity currents became relatively less important than hemipelagic deposition. As a result there was a decrease in the rate of erosion of sediments on the adjacent slopes and the rate of ponding in the Western Alboran Basin plain also declined. The sum of sedimentological data obtained favors the hypothesis that a reversal of current regime at the Strait of Gibraltar occurred at about 10,000 years B. P. Before this time, an 'estuary-type' circulation existed, with denser Atlantic water inflowing into the Alboran Sea, as opposed to the present 'lagoonal-type' circulation.

ACKNOWLEDGMENTS

We thank the SACLANT ASW Research Centre, La Spezia, Italy and the NAVOCEANO Geological Laboratory, Washington, D.C. for the use of cores, sub-bottom and bathymetric data. Bathymetric data, bottom samples and photographs were also collected on USCGC ROCKAWAY cruise $RoSm_1$ (August 1970), and appreciation is expressed to the Captain, officers and men of this ship for their help in the work at sea.

We are indebted to Mr. C. E. Gehin, SACLANT for sending us selected PGR profiles collected on the RV MARIA PAOLINA G., Dr. W. B. F. Ryan, Lamont-Doherty Geological Observatory for use of sub-bottom profiles collected on RV CONRAD 9 (1965) cruise, Mr. H. Sheng, Smithsonian Institution for plotting bathymetric data and help in processing cores, and Dr. R. Stuckenrath, Smithsonian Institution, for providing carbon-14 dates.

This study was supported by National Geographic Society grant No. 155670, Smithsonian Institution Research Awards grants 234230 and 436330 and a travel grant to La Spezia to one of us (DJS) provided in Summer 1969 by the Office of Naval Research, Washington, D.C. This paper was prepared while T. C. Huang held a Visiting Research Associateship at the Smithsonian Institution. The paper was critically read by Dr. G. Kelling, University of Wales at Swansea, Dr. D. C. Krause, University of Rhode Island, and Dr. Y. Weiler, Hebrew University of Jerusalem.

REFERENCES

Allan, T. D. 1966. Underwater photographs in the Strait of Gibraltar. *NATO Technical Memorandum.* SACLANT ASW Research Centre, La Spezia, 116 : 19 p.

Anderson, R. S. 1965. *Paleo-oceanography of the Mediterranean Sea: Some Consequences of the Würm Glaciation.* M. S. Thesis, U.S. Naval Postgraduate School, 73 p.

Auzende, J. M., J. Bonnin, J. L. Olivet, G. Pautot, and A. Mauffret 1971. Upper Miocene salt layer in the Western Mediterranean Basin. *Nature,* 230 : 82-84.

Bagnold, R. A. 1942. *Physics of Blown Sands and Desert Dunes.* Methuen and Company, London. 265 p.

Barazangi, M. and J. Dorman 1969. World seismicity maps compiled from ESSA coast and geodetic survey epicenter data 1961–1967. *Seismological Society of America Bulletin,* 59 : 369–380.

Bartolini, C., and C. E. Gehin 1970. Evidence of sedimentation by gravity-assisted bottom currents in the Mediterranean Sea. *Marine Geology,* 9 : M1–M5.

van Bemmelen, R. W. 1969. Origin of the western Mediterranean Sea. *Verhandelingen van het Koninklijk Nederlands Geologisch Mijnbouwkundig Genootschap,* 26 : 13–52. (Symposium on the Problem of Oceanization in the Western Mediterranean).

Benson, R. H. 1972. Ostracods as indicators of threshold depth in the Mediterranean during the Pliocene. In: *The Mediterranean Sea: A Natural Sedimentation Laboratory,* ed. Stanley, D. J. Dowden, Hutchinson and Ross, Inc., Stroudsburg, Pennsylvania, 63–73.

Biscaye, P. E. 1965. Mineralogy and sedimentation of recent deep-sea clay in the Atlantic Ocean and adjacent seas and oceans *Geological Society of America Bulletin,* 76:803–832.

Bouma, A. H. 1964. Notes on X-ray interpretation of the marine sediments. *Marine Geology,* 2:278–309.

Bradley, W. H. 1938. Mediterranean sediments and Pleistocene sea-level. *Science,* 88:376–379.

Broecker, W. S., M. Ewing and B. C. Heezen 1960. Evidence of abrupt change in climate close to 11,000 years ago. *American Journal of Science,* 258:429–443.

Butzer, K. W. 1964. *Environment and Archeology. An Introduction to Pleistocene Geography.* Aldine Publishers, Chicago, 524 p.

Butzer, K. W. and J. Cuerda 1962. Coastal stratigraphy of southern Mallorca. *Journal of Geology,* 70:398–416.

Caputo, M., G. F. Panza and D. Postpischl 1970. Deep structure of the Mediterranean Basin. *Journal of Geophysical Research,* 75:5919–4923.

Carey, S. W. 1958. A tectonic approach to continental drift. In: *Continental Drift, A Symposium,* ed. Carey, S. W. University of Tasmania, Hobart, 177 p.

Chave, K. E. 1952. A solid solution between calcite and dolomite. *Journal of Geology,* 60:190–192.

Drake, D. E. and D. S. Gorsline 1971. Turbid layer distribution and fine-grained sediment transport, Santa Barbara Channel, California continental borderland. *Abstracts with Programs, Annual Meeting Geological Society of America,* Cordilleran Section, 112–113.

Duplaix, S. 1958. Etude minéralogique des niveaux sableaux des carottes prélevées sur le fond de la Méditerranée. *Reports of the Swedish Deep-Sea Expedition, 1947–1948,* 8:137–166.

Emelyanov, E. M. 1972. Principal types of recent bottom sediments in the Mediterranean Sea: their mineralogy and geochemistry. In: *The Mediterranean Sea: A Natural Sedimentation Laboratory,* ed. Stanley, D. J., Dowden, Hutchinson and Ross, Inc., Stroudsburg, Pennsylvania, 355–386.

Emelyanov, E. M. and P. P. Shimkus 1972. Suspended matter in the Mediterranean Sea. In: *The Mediterranean Sea: A Natural Sedimentation Laboratory,* ed. Stanley, D. J. Dowden, Hutchinson and Ross, Inc., Stroudsburg, Pennsylvania, 417–439.

Ericson, D. B., M. Ewing, G. Wollin and B. C. Heezen 1961. Atlantic deep-sea sediment cores. *Geological Society of America Bulletin,* 72:193–286.

Ericson, D. B. and G. Wollin 1968. Pleistocene climates and chronology in deep-sea sediments. *Science,* 162:1227–1234.

Eriksson, K. G. 1961. Granulométrie des sédiments de l'île d'Alboran, Méditerranée occidentale. *Bulletin of the Geological Institute, University of Uppsala,* 15:269–284.

Eriksson, K. G. 1965. The sediment core no. 210 from the Western Mediterranean Sea. *Reports of the Swedish Deep-Sea Expedition, 1947–1948,* 8:397–594.

Ewing, J. and M. Ewing 1959. Seismic-refraction measurements in the Atlantic Ocean Basin, in the Mediterranean Sea, on the Mid-Atlantic Ridge, and in the Norwegian Sea. *Geological Society of America Bulletin,* 70:291–318.

Ewing, J. and E. M. Thorndike 1965. Suspended matter in deep ocean water. *Science,* 147:1291–1294.

Fahlquist, D. A. 1963. *Seismic Refraction Measurements in the Western Mediterranean Sea.* Ph. D. Thesis, Massachusetts Institute of Technology, Cambridge, 963 p.

Fairbridge, R. W. 1961. Eustatic changes in sea level: In: *Physics and Chemistry of the Earth,* eds. Ahrens, L. H., F. Press, K. Rankama and S. K. Runcorn, Pergamon Press, Oxford, 4: 99–185.

Frassetto, R. and N. D. Crose 1965. Observations of DSL in the Mediterranean. *Bulletin de l'Institut Océanographique, Monaco,* 65:1–16.

Gehin, C. E., C. Bartolini, D. J. Stanley, P. Blavier and B. Tonarelli 1971. Morphology and Late Quaternary fill of the Western Alboran Basin, Mediterranean Sea. *NATO SACLANT Research Centre, La Spezia, Technical Report,* 201:78 p.

Gennesseaux, M. and Y. Thommeret 1968. Datation par le radiocarbone de quelques sédiments sous-marins de la région niçoise. *Revue de Géographie Physique et de Géologie Dynamique,* 10: 375–382.

Giermann, G. 1961. Erläuterungen zur bathymetrischen Karte der Strasse von Gibraltar. *Bulletin de l'Institut Océanographique Monaco,* 1218 1:28.

Giermann, G. 1962. Meeresterrassen am Nordufer der Strasse von Gibraltar. *Bericht der Naturforschenden Gesellschaft zu Freiburg i. Br.,* 52:111–118.

Giermann, G., M. Pfannenstiel and W. Wimmenauer 1968. Relations entre morphologie, tectonique et volcanisme en mer d'Alboran (Méditerranée Occidentale), résultats préliminaires de la campagne JEAN-CHARCOT (1967). *Comptes-Rendus Sommaire des Séances de la Société Géologique de France,* 4:116–117.

Glangeaud, L. 1970. Les structures mégamétriques de la Méditerranée: la mer d'Alboran et l'"arc" de Gibraltar. *Comptes-Rendus de l'Académie des Sciences, Paris,* 271:473–478.

Glangeaud, L., C. Bobier and G. Bellaiche 1967. Tectonique-évolution néotectonique de la mer d'Alboran et ses conséquences paléogéographiques. *Comptes-Rendus de l'Académie des Sciences, Paris,* 265D:1672–1675.

Guilcher, A. 1958. *Coastal and Submarine Morphology.* Methuen & Co., Ltd. London, 274 p.

Hamblin, W. K. 1962. X-ray radiography in the study of structures in homogenous sediments. *Journal of Sedimentary Petrology,* 32:201–210.

Heezen, B. C. and G. L. Johnson 1969. Mediterranean undercurrent and microphysiography west of Gibraltar. *Bulletin de l'Institut Océanographique, Monaco,* 67 (1382):95 p.

Hersey, J. B. 1965a. Sediment ponding in the deep sea. *Geological Society of America Bulletin,* 76:1251–1260.

Hersey, J. B. 1965b. Sedimentary basins of the Mediterranean Sea. In: *Submarine Geology and Geophysics,* eds. Whittard, W. F. and R. Bradshaw, Butterworths, London, 75–91.

Hsü, K. J. 1971. Origin of the Alps and Western Mediterranean. *Nature,* 233:44–48.

Huang, T. C. and H. G. Goodell 1970. Sediments and sedimentary processes of Eastern Mississippi Cone, Gulf of Mexico. *American Association of Petroleum Geologists Bulletin,* 54:2070–2100.

Huang, T. C. and J. W. Pierce 1971. The carbonate minerals of deep-sea bioclastic turbidites, Southern Blake Basin. *Journal of Sedimentary Petrology,* 41:251–260.

Inman, D. L. 1952. Measures for describing the size distribution of sediments. *Journal of Sedimentary Petrology,* 22:125–145.

Kelling, G. and D. J. Stanley 1972. Sedimentation patterns in the vicinity of the Strait of Gibraltar. In: *The Mediterranean Sea: A Natural Sedimentation Laboratory,* ed. Stanley, D. J. Dowden, Hutchinson and Ross, Inc., Stroudsburg, Pennsylvania, 489–519.

Kermabon, A., P. Blavier, V. Cortis and H. Delauze 1966. "Sphincter" corer: A wide-diameter corer with watertight core-catcher. *Marine Geology,* 4:149–162.

Kuenen, Ph. H. 1964. The shell pavement below oceanic turbidites. *Marine Geology,* 2:236–246.

Kullenberg, B. 1947. The piston core sampler. Svenska Hydrografisk-Biologiska *Kommissionens Skrifter,* 1:25 p.

Kullenberg, B. 1952. On the salinity of the water contained in marine sediments. *Meddelanden från Oceanografiska Institutet i Göteborg,* 21:1–38.

Lacombe, H. and P. Tchernia 1960. Quelques traits généraux de l'hydrologie Méditerranéenne. *Cahiers Océanographiques,* 12: 527–547.

Lacombe, H. and P. Tchernia 1972. Caractères hydrologiques et circulation des eaux en Méditerranée. In: *The Mediterranean Sea: A Natural Sedimentation Laboratory,* ed. Stanley, D. J. Dowden, Hutchinson and Ross, Inc., Stroudsburg, Pennsylvania, 25–36.

Leclaire, L. 1970. *La Sédimentation Holocène sur le Versant Meridional du Bassin Algéro-Baléare (Pré-continent Algérien).* Thèse de Doctorat d'état. Faculté des Sciences, Paris, 1–3: 552 p.

Leclaire, L. 1972. Aspects of Late Quaternary sedimentation on the Algerian Precontinent and in the adjacent Algiers—Balearic Basin. In: *The Mediterranean Sea: A Natural Sedimentation Laboratory,* ed. Stanley, D. J., Dowden, Hutchinson and Ross, Inc., Stroudsburg, Pennsylvania, 561–582.

Lucayo, N. C. 1968. Contribucion al conocimento del Mar de Alboran I. Superficie de referencia. *Boletin Instituto Español de Oceanografia,* 135: 28 p.

McKenzie, D. P. 1970. Plate tectonics of the Mediterranean region. *Nature,* 226: 239–243.

Mars, P. 1963. Les faunes et la stratigraphie du quaternaire Méditerranéen. *Recueils des Travaux de la Station Maritime d'Endoume Bulletin,* 28 (43): 61–97.

Mikesell, M. W. 1961. *Northern Morocco, a Cultural Geography.* University of California Press, California 122 p.

Milliman, J. D., Y. Weiler and D. J. Stanley, 1972. Morphology and carbonate sedimentation on shallow banks in the Alboran Sea. In: *The Mediterranean Sea: A Natural Sedimentation Laboratory,* ed. Stanley, D. J., Dowden, Hutchinson and Ross, Inc., Stroudsburg, Pennsylvania, 241–259.

Moore, D. G. 1969. Reflection profiling studies of the California continental borderland: structure and Quaternary turbidite basins. *Geological Society of America Special Paper,* 107: 142 p.

Nesteroff, W. D. 1961. La "séquence type" dans les turbidites terrigènes modernes. *Revue de Géographie Physique et de Géologie Dynamique,* 4: 263–268.

Nesteroff, W. D. 1962. Essai d'interpretation du mécanisme des courants de turbidité. *Bulletin de la Société Géologique de France,* 7: 849–857.

Nesteroff, W. D., G. Sabatier and B. C. Heezen 1965. Les minéraux argileux dans les sédiments du bassin occidental de la Méditerranée. *Commission Internationale pour l'Exploration Scientifique de la Mer Méditerranée,* 17 (3), 1005–1007.

Nesteroff, W. D., W. B. F. Ryan, K. J. Hsü, G. Pautot, F. C. Wezel, J. M. Lort, M. B. Cita, W. Mayne, H. Stradner and P. Dumitrica 1972. Evolution de la sédimentation pendant le Néogène en Méditerranée d'après les forages JOIDES-DSDP. In: *The Mediterranean Sea: A Natural Sedimentation Laboratory,* Dowden, Hutchinson and Ross, Inc., Stroudsburg, Pennsylvania, 47–62.

Norrish, K. and R. M. Taylor 1962. Quantitative analysis by X-ray diffraction. *Mineralogical Society of Great Britain, Clay Mineral Bulletin,* 4: 109–111.

Ryan, W. B. F. 1971. Can an ocean dry up? Results of deep-sea drilling in Mediterranean. *(Abstract) Annual Meeting, Association of Petroleum Geologists and Society of Economic Paleontologists and Mineralogists, Houston, Texas,* 362.

Ryan, W. B. F. 1972. Stratigraphy of Late Quaternary sediments in the Eastern Mediterranean. In: *The Mediterranean Sea: A Natural Sedimentation Laboratory,* ed. Stanley, D. J., Dowden, Hutchinson and Ross, Inc., Stroudsburg, Pennsylvania, 149–169.

Ryan, W. B. F., D. J. Stanley, J. B. Hersey, D. A. Fahlquist and T. D. Allan 1970. The tectonics and geology of the Mediterranean Sea. In: *The Sea,* ed. Maxwell, J. C., Wiley Interscience, New York, 4: 389–492.

Scientific Staff 1970. Deep Sea Drilling Project: Leg 13. *Geotimes,* 15: 12–15.

Stanley, D. J. (Editor) 1969. *The NEW Concepts of Continental Margin Sedimentation.* American Geological Institute, Washington, D. C., 400 p.

Stanley, D. J., C. E. Gehin and C. Bartolini 1970. Flysch-type sedimentation in the Alboran Sea, Western Mediterranean. *Nature,* 228: 979–983.

Sverdrup, H. U., M. W. Johnson and R. H. Fleming 1942. *The Oceans, Their Physics, Chemistry and General Biology.* Prentice Hall, Englewood Cliffs, New Jersey, 1087 p.

Texas Instruments Incorporated 1967. 1965–67 North Atlantic Ocean, Norwegian Sea and Mediterranean Sea Area 6. *U.S. Naval Oceanographic Office Report, Contract No.* N 62306–1687, 48 p.

Todd, R. 1958. Foraminifera from Western Mediterranean deep-sea cores. *Reports of the Swedish Deep-Sea Expedition, 1947–1948,* 8 (3): 167–215.

van Straaten, L. M. J. U. 1967. Turbidites, ash layers and shell beds in the bathyal zone of the southeastern Adriatic Sea. *Revue de Géographie Physique et de Géologie Dynamique,* 9: 219–240.

van Straaten, L. M. J. U. 1970. Holocene and Late-Pleistocene sedimentation in the Adriatic Sea. *Geologische Rundschau,* 60: 106–130.

van Straaten, L. M. J. U. 1972. Holocene stages of oxygen depletion in the deep waters of the Adriatic Sea. In: *The Mediterranean Sea: A Natural Sedimentation Laboratory,* ed. Stanley, D. J., Dowden, Hutchinson and Ross, Inc., Stroudsburg, Pennsylvania, 631–643.

Vergnaud-Grazzini, C. and C. Bartolini 1970. Evolution paléoclimatique des sédiments würmiens et post-würmiens en mer d'Alboran. *Revue de Géographie Physique et de Géologie Dynamique,* 12: 325–334.

Vita-Finzi, C. 1969. *The Mediterranean Valleys.* Cambridge University Press, London, 139 p.

Vogt, P. R., R. H. Higgs and G. L. Johnson 1971. Hypotheses on the origin of the Mediterranean Basin: Magnetic data. *Journal of Geophysical Research,* 76: 3207–3228.

Warshaw, C. M. and R. Roy 1961. Classification and a scheme for the identification of layer silicates. *Geological Society of America Bulletin,* 72: 1455–1492.

Way, R. 1962. *A Geography of Spain and Portugal.* Methuen, London, 362 p.

Weibull, W. 1947. The thickness of ocean sediments measured by a reflexion method. *Meddelanden Oceanografiska Instituteti Göteborg,* 12: 17 p.

Wong, H. K., E. F. K. Zarudzki, S. T. Knott and E. E. Hays 1970. Newly discovered group of diapiric structures in Western Mediterranean. *American Association of Petroleum Geologists Bulletin,* 54: 2200–2204.

Wüst, G. 1960. Die Tietenzirkulation des Mittelländischen Meeres in den Kernschichten des Zwischen-und des Tieferwassers. *Deutsche Hydrographische Zeitschrift,* 13: 105–131.

Wüst, G. 1961. On the Vertical Circulation of the Mediterranean Sea. *Journal of Geophysical Research,* 66: 3261–3271.

Zeuner, F. E. 1952. *Dating the Past, Introduction to Geochronology.* 3rd edition, Methuen, London, 495 p.

Aspects of Late Quaternary Sedimentation on the Algerian Precontinent and in the Adjacent Algiers-Balearic Basin

Lucien Leclaire

Muséum National d'Histoire Naturelle, Paris

ABSTRACT

The textural, mineralogical and geochemical attributes of about two thousand samples of the Algerian continental shelf sediments and thirty three other samples dredged and cored from the adjacent continental rise and the abyssal plain have been investigated and the main results of this study are summarized in this paper.

The sediments of this very narrow shelf are classified into four facies: (1) detrital sands of the littoral prism with quartz and tourmaline, (2) bay muds (*vasières*) with illite and quartz, (3) algal and bryozoan calcareous sands and gravels, and (4) foraminiferal oozes of the shelf edge. The distribution of these facies is similar to the sediment distribution on broader continental shelves. The Holocene muds of *vasières* commonly blanket a large part of the old calcareous (relict) sands.

On the continental rise and the abyssal plain, pelagic sediments are interbedded with numerous detrital sands and the recent Quaternary deposits are distinctly stratified. A study of the planktonic foraminiferal assemblages shows that this stratification results from recent Quaternary climatic fluctuations.

RESUME

Environ deux mille échantillons des sédiments du plateau continental algérien, trente-trois dragages et carottages du glacis et de la plaine abyssale ont été analysés par différentes méthodes: sédimentologiques, minéralogiques et géochimiques. Les principaux résultats de cette étude, déjà développés dans une thèse, sont résumés dans cette note.

Sur le plateau continental algérien, les différents sédiments peuvent être groupés en quatre faciès: (1) les sables siliceux à quartz et tourmaline de la frange littorale, (2) les boues à illite et quartz des vasières, (3) les sables et graviers calcaires à Algues et Bryozoaires, et (4) les boues à Foraminifères du rebord. Ces sédiments se repartissent de la même façon que les dépôts de plateaux continentaux beaucoup plus vastes. Des vasières holocènes recouvrent vraisemblablement une grande partie des sables calcaires anciens (sables reliques).

Sur le glacis et dans la plaine abyssale, de nombreuses passées de sable détritique fin s'intercalent dans les sédiments pélagiques. Les dépôts du Quaternaire récent sont nettement stratifiés. Une étude des associations de Foraminifères planctoniques montre que cette stratification a pour origine les oscillations climatiques qui ont marqué le Quaternaire récent.

INTRODUCTION

The Algerian continental margin constitutes almost all the southern border of the major depression of the western Mediterranean Sea: the Algéro-Provençal Basin. The Algiers-Balearic Basin is the southward extension of this depression.

By reason of its morphology and of the nature of the sediments emanating from it, the Algerian margin contributes significantly to the filling of the basin. This paper provides some details about the main morphologic characteristics of the Algerian continental shelf, continental slope and adjacent basin plain together with a brief review of the essential sedimentary features and processes operating in these regions during the late Quaternary and especially during the Holocene.

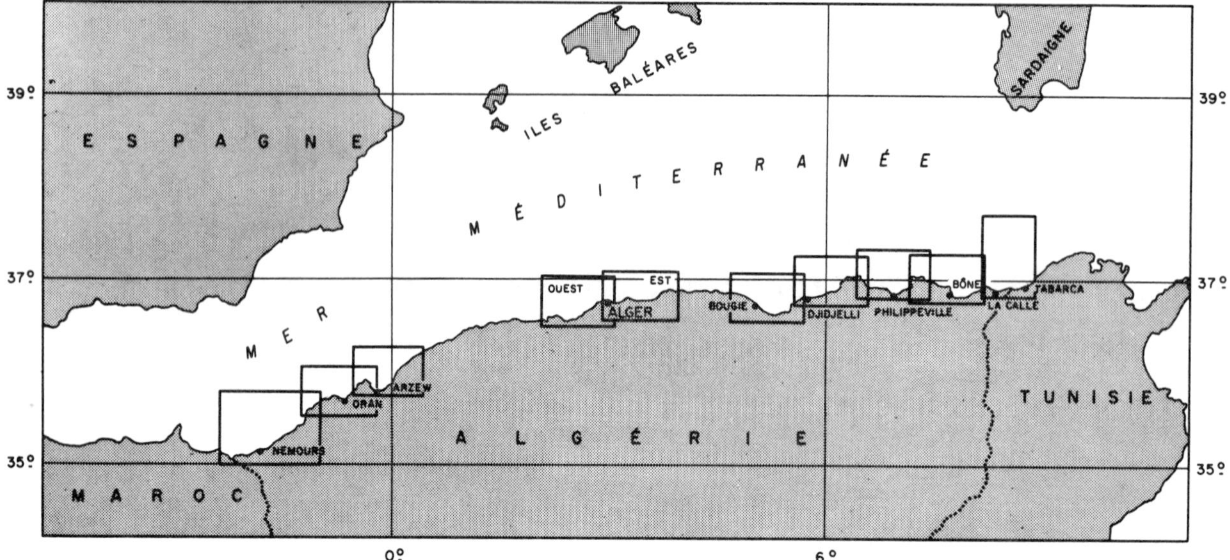

Figure 1. Chart of the western Mediterranean Sea showing the Algerian continental margin. Insets show sedimentologic maps on Figure 2, 3 and 4 and also Caulet (this volume). Modified after Leclaire (1968d and 1970).

MORPHOLOGY

Description

Detailed study of the morphology of the Algerian margin has formed the subject of previously published notes and charts (Rosfelder, 1955: 3 maps at the scale of 1/500.000; Leclaire, 1968d: 10 maps at the scale of 1/100.000, see Figure 1), which reveal this margin to be the steepest and most narrow one in the Mediterranean Sea. Indeed, from the Bay of Oran to the Gulf of Bône (700 km apart), the precontinent does not extend seawards for more than 25 to 30 km on average and the width is restricted to some ten kilometers in front of headlands (Cape Falcon near Oran, Cape Bougarouni near Collo) but increases appreciably off the Moroccan and Tunisian borders.

On the shallow part of the precontinent a break in slope usually can be seen: this is the shelf break, which forms the seaward edge of what may be broadly termed the continental shelf. The lower boundary of this *precontinent* (continental shelf plus continental slope) is sometimes difficult to delineate, but is usually situated between the 2000 and 2500 m isobaths where the continental slope merges gradually with the continental rise and the basin plain.

The Continental Shelf

The Algerian Shelf is generally very narrow (only a few kilometers wide), and resembles a series of benches or ledges. In front of headlands or high coastal mountains (Sigale Cape, Cape Aiguille near Oran, Great Kabylia mountains, Collo mountain) the shelf may be only a few kilometers in width (1 to 4 km) but it broadens to perhaps 20 km in large gulfs such as those of Arzew, Nemours, Bône (20 km). In the Bay of Castiglione and the Gulf of Philippeville, the shelf is about ten kilometers wide.

The depth of the generally prominent shelf break is also very variable: 150 m (Gulf of Nemours), 100 to 140 m in the Bay of Oran, 120 m (Gulf of Arzew), 100 m and less at the foot of Collo mountain.

Two main areas characterize this narrow shelf: (1), the inner shelf extending from 0 to 50 m depth, with a slope greater than 1° and formed either by the under-sea extension of rocky cliffs bordering the shore or by a sandy prism (or talus) overlapping the bed-rock; (2), the outer shelf, a continental platform that extends below a depth of 50 m to the shelf break and is gently sloping (less than 0.5°) and dissected by gullies, with submarine plateaus and rocky banks, all various features of a relict morphology. The submarine plateaus of the Gulf of Bône are probably formed by Tyrrhenian beach sandstones, the Djidjelli Bank by Numidian sandstone and the Toukoush Bank by eruptive rocks.

In addition to the variability in depth of the shelf break, the slope of the continental platform is also variable. Where the inner shelf merges gradually into the outer shelf (Courbet Bay and Djidjelli Bay, Figure 4), the slope is greater than 1°.

The Continental Slope

This feature also varies both in width and gradient and is often particularly steep (15°) off some coastal

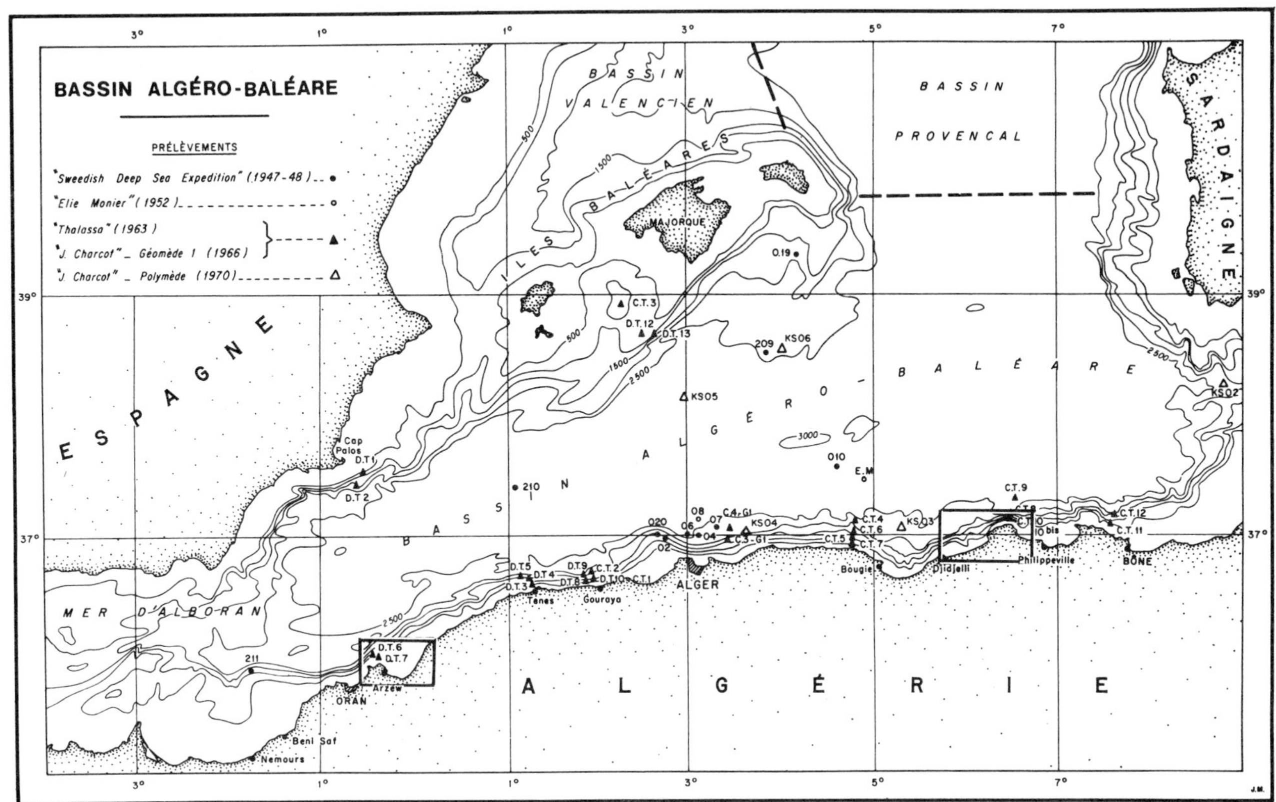

Figure 2. Chart of a part of the western Mediterranean Sea showing the main physiographic feaures of the Algiers-Balearic basin and the location of deep-sea samples and other samples cited in text.

mountains such as Collo mountain. In a general way *a precipitous continental slope and a narrow shelf both occur off a mountainous coast while a gentler continental slope generally borders the wide shelf of a low-lying coast.*

The slope is deeply dissected by numerous canyons and submarine valleys whose heads cut the shelf edge (Rosfelder, 1955). Seawards from the Gulf of Bougie, several canyons converge into broad submarine "amphitheatres."

The Courbet-Marine Canyon is the largest on the Algerian continental margin. A core (C. 3 Geom. I, Figure 2) from a depth of 2000 m near the canyon axis is believed to countain early Pleistocene or Pliocene deposits (Leclaire and Le Calvez, 1969). The Moules Canyon (Oranie) heads at a depth of 60 m, very near the sandy littoral prism, and a core from this canyon head reveals the presence of eruptive rocks. Very few of these canyons can be related, with any certainty, to river valleys on the adjacent land. Only the ravines of the precipitous continental slopes in the vicinity of high coastal mountains are often found to be continuations of land valleys or gullies.

Some outcrops have been recognized on the upper part of the continental slope: Miocene marl at a depth of about 300 m (Oranie), and metamorphic rocks at about 600 to 700 m in front of Bouzarea mountain (near Algiers). These outcrops represent submarine extensions of geological formations recognised on the adjoining land-surface.

The Continental Rise

As a general rule, the junction of the continental slope with the flat basin floor is rarely represented by a break in slope. From about 2000 m depth, a gentle slope of 1° or less leads to the basin plain, beginning at an average depth of 2500 to 2600 m. This region, several tens of kilometers wide, usually defined as the continental rise, is differentiated from the almost completely flat plain (apart from abyssal hills) by a smooth relief with low hills and undulations.

Interpretation and Discussion

As to the continental shelf, the variability of the morphology indicates that the Algerian continental margin does not include a shelf proper but comprises a set of platforms and ledges, the slope and depth of which are different. Some morphotectonic and geomorphologic studies (Bourcart and Glangeaud, 1954;

Leclaire, 1968a) have previously suggested that the heterogeneity of the Algerian Shelf is related to the late Cenozoic diastrophism that affected the hinterland. It has been established that narrow benches are always carved in the flanks of coastal mountains and intrusive rocky bodies which have undergone strong uplift (horsts of Oranie for instance, Gourinard, 1958) movements by which, locally, the earlier Quaternary (Calabrian) was raised by as much as several hundred meters (Laffitte, 1950). On the other hand, large platforms are generally situated in the axes of submarine extensions of land depressions and plains, or on surfaces of Neogene or Plio-Quaternary sublittoral faulted or subsiding basins (Bône Plain, Fetzara depression, Habra Plain, *etc.*). In a general way, the result is that "The widest shelves can be shown also to be the deepest shelves" (Curray, 1969a, p. 10).

Ledges and benches probably have been shaped by marine abrasion and are more or less buckled. Gulfs and bays result from aggradation during regressions (with subaerial erosion) and from marine erosion during transgressions. Evidence of the pre-Holocene regression and of the Holocene transgressions across the Algerian Shelf can be found in the following features: flooded landscape of the Gulf of Philippeville; features suggestive of an old shoreline at 90 to 100 m on the floor of the Gulf of Bône and between Cape Rosa and Tabarca (in Tunisia); planed surfaces and terrace levels near the shelf break of the Gulf of Bône; the gully in the Gulf of Bougie which opens on the upper part of the continental slope at about 110 m; pebbles of presumed old offshore bars at a depth of 30 to 70 m near Toukoush and Djidjelli banks and so on. In addition a relict topography, now partly obscured, can be discerned on the western part of the Gulf of Nemours and in the eastern part of the Bay of Djidjelli (Figure 4). Thus, it appears that the relief of the Algerian Shelf originates from two main processes: (1) late Cenozoic tectonic activity which varied in intensity in different areas, and (2) glacio-eustatic sea-level fluctuations.

Recent seismic surveys (Auzendre, 1969) have revealed the morphology of large gulfs such as Bougie and Philippeville, the origin of which is assigned to faulted basin-like structures. The continental slope displays considerable thicknesses of unconsolidated Plio-Pleistocene sediments only below these basins or in smaller secondary basins, more or less distorted, which have sometimes given rise to terraced surfaces. The sediment blanket seems scarcely noticeable and does not occur at all in several places where rocky substratum outcrops. The geological framework of the eastern slope essentially seems to comprise Cretaceous and Palaeocene rocks with here and there a Miocene blanket of variable thickness.

Thus the prominent features of the Algerian continental slope result mostly from the interaction of tectonic stresses producing numerous fractures in rigid formations, especially eruptive or intrusive rocks, and a variety of flexures and flexure-faults. This combination of processes produced what Bourcart (1950) named "la flexure continentale". According to Glangeaud (1962), these deformations accompanied the foundering of the Mediterranean Sea which affected the surrounding continental edges during three successive late Cenozoic phases until the Recent epoch, and possibly continue today.

This "flexuration" may have produced the submergence of canyons and former land valleys. Many of the submarine canyons are cut in very hard rocks and their origin therefore seems to be not just a result of submarine erosion. Such submarine canyons may then represent submerged traces of Cenozoic coastal drainage systems renewed by rivers at the end of the Pliocene or perhaps at the end of the early Quaternary. Perhaps this is the reason why all the canyons cannot be connected to the present drainage system; formation of the lower part of this system probably ranges back in time to the beginning of the middle Quaternary.

The continental rise consists of an accumulation of sediments ranging from about 500 to more than 1000 m in thickness (eastern part, Auzendre, 1969) whose upper layers are the usual Plio-Quaternary base-of-slope deposits (turbidites). These deposits grow thinner towards the basin plain where they overlie layers of a different nature which are affected by plications of variable magnitude. To some extent, this rise is equivalent to what American oceanologists term the continental *rise* which characterizes, for instance, the eastern margin of North America. However, the Algerian continental rise is narrower and much less thick.

Many attempts have been made to classify the different types of continental margin. The Algerian margin looks like Curray's steep-rift type (Curray, 1969b). With the Provençal margin, it constitutes a model of down-warped continental edges and an example of the flexured-type of Guilcher (1963). Moreover, this Algerian margin represents a youthful or early stage of development (Dietz, 1952) possibly maintained in this condition by tectonic effects which have played a more important role in its formation than processes of sedimentation or eustatic changes.

SEDIMENTATION ON THE CONTINENTAL SHELF

The information summarized in the following pages and in Figures 3, 4 and 5 is derived from ten detailed

Table 1. Average heavy mineral content in Algerian coastal samples.

Localities	Percentage of heavy minerals
The beach sands of Arzew (West)	0.5
The black sands of Cap-de-Garde	80.0
The Algerian shore-wide average	1.0 to 3.0

charts at the scale of 1/150.000 (Leclaire, 1968d, 1970; Figure 1), which incorporate a new sediment classification (Leclaire, 1968b, 1969b used as a base for sedimentological mapping.

Nearshore and Inner Shelf Sedimentation

An important nearshore sand prism borders the Algerian coast along a total length of more than 350 km. Generally, it forms the bottom of bays and gulfs from 0 to 20 m deep, and sometimes down to 50 m; occasionally, it goes round a headland like Djinet Cape (Algérois). This talus gives rise to numerous broad and extensive beaches (Bône, Djidjelli, Philippeville, *etc.*) and is often backed by great dune systems which have contributed material to the beaches. This sand body is interrupted by rocky peninsulas or coasts where shingle beaches are forming and where modern biogenic calcareous sands and gravels are being deposited in small amounts. These calcareous sands originate from contemporary infralittoral plant and animal associations (calcareous algae, Bryozoa, molluscs, *etc.*) and are distributed in rocky hollows or form a general veneer.

Grain-size analysis of numerous samples of sand from more than 60 beaches and from adjacent dune systems reveals the texture of beaches to be similar to the texture of dunes. Using the method of Curray (1960), it appears that cumulative curves of grain size distribution are always unimodal with small and very similar anomalies (asymmetry, skewness), while the most frequent mode of dune sands from Morocco to Tunisia is about 2.25ϕ.

Mineralogic analysis has shown quartz to be the chief component of the sandy littoral edge (50 to 70% of total weight). However, the calcareous content (15 to 30%) is also substantial, with mainly bioclastic debris (algae, shells) and also some minerals (calcite, dolomite). The heavy mineral content is highly variable (Table 1).

Some mineral species occur in all the sands of the littoral edge and constitute a common mineralogic suite with prevailing tourmaline and the following species:

- micas: biotite, chlorite,
- pyroxenes: diopside, augite,
- metamorphic silicates: andalusite, staurolite, garnet, epidote,
- accessory silicates: zircon, glauconite,
- oxides and hydroxides: titanite, goethite, magnetite, rutile, apatite, ilmenite.

Certain minerals are very scarce and confined to particular places:

- scheelite in the black sands of Cap-de-Garde,
- cinnabar found sporadically near Kabylia,
- cassiterite in the black sands of Cap-de-Garde and in Cap-de-Fer sands,
- chromite (Cap-de-Garde and Djidjelli),
- fluorite (Cap-de-Fer).

Other species, by their frequency and assemblages, characterize certain regions. These include the eastern (Constantinois)–metamorphic silicates with prevailing staurolite and garnet; western (Oranie)–amphiboles and pyroxenes with titaniferous augite and chrysolite (Gulf of Nemours); and near Cap-de-Fer or Collo mountain – actinolite and automorphic biotite, glauconite and tourmalines (blue, yellow, brown, black, bicolored) from Sidi Ferruch dune sands.

Continental Platform Sediments

Vasières (a French term used to designate broad, muddy areas in bays and gulfs) result from mud accumulation in depressions. They are sometimes very extensive as in the Gulf of Arzew (Figure 3) and Bône, and their geometry is highly variable. Thus the *vasière* of the Gulf of Bône is curiously amoeboid in shape. No data are available on the thickness of these *vasières* but they are believed to extend to several meters depth and may extend to several tens of meters where the gulf is strongly depressed. Muds accumulated in this area are fine sandy ones, gray or black-grayish or gray-brownish; off the Gulf of Nemours they are brown.

On average, these muds consist of about:

- 1% of arenite (particles with $\phi \geqslant 100\mu$),
- 45% of aleurite ($2\mu < \phi < 100\mu$),
- 55% of pelite ($\phi \leqslant 2\mu$).

Mud is generally composed of $CaCO_3$ (20%), clay (53%), siliceous silt and fine sand (27%). The calcium carbonate has been chemically analysed. The clay fraction in weight corresponds roughly to clay minerals plus pelitic quartz and other siliceous minerals or fragments (less than 10% of the pelitic part); it is calculated by subtracting the pelitic $CaCO_3$ from total pelitic part of the sediment. Siliceous silt and sand correspond to the non-carbonate arenitic and aleuritic fraction. The amounts of these three major components are expressed in relative

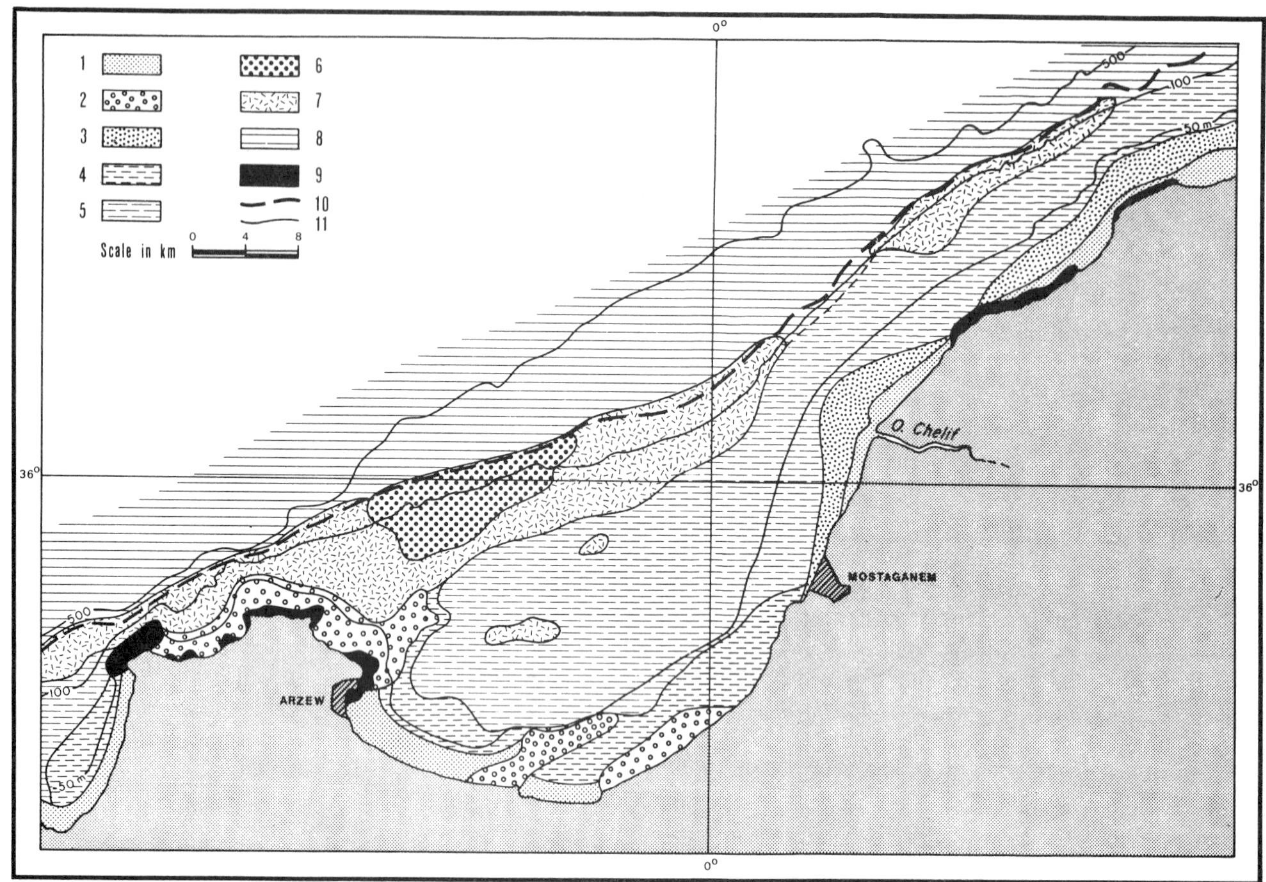

Figure 3. Chart showing the distribution of sediments in the Gulf of Arzew. The major facies include: 1, sands of the nearshore modern sand prism; 5, muds and sandy muds of mud-bands and *vasières:* 6, relict calcareous and glauconitized sands and gravels; 7, glauconitized and clayey relict sands; and 8, foraminiferal muds of the shelf edge. Other sediment facies: 2, modern biogenic sands and gravels; 3, sandy muds (transition); 4, calcareous and argillaceous muds (transition); 9, rocky outcrops; 10, shelf break; 11, isobath (m).

percentages and represent more than 96% of total dry sediment weight.

The calcium carbonate derived entirely from contemporary biocenosis is very low. Foraminifera are very scarce, especially the planktonic species. Likewise, Coccoliths seem scarce and most of them come from the Neogene marl (late Cenozoic). The greatest part of the calcium carbonate may originate either from reworking of fine calcite or dolomite grains or from Foraminifera and Coccolith fragments brought by rivers onto the shelf. In addition, some carbonate is certainly derived from shells, tests and various calcareous debris (algae, molluscs, *etc.*) along the surf and swash zones. The fine sand and silt that constitutes a notable part of the total *vasière* sediment is generally less rich in quartz than beach sand, and richer in micas and in small fragments of altered minerals.

Three groups of clay minerals are significant in these muds: illite and mixed-layer minerals like illite-montmorillonite (70%); kaolinite (20%); and chlorite (10%). This composition can change from one *vasière* to another. For instance, chlorite is sometimes absent but, locally, also may be relatively abundant (Gulf of Bougie, 20 to 30%). Montmorillonite, related to the presence of coastal volcanic formations, is relatively abundant in the Gulf of Nemours.

These muds contain also minor components such as organic matter and iron sulfides. The average organic carbon comprises from 0.5 to 0.7% of the dry sediment weight while average quantities of nitrogen vary between 0.08 and 0.10%. Values of the C/N ratio vary from one *vasière* to another but remain relatively uniform within a single *vasière*. A statistical study (Leclaire, 1968c) has shown a dependence of nitrogen content on organic carbon concentration. This dependence is practically linear, with a correlation factor often higher than 0.8. Hydrotroilite is abundant in Gulf of Bougie ($S^{--} = 0.012$ to 0.017% of dry sediment weight), with traces of pyrite. On the other hand, muds of the Gulf of Nemours are rich in ferric hydroxide ($Fe^{3+} = 3\%$, on average).

Shelf edge muds closely resemble the sediments

accumulated in the *vasières*. However, they are distinguished by the presence of a microfauna that, exceptionally, represents 40% of the total calcium carbonate. The foraminiferal assemblage is generally marked (particularly near Oranie) by the prevalence of planktonic species like *Globorotalia inflata* with *Globigerinoides rubra, Globigerinoides sacculifera, Globigerinella aequilateralis, Globorotalia truncatulinoides, Globigerina bulloides* and so on.

Among the benthic forms: *Hyalinea baltica, Bolivina, Cassidulina, Bulimina, Uvigerina, Cibicides, Elphidium, Ammonia, etc.* are the most common ones. *Bolivina* and *Bulimina* associated with *Cassidulina* are largely dominant in the outer shelf muds of Kabylia (Gulf of Bougie, Bay of Djidjelli). Coccoliths are generally scarce but sometimes constitute up to 20% of the total calcium carbonate. Therefore, these muds tend to be *foraminiferal oozes*.

The composition of the *calcareous sands and gravels* has been fully discussed by Caulet (1970 and this volume). They are of two kinds: shelly sands with encrusting or *praline shaped* calcareous algae and Bryozoa which accumulate on top of rocky banks or surround them; and calcareous algal sands with *Lithothamnium calcareum* and 'old' mixed debris forming broad elongated banks (Figure 3 and 4) near the shelf break.

The former are modern biogenic sands which are being formed today from contemporary calcareous organisms. According to Caulet (1970), the algal sands are certainly far older; this point of view is substantiated by the following observations:

1. Such sands are located in areas of relict topography.
2. The calcareous algae which constitute an important fraction of the sands, especially *Lithothamnium calcareum*, occur at abnormal depths, ranging from about 80 to more than 150 m.
3. Much of the calcareous debris is colored by limonite impregnation, colored red by iron oxide as a consequence of subaerial exposure, or blackened by iron sulphides or manganese hydroxide.
4. Within the gulfs of Nemours and Oran, calcareous sands and gravels are heavily encrusted or mixed with glauconite or limonite grains; berthierine (chamosite) may be present.
5. These sands contain foraminiferal tests filled with glauconite and goethite and scarce fossil species which are completely calcified.
6. Tests, shells and other debris tend to be abraded, broken, corroded and bored by the sponge *Clione*.
7. They are often mixed with minor amounts of siliceous sand carrying eolian quartz grains and heavy minerals similar to those of the nearshore sand prism and dunes; and
8. The texture and the nature of these sands can be compared to those of the Tyrrhenian calcareous sandstones.

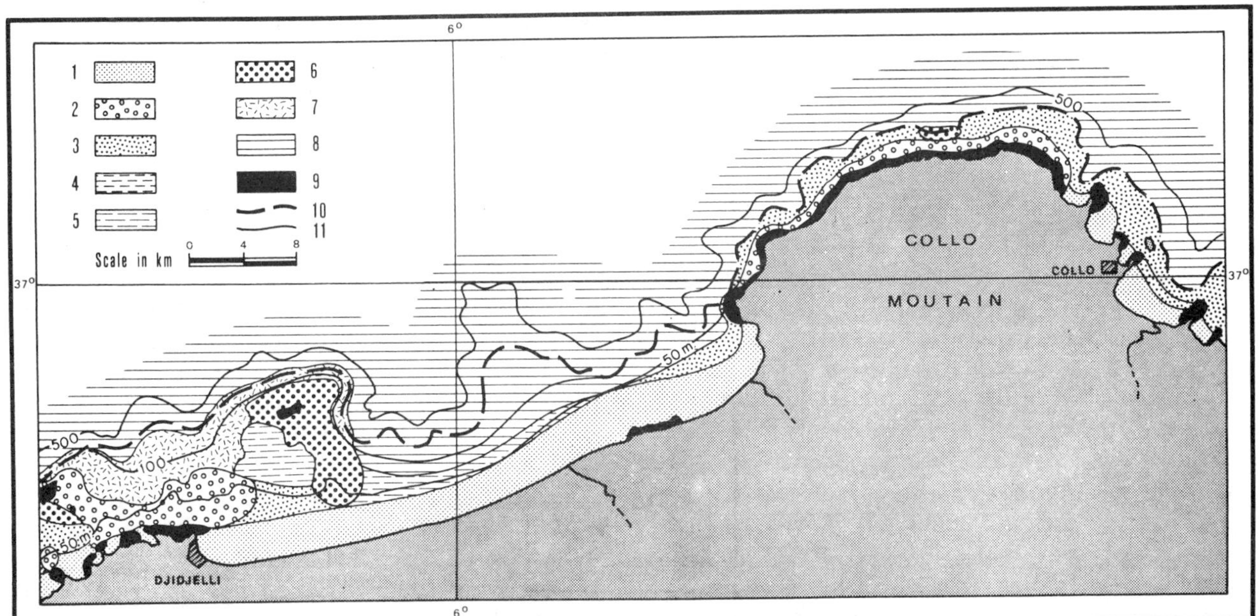

Figure 4. Chart showing the distribution of sediments in the Bay of Djidjelli. The major facies include: 1, sands of the nearshore modern sand prism; 5, muds of the *vasière* of Djidjelli; 6, calcareous and glauconitized relict sands and gravels; 7, clayey relict sands; and 8, foraminiferal muds of the outer shelf. Other sediment facies: 2, modern calcareous sands and gravels; 3, sandy muds (transition); 9, rocky outcrops; 10, shelf-break location; 11, isobath (m).

Sediment Supply on the Algerian Shelf

Few data are available concerning planktonic production in this area. Kruger (1950) counted about 30,000 to 100,000 Coccolithophorid cells per liter of Algiers Shelf waters, with an exceptional maximum of 240,000. Moreover, this author assessed 10,000 to 30,000 Diatoms per liter of similar waters. Bernard and Lecal (1953) estimated the amount of Coccolithophorids in Mediterranean waters to range from about 50,000 to 500,000 cells/liter. We have not found quantities of coccoliths or diatoms in the sediments which would agree with such amounts.

Investigating ten years of runoff and suspended load observations recorded in different gauging stations (of the "Service de l'Hydraulique et de l'Equipement rural de l'Algérie") implanted on the main rivers, we have estimated the annual discharge of suspended load to the sea at about 40 to 60 million tons. More than 2/3 of this discharge settles on the eastern continental shelf. During winter periods with great spates, certain rivers (Oued Mazafran in 1954, Oued Soummam in 1955) supply several millions of tons of sediment per month while runoff ranges from tens to one or two hundred million cubic meters. In September 1955, at the gauging station of Sidi Aïch (O. Soummam), the total runoff was approximately 38 $\times 10^6$ m^3 and the total suspended load was 2 $\times$ 10^6 tons. During this month, the concentration of sediment load, *i.e.*, the rate of turbidity was therefore, on average, greater than 50 gm/liter.

INTERPRETATION OF CONTINENTAL SHELF SEDIMENTATION

Origin of the Nearshore Sand Prism

Grain size distribution data together with morphoscopic analyses indicate that the material of the sandy shore prism is derived essentially from recent and old sand dunes. Recent fluviatile supplies seem to play a secondary role, only noticeable close to river mouths. The similarity of texture in the different dune systems is probably related to the simultaneous creation of a great part of the whole dune system by the prevailing northwest wind system which swept the exposed continental shelf during the last pre-Holocene regression or during the Holocene transgression (Hilly, 1957).

The common mineralogic suite with strongly reworked species (quartz, tourmaline, zircon) and a comparative study of sandstones of the adjoining land demonstrate that a great part of the sands of the Algerian coast originates from erosion of the huge detrital Cretaceous and Palaeogene formations. Local or regional suites with rare and fresh species are produced, in all likelihood, by recent erosion of eruptive and intrusive or metamorphic coastal rocks, uplifted during the last Plio-Quaternary tectonic phases. These local contributions constitute a very small part of the littoral sand prism.

Some previously published studies (Leclaire, 1963a, 1963b; Leclaire and Finck, 1962) have shown the modern nearshore sand prism to be intensively stirred by currents and swell-induced sea movements. These currents produce great morphological changes in the sand prism, with seasonal prograding and recession of beaches and nearshore zones.

Old Calcareous Sands and Gravels

Such calcareous sands and gravels form a particular facies (see above). Its presence near the shelf break, usually surrounded by finer sediments, shows that it is not produced by present or recent sedimentation. These old sands are no longer in their original environment and they seem to be quite similar to the so-called relict sands (Emery, 1952). More accurately, these deposits are residual sands (see Swift, 1969). It is quite probable that, as in the case of the Carolina Shelf sands (Dill, 1965; Gorsline, 1963), a part of the Algerian residual sand is derived from reworking of underlying Pleistocene or Neogene strata during low stands of sea leval, at depths ranging from about 80 to 120 m. Presumably, these relict sands were deposited during the later stages of the pre-Holocene regression or during the earlier stages of the Holocene transgression.

Relict sands are rarely clay-free on the Algerian Shelf. The amount of clay increases from the center of the calcareous banks towards the margins, especially near the *vasières*, and thus there may be distinguished some transitional sediments which may completely surround the old sands (see the Gulf of Arzew, Figure 3). This provides clear evidence of burial of the relict sands *i.e.*, of a silting process for which alternative evidence is provided elsewhere. Silting occurs on the narrow shelf adjacent to Collo Mountain, where *Corallium rubrum* stalks occur at a depth of about 100 m, fixed on bed-rock and entirely embedded in a mud blanket.

This silting is particularly obvious on the shelf off the Gulf of Oran and its approaches, which is carpeted by relict sands covered with a film of brown mud. Such a process promotes the seaward extension of the *vasières*. It can be related to the high concentration of fine sediment load in the river discharge, due to the changing climatic conditions: from a Mediterranean climate to a subarid type with strong erosion of soils. The observed silting of the continental shelf, which started at the end of the Holocene transgression, still proceeds gradually in spite of the bypassing of much of the alluvial supply (fed directly onto the slope and basin plain).

Clay deposits on top of relict sands are accompanied by deposition of fresh shells of small molluscs and Bryozoa tests that contrast with the worn and abraded old shells and glauconitized or calcified foraminifera. Thus, some old Posidonia beds include foraminifera such as *Elphidium crispum, Ammonia beccarii* and Miliolidae, all stained by iron oxide or glauconitized, which are accompanied by modern species such as *Globorotalia truncatulinoides, Globorotalia inflata, Globigerinoides ruber* (pink), *Bolivina, Uvigerina, Bulimina, Cassidulina* and so on. Initially, it was believed that this modern fauna was covering the relict sands and that the observed mixing was merely an artefact of the sampling method. However, since this mixing accompanies the lateral passage of one facies into another and is also observed in some cores, we now believe that it really exists in the sediment. Thus, there would be a sort of vertical and horizontal gradation involving both the amounts of clay and modern sediment quantities mixed with the old sands. Presumably, this gradation is comparable to that observed in Chesapeake bay by Powers and Kinsman (1953). According to Swift (1969, 1970), this phenomenon provides good evidence of modern reworking of a relict sand.

General Sediment Distribution Pattern

Apart from the nearshore sandy prism, where continuous wave-winnowing results in a lack of clay particles, one might expect a more or less thick mud blanket to cover the rest of the shelf. In fact, three different zones of pelitic sedimentation exist on the Algerian Shelf. The first (*vasières*, or mud-bands) is the site of maximum mud sedimentation. The second zone is situated further seaward, near the shelf break, and adjacent to relict sand patches: this zone is characterized by the lowest rate of pelitic sedimentation. The third zone lies at the shelf edge and at the upper part of the continental slope where mud accumulation is again very high.

Theoretical and experimental investigations (in scale model) by Lhermitte (1958) allow us to interpret varying rates of pelitic sedimentation and thus the actual manner of sediment distribution on the shelf. The principal dynamic factor results from great storm waves or large-scale wave motions. In contact with the shelf surface, such waves give rise to a laminar boundary layer which is responsible for the landward movement of a part of the muds in suspension or already deposited.

This landward movement is entirely compensated on the middle shelf (center of *vasières*) by the seaward-moving current from the surf zone, carrying fine particles. On the outer shelf, and especially on the relict sand banks, the landward movement is not completely compensated and the residual effects, added to geostrophic currents, would prevent accumulation of great quantities of river sediment. Although Lhermitte (1958) did not determine whether motion of this type is able to rework sands, it may be supposed that storm waves, in contact with the rough surface of the sands, would give rise to a turbulent rather than a laminar boundary layer, which might be capable of suspending a thin sheet of sandy sediment.

This hypothesis concerning the present distribution of sediment on parts of the Algerian Shelf is somewhat similar to the pattern proposed by Swift (1970) who, however, supposes bulk movement of storm-stirred shelf water to be more complex and disordered on the middle shelf.

The Holocene transgression and the Postglacial climatic changes have resulted in the continental shelf becoming a site of both accumulation and bypassing ever since the present sea level was reached (about 6000 to 8000 years B.P.) and since the rapid erosion of soils began. Today, the thickness of deposits overlying the pre-Holocene surface is still relatively small in spite of a high rate of pelitic sedimentation. The continental shelf has not yet reached an equilibrium profile except perhaps in certain areas such as the Bay of Djidjelli (eastern part, Figure 4) and the western part of the Gulf of Nemours. Such an equilibrium profile may be characterized by a slightly concave curve inclined seawards from the shoreline to the shelf edge. Thus, the modern sediment distribution seems to be quite ephemeral and represents only one stage of the evolution towards an equilibrium state (if present environmental conditions are maintained). In this respect, the great morphological variety of the Algerian Shelf and the differing conditions of sedimentation from east to west might explain why it is possible to distinguish here the several stages of shelf evolution defined by Curray (1969). It seems that:

1. The shelf off Oran and Bône would represent the earlier stages, with relict morphology and relict sands.

2. The Gulf of Arzew (Figure 3) and the Bay of Castiglione provide examples of the next stages (transitional), with relict morphology and relict sands partly concealed by a modern mud blanket (*vasières*, mud-bands, shelf edge muds) but discernible near the shelf edge through "windows" of the modern sediment blanket; it is at present the most widespread stage (Figure 5).

3. The Gulf of Bougie and part of the Bay of Djidjelli

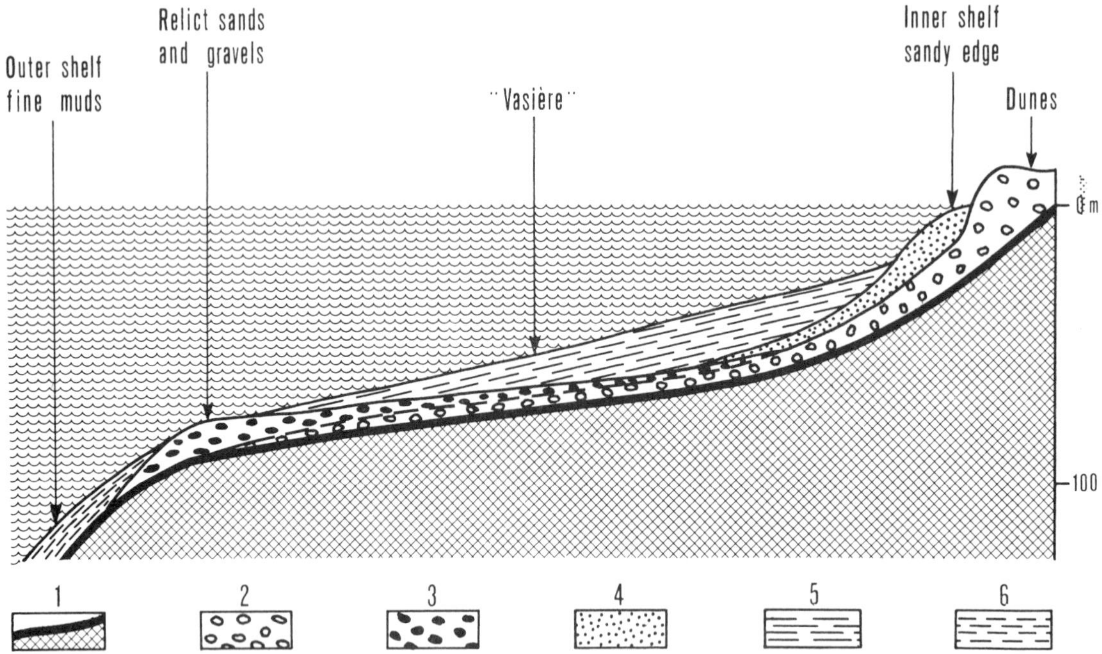

Figure 5. Idealized model of deposition of late Quaternary sediments, based mainly on study of the western part of the Algerian Shelf. 1, theoretical Pleistocene surface; 2, regression deposits at the end of Pleistocene; 3, relict sands and gravels; 4, modern nearshore sand prism; 5, *vasière*; 6, foraminiferal muds at the shelf break and outer shelf-upper slope muds.

(Figure 4) are entirely covered by a modern sediment blanket and represent the final equilibrium stage.

Such an evolution seems to be followed by an important change in the sediment blanket texture. Early stages do not show any distinct grading. Today, there exists an apparently reversed grading (nearshore sands, mud-bands, relict sands) which is due to the simultaneous presence of two different facies (modern and relict). According to Curray (1969a), the final stage would be a seaward prograding allochthonous deposit, *i.e.,* the equilibrium shelf facies with *a return to grade* (Swift, 1970).

Thus, related to a renewed high stand of the sea, there should be a textural and morphological adjustment of the shelf, corresponding logically to a dynamic equilibrium state between the unconsolidated materials and their environment. This evolution would be marked, apart from general silting or burial, by a gradual reduction of the calcareous facies. This results from burial of relict sands which formerly covered the major part of the continental shelf, and from shrinking of areas propitious to the production of modern biogenic calcareous sands, because of silting. Assuming such a hypothesis, during Pliocene and especially during the late Quaternary, there must have existed alternating phases of sedimentation directly related to the eustatic sea level changes. Finally, sedimentation on the Algerian Shelf is dependent on the morphology (related to tectonics) and on the climate, together with eustatic oscillations and accelerated soil erosion.

SEDIMENTATION ON THE ALGERIAN SLOPE

Relatively few samples are available from this region to give a comprehensive account of the sedimentation here. However the following section conveys some preliminary results and conclusions derived from the present work and previous studies.

Description of Eastern Sector

The slope off the Edough—Cap-de-Fer mountains is covered with a gray mud, brownish at the surface but darker below. Clay (50%), siliceous silt and fine sands (26%) and calcium carbonate (24%) are the major components of this deposit. It is a fine mud (pelite: 64%, aleurite: 35%, arenite: 1%) within which coarser beds with Bryozoa, shells and calcareous algal debris are allochthonous and reworked. The clay fraction consists of about 70% illite and mixed-layer clay minerals, 30% kaolinite, and traces of chlorite. Organic matter (1.24% of organic carbon, on average; 1.66% maximum), mainly plant debris, and authigenic pyrite (0.64% to 0.27% of pyritic sulfur in total dry sediment) are relatively abundant.

Further west, off Collo Mountain, at depths from about 800 to 1200 m, the slope is covered with a thin veneer of black sandy mud containing rock fragments of microgranitic nature, similar to the Collo Formation. Lower down, at about 2000 m depth, the sediment mantle seems thicker and is formed by gray muds rich in fine sand (clay: 48 %, sand: 32 %, calcium carbonate: 20%). At the base of Core C.T.8, there are calcareous bands with Bryozoa (Reteporidae) and mollusc shells originating from the shelf. Quartz, biotite, rock fragments, sponge spicules, sometimes embedded in a pyritic cement, constitute the siliceous part of this organic-rich sediment.

CORE C.T.7

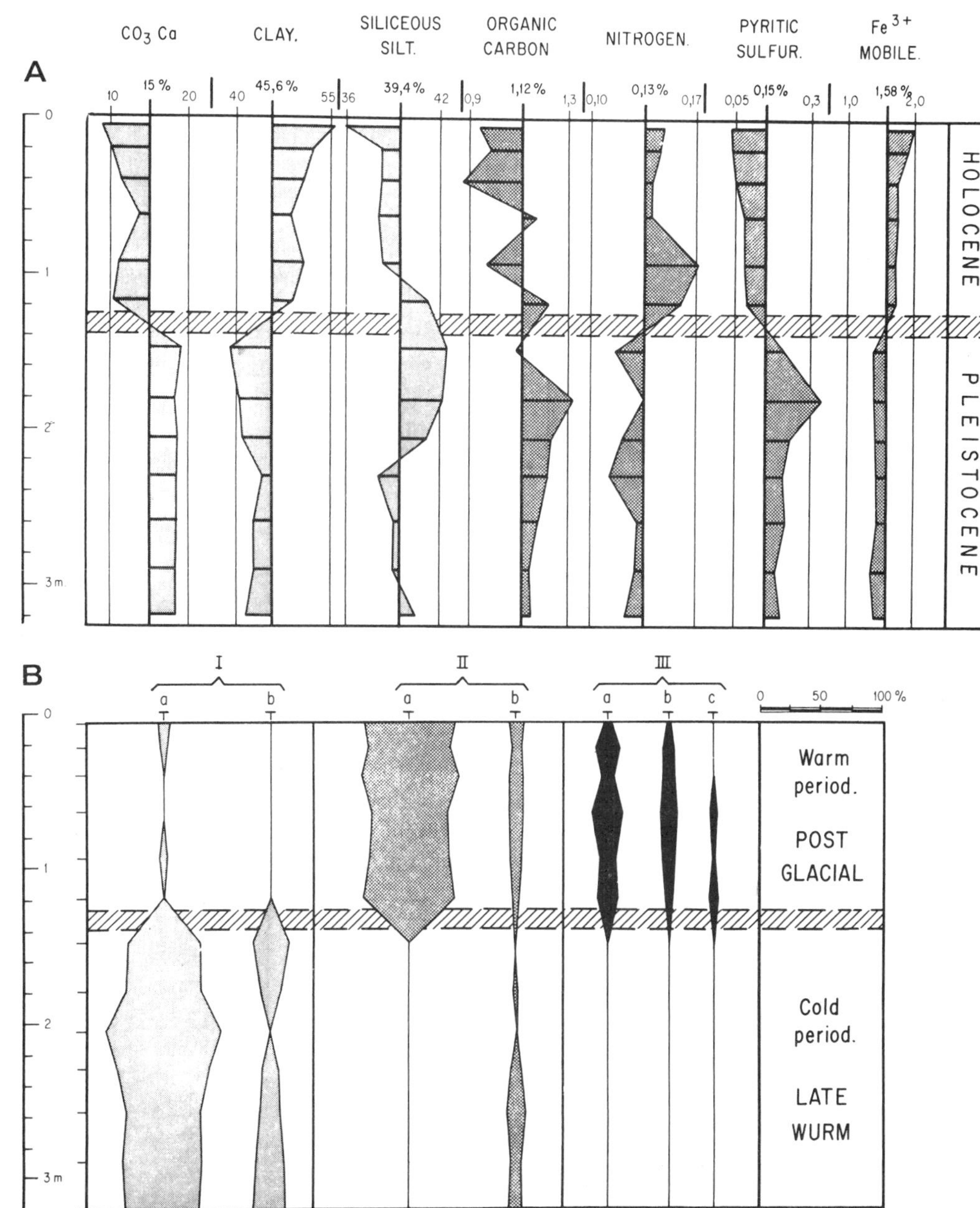

Figure 6. Lithological changes (A) and foraminiferal zonation (B) in Core C.T.7.
A. The three major components ($CaCO_3$, clay and siliceous silt) are recalculated to constitute 100 percent of the sediment (in fact they represent more than 96% of the total dry sediment weight). The other components are expressed in percentages of the dry sediment weight. The thicker vertical lines represent average amounts.
B. I, cold species: (a) *Globigerina pachyderma* (right coiling), (b) *Globorotalia scitula*; II, temperate species: (a) *Globorotalia inflata*, (b) *Globigerinoides ruber* (White); III, warm species: (a) *Globigerinoides ruber* (pink), (b) *Globorotalia truncatulinoides*, (c) *Globigerinella siphonifera*.

Further west, four cores collected approximately along the 4°45′ meridian (*e.g.*, C.T.7, Figure 6) permit assessment of the nature of sedimentation on the Great Kabylia slope. On the upper part are fine gray muds, poor in calcium carbonate (less than 10%) but rich in silt (36%), and clay (55%). The clays are composed of the usual species: illite, kaolinite and traces of chlorites. The sandy part contains: zircon, rutile, tourmaline, epidote, garnet, amphiboles, *etc.*, a heavy mineral suite also found in sandy muds of the shelf.

Organic matter (on average 1.1% of organic carbon) and authigenic pyrite (on average 0.15% of pyritic sulfur) are always abundant except at the top of the cores.

In these cores, a gradual but clear change in the sediment composition appears with a zone ranging between 0.50 and 1.50 m. This lithological sequence is characterized by:

1. A strong upward increase in the clay content (from 30 to 50%) with a concomitant but slight decrease of the sandy fraction.
2. An important flunctuation in the organic and mineral carbon percentage with a general upward decrease of organic matter and of calcium carbonate.
3. A strong upward decrease in authigenic pyrite, found near the top of the cores only in trace amounts; this decrease is accompanied by an increase in the oxidizing capacity of the sediment as evidenced by quantitative analysis of the *mobile* ferrous and ferric iron (Leclaire, 1968a).

Corresponding to this lithological sequence, a biostratigraphic zonation has been established by the use of planktonic Foraminifera. A warm species assemblage passes downward into an association of cold species (Figure 6). This zonation will be seen later to mark the late Würm-Postglacial boundary.

Central Slope Sector

Off Algiers, the color of muds at the surface of the sediment blanket is relatively pale: from dark-gray or gray-brownish to yellow. The deposit is a clayey siliceous mud with illite and quartz. However, off the Bouzarea Mountain, the upper part of the continental slope is rocky (gneiss, micaschists). Possibly, the sediment blanket is represented here by a veneer of brownish muds with shells, Bryozoa and Echinoid debris. Further seawards, in sediment collected during the Swedish Deep Sea Expedition (1947–1948), Todd (1958) observed numerous rock fragments, and abraded fossil foraminifera and benthic foraminifera originating from the shelf. The angularity of the rock fragments and the fact that the core-catcher did not deeply penetrate infer the probability of bed-rock in the vicinity. Moreover, the boundary from a warm to a cold fauna occurs at 1.60 m from the top of one core (Todd, 1958). According to Muraour (1954), planktonic Foraminifera with predominent *Globorotalia inflata* are very abundant off Algiers as within the Bay of Castiglione.

East of Algiers, at the bottom of the great Courbet-Marine Canyon (around 2,000 m depth) a blue mud, fairly well consolidated, contains *Globorotalia hirsuta aemiliana* (Colalongo and Sartoni) in relative abundance and an assemblage of Pliocene benthic foraminifera. This layer of blue mud underlies 30 cm of modern sediment. This fossil microfauna, unknown in the Mediterranean late Quaternary, suggests a Pliocene or early Pleistocene age for this sediment (Leclaire and Le Calvez, 1969). If this hypothesis is correct, there must exist a stratigraphic hiatus between the blue muds and the modern deposits.

West of Algiers, near Tenes and Gouraya, a mantle of brown or yellow muds appears, covering a gray-blue sediment which is compact and contains authigenic pyrite. In several dredgings and particularly in Core C.T. 1. (Figure 2), numerous sandy and pebbly beds are interlayered in the muds. Rock fragments are mainly derived from the flysch formations exposed on the adjacent land. These beds also contain reworked calcareous algae, bryozoa and echinoid debris together with benthic Foraminifera, all of shelf origin. Such a reworking is to be anticipated in this region when it is recognized that tremors may severely affect the continental slope and its sediment blanket. A well documented example is the Orleansville earthquake of 1954.

Western Slope Sector

Off the gulfs of Arzew and Oran, the sediments are richer in calcium carbonate than in the other regions described (26% to 30% and more). Organic matter percentages do not exceed 0.7 to 0.8% while iron sulfides are absent from the brown or gray muds. This westward trend in the sedimentation is even more marked off the Gulf of Nemours where sediments collected by the Swedish Deep Sea Expedition contain calcium carbonate in great amounts (40%) (Todd, 1958). Blue-mud boulders embedded in yellow-brownish muds have been seen in this region. Moreover, volcanic rock debris was sampled at a depth of 1000 m in the Moules Canyon; this debris lies below modern deposits some tens of centimeters thick.

SEDIMENTATION ON THE NORTHERN EDGE OF THE BASIN

Sedimentation on the Spanish sector of the Algiers-Balearic Basin appears to be quieter, and some sedimentological features from this region will be described here to furnish a basis for comparison. This northern edge of the basin seems to be covered with a mantle of yellow or mottled muds from 0.5 to 1 m thick. Calcium carbonate (40% weight of dry sediment) frequently predominates in this surficial veneer.

Radiolaria are sometimes numerous (D.T. 1, D.T. 2) and their presence is a characteristic feature of the sedimentation on this side of the basin, as compared with the Algerian edge.

On the slope of the "Emile Baudot" Bank (south of the island of Majorca, at a depth of 900 m, a calcareous ooze of exceptional composition has been sampled (D.T. 13; see Table 2).

The great amount of foraminifera should be noted (40 to 50% of $CaCO_3$) and also the abundance of coccoliths, which are always present in the deep sediments but in lesser amounts. Generally, coccoliths (Figure 7) do not constitute more than 10 to 15% in weight of the dry sediment both in the blue and in the yellow muds. So, it is somewhat inaccurate to call all the muds of the basin and especially the blue muds (that are the most deficient), coccolith oozes. More than 40% of the foraminifera are represented by *Globorotalia inflata*. Therefore, this sediment is truly a *"Globigerina ooze"*.

At about 0.5 m below the top of a core (C.T. 3) collected further north, a gradual boundary appears between a cold and a warm fauna. This boundary is also marked by an important change in the composition of the sediment *e.g.*, (1) a strong downward decrease in the amount of calcium carbonate, from more than 60% at the top to 40% near the middle of the core, (2) a corresponding downward increase in the quantity of fine sand, from 10 to 35%, (3) a conspicuous upward decrease in organic carbon and authigenic pyrite, the latter completely disappearing near the top of the core.

Further east on the Balearic slope, in Core O19, Todd (1958) also found a warm period following a cold one, the boundary being situated at 50 cm below the top. In all samples, the presence of transported material from the nearby shelf and from the top of the slope is not certain, although some benthic foraminifera (*Ammonia beccarii*) have been observed.

Preliminary investigation suggests that the sediment settling off the Algiers-Tunisian border (in the Chenal de la Galite) and on the Sardinian slope is very similar to the Spanish edge deposits. Thus the sediment blanket of the Algerian margin appears to be uniquely characterized by a predominance of detrital sedimentation and a relatively high degree of local reworking.

Table 2. Estimated composition of sample D.T. 13.

Component	%
Calcium carbonate: 67%	
Foraminifera	25%
Coccoliths	20%
Pteropods	5%
Fine grained $CaCo_3$ (microgranular)	20%
Clay: 20%	
Illite and mixed-layer minerals	15%
Kaolinite	5%
Siliceous fine sands: 12% (quartz, sponge spicules and Radiolaria)	
Organic matter and ferric hydroxide: 1%	

SEDIMENTATION ON THE CONTINENTAL RISE AND BASIN PLAIN

Both off Algiers and in front of Kabylia (Cores C. 4, Geom. 1, C.T. 4, C.T. 9, KS 03) the texture and the composition of the recent deposits do not vary appreciably. In a ten meters long core (*e.g.*,C.4, Geom. 1), 50 to 100 sandy beds are discernible to the naked eye, and are variable both in frequency and thickness. These sandy layers constitute from 15 to 30% of the total sequence.

Such layers sometimes reveal a vertical grain-size gradient (graded bedding) and are frequently interlaminated with finer sediments (Figure 8, *1*). Usually, they contain clayey sand with quartz and numerous rock fragments, together with heavy minerals such as garnet, andalusite, zircon, chlorite, biotite, tourmaline, pyroxenes, *etc.*, the same species that occur in the shelf suite. A plant-fiber felting is present, with vegetal debris and rafted fragments of wood, more or less carbonaceous, and also a mixture of displaced benthic foraminifera whose shelf assemblages are well known (see above). Calcified or silicified fossil foraminifera such as *Globotruncana* are not uncommon. The biggest foraminifera, the coarsest grains of quartz and the pteropod debris often form a pavement at the base of the sandy beds. These deposits are interpreted as turbidites.

Between those layers visible to the naked eye, within deposits that seem to be homogeneous, X-radiographs (Figure 9, *1*) often reveal a laminated texture with superposition of thin silty beds, rich in clay, and organic matter. However, a considerable thickness of sediments which are homogeneous even to X-rays may occur in the same core. These deposits have a texture similar to the Globigerina oozes found further to the north.

Turbidites form an apron at the base of the Algerian continental slope. This apron is essentially composed of detritus originating from the Algerian mainland and extends northwards for nearly 50 miles (Bourcart, 1953), underlying a thin blanket of yellow or brown

Figure 7. Scanning electron micrographs of the fine silt-clay fraction ($< 25\mu$ to $> 2\mu$) in mud samples from the deep part of the Algiers-Balearic Basin. 1, ($\times$ 1760). The base of Core C.T.3 (Balearic rim of the basin). Numerous Coccoliths with *Scyphosphaera sp.*, and variegated debris. ($CaCO_3$ content of the fraction: 50%). 2, ($\times$ 1760). Sample from 1.80 m below the top of Core C.T.3. Coccoliths (*Coccolithus huxleyi*, *Gephyrocaspsa oceanica, etc.*), clay minerals and micas with other debris. ($CaCO_3$ content of the fraction: 52%). 3, ($\times$ 1760). Sample from 3.60 m below the top of Core C.I. GEOMED 1. Coccoliths (*Gephyrocapsa oceanica*), various minerals and biogenic fragments ($CaCO_3$ content of the fraction: 43.6%). 4, (1900). Sample from 4.0 m below the top of Core C.4 GEOMED 1. Coccoliths with *Coccolithus leptoporus* and various minerals (clay, micas). ($CaCO_3$ content of the fraction: 35%). 5, ($\times$ 4200). Same sample as in (4): *Coccolithus huxleyi and Rhadbosphaera oceanica*, with undetermined debris. 6, ($\times$ 1760). Sample from 6.10 m below the top of Core C.4 GEOMED 1; the fine fraction of a sandy bed; a mixture of various Coccoliths and debris.

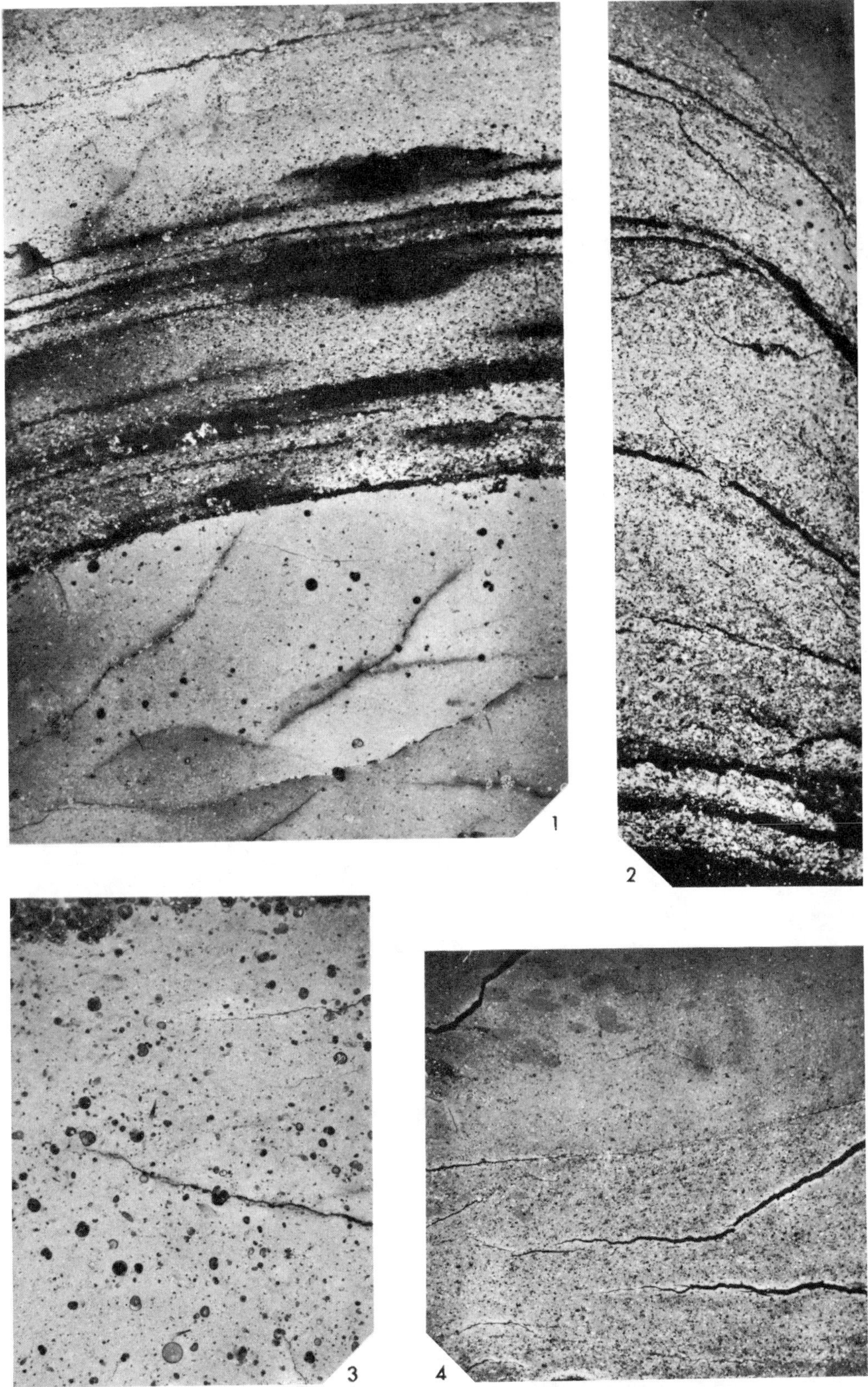

Figure 8. Some features of the texture of hemipelagic sediments in Core C.4 GEOMED 1, (Photographs of impregnated thin-sections). 1, (× 4.8). Sample from 3.72 m below the top of the core. Heterogenous hemipelagic deposit with an interbedded laminated sandy bed overlying a homogeneous pelagic sediment. Top layer is probably a redeposited sediment. 2, (× 4.8). Sample from 3.0 m below the top of the core. An interbedded sandy layer, relatively thick, gradually passing upwards into a fine pelagic deposit. 3, (× 4.8). Sample from 1.20 m below the top of the core. A pelagic-type deposit with scattered foraminifera and ungraded texture. The base of the overlying sand bed, a pavement of Foraminifera, appears at the very top of the photograph. 4, (× 4.8). Sample from 0.54 m below the top of the core. A thin sandy bed (probably resedimented) displaying normal upwards grading.

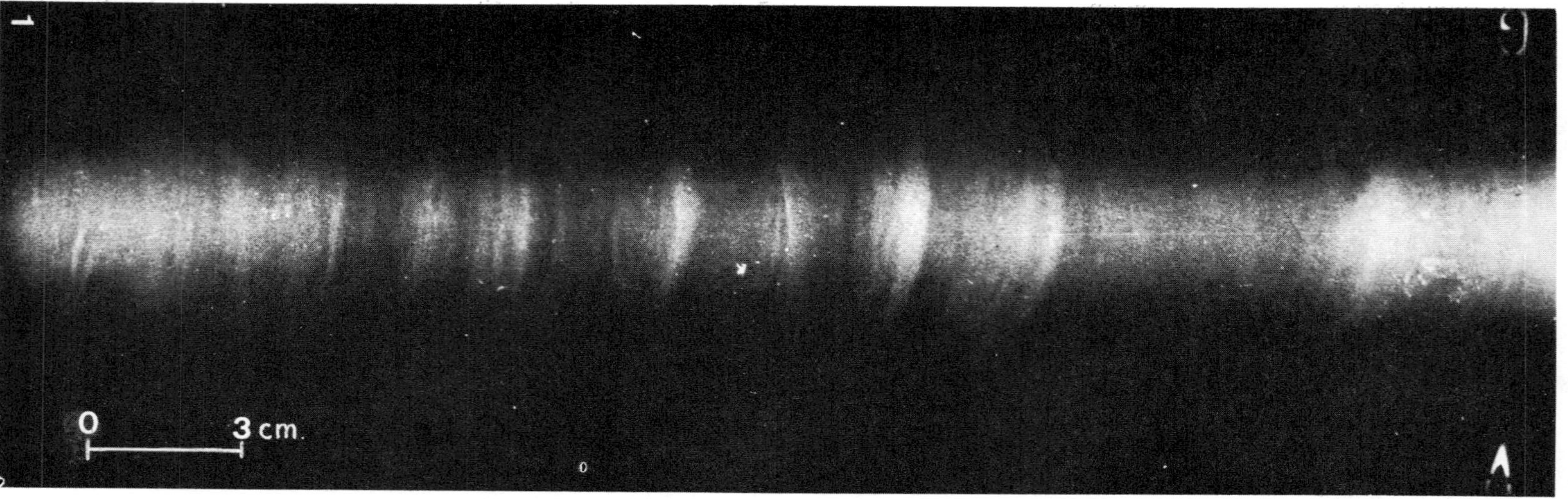

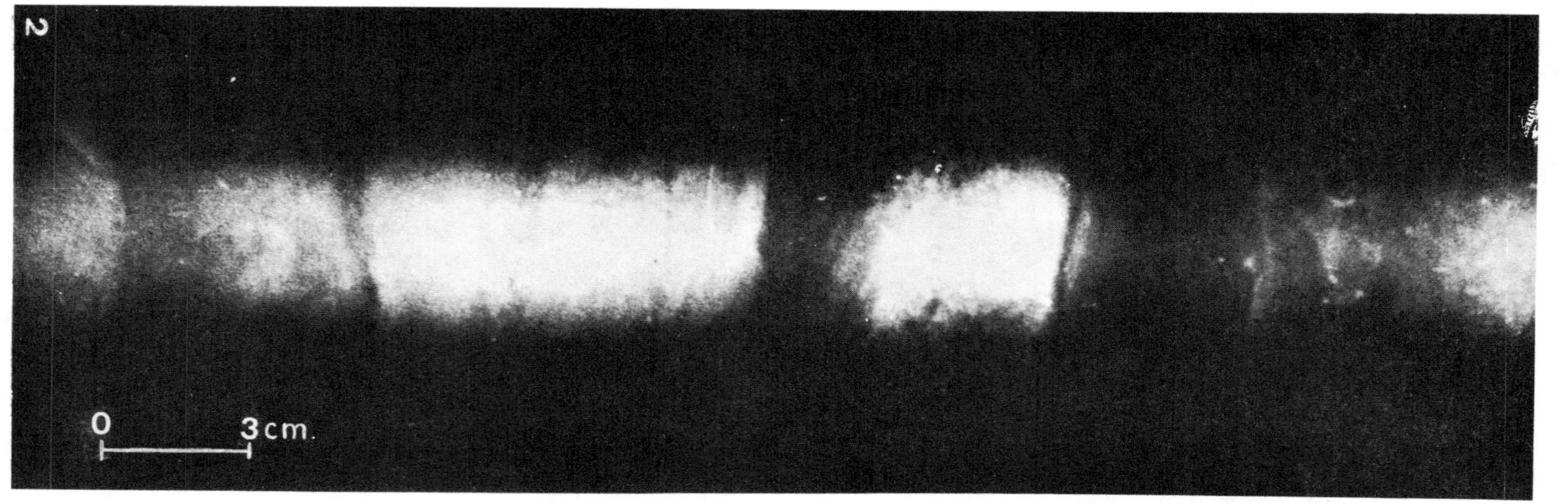

Figure 9. X-radiographs (× 1/2) of unsplit cores (diameter approximately 5 cm) collected on the continental rise off Algiers and Bougie. X-radiographs were made on board the N/O JEAN CHARCOT using apparatus provided by the C.O.B. (CNEXO).
1, Core K.S.03 (2344 m). Note distinct alternation of laminated fine sand (light) with silt and muddy layers (dark). An example of laminated texture. Depth in core: 2.40–2.70 m from top.
2, Core K.S.04 (2070 m). Thick sandy beds more or less laminated (light) interbedded with homogenous muddy strata (dark). An example of turbidite texture. Depth in core: 2.75–3.05 m from top.

muds. This apron probably ends in a bevelled edge near the basin center, south of the 38th parallel, since there is no evidence for its presence in the cores recently collected (KS 05 and 06) during the "Polymede cruise" of the N/O JEAN CHARCOT.

Homogeneous deposits, texturally similar to the globigerina oozes, occur on abyssal hills and in the eastern and northern regions of the basin. As the Core C. 1, Geom. 1 (Lat. 41°30′N, Long. 6°05′E; depth 2,418 m) reveals, similar sediments also occur in the Provençal basin. In this core, several lithologic sequences are interbedded, principally yellow muds rich in calcium carbonate, clay and limonite (but poor in organic carbon, and in iron sulfides), which alternate with blue (or black) muds having opposite characteristics (see Figure 10).

INTERPRETATION OF DEEP-SEA SEDIMENTS

Origin and Nature

As on the continental slope, sedimentation processes on the rise and in the basin plain result in deposits that are a mixture in varying proportions of terrigenous and pelagic materials. These are the so-called *hemipelagic* sediments or hemipelagites (Kuenen, 1950; Stanley, 1969a). By comparison, the muds of the *vasières* or of the shelf mud-bands would represent pure terrigenous sediments, while the *globigerina oozes* at the foot of the "Emile Baudot" Bank are essentially pelagic sediments.

Finally, two kinds of hemipelagites can be distinguished in the Algiers-Balearic basin: those with a homogeneous texture similar to pelagic ones (Figure 8, *1*) and those with a heterogeneous texture characterized by interlayered sandy beds and laminated deposits (turbidites, laminites) (Figure 8:*1,2,4,* and Figure 9). These two types of sediment are derived from two different types of process.

Terrestrial supplies derived from the Algerian mainland give rise to the deep-sea sands and the sandy part of the shelf mud-bands but also provide much if not all of the hemipelagite clays. The clay mineral suite of the basin resembles that found in soils and rocks of the mainland, especially in the marls of the late Cenozoic sublittoral basins (*e.g.,* Chelif basin).

Part of the calcium carbonate found in the hemipelagites may be provided from terrestrial sources, in addition to the contribution from the autochthonous biogenic material. About 50% of the calcium carbonate in these muds is of very fine grain-size since it is contained in the aleuritic and pelitic fraction ($<25\mu$). The origin of this finely divided calcium carbonate is still debatable. However, Eriksson and Olsson (1965) demonstrated that an important part of the finely divided calcium carbonate is older and causes a notable drift of the C^{14} datings when C is extracted from the total $CaCO_3$. Using a scanning electron microscope, fossil coccoliths and other reworked but indeterminable grains have been seen, in addition to tiny calcite rhombs. It seems probable that the older, allochthonous carbonate may be derived in two ways: (1) reworking of fine biogenic debris or calcite contained in the late Cenozoic marls, and (2) crushing of old shells and algal fragments contained in the near-shore sand prism within the surf and shore zone.

Stratigraphy and Rate of Sedimentation of the Late Quaternary Deep Sea Deposits

General stratigraphic correlation between the cores studied by the author and those of the Swedish Deep Sea Expedition has been attempted (Figure 11). Such a tentative attempt is based on the changes in planktonic foraminiferal assemblages as interpreted from a hydrological and climatic point of view and revealed by counting the following species (Figures 6B and 10B:

(1) *Globigerinoides sacculifer* (Brady), *Globigerinoides ruber* (d'Orb.) var. pink, *Globigerinella siphonifera* (d'Orb.), *Globorotalia truncatulinoides* (d'Orb.), selected as tropical to subtropical water indicators;

(2) *Globorotalia inflata* (d'Orb.) and *Globigerinoides ruber* (d'Orb.) var. white, used as subtropical to temperate water indicator species.

(3) *Globigerina quinqueloba* (Natl.), *Globorotalia scitula* (Brady), *Globigerina pachyderma* (Ehr.) right-coiling, selected as cold-temperate to subarctic water indicator species; and

(4) *Globigerina pachyderma* (Ehr.) left-coiling, used as an arctic water indicator.

Some absolute ages obtained using the C^{14} method have allowed the construction of a detailed (if hypothetical) stratigraphy for the late Quaternary (Figures 11A and B). Four broad periods seem to characterize the late Quaternary: (1) the early Würm (early Glacial, first phase of the Würm-Wisconsin), (2) the middle Würm (with glacial and *deglaciated* alternating stages), (3) the late Würm, and (4) the Postglacial (Holocene).

The two latter periods particularly held the author's attention because their boundary is conspicuous in most of the cores, being marked by a strong upward decrease in the number of *Globigerina pachyderma* (right-coiling) and by the appearance of most of the tropical species. Moreover, the Pleistocene-Holocene boundary, in the turbidites of the Algerian continental rise, seems to be marked by a decrease in the thickness and the frequency of the sandy layers: in the Core C. 4 Geom. 1 there are ten beds per meter during the late Würm, against five for the Postglacial.

Estimated from the foregoing stratigraphic scale, sedimentation rates during the Holocene range from about 30 to 60 cm/1,000 years at the Algerian base-of-slope, against 3 to 10 cm/1,000 years on the Spanish edge and 15 cm/1,000 yr (maximum) in the basin plain. The rates of deposition were even higher during the late Würm. Thus, it appears that, as Stanley (1969b) has suggested, the Algerian margin now plays a predominant part in the filling of the Algiers-Balearic basin. Much of the detritus derived from the

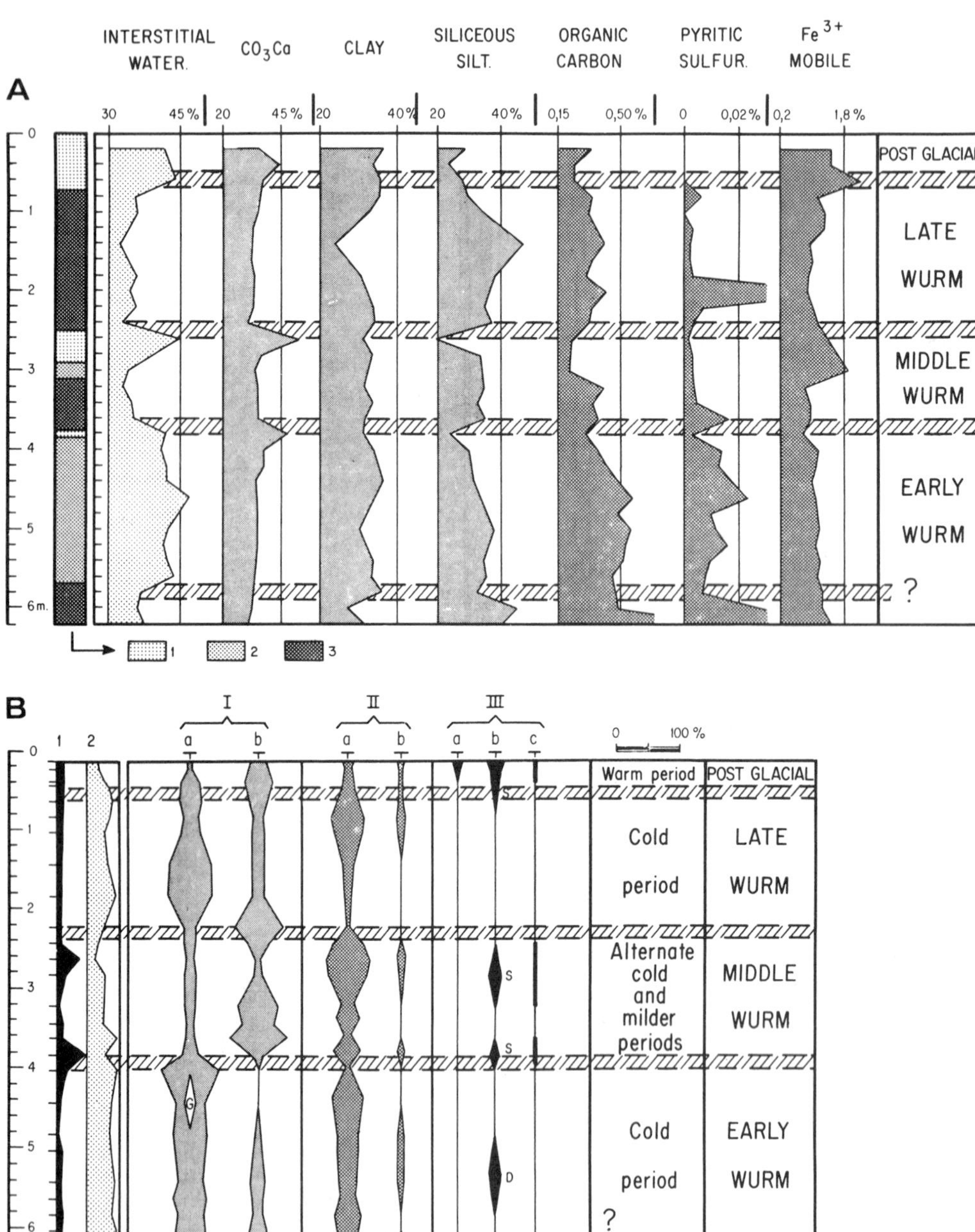

Figure 10. Lithological changes (A) and foraminiferal zonation (B) in Core C.I-GEOMED I (from the southeast Provençal Basin plain; depth: 2418 m; Latitude = 41°30′N, Longitude = 6°05′E). A, Interstitial water is expressed in per cent of dry sediment weight. The three major components ($CaCO_3$, clay, siliceous silt) are recalculated to constitute 100 per cent of the dry sediment (in fact they represent more than 97% of total weight). The amount of the other components is expressed in per cent of the dry sediment weight. 1, yellow or brown muds; 2, gray muds; 3, black muds.
B, Left hand columns: foraminiferal weight variation (1) and variation of number of *Globogerina quinqueloba* related to the total foraminiferal assemblage (2). Column I, cold species: (a) *Globigerina pachyderma* (right-coiling and left coiling), (b) *Globorotalia scitula.* Column II, temperate species: (a) *Globorotalia inflata*, (b) *Globogerinoides ruber* (white). Column III, warm species: *Globigerinoides ruber* (pink), (b) *Globorotalia truncatulinoides*, (c) *Globigerinoides sacculifera.*

North African mainland bypasses the shelf and the continental slope and accumulates at the base-of-slope and on the basin plain.

The rough correlation demonstrated above, between the lithostratigraphy and biostratigraphy of the recent deposits shows that the climatic periods of the recent Quaternary, and especially of the two last episodes, were accompanied by changes in the nature (Olausson, 1969) and in the early diagenesis of the sediments formed at that time. It is obvious that such variations in the nature of the sedimentation must be related to environmental changes which are dependent directly or indirectly on climatic fluctuations.

DISCUSSION

Terrigenous Supplies and Climatic Changes

The Holocene transgression was followed by an environmental change related to the alteration in climate. The Mediterranean or subarid present climate was preceded during the late Würm by temperate cooler conditions with less contrasted seasons. The rate of erosion of soils and the nature of the Algerian fluvial regime were both affected by these changes. It is known that maximum aggradation occurs in regions under subarid climatic conditions, where the river sediment load is frequently a hundred times higher

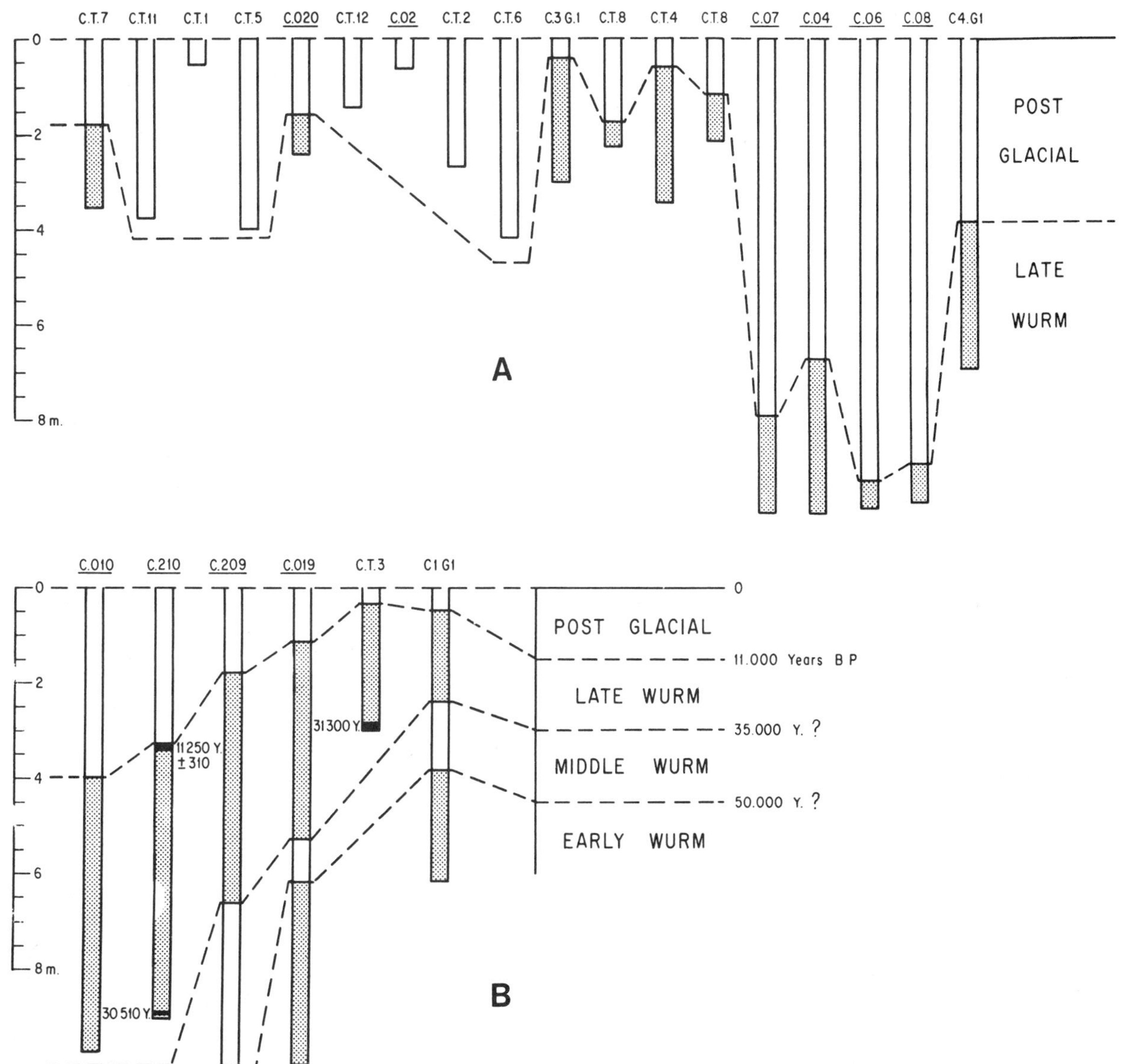

Figure 11. Tentative correlation of late Quaternary episodes in different cores collected in the Algiers-Balearic basin (except C.I. GEOMED I) based on foraminiferal zonation and some Carbon 14 dates (cores C.210 and C.T.3).

than it is in cooler zones. The appearance of subarid climatic conditions at the beginning of the Holocene probably initiated the progressive reduction of the vegetation and promoted the chemical weathering of rocks and soils, together with intermittent but strong flushing of the weathered detritus. This erosion has been locally accelerated by human deforestation. These processes have produced, and are still producing, a great deal of clayey material which is carried out to sea in great quantities, providing a thick cover of fine sediment which overlies the coarser late Würm deposits.

Moreover, during a long period of the late Würm, the shelf was almost completely uncovered by the late regression of the Pleistocene. At that time, Algerian rivers discharged their sediment load directly on to the continental slope and rise. The Holocene rise in sea level has resulted in the shelf acting as a trap, retaining a substantial proportion of the terrigenous input, particularly the coarser sediments. This may furnish an explanation for the gradual but clear decrease in frequency of the interlayered sandy beds at the Pleistocene-Holocene boundary. This factor may also account for the lower rate of basin plain sedimentation during the Holocene although the total bulk of sediment being supplied to the sea is probably higher at present. A part is trapped by the continental shelf and contributes, as previously described, to the wide distribution of the *vasières*.

Turbidites and Turbidity Currents

Turbidity currents play an important part in the progradation of the Algerian margin and are of particular importance in the vicinity of submarine canyons where they may be accompanied by sandfalls, especially where canyons head in the vicinity of the nearshore sand prism. As soon as they reach the rise and the basin plain such currents, probably, are transformed into turbidity overflow sheets, spreading fan-wise (Leclaire, 1968e). The lamination of thick sandy beds in cores may be explained by the successive passage of several of these turbidity layers proceeding from the same turbidity current.

The thin lamination whose presence is attested by X-ray investigations (Figure 9) may also be produced by turbidity flows but it is obvious that formation of such a deposit needs much less energy than deposition of sandy beds. Thus, according to several authors, it is possible to distinguish a low-density slow-moving turbidity current from other flows of higher density and velocity. These two modes of deposition possess a fundamental feature in common: they are intermittent and may be regarded as *spasmodic sedimentation processes* in the sense of Dunbar and Rodgers (1957).

The discovery of the so-called "nepheloid layers" (Ewing and Thorndike, 1965), now detected in several regions (Takashi, 1966; Eittreim *et al.*, 1969) and even postulated in the Mediterranean Sea (Bartolini and Gehin, 1970) has led several authors such as Stanley (1969a) to re-assess the role of turbidity flows in deep-sea deposit. Possibly, in the Algiers-Balearic Basin, nepheloid layers are an aspect of the *permanent sedimentation process*, designated formerly by "the rain of sediment", and are responsible both for pelagic deposits and the homogeneous part of hemipelagites.

Recent Hydrodynamic Changes and Climatic Fluctuations in the Algiers-Balearic Basin

Today, and probably since the beginning of the Holocene, the hydrological balance of the Mediterranean Sea as a whole shows a deficit due to an excess of evaporation over precipitation plus runoff. This deficit is responsible for the inflow of Atlantic water through the Strait of Gibraltar, *i.e.,* the eastward-flowing Atlantic current detected in the Western Mediterranean Sea. The dynamics of the Algiers-Balearic basin waters are largely determined by this Atlantic current and by the descent or *cascading* of cool, dense waters derived from the French coast in winter.

These surficial waters, rich in oxygen, spread down to the basin floor and extend towards the south, approaching the Algerian margin (Lacombe and Tchernia, 1960; McGill, 1962). This phenomenon permits the replenishment of oxygen in the bottom waters and thus, allied to a low rate of organic matter sedimentation, originates an oxidizing environment at the water/sediment interface, within which iron hydroxide neogenesis (limonite, goethite) occurs. Such is probably the origin of the recent brown or yellow muds carpeting the basin.

During the late Würm, the world-wide lowering of sea level and the more temperate climate probably resulted in drastic disturbance of the process of water-mass exchange through the Strait of Gibraltar. It may be assumed that the water balance of the Mediterranean Sea during this period was in excess, like the present Baltic Sea. Consequently, the current through the Gibraltar Straits was reversed (Kullenberg, 1952; Mars, 1963; Huang and Stanley, this volume), the Mediterranean surface waters flowing out into the Atlantic ocean. Moreover, these waters (rendered less saline and less dense by increased rain and ice-melting runoff) could not descend towards the basin floor and thus could not sufficiently replenish the bottom waters with oxygen. These conditions may have promoted a density stratification leading more or less rapidly to stagnation of the bottom waters. Locally, reducing conditions would then appear at the water/sediment interface; however truly euxinic conditions were not produced, since oxygen was never entirely lacking as the presence of benthic foraminifera indicates. Such conditions would give rise to the blue muds rich in iron sulfides.

Current reversals through the Strait of Gibraltar and particularly the input of great volumes of oceanic water (1,000,000 m^3/sec on average, but up to 2,600,000 m^3/sec: Lacombe and Tchernia, 1960 in Guilcher, 1963) carrying an important biomass into the Mediterranean Sea, must have had a great effect on the composition of the plankton and may be partly responsible for the changes in the foraminiferal assemblages. It can be assumed that the existence and renewal of tropical water species are also dependent on such factors.

Thus, environmental and climatic changes which took place during the recent Quaternary are reflected in the changing character of the sediments and their diagenesis.

CONCLUSIONS

The Algerian margin of the African continent provides great quantities of terrigenous material which are accumulating mainly at the base of the Algerian continental slope and in the adjacent basin plain, forming a large detrital apron. The convergent effects of factors such as morphology (related to tectonics) and glacio-eustatic and climatic changes during the late Quaternary have resulted in a pattern of sedimentation with the following broad facies: (1) on the shelf, calcareous sands and gravels, followed by muds with illite and quartz, (2) on the slope and basin floor, sandy pyritic muds (blue) followed by more argillaceous, limonitic muds (yellow) or, further north, sandy siliceous muds which are succeeded by calcareous oozes. The rapidity of these alternations (by geological standards) characterizes Quaternary sedimentation in the Algiers-Balearic basin.

ACKNOWLEDGMENTS

I wish to thank Professor R. Laffitte for his assistance and valuable comments, together with Professors G. Kelling and D. J. Stanley, whose help and advice were most useful in preparing this paper. I am also indebted to J. Evin (Laboratoire de Radiocarbone de l'Université de Lyon) who undertook the C^{14} dating on material from Core C.T. 3, and to Dr. G. Pautot (Centre Océanologique de Bretagne, CNEXO). This work was supported by the Muséum National d'Histoire Naturelle and the Centre National de la Recherche Scientifique.

REFERENCES

Auzende, J. M. 1969. *Etude par Sismique Réflexion de la Bordure Continentale Algéro-Tunisienne entre Bougie et Bizerte*. Thèse de Doctorat de 3ème cycle, Faculté des Sciences de Paris, 115 p.

Bartolini, C. and C. E. Gehin 1970. Evidence of sedimentation by gravity-assisted bottom currents in the Mediterranean Sea. *Marine Geology*, 9: M1–M6.

Bernard, F. and J. Lecal 1953. Rôle des Flagellés calcaires dans la sédimentation actuelle en Méditerranée. In: *Topographie Marine et Sédimentation Actuelle*, 19ème Congrès International de Géologie, Alger 1952, section IV, 4:11–23.

Bourcart, J. 1950. La théorie de la flexure continentale. *Congrès International de Géographie, Lisbonne 1949*, 2:167–190.

Bourcart, J. 1953 Sables néritiques à 2750 m de profondeur au large de Bougie. *Comptes Rendus des Séances de l'Académie des Sciences, Paris*, D 236:738–740.

Bourcart, J. and L. Glangeaud 1954. Morphotectonique de la marge continentale nord-africaine. *Bulletin de la Société Géologique de France*, (6) 4:751–772.

Caulet, J. P. 1972. *Les Sediments Organogènes du Précòntinent Algérien*. Mémoires Muséum Paris, 25:289 p.

Caulet, J. P. 1972. Recent biogenic calcareous sedimentation on the Algerian continental shelf. In: *The Mediterranean Sea: A Natural Sedimentation Laboratory*, ed. Stanley, D. J., Dowden, Hutchinson and Ross, Inc., Stroudsburg, Pennsylvania, 261–277.

Curray, J. R. 1960. Tracing sediment masses by size modes. *21st International Geological Congress, Copenhagen, Report Session, Norden*, 23:119–130.

Curray, J. R. 1969a. History of continental shelves. In: *The New Concepts of Continental Margin Sedimentation*, ed. Stanley, D. J., American Geological Institute, Washington, D. C., 6:1–18.

Curray, J. R. 1969b. Shallow structure of the continental margin. In: *The New Concepts of Continental Margin Sedimentation*, ed. Stanley, D. J., American Geological Institute, Washington, D. C., 12:1–12.

Dietz, R. S. 1952. Geomorphic evolution of continental terrace (continental shelf and slope). *Bulletin of the American Association of Petroleum Geologists*, 36:1802–1819.

Dill, C. E. 1965. *Formation and Distribution of Glauconite on the North Carolina Continental Shelf and Slope*. Unpublished M. Sc. thesis, Duke University, Durham, N. C. 48 p.

Dunbar, C. O. and J. Rodgers 1957. *Principles of Stratigraphy*. John Wiley and Sons, New York, 356 p.

Eittreim, S., M. Ewing and E. M. Thorndike 1969. Suspended matter along the continental margin of the North American basin. *Deep Sea Research*, 16:613–624.

Emery, K. O. 1952. Continental shelf sediments off Southern California. *Geological Society of America Bulletin*, 63:1105–1108.

Eriksson, K. G. and I. U. Olsson 1965. Some problems in connection with C 14 dating of tests of Foraminifera. *Bulletin of the Geological Institute of Uppsala*, 92:1–13.

Ewing, M. and E. M. Thorndike 1965. Suspended matter in deep ocean water. *Science*, 147:1291–1294.

Glangeaud, L. 1962. Paléogéographie dynamique de la Méditerranée et de ses bordures. Le rôle des phases ponto-plioquaternaires. In: *Océanographie Géologique et Géophysique de la Méditerranée Occidentale*, Villefranche/mer, avril 1961, Editions du C. N. R. S., 125–165.

Gorsline, D. S. 1963. Bottom sediments of the Atlantic shelf and slope off the southern United States. *Journal of Geology*, 71: 422–440.

Gourinard, Y. 1958. Recherches sur la géologie du littoral oranais. *Publication du Service de la Carte Géologique de l'Algérie*, 6: 200 p.

Guilcher, A. 1963. Continental shelf and slope (continental margin). In: *The Sea: The Earth Beneath the Sea, History*, ed. Hill, M. N., John Wiley, New York, 3:281–311.

Hilly, A. 1957. Etude géologique du massif de l'Edough et du Cap de Fer. *Publication de la Faculté des Sciences de Nancy*, 125: 408 p.

Huang, T.C. and D. J. Stanley 1972. Western Alboran Sea: sediment dispersal, ponding and reversal of currents. In: *The Mediterranean Sea: A Natural Sedimentation Laboratory*, ed. Stanley, D. J., Dowden, Hutchinson and Ross, Inc., Stroudsburg, Pennsylvania, 521–559.

Ichiye, T. 1966. Turbulent diffusion of suspended matter along the continental margin of the North-American basin. *Deep-Sea Research*, 16:613–624.

Johnson, D. W. 1919. *Shore Processes and Shoreline Development*. John Wiley, New York, 584 p.

Kruger, D. 1950. Variations quantitatives des protistes marins au voisinage du port d'Alger durant l'hiver 1949–1950. *Bulletin de l'Institut d'Océanographie de Monaco*, 978:2–20.

Kuenen, Ph. H. 1950. *Marine Geology*. John Wiley and Sons, New York, 568 p.

Kullenberg, B. 1952. On the salinity of the water contained in marine sediments. *Göteborg Klung; Vetenskaps och Vitterhts-Samhalles. Handlingar, Sjatte Folden*, (B) 6:1–37.

Lacombe, H. and P. Tchernia 1960. Quelques traits généraux de l'hydrologie méditerranéenne. *Cahiers Océanographiques*, 12: 527–547.

Laffitte, R. 1950. Sur l'existence de Calabrien dans la région oranaise. *Compte Rendus des Séances de l'Académie des Sciences, Paris*, (D) 230:217–219.

Leclaire, L. 1963a. Etudes littorales en Baie d'Alger. Zone de Fort-de-l'Eau—Ben-Mered. *Cahiers Océanographiques*, 25:109–104.

Leclaire, L. 1963b. Facteurs d'évolution d'une côte sablonneuse très ouverte. Etude préliminaire à l'implantation d'un port de pêche et de plaisance. *Cahiers Océanographiques*, 25:540–556.

Leclaire, L. 1968a. Détermination du degré d'oxydation d'un sédiment par l'étude de l'état du fer dans ses formes minérales authigènes. *Comptes Rendus des Séances de l'Académie des Sciences, Paris*, (D) 266:452–454.

Leclaire, L. 1968b. Contribution à l'étude des sédiments marins non consolidés. Les bases d'une nouvelle classification. *Comptes Rendus des Séances de l'Académie des Sciences, Paris*, (D) 266: 563–565.

Leclaire, L. 1968c. Contribution à l'étude de la relation entre le carbone et l'azote de la matière organique contenue dans les boues et vases du plateau continental algérien. *Comptes Rendus des Séances de l'Académie des Sciences, Paris*, (D) 266:2049–2051.

Leclaire, L. 1968d. Contribution à l'étude géomorphologique de la marge continentale algérienne. Note de présentation de 10 cartes topographiques du plateau continental algérien. *Cahiers Océanographiques*, 20:451–521.

Leclaire, L. 1968e. Sur la présence et la mise en place de lits sableux dans le bassin de sédimentation algéro-baléare. Analogie avec les passées gréseuses des marnes bleues néogènes du bassin de la Mitidja. *Comptes Rendus Sommaires de la Société Géologique de France*, 4:135–136.

Leclaire, L. 1969a. Sur la genèse de sulfures (pyrite) dans le bassin algéro-provençal pendant le Quaternaire récent. Relation entre l'euxinisme et la paléodynamique de ce bassin. *Comptes Rendus des Séances de l'Académie des Sciences, Paris*, (D) 268:1586–1588.

Leclaire, L. 1969b. Principe de cartographie des sédiments du plateau continental algérien. Présentation de 10 cartes sédimentologiques. *Comptes Rendus Sommaires de la Société Géologique de France*, 5:162.

Leclaire, L. 1972. *La Sédimentation Holocène sur le Versant Méridional du Bassin Algéro-Baléare (Précontinent Algérien)*. Mémoires Muséum, Paris, 24:291 p.

Leclaire, L. and C. Finck 1962. Utilisation des traceurs radio-actifs à Courbet-Marine. Etude de l'ensablement d'un port du littoral algérois. *Cahiers Océanographiques*, 24:526–542.

Leclaire, L. and Y. Le Calvez 1969. Sur la présence probable de Quaternaire ancien dans l'un des grands canyons du Précontinent nord-africain. Mise en évidence d'une lacune stratigraphique dans la série pléistocène. *Comptes Rendus des Séances de l'Académie des Sciences, Paris*, (D) 268:1252–1254.

Lhermitte, P. 1958. Contribution à l'étude de la couche limite des houles progressives. Application aux mouvements de matériaux sous l'action des houles. *Publication du Comité Central d'Océanographie et d'Etude des Côtes, Paris*, 136:171 p.

Mars, P. 1963. Les faunes et la stratigraphie du Quaternaire méditerranéen. *Recueils des Travaux de la Station Maritime d'Endoume*. 28 (43):61–97.

McGill, D. A. 1962. A preliminary study of the oxygen and phosphate distribution in the Mediterranean sea. *Woods Hole Oceanographic Institution, Contribution*, 28.

Muraour, P. 1954. Sur quelques sédiments dragués au large du littoral compris entre le cap Djinet et l'embouchure de l'Oued Sébaou. In: *Recherches sur les Fonds Chalutables de la Région d'Alger*. Publication du Comité Central d'Océanographie et d'Etude des Côtes, Paris, 90–121.

Olausson, E. 1969. On the Würm-Flandrian boundary in deep-sea cores. *Geologie en Mijnbouw*. 48:349–361.

Powers, M. C. and B. Kinsman 1953. Shell accumulations in underwater sediments and their relation to the thickness of the traction zone. *Journal of Sedimentary Petrology*, 23:229–234.

Rosfelder, A. 1955. Carte provisoire au 1/500.000 de la marge continentale algérienne. Note de présentation. *Publication du Service de la Carte Géologique de l'Algérie. Travaux des Collaborateurs*. 5:57–106.

Stanley, D. J. 1969a. Sedimentation in slope and base-of-slope environments. In: *The New Concepts of Continental Margin Sedimentation*, ed. Stanley, D. J., American Geological Institute, Washington, D. C., 8:1–25.

Stanley, D. J. 1969b. Turbidites, non-turbidites and outer continental marine paleogeography. In: *The New Concepts of Continental Margin Sedimentation*, ed. Stanley, D. J., American Geological Institute, Washington, D. C., 13:1–13.

Swift, D. J. P. 1969. Outer shelf sedimentation processes and products. In: *The New Concepts of Continental Margin Sedimentation*, ed. Stanley, D. J., American Geological Institute, 5:1–26.

Swift, D. J. P. 1970. Quaternary shelves and the return to grade. *Marine Geology*, 8:5–30.

Todd, R. 1958. Foraminifera from the western Mediterranean deep sea cores. *Reports of the Swedish Deep Sea Expedition, 1947–1948* (3):169–211.

Histoire Sédimentaire de la Région au large de la Côte d'Azur*

Guy Pautot

Centre Océanologique de Bretagne, Brest

RESUME

Dans cette étude locale, nous avons voulu montrer que la définition de paramètres aussi différents que la composition chimique et les propriétés physiques du sédiment, la stratigraphie et le cadre structural doivent participer à l'élaboration d'une synthèse à l'échelle régionale. Les relations entre le cadre structural et la stratigraphie de la couverture sédimentaire ont été étudiées avec un intérêt particulier à l'aide de la sismique réflexion, de forages et de carottages.

On aboutit à la conclusion que la couverture sédimentaire meuble de la pente continentale est Plio-Quaternaire. Elle présente de nombreuses figures de glissement dues à des mouvements tectoniques du même âge. Le substratum au Sud de Cannes est formé par le prolongement du massif cristallin des Maures qui se serait effondré au Mio-Pliocène.

La partie centrale du bassin est recouverte d'une couche salifère d'âge Messinien (6 à 9 MA) qui alimente de nombreuses structures diapiriques. Cette couche salifère se serait formée sous un régime lagunaire après la création du bassin à l'Oligo-Miocène. La partie océanique centrale s'est ensuite enfoncée par subsidence en entraînant les marges.

ABSTRACT

Detailed investigation of a portion of the sea-floor off the French Riviera serves to detail the history of the Algéro-Provençal Basin. The chemical composition and physical properties of sediments and the stratigraphy and structural framework are considered together. Both stratigraphy and structural framework of the sedimentary cover are examined by means of deep drilling, coring and seismic reflection.

It appears that the unconsolidated sedimentary cover of the continental slope is Plio-Quaternary in age. This cover displays abundant slumping structures which have resulted from contemporaneous tectonic movements. The basement south of Cannes corresponds to an extension of the crystalline Maures Massif which subsided during Mio-Pliocene time.

The central portion of the Algéro-Provençal Basin is covered by an evaporite layer of Miocene age (Messinian, 6 to 9 MY) which forms numerous diapiric structures in this region. It is likely that the evaporite layer resulted from deposition in a lagoonal environment following the formation of the basin in Oligo-Miocene time. Subsequently, the central (presently oceanic) portion of the basin and contiguous continental margins subsided.

INTRODUCTION

La région maritime s'étendant au large du massif de l'Esterel (France) est une des zones de contact entre le système pyrénéen et le système alpin. Différentes méthodes ont été utilisées pour cette étude. La *bathymétrie* met en évidence les grands traits de la marge continentale; la *sismique continue* caractérise la couverture sédimentaire et délimite son contact avec le substratum. L'analyse *stratigraphique* d'une cinquantaine de carottes trace une histoire simple de la marge continentale durant la subsidence plio-quaternaire, et les premiers forages par grands fonds ont permis de préciser cette histoire. Les mesures des *propriétés physiques* des sédiments permettent de mieux comprendre les processus de transport et de mise en place. Enfin, la *géochimie* des sédiments marins

*Contribution no. 46 du Groupe Scientifique du Centre Océanologique de Bretagne.

est en concordance avec la dynamique sédimentaire. Toutes ces approches différentes par les méthodes et par l'échelle du phénomène, permettent chacune l'apport de faits qui, une fois ordonnés, conduisent à une interprétation rationnelle du mode de formation du bassin occidental méditerranéen.

LA MARGE CONTINENTALE: CADRE STRUCTURAL ET STRATIGRAPHIE

Cadre Général

La zone étudiée s'étend dans une région complexe au carrefour du système alpin, des chaînes calcaires provençales, des massifs cristallophylliens des Maures et du Tanneron et du massif volcano-sédimentaire de l'Esterel (Figure 1).

L'analyse morphotectonique de la marge continentale permet de présenter le schéma suivant (Figure 2):

1. *Le plateau continental* est très étroit. C'est une plateforme littorale quaternaire qui se termine entre –90 et –100 m par une rupture de pente souvent très marquée. Ce ressaut correspond à la régression maximale du Würmien.

2. *La pente continentale* est abrupte. Entre 0 et 1000 m, la pente moyenne est de 9° avec un maximum de 15°. Cette valeur maximum est comparable à l'escarpement (Shepard, 1966) qui borde la plateforme

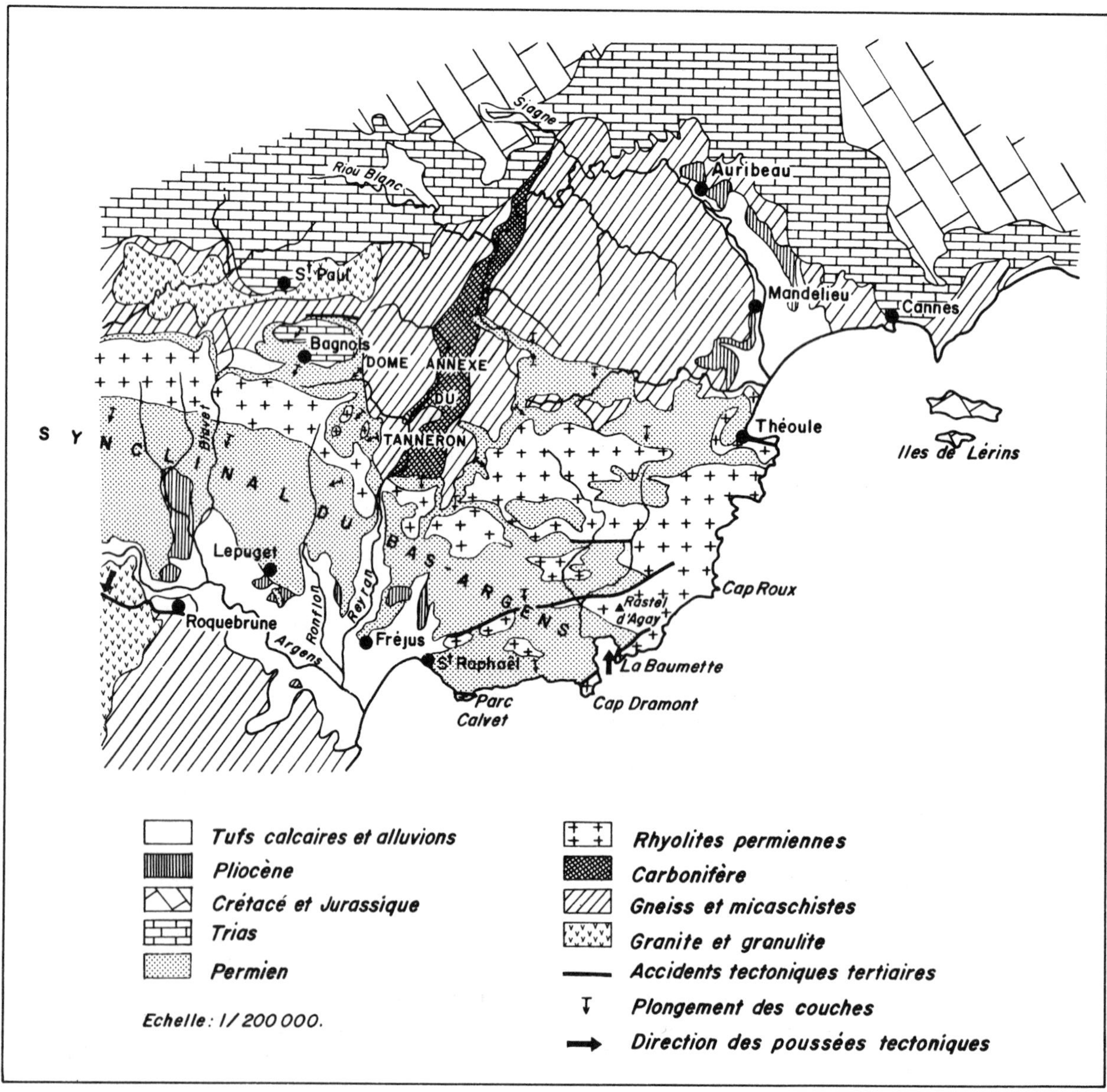

Figure 1. Schéma géologique de la bordure continentale de la région étudiée (d'après Luteaud, 1924). C'est la région de contact entre la Provence cristalline et la Provence calcaire.

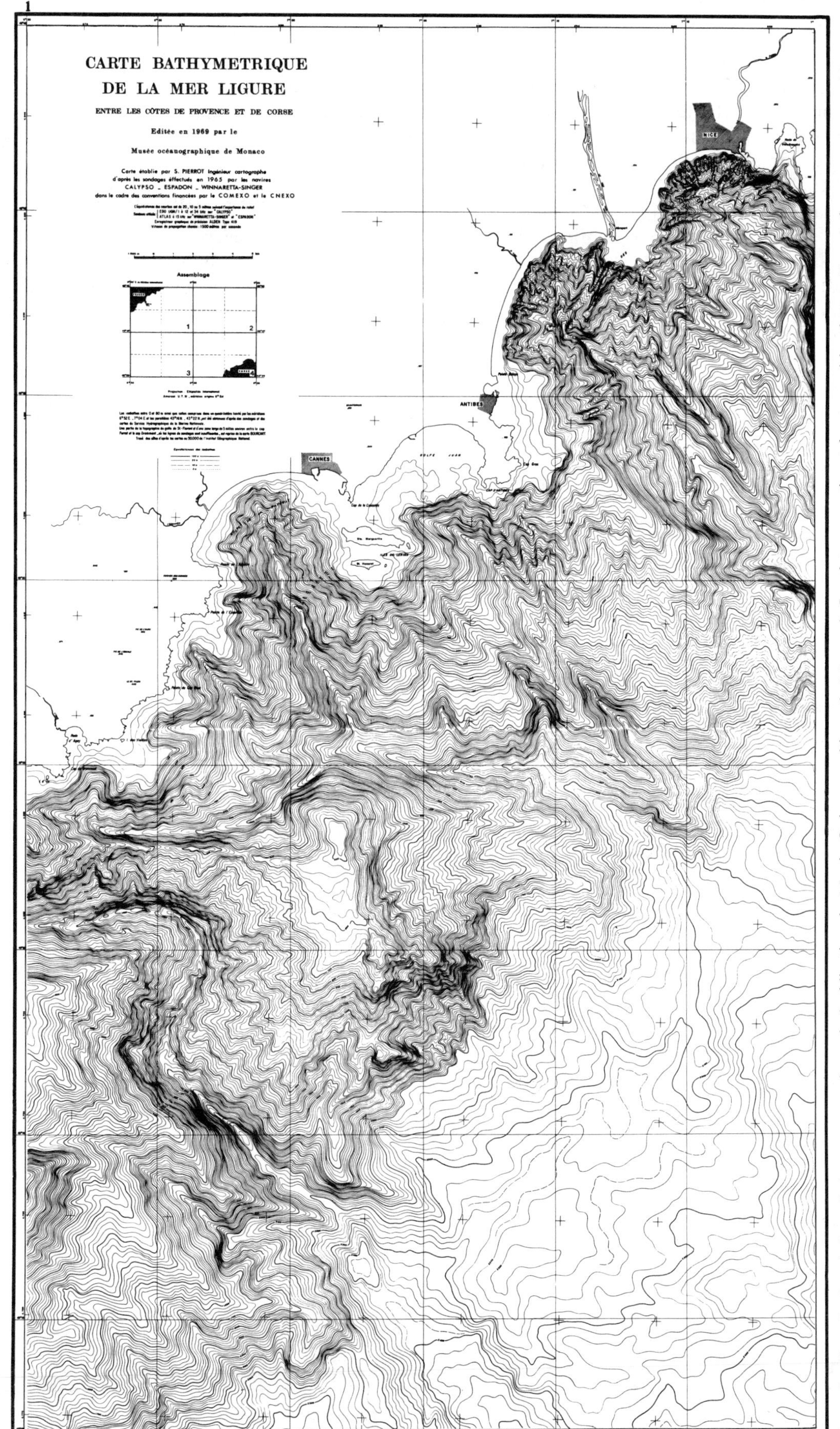

Figure 2. Carte bathymétrique de la marge continentale (publication du Musée Océanographique de Monaco). Le plateau continental est quasi-inexistant. De nombreux canyons accidentent cette pente. Le haut-fond du Méjean est visible à 20 km au Sud de Cannes.

au NE du Brésil (15 à 20°). Au-dessous de 1000 m, la pente est encore importante: entre 6° et 19°. Les canyons sont nombreux. Ils forment un véritable réseau avec de nombreux affluents; un haut-fond appelé "haut-fond du Méjean" forme un gradin important sur la pente.

3. *La limite supérieure du glacis* continental est sensiblement NE, parallèle à la direction générale du rivage, à une distance de 25 km de la côte.

4. Au niveau de la *plaine abyssale*, on distingue de nombreuses collines périabyssales d'une cinquantaine de mètres de hauteur et de 2 à 3 km de diamètre. Elles semblent plus élevées dans la partie Nord et moins marquées vers le Sud.

La Plateforme Littorale

Une étude détaillée par sismique haute fréquence et par carottages nous conduit à distinguer 3 domaines principaux (Figure 3) de l'Ouest à l'Est: le domaine de la Napoule; le domaine des îles de Lérins; et le domaine du cap d'Antibes.

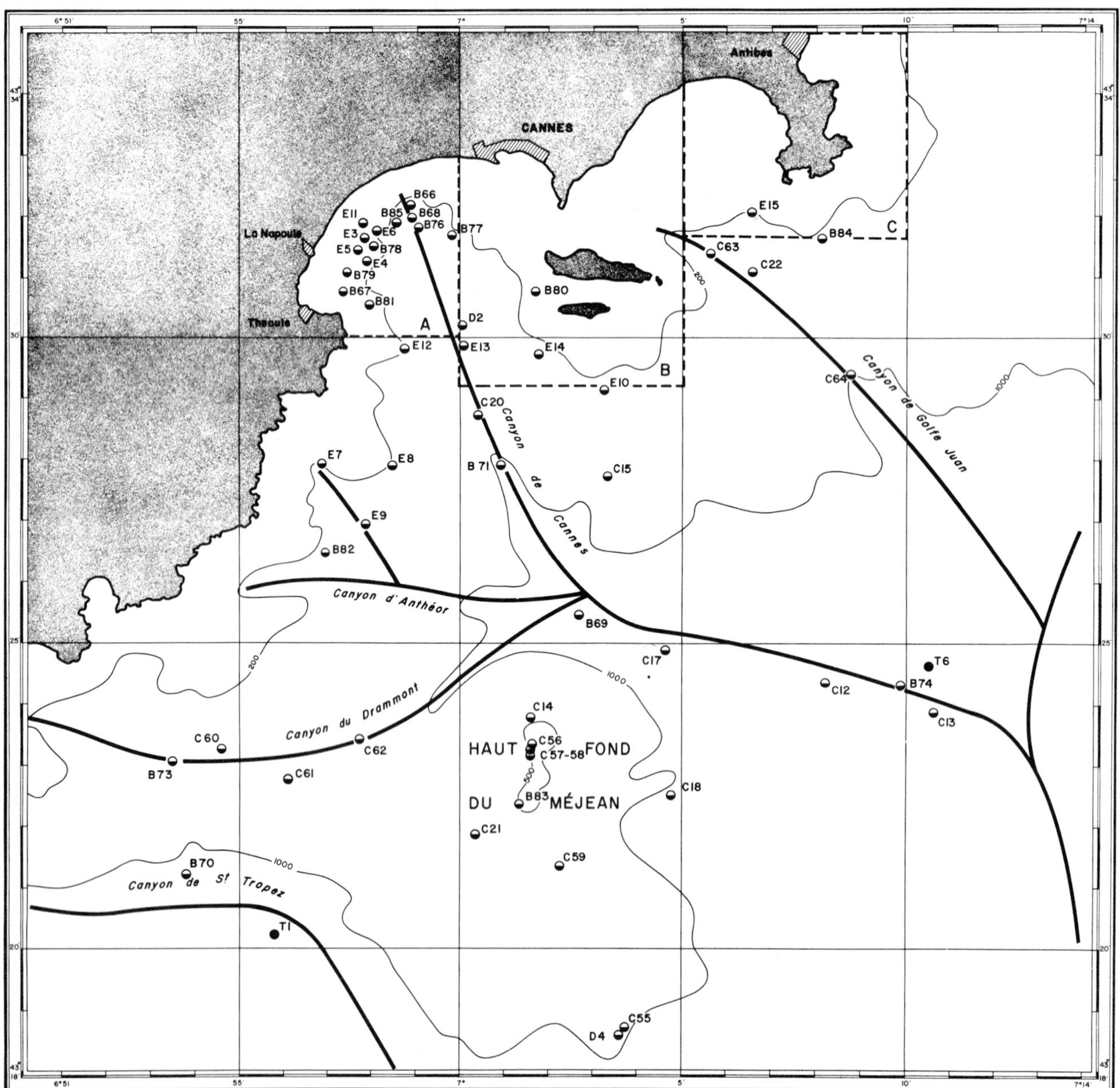

Figure 3. Cadre morphologique de la marge continentale. La plateforme continentale est divisée en trois domaines (tiretés). Les canyons sont en traits pleins. Les carottages sont représentés par des cercles noir et blanc et les forages par des cercles pleins.

Le Domaine de la Napoule

Dans ce secteur, la sismique montre que le substratum est proche. Ce substratum (Esterel) est recouvert d'une couche sédimentaire qui comble un petit bassin dont les bordures correspondent à la plage et à la rupture de pente. Plusieurs carottes de 500 cm prélevées sur la rupture de pente jusqu'à 200 mètres de profondeur montrent la stratigraphie suivante de la surface vers la base:

1. vase terrigène gris-bleu;
2. sablon à sable fin terrigène;
3. sable coquillier à gastéropodes, huîtres, serpules, turritelles et galets roulés;
4. vase consolidée.

Sur le socle formé de rhyolite ou de pyroméride se sont déposés des sédiments sableux ou vaseux qui ont été surmontés par des sédiments de plage déposés probablement au cours de la dernière régression quaternaire (Wurm). Il est à noter qu'en plus des variations eustatiques du niveau de la mer, des phénomènes de néotectonique sont visibles. On retrouve en effet sur le littoral des plages quaternaires soulevées (Luteaud 1924, et Chamley et Pastouret, communication orale): (1) un ensemble dont l'altitude oscille entre 4 et 10 m et un ensemble qui présente des variations d'altitude plus marquées de 30 à 65 m. Ceci suggère des mouvements tectoniques le long des failles anciennes qui auraient rejoué au Quaternaire.

Domaine des Iles de Lérins

Le bloc des îles de Lérins présente des parois abruptes dans son domaine maritime et une très faible extension de la plateforme littorale.

Trois carottages réalisés à 105 m, 185 m et 250 m à la limite plateforme-pente autour des îles ont atteint le substratum de calcaire dolomitique jurassique.

Dans la carotte B 80 (Figure 4) (profondeur 105 m), le sédiment superficiel est grossier à dominante organogène; il ne présente pas de granoclassement. On trouve également des fragments dolomitiques qui proviennent vraisemblablement des îles de Lérins. Leur aspect roulé et émoussé est remarquable; il peut être expliqué par l'existence d'une plage fossile ceinturant ces îles.

Domaine du Cap d'Antibes

Le Cap d'Antibes est lui aussi bordé par un système de failles, car les pentes sont abruptes et le plateau très limité.

La stratigraphie de la carotte E15 prélevée à 200 m de profondeur est la suivante, du sommet vers la base:

vase beige foncé avec de nombreux passages de fibres de posidonies;

passage graduel à un sablon;

sable vaseux avec de nombreuses coquilles de lamellibranches et des fragments rocheux; et

calcarénite.

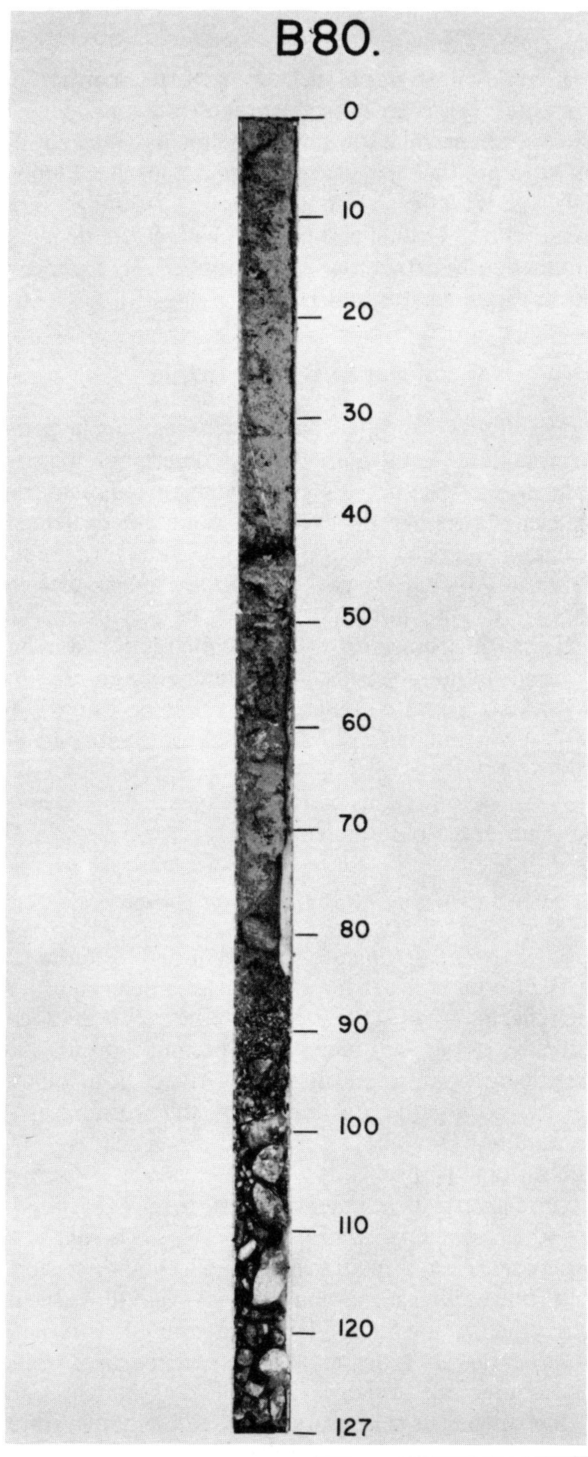

Figure 4. Photographie de la carotte B 80 (profondeur 105 m) prélevée sur la plateforme continentale à l'Ouest des îles de Lérins. Le sédiment superficiel est grossier à dominante organogène. La partie inférieure est constituée par des fragments calcaires émoussés.

On retrouve donc la même succession que dans les autres domaines.

Conclusion

Le plateau continental est très réduit dans cette région. Il est limité vers le large par une rupture de pente très nette entre 90 et 100 mètres.

L'étude stratigraphique des carottes permet de présenter le schéma de formation suivant. Sur le substratum en relief ou sur les sédiments pliocènes ou pléistocènes des parties en dépression, des dépôts de plage würmiens se sont déposés. La transgression flandrienne a ensuite entraîné un changement de mode sédimentaire caractérisé par l'apport de sédiments hémipélagiques durant l'Holocène.

La Pente Continentale

Des études sismiques ont été réalisées sur la pente continentale (Pautot, 1969) pour déterminer l'épaisseur et les structures de la couverture sédimentaire. Plusieurs dizaines de carottages ont été également réalisés (Figure 5), et enfin une mission de forages sur contrat C.N.E.X.O. par le procédé "Flexo-électroforage" mis au point par l'I.F.P. nous a permis de prélever plusieurs carottes sous 40 mètres de sédiment.

La sismique a permis de définir trois grands domaines: la pente continentale au Sud de Cannes, le prolongement du Cap d'Antibes, et le haut-fond du Méjean.

Les canyons sous-marins peuvent être classés dans un ensemble différent.

La Pente Continentale au Sud de Cannes

Sur les *profils sismiques* (Figure 6) réalisés entre les îles de Lérins et le canyon de Cannes, la détermination des réflecteurs a été faite par continuité avec la géologie terrestre et par carottages. Les premiers profils ont été effectués suffisamment près du rivage pour que la nature géologique du réflecteur soit certaine par continuité. On suit ainsi un horizon de calcaire dolomitique bathonien.

La couverture sédimentaire est formée par deux assises à caractères sismiques différents. La *couche A* superficielle est bien stratifiée, à réflecteurs nets; elle est formée d'une succession de lits de vase terrigène et de sablon (sédimentation hémipélagique). La *couche B* possède un haut degré de transparence acoustique qui indique une homogénéité réelle de sédimentation.

L'épaisseur de cette couverture sédimentaire est très variable car de nombreuses figures de glissement sont visibles, et des canyons sont entaillés dans cette assise.

Une faille N-S est bien mise en évidence à l'Ouest du Cap d'Antibes; elle présente un rejet d'environ 1000 mètres. D'autre part, des accidents E-O ennoient cette marge jusqu'au niveau du canyon de Cannes.

Les *carottages* réalisés dans cette zone, C 15 (788 m) et C 64 (1060 m) présentent un faciès hémipélagique typique (passage de lits sablonneux) mais à structure assez homogène. Les sédiments sont d'âge Holocène. La carotte E 10 (250 m) prélevée sur la partie supérieure de la pente (Figure 7) est formée d'une succession de lits de vase à sablon calcaire, de passées de sable organogène vaseux à Madréporaires. On note la présence d'encroûtements calcaires et de galets roulés de calcaire. Il faut noter ici que le calcaire du bedrock a été atteint.

Le Prolongement du Cap d'Antibes

Le cap d'Antibes se prolonge vers le SE par un relief qui présente une face abrupte vers la baie des Anges, c'est-à-dire vers le NE (Figure 2). Des profils de *sismique continue* (Figure 8) ont permis d'une part de retrouver les deux couches à caractères sismiques différents signalées plus haut et ont montré d'autre part que ce relief était formé par une accumulation sédimentaire de 600 à 800 ms d'épaisseur. Ce relief a pu être formé par les dépôts terrigènes fins du Var qui auraient été contrôlés par un courant marin assez violent autour d'un relief du socle.

Le socle est représenté par un réflecteur net; il semble affleurer sous la falaise NE.

La carotte E 1 (1590 m) prélevée au pied de ce relief présente la succession suivante: sous 8 cm de vase beige oxydée, on passe brutalement à une marne gris-bleu très rigide à faible teneur en eau (20%) et à très forte résistivité. Le forage T 8 (1775 m) réalisé par le B. F. TEREBEL sur ce même réflecteur n'a rencontré que 140 cm de sédiment meuble avant de toucher un horizon extrêmement dur. Malheureusement, ce réflecteur n'a pu être échantillonné.

Sur la dorsale sédimentaire, la carotte E 2 (1190 m) est formée de vase gris-bleu homogène, très fluide au sommet. Le forage T 4 (1275 m) sur ce même relief a prélevé 110 cm de vase gris-bleu compacte entre 39 et 40 mètres sous le fond. Au cours de l'enfoncement, plusieurs niveaux plus compacts ont été rencontrés. Ils correspondent à des réflecteurs visibles sur les enregistrements du sondeur de vase.

Ce relief sédimentaire est parcouru par des canyons qui se sont creusés dans des chenaux d'éboulement ou d'avalanche, qui rejoignent les canyons principaux: canyon de Golfe Juan et canyon de Cannes. Ces derniers ont une origine différente.

Le Haut-fond de Méjean

Ce haut-fond est situé à 15 km au Sud des îles de Lérins. C'est un ressaut important sur la pente

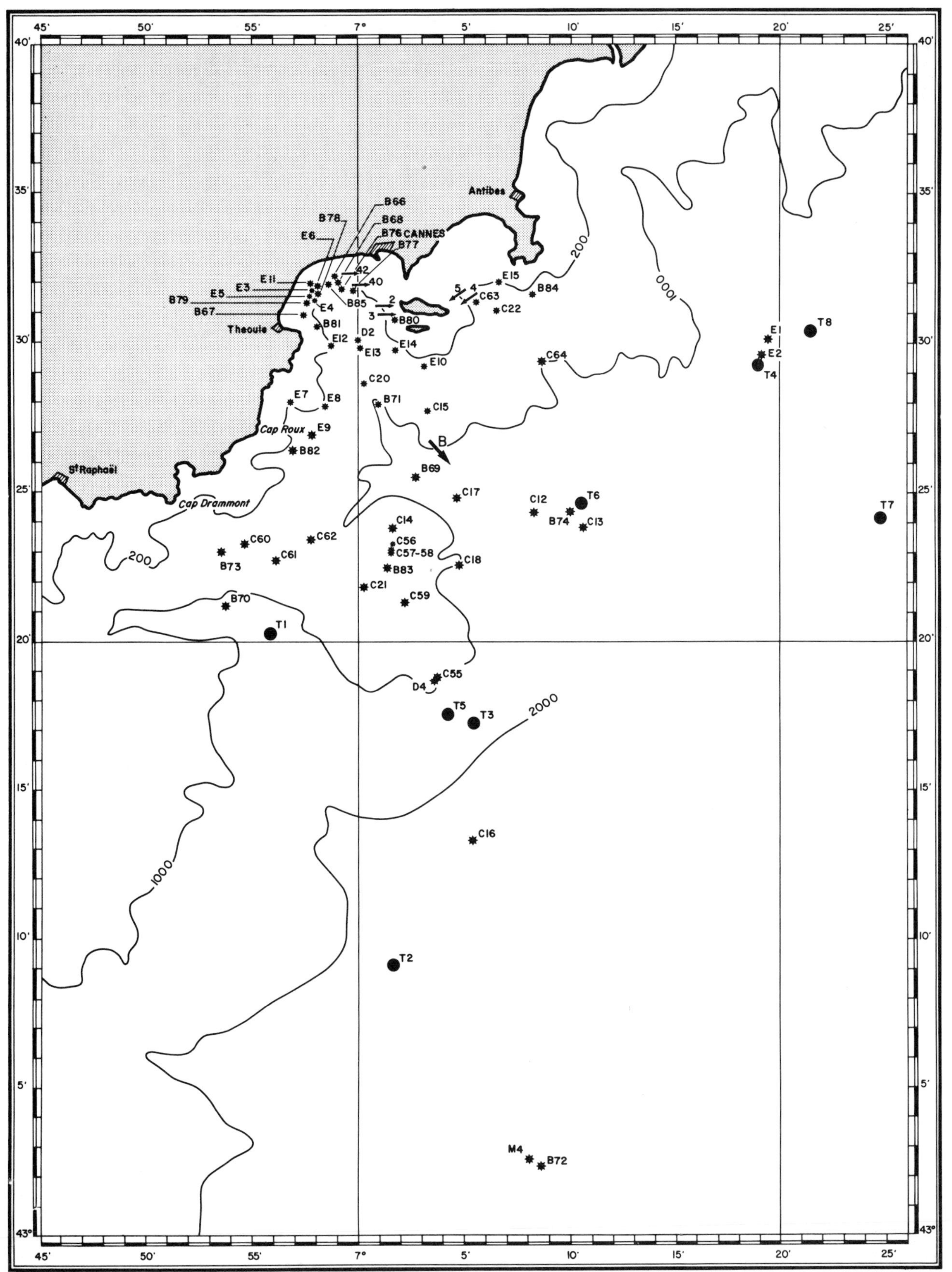

Figure 5. Cadre général de l'étude avec la position des carottages (étoile), des dragages (flèche), des forages (cercle plein) et de la plongée en bathyscaphe (flèche plus B).

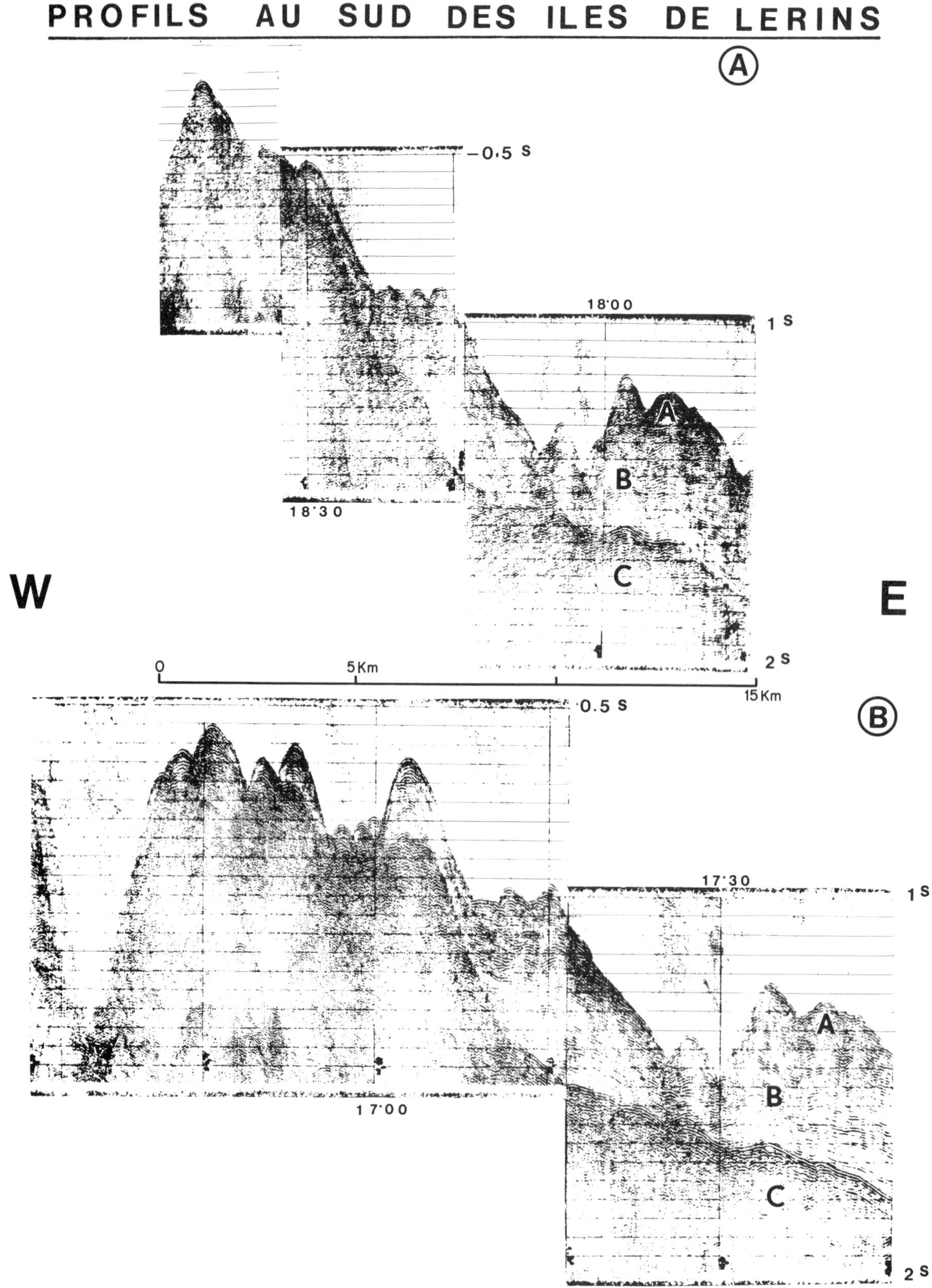

Figure 6. Profils sismiques réalisés avec un canon à air (air gun) au Sud des îles de Lérins sur la pente continentale. La position de ces profils est soulignée sur la Figure 8. Le substratum est le calcaire jurassique. La couverture sédimentaire est formée de deux assises à caractères acoustiques différents: couche A superficielle avec un litage apparent, couche B d'apparence homogène.

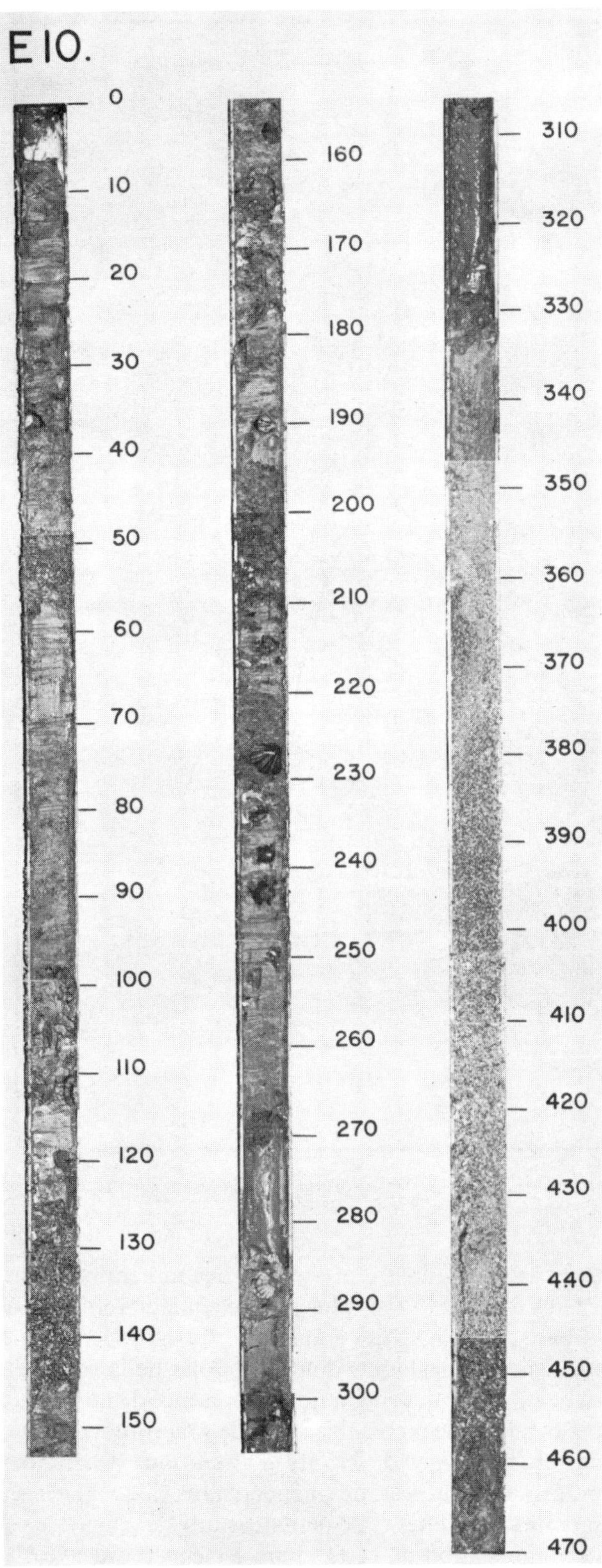

Figure 7. Photographie de la carotte E 10 (profondeur 250 m) prélevée du Sud des îles de Lérins sur la partie supérieure de la pente continentale. On note une succession de lits de vase et de sable organogène avec la présence d'encroûtements calcaires et de galets roulés de calcaire.

continentale (Figure 9). Il présente une pente abrupte vers le Nord en direction du canyon de Cannes, et vers l'Ouest en direction du canyon de Saint-Tropez. Il s'ennoie par paliers vers le Sud et brutalement par faille vers l'Est. Le point culminant est un piton qui remonte jusqu'à – 340 mètres.

Une étude détaillée à l'aide de la *sismique continue* a montré que le substratum affleurait sous forme de reliefs arrondis dans la partie Nord (Pautot, 1970) et certainement sur les reliefs SE. Ce haut-fond est recouvert dans sa partie centrale par une couverture sédimentaire ayant une puissance maximum de 600 ms. Les deux couches A et B de caractères sismiques différents sont visibles. Ce bassin suspendu ne peut s'expliquer que par des phénomènes de sédimentation différentielle.

Lowrie et Heezen (1967) ont montré le processus de formation d'une colline sédimentaire sous-marine par la décélération d'un courant chargé, due à un relief faisant obstruction à ce courant. Ce modèle peut être appliqué ici. La partie supérieure de la couverture sédimentaire (couche A) qui a un aspect lité serait due à des apports terrigènes rythmiques par flottation. La partie inférieure (couche B) est homogène mais elle se distingue des "pélagites" par une épaisseur non constante. Ils correspondent aux sédiments *homogènes* de Ewing *et al.* (1968), c'est-à-dire à une sédimentation rapide, continue, mais contrôlée par des courants de fond chargés de particules en suspension (couche "néphéloïde"). Sur ce haut-fond, seule une sédimentation hémipélagique est possible.

Si l'on applique un taux de sédimentation de 10 cm/1000 ans qui est généralement admis pour ce mode de dépôt en Méditerranée (Blanc-Vernet *et al.*, 1969), la base de la couche B serait Pliocène basal. Ceci est en accord avec un dragage qui a ramené une faune pliocène au voisinage de la couche B. Ainsi, la couche A qui présente des apports rythmiques plus grossiers liés aux glaciations serait d'âge quaternaire alors que la couche B, plus homogène sur ce type d'enregistrement serait d'âge pliocène.

Les *carottes* prélevées dans ce bassin sédimentaire sont constituées de vase gris-bleu assez homogène. On note parfois des apports d'organismes calcaires (coquilles de lamellibranches, gastéropodes, térébratules, *etc.*) qui doivent être éboulés à partir des pitons rocheux, puis transportés.

Deux échantillonnages du substratum ont été réussis. Le premier réalisé sur le *piton Nord* [carotte C 58 (435 m)] a ramené des fragments de gneiss. Le second réalisé sur la pente a ramené une carotte de 470 cm qui présente de haut en bas la stratigraphie suivante:
vase beige,
vase graveleuse avec de nombreux débris organiques:

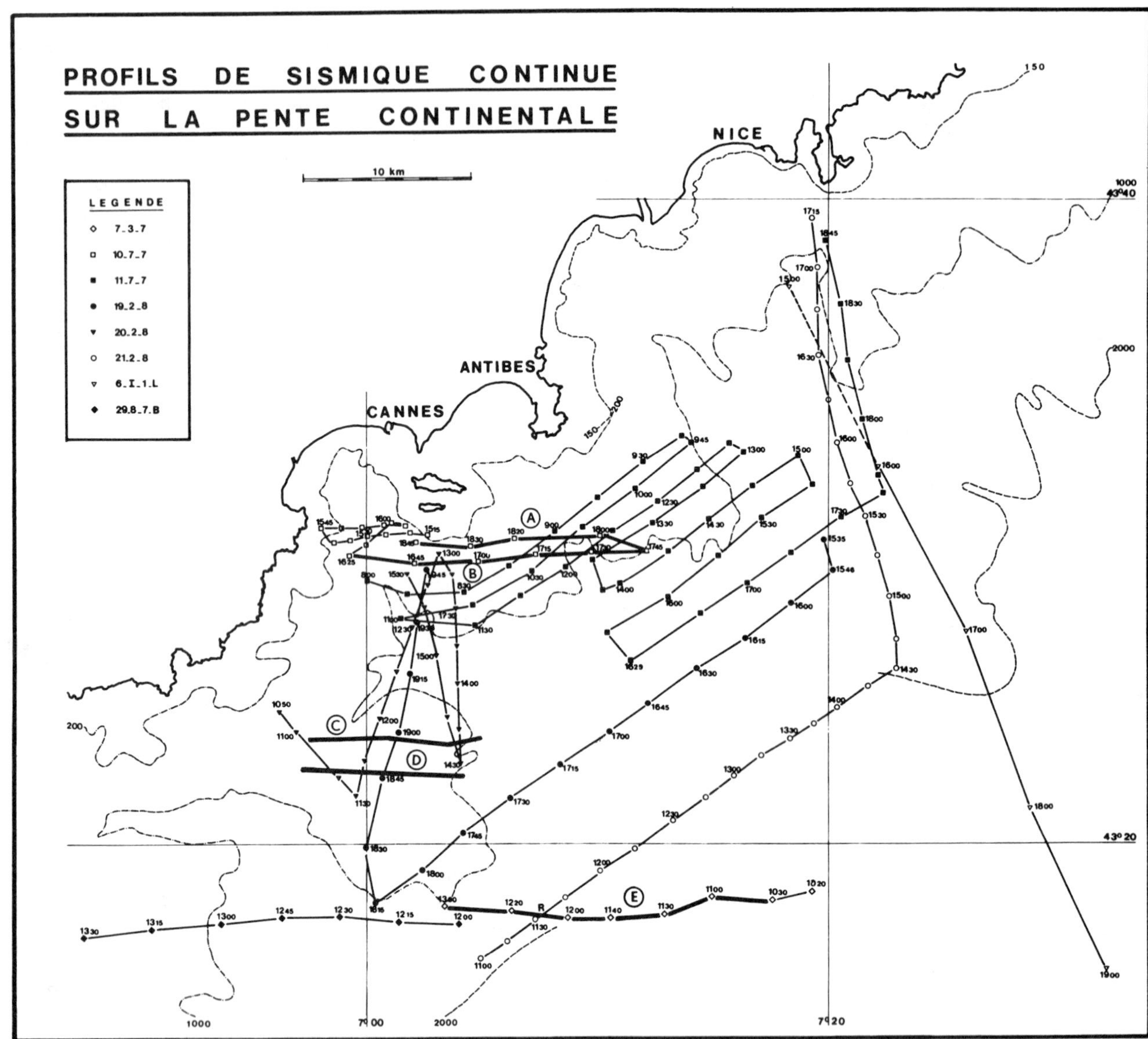

Figure 8. Position des profils de sismique continue au Sud de Cannes et sur le prolongement du cap d'Antibes (profils canon à air réalisés avec le B. O. CATHERINE-LAURENCE). Les profils présentés dans cette note sont représentés par un trait soutenu.

oursins, bryozoaires, lamellibranches, gastropodes, dentales, gros foraminifères, *etc.*

260 cm: deux fragments anguleux de roche métamorphique couverts d'une patine verte,

360 cm: nombreux polypiers dont l'âge déterminé par le carbone radioactif 14 est supérieur à 35000 ans., fragments de roches métamorphiques à patine verte et à cassure fraîche, composés de gneiss qui présente entre les feuillets de biotite une structure de granite à biotite du type anatexique (Figure 10).

Sur le *flanc S* du haut-fond, le substratum est recouvert par une pellicule sédimentaire; la couche B homogène est reconnaissable. Le carottage C 55 a été réalisé par 1300 m de profondeur. Sous 70 cm de vase gris-bleu subactuelle, on passe brutalement à une "arène" jaunâtre présentant des minéraux moins altérés et des passages plus fins de couleur verdâtre. A la base, on trouve des fragments plus grossiers d'une roche métamorphique altérée. Le fond de la roche est riche en oxyde de fer. On note la présence de nombreux petits quartz de recristallisation, de sphérolites d'oxyde de fer, de plages de calcite d'altération, de grandes biotites très altérées, de phénocristaux de plagioclases saussuritisés zonés et de petits cristaux de plagioclases. La composition de cette arène est en accord avec la pétrographie d'un gneiss à biotite très altéré.

L'analyse de cette arène (Pautot, 1967b) se trouvant actuellement à plus de 1000 mètres de profondeur, conduit à penser à une formation sédimentaire qui proviendrait de l'altération atmosphé-

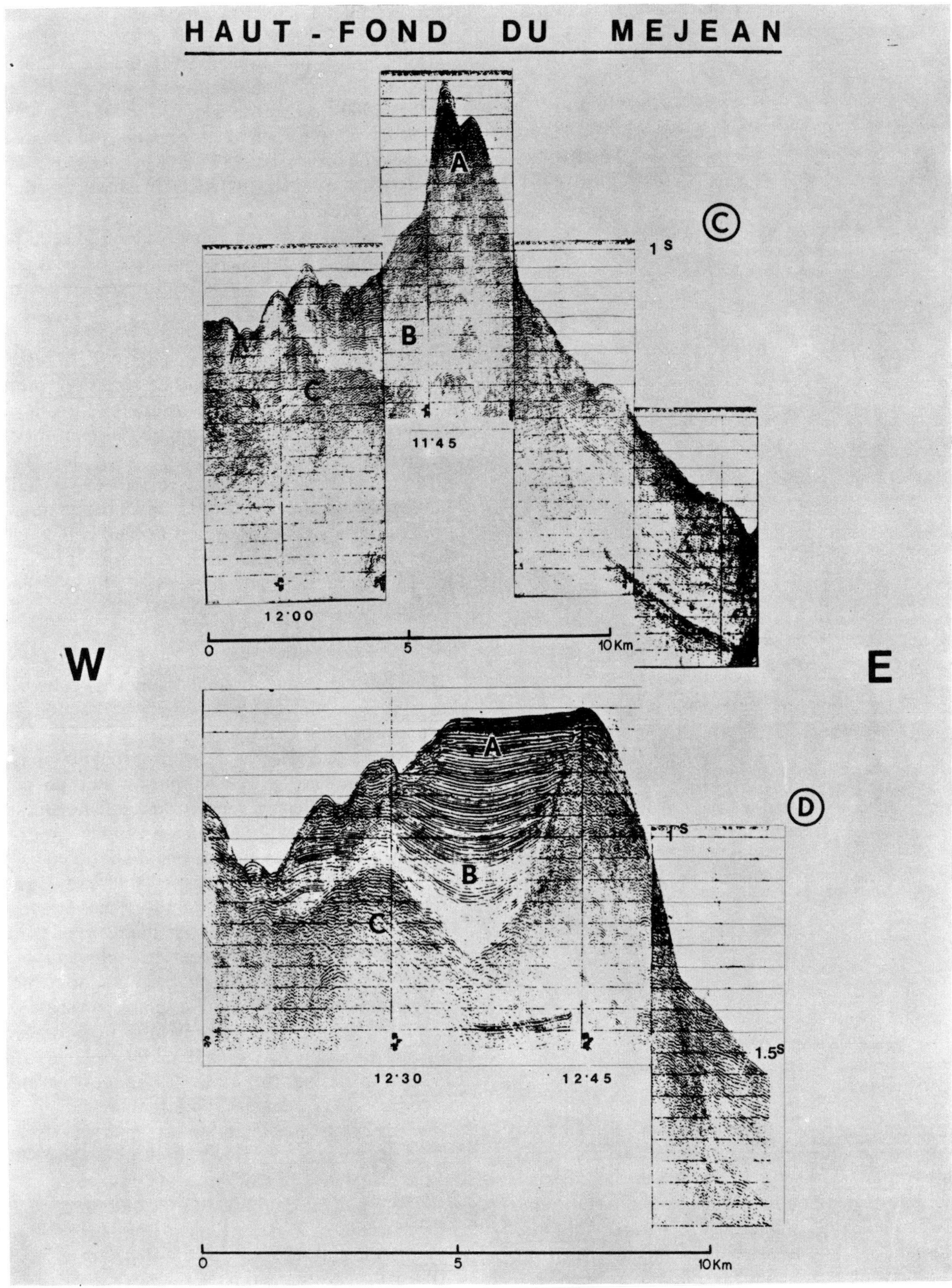

Figure 9. Profils sismiques sur le haut-fond du Méjean montrant le substratum (C) et le bassin sédimentaire suspendu (couches sédimentaires A et B; *cf.*, Figure 6) (profils canon à air réalisés avec le B. O. CATHERINE-LAURENCE). Position représentée sur la Figure 8.

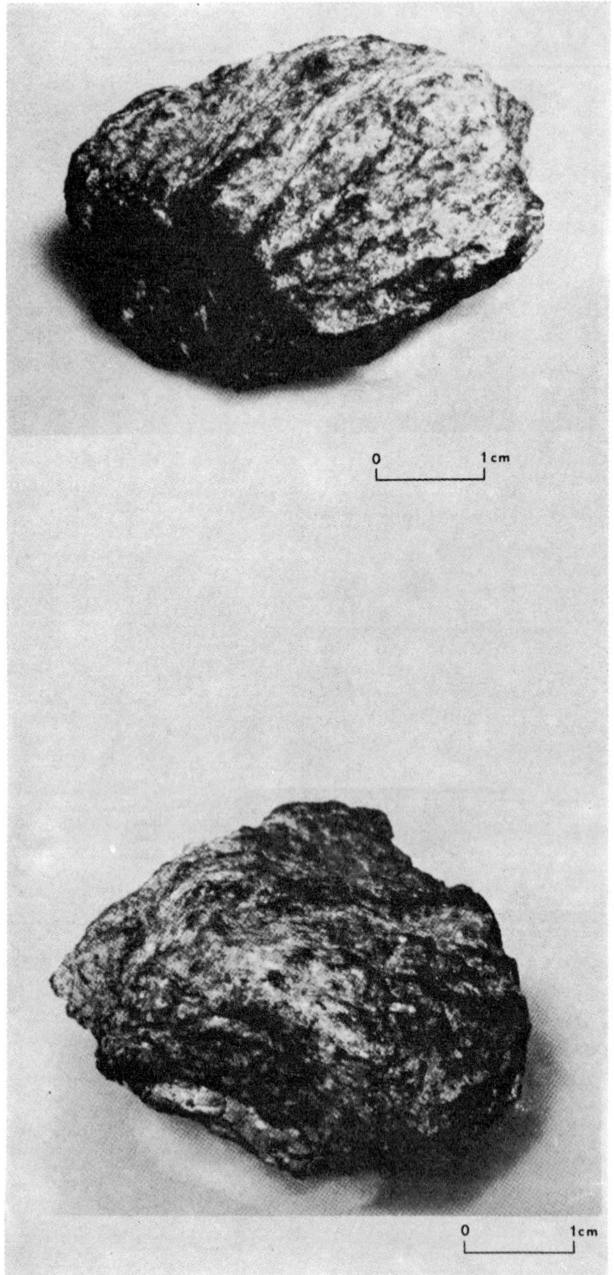

Figure 10. Echantillons de gneiss prélevés sur le haut-fond du Méjean par carottages et provenant du substratum.

rique de matériaux continentaux. Ce sédiment est absolument azoïque. En outre, l'aspect des minéraux sombres altérés et la structure de cette couche ne permet pas de penser à un transport. Il pourrait donc s'agir d'un *paléosol* en place. La roche-mère doit être proche car l'ogive du carottier a été ébrêchée.

Si l'altération a eu lieu à l'air libre, il est nécessaire d'envisager un mouvement d'ennoyage important (plus de 1000 mètres) et très rapide, afin d'expliquer que cette arène n'ait pas été lessivée et resédimentée. Ceci est une bonne illustration du phénomène de subsidence.

Deux *forages* ont été réalisés sur la falaise Sud de ce haut-fond à une profondeur plus importante, entre 1800 et 1900 m. Dans le site T3 (1840 m, sous 26 m de sédiment, on a foré 10 cm de roche qui n'a pu être prélevée. Le forage T5 (1896 m) a échoué car le carottier n'a pas déclenché, mais à 38 m un lit très dur est également présent.

Le réflecteur sismique étant continu entre le piton septentrional et la falaise méridionale, on peut penser que tout le haut-fond du Méjean a un soubassement gneissique.

Sur ce substratum, seule une couverture plio-quaternaire est visible. En tenant compte de la présence d'un paléosol continental à plus de 1000 m de profondeur recouvert par une couverture plio-quaternaire hémipélagique, on peut penser que le haut-fond du Méjean est le prolongement du massif des Maures effrondré depuis le Pliocène. Luteaud (1924), écrivait d'ailleurs: "Je crois donc que le massif des Maures . . . a continué de s'ennoyer durant la plus grande partie du Quaternaire, tandis que la région de l'Estérel et du Tanneron, au contraire, s'est progressivement relevée pendant la même période".

Les Canyons Sous-marins

De nombreux canyons sont visibles sur cette pente. Les plus importants sont le canyon de Saint-Tropez et le canyon de Cannes.

Le canyon de Saint-Tropez est au début de son cours d'abord parallèle au rivage, puis il fait un coude brusque vers le Sud en rencontrant le flanc Ouest du haut-fond de Méjean. Ce canyon semble emprunter des lignes de fractures anciennes qui affectaient la surface d'érosion Miocène émergée. Bourcart (1959) a prélevé par dragages du gneiss sur les flancs du canyon. La carotte B 70, effectuée dans le thalweg par 1420 m, est constituée de vase beige foncé plus ou moins chargée en sablon avec deux passages de sable fin terrigène micacé. Le flanc Ouest semble envasé et le forage T 1 (1395 m), réalisé à proximité du thalweg mais sur son flanc, a prélevé, sous 39 m de vase, une carotte de vase plus indurée avec des faunes mélangées provenant de glissements sur la pente.

Le canyon de Cannes présente un tracé en baïonnette: direction S-SE, puis à 1400 m de profondeur virage brutal vers l'E-SE, et à 1900 m il se dirige vers le S. Sa pente est de 6° de 10 m à 1400 m et de 3° de 1400 m à 1900 m; c'est un des canyons les plus abrupts de la Méditerranée.

Plusieurs carottages ont été réalisés dans le thalweg: de 300 à 1400 m; à partir de 1600 m; et dans la partie plus profonde à 2020 m.

De 300 à 1400 m, on ne trouve que de la vase

homogène beige à gris foncé se chargeant en sablon vers la base. Dans le prélèvement B 71, ayant 940 cm de longueur, on reste dans l'Holocène et les formes benthiques sont bien représentées avec Buliminidae, Nodosariidae, Lagenidae, Cibicidae, Miliolidae, Rotaliidae, Textulariidae et autres.

Après son confluent avec le canyon du Drammont, on note la présence de passées de sablon micacé qui proviennent de l'Esterel (Carotte B 69).

A partir de 1600 mètres de profondeur, on trouve sous quelques mètres de vase homogène une marne très indurée que nous avons prélevée dans les carottes C 13 (1600 m), C 17 (1650 m), C 12 (1750 m) et B 74 (1940 m). Cette marne présente une forte résistivité, une faible teneur en eau (18 %). La fraction grossière est plus importante et moins riche en calcaire que la vase commune. A la base de la carotte B 74 (Figures 11 et 12) le sédiment est très détritique et il contient de nombreux foraminifères remaniés du Miocène et peut-être aussi de l'Oligocène.

Ce sédiment très micacé a créé un biotope peu propice au développement des foraminifères, ce qui explique que la faune en place est rare. On note la présence de quelques foraminifères planctoniques: *Globigerina bulloides, Globigerina eggeri,* mais absence de *Orbulina* et de *Globorotalia.*

Dans la partie plus profonde de ce canyon, à 2020 m, le forage T 6 a trouvé le substratum sous 7 m de sédiments. Seuls quelques fragments de calcaire ont pu être prélevés.

Sur les flancs de ce canyon, plusieurs carottes ont été prélevées. Les flancs sont habituellement formés d'une vase beige terrigène très homogène avec de nombreuses taches de réduction. Des coquilles de lamellibranches et des fibres de Posidonies se retrouvent parfois dans cette vase.

En conclusion, le canyon de Cannes présente tout d'abord une accumulation de sédiments fins dans la partie initiale de son thalweg (direction NNO-SSE). Puis dans la seconde partie de son parcours (O-E), après avoir reçu des canyons de la pente de l'Esterel, on note la présence de passages sablonneux. A partir de 1600 m, c'est-à-dire juste après le passage de l'accident N-S du cap d'Antibes, on a prélevé à chaque carottage des marnes pliocènes. Enfin, à 2000 mètres, on trouve le substratum calcaire. La présence de cette marne à caractère peu profond (pourcentage élevé de la fraction grossière, faible teneur en calcaire, *etc.*) peut s'expliquer par des mouvements tectoniques récents qui auraient entraîné l'effondrement du compartiment à l'Est de la faille. C'est une explication voisine de celle avancée par Bourcart *et al.* (1960) dans son étude de la Baie des Anges. On doit, dans cette hypothèse, admettre un effondrement de 2000 m d'âge pliocène et post-pliocène.

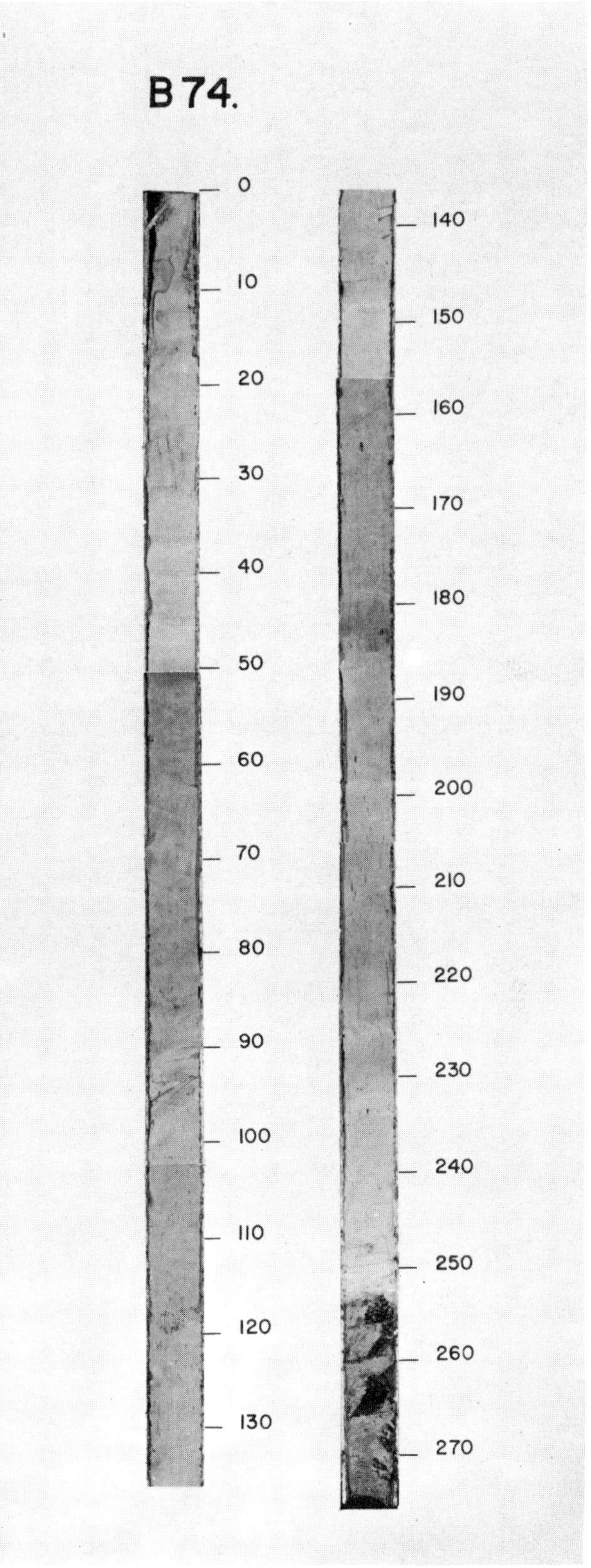

Figure 11. Photographie de la carotte B 74 (profondeur 1940 m) prélevée dans le thalweg du canyon de Cannes. Sous 255 cm de vase holocène et pléistocène on passe brutalement à une marne indurée contenant des foraminifères benthiques miocènes.

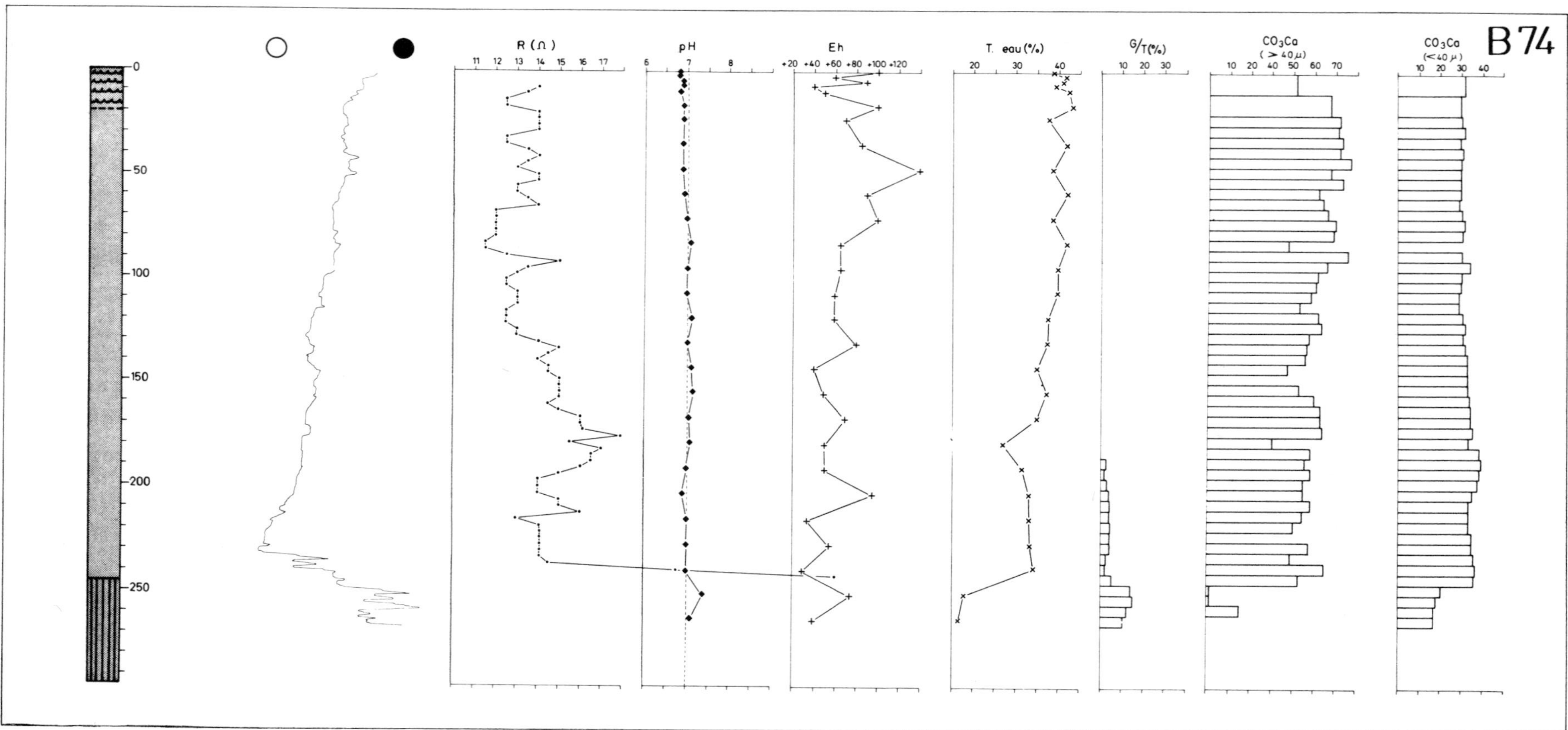

Figure 12. Exemple de log complet montrant les variations des propriétés physiques sur la carotte B 74 (Figure 11). Les cercles représentent la densité optique, **R** est la résistance électrique, G/T est le rapport de la fraction grossière sur le poids total.

Le canyon du Drammont, affluent du canyon de Cannes, semble se combler en direction de l'aval. En effet, à 750 m de profondeur, le carottage (B 73) a atteint le substratum, sous 300 cm de vase gris-bleu terrigène recouvrant du sable terrigène. Ce canyon parait actuellement inactif. Le comblement du thalweg s'effectue par apports de matériel dans l'axe mais également par glissement latéral sur les flancs. Le "galet mou" trouvé dans C 62 proviendrait du démantèlement de la couverture ancienne d'un des flancs du canyon. L'interfluve est recouvert d'une couverture sédimentaire plus épaisse.

Le canyon de l'Esquillon, autre affluent du canyon de Cannes, est parsemé de galets roulés de rhyolite et de pyroméride. Ces galets de rhyolite permienne se retrouvent aussi bien dans le thalweg du canyon (E 7 = 330 m) que sur les interfluves (E 8 = 190 m) (Figure 13). Il est donc difficile de les attribuer au seul phénomène d'écoulement dans un canyon. Une explication plus vraisemblable serait d'évoquer la présence d'une plage fossile quaternaire dans cette partie de la côte de l'Esterel. Au niveau de E 8, qui correspond à un niveau de régression de 190 m ou à un début du mouvement d'ennoyage, ces galets seraient à peu près en place. En E 7 (330 m) ils auraient glissé dans la tête du canyon et en B 82 (350 m) les passages de galets roulés interstratifiés dans la vase indiqueraient un glissement le long de la pente durant la subsidence. La couverture sédimentaire, comme dans le canyon du Drammont, augmente d'épaisseur vers l'aval (E 9).

Conclusion

Les profils bathymétriques sériés sur la pente continentale (Pautot, 1969) montrent des surfaces d'aplanissement qui descendent d'une façon presque continue du Nord au Sud par de petits accidents Est-Ouest. Au Sud de Cannes, une dénivellation importante apparaît dans la partie méridionale du haut-fond du Méjean : elle serait due à l'accentuation des failles de Roquebrune et de Saint-Tropez réunies en une faille unique.

Tandis que le Nord du Méjean reste à une altitude de 500 mètres, la partie Sud descend brutalement en-dessous de 1500 mètres, en formant une zone de "blocs éboulés". En même temps, le canyon de Saint-Tropez qui se dirigeait vers le NE prend une direction N-S. Ce virage important paraît lié à l'affaissement de la région située entre le Sud du Méjean et l'Est de Saint-Tropez d'une part, et au grand accident N-S empruntant la première partie du canyon de Cannes et se prolongeant vers le Sud d'autre part. Le faible recouvrement sédimentaire, les parcours à angle droit des canyons et leur pente excessive s'expliquent mal par une érosion sous-marine, mais beaucoup mieux

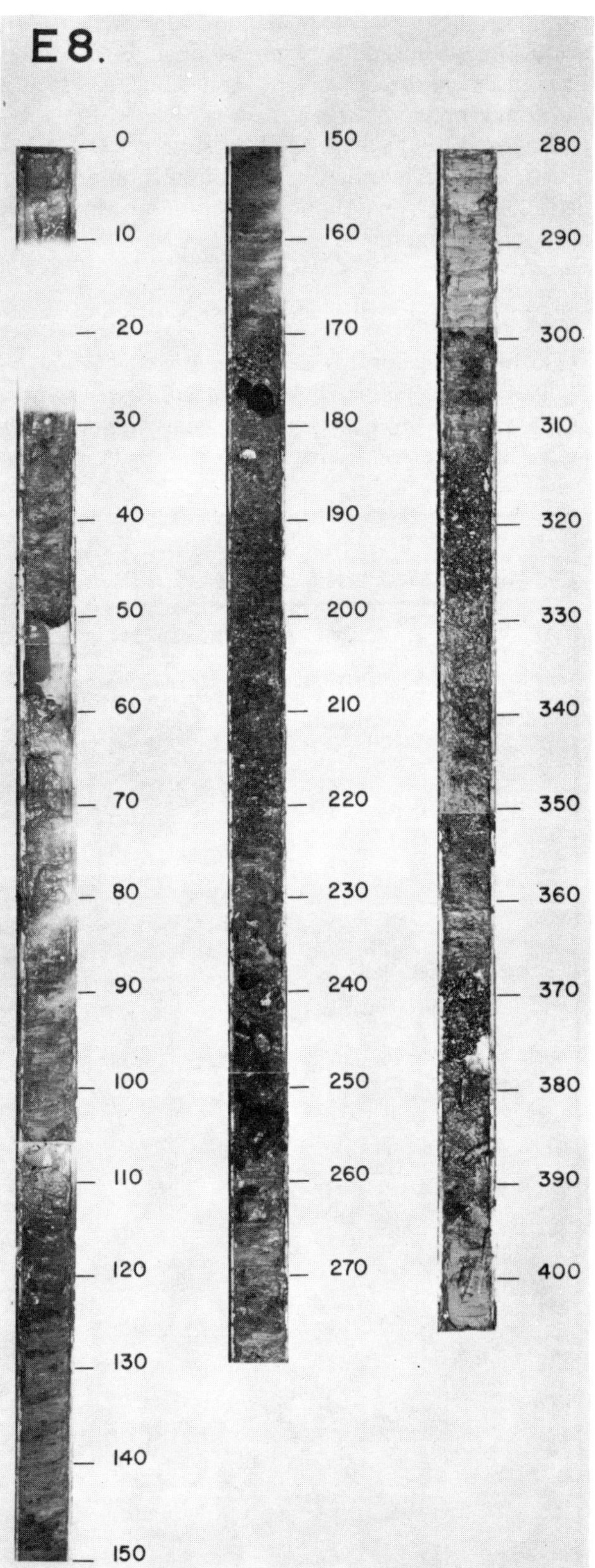

Figure 13. Photographie de la carotte E 8 (profondeur 190 m) prélevée sur un interfluve au large du massif de l'Esterel. Succession de passées de vase sablonneuse et de sable organogène avec des galets roulés de Permien de 2 à 3 cm de diamètre (niveau 176 cm). Nombreux changements de faciès.

par une phase principale d'érosion réalisée à l'*air libre*, avant l'immersion de la région au début du Pliocène. Le creusement de ces canyons serait donc, comme il a été admis pour d'autres régions (Afrique du Nord) (Glangeaud, 1961), d'âge Miocène supérieur (Pontien). A cette époque, le trajet des canyons a épousé le trajet de failles orthogonales anciennes qui auraient rejoué au Pontien (Figure 14).

La Plaine Abyssale

De nombreux profils de *sismique continue* ont été réalisés dans la plaine abyssale par le Musée Océanographique de Monaco (sparker), par la Station de Géodynamique sous-marine de Villefranche-sur-Mer (air-gun) et en 1970 par l'Institut Français du Pétrole et le Centre Océanologique de Bretagne (Flexotir) (Figure 15) (Hersey, 1965; Mauffret, 1968; et Pautot, 1969). Nous ne présenterons ici que l'étude structurale de la terminaison de la marge continentale vers le large, et l'étude sédimentologique de la zone des dômes de la plaine abyssale.

Les résultats principaux de l'interprétation de nos profils sismiques sont les suivants (Pautot, 1969):

1. Les phénomènes de glissement *subactuels* par masses importantes (olistostromes, klippes) sont communs sur la pente continentale.
2. Les deltas sous-marins des canyons de Cannes et de Saint-Tropez forment une morphologie de grande extension.

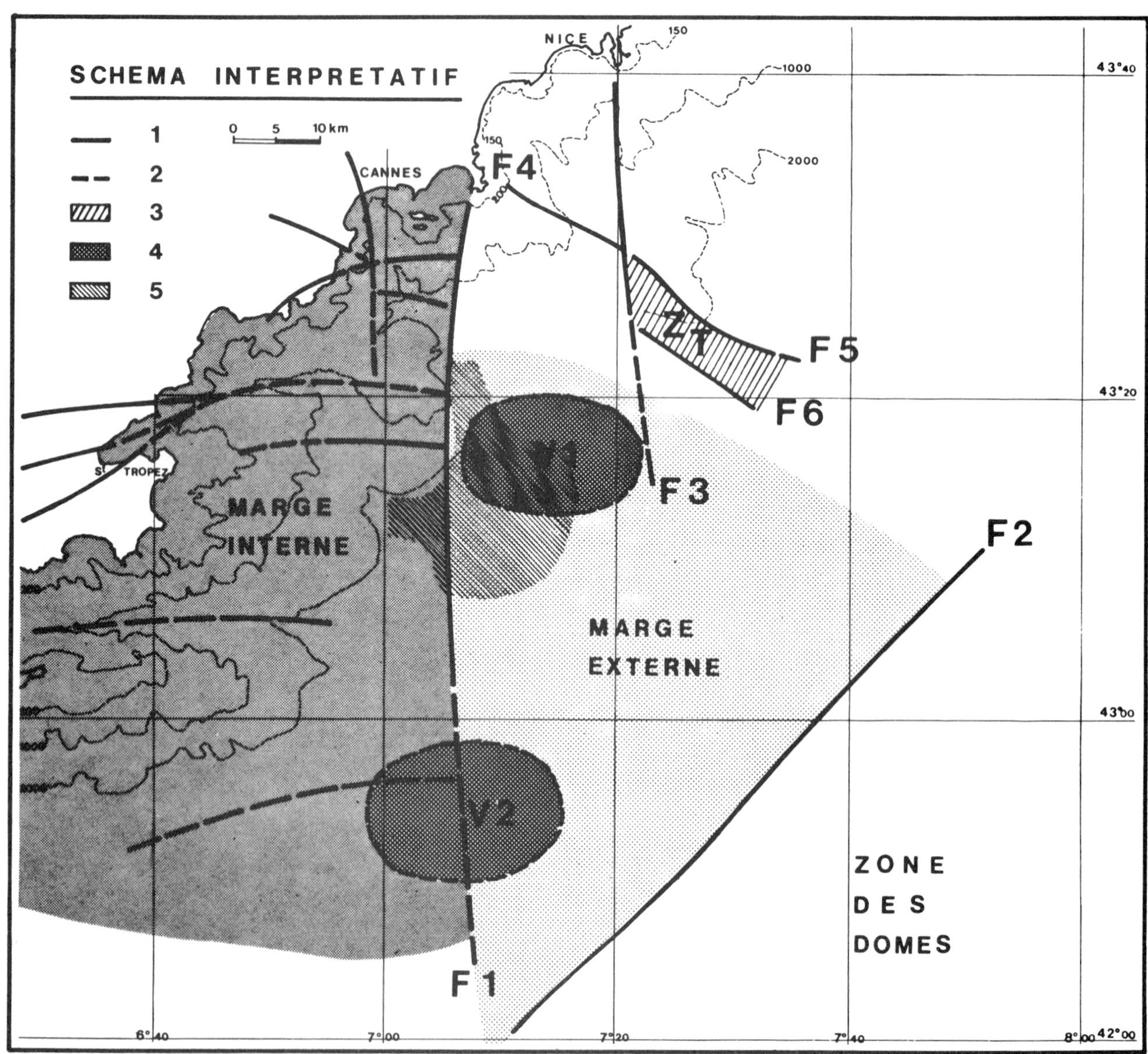

Figure 14. Schéma structural interprétatif. 1, failles; 2, failles probables; 3, zone acoustiquement transparente limitée par des failles; 4, volcans présumés d'après la sismique réfraction et le magnétisme; 5, deltas sous-marins des canyons.

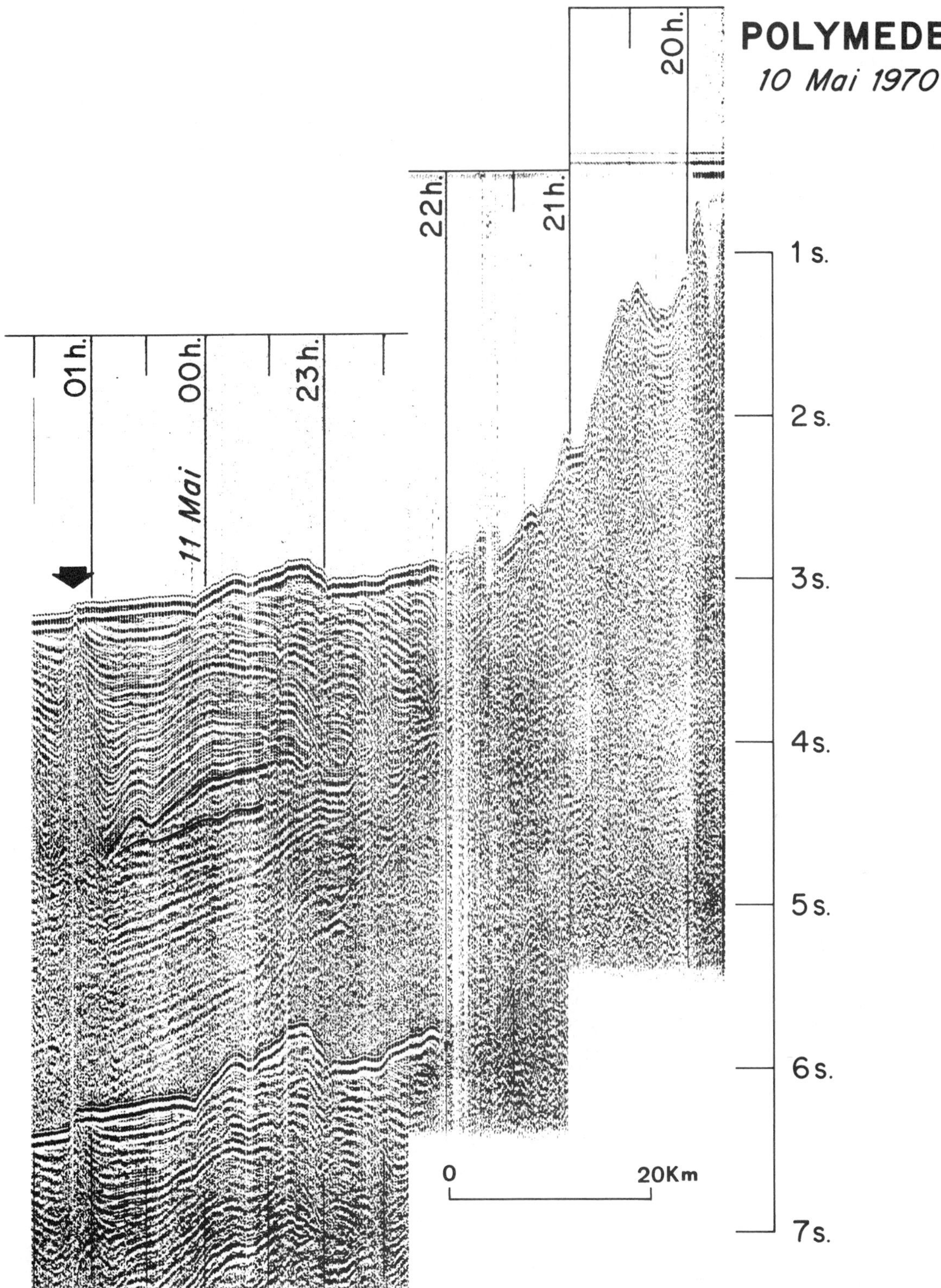

Figure 15. Profil sismique de la marge continentale et du glacis obtenu au cours de la mission POLYMEDE I du N. O. JEAN CHARCOT avec le procédé flexotir. Ce profil a été réalisé en dehors de la zone étudiée ici, à 200 kilomètres plus à l'Ouest. Ce profil a une orientation NO-SE à partir de Marseille. Vitesse: 6 noeuds. Un dôme (flèche) est visible à 1 h 15. Le toit et le plancher du sel sont également visibles.

3. On peut délimiter une zone centrale affectée par de nombreuses structures en forme de dômes.
4. Entre le pied de la pente continentale et la zone centrale parsemée de dômes, on suit un réflecteur puissant qui devient de plus en plus fracturé vers le large. Ce réflecteur est surmonté d'une couche sédimentaire de 800 ms de puissance moyenne.

Les forages JOIDES ont montré que la vitesse dans les sédiments non consolidés était plus faible que ce qui était admis jusqu'alors. Dans le Golfe de Gascogne, comme en Méditerranée, une vitesse de 1,8 km/sec peut être adoptée pour ce type de sédimentation. La couche sédimentaire envisagée ici aurait donc 700 m d'épaisseur. Le taux de sédimentation habituellement admis d'après les mesures sur les carottes (Blanc-Vernet *et al.*, 1969) ou d'après le taux de dénudation moyen des Alpes depuis l'Oligocène (Menard *et al.*, 1965) est de 10 cm/1000 ans.

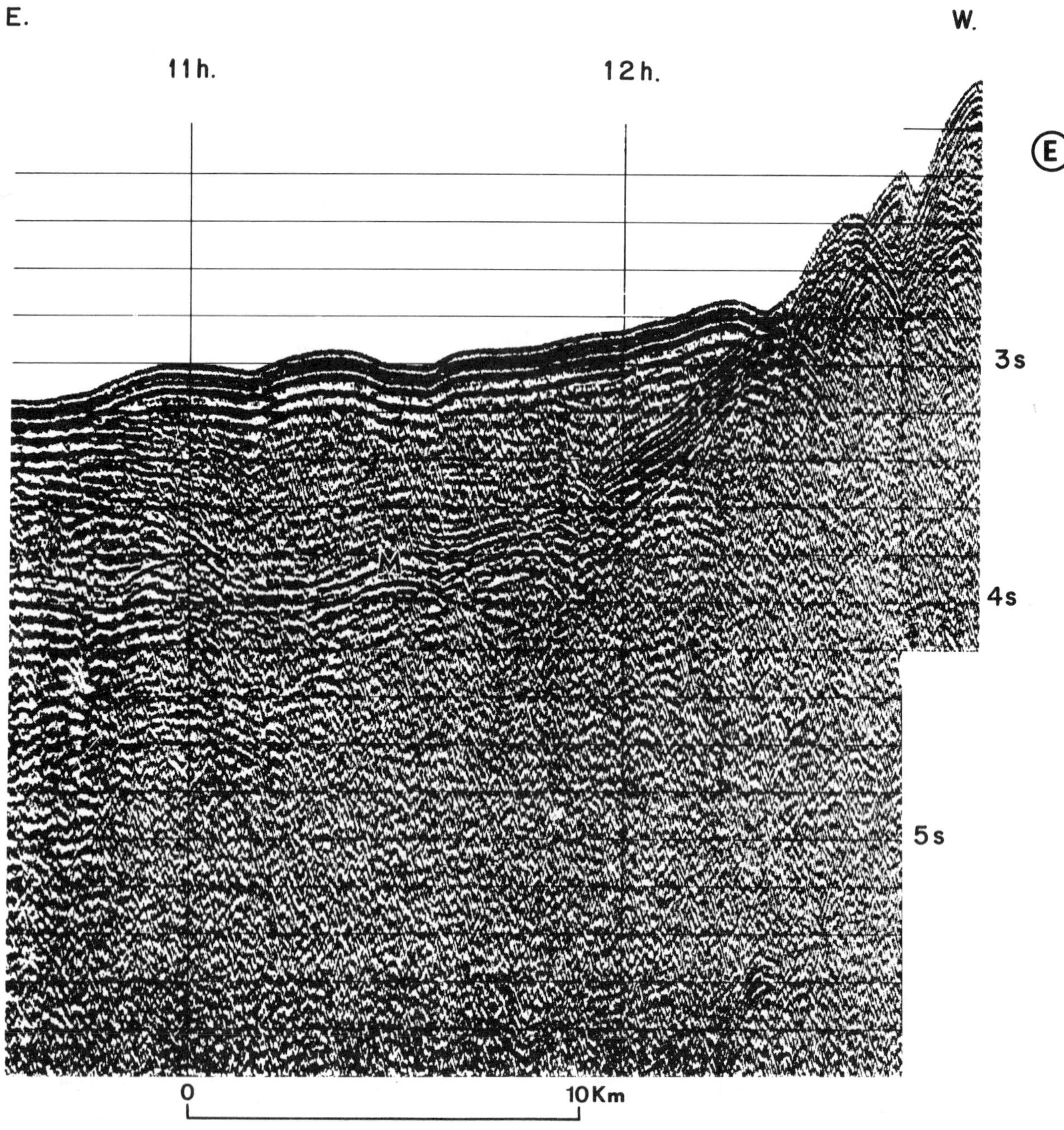

Figure 16. Profil sismique (canon à air) obtenu avec le N. O. CATHERINE-LAURENCE. La position de ce profil est soulignée sur la Figure 8. Il est situé sur la terminaison orientale du haut-fond du Méjean. La couverture sédimentaire meuble est Plio-Quaternaire. Le réflecteur puissant (M) représente probablement la surface Miocène.

La base de cette série sédimentaire non consolidée aurait 7 MA, c'est-à-dire serait Miocène terminal à Pliocène basal.

Le réflecteur puissant au bas de la marge continentale a les caractéristiques d'un horizon sédimentaire induré. Il présente des fractures et même parfois des surfaces d'érosion. Ce réflecteur représente probablement la surface miocène effondrée au cours du phénomène de subsidence.

Contact Glacis-Plaine

La carotte C 16 a été prélevée par 2200 mètres de profondeur sur le tombant Sud du haut-fond du Méjean, et à proximité du flanc Nord du canyon de Saint-Tropez. C'est une vase gris-bleu sablonneuse. A 512 cm, on trouve des coquilles de lamellibranches et à la base, de 530 à 540 cm, on note un passage franc à un sable vaseux. L'ogive a été ébréchée sur le substratum. On est certainement en présence de blocs de substratum effondrés, recouverts de sablon draîné par le canyon de Saint-Tropez et remodelé par les courants de contour puissants au pied de cette marge (observations en bathyscaphe de Bellaiche et Pautot, 1966).

La carotte B 72 provient du delta sous-marin du canyon de Saint-Tropez par 2480 m de profondeur, sur la levée Sud du canyon. Sur les 834 cm de la carotte, on note une succession de strates de vase beige et de sable fin ou de sablon. A partir de 300 cm, les passées sableuses sont plus nombreuses et leur épaisseur croît vers la base en même temps qu'apparaît un granoclassement type. On relève 6 turbidites nettes de 2 à 13 cm d'épaisseur, mais également 155 passées sablonneuses de plus de 2 mm d'épaisseur. Elles se présentent en récurrence après apport turbide et en figures de courant de contour. Ce sable fin représente 180 cm de sédiment sur la longueur totale de la carotte. Les turbidites ont une couleur foncée et une composante terrigène avec micas, pélites rouges permiennes, et quartz. La coloration s'assombrit vers la base.

La carotte C 5 prélevée par Bellaiche (1969) au large des îles du Levant par 2500 m de profondeur présente une structure différente. Cette carotte de 795 cm est constituée par 4 parties essentielles:

de 0 à 250 cm: partie riche en matériel grossier (sable et sablon)

de 250 à 592 cm: partie constituée essentiellement de vase

de 592 à 670 cm: sédiment sableux

de 670 à 795 cm: vase

La majeure partie des passées grossières de cette carotte est constituée d'un matériel organogène calcaire (foraminifères planctoniques en général). Ces passées, comme l'analyse climatique l'a montré, sont en relation avec l'abondance des foraminifères durant les périodes chaudes.

Le forage T10 réalisé dans la même zone presque au contact du socle a permis le prélèvement d'une carotte de 70 cm sous 43 m de sédiment. Elle est composée d'un sable plus ou moins vaseux. Ce sable est constitué de gros minéraux non roulés: quartz anguleux, plagioclases, hornblende et de foraminifères benthiques et planctoniques.

En conclusion, on peut évoquer *trois modes de sédimentation généraux* dans la plaine abyssale:

1. La sédimentation pélagique lorsque les îles font barrage aux apports terrigènes.
2. La sédimentation terrigène par glissement de matériel le long des parois.
3. La sédimentation par "turbidites" (apports terrigènes empruntant des canyons sous-marins) qui est souvent remodelée par les courants de fond ou de contour.

Les Dômes

Une étude sédimentologique spéciale a été réalisée sur des dômes qui perçent la plaine abyssale (Figures 17 et 18). Trois carottages ont été réalisés sur un dôme appelé structure A (42°47′N et 7°41′E) par le Musée Océanographique de Monaco:

C1: dans la dépression qui entoure ce relief (2665 m)

C2: sur la partie Sud-Est du sommet (2605 m)

C3: sur la partie Nord-Ouest du sommet (2602 m)

L'étude sédimentologique a été réalisée en collaboration avec Bellaiche et les résultats généraux ont déjà été publiés (Alinat *et al.*, 1970).

Le problème posé était la détermination de la nature de ces dômes. La présence de pyrite néoformée à la base de C 3 et l'accroissement de la concentration ionique de l'eau interstitielle dans la vase surmontant ces dômes, semblent indiquer l'existence de conditions d'oxydo-réduction liées souvent aux évaporites (Bellaiche et Pautot, 1968).

Des forages ont été réalisés sur d'autres dômes du même type au Sud de Toulon. Les carottes prélevées sous 45 m de sédiment sont habituellement des vases plus ou moins compactes; elles remonteraient à une époque comprise entre 200.000 et 300.000 ans et seraient donc d'âge Pléistocène Supérieur (datation par coccolithes en cours). Un niveau de sable a été prélevé sur un de ces dômes. Compte tenu des conditions hydrodynamiques actuelles dans ce secteur, ce sable n'a pu se déposer dans cette position. On est donc obligé d'admettre que la formation de ces figures topographiques est subactuelle. Les forages JOIDES du parcours 13 ont effectivement trouvé un horizon salifère d'âge Messinien (6 MA). Ces dômes sont donc bien des structures évaporitiques mais le cap-rock n'a pas été atteint sur ce dôme.

PROPRIETES PHYSIQUES DU SEDIMENT

Sur la cinquantaine de carottes prélevées, plus de 10.000 mesures ont été réalisées: résistivité électrique, teneur en eau, pH, potentiel d'oxydo-réduction, poinçonnement, cisaillement, densité humide, porosité, teneur en calcaire, granulométrie. La mesure de la vitesse du son n'a pu être effectuée car nous ne disposions pas de l'appareillage nécessaire. L'analyse de ces mesures a été présentée en détail (Pautot, 1969). Nous voulons donner ici les résultats généraux après traitement de ces données sur ordinateur.

Pour chaque carotte (en moyenne 50 mesures par technique) la moyenne arithmétique de chaque propriété a été calculée. Dans un premier temps, on calcule la moyenne régionale qui est la moyenne de chaque propriété physique sur une couche sédimentaire de 5 m d'épaisseur. Dans un deuxième temps, on a fait apparaître les coefficients de corrélation entre les diverses propriétés physiques.

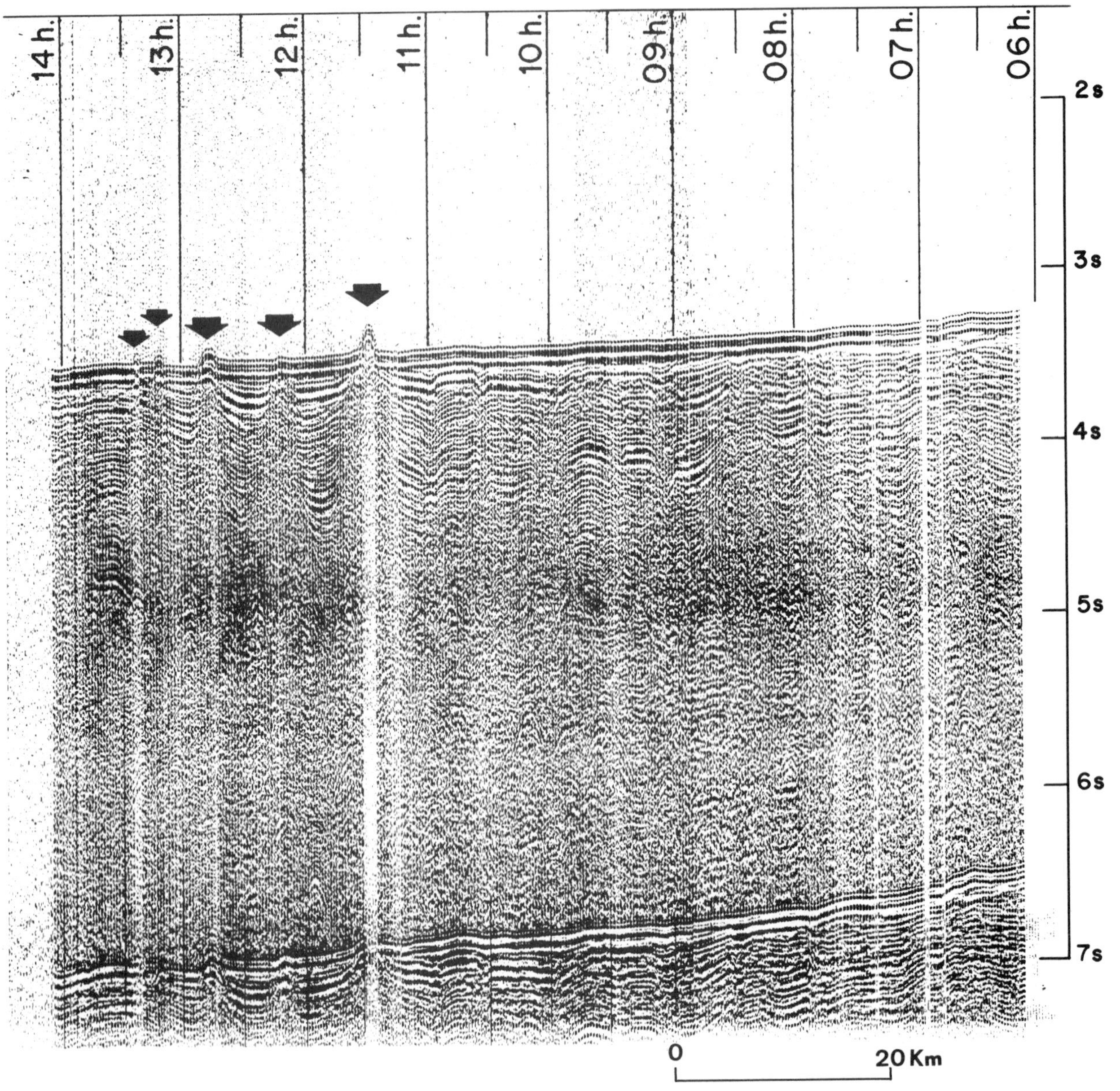

Figure 17. Exemples de structures diapiriques de nature salifère dans la plaine abyssale ligure. Cet enregistrement a été réalisé avec le système flexotir au cours de la mission POLYMEDE I avec le N. O. JEAN CHARCOT. Vitesse: 6 noeuds. Ce profil est donné comme exemple de diapirs en formation et de diapirs perçants; il est situé au Sud de Toulon.

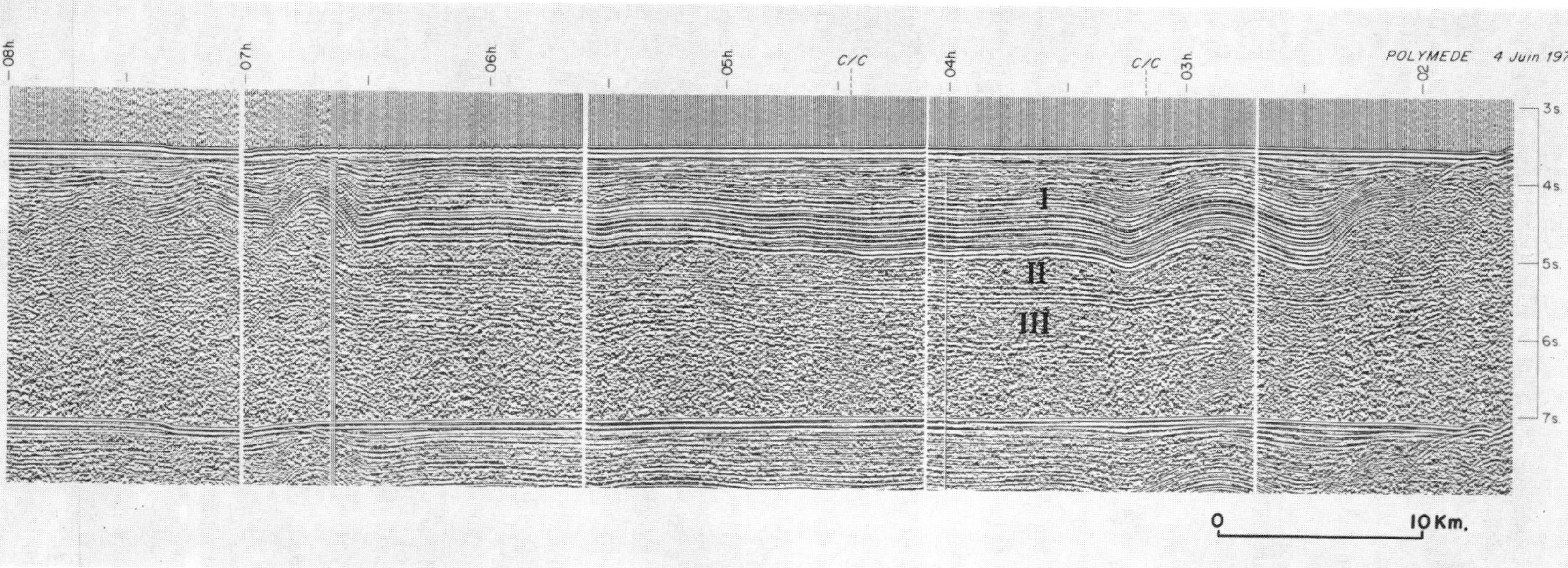

Figure 18. Exemple de profil montrant la couverture sédimentaire du bassin occidental méditerranéen. Procédé flexotir, N. O. JEAN CHARCOT, mission POLYMEDE I. Vitesse: 6 noeuds. Ce profil est orienté au N E de Minorque (Baléares). Cet exemple permet de mettre en évidence la couverture plio-quaternaire (I), la couche de sel formant des structures diapiriques (II), les niveaux sub-horizontaux anté-Miocène supérieur (III).

Propriétés Electriques

Cette méthode a été mise au point par l'auteur (Pautot, 1967a). Les sédiments marins sont composés de trois éléments: la charpente minéralogique, la matière organique et l'eau interstitielle. La partie minéralogique est formée essentiellement de quartz et d'autres minéraux non conducteurs. Seuls quelques minéraux comme la magnétite ont une conductivité appréciable. La résistivité électrique d'un sédiment marin dépend essentiellement de son eau interstitielle. Elle varie donc avec la température, la pression et la salinité. La province étudiée ayant une extension réduite, on peut l'assimiler à une zone profonde à salinité constante.

Les valeurs brutes de résistivité (Figure 19) ont été calculées et comparées aux autres propriétés physiques. Les valeurs moyennes *extrêmes* relevées dans l'étude des carottes de cette zone vont de 39 Ωcm à 94Ωcm. Ces valeurs sont comparables à celles publiées par Filloux (1967) dans sa thèse (39 Ω cm à 87 Ω cm).

La valeur moyenne régionale calculée est de 54 Ω cm. La variation avec le niveau est significative:
48 Ω cm entre la surface et le premier mètre;
52 Ω cm entre 100 et 200 cm;
53 Ω cm entre 200 et 300 cm;
56 Ω cm entre 300 et 400 cm; et
58 Ω cm entre 400 et 500 cm.

Les coefficients de corrélation entre la résistivité et les autres propriétés physiques ont été calculés. Ces coefficients de corrélation ont été obtenus carotte par carotte, et chacune de ces carottes est définie par un nombre de mesures situé entre 20 et 60. Je considère qu'un coefficient de corrélation supérieur à 0.6 dans ces conditions est significatif.

En reportant sur un graphique les valeurs des propriétés physiques qui présentent un bon coefficient de corrélation, on peut en déduire une loi générale linéaire. Les *lois générales* sont les suivantes:

1. La résistivité croît avec l'"enfouissement", la densité humide et la granulométrie. Elle augmente dans le même sens que le poinçonnement et le cisaillement.
2. La résistivité est inversement proportionnelle à la teneur en eau et à la porosité.
3. Il ne semble pas y avoir de relation entre la résistivité et le pH, le potentiel d'oxydo-réduction, la profondeur d'eau, et la pente du relief.

Horn *et al.* (1968) dans l'étude de carottes de la Mer Méditerranée et de la Mer de Norvège ont établi d'excellentes corrélations entre la vitesse du son dans le sédiment et les propriétés physiques suivantes: porosité, teneur en eau, densité humide, indice des vides, granulométrie, *etc.* (Hamilton, 1970).

L'établissement de ces lois de relation entre la résistivité et d'autres propriétés physiques devrait permettre, si elles sont univoques, *d'estimer* directement la porosité, la densité d'un sédiment, et de déterminer la vitesse du son.

Quelques carottes ne suivent pas les lois générales établies précédemment. Or, ces carottes proviennent d'une pente abrupte ou d'un thalweg de canyon. On peut donc également déterminer les zones instables actuelles ou même *fossiles* par l'analyse des propriétés physiques du sédiment.

"Enfouissement"

Pour étudier le rôle de l'"enfouissement" sur la diagenèse précoce du sédiment, on a établi des moyennes et des coefficients de corrélation mètre par mètre (Figure 20). Entre 0 et 100 cm, la disposition des points montre qu'il n'y a pas de corrélation visible entre les diverses propriétés physiques.

Par contre, entre 300 et 400 cm par exemple, les corrélations, sans être encore significatives, semblent bien se dessiner. Le début du tassement est donc visible dès les premiers mètres.

Teneur en Eau

La résistivité électrique dépend grandement de la teneur en eau du sédiment. Les lois établies pour la résistivité sont donc valables pour la teneur en eau. La moyenne régionale est de 51%. La diminution de la teneur en eau avec l'"enfouissement" est très nette après de grandes oscillations dans les dix premiers centimètres:
0 à 100 cm: 58%
100 à 200 cm: 50.5%
200 à 300 cm: 47%
300 à 400 cm: 46%
400 à 500 cm: 41.5%

Cette variation n'est pas directement linéaire.

La teneur en eau n'est pas constante après le premier mètre, comme certains auteurs le pensaient, mais elle diminue lentement avec l'enfouissement. Cette perte d'eau, rapide dans les premiers centimètres, puis plus lente, est une des premières étapes de la diagenèse.

Poinçonnement et Cisaillement

Les mesures de cisaillement et du poinçonnement sont des mesures complémentaires dont les résultats sont en bonne concordance.

γ en g/cm^2 sont les valeurs de compression simple obtenues avec un pénétromètre; γ en g/cm^2 sont des valeurs de résistance au cisaillement obtenues avec un scissomètre. La moyenne *régionale* est de 850 g/cm^2.

L'évolution avec l'"enfouissement" est très nette:
0–100 cm: $\gamma =$ 480 g/cm^2
100–200 cm: $\gamma =$ 600 g/cm^2

200–300 cm: $\gamma = 800$ g/cm^2
300–400 cm: $\gamma = 970$ g/cm^2
400–500 cm: $\gamma = 1300$ g/cm^2

Les corrélations entre τ et γ, carotte par carotte, sont bonnes à l'exception de quelques carottes qui proviennent de régions instables. Théoriquement, la "consistance" du sédiment γ et la "rigidité" du sédiment τ sont reliées dans un corps cohérent, et on peut définir la cohésion c et l'angle de frottement interne ϕ. En fait, s'il n'y a pas de relation entre τ et γ, c'est que l'on a affaire à un milieu devenant pulvérulent (cohésion nulle) ou devenant purement cohérent (angle de frottement interne nul). Par exemple, un sablon fin humide donnera une forte valeur de "consistance"

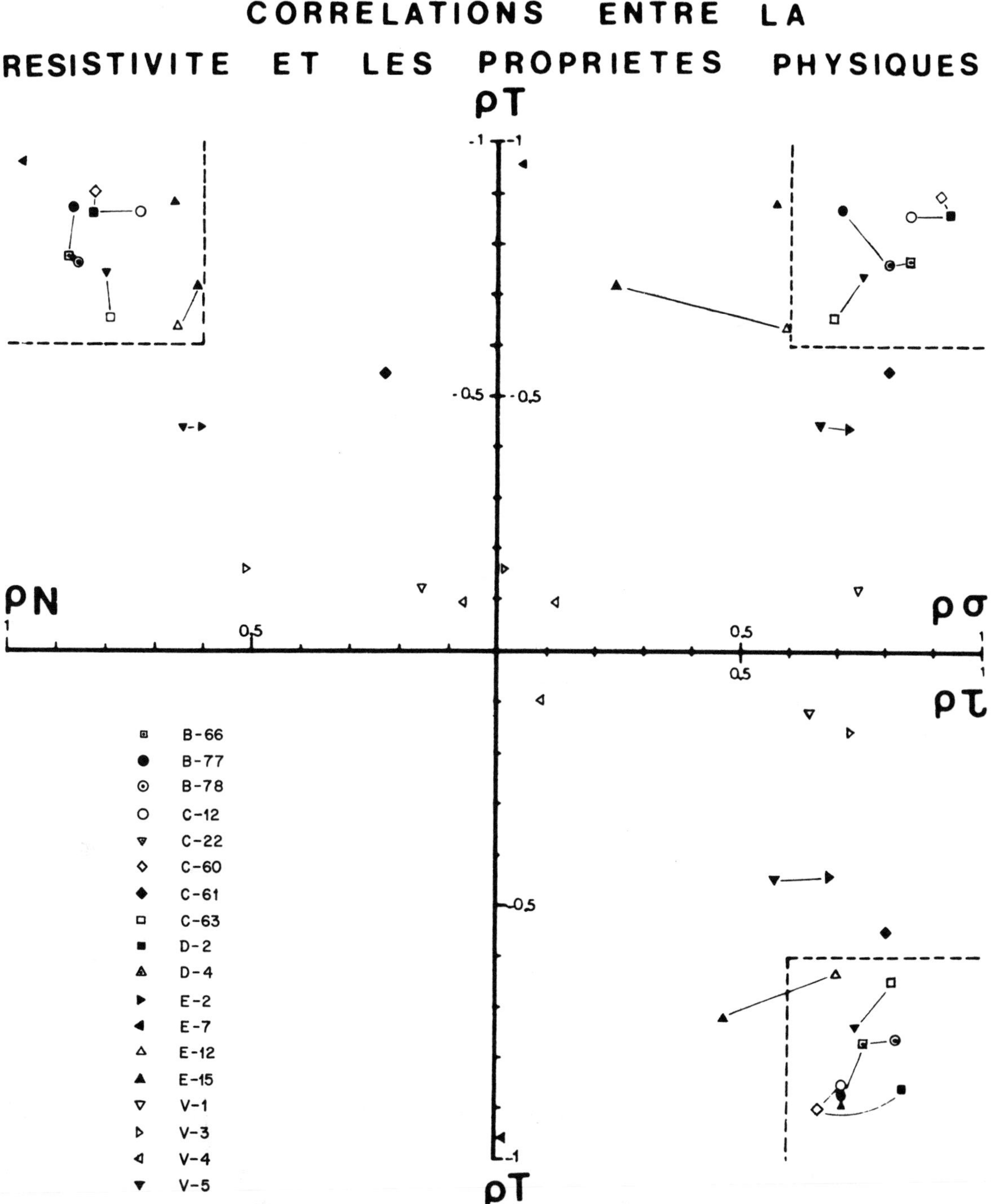

Figure 19. Corrélations entre la résistivité et diverses propriétés physiques. Chaque symbole représente la valeur moyenne d'une carotte. ρ est la résistivité; T la teneur en eau, N le niveau, γ le poinçonnement et τ le cisaillement.

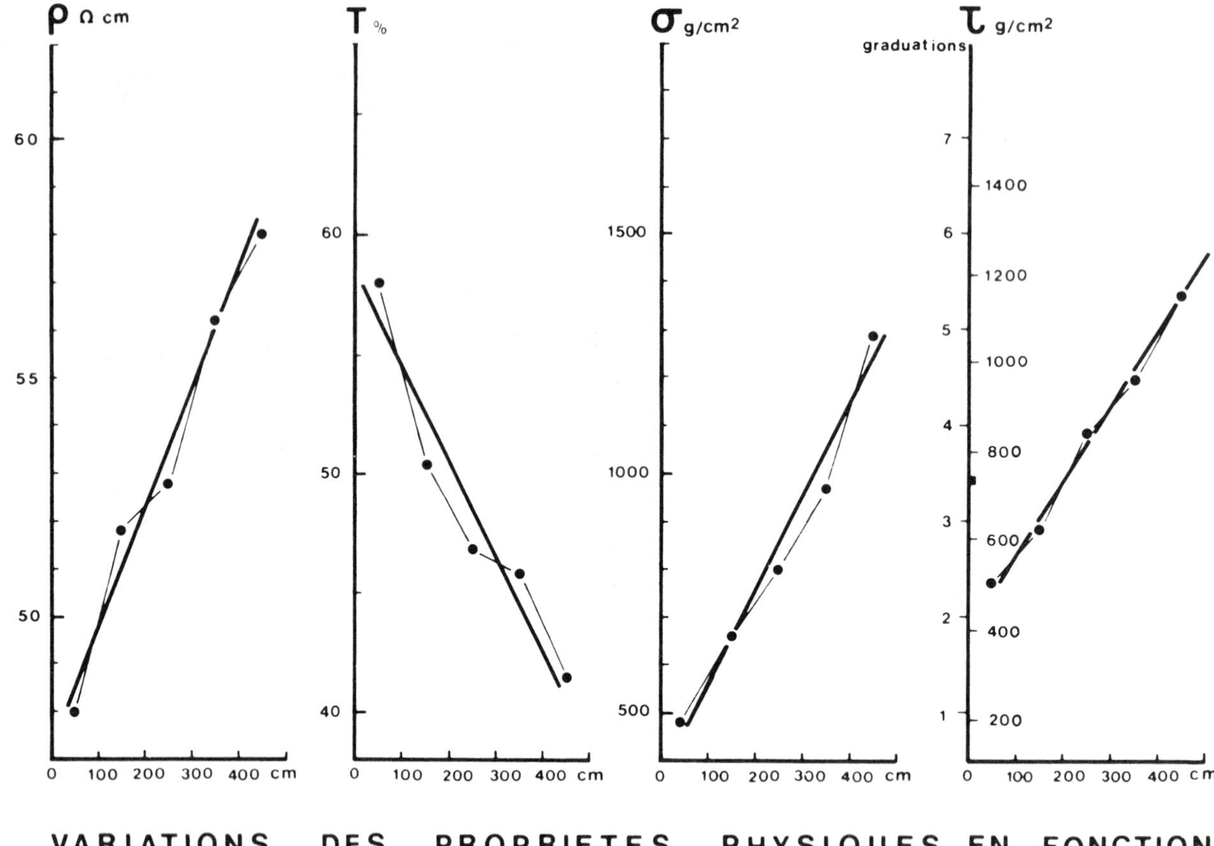

Figure 20. Variations des propriétés physiques en fonction de l'enfouissement. Chaque point représente la moyenne par mètre.

et une faible valeur de "rigidité". De la même manière, une vase chargée de coquilles ou de fibres végétales présentera une forte valeur de "consistance" pour une faible valeur de "rigidité".

Densité Humide et Porosité

Il y a un bon accord entre ces propriétés et les autres mesures mécaniques et électriques. Les valeurs extrêmes vont de 1.35 à 2.00 pour la densité et de 0.41 à 0.65 pour la porosité. La porosité est une caractéristique fondamentale en relation étroite avec la résistivité électrique et la vitesse du son (Horn *et al.*, 1968).

Conclusion: Propriétés Physiques

L'étude statistique des propriétés physiques a permis de mettre en évidence des relations plus ou moins lâches entre la granulométrie, la teneur en eau, l'"enfouissement", le coefficient de compression simple, le coefficient de torsion, la densité humide et la porosité des sédiments d'une région donnée.

En se référant aux courbes de Horn *et al.* (1968) on peut, d'après les résultats des mesures physiques, en déduire la *vitesse du son* approchée dans le sédiment étudié. La porosité moyenne donne une vitesse de 1550 m/sec; la teneur en eau et la densité humide correspondraient à une vitesse de 1500 m/sec. La partie superficielle du sédiment aurait donc une vitesse *comparable* à la couche d'eau qui recouvre le sédiment. Dans le cas de la transmission du son, les paramètres importants à connaître sont la porosité, la teneur en eau et l'indice des vides. Il ne semble pas qu'il y ait une relation étroite entre la vitesse du son et la densité sèche ou la teneur en carbonates.

Le coefficient de cisaillement reflète la granulométrie et la genèse du sédiment, mais ce n'est pas un paramètre en relation directe avec la vitesse du son. *La dimension moyenne des grains* est le paramètre le plus important pour la détermination de la vitesse du son, et cela gouverne la teneur en eau et la texture du sédiment.

La résistivité électrique est en excellent accord avec les propriétés mécaniques et physiques du sédiment. Cette méthode est précise, rapide, c'est un paramètre complémentaire important. Les mesures "in situ" permettraient de définir rapidement la lithologie du sédiment et peut-être une prospection de divers minerais.

Les propriétés physiques ont permis une approche de *la genèse* du sédiment (Figure 21). Les parties planes ou les pentes inférieures à 6° ne présentent pas d'éboulement et la vase montre de bonnes relations entre ses propriétés physiques. Dans le cas de zones érodées, de zones tectoniques instables et de pentes accentuées, les propriétés physiques sont indépendantes. Ainsi, en complément de l'analyse lithologique, les propriétés physiques peuvent montrer des pics dûs soit à une variation granulométrique, soit à une phase d'érosion et de tassement liée au passage d'une avalanche sous-marine.

PROPRIÉTÉS CHIMIQUES

Une centaine d'analyses chimiques ont été effectuées sur les carottes. La majeure partie de ces analyses a été réalisée sur la fraction fine (inférieur à 40μ).

Pour conduire et interpréter correctement l'analyse élémentaire d'une vase marine, il faut auparavant connaître les minéraux qui la composent. L'analyse minéralogique des argiles n'a pu être effectuée sur chaque échantillon, ce qui entraîne une interprétation parfois incomplète.

La moyenne des compositions chimiques (en %) de tous les échantillons étudiés ici est la suivante:

SiO_2 : 40.00
Al_2O_3 : 10.00

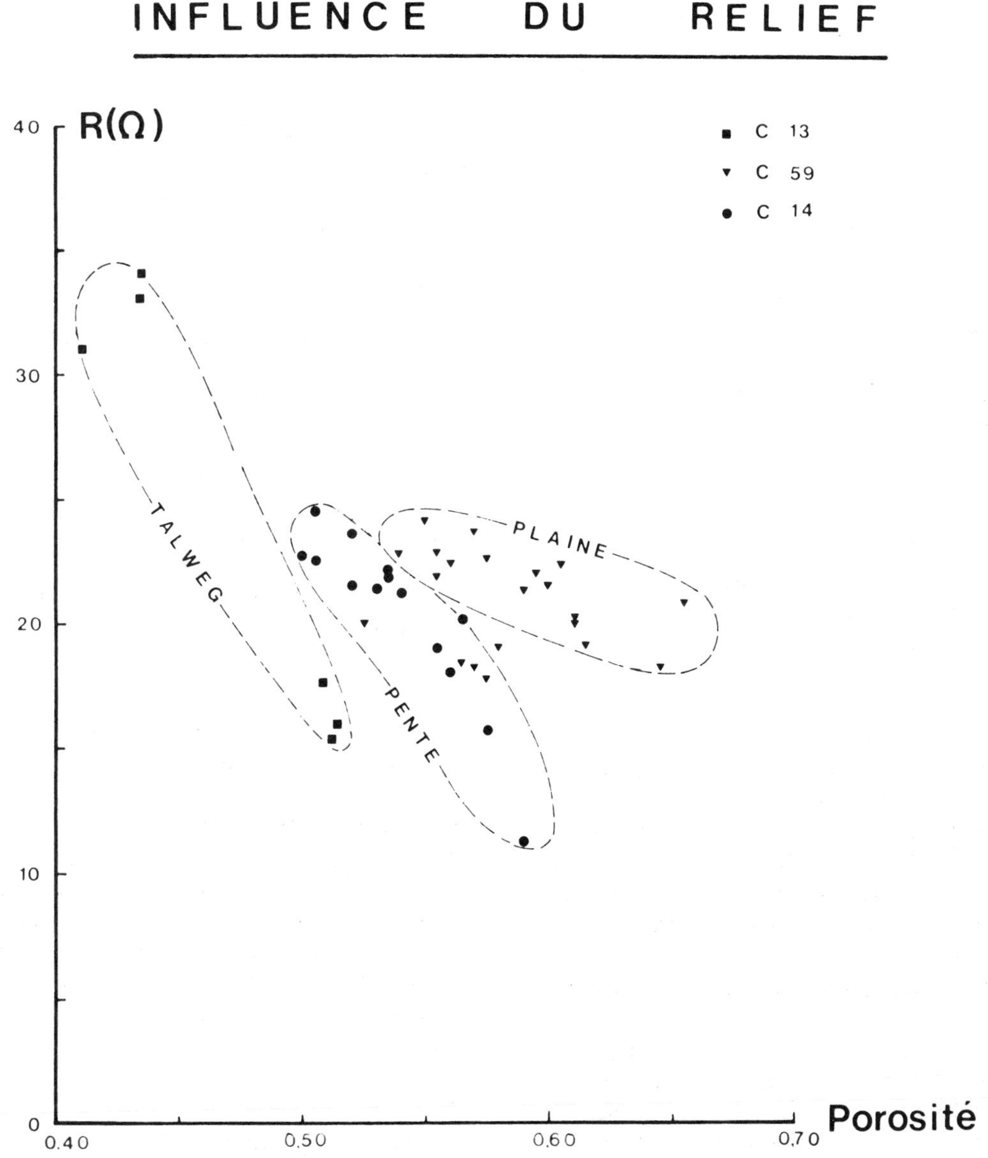

Figure 21. Influence du relief sur les relations entre la porosité et la résistivité.

Fe_2O_3 : 4.00
MnO : 0.05
MgO : 3.00
CaO : 20.00
Na_2O : 0.80
K_2O : 2.00
TiO_2 : 0.50
P_2O_5 : 0.10

C'est la composition chimique type d'une vase terrigène.

Malgré les difficultés rencontrées dans l'interprétation de l'analyse élémentaire des vases marines, il est possible dans une certaine mesure *d'estimer* quels sont les minéraux qui fournissent les éléments les plus courants du sédiment (Debyser, 1959). Le fer provient essentiellement des illites et des chlorites et de leur produit d'altération, le magnésium des chlorites et de la dolomite, le potassium des feldspaths alcalins des micas et des illites, le sodium des plagioclases. Bien qu'une telle interprétation n'ait qu'une valeur statistique et que toutes les exceptions soient possibles, elle permet d'orienter l'étude minéralogique.

Etude Minéralogique

Les travaux de Chamley (1968) et de Nesteroff *et al.*, (1963) aboutissent aux conclusions suivantes:

1. La sédimentation argileuse récente a une répartition géographique homogène en Méditerranée nord-occidentale.
2. L'illite est le minéral dominant; ceci est en relation avec la haute concentration en micas dans de nombreux types de roches et la résistance relative de ces micas à l'altération chimique.
3. La distribution verticale en Méditerranée est caractérisée par une homogénéité de l'apport au cours du temps.

Une agradation superficielle généralisée se produit.

Biscaye (1965), dans son étude des sédiments de l'Atlantique, a mis en évidence des phénomènes importants: les périodes glaciaires seraient marquées par un déficit en kaolinite et gibbsite et une augmentation de la teneur en chlorite par rapport aux sédiments actuels, et l'époque pliocène semble présenter le phénomène inverse.

L'analyse minéralogique de quelques échantillons prélevés dans le Golfe du Lion et en Provence nous conduit aux conclusions suivantes: l'illite est le principal constituant avec la chlorite et un complexe gonflant illite-montmorillonite. La kaolinite semble inexistante dans le Golfe du Lion alors qu'elle est présente en Provence (Golfe de Fréjus).

Distribution Horizontale des Eléments dans les Echantillons de Surface

CaO (Figure 22)

Autour des îles de Lérins formées de dolomie, les valeurs sont plus élevées que vers l'Estérel, de nature volcanique. La teneur en calcaire diminue avec la profondeur dans le canyon de Cannes. Le haut-fond du Méjean à soubassement gneissique a une faible teneur en calcaire.

Ces faits suggèrent que dans la région étudiée l'origine du calcaire est à dominante détritique. Plus au large, Nesteroff (1965) a montré que le pourcentage de calcaire se stabilise autour de 40% et la proportion entre le calcaire détritique et le calcaire organogène est dans ce cas certainement différente.

SiO_2

Sur le littoral, les valeurs ont un caractère assez constant (33 à 38%). Des valeurs plus élevées sont notées sur le haut-fond du Méjean et sur la partie terminale du canyon de Cannes. Dans les deux cas, il y a présence d'un substratum sous-jacent proche, gneiss dans un cas, vase ancienne dans l'autre.

Al_2O_3

Les valeurs de l'alumine sont bien groupées entre 10 et 12%. Il y a constance du rapport SiO_2/Al_2O_3 dans cette région (Figure 24).

Fe_2O_3

Les concentrations en fer varient de 3.30 à 5.87%. Il y a encore ici une relation directe entre silice, alumine et fer (Figure 24), c'est-à-dire que la teneur en fer totale du dépôt est liée à la fraction minérale silico-alumineuse.

MnO

Proportion extrêmement constante de MnO le long du littoral: 0.04%. Augmentation avec la profondeur dans le canyon de Cannes (0.08%) et sur le haut-fond du Méjean (0.19%). Cette valeur élevée ne doit pas être en relation avec le socle sous-jacent car dans la région, seules les dolérites et les andésites présentent des valeurs élevées comparables. L'augmentation de la vitesse du courant sur ce relief entraînerait plutôt des conditions d'oxygénation qui sont propices à la précipitation du manganèse (formation des "nodules").

MgO

Valeurs comprises entre 2.30 et 4.20%. Les valeurs les plus fortes sont relevées au contact de l'Esterel

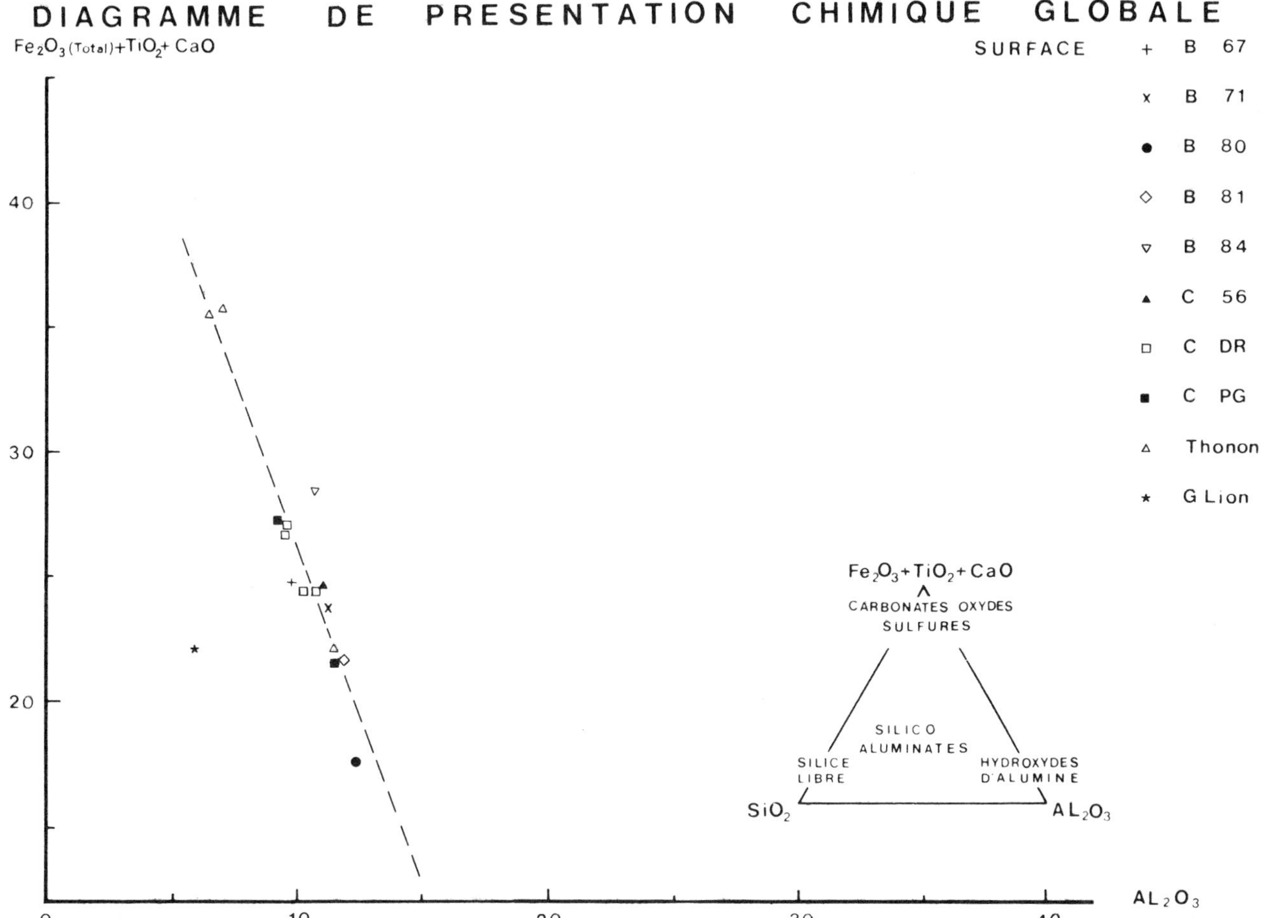

Figure 22. Diagramme de présentation chimique globale des échantillons de surface. Thonon, échantillon de vase lacustre; G. Lion provient du Golfe du Lion (ces échantillons sont présentés à titre comparatif).

alors qu'on ne note pas de concentration particulière autour des îles de Lérins dolomitiques. La concentration en magnésium de la partie Est de l'Esterel serait due aux apports terrigènes (chlorite) des rivières Siagne et Argentière qui parcourent les gneiss du Tanneron.

Na_2O

Valeurs de 0.40 à 1.40%. Le sodium provient essentiellement des plagioclases. Les valeurs élevées correspondent aux domaines éruptifs ou métamorphiques: Esterel et haut-fond du Méjean.

K_2O

Valeurs de 1.40 à 2.75%. Les valeurs les plus importantes se trouvent dans le domaine de l'Esterel et dans le cours terminal du canyon de Cannes. Le potassium dans ces sédiments récents proviendrait directement des apports terrigènes des rivières Siagne et Argentière sous forme de mica, de feldspaths alcalins et d'illite.

Distribution Verticale et Diagenèse

Sédimentation Quaternaire

L'influence du temps et de l'"enfouissement" est difficile à déterminer dans une zone accidentée où les apports sont hétérogènes comme sur cette marge continentale abrupte.

Une carotte type (B 84) a été étudiée en détail pour mettre en évidence les phénomènes de diagenèse précoce (Figure 23). Elle a été choisie à cause de sa sédimentation homogène, de sa longueur (860 cm), et de sa position. La fraction grossière, supérieure à 40μ est inférieure à 10% jusqu'à 750 cm et supérieure à 20% à partir de 800 cm. Les analyses chimiques montrent que tous les éléments marquent une discontinuité entre 810 et 820 cm, exception faite de MgO. Ce phénomène ne correspond pas à une anomalie physique notable.

Il est difficile de tirer des conclusions générales de l'étude d'une seule carotte car les variations mises en évidence peuvent être dues aussi bien au mode d'apport ou de sédimentation qu'à la diagenèse.

Toutefois les faits qui se dégagent de cette étude et qui pourront être vérifiés par d'autres travaux sont une augmentation de la teneur en CaO et Na_2O et un déficit de la teneur en SiO_2, Al_2O_3, Fe_2O_3 et K_2O avec l'enfouissement.

Le passage 810–820 cm montre un enrichissement relatif en SiO_2, Al_2O_3, Na_2O et K_2O et un déficit très sensible en CaO (25 à 17%). Ce passage correspondant à une période plus froide d'après la faune, les conclusions de Biscaye (1965) sur les sédiments atlantiques seraient applicables également en Méditerranée; en particulier l'augmentation de la teneur en chlorite expliquerait l'augmentation de la teneur en MgO.

Sédimentation Pliocène

Deux échantillons pliocènes prélevés dans le canyon de Cannes (B 74 et C 17) se distinguent des échantillons quaternaires par leur rigidité, leur faible teneur en eau et leur forte résistivité (il en est de même pour un

B84

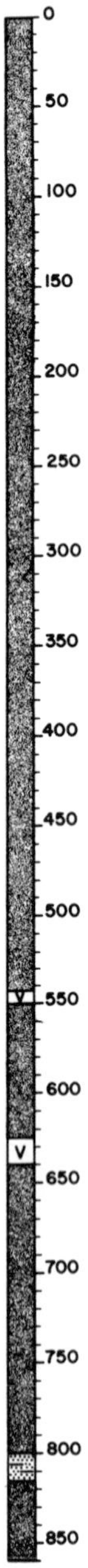

SiO_2	Al_2O_3	Fe_2O_3	MnO	MgO	CaO	Na_2O	K_2O	TiO_2	PF	TOTAL	SiO_2/Al_2O_3	Fe_2O_3/Al_2O_3
38,10	10.70	3,53	0,05	2,33	20,20	0,66	1,86	0.44	22.31	100,18	3,60	0,32
39,70	10,45	3,71	0,04	2,03	20,02	0,54	1,79	0,46	21,76	100,50	3,70	0,345
37,90	10,30	3,72	0,04	2,33	20,19	0,62	1,67	0,41	22,32	99,50	3,68	0,36
38,90	10,00	3,72	0,04	2,31	20,08	0,62	1,80	0,47	21,56	99,50	3,89	0,37
38,70	10,50	3,72	0,04	2,58	20,37	0,66	1,80	0,44	21,62	100,43	3,70	0,355
37,60	9,90	3,47	0,04	2,53	21,54	0,50	1,75	0,41	21,93	99,67	3,80	0,35
37,20	10,10	3,37	0,04	2,51	21,70	0,69	1,66	0.39	22,52	100,18	3,70	0,33
37,40	9,70	3,21	0,04	2,28	22,16	0,69	1,62	0,37	22,50	99,97	3,85	0,34
37,30	9,87	3,34	0,05	2,33	22,53	0,62	1,63	0,38	22,28	100,33	3,78	0,34
34,50	9,00	3,36	0,06	2,44	24,43	0,80	1,56	0,47	23,63	100,25	3,85	0,37
35,20	9,10	3,16	0,04	2,47	23,48	0,83	1.52	0,42	23,76	99,98	3,90	0,345
34,40	9,30	3,28	0,04	2,40	23,97	0,80	1,71	0,42	23,65	99,97	3,70	0,35
42,40	11,50	4,21	0,07	2,50	16,52	1,18	2,16	0,55	18,19	99,28	3,68	0,365
34,40	9,30	3,32	0,05	2,56	23,70	0,57	1,64	0,45	23,57	99,56	3,70	0,36
35,50	9,10	3,18	0,05	2,57	23,18	0,69	1,53	0,40	23,24	99,44	3,90	0,35

Figure 23. Analyse chimique de la carotte B 84 (350 m). Cette carotte a été prélevée au Sud du Cap d'Antibes.

échantillon pliocène prélevé dans un canyon de la marge algérienne). L'analyse chimique les caractérise très nettement. La teneur en SiO_2 est plus élevée; Al_2O_3 et Fe_2O_3 suivent cet accroissement. CaO marque un fort déficit (teneur inférieure à 10 %).

Ce faciès de Pliocène semble marqué par un apport plus important de sédiments terrigènes siliceux, sur une plateforme de faible ou moyenne profondeur (foraminifères benthiques dominants). L'apport également plus important de kaolinite à partir de sols fossiles a pu jouer dans le phénomène de concentration en Al_2O_3. La teneur en magnésium liée sans doute aux chlorites, ne montre pas de déficit car au cours de l'ennoyage des terrains métamorphiques (socle d'Afrique du Nord, Massif des Maures en Provence), les micas des gneiss ont participé aux phénomènes de sédimentation.

Conclusion: Propriétés Chimiques

La similitude de composition chimique de ces vases terrigènes pliocènes de Provence et d'Afrique du Nord est probablement liée à l'histoire tectonique du bassin occidental méditerranéen (Bourcart, 1962). Après le dépôt des évaporites dans la lagune méditerranéenne au Miocène supérieur, le Pliocène est marqué par une reprise brutale de la subsidence du bassin avec érosion intense et apports terrigènes. Au Pliocène supérieur et au Quaternaire, le mouvement de subsidence semble se poursuivre de façon moins intense (Auzende *et al.*, 1971).

Au Quaternaire, les glaciations se marquent par des variations de faune, de productivité, ce qui entraîne des variations de composition chimique.

CONCLUSION

Dans cette étude régionale, nous avons voulu définir un certain nombre de paramètres et vérifier leur importance et leurs relations dans le cadre d'une synthèse structurale.

Des méthodes indirectes comme la sismique réflexion continue sont utilisées pour connaître les

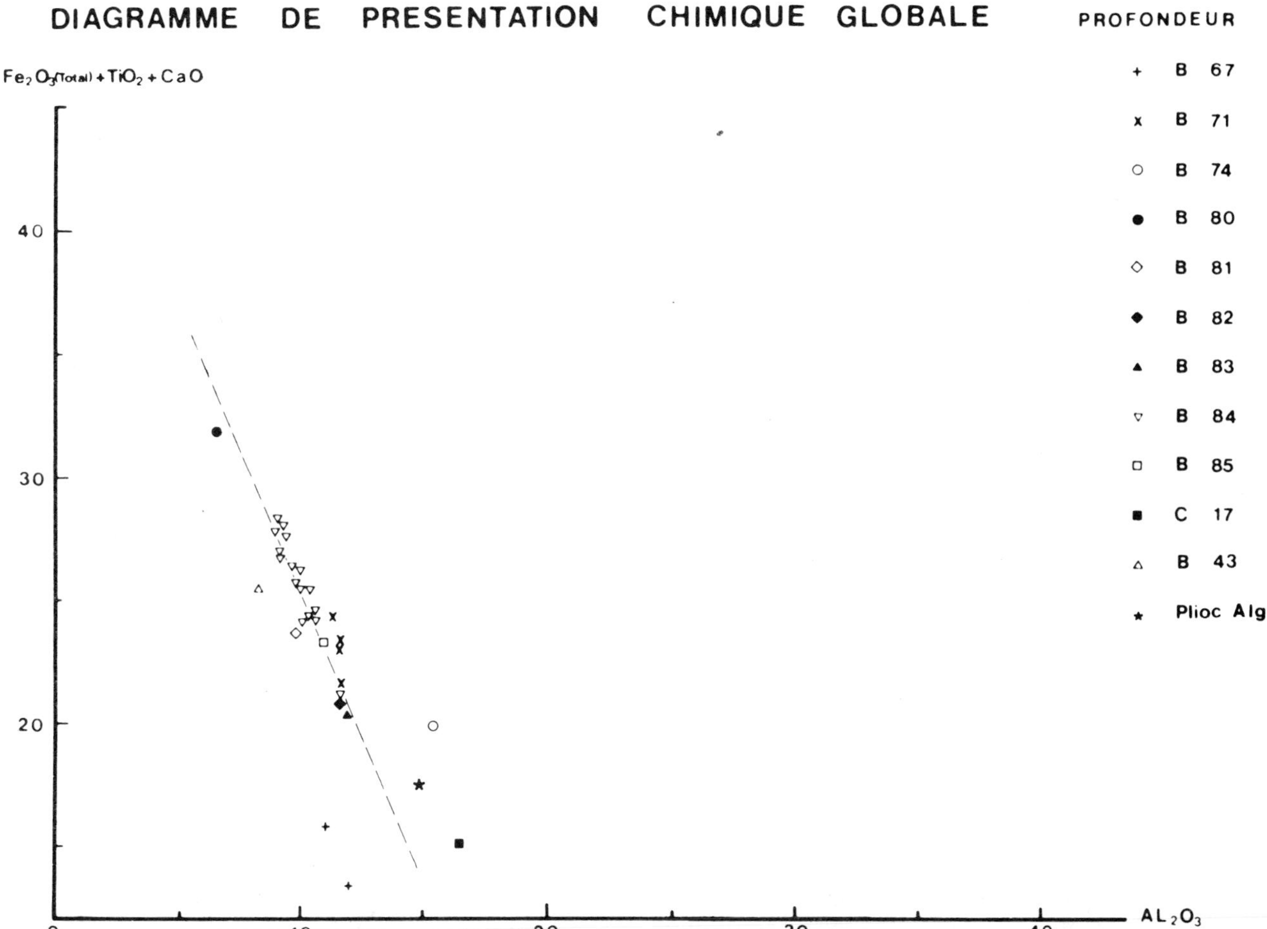

Figure 24. Diagramme de présentation chimique globale d'échantillons de niveau–5 m dans la colonne sédimentaire. Plioc Alg, correspond à un échantillon de marne pliocène provenant d'un canyon de la marge algérienne et présenté à titre comparatif.

grandes lignes de la structure sous-marine. Les forages et les carottages permettent ensuite de déterminer la nature des réflecteurs définis sur les enregistrements. La profondeur réelle des réflecteurs comme le degré de compaction des strates sédimentaires ne peuvent être abordés que par l'étude des propriétés physiques du sédiment (si l'on excepte les grands forages du type JOIDES).

Les propriétés physico-chimiques du sédiment reflètent également le mode de dépôt et la stabilité tectonique de la région. La minéralogie et la géochimie enfin montrent l'intensité des apports terrigènes, l'extension de ces apports, et soulignent donc les grandes phases d'érosion, donc d'activité tectonique.

Des moyennes régionales des divers paramètres mesurés sont présentées pour permettre une comparaison avec d'autres régions, et donc aboutir à un élargissement du champ de la synthèse.

On aboutit ainsi aux conclusions suivantes: le plateau continental est très peu étendu ou même absent par endroits. Le rebord continental est situé entre 90 et 100 m et il est dû à la régression Würmienne. La transgression pliocène n'est visible sur terre que dans les golfes pliocènes de Saint-Raphaël, La Napoule et du Var.

La pente continentale est abrupte, entre 10 et 15°. Des gradins découpent cette pente. Un gradin important au Sud de Cannes, le haut-fond du Méjean, laisse affleurer le substratum (gneiss des Maures) et montre un paléosol azoïque à 1200 mètres de profondeur. Cette pente continentale au Sud de Cannes est formée par le prolongement du massif des Maures et plus à l'Est par la couverture mésozoïque. Ce substratum est recouvert par une couverture sédimentaire meuble d'âge plio-quarternaire. Cette couverture présente de nombreuses figures de glissement dues à des mouvements contemporains du dépôt.

La plaine abyssale est formée de deux zones. La partie centrale du bassin est recouverte par une couche salifère d'âge Messinien (6 à 9 MA) qui alimente de nombreuses structures diapiriques. La couverture plio-quarternaire est extrêmement affectée par ces mouvements diapiriques.

Entre la pente continentale et cette zone centrale, on suit un réflecteur anté-pliocène qui est marqué par des entailles de canyon, des surfaces d'érosion. Ce réflecteur représente vraisemblablement une surface miocène érodée à l'air libre et amenée à sa position actuelle par un phénomène de subsidence. Ce réflecteur est très fracturé lorsqu'il arrive au contact de la zone centrale.

On propose une création du bassin à l'Oligo-Miocène par rotation du massif corso-sarde. La subsidence débute au Miocène en même temps qu'un comblement à la même vitesse. Au Miocène supérieur, fin de comblement du bassin avec le dépôt de la couche évaporitique. Une accélération de la subsidence est chronologiquement liée à la transgression pliocène. Ce mouvement entraîne l'effondrement des marges.

REMERCIEMENTS

Nous remercions MM. M. Roubault et H. de la Roche du CRPG de Nancy pour la réalisation des analyses chimiques. Mes collègues du Centre Océanologique de Bretagne ont bien voulu lire et critiquer ce texte; je les en remercie. Les carottages ont été réalisés avec le B. O. CATHERINE LAURENCE de la Station de Géodynamique sous-marine de Villefranche-sur-Mer.

BIBLIOGRAPHIE

Alinat, J., G. Bellaiche, G. Giermann, O. Leenhardt et G. Pautot 1970. Morphologie et sédimentologie d'un dôme de la plaine abyssale ligure. *Bulletin de l'Institut Océanographique de Monaco,* 1400 : (69) 22 p.

Auzende, J. M., J. Bonnin, J. L. Olivet, G. Pautot et A. Mauffret, 1971. An Upper-Miocene salt layer in the Western Mediterranean basin. *Nature, Physical Science,* 230: 82–84.

Bellaiche, G. 1969. *Etude géodynamique de la marge continentale au large du massif des Maures (Var) et de la plaine abyssale ligure.* Thèse, Paris, 221 p.

Bellaiche, G., et G. Pautot 1966. Quelques observations morphologiques et sédimentologiques effectuées à bord du bathyscaphe "Archimède" au large des Maures et de l'Esterel. *Bulletin de la Société Géologique de France,* 8 (7ème série): 769–772.

Bellaiche, G., et G. Pautot 1968. Sur la présence de niveaux à pyrite au sommet d'un dôme de la plaine abyssale ligure. *Compte-Rendu de l'Académie des Sciences, Paris,* 267 : 991–993.

Biscaye, P. E. 1965. Mineralogy and sedimentation of recent deep-sea clay in the Atlantic Ocean and adjacent seas and oceans. *Geological Society of America Bulletin,* 76 : 803–832.

Blanc-Vernet, L., H. Chamley et C. Froget 1969. Analyse paléoclimatique d'une carotte de Méditerranée nord-occidentale. Comparaison entre les résultats de trois études: foraminifères, ptéropodes, fraction sédimentaire issue du continent. *Paleogeography, Paleoclimatology, Paleoecology,* 6 : 215–235.

Bourcart, J. 1959. Morphologie du Précontinent des Pyrénées à la Sardaigne. *Colloque International du Centre National de la Recherche Scientifique, Nice-Villefranche* (5–12 mai 1958), 114–121.

Bourcart, J. 1962. La Méditerranée et la révolution du Pliocène. *Livre à la Mémoire du Professeur Paul Fallot,* 1 : 103–116.

Bourcart, J., M. Gennesseaux et E. Klimek 1960. Ecoulements profonds de sables et de galets dans la grande vallée sous-marine de Nice. *Compte Rendu de l'Académie des Sciences, Paris,* 250: 3761–3765.

Chamley, H. 1968. La sédimentation argileuse actuelle en Méditerranée nord-occidentale. Données préliminaires sur la diagenèse superficielle. *Bulletin de la Société Géologique de France,* 10: 75–88.

Debyser, J. 1959. *Contribution à l'étude géochimique des vases marines.* Thèse, Paris. Technip Editions, 210 p.

Ewing, J., M. Ewing, T. Aitken et W. J. Ludwig 1968. North Pacific sediment layers measured by seismic profiling. In: *The Crust and*

Upper Mantle of the Pacific Area (Geophysical Monograph 12), eds. Knopoff, L., C. L. Drake and P. G. Hart, American Geophysical Union, Washington, D. C., 147–173.

Filloux, J. H. 1967. *Oceanic electric currents, geomagnetic variations and the deep electrical conductivity structure of the ocean continent transition of Central California.* Thèse, San Diego, 120 p.

Glangeaud, L. 1961. Paléogéographie dynamique de la Méditerranée et de ses bordures. Le rôle des phases ponto-plio-quaternaires. *Colloque National de la Recherche Scientifique,* Villefranche, 125–161.

Hamilton, E. L. 1970. Sound velocity and related properties of marine sediments, North Pacific. *Journal of Geophysical Research,* 75 : 4423–4446.

Hersey, J. B. 1965. Sedimentary basins of the Mediterranean sea. In: *Submarine Geology and Geophysics,* eds. Whittard, W. F. and R. Bradshaw, Butterworths, London, 75–89.

Horn, D. R., B. M. Horn et M. N. Delach 1968. Correlation between acoustical and other physical properties of deep-sea cores. *Journal of Geophysical Research,* 73 : 1939–1957.

Luteaud, L. 1924. Etude tectonique et morphologique de la Provence cristalline. Revue de Géographie, 12 : 270 p.

Lowrie, A. et B. C. Heezen 1967. Knoll and sediment drift near Hudson, Canyon. *Science,* 157 : 1552–1553.

Mauffret, A. 1968. *Etude des profils sismiques obtenus au cours de la campagne GEOMEDE I au large des Baléares et en Mer Ligure.* Thèse de 3ème cycle. Paris, 90p.

Menard, H. W., S. M. Smith et R. M. Pratt 1965. The Rhone deep sea fan. In: *Submarine Geology and Geophysics,* eds. Whittard W. F. and R. Bradshaw, Butterworths, London, 271–284.

Nesteroff, W. D., G. Sabatier et B. C. Heezen 1963. Les minéraux argileux dans les sédiments du bassin occidental de la Méditerranée. *Rapport et Procès-Verbaux des Réunions du Comité International d'Etudes Scientifiques en Mer Méditerranée,* 17 (3) : 1005–1007.

Nesteroff, W. D. 1965. Recherches sur les sédiments marins actuels de la région d'Antibes. *Annales de l'Institut Océanographique,* 63:136 p.

Pautot, G. 1967a. Mesures de résistivités électriques sur des carottes de sédiments marins et lacustrees. *Bulletin de l'Institut Océanographique de Monaco,* 1376 (67):8 p.

Pautot, G. 1967b. Structure sous-marine du haut-fond du Méjean (sud de Cannes). *Comptes Rendus de l'Académie des Sciences, Paris,* 265 : 1028–1030.

Pautot, G. 1969. *Etude géodynamique de la marge continentale au large de l'Esterel.* Thèse, Paris, 269 p.

Pautot, G. 1970. La marge continentale au large de l'Esterel (France) et les mouvements verticaux pliocènes. *Marine Geophysical Research,* 1 : 61–84.

Pautot G. (éditeur) 1971. Résultats de la campagne de flexo-électro-carottage en Méditerranée nord-occidentale. *Publication du CNEXO,* en cours d'impression.

PART 10. **Sedimentation in the Central and Eastern Mediterranean**

Sedimentation in the Tyrrhenian Sea

Henry Charnock[1], Anthony I. Rees[2]
and Norman Hamilton

The University, Southampton

ABSTRACT

Gravity cores and grab samples were collected from the eastern continental borderland and the central abyssal plain of the Tyrrhenian Sea. Bathymetry and sub-bottom topography of the borderland region, studied with precision depth recorder and seismic profiler, revealed several submarine canyons and associated fan systems and a sediment dispersal pattern from the coast to the abyssal plain that is strongly influenced by structure. Several basins have trapped sediment until filled.

Sediments in selected cores from the borderland and the abyssal plain are predominantly of fine grained material but some layers of sandy sediment are observed. Much of the coarse fraction is organic or volcanic in origin. Fine silts as well as sands may have been deposited by bottom currents. Cores from the borderland show most signs of burrowing and penecontemporaneous erosion. Sedimentation on the abyssal plain appears to have been discontinuous in time and space.

Magnetic fabric measurements are of two types: primary style fabrics provide evidence of transport directions, whereas secondary fabrics give information about soft sediment deformation and allied processes. Abyssal plain cores generally have primary-style fabrics; cores from the borderland are deformed.

RESUME

Une expédition à bord du R.R.S. JOHN MURRAY a été entreprise dans la Mer Tyrrhénienne afin d'étudier la géologie et géophysique de la marge continentale dont on a prélevé des carottes et des échantillons. Les mesures bathymétriques et stratigraphiques ont été effectuées à l'aide d'enregistreurs de profondeur et de profils sismiques.

Les études bathymétriques détaillées de la partie orientale du Bassin Tyrrhénien ont révélé l'existence de plusieurs canyons et deltas sous-marins correlatifs. La bathymétrie et la stratigraphie ont montré que la structure influence fortement le transport des sédiments de la côte vers la plaine abyssale; plusieurs bassins ont été comblés par l'accumulation des sédiments.

D'après l'étude de carottes prélevées sur la marge continentale et dans la plaine abyssale on a pu obtenir des renseignements sur la sédimentologie et la microstratigraphie de ces zones. Toutes ces carottes sont principalement constituées de matériau à grain fin mais présentent des couches de sédiments sableux. La plupart de la partie grossière est d'origine organique ou volcanique. Les carottes de la marge présentent le plus grand nombre de traces organiques et d'érosion de même âge que la marge. Dans la plaine abyssale, la sédimentation semble avoir été discontinue dans le temps et l'espace. Cela semble indiquer que les silts fins de même que les sables aient été déposés par des courants de fond.

On présente les résultats des mesures du matériau magnétique. Les matériaux de style primaire donnent les directions des transports alors que les matériaux de type secondaire donnent des renseignements sur la déformation des sédiments mous et sur les processus reliés à cette déformation. Les échantillons de la plaine abyssale présentent généralement des matériaux de style primaire alors que ceux de la marge continentale sont déformés.

[1]Present address: National Institute of Oceanography, Wormley, Surrey, U. K.
[2]Present address: Natural Environment Research Council, Alhambra House, 27/33 Charing Cross Road, London, W.C., U.K.

INTRODUCTION

The Tyrrhenian Abyssal Plain forms the floor of an enclosed basin about 3500 m deep and is thought to exhibit features due to sediment ponding (Hersey, 1965). Coarse sediment is thought to be transported to the abyssal plain via submarine canyons originating near the Italian and Sardinian shores. These canyons appear to have been little studied. Volcanic debris may be carried direct to the abyssal plain area by the wind. Ryan *et al.* (1965) have studied a small number of long piston cores. The cores have layers of sand-sized sediment from a few mm to 0.62 m thick. A few of these layers are thought to persist over the whole 100 km by 60 km of the abyssal plain.

The analysis of more closely spaced cores taken from a small area of the abyssal plain (Kermabon *et al.*, 1969) does not support the notion that all sand layers are widely distributed. Some layers seem to extend from one core to another but others plainly do not. Some cores less than 1 km apart show little if any similarity in the sequence of their bedding.

The main object of our 1969 cruise aboard the R.R.S. JOHN MURRAY was to collect evidence of the movement of sediment onto and across the Tyrrhenian Abyssal Plain. A detailed bathymetric chart was to be made of part of the continental slope off Naples known to have submarine canyons and attempts made to sample the canyons and the surrounding continental margin. Our intention was to collect a set of 10 cm diameter gravity cores spaced over the abyssal plain.

Three 10 cm cores, each about 2.5 m long were obtained before the main winch failed. The coring was continued with a 6.3 cm gravity corer and 13 more cores, up to 1.8 m long, were obtained at a 20 to 30 km spacing from the abyssal plain (Figure 1). A further 23 cores and 20 grab samples were taken from the borderland off Naples.

The bathymetry of an area of about 5000 km^2 was studied by means of about 5,500 km of precision depth recorder traverse. This was supplemented by approximately 550 km of traverse with a 6000 joule E.G. & G. Sparker system.

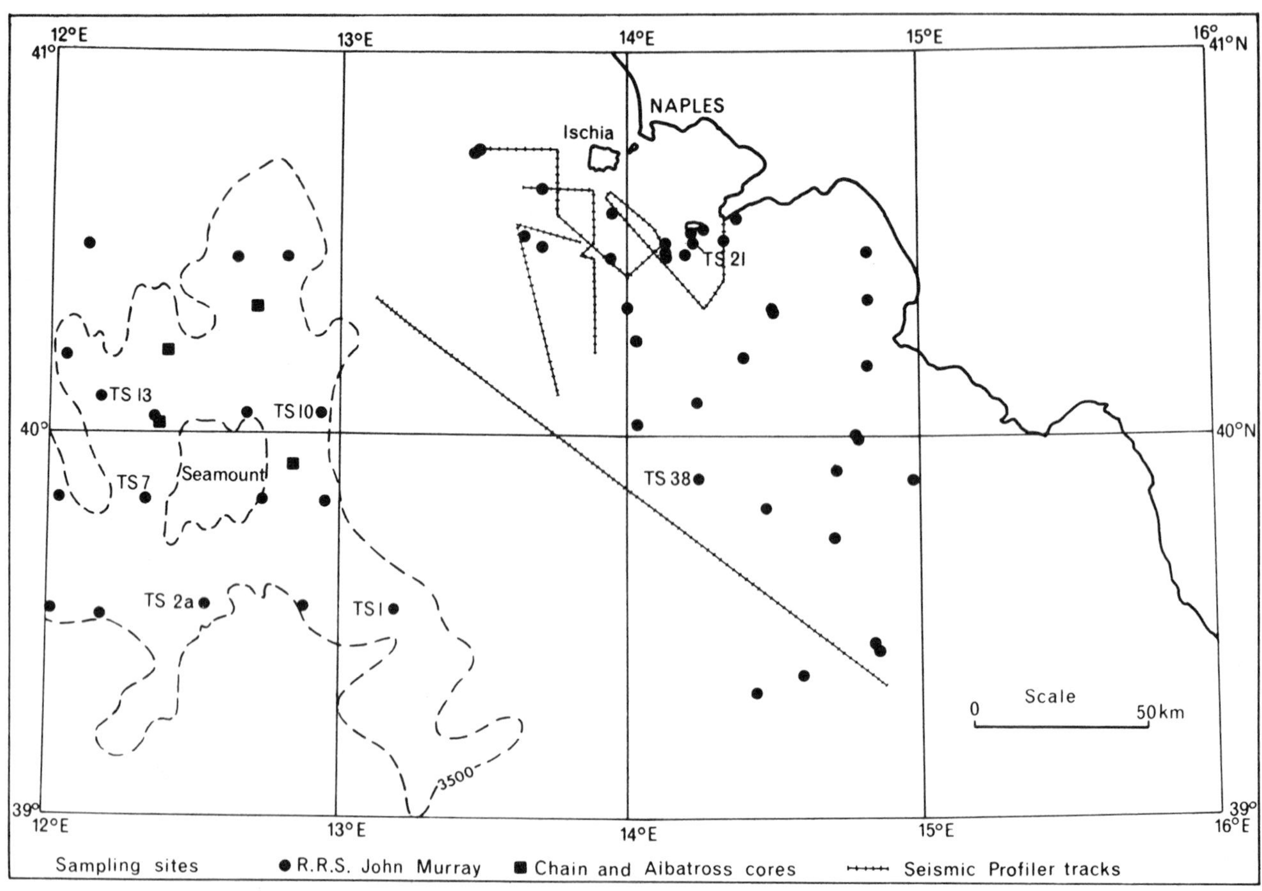

Figure 1. Location of sampling sites on the abyssal plain and continental borderland of the Tyrrhenian Sea together with seismic profiler tracks.

BATHYMETRY AND SUB-BOTTOM PROFILES

Figure 2 is a bathymetric chart of an area of the borderland off Naples. Navigation was by radar, and depths have been corrected according to the tables of Matthews (1939).

Errors in the chart may be due to errors in depth estimation or to inaccuracy of navigation. Depth estimation is usually reliable to a few meters in the depths encountered, navigation is likely to lead to errors of a few hundred meters in position. A statistical analysis of the depth differences at apparent crossovers suggests an average position fixing error of the order of 500 m. The position of the map's contours, especially in the region of the canyons where ship tracks are close together, will often be more closely defined than this and the relative positions of adjacent contours much more so.

The most obvious features of the bathymetric map are aligned on two trends—NW-SE and ENE-WSW—which are closely parallel with structural lineaments on land (Figure 3a). Two main lineaments cross the area from NW to SE forming southwestward facing scarps, the Banco delle Vedove and the 1000 m scarp. Lineaments running ENE-WSW form the southern boundaries of the Banco delle Vedove and Sorrento peninsula and the Southern Escarpment.

Other minor lineaments follow the same trends. The trends appear to control the sinuous courses of the two submarine canyons whose heads are in the Bay of Naples (Figure 3b). Profiles along the canyons (Figure 4) show steps which may be the result of movement on faults forming the minor lineaments.

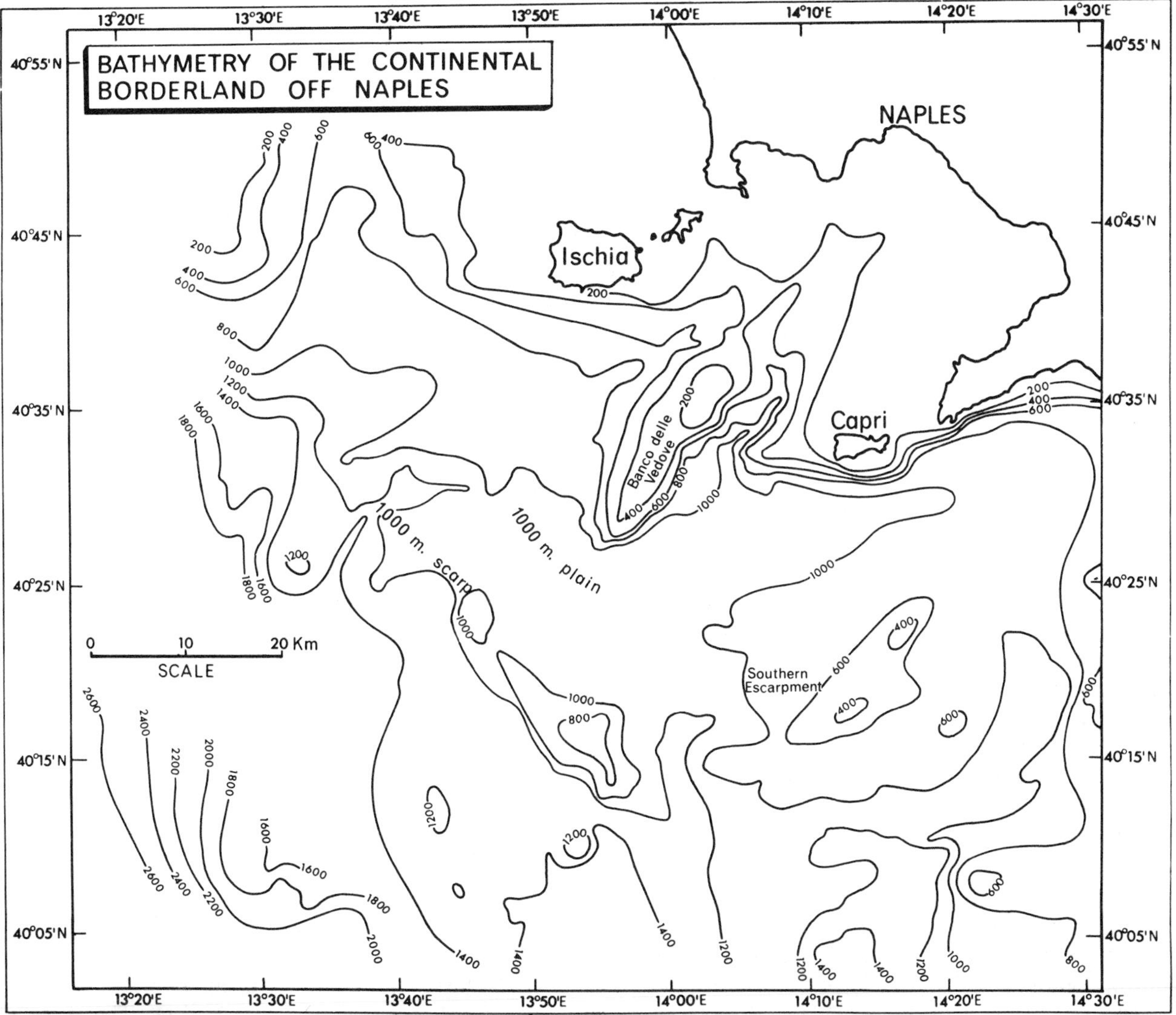

Figure 2. The bathymetry of the continental border off Naples. Contour interval 200 meters, all depths corrected according to Matthews (1939).

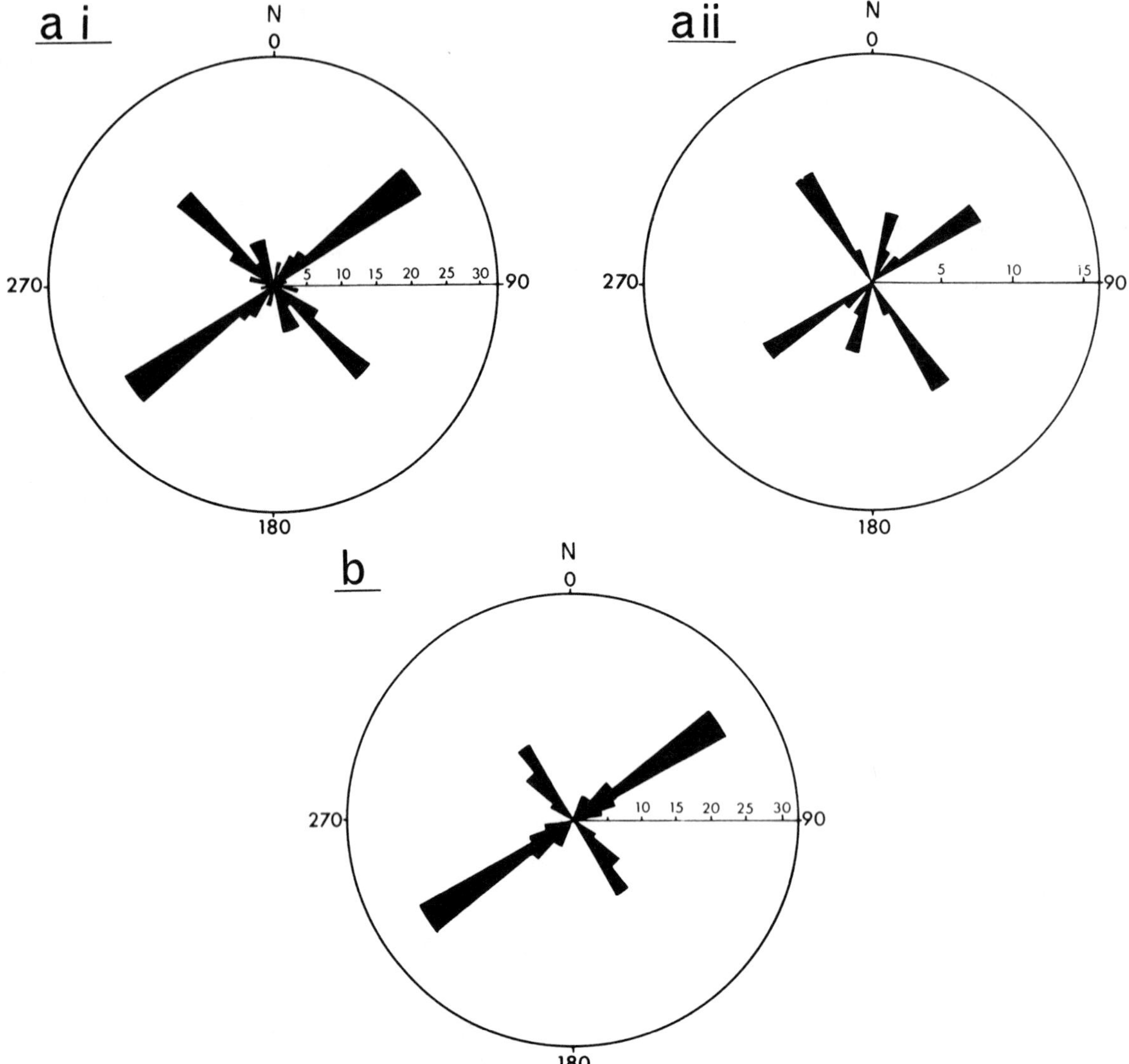

Figure 3. Rose diagrams of the directions of structural lineaments. a i, Directions from 68 faults in the Naples area; a ii, Directions from 24 submarine escarpments in the area of the bathymetric chart of Figure 2; b, Directions from 60 segments of the Naples Submarine Canyons.

Sub-bottom profiles confirm that the bottom topography is the result of recent sedimentation upon a sea floor whose configuration has been controlled by fault movement. A cross section across the Banco delle Vedove and the Southern Escarpment (Figure 5) shows that both are tilted blocks.

The rocks of the Banco delle Vedove are relatively transparent acoustically except for a comparatively thin draping of layered sediment. They may well be of the same Cretaceous sequence that forms the adjacent land of Capri and the Sorrento peninsula.

The Southern Escarpment by contrast has many reflectors which seem to indicate mildly deformed bedding, similar to the undeformed, presumably modern, sediments that have filled the depression between the two blocks. The escarpment shows many signs of slumping on both dip and scarp faces; the slumping may account for the scalloped outline of the southern boundary.

The depression to north of the southern escarpment and northeast of the 1000 m scarp has filled with sediments since major fault movement. This sediment forms a plain—the *1000 m plain*—which slopes gently upward toward the north and east. The profiles show it to be mainly horizontally parallel bedded but with some cross bedding near the mouths of the submarine canyons that open onto it (Figure 5). The thickness of the horizontally bedded sediment can be shown to be of the order of 1000 m in places.

Two submarine canyons, the Northern and Southern Naples Canyons, have their heads near the northern shore of the Bay of Naples. The northern canyon follows a more or less westerly course, ending in what appears to be a submarine fan at the 1000 m

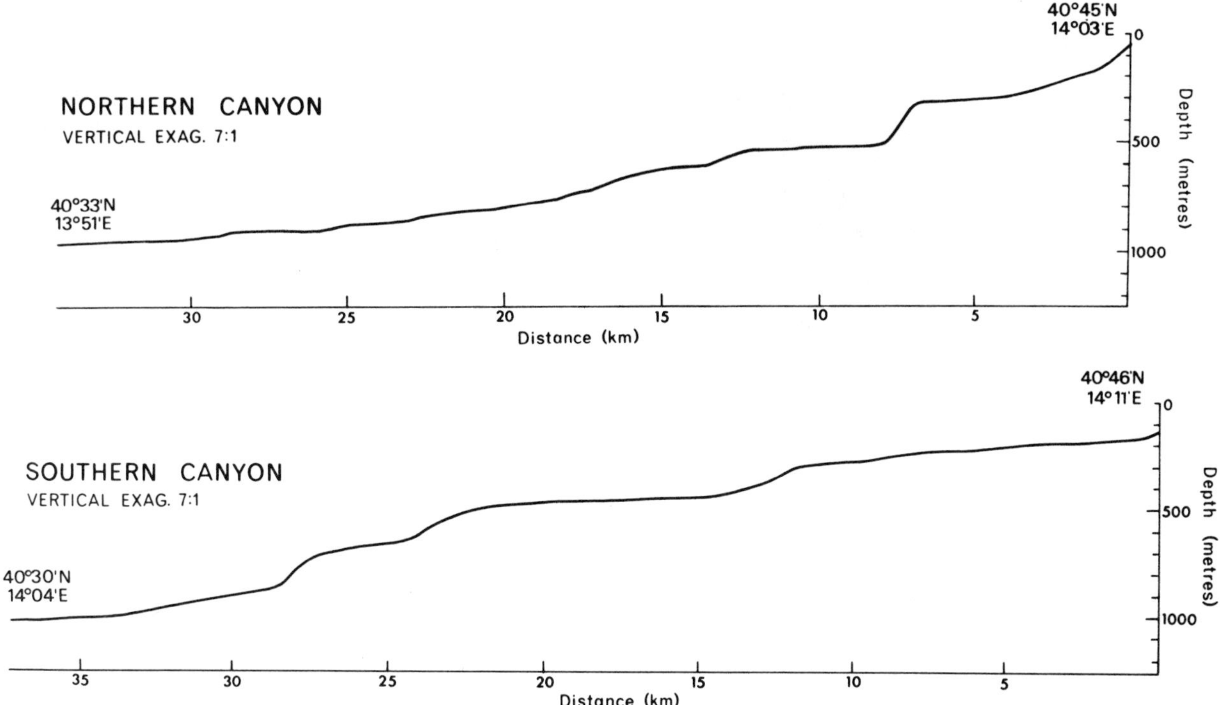

Figure 4. Profiles along the axes of the Naples Submarine Canyons.

plain. The course of the southern canyon also ends in the fan-like cross-bedded sequence to the southeast of the Banco delle Vedove that appears in Figure 5.

These canyons are assumed to have carried much of the sediment that forms the 1000 m plain, though other sources to the north and east are possible. The main positive evidence for this assumption lies in the existence of the fans. It is estimated that about 10^{11} m^3 of sediment has been transported to produce the infilling of the 1000 m plain alone. Movement during the postorogenic phase of the alpine orogeny may have been responsible for the tilting of the southern escarpment possibly during the early part of the Pleistocene. This would leave a million years or more for the filling of the 1000 m plain. Although we cannot yet make any detailed comment about the origin of the infilling, a sedimentation rate of the order of 10^5 m^3 per year does not seem likely to be beyond the capacity of local streams.

The basin of the 1000 m plain is now brimful with sediment which seems to be carried over the 1000 m scarp at several places. One of these appears to be a continuation of the northern Naples Canyon. A steep walled canyon has been cut back several kilometers into the 1000 m escarpment and has brought down sediment to a fan at a depth of about 1500 m. The bathymetry in the area of this fan has not yet been completely mapped. A reflexion profile across it (Figure 6) shows major cross bedding capping a sequence of undisturbed sediments that fill an apparently tectonic basin to a depth of about 500 m.

Evidence of similar deposition can be seen further to the west at a depth of 2300 m, approaching the edge of the Tyrrhenian Abyssal Plain (Figure 7). This figure illustrates what seems likely to be the typical sedimentary evolution of the continental borderland in areas whose topography is not such as to trap sediment. In such areas a comparatively transparent blanket of sediment of uniform thickness overlies a folded basement. The thickness of about 300 m may represent slow deposition over the period from the formation of the basin.

THE CORES

General

Seven cores have been opened, five from the abyssal plain and two from the continental borderland.

One of two 10 cm gravity cores was extruded from the core barrel into a liner, the other was taken directly into the liner as were the three 6.3 cm gravity cores from the abyssal plain and the two 6.3 cm free fall cores from the borderland. All the cores seem to be mechanically in good condition with little sign of deformation due to coring. Slight down-dragging is visible in some places near the corer walls.

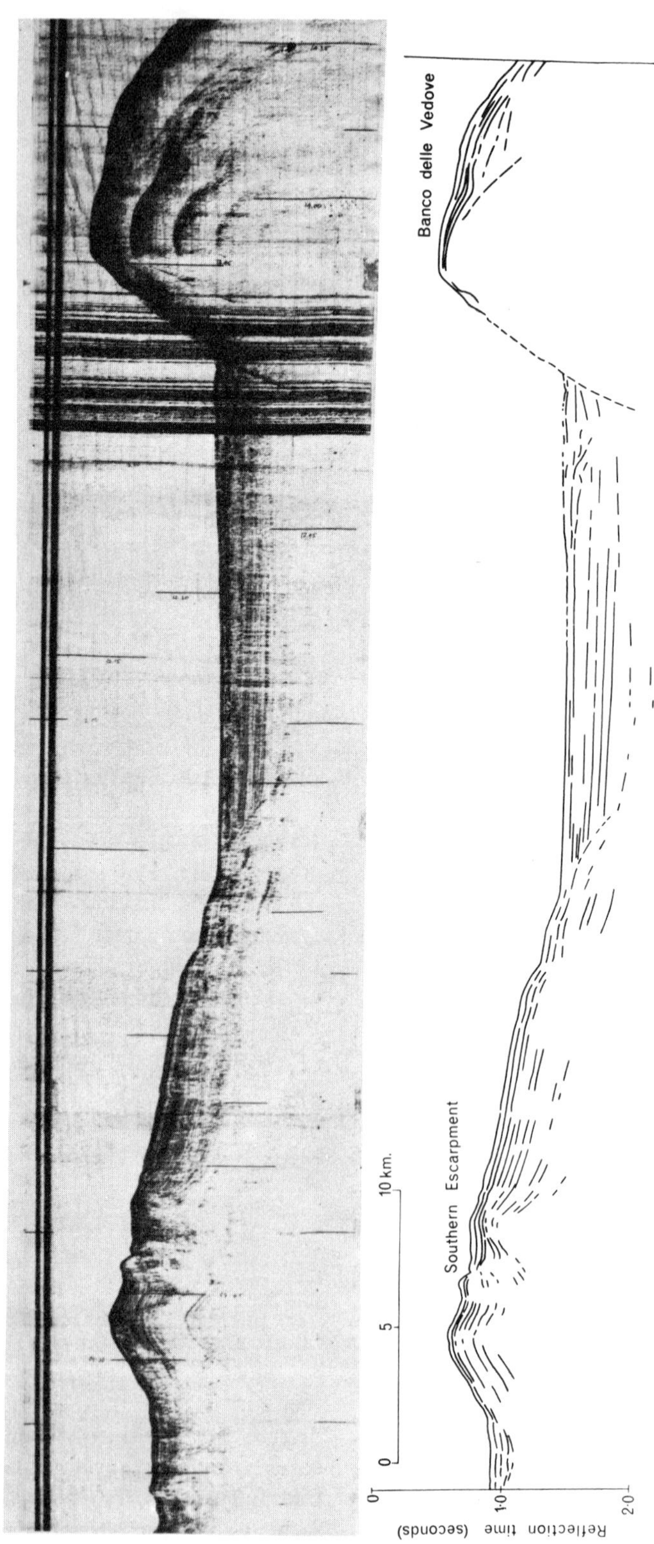

Figure 5. Profiler record showing a cross section of the Banco delle Vedove, the 1000 m plain and Southern Escarpment.

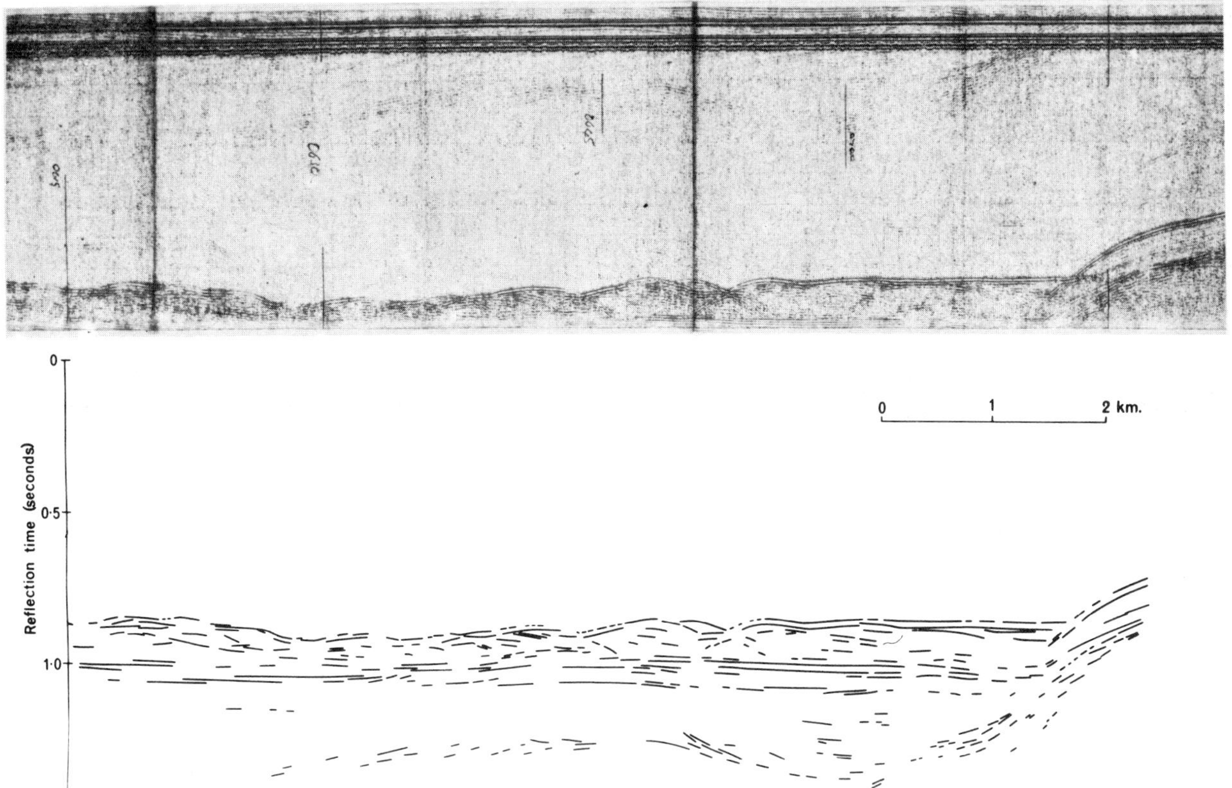

Figure 6. Profiler record showing cross-bedded sediments near the mouth of a canyon transporting sediments away from the 1000 m plain. This canyon is thought to be a continuation of the Northern Naples Canyon.

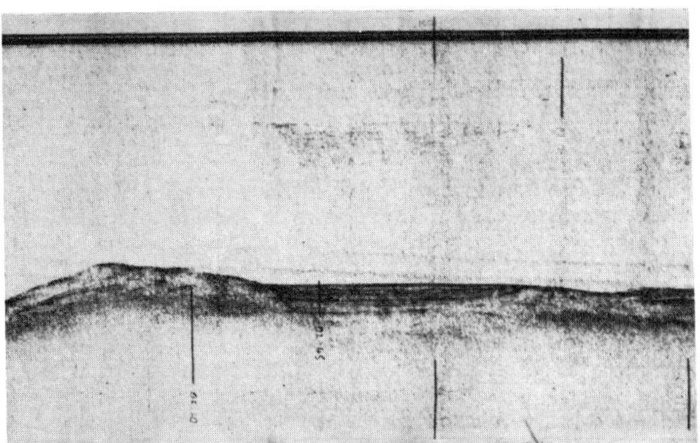

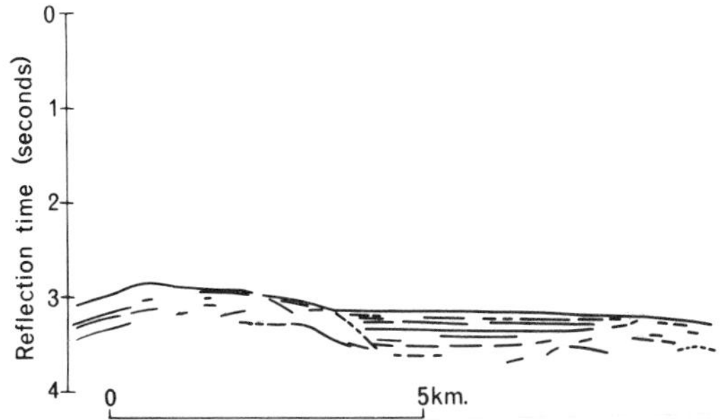

Figure 7. Profiler record showing thin sediment draping over a folded and faulted basement with horizontally bedded infilling of a depression. Water depth approximately 2500 m.

Sedimentology

The borderland and the abyssal plain cores differ markedly from each other. The former show signs of slow deposition with intermittent erosion and contain no distinct sand layers, the latter appear to have been deposited relatively rapidly.

The borderland Core TS 21 has been studied in detail. It is a free fall core, 67 cm long, from the northern edge of the 1000 m plain near Capri. The core is almost all of silt and clay but has sand sized grains scattered throughout its length and several layers in which sand grains are concentrated. The consistency is firm and the core has many small burrows filled with silt. An erosional scour-and-fill structure occurs at 45 cm (Figure 8). Fragments of the material from immediately below the channel appear in the channel filling.

Core TS 38, a free fall core 81 cm long from a depth of 1631 m, is generally similar though of softer consistency. It is silty over its whole length, with a variable proportion of sand grains and shell fragments. The core shows many signs of burrowing (Figure 9).

These cores come from gently sloping parts of the continental borderland. TS 21 is from the edge of the 1000 m plain, an area known to contain a considerable

Figure 8. Part of Core TS 21 showing a scour-and-fill structure.

Figure 9. Burrowing in Core TS 38.

thickness of horizontally bedded sediments probably laid down in a late phase of sedimentation. The condition of both TS 21 and TS 38 suggests that sedimentation during their deposition was slow and that they may have been subject to occasional erosion. This is consistent with the supposition that the area represented by TS 21 is no longer accumulating sediment at the rate it was.

Earlier workers have described abyssal plain cores containing numerous sand layers 10 cm or more thick. None of the five cores we have examined penetrated such a layer though layers up to 5 cm are common. Most of the nonorganic sand-sized material is volcanic in origin. Some layers contain many angular glass shards, others have grains which show signs of rounding which suggests transport along the sea floor.

Three of the abyssal plain cores, TS 1, TS 2A and TS 10, contain sand layers, 1 cm or more in thickness, which together make up about 5% of the total core length. The fine grained intervening material is generally soft. Cores TS 1 and TS 2A are 10 cm gravity cores and Core TS 10 is a 6.3 cm gravity core. Their lengths are 232 cm, 222 cm and 122 cm, respectively and all appear to be undisturbed.

Figure 10 shows clearly the absence of deformation in Core TS 1, except for disturbance near the top, and also shows several features common to the three cores.

The most notable of these is a rhythmic sequence of color changes which appears to be related to the state of oxidation of the sediments and is probably a result of the depositional history. The light bands at 85–90 cm, 96–98 cm and 125–138 cm for example, reflect oxidation below sediment surfaces that remained exposed at the sea floor for considerable periods. Evidence for this long exposure is found in filled burrows at the 85 cm and 96 cm surfaces and in what appears to be rippling at 125 cm.

Also noticeable is the change of dip with distance down the core. This may represent the infilling of an advancing foreset or it may reflect the influence of some minor tectonic event at a distance from the core. An even clearer example of this change of dip is seen in Core TS 2A (Figure 11).

Some of the sand layers show either normal grading—with grain size increasing from top to bottom—or reversed grading (Figure 11), and many contain a high proportion of planktonic tests.

The comparative rarity of burrowing, the cyclic color changes and the presence of appreciable quantities of sand combine to suggest that these three cores are from areas of rapid, intermittent sedimentation. This impression is also given by Core TS 13 taken to the northeast of the abyssal plain. This 6.3 cm gravity core is 170 cm long and appears undeformed (Figure 12); it contains no sand layers more than a millimeter or so thick.

The fifth core, TS 7, is conspicuously mottled (Figure 13). Though from abyssal plain depth it has a firm texture and so resembles the two borderland cores discussed above. It was taken close to the central seamount and may not belong to the true abyssal plain.

The Magnetic Fabric

The application of magnetic anisotropy measurement to the study of sediments has been discussed in detail elsewhere (Rees, 1965; Hamilton and Rees, 1970). The method provides a sensitive indicator of deformation and can also give some indication of conditions of deposition as well as providing an

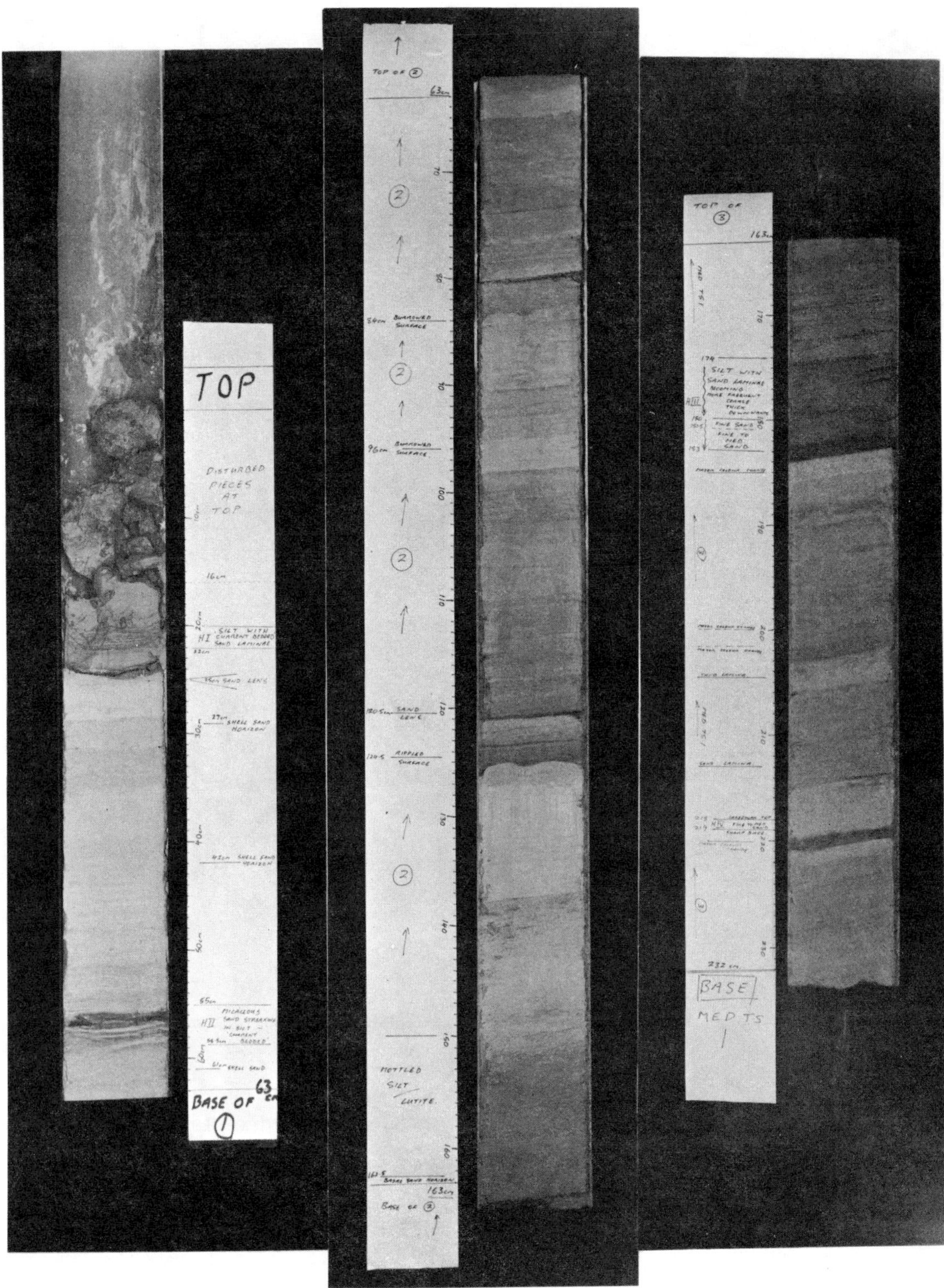

Figure 10. Core TS 1. There is little sign of deformation. The core is made up of a rhythmic sequence of beds of different colors. Signs of burrowing can be seen in a few places. The upper part of the core has horizontal bedding the lower part has beds dipping at up to 10°. In view of the apparent absence of edge effects it seems unlikely that the dips are an artefact of coring.

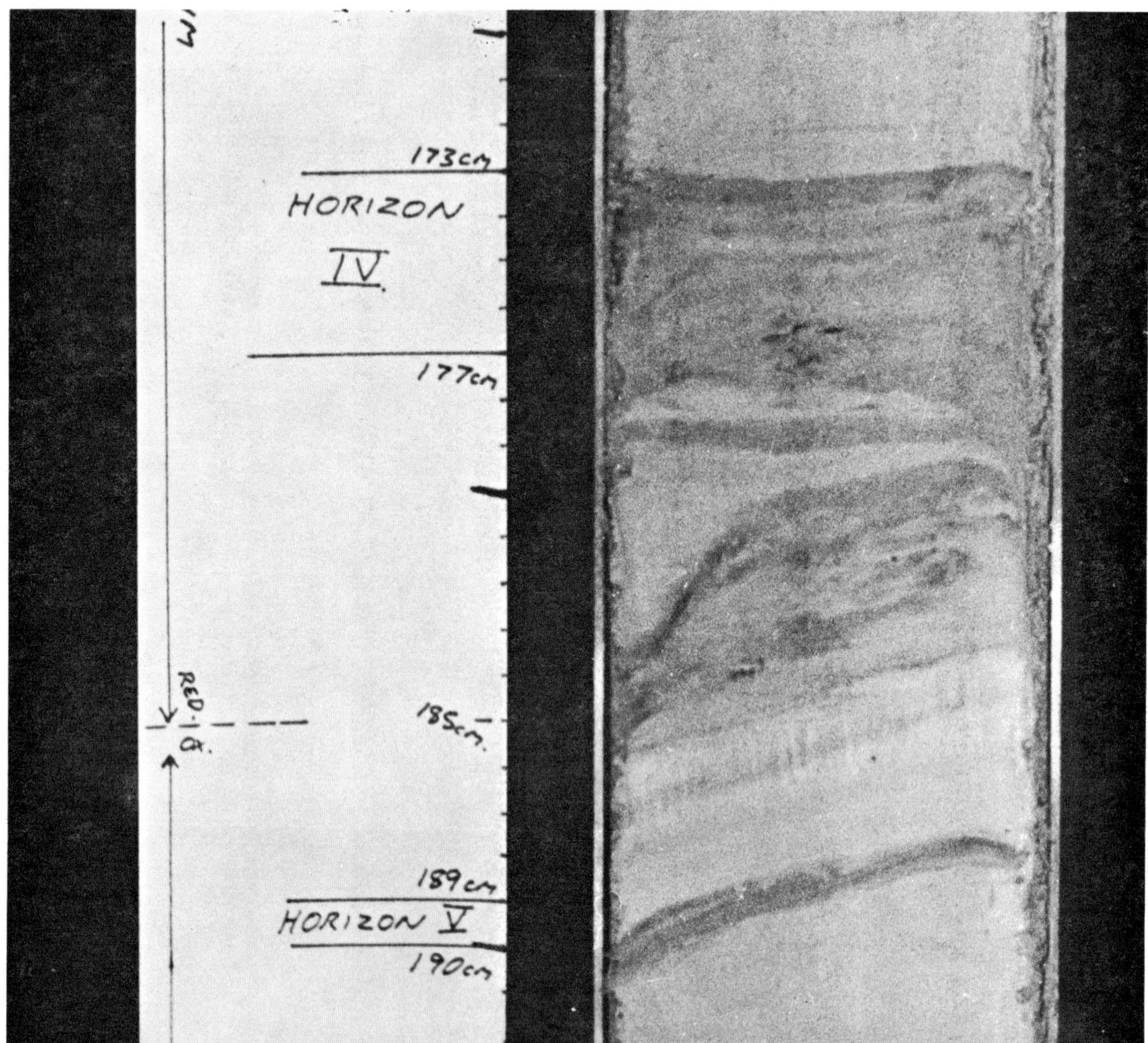

Figure 11. Part of Core TS 2A showing a wedge shaped bed, possibly the result of local tectonic movement. An inversely graded sand bed occurs just above this one.

estimate of palaeocurrent or palaeoslope direction.

The main characteristics of a primary magnetic fabric resulting from the deposition of a sediment containing magnetite are a near horizontal *magnetic foliation plane* containing principal axes of maximum and intermediate susceptibilities K_{max} and K_{int} and a value of the parameter

$$q = \frac{K_{max} - K_{int}}{\frac{1}{2}(K_{max} + K_{int}) - K_{min}}$$

in the range 0 to 0.6, K_{min} being the minimum principal susceptibility.

The magnetic fabric measurements confirm the results of the sedimentological examination. They confirm that the sediments in the four cores thought to belong to the true abyssal plain are largely undisturbed and provide evidence of deformation in the borderland and seamount cores.

All the cores so far opened show some evidence of disturbance, the least disturbed being TS 13. Figure 14 shows the directions of maximum and minimum principal susceptibility axes for this core. Those for minima are almost vertical, the action of gravity being to produce a nearly horizontal magnetic foliation, and those for maxima are well grouped in a direction resulting from the action of stresses parallel to the bed.

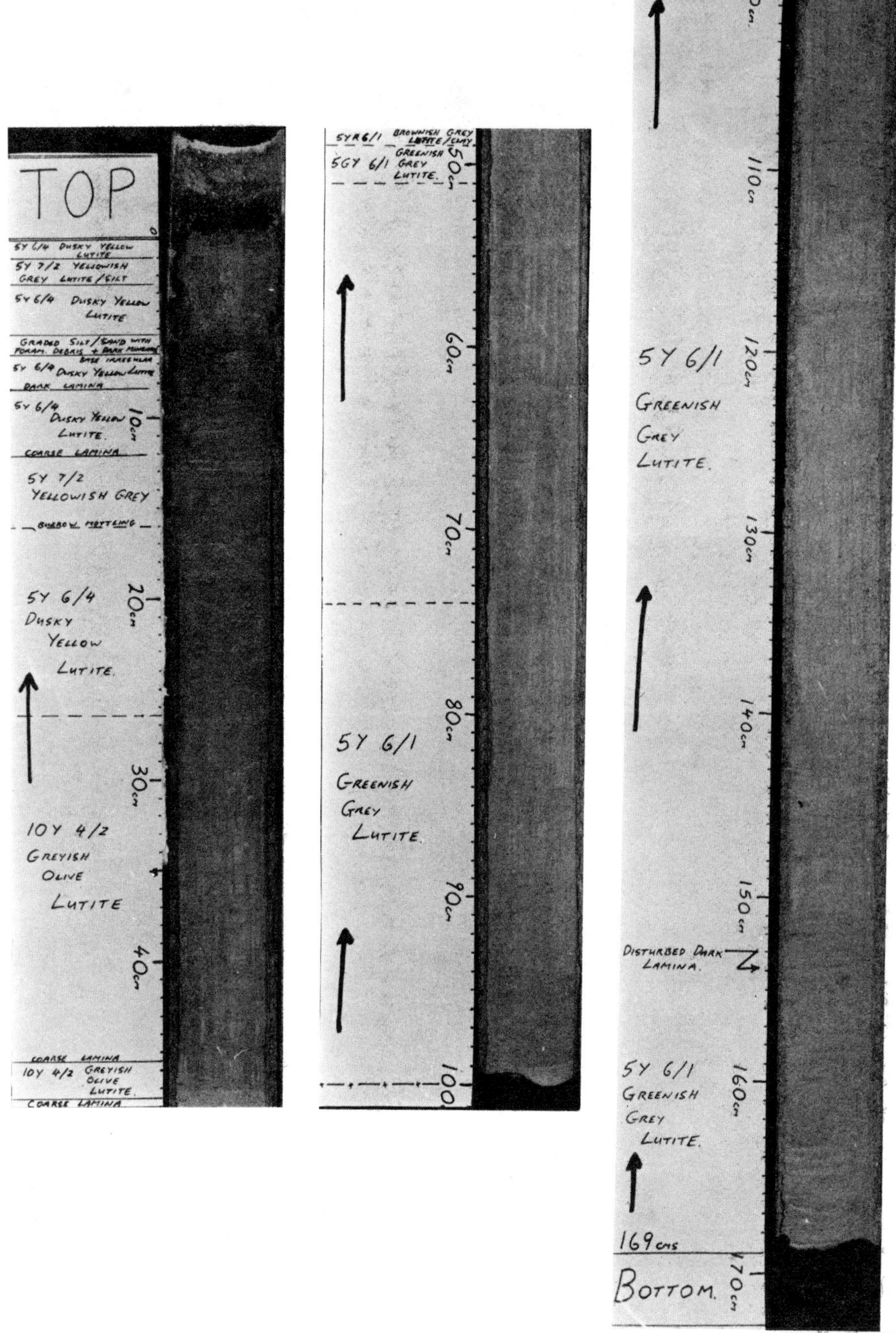

Figure 12. Core TS 13. Note absence of sand layers.

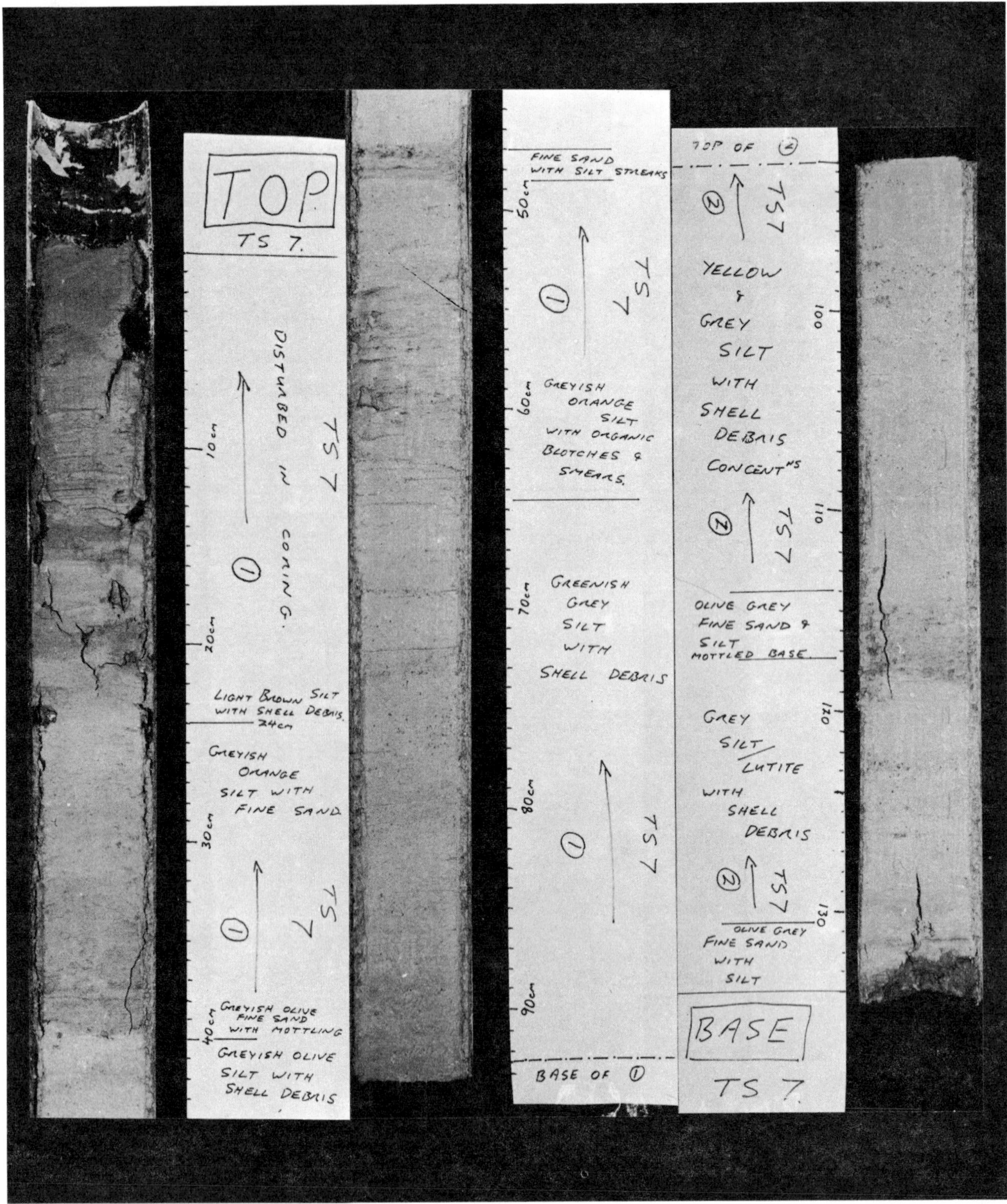

Figure 13. Mottling in core TS 7.

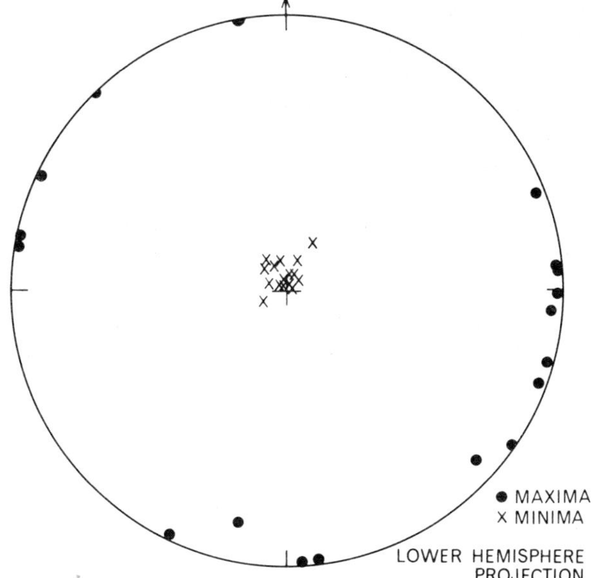

Figure 14. Magnetic fabric of specimens from core TS 13, relative to an arbitrary azimuth. The well-grouped near vertical minima and preferred orientation in the horizontal are characteristic of primary magnetic fabric.

Cores TS 7 and TS 21 show evidence of streaking of the minima away from the vertical (Figure 15). This is thought to be the result of compressive stress acting in the horizontal plane and has been considered to be due to downslope movement of sediment (Rees *et al.*, 1968).

Table 1 shows how the magnetic properties vary from core to core. Any specimen has been considered

Table 1. Magnetic fabric of the cores.

Core	TS 1	TS 2A	TS 7	TS 10	TS 13	TS 21
Location	Abyssal Plain	Abyssal Plain	Central Sea-mount	Abyssal Plain	Abyssal Plain	Border-land
No. of specimens with primary style fabric	24	30	7	10	15	2
No. of specimens with anomalous fabric	11	18	16	5	3	4

to have anomalous fabric if it does not have a nearly horizontal magnetic foliation plane and $0.6 \geqq q \geqq 0$. It is clear from Table 1 that primary style fabric is associated with abyssal plain sedimentation.

Also of interest is a correlation between the directions of maximum susceptibility axes and the directions of sedimentary dip (Figure 16). This has been found to be typical of depositional dip surfaces (Rees, 1966; Hamilton *et al.*, 1968). It implies that the dips are depositional or, if due to small scale tectonics, movement was contemporaneous with sedimentation.

DISCUSSION

Much work remains to be done on the material collected during our 1969 cruise and it is already clear

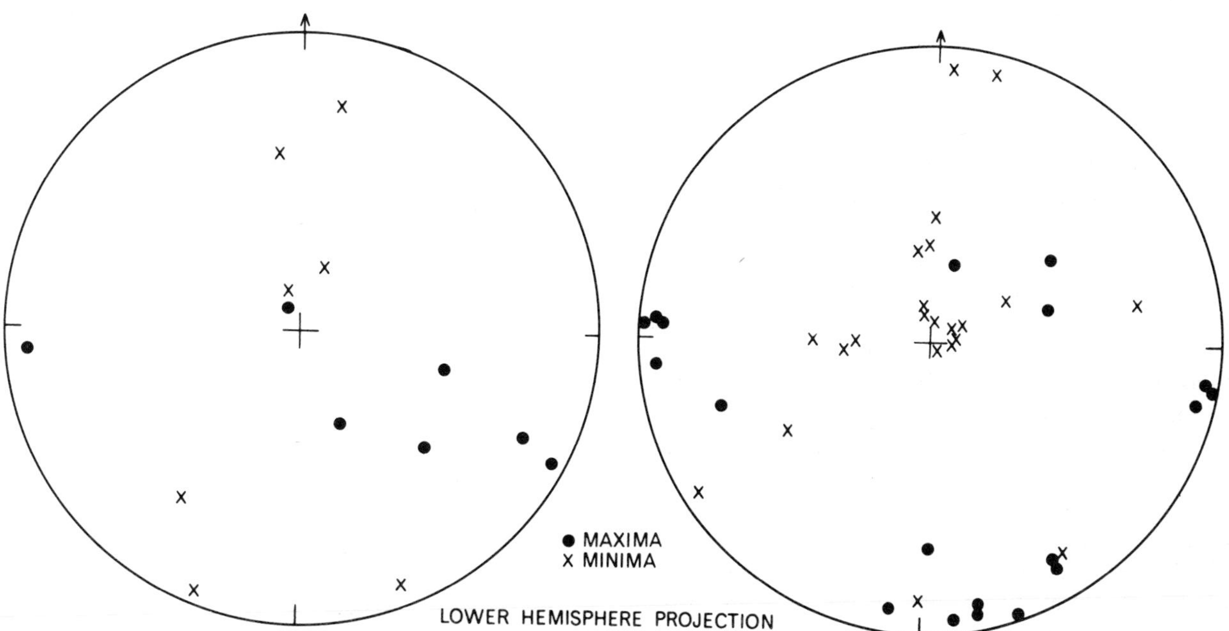

Figure 15. Magnetic fabric of specimens from Core TS 21 (left) and Core TS 7 (right). Both plots are relative to an arbitrary azimuth. Deformation is indicated by the scatter of axial directions.

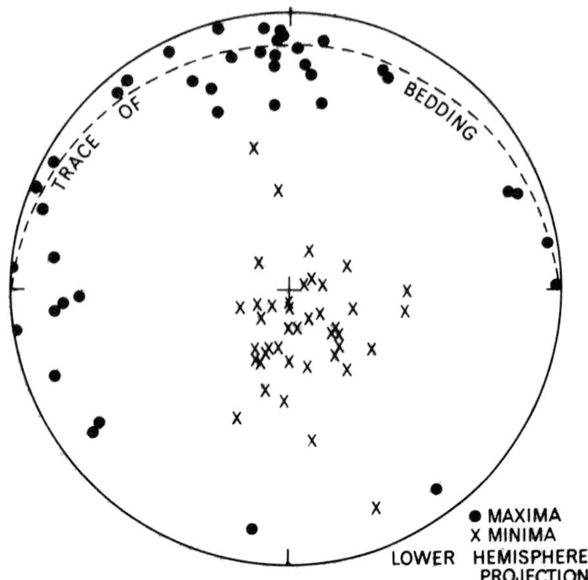

Figure 16. Magnetic fabric of specimens from Core TS 2A. A tendency for the maxima to be grouped in the direction of dip of the bedding reflects the action of stresses in that direction.

that many more observations still will be necessary before Tyrrhenian sedimentation can be completely understood. But some general conclusions can already be stated.

The first is that the movement of sediment from the land toward the abyssal plain is largely controlled by the structural configuration making up what we have called the Continental Borderland. This term is not used in exactly the same sense as Emery's (1960) description of the sea floor off California since we have no evidence of true oceanic crust anywhere in the Tyrrhenian basin. Nevertheless there are close analogies, as for example between the 1000 m plain and the San Diego Trough, both of which are structural basins filled with flat-lying sediment.

Sediment movement in the eastern Tyrrhenian appears, as it does off California, to be via submarine canyons ending in sediment fans at local baselevels, thereby achieving the infilling of structural depressions progressively away from shore. Evidence has been presented of infilling down to a water depth of 2500 m. Sub-bottom profiles show that tilting of infilled sediment blocks has taken place since the formation of the main Tyrrhenian basin and that considerable slumping has occurred as a result.

The superficial sediments of the borderland have features characteristic of slow deposition and they also show signs of soft sediment deformation. Profiler records show evidence of relatively thin draping over most topographic highs which probably represents slow deposition from suspension over a long period.

The second major conclusion is that sedimentation over the abyssal plain is discontinuous in time and over the area studied.

The lateral discontinuities of the minor sand layers are perhaps not surprising. Local deposition of sands has been deduced in similar conditions in the San Diego trough (von Rad, 1968). Similar lack of continuity in the Tyrrhenian is well known, though Ryan *et al.* (1965) claim that some layers of vitric and crystalline ash and others of terrigenous material are continuous over wide areas.

Some of the properties of the fine grained material in the abyssal plain cores are difficult to interpret in terms of the conventional view that deep sea lutites are the product of a continuous rain of particles through the whole depth of the water column.

The conspicuous and well defined vertical color

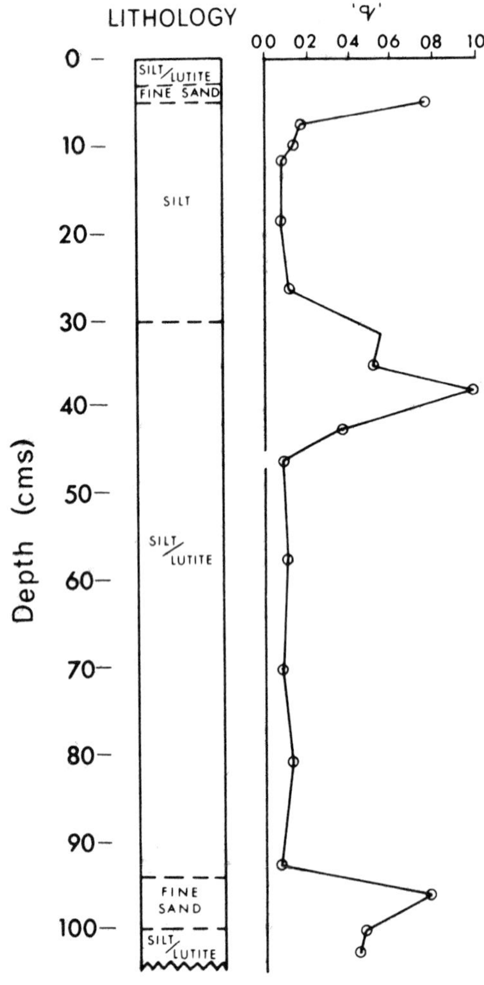

Figure 17. The parameter 'q' plotted against depth for an apparently uniform fine-grained layer in Core TS 1. The peak at 30 to 45 cm may indicate an episode of strong current activity.

changes are evidence of continual change of sedimentation conditions with time. Occasional breaks in sedimentation are indicated by organic activity at some of the surfaces where color change is most marked.

It seems likely from our preliminary study that discontinuities in sedimentation exist not only in time but also laterally over the abyssal plain. The sequence of color changes and burrowed surfaces of one core does not match that of another 20 or 30 km away.

Other evidence of discontinuity within an apparently homogeneous lutite bed comes from the magnetic measurements. The bed in question has a narrow band of high q value (Figure 17), possibly due to a stronger than usual current (Hamilton and Rees, 1970).

These forms of discontinuity are difficult to account for on a continuous rain hypothesis but could be explained by deposition from currents flowing near the bottom. Deposition from bottom currents is usually accepted to be common in deep water sands. Whether it is effective for fine grained deposits remains to be investigated.

ACKNOWLEDGMENTS

We wish to acknowledge the help of those who took part in the 1969 R.R.S JOHN MURRAY cruise. D. Frederick, R. B. Kidd and J. Wenlock have done much of the work on magnetic fabric, sedimentology and bathymetry respectively. The work was funded largely by the Natural Environment Research Council of Great Britain.

REFERENCES

Emery, K. O. 1960. *The Sea off California, a Modern Habitat of Petroleum.* Wiley, New York, 366 p.

Hamilton, N. and A. I. Rees 1970. The use of magnetic fabric in palaeocurrent estimation. In: *Palaeogeophysics,* ed. Runcorn, S.K., Academic Press, London and New York, 445–464.

Hamilton, N., W. H. Owens and A. I. Rees 1968. Laboratory experiments on the production of grain orientation in shearing sand. *Journal of Geology,* 76:465–472.

Hersey, J. B. 1965. Sediment ponding in the deep sea. *Geological Society of America Bulletin,* 76:1251–1260.

Kermabon, A., C. Gehin, P. Blavier and B. Tonarelli 1969. Acoustic and other physical properties of deep sea sediments in the Tyrrhenian Abyssal Plain. *Marine Geology,* 7:129–145.

Matthews, D. J. 1939. Tables of the velocity of sound in pure water and sea water for use in echo sounding and echo ranging. *British Admiralty Hydrographic Department Publications,* H. D. 282.

von Rad, U. 1968. Comparison of sedimentation in the Bavarian Flysch (Cretaceous) and Recent San Diego Trough (California). *Journal of Sedimentary Petrology,* 38:1120–1154.

Rees, A. I. 1965. The use of anisotropy of magnetic susceptibility in the estimation of sedimentary fabric. *Sedimentology,* 4: 257–271.

Rees, A. I. 1966. The effect of depositional slopes on the anisotropy of magnetic susceptibility of laboratory deposited sands. *Journal of Geology,* 74:856–867.

Rees, A. I., U. von Rad and F. P. Shepard 1968. Magnetic fabric of sediments from the La Jolla Submarine Canyon and Fan, California. *Marine Geology,* 6:145–178.

Ryan, W. B. F., F. Workum and J. B. Hersey 1965. Sediments on the Tyrrhenian Abyssal Plain. *Geological Society of America Bulletin,* 76:1261–1282.

Holocene Stages of Oxygen Depletion in Deep Waters of the Adriatic Sea

L. M. J. U. van Straaten

Geologisch Instituut, Groningen

ABSTRACT

New data are presented on the dark sediments of early Holocene age in the southeastern Adriatic Sea. These layers are devoid of skeleton remains or traces of benthic animals. From their depth distribution, lithology and (pelagic) mollusc shell contents the following conclusions are drawn:

1. The absence of benthic life during the formation of the dark sediments was due to lack of oxygen.
2. The lack of oxygen resulted from stagnation of the deeper water masses.
3. This stagnation was partly, perhaps mainly, the result of a rapid increase in the minimum temperatures of the upper water layers.
4. The upper boundaries of the stagnation zones lay several hundreds of meters below the sea surface.
5. Low-velocity bottom currents flowed down the entire continental slope, at least on the Italian side, during periods of normal circulation, but did not affect the bottom of the deeper parts during the stagnation stages.

RESUME

On présente de nouvelles données sur les sédiments foncés du Holocène inférieur dans le Sud-Est de l'Adriatique. Ces couches sont caractérisées par l'absence de faune benthique. L'étude de la répartition bathymétrique de ces sédiments, de leur propriétés malacologiques et de leur lithologie permet de tirer les conclusions suivantes:

1. L'absence de faune benthique pendant la formation des sédiments foncés était due à l'absence d'oxygène.
2. L'absence d'oxygène était le résultat de la stagnation de masses d'eau plus profondes.
3. Cette stagnation s'est établie sous l'influence soit légère soit dominante de l'accroissement rapide des températures minima des couches d'eau superficielles.
4. Les limites supérieures de ces zones de stagnation sont situées à plusieurs centaines de mètres de profondeur.
5. Les courants de fond de faibles vitesses ont coulé jusqu'en bas de la pente continentale, du moins sur le versant italien, pendant les périodes de circulation normale mais n'ont pas affecté le fond des zones plus profondes durant les périodes de stagnation.

INTRODUCTION

In earlier papers (van Straaten, 1966, 1967, 1970; Bottema and van Straaten, 1966) it has been shown that in the course of the Holocene two stages occurred, one shortly after the other, when the deep parts of the southeastern Adriatic Sea (Figure 1) were devoid of benthic animals. The skeletal remains found in the sediments of these stages belong exclusively to planktonic (pelagic) organisms, and the primary depositional laminations are largely or wholly undisturbed by burrowing or creeping bottom fauna. The successive laminae are alternately dark and light colored, probably as the result of seasonal differences in sedimentation.

The absence of bottom life has been ascribed to lack of oxygen in the deeper water layers, an interpretation which appears to be corroborated by the relatively high contents of organic matter, *e.g.*, 2 to 3% (by weight), or about twice as much as in the normal muds above and below. The deposits are also enriched in pyrite, which is present as isolated particles

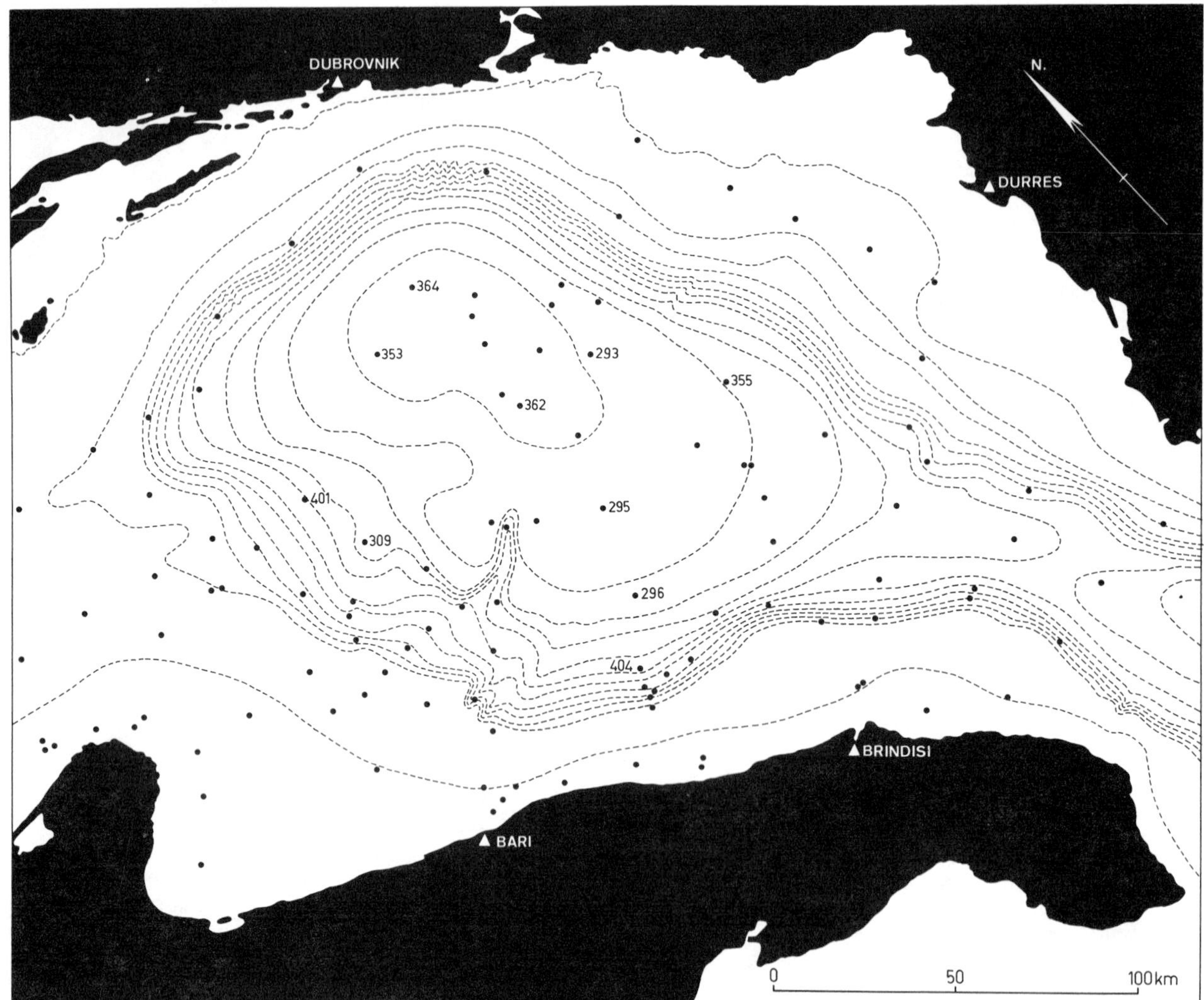

Figure 1. Chart showing sample locations in the Adriatic Sea.

and concretionary bodies, and as fillings of foraminifera tests and pteropod shells. Owing to the relative concentration of both organic matter and pyrite, the sediments as a whole contrast with the overlying and underlying deposits by their distinctly darker colors.

The lack of oxygen must have been caused by (temporary) stagnation of the deeper sea water. Since dark layers of this type are also encountered at levels above that of the sill between the Adriatic Sea and the Ionian Sea, it is likely that similar conditions must have existed in the deeper parts of the Eastern Mediterranean. In fact, dark colored deposits of presumably the same age have been mentioned previously from this eastern basin (Olausson, 1961; Ninkovich and Heezen, 1965; see also Vergnaud-Grazzini and Herman-Rosenberg, 1969). More recently they have also been described by Hesse *et al.*, (1971) from the Strait of Otranto.

It should be noted, however, that these authors refer to one layer only. This could mean that the vertical circulation, which interrupted the stagnation in the Adriatic Sea, did not penetrate to the greater depths of the Eastern Mediterranean. The dark sediment found at shallower depths in the Strait of Otranto, was present only in the lower end of the cores, so that it might correspond to the upper dark layer of the Adriatic Sea deposits.

Carbon-14 age determinations of the dark muds of the eastern Mediterranean gave the following results: 7500 years B.P. (Ninkovich and Heezen, 1965), 8830 years B.P. (Olausson, 1961) and 8210 years B.P. (Fabricius *et al.*, 1970). Preliminary palynological data (Bottema and van Straaten, 1966) seem to point to an early-Atlantic age for the upper dark layer in the Adriatic, which would correspond to an age of roughly 7500 to 9000 years for the whole zone.

Olausson (1961) suggested three factors that could have contributed to stagnation of the deeper water masses in the eastern Mediterranean: (1) increased supply of fresh water from the land, owing to melting of Würm ice-sheets, (2) increased precipitation and lowered evaporation during the Upper Dryas, and (3) a general rise in temperatures of the surface waters, lowering their density. The second factor is inapplicable, since the dark zone is younger than Upper Dryas. van Straaten (1966) believed that a rather sudden rise of the minimum (winter) temperatures of the upper water layers was mainly responsible, but that its effects might have been strengthened by increased run-off of river water from the land.

So far, no dark layers, lacking in remains of benthos organisms, are known from the Holocene deposits in the western Mediterranean (*cf.*, the core data of Norin, 1958; Todd, 1958; and Eriksson, 1965). Their absence might be due (*cf.*, Olausson, 1961) to subsurface inflow of Atlantic Ocean water through the Strait of Gibraltar, which would have prevented stagnation of the deeper water west of the sill between Sicily and Tunisia.

DEPTHS OF OXYGEN DEFICIENCY

The shallowest depth where, during the 1962 cruise in the southeastern Adriatic Sea, dark Holocene muds without benthic remains were found, was 699 m. A core from 547 m showed at its base a dark zone with partial reduction of benthic elements. This could be the result either of deposition in a transition *zone*, between oxygen-free waters below and aerated waters above, or of deposition during a transition *stage*, following a period when the waters at this depth were completely lacking in oxygen.

In the hope of determining more accurately the upper boundaries of the former oxygen-free zones, additional cores were taken during the 1968 cruise with the Italian vessel BANNOCK. Only one of these cores, from 696 m depth, contained sediment without benthic material. However, the absence of such sediment in the other cores, taken at shallower depths, cannot be used as evidence against lack of oxygen at these levels, since in all of them the Holocene deposits were very thin. Hence, if any benthos-free layers have been formed at these places, they must have become unrecognizable, owing to intermixing with normal mud by later bioturbation.

VERTICAL DISTRIBUTION OF PELAGIC MOLLUSCS

As mentioned in earlier papers (*e.g.*, van Straaten, 1966), the composition of the pelagic fauna in the Adriatic Sea has varied greatly in the course of the Holocene. Moreover, the composition of the assemblages of pelagic mollusc shells found in the cores varies with depth of deposition. Examples of both variations are given in Figure 2, which is based on the combined shell material from different cores. The statigraphic zones, represented in this diagram, have the following ranges:

I. *Oldest Holocene*, from ash fall 6 (see van Straaten, 1966, 1967) to the beginning of the first stage of oxygen deficiency (= "O.D. stage I"). The deposits include the youngest layers of the Upper Dryas stage, but these are very poor in pelagic mollusc remains.

II. *"O.D. stage II"*, corresponding to the upper dark zone in deep water, and to the normal sediments of the same age, formed in shallower depths.

III. *Later part of Cavolinia stage*, from shortly before ash fall 4 to shortly before ash fall 2.

IV. *Styliola-stage*, from shortly before ash fall 2 to the present day.

Figure 2 demonstrates that the deposits formed under normal conditions (stages I, III and IV) at moderate depths are richer in shells of *Creseis acicula*, *Creseis virgula*, *Styliola subula*, and *Cavolinia inflexa* than corresponding deposits of deeper water. Shells of *Limacina inflata* and *Clio pyramidata*, on the other hand, are most abundant in deep water sediments.

One might perhaps expect on the basis of these differences that the deep water muds, formed during the stagnation stages, would be relatively enriched in shells of the moderate-depth group. However, this is not the case. Apparently the depth of the water above the oxygen-free layers was sufficient for the "deep" species to flourish also. In fact, the shells of *Limacina inflata*, which under normal conditions are more common in deep water sediments, reach their greatest abundance in the deposits of the benthos-free stages[1]. One may probably conclude that the upper boundaries of the oxygen-free zones were situated at depths of several hundred meters below the sea surface.

VARIATIONS OF THE PELAGIC MOLLUSC FAUNA DURING THE HOLOCENE

The general trends of the stratigraphic variations in malacological composition of the deep water sedi-

[1] In this connection it is interesting to note the frequency variations of *Globigerinoides rubra*. In earlier (Pleistocene) stages, when the deeper water in the eastern Mediterranean likewise was lacking in oxygen, this species was absent, or very scarce, while in the layers formed immediately before and after, it was on the contrary common or even abundant (Olausson, 1961). During the Holocene, in the Adriatic Sea, these relations were reversed, *Globigerinoides rubra* being very abundant during the stages of oxygen deficiency (see Figure 6) and scarce or absent during the other parts of the Holocene.

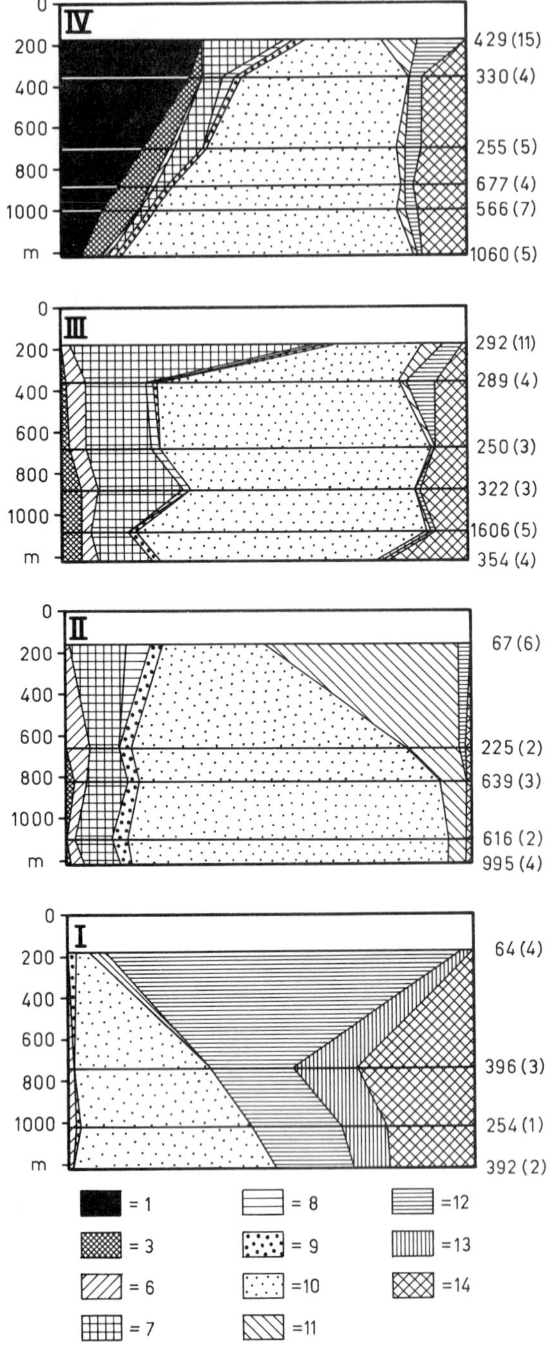

Figure 2. Average composition of pelagic mollusc shell assemblages (> 0.5 mm fraction) for different depths of water and different stages of Holocene. STAGES: IV Styliola-stage (Late Subatlantic); III. Later part of Cavolinia-stage (Subboreal?); II. Second stagnation stage (Early Atlantic?); I. Oldest Holocene (Preboreal and early Boreal?); SPECIES: 1. *Styliola subula* (Quoy et Gaymard); 3. *Hyalocylis striata* (Rang), *Peraclis reticulata* (Orb.), *Clio cuspidata* (Bosc), *Carinaria mediterranea* Lam.; 6. *Limacina bulimoides* (Orb.); 7. *Cavolinia inflexa* (Lesueur); 8. *Cavolinia gibbosa* (Orb.), *Cavolinia tridentata* (Niebuhr); 9. *Atlanta peroni* Lesueur; 10. *Limacina inflata* (Orb.); 11. *Creseis virgula* (Rang); 12. *Creseis acicula* (Rang); 13. *Diacria trispinosa* (Lesueur); 14. *Clio pyramidata* L. On right of diagram: total numbers of specimens (numbers of cores).

ments in the Adriatic Sea are shown in Figure 3. This diagram gives the average composition of samples of equal age from 6 cores, taken at depths between 929 and 1207 m. The diagram is of preliminary character. More detailed sampling of cores shows among other things (see Figure 4) that *Creseis virgula* became important only after the beginning of the first stagnation stage, and that *Clio pyramidata* became very rare during the second stagnation stage.

The variations of the pelagic fauna were due, of course, to the combined effect of many factors. Yet, from the known distribution of pelagic molluscs in the present oceans (see van der Spoel, 1967) it follows that for most species the chief basic factor must have been the water temperatures.

In Figure 3 it is seen that during the end of the Pleistocene and the first part of the Holocene the main pteropod species in the Adriatic Sea reached their maximum percentages one after the other. The order of the successive maxima is the same as the order in which the minimum water temperatures, tolerated by these species, increases:

1. *Limacina retroversa, 2°–circa 16°*
2. *Clio pyramidata, 7°–28°*
3. *Diacria trispinosa, circa 9°–27°*[1]
4. *Creseis acicula, 10°–28°*
5. *Limacina inflata, 14°–28°*
6. *Creseis virgula, 15°–28°*
7. *Cavolinia inflexa, 16°–28°*

From this sequence it may be deduced that during the above mentioned period the temperatures of the upper water layers in the Adriatic Sea, or at least the minimum temperatures, have more or less gradually risen.

During the later part of the Holocene the minimum water temperatures probably decreased slightly. Indications of this event are: (1) the decline in the proportion of *Cavolinia inflexa*, which had reached its maximum in the period directly following the second stagnation stage, and (2) the reappearance (in small numbers) of *Creseis acicula* (see Figure 6). A slight lowering of the (minimum) water temperatures is not in contradiction with the immigration of *Styliola subula* during the most recent part of the Holocene, the minimum temperature tolerated by this species being 14°C (van der Spoel, 1967). On the other hand, it is still not possible to say why this pteropod did not appear earlier.

With regard to the presence of *Clio pyramidata* in the deposits both of the cold stages of the Pleistocene

[1]The minimum temperature for *Diacria trispinosa*, according to van der Spoel, is 14°, but this value does not agree with its present day distribution, which in the Atlantic Ocean extends to Iceland and the Lofoten Islands.

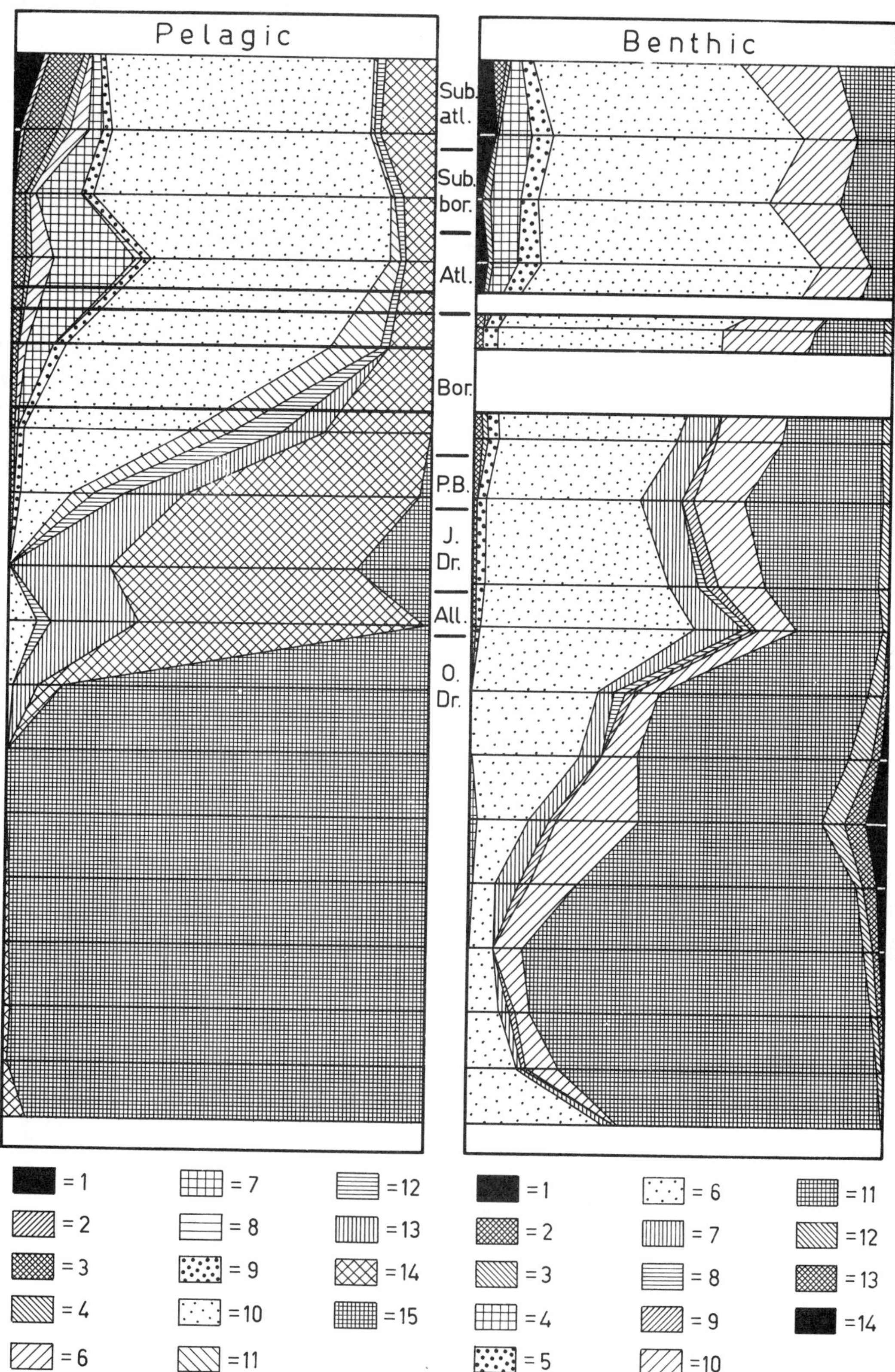

Figure 3. Average composition of pelagic and benthic mollusc shell assemblages (> 0.5 mm) of cores 293, 295, 296, 309, 353, 355 (depths: 1198, 1161, 1063, 929, 1207, 1096 m). PELAGIC SPECIES: 1. *Styliola subula* (Quoy et Gaymard); 2. *Peraclis reticulata* (Orb.); 3. *Clio cuspidata* (Bosc); 4. *Carinaria mediterranea* Lam.; 6. *Limacina bulimoides* (Orb.); 7. *Cavolinia inflexa* (Lesueur); 8. *Cavolinia gibbosa* (Orb.), *Cavolinia tridentata* (Niebuhr); 9. *Atlanta peroni* Lesueur; 10. *Limacina inflata* (Orb.); 11. *Creseis virgula* (Rang); 12. *Creseis acicula* (Rang); 13. *Diacria trispinosa* (Lesueur); 14. *Clio pyramidata L.*; 15. *Limacina retroversa* (Fleming); BENTHIC SPECIES: 1. *Leda micrometrica* Jeffr; 2. *Teretiateres* Reeve, *Pleurotoma trechii* (Phil.); 3. *Dentalium agile* M. Sars; 4. *Entalina quinquangularis* (Forb.); 5. *Abra longicallus* Scacchi; 6. *Kelliella miliaris* (Phil.); 7. *Nucula aegeensis* Forbes; 8. *Lima sarsi* (Lovén); 9. *Columbella haliaeeti* Jeffr., *Dischides bifissus* S.V. Wood, *Thyasira eumyaria* (G. O. Sars), *Lyonsia formosa* Jeffr.; 10. Other species; 11. *Benthonella tenella* (Jeffr.); 12. *Malletia obtusa* M. Sars (large variety); 13. *Dentalium striolatum* Stimpson, *Malletia obtusa* M. Sars (small variety); 14. *Yoldia lucida* Lovén, *Lima subovata* (Jeffr.).

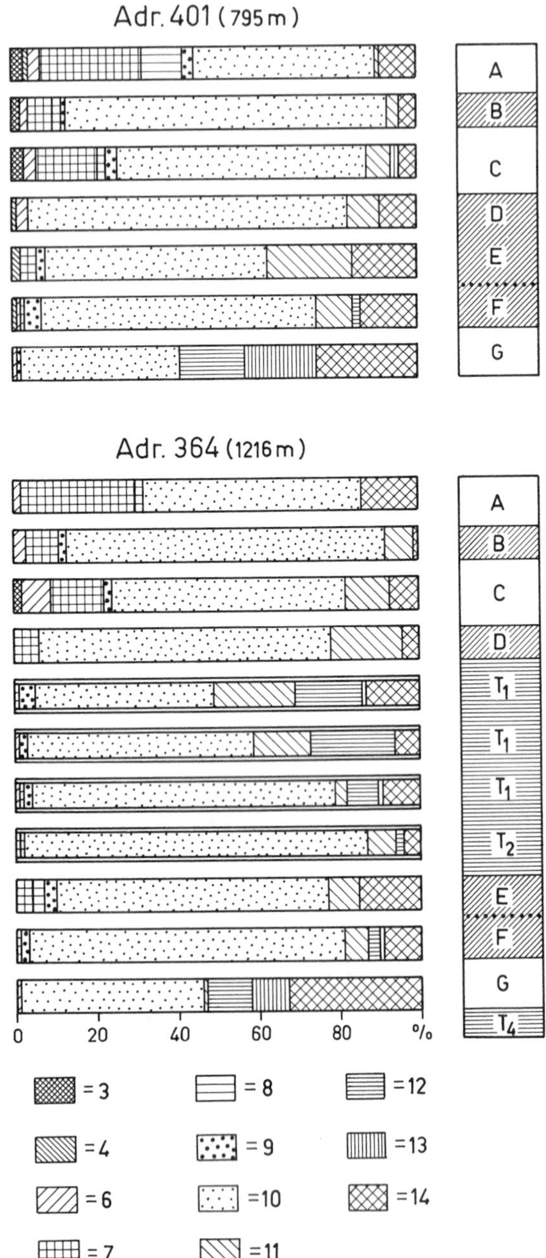

Figure 4. Composition of pelagic mollusc shell assemblages (> 0.5 mm) in cores 401 and 364. DEPTHS: 401: A, 65–82 cm; B, 82–90 cm (Upper Dark Zone); C, 90–96 cm; D, 96–101 cm; E, 101–104 cm; F, 104–113 cm (D-F = Lower Dark Zone; D-E = above ash layer 5); G, 113–126 cm. 364: A, 16–24 cm; B, 24–28 cm (Upper Dark Zone); C, 28–33 cm; D, 33–38 cm (Upper part of Lower Dark Zone); T_1–T_2 Turbidites: 75–97 cm; 97–118 cm; 118–120 cm (T_1), 136–140 cm (T_2); 155–159 cm; F, 159–172 cm (E-F = Lower part of Lower Dark Zone; E = above ash layer 5); G, 172–183 cm. SPECIES: 3. *Clio cuspidata* (Bosc); 4. *Oxygyrus keraudrenii* Rang; *Carinaria mediterranea* Lam.; 6. *Limacina bulimoides* (Orb.); 7. *Cavolinia inflexa* (Lesueur); 8. *Cavolinia gibbosa* (Orb.), *Cavolinia tridentata* (Niebuhr); 9. *Atlanta peroni* Lesueur; 10. *Limacina inflata* (Orb.); 11. *Creseis virgula* (Rang); 12. *Creseis acicula* (Rang); 13. *Diacria trispinosa* (Lesueur); 14. *Clio pyramidata* L.

and the warm stages of the Holocene, it must be remarked that this species comprises various forms. During the last parts of the Pleistocene the shells were of the type 'H' shown in Figure 5. In the early Holocene sediments they increasingly show the properties of the forma *convexa* (Boas) (Figure 5, G-D, E), then disappear during the second stage of oxygen depletion. After that, a new type, the forma *lanceolata* (Lesueur) (Figure 5, A) entered the Adriatic Sea.

The different types can be characterized quantitatively by the relation between length of the protoconch and (maximum) width of the adult shell at a certain distance (Figure 5: 350μ) from the protoconch. It is not yet known in how far the forma *convexa* indicates special temperature conditions. Possibly the temperatures may have been, in this case, of less importance than other factors, such as food or oxygen distribution in the water.

VARIATIONS OF THE BENTHIC MOLLUSC FAUNA DURING THE HOLOCENE

Comparing the benthic shell assemblages in the Older Dryas deposits with those of the late Holocene (Figure 3) one again finds great differences. It is clear that these differences are also mainly caused by changes in temperature. Many of the species composing the late Pleistocene bottom fauna are limited at the present day to the seas of northern Europe.

However, when one considers in more detail the malacological compositions of the Younger Dryas and early Holocene deposits, it is evident that in this particular period the benthic fauna changed much less than the pelagic fauna. The benthic assemblages formed just before the first stagnation stage still had practically the same composition as the Alleröd assemblages. It indicates that during the end of the Younger Dryas and the earliest parts of the Holocene the temperatures in the upper water layers increased more rapidly than in the deeper water.

Obviously this difference in the rates of warming-up must have considerably reduced the exchange of the deeper water with aerated water descending from the surface. Indeed, the total number of benthic shells in the sediments of this period gradually decrease, in contrast to the simultaneous increase of the pelagic shells (see Figure 6). Eventually the reduction of vertical circulation led to complete exhaustion of dissolved oxygen.

In this connection it may be pointed out that it is not necessary to assume complete stagnation as the explanation for the absence of oxygen. Depletion of oxygen requires only that its supply by "vertical" circulation is slower than its consumption. Hence,

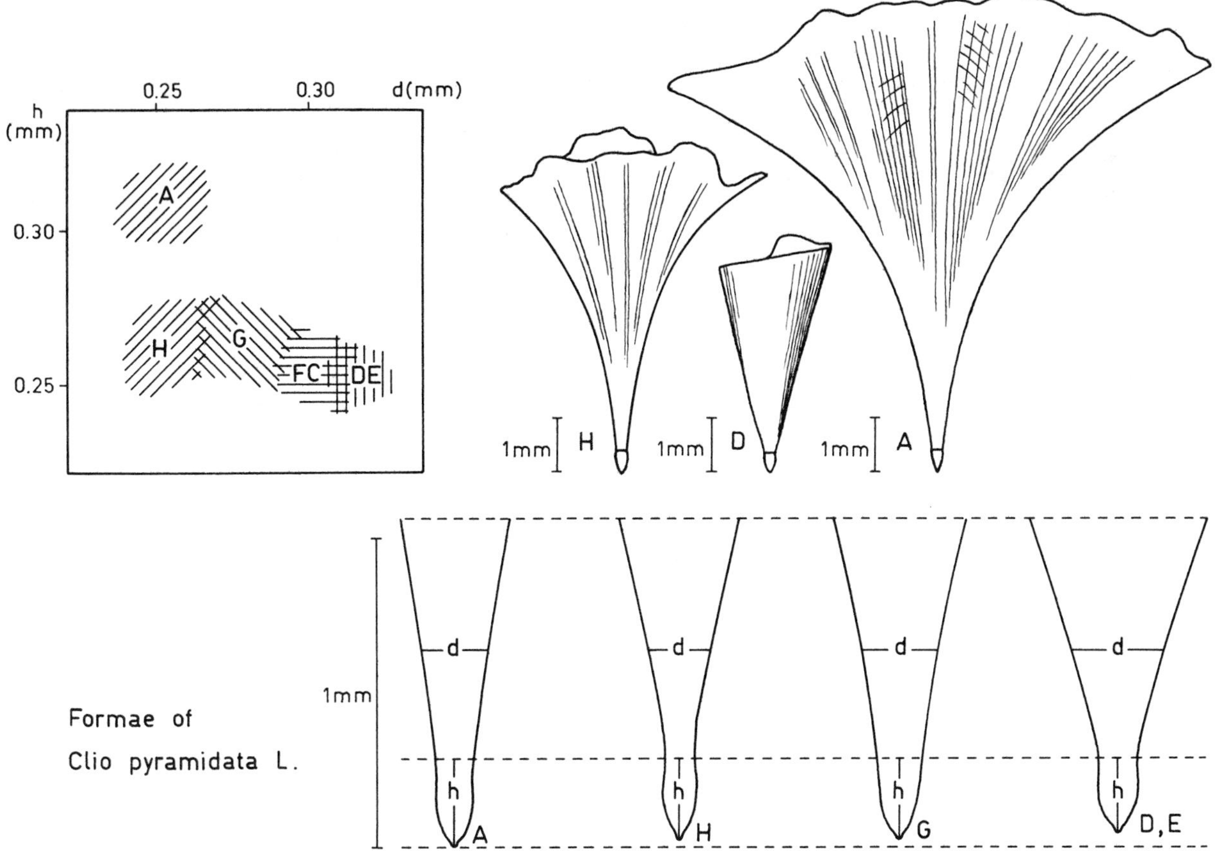

Figure 5. Formae of *Clio pyramidata* L. Letters A-H correspond to stages of Figure 4. A = Upper Holocene; C = Intercalation between two dark zones; D-F = Lower Dark Zone; G-H = Earliest Holocene. Left: relation between height (length) of protoconch (h) and maximum width at 350 μ from protoconch.

where the term stagnation is used in this paper, it is meant only to indicate that the circulation is reduced to a value below this critical limit.

The re-establishment, or re-intensification of vertical circulation down to the bottom, at the end of the stagnation stages, may have been the combined result of an increase in the maximum densities of the surface waters, and a slight warming of the deep water layers. The latter could have been caused by flow of earth heat from below, and by very slow circulation of the water (see above). A rise in density of the surface waters may have been brought about by a decrease in the supply of fresh water (precipitation and influx of river water), by a drop in the minimum (winter) temperatures, by increased evaporation, or by combinations of these factors. It is as yet impossible to say, on the basis of the pelagic mollusc shell assemblages, which among these factors may have had the greatest influence.

The benthic mollusc fauna which settled on the deepest parts of the Adriatic Sea floor at the end of the first stagnation stage had a distinctly "warmer" character than that immediately preceding this stage. Among other species, *Nucula aegeensis* suddenly had become very rare (less than 0.5%). On the other hand, the "cold" *Malletia obtusa* was still present.

After the second interruption of benthic life, the bottom fauna had approximately the same composition as to-day. *Malletia obtusa* had disappeared, and several new species had immigrated, *e.g.*, *Leda micrometrica* and *Dentalium agile,* while *Entalina quinquangularis* became more abundant.

ABSOLUTE ABUNDANCE OF MOLLUSC SHELLS

In various papers Brongersma-Sanders (1957, 1965, 1966) and other authors have called attention to the difference in organic productivity between seas with "estuarine-like" circulation (outflow at surface, inflow below surface) and seas with "anti-estuarine" circulation. At present, the Mediterranean as a whole is an example of the anti-estuarine type (*cf.*, Wüst, 1961). This fact accounts for the very low production of organic material in this sea, most of the nutrients

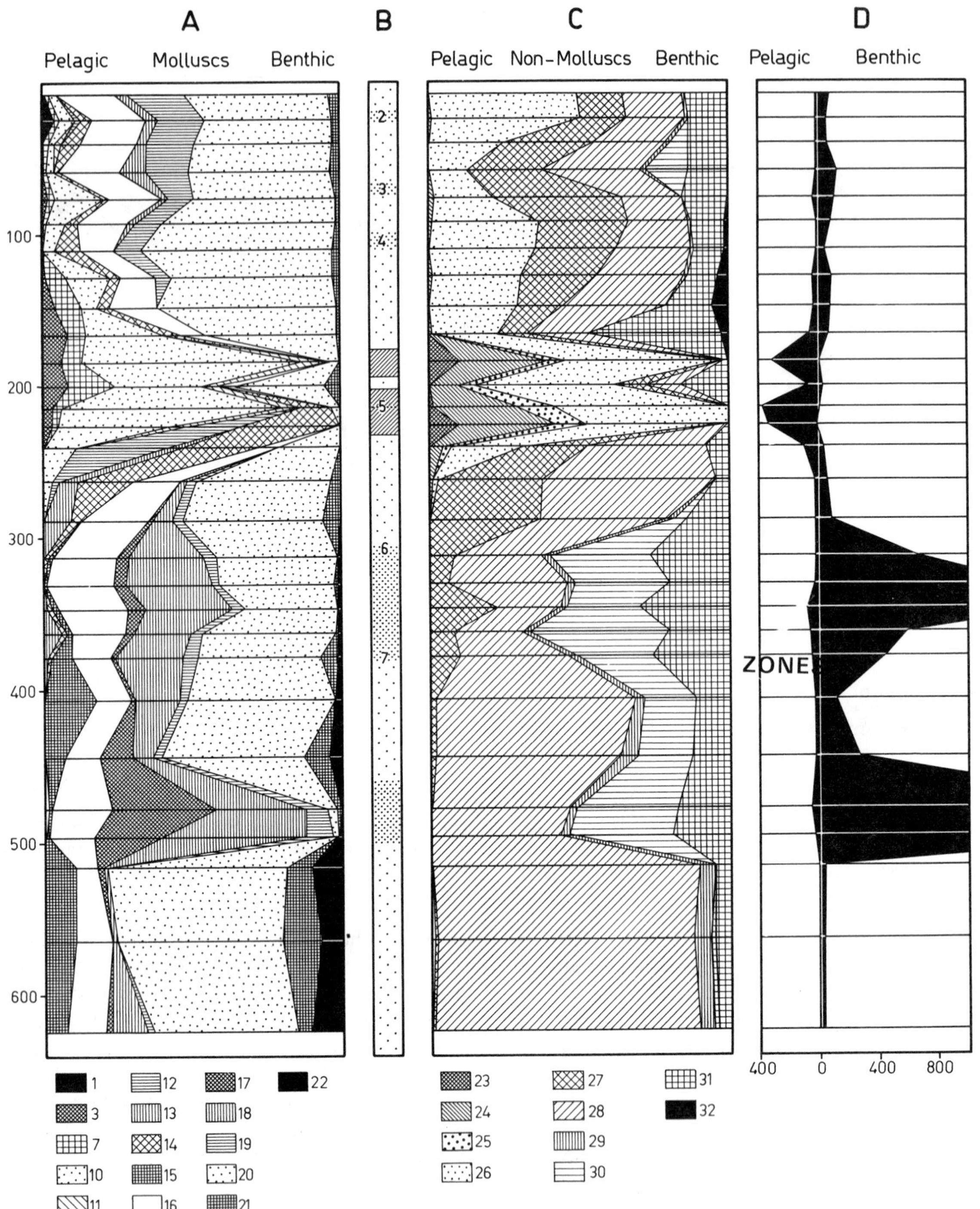

Figure 6. Detailed analysis of Core 404. A, COMPOSITION OF MOLLUSC SHELL ASSEMBLAGES (> 0.5 mm). B, LITHOLOGY. Oblique lines = dark zones; Closely stippled = ash layers 2-7 and sandy mud layers; open stippled = scattered sand grains, flakes of chlorite, etc., and displaced shallow water shells. C, COMPOSITION OF NON-MOLLUSC ORGANIC REMAINS (> 0.5 mm). D, NUMBERS OF PELAGIC AND BENTHIC MOLLUSC SHELLS (> 0.5 mm) per 20 cm of sample thickness. MOLLUSCS: PELAGIC SPECIES: 1. *Styliola subula* (Quoy et Gaymard); 3. Other species; 7. *Cavolinia inflexa* (Lesueur); 10. *Limacina inflata* (Orb.); 11. *Creseis virgula* (Rang); 12. *Creseis acicula* (Rang); 13. *Diacria trispinosa* (Lesueur); 14. *Clio pyramidata* L.; 15. *Limacina retroversa* (Fleming). BENTHIC SPECIES: 16. Various species; 17. *Bittium* cf. *deshayesi* Cirulli-Irelli; 18. *Hydrobia* sp., *Rissoa concinnata* Jeffr., *Turritella* sp., *Calyptraea chinensis* (L.), *Propeamussium incomparable* (Risso), *Pecten opercularis* (L.), *Mysella bidentata* (Mtg.), *Cardium edule* L., *Cardium minimum* Phil., *Cardium papillosum* Poli, *Venus ovata* Penn., *Spisula subtruncata* (Da C.), *Abra alba* (Wood), *Abra nitida* (Müller), *Corbula gibba* (Olivi); 19. *Anomia squamula* L., 20. *Kelliella miliaris* (Phil.); 21. *Benthonella tenella* (Jeffr.); 22. *Yoldia lucida* Loven. NON-MOLLUSCS: 23. Wood fragments, 24. Fish remains and bones in dark layers, otoliths in other layers); 25 *Globigerinoides rubra;* 26. Other pelagic foraminifera; 27. Agglutinant foraminifera; 28. Other benthic foraminifera; 29. Ostracods; 30. Crabs, Balanids, Serpulids, *Ditrupa*, Bryozoans a.o.; 31. Echinoid fragments; 32. Sponge spicules. N.B. The base of the Holocene in this core lies at a depth of circa 300 cm.

being removed by the outflowing undercurrent.

On the contrary, in seas with estuarine-type circulation a considerable part of the nutrients is brought back to the surface by upwelling. This leads to high organic productivity, and, in some areas (*e.g.*, off the coast of southwest Africa) occasionally even to mass-mortality. Under such conditions the water near the bottom may become depleted in oxygen, not by stagnation or almost-stagnation, but as a result of the exceptionally strong supply of decomposable organic matter.

In the light of these data it is necessary to consider also the temporary establishment of estuarine circulation as a possible cause of the oxygen deficiencies that occurred in the eastern Mediterranean. It is noteworthy, in this respect, that on the whole the dark zones in the sediments of the Adriatic Sea are much richer in tests of pelagic foraminifera and shells of small pteropods like *Limacina inflata,* than the normal deposits. But there are also exceptions. Thus the layer immediately overlying the upper dark layer is, at several places, likewise very rich in pelagic remains.

In the second place, large shells of larger species, like *Clio pyramidata,* are relatively scarce in the dark layers, so that in many cores the total weights of pelagic shell material, per equal quantity of sediment, are even smaller than in the normal muds.

Furthermore, if the oxygen depletion had been the result of a dense rain of organic material towards the bottom, sufficient to keep up with an active supply of oxygen by water currents, one should expect that small temporary decreases in the oxygen supply would lead immediately to strongly increased accumulation of organic matter. However, as far as available analyses indicate, the organic contents of the dark layers are nowhere exceptional, usually not greater than 3.0 %.

Hence, it appears unlikely that increased organic productivity due to estuarine circulation was the main cause of the early-Holocene stages of oxygen-depletion. Of course, it is still possible that during the benthos-free stages the circulation above the inferred zone of stagnation was indeed of the "estuarine" type.

DISTRIBUTION OF SAND AND MOLLUSC SHELLS IN CORE 404

The Holocene muds on the lower parts of the continental slope locally contain appreciable quantities of silt, as well as scattered, sand-size flakes of chlorite and muscovite, and grains of rounded quartz sand (*cf.*, van Straaten, 1970). The highest contents of such coarse material were encountered in Core 404, taken from a depth of 716 m, east of Bari.

The detrital elements are more or less equally distributed through this core, with the exception of the stagnation stage deposits in which they are much scarcer or even absent. Moderate quantities are found in the light colored layer intercalated between the two dark zones (Figure 6).

Apart from the terrigenous silt and sand grains, the mud contains small mollusc shells, up to a few millimeters in diameter, belonging to species that are characteristic of the shelf environment. They have apparently been transported down the continental slope. No such displaced shells are present in the muds of the stagnation stages.

A third peculiarity of Core 404 is its relatively high content of deep-water benthic mollusc shells, as compared to other cores from the same environment (see Table 1: A_1-A_4 and G(-H)). The numbers of pelagic shells, on the other hand, are relatively small, at least in some of the layers. Again, no such abnormal shell contents are found in the stagnation stage deposits.

BOTTOM CURRENTS AND STAGNATION ZONES

The only explanation which accounts satisfactorily for the combination of the above mentioned properties is that they are due to current action along the bottom. Although surface currents could perhaps have supplied, under exceptional conditions (*e.g.*, heavy storms), flakes of chlorite and muscovite and grains of quartz, they could not have transported the shells of shelf molluscs. The latter can only have been brought to the site of Core 404 by currents flowing downslope over the bottom.

Apart from supplying silt, sand grains, and small shells, these currents may have been responsible for the scarcity (in some of the layers) of pelagic mollusc shells, *viz.*, by preventing deposition of a proportion of the shells that sank towards the bottom. At the same time, the active supply of food and oxygen by the currents must have created very favorable living conditions for bottom fauna, and hence led to the relative abundance of benthic mollusc shells.

It may seem difficult to understand how the currents on the one hand could have hindered the deposition of pelagic mollusc shells while on the other hand so much mud was deposited at the same place. However, the flow regime may have been of intermittent character, so that clayey material could settle on the bottom during intervals of reduced current activity, to be mixed later by benthic animals with the coarser particles supplied during stages of stronger flow.

Moreover, part of the lutite material may have been transported as larger aggregates, such as fecal pellets or small granules, formed by erosion of older

Table 1. Total quantities of mollusc shells (> 0.5 mm) per horizontal area of 1 cm^2 in Holocene deposits (exclusive of turbidites) in southeastern Adriatic Sea.

	Pelagic Shells										Benthic Shells									
	Continental Slope						Turbidite Plain				Continental Slope						Turbidite Plain			
Core	349	346	284	404	401	372	353	362	363	364	349	346	284	404	401	372	353	362	363	364
Depth (m)	547	627	699	716	797	1030	1207	1216	1220	1220	547	627	699	716	797	1030	1207	1216	1220	1220
Zone A_1	38	11	16	2	31	39	46	(49)	53	—	53	21	6	16	5	4	3	(2)	1	—
A_2	8	9	4	3	3	4	5	2	0	—	13	5	7	23	1	1	1	1	0	—
A_3	1	9	10	5	4	3	4	11	15	—	6	9	13	20	2	2	2	1	2	—
A_4	10?	14?	27	7	14	15	26	35	11	—	30?	6?	6	14	5	6	3	3	2	—
O.D.II B	—	—	25	23	18	32	29	37	45	—	—	—	1	1	1	0	0	0	0	—
Interval C	—	—	14	5	7	7	3	4	8	—	—	—	4	3	2	2	1	2	1	—
O.D.I D–E	—	—	—	24	11	45	67	12	24	—	—	—	—	1	0	0	0	1	0	—
O.D.I F	—	—	—	19	21	44	38	38	—	25	—	—	—	—	0	0	2	0	—	1
G–(H)	—	—	—	13	25	31	23	32	—	7	—	—	—	12	6	7	1	5	—	3
A_1–A_4	57	43?	57	17	52	61	81	97	79	—	102?	41?	32	73	13	13	9	7	5	—
B–F	—	—	—	71	57	128	137	91	102		—	—	—	5	3	2	3	3	2	
G–H	—	—	—	13	25	31	23	32	—	7	—	—	—	12	6	7	1	5	—	3
Total	—	—	—	101	134	220	241	220	188		—	—	—	90	22	22	13	15	10	

mud layers. In fact, the upper decimeters of the mud in Core 404 show under the binocular microscope a distinctly granular (50–100 μ) structure. The absence of this granularity in most of the deeper layers may be partly primary and partly due to the welding together of these soft particles under the influence of compaction.

The currents, although able to move particles of silt and sand size and occasional small shells, cannot have been very rapid, since in that case more coarse material would have been present, as well as erosion features and residual (lag) deposits. The fact that they followed the bottom down the slope implies that the driving force must have been gravity. Hence the downflowing water must have had a higher density than the surrounding water. But, considering their low velocity, this density difference can only have been small, much smaller than in the case of normal turbidity currents, for example.

Possibly the difference in density was caused mainly by moderate quantities of fine suspended particles, and by lower temperatures. Evidence for low-velocity currents of this type has been given in recent years by many authors, *e.g*., Moore (1969). Probably they are a very common phenomenon on continental slopes in general.

The interesting point in the present case is that in the deeper parts of the Adriatic Sea this current activity was apparently interrupted during the stagnation stages. It may mean either that no such currents were generated at all in these periods, which seems rather unlikely, or that, because of the higher density of the stagnating water masses they could not descend the slope beyond a certain depth. At this limiting depth they must have flowed out horizontally and probably soon lost their original momentum.

The density stratification during the stagnation stages, of course, could not prevent the downslope continuation of normal, high-density turbidity currents. Indeed, three layers of normal turbidites were found intercalated in the stagnation stage deposits of the bathyal plain (van Straaten, 1970).

AGE OF DARK LAYERS AND SEDIMENTATION RATES

As stated in the introduction, the probable age of the dark layers has been determined by C-14 analyses and by a palynological study as late-Boreal to early-Atlantic. Theoretically this age may be a little too old, because part of the analyzed material may have been supplied by erosion of older deposits. Assuming, however, that the datings are approximately correct, it is interesting to relate them to the depths of the dark layers in the core sections. After eliminating layers of volcanic ash and turbidites, and after correcting for differences in compaction, one may in this way get an idea of variations in the rate of supply of terrigenous (and pelagic marine) material.

The influence of compaction can be estimated by determining the water content, or the bulk density of the successive layers (thereby also considering the percentages of clay material). Corrections for turbidite intercalations are easily made on account of the typical properties of these sediments (grading, granulometric composition, internal structures; or in turbidite lutites, the strongly parallel orientation of clay mineral flakes). More complicated is the correction for ash layers. Though it is easy to recognize volcanic elements of sand or coarse silt size, direct mineralogic identification becomes very uncertain for the finer grades. For the cores of the Adriatic Sea an attempt was made to estimate the amounts of volcanic admixture on the basis of the carbonate contents, assuming (1) that the percentages of carbonate in normal terrigenous and pelagic material was more or less constant, and (2) that the volcanic material was free of carbonates.

An example of a Holocene section (No. 362, *cf*., van Straaten, 1970) corrected in the above manner, is shown in Figure 7, C. Although the ash and turbidite layers are left out of this sequence, the levels (without

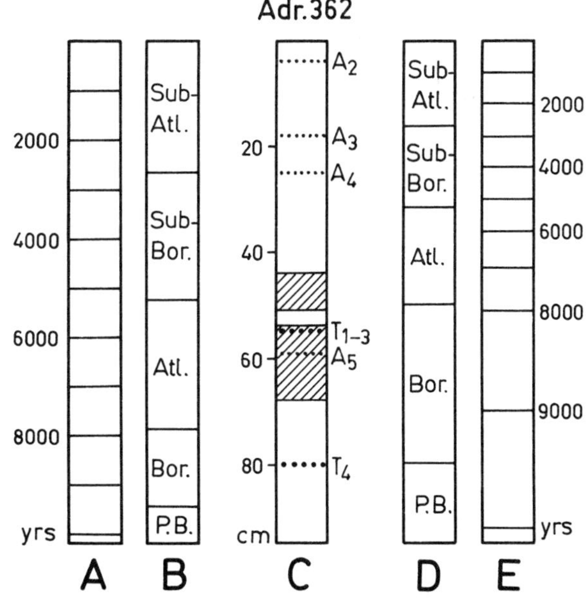

Figure 7. Analysis of Core 362. Core section, C, after elimination of volcanic ash (layers A_2–A_5) and turbidites (T_1–T_4), and after correction for differences in compaction. Oblique lines = dark zones. A, B, Time scale and stratigraphy, assuming constant deposition rate for terrigenous material and pelagic organic carbonates. D, E, Stratigraphy and time scale, assuming age of 8000 years B.P. for base of upper dark layer.

thickness) at which they occur are indicated, because they permit accurate correlation with other sections.

On the left of this diagram (Figure 7, A, B) the stratigraphic and time divisions are given as they would be if the deposition rate had been constant during the whole Holocene. The scales on the right (Figure 7, D, E) are based on the assumption that the second stage of stagnation commenced about 8000 years ago, at the beginning of the Atlantic, and that the sedimentation since that time has been approximately constant. It would imply that the rate during the early Holocene has been about 4 times higher than nowadays, which might be the result of the different positions of sea level. With a lowered sea level much more sediment can be moved over the shelf edge towards deeper water.

In an earlier paper (van Straaten, 1966) the author mentioned that the thickness of sets of one dark and one light colored lamina (in the lower dark zone) varied generally between 120 and 300 μ, and that an average thickness of 200 μ per year would correspond to the average rate of deposition of the whole Holocene section. However, at that time no quantitive data for the influence of compaction were available, while some intercalated lutite turbidites also had escaped attention. Furthermore the base of the Holocene was probably taken too low in the sections. It now appears that a rate of 200 μ per year would be much higher than the average for the whole Holocene. Moreover, reinspection of the core samples showed that the average thickness of the annual deposits in the dark zones may have been underestimated. More likely it is about 300 μ, or even more. This would indicate that the deposition rate during the stagnation stages was at least equal to, and perhaps even greater than during the preceding part of the Holocene. Further data are required before definite conclusions can be made on this point.

SUMMARY

The Holocene deposits in the deep parts of the southeastern Adriatic Sea contain two dark colored layers. Similar dark sediment (one layer only) was found in the Strait of Otranto and in other parts of the eastern Mediterranean. This material has been dated by C-14 -analysis at about 7500 to 9000 years B. P. The skeleton remains found in the dark layers in the Adriatic Sea belong exlusively to pelagic organisms. The disappearance of benthic fauna during these stages was apparently due to oxygen depletion of the deeper water masses.

From the variations in the mollusc shell assemblages in the Adriatic cores it may be deduced that the minimum (winter) temperatures of the upper water layers, just before the depletion stages, had risen much quicker than the temperatures of the deeper masses. This renders it likely that the depletion was caused, at least for a large part, by stagnation of the water below a certain depth, or at least by a strongly reduced vertical circulation. The upper limit of the depleted waters must have lain at a depth of several hundred meters, perhaps some 500 m, but less than 700 m.

Part of the cores, taken near the base of the continental slope in the deep basin of the southeastern Adriatic contain admixtures of terrigenous sand- and silt-sized elements, as well as small mollusc shells (and other organic remains), characteristic of the shelf environment. The coarse elements are common or even abundant at all depths in the cores, with the exceptions of the dark layers.

Sedimentological analysis of the cores shows that these coarse admixtures must have been supplied by low-velocity bottom currents, flowing down the continental slopes owing to a relatively increased density. Their density apparently was not high enough to enable them to break through the density stratification in the deeper water during the "stagnation" stages, in contrast to normal turbidity currents which on three occasions during this time flowed all the way down to the bathyal plain that forms the deepest part of the basin.

From the C-14-age determination of the dark layers it follows that the rate of normal terrigenous sedimentation (*i.e.*, exclusive of turbidites and volcanic ash deposits) was much higher during the early Holocene than later on.

ACKNOWLEDGMENTS

The cores mentioned in this paper were collected thanks to the financial and personal help provided by the following institutions: the Consiglio Nazionale della Ricerche at Rome, the Istituto di Geologia e Paleontologia at Bologna, the National Science Foundation at Washington D.C., the Scripps Institution of Oceanography at La Jolla, the Royal Dutch Shell Exploration and Production Laboratory at Rijswijk, the Dutch Foundation for Pure Research at the Hague, and the State University at Groningen.

REFERENCES

Bottema, S. and L. M. J. U. van Straaten 1966. Malacology and palynology of two cores from the Adriatic Sea floor. *Marine Geology*, 4:553–564.

Brongersma-Sanders, M. 1957. Mass mortality in the sea. *Geological Society of America, Memoir* 67 (1):941–1010.

Brongersma-Sanders, M. 1965. Metals of Kupferschiefer, supplied by normal sea water. *Geologische Rundschau*, 55:365–375.

Brongersma-Sanders, M. 1966. The fertility of the sea and its bearing on the origin of oil. *The Advancement of Science*, 23 (107):41–46.

Eriksson, K. G. 1965. The sediment core No. 210 from the western Mediterranean Sea, *Reports of the Swedish Deep-Sea Expedition 1947–1948*, 8 (7):395–594.

Fabricius, F. H., U. von Rad, R. Hesse, and W. Ott, 1970. Die Oberflächensedimente der Strasse von Otranto (Mittelmeer). *Geologische Rundschau*, 60:164–192.

Hesse, R., U. von Rad and F. H. Fabricius 1971. Holocene sedimentation in the Strait of Otranto between the Adriatic and Ionian seas (Mediterranean). *Marine Geology*, 10:293–355.

Lacombe, H. and P. Tchernia 1960. Quelques traits généraux de l'hydrologie méditerranéenne. *Cahiers Océanographiques*, 12: 527–547.

Moore, D. G. 1969. Reflection profiling studies of the California Continental Borderland: structure and quaternary turbidite basins. *Geological Society of America, Special Paper* 107: 142 p.

Ninkovitch, D. and B. C. Heezen 1965. Santorini Tephra. In: *Submarine Geology and Geophysics*, eds. Whittard, W. F. and R. Bradshaw, Butterworths, London, 413–453.

Norin, E. 1958. The sediments of the Central Tyrrhenian Sea. *Reports of the Swedish Deep-Sea Expedition 1947–1948*, 8 (1): 1–136.

Olausson, E. 1961. Sediment cores from the Mediterranean Sea and the Red Sea. *Reports of the Swedish Deep-Sea Expedition 1947–1948*, 8 (6):337–391.

van der Spoel, S. 1967. *Euthecosomata, a group with remarkable developmental stages (Gastropoda, Pteropada)*. Thesis, Amsterdam, Noorduyn, Gorinchem 249 p.

van Straaten, L. M. J. U. 1966. Micro-malacological investigation of cores from the southeastern Adriatic Sea. *Koninklijke Nederlandse Akademie van Wetenschappen, Amsterdam, Proceedings*, Series B, 69:429–445.

van Straaten, L. M. J. U. 1967. Turbidites, ash layers and shell beds in the bathyal zone of the southeastern Adriatic Sea. *Revue de Géographie Physique et de Géologie Dynamique* 9(2):219–239.

van Straaten, L. M. J. U. 1970. Holocene and late-Pleistocene sedimentation in the Adriatic Sea. *Geologische Rundschau*, 60:106–131.

Todd, R. 1958. Foraminifera from Western Mediterranean deep-sea cores. *Reports of the Swedish Deep-Sea Expedition 1947–1948*, 8 (3): 169–215.

Vergnaud-Grazzini, C. and Y. Herman-Rosenberg 1969. Etude paléoclimatique d'une carotte de Méditerranée Orientale. *Revue de Géographie Physique et de Géologie Dynamique* 11(2):279–292.

Wüst, G. 1961. On the vertical circulation of the Mediterranean Sea. *Journal of Geophysical Research*, 66:3261–3271.

Zore-Armanda, M. 1968. The system of currents in the Adriatic *Acta Adriatica, Institut za Oceanografiju i Ribarstvo, Split*, 10 (3):93 p.

Zore-Armanda, M. 1968. The system of currents in the Adriatic Sea. *Conseil Général des pêches pour la Méditerranée, Etudes et Revues*, 34:48 p.

Zore-Armanda, M. 1969. Temperature relations in the Adriatic Sea. *Acta Adriatica, Institut za Oceanografiju i Ribartsvo, Split*, 13 (5):51 p.

Undisturbed Large-Diameter Cores From the Strait of Otranto

Reinhard Hesse[1] and Ulrich von Rad[2]

Technische Universität, Munich

ABSTRACT

Twenty-four undisturbed cores were recovered from three profiles across the western half of the Strait of Otranto by using a 2 m long square-box corer with 15 cm side and a normal gravity corer. X-radiographs of thin slabs of the cores revealed excellently preserved bedding structures and burrowing features. The morphologic subdivision of the Strait into four major regions: (1) Apulian shelf (2) slope, (3) trough-bottom (deep northern extension of the Corfu-Kephallinia Trough of the northeastern Ionian Sea), and (4) Apulian-Ionian Ridge (southern submarine continuation of the Apulian Peninsula), is reflected within the sediments.

In the relatively coarse sediments of the shelf, penetration of the coring device was usually low. Cores from the upper slope as well as from the Apulian-Ionian Ridge show a high degree of burrowing, including laterally stoping structures, probably caused by echinoids. Cores from the lower strait slope and the Strait bottom contain graded sand layers 2 to 30 cm thick which are possibly continuous over 15 km. These layers display parallel and ripple cross lamination. Evidence of slumping is provided by one core with highly disturbed laminations. Hemipelagic sediments prevail on the Apulian-Ionian Ridge and adjacent deeper parts of the Corfu-Kephallinia Trough. Thinly laminated coccolith layers near the base of two of the cores, which are rich in organic carbon (sapropelic muds), gave C-14 ages between 8210 and 9640 years B.P.

Heavy mineral distributions show areas of different hydraulic behavior on the shelf, and prove the presence of North Adriatic (and probably Alpine) sedimentary components in the Strait of Otranto, indicating a possible multicycle marine transport of 700 km.

RESUME

On a prélevé 24 carottes intactes au long de trois profils transversaux de la partie Ouest du Détroit d'Otrante, à l'aide d'un carottier à section carrée et d'un carottier sans piston. Les radiographies de tranches minces ont révélé des structures sédimentaires et de bioturbation très bien préservées. On trouve dans les sédiments les caractéristiques des quatre zones morphologiques principales qui sont: (1) la plateforme d'Apuli, (2) la pente, (3) le fond du détroit qui est l'extension au Nord, de l'Auge de Corfou et de Céphalonie (N.O. de la mer Ionienne) et (4) la crête Apulo-Ionienne qui est le prolongement sous-marin de la péninsule d'Apulie.

La pénétration du carottier dans les sédiments grossiers de la plateforme est généralement faible. Les carottes prélevées dans la partie supérieure de la pente ainsi que dans la crête Apulo-Ionienne présentent une bioturbation intense ainsi que des structures édifiées par les échinoïdes.

Les carottes de la partie inférieure de la pente et du fond du détroit contiennent des couches de 2 à 30 cm d'épaisseur de sable granoclassé qui s'étendent probablement sur une quinzaine de kilomètres. Ces couches présentent des laminations parallèles et entrecroisées. La lamination extrêmement confuse d'une carotte indique la présence de slumping. Les sédiments hémipélagiques prédominent dans la crête Apulo-Ionienne et dans les régions plus profondes de l'Auge de Corfou et de Céphalonie. La datation au carbone 14 des couches de coccolithes à la base de 2 carottes riches en carbone organique indique un âge compris entre 8210 et 9640 ans B.P.

Les distributions de minéraux lourds marquent, sur la plateforme, des zones de régimes hydrologiques différents et prouvent l'existence de composantes sédimentaires d'origine N. adriatique et, probablement, alpine dans le Détroit d'Otrante. Ceci indique la possibilité d'un cycle multiple de transport marin sur une distance de 700 km.

Present address: [1]Department of Geological Sciences, McGill University, P. O. Box 6070, Montreal 101, Quebec, Canada. [2]Bundesanstalt für Bodenforschung, D–3, Hannover-Buchholz, Germany.

INTRODUCTION

Since the long square-box corer (Kastenlot) was introduced by Kögler (1963), this gravity-type instrument has proved to be very useful in deep-sea coring. Its large diameter (side-dimension: 15 cm) combined with a considerable core length (maximum obtainable core length: 8 m) is of particular use in the study of sedimentary structures in modern sediments. During a recent sedimentological investigation of the Strait of Otranto, the 2 m-version of the corer and a normal gravity corer (6 cm in diameter) were used for the recovery of 24 undisturbed cores from three profiles across the western half of the Strait (Figure 1).

The main objectives of this study were:

1. To reveal the distribution of the surface sediments of the Strait of Otranto, which, being the southernmost part of the otherwise well-studied Adriatic Sea, had not yet been sampled systematically;

2. to elucidate the Holocene and possibly the older sedimentary history of the Strait and

3. to grain some insight into the particular conditions of sedimentation in a strait which acts as the portal for water exchange between two adjacent, connected seas. Narrowed topography creates a bottle neck here which causes acceleration of marine currents in a jetstream-like fashion.

Whereas detailed core-description and interpretation have been published elsewhere (Hesse *et al.*, 1971), this paper emphasizes some primary and secondary structures of the sediments which were excellently revealed by the X-radiography technique. Also supplementing our previous data, new carbon-14 ages are given for various layers of the cores which will help to establish an absolute chronology of the latest sedimentary events in the Strait of Otranto.

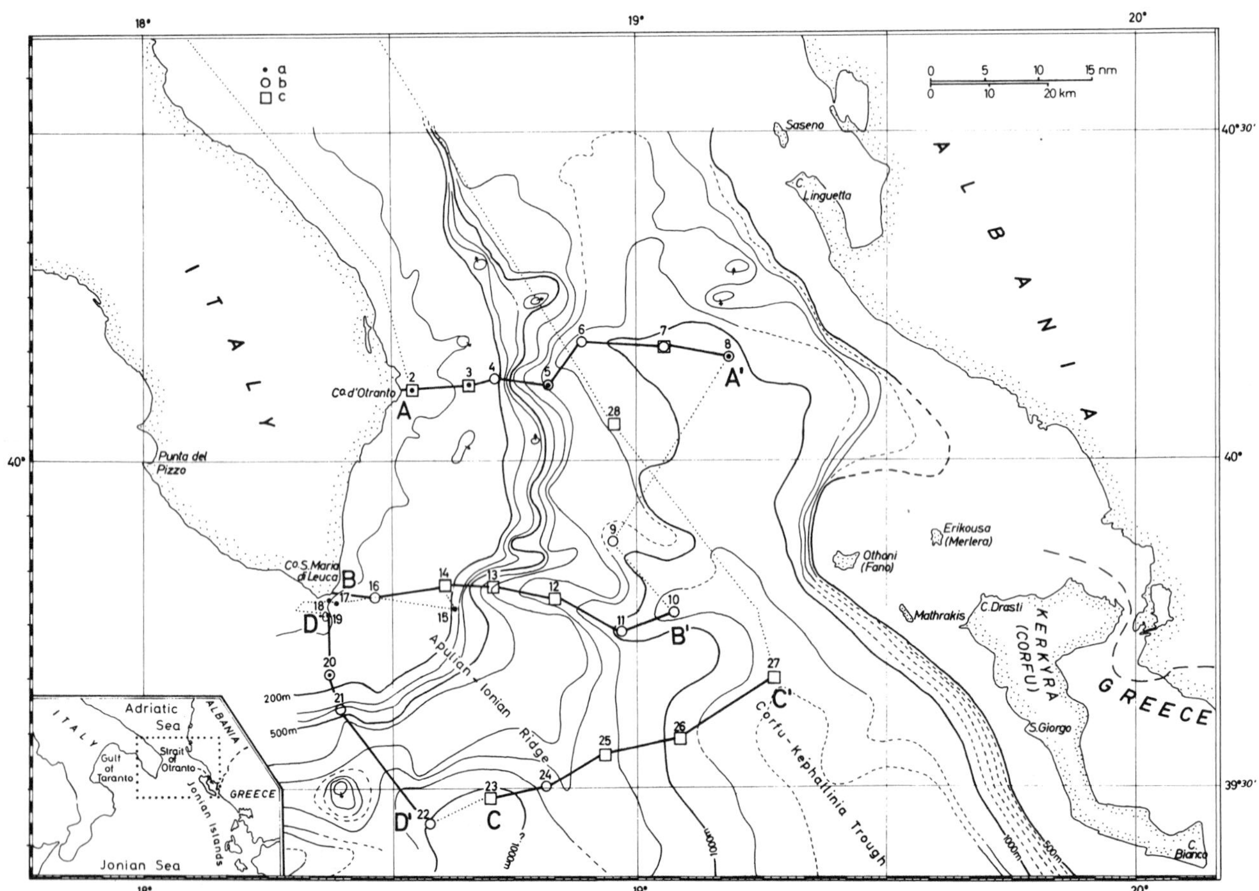

Figure 1. Bathymetry of the Strait of Otranto with ship track and core positions (expedition OTRANTO, July 1968). Depth contours in 100 m intervals, compiled from published sea charts and our own ELAC records. A-A′ to C-C′: profiles shown in Figure 2. Type of sampling device:
a = van Veen grab sampler;
b = gravity corer (6 cm in diameter);
c = square box-corer (15 × 15 cm × 2 m).

SEDIMENT DISTRIBUTION AND MORPHOLOGIC SUBUNITS OF THE STRAIT

The 70 km wide Strait of Otranto connects the Adriatic and Ionian Seas of the central Mediterranean Sea. It is morphologically subdivided into three major regions: shelf, slope and strait-bottom which are repeated more or less symmetrically on both sides of the Strait. Only the western two thirds of the Strait have been sampled, since it was not possible to enter Albanian waters on the eastern side. A few samples were taken from the Apulian-Ionian Ridge which is the southeastern submarine continuation of the Apulian Peninsula on land (Figure 1).

A close relationship exists between these major morphologic subdivisions of the Strait and sediment characteristics. According to texture and sedimentary structures, four different facies types may be distinguished on the western side of the Strait.

The 15 to 25 km wide Apulian Shelf (0 to 170 m deep) is characterized by *sandy sediments* rich in biogenic carbonates. On the inner shelf coarse-grained, very poorly sorted muddy sands (sand plus gravel content over 40%) prevail (see Hesse *et al.*, 1971, their figure 13) reflecting the effect of a strong southward directed surface current (current velocities over 2 and up to 4 knots; Mediterranean Pilot, 1957, p. 11) which prevents a major portion of the fines from settling. On the outer part of the shelf the sand fraction makes up 20 to 40%. The shelf sands may also be composed of a considerable amount of relict Pleistocene material. They contain glauconite and reworked (? Pleistocene) ooids. Oolithic grain coatings often are stained by iron oxides and hydroxides. Shelf samples are further characterized by low mica contents of the heavy mineral fraction (less than 20% phyllosilicates in samples from the inner shelf), (Hesse *et al.*, 1971, their figure 23) and by their specific organic content (Hesse *et al.*, 1971, their figures 24 and 25) which has been shown to be a useful depth indicator. Samples close to Capo Santa Maria di Leuca, which is the southern tip of the Apulian Peninsula, are composed of coarse organic debris apparently deposited in pockets of the rocky bottom. The shelf break occurs between 150 and 200 m water depth.

Proceeding to deeper waters, *fine-grained sediments* (facies type 2) were first encountered at the shelf margin (stations OT 4 and 14 in water depths of 198 and 179 m respectively) beyond which they prevail almost exclusively on the *slope* and trough bottom. Here we find sandy or silty muds with less that 13% sand-sized material. As a rule they show a high degree of burrowing. The slope (170 to 1000 m deep) is most steeply inclined (3 to 6%) on its upper part, maintains only 1 to 3% inclination on its lower part and then gradually merges with the trough bottom (Figure 2).

Facies type 3 was found on the lower portion of the slope: it is a *slump facies.* With respect to texture its presence can be seen from the high percentage of coarse material (more than 70% sand and gravel in the top layers of core OT 5, which clearly deviates in its sedimentary characteristics from all other slope cores). The original position of the sediment now found in 826 m water depth was possibly 600 to 700 m higher on the slope (Fabricius *et al.*, 1970, p. 189) as indicated by its shallow-water fauna. Its coarse grain size, low mica content, the relatively strong compaction and the bluish-gray color (compared to a normal greenish-gray and olive-gray) do not fit the surrounding sediments encountered in nearby stations. The primary laminations of the sediment are strongly contorted. The westward indented topography of the upper slope above station OT 5 further supports the idea of slumping if interpreted as a slump niche (*cf.*, Stanley and Silverberg, 1969). Carbon-14 ages give an almost uniform late-Pleistocene age for three samples taken from various depths of the core (see Table 1).

The fourth type of sediment is a *slope and trough facies* found (a) on the lower slope, (b) in the center

Table 1. Carbon-14 age determinations of samples in the Strait of Otranto area.

Core	Depth below sea-floor (in cm)	C-14 model age* years B.P. (1950)	Laboratory**	Remarks
OT 5 top	0–9	10170 ± 1120	(1)	slump facies
	27–35	10270 ± 1125	(1)	slump facies
base	102–110	10290 ± 1640	(1)	slump facies
OT 25 top	0–10	4530 ± 140	(2)	
	10–30	6885 ± 100	(1)	
	92–112	11210 ± 165	(1)	? contaminated by older material
base	170–185	9640 ± 150	(2)	below stagnation layer
OT 26 top	0–5	5035 ± 155	(1)	
base	174–178	9210 ± 185	(1)	stagnation layer
OT 28 top	0–18	8305 ± 115	(1)	
	65–75	13340 ± 180	(1)	stagnation layer below upper sand layer
	105–115	15815 ± 195	(1)	
base	130–153	21115 ± 160	(1)	

*The model age caculation is based on 85% of the modern C-14 concentration as being the starting point.

**(1) Niedersächsisches Landesamt für Bodenforschung, Hannover, Germany (Dr. M. A. Geyh); (2) Isotopes, Inc., Westwood, N.J.

of the Strait (which is the northern continuation of the Corfu-Kephallinia Trough of the northern Ionian Sea), and (c) on the eastern flank of the Apulian-Ionian Ridge. Cores from these localities contain thin sand layers as well as layers of laminated sapropelic muds. The bulk of the sediment, however, consists of homogeneous muds.

Sedimentary material from the Adriatic Basin passes into the Ionian Sea via longshore currents on the shelf, as can be seen from a regional analysis of heavy minerals which provides a typical augite and aegirine-augite suite derived from the Monte Vulture volcanic complex (southern Apennines in Apulia). Among the heavy mineral concentrates found south

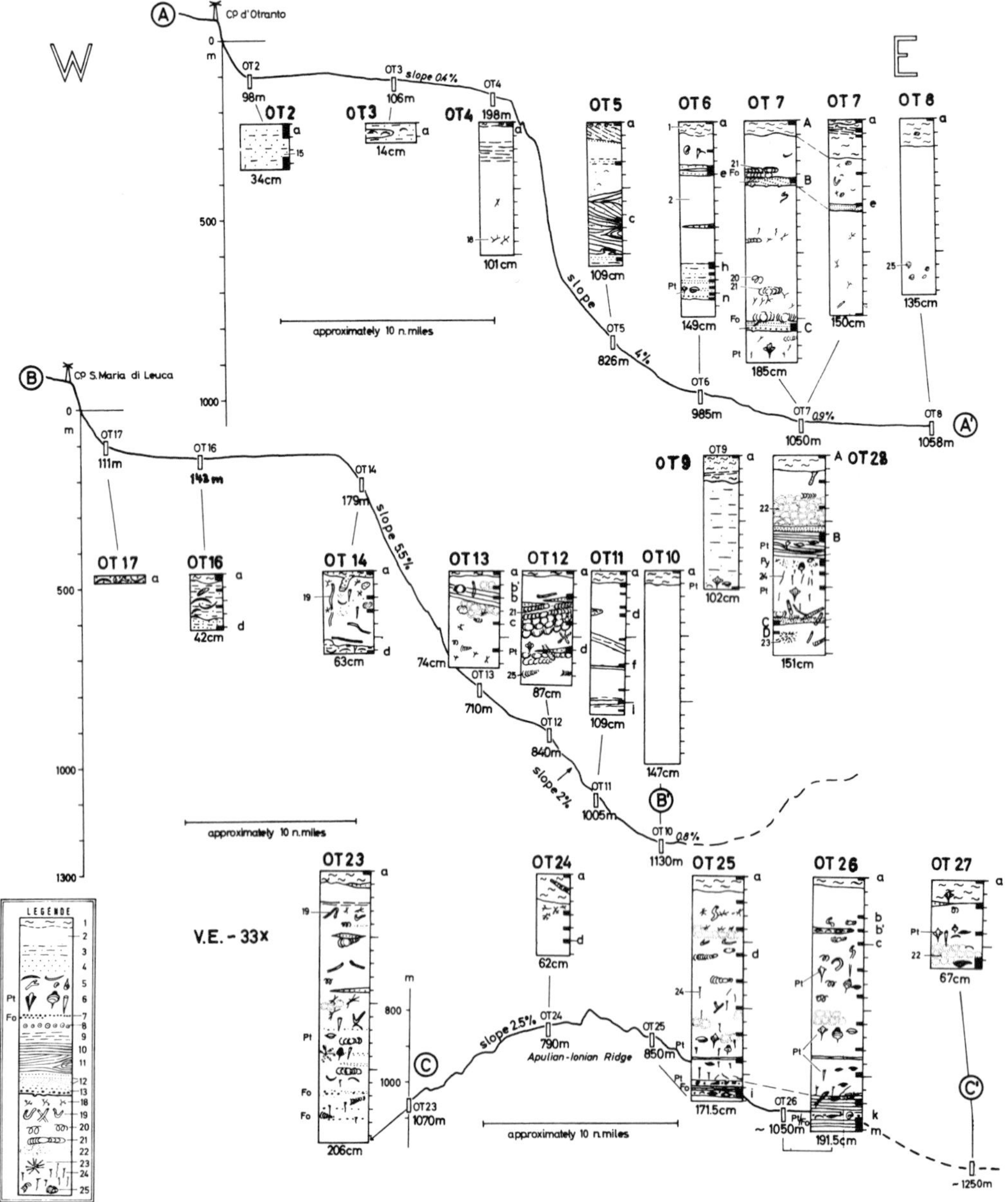

Figure 2. Bathymetric west-east profiles of the western part of the Strait of Otranto and sedimentary structures in the cores. Profiles redrawn after ELAC records. Symbols: 1, brownish oxidized top layer (mud); 2, homogeneous olive-gray mud; 3, muddy sand; 4, sand; 5, shell debris (molluscs, bryozoans, algae); 6, Pteropods (Pt); 7, concentrations of pelagic foraminifera (Fo); 8, ooids; 9 and 10, faint or distinct parallel lamination (in OT 25, 26, 28: varve-like); 11, inclined and contorted (OT 5) lamination; 12, parallel and ripple cross-lamination; 13, graded bedding; 18, tiny dendritic burrows ("fucoids"); 19, tube-shaped (?) worm burrows; 20, spiral-shaped (?) worm burrows; 21, laterally stoping structure (? *Echinocardium*); 22, mottled structure, only shown when distinct; most muds are homogeneously mottled; 23, large dendritic burrows; 24, pyritized filaments (? hyphae of marine fungi or worm tubes); 25, poorly characterized burrow structures; Py, pyrite concentrations (*e.g.*, framboidal spherules) a, b, c, A, B, *etc.*, (blank squares) = position of subsamples. Broad columns = box cores; narrow columns = gravity cores.

Figure 3. Core OT 7 showing two distinct sand layers at 42–48 and 153–160 cm depths below surface. Both layers have a sharp lower contact and are parallel laminated, the upper layer displaying a few low-angle cross-laminae. Original upper contact disturbed by lateral stoping structures. The muddy remainder of the core is completely homogenized by a kind of mottled structure with only a few spiral and other structures traceable as individual burrows. (Print of X-radiograph from 1 cm thick slabs cut from the square-box with 25 × 14 cm wide tops of plastic boxes. Mueller macrotank B X-ray unit, 35 KV, 4 mA, 4 min., distance X-ray tube to sample: 65cm; film: AGFA-GEVAERT Structurix D4. All X-ray photos by U.v.R.).

of the sill, a minor amount (normally less than 1% by grain number) of glaucophane is present, which is probably of Alpine origin. However it may also come from Tertiary flysch sediments of the northern and central Apennines, thus being derived from the Alps secondarily. Thus the marine transport over 700 km, which is implied by the occurrence of glaucophane of Alpine origin on the western side of the Strait of Otranto, may have happened in one or two steps.

SEDIMENTARY STRUCTURES

Three types of sedimentary structures deserve particular mention: burrowing structures; primary structures related to sand-layers; and laminations of the "varved zones".

As shown above, the occurrence of some of the structures is restricted to certain areas of the Strait in particular, and consequently provides another means of correlation between sediment properties and bathymetric subunits of the sea floor.

Burrowing Structures

The degree of burrowing is remarkably high in the Otranto cores. Except for the sand-layers and the laminated stagnation layers of facies type 3, the sediment has been completely reworked by burrowing and mud-ingesting animals, destroying the original structures.

Besides a kind of mottled structure (Figure 3) which is the prevailing effect of bioturbation in the cores, the outstanding feature is a layer-by-layer burrowing of the sediment. This is best seen in Core OT 12 from a water depth of 840 m (Figure 4). In the 87 cm long core, a sequence of more than 40 distinct layers of tracks can be observed. Each is between 1.0 and 2.5 cm thick and is superposed on top of the next one. The internal structure of the individual burrowing track shows concentric cusps. Similar structures have previously been described by Schäfer (1962) and Reineck (1963) from the North Sea. Following these authors we would interpret this type of burrowing, which we have called a *laterally stoping structure*, as tracks of sea-urchins. A few oblique tracks can be seen in a deeper portion of the core (not shown in Figure 4). Some echinoid spines have been isolated from the sediment. The same structure has been found in other cores, however, as isolated tracks only. In particular, the muddy layers following immediately above the two sand layers in Core OT 7 (Figure 3) display this laterally stoping structure as well. It was not observed in the cores taken above 700 m of water depth in the Strait of Otranto, though similar structures are known from shallow water sediments of the North Sea (Reineck, 1963).

Other structured burrows include tubes, 5 to 20 mm wide and up to 20 cm long, running vertically or obliquely through the sediment, in Core OT 28 penetrating the thin lower sand layer. They may tentatively be ascribed to polychaete worms or molluscs. This group also comprises smaller spiral-shaped (?) worm tubes (Figure 3). Small fucoid-like tubes branching dendritically or fan-like with a diameter less than 0.5 and 1 to 3 mm length are schematically

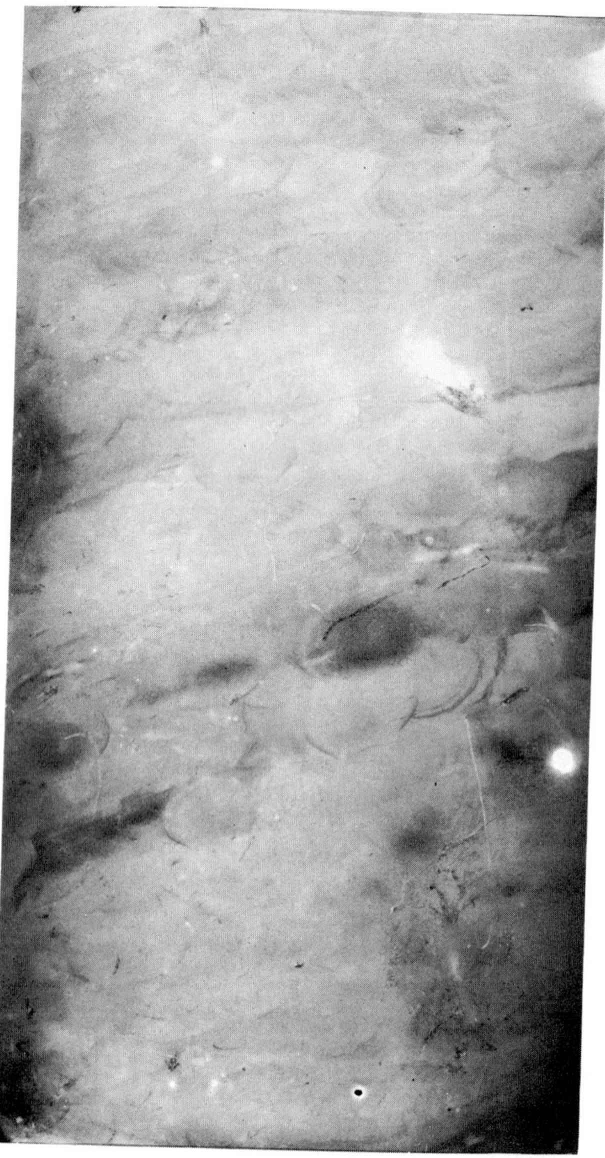

Figure 4. Lateral stoping structure caused by echinoids feeding on the sediment layer-by-layer. Inclination of the burrowed layers due to oblique impact of the corer. Core OT 12, 840 m water depth, 22 to 47 cm below surface. (Print of X-radiograph as in Figure 3; actual width of radiograph 13 cm).

shown on Figure 2 (*cf.*, Hesse *et al.*, 1971, their figure 12) as well as tiny pyritized filaments (possibly hyphae of marine fungi; personal communication, Dr. F. Werner, Kiel, 1969).

Sand Layers

Cores OT 6, 7 and 28 located in a triangular pattern 15 km apart from each other, contain two sand layers which are probably identical in each core. Correlation is suggested by their occurrence at approximately the same depth below surface and by the decrease of grain-size and thickness in an easterly direction. This correlation still has to be confirmed by absolute dating of the underlying or overlying muds. For the mud immediately below the upper sand layer in Core OT 28 a late Pleistocene age of 13340 ± 180 years B.P. was obtained. A turbidity current origin is favored for these sand layers, although other mechanisms of sand transport cannot be excluded (Hesse *et al.*, 1971). The internal structure of the upper layer in Core OT 7 is shown in Figure 3. Following the sharp base, a 5 mm thick lamina is rich in reworked shallow-water foraminifera. It is overlain by a 1 cm thick division of low-angle cross-laminae. The parallel laminated division above contains a light layer (2 mm thick) formed by a concentration of *Orbulina* sp. The upper contact of the 6 cm thick layer is disturbed by the lateral stoping structures mentioned above.

The thick lower sand layer of Core OT 6 has a similar sharp base like the other sand layer and is laminated throughout most of its thickness (Figure 5).

The sand layers are restricted to the northern part of the trough bottom and lower slope. Accordingly, a local lateral source on the near-by shelf is suggested.

Laminations of "Varved Zones"

Thinly laminated layers rich in organic carbon and planktonic organisms which occur in deep-sea cores from the eastern Mediterranean and Red Sea have been taken as evidence for stagnation of marine circulation (Olausson, 1961; Ninkovich and Heezen, 1965; van Straaten, 1966; Vergnaud-Grazzini and Herman-Rosenberg, 1969).

In the Otranto cores the number of dark-light couples per millimeter is two to five. The light layers consist mainly of coccoliths (Fabricius *et al.*, 1970, their figure 3b), an observation which has also been reported recently from similarly laminated muds of the Black Sea, slightly younger in age (Ross *et al.*, 1970).

The content of organic carbon (up to 1.3%) of the "varved zone" is higher by a factor of 2 to 3 compared to the normal hemipelagic mud (Fabricius *et al.*, 1970). The high concentration of planktonic foraminifera

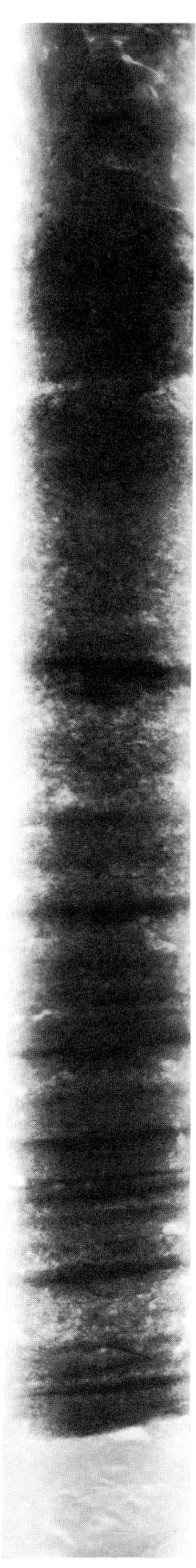

Figure 5. Lower sand layer in Core OT 6, 985 m water depth, 110–135 cm below sediment surface. (X-radiograph of a half-core, 6 cm gravity corer; actual width of radiograph, 4.5 cm).

(*Globigerina* sp., *Orbulina* sp.) and pteropods, including pyritized tests in the laminated zones (Figure 6), is remarkable.

The youngest of these stagnation layers has been dated as 8,330 years B.P. in ALBATROSS Core 194 by Olausson (1961), 7500 years in VEMA Core V10-65 (extrapolated by Ninkovich and Heezen, 1965 to Core V10-67 which correlates well with V10-65, personal communication, Y. Herman, 1970), as older than 7,935 years in VEMA Core V14-122 (Red Sea, Herman, 1968), as 8210 and younger than 9,640 years in our cores OT 26 (174–178 cm) and 25 (170–185 cm), respectively. In Core OT 28, where it occurs between interval 65 and 75 cm, the age is significantly higher (13340 ± 180 years, *s.a.*) thus excluding a correlation with cores OT 25 and 26.

For the same cores the top layers also have been dated. The results were surprising since high "zero-ages" ranging between 5000 and more than 8000 years B.P. (see Table 1) have been found. These high

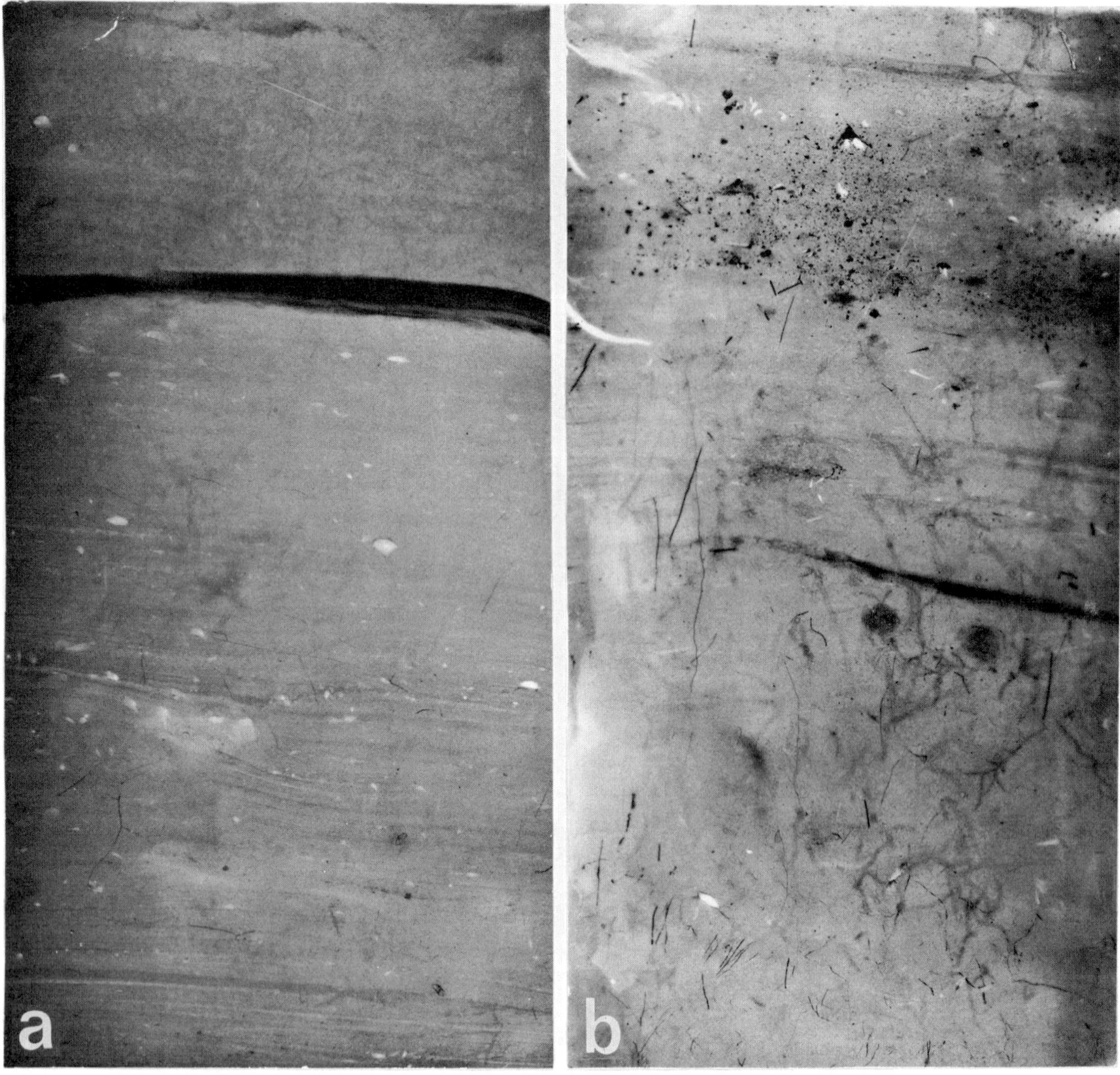

Figure 6. Core OT 28 from the Strait bottom. a, 930 m water depth. Section 50–75 cm below surface. Interval 55–57 cm: weakly ripple-laminated layer of very fine sand (= upper sand layer of Core OT 28). Interval 57–75 cm: parallel laminated mud (varve-like zone) with about 4 laminae per millimeter. Many pteropod tests and pelagic foraminifera. b, Section 75 to 100 cm below surface of the same core. The upper par of this section is thinly laminated like section 57 to 75 cm in a, it contains, however, pyrite concentrations. Lower half with many pyritized filaments, origin uncertain. Another thin silt layer in the middle of the picture (actual width of both radiographs, 13 cm).

zero ages were obtained independently by two different laboratories analysing samples from the same core, thus laboratory error is very unlikely to have caused the high ages of the top layers.

High carbon-14 ages of the surface sediment in the Strait of Otranto may result from two causes: contamination by older organic carbon-bearing material, or erosion or non-deposition. Since at the present stage of the investigation it is difficult to evaluate the relative influence of each of these factors, no sedimentation rates will be presented here. Bottom photographs and bottom-current measurements might in the future help us to arrive at a decision between the two possibilities.

The regional distribution of the stagnation layers also presents some problems. They were not observed above 850 m water depth. Not all of the cores, however, from waters deeper than 850 m, which have sufficient length, penetrated one of these layers. For instance, they are missing in the cores of the northernmost profile (A-A′ in Figure 2, except for Core OT 28 which is located somewhat south of this profile), and also on the west flank of the Apulian-Ionian Ridge. Does this indicate that the water exchange through the Strait itself was not restricted during late Pleistocene and early Holocene time and that an outgoing current from the Adriatic, which was deflected toward the west around Capo Santa Maria di Leuca in the deep phase of circulation, was still active at that time? With further core samples it might be possible to delimit more exactly the upper boundary of the Early Holocene stagnant waters. For further discussion of stagnation layers in the Adriatic Sea, see van Straaten (this volume).

CONCLUSIONS

Regional distribution of sediments in the Strait of Otranto does not show any major deviation from the normal pattern across the shelf and continental slope. Strong surface currents of outflowing Adriatic water are merely reflected in the sediments of the inner shelf. Deeper portions of the Strait (lower slope and trough bottom) received two distinct sand layers, possibly turbidites, in late Pleistocene time. Stagnation layers of Early Holocene age found in the cores are characteristic for the eastern Mediterranean in general. In the Strait of Otranto a late Pleistocene age also seems possible for one of the layers. The 750 m deep Strait of Otranto apparently has not acted as a barrier either for water or for sediment exchange between the adjacent Adriatic and Ionian seas. The exchange seems to have continued even during the time when deep waters were stagnant.

ACKNOWLEDGMENTS

Financial support made available by a grant of the Deutsche Forschungsgemeinschaft, Bad Godesberg to Drs. P. Schmidt-Thomé and F. Fabricius, Technical University of Munich, Germany (project No. Fa 46/10) is gratefully acknowledged. The C-14 determinations were carried out by the C-14/H-3 laboratory of the Niedersächsisches Landesamt für Bodenforschung, Hannover. We are very much indebted to Dr. M. A. Geyh for providing us with these data.

REFERENCES

Fabricius, F., U. von Rad, R. Hesse, and W. Ott 1970. Die Oberflächensedimente der Strasse von Otranto (Mittelmeer). *Geologische Rundschau*, 60:164–192.

Herman, Y. 1968. Evidence of climatic changes in Red Sea cores. *Proceedings of the 7th Congress of the International Association for Quaternary Research*, University of Utah Press, 8:325–348.

Hesse, R., U. von Rad, and F. Fabricius 1971. Holocene Sedimentation in the Strait of Otranto between the Adriatic and Ionian seas (Mediterranean). *Marine Geology*, 10:293–355.

Kögler, F. C. 1963. Das Kastenlot. *Meyniana*, 13:1–7.

Mediterranean Pilot, 1957. 8th Edition, Hydrographic Department, Admiralty, London 3:672 p. Supplement 6 (1967):119 p. Hydrographer of the Navy, London.

Müller, G. and R. Blaschke, 1969. Zur Entstehung des Tiefsee-Kalkschlammes im Schwarzen Meer. *Naturwissenschaften*, 56: 561–562.

Ninkovich, D. and B. C. Heezen, 1965. Santorini Tephra. In: *Submarine Geology and Geophysics*, eds. Whittard, W. F. and R. Bradshaw, Butterworths, London, 413–453.

Olausson, E. 1961. Description of sediment cores from the Mediterranean and Red Sea. *Reports of the Swedish Deep-Sea Expedition 1947–1948*, 8, (3) and (4):287–389.

Reineck, H. E. 1963. Sedimentgefüge im Bereich der südlichen Nordsee. *Abhandlungen Senckenbergische Naturforschende Gesellschaft*, 505:1–138.

Reineck, H. E. 1968. Lebensspuren von Herzigeln. *Senckenbergiana Lethea*, 49:311–319.

Ross, D. A., E. T. Degens, and J. Macilvaine 1970. Black Sea: Recent sedimentary history. *Science*, 1970:163–165.

Schäfer, W. 1962. *Aktuo-Paläontologie nach Studien in der Nordsee*. Verlag Waldemar Kramer, Frankfurt/M, 666 p.

Stanley, D. J. and N. Silverberg 1969. Recent slumping on the continental slope off Sable Island Bank, Southeast Canada. *Earth and Planetary Science Letters*, 6:123–133.

van Straaten, L. M. J. U. 1966. Micro-malacological investigations of cores from the southeastern Adriatic Sea. *Koninklijke Nederlandse Akademie Wetenschappen, Proceedings*, Series B, 69 (3): 429–445.

van Straaten, L. M. J. U. 1967. Turbidites, ash layers and shell beds in the bathyal zone of the southeastern Adriatic Sea. *Revue de Géographie Physique et Géologie Dynamique*, Séries 2, 9:219–240.

van Straaten, L. M. J. U. 1972. Holocene stages of oxygen depletion in the deep waters of the Adriatic Sea. In: *The Mediterranean Sea: A Natural Sedimentation Laboratory*, ed. Stanley, D. J., Dowden, Hutchinson and Ross, Inc., Stroudsburg, Pennsylvania, 631–643.

Vergnaud-Grazzini, C. and Y. Herman-Rosenberg 1969. Etude paléoclimatique d'une carotte de Méditerranée Orientale. *Revue de Géographie Physique et Géologie Dynamique*, Séries 2, 9: 279–292.

Submarine Canyons on the Central Continental Shelf of Lebanon

Thomas R. Goedicke

American University of Beirut, Beirut

ABSTRACT

The most prominent topographic features of the continental shelf off central Lebanon are a number of submarine canyons and sea valleys and a submarine promontory. During investigations conducted between 1968 and 1970, seven undersea valleys were surveyed using a small fishfinder type of recording fathometer. The topography of three of the submarine canyons located in the central part of the continental shelf of Lebanon has been investigated in some detail.

Four of the sea valleys investigated in the course of the present survey are located off the mouth of a river valley. Two canyons head seaward from the center of major bays, and one canyon is located offshore from the rocky promontory of Beirut. The canyon located off the Beirut Promontory appears, on the basis of preliminary evidence, to be a southern branch of the St. George's Bay Canyon. The Beirut Canyon appears to have the steepest walls of the three canyons, as well as being the most narrow and is the only one which has no apparent connection with either a river valley or a bay. It has no tributary valleys, unlike the St. George's Bay and Junieh canyons. The submarine promontory is situated just south of the Beirut headland, and projects to the west approximately 16 miles from the shore. This feature is believed to be a structural horst genetically connected to an east-west fault pattern observed southeast of Beirut.

Both St. George's and Junieh Bays show a predominance of fine silt and clay in the northern portion, grading to silty sand and sand in the southern part. In Junieh Bay, the sand fraction increases from 10 to 20% in the north to 80 to 95% in the south. Samples taken near the axes of the canyons also show mainly fine silt to clay. This sediment distribution in both bays is presumably due to similar wave and longshore current patterns.

RESUME

Canyons sous-marins, vallées marines et promontoires sous-marins sont les traits topographiques prédominants de la plateforme continentale du Liban central. Les profils de coupe de sept vallées sous-marines ont été étudiées entre 1968 et fin 1970 à l'aide d'un petit enregistreur de profondeur. On a étudié en détails la topographie de 3 canyons.

Quatre des vallées marines étudiées sont situées à l'embouchure d'une vallée fluviale. Deux têtes de canyons au centre des baies principales sont orientées vers la mer, et un canyon se situe au large du promontoire de Beyrouth; ce dernier semble, d'après des études préliminaires, être un prolongement au Sud du canyon de St. Georges. Le canyon de Beyrouth semble être celui des 3 canyons à avoir les murs les plus abrupts et être le plus étroit. C'est le seul à n'avoir aucun lien avec une vallée fluviale ou une baie; contrairement aux canyons de St Georges et de Junieh il ne semble avoir aucune vallée affluente. Le promontoire sous-marin se situe au Sud du Cap de Beyrouth et s'étend à l'Ouest jusqu'à 16 miles du rivage; c'est un horst structural de même origine qu'une faille orientée Est-Ouest au Sud-Est de Beyrouth.

Les baies de St Georges et de Junieh indiquent une distribution prédominante de silt fin à argile dans la partie Nord, changeant graduellement en sable silteux et sable dans la partie Sud. Dans la baie de Junieh, la fraction sableuse augmente de 10% à 20% au Nord et de 80% à 95% au Sud. Les échantillons prélevés le long des axes présentent également une distribution silt fin à argile. La distribution sédimentaire de ces deux baies est problement due à des types de vagues et de courants similaires.

INTRODUCTION

The most prominent topographic features on the margin of the continental shelf off Lebanon, as shown in Figure 1, are a series of submarine valleys or canyons and a prominent westward extension of the shelf edge in an area immediately south of Beirut. One of the submarine canyons, the Beirut Canyon, was described by Emery *et al.* (1966). The prominent submarine promontory south of Beirut also has been described (Pfannenstiel, 1960).

Bathymetric investigations by the author, however, have revealed the presence of a number of additional submarine canyons on the continental shelf margin. The heads of three of these have been investigated in some detail.

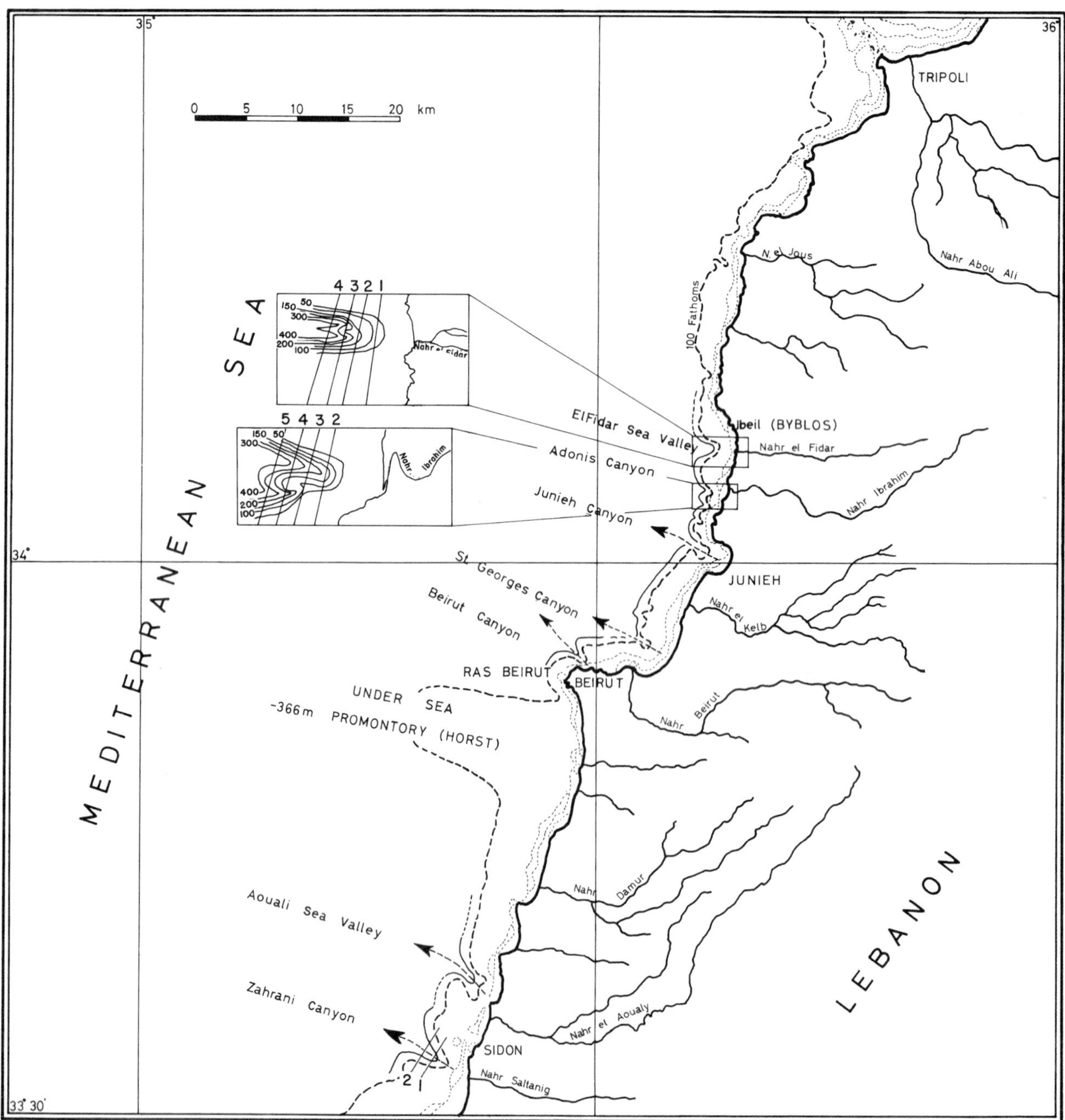

Figure 1. The coast and continental shelf off Lebanon between Sidon and Tripoli, showing location of the submarine canyons in relation to bays and river valleys.

Preliminary investigations into the distribution of sediments around the heads of two canyons have been carried out, and some short core samples have been taken near the axis of one canyon.

TOPOGRAPHY

The continental shelf off Lebanon shown in Figure 1 can be roughly divided into three sections based on its relative width. According to Renouard (1955), this can be correlated with the relative width of the adjacent coastal plain.

The widest part of the continental shelf is located in an area extending from Enfe, between Batroun and Tripoli, to the Syrian border and beyond. In this area the shelf attains a width of 18 km, and the adjacent coastal plain is over 20 km wide. Here, the coast has long expanses of sandy beaches, and few rocky headlands.

Southward from Enfe, and in the region extending to headland of Ras Beirut, the shelf narrows to 3 km or less and the coastal plain is of negligible width. The coast is rocky, with cliffs, very few bays and almost no sandy beaches.

South of Ras Beirut, towards Sidon and Tyre, the continental shelf attains an average width of 7 km and the coastal plain widens to 5 to 7 km. Here, as north of Enfe, one finds large bays, and sandy beaches separated by a few rocky promontories.

The average depth of the flat part of the continental shelf is between 20 and 40 meters. The depth at the shelf break averages between 80 and 100 m and appears to become slightly deeper from north to south.

Submarine Valleys

Prominent indentations in the edge of the continental shelf are noted at a number of points, especially immediately south and north of Sidon, off the Ras Beirut headland, in St. George's Bay, in Junieh Bay, between Junieh and Byblos, near Batroun and north of Enfe (Figure 1). It is suspected that most of these indentations are related to submarine valleys. However, the density of soundings on the shelf and the shelf margin is still insufficient to verify this hypothesis except in certain areas, some of which have been investigated in the course of the present study.

Beirut Canyon

The Beirut Canyon, situated directly offshore from the headland of Ras Beirut, is the southernmost of three canyons surveyed in some detail in the course of the present study. Bathymetry was determined by means of a small launch and a 'Furuno Fishfinder' recording fathometer having a total depth range of 480 meters. Positions of the profiles were determined by sextant angles on prominent shore features. Several aqualung dives have been made at the head of the canyon.

The Beirut Canyon was first described by Emery *et al.* (1966), on the basis of two crossings by the survey ship ARAGONESE, as being narrow and having steep, probably rocky walls, in contradistinction to the the Gaza and Akziv canyons which are wider.

Locations of fathometer profiles for the Beirut and St. George's Bay canyons are shown on Figure 2, and transverse profiles are shown on Figures 3 and 4. Bathymetry is shown on Figure 5.

The head of this canyon is situated at a distance of less than 500 m from a small bay in the rocky north shore of the Ras Beirut headland. Although the area is built over, preventing detailed examination of onshore geology, the existence of the bay and of a gully in the projected landward extension of the canyon, suggests the possible presence of a zone of weakness. A shear zone near Ras Maamelteine, Junieh Bay, also forms a bay and is aligned with a prominent tributary valley to the Junieh Bay Canyon (Figures 6 and 7).

Divers have reported evidence of overhanging walls, and slumping of unstable sediments near the head of the Beirut Canyon.

The transverse profiles of the canyon illustrated on Figure 3 show steep walls and the V-shaped cross section typical of other submarine canyons (Shepard and Dill, 1966). Because of the steep walls, no reliable fathometer data could be obtained with the present equipment beyond a depth of 290 meters.

The gradient of the axis of the canyon in the part surveyed in the present study is 138 m/km, comparable to that of a number of land valleys immediately adjacent to the coast of Lebanon.

Unlike most of the other canyons on the coast of Lebanon, the Beirut Canyon is neither in the center of a bay, nor in the vicinity of a river valley, with which it might have a genetic connection. The possible connection between the Beirut Canyon and the median course of the Nahr (River) Beirut can be postulated if it is assumed that the Nahr Beirut flowed across the present location of the horst of Ras Beirut before it was uplifted in the early Tertiary (Dubertret, 1945, 1951).

Evidence from a series of fathograms surveyed by the Wood's Hole Oceanographic Institution ship CHAIN in 1964 (Wood's Hole Oceanographic Institution, 1964, recently released to the author by Dr. C. George) indicates that the lower course of the Beirut Canyon joins that of the St. George's Bay Canyon below a depth of 1,000 meters.

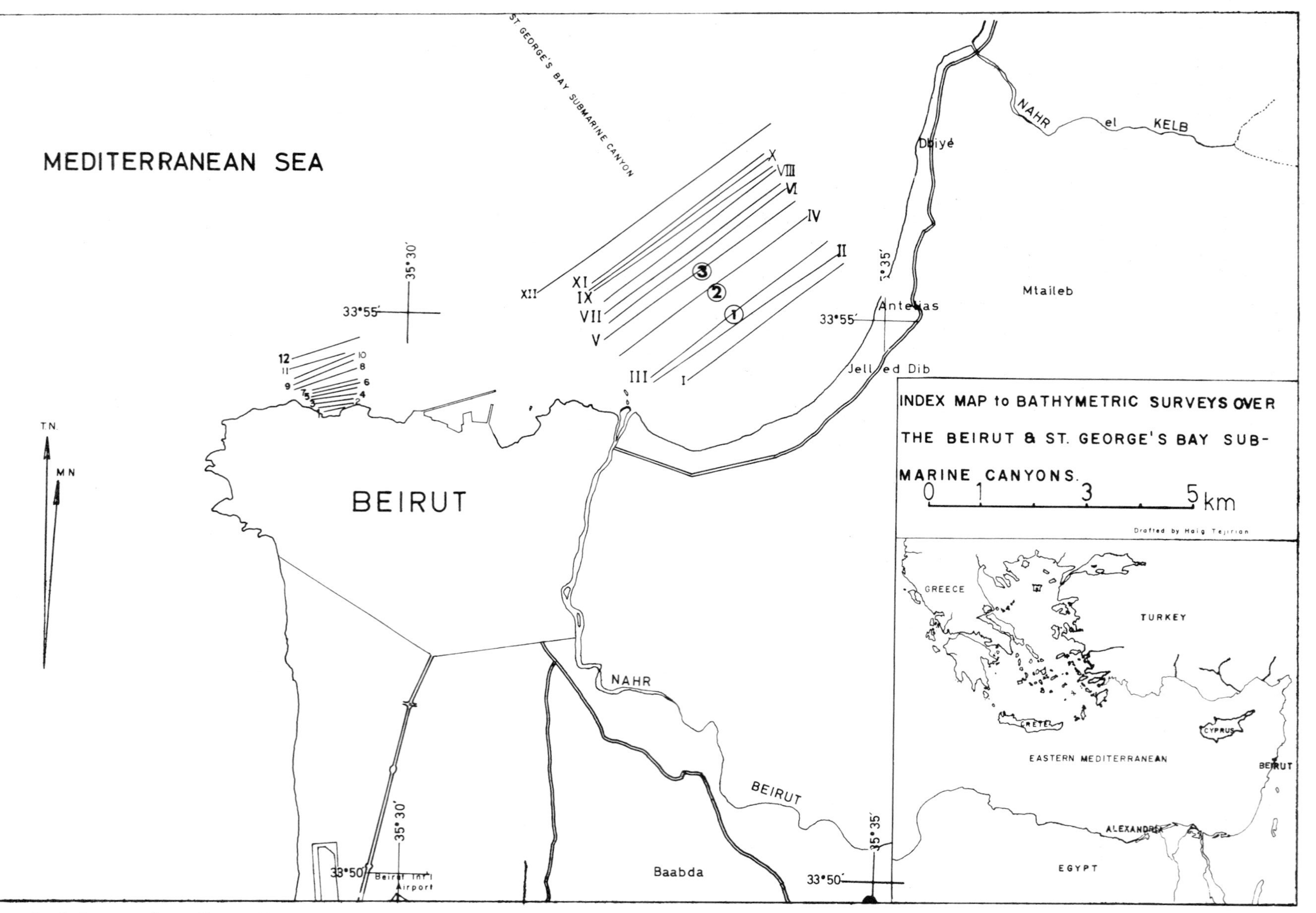

Figure 2. Index map of sounding profiles across the Beirut and St. George's Bay canyons. Numbers in circles indicate location of core samples. Inset shows location of region studied in relation to the eastern Mediterranean.

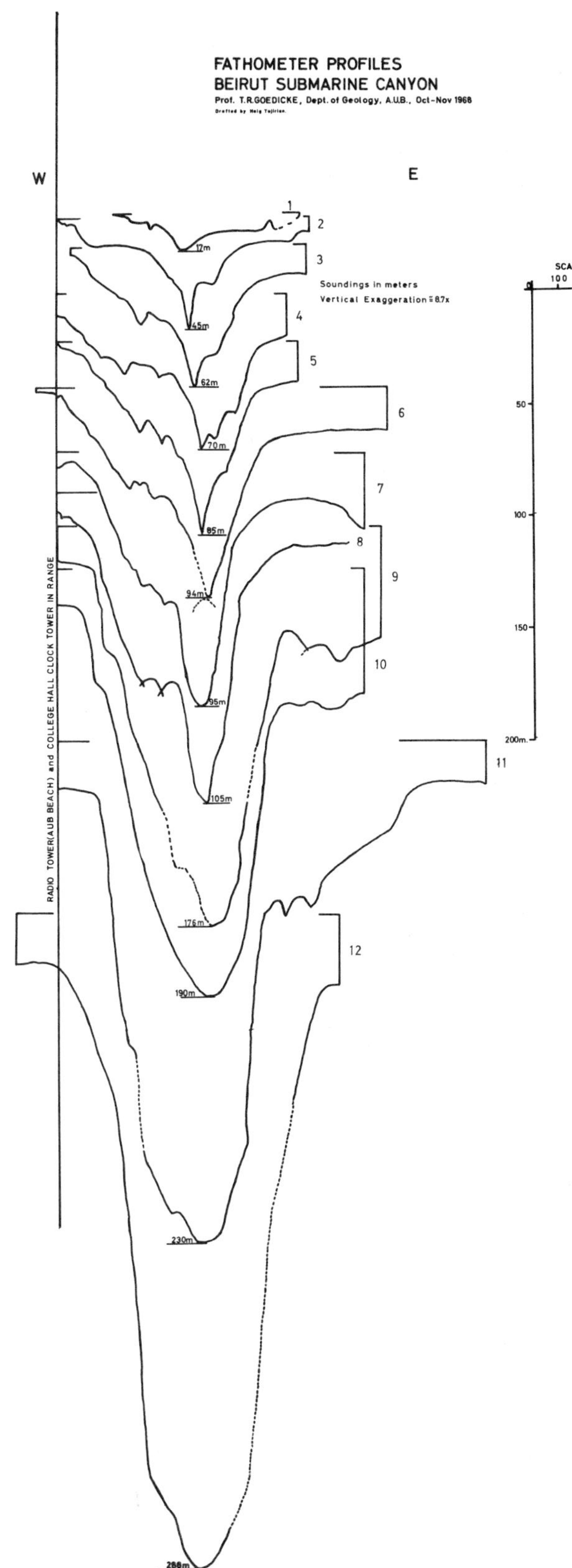

Figure 3. Transverse profiles across the head of Beirut Canyon (after Goedicke, 1969; Tejirian, 1970).

St. George's Bay Canyon

The St. George's Bay submarine canyon, whose axis is oriented in a northwesterly direction, is situated approximately in the center of St. George's Bay north of Beirut (Figures 2, 4 and 5). It is by far the widest of the three canyons explored in the present study. The axis has been followed to a depth of 482 meters, at a distance of 4.3 km from shore. The average gradient of the axis of the canyon is 154 m/km, steeper than either the Beirut Canyon or the Junieh Bay Canyon to the north.

Two tributary valleys enter the St. George's Bay Canyon from the south. These are oriented in a northerly direction parallel to the lower course of the Nahr Beirut, and in its projected extension. The axis of the main part of the canyon trends to the northwest, approximately in line with the Nahr Antilias in the center of St. George's Bay, at a distance of 2 km from the head of the canyon (Figure 5).

It is possible that the valley of the Nahr Antilias is continuous with the St. George's Bay Canyon beneath the cover of recent sediments in the near shore portions of the bay. Several of the submarine valleys on the eastern Mediterranean coast can be traced to adjacent existing or former river valleys, *e.g.*, the Gaza Canyon can be traced to a pre-Neogene erosion channel cutting through the coastal plain (Emery *et al.*, 1966; Neev, 1960; Emery and Neev, 1960).

Landward from the head of the canyon there is a flat, bowl shaped depression at depths of from 20 to 40 meters. This is similar to the depression at the head of La Jolla Canyon (Shepard and Dill, 1966).

A rocky ridge is located adjacent to the south wall of the St. George's Bay Canyon and subparallel to its axis. This ridge is oriented in a north-northwesterly direction, roughly in alignment with two small islands located off the mouth of the Nahr Beirut. No core samples could be obtained because of the hard bottom in this area. However, fragments of coralline algae have been found in dredge samples from this ridge at a depth of 140 to 150 m.

Junieh Bay Canyon

The Junieh Bay Canyon, shown in Figure 7, is situated in the center of Junieh Bay 11 km north of St. George's Bay. The axis of the canyon is oriented almost east-west, in line with a prominent land valley occupied by an intermittent stream. The canyon has been followed to a depth of 365 m and has a gradient of 136 m/km.

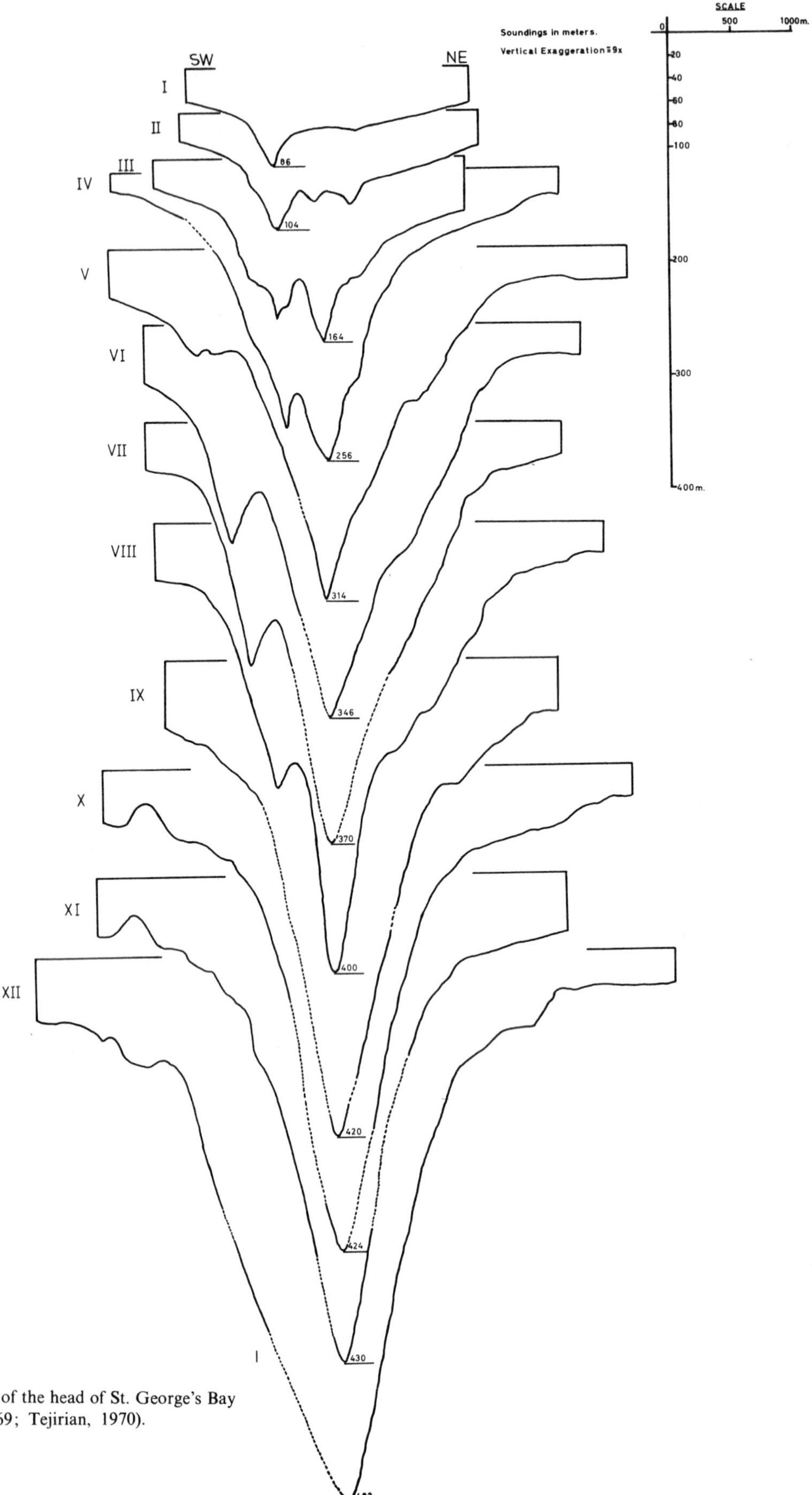

Figure 4. Transverse profiles of the head of St. George's Bay Canyon (after Goedicke, 1969; Tejirian, 1970).

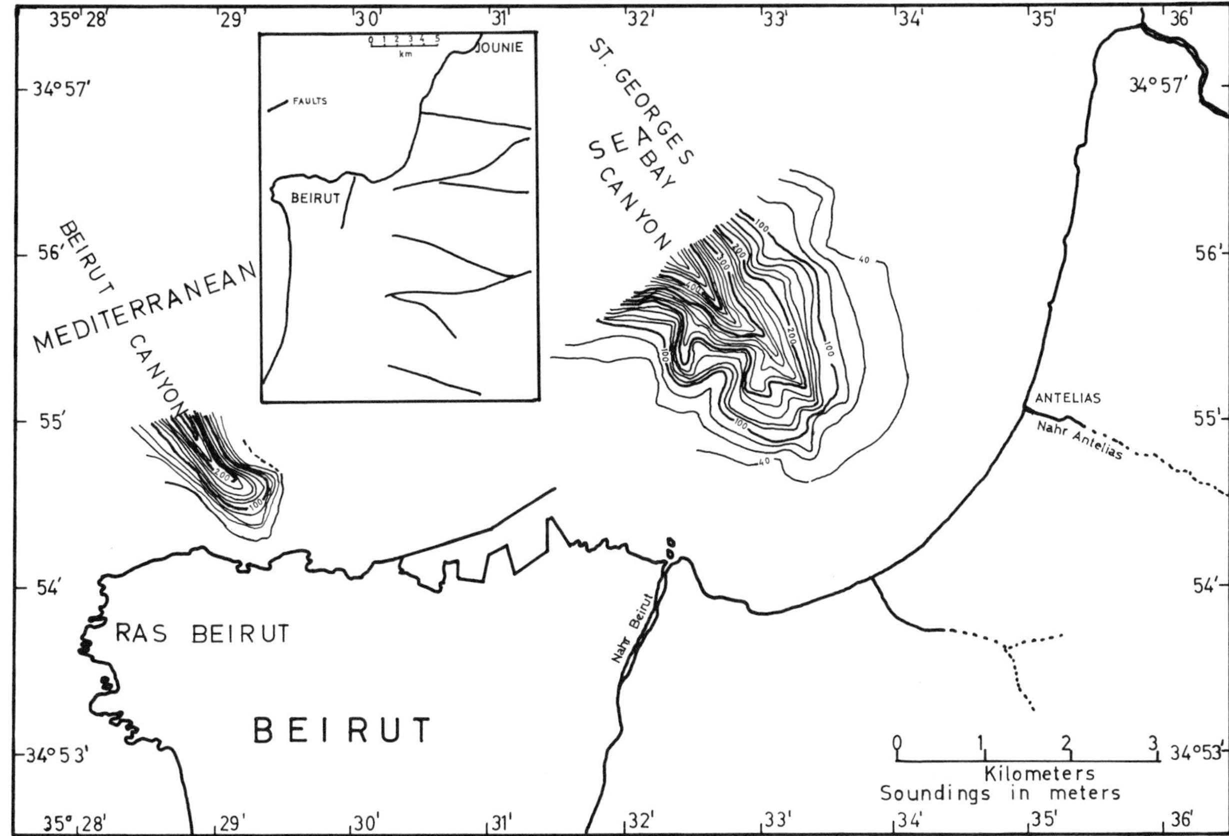

Figure 5. Generalized bathymetry of the submarine canyon heads located in St. George's Bay and off Ras Beirut. The inset geologic sketch map of Beirut and vicinity shows the major east-west directed faults which are believed to have had a major role in controlling the submarine topography of the underwater promontory south of Ras Beirut headland. Faulting may also be in part responsible for the orientation of the canyon offshore from Ras Beirut. (Chart by Tejirian, after Goedicke, 1969).

The maximum width of the canyon is about 2 km and its length to the maximum determined depth of 365 m is 2.5 km. The wall height reaches 330 meters and the transverse profiles shown on Figure 6 are predominantly V-shaped. The walls are steeper on the north side, especially near the area where two tributary valleys join the main axis.

The head of the canyon, at a distance of 300 m from the shore at the mouth of the valley, forms a depression in the north central part of Junieh Bay. Three small head branches join to form the main channel at a depth of 110 m. A fourth branch having a north-south orientation joins the axis 150 to 200 m further west, down-canyon. The junction of the tributary branches represents a local decrease in the gradient of the canyon from an average of 136 m/km to one of 57 m/km. Below this there is a slight local increase in the gradient of the axis.

This small, relatively flat, area near the junction of four tributary valleys can be interpreted as due to local accumulation of sediments supplied by the tributaries. A similar local area of low gradient is found along the axis of Scripps Canyon, off the western coast of the United States, at the junction of Sumner and North Branches (Shepard and Dill, 1966, p. 39). The gradient of Scripps Canyon, like that of the Junieh Bay Canyon, steepens below the point of junction.

Two tributary valleys are located in the north wall of the canyon about 2 km from the head. These valleys join the main axis at depths of 210 and 250 m respectively. They appear to be aligned with a prominent fault pattern observed onshore in the northern part of Junieh Bay (Figure 7).

The faults in this area are of the strikeslip, sinistral type, as shown by the presence of horizontal slickensides on small preserved sections of the fault planes, in addition to stratigraphic evidence (Tejirian, 1968 and 1970). The movement is presumed to be of post-Middle Miocene age, and to have played an important role in the creation of the present outline of the coast, especially in the formation of the headland of Ras-el-Maameltein.

The area where the fault zone reaches the coast is marked by a small bay, 80 to 100 m wide and indented into the coast over 250 m. A steep submarine scarp is in line with the north side of this bay and the

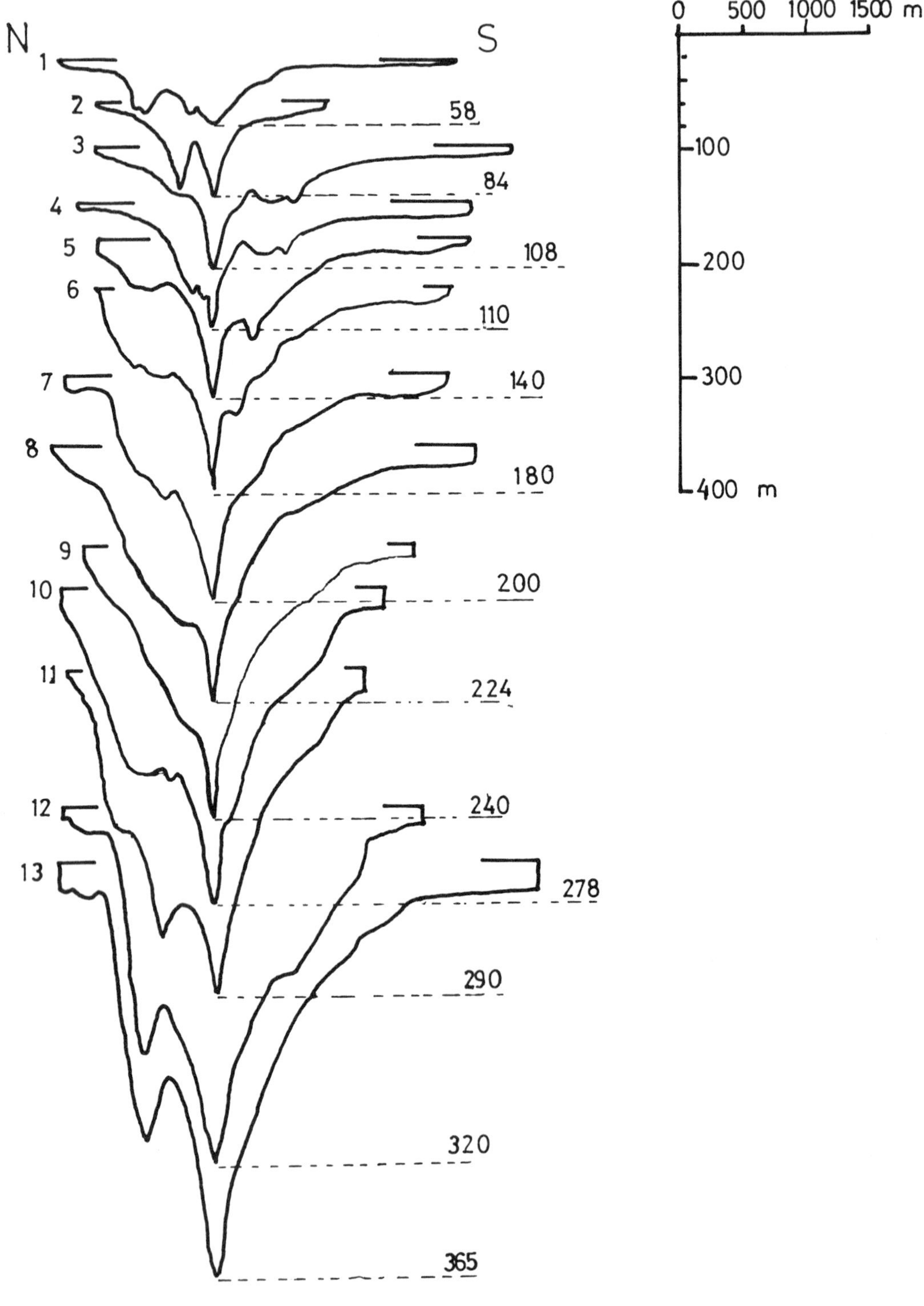

Figure 6. Transverse profiles of the head of Junieh Bay Canyon (after Goedicke 1969, Tejiran 1970).

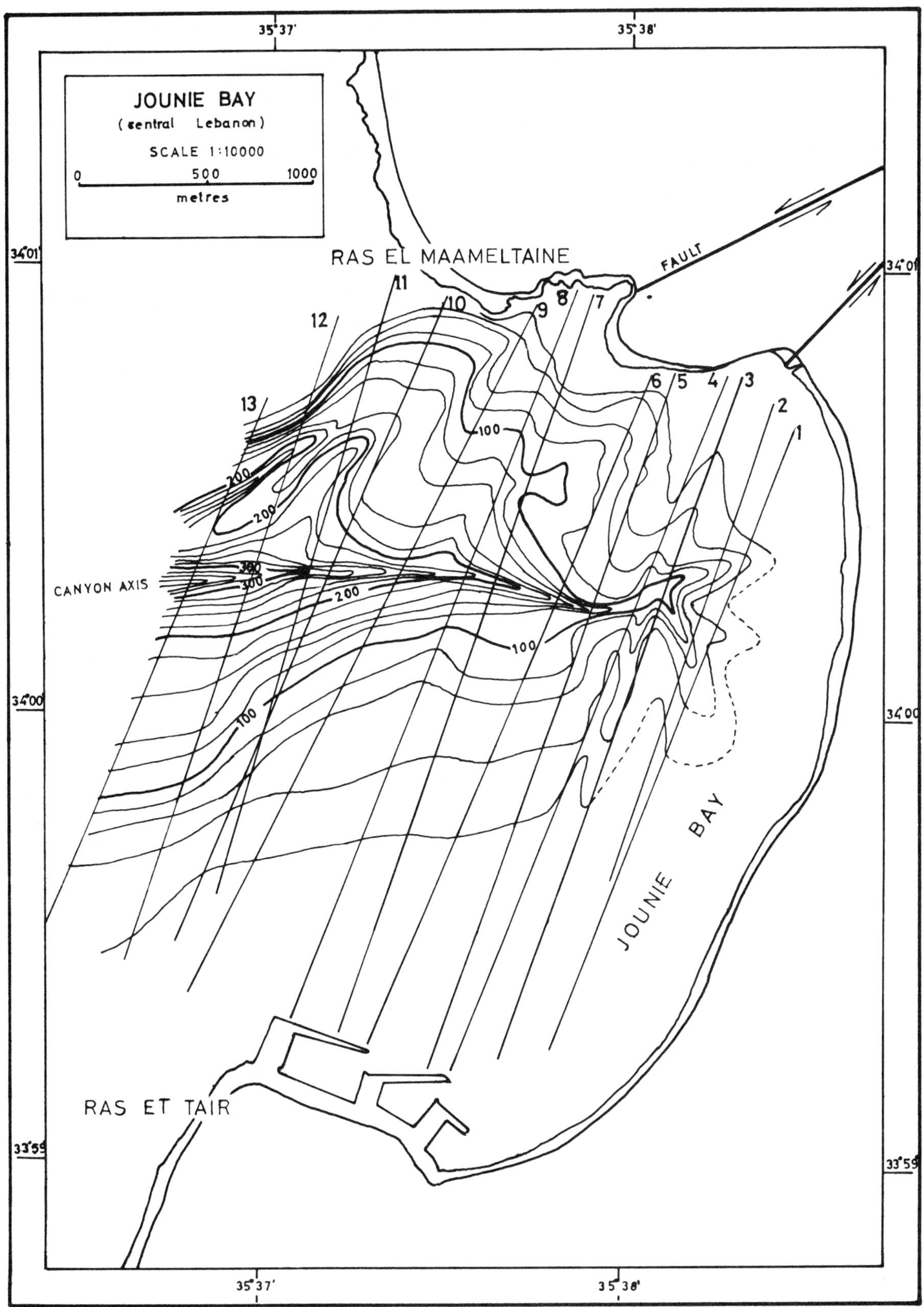

Figure 7. Submarine topography of Junieh Bay. Location of sounding profiles shown in Figure 6 is indicated. Note the alignment of the northern tributary valley of the main canyon with the shear fault near Ras el Maameltein. (Bathymetric survey made by Dening, Goedicke and Tejirian, chart after Tejirian, 1970).

fault, and at greater depth forms the north wall of the major tributary valley, whose head is at a distance of 1,100 m from the bay. The fault, the scarp and the valley have a common west-south westerly orientation, suggesting a tectonic origin for the undersea topography in this area.

The gradient of six land valleys in the immediate vicinity of Junieh Bay, measured over a distance of 2 km, averages out at 130 m/km (Tejirian, 1970). This value, when compared with the gradient of 136 m/km for the Junieh Bay Canyon, suggests, together with other evidence, the possibility that at least the upper part of the canyon was formed by subaerial erosion at a time of lower sea level. Comparable gradients of 138 m/km and 154 m/km for the Beirut and St. George's Bay canyons are significant.

Adonis Canyon and El Fidar Sea Valley

Preliminary bathymetric measurements in the area between Junieh and Byblos have revealed the presence of three submarine valleys, shown in Figure 8. Two of these, the Adonis Canyon and the El Fidar Sea Valley are in direct alignment with the lower courses of the Nahr Ibrahim (Adonis river) and the Nahr el Fidar. The third valley, observed only on two profiles, is located offshore from the rocky headland of Ras el Maameltein, which forms the north side of Junieh Bay.

The axis of the Adonis Canyon, which has been followed to a depth of 380 m, trends in an east-west direction, in line with the lower course of the Nahr Ibrahim. The head of the canyon, situated in a wide, hummocky depression at a depth of 150 to 160 m, is offset to the south from the present mouth of the Nahr Ibrahim. There is little doubt, however, that both the Adonis Canyon and the El Fidar Sea Valley have a genetic connection with their respective river valleys.

The El Fidar Sea Valley heads in a wide depression at a depth of 86 m, directly off the mouth of the Nahr el Fidar. It is oriented east-west and divides into two canyons below 160 m. The deeper of the two canyons has been followed to a depth of 480 m. At this point the south wall has a height of 420 m.

The El Fidar Sea Valley is in exact alignment with the Nahr el Fidar, which has a particularly straight course and flows in a steep walled canyon. This strongly suggests a common origin, as postulated by Shepard and Dill (1966) for canyons off Corsica and Crete.

Unlike the Beirut, St. George's Bay and Junieh Bay canyons, which show the typical V-shaped canyon profile at depths between 40 and 80 m, both the Adonis Canyon and the El Fidar Sea Valley are developed only at depths greater than 160 to 180 meters. Above this depth, there are only wide depressions having a marked hummocky topography. This may be

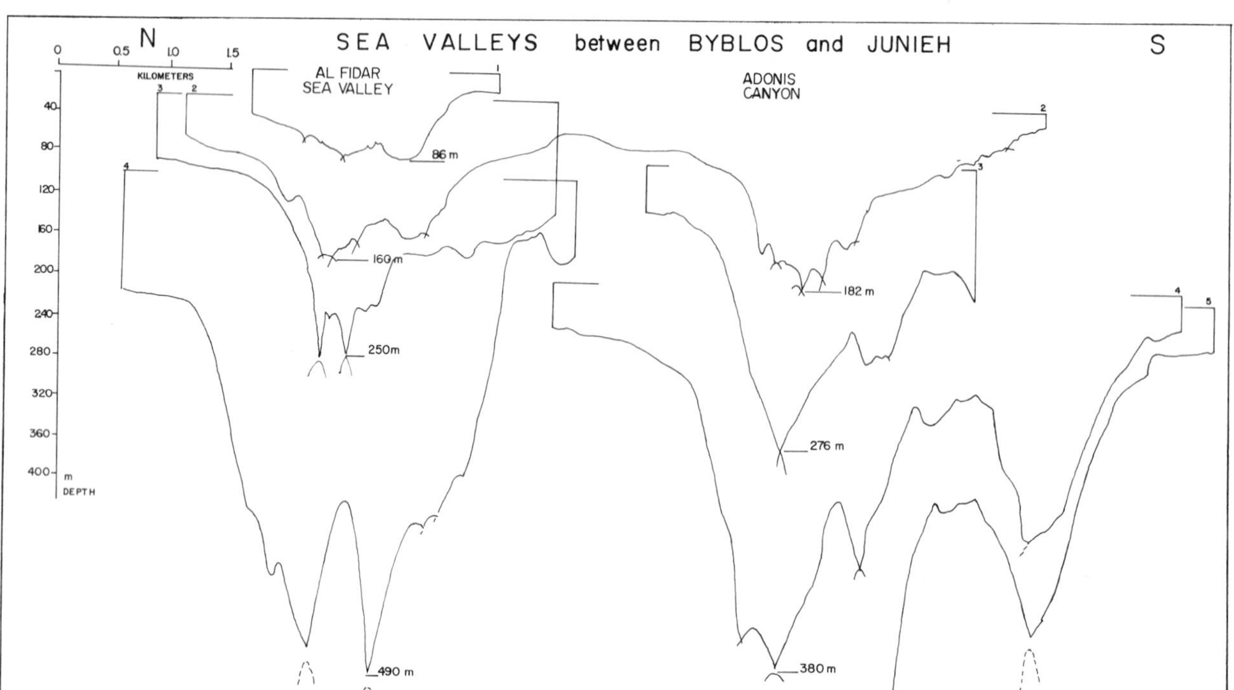

Figure 8. Fathometer profiles from an area between Byblos and Tabarja point, north of Ras el Maameltein (see Figure 1). The southernmost, Adonis Canyon, is approximately in line with the lower course of Nahr Ibrahim (Adonis River) El Fidar Sea Valley, to the north, is aligned with another major valley, Nahr El Fidar. Vertical exaggeration × 8. (Survey by Dening and Goedicke, 1969.)

due to a large supply of sediment brought onto the narrow shelf by the Nahr Ibrahim and Nahr el Fidar, both of which have large drainage basins.

Zahrani Canyon and Aouali Sea Valley

Two pronounced indentations in the edge of the continental shelf south of Ras Beirut are located north and south of the town of Sidon. These have been named the Zahrani Canyon and the Aouali Sea Valley.

Two sounding profiles across the indentation south of Sidon surveyed by the author in 1970 reveal a valley having the typical V-shaped cross section of a submarine canyon and extending to a depth of at least 396 m (Figure 9). This undersea valley, the Zahrani Canyon, is aligned with the lower channel of the Nahr Saitanig, a fact which might suggest a genetic connection (Goedicke, 1970a).

Another prominent indentation in the edge of the continental shelf north of Sidon, not explored in the course of the present study, is, like the Zahrani Can-

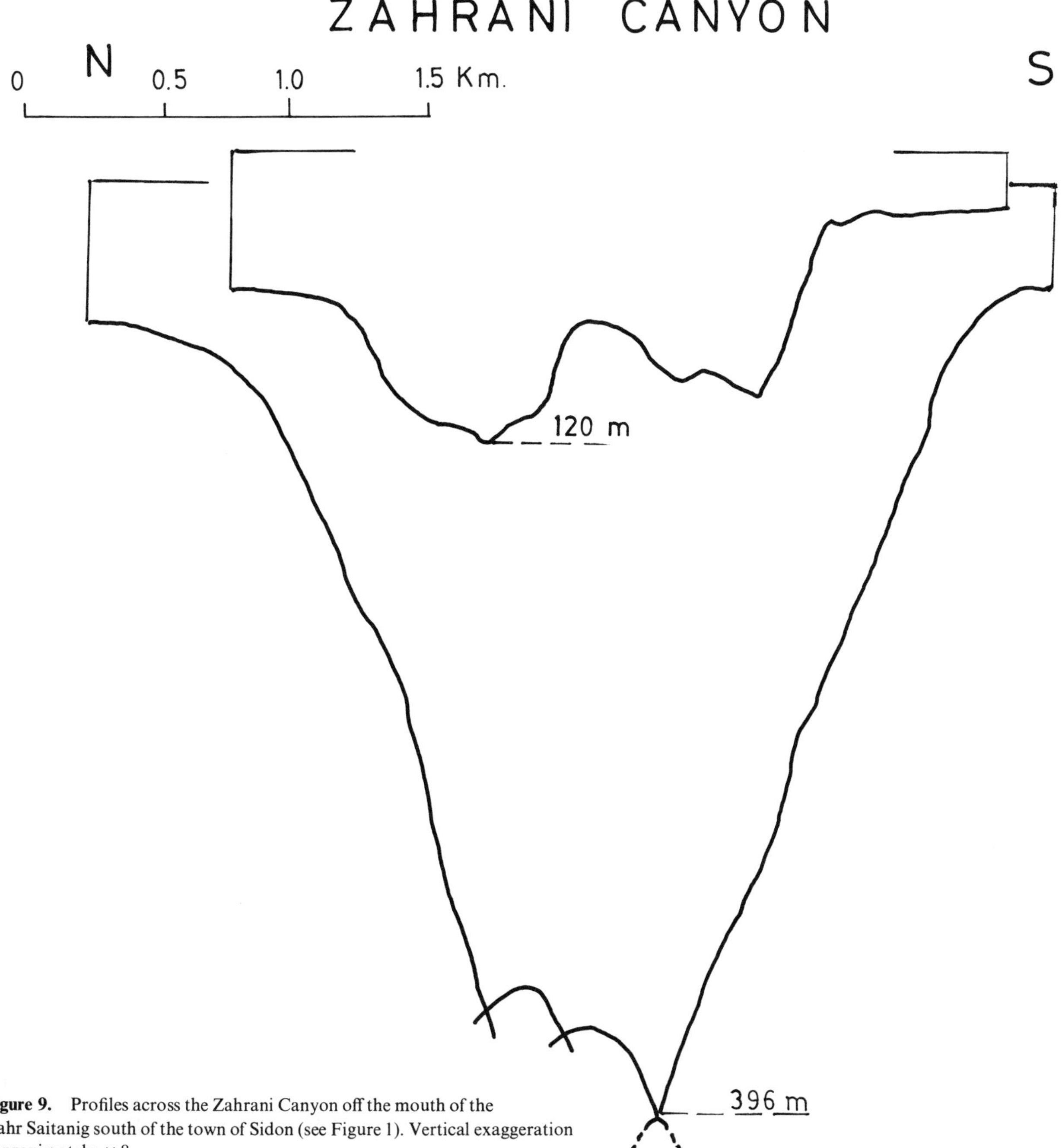

Figure 9. Profiles across the Zahrani Canyon off the mouth of the Nahr Saitanig south of the town of Sidon (see Figure 1). Vertical exaggeration approximately × 8.

yon, located in the approximate extension of a river valley, the Nahr el Aouali. By comparing the depression of the 50 m and 100 m depth contours in this area with the Zahrani Canyon and the other explored canyons it is assumed that further investigation will disclose a major submarine valley in this location.

Submarine Promontory West of Ras Beirut

The submarine promontory which extends a distance of 27 km to the west just south of the Ras Beirut headland is nearly flat at 368 m below sea-level and is surrounded by water depths in excess of 1500 m (Pfannensteil, 1960).

Figure 10 shows a precision depth recorder profile across the scarp on its northern flank. This prominent submarine scarp, which has a slope of 220 m/km, is in line with a major fault, with a throw exceeding 700 meters, which displaces the Jurassic and Lower Cretaceous strata in the valley of Nahr Beirut southeast of Beirut (Dubertret, 1945, 1951).

Because of its alignment with this east-west trending fault pattern in Lebanon, Pfannenstiel (1960) tentatively identified the undersea promontory as a horst. The coincidence of the submarine scarp with the direction of the major fault observable in the Cretaceous sediments, together with other evidence from Junieh Bay to the north, suggests a strong connection between the prominent fault patterns of the Lebanon mountains and the submarine topography of the continental shelf.

SEDIMENTS

The sediments of the coast of Lebanon were first described by Boulos (1962) and later by Emery and George (1963). Their origin is from multiple sources. Most of the detrital sand, both of calcareous and of

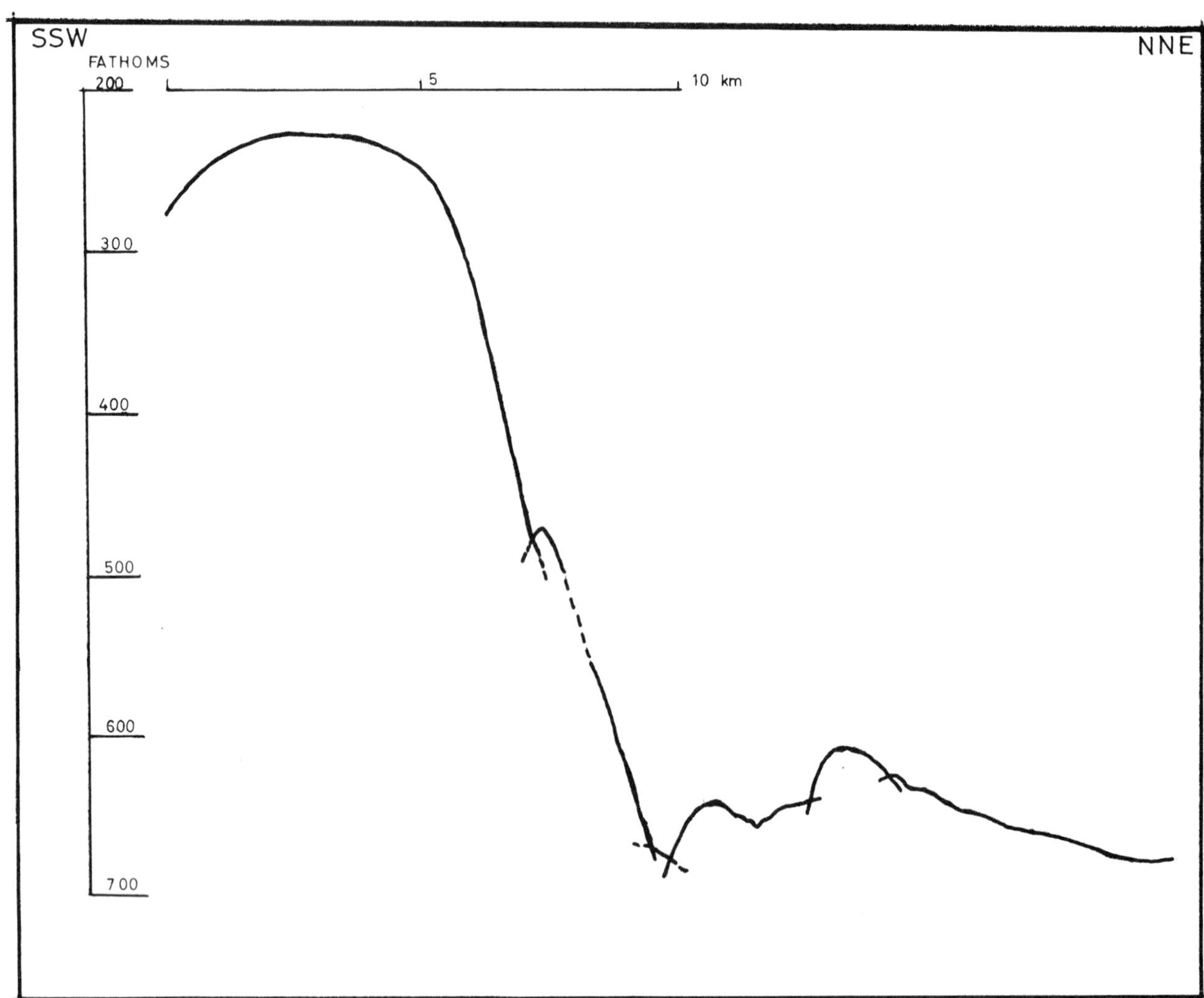

Figure 10. Section across the north flank of the submarine promontory south of the Ras Beirut headland (see Figure 1).

siliceous composition, as well as other minerals, such as hematite, is derived from erosion of the Lebanon mountain chain.

The gravel, which is mainly calcareous in composition, finds its way onto the beaches and the upper continental shelf during the period of heavy winter rains when the short, steep river valleys of the coastal zone are in flood. Most of the quartz sand and grains of hematite which are derived from the 'Grès de Base' formation of Lower Cretaceous age in the mountains are also transported by the winter floods. The heavy mineral fraction of the sediments on the coastal plain and continental shelf consists of titanaugite, augite, rutile, tourmaline, hornblende and occasionally monazite. Most of these minerals are presumed to have originated through weathering of the basalts of Upper Jurassic and Cretaceous age, outcropping in north-central Lebanon.

Some of the constituents of the fine fraction of the continental shelf sediments are presumed to have been carried north along the coast of the Levant from the Nile Delta, by the prevailing northerly longshore currents (Emery and George, 1963), However, because of observed southward sediment transport on the beaches, it is doubtful if the fraction of the silts and clays originating from the Nile constitutes a significant part of the shelf shallow sediments along the coast of Lebanon.

The most thorough sampling of continental shelf sediments has been done in Junieh Bay, on the slopes and near the axis of the Junieh Bay Canyon (Tejirian, 1970). A few samples were also collected in St. George's Bay and on the upper shelf south of Sidon. All of the Junieh Bay samples and most of the samples from St. George's Bay were collected by means of a small pipe dredge and, therefore, represent only the uppermost layer. However, three short core samples were obtained from the St. George's Bay Canyon area.

Bottom sediments found in Junieh Bay are similar to those in St. George's Bay and are mainly of terrigenous origin, consisting of four types: sand, sandy silt, silty sand and clayey silt. These grade into each other in a manner which depends on location.

The short steep valleys draining into the bay origiate mainly in areas of calcareous sediments, and are the source of most of the calcareous material found there, including an accumulation of limestone cobbles and pebbles on the beach in the north east corner of the bay. Mixed with these is some chert from the limestones of Cenomanian age outcropping around the bay (Tejirian, 1970).

Transport of terrigenous sediments into Junieh Bay is both directly from intermittent streams entering the bay and indirectly from the Nahr el Kalb entering the sea 5 km south of the bay. Most of the sediments of sand and silt grain size reach the sea via the Nahr el Kalb and are transported into Junieh Bay by longshore currents and wave action. The Nahr el Kalb is a perennial river having a large drainage basin and it transports considerable quantities of sediment during the rainy season.

An analysis of the grain size distribution of the sediments in Junieh Bay (Tejirian, 1970) shows that the sand size sediments occupy the southern and eastern shallow parts of the bay. The finest sediments, consisting of clayey silt, are found on the walls and along the axis of the Junieh Canyon.

The sorting of the sediments becomes progressively poorer with increasing water depth, increasing proximity to the axis of the submarine canyon, and decreasing grain size. This suggests possible mixing of the sediments by sliding and slumping from a semi-circular area of supply near the head of the canyon, and movement of the sediment along the axis.

The relatively good sorting of the sediment in the southern and eastern shallow areas of the bay is probably due to the action of wind, waves and swell (Tejirian, 1970).

It has been observed that at times when a heavy westerly to southwesterly swell is running, a yellow stream of turbid water extends from the headland at Ras el Tair to the northeast, past the jetties of the harbor installation (Figure 7). Furthermore, refraction causes the wave fronts to change direction as they pass the headland, curving into the southeastern part of the bay. The ends of the piers of the harbor installations act as point sources for new waves which spread out with wave fronts at more than 90° to the direction of the original swell.

These conditions, which prevail throughout almost the entire winter from October to April, together with the north trending ocean currents, are probably responsible for most of the longshore transport of terrigenous sediment from its origin at the mouth of the Nahr el Kalb into Junieh Bay. The relatively good sorting is due to the effect of the ground swell, while wave refraction is responsible for the transport of sand and silt into the newly constructed harbor installation. The innermost harbor in the southeast corner of the bay was almost completely filled with sand less than two years after its construction (Goedicke, 1969).

Distribution of sediments in St. George's Bay bears some resemblance to that in Junieh Bay, in that the sand fraction is found mainly in the southern and eastern portion and in shallow water, whereas the silt, clayey silt and silty clay fraction is found in the central and northern parts and in the areas surrounding the submarine canyon.

In the southern part of St. George's Bay, and sub-

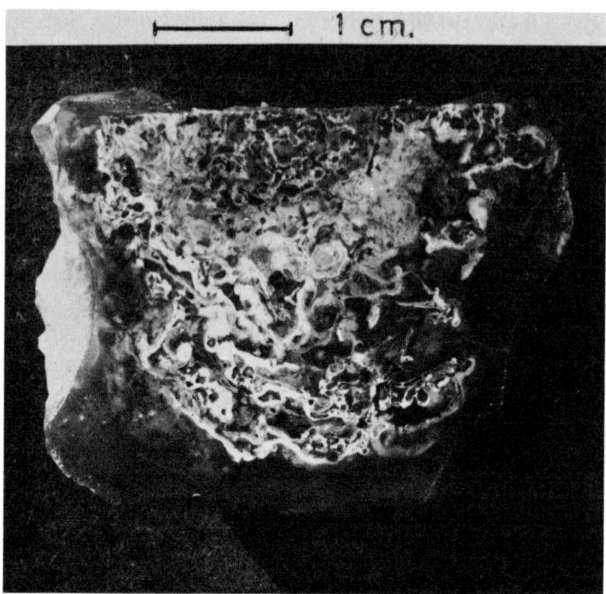

Figure 11. Sample of coralline algae dredged from the ridge adjacent to the south wall of the St. George's Bay Canyon.

parallelling the southern rim of the submarine canyon is a rocky ridge. This finds topographic expression above water in three small islands near the mouth of Nahr Beirut and can be followed for a considerable distance on the otherwise flat part of the continental shelf to the northwest. Fragments and balls of coralline algae up to 5 cm in diameter were obtained in dredge hauls from this ridge (Figure 11). The samples were obtained from depths of 140 and 150 meters and were broken off solid rock, as evidenced by fresh broken surfaces on the specimens, the hard bottom, and the fact that the dredge was caught on rock before being retrieved with the samples.

The greatest development of crustose coralline algae occurs between low tide and a depth of 25 m and the maximum depth from which they have been collected *in situ* in the Mediterranean is 120 m (Lemoine, 1940). The presence of these algae in St. Georges Bay, apparently *in situ* at 140 to 150 m is suggestive of former lower sea level stands combined with warmer water. A ridge of living coralline algae could have formed at a depth of 10 to 25 m at a time when sea level stood some 80 to 100 m below the present datum, as suggested by the 'knee' on the edge of the shelf near Monaco (Edgerton and Leenhardt, 1966) and a similar feature on some fathometer records from the continental shelf of Lebanon. The apparent continuity between the submarine and land valleys on the Lebanon coast also fits this picture.

Dredge samples from the sides and axis of the St. George's Bay Canyon were found to contain mainly fine silt and clay with some organic matter.

Four short gravity core samples were taken in the canyon at depths of 45, 153, 240 and 284 m (Goedicke, 1970a). All consist of clayey silt to silty clay and radiographs of one of the cores show cross laminations (Figure 12, *1*, *2*). Over 50 percent of the upper 30 cm of another core, taken on the axis of the canyon at a depth of 284 meters, was found to consist of matted sea grass and clay, black in color (Figure 12, *3*). This color was found to disappear overnight on exposure to air. Similar black clay and sea grass were found in dredge samples from shallow water between the mouth of the Nahr Beirut and the head of the canyon. This implies reasonably rapid down-canyon movement of material as described by Shepard and Dill (1966) and others.

SUMMARY AND CONCLUSIONS

This preliminary study of the topography and sediments of the continental shelf of Lebanon has revealed that there are several important topographic features on the upper part of the shelf. These consist of a number of submarine canyons and a submarine promontory whose origin can be linked tentatively to a prominent fault pattern observed in the Lebanon Mountains. The fact that these faults may continue under the Mediterranean, and that their direction is subparallel to a major undersea scarp on the south side of Cyprus suggests the importance of a WNW-ESE tectonic trend in the eastern Mediterranean (Giermann, 1966).

Preliminary investigations of the heads of three of the submarine canyons found along the coast suggest that at least some of these canyons are genetically related to land valleys. The importance of faulting in controlling the location of undersea valleys or their tributaries is suggested by the topography of the Junieh Bay Canyon.

The sediments of the upper continental shelf of Lebanon consist of sand, sandy silt, silt and clayey silt, derived mainly from erosion of the adjacent mountains. A small fraction of the fines may have been transported northwards from the delta of the Nile River.

Grain size analysis of dredge samples from two bays indicates a predominance of well sorted sand in the southeastern section of the bays, in relatively shallow water, grading into poorly sorted sandy silt toward the north in deeper water. Very poorly sorted clayey silt is found in the central portion of the bays near the heads and along the axes of the submarine canyons.

The presence of a deeply submerged ridge of coralline algae, together with the presence of a break or

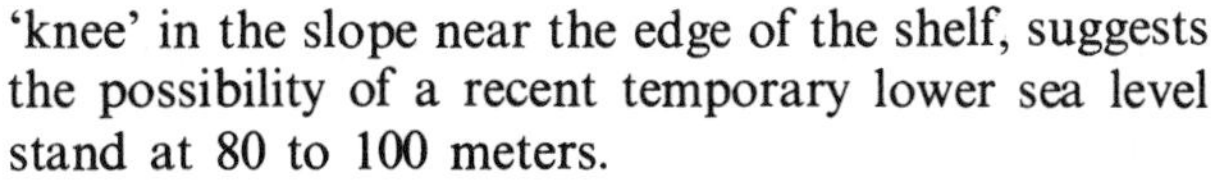

'knee' in the slope near the edge of the shelf, suggests the possibility of a recent temporary lower sea level stand at 80 to 100 meters.

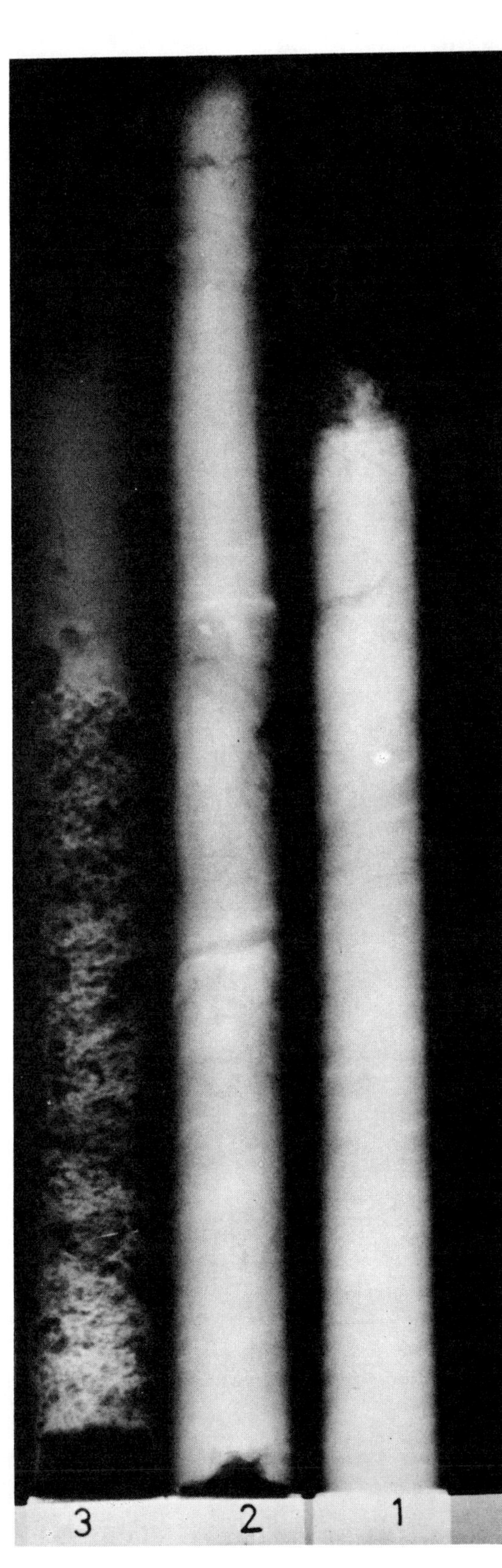

Figure 12. X-radiographs of three short gravity cores taken in St. George's Bay Canyon. Locations of the cores are shown on Figure 2. Note cross bedding and slump structure in core No. 2 and mottled appearance of core No. 3 due to air trapped amid sea grass and other organic material. Core No. 1 is from a depth of 153 m; No. 2 from 240 m; and No. 3 from 284 m.

ACKNOWLEDGMENTS

The field work for this research was supported in part by a grant from the Arts and Sciences Research Committee of the American University of Beirut. Field work near Sidon was supported by the Trans Arabian Pipe Line Company.

The following persons contributed greatly to the execution of the field and laboratory work, as well as in the preparation of illustrations: A. P. Denning; S. R. Smith, H. G. Tejirian. The constant collaboration of Mr. Tejirian and the permission to use data from his unpublished thesis are gratefully acknowledged.

REFERENCES

Boulos, I. 1962. *Carte de Reconnaissance des Cotes du Liban. 1:150.000*. Bassile Frères, Beyrouth, 2 sheets, 2nd edition.

Dubertret, L. 1945. *Géologie du Site de Beyrouth avec Carte Géologique au 1:20.000*. Délégation Générale de France au Levant, Section Géologique, Beyrouth.

Dubertret, L. 1951. *Carte Géologique au 1:50.000, Feuille de Beyrouth, avec notice explicative*. Ministère des Travaux Publics, République Libanaise, Beyrouth.

Edgerton, H. E. and O. Leenhardt 1966: Monaco: the shallow continental shelf. *Science*, 152:1106–1107.

Emery, K. O. and C. George 1963. The shores of Lebanon. *Woods Hole Oceanographic Institution Contribution*, 1385.

Emery, K. O., B. C. Heezen and T. D. Allan 1966. Bathymetry of the eastern Mediterranean Sea. *Deep-Sea Research*, 13:173–192.

Emery, K. O. and D. Neev 1960. Mediterranean beaches of Israel. *Bulletin Geological Survey of Israel*, 26:1–23.

Giermann, G. 1966: Gedanken zur Ostmediterranean Schwelle. *Bulletin de l'Institut Océanographique Monaco*, 66 (1362A); 16 p.

Goedicke, T. R. 1969. Geology of the continental shelf of Lebanon. *Lebanese Yachting Review*, 3:14–18.

Goedicke, T. R. 1970a. *Submarine topography and sediments offshore from Sidon terminal port*, Unpublished report to Trans Arabian Pipeline Company.

Goedicke, T. R. 1970b. Instruction and research in Marine Science at the American University of Beirut. *Marine Technology Society*, 6th Annual Conference, Washington, D. C. (Lecture).

Lemoine, P. 1940. Les algues calcaires de la zone néritique. *Société de Biogéographie* (Contribution à l'étude de la réparation actuelle et passée des organismes dans la zone néritique), Paris, 7:75–138.

Neev, D. 1960. A pre-Neogene erosion channel in the southern coastal plain of Israel. *Bulletin Geological Survey of Israel*, 25: 1–21.

Pfannenstiel, M. 1960. Erlauterungen zu den bathymetrischen Karten des östlichen Mittelmeeres: *Bulletin de l'Institut Océanographique Monaco*, 1192.

Renouard, G. 1955. Oil prospects of Lebanon. *Bulletin of The American Association of Petroleum Geologists*, 39:2125–2169.

Shepard, F. and R. Dill 1966. *Submarine Canyons and Other Sea Valleys*. Rand McNally and Co., Chicago, 381 p.

Tejirian, H. G. 1968. *Geological Map and Report, Maameltein Area, Lebanon*. Unpublished field course report, Department of Geology, American University of Beirut, 40 p.

Tejirian, H. G. 1970. *Preliminary Study of Submarine Topography and Sediments of Junieh Bay, Central Lebanon*. Unpublished thesis, American University of Beirut, 82 p.

Woods Hole Oceanographic Institution 1964. *Unpublished Fathograms, Cruise R/V CHAIN* No. 43, Woods Hole, Massachusetts.

PART 11. Geochemical Aspects of Mediterranean Deposits

Les Forages DSDP en Méditerranée (Leg 13): Reconnaissance Isotopique

Jean-Charles Fontes, René Létolle et
Wladimir D. Nesteroff

Université de Paris VI, Paris

RESUME

Des analyses isotopiques de reconnaissance ont porté sur les niveaux évaporitiques messiniens du leg 13 (programme DSDP-JOIDES). Elles concernent des calcites et dolomites (^{18}O, ^{13}C); des gypses (^{34}S, ^{18}O du sulfate et ^{18}O, ^{2}H de l'eau de constitution), des eaux d'imbibition de sédiments (^{18}O,^{2}H).

Sept sites (dont 4 en Méditerranée occidentale) ont été étudiés. Le faciès particulier de cette série (récurrences salines superposées allant des carbonates aux chlorures, avec intercalations argileuses) a révélé par d'autres techniques (micropaléontologie) des apports marins pélagiques discontinus. La teneur en isotopes lourds des carbonates donne à son tour la preuve d'alimentations nettement continentales; les niveaux dolomitiques dénotent le plus souvent un arrangement cristallin en présence d'une tranche d'eau réduite par évaporation. Les teneurs en ^{13}C des carbonates s'abaissent jusqu'à $-27‰$, reflétant probablement l'intervention massive de CO_2 d'origine biogénique. Les teneurs en ^{34}S et ^{18}O des 2 gypses analysés sont en accord avec les compositions des sulfates marins de l'époque. L'eau de constitution des gypses indique, toutes corrections de fractionnement faites, une recristallisation avec une eau marine de composition isotopique voisine de celle de l'eau d'imbition des sédiments. Cette eau pourrait être représentative de l'eau de mer de la fin du Miocène ou du Pliocène, car elle est différente de l'eau de mer actuelle en Méditerranée et elle n'est pas marquée par l'influence du stockage de grandes masses d'eau appauvries en ^{18}O et ^{2}H dans les calottes polaires.

Au-dessus du Messinien, le Pliocène marin pélagique a un faciès assez homogène et le carbonate est systématiquement enrichi en ^{18}O par rapport au Messinien. Il ne paraît pas y avoir, au passage avec des niveaux vraisemblablement quaternaires anciens, de changement net dans les conditions de sédimentation et en particulier dans la température.

ABSTRACT

Preliminary isotopic studies have been made of evaporitic sequences of Messinian age drilled on Leg 13 of the DSDP-JOIDES program in the Mediterranean Sea. The deposits are composed of calcite and dolomite (O^{18}, C^{13}), gypsum (S^{34}, O^{18} in sulphates, and O^{18} and H^{2} in crystallization waters). Seven sites, four of them in the western Mediterranean, are examined.

Earlier micropaleontological analyses have shown that discontinuous and sporadic marine incursions affected the saline series (carbonate to chloride sequences interbedded with clay). The isotopic composition of some carbonates provides clear evidence of the continental influence. Dolomite-rich zones generally indicate crystallization in a somewhat shallower water as a result of evaporation. The C^{13} content of some carbonates, in some cases as low as $-27‰$, indicates the importance of CO_2 of biogenic origin.

The S^{34} and O^{18} contents of gypsum are representative of marine sulphates of this period. The water of crystallization of gypsum (after correction for isotopic fractionation between mother water and crystal water) indicates complete recrystallization in marine water. The heavy isotope content of the connate water of the sediment may approximate that in which gypsum has crystallized, and may indicate the nature of sea water at the end of Miocene or in Pliocene time; this water having a significantly different isotopic content from that of modern Mediterranean water, was not modified by low O^{18} and H^{2} water masses stored in ice-caps.

The rather homogeneous pelagic Pliocene facies has a carbonate fraction (calcite) with a higher O^{18} content than in the underlying Messinian strata. There does not appear to be any notable change in the conditions of sedimentation and especially in temperature at the Pliocene-Quaternary boundary.

FACIES ET PROBLEMES

Au-dessous d'une série de vases et parfois de turbidites plio-quaternaires, le Miocène supérieur du fond de la Méditerranée se présente sous des faciès confinés à évaporites (sulfates et chlorures).

Les corrélations sont bonnes dans toute la partie occidentale du bassin (Figure 1) et démontrent la continuité des faciès évaporitiques (voir Nesteroff *et al.*, ce volume). A l'Est du détroit siculo-tunisien la situation est plus confuse. Les évaporites (gypse) ont été retrouvées à l'extrémité du sondage au site 125. Le site 126 correspond au remblaiement d'une vallée qui entaille en profondeur la Dorsale méditerranéenne jusqu'au Serravalien. Ici le remplissage quaternaire repose directement sur des niveaux de Serravalien marin. En 129, au pied de la Montagne sous-marine du Strabo, sur le flanc Nord de la fosse, le forage montre que le Plio-Quaternaire repose sur du Miocène supérieur qui est vraisemblablement remanié en "mélange tectonique".

Quoi qu'il en soit, le recensement des niveaux continus indique qu'il s'agit du plus vaste système évaporitique actuellement connu.

Les faciès sont extrêmement hétérotypiques et consistent, généralement en alternances souvent serrées, de bancs marno-calcaires et marno-dolomitiques à micro-organismes marins pélagiques, avec des couches d'évaporites. Les paragenèses salines: gypse, anhydrite, halite, témoignent de réductions de volume importantes et supérieures à 80% si on réfère les précipitations à des solutions initiales de composition ionique proche de celle de l'eau de mer actuelle.

Les principales questions concernent alors l'origine de l'apport en eaux, ions et éléments figurés, correspondant aux diverses récurrences marneuses et salines ainsi que l'épaisseur des tranches d'eau respectives.

Au stade même des premières investigations sur le matériel du programme DSDP-JOIDES en Méditerranée, l'intérêt d'une étude isotopique préliminaire est clair.

En effet, dans le cas des milieux confinés et des séquences évaporitiques, les documents, écologiques en particulier, se raréfient. Les données minéralogiques, géochimiques et isotopiques offrent en revanche des possibilités d'investigation accrues du fait même de la variété du matériel disponible.

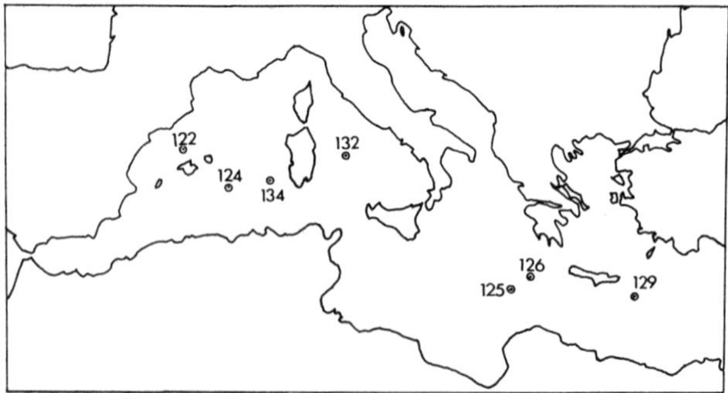

Figure 1. Localisation des sondages étudiés.

Nous présentons ici les résultats de l'étude isotopique préliminaire de carbonates (^{18}O et ^{13}C), deux analyses complètes de gypse (^{18}O et ^{34}S dans le sulfate, ^{18}O et deuterium dans l'eau de constitution) ainsi que deux analyses d'eau d'imbibition de sédiments (^{18}O et deuterium).

Nous avons examiné et analysé du matériel provenant (Figure 1) de Méditerranée occidentale: Golfe de Valence (site 122), Bassin Algéro-Provençal (site 134, 134 D et 124), Mer Tyrrhénienne (site 132), et de Méditerranée orientale: dorsale de la Mer Ionienne (sites 125 A et 126), montagne sous-marine du Strabo (site 129).

COMPOSITION ISOTOPIQUE DES CARBONATES

Le faciès minéralogique des échantillons et le contexte sédimentologique sont brièvement définis (Tableaux 1 et 2, et Figure 2).

Les Niveaux Distincts de L'Episode Evaporitique

Les résultats proviennent des sites 134 (Bassin Algéro-Provençal), 132 (Mer Tyrrhénienne) et 125 A et 126 (Mer Ionienne). Les faciès analysés sont très homogènes: boues riches en calcite (60 à 70%) avec

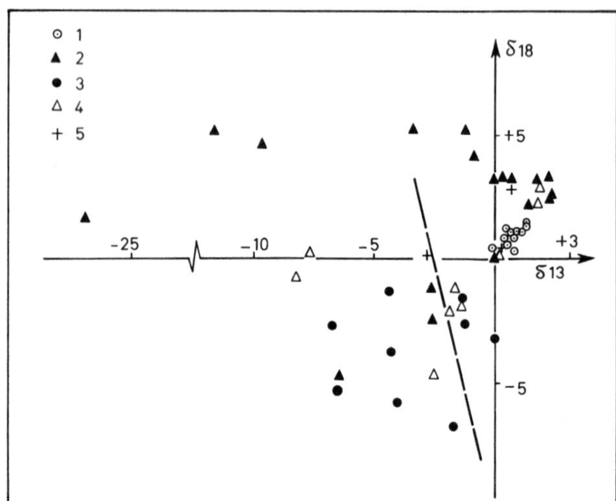

Figure 2. Diagramme $\delta 18/\delta 13$ des carbonates étudiés. I, calcites pliocènes; 2, dolomites miocènes; 3, calcites miocènes; 4, mélanges calcite-dolomite miocènes; 5, calcites serravaliennes. La droite en tireté correspond aux carbonates d'eau continentale (à gauche) et carbonates marins (à droite). (Keith et Weber, 1964).

un cortège d'éléments détritiques banals à quartz, illite souvent très ouverte et par endroits de la kaolinite. A l'Est, la phase carbonatée contient une légère surcharge dolomitique. En l'absence d'autres informations sur ce minéral qui apparaît ici en situation anormale dans l'environnement sédimentaire, nous le porterons également au compte d'un apport détritique tel qu'il a pu être observé notamment dans le Golfe Persique par Sugden (1963). Il est probable que les teneurs en ^{18}O enregistrées pour ces échantillons se trouvent ainsi "tirées" vers des valeurs plus hautes. Ainsi, en particulier, le niveau 125 A, 6-1-25 cm présente une concentration en ^{18}O assez élevée du fait de la participation d'une quantité notable de dolomite ($\sim \frac{1}{6}$) dans la phase carbonatée.

Les compositions isotopiques sont homogènes $+ 0{,}24 < \delta^{18}O < + 1{,}35$ et $- 0{,}04 < \delta^{13}C < + 1{,}25$ et se regroupent sur la Figure 2 en un domaine bien défini. Il faut y voir l'effet de la cristallisation des carbonates, essentiellement ici des foraminifères pélagiques, au sein d'une grande réserve d'eau bien tamponnée en composition isotopique.

L'échantillon qui provient du Quaternaire ancien (site 126, 1-4-90 cm) apparaît un peu plus chargé en isotopes lourds que les carbonates pliocènes, sans toutefois porter la marque de la nette augmentation que devraient entraîner les effets conjugués de l'abaissement de température et d'augmentation générale de teneur en ^{18}O des eaux marines par effet de bilan (immobilisation de grandes quantités d'eaux appauvries en isotopes lourds sous forme de glaces polaires). Des mesures complémentaires pourraient confirmer ici les observations réalisées sur le continent par Emiliani *et al.* (1961) qui ne décèlent pas de variations sensibles dans les températures isotopiques apparentes entre le Pliocène et le Calabrien.

Dans les faciès nettement moins carbonatés du Serravalien (3 à 12% de calcite), la phase argileuse est représentée par des smectites dominantes au sein d'une association qui comprend également kaolinite et illite. Ce cortège est voisin de celui des niveaux marneux de la série évaporitique messinienne (*cf.* ci-dessous). Les quelques valeurs de composition isotopique disponibles sont plus dispersées que pour les niveaux supérieurs. En l'absence d'autres indications, les teneurs en isotopes lourds peuvent être attribuées à des conditions de dépôt en milieu marin, toutefois moins homogène et moins tamponné (épicontinental?) que les masses d'eau pliocène.

Les Faciès Evaporitiques

On y trouve des associations minéralogiques à calcite ou dolomite avec souvent coexistence des deux espèces minérales. Un échantillon a révélé de la sidérite (site 132, 26-1-20 cm). Les carbonates sont en général limités à moins de 30% de l'ensemble du sédiment où les éléments détritiques sont abondamment représentés par du quartz (>10%) et parfois également des feldspaths. De même, l'association des minéraux argileux est riche en éléments hérités et se caractérise par la présence constante et majoritaire de smectites.

Lorsque les marnes contiennent de fortes proportions de sels (gypse ou anhydrite) la répartition relative des autres minéraux (carbonates, détritiques et argiles) ne semble pas varier.

Les compositions isotopiques sont très dispersées. On prendra en considération les calcites pures et les dolomites pures. On sait (Fritz et Fontes, 1966) que les teneurs en isotopes lourds attribuées aux calcites par attaque acide fractionnée des mélanges calcite-dolomite (Degens et Epstein, 1964) peuvent être largement entachées d'incertitude. En revanche, les compositions isotopiques des dolomites isolées à partir de ces mélanges cristallins par attaque acide ménagée sont représentatives. Nous les produirons dans une étude ultérieure.

Compositions Isotopiques des Calcites Pures

On relève plusieurs caractères:

1. teneurs plus faibles en ^{18}O et ^{13}C par rapport aux calcites marines du Pliocène;
2. teneurs plus faibles en ^{18}O par rapport à la plupart des dolomites messiniennes; et
3. grande dispersion des teneurs tant en ^{18}O qu'en ^{13}C.

Composition Isotopique des Dolomites Pures

Les dolomites sont en majorité caractérisées par une forte teneur en ^{18}O (moyenne voisine de $+ 4‰$).

Deux échantillons s'écartent toutefois sensiblement du domaine centré sur $\delta^{18}O = + 4‰$. Il s'agit de deux prélèvements voisins dans le sondage 132:

132 — 25 — 2(135)$\delta^{18}O = -1{,}17‰$

132 — 25 — 2(70)$\delta^{18}O = -4{,}87‰$

Les teneurs en ^{13}C montrent encore une large gamme de variations avec un groupe de valeurs proches de $+ 2‰$ et une série de teneurs plus basses jusque vers $- 12‰$. Dans le secteur de la montagne sous-marine du Strabo on relève une composition isotopique isolée très appauvrie en $^{13}C(- 26, 90‰)$.

DISCUSSIONS SUR LES CONDITIONS DE FORMATION DES CARBONATES MESSINIENS

Calcites

Le passage du Miocène au Pliocène se manifeste dans les compositions isotopiques des calcites par un

Tableau 1. Faciès minéralogiques et compositions isotopiques des échantillons DSDP-Leg XIII prélevés en Méditerranée occidentale.*

	Echantillon	Référence (cm)	Contexte Minéralogique C%	D%	Autres constituants	$\delta^{18}O$ *vs* PDB-1	$\delta^{13}C$ PDB-1
Site 132 Mer Tyrrhénienne							
Pliocène	19–3	(100)	60	0	I, K, Q	+0,24	+0,77
	20–4	(40)	80	0	I, Q	+1,33	+1,25
	20–4	(90)	60	0	I, Q	+1,05	+1,09
	21–2	(60)	30	0	I, Q	+1,12	+0,59
Miocène							
−88,25 m	21–2	(76)	6	0	M, I, Chl, G	−1,59	−1,37
	21–2	(100)	12	2	I, M, Chl, G, K	−1,09	−1,66
	22–1	(78)	4	0	n.d.	−2,67	−1,29
	22–1	(140)	+	0	G + n.d.	−3,72	−4,23
	23–1	(120)	2	0	G, Q, I, Chl	−6,37	−6,53
	25–1	(50)	0	8	G, I, Q, Chl, K, M		
	25–1	(90)	+	+	n.d.	−2,44	−2,60
	25–2	(70)	0	2	G, Q, I, Chl, M	−4,87	−6,34
	25–2	(110)	n.d.	n.d.	n.d.	−5,85	−4,00
	25–2	(135)	0	10	n.d.	−1,17^{x}	−2,66
	25–2	(140)	0	0	M, I, Q, K, Chl, F		
	26–1	(20)	0	15	I, M, Q, K, Chl, S	+0,19^{x}	0,00
	26–1	(24)	n.d.	n.d.	n.d.	−2,70	−6,66
	26–1	(110)	n.d.	n.d.	n.d.	−1,11	−4,49
Site 122 Golfe de Valence							
	Extrémité du sondage		++	T	n.d.	−3,24	+0,02
Site 134 Bassin Algéro-Provençal (SO Sardaigne)							
Pliocène	3–1	(126)	+	0	n.d.		
	3–2	(130)	+	0	n.d.	+0,68	+0,25
	3 *cc*		+	0	n.d.	+0,91	+0,55
	5 *cc*		+	0	n.d.	+0,99	+0,82
Miocène	7–5	(140)	+	+	n.d.	+0,10	+ 0,08
	7 *cc*	(24)	15	15	Q, I, M, Chl.	−1,76	−1,32
	10–1	(115)	0	++	A, H, Q	+4,60^{x}	−9,57
Site 134 D (même localisation)							
Miocène	1–1	(90)	14	2	G, M, I, Chl, Q	+2,41	+1,93
	1–1	(144)	T	T	G, Q, I, Chl, M	+2,96	+1,92
Site 124 Bassin Algéro-Provençal (centre)							
−350 m	6		0	60	I, Chl, Q, M, G	+2,14^{x}	+1,27
Miocène	7		+	+	G, Q + n.d.	−4,73	−2,43
	10–1	(113)	0	+	n.d.	+3,34^{x}	−2,18
	10 *cc*		0	+	n.d.	+2,42^{x}	+2,15
	10 *cc*		0	+	n.d.	+2,56^{x}	+2,22
	11–1	(124)	0	85	Q, I, M, K	+3,08^{x}	+1,74
	13–2	(35)	0	++	n.d.	+3,29^{x}	+0,27

**Symboles utilisés dans les tableaux numériques:* T = traces, n.d. = non dosé, C = calcite, D = dolomite, G = gypse, Q = quartz, I = illite, M = montmorillonite, Chl = chlorite, K = kaolinite, F = feldspath, A = argiles, H = halite, S = sidérite. Les constituants sont indiqués dans l'ordre d'importance relative. *cc* = core catcher; *DB* = drill bit; *CB* = center bit.
Les valeurs isotopiques pour les dolomites (x) ont été corrigées de l'effet de fractionnement isotopique signalé par Sharma et Clayton (1965).

Tableau 2. Faciès minéralogiques et compositions isotopiques des échantillons DSDP-Leg XIII prélevés en Méditerranée orientale (symboles utilisés: les mêmes que dans le Tableau 1).

	Echantillon Référence (cm)		Contexte Minéralogique			$\delta^{18}O$ *vs* $\delta^{13}C$ PDB	
			C %	D%	Autres constituants		
Site 126 Dorsale Mer Ionienne.							
Quaternaire	1–4	(90)	60	4	I, Q, K, Chl	+1,35	+0,49
Serravalien	5–1	(130)	12	0	M, K, Q, I	+0,48	+0,27
	5 *cc*		6	0	M, K, Q	+0,72	+0,33
	6–1	(130)	3	0	M, K, I, Q	+2,71	+0,65
	6 *cc*		4	0	Q, + n.d.	+0,19	−2,83
Site 125							
Plio-Quaternaire	2–2	(66)	60	0	G, Q, I	+0,37	−0,04
	5–3	(48)	64	4	I, Q, K	+1,10	+0,91
	6–1	(25)	65	10	I, Q, K	+1,48	+1,23
Miocène	6–1	(150)	0	++	n.d.	+5,20^{x}	−1,08
	7–*cc*	(140)	0	25	I, Q, K, F	+4,38^{x}	−0,92
	9–1	(113)	0	++	n.d.	+5,23^{x}	−3,28
	DB		0	++	A + n.d.	+5,09^{x}	−11,66
Site 129 Montagne sous-marine du Strabo.							
	2–*CB*	++	++	0	n.d.	−6,81	−1,92
	2–1	(136)	0	25	Q, M, Chl, I, G	+1,60^{x}	−26,90
	3–1	(49)	n.d.	n.d.	n.d.	−0,83	−8,22
	3 *cc*		15	13	M, Q, I, Chl, K	+0,17	−7,51
Site 129 A (même localisation).							
	2 *cc*		15	10	M, Q, I, Chl	−2,21	−1,97

signal très net (en ^{18}O et ^{13}C pour les sites 132 et 134, en ^{13}C seulement pour le site 125 où la présence de dolomite empêche de comparer les teneurs en ^{18}O). L'enrichissement relevé pourrait s'interpréter en termes de variation thermique dans une eau de composition isotopique comparable. Ceci implique un abaissement brusque (de l'ordre de 10°C au site 132) qui ne pourrait alors s'expliquer que par l'augmentation brutale de la tranche d'eau sur le sédiment. La logique de cette hypothèse entraîne fort loin. Peut-être prendra-t-elle cependant vigueur dans la suite des études (Benson, 1972). Par l'un de ces ironiques revirements dont l'avancement des sciences nous gratifie parfois, les moyens analytiques les plus poussés, mis au service de l'échantillonnage le plus complexe, conduiraient alors à exhumer les vieilles théories cataclysmistes enfouies sous plus de deux siècles de mépris définitif. Une autre interprétation possible consiste à attribuer une influence notable aux venues d'eaux appauvries en isotopes lourds, c'est-à-dire d'eaux continentales d'origine météorique, lors de la précipitation des calcites messiniennes. Cette dernière hypothèse rendrait par ailleurs mieux compte des variations de teneurs en ^{13}C pour lequel le fractionnement en fonction de la température est nettement plus faible que celui de l'oxygène (Mook, 1971).

D'autre part, les valeurs les plus basses en particulier se rapportent clairement à des eaux continentales. Contrairement aux dolomites peu sensibles aux processus d'échanges isotopiques postérieurs au dépôt (*cf.*, Fritz, 1971), les calcites peuvent n'avoir pas conservé leur composition isotopique originelle. Si les calcites ont subi un échange secondaire par transformation en sparite sous l'effet d'une percolation d'origine météorique, cela signifie que les sédiments se sont trouvés émergés et soumis à un certain drainage. Au cas où ces mêmes calcites auraient conservé leur teneur initiale en ^{18}O il faut admettre que le bassin lui-même était à certaines époques alimenté par des fleuves. L'analyse micrographique détaillée permettra de trancher sur la présence ou l'absence de calcisparite. Quoi qu'il en soit, les conséquences paléographiques sont dans les deux cas assez voisines et peuvent se résumer ainsi: *à plusieurs reprises le domaine évaporitique messinien s'est trouvé soustrait aux apports marins.* Cette situation s'est réalisée, soit par simple émersion, soit par rupture des communications avec le large et invasion d'eaux fluviatiles peu ou pas soumises à évaporation.

Les grandes variations relevées dans les compositions isotopiques des calcites messiniennes peuvent s'interpréter en termes de fluctuations de la teneur des eaux en isotopes lourds et dénotent le caractère sporadique des apports continentaux. Certaines valeurs voisines de 0‰ en particulier dans les échantillons du site 132 peuvent correspondre à des teneurs marines. Sur ce point l'inventaire analytique reste à compléter.

Dolomites

La teneur moyenne en ^{18}O des dolomites rend compte d'un arrangement de ce minéral en équilibre avec des solutions enrichies en isotopes lourds sous l'effet de l'évaporation. Les différences observées avec les teneurs en ^{18}O des calcites montrent une fois encore que la dolomite ne s'ordonne qu'après concentration et modification de la composition ionique initiale des eaux. Les teneurs voisines de + 4‰ sont connues pour l'origine marine (Clayton *et al.*, 1968) mais aussi dans les faciès de lacs évaporitiques (Friedman, 1966). On n'en peut donc tirer d'indication génétique. On affirmera seulement le caractère généralement surévaporé des solutions mères de la dolomite qui apparaît ici comme l'un des termes de la séquence saline. En ce qui concerne les deux valeurs atypiques on retiendra que l'une ($\delta^{18}O =$ − 1,17) ne peut procéder d'une eau de mer fortement évaporée et que l'autre ($\delta^{18}O = -4{,}87$) implique l'intervention de solutions d'origine continentale. La composition isotopique du carbone $\delta^{13}C = -6{,}34$ renforce cette conception. Une enquête au microscope électronique à balayage est ici nécessaire pour contrôler le caractère détritique éventuel de ces dolomites. Si les cristaux se sont formés *in situ*, il s'agit alors de dolosparite secondaire puisque les dolomites évaporitiques portent la marque de l'enrichissement de l'eau en isotopes lourds. Le problème des circulations d'eaux d'origine météorique se trouve à nouveau posé.

Pour l'ensemble des carbonates, les basses teneurs en ^{13}C, de l'ordre de − 3 à − 12‰ peuvent correspondre aux compositions isotopiques des carbonates du domaine continental (Keith et Weber, 1964; Mook, 1971). Pour les dolomites, les teneurs sont plus faibles que celles que l'on pourrait attendre des dolomites secondaires au sens de Fritz (1971), c'est-à-dire des cristaux organisés aux dépens d'un précurseur de protodolomite. Le groupe de valeurs à $\delta^{13}C \simeq +2$‰ peut être rapporté à l'établissement de conditions proches de l'équilibre isotopique entre le gaz carbonique atmosphérique et le carbonate en voie de cristallisation (Mook, 1971). Il en résulte l'image d'un bassin évaporitique où la tranche d'eau pouvait par endroits devenir très réduite.

La valeur très appauvrie en ^{13}C suggère au contraire l'influence de conditions réductrices avec évolution poussée de la matière organique jusqu'à libération de carbure (méthane ou éthane) qui s'est par la suite mis en équilibre isotopique avec le carbone des bicarbonates ou des carbonates par l'intermédiaire d'une réaction d'échange avec CO_2. Il n'est possible de faire ce raisonnement sur une valeur isolée que par suite de l'existence de nombreuses références à ce type de phénomène dans la série évaporitique miocène de Sicile (Dessau *et al.*, 1960; Cheeney et Jensen, 1965) ou dans des contextes géologiques analogues (*cf.*, Russel *et al.*, 1967; Hathaway et Degens, 1970; Deuser, 1970).

COMPOSITION ISOTOPIQUE DES SULFATES

Nous disposons dans ce travail préliminaire de deux analyses isotopiques complètes:

$CaSO_4$ = ^{18}O, ^{34}S

H_2O crist. = ^{18}O, ^{2}H

C'est bien trop peu pour interpréter, mais suffisant pour animer un débat qui est destiné à se poursuivre.

Sulfate

Oxygène 18 (SO_4)

Dans les comportements isotopiques respectifs de l'ion sulfate et du sel précipité, il est encore difficile d'interpréter les teneurs observées sur les échantillons naturels en fonction des données expérimentales (Longinelli et Craig, 1967; Lloyd, 1967, 1968; Misutani et Rafter, 1969).

Les évaporites précipitées à l'écart du système océanique montrent des teneurs en ^{18}O plus élevées que celles des sulfates marins (Rösler *et al.*, 1968; Rafter et Misutani, 1967; Lloyd, 1967). Les ions sulfates des masses océaniques ont une composition bien groupée et voisine de + 10‰ *vs* SMOW. Les quelques mesures disponibles sur les lacs hypersalés font état d'enrichissements en ^{18}O pouvant aller jusqu'à 23‰.

Dans ces conditions, les valeurs (+ 15,1‰ et + 20,8‰) que nous observons pour un dépôt miocène du paléobassin méditerranéen pourraient être rapportées à une alimentation mixte. Cependant les résultats obtenus sur des bassins confinés (marais salants) alimentées par de l'eau de la Méditerranée actuelle (Fontes et Schwarcz, résultats inédits) montrent que la concentration évaporitique de la solution s'accompagne d'un enrichissement de la teneur en ^{18}O des ions sulfates (+ 8 à 13%) jusqu'à précipitation d'un gypse enrichi en ^{18}O (+ 14 à 15%). Bien qu'il soit communément admis (Longinelli et Craig, 1967) que dans la gamme des températures et des pH biologiques, l'échange isotopique entre l'oxygène de l'eau et celui de l'ion sulfate est lent (plusieurs centaines à plusieurs milliers d'années), il apparaît que dans des conditions de confinement particulières (circulation univoque, forte évaporation, faible tranche d'eau, haute température et concentration des saumures), les ions sulfates peuvent rapidement amorcer leur rééquilibration avec l'eau. A titre d'hypothèse de travail, au vu de ces deux résultats, on peut estimer que les valeurs voisines de + 14% correspondent alors à une précipitation dans un bassin bien isolé du large et peu profond, mais alimenté en ion sulfate par le domaine marin.

Soufre 34 (SO_4)

Les teneurs en ^{34}S obtenues pour deux gypses (environ + 22‰) pourraient également correspondre à un sulfate précipité par concentration de l'eau de mer. On sait que cetter teneur n'a guère été modifiée par les effets de bilan depuis le Tertiaire (Nielsen, 1965; Nielsen et Rambow, 1969). En particulier, il s'agit d'une valeur supérieure à celle des dépôts de mers anciennes et en particulier du Permo-Trias. Il apparaît donc que ces terrains anciens pourtant riches en évaporites dans le pourtour méditerranéen ne sauraient constituer la source des ions sulfates qui se sont concentrés dans le bassin messinien.

Eau de Cristallisation des Gypses

Les mécanismes du fractionnement isotopique sont bien connus en ce que concerne l'intégration de l'eau au réseau de cristallisation (Gonfiantini et Fontes, 1963; Fontes et Gonfiantini, 1967).

Lors de la précipitation d'un gypse, l'eau qui s'intègre au réseau présente les caractéristiques suivantes par rapport à l'eau de la solution mère:

On définit: $\varepsilon^{18}O = +4{,}0‰$
$\varepsilon D = -16‰$
avec $\varepsilon = \delta_{\text{eau crist.}} - \delta_{\text{eau mère}}$

Dans la gamme de + 15 à + 60°C qui recouvre le domaine normal de précipitation des évaporites, ces fractionnements ne dépendent pas de la température.

Les eaux mères des gypses précipités à partir de de l'eau de la Méditerranée actuelle s'inscrivent dans un domaine proche de + 8‰ en ^{18}O et + 50 en ^{2}H (Fontes, 1966) qui correspondent à l'enrichissement dû à l'évaporation de 4/5 de la fraction liquide initiale.

Il est donc clair que l'eau de constitution des gypses miocènes étudiés ne peut correspondre à de l'eau de mer évaporée jusqu'à saturation en sulfate de calcium. En fait, il s'agit d'évaporites recristallisées. Le faciès macroscopique des échantillons (albâtrisation complète dans le cas de l'échantillon 125 A, débris de monocristaux hyalins à faces planes dans le site 122) est conforme à cette interprétation. L'eau qui s'est intégrée tardivement au réseau ne présente pas, non plus, une composition compatible avec celle de la plupart des eaux météoriques actuelles en climat semi-aride (Craig, 1961; I.A.E.A., 1969–1970). La balance générale de l'hydrosphère (au moins 98‰ de l'eau toujours immobilisée dans la partie océanique du cycle) ne permet guère de concevoir raisonnablement que les précipitations engendrées par les masses de vapeur issues des étendues océaniques aient pu avoir au Miocène une composition voisine de celles qu'indiquent les eaux de cristallisation des gypses (après correction des effets de fractionnement) (voir Figure 3).

En conséquence, une hypothèse possible est celle d'une recristallisation en présence d'eau de mer. Celle-

Tableau 3. Composition isotopique des gypses (DSDP-Leg 13).

Site	Localisation	Référence Echantillon	Faciès	Eaux crist. $\delta^{18}O$	Eaux crist. δD *vs* SMOW	Sulfate $\delta^{34}S$ *vs* C.D.	Sulfate $\delta^{18}O$ *vs* SMOW
122	Golfe de Valence	Extrémité du sondage	Débris hyalins (1 cm) Petits graviers	+ 4,07	− 16	+ 22,0	+ 15,1
125 A	Méd. orientale	Extrémité du sondage	Cristaux grossiers, parallèles, corrodés, traces Fe^{+++}	+ 3,89	− 8,5	+ 22,6	+ 20,8

Composition isotopiques des eaux d'imbibition des sédiments

Site	Localisation	Référence Echantillon	Résidu sec ‰	$\delta^{18}O$	δD
132	Mer Tyrrhénienne	25–2 (125)	53,8	− 0,78	+ 6,4
132	Mer Tyrrhénienne	26–1 (15)	55,0	− 0,88	+ 4,3

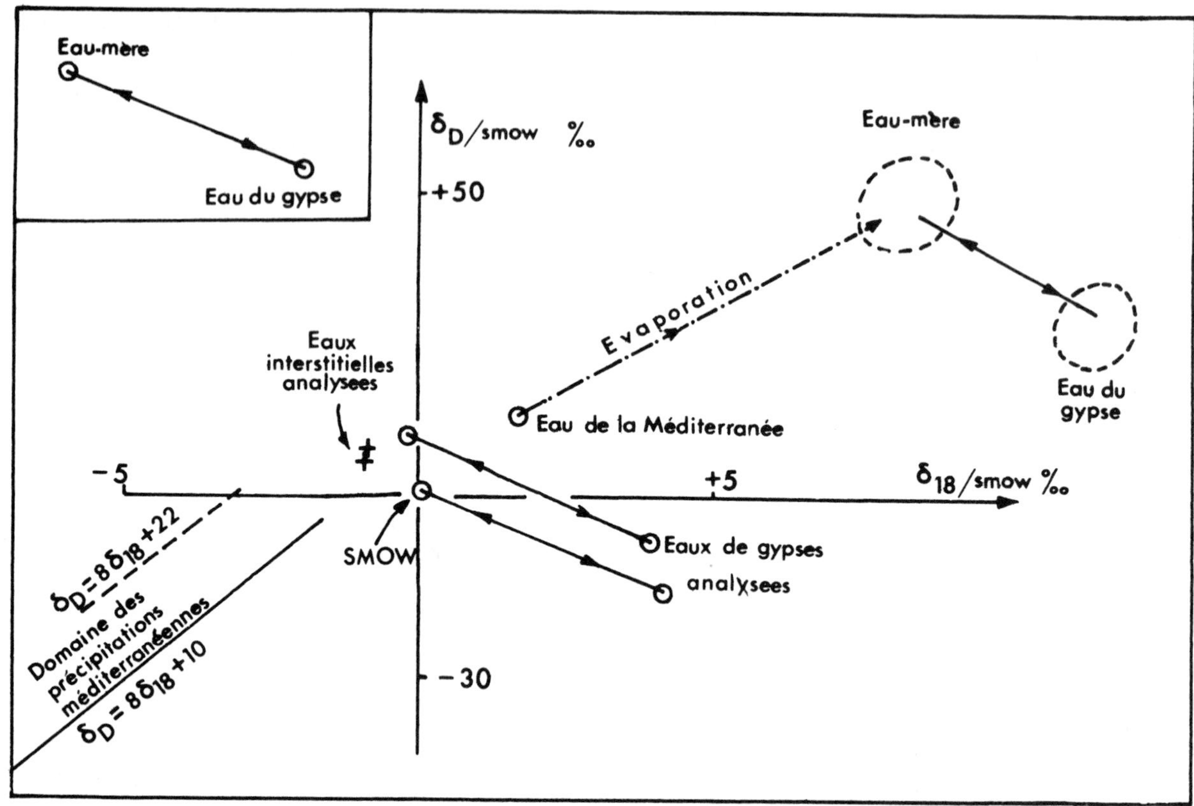

Figure 3. Diagramme $\delta D/\delta 18$ des eaux analysées. En cartouche, la relation eau-mère—eau de cristallisation des gypses; en pointillé, évolution isotopique de l'eau de la Méditerranée aboutissant au gypse d'après Fontes (1966). En bas à gauche, droites représentatives des précipitations: (1) à l'échelle du globe, (2) pour la Méditerranée orientale.

ci serait cependant bien différente de l'eau de la Méditerranée actuelle. A ce sujet, il est intéressant de constater que les teneurs en isotopes lourds des eaux interstitielles, extraites par pression des échantillons, se placent à proximité des valeurs observées pour les eaux de cristallisation corrigées du fractionnement (Figure 3).

Ces teneurs assez homogènes en ^{18}O et ^{2}H peuvent correspondre à une même masse d'eau dépourvue de traces notables d'évaporation (Craig et Gordon, 1965). Elle serait antérieure à l'enrichissement général des masses marines en isotopes lourds sous l'effet des glaciations ainsi qu'à l'isolement relatif du bassin méditerranéen. Ces arguments nous conduisent à penser que ces valeurs sont représentatives des eaux marines d'invasion sporadiques du bassin messinien ou des eaux de la mer qui ont repris possession du domaine méditerranéen au début du Pliocène.

Les conséquences de cette hypothèse sont doubles:

1. La relation des paléotempératures d'Epstein *et al.* (1953), modifiée par Craig (1965) :

$$t = 16.9 - 4.2\,(\delta - A) + 0.13\,(\delta - A)^2$$

dans laquelle on introduit les valeurs extrêmes:

δ: composition isotopique de l'oxygène du carbonate *vs* PDB $\simeq -1{,}20$ à $+0{,}2$‰

A: composition isotopique de l'oxygène de l'eau *vs* SMOW $\simeq 0$ à $-0{,}8$‰

indique une plage de variations de 8 à 13°C somme toute assez proche de la température actuelle des fonds méditerranéens ($\simeq 12$°C).

2. La valeur inférée pour la composition isotopique moyenne de la Méditerranée correspondrait à un état antérieur à celui qu'envisagent divers auteurs (Emiliani, 1966; Shackleton, 1967; Dansgaard et Tauber, 1969; Létolle *et al.*, 1971) et qui concerne plus particulièrement les variations de la composition isotopique moyenne des mers au début et au cours du Quaternaire.

L'alternative à l'hypothèse "marine" exposée ci-dessus (origine météorique de ces eaux) conduit à une discussion complexe qui sera détaillée dans une publication ultérieure. Elle présente selon nous moins de fondement que celle qui a été exposée.

CONCLUSIONS

Les premières données isotopiques relatives aux sulfates conduisent à des hypothèses de travail susceptibles d'être précisées ou même rectifiées: apports marins en ions $SO_4^{=}$, composition isotopique de l'eau de la Méditerranée à la fin du Pliocène. Il y faudra revenir.

Les interprétations de détail tirées de l'examen des teneurs en isotopes lourds peuvent également revêtir un certain caractère spéculatif: caractère détritique de la dolomite pliocène, conditions continues de sédimentation entre Pliocène et Quaternaire ancien, caractère épicontinental de la mer serravalienne, arrivée brutale d'une épaisse couche d'eau avec le Pliocène, véritable alimentation fluviatile du bassin messinien ou simple remaniement pluvial des dépôts avec échanges isotopiques et recristallisation, invasions marines sporadiques.

Nous rappellerons ci-dessous les faits essentiels et leurs conséquences immédiates:

Les carbonates de la série évaporitique (calcite et dolomite) montrent une large gamme de teneurs en ^{13}C et ^{18}O qui débordent largement du domaine de variations des carbonates marins.

$$-6{,}81 < \delta\,^{18}O < +7{,}03$$

Ces résultats parlent d'eux-mêmes: à plusieurs reprises le bassin (ou les bassins) se sont trouvés soustraits au domaine marin et largement alimentés en eau continentale. Les teneurs élevées témoignent de l'intensité de l'évaporation qui affectait les solutions et conduisait à l'organisation de la dolomite.

Le contenu en carbone 13 varie également de façon très large:

$$-26{,}90 < \delta\,^{13}C < +1{,}93$$

On y voit le reflet des fluctuations de l'activité organique liées aux épisodes de confinement, et dans certains cas l'influence de produits évolués (hydrocarbures). Dans le cas des valeurs élevées et attribuables à l'intervention du gaz carbonique atmosphérique, on décèle la diminution de la tranche d'eau et de l'activité réductrice.

L'étude de la phase carbonatée permet d'aborder une fois encore le problème de la dolomite. Sans anticiper sur les études isotopiques futures en ce domaine, nous noterons que les teneurs en ^{18}O évoquent pour la plupart des échantillons une organisation rapide et précoce (de type pénécontemporain) pour ce minéral.

REMERCIEMENTS

Nous sommes profondément redevables à la National Science Foundation des Etats-Unis d'Amérique qui a permis la réalisation du DEEP SEA DRILLING PROJECT; le Centre National d'Exploitation des Océans et le Centre National de la Recherche Scientifique ont assuré la participation française à l'état-major scientifique embarqué.

Madame Merlivat du Commissariat à l'Energie Atomique a mesuré les teneurs en deuterium. Notre collègue F. Mélières a effectué une partie des analyses diffractométriques.

Nous remercions Mesdemoiselles A. Filly et Sichère pour la partie analytique de ce travail réalisé avec le support financier du Centre National de la Recherche Scientifique (Laboratoire Associé n° 13).

BIBLIOGRAPHIE

Benson, R. H. 1972. Ostracods as indicators of threshold depth in the Mediterranean during the Pliocene. In: *The Mediterranean Sea: A Natural Sedimentation Laboratory*, ed. Stanley, D. J., Dowden, Hutchison and Ross, Inc., Stroudsburg, Pennsylvania, 63–73.

Cheeney, E. S., et M. L. Jensen 1965. Stable carbon isotopes of biogenic carbonates. *Geochimica et Cosmochimica Acta*, 29: 1331–1346.

Clayton, R., B. F. Jones et R. A. Berner 1968. Isotope studies of dolomite formation under sedimentary conditions. *Geochimica et Cosmochimica Acta*, 32:415–432.

Craig, H. 1961. Isotopic variations in meteoric waters. *Science*, 133:1702–1703.

Craig, H. 1965. The measurements of oxygen isotope paleotemperatures. In: *Stable Isotopes in Oceanographic Studies and Paleotemperatures*, ed. Tongiorgi, E., Consiglio Nazionale delle Ricerche, Spoleto, 161–182.

Craig, H. et L. I. Gordon 1965. Deuterium and oxygen 18 variations in the ocean and the marine atmosphere. In: *Stable Isotopes in Oceanographic Studies and Paleotemperatures*, ed. Tongiorgi, E., Consiglio Nazionale delle Ricerche, Spoleto, 9–130.

Dansgaard, W. et H. Tauber 1969. Glaciers oxygen 18 content and Pleistocene ocean temperatures. *Science*, 166:499–502.

Degens, E. T. et S. Epstein 1964. Oxygen and carbon isotope ratios in coexisting calcites and dolomites from recent and ancient sediments. *Geochimica et Cosmochimica Acta*, 28:23–44.

Dessau, G., R. Gonfiantini et E. Tongiorgi 1960. L'origine dei giacimenti solfiferi italiani alla luce delle indagini isotopiche sui carbonati della serie gessoso-solfifera della Sicilia. *Lo Zolfo*, 15–16:9–27.

Deuser, W. G. 1970. Extreme variations in Quaternary dolomites from the continental shelf. *Earth and Planetary Science Letters*, 8:118–124.

Emiliani, C. 1966. Paleotemperature analysis of Caribbean cores P 6304–8 and P 6304–9 and a curve for the past 425,000 years. *Journal of Geology*, 74:109–126.

Emiliani, C., T. Mayeda et R. Selli 1961. Paleotemperature analysis of the Plio-Pleistocene section at La Castella, Calabria, Southern Italy. *Geological Society of America Bulletin*, 72:679–688.

Epstein, S., R. Buchsbaum, H. A. Lowenstam et H. C. Urey 1953. Revised carbonate water isotopic temperature scale. *Geological Society of America Bulletin*, 64:1315–1326.

Fontes, J. Ch. 1966. Intérêt en géologie d'une étude isotopique de l'évaporation. Cas de l'eau de mer. *Comptes Rendus de l'Académie des Sciences, Paris*, 263:1950–1953.

Fontes, J. Ch. et R. Gonfiantini 1967. Fractionnement isotopique de l'hydrogène dans l'eau de cristallisation du gypse. *Comptes Rendus de l'Académie des Sciences, Paris*, 265:4–6.

Friedman, G. M. 1966. Occurrence and origin of Quaternary dolomites of Salt Flat, West Texas. *Journal of Sedimentary Petrology*, 36:263–267.

Fritz, P. 1971. Geochemical characteristics of dolomites and the ^{18}O content of middle Devonian oceans. *Earth and Planetary Science Letters*, 11:277–282.

Fritz, P. et J. Ch. Fontes 1966. Fractionnement isotopique pendant l'attaque acide des carbonates naturels. Rôle de la granulométrie. *Comptes Rendus de l'Académie des Sciences, Paris*, 263:1345–1348.

Fritz, P. et D. G. W. Smith 1970. The isotopic composition of secondary dolomite. *Geochimica et Cosmochimica Acta*, 34:1161–1173.

Gonfiantini, R. et J. Ch. Fontes 1963. Oxygen isotopic fractionation in the water of crystallization of gypsum. *Nature*, 200: 624–648.

Hathaway, J. C. et E. T. Degens 1968. Methane-derived marine carbonates of Pleistocene age. *Science*, 165:690–692.

I. A. E. A. 1969–1970. Environmental Isotope Data, World Survey of Isotope Concentration in Precipitation, 1 (1953–1963) and 2 (1964–1965). *International Atomic Energy Agency*, Vienne, technical reports, 96 et 117.

Keith, M. L. et J. N. Weber 1964. Carbon and oxygen isotopic composition of selected limestones and fossils. *Geochimica* et *Cosmochimica Acta*, 28:1787–1816.

Létolle, R., H. de Lumley et C. Vergnaud-Grazzini 1971. Composition isotopique de carbonates organogènes quaternaires de Méditerranée Occidentale: essai d'interprétation climatique. *Comptes Rendus de l'Académie des Sciences, Paris*, 273:2225–2228.

Lloyd. R. M. 1967. Oxygen 18 composition of oceanic sulphate. *Science*, 136:1228–1231.

Lloyd, R. M. 1968. Oxygen isotope behavior in the sulphate water system. *Journal of Geophysical Research*, 73:6099–6110.

Lloyd, R. M. et K. J. Hsü 1972. Preliminary isotopic investigations of samples from deep-sea drilling cruise to the Mediterranean. In: *The Mediterranean Sea: A Natural Sedimentation Laboratory*, ed. Stanley, D. J., Dowden, Hutchison and Ross, Inc., Stroudsburg, Pennsylvania, 681–686.

Longinelli, A. et H. Craig 1967. Oxygen 18 variations in sulphate ions in sea water and saline lakes. *Science*, 156:56–59

Misutani, Y. et T. A. Rafter 1969. Oxygen isotopic fractionation in the bisulphate ion-water system. *New Zealand Journal of Science*, 12:54–59

Mook, W. G. 1971. Paleotemperatures and chlorinities from stable carbon and oxygen isotopes in shell carbonate. *Paleogeography, Paleoclimatology, Paleoecology*, 9:245–263.

Nesteroff, W. D., W. B. F. Ryan, K. J. Hsü, G. Pautot, F. C. Wezel, J. M. Lort, M. B. Cita, W. Maync, H. Stradner and P. Dumitrica 1972. Evolution de la sédimentation pendant le Néogène en Méditerranée d'après les forages JOIDES-DSDP. In: *The Mediterranean Sea: A Natural Sedimentation Laboratory*, ed. Stanley, D. J., Dowden, Hutchison and Ross, Inc., Stroudsburg, Pennsylvania, 47–62.

Nielsen, H. 1965. Schwefelisotopen im marinen Kreislauf und das ^{34}S der früheren Meere. *Geologische Rundschau*, 55:160–172.

Rafter, A et Y. Misutani, 1967. Oxygen isotope composition of sulphates, p. 2. *New Zealand Journal of Science*, 10:816–840.

Rösler, H. J., J. Pilot, D. Harzer et P. Kruger 1968. Isotopengeochemische Untersuchungen (U, S, C) Salinar und Sapropelsedimenten Mitteleuropas. *Proceedings, 22nd International Geological Congress*, 6:89–100.

Russell, K. L.,K. S. Deffreyes, G. A. Fowler et R. M. Lloyd 1967. Marine dolomite of unusual isotopic composition. *Science*, 155: 189–191.

Shackleton, N. 1967. Oxygen isotope analyses and Pleistocene temperatures reassessed. *Nature*, 215:15–17.

Sugden, W. 1963. Some aspects of sedimentation in the Persian Gulf. *Journal of Sedimentary Petrology*, 33:355–364.

Sharma, T. et R. N. Clayton 1965. Measurements of O^{18}/O^{16} ratios of total oxygen from carbonates. *Geochimica et Cosmochimica Acta*, 29:1347–1353.

Preliminary Isotopic Investigations of Samples from Deep-Sea Drilling Cruise to the Mediterranean*

R. Michael Lloyd and K. Jinghwa Hsü

Shell Development Company, Houston, Texas, and Swiss Federal Institute of Technology, Zurich

ABSTRACT

The oxygen and carbon isotope composition of nine carbonate and three sulphate samples from the DSDP Leg XIII were analyzed to provide data for an interpretation of the environments of their deposition and diagenesis. Analyses of dolomite and anhydrite samples yielded data in support of the geological deduction by the scientific staff of the Leg XIII that the Upper Miocene Mediterranean evaporites were formed in dessicated inland basins after the Strait of Gibraltar was closed in the Late Miocene.

RESUME

Les compositions isotopiques de l'oxygène et du carbone ont été mesurées dans 9 échantillons de carbonate et 3 échantillons de sulfate provenant de la XIIIème croisière du Deep-Sea Drilling Project. Les résultats de ces analyses permettent de déduire les conditions de dépôt et de diagenèse de ces roches. Les analyses, effectuées sur les dolomites et les anhydrites, supportent l'hypothèse de l'équipe de chercheurs de cette croisière que les faciès évaporitiques du Miocène supérieur méditerranéen ont été formés par évaporation dans les bassins intérieurs créés par la fermeture du Détroit de Gibraltar vers la fin du Miocène.

INTRODUCTION

During the thirteenth cruise of the Deep-Sea Drilling Project to the Mediterranean, a number of diagenetically altered sediments and sedimentary rocks were sampled (see Nesteroff *et al.*, this volume, their Figure 1 for location map). The purpose of this investigation is to apply the stable oxygen- and carbon-isotope techniques to provide some data useful for the interpretation of the depositional and diagenetic histories of these deposits. The materials were made available to us in connection with research to assist in the preparation of the Initial Cruise Report of DSDP Leg XIII (Ryan *et al.*, 1972). Our investigations have been jointly supported by the Shell Development Company and by the Swiss Federal Institute of Technology.

MEDITERRANEAN EVAPORITES

One of the most significant results of the DSDP Mediterranean cruise is the discovery of an Upper Miocene evaporite formation, which underlies most parts of the Mediterranean (Ryan *et al.*, 1972). The origin of the evaporite is, however, a question of some

*Shell Development Company (a Division of Shell Oil Co.) EPR Publication No.608

controversy. Three alternative hypotheses have been advanced and are discussed below.

Deep-Water, Deep-Basin Model

This hypothesis assumes the deposition of evaporite minerals in a deep-water basin (Schmalz, 1969). At the time of evaporite-deposition, the Mediterranean is assumed to have been a deep-water basin, not isolated from the Atlantic, but separated from the latter by a shallow sill. However, the circulation was sufficiently reduced to cause an increasing salinization of the Mediterranean. Eventually carbonates, sulphates and halite were crystallized out of brines and accumulated on the deep basin floor to form the supposedly deep-water evaporite.

Shallow-Water, Shallow-Basin Model

This hypothesis assumes the deposition of evaporite minerals on the bottom of a shallow restricted shelf sea, which may or may not have an open connection with the Atlantic. The present depth of the Mediterranean was related to post-Miocene subsidence, subsequent to the formation of the Upper Miocene evaporites (see Ryan *et al.*, 1972).

Dessicated Deep-Basin Model

This hypothesis assumes the deposition of evaporite minerals on inland playas, whose flat basin floors were thousands of meters below the Atlantic sea level. The playas owed their origin to the dessication of the Late Miocene Mediterranean, when it was completely isolated from the Atlantic. The return of the normal marine condition in early Pliocene led to the deposition of deep-water pelagic oozes on top of the playa evaporites (see Ryan *et al.*, 1972).

The arguments, pro and con, have been discussed in a separate paper (see Ryan *et al.*, 1972). The evidence on the whole could be considered largely in favor of the dessicated deep-basin model. Isotopic analyses were carried out to throw additional light on this problem.

Oxygen and carbon isotope data for Miocene dolomites and dolomitic calcites from the Mediterranean evaporite formation and calcites from overlying Pliocene sediments are given in Table 1. These data are compared with analyses reported by Fontes *et al.* (this volume) from the Mediterranean evaporite and with values for Holocene carbonates from lacustrine and marine Miocene environments in Figures 1 and 2. Oxygen isotope data on mineral sulphates from the Miocene evaporite formation are given in Table 2.

Table 1. Isotopic composition of Upper Miocene and Lower Pliocene carbonate minerals, Mediterranean Sea.

Location	DSDP Leg XIII Sample No.	Description	Mineral	PDB δO^{18}	PDB δC^{13}
Alboran Basin	121/19/1	UM Dolomitic limestone	Calcite	−1.6	−3.4
Alboran Basin	121/19/1	UM Dolomitic limestone	Calcite	0.5	−0.4
Alboran Basin	121/21/1	UM Dense dolomite	Dolomite (Ca_{53})	1.6	−10.2
Levantine Basin	129/2/1	UM Sucrose dolomite	Dolomite (Ca_{53})	3.4	−35.8
Tyrrhenian Basin	132/21/2	Pliocene red ooze	Calcite	1.3	0.1
Tyrrhenian Basin	132/22/1	Pliocene red ooze	Calcite	1.2	0.2

These data are compared to values for marine and lacustrine sulphates in Figure 3.

MIOCENE CARBONATES

Dolomite occurs in the Mediterranean Miocene evaporite deposits in quantities varying from traces to greater than 90 weight per cent. The dolomite is fine-grained and contains excess calcium carbonate up to 55 mole per cent. It is found in both soft marly sediments and in indurated crusts. Layers of relatively pure dolomite are rare and it is usually a minor component of the sediment—of the order of 10 per cent or less (Fontes *et al.*, this volume). The remaining sediment consists of varying admixtures of calcite, quartz, feldspar, gypsum and a variety of clay minerals.

The most striking feature of the carbonate isotope data is its variability (Figure 1). This variability contrasts sharply with the rather narrow range of values for Holocene dolomites and dolomitic calcites found in tidal flat deposits associated with shallow water marine deposition. The variability compares favorably

Table 2. Isotopic composition of sulphate from Mediterranean evaporites.

Location	DSDP Leg XIII Sample No.	Description	Mineral	O^{18} SMOW
Balearic Basin	124/8/1	UM Laminated anhydrite	Anhydrite	16.7
Balearic Basin	124/10/1	UM Nodular anhydrite	Anhydrite	15.8
Tyrrhenian Basin	132/23/1	UM Laminated gypsum	Gypsum	3.8

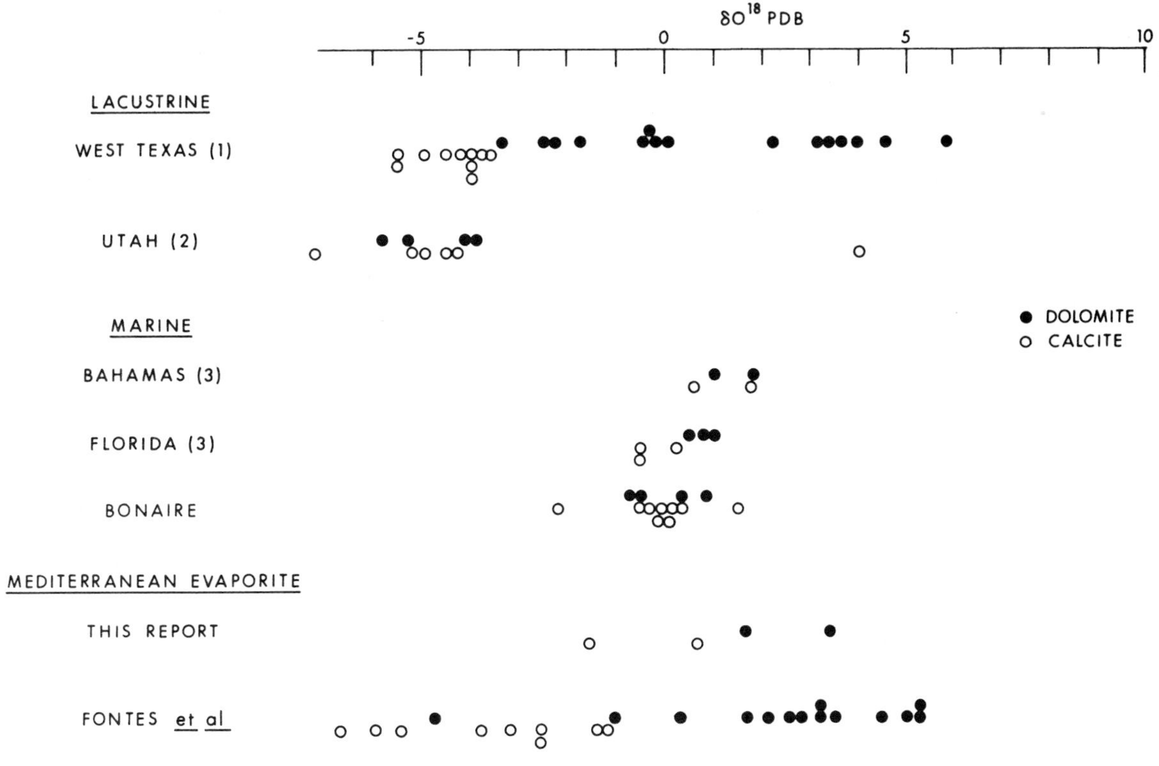

Figure 1. Oxygen isotope data for dolomite and calcite from lacustrine, marine and Mediterranean evaporite (DSDP Leg XIII) samples. Data from the literature: (1) Parry *et al.* (1970); (2) and (3) Degens and Epstein (1964); Fontes *et al.* (this volume).

with data reported for dolomites and calcites from lacustrine evaporite deposits—especially the data from a group of Pleistocene lakes from West Texas described by Parry *et al.* (1970).

The West Texas samples are from a group of isolated lake basins which formed in depressions eroded out of Pliocene deposits of the Texas high plains. The basins have maximum dimensions of the order of a few miles and are scattered over an area of 5,000 square miles. The deposits range in age from 12,000 to over 37,000 years. Dolomite and calcite are found mixed with quartz, feldspar, clays and minor amounts of celestite and gypsum. There is almost as much scatter in the isotopic data for individual lake basins as there is amongst all of the basins. The authors conclude that: "The wide scatter in isotopic compositions of the dolomites indicates that they formed from solutions of widely differing isotopic compositions and temperatures; conditions which could be expected in isolated dessicating pluvial lake systems in which evaporation is extreme" (Parry *et al.*, 1970, p. 830).

According to a recent proponent of the deep-basin deep-water evaporite formation hypothesis: "There are no active deep evaporite basins today, . . ." (Schmalz, 1969, p. 822). It is not possible, therefore, to make the simple isotopic comparisons between this environment and the Mediterranean carbonates as we did above for the shallow marine and playa lake environments. However, we may speculate on possible isotopic variations on the basis of the proposed model for deep-water evaporite formation.

The essential element of this model is the existence of a shallow sill at the rim of the basin which allows new sea water to enter in a surface layer and dense brine to escape in a lower layer in a continuous refluxing system. Though there are details to the model which are necessary to explain the sequence of salts deposited, the essential isotopic factors are:

1. The source of water and salts is sea water from the oceanic reservoir.

2. Concentration of salts is by evaporation from a free water surface.

3. Because of density stratification the floor of the basin can, at times, be stagnant.

Ocean water has a constant oxygen isotopic composition of about 0.0 per mil on the PDB scale. At earth surface temperatures ($\sim 25°C$) calcites formed in equilibrium with such water would have δO^{18} values of -2 per mil and dolomites δO^{18} values of -1 to 3 per mil[1]. Evaporation from a free water surface causes isotopic enrichment of sea water to an upper limit of about 6 per mil (Lloyd, 1967). Such enrichment would effect carbonate values directly. Thus, calcites might range from -1 to $+4$ and dolomites from 0 to $+9$ per mil[1] in a deep evaporite basin environment.

These ranges accomodate the more positive values for the Mediterranean Miocene carbonates, but fail to account for the very negative oxygen isotope values found in many of the samples from the same unit.

Some of the observed variation in carbon isotope values of the Mediterranean samples might be accommodated by the deep-water, deep-basin hypothesis if we accept the possibility of periods of stagnation of bottom waters. During such periods the normal bicarbonate carbon acquired by the water mass at the surface could be contaminated by more negative carbon from CO_2 generated by organic decay. It should be pointed out, however, that periods of stagnation were proposed by Schmalz (1969) to account for euxinic sulfide-rich deposits in ancient evaporite basins. We are unaware of similar sediments in the Mediterranean sequence.

[1]The range is given to accomodate the unresolved differences of opinion amongst isotope workers as to how much, if any, isotopic fractionation exists between dolomite and calcite.

The large negative C^{13} value of the Levantine (Leg XIII site 129) dolomite sample is very unusual. The closest match to this is one from a deep-water open marine dolomite, whose genesis was probably related to the bacterial breakdown of hydrocarbons (Russell *et al.*, 1966). Limestones associated with Sicilian sulphur deposits in the Solfifers Formation (an equivalent of the Mediterranean Evaporite) likewise have highly negative C^{13} values (-8.8 to -43); these limestones are also believed to have derived their carbonate primarily from oxidized methane enriched in C^{12} (Jensen, 1968). In any event, such extreme values speak neither for, nor against, a deep water origin.

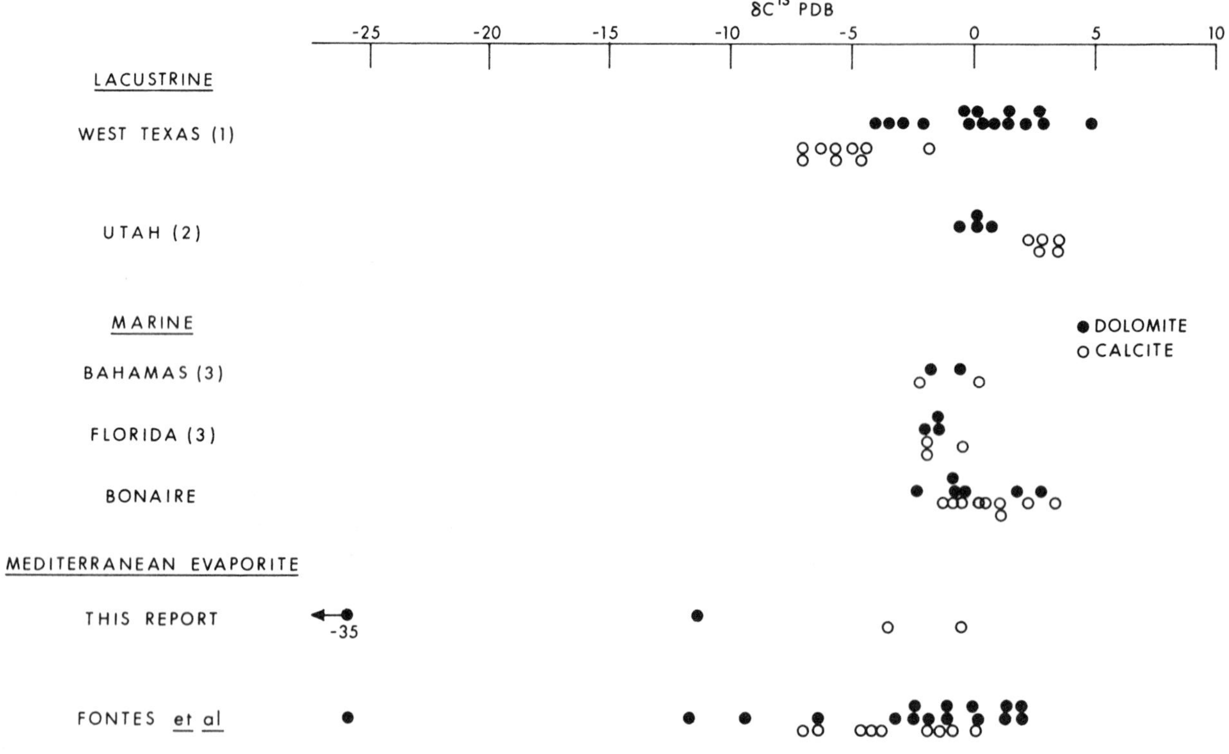

Figure 2. Carbon isotope data for dolomite and calcite from lacustrine, marine and Mediterranean evaporite (DSDP Leg XIII) samples. Data from the literature: (1) Parry *et al.* (1970); (2) and (3) Degens and Epstein (1964); Fontes *et al.* (this volume).

PLIOCENE CARBONATES

The Miocene evaporite series is directly overlain by Pliocene carbonate-rich open marine sediments with abundant pelagic organisms and a normal marine benthonic ostracod fauna. Our isotope values (Table 1) and eleven reported by Fontes *et al.* (this volume) all fall within a narrow range of 0.2 to 1.3 per mil for δO^{18} and -0.1 to 1.2 per mil for carbon relative to PDB.

The isotopic composition of the oozes could be construed to indicate formation in cool normal marine waters, say of the order of 10°C. However, partial evaporation of the Mediterranean water may be partly responsible for the positive δO^{18} values. The present-day waters in the Mediterranean have a δO^{18} value of about 1.2. If the Pliocene oozes were deposited from waters of similar isotopic composition, our result would suggest that the largely planktonic skeletons were crystallized from a water column with an average temperature of about 15°C. Benson (this volume) studied the ostracod fauna and found the ostracod assemblage "most likely to occur living in open ocean between 1,000 and 1,500 meters (bottom temperatures between 4° and 6°C)." This is consistent with the postulate that the Pliocene bottom waters were much cooler than those of the present-day Mediterranean and that the first Pliocene waters flooded deep but dessicated Mediterranean basins (Benson, this volume; Ryan *et al.*, 1972).

MIOCENE SULPHATES

The most interesting feature of the sulphate isotope values from the Miocene evaporite is again their variability (Figure 3). The fact that the values appear to avoid the range of present day marine sulphates is only an accident of sampling. Fontes *et al.* (this volume) report a value of 14.3 for a Miocene sulphate from site 122.

Our knowledge of oxygen isotope variation in sulphates is very limited. Marine oceanic sulphate has a very constant δO^{18} value of about 10 per mil relative to the SMOW standard (Longinelli and Craig, 1967; Lloyd, 1967). Lloyd (1968) has proposed that this represents a steady state value in the oxidation-reduction sulphur cycle of the ocean rather than the true equilibrium exchange value with ocean water. Exchange rates between water and sulphate are extremely slow at earth surface conditions. Therefore, mineral sulphates precipitated from evaporated marine waters would tend to reflect the constant value of marine sulphate (plus a fractionation of crystallization) rather than the conditions of temperature and

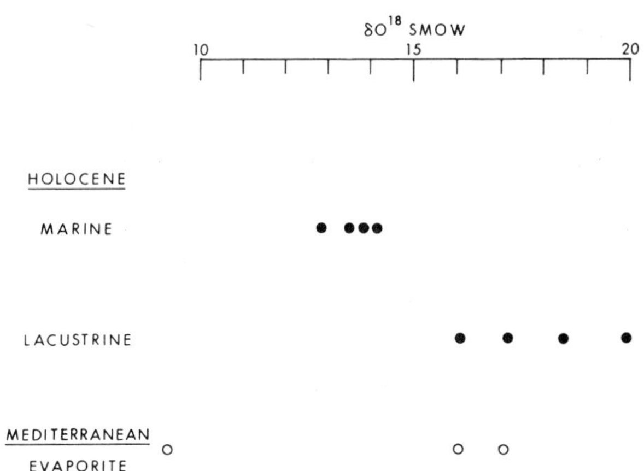

Figure 3. Oxygen isotope data for oxygen in sulphates. Marine data from Lloyd (1967). Lacustrine data from playa deposits West Texas and New Mexico.

water isotopic composition at the time of precipitation which could vary substantially from place to place (Lloyd, 1968). This accounts, then, for the fact that gypsum derived from marine water occupies a narrow range from 13 to 15 per mil relative to SMOW (Figure 3).

Gypsum samples from lacustrine environments have a wider spread of values and are, in general, more positive. This may represent inherited values from older sulphates dissolved from the outcrop (Longinelli, 1968), or perhaps, a different isotopic balance determined by local variations in the sulphur cycle oxidation-reduction system.

The fact that the Miocene sulphate samples show no tendency to group near the values for marine-derived sulphates argues against both the shallow basin-shallow water and deep basin-deep water models for evaporite genesis. While there are no Holocene data for mineral sulphates to match the one very negative Miocene sample (9 per mil) there is a report of a thermal spring water containing dissolved sulphate with a δO^{18} of 4.8 per mil (Lloyd, 1967). Gypsum precipitated from this solution would have a value near 9 per mil.

DISCUSSION

Because of the variability of the data we cannot define, in any exact sense, the environment of formation of any portion of the Mediterranean Miocene evaporite deposits. However, it is this great variability which supports the hypothesis that deposition occured in a dessicated basin consisting of playas, residual salt ponds and isolated ephemeral lakes covering a broad flat area of many thousands of square miles. That fresh water from rain and run-off was an important contributor to the basin is demonstrated by the negative carbon and oxygen isotope values for many of the carbonate samples.

The dessicated playa environment might have existed as part of a shallow basin. However, the rapid onset over the entire basin of uniform open marine conditions indicated by the Pliocene carbonate isotope values supports the conclusion of Ryan *et al.* (1972) that the basin was deep.

REFERENCES

Benson, R. H. 1972. Ostracods as indicators of threshold depth in the Mediterranean during the Pliocene. In: *The Mediterranean Sea: A Natural Sedimentation Laboratory,* ed. Stanley, D. J., Dowden, Hutchinson and Ross, Inc., Stroudsburg, Pennsylvania, 63–73.

Degens, E. T. and S. Epstein 1964. Oxygen and carbon isotope ratios in coexisting calcites and dolomites from recent and ancient sediments. *Geochimica et Cosmochimica Acta,* 28:23–44.

Fontes, J. Ch., R. Létolle and W. D. Nesteroff 1972. Les forages DSDP en Méditerranée (Leg 13): reconnaissance isotopique. In: *The Mediterranean Sea: A Natural Sedimentation Laboratory.* ed. Stanley, D. J., Dowden, Hutchinson and Ross, Inc., Stroudsburg, Pennsylvania, 671–680.

Jensen, M. L. 1968. Isotopic geology of Gulf Coast and Sicilian sulfur deposits. *Geological Society of America Special Paper,* 88:

Lloyd, R. M. 1967. Oxygen isotope composition of oceanic sulfate. *Science,* 156:1228–1231.

Lloyd, R. M. 1968. Oxygen isotope enrichment of sea water by evaporation. *Geochimica et Cosmochimica Acta,* 30:801–814.

Lloyd, R. M. 1968. Oxygen isotope behavior in the sulfate-water system. *Journal of Geophysical Research,* 73:6099–6110.

Longinelli, A. 1968. Oxygen isotopic compositions of sulfate ions in water from thermal springs. *Earth and Planetary Science Letters,* 4:206–210.

Longinelli, A. and H. Craig 1967. Oxygen-18 variations in sulfate ions in sea water and saline lakes. *Science,* 156:56–59.

Nesteroff, W. D., W. B. F. Ryan, K. J. Hsü, G. Pautot, F. C. Wezel, J. M. Lort, M. B. Cita, W. Maync, H. Stradner and P. Dumitrica 1972. Evolution de la sédimentation pendant le Néogène en Méditerranée d'après les forages JOIDES-DSDP. In: *The Mediterranean Sea: A Natural Sedimentation Laboratory,* ed. Stanley, D. J., Dowden, Hutchinson and Ross, Inc., Stroudsburg, Penssylvania, 47–62.

Parry, W. T., C. C. Reeves, Jr. and J. W. Leach 1970. Oxygen and carbon isotopic composition of West Texas lake carbonates. *Geochimica et Cosmochimica Acta,* 34:825–830.

Russell, K. L., K. S. Deffreyes, G. A. Fowler and R. M. Lloyd 1966. Marine dolomite of unusual isotopic composition. *Science,* 155:89–100.

Ryan, W. B. F., K. J. Hsü, D. Nesteroff, G. Pautot, F. C. Wezel, J. M. Lort, M. Cita, W. Maync, H. Stradner and P. Dumitrica 1972. *Initial Reports of the Deep Sea Drilling Project, Lisbon, Portugal, to Lisbon, Portugal.* U. S. Government Printing Office, Washington, D. C. 13 (in press).

Schmalz, R. F. 1969. Deep-water evaporite deposition: a genetic model. *The American Association of Petroleum Geologists Bulletin,* 53:798–823.

Etude Sédimentologique et Géochimique de Quelques Dépôts Quaternaires Récents du Plateau Continental au Large du Roussillon (Golfe du Lion)

Gustave A. Cauwet, François Y. Gadel et André A. Monaco

Centre de Recherches de Sédimentologie Marine, Perpignan

RESUME

Au cours du Quaternaire récent, le plateau continental méditerranéen a été le siège de nombreuses oscillations du niveau de la mer responsables d'une succession de faciès sédimentaires de nature et d'origine variées. Le remplissage flandrien, relativement épais et vaseux dans la portion médiane du plateau, repose sur une assise grossière, contemporaine de la dernière régression würmienne. Vers le large, ces formations passent latéralement à des sables reliques en partie remaniés au cours de la transgression. Sous les dépôts grossiers peuvent apparaître d'autres dépôts fins (silts argileux), mis en place au cours d'une période froide (faune à *Cyprina islandica*).

Les données sédimentologiques, minéralogiques et géochimiques, apportent plusieurs sortes de renseignements, suivant la position stratigraphique des horizons recoupés. Dans les niveaux récents flandriens, déposés sous des conditions climatiques et bathymétriques relativement bien connues, l'aspect dynamique et évolutif de la sédimentation se traduit par des variations verticales et horizontales des divers paramètres qui rendent compte de la progression de l'oscillation positive et des modalités d'apport. Au contraire, dans les niveaux plus anciens, l'intervention possible de plusieurs facteurs rend difficile la reconstitution des processus de sédimentation: remaniements liés à une épaisseur d'eau moins grande, effets diagénétiques plus apparents, *etc.* Les données palynologiques permettent de préciser l'origine des apports et d'établir des corrélations étroites entre les divers résultats.

ABSTRACT

Continental shelves in the Mediterranean Sea, as those elsewhere, have been modified significantly by the numerous Quaternary eustatic sea-level oscillations. These oscillations resulted in a distinct succession of sedimentary sequences. The generally fine-grained Flandrian deposits form a thick blanket on the mid-shelf off the Roussillon region of France (Gulf of Lions) and cover coarser-grained regressive Würm facies. Seaward, relict sands are encountered; these, in part, have been reworked during the last transgression. Argillaceous silts, containing *Cyprina islandica* and deposited during a "cold" phase, are found below the coarse relict sands.

This study summarizes sedimentologic, petrographic and geochemical characteristics of these facies. The properies of recent Flandrian units, deposited under a varying set of climatic and bathymetric conditions that are generally well known, show a distinct change with time. Observed vertical and horizontal lithological changes reflect changes in hydrodynamic regime and provenance in turn related to sea level oscillation. Interpretation of older units is considerably more difficult due to subsequent reworking in shallower marine conditions and diagenetic changes. A palynological study sheds light on the source of materials brought to the continental margin, and serves to insure better correlation between the different data collected.

INTRODUCTION

Les travaux antérieurs effectués sur le plateau continental roussillonnais concernent, en premier lieu, la stratigraphie et la tectonique plio-quaternaires du précontinent, ainsi que l'origine des vallées sous-marines qui l'entaillent (Bourcart, 1956; Bourcart *et al.*, 1961; Glangeaud *et al.*, 1968; Got et Monaco, 1969).

D'autres études portant sur l'analyse de thanatocoenoses ont conduit à une meilleure connaissance des assises subrécentes (Mars *et al.*, 1957; Mars, 1959).

Des recherches préliminaires, paléoécologiques et sédimentologiques, ont abouti à des corrélations stratigraphiques dans les dépôts situés à proximité des vallées sous-marines (Monaco, 1967; Levy, 1967).

Le cadre stratigraphique, établi d'après les résultats sismiques, sert de base à la présente étude, qui se propose de reconstituer les conditions de mise en place des dépôts du Quaternaire récent sur le plateau continental du Roussillon et de reconnaître les effets possibles de la diagenèse.

SITUATION DE L'ETUDE ET DONNEES GENERALES

Le plateau continental, large et développé dans le secteur septentrional au large du Roussillon, s'amenuise au contact de la côte rocheuse des Albères, prolongation de la zone pyrénéenne (Figure 1). Il est incisé vers −100 m par plusieurs vallées sous-marines creusées par étapes successives dans les séries pliocènes et quaternaires: "rechs" Lacaze-Duthiers, Pruvot et Bourcart. La carte lithologique des dépôts superficiels (Got *et al.*, 1968; Got *et al.*, 1969) fait apparaître, au-delà d'une étroite bande de sables littoraux, une vaste étendue de vases. A partir de 80 mètres de profondeur, ces vases terrigènes sont relayées par des sables reliques plus ou moins grossiers qui simulent une morphologie dunaire.

La prospection sismique superficielle (boomer et mud penetrator) et les carottages permettent de reconnaître l'épaisseur du recouvrement récent argileux (15 m maximum) et la nature de son substrat; ce dernier consiste en épandages grossiers en continuité avec les sables du large. L'extension de ces formations résiduelles donne une idée de l'ampleur de la dernière oscillation négative (Würm) dans ce secteur (− 100 m). Les estimations faites en d'autres points de la Méditerranée (Mars, 1959; Blanc, 1968; Bonifay, 1969) sont du même ordre de grandeur (−100 à −120 m).

Les carottes recoupent également sous ces assises grossières des horizons de nature variée, témoins d'autres variations du niveau de la mer.

CHOIX DES PRELEVEMENTS ET METHODES D'ETUDE

En fonction des résultats sédimentologiques et sismiques préliminaires, diverses carottes sont choisies dans les secteurs où les dépôts sont les plus caractéristiques et représentatifs des diverses étapes de la sédimentation (Figure 1).

Le sondage F 119 (36°6′E, 42°43′N; profondeur: 35 m), le plus proche de la côte actuelle, offre les changements de faciès les plus nombreux. Il se situe à proximité de hauts fonds rocheux, témoins immergés du Quaternaire ancien (Tyrrhénien ?).

Les sondages CL (3°8′E, 42°56′10″N; profonduer: 40 m), et F 116 (3°15′15″E, 42°54′20″N; profondeur: 55 m) appartiennent à la portion septentrionale et circalittorale du plateau, soumise depuis la transgression flandrienne à une assez forte sédimentation.

Les sondages F 118 (3°21′15″E, 42°50′45″N; profondeur: 90 m) et F 117 (3°28′20″E, 42°52′40″N; profondeur: 95 m) sont localisés vers la limite externe du plateau et recoupent en surface des assises sableuses ou graveleuses, reliques de la dernière régression. Ils atteignent en profondeur des horizons plus anciens, essentiellement vaseux.

L'étude sédimentologique utilise les méthodes classiques d'analyse: granulométrie des fractions fines et grossières, minéralogie des argiles par diffractométrie de rayons X.

Les teneurs en carbone sont déterminées par combustion sèche dans un four à induction et mesure pondérale du gaz carbonique produit. Le taux d'azote est déterminé volumétriquement après combustion de l'échantillon. Les carbonates sont évalués dans les pélites par méthode gazométrique.

L'analyse chimique effectuée sur le sédiment et les eaux interstitielles porte essentiellement sur quelques cations (Mg, Ca, Na, K, Fe, Mn) et les chlorures des eaux de pores. La fraction pélitique, mise en solution par une attaque nitro-fluorhydrique, est reprise par l'acide chlorhydrique à ébullition (Voinovitch *et al.*, 1962). Le dosage du sodium et du potassium se fait par photométrie de flamme, celui des autres éléments par spectrophotométrie d'absorption atomique. On prend soin d'ajouter une solution de lanthane pour le dosage du magnésium et du calcium, afin d'éliminer les interférences.

Les résultats palynologiques sont exprimés en distinguant les apports continentaux, marins et palustres, d'après le type d'associations qui les caractérisent.

L'interprétation des diverses données s'appuie également sur les analyses micropaléontologiques. L'ensemble de ces résultats fera l'objet d'une publication ultérieure.

RESULTATS

Description des prélèvements et Caractères Sédimentologiques

L'étude a porté sur cinq carottes représentatives des divers milieux rencontrés sur le plateau continental (Figure 1).

La carotte F 119 (H = 540 cm) présente une alter-

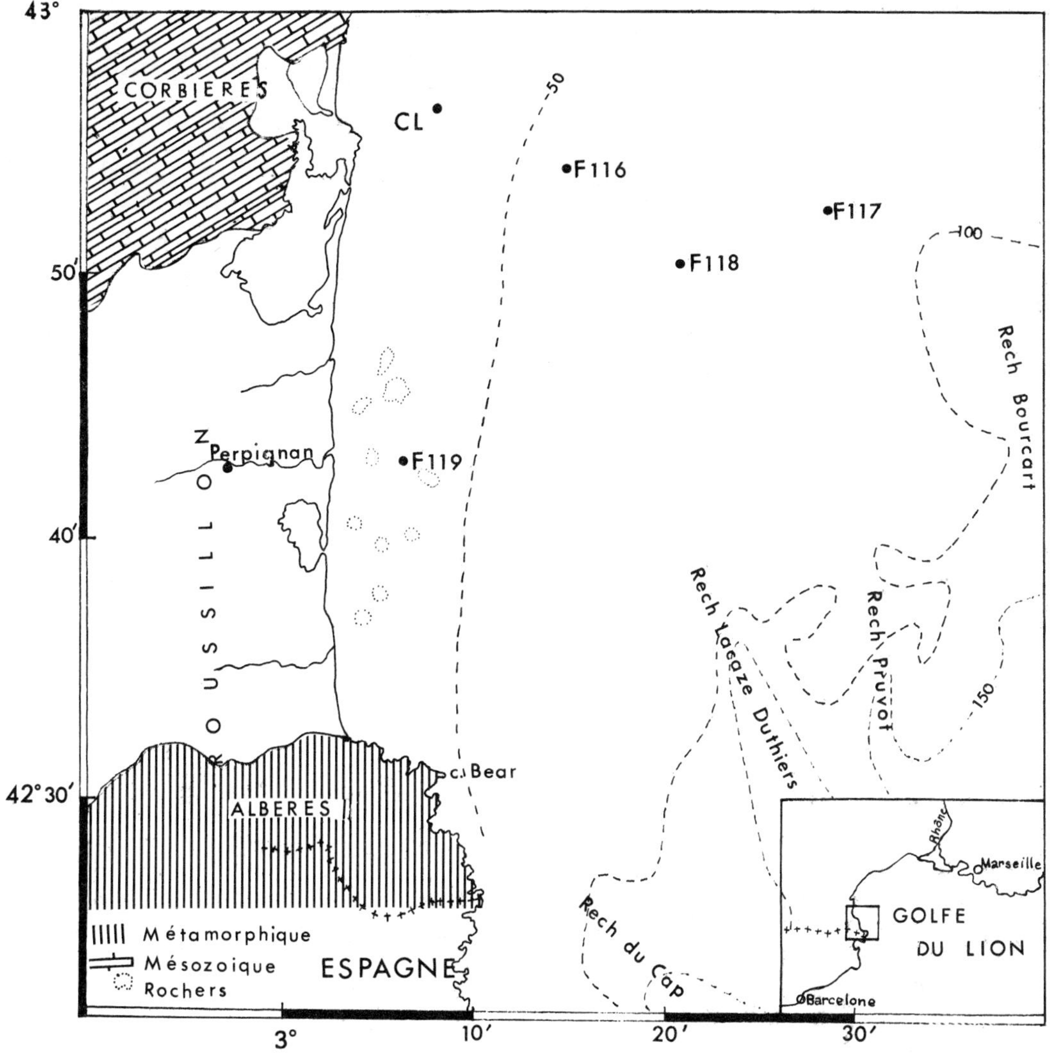

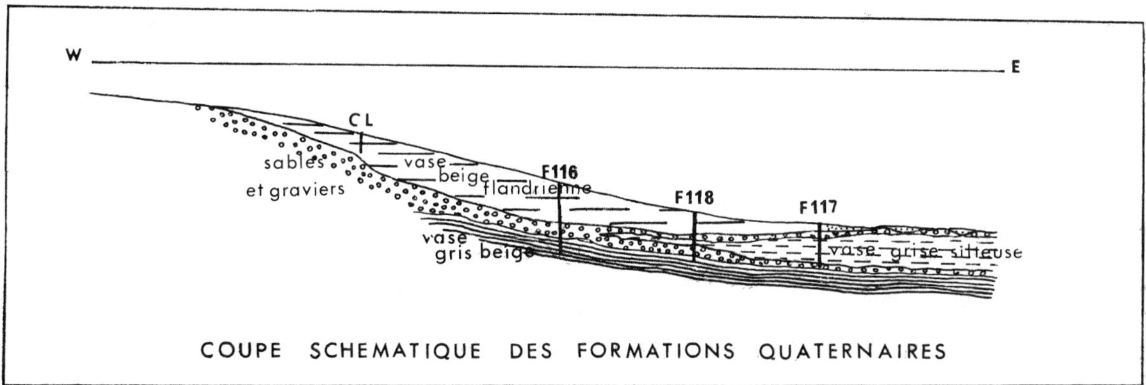

Figure 1. Situation des prélèvements. Coupe schématique des dépôts quaternaires récents.

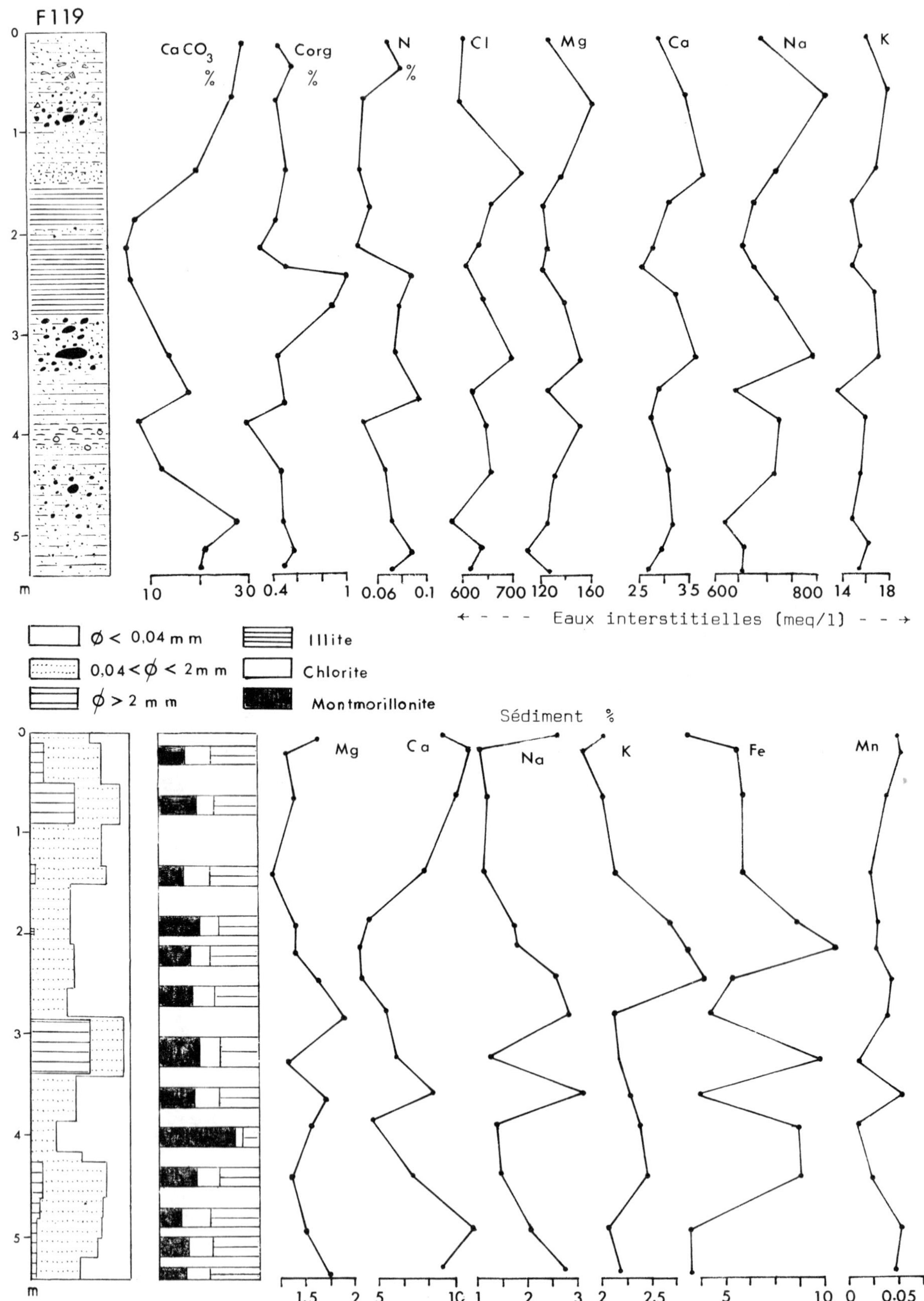

Figure 2. Carotte F 119. Analyses du sédiment et des eaux interstitielles: variations verticales.

nance de niveaux argileux et sableux, voire graveleux; elle se singularise par l'existence d'un horizon d'argiles vertes à concrétions calcaires (Figure 2). Sa position dans la zone littorale est responsable de la diversité des faciès:

—de 0 à 90 cm: vase gris-beige, sableuse et graveleuse à la base,
–de 90 à 150 cm: vase sableuse, puis sable vaseux (130 à 150 cm),
–de 150 à 280 cm: vase compacte, silteuse, vert-olive à noir, à débris végétaux,
–de 280 à 340 cm: gravelle argileuse à galets façonnés,
–de 340 à 380 cm: vase fluide légèrement sableuse,
–de 380 à 415 cm: argile verte à concrétions calcaires et débris végétaux,
–de 415 à 540 cm: vase sableuse et graveleuse.

L'analyse granulométrique confirme l'hétérogénéité de la sédimentation à laquelle participent des éléments de nature et d'origine variées. Les passées graveleuses correspondent à des épandages alluviaux remaniés par des actions marines littorales: façonnement peu poussé, débris coquilliers, encroûtement calcaire. Les horizons de vase montrent un classement médiocre des produits fins. Par contre, la vase verte (380 à 415 cm) est constituée d'une assez forte proportion de particules argileuses ($<2\mu$).

Les carottes CL (H = 457 cm) et F 116 (H = 617 cm) appartiennent aux secteurs infralittoral et circalittoral qui enregistrent depuis la transgression flandrienne le plus fort taux de sédimentation. Elles recoupent, sur une épaisseur de 400 cm, une même série de vases plus ou moins compactes, homogènes et de couleur beige à gris-beige (Figures 3 et 4). Le sondage F 116, plus profond, atteint de 390 à 530 cm, une assise graveleuse qui surmonte un horizon de vase plastique gris-beige (Figure 4).

L'analyse granulométrique des pélites fait apparaître une évolution progressive qui tend vers des faciès plus évolués en surface. La passée grossière est très hétérogène et comporte un mélange de sables frustes, de gravillons façonnés et de graviers peu roulés. La présence de galets calcaires témoigne d'apports en provenance des terrains mésozoïques des Corbières.

La carotte F 118 (H = 700 cm) présente une alternance de vases plus ou moins compactes et de niveaux grossiers graveleux ou sableux, situés notamment à 90–105 cm, 350–370 cm, et 460–500 cm (Figure 5). Ces passées grossières sont généralement hétérogènes et comportent trois fractions granulométriques: graviers, sables moyens constitués de quartz jaunâtres oxydés et sables fins micacés, riches en foraminifères et spicules d'éponges. Quelques grains de glauconie s'y trouvent mêlés. La fraction organogène offre la même hétérogénéité; elle est constituée de coquilles fraîches ou usées et de débris de taille variée. Les différents niveaux de vase ne sont pas semblables: alors que les parties supérieure et inférieure sont constituées de vase beige, identique à celle trouvée sur F 116, la passée médiane est formée de vase grise silteuse et compacte.

La carotte F 117 (H = 468 cm), la plus éloignée du rivage actuel, est caractérisée par une certaine homogénéité. Le dépôt est essentiellement constitué par une vase grise compacte, riche en débris noirs,

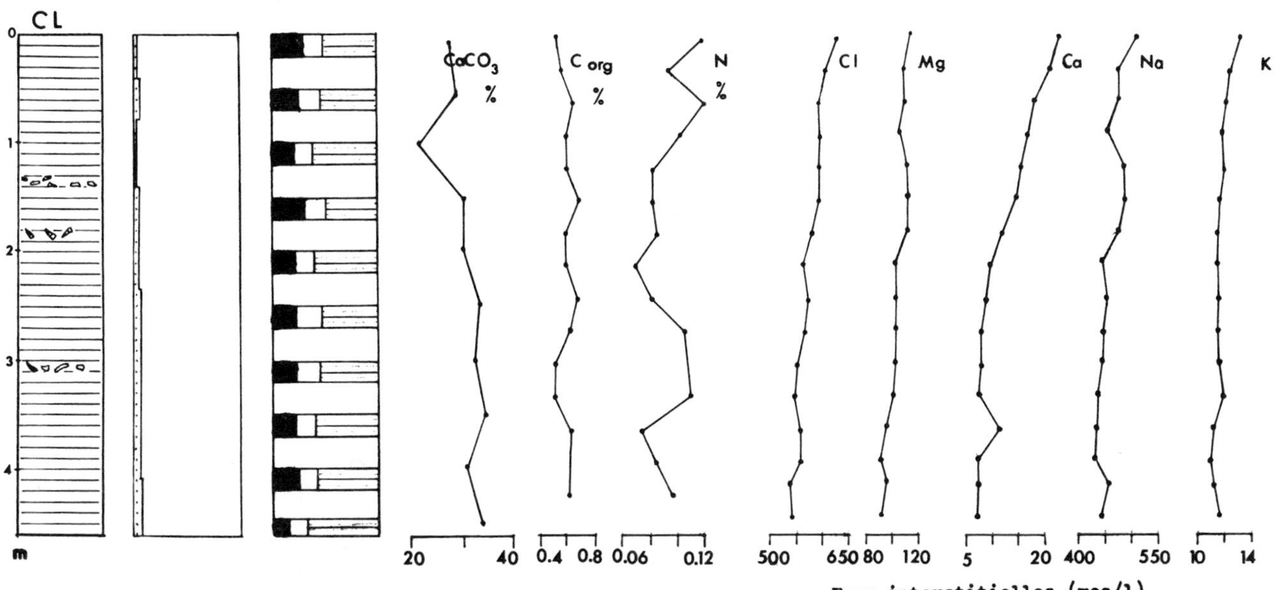

Figure 3. Carotte CL. Analyses du sédiment et des eaux interstitielles: variations verticales.

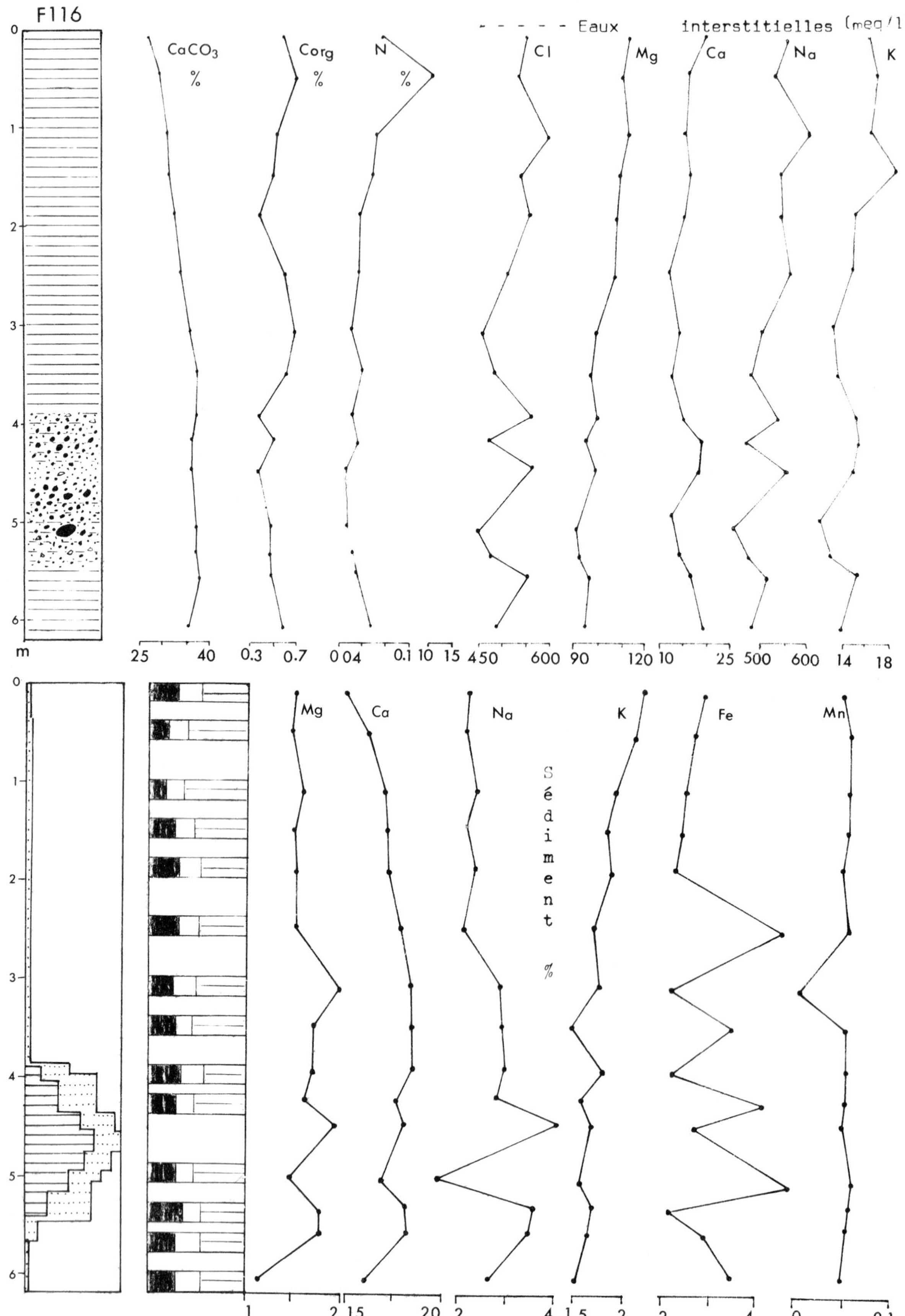

Figure 4. Carotte F 116. Analyses du sédiment et des eaux interstitielles:variations verticales.

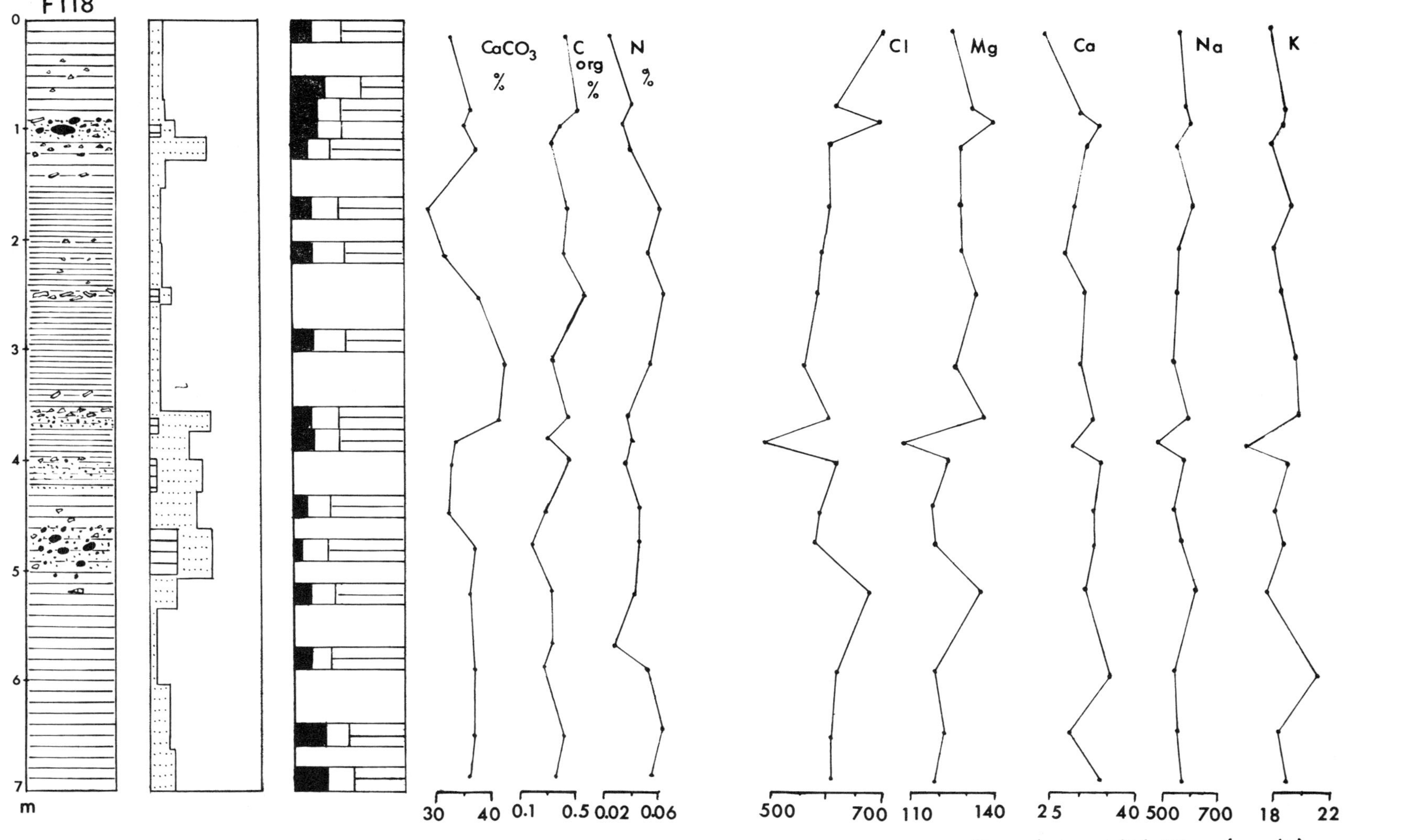

Figure 5. Carotte F 118. Analyses du sédiment et des eaux interstitielles: variations verticales.

limitée aux deux extrémités par des niveaux grossiers (0 à 10 cm et 440 à 468 cm) (Figure 6). La vase, assez uniforme, comprend une fraction silteuse importante (25%). Dans les passées grossières, les composants sont de taille et de nature variées. La fraction organogène offre un mélange de débris ou de coquilles entières plus ou moins bien conservées; on y trouve, également, quelques grains de glauconie.

Minéralogie des Argiles

Les associations minérales argileuses (illite, chlorite, montmorillonite) ne subissent pas de variations importantes. La teneur en montmorillonite, faible dans les vases grises (20%) est un peu plus abondante dans les vases beiges (jusqu'à 35%). Elle n'atteint un fort pourcentage (75%) que dans l'horizon d'argile verte de F 119. Les vases beiges à gris-beiges se caractérisent donc par un enrichissement en montmorillonite, alors que la vase grise est plus riche en illite et chlorite. Ces variations peuvent être liées au régime des apports (relation entre apports continentaux et montmorillonite), à des variations climatiques (relation illite-chlorite avec les périodes froides) ou à des effets de diagenèse (transformation montmorillonite ⟶ illite).

Géochimie du Sédiment

Les dépôts rencontrés sont riches en carbonates. Leur analyse a été effectuée sur la fraction pélitique ($< 40\mu$) afin d'éliminer l'interférence due aux accumulations de coquilles. Sauf dans les horizons argileux de F 119 (200 et 400 cm) où elle décroît jusqu'à 8%, la teneur en carbonates est toujours supérieure à 20% et dépasse souvent 30% (CL, F 116, F 118). Les faibles valeurs correspondent soit à des horizons d'argile compacte (F 119 : 200 et 400 cm), soit à des passées graveleuses ou silteuses (F 118 : 150 et 400 cm).

Les teneurs en carbone organique et en azote, dont les variations sont parallèles, sont faibles (0,3–0,6 % C; 0,03–0,05% N). Il existe pourtant des différences suivant le prélèvement envisagé et suivant les niveaux recoupés. Les dépôts littoraux sont plus riches en carbone organique (0,5 à 0,6%) que les dépôts du large (0,3%); il en est de même pour les vases beiges comparativement aux vases grises. On retrouve là l'origine continentale de la matière organique. Ce fait, déjà observé auparavant (Bordovskiy, 1965) est confirmé par les variations opposées du carbone et de la chlorinité des eaux de pores. Ce phénomène apparaît surtout sur les carottes F 116 (300 cm) et F 119 (230 cm). On note aussi un appauvrissement en matière organique au niveau des passées grossières.

La composition chimique minérale du sédiment est sous la dépendance de plusieurs facteurs liés à l'origine du dépôt et à son évolution. Ainsi, les teneurs en calcium sont directement liées au taux de carbonates, le sédiment n'ayant pas été décarbonaté avant analyse. Pour ces mêmes raisons, les teneurs en calcium et potassium sont opposées, l'enrichissement en minéraux micacés, riches en potassium, étant forcément corrélatif d'un appauvrissement en carbonates. Ce phénomène est particulièrement visible sur le sondage F 117 riche en illite.

A côté de ces remarques générales, un certain nombre de caractères particuliers à chaque dépôt se manifeste. Ainsi, dans les premiers centimètres de la carotte F 119, les teneurs en magnésium, sodium et potassium s'accroissent brusquement alors que celles en fer et calcium diminuent. On peut interpréter ce phénomène par une adsorption préférentielle de certains cations au niveau des argiles (Grim et Johns, 1953), d'autant plus probable que les eaux interstitielles s'appauvrissent conjointement en Mg, Na et K. Sur ce même sondage, les fluctuations des teneurs en fer et manganèse sont importantes à partir de 150 cm et tendent, alors, à être opposées. Les maxima en fer sont généralement liés à des horizons à accumulations grossières qui présentent des concrétions ferrugineuses et des feuillets altérés de biotite. Les valeurs les plus fortes en manganèse correspondent aux séquences argilo-sableuses où les teneurs sont comparables et de l'ordre de 0,05%.

Les variations de composition sont plus discrètes sur F 116, plus homogène. A partir de 200 cm, le fer présente une répartition très irrégulière qui peut s'interpréter par des modifications des conditions oxydo-réductrices au cours du tassement (Sevast'yanov, 1968). Le phénomène le plus général qui intervient alors, est la formation, en milieu réducteur, de sulfures qui enrichissent le sédiment en fer.

Enfin, la carotte F 117, recoupant des niveaux différents, présente une opposition entre les variations des teneurs en potassium dans le sédiment et les eaux interstitielles, qui pourrait s'interpréter comme la conséquence de phénomènes d'adsorption préférentielle et de transformation diagénétique de montmorillonite en illite (Grim et Johns, 1953; Powers, 1959).

Eaux Interstitielles

Les caractéristiques chimiques des eaux de pores traduisent des variations de composition qui coïncident souvent avec l'importance des apports marins ou continentaux. On note un accroissement de la salinité vers la surface, d'autant plus visible que les fluctuations le long de la carotte sont peu marquées (Figures 3 et 4). La teneur en ions est plus importante dans les eaux interstitielles que dans l'eau de mer sus-jacente, la compaction au niveau du sédiment ayant pour effet de concentrer les solutions initiales.

La minéralisation des eaux de pores n'a pas la même importance pour chaque sondage. Ainsi la carotte F 118 (Figure 5) présente des concentrations moyennes plus fortes (Cl: 600 à 700 meq/1; Mg: 120 à 130 meq/1; Na: 500 à 700 meq/1) que les carottes CL (Figure 3) et F 116 (Figure 4) qui sont plus littorales (Cl: 500 à 600 meq/1; Mg: 90 à 110 meq/1; Na: 450 à 550 meq/1). La vase grise recoupée par le sondage F 117, bien qu'elle corresponde à un dépôt de mer peu profonde montre également une minéralisation assez élevée des eaux de pores (Figure 6). La carotte F 119 (Figure 2) se distingue des autres pré-

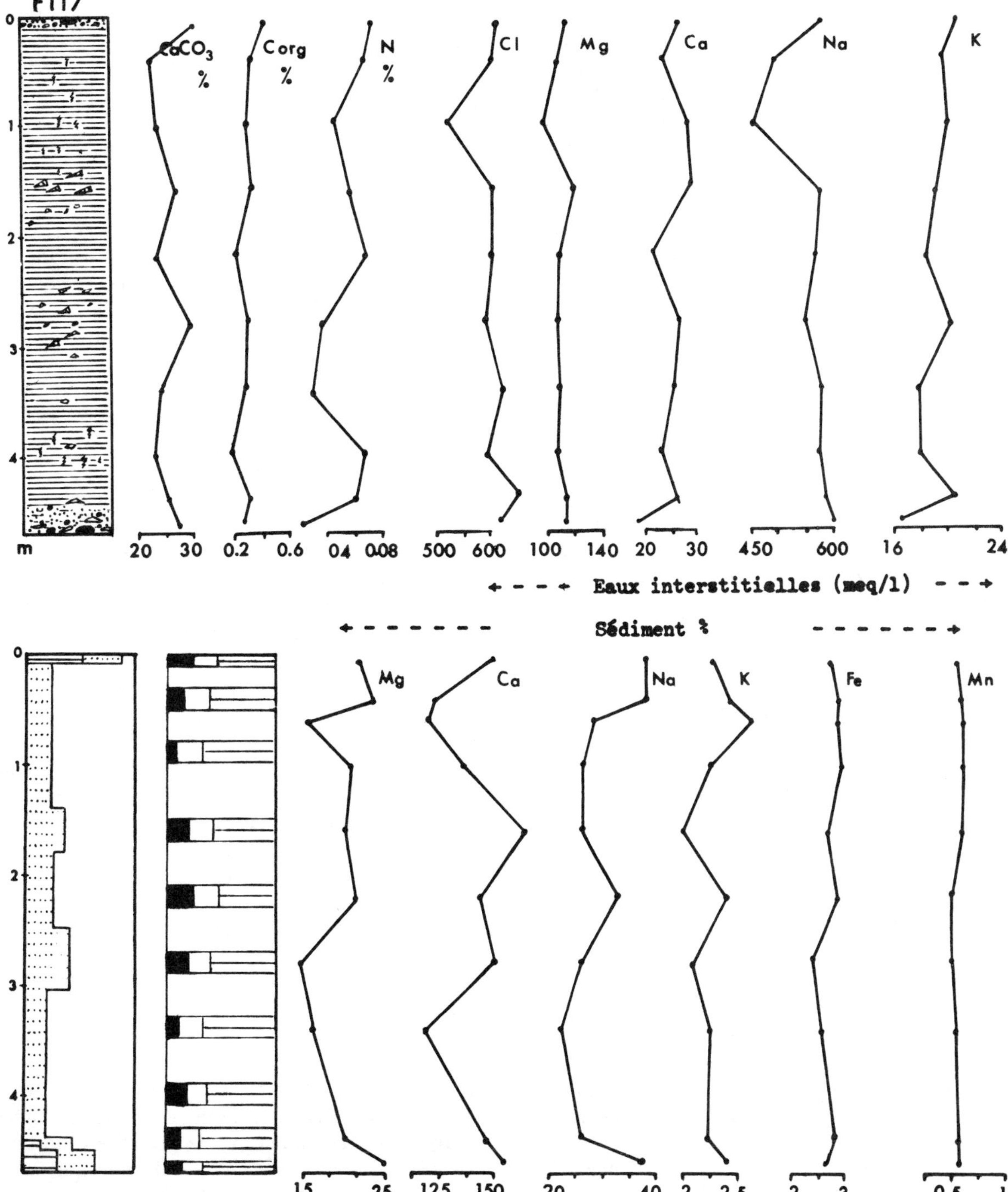

Figure 6. Carotte F 117. Analyses du sédiment et des eaux interstitielles: variations verticales.

lèvements. Bien que située à proximité immédiate de la côte, elle présente une salinité des eaux interstitielles assez élevée (Cl: 600 à 700 meq/1; Mg: 120 à 150 meq/1; Na: 650 à 750 meq/1) et surtout très variable, conséquence de l'hétérogénéité du sédiment et des conditions particulières de dépôt.

Résultats Palynologiques

L'étude palynologique des carottes à plusieurs niveaux a apporté un complément utile à la définition des faciès rencontrés. Elle permet, en effet, d'estimer l'importance relative des influences marines et continentales (Figure 7). Les résultats sont établis sur la base des teneurs relatives en espèces d'origine continentale, marine ou palustre, telles qu'elles sont définies ci-dessous:

1. Apports d'origine continentale, caractérisés par trois associations principales: association riche en pollens variés à *Pinus* dominant, association à forte proportion de pollens palustres (*Phragmites, Juncus, Potamogeton* et *Iris pseudacorus*) et association où prédominent les débris végétaux (cuticules ou débris ligneux).
2. Apports d'origine marine caractérisés par des spores d'algues ou la présence d'algues unicellulaires (*Tasmanacées* et *Leiosphéridacées*) et par des *Hystrichosphères* et des *Dinoflagellés*.

Il y a généralement enrichissement en pollens dans les passées argileuses, tandis que les niveaux grossiers comportent principalement des débris végétaux.

Sur les trois carottes situées le plus près de la côte actuelle, le caractère marin est peu prononcé, malgré une augmentation vers 90 cm (CL) et 130 cm (F 116). Les influences continentales se traduisent principalement par la présence de pollens des genres *Alnus, Quercus, Corylus, Salix*. Dans les horizons supérieurs des vases flandriennes, on enregistre un développement des espèces palustres.

Sur le sondage F 118, apparaît une distribution plus marquée des associations marines avec deux maxima vers 50 cm et 200 cm. Les pollens de *Pinus* représentent la plus grande partie des apports continentaux. Sur la coupe F 117, pourtant plus éloignée de la côte, mais qui recoupe un niveau de vase grise silteuse, la répartition pollinique fait apparaître une prédominance des apports continentaux, malgré une incursion marine vers 270 cm. Le niveau d'argile verte mis en évidence sur F 119 correspond à un milieu particulier, caractérisé par l'absence de pollens et la présence de débris végétaux (cuticules et débris ligneux). La concordance entre l'accroissement de la salinité et l'importance des apports marins, tels qu'ils sont définis par la palynologie, est assez marquée, particulièrement à 320 cm sur F 119. Elle est moins apparente sur les autres coupes, mais le nombre réduit d'analyses polliniques ne permet pas d'établir des corrélations plus étroites.

DISCUSSION

Les résultats analytiques confirment la succession de faciès variés, liée à une grande mobilité morphologique et bathymétrique du secteur littoral au cours des étapes de la sédimentation récente.

En l'absence de déterminations faunistiques, les données écologiques nous sont fournies par l'analyse palynologique. La constance des espèces rencontrées témoigne de faibles variations des associations végétales au cours de la période récente.

La pauvreté en pollens souvent constatée peut s'expliquer soit par leur dispersion plus grande dans les sédiments superficiels non encore consolidés, soit par la granulométrie du dépôt. Néanmoins, pour certains auteurs, elle est l'indice de conditions climatiques froides (Pokrovskaïa, 1950).

Les zones d'enrichissement relatif en débris végétaux coïncident le plus souvent soit avec des sédiments franchement grossiers, où leur présence peut résulter d'un classement différentiel dans les dépôts littoraux (Keller-Bonnet et Doubinger, 1963), soit avec des vases terrigènes peu évoluées, modernes (CL, F 119) ou subrécentes (F 117), déposées dans des secteurs relativement proches de la côte. Dans les deux cas, il n'y a pas obligatoirement enrichissement corrélatif en carbone organique, la fraction organique dispersée (ou soluble) pouvant être éliminée en grande partie par les phénomènes de lévigation ou par minéralisation dans des conditions oxydantes.

L'interprétation de certains résultats palynologiques et géochimiques est souvent rendue difficile par les mélanges qui interviennent au contact de niveaux correspondant à deux phases successives de mise en place.

En effet, les dépôts examinés présentent de nombreuses variations de faciès correspondant à une grande mobilité de la ligne de rivage au Quaternaire terminal. Aussi est-il souvent malaisé de faire la part des effets diagénétiques.

La succession lithologique est fonction du secteur considéré (Figure 1):

Les carottes les plus littorales sont très hétérogènes. Sur la carotte F 119, l'alternance de séquences fines et grossières, liée aux oscillations de la ligne de rivage, est interrompue par la présence d'un horizon d'argiles vertes à concrétions calcaires qui peut être interprété comme un paléosol. L'enrichissement très net en montmorillonite serait, semble-t-il, l'in-

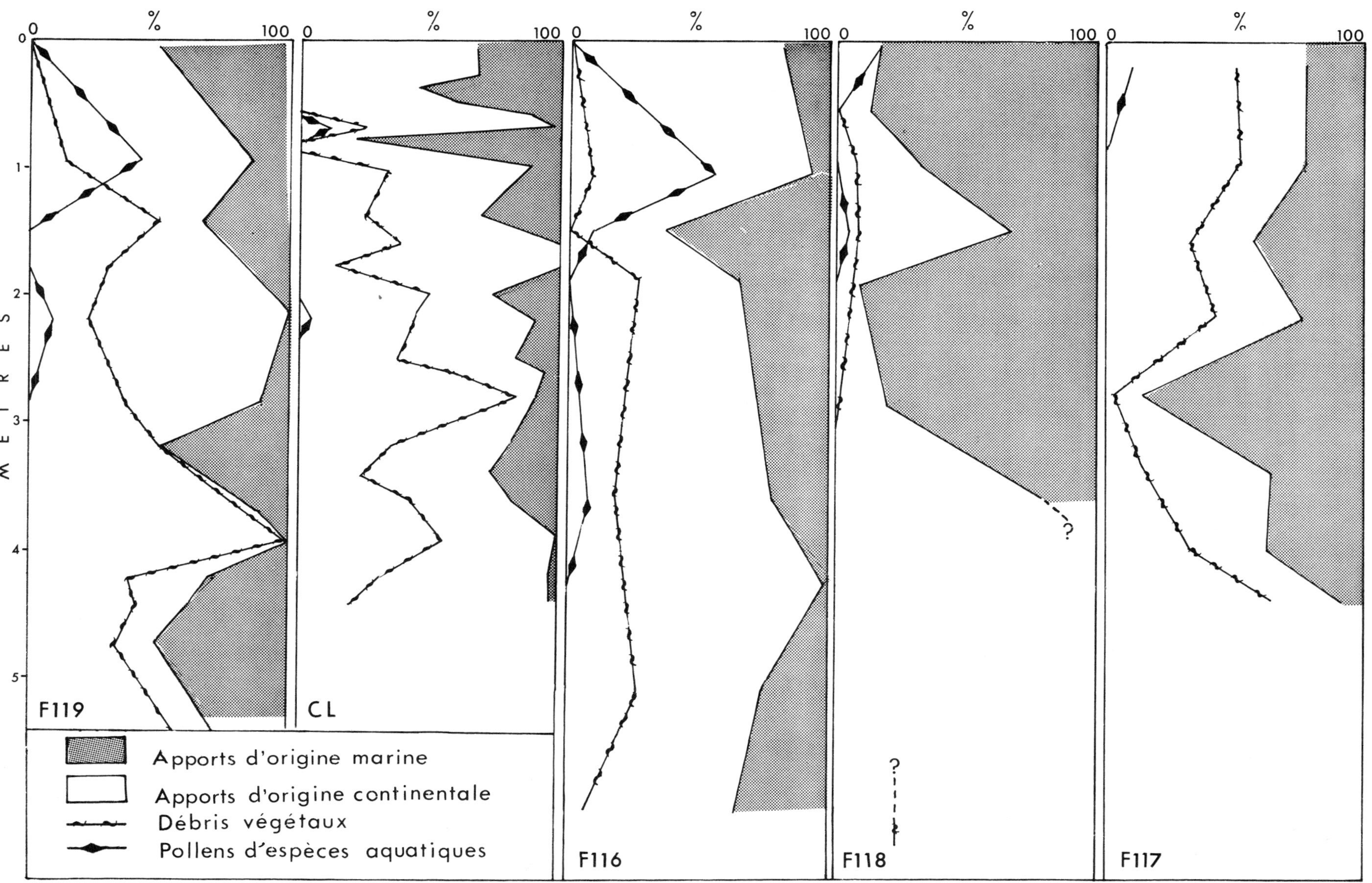

Figure 7. Diagrammes polliniques des diverses carottes.

dice d'une néogenèse de ce minéral dans des conditions de mauvais drainage, équivalentes à celles qui président à la formation des vertisols hydromorphes (Trauth *et al.*, 1967).

A ce niveau, les associations d'origine marine sont inexistantes, les apports continentaux consistent essentiellement en débris ligneux. La teneur en carbone organique est très faible (0,17%), la phase organique non particulaire est réduite.

Dans la portion médiane du plateau, les carottes CL et F 116 recoupent une épaisse série argileuse, contemporaine de la transgression flandrienne. Cette série repose sur des niveaux grossiers, témoins régressifs remaniés au cours de la remontée eustatique.

Plus au large, le faible taux de sédimentation récente, permet d'atteindre sous les sables reliques würmiens des niveaux de vase grise compacte (F 117) que l'on peut attribuer à un dépôt mis en place au cours d'une période froide. En effet, c'est dans des horizons équivalents qu'ont été trouvées, plus au Sud, les thanatocoenoses à *Cyprina islandica* (Bourcart, 1956; Mars *et al.*, 1957).

Les corrélations des résultats sur ces diverses carottes amènent à distinguer les séquences sédimentaires et à préciser leur environnement (Figure 1).

1. L'analyse de niveaux équivalents flandriens (vase homogène) sur les carottes F 119 (niveaux supérieurs), CL (0 à 457 cm), F 116 (0 à 400 cm) et F 118 (0 à 90 cm), permet d'établir une chronologie relative dans les étapes successives de la remontée du niveau marin.

De la base au sommet de ces horizons, on constate un accroissement régulier de la salinité des eaux interstitielles, lié à l'accentuation des apports d'origine marine, une évolution des faciès granulométriques, une abondance relative des associations d'origine marine et souvent une diminution du taux de carbonates en grande partie détritiques.

Le stade régressif récent se manifeste surtout au niveau des pollens par l'apparition générale des espèces aquatiques vers 100 cm, consécutif à l'extension des lagunes littorales. Ce phénomène s'accompagne quelquefois d'une légère dilution des eaux de pores.

Il semble que l'on puisse considérer les valeurs relatives de la salinité des eaux de pores comme représentatives des variations de composition de l'eau originelle (Degens, 1965; Degens et Chilingar, 1967; Bouysse *et al.*, 1966).

Cela est d'autant mieux mis en évidence que l'on se trouve dans un secteur à sédimentation rapide qui permet aux eaux initiales de conserver certaines de leurs caractéristiques majeures (Parent, 1968; Cauwet et Gadel, 1970). Le taux d'accumulation serait d'environ 10 cm par siècle dans la portion médiane du plateau. Cette vitesse explique également les faibles effets diagénétiques tels qu'ils peuvent se traduire dans l'évolution minéralogique des argiles (Chamley, 1968).

La salinité moyenne des eaux interstitielles est moins élevée sur les carottes situées à proximité de la côte (CL et F 116; Cl = 550 à 600 meq/l), que dans les dépôts plus externes (F 118; Cl = 650 à 700 meq/l).

Ce phénomène est corrélatif de l'augmentation des pollens et de la matière organique dans les sédiments littoraux.

La diminution de la matière organique vers le large (CL:C ~ 0,6%; F 118:C ~ 0,4%) s'explique principalement par la réduction des apports continentaux et l'influence certaine des transformations diagénétiques au cours du transport (matière organique plus évoluée et en partie dégradée).

Dans cette sédimentation homogène, les effets de la diagenèse se traduisent dans les premiers centimètres par la minéralisation de l'eau combinée par rapport à l'eau de surface (Friedman *et al.*, 1968).

Avec l'enfouissement, l'évolution divergente de certains éléments, notamment le magnésium et le sodium dans les eaux de pores et le sédiment, pourrait également résulter de processus diagénétiques (Shishkina, 1957; Grim et Johns, 1953; Tageeva et Tikhomirova, 1962).

Les variations de nature des argiles sont liées au régime des apports (relation entre apports continentaux et montmorillonite) et aux changements climatiques (relation illite-chlorite avec les périodes froides).

2. Dans les niveaux grossiers qui forment le substrat de la sédimentation argileuse flandrienne, l'interprétation des résultats géochimiques est rendue difficile par suite de l'hétérogénéité des dépôts qui résulte à la fois de remaniements synsédimentaires et de contaminations par la vase sus-jacente.

On enregistre dans ces horizons de fortes fluctuations, souvent anarchiques. Seules les données sédimentologiques et malacologiques peuvent fournir des renseignements d'ordre stratigraphique: caractère marin littoral du sédiment et des coquilles qui lui sont associées, présence de débris remaniés de faune fossile (*Cyprina islandica*).

Des caractères semblables se retrouvent dans les sables du large, équivalents latéraux des assises grossières.

3. La vase silteuse grise recoupée par les carottes F 117 (10–450 cm), F 118 (150–400 cm) se situe sous les affleurements graveleux et sableux de la partie externe du plateau. Elle serait l'équivalent des boues grises à thanatocoenoses würmiennes rencontrées plus au Sud, au large du Cap Creus (Bourcart, 1956; Mars *et al.*, 1957). Les divers paramètres sédimentologiques et géochimiques soulignent les particularités

de ce faciès apparemment homogène: concentration plus élevée en certains éléments (Mg, Na, K) dans les eaux interstitielles et le sédiment, faible taux de matière organique, nette prédominance de l'illite et de la chlorite.

Ces observations peuvent donner lieu à une double interprétation faisant appel soit aux facteurs climatiques, soit aux effets diagénétiques.

Ces deux hypothèses peuvent, en effet, expliquer la minéralisation des eaux et la présence de minéraux argileux résistants, résultant de conditions particulières d'érosion ou encore de transformations à partir de phyllites dégradées.

Néanmoins, malgré son état de tassement (faible teneur en eau), l'épaisseur réduite du dépôt et la plasticité des horizons inférieurs justifient difficilement un processus diagénétique avancé.

Il s'agirait plutôt de l'incidence de facteurs climatiques (période froide), apparaissant au niveau des phyllites, de la matière organique, de la microflore et de la microfaune.

Ces dépôts se seraient mis en place par une profondeur de 30 à 40 mètres, au cours d'une oscillation négative würmienne (faciès granulométrique peu évolué, abondance des apports continentaux, débris ligneux).

Certaines caractéristiques géochimiques attestent pourtant de phénomènes diagénétiques résultant de conditions réductrices: néoformation de sphérules de pyrite.

CONCLUSIONS

Les dépôts analysés sont représentatifs de la sédimentation récente du plateau continental roussillonnais. Les corrélations établies entre les données sédimentologiques, géochimiques et palynologiques permettent de préciser les modalités des apports minéraux et organiques et les mécanismes de dépôt en fonction des conditions climatiques et des fluctuations de la ligne de rivage.

Les apports continentaux se traduisent généralement par l'abondance des pollens, un accroissement de la matière organique, débris végétaux principalement; l'ensemble étant corrélatif d'une diminution de la salinité des eaux interstitielles et d'un enrichissement en certaines espèces minérales argileuses (montmorillonite).

Il y a souvent parallélisme entre l'accroissement des apports marins et la mise en place des passées grossières au cours d'oscillations positives du niveau de la mer. Ces phénomènes s'accompagnent généralement d'un accroissement de la salinité des eaux de pores.

Il est possible de distinguer diverses séquences sédimentaires dans l'ensemble recoupé par les carottages:

1. Dans les niveaux flandriens traversés dans le domaine littoral, une évolution générale apparaît, se traduisant par des valeurs croissantes de la salinité vers les horizons supérieurs, corrélatives d'un classement plus poussé des faciès granulométriques. Dans les assises plus récentes, on note un développement d'espèces palustres.

Ces phénomènes évolutifs peuvent être attribués aux effets de la transgression flandrienne ayant pour conséquence des conditions marines plus affirmées. A une époque plus récente, a succédé un stade régressif qui a vu l'extension des lagunes.

2. Les assises grossières à faune froide, formant le substrat de la sédimentation argileuse flandrienne, offrent une grande hétérogénéité. Les résultats sédimentologiques, géochimiques et palynologiques mettent en évidence de fortes variations qui peuvent s'expliquer à la fois par les remaniements ou les contaminations au contact des assises plus récentes.

3. La vase silteuse atteinte par les carottages sur la partie externe du plateau continental, sous les formations grossières, correspond à un épisode régressif würmien. Ces dépôts présentent des caractères sédimentologiques, géochimiques (minéralisation des eaux interstitielles, minéraux argileux) et paléontologiques qui traduisent des conditions peu profondes de dépôt.

Dans les niveaux équivalents, contemporains de la même phase de sédimentation, l'accroissement de salinité des eaux de pores vers le large s'accompagne d'une diminution du taux de matière organique.

Les dépôts effectués près des côtes enregistrent les fluctuations du niveau de la mer, qui se traduisent par une grande hétérogénéité dans la nature des sédiments.

La variabilité des conditions de dépôt sur le plateau continental au cours du Quaternaire terminal ne permet pas toujours de faire la part exacte des effets diagénétiques. Leur incidence est surtout sensible dans les zones profondes où les faciès sont plus constants.

REMERCIEMENTS

Nous sommes redevables à Madame M. Pi-Radondy des déterminations palynologiques. Les analyses chimiques ont été effectuées avec l'aide efficace de Monsieur H. Got. Nous tenons à leur exprimer nos vifs remerciements.

BIBLIOGRAPHIE

Blanc, J. J. 1968. Sedimentary geology of the Mediterranean Sea. *Oceanography and Marine Biology, Annual Review,* 6:377–454.

Bonifay, E. 1969. Le Quaternaire littoral et sous-marin des côtes françaises de la Méditerranée. Etudes françaises sur le Quaternaire. *VIIème Congrès International INQUA 1969*. Supplément au Bulletin de l'Association Française pour l'Etude du Quaternaire:43–55.

Bordovskiy, O. K. 1965. Transformation and diagenesis of organic matter in sediments. *Marine Geology*, 3:83–114.

Bourcart, J. 1956. Recherches sur le plateau continental de Banyuls-sur-Mer. *Vie et Milieu*, 6:435–524.

Bourcart, J., M. Gennesseaux et E. Klimek 1961. Les canyons sous-marins de Banyuls et leur remplissage sédimentaire. *Comptes Rendus de l'Académie des Sciences, Paris*, 253: 19–24.

Bouysse, P., J. Goni, C. Parent et Y. Le Calvez 1966. Recherches du B.R.G.M. sur le plateau continental. Premiers résultats sédimentologiques, micropaléontologiques et géochimiques (Baie de la Vilaine). *Bulletin du Bureau de Recherches Géologiques et Minières*, 5:1–77.

Cauwet, G. et F. Gadel 1970. Etude géologique et géochimique d'une carotte prélevée dans le Golfe du Lion. *Bulletin du Bureau de Recherches Géologiques et Minières*, 4(2):5–17.

Chamley, H. 1968. La sédimentation argileuse actuelle en Méditerranée nord-occidentale. Données préliminaires sur la diagenèse superficielle. *Bulletin de la Société Géologique de France*, 7:75–88.

Degens, E. T. 1965. *Geochemistry of Sediments. A brief survey*. Prentice Hall, Englewood Cliffs, New Jersey, 342 p.

Degens, E. T. et G. V. Chilingar 1967. Diagenesis of subsurface waters. In: *Diagenesis in Sediments*, eds. Larsen, G. et G. V. Chilingar, Elsevier, Amsterdam:477–502.

Friedman, G. M., B. P. Fabricand, E. S. Imbibo, M. E. Brey et J. N. Sanders 1968. Chemical change in interstitial waters from continental shelf sediments. *Journal of Sedimentary Petrology*, 38:1313–1319.

Glangeaud, L., G. Bellaiche, M. Gennesseaux et G. Pautot 1968. Phénomènes pelliculaires et épidermiques du rech Bourcart (Golfe du Lion) et de la mer hespérienne. *Comptes Rendus de l'Académie des Sciences. Paris*. 267:1079–1083.

Got, H., A. Guille, A. Monaco et J. Soyer 1968. Carte sédimentologique du plateau continental au large de la côte catalane française (P.-O.). *Vie et Milieu*, 19 (2B):273–290.

Got, H. et A. Monaco 1969. Sédimentation et tectonique plio-quaternaires du précontinent méditerranéen au large du Roussillon (P.-O.). *Comptes Rendus de l'Académie des Sciences, Paris*, 268:1171–1174.

Got, H., A. Monaco et D. Reyss 1969. Les canyons sous-marins de la mer catalane. Le rech du Cap et le rech Lacaze-Duthiers. *Vie et Milieu*, 20 (2B):257–278.

Grim, R. E. et W. D. Johns 1953. Clay mineral investigations of sediments in the northern Gulf of Mexico. *Clays and Clay Minerals*, 2nd National Conference: 81–103.

Keller-Bonnet, J. et J. Doubinger 1963. Etude microscopique des débris végétaux du Carbonifère et des séries infrasalifères de quelques sondages sahariens. *Bulletin du Service de la Carte Géologique Alsace-Lorraine*, 16:249–259.

Levy, A. 1967. Contribution à l'étude des foraminifères des rechs du Roussillon et du plateau continental de bordure. *Vie et Milieu*, 18 (1B):63–102.

Mars, P. 1959. Les faunes malacologiques quaternaires "froides" de Méditerranée. Le gisement du Cap Creus. *Vie et Milieu*, 9:293–309.

Mars, P., J. Mathely et J. Paris 1957. Remarques sur le gisement quaternaire du Cap Creus. *Comptes Rendus de l'Académie des Sciences, Paris*, 242:1940–1942.

Monaco, A. 1967. Etude sédimentologique et minéralogique des dépôts quaternaires du plateau continental et des rechs du Roussillon. *Vie et Milieu*, 18 (1B):33–62.

Parent, C. 1968. Réflexions sur quelques caractères des eaux interstitielles dans les sédiments actuels ou récents. Conséquences diagénétiques. *Bulletin du Bureau de Recherches Géologiques et Minières*, 4:93–94.

Pokrovskaïa, I. M. 1950. Analyses polliniques. In: *Annales du Service d'Information Géologique et Minière*, 421 p.

Powers, M. C. 1959. Adjustment of clays to chemical change and the concept of the equivalence level. *Clays and Clay Minerals*, 6th National Conference, 1957:309–326.

Sevast'yanov, V. F. 1968. Redistribution of chemical elements as a consequence of oxidation-deoxidation processes in sediments of the Mediterranean Sea. *Lithology and Mineral Resources*, 1:3–15 (Traduction de *Lithologia i Poleznye Iskopayeme)*.

Shishkina, O. V. 1957. Ooze waters of the Pacific Ocean and adjoining seas. *Akademiya Nauk S.S.S.R. Doklady*, 112:470–473.

Tageeva, N. V. et M. M. Tikhomirova 1962. Géochimie des eaux interstitielles dans la diagenèse des sédiments marins (d'après l'exemple des dépôts de la mer Caspienne). *Service d'Information Géologique du Bureau de Recherches Géologiques et Minières*, 4286:192–200. (Traduction de *Moskva Izdat Akademiya Nauk SSSR*).

Trauth, N., H. Paquet, J. Lucas et G. Millot 1967. Les montmorillonites des vertisols lithomorphes sont ferrifères: conséquences géochimiques et sédimentologiques. *Comptes Rendus de l'Académie des Sciences, Paris*, 264:1577–1579.

Voinovitch, I. A., J. Debras-Guedon et J. Louvrier 1962. *L'analyse des Silicates*. Herman, Paris, 160 p.

Submarine Iron Deposits from the Mediterranean Sea*

Enrico Bonatti, Jose Honnorez, Oiva Joensuu and Harold Rydell

University of Miami, Miami

ABSTRACT

Submarine iron-rich deposits close to the island of Thera (Aegean Sea) and Stromboli (Tyrrhenian Sea) have been studied. The Thera deposits are rich in iron hydroxides and opal, while in the Stromboli deposits manganese oxides are also present. These deposits are formed as a result of submarine volcanic emanations which introduce Fe, Mn, Si, B, Ba, P and possible other elements in sea water. In addition, these hydrothermal emanations may contribute to the extraction and concentration of some elements from sea water. Fractionation of Fe from Mn takes place during deposition, whereby Fe precipitates close to the source, while Mn has a longer residence time in solution. Si and B co-precipitate with Fe, while Ba follows Mn. Similarities between the Thera and Stomboli iron deposits and deposits from oceanic seamounts and rifts are described.

RESUME

On a étudié des dépôts sous-marins riches en fer près des îles volcaniques de Thera (Mer Egée) et de Stromboli (Mer Tyrrhénienne). Les dépôts de Thera sont principalement constitués d'hydroxyde de fer et d'opale alors que dans ceux de Stromboli s'y ajoutent des oxydes de manganèse. Ces dépôts sont le produit d'émanations volcaniques sous-marines qui introduisent du Fe, Mn, Si, B, Ba, P et probablement d'autres éléments dans l'eau de mer. La séparation du Fe et du Mn s'effectue pendant le dépôt; en conséquence, Fe précipite près de la source, alors que Mn reste plus longtemps en solution. Si et B précipitent en même temps que Fe alors que Ba précipite en même temps que Mn. On décrit les similitudes entre les dépôts de fer de Thera et de Stromboli d'une part et ceux provenant des seamounts et rifts océaniques d'autre part.

INTRODUCTION

Iron and related elements are supplied to the ocean in particulate and dissolved forms primarily by streams which drain the weathering products of rocks on the continent. An additional but locally important source of iron to the ocean is hydrothermal activity connected with submarine volcanism. Hydrothermal-volcanic iron deposits have been reported from the submarine caldera of the island of Santorin, or Thera, in the Aegean Sea (Harder, 1964; Bonatti and Joensuu, 1966; Butuzova, 1966, 1969); from the vicinity of the island of Vulcano in the Tyrrhenian Sea (Honnorez, 1969; Honnorez *et al.*, 1971); from the submarine volcano Banu Wuhu in Indonesia (Zelenov, 1964); from the flanks of some sea mounts on the East Pacific Rise (Bonatti and Joensuu, 1966). Ferruginous sediments which outcrop along the East Pacific Rise and other active oceanic ridges are also thought to be formed as a result of submarine exhalative activity (Skornyakova, 1964; Boström and Peterson, 1966, 1969; Boström, 1970).

In the present paper we report on the chemistry and mineralogy of a submarine iron deposit discovered recently on the flank of the active volcanic island of Stromboli in the Tyrrhenian Sea (Figure 1); we present new data on the deposits from the submarine caldera of Santorin; and we discuss, in general the origin of hydrothermal submarine iron deposits.

*Contribution No. 1583 from the University of Miami, Rosenstiel School of Marine and Atmospheric Science, Miami, Florida 33149.

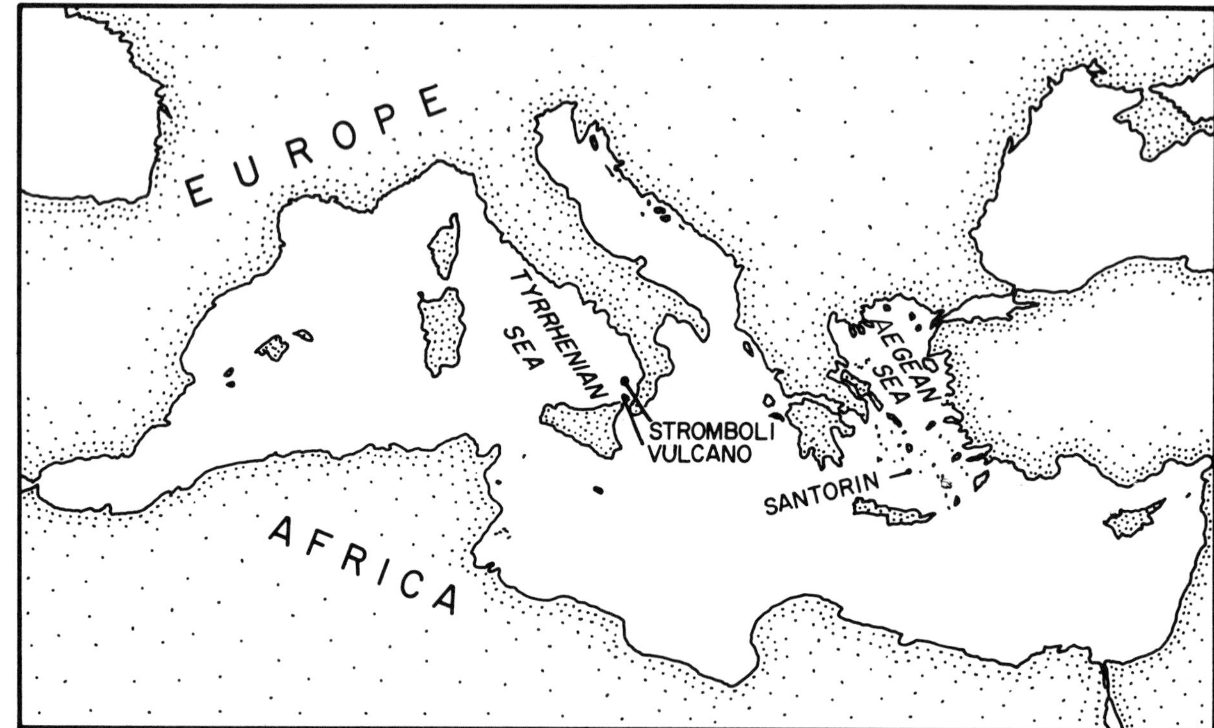

Figure 1. Location of the volcanic islands Stromboli, Vulcano and Santorin (Thera) in the Mediterranean Sea.

CHEMISTRY AND MINERALOGY OF THE STROMBOLI DEPOSIT

During a cruise of the R/V BANNOCK, operated by the Research Council of Italy, dredgings were carried out at latitude 38°49′ and longitude 15°16′, at a depth of about 250 meters (Figure 2). This site is on the northeast flank of the Stromboli volcano, off the Strombolicchio islet. Various fragments of a dark brown, powdery rock were recovered at this station. No layering was apparent on examination of the hand specimens but irregular septas display a cavernous structure.

The concentration of major and trace elements was determined in two samples of the deposit (Table 1); the samples of iron precipitates from Stromboli and Santorin contain about 23 to 37% of adsorbed and absorbed water; the analyses of Ti and all trace elements were made by optical spectroscopy, using a Bausch and Lomb dual grating spectrograph. A minimum of 10 mg of crushed sample were blended with an internal standard buffer and with graphite, in the ratio 1:1:3. The whole was burned in a graphite electrode using a 12 amp D.C. arc. The precision and accuracy of the analyses is estimated to be within ± 15% for the elements in question. Si, Al, Fe, Mn, Ca,

Table 1. Concentration of major and minor elements in two wet samples of the iron deposit from Stromboli.

Per cent	Stromboli a	Stromboli b
Si	5.1	7.9
Al	0.2	0.4
Ca	0.50	0.44
Mg	0.74	0.40
K	0.78	0.46
Na	2.2	1.7
Fe	14.3	28.0^{+}
Mn	22.4	9.2
P	0.3	0.4
Ti	0.006	0.006
H_2O^+	—	8.6
ppm		
B	87	120
Ba	1070	550
Co	30	15
Cr	<5	<5
Cu	420	160
La	< 10	< 10
Ni	15	10
Sc	<3	<3
V	55	90
Y	30	25
Zr	75	40
Fe/Mn	0.6	3.0

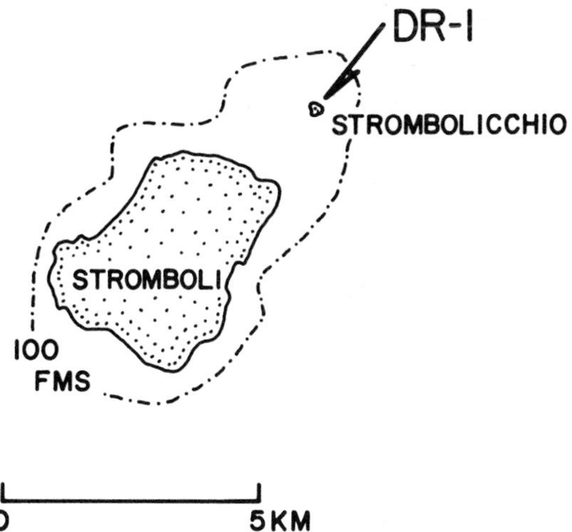

Figure 2. Location of the submarine iron deposit (DR-1) dredged NE of Stromboli, off the Strombolicchio islets.

Mg, Na, K and P were determined by atomic absorption, using a Perkin-Elmer 303 unit. Precision and accuracy are estimated to be within ±5% of the measured values. The results are reported in Table 1. Isotopic uranium and thorium analyses were done by alpha particle spectrometry, using a solid state detector and multichannel analyzer, after separation of uranium and thorium by co-precipitation, solvent extraction, ion exchange and electrodeposition. U-232 and Th-234 tracers were used to correct for analytical losses. Determined values and errors are shown in Table 6.

The mineralogy of the samples was studied by X-ray powder diffraction, using a General Electric RD-5 diffractometer; Fe K α radiation was employed for the analyses. The untreated material shows a major reflection at 7.2 Å, with minor peaks at 3.6 Å and 2.3 Å. These reflections disappear when the sample is treated with a 1 M solution of hydroxylamine sulfate at 60°C for 30 minutes (Figure 3). This treatment dissolves quantitatively MnO_2 compounds, leaving silicates and other common sedimentary minerals unaffected. On the basis of these data the X-ray reflections mentioned above were attributed to MnO_2, birnessite. The presence of other poorly crystalline manganese hydroxide and oxide phases cannot be excluded.

The samples were kept at 500°C in presence of air for 15 hours and subsequently X-rayed again. As a result, intense X-ray reflections of hematite, Fe_2O_3, and hausmannite, Mn_3O_4, appeared while the birnessite peaks disappeared (Figure 3). The fact that hematite was produced upon heating suggests that goethite, FeO(OH), is present in the untreated sample.

The presence of significant quantities of silica in the samples (Table 1), coupled with the absence of X-ray reflections of silicate minerals and with the high Si/Al ratio, suggests that silica is present in a disordered phase, probably opal.

CHEMISTRY AND MINERALOGY OF THE SANTORIN (THERA) DEPOSITS

Samples of a red, fluid mud were collected near the mouth of a submarine thermal spring in bays on the eastern shore of the islet Nea Kameni, located within the submerged caldera of Santorin, or Thera (Figure 4), during an expedition conducted in 1965 by Miami's Institute of Marine Sciences. We obtained from Professor A. Capart another sample of a similar muddy precipitate which was collected in 1967 from a bay on the northern shore of Nea Kameni (Figure 4). This islet has been formed in historic times by eruptions of andesitic-dacitic lavas within the submerged caldera of Thera. The red mud deposits were restricted to an area of a few 100 m^2 in the bays B and C and their thickness was very irregular (from a few up to 40 cm) following the bottom topography; Butuzova (1966) mentioned that "the area of the bay A is 2 to 2.5 thousand m^2" but this author did not report that the iron deposit she studied covered either completely or partly the bottom surface of this bay. Puchelt (1971) reported an area of 350 m^2 for the iron deposit in the Bay of Paleo Kameni that he had studied. These iron deposits are thus restricted to within several small bays off the shores of Nea and Paleo Kameni. Chemical

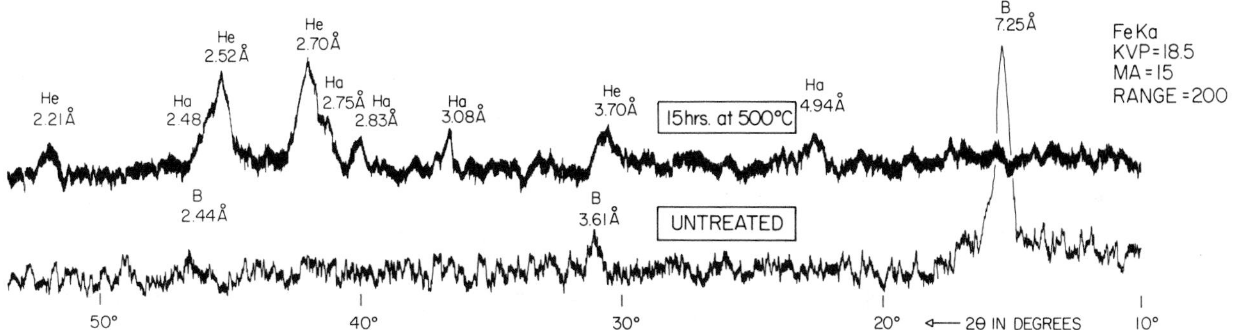

Figure 3. X-ray diffractograms for an untreated sample of the Stromboli iron deposit DR-1 and for the same sample kept at 500°C for 15 hours.

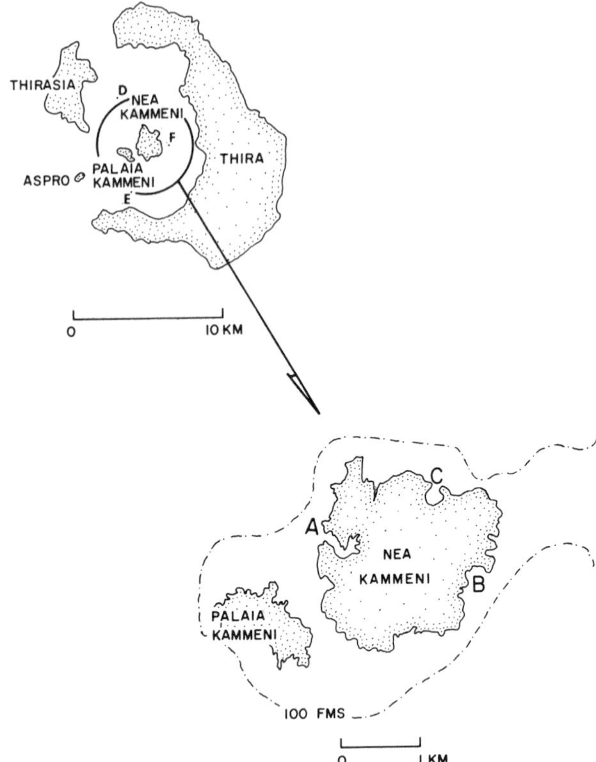

Figure 4. Location of the iron deposits (A, sample analyzed by G. Yu. Butuzova, 1966; B, sample from IMS, 1965; C, sample from A. Capart, 1967) and the sediment cores (D,E,F) from the caldera of Santorin volcano (Thera).

Table 2. Concentration of major and minor elements in various samples of iron deposits from Thera (Santorin).*

Per cent	Thera (a)	Thera (b)	Thera (c)	Thera (d)	Thera (e)	Thera (f)
Si	6.5	11.6	5.1	8.4	6.1	10.6
Al	0.3	1.2	0.5	0.03	0.3	0.7
Ca	0.66	0.57	0.38	0.10	0.91	1.18
Mg	0.83	0.34	0.49	0.12	0.63	0.84
K	0.43	0.14	0.18	—	0.49	0.56
Na	6.29	0.50	3.41	—	5.01	5.01
Fe	27.0	35.0	40.0	40.5	37.1	27.0
Mn	0.2	0.6	0.07	0.11	0.007	0.01
P	0.8	1.1	0.3	0.13	0.11	0.71
Ti	0.009	0.03	0.003	0.006	—	—
H_2O^-	—	—	—	21	13.27	9.72
H_2O^+	—	—	—		5.58	6.91
ppm						
B	130	190	135		6400	2600
Ba	80	90	88			
Co	<5	<5	<5			
Cr	<5	<5	<5			
Cu	9	30	20			
La	<10	<10	<10			
Ni	<5	<5	<5			
Sc	<3	<3	<3			
V	30	60	70			
Y	70	170	35			
Zr	<10	19	<10			
Fe/Mn	135	58	540	368	5300	2700

*a = wet sample collected in 1965 from eastern shore of Nea Kameni (see Figure 4); b = same as *a*, washed with distilled water to eliminate sea salts; c = wet sample from the northern shore of Nea Kameni; d = from Harder, 1964; e = from the western shore of Nea Kameni (Butuzova, 1966) upper layer; f = from western shore of Nea Kameni (Butuzova, 1966) lower layer.

analyses of wet samples from these two sites were carried out by the same techniques outlined in the previous section. The results are shown in Table 2 together with data on other deposits from Nea Kameni obtained from Butuzova (1966) and Harder (1964). X-ray powder diffraction patterns of samples from the two deposits show only very poorly developed, broad peaks in the 4Å region, to be attributed to the (110) planes of cryptocrystalline FeO (OH) (goethite). Upon heating for 10 hours at 500°C poor reflections of Fe_2O_3 (hematite) appear.

As in the Stromboli deposit, considerable quantities of silica are present in the Santorin samples. The absence of X-ray reflections of quartz and other silicates, and the high Si/Al ratio in the samples suggest that silica is in the form of opal.

ORIGIN OF THE STROMBOLI AND SANTORIN IRON DEPOSITS

The hot springs which debouch at the edge of the islet Nea Kameni are likely to represent recirculated waters (both meteoric and sea waters) which were heated at depth both as a result of the normal geothermal gradient and of contact with magmatic bodies. Admixture with juvenile waters cannot be excluded.

The composition of the material precipitated right at the mouth of the emanations (essentially iron hydroxides and opal) suggests that Fe and Si are among the elements introduced into sea water by the hydrothermal activity. We assume that these and other elements derive primarily from the deuteric alteration and leaching of the magmatic bodies by the acidic thermal waters.

The Fe-Mn deposit close to Stromboli probably also originated as a result of submarine hydrothermal emanations, introducing in sea water Fe, Mn and other elements leached from igneous bodies.

The behavior of various elements in the deposits under study is considered next within the framework of the before-mentioned hypothesis on the origin of the deposit.

Iron and Manganese

Experimental work by Krauskopf (1957) suggests that during leaching of basic igneous rocks by thermal waters iron and manganese are leached in about the same ratio as they are contained in the source igneous rock. Thus, the Thera thermal waters, before they debouch into sea water, are likely to contain in solution, in addition to iron, substantial quantities of manganese; both elements (and others) have been leached from the surrounding rocks; the Fe/Mn ratio of these mineralized waters should be roughly similar to the ratio in the lavas of Nea Kameni (chemical analyses of 2 samples of such lavas are reported in Table 3; other analyses can be found in Nicholls, 1971), that is, between 20 and 80 with an average of 51. According to Butuzova (1966) the temperature of the Thera hot springs measured close to the mouth reaches about 60°C, and "the volcanic solutions entering the sea contain Fe, Mn, silica and a very small quantity of phosphorus." The acidity of the Thera thermal waters is suggested by the low pH (5.2) measured by Butuzova (1966) in the sediment off the spring on the western shore of Nea Kameni, and by analogy with thermal springs from other areas. When the thermal acidic water debouches in seawater its pH gradually increases; $Fe^{2+}$and $Mn^{2+}$are gradually oxidized by seawater oxygen, possibly according to the following reactions:

$$4Fe^{2+} + 6H_2O + O_2 \leftrightharpoons 4FeO(OH) + 8H^+$$
$$2Mn^{2+} + 2H_2O + O_2 \leftrightharpoons 2MnO_2 + 4H^+$$

Since iron compounds reach the limit of solubility before manganese compounds (Krauskopf, 1957), iron will tend to precipitate first close to the source, while manganese will have longer residence time in solution. These processes will give rise to fractionation of iron and manganese, resulting in series of precipitates with various Fe/Mn ratios; these ratios will be high in the early precipitates, gradually lower in the later deposits. The longer residence time in solution of manganese will result in its tendency to be preferentially deposited away from the hydrothermal sources.

This scheme, even though admittedly oversimplified, appears to explain the quasi-absence of Mn from the material precipitated right at the mouth of the thermal springs. According to this scheme part of the Mn introduced by the thermal springs should be deposited on the sea floor both inside and outside the caldera of Thera. Partial analyses of sediment samples collected at 3 sites within the caldera are reported in Table 4. The Fe/Mn ratio of these sediments (ranging from 26 to 1057) is substantially lower than in the iron deposits close to the mouth of the hot springs (Table 2). This, however, does not prove unequivocally the preferential deposition of hydrothermal Mn in these sediments, due to their consisting prevalently of submarine and subaerial weathering products of the Thera volcanites; moreover, the Fe/Mn ratios of the sediments from the bottom of the caldera are similar to those of the lavas from Nea Kameni (Table 3) and from Thera itself (Nicholls, 1971).

Substantial quantities of Mn are present in the samples collected near Stromboli (Table 1) and the Fe/Mn ratio is quite variable, suggesting that we are

Table 3. Chemical analysis of 2 andesite-dacitic lavas from Santorin (Thera). Rock (a) was dredged in 1965 at site F (Figure 4); analysis (b) is from Butuzova (1966).

Per cent	*a*	*b*
SiO_2	60.70	63.94
Al_2O_3	20.07	15.80
Fe_2O_3	2.09	1.17
FeO	2.89	4.28
CaO	3.64	4.73
MgO	1.86	1.43
MnO	0.11	0.16
K_2O	1.80	1.87
Na_2O	4.90	4.82
TiO_2	0.92	0.84
P_2P_5	0.24	0.25
H_2O^+	0.74	0.27
Total	99.90	99.56
Fe/Mn	25	35

Table 4. Concentration of major and minor elements in sediments from the Thera caldera. Location of the samples is shown in Figure 4.

Per cent	G24	G25	Dr
Si	6.1	8.4	14.0
Al	3.3	2.9	3.6
Fe	3.6	4.8	4.6
Mn	0.14	0.15	0.08
Ti	0.09	0.15	0.18
P	<0.04	0.05	0.04
H_2O^+	—	—	—
ppm			
B	325	350	225
Ba	<200	<200	<200
Co	6	9	10
Cr	20	30	45
Cu	85	70	95
La	12	16	18
Ni	40	40	50
Sc	4	6	8
V	170	250	250
Y	18	30	30
Zr	45	70	80
Fe/Mn	26	32	57

seeing different stages in the fractionation of Mn from Fe. A similar process was suggested to have taken place in the iron deposit from the Pacific seamount Amph D2 (Table 5), where the Fe/Mn ratio of discrete samples of the various precipitates ranges from 52 to 0.1 (Bonatti and Joensuu, 1966). In a Fe-Mn deposit from the Afar Rift (Ethiopia), believed to be of submarine hydrothermal origin (Bonatti *et al.*, 1971), again evidence of fractionation of Fe and Mn was observed, the basal layers of the deposit being Fe-rich, the upper layers Mn-rich.

The iron precipitate described by Butuzova (1966) consists of 2 layers: an upper oxidized layer, from 24 to 40 cm thick, where the iron is in the form of ferric hydroxide "gel", and a lower reduced layer containing ferrous hydroxide, siderite, vivianite, "hydrotroilite", pyrite and a sulfate; both layers contain opal. The chemical composition of samples from the upper and lower layers is reported in Table 2. The redox potential of the upper layer is about + 260–270 mv; in the lower layer it is − 40 mv. The pH in both layers is 5.2. According to Butuzova, the reduced lower layer results from the reduction of the ferric hydroxides by volcanic hydrogen sulfide, but any traces of H_2S were absent from the thermal springs visited by us; thus the sulfides of the lower layer may have resulted from the activity of reducing bacteria of the type *Desulfovibrio desulfuricans* which can reduce sea water sulfates and Fe-hydroxides.

Silica

Both the Thera and the Stromboli precipitates contain substantial quantities of silica ascribed to opal; similar data (see Table 5) has been obtained from the deposits on the Amph D2 seamount (Bonatti and Joensuu, 1966) and the Banu Wuhu submarine volcano where Zelenov (1964) measured high concentrations of dissolved silica (up to 45 mg/1, that is, about 10 times that of sea water) close to the mouth of the hydrothermal sources. It is likely that considerable amounts of silica are introduced in seawater of the Thera caldera by the hydrothermal solutions. Butuzova (1966) reports silica as a major component of the thermal waters on the western shore of Nea Kameni. One of us (Bonatti) observed white, snow-like flocculating particles, probably silica, being formed in abundance in the water at the mouth of one of the Nea Kameni springs. Since the solubility of silica does not vary significantly as a function of pH up to a pH of about 9, it is unlikely that the increase in pH (to be expected when the hydrothermal liquids debouch in sea water) causes precipitation of silica. The reason for its rapid precipitation

Table 5. Concentration of major and minor elements in 6 wet samples of the iron deposits from the Amph D2 seamount (South Pacific) studied by Bonati and Joensuu (1966) and one sample of the iron deposit from the Banu Wuhu submarine volcano (Indonesia) studied by Zelenov (1964).

Per cent	$D2\text{-}a_1$	$D2\text{-}a_2$	$D2\text{-}a_3$	$D2\text{-}a_4$	$D2\text{-}_b$	$D2\text{-}_c$	Banu Wuhu
Si	8.2	6.5	5.8	6.3	3.8	5.8	up to 5.6
Al	0.5	<0.5	<0.5	0.1	0.21	1.6	up to 2.1
Ca	1.86	1.64	1.93	1.26	1.8	3.7	—
Mg	0.90	0.51	0.48	0.87	3.6	0.24	N.D.
K	—	—	—	0.27	0.92	2.8	N.D.
Na	—	—	—	2.00	—	—	N.D.
Fe	32.5	28.8	31.1	30.0	5.5	17.8	up to 35.7
Mn	1.94	2.43	0.58	1.82	38.7	19.7	up to 5.42
P	—	—	—	1.2	—	—	up to 1.18
Ti	—	—	—	0.0006	—	—	up to .150
H_2O	13.6	—	—	—	—	—	—
ppm							
B	290	210	330	250	85	230	
Ba	115	100	100	140	1700	670	
Co	35	120	32	80	290	6800	
Cr	<20	<20	<20	<5	210	87	
Cu	74	120	60	120	>500	220	
La	—	—	—	<10	—	—	
Ni	400	460	90	550	4500	3200	
Sc	—	—	—	<3			
V	90	65	180	170	210		
Y	<3	15	10	<3	23		
Zr	<10	18	<10	<10	190		
Fe/Mn	17	12	54	16	0.1	0.9	

lies in the cooling off of the hydrothermal solutions after they are injected in sea water, because the solubility of silica decreases drastically with decrease in temperature of the solution (Alexander *et al.*, 1954; Okamoto *et al.*, 1957).

Minor Elements

The concentration of minor elements in Fe-Mn deposits of the type described here is determined principally by three factors: (a) the amount of each particular element which is introduced in the bottom water by the hydrothermal source; (b) its solubility in sea water; (c) the "scavenging" of minor elements from sea water by the Fe and Mn hydroxides (Goldschmidt, 1954). The content of minor metallic elements (Ni, Co, Cu and Cr) in the Fe-rich Thera deposit (Table 2) is extremely low when compared to common deep sea Fe-Mn deposits (see Mero, 1965), possibly because the rapidity of precipitation of Fe hydroxides prevents their scavenging trace elements from sea water. The concentration of some of these elements appears to co-vary inversely to the Fe/Mn ratio; that is, the higher the content, the lower the rate of sedimentation of the deposit.

Ba does not reach 100 ppm in the Fe-rich precipitates from Thera, but is substantially higher in the Fe-Mn precipitates from Stromboli. It can be shown that Ba correlates positively with Mn and inversely with the Fe/Mn ratio (Figures 5 and 6; also included are data from the Amph D2 and Afar deposits). When Ba reaches relatively high concentrations, it is in the form of barite, as in the deposit from the Afar Rift (Bonatti *et al.*, 1971) and in deposits from the East Pacific Rise (Arrhenius and Bonatti, 1965). For the solubility product of barium sulfate to be exceeded and barite to precipitate, the concentration of Ba must exceed that of normal sea water (Hanor, 1969); thus, it is probable that Ba is among the elements introduced in sea water by hydrothermal activity. The inverse correlation of Ba with the Fe/Mn ratio suggests that the precipitation of Ba is relatively slow, resulting in its coprecipitation with the late Mn-rich phases. The introduction of Ba by hydrothermal activity along the East Pacific Rise, where barite-containing sediments are present, has been proposed by Arrhenius and Bonatti (1965) and Boström (1970).

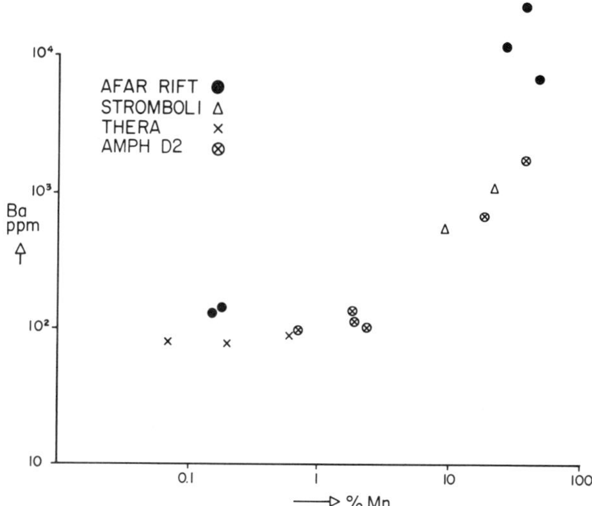

Figure 5. Plot of Ba versus Mn in iron deposit samples from Santorin (Thera), Stromboli, Afar Rift (Ethiopia), and a Pacific seamount (Amph D2).

Boron appears to rather high in the Fe-rich early precipitates from the eastern and northern shore of Nea Kameni in Thera; it reaches exceptionally high concentrations in the Fe deposits from the western shore of Nea Kameni (Butuzova, 1966; Table 2). In the Afar, Stromboli and Amph D2 deposits it appears to partition preferentially with the early, Fe-rich phases rather than with the late, Mn-rich phases. In the Fe-rich samples from seamount Amph D2 the concentration of B reaches 300 ppm. We conclude that some B is supplied by the hydrothermal sources to the deposits under study; this conclusion agrees with the common observation of abundant B in hydrothermal fields and post-volcanic emanations on

Table 6. Isotopic uranium and thorium analyses of iron deposits from Santorin, Stromboli and the Amph D2 seamount (South Pacific).*

Location	U ppm	Error +	Th ppm	Error +	U-234 / U-238	Error +	Th-230 / U-234	Error ±	Th / U ppm	Error ±	Th-230 / Th-232	Error ±
Santorin	16.12	0.25	0.19	0.06	1.14	0.02	0.01	0.002	0.01	0.003	2.27	0.82
Stromboli	11.15	0.13	0.47	0.05	1.13	0.02	0.05	0.003	0.04	0.005	3.96	0.50
Amph D2 Fe a5	1.75	0.05	0.08	0.03	1.15	0.04	0.19	0.02	0.04	0.02	15.12	5.85
Amph D2 Fe a6	2.39	0.06	0.05	0.02	1.18	0.04	0.46	0.03	0.02	0.01	77.25	38.87
Amph D2 Fe a7	1.81	0.06	0.22	0.09	1.16	0.05	0.67	0.06	0.12	0.05	19.13	8.01
Amph D2 Mn d	1.79	0.05	0.87	0.08	1.23	0.05	6.67	0.19	0.48	0.05	50.25	4.95

*Errors 1σ Radiometric Count Sea water contains: 3.3×10^{-3} ppm U; 7.0×10^{-8} ppm Th; and has a U^{234}/U^{238} activity ratio of 1.15.

land (for instance, at Larderello, Italy, and Yellowstone, USA). We also conclude that B is to a large extent precipitated rather rapidly after its introduction in sea water, thereby associating with the early, Fe-rich precipitates. It is notable that boron commonly reaches high concentrations in ancient iron ores (Goldschmidt, 1954; James, 1966).

Phosphorus may also be supplied to some extent by the hydrothermal sources in the deposits under study; in the Thera iron deposits it reaches 1% and it is almost half that much in the Stromboli deposits. It reaches almost 2% in a Fe-rich sample from Amph D2 seamount and in the iron deposit from the Banu Wuhu submarine volcano (Zelenov, 1964). These values are one order of magnitude higher than values found in deep sea sedimentary manganese nodules, where P rarely is over 0.1% (Mero, 1965). Butuzova (1966) reported vivianite $Fe_3(PO_4)_3 \cdot 8H_2O$ in the Fe deposit on the western shore of Nea Kameni in Thera. The introduction of some P by the hydrothermal sources under study is likely in view of the well-known concentration of this element in late magmatic, hydrothermal fluids (Goldschmidt, 1954). Again, phosphorus is reported in relatively high concentrations in ancient iron ores from various parts of the world (Goldschmidt, 1954; James, 1966).

The concentration of uranium in sea water is rather uniform and stands at about 3.3×10^{-6} g/l (Rona *et al.*, 1956; Spencer *et al.*, 1970). The activity ratio U^{234}/U^{238} is likewise uniform at about 1.15 (Blanchard, 1965; Koide and Goldberg, 1965). Thorium instead, is almost absent in sea water due to its insolubility and short residence time. Kaufman (1969) reports concentrations of Th232 of less than 7.0×10^{-11} g/l for surface ocean water. Th^{230}, while also rapidly removed from sea water, is constantly being produced by decay of U^{234}. As a result authigenic minerals precipitated from sea water have, at least initially, activities of Th^{230} in excess of Th^{232} and U^{234} (Goldberg and Koide, 1962; Holmes, 1968).

Less is known about uranium and thorium in waters of volcanic origin. From the data of Kuptsov and Cherdynsev (1969) on samples of vapor condensates from fumaroles in the Kamchatka and Kuril region it appears that the U content is quite variable, but generally lower than sea water, with a mean value of 1.5×10^{-7} g/l. The activity ratio U^{234}/U^{238} is less variable and shows a mean value of 1.32, that is, it is significantly higher than the sea water ratio.

In the case of the Thera and Stromboli deposits the U^{234}/U^{238} ratios are close to the sea water ratio. Removal of U from sea water by coprecipitation and/or

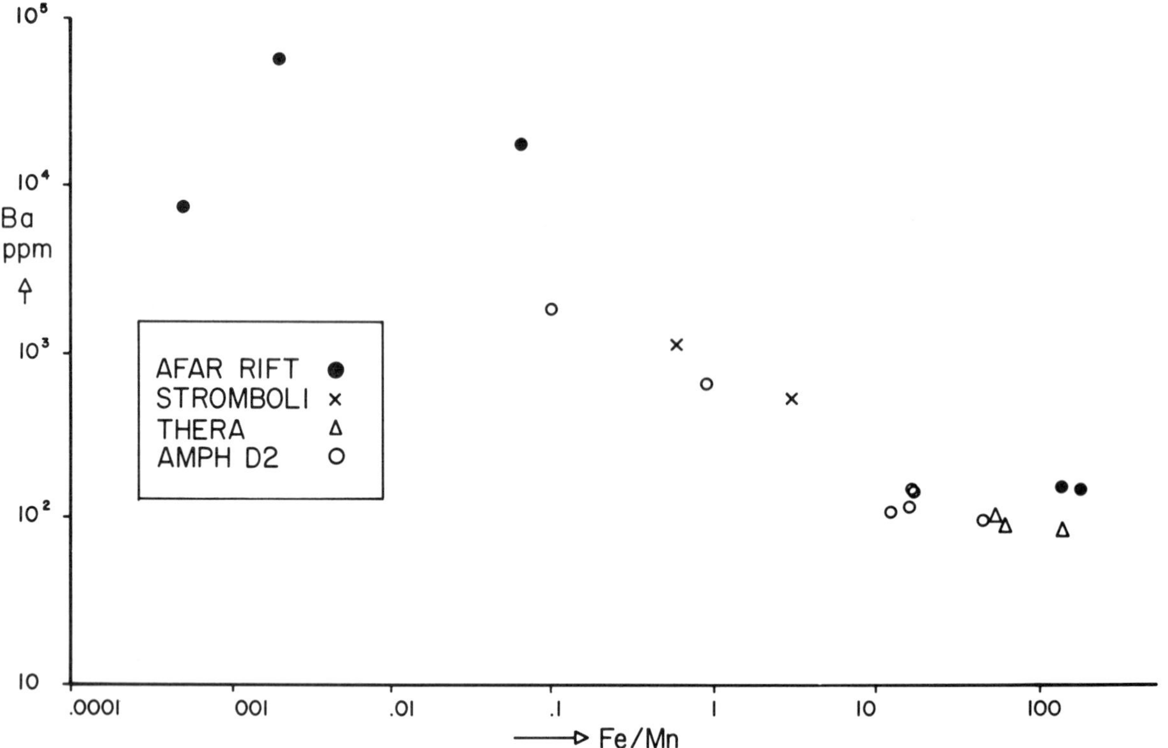

Figure 6. Plot of Ba versus Fe/Mn in iron deposit samples from Santorin (Thera), Stromboli, Afar Rift (Ethiopia), and a Pacific seamount (Amph D2).

adsorption of "volcanic" U cannot be excluded. We note that an anomalous U^{234}/U^{238} ratio has been found by us in at least one sample from the Fe-Mn deposit on seamount Amph D2 (Table 5), similarly to that found by Veeh and Boström (1971); and in an Fe-Mn deposit from the Afar Rift (Bonatti *et al.*, 1971). Such anomalous ratios suggest a volcanic source for uranium. The concentration of uranium (Table 5) appears to be rather high in both the Thera and Stromboli deposits but the isotopic analyses of Th and U (particularly the ration U^{234}/U^{238}) on these materials do not demonstrate any evidence which would allow rational choice between juvenile water or recycled sea water and/or meteoric waters.

CONCLUSIONS

1. The Thera and Stromboli submarine iron deposits were formed as a result of secondary volcanic thermal activity which supplied to sea water Fe, Mn, Si, Ba, B and P and possibly other elements. Fe precipitates rapidly close to the source, while Mn is deposited subsequently over a wider area. Si, P and B are deposited with the Fe-rich, early precipitates, while Ba coprecipitates with Mn, Ni, Co, Cu and Cr appear to be very low in the Fe-rich deposits.

2. The chemistry of the Fe deposits from Thera and Stromboli is quite similar to that of Fe deposits from seamounts and submarine volcanoes from various parts of the world (Zelenov, 1958, 1964; Bonatti and Joensuu, 1966) and from active rifts both in ocean basins (Boström and Peterson, 1966; Skornyakova, 1964; Boström, 1970) and in incipient seas (Bonatti *et al.*, 1971). The Fe-Mn rich sediments from the East Pacific Rise are enriched in Si, B (Boström, 1970), Ba (Arrhenius and Bonatti, 1965), and rather poor in Ni, Co, Cu and Cr (Boström, 1970), as are the Thera and Stromboli deposits; similar observations were made in the Afar Rift Fe-Mn deposit (Bonatti *et al.*, 1971). These similarities suggest that the Fe-Mn rich sediments from active rifts were formed by processes similar to those which gave rise to the Thera and Stromboli deposits.

3. The chemistry of various ancient Fe deposits, both Pre-Cambrian members of the "iron formation" and post-Precambrian "ironstone" types, appear to be quite similar to that of the submarine exhalative deposits discussed in this paper (James, 1966). The possibility that some such ancient Fe deposits were originated by processes of the type discussed in this paper requires further investigation.

ACKNOWLEDGMENTS

We thank Professor A. Segre for samples of the Fe-Mn deposit near Stromboli, and Professor A. Capart for a sample of Thera deposit. Research was supported by NSF Grants GA-25163 and GA-20112, and ONR Contract N00014-67-A-0201-0013.

REFERENCES

Alexander, G. B., W. M. Heston and H. K. Iler 1954. The solubility of amorphous silica in water. *Journal of Physical Chemistry*, 58:453–455.

Arrhenius, G., and E. Bonatti 1965. Neptunism and volcanism in the ocean. In: *Progress in Oceanography*, ed. Sears, M., Pergamon Press, London, 3:7–22.

Blanchard, R. L. 1965. U^{234}/U^{238} ratios in coastal marine waters and calcium carbonates. *Journal of Geophysical Research*, 70: 4055–4061.

Bonatti, E., and O. Joensuu 1966. Deep sea iron deposits from the South Pacific. *Science*, 154:643–645.

Bonatti, E., D. E. Fisher, H. Rydell and M. Beyth 1971. Iron-manganese-barium deposit from the Afar Rift (Ethiopia). *Economic Geology* (in press).

Boström, K. 1970. Submarine volcanism as a source for iron. *Earth and Planetary Science Letters*, 9:348–354.

Boström, K., O. Joensuu, M. Dalziel and A. Horowitz (in preparation). Geochemistry of barium in pelagic sediments.

Boström, K., and M. N. A. Peterson 1966. Precipitates from hydrothermal exhalations on the East Pacific Rise. *Economic Geology*, 61:1258–1265.

Boström, K., and M. N. A. Peterson 1969. The origin of aluminum-poor ferromanganoan sediments in areas of high heat flow on the East Pacific Rise. *Marine Geology*, 7:427–447.

Butuzova, G. Yu. 1966. Iron ore sediments of the fumarole field of Santorin volcano, their composition and origin (in Russian). *Akademiya Nauk SSSR Doklady*, 168:1400–1402.

Butuzova, G. Yu. 1969. Recent volcano-sedimentary iron-ore process in Santorin volcano caldera (Aegean Sea) and its effect on the geochemistry of sediments (in Russian). *Akademiya Nauk SSSR Geologicheskiy Institut, Trudy*, 194:110 p.

Goldberg, E. D. and M. Koide 1962. Geochronological studies of deep sea sediments by the ionium/thorium method. *Geochimica et Cosmochimica Acta*, 26:417–450.

Goldschmidt, V. M. 1954. *Geochemistry*. Clarendon Press, Oxford, 730 p.

Hanor, J. 1969. Barite saturation in sea water. *Geochimica et Cosmochimica Acta*, 33:894–895.

Harder, H. 1964. Kohlensäuerlinge als eine Eisenquelle der sedimentären Eisenerze. In: *Sedimentology and Ore Genesis* (Developments in Sedimentology 2), ed. Amstutz, C. C., Elsevier, Amsterdam, 107–112.

Holmes, C. W. 1968. Th^{230}/T^{232} (ionium/thorium) dating of deep sea foraminiferal ooze. In: *Means of Correlation of Quaternary Successions*, Proceedings VII Congress International Association for Quaternary Research, University of Utah Press, Press, 8:207–239.

Honnorez, J. 1969. La formation actuelle d'un gisement sous-marin de sulfures fumerolliens à Vulcano (mer Tyrrhenienne). Part I. *Mineralium Deposita*, 4:114–131.

Honnorez, J., B. M., Honnorez-Guerstein, J. Valette and A. Wauschkuhn (in press). Active crystallization of fumarolic sulfides in volcanic sediments off Vulcano (Tyrrhenian Sea). *Ores in Sediments*, VIII International Sedimentological Congress, Heidelberg, September 1971. Springer-Verlag, Berlin.

James, H. L. 1966. Chemistry of the iron-rich sedimentary rocks. *U.S. Geological Survey Professional Paper*, 440–W:61 p.

Kaufman, A. 1969. The Th-232 concentrations of surface ocean water. *Geochimica et Cosmochimica Acta*, 33:717–724.

Koide, M, and E. D. Goldberg 1965. Uranium-234/uranium-238 ratios in sea water. In: *Progress in Oceanography*, ed. Sears M., Pergamon Press, London, 173–177.

Krauskopf, K. B. 1957. Separation of manganese from iron in sedimentary processes. *Geochimica et Cosmochimica Acta*, 12: 61–84.

Kuptsov, V. M. and V. V. Cherdyntsev 1969. The decay products of uranium and thorium in active volcanism in the USSR. *Geochemistry International*, 6:532–545.

Mero, J. L. 1965. *The Mineral Resources of the Sea*. Elsevier, New York, 312 p.

Nicholls, I. A. 1971. Petrology of Santorini Volcano, Cyclades, Greece. *Journal of Petrology*, 12:120.

Okamoto, G., T. Okura and K. Goto 1957. Properties of silica in water. *Geochimica et Cosmochimica Acta*, 12:123–132.

Puchelt, H. 1971. Recent iron sediment formation at the Kaimeni Islands, Santorini, Greece (Abstract). *Program with Abstracts*, VIII International Sedimentological Congress, Heidelberg:78–79.

Rona, E., L. O. Gilpatrick and L. M. Jeffrey 1956. Uranium determination in sea water. *Transactions of the American Geophysical Union*, 37:697–701.

Skornyakova, I. S. 1964. Dispersed iron and manganese in Pacific Ocean sediments. *Lithology and Mineral Resources*, 7: 2161–2174.

Spencer, D. W., D. E. Robertson, K. K. Turekian and T. R. Folsom 1970. Trace element calibrations and profiles at the Geosecs test station in the northeast Pacific ocean. *Journal of Geophysical Research*, 75:7688–7696.

Veeh, H., and K. Böstrom 1971. Anomalous U^{234}/U^{238} on the East Pacific Rise. *Earth and Planetary Science Letters*, 10:372–374.

Zelenov, K. K. 1958. Iron contributed to the sea of Okhotsk by the thermal springs of the volcano Ebeko (Paramushir Island) (in Russian). *Akademiya Nauk SSSR Doklady*, 120:1089–1092.

Zelenov, K. K. 1964. The submarine volcano Banua Wuhu, Indonesia. *Bandung Institute of Technology, Department of Geology, Contributions*, 55:19–34.

PART 12. **Epilogue**

Sedimentology and Pollution in the Mediterranean: a Discussion

Antonio Brambati

Università degli Studi, Trieste

ABSTRACT

Solid wastes introduced into the sea are dispersed and accumulated in a manner analogous to that of natural sedimentary materials. From this it follows that sedimentological techniques may be usefully applied to the study of the transport and deposition of solid wastes in natural fluids. The present paper also indicates the possibility of utilizing some pollutants as tracers or index-markers in sedimentological research.

The Mediterranean Sea is an almost closed basin, characterized by a very low exchange of water with the Atlantic Ocean, and is therefore highly susceptible to changes in the physico-chemical regimen. Accordingly the nature and distribution of the principal types and sources of pollutants is summarized in the following paper, with particular emphasis on the petroleum products. Moreover, on the basis of modifications to the marine fauna and flora already attributable to pollution, a number of predictions are made concerning the probable fate of the Mediterranean basin in the near future. Finally, some recommendations are outlined with a view to containing and controlling the pollution threat. The most important of these visualizes a concerted, multi-national effort aimed at acquiring a more detailed knowledge of the sedimentological and oceanographic processes operating in both shallow and deep areas of the entire Mediterranean, and the utilization of this data in order to control and if possible eliminate the harmful effects of pollution.

RESUME

L'accumulation et la dispersion des détritus produits par l'homme dans la mer s'effectuent de façon analogue à celle des matériaux sédimentaires naturels. Il est donc possible d'utiliser des techniques sédimentaires pour l'étude du transport et du dépôt des détritus solides dans les eaux naturelles. Il est également possible d'utiliser des impuretés comme traceurs ou "index markers" en recherche sédimentaire.

La Mer Méditerranée est un bassin presque fermé, caractérisé par un échange d'eau très faible avec l'Atlantique et de ce fait, est très sensible aux changements de régime physico-chimique. La nature et la distribution des types appropriés d'impuretés sont énumérées en insistant particulièrement sur les dérivés du pétrole. Il est possible, en se basant sur les changements de la faune et de la flore dus à la pollution, de prévoir le sort de la Méditerranée dans un avenir proche. Des suggestions sont faites afin de circonscrire et controler la pollution. Une des plus importantes concerne la création d'un projet à l'échelle multinationale, consacré à l'accroissement et l'approfondissement de l'étude des processus sédimentaires et océanographiques aussi bien des zones profondes que des zones peu profondes sur toute l'étendue de la Méditerranée. Les résultats obtenus seraient mis en oeuvre pour controler et, si possible, éliminer les effets destructeurs de la pollution.

INTRODUCTION

For the past few decades sedimentology has been one of the most rapidly burgeoning fields of geology and much emphasis has been placed on the study of modern marine environments and processes.

Sedimentology has naturally been amalgamated with oceanography, meteorology and, finally, space research. The various types of photographs taken by satellites have, in fact, proved extremely useful for oceanographic and meteorological research, elucidating the movement of great bodies of water in the oceans or along the shelf and have also indicated their value in studying the transport of sediment in suspension both along the coast and offshore (Figure 1).

Today, however, the seas are becoming polluted. In the last fifty years man has brought about more modifications on land than in all the previous millenia. The sea has become the receptacle for every kind of waste product and the photosynthetic activity of the sea is threatened by pollution. This activity is fundamental for man because it transforms inorganic substances into organic materials, which can then be be assimilated by animals and humans, and also because it regenerates the oxygen of the atmosphere.

In several seas, but above all in the Red Sea and the Indian Ocean, large areas of coral reef are already dead or in course of rapid extinction because they represent environments particularly sensitive to a decrease in the purity of the water (Cousteau, 1971). Whole seas are almost sterile because of irrational fishing or pollution by various kinds of toxic substances.

In Japan, hundreds of people have been afflicted by the terrible disease of Minamata (Pavan, 1970a) brought on by eating fresh-water and marine fish that have absorbed and concentrated in their organisms mercury compounds derived from industrial wastes or the waters of rice-fields that have been treated with the fungicide, mercury-diphenyl.

Regarding the atmosphere, every year a quantity

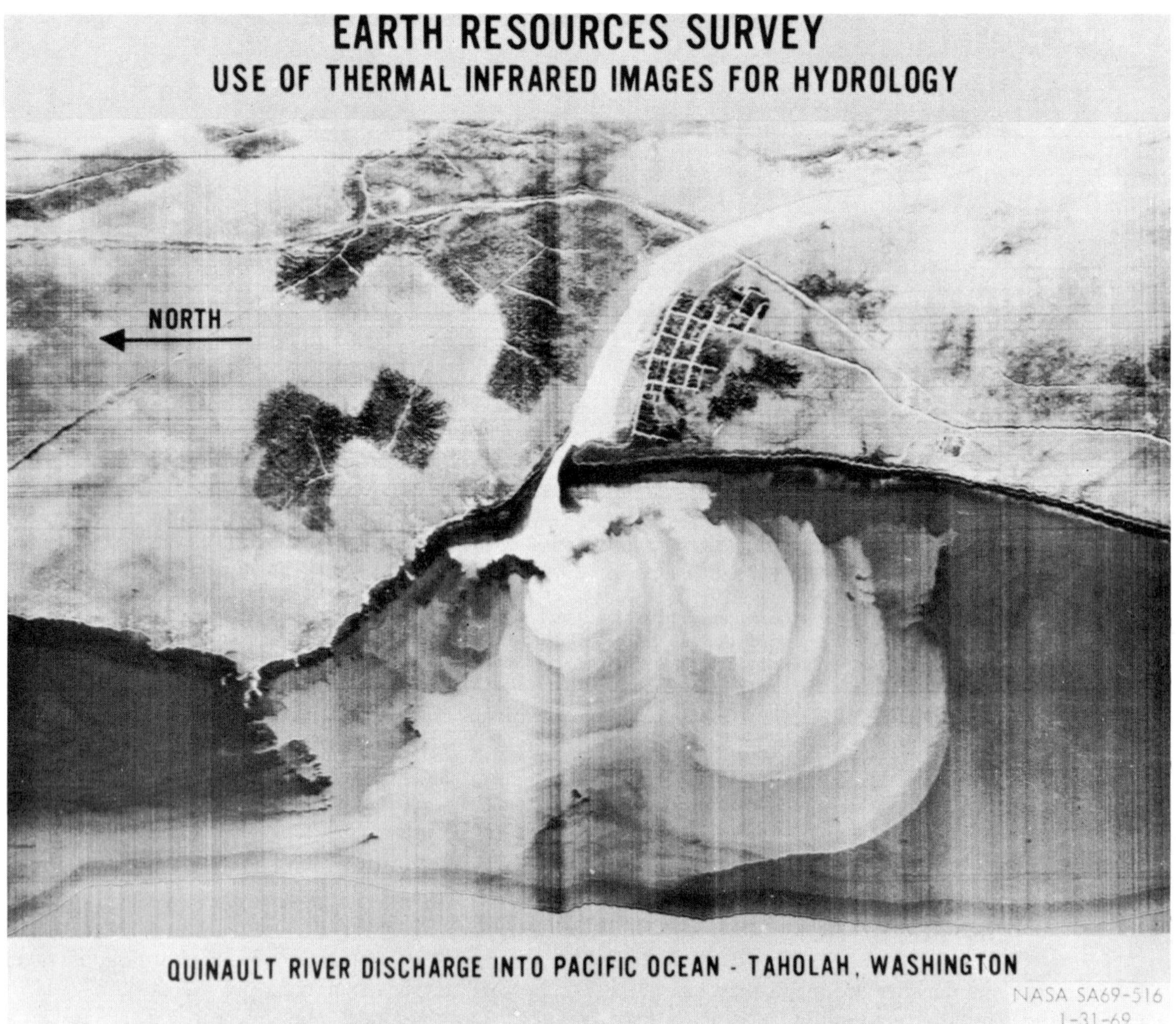

Figure 1. Quinault River discharge into the Pacific Ocean. This thermal infra-red image provides an example of the application of remote sensors to sedimentological studies and then to pollution problems involving dispersion of fluvial material into the sea. (Courtesy of N.A.S.A. and Dr. F. J. Wobber, Earth Satellite Corporation, Washington, D. C.).

of oxygen is consumed that is equal to that overlying a land area of 1,700,000 km^2, or an area as large as Italy, Switzerland, France, Spain and Western Germany combined (Pavan, 1970a). For industry and other activities that require combustion, a quantity of oxygen is consumed that would be sufficient for 43 billion people. So it seems that we, of our own free will, are beginning to create oxygen-deficient environmental conditions like those referred to by van Straaten (this volume) in the recent sediments of the Central-Southern Adriatic, except that in these the conditions are part of a natural equilibrium process.

The content of carbon dioxide increases by 0.2% annually; calculations show that since the beginning of the industrial era this increase has been approximately 10%. Such a rate of increase will cause the atmosphere to warm and thus may bring about the melting of glaciers, and therefore a decrease in marine salinity, an increase in the sea-level and will also promote strong variations in world climatic and oceanographic conditions.

The situation is alarming. It has been calculated, for example, that if the pollution in Lake Michigan were immediately checked, its waters would still take a century to be purified by nature, while for Lake Superior, it would take 500 years.

What can be done to alleviate similar catastrophes that man of his own volition has caused? Clearly all branches of science must be combined to contribute their experience towards the protection of nature and knowledge concerning marine sedimentation may therefore be of some help in the study of the behavior of marine pollution too. In fact, the solid wastes discharged in suspension into the sea behave in a similar fashion to natural sediments. Moreover, the same is true for the waste products in solution that are absorbed by "particulate matter".

This article illustrates some aspects of pollution for which sedimentology may be used advantageously, with special reference to the Mediterranean. Being an enclosed sea, surrounded by countries that are either industrialized or in the process of industrialization and that are linked to a booming tourist industry, it seems to be in danger of meeting an untimely death through pollution.

SEDIMENTATION AND POLLUTION

Sedimentology as a Technique in Pollution Studies

It is first necessary to define what is meant by pollution and which aspects of this topic require the special interest of the sedimentologist. If sedimentology may be defined as simply the science that studies the chemical and mechanical processes of sedimentation, we may define pollution as any negative modification of the environment brought about by man; modifications that are caused on a long or short term basis following the human introduction of wastes that alter the equilibrium of the ecosystem.

This being stated, it follows that any aspect of pollution which involves consideration of the type of transport, dispersion and accumulation of substances that are dangerous to the ecosystem may be tackled with sedimentological techniques, since these enable us to define (1) how transport occurs, (2) over which areas the pollutant is dispersed and also (3) in which zones the materials tend to accumulate.

Sedimentological techniques appropriate to both eolian and hydraulic processes may be utilized in the study of atmospheric pollution and in the investigation of pollution in fresh and marine waters.

In conjunction with meteorologic data, the sedimentology of eolian deposits could be applied to the problem of the transport and fall-out, on a long term basis and over long distances, of solid particles from the site of origin to the region of accumulation, so offering the possibility of predicting which areas may be affected by the precipitation of dangerous substances transported in the atmosphere.

If this may seem a limited field of application of sedimentology to pollution, the same cannot be said for marine pollution where studies of the transport and accumulation of both fine and coarse sediments in various environments are particularly appropriate.

The results may be obtained either directly or indirectly. In the former case, it is possible to follow the various changes that the material undergoes from the source to the site of accumulation; in the latter case, it is possible to define a sediment mass as the resultant of the different components of transport, and thus trace the pollutants back to the area of origin.

It is also possible to distinguish areas marked by low levels of transport energy from those marked by high levels, and therefore predict the dispersion and accumulation of different grades of material in different areas. The same techniques can be applied to processes of chemical or biochemical sedimentation, where the relationship between the means of transport and the material are chemical-physical or chemical-physical-biological.

Adopting the northern Adriatic as a typical example for the study of a shelf area, sedimentological research has been able to show the origin, dispersion and accumulation of the fluvial materials transported to the sea in suspension as well as by traction, and the role that sea currents or tides play in these processes.

In particular, the dispersion area of the terrigenous contribution of each river, the part each river plays or has played in marine sedimentation, and the relationship between the terrigenous contributions of different rivers have all been defined. The relationships between marine and lagoonal sedimentation and between fluvial and lagoonal sedimentation have also been investigated. In this way it has been ascertained that sedimentation occurs in lagoons partly through the action of rivers emptying directly into them but mainly by means of the tidal currents that, during high tides, discharge suspended materials from the open sea into the lagoons.

Moreover sedimentologic research on the transport of suspended matter along the shelf-break has shown that the finest sediments are put into suspension during high sea states and then re-dispersed towards the open sea by the large-scale, short-term movements of water masses moving back and forth across the continental margin.

The same can be said for the transport in suspension of very fine materials along canyons for which a stratiform dispersion of materials toward the open sea has been ascertained, particularly accentuated on the surface, near the bottom, and near the thermocline.

In addition, it has been possible to delimit large oceanic water masses, different because of temperature and concentration of materials in suspension in which materials are transported over long distances.

The effects of pollution can, therefore, be predicted by means of sedimentological research even in marine areas although the repercussions of pollution are still undetected, including areas not of immediate economic interest or because they are too far from coastal areas.

In the light of these results, it is obvious that problems concerning the dispersion of pollutants in seas and oceans can be solved at least partially by sedimentological principles since these can provide a rational explanation of the mechanisms of transport, dispersion, and accumulation of the materials.

In order to trace masses of sediments of different origin and different transport histories, sedimentology makes use of mineralogical and geochemical parameters as well as grain size characteristics. Heavy, light, and clay minerals or, better still, trace and/or radioactive elements form the necessary bases for research on the dispersion of sediments. Conversely, abnormally high concentrations of certain elements can be interpreted as the result of the increasing discharge into the environment of residues from industrial processes.

Thus present day sediments of the Atlantic Ocean contain abnormal concentrations of Pb, which result from the exploitation of liquid hydrocarbons and in particular from the boom in motorization. It appears that if the waste products of our age are discharged into the ecosystem in abnormal concentrations with respect to the receptive capacity of the environment, the character of the sediments will also display abnormal features.

In particular, we can say that while material in solution tends to be diluted in water, the contrary is often true for material in suspension which tends to become concentrated in special offshore zones (see Postma, 1967).

In the latter case, the accumulation of suspended material is the result of complicated transport processes for which a sound knowledge of each individual process is necessary in order to deduce the various stages of transport and, therefore, the means of accumulation.

The complexity of possible interaction between transport and depositional processes can be illustrated by the coastal regime.

Along the seashore, transport processes are controlled mainly by the motion of the waves and by the winds. Sedimentary materials are placed in suspension in a zone bounded by the coastline and a depth that corresponds to approximately half the length of the waves that break on the shore.

Ultimately, the suspended material may be transported landwards or offshore, according to whether the winds are blowing offshore or towards the land, and also dependent on the counter-currents that are set up at the bottom and on the longshore currents.

In lagoons (and tidal flats in particular) a granulometric decrease in the sediments has been recorded from the inlets towards the interior of the lagoon (pelites). The pelitic sediments become finer when traced towards the inner part of the lagoon and the accumulation of this material is ultimately dependent on the "settling and scour-lag effects" (van Straaten and Kuenen, 1957, 1958). These, in their turn, depend on the decrease in velocity of the incoming tidal currents and on the decrease in the average depth of the waters from the inlets towards the interior of the lagoons. On the basis of the settling-lag effect, the finer particles are deposited in more internal areas than those in which they should be deposited because of the currents' reduced transporting capacity. This happens because a certain period of time passes from the moment at which a current is no longer capable of transporting in suspension particles of a given size and the moment at which they reach the bottom. The scour-lag effect results from the fact that some of these particles, once deposited, cannot then be removed by the outgoing tidal flow because a greater velocity of current is required to erode such material than that required for their deposition.

Finally, very fine material can be trapped at the mouth of estuaries for a long time before it reaches the bottom. In fact, while fluvial waters are transported to the sea in the upper layers, in the lower layers there may be net transport of water upstream. In this way material transported in suspension from the fluvial waters to the sea will sink from the upper to the lower layers and then be re-transported upstream. This process can be maintained for a long time and it may be complicated by tidal currents so that while the particles of suspended material can concentrate around the estuary, the materials in solution may be dispersed offshore.

Similar models of sedimentation processes are now available for coastal and bathyal or abyssal areas where re-sedimentation processes may take place. In the same way, sedimentation processes relating to the flocculation and adsorption on clays of dissolved organic substances or trace metals can be checked and predicted.

Pollutants as Sedimentological Tools

Most elements or substances artificially discharged into the environment can be used as tracers for the study of sediment dispersion.

One example is that of pesticides, widely used in agriculture and which through erosion of the soil, are transported to the sea by way of rivers. Many other industrial waste elements can be recognized as abnormal or in abnormal concentrations in sediments. Obviously, the concept also can be extended to viral pollution arising from sewage discharged into the sea or into rivers flowing through highly urbanized regions. Pollutants, such as detergent foams and, especially, crude oils, can also be used for the study of surface currents as satellite imagery has recently demonstrated.

Furthermore, the presence of particular concentrations of elements in sediments, which correspond to the degree of industrial development (*e.g.*, Pb) may be used as a marker in studying the rate of sedimentation.

In conclusion, we may say that sedimentology and pollution are closely interdependent and complementary.

Nature and Extent of Pollution in the Mediterranean

The most obvious evidence of pollution in the Mediterranean derives from the discharge into the sea of tanker-cleaning residues and, from the processing residues of the numerous industries to be found along the northwestern shores of the Sea. The beaches and ports in this area are already endangered by oil residues.

The potential sources of pollution by oil are indicated in Figure 2 which shows the main routes followed by tankers in the Mediterranean and the major terminals for the loading and discharging of oil. During tank-cleaning operations, tankers discharge water containing a quantity of oil equal to about 1% of the cargo.

Obviously, the Mediterranean as a whole is not affected by the same intensity of traffic. In fact, there is a lack of balance between the Eastern basin, characterized by oil loading ports and only a few refineries, and the Western basin where there is a concentration of refineries, industries and main ports for oil discharging. As a result, the whole Western basin is subject, at least potentially, to a greater degree of pollution by oil.

An idea of the traffic and tonnage involved can be gained from the fact that the number of tankers which discharged their cargo in Italy in 1968 was 21,910 (Marchetti, 1970), *i.e.*, a quantity of 92.6 million tons of crude oil (Informatutto, 1970). As far as pollution is concerned, a 20 to 30% increase due to coastal traffic has to be added to the above quantity.

We can calculate that crude oil traffic in the Mediterranean, including transit towards extra-Mediterranean ports, totals about 400 million tons (Office of Oil and Gas, 1971). Since tankers discharge into the sea about 1% of their cargo in tank-cleaning, it is clear that the quantity of oil spilled every year is not less than 2,000,000 tons, as the crude oil imported by the Mediterranean countries amounts to 200 million tons, and transit traffic directed to extra-Mediterranean countries is substantially greater. Therefore, the most pessimistic estimates indicate a maximum of about 4 million tons of oil spilled into the sea.

Horn *et al.* (1970) detected quantities of oil in 75% of the 743 Mediterranean water samples taken during the cruise of ATLANTIS II (Summer, 1969) on the route Rhodes—Strait of Messina—Bonifacio Mouths—Azores. Oil lumps with a maximum concentration of 0.54 ml/m^2 were detected. The main concentrations were found in the sea between Sicily and Libya, an area where tankers can still discharge clean-up waters into the sea according to the 1967 international agreements (Figure 2).

Since oil spills into the Mediterranean amount to about 2 million tons yearly and the Basin surface is about 3 million km^2, a theoretical average oil concentration of 0.7 gm/m^2 may be calculated, *i. e.*, a concentration higher than the maximum value detected. Since Horn *et al.* (1970) proved, through chemical-biological analyses, that the maximum duration of the oil lumps found in the sea is two

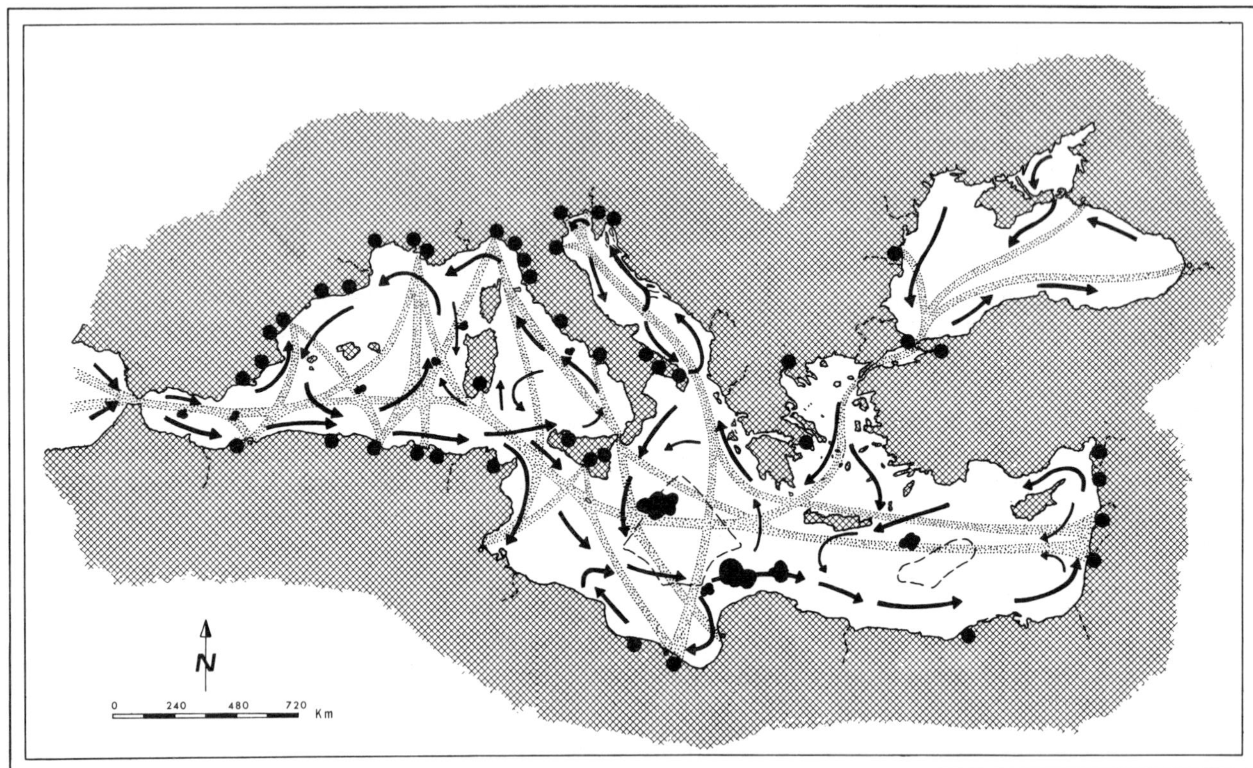

Figure 2. Oil commerce in the Mediterranean Sea: dotted tracks represent the main oil tanker routes. Arrows indicate the surface water circulation. Black dots along the coast show urban and industrial concentrations; areas enclosed in broken lines represent those regions where discharge of bilge-water from tankers is still allowed. Large black dots at sea show the distribution of petroleum lumps at the surface (after Horn *et al.*, 1970); the area of each dot is proportional to the volume of the petroleum lumps collected at that point.

months, the theoretical oil concentrations of 0.7 gm/m^2 should be reduced to about $\frac{1}{6}$, *i.e.*, to about 0.11 gm/m^2. This value corresponds to the average concentrations registered in the Mediterranean by Horn *et al.* (1970) which were calculated to be 0.10 gm/m^2. The major part of the remaining $\frac{5}{6}$, equal to about 0.6 gm/m^2 and amounting to about 1,650,000 tons of crude oil, must be carried towards the coastal areas, oxidized, or broken down by bacteriological action. One ton of crude oil can cover up to 1,200 hectares of sea surface (Leone, 1971) and in the port of Genoa maximum oil concentrations of 218 mg/l (Bianucci and Ribaldoni-Bianucci, 1970) were found. Oils (crude oil, fuel oil, heavy diesel oil, lubricating oil), if "persistent", can be carried for considerable distances by sea currents so that Figure 2 indicates the possible routes of oil residues in the Mediterranean resulting from transport by currents. Negative consequences on the coastal areas can result from current transport of hydrocarbons (Figure 3).

In fact, even if the oil film is not perceptible to the eye, it may actually persist as an invisible film. Therefore, it can be carried towards the coast, where it can stick on the shore. Fuel oils of the asphalt type, if agitated by waves and when combined with salt water and air, form very viscous and adhesive emulsions along the coasts. If these are mixed with solid substances, for example sand, they can sink to the bottom and be deposited on the shores as well, where oil residues are transformed into heavier substances by self-oxidizing processes.

The examples of the TORREY CANYON and TAPIKU MARU show the tragic consequences brought to coastal areas by a tanker's wreck. In particular they illustrate the effects due to oil pollution and indirectly to the use of detergents employed in cleaning the beaches and sea (10,000 tons). The latter proved to be more toxic than the 117,000 tons of oil which escaped from the TORREY CANYON (Pavan, 1970a).

Unfortunately, the dangerous effects of oil pollution are often complicated by the processes used for cleaning sea waters, such as treatment with pulverized solids or the employment of emulsifying substances. The former facilitate oil sinking but also pollute the bottom; the latter, as is well known, are often toxic to marine life. Their employment is not advisable near coastal areas in general, and in estuaries in particular, where a change in environmental conditions can initiate other forms of pollution. Moreover, pollution by the light oils used in the propulsion of

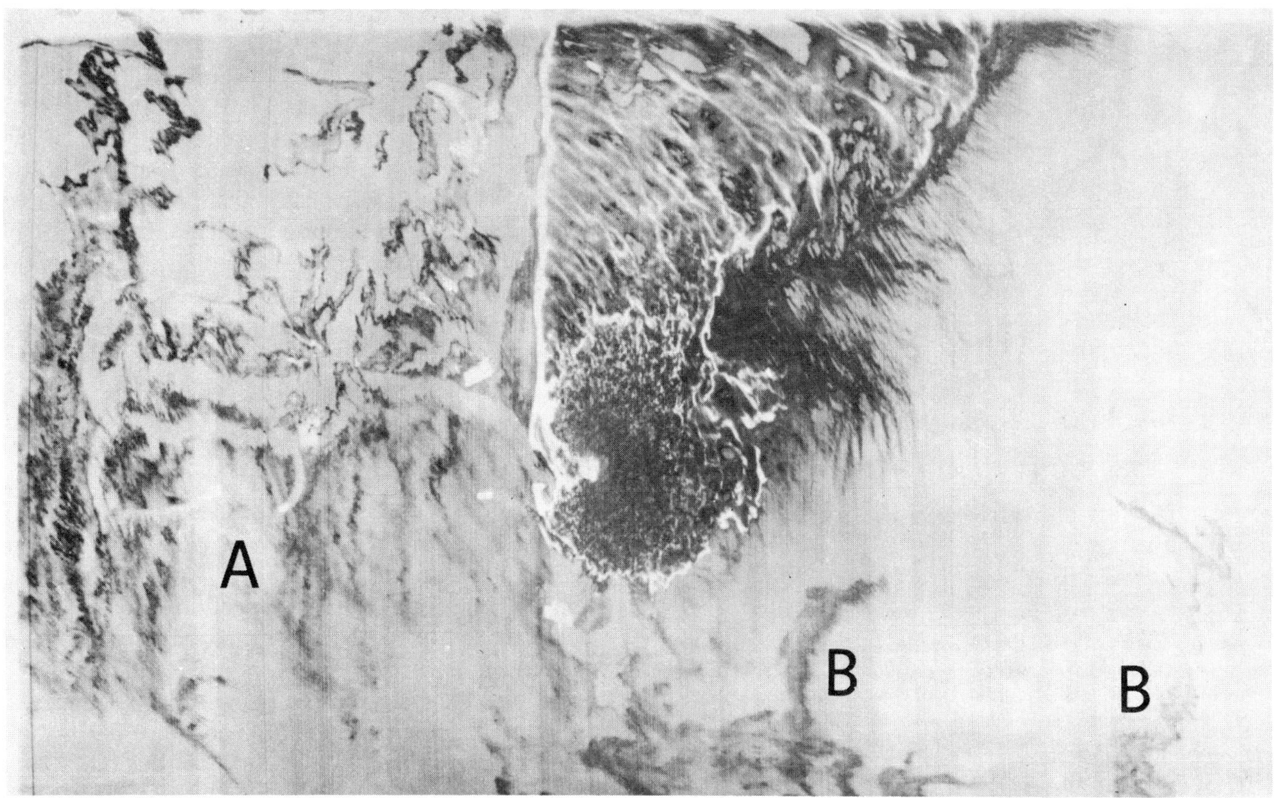

Figure 3. Infrared image of a stretch of sea near the Santa Barbara oil spill. Ship wakes. A, indicate areas of thin oil: relatively thick oil patches are imaged in dark tones, *e.g.*, B, (after Wobber, 1971). A similar situation can be envisaged in the Adriatic Sea where an intensive offshore oil exploration program has been planned.

vessels and in particular by pleasure craft must be considered.

In addition to the high degree of oil pollution in the Mediterranean in general and on the coastal areas in particular, domestic or industrial pollution, resulting from coastal urban concentrations, is to be found. For example, in Italy along 7,500 km of coasts (islands excluded) there are about 164,000 industries (Passino and Merli, 1970) and more than 16 million inhabitants out of a total population of 53 millions, tourists excluded, who permanently live along the coasts (Marchetti, 1970). Between Genoa and Barcelona there are more than 11 million coastal inhabitants (Le Group d'Experts CGPM-CIESM, 1970). We can calculate that in the whole of the Mediterranean there are about 32 million people living directly on the coast. However they are distributed, as mentioned before, in nonuniform concentrations, which are greater in the Western than in the Eastern Basin.

Taking Italy as a reference, we can determine that 59% of the coastal pollution is of domestic origin, while some 33% is due to industries and 8% due to vessels (Leone, 1970), (Figure 4).

Italian, French and Spanish coasts are the areas most endangered by domestic pollution in the Mediterranean. This already alarming situation is exacerbated by highly developed industrial infrastructure and is likely to be an extremely serious problem in the future because of increasing industrial, economic and tourist development instigated by probable increases in general population. Referring to industry the most dangerous substances known at present are: chrome salts, acids produced by steel and chemical industries, mineral oils, pesticides and radioactive materials. Omitting gases and vapors freed to the atmosphere, the dispersion in the air of solid and liquid particles is a problem which should not be underestimated. These particles include ashes, fragments of carbon, organic substances, sulphuric acid and tar substances which ultimately fall to the ground and are subsequently carried to the sea through the washing action of the water. As an example of this process 900 kg of dust are released in one day's processing, without control methods of 1,000 tons of oil. Not included in this value are 6.3 tons of oxygenated sulphur compounds (Blokker and Liedmeier, 1969).

Taking as reference the oil tonnage in the Mediter-

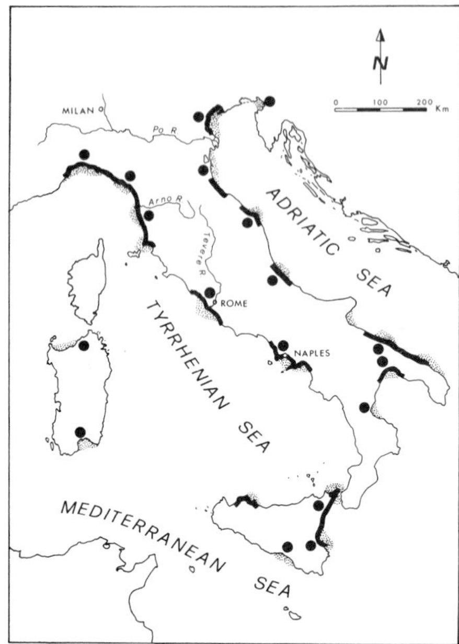

Figure 4. Zones of pollution along the Italian coast attributed to urban and industrial wastes (heavy lines). Black spots show principal oil refineries. According to Leone (1970) only 13.6% of the Italian coastline is not polluted, while 13% is very polluted and 62.5% slightly polluted. For the surface water circulation pattern, see Figure 2. (After Bianucci and Ribaldoni-Bianucci, 1970, simplified and modified).

ranean Basin (about 200 million tons), we can calculate that the refineries of the Mediterranean industrialized countries could theoretically produce 180,000 tons of dust. This inevitably tends to concentrate in the Mediterranean Sea, either falling directly on the sea surface or being transported to the sea through the surrounding drainage systems.

The quantity of solid particles that can fall on the ground in an area with a high industrial concentration is illustrated by Pittsburg, which has recorded an annual fall-out of 200 tons of industrial dust per square km; one central heating plant actually produced 28 tons of ashes per day (Pavan, 1970b).

However, urban concentrations in the coastal area are not alone responsible for pollution, which is also to be attributed to the rivers emptying into the Mediterranean. In Italy the Po, Arno and Tiber rivers together contribute to the sea pollutants corresponding to 12 million inhabitants (Le Group d'Experts CGPM-CIESM, 1970). For example, the remarkable decrease of phytoplankton registered in the northern Adriatic may be attributed principally to the effluent from the River Po. Rivers, in fact, are responsible for conveying into the sea both the waste products of urban concentrations and the pollutants resulting from the washing of agricultural soil, which is nowadays heavily treated with pesticides. Since the latter are not bio-degradable, they are a continuing menace to life in the sea.

Italy alone produced 168,000 tons of pesticides in 1967 (Le Group d'Experts CGPM–CIESM, 1970). If we consider that in the Apennines at least 50,000 km are subject to strong erosion (Pavan, 1970b), we can estimate realistically the proportional contribution to overall pollution which results from soil degradation. In Morocco, for example, it has been calculated that 45 tons of soil per hectare/year are degraded in the Ouergha Basin. This is equal to 25 million m^3 of the whole cultivable surface, which is about 450,000 hectares (Pavan, 1970a).

Unfortunately, rivers have the propensity of concentrating all the pollutants, originally spread throughout the drainage basin, into very restricted areas (river course and mouth). Hydraulic regulation of rivers can check the dispersion of pollutants into the sea as is indicated by the fact that, since the construction of the Aswan Dam, the pollution of the areas at the mouth of the Nile has considerably decreased (we have to bear in mind that about 20 million people live along the Nile River). On the other hand, this reduction produced ecological imbalance in the delta area, which resulted in a decrease in the production of sardines, so that catches which were 30,000 tons in 1964 reached only a few thousand tons in 1970 (Masini, 1970).

From the coastal areas, pollutants are subsequently dispersed into the open sea or concentrated in particular areas by sea currents, whose action depends on the geological characteristics of the coastal belt and hydrologic character of the sea. In particular, the behavior of the neritic zone is similar to that of a lake, both with regard to the accumulation of decaying organic matter and to the dangerous effects of phosphorus and nitrogen coumpounds, which cause an excessive growth of green algae. Moreover synthetic detergents hinder the absorption of atmospheric oxygen by water. In fresh water, oxygen dissolves more easily than in sea water due to the presence in the latter of salts and to a generally more elevated temperature.

In some regions of the western Mediterranean coast, particularly in the estuaries, lagoons and sheltered sea areas which are characterized by a limited water circulation, the excessive use of pesticides has been responsible for a great reduction in marine organisms of all types, from zooplankton to fish.

The most common sources of pollution which affect sea life in coastal, estuary and lagoon areas comprise the non-treated effluents of every origin and the oil residues from ships. However, it has been ascertained that oil residues are subject to bacteriological oxidation, which takes place very quickly at water temperatures between 15° and 35° (Dardel, 1969). In fact, oil oxidizing bacteria are numerous in shallow marine sediments, *i.e.*, in the lagoons, marshes and the ports where oil is always present.

Considering other bio-degradable pollutants, it is known that aerobic bacteria contained in the water attack organic substances by using oxygen and transforming nitrogen, sulphur, carbon and phosphorus into harmless oxygenated compounds. The opposite effect occurs when the oxygen supply in the water is not sufficient, thus favoring an attack on the pollutants by anaerobic bacteria and thus transforming them into very dangerous hydrogenated compounds. In this way, due to lack of oxygen, sea flora and fauna can die, starting decaying processes which imply a further consumption of oxygen (eutrophication).

An example of this is furnished by the Bay of Muggia (Trieste) which is intensely industrialized along its northern coast. Specchi and Orel (1968) noted a marked reduction in the biomass, ranging from 742.92 gm/m^2 to 2.24 gm/m^2 when passing progressively from the sparsely populated southern coast to the highly industrialized northern shore which is azoic (Figure 5). Analogous results were reported by Ghirardelli and Pignatti (1968) and Pignatti and de Cristini (1968) who pointed out the marks of pollution in the Bay of Muggia. They detected a very remarkable progressive reduction of the phaeophyceous and rhodophyceous algae towards the industrialized coast where an explosion of chlorophyceous algae was found, indicating an advanced degree of pollution. This foreshadows a collapse when all species totally disappear. About 100 years ago, on the contrary, Stossich (1876) cited the Bay of Muggia for the amount and the variety of species, most of which have now already disappeared as a result of pollutants. In fact, the sediments on the bottom now often are covered by an iridescent hydrocarbon stratum and here and there completely azoic areas are to be found.

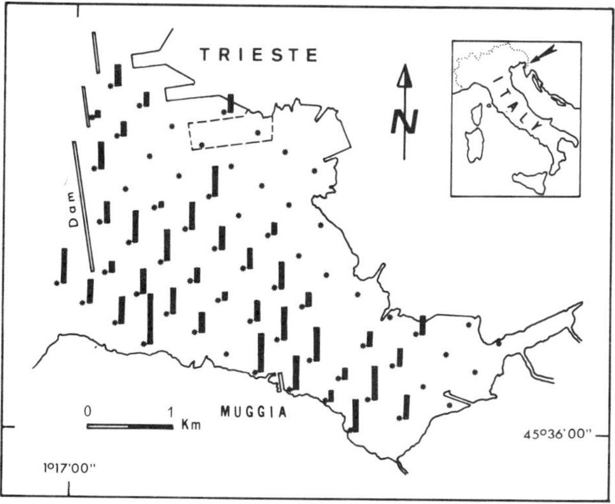

Figure 5. Muggia Bay, Trieste: distribution of the animal population/m^2 in the bottom sediments (after Specchi and Orel, 1968). The height of the black rectangles is proportional to the quantity of the animal population. Note the decrease of the biomass from the southern coast towards the northern shore which is highly industrialized and where the animal population is completely lacking. For further explanation, see text.

The lagoons of Marano and Venice provide analogous cases where the reduction of the biomass is due to the waste products of the many industries present in the immediate hinterland. The diffusion of pollutants into the sea is regulated by tidal currents, which convey the polluting substances into lagoons and inlets or redistribute them into the open sea, especially towards the west and southwest because of the prevailing currents.

Industrial dust-waste has proved to be particularly dangerous. This material varies in granulometry and is carried into the sea in suspension or by rolling movement. An example of the effects of such industrial dust is provided by the discharge of red muds at Marghera (Venice) and the bauxite residues of Marseilles. These residues resulting from alumina processing (iron oxides and silicates) have created a completely azoic area on the sea bottom in front of Marseilles (Pérès and Bellan, 1970). This azoic area is flanked by a peripheral area in which the benthonic species tend to form a biocenosis similar to that characteristic of coastal terrigenous muds (Picard, 1965).

Without further excessive cataloging we shall only recall the harmful effects due to the discharge into the sea of soda (near Leghorn), the residues from asbestos processing at Canari in Corsica, and the discharge of detrital material from the excavations involved in the construction of the new highway which occurs along the coastal waters of the Ligurian Riviera. For the future, factories planned at Scarlino (Leghorn) and Portoscuso (Sardinia) foresee a daily discharge into the sea of 5,000 tons of processing residues, in the first case of titanium dioxide and in the latter instances red muds. It is clear that whenever residues from industrial processing imply the discharge into the sea of solid particles, the techniques of sedimentology and of oceanography should be utilized in advance in order to predict the spreading of pollutants and thus prevent further and more diffused pollution.

PREDICTIONS AND RECOMMENDATIONS

The Mediterranean represents a natural laboratory. Geological and sedimentological research, carried out over the past few years, has supplied us with the patterns of the transport of sediment by traction and suspension, which demonstrates the varied sedimentological problems of this area.

On the other hand, little is yet known about the extent and nature of pollution in the Mediterranean with the exception of those coastal areas near highly industrialized or urbanized zones which have been discussed already. At this point the question arises as to what will happen to the Mediterranean. Marine life in this sea, in part is dependent on the exchange of the oxygenated Atlantic waters through the Strait of Gibraltar (about 31,600 km^3 are exchanged each year).

Oxygenated surface waters sink towards the deeper parts of the basin, ensuring sufficient oxygenation of all the large bodies of water in three preferential zones, the North Adriatic, the Provençal Basin and the Aegean sea. Sinking is due to the action of the cold winds that come from the Alps and the Turkish mountains respectively. These winds cool the upper water layers giving rise to a thermo-convective circulation and hence promote an exchange of the surface waters for the bottom layer. Unfortunately, these three zones of the Mediterranean are rapidly becoming heavily polluted and so, too, are some of the surface waters of the North Atlantic on account of the presence of substances floating on the surface which reduce the development of phytoplankton. The consequences are obvious: first, the initial link in the marine food chain (phytoplankton) is gradually being endangered; secondly, the spread of pollutants that will inevitably have repercussions on all marine life is proceeding towards the deeper parts of the basin.

Thus, it is predictable that in the course of a few decades the Mediterranean will be transformed into a dead sea unless drastic measures are taken immediately since polluted water is being introduced into the Mediterranean from the Atlantic and the three "lung areas" of the Mediterranean are becoming polluted.

The oceanographic research programs presently being undertaken by many nations in the Mediterranean present the opportunity for a working partnership to be set up to establish the extent and character of pollution, especially in the deeper areas. In this context, sedimentology can provide valuable support for applied research into the problems of pollution in the Mediterranean which, being an almost completely enclosed basin, runs the risk of becoming the most polluted sea in the world.

More sedimentological and oceanographic data for the Mediterranean are needed to provide better knowledge of the circulation of the surface and bottom waters and, above all, the nature of the processes involved in the dispersion and accumulation of materials transported either in suspension or by traction.

This knowledge will then enable predictions to be made about the diffusion of pollutants introduced into the basin and thus identify the dangers and consequences of pollution as well as the areas most likely to suffer ill-effects from this cause.

ACKNOWLEDGMENTS

Thanks are due to Dr. G. Catani and Dr. L. Carobene of the Geological Institute of Trieste for helping the author to collect data on the pollution of the Mediterranean Sea. I also wish to thank Professor D. J. Stanley of the Smithsonian Institution of Washington, Dr. G. Kelling of the University of Wales, Swansea and Professor G. A. Venzo of the Geological Institute of Trieste for helpful comments and critical revision of the manuscript. This work was carried out while the author was a Visiting Research Associate at the Smithsonian Institution, Washington, D. C. Work carried out with the financial support of the Consiglio Nazionale delle Ricerche d'Italia, Comitato per l'Oceanografia e Limnologia.

REFERENCES

Bianucci, G. and E. Ribaldoni-Bianucci 1970. In informatutto l'enciclopedia annuale. Selezione dal *Reader's Digest* 1970:458.

Blokker, P. C. and G. P. Liedmeier 1969. L'inquinamento atmosferico nella industria petrolifera. In: *Enciclopedia del Petrolio e del Gas Naturale, E.N.I.* Edizioni Carlo Colombo, Roma, 6:1–8.

Cousteau, J. Y. 1971. *Vita e Morte Nelle Scogliere di Corallo.* Edizioni Longanesi e Co., Milano, 302 p.

Dardel, W. 1969. L'inquinamento delle acque marine da prodotti petroliferi. In: *Enciclopedia del Petrolio e del Gas Naturale, E.N.I.*, Edizioni Carlo Colombo, Roma, 6:19–23.

Ghirardelli, E. and S. Pignatti 1968. Conséquences de la pollution sur les peuplements du "Vallonne de Muggia" près de Trieste.. *Revue Internationale Océanographique Méditerranée,* 10:111–112.

Horn, M. H., J. M. Teal and R. H. Backus 1970. Petroleum lumps on the surface of the sea. *Science,* 168:245–246.

Informatutto 1970. L'Enciclopedia annuale. Selezione dal *Reader's Digest*: 400 p.

Le Groupe d'Experts GGPM-CIESM de la Pollution Marine 1970. Etude sur l'état de la pollution de la mer Méditerranée. *Conférence Technique de la F.A.O. sur la Pollution des Mers et sur ses Effets sur les Ressources Biologiques et la Pêche. 9–18 Décembre 1970*, Food and Agricultural Organization Report, Rome, Italie: 17 p.

Leone, U. 1970. *L'Italia inquinata.* Edizioni Scientifiche Italiane, Roma, 134 p.

Leone, U. 1971. Piano territoriale e raffineria a Napoli. *Le Scienze,* 40:16–20.

Marchetti, R. 1970. La situazione italiana in materia di inquinamento delle acque. In: *Acque pulite,* numero speciale de *La Bo-*

nifica, Associazione Nazionale delle Bonifiche, delle Irrigazioni e dei Miglioramenti Fondiari, 11–12:557–574.

Masini, G. 1970. In pericolo la vita sulla terra. *Corriere della Sera,* 8 Dicembre 1970: 7.

Office of Oil and Gas 1971. Estimated international flow of petroleum and tanker utilisation 1970–1971. *U.S. Department of Interior*: 10 p.

Passino, R. and C. Merli 1970. Le attività dell'Istituto di Ricerca sulle Acque (I.R.S.A.) del Consiglio Nazionale delle Ricerche nel settore della protezione delle acque dall'inquinamento. In: *Acque pulite*, numero speciale de *La Bonifica*, Associazione Nazionale delle Bonifiche, delle Irrigazioni e dei Miglioramenti Fondiari, 11–12:591–602.

Pavan, M. 1970a. La difesa del suolo nella conservazione della natura. *Entre Fiera Internazionale dell'Agricoltura di Verona,* 30 p.

Pavan, M. 1970b. Che cosa vogliamo farne del pianeta Terra? Appello ai giovani. *Ministero dell'Agricoltura e delle Foreste:* 28 p.

Pérès, J. M. and G. Bellan 1970. Aperçu sur l'influence des pollutions sur les peuplements benthiques. *Conférence Technique de la F.A.O. sur la Pollution des Mers et sur ses Effets sur les Ressources Biologiques et la Pêche, 9–18 Décembre 1970,* Food and Agricultural Organization Report, Rome, Italie: 25 p.

Picard, J. 1965. Recherches qualitatives sur les biocoenoses marines des substrats meubles dragables de la région marseillaise. *Recueil des Travaux de la Station Maritime d'Entoume*, 36 (52): 5–160.

Pignatti, S. and P. de Cristini 1968. Associazioni di alghe marine come indicatori di inquinamenti delle acque nel "Vallone di Muggia" presso Trieste. *Archivi di Oceanografia e di Limnologia,* supplemento, 15: 185–191.

Postma, H. 1967. Marine pollution and sedimentology. In: *Pollution and Marine Ecology*, Interscience Publishers, New York: 225–234.

Specchi, M. and G. Orel 1968. I popolamenti dei fondi e delle rive del "Vallone di Muggia" presso Trieste. *Bollettino della Società Adriatica di Scienze Naturali* 56 (1): 137–161.

Stossich, A. 1876. Breve sunto sulle produzioni marine del Golfo di Trieste. *Bollettino della Società Adriatica di Scienze Naturali,* 2 (3): 349–371.

van Straaten, L. M. J. U. and Ph. H. Kuenen 1957. Accumulation of fine grained sediments in the Dutch Wadden Sea. *Geologie en Mijnbouw,* 19: 329–354.

van Straaten, L. M. J. U. and Ph. H. Kuenen 1958. Tidal actions as a cause of clay accumulation. *Journal of Sedimentary Petrology,* 28 (4): 406–413.

van Straaten, L. M. J. U. 1972. Holocene stages of oxygen deficiency in the deep waters of the Adriatic Sea. In: *The Mediterranean Sea: A Natural Sedimentation Laboratory*, ed. Stanley, D. J., Dowden, Hutchinson and Ross, Inc., Stroudsburg, Pennsylvania, 631–643.

Wobber, F. J. 1971. Imaging techniques for oil pollution survey purposes. *Photographic Applications in Science, Technology and Medicine,* 6 (4): 9 p.

Guidelines for Future Sediment-Related Research in the Mediterranean Sea

Daniel J. Stanley,[1] Maria B. Cita,[2] Nicholas C. Flemming,[3] Gilbert Kelling,[4] R. Michael Lloyd,[5] John D. Milliman,[6] Jack W. Pierce,[7] William B. F. Ryan,[8] and Yehezkiel Weiler.[9]

ABSTRACT

This epilogue focuses on those major problems related to sedimentation in the Mediterranean Sea that appear to warrant investigation in the near future. Studies of provenance, dispersal, post-depositional alteration and the effects of climate, eustatic oscillations and tectonic activity require multi-disciplinary research strongly supported by international and national funding.

In particular, we advocate joint studies involving physical oceanographers and geologists to determine the relation of water movement to the transport of silt and clay, the predominant sediment types in this almost enclosed sea. Deep drilling, essential for interpreting stratigraphy and paleogeography, is the only means of penetrating the thick sediment sections on basin margins and in basin plains and trenches. Drilling, however, does not preclude more intense use of conventional methods—seismic surveys coupled with dredging, coring and observation from submersibles—for retrieval of samples of the pre-Neogene ('pre-evaporite') Tethyan sea floor. Advanced technology (lasers, narrow-beam echo sounders, *etc.*) offers new means of charting the sea floor and measuring possible displacement of the North African margin relative to southern Europe. Microfossil and mineralogical variations in cores, and continued geochemical examination of trace elements, evaporites, carbonates and sapropels will serve to refine interpretation of the remarkable changes that have affected the Tethys.

Modern basins, in many respects ideal natural laboratories, offer the means to observe sedimentation on the sea floor, and to compare geologically recent deposits with ancient counterparts (flysch, molasse, carbonates, ophiolites) exposed in the circum-

RESUME

Cet épilogue est destiné à attirer l'attention sur des problèmes essentiels de la sédimentation en Méditerranée qui devraient être l'objet d'études plus approfondies. Les recherches sur la provenance, la dispersion et la diagenèse ainsi que sur les influences du climat, des oscillations eustatiques et de l'activité tectonique relèvent de diverses disciplines pour lesquelles de plus grands soutiens financiers nationaux et internationaux deviennent indispensables.

Les travaux concomitants des physiciens océanographes et des géologues sur les rapports du mouvement de l'eau et le transport du silt et de l'argile—types prédominants de sédiments—s'avèrent très prometteurs. Essentiel pour l'interprétation de la stratigraphie et de la paléogéographie, le forage de profondeur est le seul moyen qui existe pour pénétrer les épaisses couches de sédiments des marges, plaines et fausses. Néamoins le forage ne doit pas exclure l'utilisation de méthodes plus conventionelles telles que la sismique complétée par le dragage, le carottage et l'observation en submersible pour prélever des échantillons du pré-Néogène du fond de l'ancienne Téthys. La technologie perfectionnée (lasers, échosondeurs à faisceaux étroits) offre des moyens très précis pour dresser la carte du fond et pour mesurer les déplacements de la marge Nord Africaine par rapport à l'Europe méridionale. Les variations des microfossiles et des minéraux, ainsi que la recherche systématique des éléments rares, des évaporites et couches sapropéliques permettront d'affiner et d'enrichir l'interprétation des changements paléogéographiques remarquables qui ont affecté la Mer de Téthys.

Les bassins modernes—excellents laboratoires naturels en bien

[1]Smithsonian Institution, Washington; [2]University of Milan, Milan; [3]National Institute of Oceanography, Wormley; [4]University of Wales, Swansea; [5]Shell Development Company, Houston; [6]Woods Hole Oceanographic Institution, Woods Hole; [7]Smithsonian Institution, Washington; [8]Lamont-Doherty Geological Observatory, Palisades; and [9]The Hebrew University, Jerusalem.

Mediterranean region. Pollution and problems related thereto almost certainly will provide the impetus to advance Mediterranean ocean research in coming years. Sedimentologists should be prepared to participate in marine programs aimed at assessing and controlling man's influence in this closed system.

des points—permettent d'observer la sédimentation actuelle et de comparer du point de vue géologique les dépôts récents avec leurs équivalents anciens (flysch, molasse, carbonate, ophiolite) exposés dans la région méditerranéenne. De plus, la pollution et les problèmes qui s'y rattachent contriburont très certainement à accroître l'essor des recherches en Méditerranée dans les années qui viennent; les sédimentologues seront donc appelés à jouer un rôle très important dans les programmes océanographiques destinés à contrôler l'influence de l'homme dans cette région.

INTRODUCTION

It is customary to culminate a reference volume with a list of major highlights and a summary of accomplishments. However, we have decided to take a different approach. The gaps in our scientific knowledge of the Mediterranean Sea are so numerous that at this time it is more useful to focus on those aspects of the *mare internum* which require further examination and to suggest some guidelines for Mediterranean sediment and sediment-related research projects in the next decade.

This section is an outgrowth of personal interaction with the authors of this volume and with other specialists at the Mediterranean Symposium held in conjunction with the VIIIth Sedimentological Congress (1971) held in Heidelberg (see Preface and Introduction to this volume). A synthesis of the responses to a questionnaire distributed to the participants of the Symposium is summarized in the Appendix at the end of the chapter. The diversity of problems brought to light at the Heidelberg Symposium reflects the diverse backgrounds and specialties of the participants. Nevertheless, some problems of mutual concern were focused upon, and participants recognized a need for greater interdisciplinary cooperation.

Most specialists expressed an interest in participating directly in cooperative programs involving sediment studies in the Mediterranean, and many individuals appear willing to take part in team-oriented research. The following sections single out some of the problems that the authors and other individual scientists believe to be worthy of investigation in coming years. The problem of man's influence in the Mediterranean is so important that a separate chapter on this subject is included in the Epilogue (see Brambati, this volume).

PHYSICAL OCEANOGRAPHY, GEOGRAPHY AND SEDIMENTATION

The link between sedimentation and water circulation patterns in the Mediterranean remains grossly neglected although the few cooperative ventures between geological and physical oceanographers (such as the recent detection of a probable turbidity current off the French margin, Gennesseaux *et al.*, 1971) have proved so rewarding that they will undoubtedly prompt further collaboration of this type. Dispersal of sediment is unquestionably influenced by seasonal variations of fluvial flow, local sinking of denser surface water masses (Medoc Group, 1970; Stommel *et al.*, 1971), and by movement by bottom water, including possible geostrophic currents. Cores have revealed that the bulk of Recent sediment types, by far, throughout the Mediterranean are clay and silt which are supplied from fluvial and eolian sources and are most prone to long distance displacement by moving masses of water. Most urgently needed are stations monitored seasonally on land and at sea (for detection of wind blown sediments, *cf.*, Eriksson, 1961, 1965; Folger, 1970; and others). In this respect, hydrological data collected in the Straits of Sicily and along traverses in the eastern Mediterranean (Filyushkin, 1962) are of particular interest. Long-term bottom flow patterns should be recorded by continuous reading current meters, dyes (Romanovsky, 1966) and other techniques sensitive to very low flow just above the sea floor. An effort should be made to recognize the presence of nepheloid layers and (if they occur) their distribution, composition and duration. In this respect, mapping of man-made pollutants can be used to trace water movement; pollutants serve not only as dispersal indicators, but also in determining rates of deposition.

The plumes at river mouths and patterns of turbid water near fluvial point sources, and bands of suspended sediment-rich water resulting from erosion of the coast can also be monitored seasonally by aerial flights and space-vehicle photography (Figure 1). Unfortunately space flights to date have provided only partial coverage of the Mediterranean, although the developing ERTS program is designed to cover more northerly latitudes also.

In the past, most of the surveys of bottom morphology have not been sufficiently refined for marine scientists interested in the movement of sediment on the sea floor. The results of narrow beam echo sounding profilers complement those of conventional bottom echo sounding equipment (see for example

Figure 1. Photography from space showing coastal sections off North Africa (with permission of NASA). In upper photograph, offshore bed forms (arrow) near Benghazi, Libya and cloud pattern reflecting a sea-breeze system (Apollo 9, 11 March 1969). Lower photograph shows turbid water (arrows) in the vicinity of Alexandria, Egypt and the Nile (Gemini V, 22 August 1965).

the study of the Mediterranean Rise by Belderson *et al.*, 1972). The use of such narrow beam profilers in conjunction with side-scan sonar would enable charts to be greatly refined and allow better geological-tectonic interpretation of the sea floor. Much better definition is needed in mapping of trenches of the Aegean Arc and Ionian Sea, and small basins of the complex Aegean Sea (Maley and Johnson, 1971). Mapping of 'steps,' or the small breaks in topography, occasionally viewed on bottom photographs (Ryan *et al.*, 1971, their fig. 7) in areas such as the Hellenic Arc-Levantine Basin are needed to define zones of recent and sub-recent deformation of the sea floor. Measurement of displacement of the shelfbreak, canyon heads, sills and submarine terraces can be used to evaluate subtle tectonic displacement. It also would be useful to survey the above environments visually by means of submersible in order to detect ancient beach or cliff features, subaerial drainage channels, and submerged beach ridges related to Pleistocene eustatic sea level oscillations.

It is evident that the few major Mediterranean deltas are among the most significant features affecting sedimentation, yet with the exception of the Rhône Delta and its subsea fan complex, little work has been done in establishing the close relation between these major point sources, fluvial flow and hydrology and morphology of the submerged portions of the deltas.

The canyons and submarine valleys which incise the slopes throughout the Mediterranean serve as major conduits for material moving downslope to basin plains and furnish one of the keys for interpreting dispersal patterns. For instance, it is essential to take note of the valleys on the Damietta slope-and-fan (Nile Cone) complex, recorded by Carter *et al.* (this volume), in order to help define sedimentation patterns in the Levantine Basin and to understand the particular contributions from the Nile sub-sea delta, the largest single source of sediment in the eastern Mediterranean. Refined mapping of meandering patterns and subtle features such as fan-valley levees is needed to evaluate the relative importance of overbank flow. Closed-spaced surveys and detailed charting of the slope-and-base-of-slope environments, including canyons, submarine channels and subsea fans (*cf.*, charts by Bourcart, 1959; Musée Océanographique de Monaco, 1958–1960; Ryan and Heezen, 1965; and others) should be extended along all margins of the Mediterranean.

It is not too early to consider the application of laser technology to refined charts of the Mediterranean coastline. This technique may eventually be sensitive enough to measure displacement, if any, between North Africa and Southern Europe. A critical area for such measurements would be a north-south transect between Cyrenaica, Crete and the Peloponnesus.

SEDIMENTATION AS RELATED TO TECTONICS AND STRATIGRAPHY

The publication of preliminary data collected on Deep Sea Drilling Projects JOIDES Leg 13 in 1970 (Ryan *et al.*, 1972) coincident with the printing of this volume is fortunate. This, certainly the most exciting exploratory event in the Mediterranean in recent time, provides heretofore unavailable information concerning the past history and sedimentation of the Mediterranean. Examination of cores from the fourteen Leg 13 drill holes has provided a few solutions and many more surprises—enigmas which can best be resolved by future drilling. The proposed future leg in the Mediterranean by D/V GLOMAR CHALLENGER in 1974 (Deep Sea Drilling Project, Personal Communication) is clearly needed as an additional source of stratigraphic and basic geologic background information. The high cost of the drilling operation and the limited amount of time available for core retrieval requires that future operations be aimed primarily at resolving specific tectono-stratigraphic problems. Nevertheless, such a program can only increase our understanding of past sedimentation in the Mediterranean.

Several targets are of immediate interest. Penetration of the remarkable thicknesses (locally $>$ 1000 m) of unconsolidated Plio-Pleistocene clay, silt and sand units in basin plains and subsea fans (Menard *et al.*, 1965; Ryan *et al.*, 1970) can only be accomplished by deep drilling. However, considerable technological improvements are needed: sedimentologists interested in processes and primary structures require less mechanically disturbed core sections. Disturbed or not, mineralogical determination of clays and light and heavy minerals in drill cores already obtained should serve to interpret provenance patterns in Pliocene and Miocene time. Determination of source terrains and dispersal paths is needed to refine pre-Pleistocene Tethyan paleogeographic reconstructions. Interpretation of the configuration of paleo-Mediterranean basins, which appear to have been very different in Mesozoic (Figure 2) and even late Miocene and Pliocene time from those of today (Hsü, 1971; Smith, 1971; and others), would benefit from a sedimentological approach (*cf.*, Stanley and Mutti, 1968; Wezel, 1970).

One of the primary objectives of future drilling should be the recovery of 'older' (*i.e.*, 'pre-evaporite' and pre-Neogene) units. It may be appropriate to

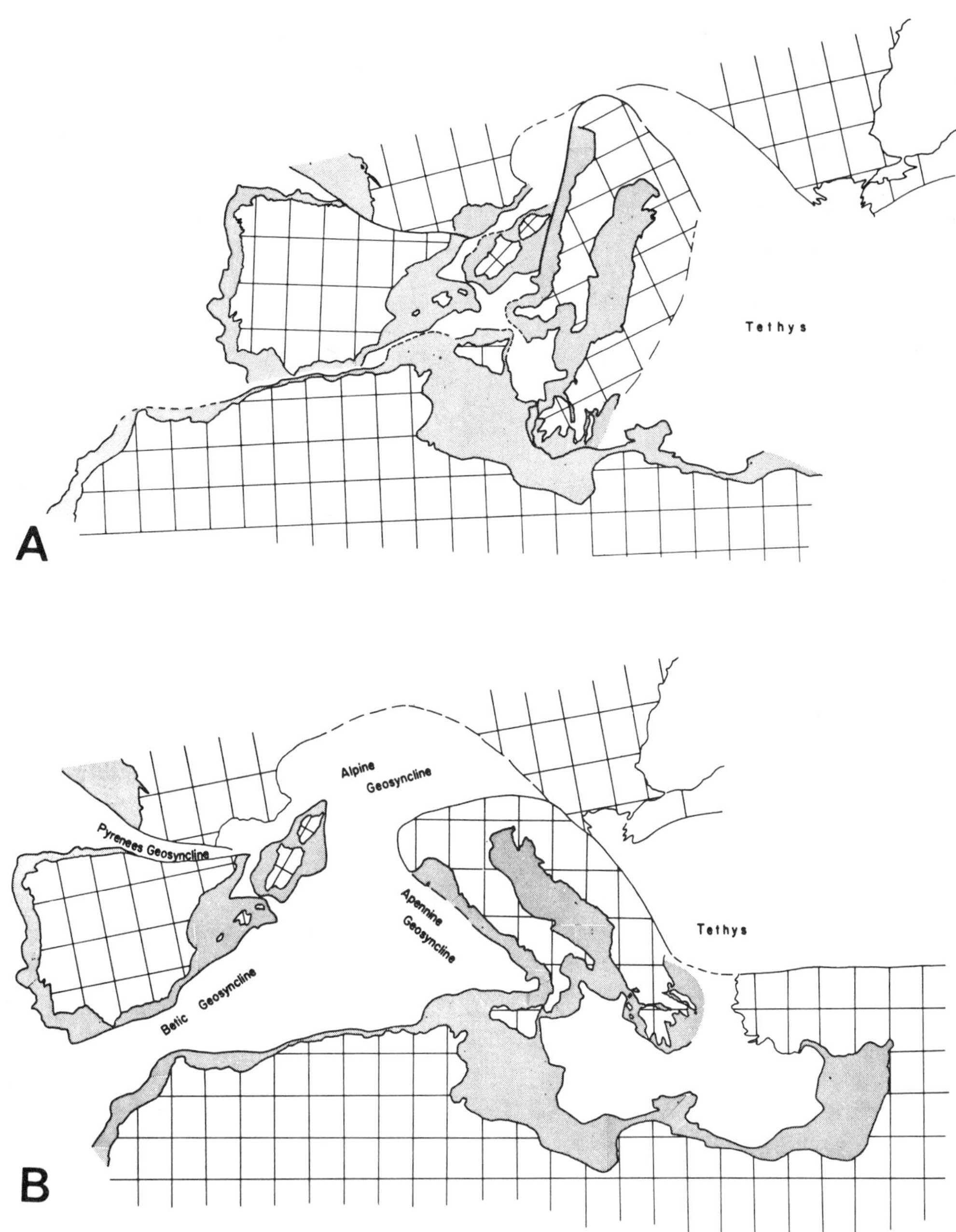

Figure 2. Paleogeographic reconstruction of the western Mediterranean region in the Mesozoic. A, in early Jurassic, before the start of movement of Africa eastward relative to Europe; B, in middle Cretaceous, when westward motion of Africa relative to Europe began. After Hsü (1971), with the author's permission.

drill in the vicinity of Crete, part of the underthrusted island arc-ridge system (Rabinowitz and Ryan, 1970), to collect material of the ancient Tethyan ocean floor. Older rock may also be cored in zones of foundering (Balearic Rise, Tyrrhenian Basin, Algerian and Provençal margins, for example) or of dislocation (sea floor off the Balearic and Corso-Sardinian platforms). It should be possible to correlate these pre-Miocene sediments with age-equivalent rocks on shore, and this would elucidate considerably the spatial framework of the Tethys and the relationship of that ancient sea with the modern Mediterranean.

Correlation of stratigraphic sections at sea with those on land are needed to shed light on the role of ancient deltas (Ebro, Nile, Ceyhan), whose influence on basin sedimentation probably predates Miocene time. Prime areas for future drilling such as the Valencia Trough, the eastern Mediterranean Trough (Saquie) and Adana-Kythrea Trough should show the influence of these ancient deltas (provenance role), and allow better paleogeographic reconstruction of the associated river basins.

There are mechanical limitations to the depth of drilling penetration (about 1000 m) and considerable problems related to penetrating the salt layer (slow drilling and the possibility of hydrocarbon pollution in an already endangered small ocean basin). As a result of these, future drilling to obtain 'old' layers will require more intense utilization of seismic subbottom data, particularly those that show pre-Pliocene targets cropping out on the sea floor (Heezen *et al.*, 1971; Fabbri and Selli, this volume). More conventional coring and dredging of rock exposures on the sea floor, apparent on seismic profiles (Figure 3), is needed before the next DSDP leg. Dredging is a faster and, in many ways, more productive means of retrieving samples of 'older' (pre-Pliocene) units; several dredge hauls can be obtained in an area in a single day, and this technique is considerably less costly and time consuming than the 'one-shot' drilling method. Targets could include young (fault?) escarpments (Emile Baudot, Mazarron, and the Malta Escarpments), the Corso-Sardinian margin, basin ridges of the type observed in the Tyrrhenian (Heezen *et al.*, 1971) and Alboran Ridge (Milliman *et al.*, this volume), and steep trench margins of the Aegean arc system.

The origin and age of isolated highs (Apulian, Cyrenian, Erathosthenes, Medina and other plateaus) in the eastern Mediterranean can best be determined by combined drilling, dredging and seismic survey. This approach could also shed light on: (a) the role of tectonics in the origin of submarine canyons (Figure 4), possibly the most important factor in canyon formation (Stanley, 1969); (b) the areal distribution of Upper Miocene evaporites; (c) the effects of separation of the eastern and western Mediterranean basins by the Straits of Sicily threshold during the Upper Miocene, Pliocene (Benson, this volume) and Pleistocene; (d) the relationship between Mediterranean evaporites and those to the south and east, including the Red Sea rift valley; and (e) possible *mélanges*, slumping and brecciation at the base of slopes and near the Mediterranean Ridge. Perhaps the use of deep-towed profiling equipment will provide more realistic records of 'cobblestone' terrain (Emery *et al.*, 1966; Ryan *et al.*, 1970) than can be obtained with standard profilers.

Global tectono-eustatic and glacial-eustatic factors have combined with local tectonics and the local effects of plate boundaries to produce significant changes in the relation between land and sea masses in the Mediterranean area. Tectono-eustatic changes and/or the closing of the Strait of Gibraltar and/or of Sicily undoubtedly lowered the level of the Mediterranean by hundreds or even thousands of meters. Following the work of Ambreseys (1965) and Flemming (1968, 1969) further studies should be undertaken to establish the vertical movement of sills in various channels, so that these can be correlated with the eustatic level at each period, and the forms of basins and sediment paths deduced. Critical sills include those of Gibraltar, Sicily, Kythera, Antikythera, Kassos, Karpathos, Rhodes, and the Bosporus passage to the Black Sea.

The direct relationship between structural activity and clastic sedimentation has been demonstrated in the western (Heezen and Ewing, 1955), central (Norin, 1958; Ryan *et al.*, 1965) and eastern (Hersey, 1965; Ryan and Heezen, 1965) Mediterranean. Special studies are now needed in particularly active areas such as the Aegean Arc. Here one can attempt to determine the effects on sedimentation in the Hellenic Trough and other depressions of the movements of Africa relative to southern Europe. In addition to maintaining laser stations at key positions on the coasts of Cyrenaica, Crete and the Aegean isles to measure movement, it also would be useful to place bottom seismograph arrays on the sea floor to record earthquake activity (particularly those of low magnitude) seaward of the Hellenic Trench. This would be used to define focal mechanisms (McKenzie, 1970; Papazachos and Comninakis, 1971) in this area. This zone is of particular interest to sedimentologists, for if flysch-type deposits are presently accumulating in the modern Mediterranean they would most likely be encountered in the Aegean arc and Cyprus area (Hersey, 1965; Stanley, 1972) where the African lithospheric plate is believed to underthrust the Eurasian plate (McKenzie, 1970; Ryan *et al.*, 1970, their fig. 4; Lort, 1971).

Sedimentation changes in the past were affected not only by tectonics but also by oscillations in sea level and differences in bottom water paleocirculation patterns. The results of these factors could be detected by a program of seismic profiling and coring. Subbottom reflectors show stratigraphic on- and off-lap patterns (Figure 3, A) reflecting alternate deposition and truncation of sedimentary deposits in the western Mediterranean (on the Balearic Rise, for instance) by water circulation in the Plio-Pleistocene. Changes in the composition of clays and trace elements accompanying such sequences can be used to determine paleodispersal patterns, including the possible influence of paleogeostrophic current activity on the ancient Mediterranean sea floor (Ryan *et al.*, 1972).

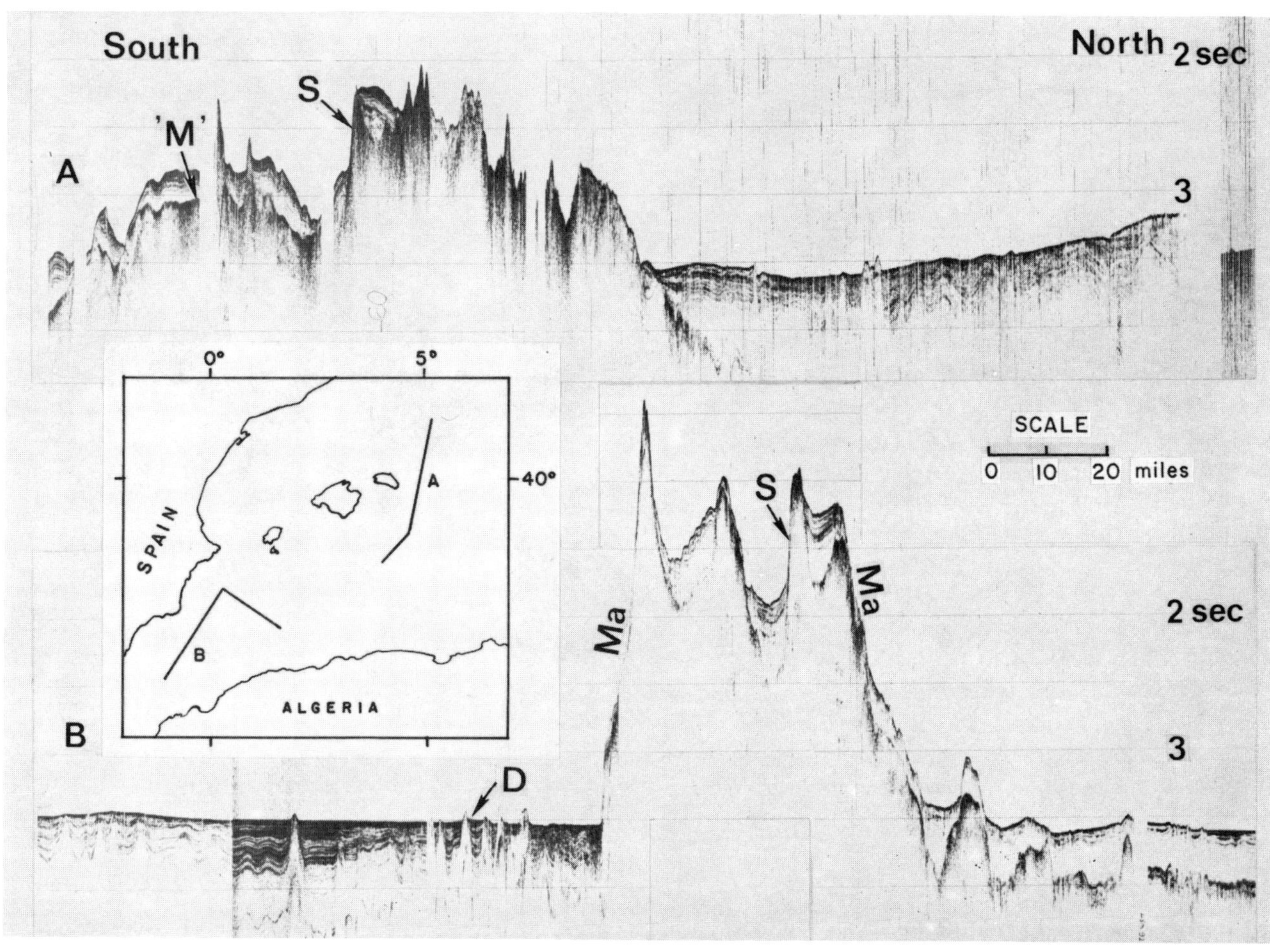

Figure 3. Air-gun seismic reflection profiles in the Western Balearic Basin (ROBERT D. CONRAD Mediterranean cruise, 1965, modified after Ryan *et al.*, 1971). Horizontal lines at 1/2 second interval (two way travel time). A, profile off Balearic Platform and crossing Balearic Rise, showing Reflector 'M' and location of scarps, S, where rocks of pre-Quaternary age can be dredged. Note onlap of Plio-Quaternary sediment on the northeast margin of the Rise. B, two lines crossing the steep Mazarron Escarpment (Ma, young fault plane along zone of foundering?). Piercement diapirs, D, in Balearic Basin plain.

MICROFOSSILS AND SEDIMENTATION

The use of organisms for interpreting stratigraphic and past sedimentary patterns in Mediterranean basins cannot be over-emphasized. Both water mass and sediments throughout much of this region contain a significantly high faunal content (Figure 5). Furthermore, both seafloor photographs and cores show that strata have been reworked (bioturbated) to varying degrees (Figure 8,C) by the activity of benthic organisms.

Pelagic oozes are dominant on some rises, ridges and other turbidite-free physiographic features, while in basins they are associated with resedimented deposits which also contain a high faunal fraction. According to available data, the rate of sedimentation in pelagic sediments in the Mediterranean ranges from 1.9 to 4 cm/1000 years (turbidite-rich sequences on the other hand accumulate at a rate of 10 to over 100 cm/1000 years). In some basins, pelagic-rich deposits comprise 60 to 70% calcareous nannofossils, 8 to 15% foraminifera, mostly planktonic. Pteropods are present at the sediment-water interface, but usually diminish at depth, while siliceous microfossils (radiolaria, silicoflagellates, diatoms, *etc.*) are rare or, in some cases, absent in normal pelagic sediments.

The species associations and assemblages of living faunas are notably different in the North Atlantic and Mediterranean basin. Mediterranean assemblages show a peculiar association of both 'warm' and 'cold' water forms (Cifelli, personal communication). Sampling programs should include collection of net samples at different depths on a seasonal basis (Glaçon *et al.*, 1971). Better definition of faunal assemblages of modern basins is needed to answer questions relative to possible reversal of current patterns during

Figure 4. Northwest coast of Corsica where the continental margin is cut by numerous canyons heading almost to shore (Bourcart, 1959). The presence of some canyons is related to major subaerial valleys extending seaward (A); others have an origin related to structural offset of the type shown in B.

the Pleistocene (Huang and Stanley, this volume) and also to Quaternary climatic oscillations (Blanc-Vernet, this volume; Herman, this volume).

The stratigraphic record, as revealed by DSDP Leg 13 cores, shows that past Mediterranean faunal abundances were comparable to those of open oceans. In some cases, abundances very clearly reflect changes in productivity while in other cases they more likely reflect changes in detrital rates and/or possibly dissolution. Evidence of a significant decrease in microfossils, especially planktonic foraminifera, has been found in Lower Pleistocene sections of the Mediterranean Ridge (DSDP Site 125; location of this and other DSDP sites are shown in Nesteroff *et al.*, this volume, their fig. 1). The decrease occurred in the latest part of the Matuyama Epoch, above the Jaramillo event. The sedimentation rate in the Tyrrhenian Basin (DSDP Site 132) also decreases in the later parts of the Pleistocene (Brunhes Epoch). This decrease of organic productivity has been interpreted as the result of progressive restriction at the Gibraltar Sill, which prior to that time was wide and deep enough to permit an oceanic type thermohaline circulation at depth (Ryan and Cita *in* Ryan *et al.*, 1972; Benson, this volume). After closure, the Mediterranean became an internal sea with elevated temperature at depth. The disappearance of psychrospheric ostracodes after this period also indicate the direct effect of paleogeographic changes at the Gibraltar portal and the Strait of Sicily on faunal abundances. Selection of a deep core in a thick sedimentary sequence free of turbidites in the western Mediterranean is needed to better define this marked faunal change.

Continuous coring in a similar thick, turbidite-free sequence in the eastern Mediterranean would also serve to define several phases of stagnation recorded in 'preglacial' Pleistocene and Late Pliocene sequences. Preliminary studies based on the presence of sapropel layers in some Leg 13 cores (Hellenic Trench, Mediterranean Ridge and Tyrrhenian Basin) show that glacio-eustatically controlled changes in sea level occurred well before the onset of glaciation in the circum-Mediterranean region (Cita and Ciaranfi, *in* Ryan *et al.*, 1972). Sapropels contain microfossils which are rare or absent in normal, well-oxygenated sediments. Siliceous microfossils are often well represented. Both benthic forms and organisms floating within the photic zone and mesopelagic forms were affected by the lack of vertical mixing in the water mass. It is noteworthy that changes in the planktonic foraminiferal populations observed in 'preglacial'

Pleistocene sapropels are different from those recorded in younger 'glacial' ones. A series of problems that have sedimentological ramifications are thus posed.

Are these earlier phases of repeated stagnation related to changes in the physiographic connections (a) between the Black Sea and the eastern Mediterranean (Olausson, 1961) and (b) between the eastern and western basins at the Straits of Sicily? Has stagnation occurred during a warming trend (near the maximum warming) of the climatic curve as suggested by Ryan (this volume)? What was the interaction between Black Sea and Mediterranean waters in the Late Pliocene to early Pleistocene? Was the shallow Bosporus Sill an effective barrier at the end of Pliocene time? Recording the microfossil and petrologic changes in a series of deep continuous cores in the sedimentary cover of the Black Sea would make it possible, perhaps, to correlate stagnant cycles in that basin with those of the Mediterranean. Furthermore, dense faunal sampling of cores on both sides of the Strait of Gibraltar should be undertaken to test the theory of Quaternary current reversals (Ryan, this volume; Huang *et al.*, 1972).

Faunal assemblages are used to help interpret depositional environments, and this is particularly useful in the case where sediment sections have undergone considerable displacement as a result of tectonic modifications. As an example, benthic faunas indicate that pelagic sediment of Pliocene age cored at the foot of the north wall of the Hellenic Trench originally accumulated in a different (non-trench) environment; the faunal analysis supports a tectonic *mélange* origin for this 3 m. y. old unit now found underlying Lower Cretaceous limestones (Ryan *et al.*, 1972).

Additional data on microfossils of the Late Miocene and lowermost Pliocene are essential to detail events related to the Pliocene 'flood', *i. e.*, the post-evaporite marine incursion. Analysis of Pliocene sections in the Tyrrhenian Sea and Ionian Basin (DSDP Leg 13 132 and 125) highlights paleontological and sedimentary changes. Recent findings (Ryan and Cita, *in* Ryan *et al.*, 1972) include: (a) a sharp sedimentary-paleontological break between pyritic marls overlying the topmost evaporite layer and the pelagic oozes, the break having been identified as the Miocene-Pliocene boundary; (b) the presence of an interval in which the most conspicuous form is represented by *Sphaeroidinellopsis* spp. (the *Sphaeroidinellopsis* Acme-zone); (c) the absence of any form of benthic life at the beginning of pelagic sedimentation (= base of the Pliocene). Benthic foraminifera appear in extremely small amounts only about 1 m above the base of the Pliocene, while ostracodes are recorded only from the next foraminiferal zone, or some 8 to 10 m above the Miocene-Pliocene boundary; (d) the pseudo-breccia some 20 cm thick immediately overlying the Miocene-Pliocene boundary; (e) red hues characterizing the *Sphaeroidinellopsis* Acme-zone; and (f) evidence of initial solution at depth (oligolithic facies) recorded in the lowermost part of the *Globorotalia margaritae margaritae* Lineage-zone at about 5 m.y., when the Mediterranean Sea probably reached its greatest depth. It is necessary to obtain a complete succession of Lower Pliocene units in the Balearic and Ionian (or Levantine) basins. Selection

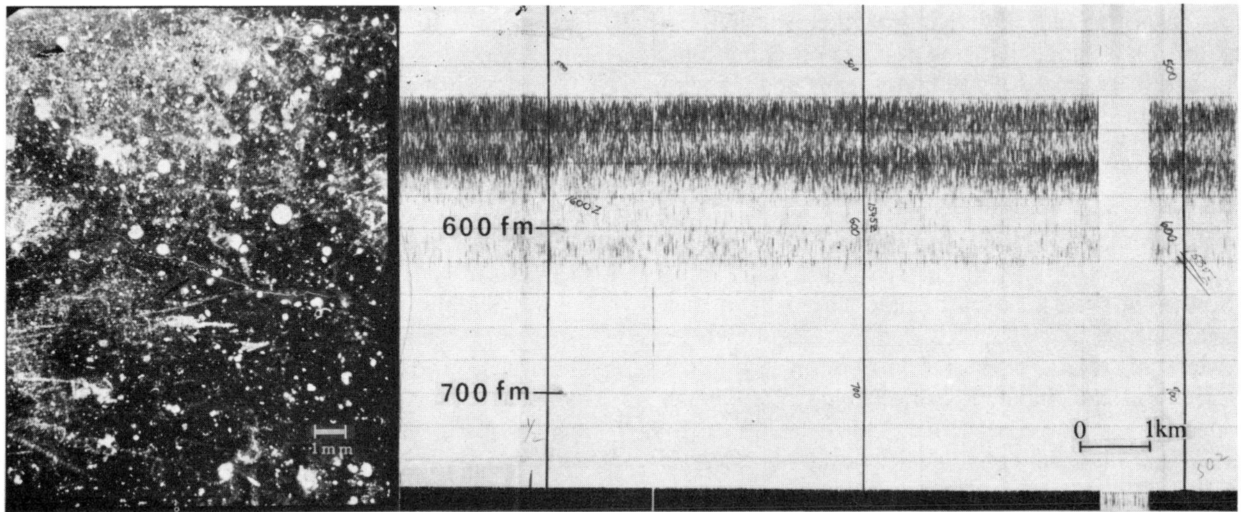

Figure 5. Deep scattering layer in near-surface water mass above the flat basin plain of the Western Alboran Sea as observed on PESR record. Thin section of impregnated Holocene hemipelagic mud from core in Western Alboran Basin plain shows abundance of microfossils, mostly planktonic foraminifera.

of suitable drill sites should take into account seismic profiles showing the location of a well-defined subbottom reflector (Reflector 'M', *in* Ryan *et al.*, 1970) which represents the top of the evaporite suite of late Miocene (Messinian) age (see Figure 3).

The retrieval and dating of Late Miocene evaporites in all of the deeper Mediterranean basins was one of the major accomplishments of deep sea drilling. However, the greatest penetration in this formation (DSDP Site 124, Balearic Basin) did not exceed 65 m which probably accounts for less than 1/10 of its total thickness. Furthermore, core recovery in this interval was poor. The evaporites are sterile, but locally the marly and clayey intercalations yield open marine and, at places, brackish faunas also. Interpretation remains controversial. Diatom epiphytes of bottom-living algae recorded at 422 m (subbottom depth) at Balearic Basin DSDP Site 124 suggest that the depth of water during deposition of this interval did not exceed the depth of the photic zone.

Open marine assemblages interbedded in the evaporite section have been recorded in the Tyrrhenian as well as the Balearic Basin, but not in the eastern Mediterranean. There, the fossil content in Late Miocene evaporites consists of diatoms and spores (Mediterranean Ridge, at Ionian Basin Site 125A), of brackish water ostracodes and benthic foraminifera (Strabo Trench Site 129A) indicating shallow deposition. Drilling deeply into eastern Mediterranean evaporites will help to reconstruct the paleogeographic setting of the area, and to determine the role of the Malta Escarpment and Strait of Sicily during periods of alternating marine incursions and dessication. These lower evaporite sections should be correlated with exposures in the circum-Mediterranean region. Intensifying microfossil-geochemical investigations of the evaporites and underlying sequences can only lead to a better understanding of paleo-oceanographic relations between the Mediterranean and eastern Atlantic during this critical period of Tethyan history.

SUSPENDED SEDIMENT INVESTIGATIONS AND DISPERSAL PATTERNS

Studies of suspended particulate matter have been too long neglected in the Mediterranean, as elsewhere, in view of the significant contribution that such information can make to sedimentology. Studies of suspended matter generally have not formed integral parts of marine sedimentological investigations despite the fact that the nature of this material ultimately determines the composition and character of the pelagic phase of bottom sediments.

Chamley (1971) points out the usefulness of suspended sediment studies in understanding the mixing of suites to derive a complex bottom sediment. Emelyanov and Shimkus (this volume) show how dispersion of material from the Nile is achieved and also how circulation patterns may be deduced from this information.

With multiple source areas contributing different suites of terrigenous material to the water masses (Venkatarathnam and Ryan, 1971) and gravitational sinking hindered at density interfaces, studies of suspended sediment should provide some indication of the relative importance of different source areas to pelagic sediments presently being deposited. Once the material emanating from a point source (river, coastal erosion or wind blown) is identified as characteristic of a specific water mass, its transport path will assist in determing paleo-circulation patterns in the Mediterranean.

Research on suspended particulate matter may also assist in the tracing of pollutants from the areas of input to the deeper basins of the Mediterranean. It is known that many pollutants are absorbed on films, on organic particles, or on sediment particles in addition to being present as dissolved species in the water. With extreme cooling in the northern portion of the Mediterranean, thermo-convective circulation (Figure 6) carries the pollutants and associated particles to deeper levels (Brambati, this volume). Tracing the paths of the suspended particles may be easier than following the pollutants themselves. Sampling of suspended material should be made in conjuntion with salinity, temperature, and real-time optical transmission measurements to define and characterize the water masses and contained particulate matter. Early investigations would best be done at the time of low terrestial run-off to minimize local perturbations.

COASTAL AND SHALLOW WATER SEDIMENTATION

Systematic mapping and classification of the circum-Mediterranean coasts coupled with definition of the near-shore dynamic processes (erosional versus accretion) should receive high priority in future research programs. Evaluation of the following can serve as basis of specific research projects: the effects of wave and current activity on the morphology and sedimentation of the inner margin in an almost tideless sea; the influence of seasonality on formation or destruction of beaches, and on the modification of surficial sediments in the shelf area; the origin of wind blown sand and movement and fixation of dunes; the origin of carbonate-cemented clastics,

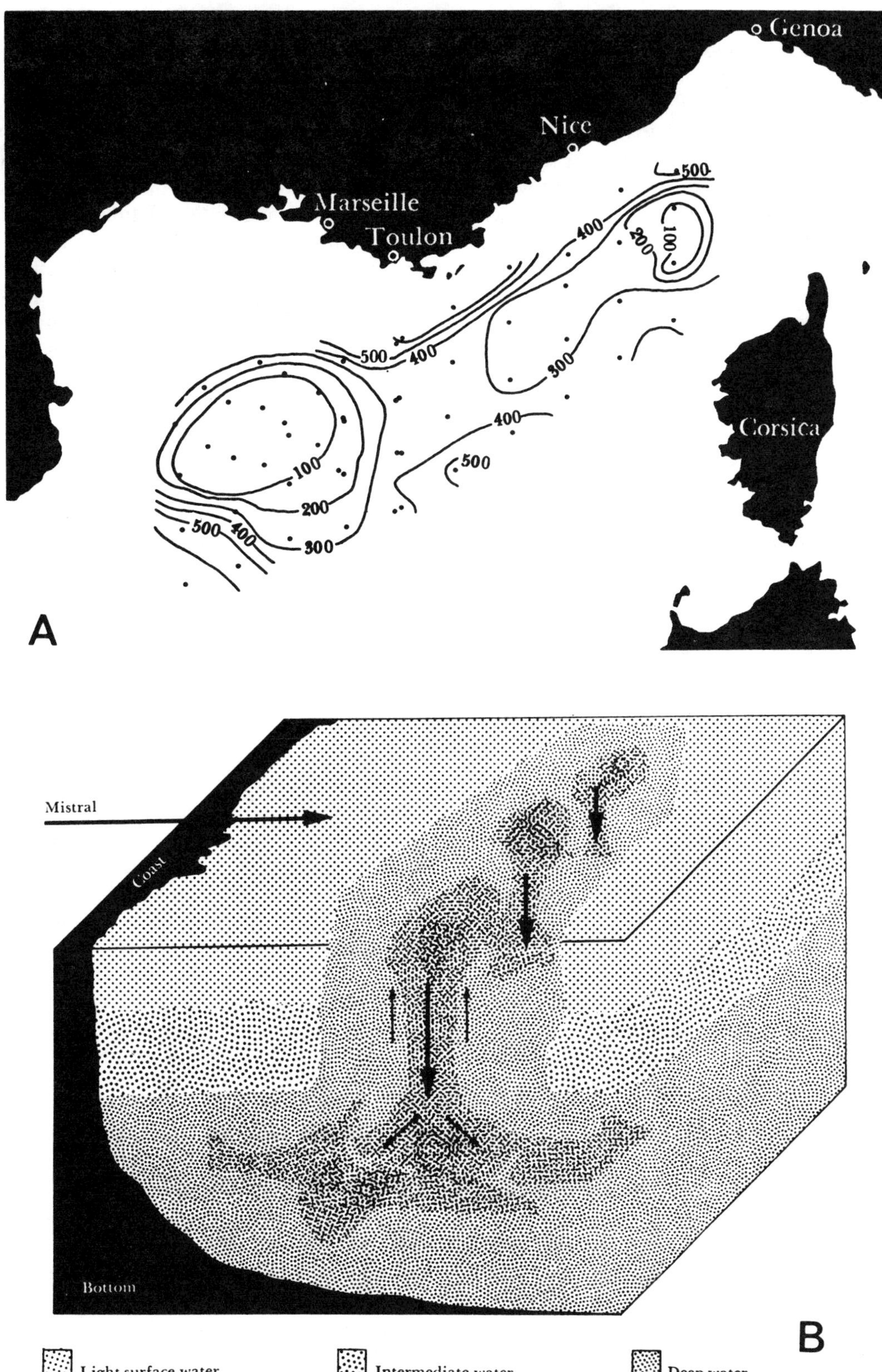

Figure 6. A, Density of surface water in the northwest Mediterranean in early January 1969. Chart shows two patches where the density between surface and deep water is less than 100 parts per million (interval between contours is 100 parts per million). Density differences between surface and deep water on the outer edges of the two patches is 500 parts per million. This is in contrast with densities found in most subtropical open oceans where density differences approximate 2000 parts per million. B, schema showing vertical flow pattern during convection. After Stommel *et al.*, 1971, with the author's permission.

submerged dunes, submerged continental and lagoonal deposits, 'drowned' and disintegrated beach rocks; and modifications of morphology and sediments due to the effect of human activity.

Rivers are the primary source of terrigenous clastics, and their importance in providing sediments to basin plains is accentuated by reason of the narrowness (and, in some cases, absence) of the shelf along much of the Mediterranean. The significance of the half dozen major rivers (including Rhône, Po, *etc.*) debouching their load in a closed system is evident, yet definition of the sediment dispersal beyond the delta mouths of such rivers as the Ebro, Nile (Figure 1) and Ceyhan, for the most part, has been neglected. With few exceptions little is known of the composition, quantities, and eventual dispersal of sediment carried to the basins seasonally (generally between autumn and spring) by the numerous short, steep rivers draining the adjacent coast; nor can much be said quantitatively of the erosion of the coast intensified in winter months, and the lateral transfer of such materials by longshore currents.

Defining provenance by mineralogical and textural analysis of grab and core samples on the shelf should be supplemented by clay studies, and these mineral suites should be correlated with fines at river mouths (Chamley, 1971). It is essential, of course, to determine if fines carried by turbid river waters to the sea at times of floods are able to pass through the salinity interfaces. High resolution profiling and coring of shelves and platforms are needed to define the nature of the Quaternary progradations and regressions and the distribution of relict sediments and the modern (Holocene) shelf blanket (Leclaire, 1970, this volume; Bellaiche, this volume; Caulet, this volume). In this regard, the wide shelf (> 300 km) east of Tunisia (a foundered lagoon or semiclosed embayment ?) would repay thorough study. Unlike the Gulf of Lion and other wide shelves in the Mediterranean, this shelf, interestingly enough, is not associated with a delta. Smaller shelves and shelf-depth areas which deserve an intensified study include those in the Straits of Sicily and off Libya, the margins off mainland Turkey and Greece and the numerous island platforms in the Aegean Sea.

Correlation of submarine terraces at the shelfbreak (Edgerton and Leenhardt, 1966; Bellaiche, this volume; Milliman *et al.*, this volume) with seismic profilers serves to measure the eustatic shifts in sea level. Sea level changes, and particularly the last glacial transgression, have resulted in extensive coastal areas being submerged. Drowned features include deltas, lagoons, sand bars, coastal dune systems, terrace and cliff erosion features, beaches, river valleys, *etc.* While features of all these types have been detected or tentatively identified by many authors (*cf.*, Breslau and Edgerton, this volume), studies have necessarily been limited in scope by available techniques. Studies have tended to be linear, that is in terms of a number of transects, rather than total analyses of an area. It would be valuable to take an area and survey it unambiguously so that there was no question of interpolation between transects or uncertainy of relationship between features. Detailed bathymetry, supported by a complete true-scale side-scan survey, would provide a complete topographic and morphological map upon which many features could be identified. The sedimentary and erosional features could then be studied with appropriate techniques to build up a complete geomorphological history of a submerged coastal strip. Of special interest are the potential land bridges which may have provided passage for palaeolithic migrations.

Climatic variations as well as changes in agricultural practices and deforestation have increased erosion in historic time and this should be reflected in the development of the modern mud blanket on the narrow shelves (Leclaire, 1970). More recently, man-made obstacles along the rivers, such as dams on the Ceyhan and the Nile, artificial diversion of flow and channelization (Var) are altering natural physical, chemical and biological processes and producing measurable depositional changes on the continental margin. To these effects should be added the dumping of industrial and municipal wastes at an ever increasing rate. The shallow marine-shelf environments of the Mediterranean are also now coming into their own as far as petroleum (Gulf of Lion, Ebro shelf, Aegean Sea, Egyptian margin, *etc*). and mineral exploration is concerned. It is desirable that the growing volume of drilling (*cf.*, Burollet and Dufaure, this volume) and sub-bottom seismic records accumulated by commercial firms be released to the scientific community after an acceptable period of time.

SEDIMENTATION IN DEEP SEA ENVIRONMENTS

The deposits in enclosed Mediterranean basins lend themselves to systematic analysis in much the same way as paleobasin deposits are examined by geologists on land. These depressions trap, or pond, sedimentary sequences that can be examined by high resolution profiling, such as 3.5 kHz records (Figure 7) where individual reflectors can be identified by means of coring surveys (*cf.*, Gehin *et al.*, 1971). In addition to isopach maps, it is possible to chart the number of reflectors (Stanley *et al.*, 1970), and lateral continuity of layers (Hersey, 1965; Ryan *et al.*,

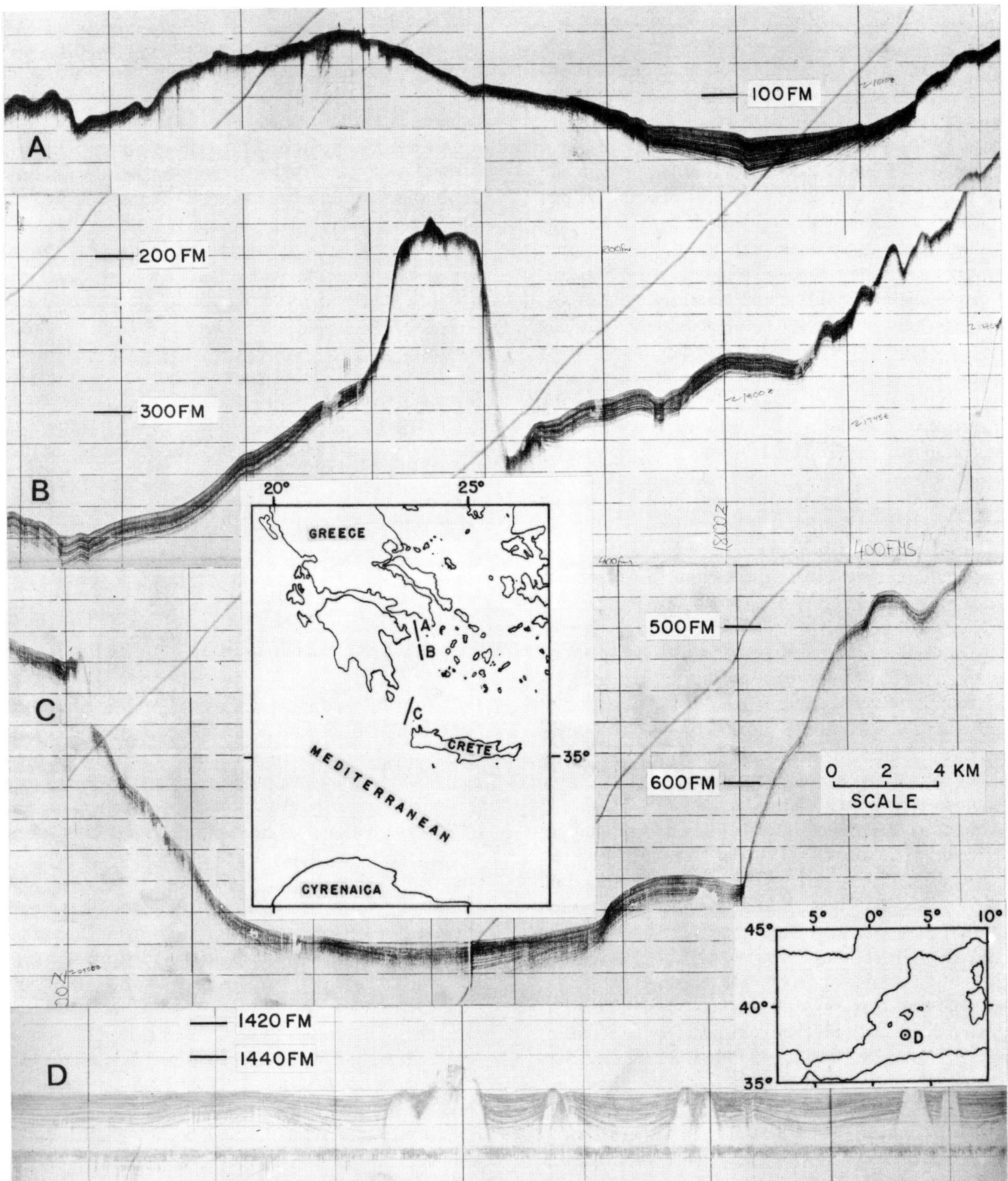

Figure 7. High resolution 3.5 kHz records off Greece and in the western Mediterranean (spacing between horizontal lines = 20 fms, or 37 m). Reflectors in profiles A and C show ponding in topographic lows; structural offset of reflectors is shown in A, B and C, a region of intense tectonic mobility. D, salt diapirs in the basin plain (Western Balearic Basin). Note disruption of reflectors (some of these are turbidites as shown in cores) by salt diapir tectonics. Truncated diapirs on right have undergone dissolution; those on left, extending above the sea floor, may be younger. Note offset and deformed reflectors extending to surface indicating recent movement.

1965). Sand/shale ratios and lateral variability of bed thickness can be determined by core study. From such patterns and the location of canyons, valleys, fans, slump deposits and slump scars it is possible to obtain a rather precise inventory of dispersal paths and basin filling. Determining the continuity of layers (Heezen and Ewing, 1955; Ryan *et al.*, 1965; van Straaten, 1967) and the presence of salt diapirs (Figure 7) can best be solved with high frequency profiling surveys coupled with a coring program.

One of the major unsolved problems concerns the origin of the clays, silts and planktonic organisms which account for the bulk of the sediment in the enclosed basins, ultimate depositional sites. X-ray mineralogy and x-radiography are needed in core studies, but much also stands to be grained by studying thin sections of core material with a standard petrographic microscope (Eriksson, 1965). It is necessary to examine the spatial relationship of microfossils such as planktonic foraminifera with the mud framework (*i.e.*, forams 'floating' or concentrated in laminae) in thin section (Figure 5). Perhaps this approach could be used as a means to distinguish graded mud turbidites from more slowly deposited hemipelagic and pelagic layers (Figure 8). This type of petrographic study can also be used to examine lateral changes in individual turbidites and help in the recognition of proximal and distal portions of gravity-induced flows. In such studies, carbon-14 age dating is needed for the correlation of layers as well as to provide a time framework and aid in the determination of sedimentation rates.

A detailed analysis of the mineralogical and paleontological composition of Quaternary core sections coupled with trace element (boron, for example) and oxygen isotope studies should provide further data to test the hypothesis of the current reversals in the Mediterranean (Huang *et al.*, 1972). In this respect, more emphasis on sapropels (Olausson, 1965; Ryan, this volume) can provide further insight on the problem of the possible stagnation of eastern basins as related to sills (Straits of Gibraltar, Sicily, Bosporus) and changes in hydrology during the Pleistocene.

Examination of the volcanics mixed with deep marine terrigenous deposits may have analogues in the fossil record, *i.e.*, the volcano-sedimentary "grès moucheté" ("salt-and-pepper") facies in the fossil record. A petrographic comparison of basin margin deposits in the Tyrrhenian Sea and in the Aegean Arc with the Tertiary Grès de Taveyannaz or Grès du Champsaur facies of the Western Alps is recommended.

The comparison of flysch facies of the fossil record with selected modern basin facies in the Mediterranean (Hersey, 1965; Stanley *et al.*, 1970) is tempting. The tectonically isolated basin plains and trenches containing thick deposits of cyclic units probably do not all contain identical facies. Plio-Quaternary strata in large, relatively quiet zones of what appears to be foundered depressions (*i.e.*, those of the Balearic Basin for instance) should be compared with those in smaller and more tectonically active settings (the Tyrrhenian basin plain) and to those in the more active compressive setting (Papazachos and Comninakis, 1971) of the Aegean arc.

Other specific problems await a solution. One of these pertains to dissolution of the salt diapirs which reach the sea-bed (Figure 7, D). Another is evaluating the role of benthic organisms that modifiy original physical structures of surficial sediments as is evident from bottom photographs. The trace fossil assemblages in each type deep sea environment need to be identified, determined and compared with possible trace fossil counterparts in the geological record.

CARBONATE, EVAPORITE AND GEOCHEMICAL RESEARCH IN THE MEDITERRANEAN

Carbonate sedimentation in the Mediterranean occurs in two general environments: shallow water (shallower than 200 m) and the deep sea. Over much of the Mediterranean, modern hermatypic organisms (primarily coralline algae) receive sufficient light at depths as great as 100 m to calcify at relatively rapid rates. The 100 to 200 m depth interval, while generally occurring below the euphotic zone, was exposed to greater light intensities during glacial lowerings of sea level, and as a result, contains many relict shallow-water organisms. In addition to this mixture of modern and relict benthonic organisms, the shelf also yields various amounts of modern and older planktonic remains. Differentiation of these various components is not easy, since it requires carbon-14 dating of hand picked samples, a process that is both expensive and time-consuming. But clearly such studies are needed if we are to understand Holocene carbonate sedimentation.

At the same time, we need to extend our knowledge about the various types of carbonate sedimentation on the Mediterranean shelves. Several excellent studies have been made off Algeria, Italy and France (some of which are reported in this volume); carbonates in other regions, particularly in the central and eastern Mediterranean, require investigation. For instance, a comparison of the shelf carbonates off Tunisia and Libya (where Pleistocene oolite deposits apparently

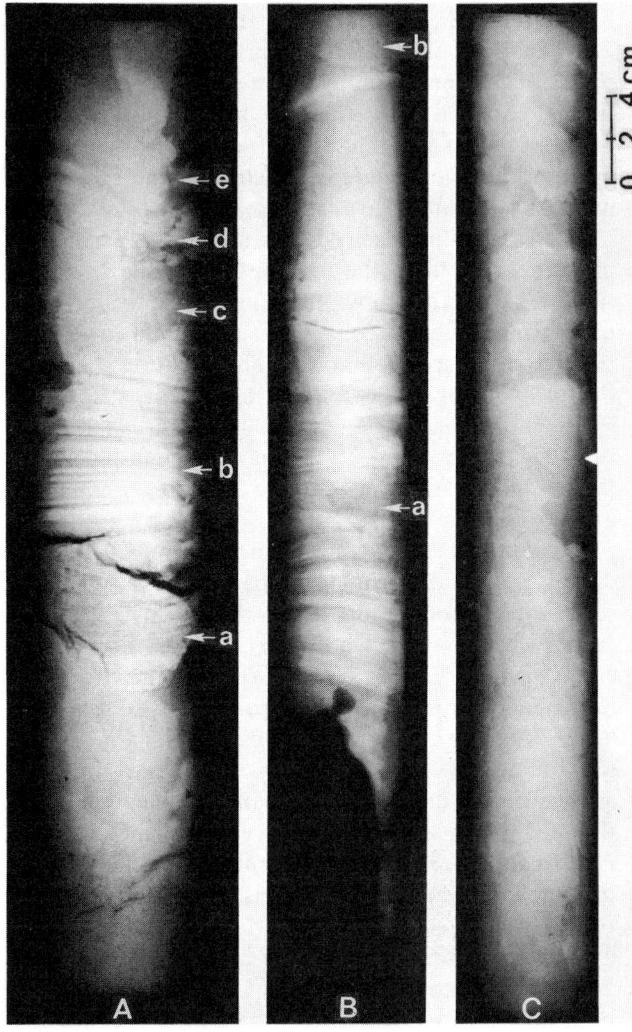

Figure 8. X-radiographs showing sedimentary and biogenic structures in cores collected on the Balearic Rise. A, section showing graded arenite above mud (note irregular erosional surface between the two units). The vertical turbidite sequence includes *a* coarse arenite, *b* well laminated, graded medium to fine grained arenite, *c* cross-laminated, very fine arenite, *d* indistinctly laminated silt and *e* lutite. B, section showing *a*, a turbidite consisting of a graded parallel- and cross-laminated fine arenite and *b*, hemipelagic mud and arenite (pteropod and foraminifera in lutite). No sharp contact between the two. C, lutite displaying mottling due to burrowing activity of benthic organisms.

are present) with those forming today in the Gulf of Suez (Sass *et al.*, this volume) would be interesting.

Reef-like bioherms composed of various types of organisms (vermetids, bryozoans, coralline algae and sponges) have been reported throughout the Mediterranean. These reefs should be studied in greater detail, with particular reference to their ecology, growth, species composition and diagenesis. This latter aspect is particularly worthwhile, since there has been little emphasis on carbonate diagenesis in such subtropical environments. A comparison of these reefs with similar features present in ancient limestones around the Mediterranean could also prove significant.

Diagenetic alteration and subsequent lithification of carbonate sediments is well documented for tropical areas, especially in the Persian Gulf and the Caribbean. Corresponding data in subtropical areas, however, are few. Preliminary studies by Alexandersson (1969, 1972) should be extended, and should especially include the diagenesis and lithification of intertidal (including beachrock) and subtidal carbonates.

The deep sea includes environments which have been essentially aphotic during most of the Quaternary. As a result, photosynthetic hermatypic organisms, such as coralline algae, are practically absent from depths greater than 200 m. In most deep-sea deposits throughout the world, deep-sea carbonates are composed primarily of calcitic planktonic foraminifera and coccoliths. The eastern Mediterranean, however, is unique in that it contains large amounts of magnesian calcite, which apparently has been precipitated *in situ*. The precipitation occurs both in the form of lutite and also as a matrix and cement within deep-sea limestones. Such limestones have been reported by many workers, but the precipitation of the magnesian calcitic lutite has only recently been documented (Milliman and Müller, in press).

Apparently precipitation was not continuous during the Pleistocene, since it is absent in layers that contain sapropel. Similarly, present evidence suggests that magnesian calcite is not important in the western Mediterranean, although further verification of this conclusion is needed. Why precipitation has been limited both geographically and stratigraphically is not known, but it probably is related to some unique oceanographic conditions present in the modern eastern Mediterranean deep sea (possibly the relatively high salinities and temperatures; Milliman and Müller, in press).

There is considerable evidence to suggest that magnesian calcite is metastable, decreasing in abundance with increasing core depth (age). In most cores the magnesian calcite is almost completely altered to low magnesium calcite within the upper 5 to 6 meters. Interstitial water studies show that some of the magnesium is present in the interstitial waters, but the rest of it is redistributed (Milliman and Müller, in press). The exact locus of redistribution, however, is not known: is it in clay minerals or has it been utilized in the formation of dolomite? If the latter alternative is not applicable, how does one explain the sporadic but prominent layers of dolomite that occur throughout many of the Mediterranean cores?

In addition to studying the process of inorganic precipitation in the deep sea, future studies should also record possible analogs within ancient limestones. It is entirely possible that magnesian calcite was a major component in the paleo-Mediterranean basins, at times when circulation was even more restricted. It is also probable that many of the fine grained limestones studied on circum-Mediterranean terrains represent ancient analogs of sediments which are presently forming in the eastern basin.

In addition to questions concerning the carbonates, the closed Mediterranean system poses a wide range of sediment related chemical problems affecting mineral genesis and diagenesis. Among those problems we may cite the relationship of pore water chemistry in sediments to water-mass chemistry and the diagenetic effects on minerals buried with the water; changes in the interstitial water chemical profiles related to the intrusion of diapiric salt structures; the existence of "fossil" pore waters which might reflect historical variations in contributions of fresher water from the Black Sea or Nile; variations in the kind and degree of reaction between metastable volcanic debris and sea water; and the mechanisms for submarine cementation in some of the Quaternary sands in the Nile Delta area.

The Upper Miocene evaporites of the Mediterranean continue to interest geologists (Figures 3, 7). More detailed sampling of the Upper Miocene salts is needed to determine their areal distribution, the relation between the eastern and western Mediterranean basins during Upper Miocene time, and the relationship to those of the Red Sea and other rift valleys. In this case, a comprehensive study of the evaporites exposed on land and drilled on the coastal plains is urgently required before any further working hypotheses are made. We need more detailed knowledge of the carbonate units associated with the evaporites: cemented layers, dolomites, pelagic limestones intercalated with salt, interbedded fresh and brackish water deposits, *etc.*

We need to come to a more precise understanding of the mechanisms responsible for the profound and rapid fluctuations in basin water chemistry as deduced from the sedimentological (van Straaten, this volume) and paleontological (Benson, this volume) evidence. Trace element (boron, for example) and geochemical analyses of sapropels are required to understand the Quaternary effects of climatic and hydrological effects on basins within the Mediterranean area.

CONCLUSIONS

The previous sections have outlined in a general manner some of the more obvious facets requiring the attention of marine scientists working in the Mediterranean. The list is by no means complete. This epilogue is an admission that only the first phase of sediment studies in the Mediterranean has been completed, and that we have now entered a period when some change of emphasis would be appropriate and could be assisted by the new techniques now available.

Deep sea drilling is needed to help resolve sediment problems related to the past history of the Mediterranean. It is, quite obviously, not the only, or even necessarily the most important, technique to resolve problems of sedimentation. The relation between fluvial hydrology, water mass movement, suspended sediment and trace element distribution can best be worked out by joint physical oceanographic-sedimentologic team research and requires immediate attention. Monitoring stations selected by groups of physical and geological oceanographers are needed on land and at sea throughout the area to map modern sediment and pollutant dispersal patterns.

The socio-political need to face an unpleasant subject—pollution—is certain to boost marine sedimentological research. The problem is so pressing in the Mediterranean that one can expect that socio-economic pressures will replace the military impetus as the primary motivator of oceanographic research in the Mediterranean. "De la chance dans la malchance." A cooperative, international multi-disciplinary effort in which marine sedimentologists must play a vital role is envisioned.

However, perhaps too little is known about the Mediterranean to terminate this volume on an alarmist note. We are not convinced that all the trends are irreversible or that man's influence necessarily need be harmful or even negative. It is not beyond reason that the addition of selected man-made materials and artificial modification of sedimentation patterns in this closed system will actually benefit life in and around the sea. In any case, challenges have been issued! The natural laboratory awaits further observation and experimentation.

ACKNOWLEDGMENTS

We thank the participants of the VIIIth International Sedimentological Congress (Heidelberg, 1971) who responded to the questionnaire cited in this chapter. Appreciation is expressed to Dr. K. J. Hsü, Geologisches Institut, Zurich, and Dr. H. Stommel, Woods Hole Oceanographic Institution, for permission to use several of their published figures to illustrate the text. Dr. R. L. Stevenson, Office of Naval Research, kindly provided the NASA Gemini and Apollo color space photographs reproduced here in black and white. Dr. R. Cifelli, Smithsonian Institution, read a draft of this chapter and made suggestions incorporated in the text.

REFERENCES

Alexandersson, T. 1969. Recent littoral and sublittoral high-Mg calcite lithification in the Mediterranean. *Sedimentology*, 12:47–61.

Alexandersson, T. 1972. Micritization of carbonate particles: process of precipitation and dissolution in modern shallow-marine sediments. *Bulletin of the Geological Institute, University of Uppsala*, New Series, 3:201–236.

Ambreseys, N. N. 1965. A note on the seismicity of the eastern Mediterranean. *Studia Geofisica et Geodaetica*, 9:405–410.

Belderson, R. H., N. H. Kenyon and A. Stride. 1972. Comparison between narrow-beam and conventional echo-soundings from the Mediterranean Ridge. *Marine Geology*, 12:M11–M15.

Bellaiche, G. 1972. Les dépôts Quaternaires immergés du Golfe de Fréjus (Var) France. In: *The Mediterranean Sea: A Natural Sedimentation Laboratory*, ed. Stanley, D. J., Dowden, Hutchinson and Ross, Inc., Stroudsburg, Pennsylvania, 171–176.

Benson, R. H. 1972. Ostracodes as indicators of threshold depth in the Mediterranean during the Pliocene. In: *The Mediterranean Sea: A Natural Sedimentation Laboratory*, ed. Stanley, D. J., Dowden, Hutchinson and Ross, Inc., Stroudsburg, Pennsylvania, 63–73.

Blanc-Vernet, L. 1972. Données micropaléontologiques et paléoclimatiques d'après des sédiments profonds de Méditerranée. In: *The Mediterranean Sea: A Natural Sedimentation Laboratory*, ed. Stanley, D. J., Dowden, Hutchinson and Ross, Inc., Stroudsburg, Pennsylvania, 115–127.

Bourcart, J. 1959. Morphologie du précontinent des Pyrénées à la Sardaigne. In: *La Topographie et la Géologie des Profondeurs Océanographiques*, Colloques Internationaux C.N.R.S., Nice-Villefranche, 33–52.

Brambati, A. 1972. Sedimentology and pollution in the Mediterranean Sea: a discussion. In: *The Mediterranean Sea: A Natural Sedimentation Laboratory*, ed. Stanley, D. J., Dowden, Hutchinson and Ross, Inc., Stroudsburg, Pennsylvania, 711–721.

Breslau, L. R. and H. E. Edgerton 1972. The Gulf of La Spezia, Italy: a case history of seismic-sedimentologic correlation. In: *The Mediterranean Sea: A Natural Sedimentation Laboratory*, ed. Stanley, D. J., Dowden, Hutchinson and Ross, Inc., Stroudsburg, Pennsylvania, 177–188.

Burrollet, P. F. and P. Dufaure 1972. The Neogene series drilled by the Mistral No. 1 well in the Gulf of Lion. In: *The Mediterranean Sea: A Natural Sedimentation Laboratory*, ed. Stanley, D. J., Dowden, Hutchinson and Ross, Inc., Stroudsburg, Pennsylvania, 91–98.

Carter, T. G., J. P. Flanagan, C. R. Jones, F. L. Marchant, R. R. Murchison, J. H. Rebman, J. C. Sylvester and J. C. Whitney 1972. A new bathymetric chart and physiography of the Mediterranean Sea. In: *The Mediterranean Sea: A Natural Sedimentation Laboratory*, ed. Stanley, D. J., Dowden, Hutchinson and Ross, Inc., Stroudsburg, Pennsylvania, 1–23.

Caulet, J. P. 1972. Recent biogenic calcareous sedimentation on the Algerian continental shelf. In: *The Mediterranean Sea: A Natural Sedimentation Laboratory*, ed. Stanley, D. J., Dowden, Hutchinson and Ross, Inc., Stroudsburg, Pennsylvania, 261–277.

Chamley, H. 1971. *Recherches sur la sédimentation argileuse en Méditerranée*. Thèse, Bibliothèque Universitaire Aix-Marseille, Section de Luminy, 401 p.

Edgerton, M. E. and O. Leenhardt 1966. Monaco: the shallow continental shelf. *Science*, 152:1106–1107.

Emelyanov, E. M. and K. M. Shimkus 1972. Suspended matter in the Mediterranean Sea. In: *The Mediterranean Sea: A Natural Sedimentation Laboratory*, ed. Stanley, D. J., Dowden, Hutchinson and Ross, Inc., Stroudsburg, Pennsylvania, 417–439.

Emery, K. O., B. C. Heezen, and T. D. Allan 1966. Bathymetry of the eastern Mediterranean Sea. *Deep-Sea Research*, 13:173–192.

Eriksson, K. G. 1961. Granulométrie des sédiments de l'île d'Alboran, Méditerranée occidentale. *Bulletin of the Geological Institute, University of Uppsala*, 40:269–284.

Eriksson, K. G. 1965. The sediment core No. 210 from the Western Mediterranean Sea. *Reports of the Swedish Deep-Sea Expedition, 1947–1948*, 8:397–594.

Fabbri, A., and R. Selli 1972. The structure and stratigraphy of the Tyrrhenian Sea. In: *The Mediterranean Sea: A Natural Sedimentation Laboratory*, ed. Stanley, D. J., Dowden, Hutchinson and Ross, Inc., Stroudsburg, Pennsylvania, 75–81.

Filyushkin, B. N. 1962. The state of oceanographical knowledge concerning the Mediterranean Sea. In: *Oceanographic Research in the Atlantic*, ed. Klenova, M. V., Academy of Sciences of the USSR. Transactions of the Institute of Oceanology, Israel Program for Scientific Translations (1967) TT 66-51029:302–315.

Flemming, N. C. 1968. Holocene earth movements and eustatic sea level change in the Peloponnese. *Nature*, 217:1031–1032.

Flemming, N. C. 1969. Archaeological evidence for eustatic change of sea level and earth movements in the Western Mediterranean in the last 2,000 years. *Geological Society of America Special Paper*, 109:125 p.

Folger, D. W. 1970. Wind transport of land-derived mineral, biogenic and industrial matter over the North Atlantic. *Deep-Sea Research*, 17:337–352.

Gehin, C., C. Bartolini, D. J. Stanley, P. Blavier and B. Tonarelli 1971. Morphology and late Quaternary fill of the Western Alboran Basin, Mediterranean Sea. *Saclantcen Technical Report*, 201:78 p.

Gennesseaux, M., P. Guibout and H. Lacombe 1971. Enregistrement de courants de turbidité dans la vallée sous-marine du Var (Alpes-Maritimes). *Comptes Rendus de l'Académie des Sciences de Paris*, 273:2456–2459.

Glaçon, G., C. Vergnaud Grazzini and M. J. Sigal 1971. Premiers résultats d'une série d'observations saisonnières des foraminifères du plancton méditerranéen. In: *Proceedings of the II Planktonic Conference Roma 1970. Volume 1*, ed. Farinacci, A., Edizioni Tecnoscienza, Rome, 555–581.

Heezen, B. C. and M. Ewing 1955. Orleansville earthquake and turbidity currents. *Bulletin of the American Association of Petroleum Geologists*, 39:2505–2514.

Heezen, B. C., C. Gray, A. G. Segre and E. F. K. Zarudzki 1971. Evidence of foundered continental crust beneath the Central Tyrrhenian Sea. *Nature*, 229:327–329.

Herman, Y. 1972. Quaternary Eastern Mediterranean sediments: micropaleontology and climatic record. In: *The Mediterranean Sea: A Natural Sedimentation Laboratory*, ed. Stanley, D. J., Dowden, Hutchinson and Ross, Inc., Stroudsburg, Pennsylvania, 129–147.

Hersey, J. B. 1965. Sediment ponding in the deep sea. *Bulletin of the Geological Society of America*, 76: 1251–1260.

Hsü, K. J. 1971. Origin of the Alps and Western Mediterranean. *Nature*, 233:44–48.

Huang, T.-C. and D. J. Stanley 1972. Western Alboran Sea: sediment dispersal, ponding and reversal of currents. In: *The Mediterranean Sea: A Natural Sedimentation Laboratory*, ed. Stanley, D. J., Dowden, Hutchinson and Ross, Inc., Stroudsburg, Pennsylvania, 521–559.

Huang, T.-C., D. J. Stanley and R. Stuckenrath 1972. Sedimentological evidence for current reversal at the Strait of Gibraltar. *Marine Technology Journal*, 6:25–33.

International Association of Sedimentologists 1971. *Program With*

Abstracts. VIII International Sedimentological Congress 1971, Laboratorium für Sedimentforschung, Heidelberg, 121 p.

Leclaire, L. 1970. *La sédimentation holocène sur le versant méridional du bassin algéro-baléare (Précontinent algérien).* Thèse, Faculté des Sciences de Paris, C.N.R.S. A.O.4492, 3 volumes, 551 p.

Leclaire, L. 1972. Aspects of late Quaternary sedimentation on the Algerian precontinent and in the adjacent Algiers-Balearic Basin. In: *The Mediterranean Sea: A Natural Sedimentation Laboratory*, ed. Stanley, D. J., Dowden, Hutchinson and Ross, Inc., Stroudsburg, Pennsylvania, 561–582.

Lort, J. M. 1971. The tectonics of the eastern Mediterranean. *Reviews of Geophysics and Space Physics*, 9:189–216.

Maley, T. S. and G. L. Johnson 1971. Morphology and structure of the Aegean Sea. *Deep-Sea Research*, 18:109–122.

McKenzie, D. P. 1970. Plate tectonics of the Mediterranean region. *Nature*, 226:239–243.

Medoc Group 1970. Observations on formation of deep water in the Mediterranean Sea. *Nature*, 227:1037–1040.

Menard, H. W., S. M. Smith and R. M. Pratt 1965. The Rhone deep-sea fan. In: *Submarine Geology and Geophysics*, eds. Whittard, W. F. and R. Bradshaw. Butterworths, London, 271–285.

Milliman, J. D. and J. Müller. Precipitation and lithification of magnesium calcite in the deep-sea sediments of the eastern Mediterranean Sea. *Sedimentology* (in press).

Milliman, J. D., Y. Weiler and D. J. Stanley 1972. Morphology and carbonate sedimentation on shallow banks in the Alboran Sea. In: *The Mediterranean Sea: A Natural Sedimentation Laboratory*, ed. Stanley, D. J., Dowden, Hutchinson and Ross, Inc. Stroudsburg, Pennsylvania, 241–259.

Musée Océanographique de Monaco. 1958–1960. *Topographic Maps* (scale 1 : 200,000), Monaco.

Nesteroff, W. D., W. B. F. Ryan, K. J. Hsü, G. Pautot, F. C. Wezel, J. M. Lort, M. B. Cita, W. Maync, H. Stradner and P. Dumitrica 1972. Evolution de la sédimentation pendant le Néogène en Méditerranée d'après les forages JOIDES-DSDP. In: *The Mediterranean Sea: A Natural Sedimentation Laboratory*, ed. Stanley, D. J., Dowden, Hutchinson and Ross, Inc., Stroudsburg, Pennsylvania, 47–62.

Norin, E. 1958. The sediments of the central Tyrrhenian Sea. In: *Reports of the Swedish Deep-Sea Expedition, 1947–1948*, 8:136 p.

Olausson, E. 1961. Studies of deep-sea cores: sediment cores from the Mediterranean Sea and the Red Sea. *Reports of the Swedish Deep-Sea Expedition, 1947–1948*, 8:337–391.

Olausson, E. 1965. Evidence of climate changes in North Atlantic deep-sea cores. In: *Progress in Oceanography*, 3, ed. Sears, M., Pergamon Press, 221–254.

Papazachos, B. and P. Comninakis 1971. Geophysical features of the Aegean Arc. *Journal of Geophysical Research*, 76:8517–8533.

Rabinowitz, P. D. and W. B. F. Ryan 1970. Gravity anomalies and crustal shortening in the Eastern Mediterranean. *Tectonophysics*, 10:585–608.

Romanovsky, V. 1966. Etudes des couches d'eau près du fond dans les grandes profondeurs méditerranéennes. In: *Disposal of Radioactive Wastes Into Seas, Oceans and Surface Waters.* International Atomic Energy Commission, Vienna, 461–469.

Ryan, W. B. F. 1972. Stratigraphy of late Quaternary sediments in the Eastern Mediterranean. In: *The Mediterranean Sea: A Natural Sedimentation Laboratory*, ed. Stanley, D. J., Dowden, Hutchinson and Ross, Inc., Stroudsburg, Pennsylvania 149–169.

Ryan, W. B. F. *et al.* 1972. *Initial Reports of the Deep Sea Drilling Project, Volume XIII*, U. S. Government Printing Office, Washington (in press).

Ryan, W. B. F. and B. C. Heezen 1965. Ionian Sea submarine canyons and the 1908 Messina turbidity current. *Bulletin of the Geological Society of America*, 76:915–932.

Ryan, W. B. F., D. J. Stanley, J. B. Hersey, D. A. Fahlquist, and T. D. Allan 1970. The tectonics and geology of the Mediterranean Sea. In: *The Sea*, Volume 4, Part II, ed. Maxwell, A. E., Wiley-Interscience, New York, 387–492.

Ryan, W. B. F., F. Workum, Jr., and J. B. Hersey 1965. Sediments on the Tyrrhenian Abyssal Plain. *Bulletin of the Geological Society of America*, 76:1261–1282.

Sass, E., Y. Weiler and A. Katz 1972. Recent sedimentation and oolite formation in the Ras Matarma Lagoon, Gulf of Suez. In: *The Mediterranean Sea: A Natural Sedimentation Laboratory*, ed. Stanley, D. J., Dowden, Hutchinson and Ross, Inc., Stroudsburg, Pennsylvania, 279–292.

Smith, A. G. 1971. Alpine deformation and the oceanic areas of the Tethys, Mediterranean, and Atlantic. *Bulletin of the Geological Society of America*, 82:2039–2070.

Stanley, D. J. (Editor) 1969. *The New Concepts of Continental Margin Sedimentation.* Short Course Lecture Notes, American Geological Institute, Washington, 400 p.

Stanley, D.J. 1972. Modern flysch sedimentation in a Mediterranean island arc setting. In: *Modern and Ancient Geosynclinal Sedimentation*, ed. Dott, R. H., Jr., University of Wisconsin, Madison, 43–44.

Stanley, D. J., C. E. Gehin and C. Bartolini 1970. Flysch-type sedimentation in the Alboran Sea, Western Mediterranean. *Nature*, 228:979–983.

Stanley, D. J. and E. Mutti 1968. Sedimentological evidence for an emerged land mass in the Ligurian Sea during the Paleogene. *Nature*, 218:32–36.

Stommel, H., A. Voorhis and D. Webb 1971. Submarine clouds in the deep ocean. *American Scientist*, 59:716–722.

van Straaten, L. M. J. U. 1967. Turbidites, ash layers and shell beds in the bathyal zone of the southeastern Adriatic Sea. *Revue de Géographie Physique et de Géologie Dynamique*, 9:219–240.

van Straaten, L. M. J. U. 1972. Holocene stages of oxygen depletion in the deep waters of the Adriatic Sea. In: *The Mediterranean Sea: A Natural Sedimentation Laboratory*, ed. Stanley, D. J., Dowden, Hutchinson and Ross, Inc., Stroudsburg, Pennsylvania, 631–643.

Venkatarathnam, K. and W. B. F. Ryan 1971. Dispersal patterns of clay minerals in the sediments of the Eastern Mediterranean Sea. *Marine Geology*, 11:261–282.

Wezel, F. C. 1970. Numidian flysch: an Oligocene-early Miocene continental rise deposit off the African Platform. *Nature*, 228: 275–276.

APPENDIX

Summary of Responses from Specialists

The following is a synthesis of responses from authors and other specialists to a questionnaire distributed at the VIIIth Sedimentological Congress in Heidelberg (International Association of Sedimentologists, 1971).

1. The following were ranked by the responders as the *most important categories to emphasize during this decade* (in order of diminishing importance): deep sea coring programs (JOIDES- and petroleum-type drilling) for geological interpretation; determination and interpretation of processes of sediment transport; relation of sedimentation to structural activity and tectonics; paleoclimatological investigations; mapping of surficial deposits; and evaluating man's influence on sediments.

2. The majority of workers feel that the eastern

Mediterranean region, including the Ionian Sea and Levantine Basin, is the area *that has been the least studied and is where major emphasis is most needed*; the Aegean Sea and Balearic Basin are the two other regions not fully investigated. In addition, the responses indicate that detailed work also should be conducted in specific areas such as the Strait of Sicily, Gibraltar, the margin off Turkey, the Nile Delta and the Tunisian Shelf.

3. The *environments requiring the most concentrated study* are the deep basin plains and submerged ridges and platforms such as the Mediterranean Ridge; submarine slope and shallow marine (beach to inner shelf) environments also have been selected for further investigation.

4. The *techniques which appear most useful to solve problems at hand* are: deep-sea drilling (JOIDES- and petroleum-type); seismic profiling (deep penetration as well as high resolution shallow penetration); current meters for use on the sea floor; and suspended sediment sampling and measuring gear. There is no agreement as to *what is a desirable spacing or density of bottom grab and core samples.* Most workers admit that the coverage should vary with the problem. For general survey purposes, however, spacing at 40 km interval is adequate; spacing of bottom samples should be reduced to less than 10 km (and less than 5 km in many instances) for specific problem solving.

5. Many marine geologists believe that ideally the *density of subbottom seismic lines* should be spaced at 10 to 20 km apart throughout the Mediterranean. Specific *areas where seismic survey nets are most needed* at present include the Mediterranean Ridge, the Aegean Sea, the region north of Crete and the Matapan Trench, the Strait of Gibraltar, the Yugoslavian sector of the Adriatic Sea, and margins off deltas (including submarine delta-fan complexes). It is emphasized that subbottom data should be made available to all workers.

6. The *need for central data banks* for sample and seismic data storage was expressed by many. The majority of scientists state that if a ten-year program involving Mediterranean sediments were initiated, they would make their data available to a central data repository. Many also would be willing to make samples and raw data available to other specialists. On the other hand, some persons expressed caution in the use of data centers and stress the necessity of intelligent management in this type of centralization.

7. The two *types of support most essential for pursuing fundamental research* are listed as increased availability of ship-time and funding by international sources (such as UNESCO). Two additional forms of research support required are cooperation through team research (*i.e.*, improving interchange between workers in different laboratories and in other specialties) and increased funding from national sources.

8. *Other aspects to be considered in planning a long-term program* cited by specialists include ease of access to laboratories and ship-time; establishing a closer liason with land-based geologists specializing on circum-Mediterranean problems in order to correlate submarine and land-exposed formations; increasing the lines of communication and scientific cooperation with states bordering the eastern Mediterranean; incorporation of physical and chemical oceanographers in sediment-related programs; and formation of multi-disciplinary groups of concerned younger as well as senior scientists to help insure that proper and equitable support be given to research endeavors.

The above generalities, gleaned from the responses of individual scientists, and not emanating from laboratories or governmental agencies, reflect the present thinking of workers concerned with the Mediterranean.

Author Index

Numbers in italics refer to pages on which complete reference is cited.

Subject Index

Page numbers which refer to illustrations or tables are in italics.

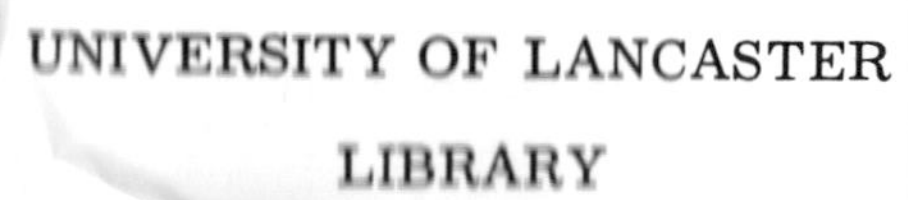

Rey de spanya
Rey de francia
Rey de tunes
Imperadore
Rey d' polonia
Soldan de babilonia
Sinay